CONVERSION TABLE[†] (MKSA TO cgs UNITS)

Quantity	Symbol	MKSA	cgs emu	cgs esu
Time	t	1 second (s)	1 s	1 s
Length	d	1 meter (m)	100 cm	100 cm
Mass	m	1 kilogram (kg)	1000 gm	1000 gm
Energy	$\mathcal{E}$	1 joule (J)	10^7 ergs	10^7 ergs
Force	F	1 newton (N)	10^5 dynes	10^5 dynes
Charge	q	1 coulomb (C)	10^{-1} abcoulomb	3×10^9 stat coulombs
Current	I	1 ampere (A)	10^{-1} abamp	3×10^9 stat amps
Potential	V	1 volt (V)	10^8 abvolts	$\frac{1}{300}$ stat volt
Electric field	E	1 V/m	10^6 abvolts/cm	$\frac{1}{3} \times 10^{-4}$ stat volt/cm
Magnetic flux	Φ	1 weber (Wb)	10^8 maxwell	$\frac{1}{300}$ stat volt-sec
Magnetic induction	B	1 tesla (T)	10^4 gauss	$\frac{1}{3} \times 10^{-6}$ stat volt-sec/cm^2
Magnetic field	H	1 ampere/m (A/m)	$4\pi \times 10^{-3}$ oersted	$12\pi \times 10^7$ stat amp/cm
Resistance	R	1 ohm (Ω)	10^9 abohm	$\frac{1}{3^2} \times 10^{-11}$ stat ohm
Conductivity	σ	1 siemens/m (S/m)	10^{-11} (abohm-cm)$^{-1}$	$3^2 \times 10^9$ (stat ohm-cm)$^{-1}$
Capacitance	C	1 farad (F)	10^{-9} abfarad	$3^2 \times 10^{11}$ stat farad
Inductance	L	1 henry (H)	10^9 abhenry	$\frac{1}{3^2} \times 10^{-11}$ stat henry
Permittivity (vacuum)	ϵ_0	$\frac{1}{3^2 \times 4\pi} \times 10^{-9}$ F/m		unity
Permeability (vacuum)	μ_0	$4\pi \times 10^{-7}$ H/m	unity	

λ (Å) $= 12{,}398/h\nu$ (eV), ν(Hz) $= 2.41804 \times 10^{14} \, h\nu$ (eV), λ^{-1}(m^{-1}) $= 8.06573 \times 10^5 h\nu$ (eV)

300 K $= 0.02585$ eV

1 bar (pressure) $= 10^6$ dynes/cm^2

1 torr (pressure) $= 1333.2 \times 10^{-6}$ bar

[†]The factor 3 should be replaced by a factor of 2.997925 if precision is required.

Fundamentals
of Semiconductor Theory
and Device Physics

Prentice Hall Series in Electrical and Computer Engineering

Leon O. Chua, Series Editor

Fundamentals

of Semiconductor Theory

and Device Physics

SHYH WANG
*University of California
at Berkeley*

PRENTICE HALL, Englewood Cliffs, New Jersey 07632

Library of Congress Cataloging-in-Publication Data

Wang, Shyh.
    Fundamentals of semiconductor theory and device physics / Shyh
Wang.
       p.   cm.—(The Prentice Hall series in electrical and
computer engineering)
    Includes index.
    ISBN   0-13-344409-0
    1. Semiconductors.   I. Title.   II. Series.
OC611.W32   1989   89-3669
621.3815'2—dc19   CIP

Editorial/production supervision and interior design: Hartley Ferguson
Cover design: Edsol Enterprises
Manufacturing buyer: Robert Anderson

Printed in the United States of America

10 9 8 7 6 5 4 3 2 1

ISBN   0-13-344409-0

PRENTICE-HALL INTERNATIONAL (UK) LIMITED, *London*
PRENTICE-HALL OF AUSTRALIA PTY. LIMITED, *Sydney*
PRENTICE-HALL CANADA INC., *Toronto*
PRENTICE-HALL HISPANOAMERICANA, S. A., *Mexico*
PRENTICE-HALL OF INDIA PRIVATE LIMITED, *New Delhi*
PRENTICE-HALL OF JAPAN, INC., *Tokyo*
SIMON & SCHUSTER ASIA PTE. LTD., *Singapore*
EDITORA PRENTICE-HALL DO BRASIL, LTDA., *Rio de Janeiro*

To my wife
Dila

Contents

PREFACE   **XV**

1 ATOMIC STRUCTURE   **1**

1.1   The Rutherford Atom   1
1.2   The Bohr Postulates and the Wilson–Sommerfeld Rule   5
1.3   The Schrödinger Wave Equation   7
1.4   The Hydrogen Atom   10
1.5   Spin and Exclusion Principle   17
1.6   The Periodic System   19
        Problems   26

2 CRYSTAL STRUCTURE   **29**

2.1   The Crystalline State   29
2.2   Bravais Lattices   31
2.3   Simple Crystal Structures   34
2.4   Crystallographic Notation   41
2.5   Reciprocal-Lattice Vectors and Plane Normal   44
2.6   Interference Phenomenon   46
2.7   Electromagnetic Waves and Matter Waves   48
2.8   Crystal Diffraction   52
2.9   Rotational Symmetry and Material-Parameter Tensor   57

2.10   Crystal Imperfections   61

2.11   Lattice Vibration   65

Problems   71

3   ATOMIC BONDING AND CRYSTAL TYPES   77

3.1   Interatomic Forces   77

3.2   Classification of Bond Types   79
The ionic bond 80
The covalent bond 80
The metallic bond 84
The van der Waals bond 84

3.3   Ionic Crystals   87

3.4   Metallic Crystals   89

3.5   The Hydrogen Molecule   92

3.6   Saturation and Directional Properties of Covalent Bonds   96

3.7   Covalent Crystals   101

Problems   105

4   FREE-ELECTRON THEORY   108

4.1   Energy Bands in Solids   108

4.2   Metals, Insulators, Semiconductors, and Semimetals   113

4.3   Free Electrons in a Box   119

4.4   Free-Electron Model and Density of States   123

4.5   The Fermi–Dirac Distribution Function   126

4.6   Fermi Energy and Contact Potential   129

4.7   Thermionic Emission   131

4.8   Electronic Conductivity and Mean Free Time   133

4.9   Electronic Contribution to Heat Capacity and Thermal Conductivity   137

Problems   140

5   THE BAND THEORY OF ELECTRONIC CONDUCTION   146

5.1   Introduction   146

5.2   The Kronig–Penney Model   147

5.3   The Bloch Wave   151

5.4   Wave Packet and Group Velocity   154

5.5   Equation of Motion and Effective Mass   157

5.6   The Brillouin Zones   158

5.7   The Zone Theory: Density of States and Origin of Energy Discontinuity   162

5.8   The Free-Electron Approximation   167

5.9   Energy-Band Structure and the Tight-Binding
        Approximation   170

5.10   Electron and Hole Conduction   176

5.11   Hall Measurement and Cyclotron Resonance
        Experiment   180
        Problems   186

6 *SEMICONDUCTOR FUNDAMENTALS*   **192**

6.1   Introduction   192

6.2   Intrinsic and Extrinsic Semiconductors   193

6.3   Free-Carrier Concentration and Fermi Level   196

6.4   Donor and Acceptor States   199

6.5   Evaluation of Fermi Level and Carrier
        Concentrations   204

6.6   Carrier Concentration and Mobility Measurements   209

6.7   Carrier Mobility and Scattering Mechanisms in Si and
        Ge   214
        Impurity scattering 214
        Lattice scattering 216

6.8   Semiconductor Materials   221
        Chemical bond 221
        Polar and piezoelectric scattering and mobility comparison 224
        Impurity states 229

6.9   Energy-Band Structure   232

6.10   Experimental Studies of Energy-Band Structure and
        Effective Masses   238
        Cyclotron-resonance experiment 239
        Density-of-state and mobility effective masses 242
        Interband magnetooptic effect 244
        Reflection experiment 245
        Magnetoresistance and piezoresistance measurements 246
        Thermoelectric effects 247

6.11   Degenerate Semiconductors   250
        Problems   255

7 *TRANSPORT AND RECOMBINATION*
   OF EXCESS CARRIERS   **265**

7.1   Diffusion Current   265

7.2   The Thermal Equilibrium Condition   266

7.3   Excess Carriers and Quasi-Fermi Levels   269

7.4   The Charge-Neutrality Condition   272

7.5   Carrier Recombination Processes   274
The intrinsic recombination process (band-to-band transition) 275
The extrinsic recombination process (recombination via impurities) 277

7.6   The Continuity Equation and the Time-Dependent Diffusion Equations   283

7.7   Discussion of the Solutions of the Time-Dependent Diffusion Equation   287

7.8   The Haynes–Shockley Experiment   295

7.9   Surface States and Field-Effect Experiments   301

7.10   Fast States and Surface-Recombination Velocity   309

7.11   Effects of Surface Condition, Particle Bombardment, and Impurity Incorporation on Carrier Lifetime   312
Problems   322

8 SIMPLE SEMICONDUCTOR JUNCTION DEVICES: THEORY OF p-n HOMOJUNCTION AND METHODS OF JUNCTION FORMATION   330

8.1   *p-n* Junction Diodes   330

8.2   Space-Charge Region and Junction Capacitance   333

8.3   Minority-Carrier Injection and Ideal Diode Characteristics   338

8.4   Carrier Storage and Diffusion Capacitance   343

8.5   Switching Response and Recovery Time   346

8.6   Junction-Formation and Film-Growth Techniques   354
Alloying process 355
Liquid-phase epitaxy 356
Diffusion 359
Chloride vapor-phase epitaxy 365
Molecular-beam epitaxy 366
Metalorganic chemical-vapor deposition 367
Ion implantation 369

8.7   Fabrication Techniques for Planar Structures   371

8.8   Transition-Region Capacitance of Diffused Diodes   375

8.9   Generation and Recombination Current   377

8.10   Junction Breakdown   381
Problems   384

9 MULTIJUNCTION AND INTERFACE DEVICES   390

9.1   Bipolar Transistors with Abrupt Doping Profiles: Small-Signal Equivalent Circuits   390

9.2   Diffused Transistors and Design Considerations   397

9.3   Design Considerations for Power Transistor, Microwave
        Transistor, and Switching Transistor   403
        Power transistor 403
        Microwave transistors 406
        Switching transistor 408

9.4   Large-Signal Analytical Models   411
9.5   The *p-n-p-n* Structure   416
9.6   Junction Field-Effect (Unipolar) Transistor   420
9.7   Metal–Semiconductor Contact and Measurement of Barrier
        Height   422
9.8   Current Conduction in Metal–Semiconductor Barrier   426
9.9   Metal–Insulator–Semiconductor Structure   428
9.10   Metal–Oxide–Semiconductor Field-Effect Transistor   432
9.11   Refined MOSFET Analyses: Important Device
        Parameters   434
9.12   Gate Structures, Ohmic Contacts, and Electrode
        Interconnects   439
9.13   Metal–Semiconductor Field-Effect Transistor   444
9.14   Circuit Integration and Thin-Film Work   449
        Problems   453

**10   HIGH-FIELD PHENOMENA AND HOT-ELECTRON
       EFFECTS**   **462**

10.1   Introduction   462
10.2   High-Field Drift Velocity of Carriers   464
10.3   The Electron-Transfer Effect   470
10.4   Impact Ionization and Carrier Multiplication
        Phenomena   474
10.5   Tunneling   481
10.6   Analysis of Junction Breakdown   486
10.7   Hot-Electron Effects in MOSFET   492
10.8   Analysis of Velocity Saturation by Transport Equations   500
10.9   Electron Transfer and Velocity-Field Characteristics in
        Two-Valley Semiconductors   503
        Problems   510

11   PROPERTIES OF HETEROSTRUCTURES   **516**

11.1   Material Requirements   516
11.2   The (GaAl)As/GaAs and (GaIn)(AsP)/InP System   521
11.3   Band-Edge Discontinuity: Capacitance–Voltage and
        Current–Voltage Measurements in SIS Structures   526

Contents

11.4   Two-Dimensional Electron (Hole) Gas   530

11.5   Magnetoresistance and Quantum Hall Effect   535

11.6   Semiconductor-Heterostructure Interface   539

11.7   Modulation-Doped Heterostructure   542

11.8   Hot Electrons and Ballistic Electrons In Heterostructures   547

11.9   Quantum Wells and Superlattices   555

11.10   Selected Properties of GaAs and (GaAl)As   561

11.11   Heterostructures and Superlattices of Different Kinds   565

Problems   570

12 DIELECTRIC AND OPTICAL PROPERTIES   **577**

12.1   Polarization and Dipole Moment   577

12.2   Electronic Polarizability   580

12.3   Ionic Polarizability   583

12.4   Orientational Polarizability   584

12.5   Chemical Bond, Molecular Structure, and Dielectric Properties of Materials   586

12.6   Internal Fields: Depolarization Field and Lorentz Field   587

12.7   Polarizability and Dielectric Constant   591

12.8   Optical Transitions in Solids   594

12.9   Absorption and Emission of Radiation   597

12.10   Transition Probability and Selection Rules   598
Ionic crystals 600
Semiconductors 602

12.11   Absorption Coefficient, Oscillator Strength, and Spontaneous Lifetime   603

12.12   Lyddane–Sachs–Teller and Kramers–Kronig Relations   607

12.13   Optical Properties of GaAs   610

12.14   Franz–Keldysh, Pockels, and Stark Effects   618

12.15   II–VI Compound and Diluted Magnetic Semiconductors   624

Problems   627

13 HIGH-FREQUENCY AND HIGH-SPEED DEVICES   **636**

13.1   Introduction   636

13.2   Varactor Diodes and Parametric Interaction   636

13.3   Negative Differential Conductance: Esaki Diode and Resonant-Tunneling Barrier Structure   641

13.4   IMPATT and TRAPATT Diodes   649

13.5   Transit-Time Effects   654

13.6   Growth and Propagation of Carrier Waves in Two-Valley Semiconductors   661

13.7   Domain Formation   664

13.8   Dynamics of Dipole Domains   668

13.9   Modes of Operation of Transferred Electron Devices   672

13.10   Considerations for Ultrahigh-Speed Operation of Field-Effect Transistors   676
Characteristic frequency f_t and switching delay τ_t 676
Power–gain cutoff frequency f_{max} 679
Power–delay product $P\tau_d$ 681

13.11   Issues for Integration of Devices of Ultrasmall Dimensions   683
Self-aligned gate 684
Defects and threshold-voltage variation 685
Backgating effect 686
Intrinsic defects in GaAs and Si 688
Graded drain structure 692
Oxide degradation 693

13.12   MESFET and MODFET   695
Analysis of short-channel MODFETs 698
Model for device simulation 702
Nonstationary carrier dynamics—velocity overshoot 704
Inverted interface and DX center 707

13.13   MOSFET   708
Scaling rules for MOSFET miniaturization 708
Short-channel and narrow-channel effects on threshold voltage 710
Proposed design for ¼-μm MOSFET 713
High-field effects 714
Channel mobility 717

13.14   Alternative Approaches for Performance Enhancement   719
Scaling 720
GaAs versus Si 720
Cryogenic operation 722
Three-dimensional architecture 723

13.15   Bipolar and Heterostructure Bipolar Transistors   723
Problems   729

14 OPTICAL DEVICES   739

14.1   Introduction   739

14.2   Luminescence in Large-Gap Phosphors   740

14.3   Luminescence in Sulfide (Medium-Gap) Phosphors   745

Contents   **xiii**

14.4   Junction Luminescence and Light-Emitting Diodes   747

14.5   Spontaneous Emission and Carrier Lifetime for Band-to-Band Transitions in Direct-Gap Semiconductors   755

14.6   Effects of Heavy Dopant and Carrier Concentrations: Band Tailing and Gap Shrinkage   760

14.7   Stimulated Emission: Population Inversion and Gain Spectrum   766

14.8   Injection Lasers   771

14.9   Rate Equations   779

14.10   Photodetectors   782

14.11   Photoconductive Devices   791

14.12   PIN and Schottky Photodiodes   795

14.13   Phototransistors and Avalanche Photodiodes   798

        Problems   800

APPENDIX 1   **BOLTZMANN STATISTICS**   **805**

APPENDIX 2   **REVIEW OF STATISTICAL THERMODYNAMICS: THE BOLTZMANN RELATION AND THERMODYNAMIC ENERGY FUNCTIONS**   **809**

APPENDIX 3   **USEFUL INTEGRALS FOR THE EVALUATION OF CONDUCTIVE PROPERTIES OF METALS AND SEMICONDUCTORS**   **813**

APPENDIX 4   **MOMENTUM, VELOCITY, AND CURRENT IN BLOCH-WAVE FORMULATION**   **816**

APPENDIX 5   **BOLTZMANN EQUATION**   **819**

APPENDIX 6   **CALCULATION OF HALL COEFFICIENT AND MAGNETORESISTANCE BY BOLTZMANN TRANSPORT EQUATION**   **823**

APPENDIX 7   **SOLUTIONS OF THE TIME-DEPENDENT DIFFUSION EQUATION**   **828**

APPENDIX 8   **TRANSIENT RESPONSE OF JUNCTION DIODES**   **832**

APPENDIX 9   **TRANSPORT EQUATIONS FOR ONE-VALLEY SEMICONDUCTORS**   **835**

APPENDIX 10   **TRANSITION PROBABILITY AND TIME-DEPENDENT PERTURBATION CALCULATION**   **840**

INDEX   **845**

Preface

This book has been designed to serve as a text for a one-year graduate course in semiconductor theory and device physics for electrical-engineering and applied-physics students. During the seventies and eighties, we have witnessed remarkable advances in our understanding of semiconductor properties and rapidly expanding uses of semiconductor devices in industrial and consumer products. The phenomenal growth of the semiconductor industry has resulted in a gradual but unmistakable tendency toward specialization. This tendency is reflected in the dichotomy of graduate texts into one group of application-driven and circuit-oriented books and a much smaller group of theory-based and physics-oriented books. Therefore, there is a need for a textbook to bridge the gap between the two groups of texts—which can serve both electrical engineers and applied physicists. This manuscript represents an attempt to provide first-year graduate students with basic knowledge about semiconductor theory and device physics. The combination of subject matter in a single book enables one to present the materials from a broader perspective and in a more coherent manner.

Semiconductor theory is basic to modern electronic and optoelectronic devices, especially those employing microstructures and heterostructures. The rapid advance in semiconductor-device technology is made possible through the collaborative efforts of many disciplines: electrical engineering, applied physics, material science, and chemical engineering. Even though the main focus of this book is on aspects of semiconductor theory and device physics germane to electrical engineering, aspects that have been considered outside the traditional realm of electrical engineering, but which have become an essential part of high technology, are also discussed. For example, answers to questions regarding degradation of device performance and ultimate limit of device physical dimensions rely heavily on our basic knowledge in applied sciences, such as that related to carrier transport, defect mechanism, and bond formation. A considerable portion of the discussion is devoted to these nontraditional subjects, important to extending further the functional capabilities of existing devices and to exploring possibilities for new device concepts. The wealth of research accomplishments in semiconductor science and technology has resulted from fruitful interdisciplinary cross fertilization. A strong motivation for writing a book of this kind is to reinstate, in a first-year graduate course in semiconductors, close ties between electrical engineering and applied physics.

The first year of graduate study is an important transition for a student from the phase of following textbook treatments to the phase of forming one's own thoughts and generating new ideas. To help students make this transition, this book draws substantially from current literature. It analyzes experimental and theoretical results selected from journal articles to build a student's understanding to a level that will allow him or her to read and digest much of the current literature. Analyses generally begin at the level of an intuitive understanding of the physics of subject matter under discussion before mathematical treatments are presented. The analytical results are then interpreted on the basis of underlying physics to reinforce our understanding of the subject. At the end of each chapter, two types of problems are provided, one serving as exercises and the other serving as extensions of or supplements to discussions in the text. The problems are ordered according to the subject matter presented in the text. One mark that distinguishes graduate from undergraduate teaching is a shift from heavy reliance on textbooks to increased usage of current literature. References to journal articles are cited in the text for further probing into the subject matter discussed. It is hoped that this book can serve not only as a basis for understanding the underlying principles of semiconductor devices but also as a vehicle to instill confidence in students in the thought-generating process.

The materials presented have been used by the author at various times in a one-year graduate-course sequence on semiconductor theory and device physics at the University of California, Berkeley. Because of the diverse background of students interested in these subjects, some introductory materials are included and used as supplementary reading to help bring all students to the same starting point. In writing, as in teaching, one often finds that the coverage is limited either by space or by time. Many important subjects are not covered in this book. These include, for example, amorphous semiconductors, noise problems, material-characterization techniques, and processing technologies. To provide an adequate discussion on these subjects would make the text overly long.

The author wishes to express his deep gratitude to reviewers for their generous encouragement and valuable comments. In view of their suggestions, some parts of the text have been reorganized and new materials have been added. He is also deeply appreciative of the efforts of many graduate students, particularly Jim Moon, Inho Kim, and Leslie Field, who have helped eliminate some of the obscurities and errors in the class notes on which this book is based. The author, of course, accepts all responsibility for the final text. He also wishes to express his appreciation of the work of Bettye Fuller in skillfully typing the various versions of the manuscript.

Shyh Wang
Berkeley, California

1

Atomic Structure

1.1 THE RUTHERFORD ATOM

This book is concerned primarily with the physical principles of modern electronic devices that utilize the conductive, dielectric, and optical properties of materials. For a systematic and coherent discussion of these properties, a clear understanding of the structure of atoms and the chemical bonding between them is always helpful and often essential. To provide this requisite background, the first two chapters are devoted to a general review of the elements of modern physics and the fundamentals of crystals.

Since the turn of the century, a number of experiments, notably the scattering experiments of Lenard and Rutherford, have yielded definite information about the structure of atoms. (A thorough discussion of the subject can be found in textbooks on atomic physics. See, for example, M. Born, *Atomic Physics,* Hafner Publishing Company, New York, 1957, p. 60; R. L. Sproull, *Modern Physics,* John Wiley & Sons, Inc., New York, 1963, Chap. 3.) According to Rutherford, the constitution of an atom can be described as follows. In the center of the atom, there is the *nucleus,* which is composed of z protons and n neutrons. The *protons* contribute to both the charge (positive) and the mass of the nucleus, whereas the *neutrons* contribute only to the mass of the nucleus. To keep the atom electrically neutral, there are z electrons moving around the nucleus. (If atoms were not electrically neutral, an enormous electric field would be exhibited by objects of a finite physical dimension. This is contrary to facts.)

Insofar as the material properties (mentioned above) are concerned, the significant number is the number of electrons z, not the number of protons and neutrons, $z + n$. It is the electrons that determine the chemical behavior of an element. The number of electrons that an atom has can be determined from a scattering experiment. A beam of charged particles (say, protons carrying a charge $+e$) is incident on a foil of a chosen target material (Fig. 1.1a). We note that the diameter of an atom is on the order 10^{-10} m, whereas the nuclear dimension is on the order of 10^{-14} m. Since the electron distribution in an atom is well dispersed compared to the nuclear charge distribution, its effect on the incident particle can be ignored. Therefore, the scattering problem is re-

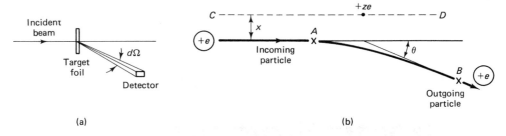

Figure 1.1 Schematic diagram showing the Rutherford scattering experiment: (a) the experimental arrangement and (b) the trajectory of a moving proton in the field of a stationary nucleus having z protons.

duced to a simple electrostatic interaction between a moving charged particle of charge $+e$ and a stationary charged particle of charge $+ze$ (Fig. 1.1b).

The trajectory of the moving particle can be obtained from classical particle mechanics. At large distances from the nucleus, the motion of the particle is little affected by the nucleus; hence the trajectory approaches asymptotically two straight lines at the two ends away from the nucleus. The angle θ formed by these two lines (Fig. 1.1b) defines the angle of scattering (i.e., the angle of deflection of the direction of the charged particle). The relation between θ and x (the distance of the nuclear charge from the asymptotic straight line) is found to be

$$\tan \frac{\theta}{2} = \frac{ze^2}{8\pi\epsilon_0 Kx} \qquad (1.1.1)^{\dagger}$$

where ϵ_0 is the dielectric constant of free space and $K = Mv^2/2$ is the kinetic energy of the moving particle. (The relation can be derived from ordinary mechanics. See, for example, M. Born, *Atomic Physics,* Hafner Publishing Company, New York, 1956, Appendix IX; R. M. Eisberg, *Fundamentals of Modern Physics,* John Wiley & Sons, New York, 1961, pp. 100–106.) The dependences of θ on K and x can be understood qualitatively from the following observations. Most of the deflection occurs in a portion AB of the trajectory when the particle is nearest to the nucleus. The smaller the distance x, the closer the portion AB is to the nucleus and hence the stronger is the electrostatic repulsion. The time during which a particle travels the distance AB depends on the velocity of the particle. A particle having a lower kinetic energy spends more time in the region AB, and hence the overall effect of the nuclear potential on the particle motion is greater.

In the actual scattering experiment, the significant quantity is the ratio of the number of particles collected at a fixed angle θ by the detector to the incident particles. Let $d\Omega$ be the solid angle extended to the detector surface from the target and $S(\theta)$ be the number of particles collected per solid angle. Note that the problem at hand is symmetrical about the axis CD. Particles whose incoming path is inside a circular ring

---

†An alternate terminology for ϵ is permittivity. In this text, dielectric constant and permittivity are used interchangeably.

about the axis CD have the same deflection angle θ. If dA is the area of this ring and ρ is the density of incident particles per unit area, the total number of particles within this ring is $\rho\, dA$. Since these particles are all scattered into the same angle θ,

$$\rho\, dA = S(\theta)\, d\Omega \qquad (1.1.2)$$

Realizing that in polar coordinates

$$dA = 2\pi x\, dx \quad \text{and} \quad d\Omega = 2\pi \sin\theta\, d\theta \qquad (1.1.3)$$

we find the ratio of $S(\theta)/\rho$, called the *differential cross section of scattering* $\sigma(\theta)$, to be

$$\sigma(\theta) = \frac{S(\theta)}{\rho} = \frac{2\pi x}{2\pi \sin\theta}\frac{dx}{d\theta} = \left(\frac{ze^2}{16\pi\epsilon_0 K}\right)^2 \csc^4\frac{\theta}{2} \qquad (1.1.4)$$

[In the original experiment of Rutherford, α particles (He^{2+} ions) were used. In that case, 16 should be replaced by 8.] Equation (1.1.4) is known as the *Rutherford scattering law*.

Now let us apply Eq. (1.1.4) to the scattering experiment. If N is the number of nuclei per square meter of the cross-sectional area of the target foil, each nucleus occupies an effective cross-sectional area of N^{-1} m^2. Consider an incident beam of n_0 particles having a cross-sectional area A_0. Imagine that the beam is divided into many minute beams, each having a cross-sectional area N^{-1} m^2 and each interacting with a different nucleus. The particle density in each minute beam is equal to n_0/A_0. However, there are all together $A_0 N$ such minute beams. Summing the effects of all these minute beams as if the beams were on top of each other, we obtain an effective particle density

$$\rho = \frac{n_0}{A_0}\frac{A_0}{N^{-1}} = n_0 N \qquad (1.1.5)$$

Substituting Eq. (1.1.5) into Eq. (1.1.4), we obtain

$$S(\theta) = n_0 N \sigma(\theta) \qquad (1.1.6)$$

If a detector having an exposure area A_1 normal to the deflected beam is placed at a distance R from the target, we have $A_1 = R^2\, d\Omega$. Therefore, the number n_1 of particles collected at the detector is given by

$$n_1 = S(\theta)\, d\Omega = \frac{n_0 N A_1}{R^2}\left(\frac{ze^2}{16\pi\epsilon_0 K}\right)^2 \csc^4\frac{\theta}{2} \qquad (1.1.7)$$

Knowing N, K, A_1, and R, we can determine the value of z from the plot of n_1/n_0 as a function of θ.

Long before the modern theory of atomic structure was firmly established, it was found that if chemical elements are arranged in a periodic table as shown in Table 1.1, those elements occupying similar places have similar chemical properties. The arrangement of elements in the table was first in the order of increasing atomic weight; however, it was later discovered that elements should be arranged according to the number of electrons that an element possesses. The number z is now called the *atomic number*, as it determines the position that an element occupies in the periodic table. For example, carbon has an atomic number $z = 6$, meaning that carbon has six electrons and occupies the sixth place in the periodic table. Different elements have different atomic numbers.

TABLE 1.1 PERIODIC TABLE OF THE ELEMENTS[a]

Period	I a	I b	II a	II b	III a	III b	IV a	IV b	V a	V b	VI a	VI b	VII a	VII b	VIII a	VIII a	VIII a	VIII b		
I	1 H 1.0079																	2. He 4.003		
II	3 Li 6.94		4 Be 9.02			5 B 10.82		6 C 12.01		7 N 14.01		8 O 16.00		9 F 19.00					10 Ne 20.18		
III	11 Na 22.99		12 Mg 24.32			13 Al 26.97		14 Si 28.06		15 P 30.98		16 S 32.06		17 Cl 35.45					18 Ar 39.94		
IV	19 K 39.09	29 Cu 63.54	20 Ca 40.08	30 Zn 65.38	21 Sc 44.96	31 Ga 69.72	22 Ti 47.90	32 Ge 72.60	23 V 50.95	33 As 74.91	24 Cr 52.01	34 Se 78.96	25 Mn 54.93	35 Br 79.91	26 Fe 55.85	27 Co 58.94	28 Ni 58.69	36 Kr 83.7		
V	37 Rb 85.48	47 Ag 107.88	38 Sr 87.63	48 Cd 112.41	39 Y 88.92	49 In 114.76	40 Zr 91.22	50 Sn 118.70	41 Nb 92.91	51 Sb 121.76	42 Mo 95.95	52 Te 127.61	43 Tc 99	53 I 126.92	44 Ru 101.7	45 Rh 102.91	46 Pd 106.4	54 Xe 131.3		
VI	55 Cs 132.91	79 Au 197.2	56 Ba 137.36	80 Hg 200.61	57–71 *Rare earths*	81 Tl 204.39	72 Hf 178.6	82 Pb 207.21	73 Ta 180.88	83 Bi 209.00	74 W 183.92	84 Po 210	75 Re 186.31	85 At 211	76 Os 190.2	77 Ir 193.1	78 Pt 195.2	86 Rn 222		
VII	87 Fr 223		88 Ra 226.05		89 Ac 227		90 Th 232.12		91 Pa 231		92 U 238.07	93 Np 237	94 Pu 239	95 Am 241	96 Cm 242	97 Bk 246	98 Cf 249	99 Es 254	100 Fm 256	101 Md 256

Rare Earths

VI 57–71															
57 La 138.92	58 Ce 140.13	59 Pr 140.92	60 Nd 144.27	61 Pm 147	62 Sm 150.43	63 Eu 152.0	64 Gd 156.9	65 Tb 159.2	66 Dy 162.46	67 Ho 164.90	68 Er 167.2	69 Tm 169.4	70 Yb 173.04	71 Lu 174.99

[a]The numbers in front of the symbols of the elements denote the atomic numbers; the numbers underneath are the atomic weights.

Although more accurate methods of determining the atomic number were developed later, the Rutherford scattering experiment is of historical importance because it led to the discovery by Rutherford of the constitution of the atom.

1.2 THE BOHR POSTULATES AND THE WILSON–SOMMERFELD RULE

In Section 1.1 we discussed the constitution of an atom as conceived by Rutherford. The Rutherford atom is composed of a small, heavy, positively charged nucleus and one or more extranuclear electrons. This model of an atom has been verified by experiments. Our next task is to see how the classical theories can be applied to analyze the motion of electrons in an atom. For this purpose we use the hydrogen atom as an example. Hydrogen with atomic number $z = 1$ has a nucleus of charge $+e = 1.6 \times 10^{-19}$ C and of mass $M = 1.67 \times 10^{-24}$ g. Since the mass of the electron $m = 9.11 \times 10^{-28}$ g is much smaller than M, the nucleus can be considered as stationary in space. According to the classical theory, the electron that is attracted toward the nucleus by the Coulomb force would describe either an elliptical or a circular orbit about the nucleus, similar to the orbit of the earth about the sun.

For simplicity, we treat only the circular orbit (Fig. 1.2) in the following discussion. A balance of the centrifugal force and the Coulomb force gives the following equation:

$$\frac{mv^2}{r} = \frac{e^2}{4\pi\epsilon_0 r^2} \tag{1.2.1}$$

where r is the radius of the circular orbit and ϵ_0 is the dielectric constant of free space. However, consideration of the balance of mechanical forces alone is not enough. According to the classical electromagnetic theory, the acceleration of a charged particle would lead to the emission of light (electromagnetic radiation). Such a radiation would have a frequency v equal to the frequency of the electron motion, that is, $v = v/2\pi r$.

The total energy $\mathscr{E}$ of the electron is equal to the sum of its kinetic and potential energy, or

$$\mathscr{E} = \frac{mv^2}{2} - \frac{e^2}{4\pi\epsilon_0 r} = -\frac{e^2}{8\pi\epsilon_0 r} \tag{1.2.2}$$

The negative sign in Eq. (1.2.2) accounts for the fact that the electron in the hydrogen atom is in the bound state. The last step in Eq. (1.2.2) is performed through the use of Eq. (1.2.1) in eliminating v. With the loss of energy through radiation, the electron would become more tightly bound to the nucleus. From Eqs. (1.2.2) and (1.2.1) we see that a more negative $\mathscr{E}$ would require a smaller r, which in turn would result in a higher angular frequency ω. Therefore, according to the classical theory, the emitted radiation should show a wide (continuous) range of frequencies. This prediction, however, is not borne out by experiments.

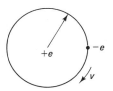

Figure 1.2 Classical model of a hydrogen atom. An electron moves in a circular orbit around a proton.

The spectrum of the radiation emitted by hydrogen consists of sharply defined (discrete) frequencies. It is clear that classical theories alone cannot explain the electron motion in an atom. The first attempt to resolve this difficulty is to seek further restrictions that must be imposed on the electron motion. One of the most important steps toward an explanation of the discrete nature of the radiation spectrum was taken by Bohr, who postulated the existence of stationary states and the Bohr frequency rule. The two Bohr postulates can be stated as follows. An atom can exist only in certain stationary states, each state having a definite energy. An atom can change from a lower-energy state (with energy $\mathscr{E}_1$) to a higher one (with energy $\mathscr{E}_2$) by absorbing a photon of energy $h\nu$ such that

$$h\nu = \mathscr{E}_2 - \mathscr{E}_1 \qquad (1.2.3)$$

Similarly, emission of a photon of the same energy results if the atom returns from the higher ($\mathscr{E}_2$) to the lower ($\mathscr{E}_1$) energy state.

In Eq. (1.2.3), ν is the frequency of the absorbed (or emitted) radiation and h is Planck's constant. In his theoretical analysis of the distribution law for blackbody radiation (which is discussed in detail in Chapter 12), Planck showed that the experimentally observed distribution can be accounted for by postulating that matter (in this case, the blackbody) absorbs or emits energy in units of $h\nu$. The proportionality constant h, which is of dimension energy-time, has a value 6.625×10^{-34} J-s. It is fitting to say that the quantum theory was born with the introduction of Planck's constant h in 1900.

Soon after the successful formulation by Bohr of the two postulates in 1913, Wilson and Sommerfeld independently discovered in 1915 a powerful method of finding the stationary states. They defined stationary states as those for which the following conditions are satisfied:

$$\int p\, dq = nh \qquad (1.2.4)$$

where p is the linear momentum, q the space coordinate of the electron, and n an integer. The condition stated in Eq. (1.2.4) is known as the *Wilson–Sommerfeld rule of quantization*. The validity of Eq. (1.2.4) was tested by Sommerfeld in his analysis of the emission spectra of hydrogen and ionized helium.

Let us now apply Eq. (1.2.4) to the electron motion in a hydrogen atom (Fig. 1.2). The linear momentum p is equal to

$$p = mv \qquad (1.2.5)$$

and the elementary path dq of the electron orbit is given by

$$dq = r\, d\phi \qquad (1.2.6)$$

Substituting Eqs. (1.2.5) and (1.2.6) into Eq. (1.2.4) and integrating the resultant expression over $d\phi$ from 0 to 2π, we find that

$$mvr = \frac{nh}{2\pi} = n\hbar \qquad (1.2.7)$$

The quantity $\hbar = h/2\pi$ is called "h-bar" and is used as often as h in modern physics. Elimination of v from Eqs. (1.2.1) and (1.2.7) yields

$$r = n^2 a \qquad (1.2.8)$$

where a is an atomic unit for distance known as the *Bohr radius* and has the value

$$a = \frac{4\pi\epsilon_0\hbar^2}{me^2} = 0.529 \text{ Å} = 5.29 \times 10^{-11} \text{ m} \qquad (1.2.9)$$

Using the value of r from Eq. (1.2.8) in Eq. (1.2.2), we obtain

$$\mathcal{E} = -\frac{me^4}{2(4\pi\epsilon_0\hbar)^2} \frac{1}{n^2} \qquad (1.2.10)$$

Thus for transitions between any two states n_1 and n_2, the energy of the emitted (or absorbed) radiation is given by

$$h\nu = \mathcal{E}_{n_1} - \mathcal{E}_{n_2} = \frac{me^4}{2(4\pi\epsilon_0\hbar)^2} \left(\frac{1}{n_1^2} - \frac{1}{n_2^2}\right) \qquad (1.2.11)$$

with $n_2 > n_1$. The many emission lines of hydrogen can indeed be identified by assigning different integers for n_1 and n_2 in Eq. (1.2.11). One simple check of Eq. (1.2.11) is to calculate the ionization energy of a hydrogen atom. For the ground state, $n_1 = 1$. For the ionized state, $n_2 = \infty$. Using these values in Eq. (1.2.11), we find a value of $h\nu = 13.6$ electron volts (eV), which agrees well with the experimentally determined value.

1.3 THE SCHRÖDINGER WAVE EQUATION

In Section 1.2 we discussed inadequacy of classical theories in dealing with the electron motion in an atom. The old quantum theory, which is based on the Bohr postulates and the Wilson–Sommerfeld quantization rule, helped greatly to overcome some of the difficulties with classical theories; however, many difficulties remained. For example, although the frequency of the emitted radiation can be predicted from Eq. (1.2.11), the intensity of the radiation cannot be calculated from the old quantum theory. The next giant step in the development of modern physics is the introduction of quantum mechanics.

As classical mechanics furnishes the description of the behavior of large objects, quantum mechanics provides the description of the behavior of matter on an atomic scale. Because all our direct experience and hence our intuition apply only to large objects, it is not surprising that we find quantum mechanics more difficult to understand than classical mechanics. For example, we cannot learn the Wilson–Sommerfeld rule of quantization by association with our direct experience as we learn Newton's law of motion. For this reason, the rule would appear to us to be merely an abstract concept. The correctness of such an abstract concept can be proved only by observing the consequences that result from the imposition of the concept. In the same way, we have to learn quantum-mechanical principles as abstract concepts. These concepts are accepted as valid principles because they give a consistent and correct description of the behavior of matter on an atomic scale. It is in this spirit that we introduce the Schrödinger wave equation in the following discussion.

Quantum mechanics differs from classical mechanics at its outset in the fundamental premises underlying them. Classical mechanics assumes that given the equation of motion, we can have a precise knowledge of the state of a particle in question. For example, we learn in kinetics that we can predict the trajectory of a falling body provided that its initial conditions (regarding its position and velocity) and the force acting on it are known. Quantum mechanics, on the other hand, does not presuppose a complete and precise knowledge of the particle. As a matter of fact, quantum mechanics is built on the premise that it is not possible to predict exactly what will happen in a definite circumstance. Instead of specifying the state of a particle in motion by its position and velocity as we did in Eq. (1.2.1), we define a wave function $\psi(x, y, z)$ such that the probability dW of finding a particle in a volume element $d\tau = dx\, dy\, dz$ is equal to

$$dW = |\psi|^2\, d\tau \qquad (1.3.1)$$

The equation from which information concerning ψ can be derived is the *Schrödinger wave equation*:

$$\left(-\frac{\hbar^2}{2m}\nabla^2 + V \right)\psi = \mathscr{E}\psi \qquad (1.3.2)$$

where the Laplacian operator ∇^2 stands for

$$\nabla^2 = \frac{\partial^2}{\partial x^2} + \frac{\partial^2}{\partial y^2} + \frac{\partial^2}{\partial z^2} \qquad (1.3.3)$$

We consider the Schrödinger wave equation [Eq. (1.3.2)] and the probabilistic interpretation of wave function [Eq. (1.3.1)] as fundamental postulates, similar to the Bohr postulates and the Wilson–Sommerfeld rule of quantization. As postulates, these two equations need no derivation from other principles to justify their existence. As stated earlier, the justification of their existence can only be supported by experiments, which are the sole test of the validity of these postulates. The following observation, however, may be useful to see the physical origin of the terms in Eq. (1.3.2). In the classical expression for energy,

$$\mathscr{E} = K + V = \frac{1}{2m}\left(p_x^2 + p_y^2 + p_z^2 \right) + V \qquad (1.3.4)$$

the term K is the kinetic energy and the term V is the potential energy of a particle. If we replace the classical momentum by a quantum-mechanical operator such that

$$p_x \rightarrow \frac{\hbar}{i}\frac{\partial}{\partial x} \quad , \qquad p_y \rightarrow \frac{\hbar}{i}\frac{\partial}{\partial y} \quad , \qquad p_z \rightarrow \frac{\hbar}{i}\frac{\partial}{\partial z} \qquad (1.3.5)$$

with $i = \sqrt{-1}$, and if we let these operators operate on ψ, we obtain

$$\mathscr{E}\,\psi = (K + V)\psi = \left(-\frac{\hbar^2}{2m}\nabla^2 + V \right)\psi \qquad (1.3.6)$$

which is identical to Eq. (1.3.2).

Let us now recapitulate the quantum-mechanical treatment in contrast to the classical method. It is obvious from Eq. (1.3.2) that the wave function ψ is, in general, a function of the coordinates (x, y, z). If the particle under consideration is an electron (e.g., the same electron as in Fig. 1.2), the electron motion can no longer be defined as precisely as in Eqs. (1.2.1) and (1.2.8), and thus the probability of finding the electron outside the prescribed orbit is zero. The wave function ψ, on the other hand, spreads over a finite region, and hence the probability of finding an electron is nonzero even in regions outside the classically allowed orbit. This smearing of electron orbit is a direct consequence of the notion that the electron orbit cannot be predicted in exact detail. The acceptance of the quantum-mechanical treatment, that is, the two postulates stated in Eqs. (1.3.1) and (1.3.2), necessitates our renunciation of the hope of exact prediction.

At first, the wave function ψ seems to be a rather poor substitute for knowing exactly where the electron is. However, there are many phenomena in nature, about which only the average properties are known and of interest to us. The thermionic emission from a vacuum tube is one example. The current density J resulting from thermionic emission is given by the *Dushman equation:*

$$J = \frac{4\pi m e}{h^3} (kT)^2 \exp\left(-\frac{\mathscr{E}_w}{kT}\right) \tag{1.3.7}$$

where $k = 1.38 \times 10^{-23}$ J/K is the Boltzmann constant, T the temperature (in Kelvin) of the cathode, $\mathscr{E}_w$ the work function (or barrier potential) of cathode surface, and m the electron mass. The fact that we cannot predict exactly when the emission of an electron takes place does not prevent us from deriving Eq. (1.3.7), which is based on the average rate of emission. A similar situation applies to the quantum-mechanical treatment of the atomic system. Even though we do not know in exact details the behavior of the system, we can derive useful and definite information about the average properties of the system.

To illustrate how the average value of a physical quantity is found from a given probability distribution, we use the familiar example of computing the class average in an examination. The average score is computed from the data sheet of Fig. 1.3 as follows:

$$\frac{1}{5 + 6 + 7 + 8 + 10} (6 \times 45 + 8 \times 55 + 10 \times 65$$

$$+ 7 \times 75 + 5 \times 85) = 64 \tag{1.3.8}$$

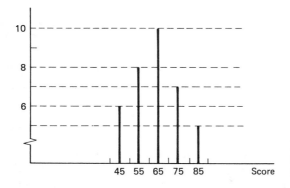

Figure 1.3 Score sheet in an examination. There are five students having a score of 85, and so on.

Note that the probability of finding a student having a score of 65 is $10/(5 + 6 + 7 + 8 + 10)$. Thus if W_i is the probability of finding a student having a score f_i, the average score $\langle f \rangle$ of the class is simply equal to

$$\langle f \rangle = \sum_i f_i W_i \qquad (1.3.9)$$

Equation (1.3.9) applies only to a discrete distribution. (This means that f_i takes discrete values of 45, 55, and so on.) For a continuous distribution, Eq. (1.3.9) must be replaced by

$$\langle f \rangle = \int f \, dW \qquad (1.3.10)$$

The quantity $\langle f \rangle$ is generally known as the *expectation value* of f.

Let us now illustrate the use of Eq. (1.3.10) in the description of an atomic system. First, we can define an electron density function

$$\rho(x, y, z) = e|\psi(x, y, z)|^2 \qquad (1.3.11)$$

If only a single electron is involved, the expectation value of the charge must be e. In other words,

$$\iiint_{-\infty}^{\infty} \rho(x, y, z) \, dx \, dy \, dz = e \quad \text{or} \quad \iiint_{-\infty}^{\infty} |\psi(x, y, z)|^2 \, dx, \, dy \, dz = 1 \qquad (1.3.12)$$

Another quantity of interest is the position of an electron. Although we do not know exactly where an electron is in an atom, we still can find the average distance $\langle r \rangle$ away from the nucleus as follows:

$$\langle r \rangle = \iiint_{-\infty}^{\infty} r|\psi(x, y, z)|^2 \, dx \, dy \, dz \qquad (1.3.13)$$

We should point out that the average value of f^{-1} is, in general, not equal to $\langle f \rangle^{-1}$. Thus the average value of the potential energy of an electron in a nuclear potential of $e^2/4\pi\epsilon_0 r$ is

$$\langle V \rangle = \iiint_{-\infty}^{\infty} \frac{e^2}{4\pi\epsilon_0 r} |\psi(x, y, z)|^2 \, dx \, dy \, dz \qquad (1.3.14)$$

and not $e^2/4\pi\epsilon_0\langle r \rangle$. To find the expectation values of the momentum and the kinetic energy of an electron is much more involved, because as indicated in Eq. (1.3.5), both quantities are represented by differential operators. We postpone the discussion of how the values of $\langle p \rangle$ and $\langle K \rangle$ can be found until we have had chances of getting a little more acquainted with the Schrödinger wave equation and its solution.

1.4 THE HYDROGEN ATOM

In this section we analyze the electron motion in a hydrogen atom based on the Schrödinger wave equation. The analysis is a necessary step toward understanding the structure of atoms. Let us refer to Eq. (1.3.2). Because the potential-energy term $V \, (= -e^2/$

$4\pi\epsilon_0 r$) has a $1/r$ dependence, it is desirable to express Eq. (1.3.2) in spherical coordinates r, θ, and ϕ (Fig. 1.4) as follows:

$$\frac{\hbar^2}{2m}\left[\frac{1}{r^2}\frac{\partial}{\partial r}\left(r^2\frac{\partial}{\partial r}\right) + \frac{1}{r^2\sin\theta}\frac{\partial}{\partial\theta}\left(\sin\theta\frac{\partial}{\partial\theta}\right)\right.$$

$$\left. + \frac{1}{r^2\sin^2\theta}\frac{\partial^2}{\partial\phi^2}\right]\psi + \left(\mathscr{E} + \frac{e^2}{4\pi\epsilon_0 r}\right)\psi = 0 \tag{1.4.1}$$

Equation (1.4.1) is a partial differential equation with ψ as the dependent variable and r, θ, and ϕ as the independent variables. For those who have not been exposed to partial differential equations, Eq. (1.4.1) may look formidable indeed. Fortunately, Eq. (1.4.1) belongs to a class of partial differential equations that occur very frequently in physics and engineering problems and of which the solutions have been worked out by mathematicians.

The standard way of finding the solution of Eq. (1.4.1) is by the method of separation of variables. We propose a solution of the following form:

$$\psi(r, \theta, \phi) = R(r)\Theta(\theta)\Phi(\phi) \tag{1.4.2}$$

which is a product of three separate functions, each of which depends only on one independent variable. Substituting Eq. (1.4.2) into Eq. (1.4.1), we obtain

$$\frac{\Theta\Phi}{r^2}\frac{\partial}{\partial r}\left(r^2\frac{\partial R}{\partial r}\right) + \frac{R\Phi}{r^2\sin\theta}\frac{\partial}{\partial\theta}\left(\sin\theta\frac{\partial\Theta}{\partial\theta}\right) + \frac{R\Theta}{r^2\sin^2\theta}\frac{\partial^2\Phi}{\partial\phi^2}$$

$$+ \frac{2m}{\hbar^2}\left(\mathscr{E} + \frac{e^2}{4\pi\epsilon_0 r}\right)R\Theta\Phi = 0 \tag{1.4.3}$$

In Eq. (1.4.3) we take outside the differentiation the part of ψ that does not depend on the independent variable with respect to which the differentiation is taken.

Dividing Eq. (1.4.3) by $R\Theta\Phi$, multiplying Eq. (1.4.3) by $r^2\sin\theta$, and rearranging terms, we have

$$\frac{\sin^2\theta}{R}\frac{d}{dr}\left(r^2\frac{dR}{dr}\right) + \frac{\sin\theta}{\Theta}\frac{d}{d\theta}\left(\sin\theta\frac{d\Theta}{d\theta}\right) + \frac{2m}{\hbar^2}r^2\sin^2\theta\left(\mathscr{E} + \frac{e^2}{4\pi\epsilon_0 r}\right)$$

$$= -\frac{1}{\Phi}\frac{d^2\Phi}{d\phi^2} \tag{1.4.4}$$

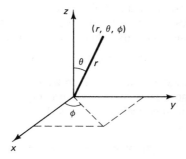

Figure 1.4 Spherical coordinates. The variable ϕ takes values from 0 to 2π, the variable θ from 0 to π (*not* from 0 to 2π), and the variable r from 0 to ∞ (*not* from $-\infty$ to ∞).

By virtue of the assumption that $\Phi(\phi)$ is a function of ϕ only, the right-hand side of Eq. (1.4.4) depends on ϕ only. Applying similar reasoning to the left-hand side, we find that it must depend on r and θ only. The only way that both conditions can be satisfied is to set both sides equal to a constant. Letting this constant be m^2, we have

$$\frac{d^2\Phi}{d^2\phi} = -m^2\Phi \qquad (1.4.5)$$

$$\frac{1}{R}\frac{d}{dr}\left(r^2\frac{dR}{dr}\right) + \frac{2m}{\hbar^2}r^2\left(\mathscr{E} + \frac{e^2}{4\pi\epsilon_0 r}\right)$$

$$= -\frac{1}{\Theta\sin\theta}\frac{d}{d\theta}\left(\sin\theta\frac{d\Theta}{d\theta}\right) + \frac{m^2}{\sin^2\theta} \qquad (1.4.6)$$

Note that the left-hand side of Eq. (1.4.6) depends only on r, whereas the right-hand side of Eq. (1.4.6) depends only on θ. Therefore, we must set both sides equal to a constant. Letting this new constant be $l(l + 1)$, we obtain

$$\frac{1}{\sin\theta}\frac{d}{d\theta}\left(\sin\theta\frac{d\Theta}{d\theta}\right) - \frac{m^2}{\sin^2\theta}\Theta = -l(l + 1)\Theta \qquad (1.4.7)$$

$$\frac{1}{r^2}\frac{d}{dr}\left(r^2\frac{dR}{dr}\right) + \frac{2m}{\hbar^2}\left(\mathscr{E} + \frac{e^2}{4\pi\epsilon_0 r}\right)R = \frac{l(l + 1)}{r^2}R \qquad (1.4.8)$$

Equations (1.4.5), (1.4.7), and (1.4.8) are three ordinary differential equations from which the solutions are to be found. It may seem mysterious that out of the infinite possible values for the constants in Eqs. (1.4.4) and (1.4.6), we choose m^2 and $l(l + 1)$. Equations (1.4.5), (1.4.7), and (1.4.8) would have different solutions if these constants were set in different forms. Only a certain type of solution yields a wave function ψ that is physically meaningful. The choice of m^2 and $l(l + 1)$ is to ensure a physically reasonable solution. We explain what we mean by *reasonable solution* in the following discussion.

Let us refer to Eq. (1.3.1) for the physical interpretation of the wave function. If $|\psi|^2 \, d\tau$ indeed represents the probability of finding an electron in a volume element $d\tau$, the integral $\int |\psi|^2 \, d\tau$ over the whole volume must be equal to the number of electrons involved and hence must be finite. Furthermore, a probability by its very definition can take one and only one value for the occurrence of a specific event. Thus the wave function ψ must be single valued. These two conditions on ψ are auxiliary postulates that must be taken into account together with the two fundamental postulates of quantum mechanics stated in Eqs. (1.3.1) and (1.3.2). These two auxiliary postulates, in essence, replace the Wilson–Sommerfeld rule of quantization in the old quantum theory.

The two auxiliary postulates stated above limit our selection of the two constants in separating the two sides of Eqs. (1.4.4) and (1.4.6). Let us now examine the ϕ dependence of the wave function. The solution of Eq. (1.4.5) is

$$\Phi = A \exp(\pm im\phi) \qquad (1.4.9)$$

where A is a proportionality constant to be fixed later. To examine whether the function Φ satisfies the auxiliary postulates or not, we use Eq. (1.4.2) and examine each part separately. Realizing that in spherical coordinates (Fig. 1.4) the volume elements $d\tau$ are given by

$$d\tau = r^2 \, dr \sin\theta \, d\theta \, d\phi \qquad (1.4.10)$$

we can write

$$\int |\psi|^2 \, d\tau = \int_0^\infty |R(r)|^2 r^2 \, dr \int_0^\pi |\Theta(\theta)|^2 \sin \theta \, d\theta \int_0^{2\pi} |\Phi(\phi)|^2 \, d\phi \quad (1.4.11)$$

Using Eq. (1.4.9) in Eq. (1.4.11), we can easily show that $\int |\Phi|^2 \, d\phi$ is finite. For the function $\Phi(\phi)$ to be single valued at $\phi = 0$ (which is identical to $\phi = 2\pi$), that is, $\Phi(0) \equiv \Phi(2\pi)$, the parameter m must be an integer. Had we changed $-m^2$ to m^2 in Eq. (1.4.5), the solution would not be a periodic function and hence would not satisfy the condition of single valuedness.

Equation (1.4.7) is known in mathematics as the *associated Legendre equation*. The solution of Eq. (1.4.7) can be expressed in terms of an infinite series that diverges at $\theta = 0$ and π if l is a noninteger. This type of solution is, of course, not acceptable. However, the infinite series terminates and thus becomes a polynomial if l is an integer such that

$$l = |m|, \quad |m| + 1, \quad |m| + 2, \quad \ldots \quad (1.4.12)$$

The polynomial is known as the *associated Legendre polynomial,* which is designated as $P_l^{|m|}(\cos \theta)$. Thus we write

$$\Theta(\theta) = P_l^{|m|} (\cos \theta) \quad (1.4.13)$$

The superscript $|m|$ and the subscript l specify the two integers in Eq. (1.4.7). Only for integral l and m such that $l \geq |m|$, the solution of Eq. (1.4.7) behaves properly at $\theta = 0$ and $\theta = \pi$. Therefore, Eq. (1.4.13) is the only acceptable solution of Eq. (1.4.7).

Similarly, the radial function of Eq. (1.4.8) has an acceptable solution only under certain conditions. First, the energy $\mathscr{E}$ can only take certain discrete values:

$$\mathscr{E} = - \frac{me^4}{2(4\pi\epsilon_0\hbar)^2} \frac{1}{n^2} \quad (1.4.14)$$

where n is a positive integer. Second, the integer n must be either equal to or greater than $l + 1$. If these conditions are met, the solution of Eq. (1.4.8) can be written as

$$R(r) = \left(\frac{r}{na} \right)^l \exp \left(- \frac{r}{na} \right) L_{n+l}^{2l+1} \left(\frac{2r}{na} \right) \quad (1.4.15)$$

where a is the Bohr radius. The function $L_{n+l}^{2l+1}(2r/na)$ is the *associated Laguerre polynomial* of order $2l + 1$ and of degree $n - l - 1$. In some books, the notation L_{n-l-1}^{2l+1} instead of L_{n+l}^{2l+1} is used so that the subscript $n - l - 1$ denotes directly the degree of the polynomial. Here we follow the notation used by Pauling and Wilson and by Condon and Shortley. (L. Pauling and E. B. Wilson, *Introduction to Quantum Mechanics,* McGraw-Hill Book Company, New York, 1935; E. U. Condon and G. H. Shortley, *The Theory of Atomic Spectra,* Cambridge University Press, Cambridge, 1959). In their notation, the degree of the polynomial is equal to the difference between the subscript $(n + l)$ and the superscript $(2l + 1)$.

Let us summarize the essential points in the discussion above. The wave function $\psi(r, \theta, \phi)$ and hence the electron motion in a hydrogen atom are specified by three numbers, n, l, and m. Symbolically, we write

$$\psi_{n,l,m} = R_{n,l}(r)\Theta_{l,m}(\theta)\Theta_m(\phi) \quad (1.4.16)$$

Since $|\psi|^2$ represents a probability distribution function, the wave function ψ must be single valued and bounded. These two requirements limit the numbers, n, l, and m to integers. Furthermore, for a fixed integer n, the allowed values of l are

$$l = 0, 1, 2, \ldots, (n - 1) \tag{1.4.17}$$

Both n and l are positive integers. For a fixed value of l, the allowed values of m are

$$m = -l, -(l - 1), \ldots, 0, \ldots, (l - 1), l \tag{1.4.18}$$

The positive and negative values of m account for the $\pm$ signs in Eq. (1.4.9).

The integer n is known as the *principal quantum number;* the integer l, the *azimuthal quantum number;* and the integer m, the *magnetic quantum number.* The following letter code is generally used by spectroscopists for the values of l:

$$l = 0, 1, 2, 3, 4, 5, 6, 7 \tag{1.4.19}$$
$$s, p, d, f, g, h, i, k$$

Further, the number that precedes the letter is the value of n. Thus, if we refer to a $3d$ electron, we mean an electron that is in the state with $n = 3$ and $l = 2$.

The reader who has not been exposed to the associated Legendre and Laguerre polynomials may be anxious to know what they look like. We list here a selected number of these polynomials:

$$\Theta_{0,0} = 1$$

$$\Theta_{1,0} = \cos\theta, \qquad\qquad \Theta_{1,\pm 1} = \sin\theta \tag{1.4.20}$$

$$\Theta_{2,0} = 3\cos^2\theta - 1, \qquad \Theta_{2,\pm 1} = \sin\theta\cos\theta, \qquad \Theta_{2,\pm 2} = \sin^2\theta$$

$$R_{1,0} = \exp\left(-\frac{r}{a}\right)$$

$$R_{2,0} = \left(2 - \frac{r}{a}\right)\exp\left(-\frac{r}{2a}\right), \qquad R_{2,1} = \frac{r}{a}\exp\left(-\frac{r}{2a}\right)$$

$$R_{3,0} = \left(27 - \frac{18r}{a} + 2\frac{r^2}{a^2}\right)\exp\left(-\frac{r}{3a}\right) \tag{1.4.21}$$

$$R_{3,1} = \frac{r}{a}\left(6 - \frac{r}{a}\right)\exp\left(-\frac{r}{3a}\right), \qquad R_{3,2} = \left(\frac{r}{a}\right)^2\exp\left(-\frac{r}{3a}\right)$$

For a tabulation of these polynomials with indices n, l, and m other than those given above, the reader is referred to the book by Pauling and Wilson. The reader can readily verify that the polynomials of Eq. (1.4.20) and those of Eq. (1.4.21) indeed satisfy, respectively, Eqs. (1.4.7) and (1.4.8). Furthermore, these polynomials do not have any singularities, and thus their values are finite.

Let us now illustrate how we obtain the wave function for a specific atomic state. For the $1s$ state, $n = 1$, $l = 0$, and thus m must be zero as required by Eq. (1.4.18). Combining Φ_0 of Eq. (1.4.9), $\Theta_{0,0}$ of Eq. (1.4.20), and $R_{1,0}$ of Eq. (1.4.21) gives

$$\psi_{1,0,0} = \frac{1}{\sqrt{\pi a^3}}\exp\left(-\frac{r}{a}\right) \tag{1.4.22}$$

which is the only wave function for the $1s$ state. The reason for choosing $A = (\pi a^3)^{-1/2}$ in Eq. (1.4.22) will be explained shortly. For the $2p$ state, there exist three wave functions:

$$\psi_{2,1,1} = \frac{-1}{\sqrt{64\pi a^3}} \frac{r}{a} \exp\left(-\frac{r}{2a}\right) (\sin\theta) \exp(i\phi)$$

$$\psi_{2,1,0} = \frac{1}{\sqrt{32\pi a^3}} \frac{r}{a} \exp\left(-\frac{r}{2a}\right) (\cos\theta) \qquad (1.4.23)$$

$$\psi_{2,1,-1} = \frac{1}{\sqrt{64\pi a^3}} \frac{r}{a} \exp\left(-\frac{r}{2a}\right) (\sin\theta) \exp(-i\phi)$$

because of three possible choices (1, 0, or -1) for the value of m. Note that since $l = 1$, the p state must start with $n = 2$ according to Eq. (1.4.17). This means that we have $2p, 3p, \ldots$ states but no $1p$ state. Similarly, the d state must start with $n = 3$. The d state, irrespective of whether the state is $3d$, $4d$, or $5d$, has five different wave functions because of five possible choices (2, 1, 0, -1, or -2) for the value of m.

Other important properties of the wave function are that they are normalized and orthogonal to each other. According to Eq. (1.3.1), the total probability of finding an electron in a given state is equal to

$$W = \int_0^\infty \int_0^\pi \int_0^{2\pi} \psi_{n,l,m}^* \psi_{n,l,m} r^2 \, dr \, \sin\theta \, d\theta \, d\phi \qquad (1.4.24)$$

where the asterisk indicates the complex conjugate. Note that the wave function is in general complex, and thus $|\psi|^2 = \psi^*\psi$. As required by one of the Bohr postulates, an electron can occupy only one of the quantum states. Once the electron is in that state, the total probability of finding the electron over all space coordinates of that state must be unity. In other words,

$$\int_0^\infty \int_0^\pi \int_0^{2\pi} |\psi|^2 r^2 \, dr \, \sin\theta \, d\theta \, d\phi = 1 \qquad (1.4.25)$$

A wave function that satisfies Eq. (1.4.25) is said to be *normalized*. The constants in Eqs. (1.4.22) and (1.4.23) are chosen so as to make the wave functions normalized. By the same Bohr postulate, an electron is forbidden to change back and forth freely between any two atomic states without the absorption or emission of radiation. Therefore, it seems reasonable to expect that

$$\int_0^\infty \int_0^\pi \int_0^{2\pi} \psi_{n,l,m}^* \psi_{n',l',m'} r^2 dr \, \sin\theta \, d\theta \, d\phi = 0 \qquad (1.4.26)$$

unless $n' = n$, $l' = l$, and $m' = m$. The two wave functions that satisfy Eq. (1.4.26) are said to be *orthogonal* to each other. The solutions of Eq. (1.4.1) and hence all the atomic wave functions have this property.

Finally, let us compare the results based on quantum mechanics with those derived from old quantum theory. First, we see that the energy expression [Eqs. (1.2.10) and

(1.4.14)] are identical. Next we compute the probability distribution of the electron. Within a spherical shell of volume $4\pi r^2 \, dr$, the probability of finding the electron is

$$dW = \int_0^\pi \int_0^{2\pi} |\psi|^2 r^2 \, dr \sin\theta \, d\theta \, d\phi \qquad (1.4.27)$$

For simplicity, we choose the ground state, that is, the lowest-energy state with $n = 1$, $l = 0$, and $m = 0$. Substituting Eq. (1.4.22) into Eq. (1.4.27) and defining a density function $D(r) = dW/dr$, we find that

$$D(r) = 4\pi r^2 |\psi_{1,0,0}|^2 = \frac{4r^2}{a^3} \exp\left(-\frac{2r}{a}\right) \qquad (1.4.28)$$

It is obvious that $D(r) \, dr$ is the probability that the electron lies within the spherical shell of volume $4\pi r^2 \, dr$.

The top figure of Fig. 1.5 is a plot of Eq. (1.4.28) as a function of r/na with $n = 1$. The other two curves are the corresponding plots for the 2s and 3s states. The vertical dashed lines represent the radii of the various orbits as calculated from Eq. (1.2.8). By comparison, the solid lines are the average distances of the electron from the nucleus for these states as calculated from Eq. (1.3.13), or

$$\langle r_{n,0,0} \rangle = \int_0^\infty \int_0^\pi \int_0^{2\pi} r |\psi_{n,0,0}|^2 r^2 \, dr \sin\theta \, d\theta \, d\phi \qquad (1.4.29)$$

It is to be noted that at distances far from the nucleus, the electron distribution drops exponentially with distance. Thus, from Fig. 1.5, we see that quantum mechanics predicts an atom size of the order of 10^{-10} m, a result not far from the semiclassical result. The significant difference is that the quantum-mechanical electron distribution spreads diffusely in space instead of being confined to a certain orbit.

As stated earlier, one of the mysteries challenging the classical theory is what prevents the electron from falling on top of the nucleus. This question is answered in quantum mechanics by another postulate, known as the *uncertainty principle*, which says that the position and the momentum of a particle cannot be determined simultaneously. More precisely, the uncertainty Δx in the value of the position and the uncer-

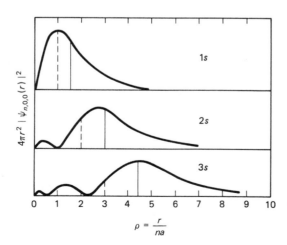

Figure 1.5 Electron distribution functions $D(r) = 4\pi r^2 |\psi_{n,0,0}|^2$ for a hydrogen atom. The abscissa is $\rho = r/na$, where a is the Bohr radius and n is the principal quantum number. The three curves are for $n = 1$, $n = 2$, and $n = 3$. The dashed lines represent the orbital radii based on the old quantum theory, and the solid lines represent the expectation values $\langle r \rangle$ based on quantum mechanics.

tainty Δp in the value of the corresponding momentum are such that the value of their product has a lower bound $\hbar/2$, which means that

$$\Delta p \, \Delta x \geq \frac{\hbar}{2} \tag{1.4.30}$$

If an electron fell into the nucleus, the position of the electron would be of nuclear dimension (of the order of 10^{-14} m, in contrast to the atomic dimension of 10^{-10} m). In other words, Δx would become so small as to require a tremendously large Δp in Eq. (1.4.30). This large value of p then would make the electron to have sufficient kinetic energy to break away from the nucleus. The spreading of the electron distribution in Fig. 1.5 can be understood on the same basis. Based on Eq. (1.4.30), an electron cannot be confined to a certain orbit with precisely defined position and momentum.

1.5 SPIN AND EXCLUSION PRINCIPLE

In Section 1.4 we discussed the motion of the electron in a hydrogen atom. However, experimental evidences indicate that the quantum description of an electron state based on three quantum numbers n, l, and m is still incomplete. For example, in the Stern–Gerlach experiment, a beam of neutral hydrogen atom is sent into an inhomogeneous magnetic field B_z. If an atom possesses a magnetic dipole of moment u, the magnetic energy of the dipole is

$$\mathscr{E} = -\mathbf{B} \cdot \mathbf{u} = -B_z u_z \tag{1.5.1}$$

In the presence of an inhomogeneous field, the atom experiences a force

$$F_z = -\frac{\partial \mathscr{E}}{\partial z} = u_z \frac{dB_z}{dz} \tag{1.5.2}$$

The magnetic dipole moment u is related to the angular momentum L of an electron. Thus, to understand the physical significance of the experiment, we must know how the angular momentum of an electron can be found.

Classically, the angular momentum vector $\mathbf{L}$ is defined as

$$\mathbf{L} = \mathbf{r} \times \mathbf{p} \tag{1.5.3}$$

Thus the z component of the angular momentum is equal to

$$L_z = x p_y - y p_x \tag{1.5.4}$$

According to our discussion in Section 1.3, the quantum-mechanical counterpart of L_z is

$$\langle L_z \rangle = \int\!\!\!\int\!\!\!\int_{-\infty}^{\infty} \psi^* \frac{\hbar}{i} \left(x \frac{\partial}{\partial y} - y \frac{\partial}{\partial x} \right) \psi \, dx \, dy \, dz \tag{1.5.5}$$

In obtaining Eq. (1.5.5), the classical momenta p_y and p_x of Eq. (1.5.4) are replaced by the corresponding momentum operator. Similar expressions are also obtained for L_x and L_y. Evaluation of Eq. (1.5.5) and similar expressions yields the following results:

$$\langle L_z \rangle = m\hbar \tag{1.5.6}$$

$$\langle L^2 \rangle = \langle L_x^2 + L_y^2 + L_z^2 \rangle = \hbar^2 l(l + 1) \tag{1.5.7}$$

For hydrogen atoms in the ground state, $l = 0$ and $m = 0$. If the angular momentum of the electron were totally accounted for by the orbital motion (i.e., the motion around the nucleus), we would have $u_z = 0$ in Eq. (1.5.2), and hence no deflection of the beam would be expected. This is contrary to the experimental finding.

Actually, the original beam is split into two separate beams in an inhomogeneous magnetic field. The answer to this puzzling result was given by Uhlenbeck and Goudsmit, who postulated the following hypothesis. In addition to its motion around the nucleus, an electron also possesses rotational freedom about an axis through its charged body. Such an additional property of an electron is called the *spin*. From the deflection of the beam, the magnetic moment u_z and in turn the amount of angular momentum associated with the spinning motion can be determined.

The angular momentum associated with the spinning motion of an electron can be expressed, in a manner similar to Eqs. (1.5.6) and (1.5.7), in terms of two quantum numbers s and s_z as follows:

$$\langle L^2 \rangle = \hbar^2 s(s + 1) \tag{1.5.8}$$

$$\langle L_z \rangle = s_z \hbar \tag{1.5.9}$$

The value of s is found experimentally to be $\frac{1}{2}$. The relation between s_z and s is also similar to that between m and l. For a given l, the maximum and minimum values of m are $\pm l$, and the changes between successive m's are ± 1. Applying the same principle to s_z, we find that the only possible values for s_z are $s_z = +\frac{1}{2}$ and $s_z = -\frac{1}{2}$. It is the existence of two possible values of s_z that gives rise to two different values of u_z in Eq. (1.5.2) and thus accounts for the splitting of the hydrogen beam.

In summary, the motion of an electron in an atom consists of two parts: the orbital motion around the nucleus and the spinning motion. It has been shown by Dirac that the electron spin is a purely relativistic property of an electron. For our present discussion of atomic theory, however, the classical concept of an electron having a spinning charge distribution can still be retained without special inquiry into the origin of spin. In other words, an electron can be visualized as a small solid body rather than a point, carrying out a rotation about its own axis similar to that of a spinning top. The angular momentum of the spinning motion is specified by the quantum number s_z ($= \pm\frac{1}{2}$). Since the value of s is always $\frac{1}{2}$, it need not be specifically stated. Therefore, the total motion of an electron is specified by four quantum numbers n, l, m, and s_z.

Now we have four quantum numbers; our next task is the assignment of these numbers to each electron in an atom. The law that regulates the assignment is known as the *Pauli exclusion principle*, which states that no two electrons can have the same set of four quantum numbers. As we discuss in Section 1.6, the shell structure of atoms can be understood only in the light of this principle.

For our present discussion, let us apply the exclusion principle to a helium atom, which is next to hydrogen in the periodic table (Table 1.1). In the ground state (the lowest-energy state), the two electrons of helium belong to the $1s$ configuration. Since these two electrons have the same orbital quantum numbers $n = 1$, $l = 0$, and $m = 0$, their spin quantum numbers must be different, that is, one having $s_z = -\frac{1}{2}$ and the other having $s_z = +\frac{1}{2}$. Note that the two combinations

$$n = 1, \quad l = 0, \quad m = 0, \quad s_z = +\tfrac{1}{2}$$

$$n = 1, \quad l = 0, \quad m = 0, \quad s_z = -\tfrac{1}{2}$$

are the only possible combinations of the four quantum numbers. Therefore, the maximum number of electrons that can be assigned to the $1s$ level is two. The same applies to other ns levels. When this maximum occupancy is achieved in an s level, no more electrons can be assigned to that level.

The discussion above can be extended to other atomic states, such as p and d states. For p states, $l = 1$. Out of three possible choices (1, 0, or -1) for m and two possible choices ($+\frac{1}{2}$ or $-\frac{1}{2}$) for s_z, there are, all together, six combinations. Therefore, the maximum number of electrons that can be accommodated in a p level is 6. In general, for a given l, there are $(2l + 1)$ values of m. Counting the two possible spin orientations, there are all together $2(2l + 1)$ different modes of electron motion in an atom. On the basis of the exclusion principle, only one electron can be assigned to a particular mode. The state of maximum occupancy is reached if all the $2(2l + 1)$ different modes have been assigned to $2(2l + 1)$ electrons. Therefore, for the nd level, which starts with $n = 3$, the maximum number of electrons is 10.

1.6 THE PERIODIC SYSTEM

The periodic table of chemical elements (Table 1.1) has long been useful in studying material properties. Elements occupying similar places, to a large extent, have similar chemical properties. Notable examples of these elements are the alkali metals, the halogens, and the inert gases. The periodic table has also aided in the search for new materials by the semiconductor industry. Since the early development of semiconductor diodes and transistors, Si, GaAs, InSb, and many other compounds have joined Ge to appear on the list of useful semiconductor materials. Referring to Table 1.1, we see that Ge and Si both belong to the group IVb of elements, whereas GaAs and InSb are compounds of group IIIb and group Vb elements. As discussed later, all these materials have similar electrical properties. To explain this fact as well as to understand the origin of the periodic arrangement of elements, we must study the periodic table in conjunction with the electronic structure of atoms given in Table 1.2.

As discussed in Section 1.1, the Rutherford atom consists of a nucleus carrying a charge of $+ze$ and a revolving electron cloud carrying a charge of $-ze$ where z is the atomic number. If we neglect the electrostatic (Coulomb) interaction between the electrons, each electron can be considered as moving independently in a nuclear potential of $ze/4\pi\epsilon_0 r$. In other words, the result of our analysis in Section 1.4 for a hydrogen atom is also applicable to other atoms provided that we replace e^2 by ze^2 in the potential-energy V term in Eq. (1.4.1).

Many physics as well as engineering problems defy an exact solution simply because the problems are too complicated. Also, if we attempt to solve such problems exactly, we find ourselves deeply involved in mathematics and thus often lose physical insight. The general approach for treating problems of this kind is known as *successive approximation*. The important terms in an equation are treated first. In the present case, since the electron–electron interaction is much weaker than the electron–nucleus interaction, the terms representing the former are ignored in the Schrödinger equation. The solution thus obtained is the first-order approximate solution. Improvement of the results can be made later, when necessary, by taking into account the electron–electron interaction. Such a refinement, however, is not needed in our present discussion of the periodic system.

TABLE 1.2 ATOMIC STRUCTURE OF THE ELEMENTS

z	Element		$\mathscr{E}_i$ (eV)	Inner shells	$1s$ / $2s$ / $3s$	Outer shells $2p$ / $3p$	Atomic radius (Å)	Electro-negativity (Pauling)
1	Hydrogen	(H)	13.59		1		0.79	2.20
2	Helium	(He)	24.56		2		0.49	—
3	Lithium	(Li)	5.40	←—— (He) ——→	1	—	2.05	0.98
4	Beryllium	(Be)	9.32		2	—	1.40	1.57
5	Boron	(B)	8.28		2	1	1.17	2.04
6	Carbon	(C)	11.27		2	2	0.91	2.55
7	Nitrogen	(N)	14.55		2	3	0.75	3.04
8	Oxygen	(O)	13.62		2	4	0.65	3.44
9	Fluorine	(F)	17.43		2	5	0.57	3.98
10	Neon	(Ne)	21.56		2	6	0.51	—
11	Sodium	(Na)	5.14	←—— (Ne) ——→	1	—	2.23	0.93
12	Magnesium	(Mg)	7.64		2	—	1.73	1.31
13	Aluminum	(Al)	5.97		2	1	1.82	1.61
14	Silicon	(Si)	8.15		2	2	1.46	1.90
15	Phosphorus	(P)	10.49		2	3	1.23	2.19
16	Sulfur	(S)	10.36		2	4	1.09	2.58
17	Chlorine	(Cl)	12.90		2	5	0.97	3.16
18	Argon	(Ar)	15.76		2	6	0.88	—

#	Element		IE	Core	3d	4s	4p			EN
19	Potassium	(K)	4.34	(Ar)	—	1			2.77	0.82
20	Calcium	(Ca)	6.11	(Ar)	—	2			2.23	1.00
21	Scandium	(Sc)	6.54	(Ar)	1	2			2.09	1.36
22	Titanium	(Ti)	6.84	(Ar)	2	2			2.00	1.54
23	Vanadium	(V)	6.71	(Ar)	3	2			1.92	1.63
24	Chromium	(Cr)	6.74	(Ar)	5	1			1.85	1.66
25	Manganese	(Mn)	7.43	(Ar)	5	2			1.79	1.55
26	Iron	(Fe)	7.83	(Ar)	6	2			1.72	1.83
27	Cobalt	(Co)	7.84	(Ar)	7	2			1.67	1.88
28	Nickel	(Ni)	7.63	(Ar)	8	2			1.62	1.91

#	Element		IE	Core	3d	4s	4p			EN
29	Copper	(Cu)	7.72	(Ar)	10	1	—		1.57	1.90
30	Zinc	(Zn)	9.39	(Ar)	10	2	—		1.53	1.65
31	Gallium	(Ga)	5.97	(Ar)	10	2	1		1.81	1.81
32	Germanium	(Ge)	7.90	(Ar)	10	2	2		1.52	2.01
33	Arsenic	(As)	9.81	(Ar)	10	2	3		1.33	2.18
34	Selenium	(Se)	9.73	(Ar)	10	2	4		1.22	2.55
35	Bromine	(Br)	11.81	(Ar)	10	2	5		1.12	2.96
36	Krypton	(Kr)	14.00	(Ar)	10	2	6		1.03	—

#	Element		IE	Core	4f	4d	5s			EN
37	Rubidium	(Rb)	4.17	(Kr)	—	—	1		2.98	0.82
38	Strontium	(Sr)	5.69	(Kr)	—	—	2		2.45	0.95
39	Yttrium	(Y)	6.5	(Kr)	—	1	2		2.27	1.22
40	Zirconium	(Zr)	6.8	(Kr)	—	2	2		2.16	1.33
41	Niobium	(Nb)	6.9	(Kr)	—	4	1		2.08	1.60
42	Molybdenum	(Mo)	7.1	(Kr)	—	5	1		2.01	2.16
43	Technetium	(Tc)	7.3	(Kr)	—	6	1		1.95	1.9
44	Ruthenium	(Ru)	7.4	(Kr)	—	7	1		1.89	2.2
45	Rhodium	(Rh)	7.5	(Kr)	—	8	1		1.83	2.28
46	Palladium	(Pd)	8.3	(Kr)	—	10	—		1.79	2.20

TABLE 1.2 Continued

First group (z = 47–54):

z	Element	$\mathscr{E}_i$ (eV)	Inner shells	4f	5s	5p	Atomic radius (Å)	Electro-negativity (Pauling)
47	Silver (Ag)	7.58		—	1	—	1.75	1.93
48	Cadmium (Cd)	8.99		—	2	—	1.71	1.69
49	Indium (In)	5.79	(Kr) + $4d^{10}$	—	2	1	2.00	1.78
50	Tin (Sn)	7.30		—	2	2	1.72	1.96
51	Antimony (Sb)	8.64		—	2	3	1.53	2.05
52	Tellurium (Te)	8.96		—	2	4	1.42	2.10
53	Iodine (I)	10.44		—	2	5	1.32	2.66
54	Xenon (Xe)	12.13		—	2	6	1.24	—

Second group (z = 55–71):

z	Element	$\mathscr{E}_i$ (eV)	Inner shells	4f	5d	6s	Atomic radius (Å)	Electro-negativity (Pauling)
55	Cesium (Cs)	3.89		—	—	1	3.34	0.79
56	Barium (Ba)	5.21		—	—	2	2.78	0.89
57	Lanthanum (La)	5.61		—	1	2	2.74	1.10
58	Cerium (Ce)	(5.54)		2	—	2	2.70	1.12
59	Praseodymium (Pr)	(5.46)		3	—	2	2.67	1.13
60	Neodymium (Nd)	(5.53)		4	—	2	2.64	1.14
61	Promethium (Pm)	(5.55)		5	—	2	2.62	1.13
62	Samarium (Sm)	(5.6)	(Xe)	6	—	2	2.59	1.17
63	Europium (Eu)	5.64		7	—	2	2.56	1.20
64	Gadolinium (Gd)	6.2		7	1	2	2.54	1.20
65	Terbium (Tb)	(5.86)		8	—	2	2.51	1.2
66	Dysprosium (Dy)	(5.94)		9	—	2	2.49	1.22
67	Holmium (Ho)	6.0		10	—	2	2.47	1.23
68	Erbium (Er)	6.1		11	—	2	2.45	1.24
69	Thulium (Tm)	6.2		12	—	2	2.42	1.25
70	Ytterbium (Yb)	6.2		13	—	2	2.40	1.1
71	Lutetium (Lu)	5.4		14	1	2	2.25	1.27

					5d	5f	6s	6p		
72	Hafnium	(Hf)	6.65		2	—	2	—	2.16	1.3
73	Tantalum	(Ta)	7.89		3	—	2	—	2.09	1.5
74	Tungsten	(W)	7.94		4	—	2	—	2.02	2.36
75	Rhenium	(Re)	7.87		5	—	2	—	1.97	1.9
76	Osmium	(Os)	8.7	(Xe)	6	—	2	—	1.92	2.2
77	Iridium	(Ir)	9.2	+	9	—	0	—	1.87	2.20
78	Platinum	(Pt)	8.96	$4f^{14}$	9	—	1	—	1.83	2.38
79	Gold	(Au)	9.23		9	—	1	—	1.79	2.54
80	Mercury	(Hg)	10.44		10	—	2	—	1.76	2.00
81	Thallium	(Tl)	6.12		10	—	2	1	2.08	2.04
82	Lead	(Pb)	7.42		10	—	2	2	1.81	2.33
83	Bismuth	(Bi)	7.29		10	—	2	3	1.63	2.02
84	Polonium	(Po)	8.42		10	—	2	4	1.53	2.0
85	Astatine	(At)	(9.6)		10	—	2	5	1.43	2.2
86	Radon	(Rn)	10.75		10	—	2	6	1.34	—

					5f	6d	7s		
87	Francium	(Fr)	(4.0)		—	—	1		0.7
88	Radium	(Ra)	5.27		—	—	2		0.9
89	Actinium	(Ac)	5.17		—	1	2		1.1
90	Thorium	(Th)	6.1		1(—)	1(2)	2		1.3
91	Protactinium	(Pa)	5.9		2(1)	1(2)	2		1.5
92	Uranium	(U)	6.05	(Rn)	3	1	2		1.38
93	Neptunium	(Np)	6.2		5(4)	—(1)	2		1.36
94	Plutonium	(Pu)	6.1		6(5)	—(1)	2		1.28
95	Americium	(Am)	6.0		7	—	2		1.28
96	Curium	(Cm)	6.0		7	1	2		1.3
97	Berkelium	(Bk)	6.2		9	—	2		1.3
98	Californium	(Cf)	6.3		10	—	2		1.3

Let us return to our original task concerning the electron structure of an atom having z electrons. Note that replacing e^2 by ze^2 in Eq. (1.4.1) will not change the results presented in Section 1.4 except that e^4 should be replaced by z^2e^4 in the energy expression of Eq. (1.4.14). The electron motion is again specified by four quantum numbers, n, l, m, and s_z. Therefore, our task now reduces to assigning a different set of four quantum numbers to each electron. For such an assignment, we need to know the energy associated with each atomic level. A modification of Eq. (1.4.14) to account for the difference in nuclear charge and to include relativistic variation of the mass leads to the following energy expression:

$$\mathcal{E} = -\frac{mz^2e^4}{2(4\pi\epsilon_0\hbar)^2}\frac{1}{n^2}\left[1 + \frac{\alpha^2 z^2}{n^2}\left(\frac{n}{l+\frac{1}{2}} - \frac{3}{4}\right)\right] \qquad (1.6.1)$$

where $\alpha = e^2/4\pi\epsilon_0\hbar c$ is the fine structure constant, and c is the velocity of light.

From the dependence of $\mathcal{E}$ on l in Eq. (1.6.1), we see that the $2p$ level should be slightly above the $2s$ level in energy and that the $3d$ level should be slightly above the $3p$ level (Fig. 1.6). Now we redraw part of Fig. 1.6 as Fig. 1.7, and distribute electrons of a selected number of elements among the various available energy states. For helium it is obvious that the configuration of lowest energy has two electrons in the $1s$ level. Since two is the maximum number possible for the $1s$ level, the level is full. A *closed shell* is said to have formed if electrons fill completely a given atomic level. Hence the first shell is completed in helium.

After He comes Li. The third electron of Li must go to a higher energy level, that is, the $2s$ level, as illustrated in Fig. 1.7. In Be, both the levels $1s$ and $2s$ are full. The $2p$ level begins to become occupied in boron, and it is completely occupied in Ne. Because the energy separation between $2s$ and $2p$ levels is not very large, we usually consider them together as constituting a single shell and speak of the next shell when both are complete. The second closed shell, therefore, is completed in neon. This definition of a closed shell also applies to $3s$ and $3p$ electrons and to $4s$ and $4p$ electrons. Thus, after He and Ne, a closed-shell configuration is again reached in argon. Elements having their shells completed are chemically inert, such as He, Ne, Ar, and Kr.

The one-electron scheme (i.e., the scheme considering each electron as moving independently of other electrons) on which Eq. (1.6.1) is based works well qualitatively up to argon in explaining the distribution of electrons among the different atomic levels. But potassium shows a deviation from the normal pattern (Table 1.2). A state of lower energy is achieved by filling the $4s$ level first instead of the $3d$ level, which has a smaller principal quantum number. For electrons in an outer orbit, $4s$ orbit or $3d$ orbit, for example, the Coulomb field of the nucleus is partly shielded by the charge cloud of the inner-core electrons, $1s$ and $2s$ electrons, for example. Therefore, in Eq. (1.6.1), the value of ze should be the effective nuclear charge whose influence is actually felt by the electron. The value of ze will be different even in the same atom for electrons in differ-

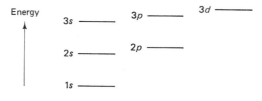

Figure 1.6 Energy diagram for the few lowest-energy states. The energy separations between the levels are not drawn to scale.

Atomic Structure   Chap. 1

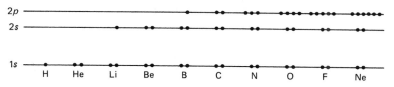

Figure 1.7 Distribution of electrons among available quantum states in a selected number of chemical elements. Lower-energy states are filled first.

ent atomic orbits because the degree of shielding and hence the value of ze depend on the depth of penetration of the electron wave function.

For small values of n, the energy separation between levels with different n is so large that the effect due to different degrees of shielding is secondary. For large values of n as in the case of potassium, the $4s$ wave function penetrates much more deeply than the $3d$ wave function into the region of low potential energy near the nucleus. The effect on energy caused by the difference in the effective ze for the $4s$ and $3d$ wave functions actually overwhelms that caused by the difference in n in Eq. (1.6.1). A similar situation occurs in the N shell (X-ray nomenclature with $n = 4$). Electrons begin to settle in the $5s$ and $5p$ levels of the O shell before they fill up the $4d$ and $4f$ levels of the N shell.

The reader may also be aware of the fact that in Table 1.2 we have assigned to electrons only the values of n and l. The existence of different values for the other two quantum numbers m and s_z can only be inferred from the maximum number of electrons allowed in Table 1.2 and Fig. 1.7 for a given set of n and l. The reason is simple for not mentioning explicitly the values of m and s_z. To be able to assign the values of m and s_z to the electron, we must consider interactions of second-order importance, such as the electron–electron interaction, in the energy expression.

Let us now discuss how we can use the information derived from the electronic structure of atoms. As mentioned earlier, elements having their shells completed are chemically inert. The same can be said about electrons in an atom. Those electrons that belong to closed shells are core electrons. Because it requires considerable energy to break up a closed shell, the core electrons are inactive in a chemical reaction and are not affected by small external disturbances, for example the application of a small electric field in a conductance measurement. Therefore, in our discussion of material properties, we can ignore the core electrons and speak of only the electrons outside the closed shells. The closed-shell configuration in an atom can easily be identified in Table 1.2 by comparing the given electron configuration with that of a nearest inert gas (He, Ne, Ar, Kr, or Xe).

The alkali metals have a single electron outside closed shells from $2s$ in Li to $6s$ in Cs. The halogens, on the other hand, are one electron short of completing the s–p shell (from the $2s$–$2p$ shell in F to the $6s$–$6p$ shell in At). The alkali metals and the halogens form salts of the type XY. As discussed in Chapter 2, crystals of alkali halides are electrical insulators. In contrast to alkali halides, the III–V compounds are semiconductors. The electron configurations in elemental semiconductors Ge and Si and those in III–V compound semiconductors are quite similar. By *III–V compounds* we mean compounds made of group IIIb elements and group Vb elements. The group IIIb elements such as Ga have two s electrons and one p electron, whereas the group Vb elements such as As have two s electrons and three p electrons. If we suppose that these electrons are shared equally between the atoms in a compound, we would have a similar electron configuration in III–V compound semiconductors as that in Ge and Si. Therefore, it is not surprising that III–V compounds are also semiconductors having similar electrical properties as Ge and Si. The differences as well as the similarities between the properties of these materials are explored further in subsequent chapters.

PROBLEMS

1.1. According to Rutherford, an atom was thought of as composed of protons and electrons. Neutrons were later found also to be a constituent part of the nucleus. From the Rutherford scattering experiment, what qualitative information can be derived about the relative mass and physical dimension of the proton and the electron? Would the experiment be equally successful if electrons were used instead of α particles? Which quantity, the atomic number z or the atomic mass number A, can be obtained from the experiment? Explain why the two numbers are different.

1.2. Consider the motion of an electron in a solid. Apply what we have learned about the Rutherford experiment to the problem and present a qualitative picture of what happens to the electron motion. Does the electron feel the nuclear charge alone or the total charge of an atom? Is the scattering effect stronger in a solid made of neutral atoms such as in Ge than in a solid made of charged ions such as in NaCl?

1.3. (a) It was found by Balmer in 1885 that the frequency of a series of emission lines of hydrogen can be fitted in the visible region by

$$ \nu = cR\left(\frac{1}{4} - \frac{1}{n^2}\right) $$

where R is the Rydberg constant and c is the velocity of light. Find the value of R (expressed in cm^{-1}).

(b) One of the emission lines in the series has a wavelength $\lambda = 6563$ Å ($\lambda = c/\nu$). Identify the principal quantum numbers of the two states in Eq. (1.2.11).

1.4. The first four emission lines of hydrogen in the Balmer series have the following wavelengths: $\lambda = 6563$ Å, 4861 Å, 4340 Å, and 4102 Å. Show that the observed emission can be accounted for by the old quantum theory.

1.5. In the Franck–Hertz experiment, a mercury discharge tube was used, and two important observations were made: the emission of radiation at $\lambda = 2536$ Å and a sharp current drop when the applied grid voltage exceeds 4.9 V. Determine whether there is any correlation between the two observations and offer a plausible explanation for the current drop. The anode is negatively biased with respect to the grid.

1.6. (a) Consider a one-dimensional, mechanical harmonic oscillator that obeys

$$ M\frac{d^2x}{dt^2} = -\gamma x $$

where γ is the spring (or force) constant. Show that the energy of the oscillator is

$$ \mathscr{E} = \frac{p^2}{2M} + \frac{\gamma x^2}{2} $$

(b) Applying the Wilson–Sommerfeld quantization rule, show that the energy $\mathscr{E}$ is quantized and equal to

$$ \mathscr{E} = n\hbar\omega $$

where $\omega = \sqrt{\gamma/M}$ and n is an integer. (*Hint:* Let $x = \sqrt{2\mathcal{E}/\gamma} \sin \theta$ in performing the phase integral.)

1.7. (a) Find the Schrödinger equation for the one-dimensional harmonic oscillator of Problem 1.6 by replacing p by an appropriate operator.

(b) In the quantum-mechanical treatment, the energy of a harmonic oscillator is

$$\mathcal{E} = \left(n + \frac{1}{2} \right) \hbar\omega$$

Show that the wave function

$$\psi_1 = A_1 x \exp \left(-\frac{M\omega}{2\hbar} x^2 \right)$$

is a solution of the Schrödinger equation for $n = 1$.

1.8. (a) Apply the Wilson–Sommerfeld quantization rule to an electron of mass m which moves with a constant velocity in one dimension between $x = -a$ and $x = a$. The electron undergoes perfectly elastic collisions with rigid walls at $x = -a$ and $x = a$. Show that the energy of the electron is

$$\mathcal{E} = \frac{n^2 h^2}{32 m a^2}$$

(b) Consider a quantum-mechanical problem in which an electron moves in a one-dimensional potential well (Fig. P1.8). For simplicity, we assume that $V = 0$ for $-a < x < a$ and $V = \infty$ for $x > a$ and $x < -a$. Find the wave function for the electron motion and the quantized energy of the electron.

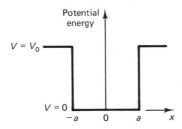

Figure P1.8

1.9. (a) Consider a problem in which an electron faces a one-dimensional potential barrier (Fig. P1.9). The electron energy $\mathcal{E}$ is such that $\mathcal{E} < V_0$. Write down the Schrödinger equation and its solution for the three regions.

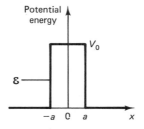

Figure P1.9

(b) Classically, we know that the electron can never penetrate the barrier for $\mathscr{E} < V_0$. Do you reach the same conclusion from the quantum-mechanical treatment? (*Hint:* Apply the probabilistic interpretation of $|\psi|^2$ and note that

$$\frac{d^2\psi}{dx^2} = \alpha\psi$$

has an exponential solution if α is positive and a sinusoidal solution if α is negative.)

1.10. Show that the average value of r for the $1s$ electron in a hydrogen atom is $\langle r_{1,0,0}\rangle = 1.5a$, where a is the Bohr radius.

1.11. The angular-momentum operator L_z is

$$L_z = \frac{\hbar}{i}\frac{\partial}{\partial\phi}$$

in spherical coordinates. Show that

$$\langle L_z\rangle_{2,1,1} = \hbar, \qquad \langle L_z\rangle_{2,1,0} = 0, \qquad \langle L_z\rangle_{2,1,-1} = -\hbar$$

for the $2p$ electrons. [*Hint:* $\int_0^\infty v^n \exp(-v)\, dv = n!$.]

1.12. Show that $\Theta_{2,0}$ of Eq. (1.4.20) satisfies Eq. (1.4.7) for $l = 2$ and $m = 0$ and that $R_{2,0}$ of Eq. (1.4.21) satisfies Eq. (1.4.8) for $n = 2$ and $l = 0$. (*Hint:* Express the terms $\mathscr{E}$ and $e^2/4\pi\epsilon_0 r$ in terms of the Bohr radius a.)

1.13. (a) Use the chemical elements Na, Cl, Ca, and O as examples. Show how these elements can achieve closed-shell configuration by becoming ions.

(b) How many distinct orbital wave functions are there for d electrons? What is the maximum number of electrons for the d state?

2

Crystal Structure

2.1 THE CRYSTALLINE STATE

In Chapter 1 we dealt with the structure of free atoms. However, materials in nature do not exist in the form of isolated atoms. As we all know, atoms are aggregated together in either one of the three natural states: the gas state (as molecules), the liquid state, and the solid state. A large majority of components and devices used in electrical and electronic engineering is made of solid materials. Compared to gases and liquids, solids have many desirable attributes. Rigidity, mechanical strength, and large density of atoms are the obvious advantages. There are between 10^{22} and 10^{23} atoms/cm^3 in a solid. For comparison, we calculate the density D of molecules per cubic centimeter of water vapor at 373 K and 1 atm of pressure from its specific volume

$$D = \frac{N}{1670 \times w} = \frac{6.02 \times 10^{23}}{1670 \times 18} = 2 \times 10^{19} \text{ molecules/cm}^3 \qquad (2.1.1)$$

where N is Avogadro's number and w is the molecular weight.

The solid state differs from the liquid state not so much in the density of atoms but in the arrangement of atoms. The fact that a liquid can be poured from one container into another indicates that the arrangement of atoms in the liquid state cannot be rigid and hence must be disorderly. When a material solidifies upon cooling from the liquid state, the atoms generally assume orderly arrangements to form a three-dimensional pattern. The solid thus formed is said to be in a *crystalline state*. If the cooling is sudden, however, atoms may lose greatly their freedom of movement before they can rearrange themselves in orderly fashion. This type of solid is not in a crystalline state but in a glassy state, and the solid is known as an *amorphous solid*. In this book we are concerned only with solids in crystalline form. As we discuss in Chapter 3, atoms are held together in a solid by the attractive forces. The number of neighboring atoms is maximized and the energy of a solid is minimized by an orderly arrangement of atoms. Therefore, the crystalline state is the natural state of a solid.

The most easily recognizable feature of a crystalline solid is the periodicity of its pattern. First, let us consider the two-dimensional pattern shown in Fig. 2.1a. In an

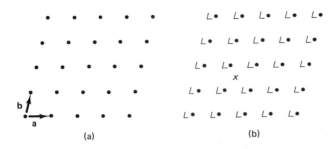

(a) (b)

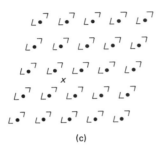

(c)

Figure 2.1 Two-dimensional lattice. (a) Each lattice point is a translation of another lattice point by **T** of Eq. (2.1.2). (b) In a complex solid, each lattice point may be occupied by a molecule denoted by $\angle$. (c) The pattern is used to illustrate that a crystal may have rotational symmetry if the arrangement of atoms is symmetrical with respect to the lattice point.

actual situation, each point in the diagram represents either an atom or the center of a molecule, and Fig. 2.1a describes the arrangement of these atoms or molecules over a given cross section of a solid. Note that each point is displaced from other points by a translation vector **T** such that

$$\mathbf{T} = n_1 \mathbf{a} + n_2 \mathbf{b} \qquad (2.1.2)$$

where **a** and **b** are two vectors defined in Fig. 2.1 and n_1 and n_2 are two integers. If translations by **a** and **b** are applied repeatedly to any point, the entire pattern can be generated. Such a periodic arrangement of points in space is called a *lattice*.

Besides the translation operation, there may exist two other operations, the rotation about an axis and the reflection in a mirror plane, which can bring a crystal into itself. However, in discussing the latter two operations, the symmetry of the molecule involved must be considered in conjunction with the symmetry of the lattice. Two different situations are illustrated schematically in Fig. 2.1b and c. In Fig. 2.1b the symbol $\angle$ represents an asymmetric molecule. A rotation of 180° about a point marked $\times$ obviously will not carry the crystal into itself. If another molecule having an inverted $\angle$ structure is placed opposite the $\angle$ molecule, the combination is symmetric with respect to the lattice point. It is evident that the pattern shown in Fig. 2.1c has a rotational symmetry about the point marked $\times$. This consideration of the symmetry of the molecule or a group of molecules further distinguishes the two cases even though both are represented by the same lattice.

The study of different ways of arranging atoms or molecules in a periodic manner is called *crystallography*. However, it is not the purpose of this book to give an exhaustive discussion of the different periodic structures of crystals. Instead, the aim of this chapter is to acquaint the reader with simple facts about crystals. Insofar as crystal symmetry is concerned, we shall focus our attention on the effects of crystal symmetry on material properties. In many engineering problems, we sometimes face the situation that material properties, such as the dielectric constant ϵ, are different in different directions in certain crystals. Therefore, considerations of crystal symmetry provide a basis for our later discussion of the use of these materials in engineering applications.

2.2 BRAVAIS LATTICES

Our discussion in Section 2.1 concerning a two-dimensional lattice can easily be extended to a three-dimensional lattice by introducing a third translation vector $\mathbf{c}$ for the third dimension. The directions of the three vectors $\mathbf{a}$, $\mathbf{b}$, and $\mathbf{c}$ define the three axes of a crystal, whereas the lengths of these vectors are the distances between neighboring atoms along the three axes. In Fig. 2.2 a parallelepiped is shown, with the length of each side designated by a, b, and c and the angle between the axes by α, β, and γ. Depending on the specifications about the lengths and the angles, crystals are grouped into seven systems, as shown in Table 2.1.

The simplest representative unit of each system has lattice points at the corners of the parallelepiped only. These units are referred to in Fig. 2.3 as primitive units. Bravais showed, however, that various symmetry operations possible for a three-dimensional point lattice require 14 different space lattices, now known as *Bravais lattices*. The other seven units shown in Fig. 2.3 have additional lattice points at the base centers, the face centers, or the body center of the basic unit. The following symbols are used together with crystal systems to specify the basic lattice unit: P for primitive unit, F for face-centered unit, C for base-centered unit, I (German *Innerzentrierte*) for body-centered unit, and R for rhombohedral unit.

All 14 Bravais lattices can be generated from the representative units of Fig. 2.3 by translations of the various units along the three crystal axes. These representative units are called *unit cells*. The translation vector $\mathbf{T}$ for a three-dimensional lattice is

$$\mathbf{T} = n_1 \mathbf{a} + n_2 \mathbf{b} + n_3 \mathbf{c} \qquad (2.2.1)$$

where n_1, n_2, and n_3 are integers. The reader may ask whether the three axes chosen in Fig. 2.3 represent a unique choice and, if not, what determines the choice of a particular set of axes. To answer this question, we must count the number of lattice points in each cell. The cell that possesses only one lattice point is the smallest unit possible to fill all space through proper translation operations. The cell of minimum size is known as a *primitive cell*.

To count the number of lattice points in each cell, we note that each lattice point at the corner of the parallelepipedon is shared by the eight cells which meet there and

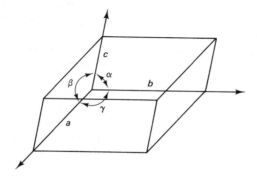

Figure 2.2 A parallelepiped that defines three lengths a, b, and c and three angles α, β, and γ. Crystal structures are classified into seven systems depending on the specifications about the lengths and the angles.

TABLE 2.1 CRYSTAL SYSTEMS

Triclinic	$a \neq b \neq c$	$\alpha \neq \beta \neq \gamma$
Monoclinic	$a \neq b \neq c$	$\alpha = \beta = 90° \neq \gamma$
Orthorhombic	$a \neq b \neq c$	$a = \beta = \gamma = 90°$
Tetragonal	$a = b \neq c$	$\alpha = \beta = \gamma = 90°$
Cubic	$a = b = c$	$\alpha = \beta = \gamma = 90°$
Hexagonal	$a = b \neq c$	$\alpha = \beta = 90°, \gamma = 120°$
Trigonal	$a = b = c$	$\alpha = \beta = \gamma \neq 90°$

that each face point is shared by two cells. The seven primitive units shown in Fig. 2.3 possess only one lattice point per cell, and hence the sizes of these cells cannot be further reduced. That is not true, however, with the units labeled C, F, and I. For example, the base-centered monoclinic structure and the face-centered cubic structure have per cell two and four lattice points, respectively. Therefore, it should be possible to reduce the sizes of these cells.

For the base-centered monoclinic structure, if the two base diagonals are chosen as two new vectors $\mathbf{a}'$ and $\mathbf{b}'$ replacing vectors $\mathbf{a}$ and $\mathbf{b}$ in Eq. (2.2.1), a primitive cell is formed which possesses only corner atoms and thus only one lattice point per cell. The relation between the primitive and the unit cell is illustrated in Fig. 2.4a. For the face-centered cubic structure, a primitive cell is formed by using the three face diagonals as the three axes (Fig. 2.4b). The volume V of a parallelepipedon shown in Fig. 2.2 is

$$V = |\mathbf{a} \cdot (\mathbf{b} \times \mathbf{c})| \tag{2.2.2}$$

Using Eq. (2.2.2), the reader can easily show that the volumes of the primitive cells shown in Fig. 2.4a and b are, respectively, one-half and one-fourth of the volume of the corresponding unit cell shown in Fig. 2.3

The three vectors used as basis vectors in defining a primitive cell are known as *primitive vectors*. The entire lattice can be generated from a lattice point by suitable translation operations only if primitive vectors are used in Eq. (2.2.1). As the examples of Fig. 2.4 show, the primitive vectors in general will be different from the three vectors defining the crystal axes. It is true that in the example of Fig. 2.4a the corresponding lattice points in the two neighboring cells are related through a translation by the vector $\mathbf{a}$. However, it is not possible to go from a corner point to a base point, or vice versa, through a translation by $\mathbf{a}$. Such a translation is only possible by the primitive vectors. The same applies to the lattice points in Fig. 2.4b. Therefore, for the body-centered, the face-centered, and the base-centered structures of Fig. 2.3, primitive vectors must be used in defining the periodicity of a crystal structure.

The discussion above does not mean, however, that primitive vectors should take the place of crystal axes in all discussions. The conventional crystal axes will be used as reference axes in later discussions. The reason for doing this is that the unit cells (Fig. 2.3) in each crystal system have similar symmetry properties. For example, the fact that a face-centered cubic structure belongs to the cubic system is more easily recognized in the system of conventional axes $\mathbf{a}$, $\mathbf{b}$, and $\mathbf{c}$ than in the system of primitive axes $\mathbf{a}'$, $\mathbf{b}'$, and $\mathbf{c}'$. Therefore, the conventional axes will be kept for general references, whereas the primitive axes will be used for specific references to the periodic arrangement of atoms in a structure.

Crystal Structure   Chap. 2

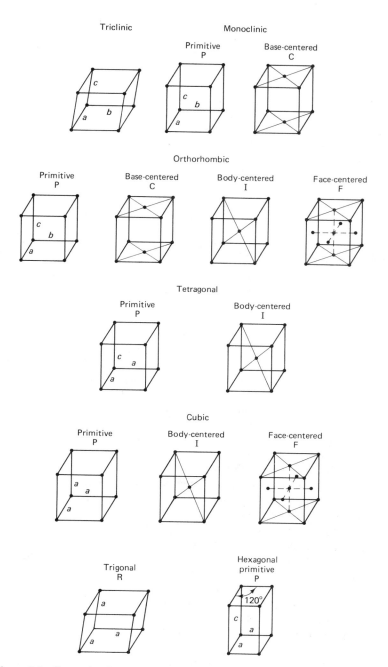

Figure 2.3 Conventional unit cells of the 14 Bravais lattices. The primitive units in each system have atoms at the corners of the parallelepiped only and are designated by the letter P. The letter C indicates that the structure has extra atoms at the centers of its bases. The letters I (German *Innenzentrierte*) and F refer, respectively, to body-centered and face-centered structures.

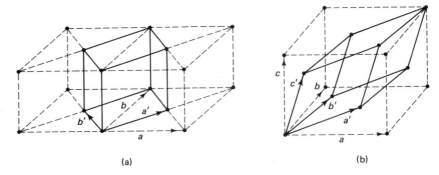

Figure 2.4 Diagrams showing the relationship between a primitive cell (heavy lines) and a unit cell for (a) the monoclinic structure and (b) the face-centered cubic structure.

2.3 SIMPLE CRYSTAL STRUCTURES

In this section a selected number of simple crystal structures of general interest are described in terms of their unit cells. Before we introduce these structures, it may be helpful that we discuss certain rationalizations regarding the arrangement of atoms in a crystal. If the bonding force between atoms is nondirectional, the arrangement of atoms is simply a matter of stacking them in as closely packed an arrangement as possible. In this way the number of bonds per unit volume is maximized, and hence the bonding energy (which is negative on account of the attractive nature of the bonding force) per unit volume is minimized. Therefore, for nondirectional bonding, close atomic packing is the primary consideration. On the other hand, if the bonding is directional, atoms will naturally join in the direction of the bonding force. In this case, the orientation of the neighboring atoms with respect to the central atom is important rather than the number of these atoms surrounding it.

Most metals have simple crystal structures. The alkali metals (Li, Na, K, Rb, Cs) form body-centered cubic lattices, while the noble metals form face-centered cubic lattices. The number of nearest neighbors in a crystal structure is called the *coordination number*. As we discuss in Chapter 3, purely metallic bonding is nondirectional. Thus we expect a high coordination number for metals. As we can see from Fig. 2.3, the coordination number is 8 for the bcc (body-centered cubic) structure and is 12 for the fcc (face-centered cubic) structure.

In Fig. 2.5 the arrangement of atoms in the face-centered cubic structure is shown in detail. If we compare the arrangements of atoms in three different horizontal planes viewed from the top (Fig. 2.5a) with those in three different vertical planes viewed from the front (Fig. 2.5c), we find that not only are the patterns the same, but atoms 12, 3, 8, and 6 in the front view and atoms 2, 10, 5, and 9 in the side view are similarly situated with respect to atom 0 as atoms 1, 7, 4, and 11 in the top view. Further, we see that there is very little space between the atoms.

Another way of arranging the atoms to minimize the interstitial space leads to a hexagonal close-packed (hcp) structure of Fig. 2.6a, in which a middle layer is added to the regular hexagonal structure. The atoms in the middle layer rest in the depression against three atoms in the basal plane (e.g., atom 7 against atoms 0, 2, and 3). If circles

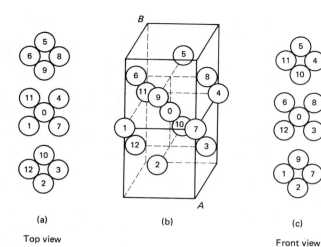

(a)

Top view

(b)

(c)

Front view

Figure 2.5 The face-centered (close-packed) cubic structure with (a) the arrangement of atoms in three different horizontal planes as viewed from the top and (b) that in three different vertical planes as viewed from the front.

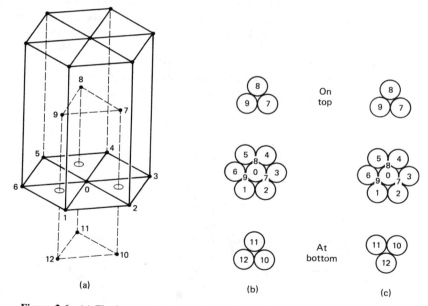

(a)

(b)

(c)

Figure 2.6 (a) The hexagonal close-packed structure. A middle layer of atoms 7, 8, and 9 is added to the regular hexagonal structure. (b) The arrangement of atoms in three consecutive layers in the hcp structure. (c) The corresponding arrangement in the cubic close-packed structure. The difference between the two arrangements is in the third layer of atoms 10, 11, and 12.

Sec. 2.3   Simple Crystal Structures

are drawn in a two-dimensional space to represent the atoms, it is clear that there is only room for three atoms, shown as atoms 7, 8, and 9 in Fig. 2.6b. However, there are two sides to the basal plane. Atoms 10, 11, and 12 represent atoms from the bottom, which have a similar arrangement with respect to atoms in the basal plane as atoms 7, 8, and 9. Thus the hcp structure, like the fcc structure, has a coordination number of 12. These two structures are the most densely packed structure. Metals such as Be, Mg, Zn, and Cd have the hexagonal close-packed structure. Since a high coordination number is the characteristic of a metallic crystal, most metals crystallize in either one of the following three structures: the bcc structure, the fcc structure, or the hcp structure (W. Hume-Rothery, *Electrons, Atoms, Metals and Alloys*, Philosophical Library, Inc., New York, 1955, p. 192).

Let us compare the arrangements of atoms in the two close-packed structures. If a body-diagonal plane passing through $A14B$ is drawn in Fig. 2.5b, atoms 1, 2, 3, 4, 5, and 6, which lie in this diagonal plane, are arranged in a similar manner as the corresponding atoms in Fig. 2.6a. The same applies to atoms 7, 8, and 9 in both figures. However, atoms 10, 11, and 12 in Fig. 2.5a, instead of taking the positions directly opposite atoms 7, 8, and 9 as shown in Fig. 2.6a, have their positions rotated by 60° with respect to those of atoms 7, 8, and 9, so that atom 10, for example, now rests against atoms 0, 3, and 4. Therefore, the hcp structure and the fcc structure differ only in the third layers. This difference is illustrated in Fig. 2.6b and c.

In many ionic crystals, the size of the anion is usually much larger than that of the cation. The size ratio is important because it limits the number of anions that can make contacts with any single cation. The two common structures for ionic crystals are the sodium chloride (NaCl) structure (Fig. 2.7a) and the cesium chloride (CsCl) structure (Fig. 2.7b). The NaCl structure consists of two face-centered cubic sublattices, one occupied by Na^+ and the other by Cl^- ions. The two sublattices are displaced from each other by a distance $a/2$ along either one of the three cube axes. The CsCl structure looks very much like a body-centered cubic structure except that one element occupies the eight corners and the other the center of the cube. Actually, the structure consists of two interpenetrating simple cubic sublattices, displaced ($a/2$, $a/2$, $a/2$) apart and oc-

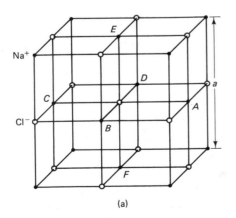

(a)

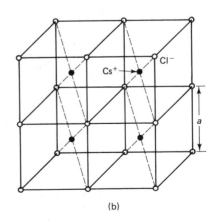

(b)

Figure 2.7 Common structures for ionic crystals. (a) NaCl structure. (b) CsCl structure.

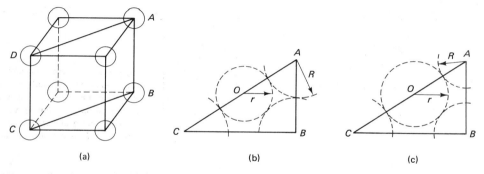

Figure 2.8 Figures showing (a) a simple cubic structure, (b) the void in the center of the cube with $r = 0.732R$, and (c) the void in the center of the cube with $r = R$.

cupied by two different elements. Among ionic crystals, the alkali halides have the NaCl structure except for CsCl, CsBr, and CsI, which have the CsCl structure.

To see why atom sizes are important in determining the crystal structure, we examine the size of the void between atoms in various forms of atomic packing. First, we consider the simple cubic structure of Fig. 2.8a. Let us draw a face-diagonal plane ABC so that we can have a cross-sectional view of the arrangement of atoms around the center O of the cube. The atoms are represented as spheres. Even if the nearest-neighbor atoms are in direct contact (e.g., atoms A and B), there still exists a void in the center of the cube, as illustrated in Fig. 2.8b. This void can accommodate a sphere of radius r. Realizing that the distance $AC = \sqrt{3}\,AB$, we find that

$$r = \tfrac{1}{2}AC - R = (\sqrt{3} - 1)R = 0.732R \qquad (2.3.1)$$

The results of our calculation for the cubic packing and of similar calculations for various other packing forms are summarized in Table 2.2. As discussed earlier, most metals take either one of the three structures: the bcc structure, the fcc structure, or the hcp structure. Since only one kind of atom is involved, the ratio of atom sizes is automatically 1. Both the bcc structure and the CsCl structure belong to cubic packing. The only difference between the two structures is that the space in the center of the cube is occupied by the same kind of atom in the bcc structure and by a different kind of atom in the CsCl structure. The other common types of atomic packing are the octahedral arrangement of Fig. 2.9 and the tetrahedral arrangement of Fig. 2.10. The reader is

TABLE 2.2 ATOM-SIZE RATIO IN VARIOUS PACKING FORMS

Ratio r/R of atom sizes	Coordination number	Packing form	Examples
$0.155 < r/R < 0.225$	3	Triangular	B_2O_3
$0.225 < r/R < 0.414$	4	Tetrahedral	silica (SiO_2)
$0.414 < r/R < 0.732$	6	Octahedral	NaCl
$0.732 < r/R < 1$	8	Cubic	CsCl, alkali metals
$r/R = 1$	12	fcc	Noble metals
$r/R = 1$	12	hcp	Be, Mg

Sec. 2.3   Simple Crystal Structures

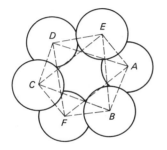

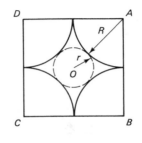

Figure 2.9 (a) An octahedral site that is surrounded symmetrically by six atoms. (b) A view of the atomic arrangement in the horizontal plane. Atoms E and F sitting in the front and back depressions between atoms A, B, C, and D are not shown.

asked to show that the minimum value of r/R is 0.414 for the octahedral packing and 0.225 for the tetrahedral packing. Referring to Fig. 2.7a, we see that each Na^+ ion is surrounded by six Cl^- ions. These six Cl^- ions form an octahedron like the one shown in Fig. 2.9a, and the Na^+ ion takes the place in the center of the octahedron. Because of the difference in the sizes of the cation and the anion, most ionic crystals take the cubic, the octahedral, or the tetrahedral packing.

In the discussion above we have treated atoms as if they were spheres. We can do this because the bonding force between atoms does not depend on their relative orientation. Atoms in nondirectional-bonding crystals are so arranged as to achieve maximum atomic packing per unit volume. This principle no longer holds in directional-bonding crystals. As we discuss in Chapter 3, the bonding between atoms in a covalent crystal is directional. One covalent structure with which the reader may be familiar is the diamond structure. Common semiconductors such as germanium and silicon crystallize in this structure.

Referring to Fig. 2.11, we see that the diamond structure is a combination of two interpenetrating face-centered cubic lattices, one sublattice shown in dots and the other in open circles. The two sublattices are displaced from each other by $(-\frac{1}{4}, \frac{1}{4}, \frac{1}{4})$ in unit of the lattice constant a, as represented, for example, by the displacement of atom E from D. To help see that atoms shown in circles do form a face-centered cubic structure, the coordinates of atoms E, F, G, and H are given in units of a in Fig. 2.11. The displacement $\mathbf{d}$ of atoms from their nearest-neighbor atoms in a fcc structure is given by one of the following expressions:

$$\mathbf{d} = \pm\hat{\mathbf{x}}\frac{1}{2} \pm \hat{\mathbf{y}}\frac{1}{2}, \quad \pm\hat{\mathbf{x}}\frac{1}{2} \pm \hat{\mathbf{z}}\frac{1}{2}, \quad \text{or} \quad \pm\hat{\mathbf{y}}\frac{1}{2} \pm \hat{\mathbf{z}}\frac{1}{2} \tag{2.3.2}$$

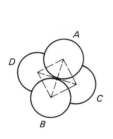

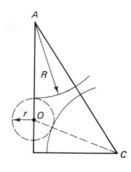

Figure 2.10 (a) A tetrahedral site that is surrounded symmetrically by four atoms in a pyramidal arrangement with atoms BCD at the base and atom A on top. (b) A view in the vertical plane containing atoms A and C and the center O of the tetrahedron.

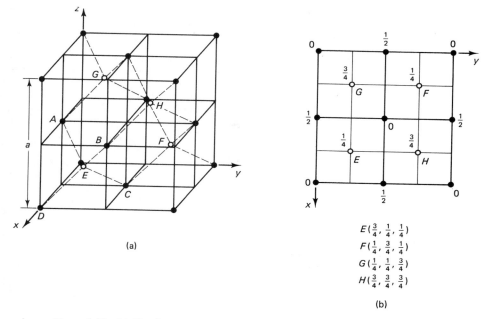

$$E(\tfrac{3}{4}, \tfrac{1}{4}, \tfrac{1}{4})$$
$$F(\tfrac{1}{4}, \tfrac{3}{4}, \tfrac{1}{4})$$
$$G(\tfrac{1}{4}, \tfrac{1}{4}, \tfrac{3}{4})$$
$$H(\tfrac{3}{4}, \tfrac{3}{4}, \tfrac{3}{4})$$

(b)

Figure 2.11 (a) The diamond structure. The numbers in parentheses (x, y, z) indicate the coordinates of an atom expressed in units of a. (b) A projection of the positions of atoms in a horizontal plane. The numbers represent the z coordinates of the atoms.

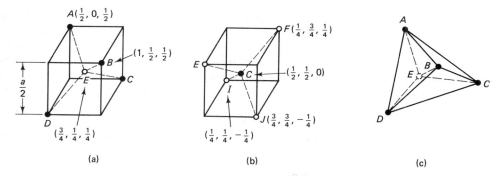

Figure 2.12 The building blocks in a diamond structure. (a) The block containing atoms *ABCDE* of Fig. 2.11. (b) The block containing atoms *EFIJC*. Atoms *I* and *J*, which are not shown in Fig. 2.11, are directly below atoms *G* and *H* by a full lattice constant. (c) Atomic bonding (tetrahedral) in the diamond structure. If the corner atoms are joined by lines, a tetrahedron is formed.

TABLE 2.3 LATTICE CONSTANTS AND STRUCTURES OF A SELECTED NUMBER OF METALLIC, IONIC, COVALENT, AND MOLECULAR CRYSTALS

Substance	Structure	Lattice constant (Å)		Nearest-neighbor distance (Å)
Carbon	Diamond	3.56		1.54
Cesium	bcc	6.05	(92 K)	5.24
Copper	fcc	3.61		2.55
Germanium	Diamond	5.65		2.44
Iron	bcc	2.86		2.48
Magnesium	hcp	3.20	5.20 (c axis)	3.20
Silicon	Diamond	5.43		2.35
Sodium	bcc	4.28		3.71
Xenon	fcc	6.24	(92 K)	4.41
Zinc	hcp	2.66	4.94 (c axis)	2.66
NaCl	NaCl	5.63		2.82
MgO	NaCl	4.20		2.10
CsCl	CsCl	4.11		3.56
NH$_4$Cl	CsCl	3.87		3.35
ZnS	Zinc blende	5.41		2.34
CdS	Zinc blende	5.82		2.51
InSb	Zinc blende	6.46		2.79

where $\hat{x}$, $\hat{y}$, and $\hat{z}$ are vectors in units of a along the x, y, and z directions, respectively. The reader can easily show that the relation given by Eq. (2.3.2) holds for both sublattices, that is, separately for lattice points in dots and lattice points in open circles.

The zinc-blende (sphalerite) structure differs from the diamond structure only in that the two sublattices are occupied by two different elements in the former structure and by the same element in the latter stucture. Thus element semiconductors Ge and Si have the diamond structure, whereas compound semiconductors such as GaAs, InSb, and GaSb have the zinc-blende structure. In both structures the basic unit is a small cube (Fig. 2.12) with the lattice points of one sublattice sitting at the center of the cube and the lattice points of the other sublattice sitting at alternate corners of the cube. Two such typical cubes are shown in Fig. 2.12, one with atom E (atom drawn in open circle) and the other with atom C (atom drawn in dot) at the center.

The important feature of the diamond and zinc-blende structure is more easily recognizable if corner atoms in Fig. 2.12 are joined together to form a tetrahedron. The bond between atoms is directed from the center toward each corner of the tetrahedron. Thus each atom forms a bond with four nearest-neighbor atoms. In covalent-bond crystals, arrangement of atoms is dictated by the direction of the bond and not by atomic packing. To make our point clear, we take the small cube (of edge $a/2$) to the right of the cube formed by atoms A, B, C, and D. These two cubes have identical volume, yet the former does not have an atom at its center. The reason for this difference in their arrangements can be understood if we examine the bond formed by atom C. Note that in the tetrahedron (Fig. 2.12c) specifying the bonding directions, once the direction of one bond is chosen, the directions of the other three bonds are automatically fixed. Now let us refer to Fig. 2.12b. Since we have already assigned the bond between atoms E and C, the other three bonds of atom C must be directed toward atoms F, I, and J as

Crystal Structure   Chap. 2

shown. None of the three atoms fall in the center of the cube to the right of the cube *ABCD*.

In summary, arrangement of atoms in a crystal depends on the nature of the bonding force. If the bonding is nondirectional, atoms are so arranged as to achieve maximum packing. On the other hand, if the bonding is directional, atoms are arranged to satisfy the directionality requirement. The examples chosen for illustration in this section represent the simplest and probably the most common structures in each category (metals, ionic crystals, and covalent crystals). More complicated structures are discussed in later chapters when appropriate. The lattice constants and the structures of a selected number of common substances are given in Table 2.3.

2.4 CRYSTALLOGRAPHIC NOTATION

In Section 2.3 we presented several simple crystal structures. The description of a structure by the unit cell is useful because a structure is completely specified by atom positions in the cell. However, knowing the unit cell is not enough. In many applications we want to examine the properties of a crystal along a specific direction or in a specific plane. For example, the dielectric constant of an insulator may be different if measured along different directions. It is also found that in making semiconductor junctions, crystals which are cut in certain specific planes generally produce better junctions. To specify directions and planes in crystals, a system of crystallographic indices has been adopted.

Before we can define a direction or a plane, we must have a unique set of reference axes. It is the general practice that the conventional crystal axes, but not the primitive (basis) vectors, are chosen for the purpose of indicating specific directions and planes. Referring to Fig. 2.2, any vector $\mathbf{r}$ starting from the origin can be expressed in terms of the three vectors $\mathbf{a}$, $\mathbf{b}$, and $\mathbf{c}$ defining the three crystals axes as

$$\mathbf{r} = u'\mathbf{a} + v'\mathbf{b} + w'\mathbf{c} \qquad (2.4.1)$$

Since important directions in a crystal are those which aim at lattice points, the three constants u', v', and w' are integers. Also, since any multiplication of $\mathbf{r}$ by a number does not change the direction of $\mathbf{r}$, the indices of a direction are given by the smallest set of integers. The three indices are conventionally arranged in brackets as $[u'v'w']$. We should point out that a parallel translation of a vector does not change the vector direction. Thus we can always make $\mathbf{r}$ to start from the origin by a parallel translation.

Let us now use the monoclinic structure and the face-centered cubic structure of Fig. 2.4 as examples. The direction indices of the three crystal axes are, of course, [100], [010], and [001]. In the cubic structure, however, it is really immaterial as to which direction is chosen for the $\mathbf{a}$, $\mathbf{b}$, or $\mathbf{c}$ axis. Thus the three directions [100], [010], and [001] are equivalent in a cubic structure. A full set of equivalent directions is denoted in carats as $\langle 100 \rangle$. The three face diagonals a', b', and c' of the cubic structure are also equivalent, and hence can be designated as a group by $\langle 110 \rangle$. On the other hand, the two face diagonals a' and b' of the monoclinic structure are not equivalent and they must be designated separately by [110] and [$\bar{1}$10]. The minus sign above an index means that the index has a negative value.

The orientation of a crystal plane can similarly be determined by three numbers called the *Miller indices* of a plane. These three numbers are determined from the intercepts of the plane (Fig. 2.13a) in the following manner. First, we express u, v, and w

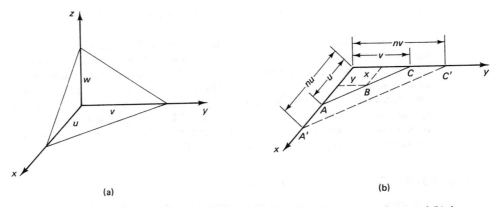

Figure 2.13 Figures showing (a) the intercepts of a plane along the three crystal axes and (b) the relation between the intercepts of two parallel planes. Note that the ratio of the intercepts remains unchanged by a parallel shift of the plane.

in units of lattice constants and take the reciprocals of these numbers to obtain a set of numbers h', k', and l' such that

$$h' = \frac{a}{u,} \qquad k' = \frac{b}{v,} \qquad l' = \frac{c}{w} \qquad (2.4.2)$$

Then we reduce the set of numbers h', k', and l' to the smallest set of three integers h, k, and l. The Miller indices of a plane are written in parentheses as (hkl). For example, to obtain the Miller indices of a plane that has intercepts $3a$, $6b$, and $2c$, we take the reciprocals of 3, 6, and 2 to obtain $\frac{1}{3}$, $\frac{1}{6}$, and $\frac{1}{2}$. The resultant fractions are then reduced to the smallest set of three integers having the same ratio. These three integers are 2, 1, and 3. Thus the Miller indices of the plane are (2 1 3).

We should point out that multiplication of the three indices by a common factor does not change the orientation of a plane because the two planes, (2 1 3) and (4 2 6), for example, are parallel. The plane shown in Fig. 2.13a satisfies the equation

$$\frac{x}{u} + \frac{y}{v} + \frac{z}{w} = 1 \qquad (2.4.3)$$

From geometry we know that if u, v, and w in Eq. (2.4.3) are replaced, respectively, by nu, nv, and nw, the new equation represents a new plane parallel to the old plane. This relation can easily be seen in a two-dimensional space. Referring to Fig. 2.13b, we have

$$\frac{x}{u} = \frac{BC}{AC,} \qquad \frac{y}{v} = \frac{AB}{AC} \qquad (2.4.4)$$

Since $z = 0$ in the xy plane, combining the two equations leads to Eq. (2.4.3). Let us return to the two-dimensional space. The two lines represented by

$$\frac{x}{u} + \frac{y}{v} = 1 \quad \text{and} \quad \frac{x}{nu} + \frac{y}{nv} = 1 \qquad (2.4.5)$$

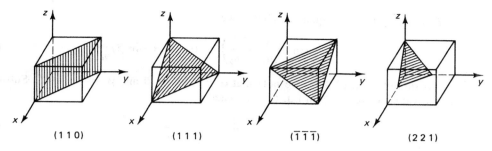

(1 1 0)   (1 1 1)   ($\bar{1}$ $\bar{1}$ $\bar{1}$)   (2 2 1)

Figure 2.14 Illustrations showing several planes and their Miller indices.

are parallel because they have the same slope. For example, the line $A'C'$ is parallel to AC in Fig. 2.13b. Like the direction indices, the plane indices are not limited to a specific plane, but represent a set of parallel planes.

As examples, several planes (110), (111)), ($\overline{111}$), and (221) in a cubic crystal are shown in Fig. 2.14. As can be seen from Fig. 2.14, the two planes (111) and ($\overline{111}$) are parallel to each other. Furthermore, all the diagonal planes in a cubic crystal, even though not parallel, are equivalent in the sense that the arrangements of atoms in these planes are the same. Therefore, the planes (111), ($\bar{1}$11), (1$\bar{1}$1), and (11$\bar{1}$) are said to belong to the set {111}. Two planes (0001) and (10$\bar{1}$0) in a hexagonal crystal are shown in Fig. 2.15a. For hexagonal systems, it is conventional to use four indices (hkjl). Instead of using two vectors **a** and **b** in the basal plane as we did in Fig. 2.3, we introduce a third axis. The three axes a_1, a_2, and a_3 form an angle of 120° with each other. The three indices h, k, and j are determined by the intercepts of the plane with the three axes. Referring to Fig. 2.15b and applying the law of sine, we find that

$$\frac{\sin \theta}{u_1} = \frac{\sin \alpha}{u_3}, \qquad \frac{\sin (180° - \theta)}{u_2} = \frac{\sin \beta}{u_3} \tag{2.4.6}$$

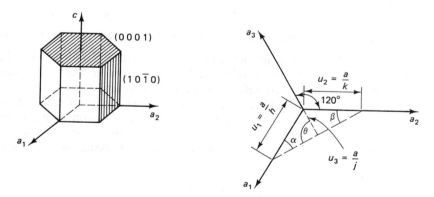

Figure 2.15 Illustrations showing two planes and their Miller indices in a hexagonal crystal. Three axes a_1, a_2, and a_3 are used to specify the three intercepts in the basal plane. However, information concerning the third intercept u_3 is redundant since a relation exists between u_1, u_2, and u_3.

Sec. 2.4   Crystallographic Notation

43

Combining the two equations above, we obtain

$$\sin\theta\left(\frac{1}{u_1} + \frac{1}{u_2}\right) = (\sin\alpha + \sin\beta)\frac{1}{u_3} \tag{2.4.7}$$

To simplify Eq. (2.4.7), we note that $\alpha = 120° - \theta$ and $\beta = \theta - 60°$. Substituting the values of α and β in Eq. (2.4.7) leads to

$$\frac{1}{u_1} + \frac{1}{u_2} = \frac{1}{u_3} \tag{2.4.8}$$

Since one intercept (in the present case, the intercept on the $\mathbf{a}_3$ axis) is negative, Eq. (2.4.8) gives

$$h + k + j = 0 \tag{2.4.9}$$

Thus the added index j serves as a check and does not provide any new information about the plane.

2.5 RECIPROCAL-LATTICE VECTORS AND PLANE NORMAL

The concept of the reciprocal lattice is sometimes extremely useful in describing the periodic property of a crystal. Here we introduce the three basis vectors defining a reciprocal lattice, but leave the use of the reciprocal-lattice concept to later discussions. If $\mathbf{a}$, $\mathbf{b}$, and $\mathbf{c}$ are the basis vectors of the primitive cell, we define three new vectors, $\mathbf{a}^*$, $\mathbf{b}^*$, and $\mathbf{c}^*$, such that

$$\mathbf{a}^* = \frac{\mathbf{b} \times \mathbf{c}}{\mathbf{a} \cdot (\mathbf{b} \times \mathbf{c})}, \qquad \mathbf{b}^* = \frac{\mathbf{c} \times \mathbf{a}}{\mathbf{a} \cdot (\mathbf{b} \times \mathbf{c})}, \qquad \mathbf{c}^* = \frac{\mathbf{a} \times \mathbf{b}}{\mathbf{a} \cdot (\mathbf{b} \times \mathbf{c})} \tag{2.5.1}$$

From Eq. (2.5.1), it is obvious that the vectors $\mathbf{a}^*$, $\mathbf{b}^*$, and $\mathbf{c}^*$ have the following properties:

$$\mathbf{a}^* \cdot \mathbf{a} = 1, \qquad \mathbf{b}^* \cdot \mathbf{b} = 1, \qquad \mathbf{c}^* \cdot \mathbf{c} = 1 \tag{2.5.2}$$

$$\mathbf{a}^* \cdot \mathbf{b} = \mathbf{a}^* \cdot \mathbf{c} = 0, \qquad \mathbf{b}^* \cdot \mathbf{a} = \mathbf{b}^* \cdot \mathbf{c} = 0, \qquad \mathbf{c}^* \cdot \mathbf{a} = \mathbf{c}^* \cdot \mathbf{b} = 0 \tag{2.5.3}$$

The set of three vectors $\mathbf{a}^*$, $\mathbf{b}^*$, and $\mathbf{c}^*$ defines a new primitive cell. The new lattice based on the new primitive cell is called the *reciprocal lattice,* and hence the three vectors $\mathbf{a}^*$, $\mathbf{b}^*$, and $\mathbf{c}^*$ are referred to as *reciprocal-lattice vectors.*

To each lattice there exists a unique reciprocal lattice. The actual construction of a reciprocal lattice is described later when we discuss the conductive properties of metals and semiconductors. The purpose of our present discussion is merely to illustrate the use of the reciprocal-lattice vectors in the analysis of crystal geometry. Consider a vector $\mathbf{r}^*$ of the form

$$\mathbf{r}^* = h\mathbf{a}^* + k\mathbf{b}^* + l\mathbf{c}^* \tag{2.5.4}$$

This vector $\mathbf{r}^*$ is a normal to the (hkl) plane. To prove this statement, we refer to Fig. 2.16. If the intersections of the (hkl) plane with xy and yz planes are represented, respectively, by $\mathbf{u} - \mathbf{v}$ and $\mathbf{v} - \mathbf{w}$, then using Eqs. (2.5.1) to (2.5.4), we have

$$\mathbf{r}^* \cdot (\mathbf{u} - \mathbf{v}) = h\mathbf{a}^* \cdot \mathbf{u} - k\mathbf{b}^* \cdot \mathbf{v} = 0, \qquad \mathbf{r}^* \cdot (\mathbf{v} - \mathbf{w}) = 0 \tag{2.5.5}$$

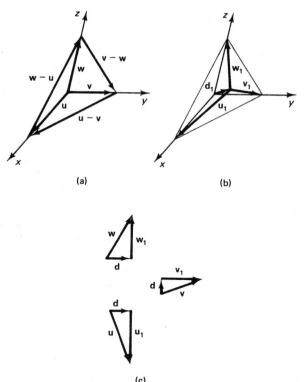

(a)

(b)

(c)

Figure 2.16 Diagrams showing (a) the intercepts of the (*hkl*) plane with the crystal axes, (b) the vector **d** normal to the (*hkl*) plane, and (c) the vectorial relation between the vector **d** and the other vectors.

Since any vector in the (*hkl*) plane can be expressed as a linear combination of $\mathbf{u} - \mathbf{v}$ and $\mathbf{v} - \mathbf{w}$, we have shown that the vector **r*** is indeed a normal to the (*hkl*) plane. Any vector **r** with direction indices [*u'v'w'*] is parallel to the (*hkl*) plane if

$$\mathbf{r} \cdot \mathbf{r^*} = u'h + v'k + w'l = 0 \tag{2.5.6}$$

The distance d separating two adjacent planes can also be found from **r***. As illustrated in Fig. 2.16b and c, the distance d is the projection of any one of the three intercepts in the normal direction. This means that

$$d = \mathbf{u} \cdot \frac{\mathbf{d}}{d} = \mathbf{u} \cdot \frac{\mathbf{r^*}}{|r^*|} = \frac{\mathbf{a}}{h} \cdot \frac{\mathbf{r^*}}{|r^*|} = \frac{1}{|r^*|} \tag{2.5.7}$$

Equations (2.5.4) and (2.5.7) are especially useful for cubic crystals because they yield extremely simple results. In a simple cubic lattice, the three primitive vectors **a**, **b**, and **c** are mutually perpendicular and equal in magnitude. Using this information in Eq. (2.5.1), we find that

$$\mathbf{a^*} = \frac{\mathbf{a}}{a^2}, \qquad \mathbf{b^*} = \frac{\mathbf{b}}{a^2}, \qquad \mathbf{c^*} = \frac{\mathbf{c}}{a^2} \tag{2.5.8}$$

where a is the magnitude of the three primitive vectors. After substituting Eq. (2.5.8) into Eq. (2.5.4), we can see that the normal to the (*hkl*) plane is in the [*hkl*] direction.

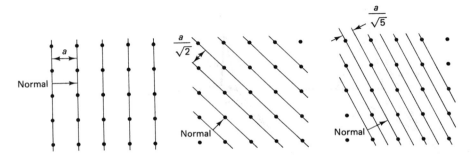

Figure 2.17 A projection of a cubic lattice in an *xy* plane. The lines in the three diagrams represent the intersections of the three crystal planes (100), (110), and (210) with the *xy* plane.

Thus in a simple cubic crystal, all planes and directions with identical indices are perpendicular. For example, the [110] direction is normal to the (110) plane, as illustrated in Fig. 2.17 together with two other examples.

From Eq. (2.5.8) we also note that the three reciprocal-lattice vectors in Eq. (2.5.4) are mutually perpendicular. Thus Eq. (2.5.7) simply reduces to

$$d = \frac{1}{|r^*|} = \frac{a}{(h^2 + k^2 + l^2)^{1/2}} \qquad (2.5.9)$$

which holds for all cubic crystals if we use the conventional crystal axes of a simple cubic structure as reference axes in defining the indices (*hkl*). Figure 2.17 shows the projection of a simple cubic lattice in the *xy* plane. Any line thus represents a (*hk*0) plane in a three-dimensional drawing. Using Fig. 2.17, the reader is asked to verify that the distances between (100) planes, between (110) planes, and between (210) planes are, respectively, a, $a/\sqrt{2}$, and $a/\sqrt{5}$, as obtained from Eq. (2.5.9). We also see from Fig. 2.17 that planes with lower indices not only have a wider interplanar spacing but also a higher density of atoms in the plane.

2.6 INTERFERENCE PHENOMENON

In Section 2.3 we showed the distribution of atoms in the unit cell of a selected number of simple crystal structures. Such information concerning the distribution of atoms in a crystal can be obtained from diffraction experiments. In a diffraction experiment a wave having a wavelength comparable to atom spacing is incident on a crystal, and the wave diffracted from the crystal is measured as a function of the orientation of the sample with respect to the incident wave. The physical phenomenon important to the experiment is the interference of two waves. It is the purpose of this and the next section to discuss the interference phenomenon and to introduce waves of different types suitable for use in a diffraction experiment. The actual diffraction experiment is described in Section 2.8.

Before we treat the interference of waves, we first consider a simpler problem, the superposition of two ac voltages of equal amplitude:

$$V = V_0 \cos (\omega t + \beta_1) + V_0 \cos (\omega t + \beta_2) \qquad (2.6.1)$$

It is obvious from Eq. (2.6.1) that the amplitude of the resultant voltage V depends on the relative phase $\beta_1 - \beta_2$ of the two component voltages. The reader can easily show that $V = 0$ for $\beta_1 - \beta_2 = \pi$ and that the amplitude of V is a maximum and equal to $2V_0$ for $\beta_1 = \beta_2$. To make our analysis relevant to our subsequent discussion of diffraction experiments, we consider the superposition of n voltages, whose phase angles are shifted progressively by β. Thus the resultant voltage takes the form

$$V_0 = V_0\{\cos(\omega t) + \cos(\omega t + \beta) + \cdots + \cos[\omega t + (n-1)\beta]\} \quad (2.6.2)$$

For convenience, the phase angle of one component is chosen to be the reference phase angle and hence is zero in Eq. (2.6.2).

The problem of adding voltages should be familiar to electrical engineers. Each component can be considered in a complex plane as the projection of a vector upon the real axis. Analytically, we have

$$V = \mathrm{Re}\, V_0 \exp(i\omega t) \sum_{m=0}^{n-1} \exp(im\beta) \quad (2.6.3)$$

where Re means to take the real part. Figure 2.18a shows the position of six vectors in the complex plane. To find the summation in Eq. (2.6.3), we add these vectors in a manner illustrated in Fig. 2.18b. The amplitude of the resultant voltage V is given by the length OA. Were β equal to zero, the amplitude V would have the length OB. As we can see, the phase difference causes a partial cancellation of the combined effects of the voltages. This phenomenon is called *interference*.

The sum in Eq. (2.6.3) can be evaluated by multiplying and then dividing by a factor $1 - \exp(i\beta)$. Realizing that

$$\sum_{m=0}^{n-1} x^m = \frac{1 - x^n}{1 - x} \quad (2.6.4)$$

and using $\exp(i\beta)$ for x in Eq. (2.6.4), we have

$$V = \mathrm{Re}\, V_0 \frac{\exp(i\omega t + in\beta/2)}{\exp(i\beta/2)} \frac{\sin(n\beta/2)}{\sin(\beta/2)} \quad (2.6.5)$$

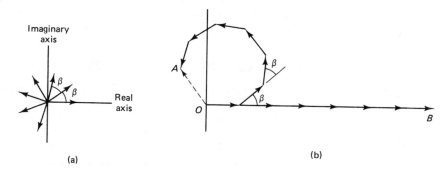

(a)

(b)

Figure 2.18 (a) Six vectors denoting six voltages in phasor representation at $t = 0$. The instantaneous value of a voltage is given by the projection of the vector upon the real axis. (b) The vectorial addition of the six voltages. The amplitude of the resultant voltage is given by the length OA.

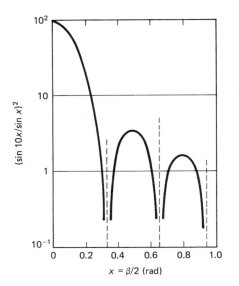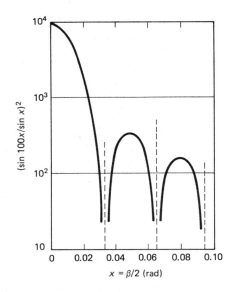

Figure 2.19 The values of $\sin^2(nx)/\sin^2 x$ as a function of x for (a) $n = 10$ and (b) $n = 100$. Note that as n increases, the curve becomes sharper.

Let us define an amplitude factor A, which is the ratio of the amplitude of V to V_0. From Eq. (2.6.5) we find that

$$A = \frac{\sin (n\beta/2)}{\sin (\beta/2)} \qquad (2.6.6)$$

The quantity A^2 is plotted in Fig. 2.19 as a function of $\beta/2$ for two values of $n = 10$ and 100. The function drops very sharply as β deviates from zero. Note the changes in scale from the case $n = 10$ to the case $n = 100$. As n increases, the range of β in which the value of A^2 remains nearly constant becomes smaller. Within this small range of β, the amplitude of V can be approximated by nV_0. For sufficiently large n, the condition for $A^2 \cong n^2$ practically demands $\beta = 0$ in Eqs. (2.6.6) and (2.6.5). This condition is known as the *condition for constructive interference*.

2.7 ELECTROMAGNETIC WAVES AND MATTER WAVES

Let us now discuss the sources of incident waves that are used in diffraction experiments. Figure 2.20a illustrates the span of the electromagnetic wave spectrum expressed in terms of both the frequency and the wavelength. Referring to Table 2.3, we see that atom spacing in crystals is on the order of 3 Å or 3×10^{-10} m. The form of electromagnetic wave having its wavelength comparable to atom spacing in crystals is the X ray. At low frequencies, we are accustomed to the use of voltage or current. At high frequencies, however, an electric signal is generally described in terms of an electric field rather than a voltage. As a matter of fact, the voltage concept is no longer useful because we must consider the effects associated with the propagation of an electric signal. A set of equations that describe the propagation of an electromagnetic wave is known as *Maxwell's equations*. In terms of electric field E and magnetic field H, Max-

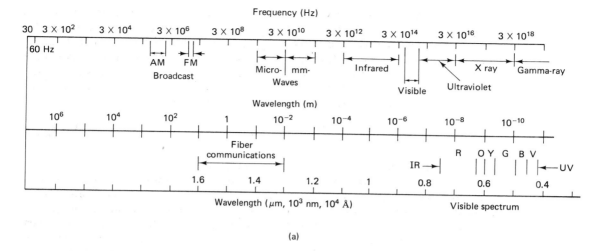

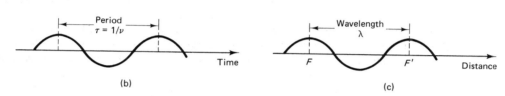

Figure 2.20 (a) The electromagnetic wave spectrum expressed in frequency (top scale) and in wavelength (in meters, middle scale and in micrometers, bottom scale). The letterings for the bottom scale have the following meaning: IR (infrared), R (red), O (orange), Y (yellow), G (green), B (blue), V (violet), and UV (ultraviolet). (b) Variation of electromagnetic wave as a function of time at a fixed distance. (c) Propagation of electromagnetic wave as a function of distance taken at a fixed time. The electric field varies both in time and in space. In part (a), the top and middle scales, related by $c = \nu\lambda$, are logarithmic and the bottom scale is linear.

well's equations can be derived from fundamental laws such as Faraday's law, Biot's law, and the laws of conservation of electric and magnetic fluxes.

A familiar example, the analysis of which must be based on Maxwell's field equations, is the propagation of radio waves. Since an analysis of the solution of Maxwell's equations is outside the scope of this book, we mention only the pertinent results from such an analysis. Suppose that a radio broadcasting station is sending out a signal at $t = 0$, and the electric field set up by the signal is $E_0 \cos \omega t$. If a receiver is placed at a distance d from the station to detect the electric field, there will be a time delay of d/v second in the detected signal on account of the finite time required for the propagation of the signal. In other words, the electric field at the receiver is

$$E = E_1 \cos \left[\omega \left(t - \frac{d}{v} \right) \right] = E_1 \cos (\omega t - \beta_1) \qquad (2.7.1)$$

where v is the velocity of propagation of the signal. Equation (2.7.1) is plotted in Fig. 2.20b as a function of t at a fixed d.

Figure 2.20c shows the variation of E with distance at a fixed time t. Within the short distance of travel illustrated in Fig. 2.20c, the amplitude E_1 stays constant. We define a wavelength λ as the distance traveled by the field in order for the field of Eq. (2.7.1) to go through a complete cycle. Suppose that the electric field is a maximum at point F, and again reaches a maximum at F'. The argument of the cosine function changes by 2π from F to F'. Since the time t is fixed in Eq. (2.7.1), the change in the argument is entirely due to the change in d. Thus we obtain

$$\frac{\omega\lambda}{v} = 2\pi \quad \text{or} \quad v = \nu\lambda \qquad (2.7.2)$$

where $\nu = \omega/2\pi$ is the frequency of the electromagnetic radiation.

As mentioned earlier, X rays, visible lights, and microwave and radio waves are electromagnetic in nature. The disturbances caused by these waves can be described in terms of an electric field of Eq. (2.7.1) and a related expression for the magnetic field, despite the tremendous difference in their frequency and wavelength. In the late seventeenth century, two drastically different theories concerning the nature of light were proposed, the corpuscle theory of Newton and the wave theory of Huygens. The controversy was not resolved until the wave nature was confirmed in Young's interference experiment in the early nineteenth century. As shown in Fig. 2.21, a beam of monochromatic light is converted into parallel beams by means of a lens. On the screen a series of bright and dark stripes appears as a result of interference of the two beams diffracted by the two slits. When $d_1 - d_2 = n\lambda$, where n is an integer, $\beta_1 - \beta_2 = n2\pi$ in Eq. (2.7.1), the two waves are in phase and bright stripes appear. On the other hand, a complete cancellation of the two waves results when $d_1 - d_2 = (n + \frac{1}{2})\lambda$.

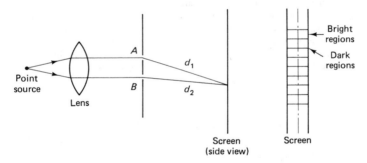

Figure 2.21 Schematic illustration of Young's interference experiment. Note that the light beams passing through the two slits A and B have different optical path lengths.

The main points in the discussion above are the following. First, light beams and X rays, being an electromagnetic wave, can be described by an electric field E and a companion magnetic field H. For the present discussion, E alone suffices. The superposition of two light beams or X rays can be treated in a similar fashion as the superposition of two ac voltages. The relative phase of two waves originating from the same source is determined by the difference in the path lengths of the two waves. If we have n wave components differing progressively in phase by β, the intensity I of the resultant wave is

$$I = A^2 I_0 = \left| \frac{\sin (n\beta/2)}{\sin (\beta/2)} \right|^2 I_0 \qquad (2.7.3)$$

where I_0 is the intensity of each individual wave. At high frequencies a wave is measured by its power, not by its amplitude. This is the reason why we have plotted the quantity A^2 in Fig. 2.19. Using $n = 2$ in Eq. (2.7.3), we see that $I = 4I_0$ for $\beta = 0$ and $I = 0$ for $\beta = (n + \frac{1}{2})2\pi$, where n is an integer. These predictions have been proved correct in Young's interference experiment.

An X ray of sufficient intensity to be of practical use is produced in a cathode ray tube by the impact of high-velocity electrons with matter. As a result of the impact, certain atoms of the anode are ionized. This ionization involves the expulsion of inner core electrons. The place thus evacuated is then reoccupied by an electron from an outer orbit. During the electron transition, an electromagnetic radiation is emitted. For example, the 2.29-Å line in chromium is caused by the transition of an electron from the $2p$ shell to the $1s$ shell which is made available during bombardment by high-energy electrons. The energy released on the transition of an electron from one atomic level to another is in the form of electromagnetic waves. Using Eqs. (1.2.3) and (2.7.2), we find the wavelength λ of the electromagnetic wave to be

$$\lambda = \frac{c}{\nu} = \frac{hc}{\mathscr{E}_1 - \mathscr{E}_2} = \frac{12{,}400}{\Delta\mathscr{E}(\text{eV})} \quad \text{Å} \qquad (2.7.4)$$

In Eq. (2.7.4), λ is the free-space wavelength expressed in angstroms, v is set equal to the velocity of light $c = 3 \times 10^8$ m/s, and the energy difference $\Delta\mathscr{E}$ is expressed in electron volts (1 eV = 1.6×10^{-19} J).

Besides electromagnetic waves, we can use neutrons and electrons in a diffraction experiment. It might seem highly improbable that particles were capable of producing interference phenomenon. Before quantum mechanics was introduced, de Broglie proposed that a particle in motion may have wavelike properties, and suggested that electrons having a momentum p should be diffracted as if they had a wavelength

$$\lambda = \frac{h}{p} = \frac{h}{(2\mathscr{E}m)^{1/2}} \qquad (2.7.5)$$

where $\mathscr{E}$ is the kinetic energy of an electron. Equation (2.7.5) was later verified in the electron diffraction experiment of Davisson and Germer.

The *Davisson–Germer experiment* is of historical importance because it confirmed that matter may have wavelike properties. It is this wave–particle dualism that led to the development of quantum mechanics. Therefore, the wavelike behavior of matter can be explained only on the basis of quantum mechanics. According to *Heisenberg's uncertainty principle* (Section 1.4), we must abandon the idea that the position of a particle can be specified precisely. On the atomic scale, the position of a particle must be described by a wave function ψ of the form

$$\psi(x, t) = A(x) \cos \left[2\pi \left(\nu t - \frac{x}{\lambda} \right) \right] \qquad (2.7.6)$$

where λ is given by Eq. (2.7.5). For simplicity, we present a one-dimensional case in Eq. (2.7.6).

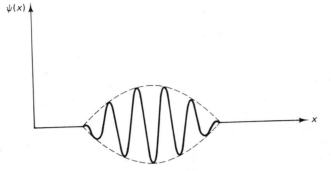

$\psi(x)$

x

Figure 2.22 Schematic representation of a one-dimensional wave packet. Such a localized wave is used to describe the behavior of a particle on the atomic scale. It accounts for the localized nature of a particle without, at the same time, violating the uncertainty princple.

As discussed in Section 1.3, the wave function ψ has the physical meaning that $|\psi|^2\, dx$ is the probability of finding a particle in space element dx. Macroscopically, a particle must behave like a particle. Thus the amplitude function $A(x)$ in Eq. (2.7.6) cannot extend far in distance and must be localized. Such a localized wave is called a *wave packet* and is illustrated schematically in Fig. 2.22. The range in which $A(x)$ is nonvanishing is determined by the uncertainty principle $\Delta p\, \Delta x \geq \hbar/2$. In dealing with macroscopic phenomena, for example the motion of an electron in electric and magnetic fields, we still can treat the electron as if it were a point charge moving according to Newton's laws because the uncertainty in distance Δx is by many orders of magnitude smaller than the physical dimension of any laboratory-size object. In dealing with the interaction of electrons with atoms in a lattice, however, the uncertainty Δx is significant because it is comparable in magnitude to atom spacing in crystals. Electrons interfere with each other in a diffraction experiment on account of the phase factor $\cos(\omega t - 2\pi x/\lambda)$ in Eq. (2.7.6). We shall use Eq. (2.7.6) again when we discuss the conductive properties of metals and semiconductors.

Wave properties are not confined to electrons, but are a general attribute of matter. Neutrons are now extensively used in the study of crystal structures. Insofar as the diffraction experiment is concerned, two important differences should be noted. First, electrons are charged and interact strongly with matter. As a result, electrons can penetrate only a relatively short distance into a crystal. This fact limits electron-diffraction studies to the structure of thin crystals and crystal surfaces. Second, neutrons have a magnetic moment and hence interact with electrons in a magnetic substance. Neutron-diffraction studies have been extremely useful in ascertaining the magnetic structures of antiferromagnetic and ferrimagnetic materials. Finally, we note that neutrons have a much larger mass ($M = 1.675 \times 10^{-27}$ kg) than electrons. To make $\lambda = 1$ Å in Eq. (2.7.5), we need $\mathscr{E} = 0.079$ eV for neutrons and $\mathscr{E} = 150$ eV for electrons. Neutrons from atomic reactors have to be slowed down considerably, as a matter of fact, almost to the thermal velocity for use in a diffraction study of crystal structures.

2.8 CRYSTAL DIFFRACTION

Let us apply the results of our analyses in Sections 2.6 and 2.7 to a diffraction experiment. When an X ray impinges on a crystal, the motion of electrons inside each atom is perturbed by the ray. The perturbed motion creates an oscillating electric dipole and thus causes the atom to radiate on the transition returning to its normal state. This

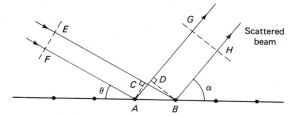

radiation, which we shall refer to as the scattered X ray, is nearly isotropic in all directions. Our task is to examine the intensity of the scattered radiation.

Figure 2.23 shows the scattering of an X ray by a plane of atoms. For clarity, we refer specifically to the scattered rays from atoms A and B. Let us examine the relative phase of the two scattered rays. In Fig. 2.23, the two dashed lines EF (perpendicular to the incident rays) and GH (perpendicular to the scattered rays) represent, respectively, the wavefronts of the incident and scattered rays. All the points on the incident wavefront have the same phase and all the points on the scattered wavefront have the same phase. Thus the difference in the path lengths of the two rays is given by

$$d_A - d_B = FA + AG - EB - BH = AD - BC \qquad (2.8.1)$$

Since the two rays are parallel, we have

$$d_A - d_B = AB(\cos\alpha - \cos\theta) \qquad (2.8.2)$$

To make $\beta_A = \beta_B$ in Eq. (2.7.1), α must equal θ in Eq. (2.8.2). Thus for constructive interference (i.e., for scattered rays from all atoms in a plane to be in phase), the angle of incidence equals the angle of reflection.

It is clear from the great penetrating power of X rays that a single plane of atoms can diffract only a tiny fraction of the incident rays. Therefore, we must add contributions to the diffracted beam from many planes that are parallel. In Fig. 2.24, the diffraction by two adjacent parallel planes is shown. The difference in the path lengths of the two diffracted rays is

$$d_B - d_A = FBH - EAG = BC + BD = 2d_{hkl}\sin\theta \qquad (2.8.3)$$

where d_{hkl} is the spacing between two adjacent (hkl) planes and is given by Eq. (2.5.7).

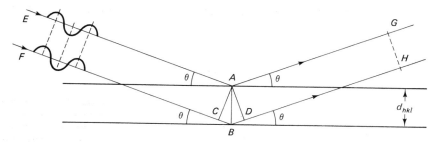

Figure 2.24 Diffraction of an X ray by atoms in two parallel planes (hkl). For constructive interference of the diffracted rays AG and BH, the incident angle θ must satisfy the Bragg condition expressed in Eq. (2.8.5).

We note from Eq. (2.6.6) that the quantity A^2 is periodic and that $|A(\beta)|^2 = |A(\beta + n2\pi)|^2$, where n is an integer. For $|A|^2$ to be a maximum, β must be either zero or any integral multiple of 2π. Thus the condition for constructive interference of the two rays is

$$\beta = \beta_B - \beta_A = \frac{\omega}{v}(d_B - d_A) = n2\pi \qquad (2.8.4)$$

where n is an integer. Substituting Eq. (2.8.3) into Eq. (2.8.4) and using Eq. (2.7.2) for the value of ω/v, we find the condition for constructive interference to be

$$2d_{hkl}\sin\theta = n\lambda \qquad (2.8.5)$$

Equation (2.8.5) is the *Bragg law of diffraction*.

Although we use X rays as the incident beam in the discussion above, Eq. (2.8.5) applies equally well to electron- and neutron-diffraction experiments. Both electrons and neutrons are preferentially scattered by a crystal into those directions specified by Eq. (2.8.5). As a matter of fact, it is the verification of Eq. (2.8.5) in electron- and neutron-diffraction experiments that proves the wave–particle dual nature of electrons and neutrons.

Let us discuss briefly experimental methods of crystal diffraction. These are the powder method, the rotating-crystal method, and the Laue method. In the powder method illustrated in Fig. 2.25, a monochromatic X ray or neutron beam is incident on a finely powdered specimen. Because a powder is a collection of many randomly oriented crystallites, we can find in some, but not all, crystallites (hkl) planes that satisfy the Bragg law of diffraction. However, the diffracted beams will not have a single direction but will form a cone that makes an angle θ with the incident beam. If a photographic film is placed along a circle around the powder, diffracted rays intercepted by the film are marked by dark concentric rings. We can also place a detector (at A) and the specimen (at C) on a table that rotates about C. The intensity of the diffracted beam is plotted as a function of 2θ, and a fit of Eq. (2.8.5) to the observed curve gives the value of the lattice constant.

Let us use diamond as an example. The lattice constant of diamond is 3.56 Å (Table 2.3). The interplanar spacing is found from Eq. (2.5.9) to be

$$d_{111} = \frac{3.56}{\sqrt{3}} = 2.06 \text{ Å}, \qquad d_{220} = \frac{3.56}{2\sqrt{2}} = 1.26 \text{ Å}, \qquad \text{and so on} \qquad (2.8.6)$$

If a neutron beam having an energy 0.069 eV is used, we obtain from Eq. (2.7.5) a

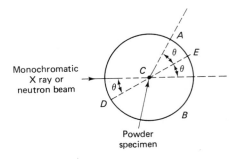

Monochromatic
X ray or
neutron beam

Powder
specimen

Figure 2.25 The powder method used in X ray or neutron diffraction. The specimen is a fine powder. Some microcrystals of the powdered specimen have a plane, such as *DE*, which is oriented at the correct diffraction angle θ. In the powder method, the incident beam must be monoenergetic.

Crystal Structure   Chap. 2

value of $\lambda = 1.07$ Å. Using $n = 1$ in Eq. (2.8.5), we find the angle of scattering to be

$$\theta_{111} = \sin^{-1}\frac{1.07}{4.12} = 15°, \qquad \theta_{220} = \sin^{-1}\frac{1.07}{2.52} = 25° \qquad (2.8.7)$$

The observed angles (Fig. 2.26) of the scattered neutron beam indeed agree with the calculated values. The reader is asked to show that the other two observed angles in Fig. 2.26 can also be accounted for. In the calculation above, we use the crystal axes of a simple cube to define the plane indices (*hkl*). How the difference between a simple and a face-centered cubic structure will affect the result of diffraction experiments will be discussed shortly.

The rotating-crystal method is used for structure determination in single crystals. The experimental arrangement is similar to that shown in Fig. 2.25 except that the specimen is a single crystal and is rotating. Suppose that the crystal is oriented with the *c* axis parallel to the axis of rotation. Then all planes with $l = 1$ and $l = -1$, for example, will reflect in layers above and below the horizontal plane. The separation between the layers recorded on the film determines the spacing between crystal planes. In the Laue method, a white X ray is incident on a stationary crystal and the diffracted beam is recorded as darkened spots on a photographic film. Since the wavelength λ of the beam covers a wide range, many planes will satisfy the Bragg equation. The positions of the darkened spots on the Laue pattern will change as the orientation of the crystal with respect to the incident beam changes. Known ways for indexing the Laue pattern are available. The method is most extensively used to orient crystals. If a crystal is oriented in one of the principal directions, the pattern will show the crystal symmetry with respect to that axis. For example, if the $\langle 100 \rangle$ axis of a cubic crystal is oriented parallel to the beam, the pattern will show a fourfold symmetry.

Now we return to Fig. 2.26. We find that the diffracted beam from the (100) plane is not shown. As a matter of fact, many other lines that would be expected from Eq. (2.8.5) do not appear in the figure. As mentioned earlier, the crystal axes of a simple cubic structure are used as reference axes in defining the plane indices. This is equivalent to neglecting all other atoms in the unit cell except the corner atoms. To

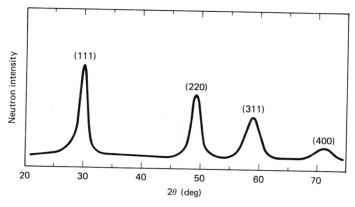

Figure 2.26 Observed neutron-diffraction pattern for diamond.

explain the diffraction pattern of crystals other than the primitive types, we must include the contributions to the diffracted beam from all atoms in the unit cell. Let f_j be the scattering power of the jth atom. Then the total scattered wave can be written as

$$\sum_j E_j = \sum \text{Re } f_j E_0 \exp\ (i\omega t - i\beta_j) \qquad (2.8.8)$$

where E_0 is the amplitude of the incident wave and β_j is the phase angle. The symbol Re stands for taking the real part.

For different atoms in the unit cell, the phase angle β_j in Eq. (2.8.8) is different. The value of β_j can be evaluated if the path of the diffracted beam is known. If a beam is diffracted from the (hkl) plane, the phase angle is

$$\beta_j = 2\pi(u_j h + v_j k + w_j l) \qquad (2.8.9)$$

for an atom situated at a distance $\mathbf{r} = u_j\mathbf{a} + v_j\mathbf{b} + w_j\mathbf{c}$ from the origin of the unit cell. A proof of Eq. (2.8.9) may be found in Problem 2.14 by letting $2d \sin \theta = \lambda$. The geometrical structure factor S is defined as

$$S = \sum_j f_j \exp\ [-2\pi i(u_j h + v_j k + w_j l)] \qquad (2.8.10)$$

In terms of S, the intensity I of a diffracted beam is

$$I \sim \left| \sum E_j \right|^2 \sim I_0\ |S|^2 \qquad (2.8.11)$$

where I_0 is the intensity of the incident beam.

Let us apply Eq. (2.8.10) to the diffraction curve of Fig. 2.26. For diffraction from the (100) plane, Eq. (2.8.10) reduces to

$$S = f \sum_j \exp\ (-2\pi i u_j) \qquad (2.8.12)$$

Referring to the diamond structure shown in Fig. 2.11, we find that

$$u_j = 0, 0, 0, 0, 1, 1, 1, 1 \qquad \text{for eight corner atoms}$$

$$0, \tfrac{1}{2}, \tfrac{1}{2}, \tfrac{1}{2}, \tfrac{1}{2}, 1 \qquad \text{for six face atoms}$$

$$\tfrac{1}{4}, \tfrac{1}{4}, \tfrac{3}{4}, \tfrac{3}{4} \qquad \text{for four inside atoms}$$

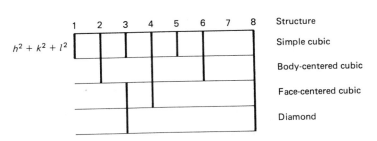

Figure 2.27 Diffraction patterns from various cubic structures. The plane indices (hkl) refer to the crystal axes of a simple cubic structure. The existence of a diffracted wave from a given plane is indicated by a vertical line.

In evaluating S, we should again note that corner atoms are shared by eight cells and face atoms are shared by two cells. Thus we have

$$S_{100} = \frac{f}{8}8 + \frac{f}{2}(2 - 4) + f(-2i + 2i) = 0 \tag{2.8.13}$$

Equation (2.8.13) shows that the contribution from the face atoms and the inside atoms completely cancels out that from the corner atoms for diffraction from the (100) plane. The experimental result of Fig. 2.26 is in agreement with this prediction. Similar calculations can be made for diffraction from other planes. The results of such calculations for several cubic structures are shown in Fig. 2.27. Those diffractions with a nonvanishing S are indicated by vertical lines.

2.9 ROTATIONAL SYMMETRY AND MATERIAL-PARAMETER TENSOR

In Section 2.2 we showed that starting with a unit cell, we can generate the whole lattice by a translation vector **T** defined in Eq. (2.2.1). Depending on the specifications about the directions and the lengths of the three crystal vectors **a**, **b**, and **c**, crystals are classified into seven distinct systems given in Table 2.1. Besides translation, there are three other operations by which the periodic pattern of a lattice can be reproduced: (1) rotation through an axis, (2) mirror reflection in a plane, and (3) inversion through a point. These operations are illustrated in Fig. 2.28, using a simple cubic structure as an example.

The purpose of our present discussion is to see how material properties are affected by crystal structures. Take the material parameter ϵ, the dielectric constant, as an example. The electric flux density **D** is related to the electric field **E** through

$$\mathbf{D} = \epsilon\mathbf{E} \tag{2.9.1}$$

Since **D** and **E** are vectors, the dielectric constant ϵ is, in general, not a scalar but a tensor of rank 2. In other words, Eq. (2.9.1) takes the general form

$$\begin{bmatrix} D_x \\ D_y \\ D_z \end{bmatrix} = \begin{bmatrix} \epsilon_{11} & \epsilon_{12} & \epsilon_{13} \\ \epsilon_{21} & \epsilon_{22} & \epsilon_{23} \\ \epsilon_{31} & \epsilon_{32} & \epsilon_{33} \end{bmatrix} \begin{bmatrix} E_x \\ E_y \\ E_z \end{bmatrix} \tag{2.9.2}$$

To have a complete knowledge about ϵ, we must know how many independent parameters ϵ_{ij} are there and what are the values of these parameters. In the following discussion, we consider only the rotational symmetry of a crystal because such considerations

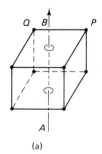

(a)

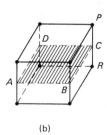

(b)

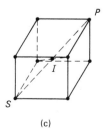

(c)

Figure 2.28 Symmetry operations in a crystal: (a) rotations through an axis, (b) reflection in a plane, and (c) inversion through a point. A simple cubic cell is used for illustration. Starting with lattice point P, we get lattice point Q by a 90° rotation about the vertical axis, lattice point R by a reflection in a horizontal plane, and lattice point S by an inversion through the point I marked by $\times$.

will tell us how many nonzero ϵ_{ij}'s are there. This applies also to any material-parameter tensor of rank 2. For material parameters described by a tensor of rank 3, we must consider rotational symmetry together with reflection and inversion operations. Examples involving rank 3 tensor include electrooptic and nonlinear optical effects.

Let us now discuss the rotational symmetry of a crystal. The axis of rotation is called an n-fold axis if the angle ϕ of rotation to carry a unit cell into itself is equal to $\phi = 2\pi/n$. Obviously, the number n must be an integer. In other words, the arrangement of similar atoms about the rotation axis must take in a two-dimensional space the shape of a regular polygon, for example, an equilateral triangle for a threefold axis, a square for a fourfold axis, and a hexagon for a sixfold axis. These polygons are shown in Fig. 2.29 and the centers of polygons through which passes the rotation axis are marked by $\times$. However, this is not the whole story. The pattern of atomic arrangement repeats itself throughout the space. That means that these regular polygons must be able to fill the two-dimensional space by joining other polygons of the same shape at the corners where they meet. For example, at the corner point A, the angle θ of the polygon must be such that 2π be divisible by θ. The reader can easily show that

$$\theta = \pi - \frac{2\pi}{n} \tag{2.9.3}$$

For 2π to be divisible by θ, the number n is limited to the following values: 1, 2, 3, 4, or 6. It can be seen from Fig. 2.29 that the case $n = 5$ cannot meet the requirement.

Information concerning the rotational symmetry of seven crystal systems is summarized in Table 2.4. It is obvious that we have to rotate a full 2π in order to bring a triclinic cell into itself. Therefore, $n = 1$ for triclinic crystals. The parallelogram base in a monoclinic crystal has a twofold symmetry about its center. A rotation of π about an axis passing through the two base centers will bring a monoclinic cell into itself. Thus for monoclinic crystals $n = 2$. In the orthorhombic system, all three faces are perpendicular to one another. Thus we have three twofold axes. These axes pass through the centers of two opposite faces. In the tetragonal crystal system, because the base is a square, the axis passing through the centers of the two square bases is an axis of fourfold symmetry.

In the cubic system, all three crystal axes are equivalent. This fact is shown clearly along any body diagonal of a cubic crystal. As illustrated in Fig. 2.30, atoms B, C, and D are symmetrically situated with respect to the body diagonal EA. A clockwise rotation of $2\pi/3$ about this body diagonal EA brings atom B to C, C to D, and D to B. If the coordinate axes x, y, and z are chosen, respectively, to be along the directions

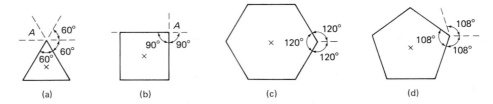

Figure 2.29 Polygons of (a) threefold, (b) fourfold, (c) sixfold, and (d) fivefold symmetry. The centers of the polygons through which the respective rotation axis passes are marked by $\times$. To form a lattice, however, the polygons must fit together to fill all the space. Therefore, it is not possible for a crystal to have a fivefold rotation axis.

TABLE 2.4 ROTATIONAL SYMMETRY OF CRYSTALS

System	Rotational symmetry
Triclinic	1
Monoclinic	2 (or a mirror plane $\perp$ to the c axis)
Orthorhombic	222
Tetragonal	4
Cubic	Four threefold axes
Hexagonal	6
Trigonal	3

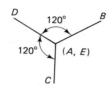

Figure 2.30 Diagrams showing the three-fold axis (body diagonal) in a cubic crystal. Since there are four body diagonals, there are altogether four threefold axes.

joining atoms A and B, A and C, and A and D, a clockwise rotation of $2\pi/3$ changes $x \rightarrow y$, $y \rightarrow z$, and $z \rightarrow x$. All four body diagonals, [111], [$1\bar{1}1$], [$\bar{1}1\bar{1}$], and [$\bar{1}\bar{1}1$], have this threefold symmetry. In the hexagonal system, the c axis is the axis of sixfold symmetry. In the trigonal system, the body diagonal joining corners at which the three angles α, β, and γ are equal is the axis of threefold symmetry. There is one and only one such threefold axis.

Before we apply symmetry considerations to the dielectric-constant tensor, we shall first prove the relation $\epsilon_{ij} = \epsilon_{ji}$ for $i \neq j$. In other words, the ϵ tensor is a symmetric tensor. The work dW done to raise electric fields in a dielectric system is

$$dW = \mathbf{E} \cdot d\mathbf{D} = \sum \epsilon_{ij} E_i \, dE_j + \sum \epsilon_{ji} E_j \, dE_i \qquad (2.9.4)$$

For dW to be a total differential, that is, W to be independent of the previous history of a sample (or the path of integration in mathematical terms), the following conditions must be met:

$$\frac{\partial W}{\partial E_j} = \sum_i \epsilon_{ij} E_i + \epsilon_{jj} E_j \qquad i \neq j \qquad (2.9.5)$$

$$\frac{\partial W}{\partial E_i} = \sum_j \epsilon_{ji} E_j + \epsilon_{ii} E_i \qquad i \neq j \qquad (2.9.6)$$

Differentiating Eq. (2.9.5) with respect to E_i and Eq. (2.9.6) with respect to E_j, we find the condition for

$$\frac{\partial^2 W}{\partial E_j \, \partial E_i} = \frac{\partial^2 W}{\partial E_i \, \partial E_j} \qquad (2.9.7)$$

to be

$$\epsilon_{ij} = \epsilon_{ji} \qquad \text{for} \quad i \neq j \qquad (2.9.8)$$

Sec. 2.9   Rotational Symmetry and Material-Parameter Tensor   **59**

Let us now apply symmetry considerations to the ϵ tensor. First we consider a cubic crystal, and let x, y, and z be the three mutually perpendicular crystal axes. A clockwise rotation of $2\pi/3$ about the [111] diagonal, as described earlier, brings $x \rightarrow y$, $y \rightarrow z$, and $z \rightarrow x$ in Eq. (2.9.2). Thus Eq. (2.9.2) becomes

$$\begin{bmatrix} D_y \\ D_z \\ D_x \end{bmatrix} = \begin{bmatrix} \epsilon_{11} & \epsilon_{12} & \epsilon_{13} \\ \epsilon_{12} & \epsilon_{22} & \epsilon_{23} \\ \epsilon_{13} & \epsilon_{23} & \epsilon_{33} \end{bmatrix} \begin{bmatrix} E_y \\ E_z \\ E_x \end{bmatrix} \tag{2.9.9}$$

Equation (2.9.9) should be identical to Eq. (2.9.2) on account of cubic symmetry. Comparing coefficients of E_x, E_y, and E_z in Eqs. (2.9.2) and (2.9.9), we obtain

$$\epsilon_{11} = \epsilon_{22} = \epsilon_{33} \tag{2.9.10}$$

$$\epsilon_{12} = \epsilon_{13} = \epsilon_{23} \tag{2.9.11}$$

The reader is asked to show that a rotation of $2\pi/3$ about the $[1\bar{1}\bar{1}]$ diagonal brings $x \rightarrow -y$, $-y \rightarrow -z$, and $-z \rightarrow x$. Thus Eq. (2.9.2) transforms into

$$\begin{bmatrix} -D_y \\ D_z \\ -D_x \end{bmatrix} = \begin{bmatrix} \epsilon_{11} & \epsilon_{12} & \epsilon_{13} \\ \epsilon_{12} & \epsilon_{22} & \epsilon_{23} \\ \epsilon_{13} & \epsilon_{23} & \epsilon_{33} \end{bmatrix} \begin{bmatrix} -E_y \\ E_z \\ -E_x \end{bmatrix} \tag{2.9.12}$$

Identifying coefficients in Eqs. (2.9.2) and (2.9.12), we find that Eq. (2.9.10) still holds. However, for the off-diagonal matrix elements, another set of equations

$$\epsilon_{12} = -\epsilon_{23}, \qquad \epsilon_{12} = -\epsilon_{13}, \qquad \epsilon_{13} = -\epsilon_{23} \tag{2.9.13}$$

must also be obeyed. The only way to satisfy both Eqs. (2.9.11) and (2.9.13) is to demand

$$\epsilon_{ij} = 0 \qquad \text{for} \quad i \neq j \tag{2.9.14}$$

We can apply symmetry considerations to other six crystal systems. The results of such analyses are summarized as follows. Insofar as the second-rank material-parameter tensor is concerned, crystal systems can be grouped into three categories: (1) cubic crystals; (2) tetragonal, trigonal, and hexagonal crystals; and (3) orthorhombic, monoclinic, and triclinic crystals. For cubic crystals, the parameter describing material property is isotropic. We have one and only one dielectric constant.

For tetragonal, trigonal, and hexagonal crystals, the second-rank tensor takes the form

$$(\epsilon) = \begin{bmatrix} \epsilon_{11} & 0 & 0 \\ 0 & \epsilon_{11} & 0 \\ 0 & 0 & \epsilon_{33} \end{bmatrix} \tag{2.9.15}$$

if the z axis is chosen to be the axis of fourfold, threefold, and sixfold symmetry, respectively, and the x and y axes are two perpendicular axes in the basal plane. Equation (2.9.15) can be deduced from Eq. (2.9.2) by considering a rotation of θ about the n-fold axis. The new (primed) coordinates are related to the old coordinates through

$$x' = x \cos \theta + y \sin \theta \tag{2.9.16a}$$

$$y' = y \cos \theta - x \sin \theta \tag{2.9.16b}$$

where $\theta = \pi/2$ for tetragonal crystals, $\theta = 2\pi/3$ for trigonal crystals, and $\theta = 2\pi/6$ for hexagonal crystals. Again, the two tensors in the old and new coordinate systems should be identical. Tetragonal, trigonal, and hexagonal crystals are generally called *optically uniaxial crystals,* and they have two principal dielectric constants: ϵ_{33} along the c axis and ϵ_{11} in the basal plane.

Orthorhombic, monoclinic, and triclinic crystals have three principal dielectric constants and they are called *optically biaxial crystals.* The dielectric-constant tensor of these crystals takes the form

$$(\epsilon) = \begin{bmatrix} \epsilon_{11}' & 0 & 0 \\ 0 & \epsilon_{22}' & 0 \\ 0 & 0 & \epsilon_{33}' \end{bmatrix} \tag{2.9.17}$$

From the theory of matrices, we know that we can always diagonalize a symmetric matrix (which the material-parameter matrix is) by a suitable transformation of coordinates. The relation between the new set of coordinates in which Eq. (2.9.17) applies and the old set depends on the values of ϵ_{ii} and ϵ_{ij} in Eq. (2.9.2). In triclinic crystals, none of the ϵ_{ij}'s vanishes. Therefore, none of the three principal axes is fixed. In other words, the coordinate axes, to which Eq. (2.9.17) refers, not only differ from the crystal axes but also change from one triclinic crystal to another.

Monoclinic crystals possess either a twofold c axis or a mirror-reflection plane perpendicular to the c axis. A mirror reflection makes $x \to x$, $y \to y$, and $z \to -z$, whereas a rotation of π about the c axis makes $x \to -x$, $y \to -y$, and $z \to z$. These two symmetry operations produce the same effect on the ϵ tensor and make $\epsilon_{13} = \epsilon_{23} = 0$. In other words,

$$(\epsilon) = \begin{bmatrix} \epsilon_{11} & \epsilon_{12} & 0 \\ \epsilon_{12} & \epsilon_{22} & 0 \\ 0 & 0 & \epsilon_{33} \end{bmatrix} \tag{2.9.18}$$

Therefore, in monoclinic crystals, only one of the three principal axes, to which Eq. (2.9.17) refers, is fixed, and this axis is the c axis. In orthorhombic crystals, the existence of two other twofold axes perpendicular to the c axis makes $\epsilon_{12} = 0$. Thus all the three principal axes coincide with the three mutually perpendicular crystal axes.

In summary, we have shown that symmetry considerations help us find the relations that may exist between the matrix elements in a material-parameter tensor such as the dielectric-constant tensor. The discussion above applies equally well to other material-parameter tensors of rank 2. However, the values of these parameters can only be determined through experiments or calculated theoretically by analyzing the physical processes regulating these parameters.

2.10 CRYSTAL IMPERFECTIONS

In our discussion thus far, we have assumed an ideal lattice in which not only are the atoms arranged in a regular and repetitive pattern but also fixed in their positions. In a real crystal, however, departure from an ideal structure can occur. First, the atoms are not stationary but vibrate about their equilibrium positions. The vibrational motion of atoms in a periodic lattice is discussed in Section 2.11. Second, there are defects in the arrangement of atoms. The three principal kinds of defects are (1) point defects, such as interstitials, vacancies, and substitutional impurities; (2) line defects, commonly known as dislocations; and (3) plane defects, such as the grain boundary in a polycrystalline material. In this section we give a brief general discussion of the various lattice defects and their effects on material properties.

Vacancies and interstitials (Fig. 2.31a) represent, respectively, a vacant lattice site which is normally occupied and an atom occupying an interstitial space which is normally vacant. A vacant site is created when an atom is removed from the interior to the surface of a crystal (Fig. 2.31b). During the process, old bonds are broken and new bonds are formed, but the number of broken bonds (in the interior) always exceeds that of bonds created at the new site (at the surface). The type of defect in which only vacancies are involved is called a *Schottky defect*. An energy $\mathscr{E}_v$ is required to create a Schottky defect, and the number of such defects is determined on thermodynamic considerations.

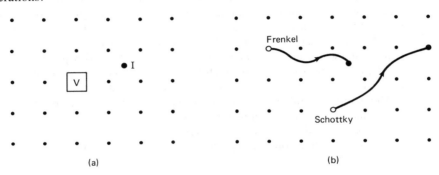

(a)   (b)

Figure 2.31 Illustrations showing (a) a vacancy (marked by V) and an interstitial (marked by I) and (b) Frenkel and Schottky defects in a two-dimensional lattice. A Schottky defect is generated by removing an atom from the interior to the surface of a crystal. In the case of a Frenkel defect, the removed atom remains in the interior of the crystal and occupies an interstitial position. Therefore, a Frenkel defect is made of a pair of vacancy and interstitial. The arrows indicate the movement of the atoms.

Consider the arrangement of n vacancies among N lattice sites. The total number of ways W is $N!/n! \, (N - n)!$; thus according to statistical thermodynamics, the entropy S is given by the Boltzmann relation (Appendix A2), that is [Eq. (A2.11)],

$$S = k \ln \frac{N!}{n! \, (N - n)!} \qquad (2.10.1)$$

The free energy of the system is given by Eq. (A2.15), or

$$F = n\mathscr{E}_v - TS \qquad (2.10.2)$$

The most probable distribution of Schottky defects is one that minimizes F. Applying the Stirling approximation $\log x! = x \log x$ to Eq. (2.10.2) and minimizing F with respect to n, we find that

$$\frac{dF}{dn} = 0 = \mathscr{E}_v + kT \ln \frac{n}{N - n} \qquad (2.10.3)$$

For $n \ll N$, $N - n$ can be approximated by N. Thus we have

$$n \cong N \exp\left(\frac{-\mathscr{E}_v}{kT}\right) \tag{2.10.4}$$

Typical values of $\mathscr{E}_v$ are around 1 eV. At a temperature $T = 1000$ K, a value of n/N of about 10^{-5} is obtained.

Another type of lattice defect is caused by an interstitial atom. Because the interatomic distance is very much reduced by the interstitial atom, the energy $\mathscr{E}_i$ required in the formation of an interstitial is very large. For example, in aluminum, the formation energies are, respectively, 0.75 eV and 3 eV for vacancies and interstitials. This large difference makes interstitials much less likely than vacancies in most crystals. However, an interstitial can be formed by transferring an atom from a lattice site to an interstitial position, leaving a vacancy behind. This type of lattice defect, known as a *Frenkel defect* (Fig. 2.31b), has a formation energy about equal to the sum of vacancy and interstitial energies, or $\mathscr{E} = \mathscr{E}_v + \mathscr{E}_i$. The number of ways of arranging n pairs of vacancies and interstitials among N lattice sites and N' interstitial positions is given by

$$W = \frac{N}{n!\,(N-n)!} \frac{N'!}{n!\,(N'-n)!} \tag{2.10.5}$$

Minimizing the free energy yields

$$\mathscr{E} = kT \ln \frac{(N-n)(N'-n)}{n^2} \quad \text{or} \quad n \cong (NN')^{1/2} \exp\left(\frac{-\mathscr{E}}{2kT}\right) \tag{2.10.6}$$

for $n \ll N, N'$. Because the defects (vacancies and interstitials) are formed in pairs, the activation energy is only one-half of the sum $\mathscr{E}_v + \mathscr{E}_i$, that is, $\mathscr{E}' = (\mathscr{E}_v + \mathscr{E}_i)/2$.

One way of determining whether Schottky defects or Frenkel defects are the more common type is to measure the density of a crystal before and after these defects are introduced. In the production of Frenkel defects, atoms simply shift from a lattice site to an interstitial position, so the volume of the crystal is expected to remain unchanged. This is not the case with Schottky defects. One of the most common methods of producing lattice defects over and above the thermal equilibrium concentrations given by Eqs. (2.10.4) and (2.10.6) is by quenching a crystal from a high temperature. Other methods include bombardments by high-energy particles and undergoing severe deformation by mechanical stress.

Lattice defects can also be produced by deviations from the stoichiometric composition. The most common and best-known defects of this type are the color centers in alkali halides. If excess alkali atoms exist and take the normal positive-ion positions, there are not enough halogen atoms to fill in all the positions reserved for negative ions. Therefore, negative ion vacancies are created. The vacancy, being surrounded by positive ions, attracts an electron. The electron bound to a negative-ion vacancy is known as an *F center*. By the same token, if excess halogen atoms are present, positive-ion vacancies are produced. The vacancy, being surrounded by negative ions, will repel an electron or, in semiconductor language, attract a hole. The complex formed by a hole and a positive-ion vacancy is called a *V center*. Because of the large gap energy, pure alkali-halide crystals are transparent to visible light. However, these crystals can be colored by the introduction of color centers, which absorb light by exciting an electron from the valence band to the V center or from the F center to the conduction band (Fig. 2.32).

Sec. 2.10   Crystal Imperfections

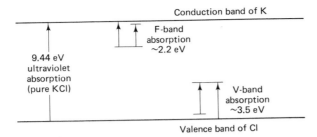

Conduction band of K

F-band
absorption
~2.2 eV

9.44 eV
ultraviolet
absorption
(pure KCl)

V-band
absorption
~3.5 eV

Valence band of Cl

Figure 2.32 Energy levels of color centers in KCl. An F center is created when an electron is bound to a negative-ion vacancy. A V center is formed when a hole is trapped at a positive-ion vacancy. In the presence of color centers, a KCl crystal will absorb light in the 2.2 eV and 3.5 eV regions of the visible spectrum. The arrows indicate the excitation of electrons as a result of the absorption.

Now we turn our attention to line defects. There are two types of dislocations, the edge and screw types (Fig. 2.33). The distortion due to an edge dislocation may be considered as caused by the insertion of an extra plane of atoms into the upper half of the crystal. In Fig. 2.33a, this half-plane is parallel to the *yz* plane and stops above the origin at a distance equal to one-half of the lattice spacing. The screw dislocation may be imagined as being produced by cutting the crystal partway through and pushing the upper part one lattice spacing over. The distortion in crystal structure generates a shear stress near a dislocation line. Under an externally applied stress of sufficient strength, the dislocation line moves and one part of the crystal slides across an adjacent part. This phenomenon, known as *slip,* greatly reduces the mechanical strength of a crystal. (For further discussion of the effect of dislocations on the mechanical strength, the following references may be helpful: C. Kittel, *Introduction to Solid-State Physics,* 3rd ed., John Wiley & Sons, Inc., New York, 1966, Chap. 19; A. J. Dekker, *Solid-State Physics,* Prentice-Hall, Inc., Englewood Cliffs, N.J., 1957, Chap. 3.)

Most crystals have dislocations except in tiny whiskers which are grown under high supersaturation (pressure/equilibrium vapor pressure) conditions. It is generally believed that the presence of dislocations greatly accelerates the rate of crystal growth. (J. Friedel, *Dislocations,* Addison-Wesley Publishing Company, Inc., Reading, Mass., 1964, Chap. 7.) Dislocations can be produced in a number of ways for experimental study, including plastic deformation of a crystal by bending or severe cold work. The low-angle grain boundary between two crystals is another example; it consists of an array of edge dislocations. The density of dislocations, which is defined as the number

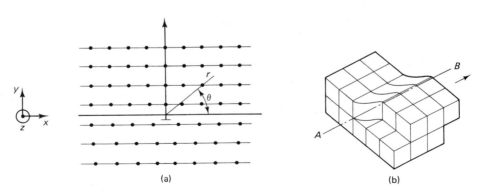

(a)

(b)

Figure 2.33 Schematic illustrations showing line defects: (a) an edge dislocation and (b) a screw dislocation.

of dislocation lines intersected by a plane of unit area in the crystal, can be measured by established methods. Among these is the *etch pit technique*. When a highly polished germanium or silicon surface is etched in an acid, conical pits are formed at places where dislocation lines intersect the surface. A precise measurement of the dislocation density is obtained by counting the number of etch pits. It is also possible to estimate dislocation densities from X-ray diffraction measurements. The observed spread of the Bragg angle θ in Eq. (2.8.5) and Fig. 2.26 is a measure of the dislocation density.

Many of the effects associated with a dislocation line are produced directly or indirectly by the stress field. Referring to Fig. 2.33a, the normal stresses, T_{rr} and $T_{\theta\theta}$ in the direction of r and θ, respectively, and the shear stress $T_{r\theta}$ are found to be

$$T_{rr} = T_{\theta\theta} = - \frac{Gb}{2\pi(1 - \nu)} \frac{\sin \theta}{r}$$

$$T_{r\theta} = \frac{Gb}{2\pi(1 - \nu)} \frac{\cos \theta}{r} \tag{2.10.7}$$

where G is the shear modulus, ν the Poisson ratio, and b the magnitude of the slip. For most crystals, $\nu \cong 0.3$; therefore, in the vicinity of the dislocation with $r \cong b$, a stress of the order of $0.1G$ is expected. In aluminum $G = 2.5 \times 10^{11}$ dyn/cm^2. Under such a large stress field, impurity atoms may be attracted and precipitated along the dislocation line. In a semiconductor, the energy gap may be affected substantially by the stress field, so that the electrical properties in regions near a dislocation line differ from those in dislocation-free regions. For devices whose electrical characteristics are very sensitive to material properties, it is essential that the material has a very low dislocation density. Such is the case with semiconductor junction devices, especially near the junction region.

Finally, we should mention that many useful materials have foreign or impurity atoms purposely introduced into the material. Examples include Cr in Al_2O_3 used for laser action, Sb or P in Si to make electrons available for conduction, and Cu and Cl in ZnS to act as luminescent centers. The useful impurities are either incorporated into the host crystal during the crystal-growing process or introduced into the crystal through the diffusion or alloying process. Since impurities play a very important part in device applications, their role will be discussed separately when we describe the operation of a particular device.

2.11 LATTICE VIBRATION

At ordinary temperatures, atoms in a solid possess a fair amount of thermal energy. This random motion of atoms has many important effects on the physical properties of a solid. For example, most conductors have a finite electrical conductivity, meaning that the electron motion in a lattice is not without impediment. The thermal motion of atoms is one of the causes that limit the mobility of electrons. We know that a solid conducts and stores heat. The same thermal motion of atoms also contributes to the thermal properties of a solid. In this section we discuss how the thermal motion of atoms can be analyzed.

Figure 2.34a shows schematically the energy $\mathscr{E}$ of an atom in a solid as a function of the interatomic distance R. The position at which $\mathscr{E}$ is a minimum determines the lattice spacing of the solid. As the atom moves away from its equilibrium position, its

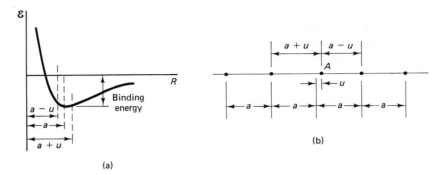

Figure 2.34 Schematic diagrams showing (a) the energy of an atom as a function of the interatomic distance R and (b) a linear array of atoms. Diagram (b) represents a hypothetical case in which only one atom is assumed to have been displaced from its equilibrium position.

energy increases. The average incremental energy above the minimum energy constitutes the thermal energy of the atom. Consider a one-dimensional array of atoms in which one atom (A) is displaced from its equilibrium position by a small displacement u (Fig. 2.34b). For simplicity, we assume that the positions of the other atoms are held rigid and only the position of A is subject to change. In terms of u, the change in energy is

$$\Delta\mathcal{E} = \mathcal{E}(a + u) + \mathcal{E}(a - u) - 2\mathcal{E}(a) \tag{2.11.1}$$

An expansion of Eq. (2.11.1) into a Taylor series gives

$$\Delta\mathcal{E} = \frac{\partial^2\mathcal{E}}{\partial u^2}u^2 = \gamma u^2 \quad \text{with} \quad \gamma = \frac{\partial^2\mathcal{E}}{\partial u^2} \tag{2.11.2}$$

From Eq. (2.11.2) it is obvious that the displaced atom experiences a force $F = -2\gamma u$ which is known as *Hooke's law*. Thus the equation describing the motion of the atom is

$$M\frac{d^2u}{dt^2} = -2\gamma u \tag{2.11.3}$$

where M is the mass of the atom. Equation (2.11.3) predicts a harmonic motion for the displaced atom with an angular frequency $\omega = \sqrt{2\gamma/M}$. For a simple cube, the force constant γ is related to Young's modulus Y by $\gamma = aY$, a being the lattice constant. Taking $a = 2.5$ Å, $Y = 10^{11}$ N/m^2, and $M = 10^{-25}$ kg, we find the frequency $\nu = \omega/2\pi$ to be of the order of 3×10^{12} Hz, which is in the far-infrared region (Fig. 2.20a). The amplitude of oscillation can be found from the average energy of a harmonic oscillator. At high temperatures, $\gamma u^2 = 3kT$, giving a value for u of about 0.2 Å at room temperature. These values of ν and u fall in the range of values observed experimentally. Since u/a is quite appreciable even at room temperature, we expect that lattice vibrations will affect material properties. This is indeed the case.

The analysis above is rather simple but not quite correct. First, the neighboring atoms are also displaced from their equilibrium positions. Second, the restoring force depends not on the absolute but on the relative displacement of an atom with respect to its neighbors. Figure 2.35a shows a linear chain consisting of identical atoms with mass M and spaced at a distance a apart from each other. If u_n denotes the displacement of the nth atom from its equilibrium position, the force acting on the nth atom is given by

$$F_n = \gamma(u_{n+1} - u_n) - \gamma(u_n - u_{n-1}) \qquad (2.11.4)$$

where γ is the same force constant as the one used in Eq. (2.11.3). For simplicity, only interactions between nearest neighbors are considered. In obtaining Eq. (2.11.4), we note that the force acting on the nth atom from the $(n-1)$th atom is opposite to that from the $(n+1)$th atom. If $u_n > u_{n-1}$, the force pulls the nth atoms toward the $(n-1)$th atom. From Eq. (2.11.4), the equation of motion of the nth atom is

$$M \frac{d^2 u_n}{dt^2} = \gamma(u_{n+1} + u_{n-1} - 2u_n) \qquad (2.11.5)$$

Similar equations are obtained for other atoms.

It is obvious from Eq. (2.11.5) that the motion of an atom is coupled to that of other atoms in the linear chain. If there are N atoms in the chain, there will be N simultaneous differential equations. It seems highly desirable that a special technique be developed to solve these equations. One of the useful methods is the method of Fourier analysis. For a particular Fourier component of lattice vibration having angular frequency ω and wavelength λ, the displacement must follow the wave motion, or

$$u_n = A \exp[i(\omega t + kna + \phi)] \qquad (2.11.6)$$

where A is the amplitude of the Fourier component, na the equilibrium position of the nth atom, and $k = 2\pi/\lambda$. The phase angle ϕ in Eq. (2.11.6) is random because the thermal motion of atoms is random.

For the $(n+1)$th atom and $(n-1)$th atom, u_{n+1} and u_{n-1} can be found from Eq. (2.11.6) by changing n to $n+1$ and $n-1$, respectively. Substituting Eq. (2.11.6) into Eq. (2.11.5) gives

$$-\omega^2 M = \gamma[\exp(ika) + \exp(-ika) - 2] \qquad (2.11.7)$$

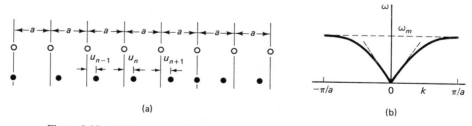

(a)

(b)

Figure 2.35 (a) A monatomic linear chain which corresponds to the situation in a solid possessing one atom per primitive cell and (b) its vibrational dispersion (ω versus k) curve. The open circles represent the equilibrium positions, whereas the dots represent the displaced particles. The dashed lines, which predict a constant phase velocity ω/k, correspond to the continuum approximation. In the approximation an elastic medium is considered to be made of continuous substance instead of discrete atoms.

Combining the exponential terms in Eq. (2.11.7), we find that

$$\omega = \pm \left(\frac{4\gamma}{M}\right)^{1/2} \sin \frac{ka}{2} \tag{2.11.8}$$

which is plotted in Fig. 2.35b. For $ka = \pi$, ω reaches a maximum value of $\sqrt{4\gamma/M}$, which is $\sqrt{2}$ of the value found from Eq. (2.11.3). The factor $\sqrt{2}$ comes from the fact that the displacements of neighboring atoms are completely out of phase, that is, $u_{n-1} = -u_n = u_{n+1}$. For small values of ka, ω/k is a constant

$$\frac{\omega}{k} = a \left(\frac{\gamma}{M}\right)^{1/2} \tag{2.11.9}$$

which is the acoustic velocity in the solid.

For crystals consisting of two different kinds of atoms or having two or more atoms per primitive cell, the analysis just presented is no longer adequate. Consider a diatomic linear chain (Fig. 2.36a) in which the even-numbered sites are occupied by atoms with mass M_2 and the odd-numbered sites by atoms with mass M_1. For such a diatomic chain, two equations of motion are needed:

$$M_2 \frac{d^2 u_{2n}}{dt^2} = \gamma(u_{2n+1} + u_{2n-1} - 2u_{2n}) \tag{2.11.10a}$$

$$M_1 \frac{d^2 u_{2n+1}}{dt^2} = \gamma(u_{2n+2} + u_{2n} - 2u_{2n+1}) \tag{2.11.10b}$$

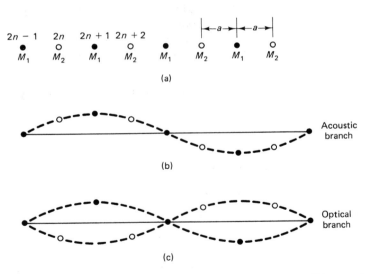

(a)

(b)

(c)

Figure 2.36 (a) A diatomic linear chain that is used to represent a solid possessing two atoms per primitive cell, (b) the transverse acoustical mode of vibration, and (c) the transverse optical mode of vibration. The main difference between (b) and (c) is that the two atoms M_1 and M_2 move in the same direction in (b) and move in opposite directions in (c).

Crystal Structure   Chap. 2

We propose that the Fourier components of the lattice displacement be of the following form:

$$u_{2n} = A \exp [i(\omega t + k2na) + i\phi] \qquad (2.11.11a)$$

$$u_{2n+1} = B \exp \{i[\omega t + k(2n + 1)a] + i\phi\} \qquad (2.11.11b)$$

Substituting Eq. (2.11.11) into Eq. (2.11.10) gives

$$-\omega^2 M_2 A = 2B\gamma \cos ka - 2A\gamma \qquad (2.11.12a)$$

$$-\omega^2 M_1 B = 2A\gamma \cos ka - 2B\gamma \qquad (2.11.12b)$$

For a nontrivial solution, the determinant of the coefficients of A and B must vanish or

$$\omega^2 = \gamma\left(\frac{1}{M_1} + \frac{1}{M_2}\right) \pm \gamma\sqrt{\left(\frac{1}{M_1} + \frac{1}{M_2}\right)^2 - \frac{4\sin^2 ka}{M_1 M_2}} \qquad (2.11.13)$$

Figure 2.37 shows a plot of ω as a function of k. The main feature is that the curve consists of two branches corresponding to the positive and negative signs in Eq. (2.11.13). The physical origin of these two branches lies not in the fact that two different masses are involved but in the fact that we allow two different modes of vibration in the proposed solution of the form given by Eq. (2.11.11). To show the two different modes of vibration, we consider the motion of two neighboring atoms. For clarity, we choose $ka = 0$. For the lower branch, $\omega = 0$; thus Eq. (2.11.12) leads to $A = B$. From Eq. (2.11.11), we see that the wave motions associated with the odd-numbered and even-numbered atoms are in phase or, in other words, they move together with the same sense of direction as illustrated in Fig. 2.36b.

For the upper branch of Fig. 2.37, we take the plus sign in Eq. (2.11.13). Thus $\omega^2 = 2\gamma(M_1^{-1} + M_2^{-1})$ at $ka = 0$. Using the value of ω^2 and the value of $ka = 0$ in Eq. (2.11.12), we find that $M_2 A = -M_1 B$. The negative sign means that the odd-numbered and even-numbered atoms vibrate in opposite sense. The opposite motion (Fig. 2.36c) results in a wider relative displacement and hence involves a larger energy than the motion in the same direction (Fig. 2.36b). Note that $\hbar\omega$ has the dimension of energy. The two branches of Fig. 2.37 are different because the mechanical energies associated with the two modes are different. The upper and lower branches of the dis-

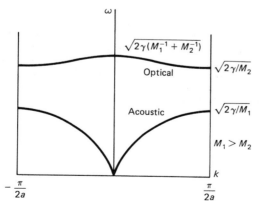

Figure 2.37 The dispersion relation (ω versus k curve) for the acoustical and optical modes of lattice vibration in a diatomic linear lattice.

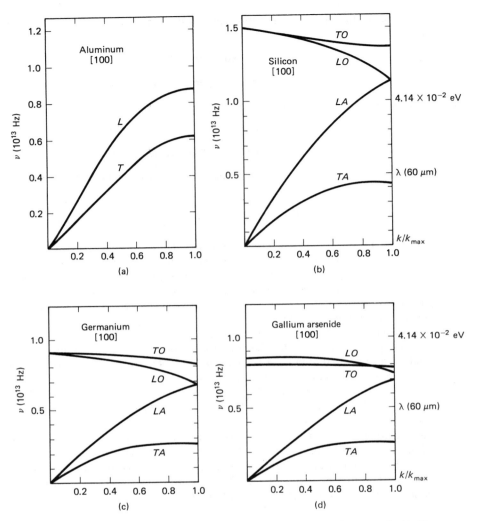

Figure 2.38 Experimental dispersion (ω versus k) curve for lattice waves in (a) Al, (b) Si, (c) Ge, and (d) GaAs. The acoustical and optical branches are denoted, respectively, by A and O, whereas the longitudinal and transverse waves are indicated, respectively, by L and T. All the curves are for lattice waves propagating in the [100] direction. (C. B. Walker, *Phys. Rev.*, Vol. 103, p. 547, 1956; B. N. Brockhouse, *Phys. Rev. Lett.*, Vol. 2, p. 256, 1959; B. N. Brockhouse and P. K. Iyengar, *Phys. Rev.*, Vol. 111, p. 747, 1958; J. L. T. Waugh and G. Dolling, *Phys. Rev.*, Vol. 132, p. 2410, 1963.)

persion curve (ω versus k curve) are generally known as the optical and acoustical branches, for the energies associated with these two modes of vibrations are, respectively, in the optical and acoustical range of the frequency spectrum.

The actual dispersion curves in Al, Si, Ge, and GaAs are shown in Fig. 2.38. These curves are obtained from measurements of the diffuse scattering of monochromatic X rays and the energy distribution of coherently scattered monoenergetic neutrons. In a three-dimensional lattice, the displacement of atoms can be either transverse or longitudinal to the direction of wave propagation and these modes are marked by T and L in Fig. 2.38. The face-centered cubic lattice of Al has only one atom per primitive cell, so only the acoustical branch exists, as in Fig. 2.38a. The diamond structure, on the other hand, consists of two interpenetrating face-centered cubic structures, and hence it contains two atoms per primitive cell, allowing two different modes of vibrations. This explains the appearance of the optical branch in Fig. 2.38b and c. The same explanation applies to the zinc-blende structure of GaAs. We should point out that the dispersion curves depend on the direction of the k vector in the reciprocal lattice. For other directions we expect the curves to be quantitatively different from those shown in Fig. 2.38.

PROBLEMS

2.1. How many atoms does the unit cell of a face-centered cubic structure (Fig. 2.3) possess? Using the three face diagonals of half length (a', b', c' of Fig. 2.4), construct a primitive cell. Find the volume of the primitive cell and the number of atom or atoms in it.

2.2. (a) How many atoms are there in the unit cell of a body-centered cubic structure (Fig. 2.3)?

(b) Refer to Fig. P2.2. Use the three basis vectors a', b', and c' as shown to find the coordinates of the eight corner atoms of the primitive cell, and then join the eight atoms to construct the primitive cell. (*Hint:* The opposite corners are separated by $a' \pm b' \pm c'$.)

(c) Find the volume of the primitive cell and the number of atom or atoms in it.

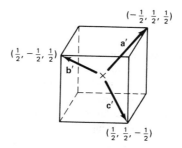

Figure P2.2

2.3. Find the distance between the nearest-neighboring atoms in each of the following structures (Fig. 2.3): the bcc, the fcc, the hcp, the diamond, the NaCl, and the CsCl structure. You can use the data given in Table 2.3 to check your results.

2.4. (a) Using Figs. 2.9 and 2.10, show that the minimum ratio of r/R is 0.414 for the octahedral packing and 0.225 for the tetrahedral packing.

(b) Refer to Table 2.3 for the structures of the following crystals: NaCl, CsCl, and ZnS.

What packing form are the atoms in for each crystal? The constituent atoms in these crystals appear as ions and their ionic radii are as follows:

$$\text{Na}^+ \qquad \text{Cs}^+ \qquad \text{Zn}^{2+} \qquad \text{Cl}^- \qquad \text{S}^{2-}$$
$$0.95\ \text{Å} \qquad 1.69\ \text{Å} \qquad 0.74\ \text{Å} \qquad 1.81\ \text{Å} \qquad 1.84\ \text{Å}$$

Calculate the ratio r/R for each crystal, and then use Table 2.2 to determine the packing form.

2.5. Refer to the hexagonal close-packed structure shown in Fig. 2.6. Show that the ratio of c/a in an ideal structure is $c/a = \sqrt{8/3} = 1.633$.

2.6. Show that any vector $\mathbf{r} = u\mathbf{a} + v\mathbf{b} + w\mathbf{c}$ lying in a plane (hkl) must satisfy the relation

$$uh + vk + wl = 0$$

Take the (110) plane in a cubic crystal as an example. Draw the directions of the following vectors: [001], [1$\bar{1}$0], [1$\bar{1}$1], and [110]. Identify the vector perpendicular to the plane and the vectors in the plane by inspection and by using the appropriate equation.

2.7. (a) Find the ratio of the lengths OA/OB in Fig. 2.18 for (1) $n = 10$ and $\beta = 30°$ and (2) $n = 1000$ and $\beta = 30°$.
   (b) Supposing that the amplitude of each component is $V_0 = 0.1$ V, find the values of OA and OB for each case. Explain why the value of OA hardly changes as n increases from 10 to 1000.

2.8. Consider an optical experiment in which a coherent light beam is split into two beams at an optical plate A and then reflected at two mirrors B and C (Fig. P2.8). Finally, the two beams are combined at another optical plate D. The electric field of a coherent light beam can be represented by Eq. (2.7.1). Suppose now that a dielectric slab of thickness d is inserted in one of the two optical paths. The light beam propagates with a velocity c in air and a velocity v in the dielectric slab. What phenomenon do you expect to observe as you turn the dielectric slab? Explain how the velocity v can be measured, and find an expression relating v to θ. (*Hint:* Find the path length EF.)

Figure P2.8

2.9. Describe the arrangement of apparatus in electron and neutron diffraction experiments (see, for example, R. L. Sproul, *Atomic Physics*, John Wiley & Sons, Inc., New York, 1963, pp. 115–122). Also discuss the essence of the experiments.

2.10. Find the de Broglie wavelength for
   (a) an electron having a kinetic energy of 60 eV and
   (b) a thermal neutron having a kinetic energy of $3\ kT/2$ at room temperature.

2.11. Calculate the Bragg angles for neutron diffraction from the (311) plane and from the (400) plane in Fig. 2.26. If we replace the neutron beam by an X ray, what should be the photon energy (expressed in eV) of the X ray in order to observe the same diffraction pattern?

2.12. (a) Determine the Bragg angles at which the (111), (200), and (220) planes of a nickel crystal will diffract X rays of wavelength $\lambda = 1.54$ Å. Nickel has a face-centered cubic structure with a lattice constant $a = 3.52$ Å. The indices (hkl) refer to the crystal axes of a simple cubic crystal.
   (b) In Fig. 2.26, no diffraction from the (200) plane in diamond is observed. Do you expect this to happen in nickel? Give a reason for the expected behavior.

2.13. The basic arrangements in the three diffraction experiments are: (1) a monochromatic source and a powder specimen for the powder method, (2) a monochromatic source and a rotating single crystal for the rotating-crystal method, and (3) a radiation source of continuous wavelength and a stationary crystal for the Laue method. Describe what you expect to see from the diffracted beam in each case. Will it appear as a spot, a line, or something else? Explain why the combination of arrangements as stated above is necessary for each method.

2.14. In a general derivation of the Bragg condition (Laue derivation), we consider any two points A and B in the (hkl) planes as shown in Fig. P2.14. We let $\mathbf{p_0}$, $\mathbf{p}$, and $\mathbf{n}$ be three unit vectors pointing in the directions of the incident ray, the diffracted ray, and the plane normal, respectively.

Figure P2.14

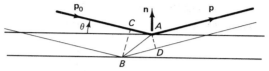

(a) Show that the difference in the two path lengths is

$$AC - BD = \mathbf{r} \cdot \mathbf{p_0} - \mathbf{r} \cdot \mathbf{p} = (2 \sin \theta)(\mathbf{r} \cdot \mathbf{n})$$

(b) Next we let $\mathbf{r} = u\mathbf{a} + v\mathbf{b} + w\mathbf{c}$. Show that the scalar product of $\mathbf{n} \cdot \mathbf{r}$ is

$$\mathbf{n} \cdot \mathbf{r} = (uh + vk + wl)d$$

(c) Show that the phase difference between the two diffracted rays is

$$\beta = \frac{2\pi}{\lambda} (2d \sin \theta)(uh + vk + wl)$$

2.15. Verify that diffraction from the (111) plane in a body-centered cubic structure will not be present.

2.16. Verify that diffraction from the (110) plane in a face-centered cubic structure will not appear.

2.17. Verify that diffraction from the (210) plane in a diamond structure will not be present.

2.18. Using the unit cell shown in Fig. 2.7a as the reference cell, we find that diffraction from the (111) plane is missing in KCl but present in KBr. Both crystals have the NaCl structure. Apply Eq. (2.8.10) to the two cases and explain why the difference in the observed behavior may be expected.

2.19. **(a)** Figure P2.19a shows a trigonal cell. At corner A, the three angles are equal, that is, $\alpha = \beta = \gamma$. Prove that the body diagonal AB is the only threefold axis in the structure.

Figure P2.19

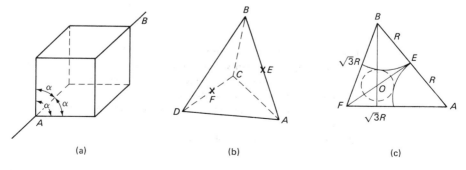

| (a) | (b) | (c) |

(*Hint:* Show that the three angles are not all equal at the other corners except A and B.)

(b) The CH_4 molecule has a tetrahedral symmetry with the hydrogen atoms situated at A, B, C, and D and the carbon atom at the center O. In Fig. P2.19b and c, E and F are midway points between AB and between CD. Show that OB is a threefold rotation axis and OE is a twofold rotation axis.

2.20. An octahedron has six twofold axes, four threefold axes, and three fourfold axes (Fig. P2.20a). Verify the statement by showing that the structure remains unchanged under the said rotations. (*Hint:* The six atoms A, B, C, D, E, and F can be considered as occupying the face centers of a cube as shown in Fig. P2.20b. The threefold axes are the body diagonals of the cube, whereas the twofold axes are parallel to the face diagonals and pass through the center of the cube.)

Figure P2.20

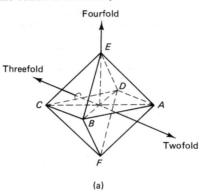
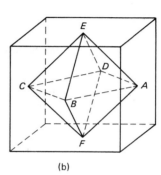

(a)　　　　　　　　　　　　　　　　(b)

2.21. Consider crystals possessing an n-fold rotation axis which is chosen to be the z axis. By applying a rotation of $2\pi/n$ to $\mathbf{D}$ of Eq. (2.9.2) about the n-fold axis, show that

$$\epsilon_{13} = 0, \qquad \epsilon_{23} = 0 \qquad \text{from } D_z$$
$$\epsilon_{12} = 0, \qquad \epsilon_{11} = \epsilon_{22} \qquad \text{from } D_x \text{ and } D_y$$

Note that if $[T]$ is the transformation matrix relating $\mathbf{D}$ before rotation to $\mathbf{D}'$ after rotation, the requirement for invariance of $[\epsilon]$ under the rotation operation is

$$[\epsilon] = [T][\epsilon][T]^{-1}$$

You may choose any one of the three uniaxial crystals (trigonal, tetragonal, and hexagonal) for the proof.

2.22. Show that Eq. (2.9.18) can be transformed to Eq. (2.9.17) if we choose

$$x' = x \cos \theta + y \sin \theta$$
$$y' = -x \sin \theta + y \cos \theta$$

Express θ, ϵ'_{11}, and ϵ'_{22} in terms of ϵ_{11}, ϵ_{12}, and ϵ_{22}.

2.23. **(a)** Show that the ϵ matrix given in Eq. (2.9.15) for a uniaxial crystal is invariant under any rotation θ about the z axis. Therefore, the c axis is the only axis specified for a uniaxial crystal insofar as the validity of Eq. (2.9.15) is concerned. The other two axes, x and y are arbitrary as long as the three axes x, y, and z are mutually perpendicular.

(b) Show that for cubic crystals the ϵ matrix

$$(\epsilon) = \begin{bmatrix} \epsilon & 0 & 0 \\ 0 & \epsilon & 0 \\ 0 & 0 & \epsilon \end{bmatrix}$$

is independent of the choice of the rectangular coordinates.

2.24. Show that Eq. (2.9.18) applies to a monoclinic crystal.

2.25. Calculate the number of vacancies per atom (i.e., the ratio n/N) in a crystal at $T = 300$ K, assuming that the energy required to remove an atom from the interior to the surface is 1 eV.

2.26. Show that the number of Frenkel defects in a solid in thermal equilibrium at a temperature T is given by Eq. (2.10.6).

2.27. In ionic crystals it is energetically favorable to form roughly equal numbers of positive and negative ion vacancies. The formation of vacancies in pairs keeps the crystal electrically neutral. Show that the number n of pair vacancies is given approximately by

$$n = N \exp\left(\frac{-\mathscr{E}}{2kT}\right)$$

where $\mathscr{E} = \mathscr{E}_+ + \mathscr{E}_-$ is the formation energy.

2.28. (a) The mechanical wave equation that describes the motion of an elastic wave in a one-dimensional space is

$$\frac{1}{v_s^2} \frac{\partial^2 u}{\partial t^2} = \frac{\partial^2 u}{\partial x^2}$$

where v_s is the sound velocity. The wave equation is applicable for low-frequency or long-wavelength oscillations where $\lambda \gg a$, and can be derived from Eq. (2.11.5).

(b) In the long-wavelength limit, we may express u_{n+1} and u_{n-1} in terms of u_n in a Taylor's series expansion. Derive the wave equation by following such a procedure. Show that the sound velocity v_s is given by Eq. (2.11.9), or

$$v_s = a\left(\frac{\gamma}{M}\right)^{1/2}$$

2.29. (a) Show that for a simple cubic crystal, the ratio γ/M is equal to $Y/\rho a^2$, where ρ is the mass density (kg/m³) of the crystal and Y is Young's modulus.

(b) Given $Y = 1.14 \times 10^{11}$ N/m², $\rho = 2.73 \times 10^3$ kg/m³, and $a = 4.03$ Å, calculate the values of ω_M in Fig. 2.35 and the sound velocity v_s in Problem 2.28. Also find the average displacement of atoms due to thermal motion.

2.30. In Fig. 2.35 and Eq. (2.11.6), the displacement of atoms is in the same direction as the direction of wave propagation. Such a wave is called the *longitudinal wave*. In two- and three-dimensional cases, we can also have transverse waves for which the displacement of atoms is transverse to the direction of wave propagation.

(a) Draw a two-dimensional lattice. Show the displacement of atoms for longitudinal and transverse lattice waves propagating in the x direction and having a wavelength $\lambda = 8a$. It is sufficient to show three rows of atoms in the y direction.

(b) Find an expression similar to Eq. (2.11.5) for the transverse lattice wave.

2.31. (a) Analyze the motion of the two kinds of atoms in Fig. 2.36 for $ka = \pi/2$. From Fig. 2.37, we see that ω is determined by one mass alone, either M_1 or M_2. Therefore, we expect that atoms of one kind would remain stationary for one branch of wave motion. Show that our physical reasoning is borne out by an analysis.

(b) Draw diagrams similar to Fig. 2.36b and c to show the movements of atoms for (1) $ka = \pi/2$, (2) $0 < ka < \pi/2$, and (3) $ka = 0$.

2.32. Consider a solid which has three atoms per primitive cell. We represent the solid by a linear chain made of three atoms as shown in Fig. P2.32, which is an obvious extension of Fig. 2.36a. How many distinct modes of lattice vibrations are there? How many modes are in the acoustical and optical branches and how many modes are associated with longitudinal and transverse waves? Draw diagrams similar to Fig. 2.36b and c to show the movement of atoms for the transverse wave.

Figure P2.32

Crystal Structure   Chap. 2

3

Atomic Bonding and Crystal Types

3.1 INTERATOMIC FORCES

When substances are cooled down to sufficiently low temperatures, they all become solids. Obviously there must be bonding forces of some kind to keep atoms together to form a solid material. On the other hand, the atoms in a solid maintain a finite distance from each other. Therefore, there must be two opposite types of forces: an attractive force F_{att} and a repulsive force F_{rep}. Both forces are a function of the interatomic distance R. At large distances, the attractive force dominates; hence atoms are drawn toward each other. At small distances, the repulsive force becomes much larger than the attractive force, and atoms are held back from getting too close to each other. At equilibrium, the two opposing forces balance out, and we have

$$F_{att}(R) = F_{rep}(R) \tag{3.1.1}$$

Equation (3.1.1) determines the atomic spacing R in a crystal.

In dealing with the interactions between atoms, it is more convenient that the mathematical formulation is based on energies rather than forces. Realizing that the force is the derivative of energy, we can rewrite Eq. (3.1.1) in terms of two energies, one attractive and one repulsive, as

$$\frac{d}{dR} [\mathscr{E}_{rep}(R) - \mathscr{E}_{att}(R)] = 0 \tag{3.1.2}$$

As we shall see shortly, the attractive energy is negative and the repulsive energy is positive. The total energy $\mathscr{E}_{tot}$ is thus equal to $\mathscr{E}_{rep} - \mathscr{E}_{att}$. In terms of the total energy, Eq. (3.1.2) becomes

$$\frac{d}{dR} \mathscr{E}_{tot} = 0 \tag{3.1.3}$$

Equation (3.1.3) says that at equilibrium the total energy of a system has a minimum value. This minimum-energy law is a fundamental law of nature.

As we shall see in discussions of subsequent sections, the nature of the attractive and repulsive energy is different in different types of crystals. In general, we consider an atom as consisting of an ion core (nucleus plus core electrons) and some valence electrons. The nature of the interaction between neighboring atoms depends on whether the orbits of neighboring core electrons overlap or not. For example, Na in metallic sodium has an ion core of charge $+e$ and one valence electron of charge $-e$. This valence electron keeps the Na^+ ion core from coming into contact with neighboring Na^+ ion cores. However, the situation in NaCl is different. In NaCl crystal, a sodium atom loses its valence electron and becomes a Na^+ ion. The Na^+ ion core now makes direct contact with neighboring Cl^- ion cores, and thus overlapping of these two ion cores may occur.

In cases where there are valence electrons to keep the orbits of neighboring core electrons from overlapping, the principal interaction is of electrostatic nature. Consider the Coulomb interaction between two charged particles, one with charge q_1 and the other with charge q_2, separated at a distance r_{12}. The potential energy $\mathscr{E}_P$ is

$$\mathscr{E}_P = \frac{q_1 q_2}{4\pi\epsilon_0 r_{12}} \tag{3.1.4}$$

as can be derived from Coulomb's law. Between charges of like polarity, the force is repulsive. Since work must be done to bring the two charges together, $\mathscr{E}_P$ is positive. Between charges of opposite polarities, the force is attractive and hence $\mathscr{E}_P$ is negative.

If the orbits of core electrons of two neighboring atoms overlap, interaction of a different kind must also be considered. Since core electrons have closed-shell configurations, all quantum states in the shell (e.g., the $2s2p$ shell in Na^+ and Ne) are occupied, and hence no quantum state in the shell is available for an additional electron. When the orbits of core electrons begin to overlap, some of the core electrons must go to unoccupied shells of higher energy. The lifting of an electron from a lower-energy state to a higher-energy state raises the energy of the entire system. This increase in energy is known as the *core repulsive energy* $\mathscr{E}_c$. The cause of this energy can be understood only on the basis of quantum mechanics (the exclusion principle).

Besides $\mathscr{E}_P$ and $\mathscr{E}_C$, there are two other kinds of energy that may be important in certain types of crystals. As we all know, metals conduct electric current with very little resistance. This high electrical conductivity is generally attributed to the fact that the valence electrons of metals move rather freely in a crystal. Associated with this freedom of movement, there is a kinetic energy $\mathscr{E}_K$. Since the kinetic energy tends to make the volume of a crystal expand, the term $\mathscr{E}_K$ is repulsive in nature. The fourth interaction energy $\mathscr{E}_W$ comes from electrostatic interaction between two charge pairs. The charge pair, known as an *electric dipole,* consists of two charges of opposite polarities kept apart at a small distance. The energy $\mathscr{E}_W$, known as the *van der Waals energy* (attractive), is discussed in more detail in Section 3.2.

Summarizing the discussion above, we may express the total energy $\mathscr{E}_{total}$ of a system as

$$\mathscr{E}_{total} = \sum \mathscr{E}_K + \sum \mathscr{E}_C - \sum \mathscr{E}_P - \sum \mathscr{E}_W \tag{3.1.5}$$

In Eq. (3.1.5) it is assumed that the electrostatic interaction between charges of opposite polarities predominates and hence the sign of $\mathscr{E}_P$ is chosen negative. As discussed in

Atomic Bonding and Crystal Types   Chap. 3

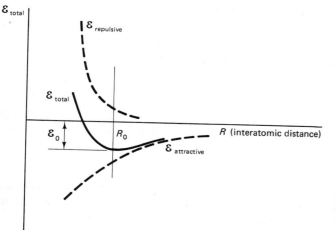

Figure 3.1 Schematic diagram showing the total energy per atom as a function of the interatomic distance. The total energy is a sum of the repulsive and attractive energies. The position at which $\mathcal{E}_{total}$ is a minimum determines the interatomic spacing R_0 in a crystal. The amount of energy $\mathcal{E}_0$ lowered by forming a crystal is the bonding energy of the crystal.

Section 3.2, crystals can be classified into four general groups. For each crystal group, different terms in Eq. (3.1.5) are important. It is the purpose of our discussions in subsequent sections to show the important differences between different crystal types.

Figure 3.1 shows the general behavior of the attractive and repulsive energies as functions of the interatomic distance R. The equilibrium separation R_0 in a crystal is determined by the condition $d\mathcal{E}_{tot}/dR = 0$ or by the position of minimum $\mathcal{E}_{tot}$. In Fig. 3.1 we choose the energy of isolated atoms as the reference energy. The energy of a system is lowered by an amount $\mathcal{E}_0$ if the atoms come to the proximity of one another. This lowering of energy constitutes the bonding between the atoms in a crystal.

3.2 CLASSIFICATION OF BOND TYPES

In Section 1.6 we discussed the periodic structure of chemical elements. One important observation is that elements having their shells completed are chemically inert. This means that a stable electron configuration is one which has its shells completed. The attractive forces differ in nature from one class of materials to another, and this difference can be understood on the basis of the shell structure of atoms. The general tendency is that atoms will give up electrons, receive electrons from other atoms, or share electrons with other atoms so that they may achieve individually or collectively a stable electron configuration. According to the different ways in which atoms achieve a stable electron configuration, the bonding between atoms in a crystal is commonly classified into four distinct types: the ionic bond, the covalent bond, the metallic bond, and the van der Waals bond.

The strength of a bond can be measured by the energy required to dissociate a solid into isolated neutral atoms or molecules. This energy is called the *cohesive energy* of a solid. Values for a selected number of materials representing each bond type are listed in Table 3.1. The bond strength in ionic, covalent, and metallic crystals is much stronger than that in molecular crystals. Hence the three strong bonds, ionic, covalent, and metallic, are termed *primary bonds*. In molecular crystals, the constituent atoms or molecules already have a stable electron configuration. The bond between such atoms

TABLE 3.1 COHESIVE ENERGIES[a] (eV) OF DIFFERENT CRYSTAL TYPES

Ionic		Covalent		Metallic		Molecular	
Material	Energy	Material	Energy	Material	Energy	Material	Energy
LiCl	8.9	Diamond	7.4	Na	1.13	Ar	0.08
NaCl	8.0	Si	4.6	Au	3.78	Kr	0.11
KI	6.6	Ge	3.9	Mg	1.56	CH_4	0.10

[a]Energy required to dissociate a solid into monatomic gas in the case of element solids and into molecular gas in the case of compound solids.

or molecules comes from a much weaker interaction. In the following discussion we describe the bonding mechanisms in crystals of the four distinct types: the ionic, the covalent, the metallic, and the molecular type.

The Ionic Bond

Alkali metals such as sodium have a single valence electron outside closed shells, whereas halogens such as chlorine are one electron short of having a complete outer shell (Fig. 3.2). An electron transfer from the alkali metal X to the halogen Y results in closed-shell configurations in both X^+ and Y^- ions. For example, Na^+ has an electron configuration $1s^2 2s^2 2p^6$ (the numbers in the superscripts indicating the number of electrons), which is the same as that of neon, whereas Cl^- has an electron configuration $1s^2 2s^2 2p^6 3s^2 3p^6$, which is the same as that of argon. The Coulomb attraction between X^+ and Y^- ions keeps the ions together in an ionic XY molecule or crystal. Besides alkali halides, ionic bonds can also be formed between divalent and trivalent elements. Selected examples are CaO, $MgCl_2$, Al_2O_3, and ZnO.

The Covalent Bond

Besides electron transfer, an incomplete shell can be filled if atoms in a molecule or solid share their electrons. The bond thus formed between atoms is called the covalent bond. To illustrate the essential difference between an ionic and a covalent bond, we use a nitrogen molecule N_2 as an example. From Table 1.2 we find that the $2p$ level with only three electrons in nitrogen is incomplete. If one nitrogen atom transfers its three $2p$ electrons to the other nitrogen atom, we have complete $1s$ and $2s$ levels in the N^{3+} ion, and complete $1s$, $2s$, and $2p$ levels in the N^{3-} ion (Fig. 3.3a). This arrange-

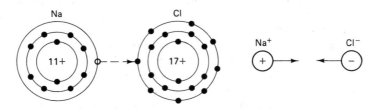

Figure 3.2 Electron configuration in NaCl. An electron transfer from Na to Cl (indicated by a broken arrow) results in closed-shell configurations in both Na^+ and Cl^- ions. The bonding in NaCl is due to Coulomb attraction between Na^+ and Cl^- ions.

Atomic Bonding and Crystal Types   Chap. 3

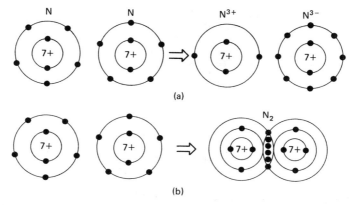

N   N   N³⁺   N³⁻

(a)

N₂

(b)

Figure 3.3 Schematic diagram illustrating the difference between (a) ionic bonding and (b) covalent bonding in a nitrogen molecule. Electron transfer is needed in forming an ionic bond. A closed-shell configuration is formed in the covalent bond by sharing electrons, the $2p$ electrons in the case of N_2, between two atoms. The actual bonding in N_2 is covalent.

ment of electrons is similar to that in NaCl and hence results in an ionic bond. In the covalent bond, the six $2p$ electrons do not really belong to a particular nitrogen atom but are shared equally between the two atoms. In other words, the charge cloud formed by the six $2p$ electrons is orbiting around and hence surrounds both nitrogen atoms (Fig. 3.3b).

If the six $2p$ electrons are shared equally by the two atoms, each atom does not lose or gain any electron, and hence both atoms should remain electrically neutral. On the basis of classical physics, it is quite impossible to understand why two neutral nitrogen atoms form a nitrogen molecule. The physical origin as well as the basic properties of a covalent bond can be understood only on the basis of quantum mechanics. The previous statement about electron sharing in a covalent bond is an oversimplified description of a quantum-mechanical phenomenon known as *electron-pair bond*. The six $2p$ electrons in a nitrogen molecule are bound in pairs with spins aligned antiparallel in each pair (Fig. 3.4a). A similar bond is formed between the two $1s$ electrons in a hydrogen molecule (Fig. 3.4b).

To describe the physical situation in the formation of an electron-pair bond, we may consider first the case of a singly ionized H_2^+ molecule, in which the two protons are held together by one electron. This problem can be solved exactly by quantum mechanics. The result of quantum-mechanical calculation shows that the two lowest-energy states have spatially symmetric and antisymmetric wave functions. As illustrated

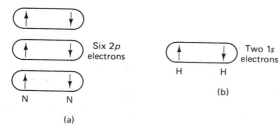

Six $2p$ electrons

N   N

(a)

Two $1s$ electrons

H   H

(b)

Figure 3.4 Electron-pair bond in (a) N_2 and (b) H_2. The bond is formed between a pair of electrons with antiparallel spins.

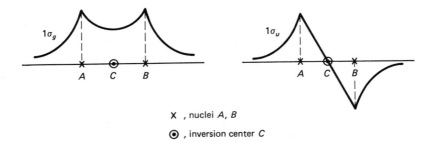

X , nuclei A, B

⊙ , inversion center C

Figure 3.5 (a) Symmetrical and (b) antisymmetric wave function of a hydrogen molecule-ion H_2^+ along the line passing through the nuclei.

in Fig. 3.5, the symmetric (even) wave function, denoted by a subscript g, is unchanged and the antisymmetric (odd) wave function, denoted by a subscript u, merely changes sign when we perform an inversion operation with respect to the midpoint between the two nuclei. The wave function that describes the electron motion in a molecule is called the *molecular orbital*. It is customary to use Greek σ, π, δ in reference to the various molecular orbitals.

The variation of the nuclear potential along the line joining the two nuclei is illustrated schematically in Fig. 3.6. Note that the Coulomb potential in the region AB between the two nuclei is depressed as compared to the potential in a single atom (represented by the dashed curves in Fig. 3.6). Using the expression of the electron density defined in Eq. (1.3.11), we can write the potential energy of a H_2^+ molecule ion as

$$\mathscr{E}_P = - \frac{e}{4\pi\epsilon_0} \iiint_{-\infty}^{\infty} \left(\frac{e}{r_a} + \frac{e}{r_b} \right) |\psi|^2 \, dx \, dy \, dz + \frac{e^2}{4\pi\epsilon_0 R} = -V_{en} + V_{nn} \quad (3.2.1)$$

where r_a and r_b are the electron distance from the two nuclei. In Eq. (3.2.1) the term V_{en} represents the electron–nucleus interaction energy and is attractive, whereas the term V_{nn} represents the nucleus–nucleus interaction energy and is repulsive. Evidently, the two states *gerade* and *ungerade* (German words for *even* and *odd*, respectively) will have different potential energies because they have different electron distributions (Fig. 3.5), especially in the region AB between the two nuclei.

In the gerade state, the electron density (i.e., the probability of finding the electron) has an appreciable value in the region AB of low potential energy. This means

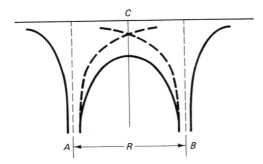

Figure 3.6 Potential energy (solid curve) of an electron in the field of two hydrogen nuclei, along the line passing through the nuclei A and B. The dashed curve is the potential energy in the field of a single hydrogen nucleus.

that the term V_{en} in Eq. (3.2.1) is expected to be lowered by the accumulation of negative charge in the low-potential-energy region. This lowering of the potential energy produces attraction in a H_2^+ molecule ion. The energy of a H_2^+ molecule ion can be calculated through the use of quantum mechanics, and the result of such a calculation is shown in Fig. 3.7 as a function of the internuclear distance R. At large R, the term V_{en} in Eq. (3.2.1) dominates. As R decreases, not only is the Coulomb potential (Fig. 3.6) further depressed but the amplitude of $|\psi|^2$ is increased in region AB. Therefore, the energy of the gerade state decreases with decreasing R until it reaches a minimum. For further decrease of R, the energy rises as the electrostatic interaction between the two nuclei (represented by the V_{nn} term) becomes the predominant term in Eq. (3.2.1). The situation for the ungerade state is entirely different. The electron cloud in the ungerade state is not concentrated in the region of low potential energy. As the two nuclei approach each other, the term V_{en} no longer decreases. Since there is nothing in Eq. (3.2.1) to compensate the nuclear repulsion, the energy increases continuously with decreasing R.

The discussion above can be extended to a neutral hydrogen molecule by considering the two electrons of H_2 separately. Each of the two electrons moves in the field of the two nuclei and in an average field (averaged over the electron distribution) of the other electron. The average electron field, of course, compensates the nuclear field. However, the main features concluded from the quantum-mechanical treatment of a H_2^+ molecule-ion can still be used to describe the situation in a neutral hydrogen molecule. The electron wave function will still be either symmetric (gerade) or antisymmetric (ungerade). Again it is the gerade state that shows a minimum in the total-energy curve. Therefore, the ground (lowest energy) state of a hydrogen molecule is the one in which both electrons have the symmetric wave function. Since the two electrons have the same orbital wave function, they must have different spin orientations.

Note that the electron in a H_2^+ molecule-ion does not belong to a particular nucleus. Similarly, the two electrons in a H_2 molecule are shared equally by the two nuclei. The bond formed between two atoms by sharing two electrons with antiparallel spins is known as the electron-pair bond. The two electrons have antiparallel spins not on account of any magnetic interaction between the spins but because of the Pauli exclusion principle.

In summary we have used the hydrogen molecule to illustrate how the potential energy can be lowered in a molecule. In atomic and molecular physics few simple problems have either exact solutions or simple enough, approximate solutions. Hydro-

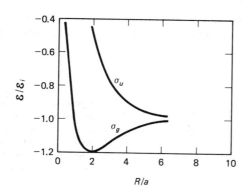

Figure 3.7 Total energy $\mathscr{E}_P$ of symmetric (σ_g) and antisymmetric (σ_u) states of the hydrogen molecule-ion as a function of the interatomic distance. The abscissa is expressed in units of the Bohr radius and the ordinate in units of the ionization energy of a hydrogen atom.

gen atom and hydrogen molecule are such problems. The principles that we learn from these examples are general and applicable to more complicated atoms and molecules. The principal features of a covalent bond, so far discussed, are electron sharing and spin antiparallel arrangement. Other important features are the directional and saturation properties. These properties are discussed in Section 3.6.

The Metallic Bond

Metallic bonds differ from ionic and covalent bonds in many respects. First, a metallic bond exists only in the condensed state. In the gaseous state, metallic elements appear as monatomic vapors. Although molecules formed by metallic elements, for example Li_2 and Na_2, have been observed, these molecules are not very stable. In our present discussion, we are not concerned with the formation of such molecules. Instead, we describe the bonding mechanism in a metallic crystal.

The outstanding characteristic of a metal is its high electrical conductivity. It can be inferred from this property that the valence electrons, such as the $2s$ electrons in Li and the $3s$ electrons in Na, must be free to move about. Because alkali metals have only a single valence electron outside closed shells, it is possible to find reasonably accurate solutions of the Schrödinger wave function. In the classical theory of metals, we think of valence electrons as forming an electron sea and the positive ions as being embedded in this sea. The positive ion is made of the nucleus and the core electrons in the closed-shell configuration. Later, quantum-mechanical calculations on alkali metals have generally supported this picture.

In Fig. 3.8 we compare the wave function of the valence electron in sodium crystals (solid curve) with the $3s$ wave function of a sodium atom (dashed curve). The wave function is not disturbed inside the ion-core region. However, in the outer region the crystal wave function is considerably flattened and squeezed in by neighboring valence electrons. It should be pointed out that the electron distribution in a volume element $4\pi r^2\,dr$ (in spherical coordinates) is, according to Eq. (1.3.11), given by $|\psi|^2 4\pi r^2\,dr$. From the shape of ψ shown in Fig. 3.8, we can see that over 90% of the electron distribution is in the flat region. The potential energy of the electron is obviously lowered because the average electron distribution in a crystal is closer to the nucleus. The kinetic energy, being proportional to $|\hbar\,\partial\psi/\partial r|^2$, is also reduced because of a smaller $\partial\psi/\partial r$ in the flattened region. This reduction in the total energy constitutes the bond in a metallic crystal.

The van der Waals Bond

When inert gases are cooled down to sufficiently low temperatures, they also become solids. The bonding between inert-gas atoms in a crystal comes from interactions of

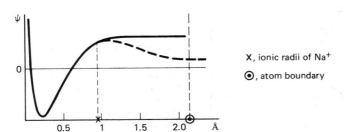

X, ionic radii of Na$^+$

⊙, atom boundary

Figure 3.8 Electron wave function in a sodium crystal (solid curve) as compared to that in an isolated sodium atom (dashed curve). One vertical line indicates the boundary between neighboring atoms in a sodium crystal, and the other indicates the region of the Na$^+$ ion core.

electric dipoles. An electric dipole is made of two electric charges of opposite polarities placed at a small but finite distance $2d$ apart (Fig. 3.9). The electric field E at point P situated at a distance R from the center of the charge distribution is

$$E = \frac{-e}{4\pi\epsilon_0(R + d)^2} + \frac{e}{4\pi\epsilon_0(R - d)^2} = \frac{2p}{4\pi\epsilon_0 R^3} \qquad (3.2.2)$$

for $R \gg d$. Equation (3.2.2) is derived for the configuration of Fig. 3.9a. In Equation (3.2.2), $p = 2ed$ is the *electric dipole moment,* which is defined as the product of the charge and the separation.

We all know that an insulator can become polarized in an electric field. A familiar example of polarizable material is a dielectric used in a capacitor. The origin of electric polarization lies in the fact that charges of opposite polarities are displaced in opposite directions in an electric field. This relative displacement of opposite charges produces an electric dipole. If an atom is placed at point P in Fig. 3.9, the atom will become polarized and will have an electric dipole moment p' induced by the field of Eq. (3.2.2). The induced dipole moment p' is related to E through

$$p' = \alpha E \qquad (3.2.3)$$

where the proportionality constant α is called the *polarizability,* and further, the potential energy of the induced dipole is given by

$$\mathscr{E}_W = -p'E = -\frac{2pp'}{4\pi\epsilon_0 R^3} = -\alpha\left(\frac{2p}{4\pi\epsilon_0 R^3}\right)^2 \qquad (3.2.4)$$

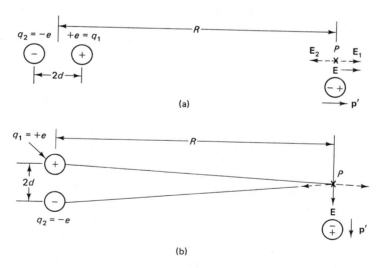

Figure 3.9 Illustration showing the effect of an electric dipole, which consists of a pair of charges of opposite polarities separated at a distance $2d$ apart. Two situations are considered: (a) The point P is along the line passing through the dipole and (b) the point P is along the line perpendicular to the dipole. If a molecule is placed at P, the dipole p' induced in the molecule will point in the same direction as the direction of **E** which is horizontal in (a) and vertical in (b).

It is clear from Fig. 3.9 that the two dipoles p and p' attract one another. This accounts for the negative sign in Eq. (3.2.4).

The interaction of the type described in Eq. (3.2.4) is known as the *van der Waals interaction* and is used for the attractive term in the gas equation of van der Waals. The van der Waals interaction is responsible for the cohesion not only in inert-gas crystals but also in other molecular crystals such as H_2 and CH_4. The molecules H_2 and CH_4 are formed by covalent bonds. However, all the bonds in H_2 and CH_4 have been used up in forming the molecules H_2 and CH_4; thus no additional bond pairs can be formed with neighboring H_2 or CH_4 molecules. The attraction between molecules in solid H_2 and CH_4 comes from forces of the van der Waals type. We should point out that the van der Waals interaction is a relatively weak interaction. The molecules H_2 and CH_4 have already achieved a stable configuration through electron sharing, whereas the inert-gas elements have a closed-shell configuration to begin with. Hence the dipole electric field can only cause a slight distortion of the electron distribution. The fact that all molecular and inert-gas crystals have very low melting and boiling temperatures is a manifestation of the very weak bond.

The reader may ask what gives rise to a nonvanishing dipole moment p in Eq. (3.2.4) in the first place. The electron distribution associated with a closed-shell configuration is known to be spherically symmetrical with respect to the nucleus. From electrostatics, we know that we can replace a spherical charge distribution by an equivalent charge at the center of the sphere. Consequently, no electric dipole is expected. Here we must distinguish between time-average and instantaneous dipole moment. It is true that the average dipole moment of an inert-gas atom is zero, as obtained from the electron wave function. However, at a given instant, an atom may possess a nonvanishing dipole moment because of the motion of the electron around the nucleus.

To prove our point about instantaneous and time-average values, we consider a one-dimensional harmonic oscillator which consists of a stationary positive charge $+e$ and a moving electron of charge $-e$. For harmonic motion, the instantaneous displacement x changes with time as

$$x = x_0 \sin (\omega_0 t + \phi) \qquad (3.2.5)$$

Although the average value of x (over a period) is zero, both the instantaneous value and the mean-square value of x are nonzero. In Eq. (3.2.4) it is the average value of pp' that is to be evaluated. Since the motions of electrons in neighboring atoms (Fig. 3.9) are in harmony with one another, the time-average value of pp' is equal to the time-average value of p^2 and hence is nonzero.

In summary, we have given in this section some general characterizations of the four distinct bond types. In ionic crystals, electron transfer from atoms of one kind to atoms of another kind is involved. In covalent crystals, the bond is formed between neighboring atoms by sharing a pair of electrons. In metallic crystals, the valence electrons are essentially free. The different arrangements in these crystals are designed such that the atoms or ions involved may achieve a closed-shell configuration. In molecular and inert-gas crystals, the constituent molecule or atom already has a stable configuration, and hence the bond is much weaker. The weak bond comes from attraction between electric dipoles. In some cases we may even have to consider attraction arising from electric quadrupoles (van der Waals interaction having a $1/R^8$ dependence). In making the foregoing general characterization of the four bond types, we must emphasize that the transition between crystal types is not at all sharp. Many materials have two bonding mechanisms operating simultaneously. For example, ZnS can be considered as having an admixture of ionic and covalent bonding, with the character of the former being stronger than that of the latter. The situation in SiC is just the opposite. Some metal alloys such as Mg_3Sb_2 and Zn_3As_2 can be considered partly metallic and partly ionic, and some element crystals such as S and P can be considered partly molecular and partly covalent.

3.3 IONIC CRYSTALS

In Section 3.2 we presented a general discussion concerning the bonding mechanisms in four principal crystal types: the ionic, the covalent, the metallic, and the molecular crystals. In discussing material properties, especially the conductive and dielectric properties, we must know how tightly electrons are bound to the parent atoms or ions in a crystal. More specifically, electrons are known to be free to move in metals, but not so in ionic crystals. The foundations of the classical theory of ionic and metallic crystals were laid before the quantum theory was advanced. The classical theory provides us with a physical model that is simple to understand and easy to work with. In this section and in Section 3.4, we explore the physical models for these crystals and see how accurate are the predictions based on the classical model.

The idealized model of an ionic crystal assumes that the constituent ions have closed-shell configurations. Hence it is expected that these ions have spherically symmetric charge distributions as inert-gas atoms. X-ray study of ionic crystals has indeed confirmed this picture of spherical electron distribution except near the region where neighboring ions come in contact. This spherical model makes the calculation of the cohesive energy of a crystal tractable.

According to the theory of electrostatics, the Coulomb energy between two non-overlapping spherical charge distributions is the same as that between two point charges. Thus the total electrostatic energy of an ionic crystal is given by

$$\mathscr{E}_P = \sum (\pm) \frac{e^2}{4\pi\epsilon_0 r_{ij}} \tag{3.3.1}$$

where r_{ij} is the distance between the two ions involved, ϵ_0 the dielectric constant of free space, and the positive and negative signs are for interactions between charges of like and opposite polarities, respectively. The summation in Eq. (3.3.1) extends over all ion pairs.

To illustrate how the summation can be evaluated, we use NaCl as an example, the structure of which is shown in Fig. 2.7a. First, we choose the center Cl ion as the reference ion. In other words, we let the index j in Eq. (3.3.1) be fixed. The contribution to Eq. (3.3.1) from the six nearest Na ions is $6e^2/R$, R being the distance between the nearest neighbors. Next are the 12 nearest Cl ions, which are $\sqrt{2} R$ apart from the center Cl ion. Then there are eight next to the nearest Na ions at a distance $\sqrt{3} R$ apart and six next to the nearest Cl ions at a distance $\sqrt{4} R$ apart. If there are $2N$ ions or N pairs of NaCl ions in a crystal, summing Eq. (3.3.1) over the index j simply means repeating the same process of calculation N times. Thus Eq. (3.3.1) can be written as

$$\mathscr{E}_P = \frac{Ne^2}{4\pi\epsilon_0 R} \left(\frac{6}{1} - \frac{12}{\sqrt{2}} + \frac{8}{\sqrt{3}} - \frac{6}{\sqrt{4}} + \frac{24}{\sqrt{5}} - \cdots \right) = \frac{ANe^2}{4\pi\epsilon_0 R} \tag{3.3.2}$$

In Eq. (3.3.2) the constant A is a numerical constant known as the *Madelung constant*. Since the series in parentheses converges very slowly, many terms are needed to make a fair estimate of A. For the NaCl structure the value of A is found to be 1.75, and for the CsCl structure, $A = 1.76$.

Equation (3.3.2) accounts for the attractive energy $\mathscr{E}_{att}$ in Eq. (3.1.2). Let us now turn to the repulsive energy. Because the ions, the Na^+ and Cl^- ions in NaCl for example, have closed-shell configurations, the repulsive energy is due to the ion-core repulsion energy $\mathscr{E}_C$, which increases very rapidly with decreasing interatomic distance R_0. The actual dependence of this repulsive energy can be calculated quantum mechanically as a function of R. However, we are interested only in the variation of this repulsive energy near R_0, the equilibrium value of R (Fig. 3.1). For this purpose, the repulsive energy can be approximated by

$$\mathscr{E}_C = \frac{BN}{R^n} \tag{3.3.3}$$

Since the two energies $\mathscr{E}_P$ and $\mathscr{E}_C$ are opposite in nature, the total energy of a NaCl crystal is given by $\mathscr{E}_C - \mathscr{E}_P$.

At the equilibrium separation R_0, $\mathscr{E}_C - \mathscr{E}_P$ is a minimum. Using Eq. (3.3.3) and setting $R = R_0$, we obtain

$$\frac{-nBN}{R_0^{n+1}} + \frac{ANe^2}{4\pi\epsilon_0 R_0^2} = 0 \quad \text{or} \quad R_0^{n-1} = \frac{4\pi\epsilon_0 Bn}{Ae^2} \tag{3.3.4}$$

Substituting R_0 from Eq. (3.3.4) into Eqs. (3.3.2) and (3.3.3), we find the binding energy $\mathscr{E}_0$ (Fig. 3.1) to be

$$\mathscr{E}_0 = (\mathscr{E}_C - \mathscr{E}_P)_{R=R_0} = -\frac{ANe^2}{4\pi\epsilon_0 R_0}\left(1 - \frac{1}{n}\right) \tag{3.3.5}$$

To compare the theoretical and experimental values of $\mathscr{E}_0$, we need to know the value of n in Eq. (3.3.5). Such information can be obtained from the compressibility K. By definition,

$$K = \frac{-1}{V}\frac{dV}{dp} \tag{3.3.6}$$

where V is the volume of a crystal and p is the applied pressure. If we assume that each ion occupies a small cube of volume R^3, the volume of the crystal is given by $V = 2NR^3$. Noting that under constant p, $d\mathscr{E} = -p\,dV$, we obtain from Eqs. (3.3.2) and (3.3.3),

$$\frac{1}{K} = V\frac{d^2\mathscr{E}}{dV^2} = \frac{(n-1)e^2A}{72\pi\epsilon_0 R_0^4} \tag{3.3.7}$$

Take NaCl as an example of an ideal ionic crystal. The value of K is 4.1×10^{-12} cm^2/dyn, and the value of R_0 is 2.82 Å (Table 2.3). Using these values in Eq. (3.3.7) gives a value of $n = 7.9$. The binding energy of NaCl crystal (Table 3.1) is found experimentally to be 183 kcal/mol or 7.95 eV/molecule. Using $A = 1.75$, $R_0 = 2.82$ Å, and $n = 7.9$ in Eq. (3.3.5), we find the value of $\mathscr{E}_0/N = 7.85$ eV. The small discrepancy between the calculated and measured values is due to the fact that we have neglected in our calculation a small contribution from van der Waals forces to the total energy. A good agreement between the two values is also found in other alkali-halide crystals. This general agreement confirms that the physical model of an ionic crystal as being made of X^+ and Y^- ions is a correct one. We further note from Eq. (3.3.5) that the repulsive energy is smaller than the Coulomb energy by the factor $1/n$. Therefore, the cohesive energy of an ionic crystal is due mostly to Coulomb attraction between the ions.

3.4 METALLIC CRYSTALS

According to the classical description mentioned in Section 3.2, metallic crystals are generally considered as being composed of arrays of positive ions embedded in a uniform sea of conduction electrons. In the following discussion we use this classical model in the calculation of the cohesive energy of a metallic crystal. For the calculation, we take sodium as an example. As shown in Fig. 3.8, the distribution of valence electrons in sodium crystals is essentially uniform over 90% of the volume occupied by electrons. Therefore, in the calculation of the electrostatic interaction energy $\mathscr{E}_P$, it seems reasonable to treat a sodium atom as being made of a point charge of $+e$ (which represents the nucleus plus core electrons) and an electron cloud of charge $-e$ (which represents the valence electron) uniformly distributed over a sphere of volume V.

Referring to Fig. 3.10, we see that the charge dq carried by the electron cloud in a volume element $4\pi r^2\, dr$ is given by

$$dq = -e\, \frac{4\pi r^2\, dr}{V} \tag{3.4.1}$$

This charge dq experiences a positive Coulomb potential $e/4\pi\epsilon_0 r$ due to the positive point charge and a negative Coulomb potential $-q/4\pi\epsilon_0 r$ due to the electron cloud, where q is the charge carried by the electron cloud. From electrostatics, we know that to evaluate the potential at a point r from the center of a spherical charge distribution, we count only the charge within a sphere of radius r. Thus $q = e4\pi r^3/3V$. The total Coulomb energy per atom is therefore given by

$$\mathscr{E}_P = \int_0^R \left(\frac{e}{4\pi\epsilon_0 r} - \frac{er^2}{3\epsilon_0 V} \right) \left(-e\, \frac{4\pi r^2\, dr}{V} \right) = \frac{-9e^2}{40\pi\epsilon_0 R} \tag{3.4.2}$$

For alkali metals, the other important term in the calculation of cohesive energy is the kinetic energy $\mathscr{E}_K$ in Eq. (3.1.5). To calculate this energy, we must know the energy distribution of electrons. Figure 3.11a is the X-ray emission spectrum of sodium caused by the $3s \rightarrow 2p$ transition (Fig. 3.11b). As a result of the impact of high-energy electrons (Section 2.7), the $2p$ electrons are liberated from certain sodium atoms. The

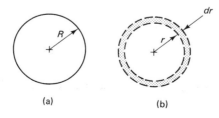

(a)   (b)

Figure 3.10 Diagrams used for the calculation of electrostatic energy in a metallic crystal. Consider the electrostatic potential experienced by charge dq in a spherical shell (shaded area). The potential consists of two parts: (a) due to the positive ion core and (b) due to the electron cloud inside the spherical shell.

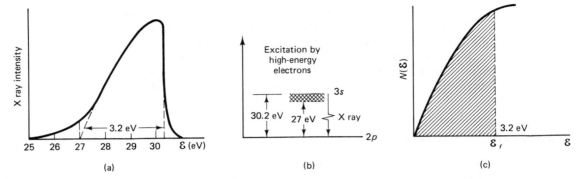

Figure 3.11 Information used in the calculation of kinetic energy in a sodium crystal: (a) X-ray emission spectrum, (b) energy levels involved in the X-ray emission, and (c) energy distribution of the 3s electrons (valence electrons) as derived from the X-ray emission spectrum. Note the similarity between the curves in (a) and (c). In curve c, the abscissa represents the kinetic energy only.

2p levels thus vacated are ready for reoccupation by electrons from the 3s levels. The energy of the 3s electrons in a sodium crystal, unlike the 3s electron in a free sodium atom, is not discrete but forms a band. Therefore, the emission spectrum of the X ray $(\mathscr{E}_{3s} - \mathscr{E}_{2p})$ covers a range from approximately 27 to 30.2 eV because the energy of the 3s electrons spreads over a range of 3.2 eV.

As discussed in Chapter 4, the energies of atomic states broaden in a crystal to form energy bands. Since the 3s (valence) electrons in Na are free to move, the spread in energy is due to different kinetic energies. We define a distribution function $D(\mathscr{E})$ such that the density of electrons dn within an energy interval $d\mathscr{E}$ is

$$dn = D(\mathscr{E}) \, d\mathscr{E} \qquad \text{electrons/m}^3 \qquad (3.4.3)$$

The intensity of the X ray shown in Figure 3.11a is proportional to $D(\mathscr{E})$ of Eq. (3.4.3). It is shown in Chapter 4 that the function $D(\mathscr{E})$ is given by

$$D(\mathscr{E}) = \frac{4\pi(2m)^{3/2}}{h^3} \mathscr{E}^{1/2} \qquad (3.4.4)$$

The $\mathscr{E}^{1/2}$ dependence is shown in Fig. 3.11c. Since $\mathscr{E}_K = 0$ for the 3s electron having the lowest energy, we move the origin to $\mathscr{E}_K = 0$ in Fig. 3.11c. Otherwise, Fig. 3.11c is a reproduction of Fig. 3.11a.

We define a quantity $\mathscr{E}_f$, generally called the *Fermi energy,* as the topmost kinetic energy. From Eqs. (3.4.3) and (3.4.4) we find the total number of 3s electrons to be

$$n = \int_0^{\mathscr{E}_f} D(\mathscr{E}) \, d\mathscr{E} = \frac{8\pi}{3} \frac{(2m)^{3/2}}{h^3} \mathscr{E}_f^{3/2} \qquad \text{electrons/m}^3 \qquad (3.4.5)$$

The total kinetic energy of all electrons is

$$\sum \mathscr{E}_K = \int_0^{\mathscr{E}_f} \mathscr{E} D(\mathscr{E}) \, d\mathscr{E} = \frac{8\pi}{5} \frac{(2m)^{3/2}}{h^3} \mathscr{E}_f^{5/2} \qquad (3.4.6)$$

Thus the average kinetic energy $\overline{\mathscr{E}_K}$ per atom (the same as per 3s electron in the case of sodium) is

$$\overline{\mathscr{E}}_K = \frac{\Sigma \mathscr{E}_K}{n} = \frac{3}{5}\mathscr{E}_f = \frac{3h^2}{10m}\left(\frac{3n}{8\pi}\right)^{2/3} \tag{3.4.7}$$

Referring to Fig. 3.10, we see that each Na atom occupies a volume $4\pi R^3/3$. Since there is one $3s$ electron per atom, the density of electrons is equal to $n = 3/(4\pi R^3)$. Using this value of n in Eq. (3.4.7), we obtain

$$\overline{\mathscr{E}}_K = \frac{3h^2}{10m}\left(\frac{9}{32\pi^2}\right)^{2/3}\frac{1}{R^2} \tag{3.4.8}$$

To calculate the value of R from the lattice constant, we note the following fact. In a body-centered cubic structure, the eight corner atoms are shared by eight cubes; hence the cube is occupied by two atoms, one being the center atom and the other being contributed by the eight corner atoms of one-eighth each. If a is the lattice constant, the volume occupied by each Na atom is $a^3/2$, which should be set equal to $4\pi R^3/3$. For sodium, $a = 4.28$ Å (Table 2.3) which gives a value of $R = 2.11$ Å. Using this value of R in Eqs. (3.4.2) and (3.4.7), we find that $\mathscr{E}_P = -6.13$ eV and $\overline{\mathscr{E}}_K = 1.92$ eV.

Now we refer to Fig. 3.12 in calculating the cohesive energy. The energy $\mathscr{E}_P$ is the potential energy required to remove the valence ($3s$) electron from Na in the crystal state. However, the cohesive energy is the energy difference between crystalline sodium and atomic sodium. The energy difference between atomic sodium and sodium in the ionized state (Na$^+$ + free electron) is just the ionization energy ($\mathscr{E}_i = 5.18$ eV) of the $3s$ electron. From Fig. 3.12 we have

$$\mathscr{E}_0 = |\mathscr{E}_P - \overline{\mathscr{E}}_K - \mathscr{E}_i| \tag{3.4.9}$$

Before we substitute the values in Eq. (3.4.9), we note that if Eqs. (3.4.2) and (3.4.8) are used in Eq. (3.1.2), we obtain an equilibrium value of $R = 1.3$ Å quite different from the value of 2.11 Å. Apparently, either Eq. (3.4.2) or (3.4.8) must be in error. We may find the clue from the ionization energy. For the $3s$ electron, the principal and orbital quantum numbers are $n = 3$ and $l = 0$. If the $3s$ electron felt an effective nuclear charge of $+e$ as our model assumes, Eq. (1.6.1) would predict an ionization energy of $13.6/9 = 1.5$ eV. The fact that the measured value (5.18 eV) is considerably larger than this value indicates that the $3s$ electron sees an effective nuclear charge (nucleus plus core electrons) appreciably larger than $+e$.

The classical model, on which Eq. (3.4.2) is based, treats electrons of closed shells as if they form a spherical core. The model further assumes that the valence electron moves completely outside this core. A quantum-mechanical calculation of the $3s$ electron wave function (Fig. 3.8) shows that the $3s$ electron wave function penetrates the ion core. When the $3s$ electron is inside the core region, that is, for $r < 0.9$ Å in Fig. 3.8, the shielding effect of the core electrons is much reduced. Therefore, it is Eq. (3.4.2), which is in error.

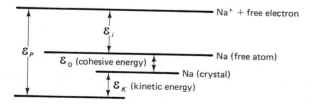

Figure 3.12 Diagram used for the calculation of cohesive energy of sodium. The difference in energy between a crystalline sodium atom and an ionized sodium atom is $\mathscr{E}_P - \overline{\mathscr{E}}_K$. The energy difference between a free and ionized sodium atom is $\mathscr{E}_i$. The cohesive energy $\mathscr{E}_0$ is, therefore, given by Eq. (3.4.9).

The correct value of $\mathscr{E}_P$ is found quantum mechanically to be -8.40 eV for $R = 2.11$ Å. Using this value, $\overline{\mathscr{E}}_K = 1.98$ eV, and $\mathscr{E}_i = 5.18$ eV in Eq. (3.4.9), we obtain a value of cohesive energy per atom $\mathscr{E}_0 = 1.30$ eV, in reasonable agreement with the observed value of 1.13 eV (Table 3.1). The agreement is a confirmation of the validity of Eq. (3.4.7). As shown in Chapter 4, Eq. (3.4.4) is derived on the basis of the relation

$$\mathscr{E} = \frac{p^2}{2m} = \frac{\hbar^2 k^2}{2m} \tag{3.4.10}$$

between the energy $\mathscr{E}$ and the wave number k of an electron. Equation (3.4.10), in turn, is based on the assumptions that valence electrons in a metal are essentially free and that the wave number $k (= 2\pi/\lambda)$ is related to the momentum p of an electron by the de Broglie relation $\lambda = h/p$. The success that we have had with Eq. (3.4.4) in predicting the value of $\overline{\mathscr{E}}_K$ proves the correctness of these assumptions. As we shall see in Chapters 4 and 5, these assumptions lead to a great simplification of the treatments of the conductive properties of metals and semiconductors.

We should point out that the relative importance of $\mathscr{E}_K$ and $\mathscr{E}_P$ varies from metal to metal; therefore, sodium should not be taken as a typical example for other metals. If the size of the inner shell is large, ion-core repulsive energy $\mathscr{E}_C$ may become significant. The wide variation of cohesive energy from 0.8 eV in Cs (alkali metal) to 4.2 eV in Fe (transition metal) is an indication that bonding forces of different physical origins may be involved. It is generally believed that covalent bonding plays a significant part in transition metals with incomplete $3d$ shells. However, for alkali metals, the binding energies calculated from Eq. (3.4.9) are in good agreement with the measured values. Because the orbits of core electrons are relatively small compared with the interatomic distances, the simple model (Fig. 3.10) works well for these metals. For this reason alkali metals are considered the closest to being ideal metals.

3.5 THE HYDROGEN MOLECULE

In the study of atoms the hydrogen atom is generally used as a model in understanding the structure of other atoms. Similarly, a great deal can be learned from the hydrogen molecule. As discussed in Section 3.2, the bond between two hydrogen atoms is formed by sharing a pair of electrons with antiparallel spins. Many important properties of covalent bonds are illustrated in this simple case. We outline below the methods used in the study of hydrogen molecules. The material presented in this section supplements our discussion in Section 3.2.

The Schrödinger equation for a hydrogen molecule is

$$\frac{\hbar^2}{2m}(\nabla_1^2 \psi + \nabla_2^2 \psi) + \left[\mathscr{E} + \frac{1}{4\pi\epsilon_0} \left(\frac{e^2}{r_{1a}} + \frac{e^2}{r_{1b}} \right. \right.$$
$$\left. \left. + \frac{e^2}{r_{2a}} + \frac{e^2}{r_{2b}} - \frac{e^2}{r_{12}} - \frac{e^2}{r_{ab}} \right) \right] \psi = 0 \tag{3.5.1}$$

where the subscripts a and b refer to the two nuclei, the subscripts 1 and 2 refer to the two electrons, and r is the distance between them, as illustrated in Fig. 3.13. In Eq. (3.5.1), ∇^2 is the Laplacian operator. Because of the various mutual interaction terms, it is difficult to find the solution of Eq. (3.5.1). Therefore, we must resort to approxi-

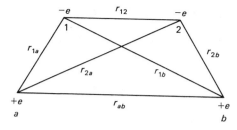

Figure 3.13 Coordinates of electrons and nuclei in a hydrogen molecule. Points a and b represent the positions of the two hydrogen nuclei, whereas points 1 and 2 represent the positions of the two electrons.

mate methods. There are two approaches that are taken to find an approximate solution of Eq. (3.5.1): the method of molecular orbitals and the method of Heitler and London.

As pointed out in Section 3.2, the Schrödinger equation for the $H_2{}^+$ molecule ion [Eq. (3.5.1) without electron 2, that is, without the r_{2a}, r_{2b}, and r_{12} terms] can be exactly solved. The wave function obtained from such a solution is called the *molecular orbital*. The wave functions σ_g and σ_u used in the computation of the two curves in Fig. 3.7 are molecular orbitals. If we neglect the mutual interaction between the two electrons in H_2, we can treat the neutral H_2 molecule as if the two electrons move independently in the Coulomb potential of $H_2{}^+$ ion. For the lowest-energy state of a neutral H_2 molecule, both electrons have the σ_g wave function. Since the orbital wave functions of the two electrons are the same, their spin wave functions must be different because of the exclusion principle. This treatment of a H_2 molecular, known as the *method of molecular orbitals*, is used in Section 3.2 to explain the bonding property and spin antiparallel arrangement of an electron-pair bond.

The other approach to the H_2 problem is the method of Heitler and London. As discussed in Section 1.4, the Schrödinger equation for a hydrogen atom can be solved exactly. If we neglect the mutual interaction terms r_{1b}, r_{2a}, r_{12}, and r_{ab} in Eq. (3.5.1), we can write Eq. (3.5.1) as

$$\frac{\hbar^2}{2m}(\nabla_1^2\psi + \nabla_2^2\psi) + \left[\mathscr{E} + \frac{1}{4\pi\epsilon_0}\left(\frac{e^2}{r_{1a}} + \frac{e^2}{r_{2b}}\right)\right]\psi = 0 \qquad (3.5.2)$$

Let ψ_{1a} and ψ_{2b} be two atomic wave functions of a hydrogen atom. In associating the subscript 1 with a and the subscript 2 with b, we assign electron 1 to nucleus a and electron 2 to nucleus b. Since electron 1 belongs to nucleus a, the wave function must satisfy the equation

$$\frac{\hbar^2}{2m}\nabla_1^2\psi_{1a} + \left(\mathscr{E}_1 + \frac{e^2}{4\pi\epsilon_0 r_{1a}}\right)\psi_{1a} = 0 \qquad (3.5.3)$$

Equation (3.5.3) is simply the Schrödinger equation for a hydrogen atom. A similar equation for electron 2 and nucleus b can be obtained from Eq. (3.5.3) by substituting the subscript 2 for 1 and the subscript b for a.

Using separation of variables, we can express the solution of Eq. (3.5.2) as

$$\Psi = \psi_{1a}\psi_{2b} \qquad (3.5.4)$$

if we let $\mathscr{E} = \mathscr{E}_1 + \mathscr{E}_2$. Now let us return to Eq. (3.5.1). We see that Eq. (3.5.1) remains the same if the subscripts 1 and 2 are interchanged. Since $|\Psi|^2$ is the joint probability function, there is no a priori reason why electron 1 should belong to nucleus a and electron 2 should belong to nucleus b. If the function $\psi_{1a}\psi_{2b}$ is an approximate solution to Eq. (3.5.1), the function $\psi_{2a}\psi_{1b}$ with electron 1 belonging to nucleus b and electron 2 belonging to nucleus a is just as good an approximate solution. Both these functions must be considered in finding the approximate solution of Eq. (3.5.1).

Let $\Psi(1, 2)$ be the joint wave function of the two electrons in a hydrogen molecule. The indistinguishability of electrons requires that the probability of finding electron 1 at r_{1a} and electron 2 at r_{2b} be the same if the labels of the two electrons are interchanged. In other words,

$$\Psi^*(1, 2)\Psi(1, 2) = \Psi^*(2, 1)\Psi(2, 1) \tag{3.5.5}$$

Equation (3.5.5) is satisfied if

$$\Psi(1, 2) = \Psi(2, 1) \tag{3.5.6}$$

or if

$$\Psi(1, 2) = -\Psi(2, 1) \tag{3.5.7}$$

The wave function that satisfies Eq. (3.5.6) is the symmetric wave function, whereas the wave function that satisfies Eq. (3.5.7) is the antisymmetric wave function. The functions

$$\Psi_{\text{S-Or}} = \psi_{1a}\psi_{2b} + \psi_{2a}\psi_{1b} \tag{3.5.8}$$

$$\Psi_{\text{A-Or}} = \psi_{1a}\psi_{2b} - \psi_{2a}\psi_{1b} \tag{3.5.9}$$

certainly satisfy the symmetric and antisymmetric requirement of Eqs. (3.5.6) and (3.5.7), respectively. The subscripts S-Or and A-Or are used to denote, respectively, symmetric and antisymmetric orbital wave functions.

In the treatment of a hydrogen molecule by Heitler and London, products of atomic wave functions expressed in symmetric and antisymmetric forms as in Eqs. (3.5.8) and (3.5.9) are used as the approximate solution to Eq. (3.5.1). The electron density for the symmetric wave function tends to be concentrated in the region of low potential energy between the two nuclei. Using the symmetric and antisymmetric wave functions of Eqs. (3.5.8) and (3.5.9), the energy of a hydrogen molecule can be calculated quantum mechanically. The results of such a calculation are shown in Fig. 3.14 as a function of the internuclear distance r_{ab}. As expected, the state represented by the symmetric wave function is the bonding state, whereas the state represented by the antisymmetric wave function is the antibonding state.

Besides the orbital wave function, we must consider the spin wave function. Like the orbital wave function, the spin wave function must also be either symmetric (denoted by $\Psi_{\text{S-Sp}}$) or antisymmetric (denoted by $\Psi_{\text{A-Sp}}$). The symmetric spin wave function can be expressed in either one of the following three combinations:

$$\Psi_{\text{S-Sp}} = \begin{cases} \alpha_1\alpha_2 \\ \beta_1\beta_2 \\ \alpha_1\beta_2 + \beta_1\alpha_2 \end{cases} \tag{3.5.10}$$

whereas the antisymmetric spin wave function is given by

$$\Psi_{\text{A-Sp}} = \alpha_1\beta_2 - \beta_1\alpha_2 \tag{3.5.11}$$

where α and β denote the one-electron spin wave function having $s_z = +\frac{1}{2}$ and $s_z = -\frac{1}{2}$, respectively.

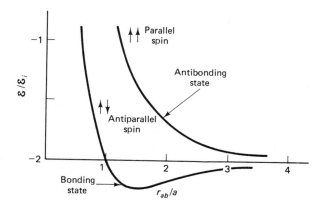

Figure 3.14 Energy curves for the hydrogen molecule plotted as a function of the internuclear distance. The ordinate and abscissa are normalized with respect to the ionization energy and the Bohr radius of a hydrogen atom. The lower curve has an energy lower than the energy of two separate hydrogen atoms. Therefore, the state represented by the antisymmetric spin wave function is the bonding state.

The total wave function in a hydrogen molecule is a product of the orbit and spin wave functions, or

$$\Psi_{total} = \begin{matrix} \Psi_{S-Or} \\ \Psi_{A-Or} \end{matrix} \times \begin{cases} \Psi_{A-Sp} \\ \Psi_{S-Sp} \end{cases} \qquad (3.5.12)$$

The question as to which combinations in Eq. (3.5.12) occur in nature can only be answered by experiments. Analyses of the atomic and molecular spectra show that only the combinations of $\Psi_{S-Or}\Psi_{A-Sp}$ and $\Psi_{A-Or}\Psi_{S-Sp}$ are allowed. In other words, the total wave function Ψ_{total} must be antisymmetric. This condition is consistent with the Pauli exclusion principle. Let us take the case of H^-, that is, a hydrogen ion with two electrons. Since we have a single nucleus, $a = b$ in Eqs. (3.5.8) and (3.5.9). The Pauli exclusion principle demands that the probability of finding two electrons in the same state must be zero. This means that if we change α to β or β to α in Eqs. (3.5.10) and (3.5.11), Ψ_{total} must be zero. The antisymmetric Ψ_{total} indeed satisfies this condition.

In Section 1.5, the exclusion principle is stated as follows. No two electrons can have the same set of four quantum numbers, n, l, m, and s_z. So that the principle can easily be applied to electrons in molecules, this principle has been changed to the following statement: If any two electrons are interchanged, the proper wave function must remain unchanged except for the factor -1. This antisymmetric requirement on the total wave function demands that the bonding state have antiparallel spins. It can be shown quantum mechanically that the resultant spin of the two electrons as represented by Ψ_{A-Sp} of Eq. (3.5.11) is zero and thus the spins must be antiparallel.

Although in the two methods (the method of molecular orbitals and the method of Heitler and London), different wave functions are used as the starting approximate solution of Eq. (3.5.1), the conclusions drawn from the two analyses are qualitatively the same. The energy of the bonding state is lowered by an accumulation of negative charges in the region of low potential energy between the two nuclei. The two electrons do not specifically belong to a particular nucleus, but are shared equally between the two nuclei. The spins of the two electrons must have an antiparallel arrangement because of the exclusion principle. These properties as derived from the analysis of a hydrogen molecule are the general properties of a covalent bond.

Sec. 3.5   The Hydrogen Molecule

3.6 SATURATION AND DIRECTIONAL PROPERTIES OF COVALENT BONDS

In Sections 3.2 and 3.5 we described a covalent bond as an electron-pair bond which is formed between two atoms as a result of minimizing the energy by sharing a pair of electrons with antiparallel spins. On the other hand, the wave function of the antibonding state (Fig. 3.14) is represented by $\Psi_{A\text{-}Or}\Psi_{S\text{-}Sp}$, a product of antisymmetric orbit wave function and symmetric spin wave function. It can be shown quantum mechanically that the resultant spin of the two electrons as represented by $\Psi_{S\text{-}Sp}$ of Eq. (3.5.10) is 1 $(= \frac{1}{2} + \frac{1}{2})$, and thus the two spins are parallel. The property of an electron-pair bond can be stated in terms of spin orientations as follows. Sharing a pair of electrons with antiparallel spins results in an attraction of the two atoms (the bonding state), whereas sharing a pair of electrons with parallel spins results in a repulsion of the two atoms (the antibonding state).

The requirement on the spin orientation leads to an important property of a covalent bond, the saturation property. Consider the interaction of a hydrogen atom with a helium atom. As illustrated in Fig. 3.15, there are only two possible interactions: sharing a pair of electrons with antiparallel spins or sharing a pair of electrons with parallel spins. If the He atom and H atom share electrons with antiparallel spins, the other electron of the He atom will have the same spin orientation as the electron from the hydrogen atom (Fig. 3.15a). This situation clearly violates the Pauli exclusion principle unless either one of the two electrons is promoted to an unoccupied shell of higher energy. On the other hand, if the He atom and H atom share electrons with parallel spins, our previous discussion of a hydrogen molecule tells us that the state will be an antibonding state. In either case, the energy of the hypothetic H—He molecule would be higher than the combined energy of two separate atoms. In other words, a bond will not form between a H atom and a He atom.

The saturation property of a covalent bond, as illustrated by the foregoing example, can be stated as follows. An electron-pair bond can be formed only by those electrons of an atom that have unpaired spins. The spins of the two electrons in He are already paired (one with $s_z = +\frac{1}{2}$ and the other with $s_z = -\frac{1}{2}$), and hence no electron-pair bond will be formed between either one of the two electrons and a third electron from another atom.

To illustrate the saturation property of a covalent bond further, we examine the situation in H_2O, NH_3, and CH_4. Referring to Table 1.2, we see that oxygen has an electron configuration $1s^2 2s^2 2p^4$ and nitrogen has an electron configuration $1s^2 2s^2 2p^3$. The s electrons already have their spins paired (one up and the other down), so only the p electrons need to be considered. For $l = 1$, there are three possible values for m, the

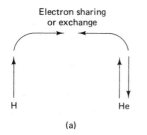

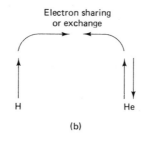

Figure 3.15 Interaction between a hydrogen and a helium atom. The arrows indicate the direction of spin. There are two possible situations: (a) sharing electrons with antiparallel spins, and (b) sharing electrons with parallel spins.

magnetic quantum number. Therefore, there are altogether six possible arrangements for the p electrons:

m	1	0	-1	1	0	-1
Spin	Up	Up	Up	Down	Down	Down

For a nitrogen atom we can have a maximum of three electrons that have unpaired spins, whereas for an oxygen atom, at least two of the four electrons must have paired spins, leaving a maximum of two electrons having unpaired spins. Therefore, a nitrogen atom is capable of making three electron-pair bonds with three hydrogen atoms in NH_3, and an oxygen atom is capable of forming only two electron-pair bonds in H_2O even though it has four $2p$ electrons. The situation is illustrated in Fig. 3.16.

Let us turn to CH_4. The ground state of carbon has the electron configuration $1s^2 2s^2 2p^2$. For the ground state, we would expect a maximum of two unpaired spins. As pointed out in Section 1.6, the energy of the $2p$ state is not too far above the energy of the $2s$ state. In CH_4, the carbon atom actually has an electron configuration $1s^2 2s 2p^3$. In this configuration, the carbon atom can have a maximum of four unpaired spins. In CH_4, all four unpaired spins of the carbon atom become paired with spins of the four hydrogen atoms; therefore, the state of saturation is reached in forming four electron-pair bonds. Energy is gained in promoting one electron of the carbon atom from the $2s$ level to the $2p$ level, but at the same time, energy is reduced by forming electron-pair bonds. The CH_4 molecule is a stable one because the reduction in energy is bigger than the gain in energy.

Typical examples of crystals having purely covalent bonds are diamond, silicon, and germanium. All these elements crystallize in the diamond structure. As discussed in Section 2.3, each atom in this structure (Figs. 2.11 and 2.12) has four nearest neighbors. Again, one electron must be promoted from the s to the p level ($2s$ to $2p$ in diamond, $3s$ to $3p$ in silicon, and $4s$ to $4p$ in germanium) in order that a maximum of four bonds can be achieved. Thus the four outmost electrons have the following configurations: $2s2p^3$ in diamond, $3s3p^3$ in silicon, and $4s4p^3$ in germanium. These four electrons are said to form sp^3 hybrid orbitals.

The reader may wonder why CH_4 does not crystallize in the diamond structure. The difference in the bond properties of CH_4 and Ge, for example, can be understood on the basis of the saturation property of the covalent bond. Figure 3.17a is a schematic representation of the bond pairs of Ge in a diamond structure. Note that germanium

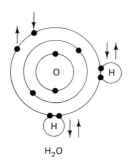

H_2O

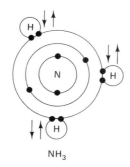

NH_3

Figure 3.16 Electron-pair bonds in (a) H_2O and (b) NH_3. Oxygen has two free (unpaired) spins and nitrogen three free spins. The number of free spins determines the maximum number of hydrogen bonds which an atom can make.

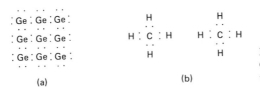

Figure 3.17 Electron-pair bonds in (a) Ge and (b) CH_4. The bond is indicated by two dots.

atoms can always be added to the existing structure because there exist unpaired bonds at the boundary of the structure. Therefore, covalent bonding is all that is needed for Ge atoms to form a Ge crystal. This is not the case with CH_4. The bonds in CH_4 are all used up (Fig. 3.17b); thus no additional bonds can be formed with neighboring CH_4 molecules. In Table 3.1 the CH_4 crystal is classified as a molecular crystal. It is true that the bonding between atoms in a CH_4 molecule is of the covalent type. However, the bonding between two CH_4 molecules in a CH_4 crystal is of the van der Waals type. Because CH_4 forms a stable molecule, they stay together as a molecule in the crystal. In other words, each lattice point in a molecular crystal (e.g., the lattice point in Fig. 2.1) actually represents a whole molecule. The same situation applies to other covalent-bond molecules, such as H_2.

Now let us discuss the directional property of a covalent bond. It is known from experiments that the H_2O molecule forms a triangle and the NH_3 molecule is in the form of a pyramid (Fig. 3.18). We also know that the angle θ between O—H bonds in H_2O and between any pair of the N—H bonds in NH_3 is close to and slightly larger than $90°$. As discussed in Sections 3.2 and 3.5, the lowering of energy is achieved in a covalent-bonding molecule by an accumulation of negative charges in the region between the two nuclei. In the language of quantum mechanics, the accumulation of negative charges is the result of the overlapping of two atomic wave functions. A strong bond is formed if there is a maximum of overlapping. Therefore, to understand the directional property of a covalent bond, we must examine the directional property of the atomic wave functions that take part in forming the bond.

Referring to the discussion in Section 1.4, we find that the wave function ψ_s for s electrons ($l = 0$) is spherically symmetrical; that is, it has no θ and ϕ dependence. Thus we can write ψ_s as

$$\psi_s = g(r) \qquad (3.6.1)$$

For the p electrons ($l = 1$), there are three wave functions whose angular dependences are of the following form: $\Theta(\theta)\Phi(\phi) \sim \sin \theta \exp (i\phi)$, $\sin \theta \exp (-i\phi)$, $\cos \theta$. Since the Schrödinger equation is a linear differential equation, any linear combination of the solutions is also a solution to the differential equation. Combining the exponential functions to form $\sin \phi$ and $\cos \phi$, we have $\Theta(\theta)\Phi(\phi) \sim \sin \theta \cos \phi$, $\sin \theta \sin \phi$, $\cos \theta$.

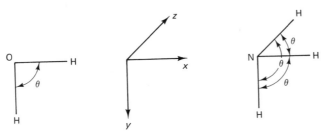

Figure 3.18 Examples used to illustrate the directional property of a covalent bond. (a) The H_2O molecule is in the form of a triangle and (b) the NH_3 molecule is in the form of a pyramid. The angle between any pair of bonds is close to $90°$. The characteristic is a result of the directional property of the p orbitals (Fig. 3.19).

Atomic Bonding and Crystal Types   Chap. 3

Realizing that

$$x = r \sin \theta \cos \phi, \quad y = r \sin \theta \sin \phi, \quad \text{and} \quad z = r \cos \theta \qquad (3.6.2)$$

and letting $R(r) = rf(r)$, we can express the three p wave functions as

$$\psi_{px} = xf(r), \quad \psi_{py} = yf(r), \quad \text{and} \quad \psi_{pz} = zf(r) \qquad (3.6.3)$$

The electron distributions as represented by the s wave function of Eq. (3.6.1) and the p wave functions of Eq. (3.6.3) are shown in Fig. 3.19. We see that each of the three p orbitals extends along a certain coordinate axis, for example, the x axis for the ψ_{px} orbital. It is expected, therefore, that the covalent bond will be formed along one of these axes to achieve a bond of maximum strength. As stated earlier, a stable bond is formed if there is a maximum overlapping of electron distribution.

In the H_2O molecule, the two electrons of the oxygen atom whose spins are unpaired must have different orbital wave functions because of the Pauli exclusion principle. Let us say that the two electrons have ψ_{px} and ψ_{py}. The condition for maximum bond strength requires that the two hydrogen atoms be located along the x and y axes. Thus we have a H_2O molecule with an angle of 90° between the bonds. The actual bond angle θ (Fig. 3.18a) in H_2O is 109°. Since the two hydrogen atoms have already used up their two electrons in forming bonds with the oxygen atom, they will not form an electron-pair bond between themselves. However, the repulsive force is still there, and the repulsion between the two hydrogen atoms increases the bond angle from 90° to 109°. The situation in NH_3 is quite similar. The three electrons of the nitrogen atom

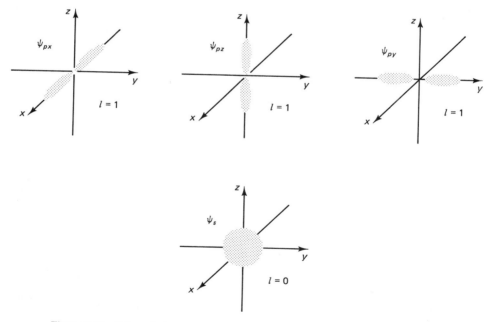

Figure 3.19 Schematic illustration showing the electron distributions for the s and p atomic wave functions. The s wave function is spherically symmetrical, whereas each of the three p wave functions extends along a specific axis, ψ_{px} along the x axis, ψ_{py} along the y axis, and ψ_{pz} along the z axis.

take three different p wave functions and form bonds with the three hydrogen atoms located separately along the x, y, and z axes. The bond angle θ (Fig. 3.18b) between them is slightly greater than $90°$ because of the mutual repulsion between the three hydrogen atoms.

Let us discuss the situation in CH_4. The four electrons of the carbon atom have wave functions which are an admixture of the s and p wave functions called the sp^3 *hybrid orbitals*. We learn from the theory of linear differential equations that the number of independent solutions is not altered by any linear combination of the independent solutions. Referring to the p state, we have three independent solutions irrespective of whether the solutions are expressed in terms of $\sin \phi$ and $\cos \phi$ or in terms of $\exp (-i\phi)$ and $\exp (i\phi)$. For the carbon atom, we have one s orbital and three p orbitals to begin with. The total number is four. Therefore, we expect to have four different sp^3 orbitals.

The four sp^3 orbitals can be expressed in terms of the s orbital of Eq. (3.6.1) and the p orbitals of Eq. (3.6.3) as

$$\psi_{111} = \tfrac{1}{2} [g(r) + xf(r) + yf(r) + zf(r)] \qquad (3.6.4)$$

$$\psi_{1\bar{1}\bar{1}} = \tfrac{1}{2} [g(r) + xf(r) - yf(r) - zf(r)] \qquad (3.6.5)$$

$$\psi_{\bar{1}1\bar{1}} = \tfrac{1}{2} [g(r) - xf(r) + yf(r) - zf(r)] \qquad (3.6.6)$$

$$\psi_{\bar{1}\bar{1}1} = \tfrac{1}{2} [g(r) - xf(r) - yf(r) + zf(r)] \qquad (3.6.7)$$

The factor $\tfrac{1}{2}$ in front of the brackets is to make the hybrid orbitals normalized. Each hybrid orbital represents an electron distribution which extends along one of the four body diagonals [111], [1$\bar{1}\bar{1}$], [$\bar{1}$1$\bar{1}$], and [$\bar{1}\bar{1}$1] of a cube as illustrated in Fig. 3.20. This fact can easily be recognized by a rotation of $120°$ about one of the body diagonals. As

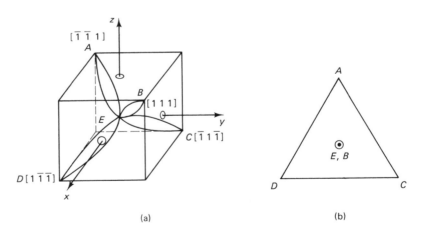

(a)   (b)

Figure 3.20 (a) The four sp^3 hybrid orbitals of carbon in CH_4, each having an electron distribution directed toward one of the four alternate corners of a cube. The center of the cube is chosen as the origin of the coordinates. (b) The tetrahedron formed by the CH_4 molecule with the four hydrogen atoms, A, B, C, and D, occupying the four corners and the carbon atom E occupying the center of the tetrahedron. The view is taken along the direction BE.

discussed in Section 2.9, a counterclockwise rotation of $2\pi/3$ about the [111] diagonal brings $x \rightarrow y$, $y \rightarrow z$, and $z \rightarrow x$. Making these changes in Eqs. (3.6.4) to (3.6.7) results in the following transformations:

$$\psi_{111} \rightarrow \psi_{111}, \quad \psi_{1\bar{1}\bar{1}} \rightarrow \psi_{\bar{1}1\bar{1}}, \quad \psi_{\bar{1}1\bar{1}} \rightarrow \psi_{\bar{1}\bar{1}1}, \quad \psi_{\bar{1}\bar{1}1} \rightarrow \psi_{1\bar{1}\bar{1}} \qquad (3.6.8)$$

In Fig. 3.20a, the electron distribution associated with each hybrid orbital is represented by a lobe extending along the four body diagonals. In CH_4, the carbon atom occupies the center of the cube, whereas the four hydrogen atoms occupy the alternate corners of the cube. As illustrated in Fig. 3.20b, a counterclockwise rotation of 120° about an axis joining atoms E and B brings atom $B \rightarrow B$, $D \rightarrow C$, $C \rightarrow A$, and $A \rightarrow D$. This is exactly what Eq. (3.6.8) says. We can apply similar rotations about other body diagonals. For example, a rotation of 120° about the [$\bar{1}\bar{1}1$] diagonal interchanges atoms B, C, and D, or hybrid orbitals $\psi_{1\bar{1}\bar{1}}$, $\psi_{\bar{1}1\bar{1}}$ and ψ_{111}. Note that a regular tetrahedron is formed by joining the four hydrogen atoms A, B, C, and D with the carbon atom sitting at the center of the tetrahedron. The C—H bonds are thus directed from the center toward each corner of the tetrahedron.

In summary, we have described two important properties of a covalent bond, the saturation and the directional property. First, an electron-pair bond can be formed only between electrons having unpaired spins. An oxygen atom has four p electrons. However, only two of the four electrons can have unpaired spins because of the Pauli exclusion principle. In carbon, one of the two s electrons must be promoted to become a p electron, so that carbon may have a maximum of four unpaired spins and thus form a maximum of four bonds. Once an atom has used up all its electrons having unpaired spins, no additional bond can be formed. The state of saturation is said to have been reached. Second, the strength of a bond depends on the overlapping of the electron wave function. Since each atomic orbital represents a specific charge distribution in space, maximum overlapping of the charge distributions is possible only if atoms are oriented with respect to one another in specific directions. The arrangements of atoms in covalent-bond molecules such as H_2O, NH_3, and CH_4 are such as to achieve maximum bond strength (or minimum energy) in these molecules.

3.7 COVALENT CRYSTALS

The elements in the periodic table (Table 1.1) can be grouped into several categories. The group VIIIb elements from He to Xe all have closed-shell configurations. Upon solidification, they form crystals of the molecular type in which bonding is due to the van der Waals force. The group I, II, and III elements are all metals. The alkali metals (from Li to Cs) and group Ib metals (Cu, Ag, and Au) are univalent. The alkali-earth metals (from Be to Ba) and the group IIb metals (Zn, Cd, and Hg) are divalent. The group IIIb metals (from Al to Tl) are trivalent. (Boron is a special case. It has been reported that the crystal may be of the covalent type.) The transition-metal group consists of elements having the incomplete $3d$ shell (elements of the iron group from Sc to Ni), elements having the incomplete $4d$ shell (from Y to Pd), elements having the incomplete $4f$ shell (elements of the rare-earth group from La to Lu), and elements having the incomplete $5d$ shell (from Hf to Pt). Almost all metals crystallize in either one of the three structures: the hcp, the fcc or the bcc structure.

If we remove the aforementioned elements from the periodic table, we are left with the group IVb–VIIb elements. These elements form covalent crystals. In the following discussion, we describe the structure and the bonding typical of each group. Let us start with the group IVb elements. Diamond, silicon, germanium, and gray tin all crystallize in the diamond structure. The bond between atoms is formed by the sp^3 hybrid orbitals. As discussed in Section 3.6, there are four sp^3 orbitals. Therefore, each atom needs four nearest neighbors to complete the covalent bond. It is not surprising to find that the basic unit in a diamond structure (Fig. 2.12) has a structure similar to that of CH_4 (Fig. 3.20). In the tetrahedral unit of Figs. 2.12 and 3.20, the center atom is surrounded symmetrically by four atoms, and hence the atom arrangement satisfies the bonding requirement of the sp^3 orbitals. We should mention that not all group IVb elements have structures in which all the bonds are covalent. Lead, which is metallic, has a fcc structure. Metallic white tin, which is stable at room temperature, takes the body-centered tetragonal structure having two atoms to each lattice point. Graphite, which is the other crystalline form of carbon, has a strong covalent bonding between atoms in the plane of a hexagon but a weak metallic bonding between the planes.

The atoms of the group Vb elements (P, As, Sb, and Bi), like the nitrogen atom, have three unpaired electrons, and the bond between the atoms is formed by the p^3 orbitals. The structure suitable for the p^3 orbitals must have a coordination number of 3. Although the four aforementioned elements crystallize in a trigonal structure, the characteristic feature of the structure is best seen in planes perpendicular to the trigonal axis. Figure 3.21 shows the double atomic layer structure of an arsenic crystal, with the atoms in the top layer drawn as dots and the atoms in the bottom layer drawn as circles. The basic unit in the structure consists of four atoms 1, 2, 3, and 4 in the form of a pyramid. From our discussion of the NH_3 molecule in Section 3.6, we expect that the angle between two As—As bonds should be close to 90°. Table 3.2 lists some experimentally measured bond angles in materials whose atoms are held together by p^3 and p^2 bonds. As can be seen, the measured angles are indeed close to 90°.

Whereas the atoms of P, As, Sb, and Bi crystallize in puckered sheets to satisfy the p^3 bond requirement, the atoms of Se and Te form spiral chains (Fig. 3.22) so that each atom has two close neighbors in fulfilling the p^2 bond requirement. Since the two p electrons must have different p orbitals (e.g., one has ψ_{px} and the other has ψ_{py}), the bond angle is again close to 90°. One arrangement of atoms to have a coordination number of two is a square ring. However, this arrangement requires a bond angle of exactly 90°. Furthermore, the ring is a closed structure and hence does not allow any extension of the structure. On the other hand, atoms can always be added to the open ends of the spiral chain so that the chain structure can be repeated while it propagates.

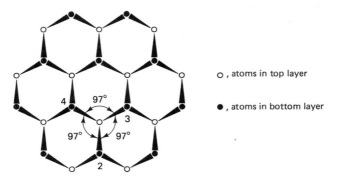

○ , atoms in top layer

● , atoms in bottom layer

Figure 3.21 The layer structure of arsenic crystals as viewed in a direction perpendicular to the plane of the layer structure. There are two layers in the diagram. Atoms in the top layer are shown as dots, whereas atoms in the bottom layer are shown as open circles. The two layers of atoms should be considered as forming a puckered sheet because the p^3 bond requirement is satisfied by atoms in the sheet. Bond between atoms in the same sheet is strong, whereas bond between atoms in two adjacent sheets is weak.

TABLE 3.2 BONDING ORBITAL AND BOND ANGLE IN A SELECTED NUMBER OF COVALENT SUBSTANCES

Substance	Bonding orbital	Bond angle
P	p^3	99°
As	p^3	97°
Sb	p^3	96°
Bi	p^3	94°
NH_3	p^3	107°
S	p^2	106°
Se	p^2	105°
Te	p^2	102°
H_2O	p^2	109°

In both the trigonal structure of P, As, Sb, and Bi and the hexagonal structure of Se and Te, there is a weak secondary bond either of the van der Waals or metallic type. Adjacent sheets and adjacent chains in the two structures are loosely bound together by this weak force.

The group VIIb atoms Cl, Br, and I possess seven electrons (two s and five p electrons). Since these atoms have one unpaired spin, they need only another atom to complete the covalent bond. In the crystalline state, each lattice point thus represents a molecule Cl_2, Br_2, or I_2, and the bond between the molecules is due to the van der Waals forces. Figure 3.23 shows the orthorhombic structure that the elements Cl, Br, and I take.

In summary, we have described the crystal structures of group IVb–VIIb elements. The number of unpaired spins in these elements is equal to $8 - N$, where N is the number of the group. Therefore, to fulfill the covalent-bond requirement, each atom

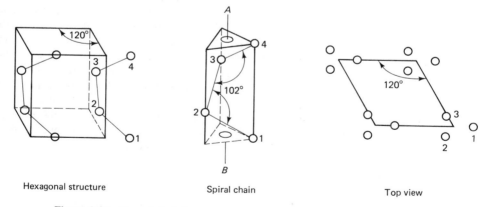

Hexagonal structure   Spiral chain   Top view

Figure 3.22 The spiral chain structure of selenium and tellurium crystals. Although both crystals have the hexagonal structure, the basic structure that satisfies the p^2 bond requirement is the spiral chain (formed by atoms 1, 2, 3, and 4, for example). Bonding among atoms in the same chain is strong, whereas bonding among atoms in two adjacent chains is weak.

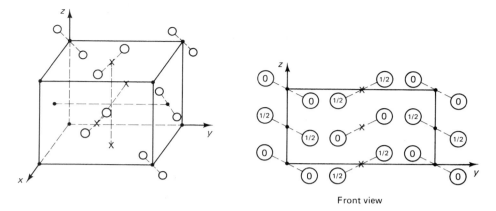

Front view

Figure 3.23 The face-centered orthorhombic structure of chlorine, bromine, and iodine. Each lattice point is occupied by a molecule. The numbers appearing inside circles are the x coordinates of the atoms, expressed in units of the lattice constant a.

must have $8 - N$ close neighbors. The characteristic feature of a covalent-crystal structure is such that both this $8 - N$ rule and the directional requirement of the bond can be satisfied. To meet the two conditions, covalent-bond crystals generally have fairly complicated structures. In contrast, metallic- and ionic-crystal structures are comparatively simple. We also see that only for the group IVb elements do the covalent bonds bind the whole structure. For the group Vb, VIb, and VIIb elements, the crystal structures involve both the primary covalent bond and a secondary weak bond. If the secondary bond is metallic, the crystal is partly covalent and partly metallic. If the secondary bond is of the van der Waals type, the crystal is partly covalent and partly molecular.

The reader may ask why covalent bonding starts with crystals of group IVb elements and not with crystals of group IIIb elements. One way to explain the difference is to say that metals have partially filled energy bands, whereas covalent crystals have fully occupied valence bands. The explanation based on the energy band is discussed further in Chapter 4. Here we give a more straightforward explanation based on the total ionization energy of an atom. In metals, the valence electrons are free from the attraction of the nucleus to form an electron sea. To free the valence electron (e.g., $3s$ electron in Na, $3s^2$ electrons in Mg, and $3s^23p$ electrons in Al) from the nucleus requires energy. Referring to Fig. 3.12, we see that a large part of the total energy is spent in raising a Na atom to the ionized state (Na^+ + free electron). As the number of the valence electrons increases, the total ionization energy (e.g., the energy required to make $\text{Al} \rightarrow \text{Al}^{3+}$ + 3 free electrons) increases accordingly. If this energy becomes too large, valence electrons will not form an electron sea but instead, will form electron-bond pairs.

Table 3.3 lists the total ionization energies and the crystal structures of the group Ib–VIIb elements except for Li and Na, which belong to group Ia, and Be and Mg, which belong to group IIa. The boundary line between metallic and covalent bonding seems to be around 40 eV for the value of the total ionization energy. The atoms having the total ionization energy below this value tend to crystallize in close-packed or nearly close-packed structures. Boron, tin, and lead seem to be borderline cases. Boron has

TABLE 3.3 CRYSTAL STRUCTURES AND TOTAL IONIZATION ENERGIES (IN eV) OF THE GROUP Ib–VIIb ELEMENTS[a]

Li	Be	B	C	N	O	F
bcc 5.4	hcp 18.1	Tetrahedral 37.7	D 64.2	— 97.4	— 137	— 184
Na	Mg	Al	Si	P	S	Cl
bcc 5.2	hcp 15.0	fcc 28.3	D 44.9	8 − N 64.7	8 − N 87.6	8 − N —
Cu	Zn	Ga	Ge	As	Se	Br
fcc 7.7	hcp 18.0	Orthorhombic 30.6	D 45.5	8 − N 62.5	8 − N —	8 − N —
Ag	Cd	In	Sn	Sb	Te	I
fcc 7.5	hcp 16.8	fcc 27.9	D (gray) 39.4	8 − N 55.5	8 − N (72)	8 − N —
Au	Hg	Tl	Pb	Bi		
fcc 9.2	Trigonal 18.6	hcp 29.7	fcc 42.1	8 − N 55.7		

[a]Covalent bonding is favored over metallic bonding in elements with ionization energy exceeding 40 eV to the right of the heavy line.

been reported to be partly covalent, whereas lead is known to be metallic. Tin can exist in two forms, with gray tin being covalent and white tin being metallic.

PROBLEMS

3.1. Discuss the essential differences between ionic and covalent bonding. Use KCl and H_2 to illustrate the differences.

3.2. The bonding in NH_3 and O_2 is known to be covalent. Use illustrations similar to Figs. 3.3b and 3.4 to explain how electron-pair bonds are formed in them.

3.3. Based on our discussion of the hydrogen (H_2^+) molecule-ion, explain the bonding mechanism in a hydrogen molecule H_2.

3.4. Find the interaction energy between a pair of collinear electric dipoles located at a distance R apart. Show that the energy is attractive. Explain the difference, if there is any, between your result and Eq. (3.2.4).

3.5. Consider a two-dimensional square lattice with alternate positive and negative ions. Choose any ion as the reference ion. Find the distances between this reference ion and its neighbors from the nearest neighbors up to the ninth nearest neighbors. Calculate the Modelung constant with terms up to the ninth nearest neighbors.

3.6. Verify Eq. (3.3.7). Explain why it is incorrect to use $\mathscr{E}_0$ of Eq. (3.3.5) as $\mathscr{E}$ in Eq. (3.3.7).

3.7. Find the values of n (the repulsive-energy parameter) and $\mathscr{E}_0/N$ (the cohesive energy) in a KCl crystal, given $K = 5.1 \times 10^{-12}$ cm²/dyn and $R_0 = 3.14$ Å. It is also known that the KCl crystal has an NaCl structure.

3.8. Describe the classical model used in the calculation of cohesive energy in a metallic crystal. Derive Eq. (3.4.2) and explain clearly the steps taken in the derivation.

3.9. (a) Assume that the density of electrons dn in an energy interval $d\mathscr{E}$ is $dn = B\mathscr{E}^{1/2}\, d\mathscr{E}$, where B is a proportionality constant. Express the average kinetic energy of an electron in terms of the Fermi energy $\mathscr{E}_f$.

(b) Show that the Fermi energy in a metal having a body-centered cubic structure, such as potassium, can be expressed in terms of the lattice constant a as

$$\mathscr{E}_f = \frac{h^2}{2ma^2}\left(\frac{3}{4\pi}\right)^{2/3}$$

(c) Given $a = 5.33$ Å in a potassium crystal, find the values of the Fermi energy and the average kinetic energy of the $4s$ electrons.

3.10. Review the models used for the discussion of ionic and metallic crystals in Sections 3.3 and 3.4, and then refer to the general discussion in Secton 3.1. Comment on the essential differences between an ionic and a metallic crystal by pointing out the principal energy terms in Eq. (3.1.5) for each type. Describe the physical origin, and explain the attractive or repulsive nature, of the principal energy terms in each case.

3.11. The $1s$ electron wave function in a hydrogen atom is given by

$$\psi_{1s} = \frac{1}{\sqrt{\pi}\, a^{3/2}} \exp\left(-\frac{r}{a}\right)$$

where a is the Bohr radius. Find the expression for the symmetric wave function $\psi_{S\text{-}Or}$ in a hydrogen molecule. The electron coordinates are (x_1, y_1, z_1) and (x_2, y_2, z_2), whereas the nuclear coordinates are $(d, 0, 0)$ and $(-d, 0, 0)$. Find the ratio of the probabilities of finding both electrons at $(0, 0, 0)$ and at $(2b, 0, 0)$.

3.12. The electron density associated with Eqs. (3.5.8) and (3.5.9) is given by

$$\rho(\mathbf{r}_1, \mathbf{r}_2) = e(\psi_{1a}^2\psi_{2b}^2 + \psi_{2a}^2\psi_{1b}^2) \pm 2e\psi_{1a}\psi_{1b}\psi_{2a}\psi_{2b}$$

where the $+$ and $-$ signs apply to the symmetric and antisymmetric wave functions, respectively. The part of the electron density

$$\rho_{ex} = e\psi_{1a}\psi_{1b}\psi_{2a}\psi_{2b}$$

is shared by the two nuclei and is known as the exchange charge density.

(a) Using the $1s$ electron wave function given in Problem 3.11, sketch the four wave functions ψ_{1a}, ψ_{1b}, ψ_{2a}, and ψ_{2b} in a one-dimensional space for the case where the two hydrogen nuclei are at a distance $2d$ apart. Show that ρ_{ex} has an appreciable value only in the region between the two nuclei.

(b) Based on Coulomb interaction between the electron cloud and nuclei, which state, the symmetric or the antisymmetric state, do you expect to be the bonding state?

3.13. Chemical elements of the first period form molecules of the type AH_n, where n is the maximum number of hydrogen atoms that can be bound. The following molecules are known to exist in chemistry: H_2, LiH, BH, CH_4, NH_3, OH_2, and FH. Show the orbital and spin states of all the electrons in the element A in forming the molecule AH_n. In OH_2, for example, the eight electrons of oxygen have the following arrangement: $1s \downarrow\uparrow\ 2s\downarrow\uparrow$ $2p \uparrow\uparrow\uparrow\downarrow$, with arrows indicating the direction of the spin.

3.14. (a) In ionic crystals such as NaCl, the attractive force between the X^+ and Y^- ions is nondirectional because the ions which have attained closed-shell configurations have spherically symmetric charge distributions. How would the atoms be arranged in H_2O

Atomic Bonding and Crystal Types   Chap. 3

and NH_3 if the bonding were ionic in nature in these molecules? The proposed arrangement should take into account the repulsive force between hydrogen atoms. Would the arrangement shown in Fig. 3.18 be stable?

(b) On the other hand, if the atoms are covalently bonded, the O—H and N—H bonds are directional. What are the principal factors that make covalent bonding directional? Explain why it is not possible to have a collinear H_2O molecule and a T-shaped NH_3 molecule.

3.15. A tetrahedron has three twofold axes (the x, y, and z axes in Fig. 3.20) besides the four threefold axes (the body diagonals in Fig. 3.20). Consider a 180° rotation about the z axis. Comment on how the four sp^3 orbitals of carbon in CH_4 will transform under the said rotation. Check your answer by comparing the results obtained from a visual inspection of Fig. 3.20 and from a transformation of coordinates in Eqs. (3.6.3) to (3.6.7).

3.16. Comment on the essential features of the structures of arsenic and selenium crystals in reference to satisfying the covalent-bond requirement. Explain why a weak secondary bond is needed in forming a crystal for both As and Se.

3.17. What are the two essential properties of a covalent bond? Use Ge and Te crystals as examples to illustrate how the atoms are arranged to satisfy the covalent-bond requirements. Explain why a weak secondary bond is needed in Te but not in Ge.

4

Free-Electron Theory

4.1 ENERGY BANDS IN SOLIDS

In Chapter 3 we presented a general discussion on the nature of the attractive force that holds atoms together in a solid. Atoms in different types of crystals redistribute their electrons differently so as to achieve a stable electron configuration. Since the properties of a material are determined by the behavior of electrons in the outer shell or shells, it is to be expected that this redistribution of electrons will have a profound effect on material properties. We begin with the conductive property.

It is known that sodium crystals are good electrical conductors and sodium chloride crystals are good electrical insulators. This statement should not surprise us, as we may have anticipated this from our discussion in Sections 3.2 to 3.4. The Na atoms in both sodium and sodium-chloride crystals give up their $3s$ electrons in order to achieve the stable electron configuration of neon. In a sodium-chloride crystal, the electron freed from a Na atom is immediately captured by a Cl atom to form a Cl^- ion. Therefore, all the electrons in NaCl are tied down and thus unable to participate in the conduction of electric current. On the other hand, the freed electron in a sodium crystal remains in the state of nonattachment, and hence is free to conduct electric current.

The foregoing discussion qualitatively explains the essential difference between an electrical conductor and nonconductor. However, the discussion is limited in its usefulness. Figure 4.1 shows the resistivity of some typical materials, ranging from 10^{-8} Ω-m for good conductors to 10^{12} Ω-m for good insulators. The quantitative difference in resistivity in different materials can only be understood on the basis of band theory. It is the purpose here to present an introductory discussion on the formation of energy bands in solids.

It is well known from circuit analysis that if two identical LC circuits are coupled together through a mutual inductance M, the two circuits will oscillate as a whole unit. Since the degree of freedom of the system is increased by a factor of 2, there will be two different modes and hence two different frequencies of oscillation with

$$\omega_1 = \frac{\omega_0}{\sqrt{1 + M/L}} \quad \text{and} \quad \omega_2 = \frac{\omega_0}{\sqrt{1 - M/L}} \quad (4.1.1)$$

108

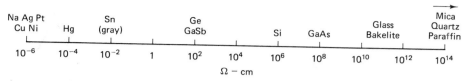

Figure 4.1 The range of resistivity observed in different materials at room temperature.

where $\omega_0 = (LC)^{-1/2}$. Physically, the existence of two distinct modes can be understood as follows. The kinetic energy (i.e., magnetic energy) of two coupled circuits is

$$\text{KE} = \tfrac{1}{2} LI_1^2 + \tfrac{1}{2} LI_2^2 + MI_1I_2$$
$$= \tfrac{1}{2} (L + M)(I_1^2 + I_2^2) - \tfrac{1}{2} M(I_1 - I_2)^2 \tag{4.1.2}$$

where I_1 and I_2 are the current flowing through each coil. Thus the kinetic energy depends on the relative phase of the two currents. The two resonant modes of Eq. (4.1.1) correspond to the situation in which the currents I_1 and I_2 are in phase and the situation in which I_1 and I_2 are out of phase.

In the case of two atoms, the mutual interaction comes from the Coulomb potential. When the distance between two atoms is large compared to the Bohr radius, the electron wave functions are represented by two distinct atomic orbitals, ψ_1 and ψ_2, one for each atom. As the two atoms approach each other, the potential energy of the system depends on the relative phase of the two wave functions. Analogous to the situation in two identical coupled LC circuits, the two normal modes of an atomic system consisting of two identical atoms have ψ_1 and ψ_2 in phase (Fig. 4.2a) and ψ_1 and ψ_2 out of phase (Fig. 4.2b). Note the similarities between the wave functions shown in Fig. 4.2 and those shown in Fig. 3.5. From our discussion in Section 3.2, we see that the two states represented by the wave functions $\psi_1 + \psi_2$ and $\psi_1 - \psi_2$ will have different potential energies. This means that the atomic energy level, which is analogous to the frequency in coupled LC circuits, is now split into two energy levels, as shown in Fig. 4.3. As the separation d between the two atoms changes, the potential energies of the two states change by different amounts. Thus the splitting between the energy levels is a function of the interatomic distance d.

The discussion above can be extended to cases involving more than two atoms. First, let us consider a linear chain consisting of N atoms. If ψ_j is the wave function associated with the jth atom, the various normal modes of the system can be represented by different wave functions of the following form:

$$\Psi = \psi_1 \exp(i\theta_1) + \cdots + \psi_j \exp(i\theta_j) + \cdots + \psi_N \exp(i\theta_N) \tag{4.1.3}$$

where the factor $\exp(i\theta_j)$ accounts for the phase difference between the components. In circuit analysis we are familiar with the practice that for easy handling, a sine or cosine function is expressed as the imaginary or real part of an exponential function. To get back a physical quantity from the complex representation, we take either the imaginary or the real part of the complex exponential function. Taking the imaginary part of Eq. (4.1.3), we have

$$\Psi = \psi_1 \sin \theta_1 + \cdots + \psi_j \sin \theta_j + \cdots + \psi_N \sin \theta_N \tag{4.1.4}$$

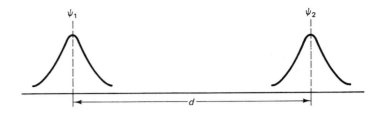

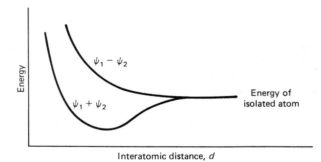

(a)   (b)

Figure 4.2 Diagram showing two possible modes of interaction between two atomic orbitals. (a) The two wave functions are in phase and (b) the two wave functions are out of phase. In (a) the resultant wave function is $\psi_1 + \psi_2$, whereas in (b) the resultant wave function is $\psi_1 - \psi_2$.

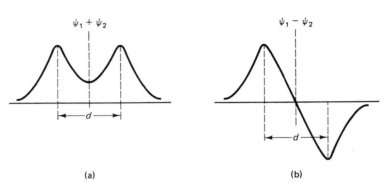

Figure 4.3 Energy as a function of the interatomic distance in a system consisting of two identical atoms. At large distances, the energies of the two atoms are equal. As the two atoms approach each other, the atomic level splits into two energy levels as a result of mutual interaction between the two atoms.

As discussed in Chapter 5, the phase angle θ_j can be set equal to

$$\theta_j = \left(j - \frac{1}{2}\right)\frac{q\pi}{N} \qquad (4.1.5)$$

where q is a positive integer that may take any value from 1, 2, ... to N. Let us apply Eq. (4.1.5) to the system involving two atoms. For $N = 2$ we have two possible choices for q, that is, $q = 1$ or 2. Thus we find $\theta_1 = \pi/4$ and $\theta_2 = 3\pi/4$, or $\theta_1 = \pi/2$ and $\theta_2 = 3\pi/2$. The variation of ψ with distance x is illustrated in Fig. 4.4a. The

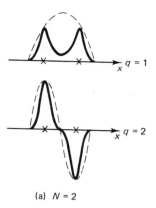

(a) $N = 2$

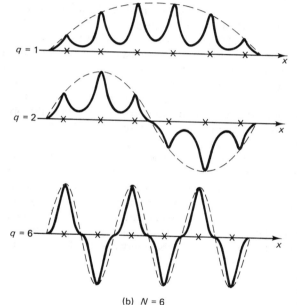

(b) $N = 6$

Figure 4.4 The various modes of the composite wave function in a system of (a) two identical atoms and (b) six identical atoms. In (a) there are two possible modes, whereas in (b) there are six possible modes. The other three modes with $q = 3$, 4, and 5 are also possible but are not shown in (b).

dashed curve represents the spatial variation of the phase factor $\sin \theta_j$ from one atom to the other. Since $\sin (\pi/4) = \sin (3\pi/4)$ and $\sin (\pi/2) = -\sin (3\pi/2)$, the top curve ($q = 1$) in Fig. 4.4a corresponds to the situation shown in Fig. 4.2a, whereas the bottom curve ($q = 2$) in Fig. 4.4a corresponds to that shown in Fig. 4.2b.

Figure 4.4b illustrates the situation in a system having six identical atoms. For $N = 6$ there are all together six different situations, each corresponding to a specific value of q ($q = 1, 2, \ldots$ or 6). In Fig. 4.4 the positions of the nuclei are indicated by $\times$. As we can see, the electron distribution in the region between two neighboring nuclei depends on the number q in Eq. (4.1.5). For $q = 1$, nonzero electron distribution occurs in all five internuclear regions. For $q = 2$, the number of internuclear regions that have a nonzero electron distribution decreases to four. This number reduces to zero for $q = 6$.

Referring to our discussions in Sections 3.2 and 3.5, we recall that the energy level of two hydrogen atoms is split into two separate levels when the two atoms form a hydrogen molecule (a two-atom system). The splitting of the energy is due to the fact that there are two different electron distributions in the internuclear region. In the case of six atoms, the atomic level will be split into six different energy levels because there exist six different normal modes of electron distributions. The situation is illustrated in Fig. 4.5. Since an accumulation of electrons in the internuclear region lowers the total energy of the system, we expect that the normal mode with $q = 1$ will be the lowest-energy state and that with $q = 6$ will be the highest-energy state, as shown in Fig. 4.5.

From the discussion above, it is not difficult to see that in a system involving N atoms, each atomic level will be split into N different energy levels. This splitting is caused by the difference in the electrostatic energy. There are about 10^{22} atoms/cm^3 in a solid. For a laboratory-size sample of dimension 2 cm $\times$ 5 mm $\times$ 1 mm, we have a total of 10^{21} atoms. The difference in energy between the highest energy level ($q = N$) and the lowest energy level ($q = 1$) in Fig. 4.5 is of the order of 10 eV. Therefore, the spacing between two successive levels (say, between $q = k$ and $q = k + 1$ levels) is of the order of 10^{-20} eV. Thus, for all practical purposes, we may consider the group of energy levels (e.g., the group formed by the $2s$ electrons in Fig. 4.5) as a band and treat the band as if the distribution of allowed energy levels were continuous.

Figure 4.6a shows the distribution of energy levels in a sodium crystal as a function of the interatomic distance d. When atoms are brought together, the orbits of the valence electrons will overlap first simply because these orbits extend to a greater distance from the nucleus than the orbits of core electrons. As discussed in Section 3.1, a strong repulsion results between two ion cores when the orbits of core electrons begin to overlap. Therefore, at the equilibrium separation R_0 (Fig. 4.6), the overlapping of the orbits of core electrons cannot be appreciable; consequently, the band formed by the core electrons has negligible width. Figure 4.6b shows the energy bands in sodium taken at $d = R_0$ in Fig. 4.6a. The $3p$, $3s$, and $2p$ bands are called *allowed bands* because there are quantum states available to electrons if the energy of an electron happens to fall in one of the bands. On the other hand, in the energy interval between the allowed

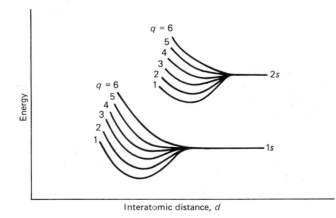

Interatomic distance, d

Figure 4.5 Splitting of the atomic levels for the states $1s$ and $2s$ in a system of six identical atoms. As the atoms approach one another, energy splitting takes place first between electrons in the outmost shell ($2s$ electrons in the present case). Energy splitting for core electrons ($1s$ electrons) is not significant until the interatomic distance approaches the dimension of the ion core.

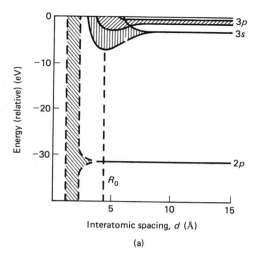

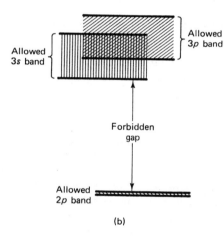

Figure 4.6 (a) Energy-band diagram of sodium as a function of the interatomic distance. The distance R_0 represents the separation between nearest neighbors in a sodium crystal. (b) Energy bands (taken at $d = R_0$) in a sodium crystal. Note that there is an overlap between the $3s$ and $3p$ bands.

bands, no quantum states exist, and hence it is not possible for electrons to have energies in this energy gap. Thus the energy gap between the allowed bands is called the *forbidden energy gap*.

4.2 METALS, INSULATORS, SEMICONDUCTORS, AND SEMIMETALS

Consider a free electron gas with a density n (electrons/m^3). Under an electric field **E**, electrons acquire, on the average, a drift velocity $\mathbf{v}_d$ (m/s) opposite to the direction of **E**. The drift motion of electrons constitutes a current of density J (A/m^2). Figure 4.7 illustrates the movement of a sample column of the electron gas. At $t = 0$, the column occupies the space BC. If d is the length of the column, it will take a time $t = d/v_d$ to move the entire column BC to a new position CD. The total number of electrons passing

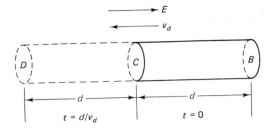

Figure 4.7 Diagram used for the calculation of conduction current density. At time $t = 0$, the column of electrons occupies the space BC. At time t, the whole column drifts to a new space CD.

through C is $N = nAd$, where A is the cross-sectional area. Therefore, the current density J is

$$J = \frac{Q}{At} = \frac{-eN}{At} = -env_d \qquad (4.2.1)$$

At low fields, the drift velocity is linearly proportional to the field. It is customary to write

$$\mathbf{v}_d = -\mu\mathbf{E} \qquad (4.2.2)$$

where μ (m^2/V-s) is called the *drift mobility*. Substituting Eq. (4.2.2) in Eq. (4.2.1), we obtain

$$J = e\mu nE = \sigma E \qquad (4.2.3)$$

The proportionality constant in front of E is the conductivity of the medium with $\sigma = e\mu n$ and σ has a dimension of siemens ($= \Omega^{-1}$). Equation (4.2.3) is merely a statement of *Ohm's law*.

The enormous variation of the electrical resistivity (Fig. 4.1) from one type of substance to another is due mainly to the difference in the density n of free electrons. To illustrate the point, we take silicon and sodium as examples. Silicon has an electron configuration $1s^2 2s^2 2p^6 3s^2 3p^2$, whereas sodium has $1s^2 2s^2 2p^6 3s^1$. The core electrons $1s^2$, $2s^2$, and $2p^6$ in both silicon and sodium are tightly bound to the nucleus and hence they do not contribute to electrical conductivity. If we simply compare the number of valence electrons, for example, 4 for silicon and 1 for sodium, we would conclude that silicon should be more conductive than sodium. This is contrary to facts. The difference in the electrical behavior of a semiconductor such as silicon and that of a conductor such as sodium is best described in terms of energy bands.

Figure 4.8 shows the energy bands of carbon in the diamond structure. The core $1s^2$ electrons of carbon are tightly bound to the nucleus, and hence unimportant in the present discussion. Associated with the s level ($l = 0$), there are two atomic states. The number 2 accounts for the two different spin orientations. For the p level ($l = 1$), there are six atomic states (three different values of m and two different spin orientations). If there are N carbon atoms in a diamond crystal, the number of available quantum states associated with the $2s$ and $2p$ states is $2N$ and $6N$, respectively. At large interatomic distances, the four valence electrons are divided such that the $2s$ band is completely filled and the $2p$ band only one-third filled. At smaller distances, however, the $2s$ and $2p$ states are admixed to form sp^3 hybrid orbitals and the total $8N$ quantum states are split equally between the lower and upper bands (Fig. 4.8b). The lower band is called the *valence band* and the upper band is called the *conduction band*.

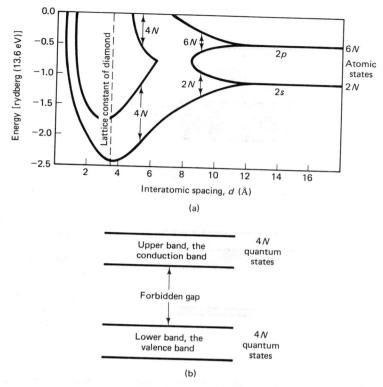

Figure 4.8 (a) Energy-band diagram of diamond as a function of the interatomic distance. There are six quantum states associated with the 2p level and two quantum states associated with the 2s level. In a crystal of N atoms, there are altogether 8N available quantum states. At small interatomic distances, the 8N states form two groups of 4N states each. (b) Energy bands of diamond crystal. The two groups of 4N states each are separated in energy by a forbidden gap.

The situation in silicon and germanium is qualitatively similar to that in diamond except that the 2s and 2p states are changed to 3s and 3p states, and 4s and 4p states, respectively, and that the energy scale is different. The filling of the bands follows a simple rule: states of lowest energy are filled first. This rule is observed strictly at absolute zero temperature. At a finite temperature, however, there is a slight deviation from this rule on account of the thermal energy of electrons. The law that governs the electron distribution is discussed in Section 4.5. For our present discussion, we assume that the temperature is so low that the simple rule is observed. Therefore, the lower band (i.e., the valence band) of silicon is completely filled by the valence electrons and the upper band (the conduction band) is completely empty, as shown in Fig. 4.9a. The situation in sodium is different. The 3s band of sodium has 2N available states, whereas there are only N valence electrons in sodium; hence the band is only half filled, as Fig. 4.9b shows.

The difference in the electrical behavior of silicon and sodium can now be understood. As indicated in Eq. (4.2.1), an electron acquires a drift velocity under an applied

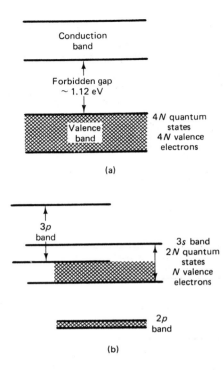

Figure 4.9 Diagrams showing the occupancy of energy bands in (a) silicon and (b) sodium. In Si, the $4N$ available quantum states in the valence (lower) band are completely occupied by the $4N$ valence electrons. In Na, on the other hand, the $2N$ available quantum states are only half filled because there are only N valence electrons. The occupied portion of a band is crosshatched.

electric field. When an electron acquires energy from the applied field, it moves from a lower to a higher energy state. However, the movement of an electron from one state to another is regulated by the exclusion principle. The final state must be vacant (i.e., not occupied by an electron). In sodium, there are vacant states immediately above the occupied states. After acquiring energy from an applied electric field, the valence electrons can readily move into a vacant higher energy state and thus leave behind a new vacant state for other electrons to move into. Therefore, a simultaneous and continuous change of electronic state is possible, ensuring electrical conduction. This applies to other materials with a partially filled band. In a semiconductor such as silicon, the lower band is completely filled at absolute zero temperature, and hence the valence electrons find themselves unable to move anywhere in the same band because of the Pauli exclusion principle.

Let us now describe the situation in ionic crystals. Take NaCl as an example. In the atomic picture, the eight valence electrons (1 from Na and 7 from Cl) occupy the outer $3s$ and $3p$ shells of the Cl^- ion. In a crystal, the atomic levels broaden to form bands. Figure 4.10a illustrates broadening of atomic levels into bands as the ions are brought together. The $3s$ level of Na is raised while the $3p$ level of Cl^- is lowered because of the Madelung potential of Eq. (3.3.2). If an electron is lifted from the $3p$ band, that is, if it is taken away from one Cl^- ion, this electron is taken by a Na^+ ion, thus becoming a $3s$ electron of Na. Since it takes an energy about equal to the Madelung energy to convert Na^+ and Cl^- ions to neutral Na and Cl atoms in NaCl crystal, the $3p$ band of Cl^- must be lower than the $3s$ band of Na by an energy of that amount.

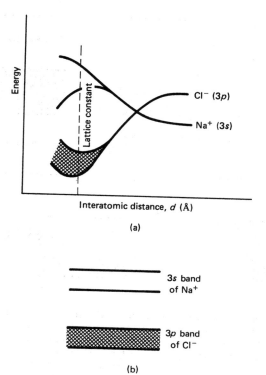

Interatomic distance, d (Å)

(a)

3s band
of Na$^+$

3p band
of Cl$^-$

(b)

Figure 4.10 (a) Energy-band diagram of NaCl as a function of the interatomic distance and (b) energy bands taken at normal lattice spacing.

For a NaCl crystal containing N sodium and chlorine ions, there are all together $2N$ quantum states in the 3s band of Cl$^-$ and 6N quantum states in the 3p band of Cl. In the atomic picture, we say that an electron transfer from Na to Cl makes Na and Cl atoms become Na$^+$ and Cl$^-$ ions. In the band scheme, we say that the 3p band of Cl$^-$ is completely occupied while the 3s band of Na$^+$ is completely empty. This explains why NaCl crystals are good insulators.

From the discussions above, we can draw the following conclusions. A substance having a partly filled band is a metal. The situation is illustrated in Fig. 4.11a. Since all the electrons in this partly filled band may participate in the conduction of electric current, the density n of free electrons in Eq. (4.2.3) is exceedingly large. For example, in sodium, n is just equal to density of sodium atoms per cubic centimeter because there is one valence electron to each atom. In an ideal insulator, the value of n should be zero. As illustrated in Fig. 4.11b, the upper band is completely empty and the lower band is completely occupied.

Theoretically, the state of ideal insulator (in which $n = 0$) can be reached only at absolute zero temperature. As mentioned earlier, the simple rule of filling the energy states is not strictly observed at finite temperatures. Because of thermal energy, electrons have a finite, even though exceedingly small, probability to be in a higher energy state than that allowed by the simple rule. In other words, there may be some empty states in the lower band and some occupied states in the upper band. Even though the

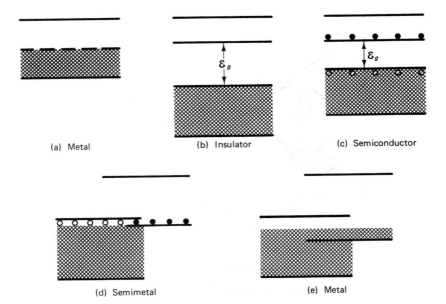

Figure 4.11 Diagrams showing five different situations regarding the occupancy of energy bands: (a) a metal with a partially filled band, (b) an insulator with a completely filled lower band and a completely empty upper band, (c) a semiconductor in which there are few occupied states (dots) in the upper band and few empty states (open circles) in the lower band, (d) a semimetal in which there are equal numbers of occupied states (dots) in the upper band and empty states (open circles) in the lower band, and (e) a metal in which two bands are partially occupied as a result of considerable overlap between the two bands.

percentage of these states is exceedingly small, the absolute value of these states can be appreciable. There are approximately 10^{22} atoms/cm^3. This means that in silicon we have a free electron density of about 4×10^{16} per cubic centimeter if 1 part in a million of the conduction-band states is occupied. For mobility of 10^3 cm^2/V-s, we find from Eq. (4.2.3) value of $\sigma = 6.4$ (Ω-cm)$^{-1}$.

The name *semiconductor* applies to materials whose resistivity lies somewhere from 10^{-2} to 10^6 Ω-cm. At absolute-zero temperature, all semiconductors become insulators. At finite temperatures, however, a semiconductor has an appreciable density of occupied states in the upper band or an appreciable density of unoccupied states in the lower band, or both. The situation is illustrated in Fig. 4.11c. Theoretically, all insulators will become semiconductors at sufficiently high temperatures. Therefore, the difference between a semiconductor and an insulator is a matter of degree. Since all practical semiconductors have an energy gap $\mathscr{E}_g$ of about 3.5 eV or smaller, we may use this value as a rough demarcation line between semiconductors and insulators. For $\mathscr{E}_g > 3.5$ eV, we have insulators, and for $\mathscr{E}_g < 3.5$ eV, we have semiconductors.

According to the classification above, examples of semiconductors (with $\mathscr{E}_g$ given in parentheses) are: Ge (0.67), Si (1.11), GaAs (1.42), InSb (0.18), CdS (2.58), ZnO (3.35), PbS (0.41), Te (0.35), and many others; and examples of insulators are: diamond ($\sim$6), MgO ($\sim$7), and NaCl ($\sim$8). We should point out that not all semiconductors are covalently bound. In most cases the bonding is an admixture of covalent and ionic bonding. How the bonding mechanism influences the electrical properties of a semiconductor is discussed in Chapter 6.

Figure 4.11d shows the energy bands in semimetals. By *semimetals* we mean those elements or materials whose conduction and valence bands overlap slightly. Such is the case with Bi, As, Sb, Se, Te, and some of their alloys with transition metals. Had it not been for the slight overlap of the energy bands, the valence band would be completely filled and the conduction band completely empty at absolute zero temperature. Because of the slight overlap, a small number of electrons go from the valence band into the conduction band, creating an equal number of free holes and free electrons, respectively, in the two bands. By *free holes* we refer to the unoccupied states in the valence band; and by *free electrons* we refer to the occupied states in the conduction band. As mentioned in Section 3.7, graphite, Vb elements, and VIb elements crystallize in complicated structures which have a primary covalent bonding and a weak metallic bonding. Therefore, the electrical properties of these materials can be highly anisotropic. They may exhibit semiconductor properties (i.e., they have an energy gap) along a certain direction or in a certain plane, and show semimetal properties along a different direction.

Finally, when there is a considerable overlapping between the bands, the material becomes a metal. The situation is illustrated in Fig. 4.11e. Take magnesium as an example. A free magnesium atom has the electron configuration of neon plus $3s^2$ valence electrons. At first we might think that the valence electrons would completely fill the $3s$ band in a solid. However, as Fig. 4.6 shows, the $3p$ band overlaps the $3s$ band. As a result, a considerable number of electrons occupy states in the $3p$ band instead of filling the $3s$ band. Both the situation illustrated in Fig. 4.11a and that in Fig. 4.11e are common in metals.

4.3 FREE ELECTRONS IN A BOX

In Sections 4.1 and 4.2 we presented a qualitative discussion of the formation of energy bands in a solid and explained the difference in the electrical behavior of materials on the basis of energy bands. Now that we have acquired this background knowledge, we proceed to analyze the problem of electric conduction on a quantitative basis. Again we start with the simplest case first, that is, the case of free electrons. As we recall our discussion in Sections 4.1 and 4.2, the central and crucial factors that influence electric conduction are the total number of available states and the distribution of the valence electrons among these available states. The concept state is purely a quantum-mechanical one, and hence the electron must be treated as a quantum particle.

For a free particle, the energy is entirely made of the kinetic energy, which is equal to $p^2/2m$. Letting $V = 0$ in Eq. (1.3.6), we obtain the following wave equation for a free electron:

$$-\frac{\hbar^2}{2m}\left(\frac{\partial^2\psi}{\partial x^2} + \frac{\partial^2\psi}{\partial y^2} + \frac{\partial^2\psi}{\partial z^2}\right) = \mathscr{E}\psi \qquad (4.3.1)$$

Equation (4.3.1), like Eq. (1.4.1), can be solved by separation of variables. Let

$$\psi(x, y, z) = X(x)Y(y)Z(z) \qquad (4.3.2)$$

Following a similar procedure as used in Section 1.4, we first substitute Eq. (4.3.2) into Eq. (4.3.1) and then divide both sides by ψ. Thus Eq. (4.3.1) transforms into

$$-\frac{\hbar^2}{2m}\left(\frac{1}{X}\frac{\partial^2 X}{\partial x^2} + \frac{1}{Y}\frac{\partial^2 Y}{\partial y^2} + \frac{1}{Z}\frac{\partial^2 Z}{\partial z^2}\right) = \mathscr{E} \qquad (4.3.3)$$

The first term on the left-hand side of Eq. (4.3.3) depends only on x, the second term only on y, and the third term only on z. However, the sum of the three terms must equal a constant $\mathscr{E}$. The only way to satisfy this condition is to set each term equal to a constant. Thus we let

$$-\frac{\hbar^2}{2m}\frac{\partial^2 X}{\partial x^2} = \mathscr{E}_1 X \qquad (4.3.4)$$

Similar equations are obtained for Y and Z. The solution of Eq. (4.3.4) is given by

$$X = A \sin k_x x + B \cos k_x x \qquad (4.3.5)$$

where A and B are two constants to be determined later and $k_x = \sqrt{2m\mathscr{E}_1}/\hbar$.

To help us see how the analysis above can be applied to conduction electrons in a metal, we draw the potential-energy curves together with the energy bands in Fig. 4.12a. The potential-energy curves are obtained by adding together the nuclear potentials due to each individual atom. The procedure is the same as that followed in obtaining Fig. 3.6 except that we now have a chain of atoms instead of just two atoms. Since there are no nuclei to the right of the solid boundary indicated by the vertical dashed lines, the potential energy does not turn downward but approaches a constant value. As

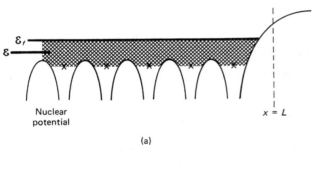

(a)

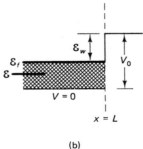

(b)

Figure 4.12 (a) Energy of valence electrons (shaded area) near the surface of a metal and (b) an idealized model (the free-electron model) of a metal surface.

Free-Electron Theory   Chap. 4

a result, there is a difference between the potential energy of an electron inside the metal and that outside the metal.

For the present discussion, let us ignore the effect of the nuclear potential except at the boundary of the solid. In other words, we assume that the potential energy inside the metal is uniform and chosen to be zero (i.e., chosen to be the reference energy). However, at the boundary, there is a jump in the potential energy by an amount V_0. Thus we may replace Fig. 4.12a by Fig. 4.12b. In both diagrams the energy band is that of valence electrons (e.g., the $3s$ band in sodium). The difference in energy between electrons in different states is entirely due to the difference in the kinetic energy. This simplified treatment of valence electrons is known as the *free-electron theory*.

Let us temporarily forget about the energy band and inquire as to what determines the energy of an electron. Consider an electron in a potential well (Fig. 4.13a). Outside the potential well, the wave equation in one dimension becomes

$$-\frac{\hbar^2}{2m}\frac{\partial^2 X}{\partial x^2} + V_0 X = \mathscr{E}X \tag{4.3.6}$$

For an electron having energy $\mathscr{E} < V_0$, the solution of Eq. (4.3.6) is

$$X = C\exp(-\alpha x) + D\exp(\alpha x) \tag{4.3.7}$$

where $\alpha = [2m(V_0 - \mathscr{E})/\hbar^2]^{1/2}$.

The complete solution of the potential-well problem is illustrated in Fig. 4.14a. Inside the well, the solution is in the form of sine and cosine functions; and outside the well, the solution is in the form of exponential functions. At the boundary ($x = 0$ and $x = L$), the two solutions must have their values and the values of their derivatives matched. However, we shall not go through with the mathematical analysis. Instead, we present a qualitative description of the situation. We know classically that an electron having energy $\mathscr{E}$ less than the potential barrier V_0 is a bound electron. For X to be zero at $x = \pm\infty$, only one term in Eq. (4.3.7) survives. For $x > L$ we have

$$X = C\exp(-\alpha x) \tag{4.3.8}$$

As discussed in Section 4.7, the difference $V_0 - \mathscr{E}$ for electrons having the topmost energy in the band (Fig. 4.12b) is called the *work function* $\mathscr{E}_w$. For metals, the

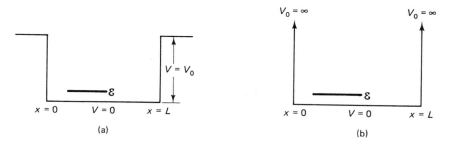

(a)

(b)

Figure 4.13 Electrons (with energy $\mathscr{E}$) in a potential well with a barrier (a) of finite height V_0 and (b) of infinite height.

Sec. 4.3   Free Electrons in a Box

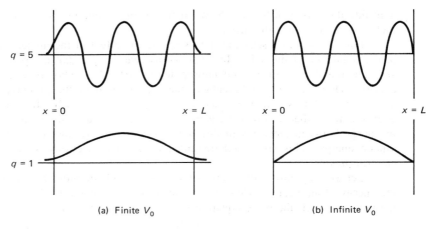

$q = 5$

$x = 0$ $x = L$ $x = 0$ $x = L$

$q = 1$

(a) Finite V_0 (b) Infinite V_0

Figure 4.14 Electron wave functions (for $q = 5$ and $q = 1$) inside a potential well with (a) a finite barrier and (b) an infinite barrier.

value of $\mathcal{E}_w$ ranges from 1.93 eV in Cs to 5.36 eV in Pt. For our discussion, we take $V_0 - \mathcal{E}$ to be 2 eV. The value of α is found to be 7.25×10^7 cm^{-1}. In other words, the value of X drops to 1/2.72 of its value at the boundary if x is away from the boundary for a distance of 1.38 Å. Thus, for all practical purposes, we may set $X = 0$ at the boundary, which is equivalent to assuming that $V_0 = \infty$ in Eq. (4.3.6). The situation is illustrated in Figs. 4.13b and 4.14b. For a one-dimensional potential well having sides of infinite heights, the solution is

$$X = A \sin (k_x x) \tag{4.3.9}$$

for $0 \leq x \leq L$, and $X = 0$ for $x < 0$ and $x > L$. The sine function is chosen so that X satisfies the boundary condition $X = 0$ at $x = 0$. The other boundary condition $X = 0$ at $x = L$ requires that

$$k_x = \frac{q}{L} \pi \tag{4.3.10}$$

where q is a positive integer $(0, 1, 2, \ldots)$.

The discussion above can be extended to a three-dimensional potential well. In the free-electron theory of metals, we consider electrons as being confined to a three-dimensional potential well having sides of infinite heights. As our calculation of α shows, the assumption that $V_0 = \infty$ and hence $\psi = 0$ outside the potential well is a good one. For simplicity, we consider a solid of cubic shape having dimensions L^3 (Fig. 4.15). The electron wave function inside the solid is given by

$$\psi(x, y, z) = A \sin (k_x x) \sin (k_y y) \sin (k_z z) \tag{4.3.11}$$

Again, the sine functions are chosen to satisfy the boundary conditions at $x = 0$, $y = 0$, and $z = 0$. To satisfy the condition $\psi = 0$ at the other sides of the boundary, that is, at $x = L$, $y = L$, and $z = L$, the values of k_x, k_y, and k_z must be such that

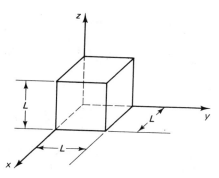

Figure 4.15 A solid of cubic shape having dimensions L^3.

$$k_x = q_1 \frac{\pi}{L}, \qquad k_y = q_2 \frac{\pi}{L}, \qquad k_z = q_3 \frac{\pi}{L} \qquad (4.3.12)$$

where q_1, q_2, and q_3 are three positive integers (0, 1, 2, . . .).

4.4 FREE-ELECTRON MODEL AND DENSITY OF STATES

Let us restate the physical model on which the free-electron theory of metal is based. In a metallic crystal, the constituent atoms consist of an ion core and one or more valence electrons. The valence electrons are weakly bound to the ion core. In the first-order approximation, that is, the free-electron approximation, we completely neglect the forces between the valence electrons and the ion core. In doing so, we treat the valence electrons as if they were a free-electron gas. These electrons are allowed to move about freely inside the metal, but their motions are confined within the metal. These two physical conditions are incorporated in the mathematical formulation of the free-electron theory by setting $V = 0$ inside the cube box and demanding $\psi = 0$ outside the cube box.

In specifying the orbital motion of an electron around a nucleus, we use three quantum numbers, n, l, and m. As we recall, because of the $1/r$ dependence of the potential energy the solution of the Schrödinger equation is separable only in the spherical coordinate system (r, θ, ϕ). The three quantum numbers n, l, and m completely specify the dependence of the electron wave function on r, θ, ϕ. In the case of a free electron with $V = 0$, it is simplest to use the rectangular coordinates (x, y, z). The three integers q_1, q_2, and q_3 are three quantum numbers which specify the dependence of the electron wave function on x, y, and z. In the free-electron theory, therefore, the state of an electron is specified by four quantum numbers: q_1, q_2, and q_3, which specify the translational motion of the electron, and s_z, which specifies the spinning motion of the electron.

Let us pursue our analysis further to find the energy of an electron. The solution of Eq. (4.3.11) has the following property:

$$\frac{\partial^2 \psi}{\partial x^2} = -k_x^2 \psi, \qquad \frac{\partial^2 \psi}{\partial y^2} = -k_y^2 \psi, \qquad \frac{\partial^2 \psi}{\partial z^2} = -k_z^2 \psi \qquad (4.4.1)$$

Substituting Eq. (4.3.11) into Eq. (4.3.1), we find

$$-\frac{\hbar^2}{2m}\left(\frac{\partial^2 \psi}{\partial x^2} + \frac{\partial^2 \psi}{\partial y^2} + \frac{\partial^2 \psi}{\partial z^2}\right) = \frac{\hbar^2}{2m}(k_x^2 + k_y^2 + k_z^2)\psi = \mathscr{E}\,\psi \qquad (4.4.2)$$

Equation (4.4.2) is satisfied only if

$$\mathscr{E} = \frac{\hbar^2}{2m}(k_x^2 + k_y^2 + k_z^2) = \frac{\hbar^2}{2m}\left(\frac{\pi}{L}\right)^2 (q_1^2 + q_2^2 + q_3^2) \qquad (4.4.3)$$

The last step in Eq. (4.4.3) is arrived at through the use of Eq. (4.3.12).

In the classical theory, the kinetic energy of a particle is given by $p^2/2m$. For a free particle with $V = 0$, the total energy $\mathscr{E}$ is entirely made of the kinetic energy. Hence

$$\mathscr{E} = \text{KE} = \frac{p^2}{2m} = \frac{1}{2m}(p_x^2 + p_y^2 + p_z^2) \qquad (4.4.4)$$

where p_x, p_y, and p_z are the components of the electron momentum $\mathbf{p}$. Comparing Eq. (4.4.3) with Eq. (4.4.4), we see that the momentum of an electron in the quantum theory is equal to

$$\mathbf{p} = \hbar\mathbf{k} \qquad (4.4.5)$$

and that the k_x, k_y, and k_z in Eq. (4.4.3) represent the three components of $\mathbf{k}$ such that

$$k^2 = k_x^2 + k_y^2 + k_z^2 \qquad (4.4.6)$$

Since $\mathbf{p}$ is a vector, so must be $\mathbf{k}$.

We further note that the spatial variation of ψ shown in Fig. 4.14b is similar to that of E of an electromagnetic wave shown in Fig. 2.20c. In our present discussion, we have not considered the time dependence of ψ. If we include the time variation of ψ in the solution of the Schrödinger equation, we indeed obtain a wavelike solution similar to that of Eq. (2.7.1). In other words, the electrons in a potential well have wavelike motions. Referring to Fig. 4.14b, we see that for $q = 1$, the wavelength λ of the wave motion is equal to $2L$. Since $k = \pi/L$ for $q = 1$, we have

$$\lambda = \frac{2\pi}{k} \qquad (4.4.7)$$

From Eq. (4.4.3), $2m\,\mathscr{E} = \hbar^2 k^2$. In terms of $\mathscr{E}$, Eq. (4.4.7) becomes

$$\lambda = \frac{h}{\sqrt{2m\,\mathscr{E}}} = \frac{h}{p} \qquad (4.4.8)$$

Equation (4.4.8) is the celebrated de Broglie relation expressed in Eq. (2.7.5), whose validity has been proved in the electron diffraction experiment of Davisson and Germer.

One function that is essential in the calculation of the electrical property of an electric conductor is the density of state $D(\mathscr{E})$. To find this function, we must learn how to count the number of proper modes of electron motion. First let us consider a two-dimensional case with $q_3 = 0$ and draw a two-dimensional lattice as shown in Fig. 4.16a. Each lattice point represents a definite set of integers for the values of q_1 and q_2, which means a specific normal mode of oscillation. Therefore, to find the number of

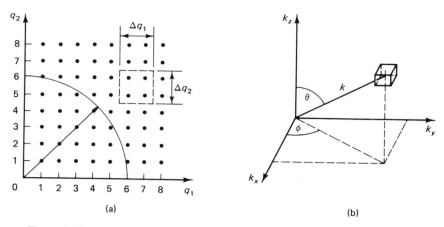

Figure 4.16 (a) Diagram used in counting the normal modes of electron motion in a two-dimensional box. Each lattice point represents a set of integers q_1 and q_2 in Eq. (4.3.12) with $q_3 = 0$. The number of normal modes in a rectangular box (or a circular ring) is equal to the total number of lattice points enclosed by the box (or the ring). (b) A three-dimensional k space that is obtained from the two-dimensional q space by adding a third dimension and by converting q to k through the use of Eq. (4.3.12).

normal modes inside a rectangular box with dimension $\Delta q_1 \Delta q_2$, we simply count the number of lattice points inside the box. If we choose $\Delta q_1 = 2$ and $\Delta q_2 = 2$ as illustrated in Fig. 4.16a, the number of normal modes is four. This is exactly equal to the area of the box. For a three-dimensional box, the number of normal modes $\Delta N'$ is equal to

$$\Delta N' = \Delta q_1 \, \Delta q_2 \, \Delta q_3 \qquad (4.4.9)$$

Let us now estimate the value of q in Eq. (4.4.3). For an electron having $\mathscr{E} = 3.2$ eV (topmost energy in Na as illustrated in Fig. 3.11), we find from Eq. (4.4.8) a value of $\lambda = 6.9 \times 10^{-8}$ cm. From Eqs. (4.4.7) and (4.3.10), the value of q is calculated to be $q = 2.9 \times 10^7$ for $L = 1$ cm. In other words, the value of q can be any integer from 0 to 2.9×10^7. For such a large value of q, we can treat the discrete variables q_1, q_2, and q_3 as if they were continuous, and replace Eq. (4.4.9) by

$$dN' = dq_1 \, dq_2 \, dq_3 \qquad (4.4.10)$$

From Eq. (4.4.3) we see that the constant energy surface in the q space is spherical. All the lattice points that represent states having energy $\mathscr{E} < \mathscr{E}_0$ lie inside a sphere with radius q such that

$$q = \frac{L}{\pi} \frac{(2m\mathscr{E}_0)^{1/2}}{\hbar} \qquad (4.4.11)$$

However, q_1, q_2, and q_3 are positive integers. Therefore, we should count only lattice points in one-eighth of the sphere. From Eq. (4.4.10) the number of normal modes having $\mathscr{E} < \mathscr{E}_0$ is given by

$$N'(\mathscr{E} < \mathscr{E}_0) = \frac{1}{8} \frac{4\pi}{3} q^3 = \frac{4\pi}{3} \left(\frac{L}{2\pi}\right)^3 \left(\frac{2m\mathscr{E}_0}{\hbar^2}\right)^{3/2} \qquad (4.4.12)$$

Sec. 4.4   Free-Electron Model and Density of States

125

Counting the two possible spin orientations, the total number N of electron states is

$$N = 2N' = \frac{8\pi}{3} L^3 \left(\frac{2m\mathscr{E}}{h^2} \right)^{3/2} \qquad (4.4.13)$$

Thus, within an energy interval $d\mathscr{E}$ between $\mathscr{E}$ and $\mathscr{E} + d\mathscr{E}$, the number of electron state is

$$dN = 4\pi L^3 \left(\frac{2m}{h^2} \right)^{3/2} \mathscr{E}^{1/2} \, d\mathscr{E} \qquad (4.4.14)$$

The *density-of-state function* $D(\mathscr{E})$ is defined as the density of electron states (per unit volume of the sample) per unit energy interval, or

$$D(\mathscr{E}) = \frac{1}{L^3} \frac{dN}{d\mathscr{E}} = 4\pi \left(\frac{2m}{h^2} \right)^{3/2} \mathscr{E}^{1/2} \qquad (4.4.15)$$

Equation (4.4.15) is identical to Eq. (3.4.4), which we used in the calculation of the average kinetic energy of electrons in sodium.

The density-of-state function can be written in different forms. In the k space and in the momentum p space, Eq. (4.4.10) becomes

$$dN = 2 \, dN' = \frac{2}{8} \left(\frac{L}{\pi} \right)^3 dk_x \, dk_y \, dk_z = \frac{2}{8} \left(\frac{2L}{h} \right)^3 dp_x \, dp_y \, dp_z \qquad (4.4.16)$$

Again the factor 2 accounts for two spin orientations and the factor $\frac{1}{8}$ accounts for the fact that the values of q_1, q_2, and q_3 and hence those of k_x, k_y, and k_z in Eq. (4.3.12) are positive, whereas the values of k_x, k_y, and k_z and those of p_x, p_y, and p_z in Eq. (4.4.16) are allowed to be either positive or negative. Realizing that in polar coordinates (Fig. 4.16b),

$$dk_x \, dk_y \, dk_z = (dk)(k \, d\theta)(k \sin \theta \, d\phi) \qquad (4.4.17)$$

we can rewrite Eq. (4.4.16) as

$$\frac{dN}{L^3} = \frac{2}{(2\pi)^3} k^2 \, dk \sin \theta \, d\theta \, d\phi \qquad (4.4.18)$$

Thus the density-of-state function $D(k)$ in k space is

$$D(k) = \frac{1}{L^3} \frac{dN}{dk} = \frac{2}{(2\pi)^3} k^2 \sin \theta \, d\theta \, d\phi \qquad (4.4.19)$$

The reader is asked to find the expression for $D(p)$ in the momentum p space.

4.5 THE FERMI–DIRAC DISTRIBUTION FUNCTION

In determining the electrical property of a solid, we need to know how valence electrons are distributed among the available energy states. At absolute-zero temperature, the filling of the available states follows a simple rule: States of lowest energy are filled first. At a finite temperature, however, this simple rule is no longer strictly obeyed. The probability that a given energy state of energy $\mathscr{E}$ is occupied by an electron is given by the Fermi–Dirac distribution function $f(\mathscr{E})$:

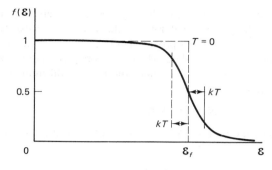

Figure 4.17 The Fermi function $f(\mathcal{E})$ of Eq. (4.5.1) plotted as a function of energy $\mathcal{E}$. At the absolute zero temperature, $f(\mathcal{E})$ drops abruptly from 1 to 0 at $\mathcal{E} = \mathcal{E}_f$ (dashed curve). At a finite temperature, $f(\mathcal{E})$ changes continuously from 1 to 0 (solid curve).

$$f(\mathcal{E}) = \frac{1}{1 + \exp\left[(\mathcal{E} - \mathcal{E}_f)/kT\right]} \qquad (4.5.1)$$

In Eq. (4.5.1) the energy parameter $\mathcal{E}_f$ is called the *chemical potential,* but better known to electrical engineers as the *Fermi level,* and k is the Boltzmann constant.

The function $f(\mathcal{E})$ has the general shape shown in Fig. 4.17. At absolute-zero temperature, the value of $(\mathcal{E} - \mathcal{E}_f)/kT$ changes abruptly from $-\infty$ to $+\infty$ as $\mathcal{E}$ increases from a value infinitesimally smaller than $\mathcal{E}_f$ to a value infinitesimally larger than $\mathcal{E}_f$. This abrupt change makes $f(\mathcal{E})$ discontinuous, that is, $f(\mathcal{E}) = 1$ for $\mathcal{E} < \mathcal{E}_f$ and $f(\mathcal{E}) = 0$ for $\mathcal{E} > \mathcal{E}_f$. Note that a value of 1 for $f(\mathcal{E})$ means total occupancy of the energy state. Therefore, the step function in Fig. 4.17 is a manifestation of the physical situation that at $T = 0$ K, the states with $\mathcal{E} < \mathcal{E}_f$ are completely occupied by electrons, whereas the states with $\mathcal{E} > \mathcal{E}_f$ are empty. At a finite temperature, the function $f(\mathcal{E})$ changes continuously from 1 to 0, meaning that the occupancy of energy states changes smoothly from full occupancy to partial occupancy and then to nonoccupancy as the energy increases. For $\mathcal{E} = \mathcal{E}_f$, $f(\mathcal{E}) = \frac{1}{2}$; thus the states with $\mathcal{E} = \mathcal{E}_f$ are half-filled.

The formal derivation of Eq. (4.5.1) based on the theory of probability can be found in many textbooks and will not be given here. Among the three distribution functions, the Maxwell–Boltzmann, the Fermi–Dirac, and the Bose–Einstein distributions, only the Maxwell–Boltzmann distribution law is derived (see Appendix A1). The Fermi–Dirac distribution differs from the Maxwell–Boltzmann distribution because of the Pauli exclusion principle. In the following discussion, we use different examples to illustrate the role of the exclusion principle.

First we consider a Boltzmann system of particles having discrete and nondegenerate energies. Figure 4.18 shows a collision process involving two particles that have energies $\mathcal{E}_1$ and $\mathcal{E}_4$ before collision and energies $\mathcal{E}_2$ and $\mathcal{E}_3$ after collision. According to the *law of mass action,* the total rate R of the number of such collisions per second is equal to

$$R = CN_1N_4 \qquad (4.5.2)$$

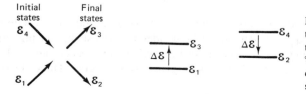

Figure 4.18 Collision process between two particles. The result of the collision is that one particle changes its energy from $\mathcal{E}_1$ to $\mathcal{E}_3$, whereas the other particle changes its energy from $\mathcal{E}_4$ to $\mathcal{E}_2$. For conservation of energy, $\mathcal{E}_3 - \mathcal{E}_1 = \mathcal{E}_4 - \mathcal{E}_2$.

Sec. 4.5   The Fermi–Dirac Distribution Function

where N_1 and N_4 are the number of particles having energy $\mathscr{E}_1$ and $\mathscr{E}_4$, and C is a proportionality constant. Similarly, the total rate of the reverse process $(\mathscr{E}_2, \mathscr{E}_3) \rightarrow (\mathscr{E}_1, \mathscr{E}_4)$ is $R = CN_2N_3$. Since the collision process between two particles is elastic, no reaction energy is either liberated or absorbed. Hence the proportionality constant C is the same for the two processes. For a system under thermal equilibrium, the two rates must be equal, yielding the relation

$$N_1N_4 = N_2N_3 \qquad (4.5.3)$$

Let $p(\mathscr{E}_i)$ be the ratio of the number N_i in the state of $\mathscr{E}_i$ to the total number N of particles. Thus we have

$$N_i = Np(\mathscr{E}_i) \qquad (4.5.4)$$

Substituting Eq. (4.5.4) into Eq. (4.5.3) gives

$$p(\mathscr{E}_1)p(\mathscr{E}_4) = p(\mathscr{E}_2)p(\mathscr{E}_3) \qquad (4.5.5)$$

In view of the fact that $\mathscr{E}_3 - \mathscr{E}_1 = \mathscr{E}_4 - \mathscr{E}_2$ for conservation of energy, Eq. (4.5.5) is satisfied if

$$p(\mathscr{E}_i) = B \exp\left(\frac{-\mathscr{E}_i}{kT}\right) \qquad (4.5.6)$$

where B is a constant. The reader is asked to show that Eq. (4.5.6) is indeed a solution of Eq. (4.5.5). Note that Eq. (4.5.6) is the Boltzmann distribution law derived in Appendix A1.

Now we consider the interaction between particles in a Fermi system with particles in a Boltzmann system. For such a discussion, we assume that the two energy levels $\mathscr{E}_2$ and $\mathscr{E}_4$ in Fig. 4.17 are energy levels of the Fermi system. The transition from the state $\mathscr{E}_4$ to the state $\mathscr{E}_2$ must observe the Pauli exclusion principle. For such a transition to occur, the initial state must be occupied and the final state must be vacant. The probability that the state $\mathscr{E}_4$ is occupied is given by $f(\mathscr{E}_4)$, whereas the probability that the state $\mathscr{E}_2$ is empty is $1 - f(\mathscr{E}_2)$. A similar consideration applies to the reverse process. Thus Eq. (4.5.5) must be replaced by

$$p(\mathscr{E}_1)f(\mathscr{E}_4)[(1 - f(\mathscr{E}_2)] = p(\mathscr{E}_3)f(\mathscr{E}_2)[1 - f(\mathscr{E}_4)] \qquad (4.5.7)$$

Using the following relation from Eq. (4.5.6),

$$\frac{p(\mathscr{E}_1)}{p(\mathscr{E}_3)} = \exp\left(\frac{\mathscr{E}_3 - \mathscr{E}_1}{kT}\right) = \exp\left(\frac{\mathscr{E}_4 - \mathscr{E}_2}{kT}\right) \qquad (4.5.8)$$

for the two Boltzmann states, we rewrite Eq. (4.5.7) as

$$\frac{f(\mathscr{E}_4)}{1 - f(\mathscr{E}_4)} \exp\left(\frac{\mathscr{E}_4}{kT}\right) = \frac{f(\mathscr{E}_2)}{1 - f(\mathscr{E}_2)} \exp\left(\frac{\mathscr{E}_2}{kT}\right) \qquad (4.5.9)$$

In Eq. (4.5.9) the left side depends on $\mathscr{E}_4$ and the right side depends on $\mathscr{E}_2$. Obviously, the only way to satisfy Eq. (4.5.9) is to set both sides to a constant. Let this constant be $C = \exp(\mathscr{E}_f/kT)$. Using this constant C in Eq. (4.5.9), the reader can easily show that the function $f(\mathscr{E}_i)$ in Eq. (4.5.9) is indeed given by the Fermi–Dirac distribution function of Eq. (4.5.1).

The examples above serve to show the essential difference between the Maxwell–Boltzmann and the Fermi–Dirac distribution functions. The two distribution functions [Eqs. (4.5.1) and (4.5.6)] are different because the laws governing the particles are different. Whenever it is necessary to consider the Pauli exclusion principle, the Fermi–Dirac distribution function is to be used. The distribution of electrons obeys the Fermi–Dirac statistics in systems where the density of electrons is high and hence mutual interaction between electrons is strong. Such is the case with metals and semiconductors. In the following sections we illustrate with examples the use of the Fermi–Dirac distribution law.

4.6 FERMI ENERGY AND CONTACT POTENTIAL

In Section 4.3 we treated electrons inside a metal as free electrons in a box. This free-electron model is useful for the understanding of many simple physical phenomena. Figure 4.19a is a reproduction of Fig. 4.13a showing electrons in a potential well. Two quantities of importance are the Fermi energy $\mathscr{E}_f$ and the work function $\mathscr{E}_w$.

The number of electrons dQ in a metal is equal to the number of available energy states dN times the probability $f(\mathscr{E})$ that a given state is occupied by an electron. According to Eqs. (4.4.14) and (4.5.1),

$$Q = \int dQ = \int f(\mathscr{E})\, dN = \int_0^\infty \frac{4\pi V}{1 + \exp\left[(\mathscr{E} - \mathscr{E}_f)/kT\right]} \left(\frac{2m}{h^2}\right)^{3/2} \mathscr{E}^{1/2} d\mathscr{E} \qquad (4.6.1)$$

where V is the volume of the crystal. The situation at absolute-zero temperature is particularly simple and is treated here.

At 0 K, the Fermi function $f(\mathscr{E})$ can be replaced by a step function (Fig. 4.17). Since the energy states with $\mathscr{E} > \mathscr{E}_f$ are not occupied, that is, $f(\mathscr{E}) = 0$ for $\mathscr{E} > \mathscr{E}_f$, the integration limits run from zero to $\mathscr{E}_f$. Further, $f(\mathscr{E}) = 1$ for $\mathscr{E} < \mathscr{E}_f$, hence the total number Q of electrons in a volume V is equal to

$$Q = \int_0^{\mathscr{E}_f} 4\pi V\left(\frac{2m}{h^2}\right)^{3/2} \mathscr{E}^{1/2}\, d\mathscr{E} = \frac{8\pi V}{3}\left(\frac{2m}{h^2}\right)^{3/2} \mathscr{E}_f^{3/2} \qquad (4.6.2)$$

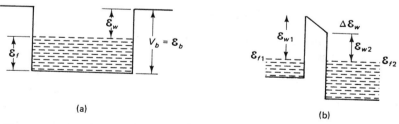

(a)   (b)

Figure 4.19 Free-electron model representing (a) a metal and (b) two metals in contact. In the free-electron model, valence electrons of a metal can be treated as electrons in a box (Section 4.4). Within the confines of the box (i.e, inside the metals), the electrons are free. At a finite temperature, almost all electrons have kinetic energies $\mathscr{E} \leq \mathscr{E}_f$ and only a few electrons have $\mathscr{E} > \mathscr{E}_f$ (Fig. 4.17). The energy $\mathscr{E}_b$ represents the potential barrier at the surface of the metal.

In terms of electron density $n = Q/V$, the Fermi level is

$$\mathscr{E}_{f0} = \frac{h^2}{2m} \left(\frac{3n}{8\pi} \right)^{2/3} \quad J \qquad (4.6.3)$$

The subscript 0 is added because Eq. (4.6.3) holds only at 0 K. Calculated values from Eq. (4.6.3) are: 3.1 eV for Na, 1.5 eV for Cs, 7.0 eV for Cu, 3.8 eV for Ba, and 11.7 eV for Al. Experimentally determined values for a selected number of metals are given in Table 4.1.

TABLE 4.1 EXPERIMENTALLY DETERMINED VALUES OF FERMI ENERGY $\mathscr{E}_f$ FOR A SELECTED NUMBER OF METALS

	Element										
	Li	Na	K	Rb	Cs	Cu	Ag	Au	Mg	Be	Al
$\mathscr{E}_f$ (eV)	4.76	3.20	2.12	1.81	1.57	7.04	5.51	5.51	7.50	14.8	13.0
a (Å)	3.47	4.28	5.20	5.62	6.05	3.61	4.08	4.07	3.20	2.23	4.04
Structure[a]	bcc	bcc	bcc	bcc	bcc	fcc	fcc	fcc	hcp	hcp	fcc

[a]bcc, body-centered cubic; fcc, face-centered cubic; hcp, hexagonal close-packed with $c = 5.20$ Å in Mg and 3.59 Å in Be.
Source: H. W. B. Skinner, *Rep. Progr. Phys.*, Vol. 5, p. 27, 1938.

Next we consider the situation in which two different metals are brought in contact with each other. Since the kinetic energies of electrons are different in the two metals, the potential energies must adjust themselves so that thermodynamic equilibrium may be established. The Gibbs potential G (Appendix A2) of a composite system under infinitesimal changes is

$$dG = -S\,dT + V\,dp + \sum g_i\,dn_i \qquad (4.6.4)^\dagger$$

where S, T, V, and p stand for entropy, temperature, volume, and pressure, respectively, while g_i is the chemical potential known as the Fermi level to electrical engineers and physicists and n_i is the concentration of particles in the subsystem i. Under constant T and p, the change in G as a result of transferring dn electrons from metal 1 to metal 2 is

$$dG = (g_2 - g_1)\,dn = (\mathscr{E}_{f2} - \mathscr{E}_{f1})\,dn \qquad (4.6.5)^\dagger$$

Under thermal equilibrium, G must be a minimum. This means that $dG/dn = 0$, requiring that

$$\mathscr{E}_{f2} = \mathscr{E}_{f1} \qquad (4.6.6)$$

Figure 4.19b shows the situation under thermal equilibrium between two different metals in contact as required by Eq. (4.6.6). As we can see, because of the difference in the two work functions, a potential difference is established between the two vacuum levels. The difference, usually expressed in eV,

---

†See Appendix A2. Equations (4.6.4) and (4.6.5) are the same as Eqs. (A2.26) and (A2.31), respectively.

TABLE 4.2 WORK FUNCTIONS OF METALS AND SEMICONDUCTORS (eV)

Li	2.48	Cu	4.45	W	4.54	Si	4.02
Na	2.28	Ag	4.46	Ta	4.13	Ge	4.50
K	2.22	Au	4.89	Zn	4.29	Ti	4.17
Cs	1.80	Pt	5.36	Cd	4.10	Ni	5.03
Rb	2.09	Mo	4.20	Al	4.08	Co	4.40

$$\Delta \mathcal{E}_w = \mathcal{E}_{w1} - \mathcal{E}_{w2} \quad \text{eV} \tag{4.6.7}$$

is called the *contact potential*. The work function $\mathcal{E}_w$ is defined as the energy required to bring an electron from inside the metal lying at the Fermi energy level to the vacuum level outside the metal.

In Table 4.2 the values of work function in a selected number of metals are given. In a photoelectric experiment, electrons are liberated from the cathode of a vacuum tube by absorbing a photon of energy $h\nu$. Electron emission starts when $h\nu \geq \mathcal{E}_w$. Therefore, the threshold frequency ν_{th} for the photoelectric current to occur gives the value of $\mathcal{E}_w$. As discussed in Section 4.7, the thermionic current from a cathode is dependent on $\mathcal{E}_w$ of the cathode. Values of $\mathcal{E}_w$ can also be obtained from an analysis of the thermionic emission data.

The contact potential can be measured in an experiment described below (W. A. Zisman, *Rev. Sci. Instrum.*, Vol. 3, p. 367, 1932). If two metal plates are connected electrically to form a plane-parallel capacitor, the two surfaces of the plates facing each other will become charged because of the contact potential. A current pulse is generated when the distance between the plates is made to change suddenly. The magnitude of this current is proportional to the contact potential. The reader is asked to consult the literature for further references to experimental data on work function and contact potential (see, for example, C. Herring and M. H. Nichols, *Rev. Mod. Phys.*, Vol. 21, p. 185, 1949; G. Hermann and S. Wagener, *The Oxide Coated Cathode*, Chapman & Hall Ltd., London, 1951).

4.7 THERMIONIC EMISSION

The process of thermionic emission consists of providing sufficient thermal energy of electrons in a solid so that the electrons near the surface can be emitted from the solid. Let us define a quantity $\rho(\mathcal{E})$ such that $\rho(\mathcal{E}) \, d\mathcal{E}$ represents the density of electrons having energy between $\mathcal{E}$ and $\mathcal{E} + d\mathcal{E}$. Following a similar procedure as used in deriving Eq. (4.6.1), we have

$$\rho(\mathcal{E}) = D(\mathcal{E})f(\mathcal{E}) = \frac{4\pi}{1 + \exp\left[(\mathcal{E} - \mathcal{E}_f)/kT\right]} \left(\frac{2m}{h^2}\right)^{3/2} \mathcal{E}^{1/2} \tag{4.7.1}$$

where $D(\mathcal{E})$ is the density of state function given by Eq. (4.4.15). Figure 4.20 shows the electron distribution $\rho(\mathcal{E})$ as a function of energy near the surface of a metal. Only electrons in the tail end of $\rho(\mathcal{E})$ with $\mathcal{E} > \mathcal{E}_b$ have sufficient energy to escape from the solid, and hence contribute to thermionic emission. As the temperature is raised, more and more electrons will have sufficient energy for emission.

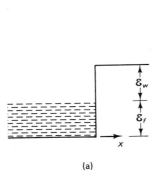

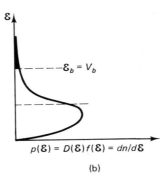

(a)   (b)

Figure 4.20 Diagrams used in the calculation of thermionic current: (a) energy diagram at the surface of a metal and (b) electron distribution $\rho(\mathscr{E})$ as a function of energy. The curve on the left is so arranged that the two energies corresponding to $\mathscr{E}_f$ and $\mathscr{E}_b$ in (a) can easily be marked on the curve in (b). The electrons in the tail end of the distribution [dark area under the $\rho(\mathscr{E})$ versus $\mathscr{E}$ curve] will be able to escape from the solid.

The number of electrons striking the surface per unit area per unit time is equal to $v_x\rho(\mathscr{E})\,d\mathscr{E}$, where v_x is the component of electron velocity normal to the surface of the emitting solid. If it is assumed that every electron striking the surface with $\mathscr{E} > \mathscr{E}_b$ escapes from the solid, the current density J_x resulting from thermionic emission is given by

$$J_x = \int_{\mathscr{E}_b}^{\infty} ev_x\rho(\mathscr{E})\,d\mathscr{E} \qquad (4.7.2)$$

To evaluate the integral in Eq. (4.7.2), we make the following observation about $f(\mathscr{E})$. The smallest value for $(\mathscr{E} - \mathscr{E}_f)$ is $\mathscr{E}_w$, which is of the order of 2 eV (Table 4.2). In comparison, the value of $kT = 0.11$ eV at 1300 K. Therefore, we can approximate $f(\mathscr{E})$ by $\exp\left[-(\mathscr{E} - \mathscr{E}_f)/kT\right]$ and rewrite Eq. (4.7.2) as

$$J_x = \int_{\mathscr{E}_b}^{\infty} 4\pi ev_x\left(\frac{2m}{h^2}\right)^{3/2}\mathscr{E}^{1/2}\exp\left(-\frac{\mathscr{E} - \mathscr{E}_f}{kT}\right)d\mathscr{E} \qquad (4.7.3)$$

Comparing Eqs. (4.4.14) and (4.4.16), we obtain

$$4\pi(2m)^{3/2}\mathscr{E}^{1/2}\,d\mathscr{E} = 2\,dp_x\,dp_y\,dp_z \qquad (4.7.4)$$

Further, from Eq. (4.4.4), we find that

$$\frac{\partial\mathscr{E}}{\partial p_x} = \frac{p_x}{m} = v_x \qquad (4.7.5)$$

Using Eqs. (4.7.4) and (4.7.5) in Eq. (4.7.3), we have

$$j_x = \iiint \frac{2e}{h^3}\exp\left(-\frac{\mathscr{E} - \mathscr{E}_f}{kT}\right)\frac{\partial\mathscr{E}}{\partial p_x}\,dp_x\,dp_y\,dp_z \qquad (4.7.6)$$

Note that since the surface barrier potential exists only in the x direction, p_y and p_z remain unchanged. If we let

$$\mathscr{E} = \frac{1}{2m}(p_x^2 + p_y^2 + p_z^2) = \mathscr{E}_1 + \frac{1}{2m}(p_y^2 + p_z^2) \qquad (4.7.7)$$

the change in energy must come from the p_x^2 term, which is represented by $\mathscr{E}_1$. In other words, $d\mathscr{E} = d\mathscr{E}_1$.

In terms of $\mathscr{E}_1$, Eq. (4.7.6) becomes

$$J_x = \frac{2e}{h^3} \int_{\mathscr{E}_b}^{\infty} \exp\left(-\frac{\mathscr{E}_1 - \mathscr{E}_f}{kT}\right) d\mathscr{E}_1 \int_{-\infty}^{\infty}$$

$$\exp\left(-\frac{p_y^2}{2mkT}\right) dp_y \int_{-\infty}^{\infty} \exp\left(-\frac{p_z^2}{2mkT}\right) dp_z \tag{4.7.8}$$

Using Eq. (A3.1) given in Appendix A3, we obtain

$$J_x = \frac{4\pi em}{h^3} (kT)^2 \exp\left(-\frac{\mathscr{E}_w}{kT}\right) \tag{4.7.9}$$

In the derivation of Eq. (4.7.9) we have assumed that every electron striking the surface with sufficient energy escaped. In reality, part may be reflected. Letting r be the reflection coefficient and $A = 4\pi emk^2/h^3$, we find the thermionic current density to be

$$J = (1 = r)J_x = A(1 - r)T^2 \exp\left(-\frac{\mathscr{E}_w}{kT}\right) \tag{4.7.10}$$

which is known as the *Dushman–Richardson equation*. The quantity k in the expression of A and in Eq. (4.7.10) is the Boltzmann constant.

4.8 ELECTRONIC CONDUCTIVITY AND MEAN FREE TIME

In Section 4.2 we showed that Ohm's law is based on the linear relationship between the drift velocity v_d and the applied field E as assumed in Eq. (4.2.2). The simple relation $v_d = -\mu E$ is conceptually more involved than it appears to be. In this section we present a derivation of this relation so that we may appreciate the underlying physics. First, we must realize that v_d represents only a small perturbation on the velocity which an electron already possesses. In metals, the mobility μ is of the order 20 cm^2/V-s. For $E = 10$ V/cm, $v_d = 200$ cm/s. For electrons at $\mathscr{E}_{f0}$, we find from Eq. (4.6.3) a velocity v_f equal to

$$v_f = \left(\frac{2\mathscr{E}_{f0}}{m}\right)^{1/2} = \frac{h}{m}\left(\frac{3n}{8\pi}\right)^{1/3} \tag{4.8.1}$$

Equation (4.8.1) applies to materials whose electron distribution obeys the Fermi–Dirac statistics. Using Na with $\mathscr{E}_f = 3.1$ eV as an example, we find that $v_f = 1.05 \times 10^6$ m/s.

Let us now turn to the more familiar situation in semiconductors. As we discuss in Chapter 6, electron distribution in nondegenerate semiconductors can be approximated by the Boltzmann distribution. According to the equipartition theorem in thermodynamics, the average kinetic energy is equal to $3kT/2$. Thus the average thermal velocity of electrons is

$$v_{th} = \left(\frac{3kT}{m}\right)^{1/2} \tag{4.8.2}$$

At $T = 300$ K, we find that $v_{th} = 1.17 \times 10^5$ m/s. Take Ge as an example, with an electron mobility $\mu = 3900$ cm^2/V-s. At a field $E = 10$ V/cm, $v_d = 390$ m/s, which is much smaller than the thermal velocity.

The question is raised as to what prevents the electrons from acquiring a much larger drift velocity. The answer lies in the fact that electrons constantly suffer collisions with atomic movements (i.e., lattice vibrations), with impurities, and with other electrons. The collisions that keep electrons in thermal equilibrium with the environment tend to randomize the motion of electrons. Figure 4.21 shows (a) the random motion of an electron, (b) the drift motion of the electron due to the applied field, and (c) the actual motion of the electron, which is the combination of (a) and (b). Two important features are that the direction of motion changes randomly after each collision at various times, t_2, t_3, and so on, and that the time interval between successive collisions, that is, $t_4 - t_3$, $t_5 - t_4$, and so on, changes from time to time.

To treat the problem of random collisions, we first define a time constant τ such that the probability that an electron makes a collision in a time interval dt is dt/τ. Let N_0 be the number of electrons that suffered a collision at time t_0. We assume that of these, N electrons have not suffered subsequent collisions between t_0 and t. During any time interval dt, the number of uncollided electrons decreases by $N \, dt/\tau$. That means that

$$dN = -N \frac{dt}{\tau} \quad \text{or} \quad N = N_0 \exp\left(-\frac{t - t_0}{\tau}\right) \qquad (4.8.3)$$

Therefore, the probability w that an electron is uncollided at time t after a collision at t_0 is

$$w = \frac{N}{N_0} = \exp\left(-\frac{t - t_0}{\tau}\right) \qquad (4.8.4)$$

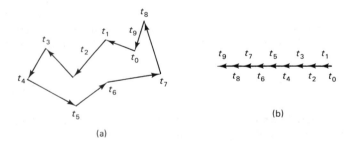

(a)

(b)

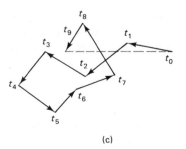

(c)

Figure 4.21 Diagrams showing (a) the random motion of an electron, (b) the drift motion of the electron under an electric field, and (c) the superposition of these two motions. Collisions that tend to randomize the electron motion are assumed to occur at t_1, t_2, t_3, and so on.

To proceed to the next step, we must realize that the collision time t_0 varies continuously from electron to electron. Let dn_0 be the number of electrons that suffered collision between t_0 and $t_0 + dt_0$. Obviously, dn_0 is proportional to $N_0\, dt_0$. Thus we assume that $dn_0 = AN_0\, dt_0$, where A is a proportionality constant. Similarly, we let dn be the number of electrons that suffered the last collision between t_0 and $t_0 + dt_0$ and have not suffered another collision at t. Since dn must bear the same relation to dn_0 as N to N_0, we have

$$dn = w\, dn_0 = AN_0 w\, dt_0 = Bw\, dt_0 \qquad (4.8.5)$$

where $B = AN_0$ is a different proportionality constant.

If there are n electrons in total, then integrating dn over all t_0 from $-\infty$ to t (the observation time)

$$\int dn = \int_{-\infty}^{t} Bw\, dt_0 = n \qquad (4.8.6)$$

must be equal to n. Substituting the value of B from Eq. (4.8.6) into Eq. (4.8.5), we find that

$$dn = n \exp\left(-\frac{t - t_0}{\tau}\right) \frac{dt_0}{\tau} \qquad (4.8.7)$$

Equation (4.8.7) is a statement of the composition of electrons. Of all n electrons, some suffered the last collision at t_{01}, some at t_{02}, and so on. If we divide the collision events in time intervals dt_0, the number of those electrons whose collision event falls in between t_0 and $t_0 + dt_0$ is given by Eq. (4.8.7).

Let us now apply Eq. (4.8.7) to the drift motion of electrons. The velocity of an electron that suffered the last collision at t_0 is given by

$$\mathbf{v} = \mathbf{v}_0 - \frac{e}{m} \mathbf{E}(t - t_0) \qquad (4.8.8)$$

where the first term represents the initial velocity after collision and the second term represents the velocity acquired under the applied field. Since the velocity $\mathbf{v}_0$ after collision is completely random in its direction, the average of $\mathbf{v}_0$ over many electrons is zero. Therefore, the average velocity of electrons is equal to

$$\langle \mathbf{v} \rangle = \frac{1}{n} \int -\frac{e\mathbf{E}}{m}(t - t_0)\, dn \qquad (4.8.9)$$

Using Eq. (4.8.7) in Eq. (4.8.9) and setting the integration limits from $-\infty$ to t for dt_0, we obtain

$$\mathbf{v}_d = \langle \mathbf{v} \rangle = -\frac{e\tau}{m} \mathbf{E} \qquad (4.8.10)$$

Comparing Eq. (4.8.10) with Eq. (4.2.2), we have

$$\mu = \frac{e\tau}{m} \quad (\text{m}^2/\text{V-s}) \qquad (4.8.11)$$

According to Eq. (4.8.10), the average time during which an electron acquires a unidirectional velocity v_d is τ. Since this can occur only between successive collisions, the time τ is called the *mean free time*. This statement can be proved directly from Eq. (4.8.4). There are $N\,dt/\tau$ electrons suffering collisions in time interval dt. For these electrons, the free time between two collisions is $t_f = t - t_0$. The reader is asked to show that $\langle t_f \rangle$ averaged over the N_0 electrons is indeed equal to τ.

Finally, we should mention experimental evidences that support the free-electron model. For electrons at $\mathscr{E}_f$ in a metal, the average distance traveled between successive collisions is

$$l = v_f \tau \tag{4.8.12}$$

where v_f is given by Eq. (4.8.1). The distance l is called the *mean free path*. Reynold and Stiwell (F. W. Reynolds and G. R. Stiwell, *Phys. Rev.*, Vol. 88, p. 418, 1952) measured the resistivity of evaporated copper and silver film as a function of the thickness of the film. As the film thickness approaches, or is smaller than, the mean free path, the scattering of electrons at the surface becomes significant and the resistivity becomes dependent on thickness. By fitting the experimental result to a theoretical expression, they found that $l = 450$ Å in Cu and 520 Å in Ag at 25°C.

For Cu, $\sigma = 6 \times 10^7$ $(\Omega\text{-m})^{-1}$ at room temperature. Using $a = 3.61$ Å in Fig. 2.3 for a face-centered cubic structure and assuming one electron per copper atom, we find that $n = 8.5 \times 10^{28} \text{m}^{-3}$. From Eq. (4.2.3) we find that $\mu = 4.42 \times 10^{-3}$ m²/V-s. Using $m = 9.1 \times 10^{-31}$ kg in Eq. (4.8.11), we obtain a value of $\tau = 2.5 \times 10^{-14}$ s. The value of v_f can be calculated from $v_f = \sqrt{2\mathscr{E}_f/m}$. For $\mathscr{E}_f = 7.0$ eV in Cu, $v_f = 1.56 \times 10^6$ m/s. Using the values of τ and v_f in Eq. (4.8.12), we find $l = 390$ Å, in reasonable agreement with the experimentally determined value of 450 Å.

The value of τ can also be found from measuring the conductivity as a function of frequency. For a time-varying field $E = E_1 \exp(i\omega t)$, Eq. (4.8.8) should be replaced by

$$\mathbf{v} = \mathbf{v}_0 - \int_{t_0}^{t} \frac{eE_1}{m} \exp(i\omega t')\,dt' = \mathbf{v}_0 - \frac{eE_1}{i\omega m}[\exp(i\omega t) - \exp(i\omega t_0)] \tag{4.8.13}$$

The reader is asked to show that

$$\mathbf{v}_d = \langle \mathbf{v} \rangle = \frac{1}{n}\int \mathbf{v}\,dn = -\frac{e}{m}\frac{\tau}{1 + i\omega\tau}\mathbf{E}_1 \exp(i\omega t) \tag{4.8.14}$$

through the use of Eq. (4.8.7) for dn. The reader is also asked to verify that the results stated in Eqs. (4.8.10) and (4.8.14) can be obtained from the following equation of motion:

$$\frac{d\langle \mathbf{v}\rangle}{dt} = -\frac{\langle \mathbf{v}\rangle}{\tau} + \frac{\mathbf{F}}{m} \tag{4.8.15}$$

where $\mathbf{F}$ is the force acting on electrons. The term $-m\langle \mathbf{v}\rangle/\tau$ represents phenomenologically the restoring force due to collisions.

According to Eq. (4.8.15), the conductivity σ should vary with frequency as

$$\sigma = \frac{\sigma_0}{1 + i\omega\tau} \tag{4.8.16}$$

where σ_0 is the dc conductivity. In n-type germanium, $\mu = 3900$ cm^2/V-s at 300 K and 4×10^4 cm^2/V-s at 77 K, corresponding to a value of $\tau = 2.2 \times 10^{-12}$ s and 2.3×10^{-11} s, respectively. Therefore, a significant change in σ is expected at 77 K, for $\nu = 9 \times 10^9$ Hz. Such a change has indeed been observed in the microwave region. At still higher frequencies such that $\omega\tau \gg 1$, Eq. (4.8.16) becomes

$$\sigma = -i\frac{\sigma_0}{\omega\tau} = -i\frac{e^2 n}{m\omega} \qquad (4.8.17)$$

Note that an imaginary σ is equivalent to a dielectric constant ϵ with $i\omega\epsilon = \sigma$. From Eq. (4.8.17), the free-electron contribution to dielectric constant is

$$\epsilon_e = -\frac{e^2 n}{m\omega^2} = -\frac{\omega_p^2}{\omega^2}\epsilon_0 \qquad (4.8.18)$$

where the quantity

$$\omega_p = \left(\frac{ne^2}{m\epsilon_0}\right)^{1/2} \qquad (4.8.19)$$

is called the *plasma frequency*. The effect represented by Eq. (4.8.18) has indeed been observed in metals in the ultraviolet region. The reader is referred to Problem 4.25 for a discussion of the effect.

4.9 ELECTRONIC CONTRIBUTION TO HEAT CAPACITY AND THERMAL CONDUCTIVITY

The heat capacity of metals is found experimentally to vary as

$$C = aT + bT^3 \qquad (4.9.1)$$

at sufficiently low temperatures (i.e., at temperatues below the Debye temperature). In Eq. (4.9.1), a and b are two proportionality constants. The first term represents the contribution from electrons and the second term the contribution from the lattice. At low temperatures, the electronic contribution is dominant (see, for example, W. H. Lien and N. E. Phillips, *Phys. Rev.*, Vol. 133, A1370, 1964). Our present discussion concerns the electronic part.

To understand the physical origin of electronic heat capacity, we refer to Fig. 4.22, in which the electron distribution function $\rho(\mathscr{E})$ of Eq. (4.7.1) at a finite temperature is compared with that at absolute zero temperature. As the temperature increases, electrons move from the lower-energy region R_1 into the higher-energy region R_2. This movement increases the energy of the system. For a metal of volume V, the total electronic energy is

$$\mathscr{E}_{\text{el}} = V \int_0^\infty \mathscr{E}D(\mathscr{E})f(\mathscr{E})\, d\mathscr{E} \qquad (4.9.2)$$

The evaluation of Eq. (4.9.2) is discussed in Appendix A3. Using Eq. (A3.11), we obtain

$$\mathscr{E}_{\text{el}} = VG\mathscr{E}_f^{5/2}\left[\frac{2}{5} + \frac{\pi^2}{4}\left(\frac{kT}{\mathscr{E}_f}\right)^2\right] \qquad (4.9.3)$$

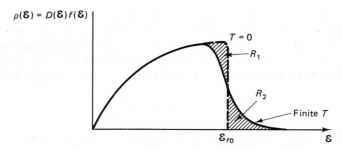

Figure 4.22 Comparison of the electron distribution function $\rho(\mathscr{E})$ at a finite temperature (solid curve) with that at $T = 0$ (dashed curve). The area under the respective curve is equal to the total number of electrons. Note that there is a transfer of electrons from the lower-energy region R_1 to the higher-energy region R_2 at a finite temperature.

where $G = 4\pi(2m/h^2)^{3/2}$ and k is the Boltzmann constant.

The value of $\mathscr{E}_f$ also depends on temperature. According to Eq. (A3.14),

$$\mathscr{E}_f^{5/2} = \mathscr{E}_{f0}^{5/2}\left[1 - \frac{5\pi^2}{24}\left(\frac{kT}{\mathscr{E}_{f0}}\right)^2\right] \tag{4.9.4}$$

Substituting Eq. (4.9.4) in Eq. (4.9.3) and letting $\mathscr{E}_f = \mathscr{E}_{f0}$ for the term $(kT/\mathscr{E}_f)^2$, we have

$$\mathscr{E}_{el} = VG\mathscr{E}_{f0}^{5/2}\left[\frac{2}{5} + \frac{\pi^2}{6}\left(\frac{kT}{\mathscr{E}_{f0}}\right)^2\right] \tag{4.9.5}$$

Since $kT < \mathscr{E}_{f0}$, we neglect the term involving $(kT/\mathscr{E}_{f0})^4$. From Eq. (4.9.5), the electronic heat capacity is

$$C_{el} = \frac{1}{V}\frac{d\mathscr{E}_{el}}{dT} = \frac{2\pi^2k^2m}{h^2}\left(\frac{8\pi^2}{9}n\right)^{1/3}T \tag{4.9.6}$$

A comparison of Eqs. (4.9.6) and (4.9.1) shows that

$$a = \frac{2\pi^2k^2m}{h^2}\left(\frac{8\pi^2}{9}n\right)^{1/3} \qquad \text{J/m}^3\text{-K}^2 \tag{4.9.7}$$

To calculate the thermal conductivity of a metal from its heat capacity, we refer to Fig. 4.23. Consider the motion of electrons across the boundary line BB. The power flux density P/A associated with the electron motion is equal to $v\mathscr{E}_{el}/V$, where v is the electron velocity and A is the cross-sectional area. Therefore, the net power flow per unit area across the boundary is equal to

$$\frac{P}{A} = \frac{1}{2V}v_x[\mathscr{E}_{el}(F) - \mathscr{E}_{el}(D)] \tag{4.9.8}$$

the factor $\frac{1}{2}$ enters in Eq. (4.9.8) because electrons can move either toward or away from the boundary, and only those moving toward the boundary should be counted.

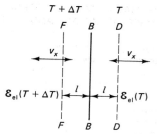

Figure 4.23 Diagram used for the calculation of thermal conduction in a solid.

Note that on the average only electrons within a time τ (mean free time) away from the boundary will be able to move across the boundary without suffering collisions. Therefore, the lines FF and DD in Fig. 4.23 are chosen to be one mean free path $l = v_x\tau$ from the line BB, and

$$\mathscr{E}_{el}(F) - \mathscr{E}_{el}(D) = \frac{d\mathscr{E}_{el}}{dx}(-2l) = \frac{d\mathscr{E}_{el}}{dx}(-2v_x\tau) \qquad (4.9.9)$$

Suppose that a temperature difference ΔT exists between lines FF and DD. We can rewrite Eq. (4.9.9) as

$$\mathscr{E}_{el}(F) - \mathscr{E}_{el}(D) = \frac{d\mathscr{E}_{el}}{dT}\frac{dT}{dx}(-2v_x\tau) \qquad (4.9.10)$$

Substituting Eq. (4.9.10) in Eq. (4.9.8) and using Eq. (4.9.6) for C_{el}, we find that

$$\frac{P}{A} = -C_{el}v_x^2\tau\frac{dT}{dx} = -K\frac{dT}{dx} \qquad (4.9.11)$$

By definition, the thermal conductivity is

$$K = C_{el}v_x^2\tau \qquad (4.9.12)$$

The value of v_x^2 in Eq. (4.9.12) is not arbitrary. From Fig. 4.22 we see that only electrons having energies close to $\mathscr{E}_{f0}$ are affected. In other words, $v_x^2 + v_y^2 + v_z^2 = v_f^2$. Since on the average $v_x^2 = v_y^2 = v_z^2$, we have

$$v_x^2 = \frac{1}{3}v_f^2 = \frac{2}{3}\frac{\mathscr{E}_f}{m} \cong \frac{2}{3}\frac{\mathscr{E}_{f0}}{m} \qquad (4.9.13)$$

Substituting Eqs. (4.9.13) and (4.9.6) in Eq. (4.9.12) and using Eq. (4.6.3) for $\mathscr{E}_{f0}$, we obtain

$$K = \frac{\pi^2k^2n\tau}{3m}T \qquad \text{W/m-K} \qquad (4.9.14)$$

The ratio of the thermal to electrial conductivity in metals obeys the relation

$$\frac{K}{\sigma} = \frac{\pi^2}{3}\left(\frac{k}{e}\right)^2 T \qquad \text{W/S-K} \qquad (4.9.15)$$

TABLE 4.3 THERMAL DATA AND LORENZ NUMBER FOR A SELECTED NUMBER OF METALS AND SEMICONDUCTORS

Metal	Debye temperature, T_D (K)	Electronic heat capacity constant, a (10^{-3} J/mol–K^2)	Lorenz number at 0°C, $L(10^{-8}$ W–Ω/K^2)
Li	363–400	1.63	
Na	150–160	1.38	2.15
K	90–100	2.08	
Rb	52–68	2.41	
Cs	40–54	3.20	
Cu	310–343	0.695	2.23
Ag	210–226	0.646	2.31
Au	162–185	0.729	2.35
Zn	190–310	0.64	2.31
Cd	120–300	0.688	2.42
Hg	69–100	1.79	
Al	380–428	1.35	2.19
Ga	125–320	0.597	
In	108–129	1.69	
Si	625–640	—	
Ge	360–370	—	
Fe	420–467	4.98	2.47
Co	385–445	3.73	

Source: Data from C. Kittel, *Introduction to Solid-State Physics,* 3rd ed., John Wiley & Sons, Inc., New York, 1966 and F. J. Blatt, *Physics of Electronic Conduction in Solids,* McGraw-Hill Book Company, New York, 1968.

which is known as the *Wiedemann–Franz law* and can be obtained from Eqs. (4.9.14), (4.2.3), and (4.8.11). The Lorenz number L is defined as

$$L = \frac{K}{\sigma T} \qquad \text{W}\Omega/\text{K}^2 \qquad (4.9.16)$$

which has a theoretical value of 2.45×10^{-8} W–Ω/K^2 from Eq. (4.9.15).

In Table 4.3 we list the thermal data and the Lorenz number for a selected number of metals and semiconductors. A general agreement between the observed and theoretical values of L is obtained. We should point out that agreement is expected only in metals, where the heat capacity is dominated by the electronic part in Eq. (4.9.1). According to Debye's theory, the lattice specific heat, the bT^3 term in Eq. (4.9.1), is equal to $2.4\pi^4 R(T/T_D)^3$ for $T \ll T_D$, where R is the gas constant and T_D is the Debye temperature. The range of T_D listed in the table represents the range of values measured at different temperatures and by different methods.

PROBLEMS

4.1. Consider two identical *LC* circuits coupled by a mutual inductance M. Verify that the two resonant frequencies of the coupled circuits are given by Eq. (4.1.1). Also show that the currents I_1 and I_2 in the two circuits are in phase ($I_1 = I_2$) for one frequency and are out of phase ($I_1 = -I_2$) for the other frequency. Finally, evaluate the magnetic energy stored in the circuits.

4.2. **(a)** Consider three coupled LC circuits with mutual inductance M between circuit 1 and circuit 2 and between circuit 2 and circuit 3, as shown in Fig. P4.2. Show that the characteristic frequency equation is

$$x^3 - 2xy^2 = 0$$

where $x = 1 - \omega^2 LC$ and $y = \omega^2 MC$.

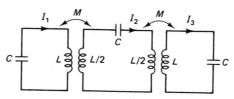

Figure P4.2

(b) Find the three resonant frequencies of the coupled circuits. Also evaluate the ratio $I_1:I_2:I_3$ at the three frequencies from the cofactors of the determinant.

4.3. Now we generalize Problem 4.2 to N coupled circuits. Show that the characteristic equation is given by

$$\begin{vmatrix} x & y & 0 & \cdots & 0 & 0 & 0 \\ y & x & y & \cdots & 0 & 0 & 0 \\ 0 & y & x & \cdots & 0 & 0 & 0 \\ \cdot & \cdot & \cdot & \cdot & \cdots & & \\ & & & \cdots & y & x & y \\ 0 & 0 & 0 & \cdots & 0 & y & x \end{vmatrix} = 0$$

Its solution can be found by noting that

$$D_N = xD_{N-1} - y^2 D_{N-2}$$

where D_N is the determinant of rank N. Also note that $D_0 = 1$ and $D_1 = x$. Next, we propose a solution of the form $D_N = z^N$. By letting $x = 2y \cos \theta$, show that the solution for D_N is of the form

$$D_N = ay^N \exp (jN\theta) + by^N \exp (-jN\theta)$$

Then we use the condition $D_0 = 1$ and $D_1 = x$ to express a and b in terms of x and y. Finally, we proceed to find the proper values for ω by setting $D_N = 0$. Express ω in terms of $\omega_0 = (LC)^{1/2}$ and $\omega_1 = (MC)^{1/2}$. How many ω's are there, and what are the proper values for θ? Show that the currents in the N coupled circuits obey a certain relation

$$I_n = I \sin n\phi$$

Find the relation between ϕ and θ.

4.4. Evaluate the values of θ_j in Eq. (4.1.5) and draw diagrams similar to Fig. 4.4b for a system of six identical atoms to show the spatial variation of ψ of Eq. (4.1.4) for the modes with $q = 1, 3,$ and 4.

4.5. Use energy bands to explain the difference in the electrical properties of Na and NaCl. Is it correct to treat energy bands as if the distribution of allowed energy levels were continuous within the bands? Is this feature (the quasi-continuous distribution of allowed energy levels) important to electrical conduction?

4.6. **(a)** Find an expression for the energy of an electron in a one-dimensional potential well with infinite sides and $L = 1$ cm (Fig. 4.13b).

(b) Suppose that the topmost electron has an energy of 3.1 eV as in sodium. Find the value of q in Eq. (4.3.10).

(c) Calculate the energy separation (in eV) between the topmost level and the level just below it. Also find the energy difference between the two levels q and $q - 1$ with $q = 10^4$. For comparison, calculate the thermal energy $\mathscr{E}_T = kT$ (in eV) at $T = 1$ K.

4.7. (a) Consider the motion of an electron in a three-dimensional well with infinite sides (Fig. 4.13b). The dimension of the well is L^3 (Fig. 4.15) with $L = 1$ cm. Find an expression for the energy of the electron.

(b) Calculate the energy (in eV) for the following sets of numbers (q_1, q_2, q_3): (1, 0, 0), (0, 1, 0), (0, 0, 1), (1, 1, 0), (1, 1, 1), and (2, 0, 0). The first three states have the same energy. Can three electrons occupy these three states simultaneously? Is this possibility forbidden by the exclusion principle?

4.8. (a) Find an expression for the density of states existing in a spherical shell with an inner radius k and an outer radius $k + dk$ in k space.

(b) Since the total number of available quantum states dN per unit volume is $dN = D(\mathscr{E})$ $d\mathscr{E} = D(k) \, dk$, the two density-of-state functions $D(\mathscr{E})$ and $D(k)$ are related. Obtain $D(k)$ from $D(\mathscr{E})$ of Eq. (4.15) through use of the relation

$$D(k) = D(\mathscr{E}) \frac{d\mathscr{E}}{dk}$$

Explain the difference between $D(\mathscr{E})$ thus obtained and Eq. (4.4.19).

(c) Sketch $D(\mathscr{E})$ and $D(k)$ as functions of $\mathscr{E}$ and k, respectively.

4.9. (a) Show that for an infinitesimal energy change $\Delta\mathscr{E}$, Eq. (4.5.5) can be satisfied only if the following relation holds,

$$\frac{1}{p} \frac{dp}{d\mathscr{E}} = \alpha$$

where α is a constant. Find the explicit dependence of p on $\mathscr{E}$.

(b) The two important features in the exponent of Eq. (4.5.6) are the negative sign in front of $\mathscr{E}$ and the dependence of $1/T$. Apply Eq. (4.5.6) to some natural phenomena with which you are familiar. Use intuitive arguments to show that Eq. (4.5.6) is consistent with common facts.

4.10. Discuss the essential differences between Eqs. (4.5.7) and (4.5.5). Show that Eq. (4.5.7) can be satisfied if Eqs. (4.5.1) and (4.5.6) are true.

4.11. Derive a rate equation, similar to Eq. (4.5.7), for collision processes (Fig. 4.18) in which all the particles involved in the processes obey the Pauli exclusion principle. Show that the rate equation can be satisfied if Eq. (4.5.1) is true.

4.12. In Problem 4.11 we obtained the following equation for the probability function for Fermi particles:

$$\frac{f(\mathscr{E}_1)}{1 - f(\mathscr{E}_1)} \frac{f(\mathscr{E}_4)}{1 - f(\mathscr{E}_4)} = \frac{f(\mathscr{E}_2)}{1 - f(\mathscr{E}_2)} \frac{f(\mathscr{E}_3)}{1 - f(\mathscr{E}_3)}$$

in an elastic collision $\mathscr{E}_1 + \mathscr{E}_4 = \mathscr{E}_2 + \mathscr{E}_3$. By letting $\mathscr{E}_4 = \mathscr{E}_2 + \Delta\mathscr{E}$ and $\mathscr{E}_3 = \mathscr{E}_1 + \Delta\mathscr{E}$, show that

$$\frac{1 - f}{f} \left(\frac{d}{d\mathscr{E}} \right) \frac{f}{1 - f} = -\alpha$$

and that the solution for f is given by

$$f = \frac{1}{1 + \exp{(\alpha \mathscr{E} + \beta)}}$$

4.13. Plot the Fermi–Dirac function $f(\mathscr{E})$ as a function of energy at four different temperatures: $T = 0$ K, $T = 4.2$ K (liquid helium temperature), $T = 77$ K (liquid nitrogen temperature), and $T = 300$ K. Find the value of $\Delta\mathscr{E}$ such that $f(\mathscr{E})$ drops from 0.99 at $\mathscr{E}_f - \Delta\mathscr{E}$ to 0.01 at $\mathscr{E}_f + \Delta\mathscr{E}$ at the various temperatures. What is the physical meaning of $f(\mathscr{E})$?

4.14. Show that the atmospheric pressure p obeys the relation

$$p = N_0\, kT \exp\left[\frac{-Mg(z - z_0)}{kT}\right]$$

where N_0 is the density of molecules (in m^{-3}) at sea level, M the molecular mass, g the acceleration of gravity, and $z - z_0$ the elevation. The gaseous molecules obey the Boltzmann distribution law.

4.15. Apply the Boltzmann distribution law to an ideal gas. The molecules are assumed to possess translational kinetic energy only. The number of states in the velocity space is proportional to $4\pi v^2\, dv$. Show that the average energy $\langle\mathscr{E}\rangle = 3kT/2$.

4.16. The average kinetic energy is given by

$$\langle KE \rangle = \frac{\int (\hbar^2 k^2/2m^*)\, dn}{\int dn}$$

where

$$dn = 2\, \frac{4\pi k^2\, dk}{(2\pi)^3} \frac{1}{1 + \exp{(\mathscr{E} - \mathscr{E}_f)/kT}}$$

Find $\langle KE \rangle$ of electrons in degenerate semiconductors and metals in terms of the Fermi energy $\mathscr{E}_f$ at $T = 0$ K. In both cases, take $\mathscr{E} = 0$ at the bottom of the band.

4.17. For nondegenerate semiconductors, $(\mathscr{E}_f < \mathscr{E}_c)$ the Fermi function can be approximated by

$$f \cong \exp\left(-\frac{\mathscr{E} - \mathscr{E}_f}{kT}\right)$$

Show that $\langle KE \rangle = \dfrac{3}{2}\, kT$.

4.18. Find the value of the Fermi energy in Na and Cu. Sodium has a bcc structure with a lattice constant 4.28 Å, whereas copper has a fcc structure with a lattice constant 3.61 Å. It is assumed that each Na or Cu atom contributes one valence electron to the conduction band.

4.19. From the definition of the electrochemical potential, show that $g_2 - g_1 = \mathscr{E}_{f2} - \mathscr{E}_{f1}$, using Eqs. (A2.23) and (A2.28). Find the contact potential between silver and platinum.

4.20. Check the dimension and find the value of the constant A in Eq. (4.7.10). The value of thermionic current from tungsten is found to be 3×10^3 A/m^2 at a cathode temperature $T = 2500$ K and 9×10^{-7} A/m^2 at $T = 1250$ K. Find the work function of tungsten.

4.21. It is apparent from Eq. (4.7.9) that a cathode having a lower work function would yield a much larger thermionic current than a cathode with a higher work function. The value of $\mathscr{E}_w$ is lowered from 4.54 eV to 1.36 eV when cesium is deposited on tungsten. Find the ratio of the thermionic currents in the two cases. The cathode temperature is assumed to be 1500 K. Find in each case the probability that the state with $\mathscr{E} = \mathscr{E}_b$ (Fig. 4.20) is occupied by an electron.

4.22. Derive Eq. (4.8.7) from Eq. (4.8.5). Also show that the mean free time $\langle t_f \rangle$ is equal to τ.

4.23. (a) Derive Eq. (4.8.14) from Eq. (4.8.13). Show that Eq. (4.8.15) yields the same result as that expressed in Eq. (4.8.14).

(b) Given $\mu = 8000$ cm^2/V-s and $m^* = 0.067m_0$ for electrons in GaAs, calculate the value of τ. Also find the value of $\sigma + i\omega\epsilon$ in a GaAs sample with σ_0 (dc) $= 1$ $(\Omega -$ cm$)^{-1}$ and $\epsilon = 13.2\epsilon_0$ at frequency $\nu = 10^{10}$ Hz.

4.24. The following values are known for copper:

$$\sigma = \text{conductivity} = 6 \times 10^7 \; (\Omega\text{-m})^{-1}$$
$$n = \text{electron concentration} = 8.5 \times 10^{28} \; \text{m}^{-3}$$
$$l = \text{mean free path} = 450 \; \text{Å}$$

Find the mobility value μ and the mean free time τ from σ and n. Also find the mean free time τ from $l = v_f\tau$, where $v_f = \sqrt{2\mathscr{E}_f/m}$ is the Fermi velocity.

4.25. Consider an optical experiment in which a light beam is incident on a solid. At the air–solid boundary, reflection of the beam takes place. The ratio of the electric field of the reflected beam to that of the incident beam is

$$\frac{E_r}{E_i} = \frac{\sqrt{\epsilon} - \sqrt{\epsilon_0}}{\sqrt{\epsilon} + \sqrt{\epsilon_0}}$$

where ϵ is the effective dielectric constant of the solid. For a solid possessing free electrons,

$$\epsilon = \epsilon_0 - \epsilon_e = \epsilon_0 - \epsilon_0 \frac{\omega_p^2}{\omega^2}$$

where ϵ_0 is the dielectric constant of free space. For a negative ϵ, the magnitudes of E_r and E_i are equal, and only their phases differ. Therefore, the light beam is totally reflected at the boundary and no transmitted light is observed. For a positive ϵ, the light beam is partly reflected and partly transmitted. Calculate the wavelength at which the transition from a total reflection to a partial transmission takes place in sodium, given that the free electron concentration is $n = 2.5 \times 10^{28}$ m^{-3} in Na.

4.26. (a) The average energy $\langle \mathscr{E} \rangle$ associated with the translational motion of particles is $3kT/2$ according to the calculation in Problem 4.15, which is based on the Boltzmann statistics. Find the heat capacity (expressed in J/mol-deg) of an ideal gas. Can we use this value for electrons in a metal?

(b) Calculate the heat capacity (in J/mol-deg) of sodium at 4 K. The measured value of C_{el}/T is 1.4×10^{-3} J/mol-deg^2. Given: $N/V = 2.5 \times 10^{28}$ atoms/m^3 in Na, and Avogadro's number is 6.025×10^{23} atoms (or molecules)/mol.

4.27. (a) In semimetals, semiconductors, and insulators, the thermal conductivity due to the lattice may be comparable to or greater than that due to conduction electrons. Do we expect the Wiedemann–Franz law to hold in these materials?

(b) The resisitivity of a metal may be expressed as

$$\rho = \rho_l + \rho_i$$

where ρ_l is the part caused by scattering of electron motion by lattice vibration and ρ_i is the part caused by scattering of electron motion by impurity atoms. We also know that ρ_i is independent of temperature and

$$\rho_l \sim T \quad \text{for } T > T_D, \qquad \rho_l \sim T^5 \quad \text{for } T < T_D$$

where T_D is the Debye temperature (Table 4.3). Sketch the temperature variations of resistivity and thermal conductivity.

(c) Deviations from the Wiedemann–Franz law have been found in the intermediate-temperature region. What assumptions are made in obtaining Eq. (4.9.15) from Eqs. (4.9.14) and (4.2.3)? Give a possible reason or reasons why Eq. (4.9.15) may be in error.

5

The Band Theory of Electronic Conduction

5.1 INTRODUCTION

The free-electron theory of metals introduced in Chapter 4 presents a simple physical model which is useful for an understanding of some physical phenomena, such as electrical and thermal conduction. The free-electron theory, however useful it may be, is open to critical examination. One of the most fundamental questions arising from this examination concerns the magnitude of k, the wave number associated with the electron motion. According to Eq. (4.4.3), the energy of an electron increases indefinitely with k. The $\mathscr{E}$–k diagram based on the free-electron theory is shown in Fig. 5.1a.

To refresh our memory, we return to Eq. (4.3.12). The values of k are given by

$$k_x = q_1 \frac{\pi}{L}, \qquad k_y = q_2 \frac{\pi}{L}, \qquad k_z = q_3 \frac{\pi}{L} \qquad (5.1.1)$$

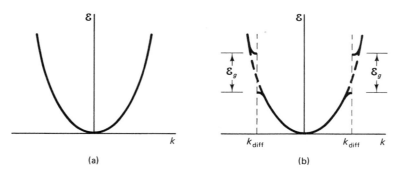

(a) (b)

Figure 5.1 Plot of energy $\mathscr{E}$ versus wave vector k: (a) for free electrons and (b) for electrons in a periodic lattice. One important difference is that reflection of electron motion occurs in a periodic lattice when the Bragg condition is satisfied. At k_{diff} where reflection takes place, the energy curve is no longer continuous and an energy gap is created.

where q_1, q_2, and q_3 are positive integers and L is the linear dimension of a solid of cubic shape (Fig. 4.15). The fundamental question is whether there exists an upper limit for the value of q_1, q_2, and q_3. In the free-electron theory, we never raised this question. To answer this question, we refer to our discussion in Sections 2.7 and 2.8. The relevant point is that electron motion has wavelike properties. Like other waves, an electron wave obeys the Bragg law of diffraction. This means that at certain specific values of k as determined from Eq. (2.8.5), the electron motion in a solid will be diffracted. In Fig. 5.1b, k_{diff} is the value of k at which the Bragg condition is satisfied. Around k_{diff}, the $\mathscr{E}$–k curve is significantly modified due to strong coupling between the electron wave and its Bragg-scattered wave. As we discuss in Sections 5.2 and 5.7, the energy curve is no longer continuous at k_{diff}, and consequently an energy gap is created.

The fact that there exist in solids bands of allowable energy states separated by a forbidden gap is central to the mechanism of electronic conduction in solids. In Section 4.2 we explained the difference in the electrical behavior of insulators, metals, semiconductors, and semimetals on the basis of energy bands (Fig. 4.11). In this chapter we examine in some detail the origin of the energy gap, and discuss the consequences of energy bands on electronic conduction.

5.2 THE KRONIG–PENNEY MODEL

The free-electron model presented in Section 4.3 and illustrated in Fig. 4.12 is a much oversimplified model. The most important feature of a solid that the model has ignored is the periodicity of the lattice. In this section we discuss the Kronig–Penney model, which takes into account the lattice periodicity. The model is useful because we may be able to see the essential features of the behavior of electrons in a periodic potential.

If we were to draw a line passing through the centers of atoms in a solid, we would see a periodic potential as illustrated in Fig. 5.2a. The potential energy has a singularity at the position of each nucleus. To make our analysis mathematically tract-

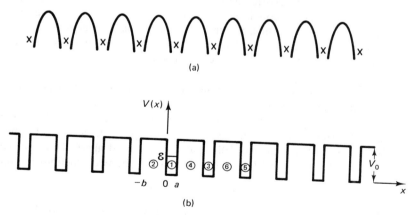

Figure 5.2 (a) The periodic lattice potential in a real crystal. The Xs represent the positions of the nuclei. (b) One-dimensional periodic potential used in the Kronig–Penney model. The central question is whether an electron with energy $\mathscr{E}$ will be able to propagate from one lattice cell to another.

able, we replace the actual potential by an approximate one as shown in Fig. 5.2b. Further, the lattice is taken to be one-dimensional. This one-dimensional, rectangular potential was used by Kronig and Penney in their analysis (R. de L. Kronig and W. G. Penney, *Proc. R. Soc. London*, Vol. A130, p. 499, 1930).

Consider an electron with energy $\mathscr{E}$ in the periodic lattice of Fig. 5.2b. The Schrödinger equation reads

$$\frac{\hbar^2}{2m}\frac{d^2\psi}{dx^2} + [\mathscr{E} - V(x)]\psi = 0 \tag{5.2.1}$$

In the region $0 < x < a$, $V(x) = 0$; hence the solution takes the oscillatory form

$$\psi_1(x) = A \exp(i\beta x) + B \exp(-i\beta x) \tag{5.2.2}$$

In the region $-b < x < 0$, $V(x) = V_0$ and $\mathscr{E} - V_0 < 0$; hence the solution takes the exponential form

$$\psi_2(x) = C \exp(\alpha x) + D \exp(-\alpha x) \tag{5.2.3}$$

In Eqs. (5.2.2) and (5.2.3), the two constants β and α are

$$\beta = \frac{(2m\mathscr{E})^{1/2}}{\hbar}, \qquad \alpha = \frac{[2m(V_0 - \mathscr{E})]^{1/2}}{\hbar} \tag{5.2.4}$$

The coefficients A, B, C, and D are related to one another through the boundary conditions. Since the quantity $\mathscr{E} - V(x)$ is finite at the boundary, the derivative $d\psi/dx$ and the function ψ itself must be continuous at the boundary $x = 0$. In other words,

$$A + B = C + D \tag{5.2.5}$$

$$i\beta(A - B) = \alpha(C - D) \tag{5.2.6}$$

Note that we have only two equations but four unknown coefficients A, B, C, and D. To find the characteristic energy equation, we need two additional equations. Applying the other boundary conditions of region 1 at $x = a$ requires the use of ψ_4 in region 4 ($a < x < a + b$), which has the same functional dependences on x but introduces two new unknown coefficients. Therefore, it is imperative that a proper relation between ψ_4 and ψ_2 be found so that the new coefficient in ψ_4 can be expressed in terms of the old coefficients in ψ_2.

An important clue to finding this relation comes from Eq. (4.1.3). In a periodic structure, the cell function as represented by ψ_j in Eq. (4.1.3) is the same from cell to cell, but the phase factor $\exp(i\theta_j)$ is different for each cell. Therefore, $\psi_4(x)$ in cell 4 should differ from $\psi_2(x)$ in cell 2 only by a phase factor. We let the phase difference be $k(a + b)$ between two cells spaced one period apart. Thus we have, for example, $\psi_4(a) = \psi_2(-b) \exp[ik(a + b)]$. In general, the two functions are related by

$$\psi_4(x_4) = \psi_2(x_2) \exp[ik(a + b)] \tag{5.2.7}$$

if the two coordinates x_4 and x_2 are separated by one period, that is, $x_4 - x_2 = a + b$. The boundary conditions at $x = a$ thus become

$$A \exp(i\beta a) + B \exp(-i\beta a) =$$
$$[C \exp(-\alpha b) + D \exp(\alpha b)] \exp[ik(a + b)] \tag{5.2.8}$$

$$i\beta[A \exp (i\beta a) - B \exp (-i\beta a)] =$$
$$\alpha[C \exp (-\alpha b) - D \exp (\alpha b)] \exp [ik(a + b)] \qquad (5.2.9)$$

For a nontrivial solution, the determinant of the coefficients must be zero, that is,

$$\begin{vmatrix} 1 & 1 & -1 & -1 \\ i\beta & -i\beta & -\alpha & \alpha \\ F & F^{-1} & -G^{-1}H & -GH \\ i\beta F & -i\beta F^{-1} & -\alpha G^{-1}H & \alpha GH \end{vmatrix} = 0 \qquad (5.2.10)$$

In Eq. (5.2.10), the three quantities F, G, and H are

$$F = \exp (i\beta a), \qquad G = \exp (\alpha b), \qquad H = \exp [ik(a + b)] \qquad (5.2.11)$$

Next we form a new determinant by adding and subtracting the columns in the following manner:

$$I' = -i(I - II) \qquad II' = I + II$$
$$III' = III + IV \qquad IV' = IV - III \qquad (5.2.12)$$

where the primed and unprimed roman numerals represent, respectively, the columns of the new and the old determinant. Following this procedure, we obtain

$$\begin{vmatrix} 0 & 2 & -2 & 0 \\ 2\beta & 0 & 0 & 2\alpha \\ -i(F - F^{-1}) & F + F^{-1} & -H(G + G^{-1}) & -H(G - G^{-1}) \\ \beta(F + F^{-1}) & i\beta(F - F^{-1}) & \alpha H(G - G^{-1}) & \alpha H(G + G^{-1}) \end{vmatrix} = 0 \qquad (5.2.13)$$

$$(5.2.13)$$

Expanding Eq. (5.2.13) and collecting the terms with α^2, β^2, and $\alpha\beta$, we find

$$(\alpha^2 - \beta^2)K + 2\alpha\beta L = 0 \qquad (5.2.14)$$

where the two coefficients K and L are

$$K = 2i(F - F^{-1})H(G - G^{-1}) \qquad (5.2.15)$$
$$L = 4(H^2 + 1) - 2H(F + F^{-1})(G + G^{-1}) \qquad (5.2.16)$$

Dividing Eqs. (5.2.15) and (5.2.16) by H and using Eq. (5.2.11), we have

$$\frac{K}{H} = -8 \sin \beta a \sinh \alpha b \qquad (5.2.17)$$

$$\frac{L}{H} = 8 \cos [k(a + b)] - 8 \cos \beta a \cosh \alpha b \qquad (5.2.18)$$

Using Eqs. (5.2.17) and (5.2.18) in Eq. (5.2.14), we obtain

$$\frac{\alpha^2 - \beta^2}{2\alpha\beta} \sinh \alpha b \sin \beta a + \cosh \alpha b \cos \beta a = \cos [k(a + b)] \qquad (5.2.19)$$

In order that we may derive some meaningful results from Eq. (5.2.19), the following assumptions are made. We let $V_0 \to \infty$ and $b \to 0$ in such a way as to maintain a finite value of $\alpha^2 b = 2P/a$, P being a dimensionless quantity. Note that $\alpha^2 b^2 = 2Pb/a$ < 1. Thus we may approximate sinh αb by αb and cosh αb by 1, and Eq. (5.2.19) becomes

$$P \frac{\sin \beta a}{\beta a} + \cos \beta a = \cos ka \qquad (5.2.20)$$

The left side of Eq. (5.2.20) is plotted as a function of βa in Fig. 5.3 for $P = 3\pi/2$. Since $|\cos ka| \leq 1$, only those values of βa are allowed for which the left-hand side of Eq. (5.2.20) falls between $+1$ and -1. Within this range, which is marked with heavy lines in Fig. 5.3, for one value of βa there is a corresponding value of ka. Since β is directly related to the energy $\mathscr{E}$ of the electron, a plot of $\mathscr{E}$ against ka can be obtained from Fig. 5.3 and is shown in Fig. 5.4.

The important question to which we hope to find an answer from the analysis is whether an electron will be able to move to neighboring cells, say cells 3 and 5, if it starts out in one of the cells, say cell 1 (Fig. 5.2b). Since $|\psi(x)|^2 \, dx$ represents the probability of finding the electron with coordinate between x and $x + dx$, the question now becomes whether it is possible to find a nondecaying or propagating mode for $\psi(x)$. The answer depends on the value of k. For a real k, the wave function ψ changes from one cell to another only by a phase factor, but its amplitude remains the same. In other words, an electron can propagate from one cell to another without suffering an amplitude change. This situation corresponds to a propagating (or allowed) mode of the solid. On the other hand, if k is imaginary, say $k = i\gamma$, the amplitude of $\psi(x)$ decays by a factor exp $(-\gamma a)$ from one cell to a neighboring cell. This situation corresponds to a nonpropagating (or forbidden) mode.

The regions AB, CD, and EF in Fig. 5.4 are the allowed energy bands because electrons having energies in these ranges can propagate from one cell to another quite freely. On the other hand, electrons having energies in the range BC, DE, and so on (Fig. 5.3), must have an imaginary k in order to satisfy Eq. (5.2.20). Note that for

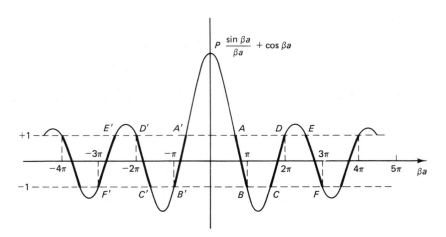

Figure 5.3 Plot of the function $Px^{-1} \sin x + \cos x$ where $x = \beta a$. The allowed values of energy are given by those ranges (marked by heavy lines on the curve) of β $= (2m\mathscr{E}/\hbar^2)^{1/2}$ for which the function lies between $+1$ and -1.

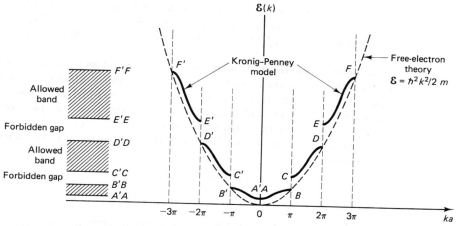

Figure 5.4 Plot of energy versus wave number for the Kronig–Penney model. The solid curves represent the allowed energy bands *AB*, *CD*, and *EF*. The energy intervals *BC* and *DE* are the forbidden zones. For comparison the $\mathscr{E}$–k relation based on the free-electron model is drawn as the dashed curve.

$k = i\gamma$, $\cos ka$ becomes $\cosh \gamma a$ whose value exceeds 1. This means that in the energy ranges *BC* and *DE* the electron wave motion is highly attenuated; therefore, these energy ranges are referred to as the *forbidden zones* where only localized states can exist.

5.3 THE BLOCH WAVE

In our discussion of the Kronig–Penney model in Section 5.2, we introduced a phase factor exp $[ik(a + b)]$ to account for the phase difference of the wave function in different crystal cells. This phase factor can be incorporated directly in the wave function. In other words, we can write the solution of Eq. (5.2.1) as

$$\psi(x, k) = u(x, k) \exp (ikx) \tag{5.3.1}$$

where the function $u(x, k)$ is equal to $\psi_1(x)$ of Eq. (5.2.2) for $0 < x < a$ and $u(x, k) = \psi_2(x)$ of Eq. (5.2.3) for $-b < x < 0$. Since the coefficients *A*, *B*, *C*, and *D* in Eqs. (5.2.2) and (5.2.3) depend on the value of k, the function $u(x, k)$ is a function of both x and k. Furthermore, the functions $\psi_1(x)$ and $\psi_2(x)$ are periodic in terms of the cell dimension $(a + b)$ and so is the function $u(x, k)$.

The solution of the type expressed in Eq. (5.3.1) is applicable not only to the Kronig–Penney model but to a general class of problems concerning electron motion in a one-dimensional periodic potential. The stipulation for its applicability is that the potential function in Eq. (5.2.1) must be periodic, that is,

$$V(x + c) = V(x) \tag{5.3.2}$$

where $c = a + b$ is the period. Under this condition, Eq. (5.2.1) has a solution of the form shown in Eq. (5.3.1) with

$$u(x + c) = u(x) \tag{5.3.3}$$

This theorem is known as the *Bloch theorem* in solid-state physics (F. Bloch, *Z. Physik*, Vol. 52, p. 555, 1928) and as the *Floquet theorem* in mathematics (see, for example, A. H. Wilson, *Theory of Metals*, Cambridge University Press, Cambridge, 1953, p. 21).

The proof of the Bloch theorem may proceed as follows. (An alternative proof is given in Problem 5.5.) Since Eq. (5.2.1) is a second-order differential equation, it has two independent solutions. Let the two solutions be $f(x)$ and $g(x)$. For simplicity in arguments, we choose $f(x)$ and $g(x)$ to be real functions. For example, we use $\sin \beta x$ and $\cos \beta x$ instead of $\exp(i\beta x)$ and $\exp(-i\beta x)$. Since all other solutions of Eq. (5.2.1) can be expressed as a linear combination of the two independent solutions, we must have the relations

$$f(x + c) = \alpha_1 f(x) + \alpha_2 g(x) \tag{5.3.4a}$$

$$g(x + c) = \beta_1 f(x) + \beta_2 g(x) \tag{5.3.4b}$$

where c is the period and $\alpha_{1,2}$ and $\beta_{1,2}$ are expansion coefficients totally unrelated to α and β of Eq. (5.2.4).

The general solution of Eq. (5.2.1) is

$$\psi(x) = Af(x) + Bg(x) \tag{5.3.5}$$

where A and B are two arbitrary constants. What is of interest to us is the relation between $\psi(x + c)$ and $\psi(x)$. We let

$$\psi(x + c) = \lambda \, \psi(x) = A\lambda f(x) + B\lambda g(x) \tag{5.3.6}$$

where λ is a quantity to be determined. According to Eqs. (5.3.4) and (5.3.5), we have

$$\psi(x + c) = Af(x + c) + Bg(x + c)$$
$$= (A\alpha_1 + B\beta_1)f(x) + (A\alpha_2 + B\beta_2)g(x) \tag{5.3.7}$$

Comparing Eqs. (5.3.6) and (5.3.7), we obtain

$$A\alpha_1 + B\beta_1 = A\lambda \tag{5.3.8}$$
$$A\alpha_2 + B\beta_2 = B\lambda$$

For nontrivial solutions of A and B, the determinant must be zero, yielding

$$\lambda^2 - (\alpha_1 + \beta_2)\lambda + \alpha_1\beta_2 - \alpha_2\beta_1 = 0 \tag{5.3.9}$$

To see what values of λ are permissible, we examine the function $fg' - gf'$. Note that

$$\frac{d}{dx}(fg' - gf') = f\frac{d^2g}{dx^2} - g\frac{d^2f}{dx^2}$$
$$= \frac{2m}{\hbar^2}[-f(\mathcal{E} - V)g + g(\mathcal{E} - V)f] = 0 \tag{5.3.10}$$

through the use of Eq. (5.2.1). From Eq. (5.3.10) we conclude that

$$fg' - gf' = \text{constant} \tag{5.3.11}$$

From Eq. (5.3.4) we have

$$f(x + c)g'(x + c) - g(x + c)f'(x + c)$$

$$= (\alpha_1\beta_2 - \alpha_2\beta_1)[f(x)g'(x) - g(x)f'(x)] \qquad (5.3.12)$$

In view of the constancy of $(fg' - gf')$ expressed in Eq. (5.3.11), we must have

$$\alpha_1\beta_2 - \alpha_2\beta_1 = 1 \qquad (5.3.13)$$

so that $fg' - gf'$ is the same at x and $x + c$.

The condition stated in Eq. (5.3.13) imposes a relation between two solutions λ_1 and λ_2 of Eq. (5.3.9). This relation is

$$\lambda_1\lambda_2 = 1 \qquad (5.3.14)$$

Furthermore, from Eq. (5.3.9),

$$\lambda_1 + \lambda_2 = \alpha_1 + \beta_2 = \text{real} \qquad (5.3.15)$$

because we have chosen the functions of f and g in Eq. (5.3.4) to be real. The conditions stated in Eqs. (5.3.14) and (5.3.15) are both satisfied if

$$\lambda_1 = \exp(ikc), \qquad \lambda_2 = \exp(-ikc) \qquad (5.3.16)$$

where ik is a purely imaginary number, or if

$$\lambda_1 = \exp(\gamma c), \qquad \lambda_2 = \exp(-\gamma c) \qquad (5.3.17)$$

where γ is a real number. If Eq. (5.3.17) is used, the solutions will become unbounded at either $+\infty$ or $-\infty$. This corresponds to the situation in the forbidden zones. On the other hand, the solutions will remain finite if Eq. (5.3.16) is used.

Using λ_1 of Eq. (5.3.16) in Eq. (5.3.6), we obtain

$$\psi(x + c) = \psi(x) \exp(ikc) \qquad (5.3.18)$$

Let us now examine whether the solution stated in Eq. (5.3.1) is consistent with Eq. (5.3.18). We find that

$$\psi(x + c, k) = u(x + c, k) \exp[ik(x + c)] = \psi(x, k) \exp(ikc) \qquad (5.3.19)$$

if Eq. (5.3.3) is satisfied. This proves the Bloch theorem. We also note that the relation stated in Eq. (5.2.7) is the same as Eq. (5.3.18) with $a + b = c$. The reason for choosing Eq. (5.2.7) in the treatment of the Kronig–Penney model is that we have anticipated what is the correct form to use.

Summarizing the foregoing discussion, we have shown that in the allowed bands, the solution of the Schrödinger equation for a one-dimensional periodic potential takes the form of Eq. (5.3.1). Since the cell function $u(x, k)$ is the same from one cell to another, any description about the motion of an electron must be contained in the phase factor $\exp(ikx)$. Furthermore, the probability dw of finding an electron in any cell between x and $x + dx$ is

$$dw = |\psi(x)|^2 dx = |\psi(x + c)|^2 dx \qquad (5.3.20)$$

and is independent of the cell. This means that the electron motion is not confined to any particular cell, but spreads over the whole crystal. Obviously, the solution of the type given in Eq. (5.3.1) is applicable only to materials where electrons are relatively free. Such is the case with electrons in metals and semiconductors.

In a three-dimensional lattice, a generalization of Eq. (5.3.1) leads to a solution of the form

$$\psi(\mathbf{r}, \mathbf{k}) = u(\mathbf{r}, \mathbf{k}) \exp(i\mathbf{k} \cdot \mathbf{r}) \qquad (5.3.21)$$

We should point out that Eq. (5.2.1) is the time-independent Schrödinger equation. To incorporate the time dependence in Eq. (5.2.1), we have the time-dependent Schrödinger equation

$$-\frac{\hbar^2}{2m}\nabla^2\psi + V(\mathbf{r})\psi = i\hbar\frac{\partial\psi}{\partial t} \qquad (5.3.22)$$

Equation (5.3.22) reduces to Eq. (5.2.1) if we let

$$\psi(\mathbf{r}, t) = \psi(\mathbf{r}, \mathbf{k})\exp\left(\frac{-i\mathscr{E}t}{\hbar}\right) \qquad (5.3.23)$$

Substituting Eq. (5.3.21) in Eq. (5.3.23), we have

$$\psi(\mathbf{r}, t) = u(\mathbf{r}, \mathbf{k})\exp[i(\mathbf{k} \cdot \mathbf{r} - \omega t)] \qquad (5.3.24)$$

where $\omega = \mathscr{E}/\hbar$. Equation (5.3.24) represents an electron wave propagating in $\mathbf{k}$ direction. If λ_2 is chosen, we have the sign of the $\mathbf{k} \cdot \mathbf{r}$ term changed in the exponential factor. This represents a wave propagating in the opposite direction. The wave represented in Eq. (5.3.24) with either $+\mathbf{k}$ or $-\mathbf{k}$ is called a Bloch wave. All one-electron wave functions in an ideal crystal are of the Bloch form.

5.4 WAVE PACKET AND GROUP VELOCITY

Consider a Bloch wave propagating in the x direction,

$$\psi(x, k) = u(x, k) \exp[i(kx - \omega t)] \qquad (5.4.1)$$

with $\omega = \mathscr{E}/\hbar$. If a wave is to be associated with an electron, it seems reasonable to demand that the wavelike property be confined to the vicinity of the electron. The monoenergetic wave represented by Eq. (5.4.1) does not have this property. A wave packet, in contrast to a monoenergetic wave, embodies both the wavelike and particle-like properties of an electron. In this section we introduce the concept of wave packet and discuss its properties in connection with electronic conduction.

In circuit analysis we are familiar with the fact that a short pulse in the time t domain can be expressed in terms of a distribution, not a single frequency, in the frequency ν domain through the Fourier transform integral as illustrated in Fig. 5.5. In a similar manner, the wave motion of a localized (in space coordinate x) electron is related to a distribution function $\phi(k)$ in the k space by the transform

$$\psi(x, t) = \sqrt{\frac{\hbar}{2\pi}}\int_{-\infty}^{\infty}\phi(k)\exp[i(kx - \omega t)]\,dk \qquad (5.4.2)$$

To see the important concepts conveyed in Eq. (5.4.2), we use a simplified example by assuming $\phi(k)$ to be real. We assume further that both k and ω are centered around k_0 and ω_0. Thus we let $\omega = \omega_0 + \omega'$ and $k = k_0 + k'$, and Eq. (5.4.2) becomes

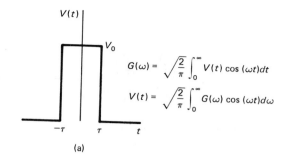

$$G(\omega) = \sqrt{\frac{2}{\pi}} \int_0^\infty V(t) \cos{(\omega t)} dt$$

$$V(t) = \sqrt{\frac{2}{\pi}} \int_0^\infty G(\omega) \cos{(\omega t)} d\omega$$

(a)

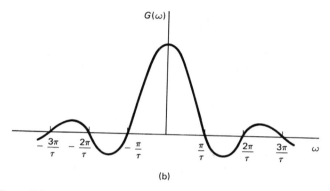

(b)

Figure 5.5 (a) An impulse function $V(t)$ in the time domain and (b) the frequency distribution function $G(\omega)$ in the frequency domain. The two functions are related to each other by the Fourier transform. One important property is that a pulse of finite duration $\Delta\tau$ has a finite bandwidth $\Delta\omega$.

$$\psi(x,\ t)\ =\ \sqrt{\frac{\hbar}{2\pi}}\ \exp{[i(k_0 x\ -\ \omega_0 t)]} \int_{-\infty}^\infty \phi(k') \exp{[i(k'x\ -\ \omega't)]}\ dk' \qquad (5.4.3)$$

At a given time t, the amplitude of $\psi(x,\ t)$ reaches a maximum when $k'x\ -\ \omega't\ =\ 0$ and hence all the frequency components are in phase. Therefore, the wave function ψ of Eq. (5.4.3), unlike that of Eq. (5.4.1), is localized in space (which is a particle-like property). The localization of $\psi(x,\ t)$ is a result of constructive interference among the frequency components $\phi(k')$. Therefore, a collection of waves can be constructed to describe the dynamic behavior of a particle even though each individual component alone behaves like a wave. Such a collection of waves as represented by Eq. (5.4.2) is called a *wave packet*.

In Fig. 5.6 we illustrate schematically the waveform of a wave packet at two particular instants in space. The rapid spatial variation of the wave packet is given by the phase factor $\exp{[i(k_0 x\ -\ \omega_0 t)]}$ and this spatial variation is represented by the solid curve. In the mean time, however, there are phase differences among the various components. The phase differences produce an interference phenomenon, resulting in an amplitude variation of the wave. The envelope of the wave is represented by the dashed

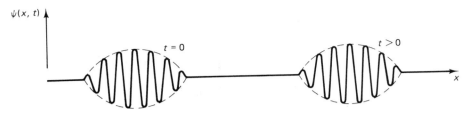

Figure 5.6 Schematic representation of a wave packet at two instants $t = 0$ and $t > 0$. The whole packet moves in space as time progresses.

curve in Fig. 5.6. Because the phase differences change with time, the envelope travels in space as time progresses.

To expound the concept of wave packet, we consider the superposition of two plane waves of equal amplitude but with slightly different frequency and wave number. Mathematically, we have

$$
\begin{aligned}
\psi_1 + \psi_2 &= A \exp\{i[(k + \Delta k)x - (\omega + \Delta\omega)t]\} + \\
&\qquad A \exp\{i[(k - \Delta k)x - (\omega - \Delta\omega)t]\} \\
&= 2A \cos(\Delta k x - \Delta\omega t) \exp[i(kx - \omega t)]
\end{aligned}
\tag{5.4.4}
$$

From Eq. (5.4.4), two velocities can be defined. The *phase velocity* v_p is the speed with which a point moves in space in order to maintain a constant phase angle as the wave propagates. Thus we have

$$
kx - \omega t = \text{constant} \quad \text{or} \quad v_p = \frac{dx}{dt} = \frac{\omega}{k}
\tag{5.4.5}
$$

The envelope of the wave, however, moves with a different velocity called the *group velocity* v_g. In the limit Δk goes to zero,

$$
v_g = \lim_{\Delta k \to 0} \frac{\Delta\omega}{\Delta k} = \frac{d\omega}{dk}
\tag{5.4.6}
$$

To see which velocity we should use for electron velocity, we apply the foregoing analysis to a free electron. According to the theory of relativity, the energy and momentum of an electron with velocity v and rest mass m_0 are given by

$$
\mathscr{E} = \frac{m_0 c^2}{\sqrt{1 - \beta^2}} \quad \text{and} \quad p = \frac{m_0 c \beta}{\sqrt{1 - \beta^2}}
\tag{5.4.7}
$$

where $\beta = v/c$ and c is the velocity of light. The phase velocity of the electron is given by

$$
v_p = \frac{\omega}{k} = \frac{\mathscr{E}}{p} = \frac{c^2}{v}
\tag{5.4.8}
$$

and the group velocity by

$$
v_g = \frac{\partial\mathscr{E}}{\partial p} = \frac{\partial\mathscr{E}}{\partial\beta} \bigg/ \frac{\partial p}{\partial\beta} = v
\tag{5.4.9}
$$

Apparently, the phase velocity which is greater than c for $v < c$ has no physical significance except for the fact that it is the velocity of propagation of points of constant phase On the other hand, the group velocity of a wave packet is actually associated with the velocity of an electron.

In summary, we have shown that a wave packet has the propagation property of a wave and the group behavior of a particle. Further, the motion of an electron is described by the group velocity v_g of Eq. (5.4.6), not by the phase velocity v_p of Eq. (5.4.5).

5.5 EQUATION OF MOTION AND EFFECTIVE MASS

In this section we apply the results of our analyses in Sections 5.3 and 5.4 to the problem of electron motion in a perfect crystal. Let us refer to Eq. (5.3.24). The function $u(\mathbf{r}, \mathbf{k})$ is the same from one lattice cell to another; therefore, the property of wave propagation is contained in the exponential factor $\exp(i(\mathbf{k} \cdot \mathbf{r} - \omega t))$. Since we let $\omega = \mathscr{E}/\hbar$, the group velocity of a wave packet in the jth direction containing electrons with energy $\mathscr{E}$ and wave vector $\mathbf{k}$ is given by

$$(v_g)_j = v_j = \frac{1}{\hbar}\frac{\partial \mathscr{E}}{\partial k_j} \qquad (5.5.1)$$

where $j = x, y,$ or z. We should point out that the expectation value of the momentum $\mathbf{p}$ associated with a wave packet is defined quantum mechanically as

$$\mathbf{p} = \int \frac{\hbar}{i} \psi^*(\mathbf{r}, \mathbf{k}) \nabla \psi(r, k)\, dx\, dy\, dz \qquad (5.5.2)$$

It can be shown through mathematical manipulation (Appendix A4) that Eq. (5.5.2) reduces to

$$p_j = mv_j = \frac{m}{\hbar}\frac{\partial \mathscr{E}}{\partial k_j} \qquad (5.5.3)$$

Therefore, Eq. (5.5.1) is rigorous (see, for example, E. Spenke, *Electronic Semiconductors*, McGraw-Hill Book Company, New York, 1958, pp. 201–206; W. Shockley, *Electrons and Holes in Semiconductors*, D. Van Nostrand Company, Inc., New York, 1950, pp. 424–435).

Next we proceed to find the equation of motion for the electron wave packet. Since the state of an electron is represented by a given set of $\mathbf{k}$, any electron motion must be accompanied by a change of state in $\mathbf{k}$ space. Therefore, the equation of motion must involve $d\mathbf{k}/dt$. If $\mathbf{F}$ is the force acting on electrons due to an applied field, the work done on an electron by the force must be equal to the gain in energy by the electron, that is,

$$\sum F_j v_j\, dt = d\mathscr{E} = \sum \frac{\partial \mathscr{E}}{\partial k_j}\, dk_j = \sum \hbar v_j\, dk_j \qquad (5.5.4)$$

through the use of Eq. (5.5.1) for $\partial \mathscr{E}/\partial k_j$. Equating the coefficients of v_j in Eq. (5.5.4), we obtain $F_j\, dt = \hbar\, dk_j$, or vectorially,

$$\frac{d\mathbf{k}}{dt} = \frac{\mathbf{F}}{\hbar} \qquad (5.5.5)$$

Except for an integration constant, the momentum $\mathbf{p}$ of an electron wave packet is

$$\mathbf{p} = \hbar\mathbf{k} \tag{5.5.6}$$

which is called the *crystal momentum* of the electron. In problems where momentum conservation must be considered, the electron momentum is represented by $\mathbf{p}$ of Eq. (5.5.6).

Differentiating Eq. (5.5.1) with respect to time, we have

$$\frac{dv_j}{dt} = \frac{d}{dt}\left(\frac{1}{\hbar}\frac{\partial \mathscr{E}}{\partial k_j}\right) = \frac{1}{\hbar}\sum_i \frac{\partial}{\partial k_i}\left(\frac{\partial \mathscr{E}}{\partial k_j}\right)\frac{dk_i}{dt} \tag{5.5.7}$$

Using Eq. (5.5.5) for $d\mathbf{k}/dt$, we obtain

$$\frac{dv_j}{dt} = \frac{1}{\hbar^2}\sum_i \frac{\partial^2 \mathscr{E}}{\partial k_i\,\partial k_j} F_i \tag{5.5.8}$$

The quantity $F_i^{-1}\,dv_j/dt$ has a dimension of mass^{-1}. Therefore, we let

$$\frac{1}{m_{ij}} = \frac{1}{m_{ji}} = \frac{1}{\hbar^2}\frac{\partial^2 \mathscr{E}}{\partial k_i\,\partial k_j} \tag{5.5.9}$$

The equation of motion then becomes

$$\frac{d}{dt}\begin{bmatrix} v_x \\ v_y \\ v_z \end{bmatrix} = \begin{bmatrix} m_{xx}^{-1} & m_{xy}^{-1} & m_{xz}^{-1} \\ m_{yx}^{-1} & m_{yy}^{-1} & m_{yz}^{-1} \\ m_{zx}^{-1} & m_{zy}^{-1} & m_{zz}^{-1} \end{bmatrix}\begin{bmatrix} F_x \\ F_y \\ F_z \end{bmatrix} \tag{5.5.10}$$

The quantity m_{ij} is called the *effective mass* of an electron wave packet. In general, m_{ij} is a tensorial quantity.

As Fig. 5.4 shows, the energy $\mathscr{E}$ of an electron depends on the wave number k. This energy dependence can be expanded into a Taylor's series around a particular point of interest (k_{x0}, k_{y0}, k_{z0}). In other words,

$$\mathscr{E} = \mathscr{E}_0 + \sum \frac{\hbar^2}{2m_{ii}}(k_i - k_{i0})^2 + \sum \frac{\hbar^2}{m_{ij}}(k_i - k_{i0})(k_j - k_{j0}) \tag{5.5.11}$$

For example, in semiconductors the point of interest $\mathscr{E}_0$ is either at the bottom of the conduction band or at the top of the valence band. As we can see from Eq. (5.5.11), the various effective masses are the coefficients in the Taylor's series expansion. Equation (5.5.11) for a constant $\mathscr{E}$ describes a surface in k space. The constant-energy surface is said to be spherical if

$$m_{xx} = m_{yy} = m_{zz} = m^* \quad \text{and} \quad m_{xy}^{-1} = m_{yz}^{-1} = m_{zx}^{-1} = 0 \tag{5.5.12}$$

5.6 THE BRILLOUIN ZONES

As pointed out in Section 5.1, there exist certain critical values of k at which Bragg diffraction of electron wave occurs. According to Eq. (2.8.5), the Bragg condition for diffraction is

$$2d_{hkl}\sin\theta = n\lambda \tag{5.6.1}$$

where d_{hkl} is the distance between two adjacent planes which belong to the set (hkl). In terms of the wave number $k = 2\pi/\lambda$, Eq. (5.6.1) becomes

$$kd_{jkl} \sin \theta = n\pi \tag{5.6.2}$$

where n is an integer. As defined in Fig. 5.7, the angle θ is the incident angle that the electron motion makes with the plane (hkl).

In the discussion of the Kronig–Penney model, we showed that discontinuities in the $\mathscr{E}$–k curve (Fig. 5.4) occur at $\pm n\pi/a$. For a one-dimensional lattice, an electron wave packet can propagate only along the chain. This means that $\theta = 90°$ in Eq. (5.6.2). Letting $d_{hkl} = a$, which is the periodicity of the one-dimensional lattice, we find that the discontinuities indeed occur at those values of $k = k_{\text{diff}}$ where the Bragg condition is satisfied.

To extend our discussion to two-dimensional and three-dimensional lattices, we use the concept of reciprocal lattice introduced in Section 2.5. If $\mathbf{a}$, $\mathbf{b}$, and $\mathbf{c}$ are the basis vectors of the primitive cell, the three basis vectors for the reciprocal lattice are defined by Eq. (2.5.1) as

$$\mathbf{a}^* = \frac{\mathbf{b} \times \mathbf{c}}{\mathbf{a} \cdot (\mathbf{b} \times \mathbf{c})}, \qquad \mathbf{b}^* = \frac{\mathbf{c} \times \mathbf{a}}{\mathbf{a} \cdot (\mathbf{b} \times \mathbf{c})}, \qquad \mathbf{c}^* = \frac{\mathbf{a} \times \mathbf{b}}{\mathbf{a} \cdot (\mathbf{b} \times \mathbf{c})} \tag{5.6.3}$$

The normal to a plane (hkl) is given by Eq. (2.5.4) as

$$\mathbf{r}^* = h\mathbf{a}^* + k\mathbf{b}^* + l\mathbf{c}^* \tag{5.6.4}$$

and the distance between two adjacent (hkl) planes is shown in Eq. (2.5.7) to be

$$d_{hkl} = \frac{1}{|\mathbf{r}^*|} \tag{5.6.5}$$

Noting that the angle between $\mathbf{k}$ and $\mathbf{r}^*$ is $90° - \theta$ (Fig. 5.7), we can express Eq. (5.6.2) in terms of $\mathbf{r}^*$ as

$$\mathbf{k} \cdot \mathbf{r}^* = \pi |r^*|^2 \tag{5.6.6}$$

Since there are integers h, k, and l in the expression of r^*, the integer n in Eq. (5.6.2) is no longer needed here. It is often convenient to introduce a new vector defined as

$$\mathbf{G} = 2\pi\mathbf{r}^* \tag{5.6.7}$$

In terms of G, the Bragg condition of Eq. (5.6.6) becomes

$$\mathbf{k} \cdot \mathbf{G} = \tfrac{1}{2} G^2 \tag{5.6.8}$$

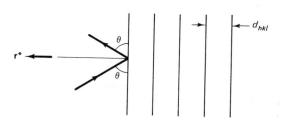

Figure 5.7 Reflection of electron wave by crystal planes $(h\ k\ l)$. The plane normal is denoted by the vector $\mathbf{r}^*$. The angle θ is the Bragg angle.

Sec. 5.6   The Brillouin Zones

The application of Eq. (5.6.8) to a simple cubic crystal is straightforward, and for this reason it is considered here. According to Eq. (2.5.8), the three reciprocal-lattice vectors are

$$\mathbf{a^*} = \frac{\mathbf{a}}{a^2}, \qquad \mathbf{b^*} = \frac{\mathbf{b}}{a^2}, \qquad \mathbf{c^*} = \frac{\mathbf{c}}{a^2} \qquad (5.6.9)$$

Hence the vector G is

$$\mathbf{G} = \frac{2\pi}{a^2} (h\mathbf{a} + k\mathbf{b} + l\mathbf{c}) \qquad (5.6.10)$$

First, we consider a two-dimensional, square lattice (Fig. 5.8a). The value of k_{diff} for the first Bragg diffraction can be found as follows. The lowest set of numbers (hkl) in Eq. (5.6.10) consists of (1, 0, 0), (0, 1, 0), and (0, 0, 1). For a two-dimensional lattice, we have

$$\mathbf{G}_{1\alpha} = \frac{2\pi}{a^2} \mathbf{a}, \qquad \mathbf{G}_{1\beta} = \frac{2\pi}{a^2} \mathbf{b} \qquad (5.6.11)$$

These two vectors are shown in Fig. 5.8b. The subscript 1 refers to the vectors used for the first Bragg diffraction. If we draw lines that bisect $\mathbf{G}_{1\alpha}$ and $\mathbf{G}_{1\beta}$ in a perpendicular direction, a zone boundary is formed. This boundary is indicated by $ABCD$ in Fig. 5.8b.

Equation (5.6.8) defines a boundary. To see whether the Bragg condition of Eq. (5.6.8) is satisfied or not along the boundary, we choose as an example $\mathbf{k}$ to be in the direction of OB and of length OB. The projection of OB upon the $\mathbf{G}_{1\alpha}$ vector is OJ. Thus Eq. (5.6.8) reduces to

$$\mathbf{k} \cdot \mathbf{G} = OJ \times OF = \tfrac{1}{2} (OF)^2 = \tfrac{1}{2} G^2 \qquad (5.6.12)$$

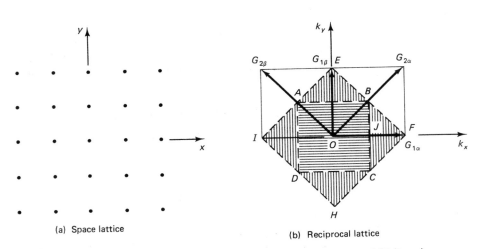

(a) Space lattice

(b) Reciprocal lattice

Figure 5.8 (a) A two-dimensional square lattice in coordinate space and (b) its reciprocal lattice in k space. The boundary of the first Brillouin zone is $ABCD$, whereas the boundary of the second zone is $EFHI$. The Bragg condition is satisfied if the k vector terminates on the zone boundary.

The Band Theory of Electronic Conduction   Chap. 5

which is obviously true. The reader can easily show that Eq. (5.6.8) is satisfied as long as the k vector terminates on lines AB, BC, CD, or DA. This proves that the line $ABCD$ forms the boundary of the first energy zone. Such an energy zone is called the *Brillouin zone* in k space. The word "first" is used because the vector $\mathbf{G}$ in defining the zone boundary is of the lowest order.

The set next to the lowest set of numbers consists of $(\pm 1, \pm 1, 0)$, $(\pm 1, 0, \pm 1)$, and $(0, \pm 1, \pm 1)$. For a two-dimensional lattice, the $\mathbf{G}$ vectors are

$$\mathbf{G}_{2\alpha} = \frac{2\pi}{a^2}(\mathbf{a} + \mathbf{b}), \qquad \mathbf{G}_{2\beta} = \frac{2\pi}{a^2}(\mathbf{a} - \mathbf{b}) \qquad (5.6.13)$$

Again, the boundaries are formed by lines EF, FH, HI, and IE, which bisect in a normal direction the vectors $\mathbf{G}_{2\alpha}$ and $\mathbf{G}_{2\beta}$, where the subscript 2 refers to the second zone. The reader is asked to show that Eq. (5.6.8) is satisfied along the zone boundary. In Fig. 5.8b, the second Brillouin zone occupies in k space an area marked with vertical lines, in contrast to the first zone, which is marked with horizontal lines.

The foregoing discussion can easily be extended to three-dimensional lattices. For a two-dimensional lattice, the zone boundary is made of lines. For a three-dimensional lattice, the zone boundary is formed by planes. Obviously, the first Brillouin zone for a simple cubic crystal is a cube. The first zones for the face-centered cubic, body-centered cubic, and primitive hexagonal structures are shown in Fig. 5.9. For the face-centered cubic (fcc) structure, we use the face diagonals

$$\mathbf{a}' = \frac{\mathbf{b} + \mathbf{c}}{2}, \qquad \mathbf{b}' = \frac{\mathbf{c} + \mathbf{a}}{2}, \qquad \mathbf{c}' = \frac{\mathbf{a} + \mathbf{b}}{2} \qquad (5.6.14)$$

as the basis vectors for the primitive cell, and obtain

$$\mathbf{a}^* = \frac{-\mathbf{a} + \mathbf{b} + \mathbf{c}}{a^2}, \qquad \mathbf{b}^* = \frac{\mathbf{a} - \mathbf{b} + \mathbf{c}}{a^2}, \qquad \mathbf{c}^* = \frac{\mathbf{a} + \mathbf{b} - \mathbf{c}}{a^2} \qquad (5.6.15)$$

as the three reciprocal lattice vectors. Thus, from Eqs. (5.6.4) and (5.6.7), we find that

$$\mathbf{G}_{\alpha}(\text{fcc}) = \frac{2\pi}{a}\langle 111 \rangle, \qquad \mathbf{G}_{\beta}(\text{fcc}) = \frac{4\pi}{a}\langle 100 \rangle \qquad (5.6.16)$$

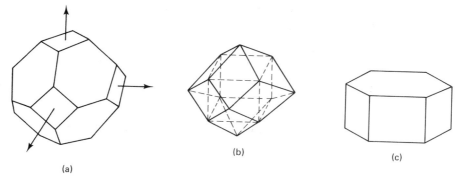

(a)   (b)   (c)

Figure 5.9   The first Brillouin zone of (a) the face-centered cubic lattice, (b) the body-centered cubic lattice, and (c) the hexagonal close-packed lattice.

along the body diagonal and along the principal axis, respectively. Therefore, the zone boundary for a fcc structure is at $\mathbf{k} = (\pi/a)\langle 111 \rangle$ and $\mathbf{k} = (2\pi/a)\langle 100 \rangle$ along the two directions.

For a body-centered cubic (bcc) structure, the primitive cell has three body diagonals

$$\mathbf{a}' = \frac{\mathbf{a} + \mathbf{b} + \mathbf{c}}{2}, \qquad \mathbf{b}' = \frac{\mathbf{a} - \mathbf{b} + \mathbf{c}}{2}, \qquad \mathbf{c}' = \frac{\mathbf{a} + \mathbf{b} - \mathbf{c}}{2} \qquad (5.6.17)$$

as its basis vectors. Using Eq. (5.6.17) in Eq. (5.6.3), we obtain

$$\mathbf{a}^* = \frac{\mathbf{b} + \mathbf{c}}{a^2}, \qquad \mathbf{b}^* = \frac{\mathbf{a} - \mathbf{b}}{a^2}, \qquad \mathbf{c}^* = \frac{\mathbf{a} - \mathbf{c}}{a^2} \qquad (5.6.18)$$

The vector $\mathbf{G}$, therefore, is given by

$$\mathbf{G}_\alpha(\text{bcc}) = \frac{2\pi}{a}\langle 110 \rangle, \qquad \mathbf{G}_\beta(\text{bcc}) = \frac{4\pi}{a}\langle 100 \rangle \qquad (5.6.19)$$

and the zone boundary is located at $\mathbf{k} = (\pi/a)\langle 110 \rangle$ and $\mathbf{k} = (2\pi/a)\langle 100 \rangle$ along the two directions. The reader is asked to show that the first Brillouin zone for a simple (primitive) hexagonal structure (Fig. 2.3) is a regular hexagonal prism. (A. H. Wilson, *Theory of Metals,* Cambridge University Press, Cambridge, 1953; M. J. Sinnott, *The Solid State for Engineers,* John Wiley & Sons, Inc., New York, 1958; W. Hume-Rothery, *Atomic Theory,* The Institute of Metals, London, 1962.) For their construction, the reader is referred to the books by Hume-Rothery and by Kittel (C. Kittel, *Introduction to Solid-State Physics,* John Wiley & Sons, Inc., New York, 1956).

5.7 THE ZONE THEORY: DENSITY OF STATES AND ORIGIN OF ENERGY DISCONTINUITY

One direct consequence of the zone theory is that the number of available states in a given zone is finite. Let us apply the Bloch wave to a one-dimensional lattice. If the lattice is taken to be a representative segment of a physical material, its property must repeat the same pattern when another segment is connected to it. The condition of repeatability is automatically satisfied by letting the linear lattice to form a closed ring (Fig. 5.10). If the electron wave function for the first atom is $\psi(x)$, the electron wave function for the $(n + 1)$th atom is

$$\psi(x + nc) = \psi(x) \exp{(iknc)} \qquad (5.7.1)$$

Since there are N atoms in the ring,

$$\psi(x + Nc) = \psi(x) \qquad (5.7.2)$$

must be true for a single-valued wave function. The condition of Eq. (5.7.2) requires that

$$kNc = q2\pi \quad \text{or} \quad k = q\,\frac{2\pi}{L} \qquad (5.7.3)$$

where $L = Nc$ and q is an integer.

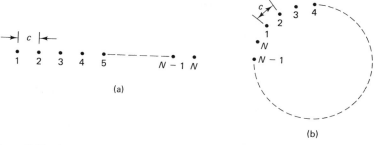

Figure 5.10 (a) A one-dimensional periodic lattice. If this lattice is considered to be a representative segment of a solid, we may extend the lattice by repeating the lattice at both ends. At the right end we have atom 1 following atom N, whereas at the left end we have atom N preceding atom 1. (b) A ring structure used to represent a finite segment of an infinite lattice.

The condition in Eq. (5.7.2) is called the periodic boundary condition. Under this condition, the wave number k takes discrete values from

$$k = 0, \quad \pm \frac{2\pi}{L}, \quad \pm \frac{4\pi}{L}, \quad \cdots, \quad + \frac{\pi}{c} \left(\text{or} - \frac{\pi}{c}\right) \tag{5.7.4}$$

The points $+\pi/c$ and $-\pi/c$ are on the zone boundary; therefore, only one of the two points should be counted. Since $L = Nc$, there are altogether N available states in the first zone. In other words, there are as many independent k states in the zone as the number of primitive cells in the linear dimension L.

The argument can be extended to three-dimensional zones. For a simple cubic structure, the number of primitive cells in a cube of volume L^3 is

$$N' = N_1 N_2 N_3 = \frac{L}{a}\frac{L}{b}\frac{L}{c} = \left(\frac{L}{c}\right)^3 \tag{5.7.5}$$

where N_i is the number of periods along a given direction of the primitive-cell basis vector. To find the number of states in the k space, we apply Eq. (5.7.4) separately to k_x, k_y, and k_z. There are $N_1 = (L/a)$ independent values of k_x, $N_2 = (L/b)$ independent values of k_y and $N_3 = (L/c)$ independent values of k_z. Therefore, the total number of independent combinations of k_x, k_y, and k_z is $N' = N_1 N_2 N_3 = (L/c)^3$, which is identical to N' of Eq. (5.7.5). The result again leads to the conclusion that each primitive cell contributes exactly one independent value of $\mathbf{k}$ (meaning a combination of k_x, k_y, and k_z) in the zone.

To calculate the density of states in k space, we refer to Fig. 5.8b. The area of $ABCD$ is $(G_1)^2$. For a three-dimensional cubic lattice, the volume of the first zone is

$$V_{\text{zone}} = (G_1)^3 = \left(\frac{2\pi}{a}\right)^3 = \frac{8\pi^3}{a^3} \tag{5.7.6}$$

If N' is the total number of states in the zone, the number of states in a volume dk_x dk_y dk_z is

$$dN' = \frac{N'}{V_{\text{zone}}} dk_x \, dk_y \, dk_z \tag{5.7.7}$$

In a volume V of the space lattice (Fig. 5.8a), there are V/a^3 primitive cells. According to the foregoing discussion, this number must be equal to the number of independent k values. In other words,

$$N' = \frac{V}{a^3} \tag{5.7.8}$$

Using Eqs. (5.7.6) and (5.7.8) in Eq. (5.7.7), we find that

$$dN' = \frac{V}{(2\pi)^3} \, dk_x \, dk_y \, dk_z \tag{5.7.9}$$

Counting two possible spin orientations, the number of available states in a volume V of the crystal is

$$dN = 2 \, dN' = \frac{2V}{(2\pi)^3} \, dk_x \, dk_y \, dk_z \tag{5.7.10}$$

which is the same as Eq. (4.4.16). However, two different physical situations are assumed in deriving Eqs. (4.4.16) and (5.7.10). In Section 4.4 the dimensions of the crystal are finite, and hence standing-wave solutions are used in Eq. (4.3.11). In Fig. 5.10 we assume that the crystal is considered to be a part of an infinite periodic structure. In this treatment, traveling-wave solutions are used. The result expressed in Eqs. (4.4.16) and (5.7.10) is independent of the physical model we choose.

Sometimes it is convenient to restrict the value of $k_i a$ to the interval between $-\pi$ and π, where $i = x$, y, or z. This can be done by either subtracting from or adding to

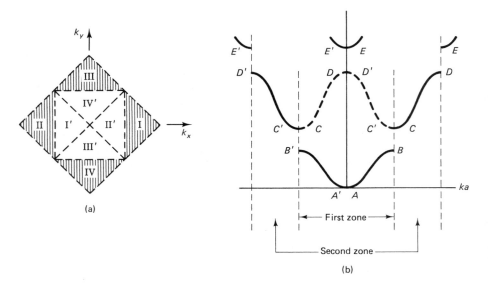

(a)

(b)

Figure 5.11 (a) The second Brillouin zone (shown as areas with vertical lines) of a square lattice. In the reduced-zone scheme, we translate I, II, III, and IV by either $\mathbf{G}_{1\alpha}$ or $\mathbf{G}_{1\beta}$ (Fig. 5.8b) to form I', II', III', and IV'. (b) The $\mathscr{E}$–k diagram in the extended (Brillouin) zone scheme (solid curves) and in the reduced zone scheme (dashed curves).

k an integral multiple of **G**. This representation is called the *reduced zone scheme*. In Fig. 5.11a, the second Brillouin zone of a square lattice shown in Fig. 5.8b is redrawn for illustration. If the portion I of the second zone is translated along the k_x direction by $-2\pi/a$, it is relocated at I'. Through this process, we can show that the area occupied by the second zone is equal to that occupied by the first zone. In Fig. 5.11b, a typical energy versus k diagram is shown. The dashed curve shows the energy in the second zone in the reduced zone scheme. There are as many available states in the zone *CDC* as in the zone *BAB* because the two zones have the same area. The statement is true in general, although its proof is given here for a two-dimensional, square lattice. Therefore, Eq. (5.7.10) is also a general expression.

Another important consequence of the zone theory is the existence of a discontinuity in energy (energy gap) at the zone boundary. In Section 5.6 we defined a zone boundary as the boundary at which the Bragg condition for diffraction is satisfied. Figure 5.7 illustrates that an incident wave is diffracted into a reflected wave by planes of atoms. At zone boundaries where diffraction occurs, it is necessary that we consider both the incident and the reflected waves. In the following discussion we see that the energy of an electron is split into two levels, depending on the admixture of the reflected and incident waves. An energy discontinuity is a direct result of this energy split.

For simplicity, a one-dimensional lattice is used to expound the underlying physics. Figure 5.12a shows the periodic potential associated with the nuclear potential in

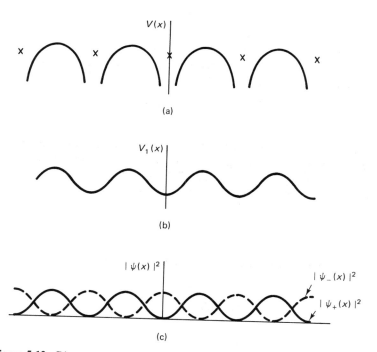

(a)

(b)

(c)

Figure 5.12 Diagrams used to show the physical origin of an energy gap: (a) nuclear potential in a one-dimensional lattice, (b) the first Fourier component of the nuclear potential, and (c) the two normal modes of electron distribution for electrons having values of k at the zone boundaries.

one dimension. Since the potential is periodic, we can express $V(x)$ in a Fourier series as

$$V(x) = V_0 - \sum_n 2V_n \cos\left(\frac{2\pi}{a} nx\right) \tag{5.7.11}$$

where n is an integer. Only cosine functions are included because $V(x)$ is an even function in the coordinate system of Fig. 5.12. At the first zone boundary, $k = \pm\pi/a$ according to Eq. (5.7.4). Let us choose $k = +\pi/a$. Thus the incident wave ψ_i has an exp $(i\pi x/a)$ dependence.

Now we consider the interaction of the incident wave with the lattice potential, especially the V_1 term. We write

$$V_1(x) = -V_1\left[\exp\left(i\frac{2\pi}{a}x\right) + \exp\left(-i\frac{2\pi}{a}x\right)\right] \tag{5.7.12}$$

which is shown in Fig. 5.12b. We see that the product of ψ_i and V_1 produces a term which has an exp $(-i\pi x/a)$ dependence. When the Bragg condition is satisfied, the interactions of the wave with atoms in consecutive planes become constructive, and a large amplitude of the reflected wave results. Therefore, at the zone boundary, the wave must be expressed as a combination of incident and reflected waves.

We let

$$\psi(x) = A_1 \exp(ikx) + A_2 \exp(-ikx) \tag{5.7.13}$$

where $k = \pi/a$. Strictly speaking, the two coefficients A_1 and A_2 should contain the cell function $u(x)$ in the Bloch-wave representation. However, our present discussion is qualitative in nature. Therefore, we use the free-electron wave function. In Eq. (5.7.13), A_1 and A_2 are two constants. Substituting Eq. (5.7.13) in Eq. (5.2.1), we have

$$(\mathscr{E} - \mathscr{E}_k)[A_1 \exp(ikx) + A_2 \exp(-ikx)]$$
$$= [V_0 + V_1(x)][A_1 \exp(ikx) + A_2 \exp(-ikx)] \tag{5.7.14}$$

where $\mathscr{E}_k = \hbar^2 k^2/2m$. In Eq. (5.7.14) we keep only those Fourier components of $V(x)$ that contribute to the coupling of the exp (ikx) term to the exp $(-ikx)$ term.

Using Eq. (5.7.12) in Eq. (5.7.14) and collecting the terms with the exp $(i\pi x/a)$ and the exp $(-i\pi x/a)$ dependence, respectively, we obtain

$$(\mathscr{E} - \mathscr{E}_k - V_0)A_1 + V_1 A_2 = 0 \tag{5.7.15}$$

$$(\mathscr{E} - \mathscr{E}_k - V_0)A_2 + V_1 A_1 = 0 \tag{5.7.16}$$

For nontrivial solutions, the determinant must be zero. Thus we have

$$(\mathscr{E} - \mathscr{E}_k - V_0)^2 = V_1^2 \tag{5.7.17}$$

Equation (5.7.17) has two solutions:

$$\mathscr{E}_+ = \mathscr{E}_k + V_0 + V_1 \tag{5.7.18}$$

$$\mathscr{E}_- = \mathscr{E}_k + V_0 - V_1 \tag{5.7.19}$$

From Eq. (5.7.15), the relation between A_1 and A_2 is found to be

$$A_1 = -A_2 \quad \text{for} \quad \mathscr{E}_+ \tag{5.7.20a}$$

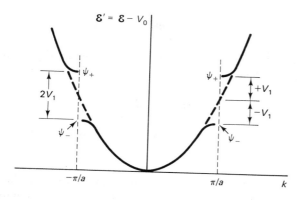

$$\mathcal{E}' = \mathcal{E} - V_0$$

$-\pi/a$ π/a k

Figure 5.13 Energy split at zone boundaries. For comparison, the free-electron energy is shown as a dashed curve.

$$A_1 = +A_2 \quad \text{for} \quad \mathcal{E}_- \tag{5.7.20b}$$

Figure 5.13 shows the energy split at the zone boundary according to Eqs. (5.7.18) and (5.7.19). To understand physically why such a split exists, we refer to Fig. 5.12c. The two energy states $\mathcal{E}_+$ and $\mathcal{E}_-$ have two different wave functions:

$$\psi_+ = 2iA \sin \frac{\pi x}{a} \quad \text{for} \quad \mathcal{E}_+ \tag{5.7.21a}$$

$$\psi_- = 2A \cos \frac{\pi x}{a} \quad \text{for} \quad \mathcal{E}_- \tag{5.7.21b}$$

through the use of Eqs. (5.7.20a) and (5.7.20b). The electron distribution represented by $|\psi(x)|^2$ has a maximum at the potential minimum for ψ_- and a minimum at the potential minimum for ψ_+. The spatial relation between the electron distribution and the lattice potential is shown in Fig. 5.12b and c. Since $|\psi_-(x)|^2$ spends more time near the potential minimum, it is the lower-energy state.

In brief, we have shown that the solution at the zone boundary actually consists of two traveling waves of equal amplitude, one with $k = +\pi/a$ and the other with $k = -\pi/a$. Depending on their relative phase, the two traveling waves combine to give two different standing waves. These two standing waves represent two different electron distributions with respect to the lattice potential, and hence have different potential energies. The difference in potential energy leads to a splitting of the energy bands at the zone boundary.

5.8 THE FREE-ELECTRON APPROXIMATION

In Section 5.7 we presented a simplified physical model to show the origin of energy gap in a periodic lattice. In the treatment, only the first Fourier components of the lattice potential [see Eq. (5.7.12)] and the free-electron wave function [see Eq. (5.7.13)] are considered. In this section we extend this treatment by including the higher-order Fourier components for the lattice potential and by using the Bloch wave function for electrons. Since the cell function is periodic [see Eq. (5.3.3)], it can also be expanded into a Fourier series. Thus Eqs. (5.7.12) and (5.7.13), respectively, become

$$V(x) = V_0 - \sum_n V_n \exp(inGx) \tag{5.8.1}$$

$$\psi(x) = [B_0 + \sum_m B_m \exp(imGx)] \exp(ikx) \qquad (5.8.2)$$

where $G = 2\pi/a$, and n and m are two nonzero integers. Substituting Eqs. (5.8.1) and (5.8.2) into the one-dimensional Schrödinger equation, Eq. (5.2.1), we obtain

$$(\mathscr{E}_k - \mathscr{E}')B_0 + \sum_n \left\{ \left[\mathscr{E}_k \left(1 + \frac{nG}{k} \right)^2 - \mathscr{E}' \right] B_n - V_n B_0 \right\} \exp(inGx)$$

$$- \sum_n \sum_m V_n B_m \exp[+iG(n+m)x] = 0 \qquad (5.8.3)$$

where $\mathscr{E}_k = \hbar^2 k^2/2m$ and $\mathscr{E}' = \mathscr{E} - V_0$.

Note that here the wave function we are dealing with involves only a single electron. Even so, an exact solution of Eq. (5.8.3) is impractical. Therefore, we must resort to approximation methods. The method we follow in this section is called the *free-electron approximation*. In this approximation we start out with free electrons by assuming that $V_n = 0$ and $B_n = 0$ as the zeroth-order solution and thus obtain $\mathscr{E} = \mathscr{E}_k = \hbar^2 k^2/2m$ which is the energy for free electrons. For electrons in a periodic lattice with k values away from the zone boundary, the terms linear in V_n and B_n are considered small and treated as perturbations. The first-order solution is obtained by neglecting the second-order terms involving the product $V_n B_m$ in Eq. (5.8.3). Thus we find that

$$(\mathscr{E}_k - \mathscr{E}')B_0 + \sum_n \left\{ \left[\mathscr{E}_k \left(1 + \frac{nG}{k} \right)^2 - \mathscr{E}' \right] B_n - V_n B_0 \right\} \exp(inGx) = 0 \qquad (5.8.4)$$

For a given crystal, the lattice potential $V(x)$ and hence its Fourier coefficients V_n are known. Our task is to find the Fourier coefficients B_n for the wave function and the proper value of the energy $\mathscr{E}$.

To solve Eq. (5.8.4), we take note of the following property of the Fourier integral:

$$\int_0^a \exp(ipGx)\, dx = \begin{cases} 0 & \text{if } p \neq 0 \\ a & \text{if } p = 0 \end{cases} \qquad (5.8.5)$$

Multiplying Eq. (5.8.4) by $\exp(-inGx)$ and then integrating the product over a lattice cell, we obtain

$$\left[\mathscr{E}_k \left(1 + \frac{nG}{k} \right)^2 - \mathscr{E}' \right] B_n - V_n B_0 = 0 \qquad (5.8.6)$$

Note that Eq. (5.8.6) can be obtained by simply equating the coefficients of $\exp(inGx)$ on the two sides of Eq. (5.8.4). This procedure is valid on account of Eq. (5.8.5). From Eq. (5.8.6) we find

$$B_n = \frac{-V_n}{\mathscr{E}' - \mathscr{E}_k(1 + nG/k)^2} B_0 \qquad (5.8.7)$$

Using B_n of Eq. (5.8.7) in Eq. (5.8.2), we have the first-order solution $\psi_1(x)$ for the one-electron wave function. The first-order correction in the wave function, however, does not introduce a corresponding first-order correction in energy because the coefficient of the constant term in Eq. (5.8.4) yields $\mathscr{E}' = \mathscr{E}_k$. To find the correction in energy which is second order, we must go back to Eq. (5.8.3). Equating the coefficients of the constant terms on the two sides, we obtain

$$\mathscr{E}' = \mathscr{E}_k - \sum_n \frac{V_{-n}B_n}{B_0} = \mathscr{E}_k + \sum_n \frac{V_n V_{-n}}{\mathscr{E}' - \mathscr{E}_k(1 + nG/k)^2} \tag{5.8.8}$$

To simplify Eq. (5.8.8), we set $\mathscr{E}' = \mathscr{E}_k$ in the correction term, and thus obtain

$$\mathscr{E} = V_0 + \frac{\hbar^2 k^2}{2m} + \sum_{n \neq 0} \frac{|V_n|^2}{(\hbar^2 k^2/2m) - (\hbar^2/2m)(k + n2\pi/a)^2} \tag{5.8.9}$$

Note that the periodic potential $V(x)$ is a real function. Therefore, its Fourier coefficients must be related by $V_{-n} = V_n^*$. The substitution $\mathscr{E}' = \mathscr{E}_k$ and hence Eq. (5.8.9) are satisfactory only for k values away from the zone boundary. Rearranging terms in the denominator of Eq. (5.8.9) yields

$$\mathscr{E} = V_0 + \frac{\hbar^2 k^2}{2m} - \sum_{n \neq 0} \frac{|V_n|^2}{(\hbar^2 n^2 G^2/2m)(1 + 2k/nG)} \tag{5.8.10}$$

From Eq. (5.8.10), the effective mass m^* can be found and is given by

$$\frac{1}{m^*} = \frac{1}{m}\left[1 - \sum_n \frac{|V_n|^2}{(\hbar^2 n^2 G^2/2m)^2}\frac{2}{(1 + 2k/nG)^3}\right] \tag{5.8.11}$$

where k is the wave number in the extended zone. Therefore, the value of m^* depends not only on the lattice potential $|V_n|^2$, but also on the particular zone. In Eq. (5.8.11) the summation is over all integers extending from $-\infty$ to $+\infty$ except $n = 0$.

As the value of k approaches $nG/2$, that is, the value at a zone boundary, the denominator of Eq. (5.8.7) becomes smaller and smaller, and hence the Fourier coefficient B_n becomes larger and larger. As a matter of fact, the treatment in Section 5.7 shows that at $k = G/2$, $|A_1| = |A_2|$. This means that the incident wave and the Bragg-reflected wave are of equal strength. If we consider the term A_1 in Eq. (5.7.13) and the term B_0 in Eq. (5.8.2) to be the incident wave, the term A_2 in Eq. (5.7.13) and the term B_{-1} in Eq. (5.8.2) represent the Bragg-reflected wave. In our previous treatment of Eq. (5.8.3) earlier in this section, we considered the terms involving $V_n B_m$ to be of second-order importance. This approximation leads to Eq. (5.8.4). Near a zone boundary where the Bragg condition $k = qG/2$ is approximately satisfied, one Fourier coefficient B_{-q} attains a magnitude comparable to that of B_0, the amplitude of the incident wave.

In view of the discussion above, we must modify our treatment of Eq. (5.8.3). First, the product $V_n B_{-q}$ now becomes a first-order correction, whereas the other products $V_n B_m$ with $m \neq -q$ remain second-order corrections. Second, for $n = -m = q$, the dependence $\exp[iG(n + m)x]$ becomes a constant. Therefore, equating the coefficients of the constant terms in Eq. (5.8.3) yields

$$(\mathscr{E}_k - \mathscr{E}')B_0 - V_q B_{-q} = 0 \tag{5.8.12}$$

Equating the coefficients of the terms with $\exp(-iqGx)$ dependence in Eq. (5.8.3) yields another equation

$$\left[\mathscr{E}_k\left(1 - \frac{qG}{k}\right)^2 - \mathscr{E}'\right]B_{-q} - V_{-q}B_0 = 0 \tag{5.8.13}$$

which is similar to Eq. (5.8.6). From Eqs. (5.8.12) and (5.8.13), we obtain the following characteristic equation for energy:

$$(\mathscr{E}_k - \mathscr{E}')\left[\mathscr{E}_k\left(1 - \frac{qG}{k}\right)^2 - \mathscr{E}'\right] = |V_q|^2 \tag{5.8.14}$$

Solving for $\mathscr{E}'$, we obtain

$$\mathscr{E}' = \tfrac{1}{2}\mathscr{E}_k\left[1 + \left(1 - \frac{qG}{k}\right)^2\right] \pm \tfrac{1}{2}\sqrt{\mathscr{E}_k^2\left[1 - \left(1 - \frac{qG}{k}\right)^2\right]^2 + 4|V_q|^2} \tag{5.8.15}$$

where $\mathscr{E}_k = \hbar^2 k^2/2m$, $\mathscr{E}' = \mathscr{E} - V_0$, and q is an integer indicating the zone number.

If we start with $\exp(-ikx)$ as the incident wave, an equation similar to Eq. (5.8.15) is obtained except that the term $-qG/k$ is changed to $+qG/k$. Therefore, what happens to the $\exp(+ikx)$ wave at $k = qG/2$ applies equally well to the $\exp(-ikx)$ wave at $k = -qG/2$. In the following discussion, we focus our attention on Eq. (5.8.15). At $k = qG/2$, Eq. (5.8.5) becomes

$$\mathscr{E} = V_0 + \mathscr{E}_k \pm |V_q| = V_0 + \frac{\hbar^2 k^2}{2m} \pm |V_q| \tag{5.8.16}$$

which is similar to Eqs. (5.7.18) and (5.7.19). From Eq. (5.8.16) the energy gap $\Delta\mathscr{E}$ at the zone boundary $k = qG/2$ is given by

$$\Delta\mathscr{E} = 2|V_q| \tag{5.8.17}$$

We further let $k = (qG/2) - k'$. Near the zone boundary, k' is a very small quantity and $\hbar^2 k' qG/2m$ is also small compared to $|V_q|$. Expanding Eq. (5.8.15) in a power series in k', we find

$$\mathscr{E} = V_0 + \frac{\hbar^2}{2m}\left[\left(\frac{qG}{2}\right)^2 + k'^2\right] \pm \left[\frac{\Delta\mathscr{E}}{2} + \frac{2}{\Delta\mathscr{E}}\left(\frac{\hbar^2 Gqk'}{2m}\right)^2\right] \tag{5.8.18}$$

From Eq. (5.8.18), the effective mass at the zone boundary $k = qG/2$ is given by

$$\frac{1}{m^*} = \frac{1}{\hbar^2}\frac{\partial^2\mathscr{E}}{\partial k'^2} = \frac{1}{m}\left(1 \pm \frac{\hbar^2 q^2 G^2}{m\Delta\mathscr{E}}\right) \tag{5.8.19}$$

5.9 ENERGY-BAND STRUCTURE AND THE TIGHT-BINDING APPROXIMATION

In Sections 5.7 and 5.8 we presented a physical model to show the origin of energy gap in a periodic lattice. In the model, the electron wave function was represented by plane waves of Eq. (5.7.13) or (5.8.2). Therefore, as the starting point, electrons were considered as free electrons and the lattice potential was treated as a weak perturbation except at the zone boundary. This method of approximate treatment is called the nearly-free-electron method or the free-electron approximation. It applies to electrons with relatively high energy within the nuclear potential funnel (Fig. 4.12a). In this section we present an entirely different scheme for electrons with relatively low energy. These electrons are strongly bound to an atomic core. Therefore, as a starting point for these electrons, atomic orbitals are most suitable. Then we combine the atomic orbitals to represent a state running throughout the crystal, with each orbital localized on a particular atom. This scheme is called the *tight-binding approximation.*

Refer to our discussion in Section 4.1 and specifically to Eq. (4.1.4) and Fig. 4.4. In the Bloch representation, we may rewrite Eq. (4.1.4) as

$$\psi(x, k) = \sum_n u_a(x - nc) \exp(iknc) \tag{5.9.1}$$

where $u(x - nc)$, the same as ψ_j in Eq. (4.1.4), is the atomic orbital of the nth atom and $\theta = knc$, the same as θ_j in Eq. (4.1.4), is the phase angle. The only difference between Eqs. (5.9.1) and (4.1.4) is that complex notation is used for the phase factor in Eq. (5.9.1). This wave function, which is illustrated in Fig. 4.4, looks like a series of localized atomic orbitals multiplied by a wavy phase factor. We further note that

$$\psi(x + c, k) = \exp(ikc) \sum_n u_a(x - n'c) \exp(ikn'c) = \psi(x, k) \exp(ikc) \tag{5.9.2}$$

where $n' = n - 1$. Since $\psi(x, k)$ satisfies the Bloch condition of Eq. (5.3.18), it is in the Bloch form.

For simplicity, we again consider the one-dimensional case. The Schrödinger equation is given by Eq. (5.2.1). In Fig. 5.14, the nuclear potential $V_0(x - nc)$ of the nth atom is drawn for comparison with the lattice potential $V(x)$ of Eq. (5.2.1). Let

$$\mathcal{H}_{0n} = -\frac{\hbar^2}{2m} \frac{\partial^2}{\partial x^2} + V_0(x - nc) \tag{5.9.3}$$

$$\mathcal{H}_{1n} = V_0(x - nc) - V(x) \tag{5.9.4}$$

be the unperturbed and perturbation Hamiltonian, respectively, for the nth atom. Then we have

$$\mathcal{H} = -\frac{\hbar^2}{2m} \frac{\partial^2}{\partial x} + V(x) = \mathcal{H}_{0n} - \mathcal{H}_{1n} \tag{5.9.5}$$

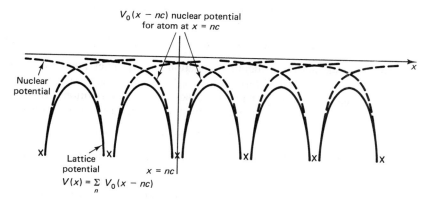

Figure 5.14 Comparison of the lattice potential $V(x)$ (solid curves) and the nuclear potential $V_0(x-nc)$ (dashed curves).

and we can write Eq. (5.2.1) as

$$(\mathcal{H}_{0n} - \mathcal{H}_{1n})\,\psi = \mathcal{E}\psi \tag{5.9.6}$$

Since we are dealing with the band formed by a given atomic state, for example the $3s$ state of sodium, the atomic orbitals u_a are the same in Eq. (5.9.1) for all atoms except that the reference points nc are displaced by an amount c from one atom to a neighboring atom. Thus we expect u_a to satisfy the local Schrödinger equation:

$$\mathcal{H}_{0n}u_a(x - nc) = \mathcal{E}_a u_a(x - nc) \tag{5.9.7}$$

where $\mathcal{E}_a$ is the eigenenergy of the atomic state. Substituting Eq. (5.9.1) in Eq. (5.9.6) and using Eq. (5.9.7), we find that

$$\mathcal{E}\psi(x, k) = \mathcal{E}_a\psi(x, k) - \mathcal{H}_{1n}\psi(x, k) \tag{5.9.8}$$

The expectation value of $\mathcal{E}$ is equal to

$$\langle\mathcal{E}\rangle = \mathcal{E}_a - \frac{\int \psi^*\mathcal{H}_{1n}\psi \, dx \, dy \, dz}{\int \psi^*\psi \, dx \, dy \, dz} \tag{5.9.9}$$

Equation (5.9.9) is obtained by multiplying Eq. (5.9.8) by $\psi^*(x, k)$ and then integrating the product over space coordinates.

A typical term in the numerator of Eq. (5.9.9) is

$$I_m = \exp\,[(ikc)(n - m)] \int u_a^*(x - mc)\mathcal{H}_{1n}u_a(x - nc) \, d\tau \tag{5.9.10}$$

where $d\tau$ is the volume element and the integral is over the volume of the crystal. The running index n in Eq. (5.9.4) goes together with that in Eq. (5.9.1). Therefore, the perturbation potential $\mathcal{H}_{1n}$ changes with $u_a(x - nc)$ in Eq. (5.9.10). Since the atomic orbital $u_a(x - nc)$ falls very rapidly as x moves away from the nuclear position $x = nc$, the integral I_m has an appreciable magnitude only if there is sufficient overlap between $u_a(x - nc)$ and $u_a(x - mc)$. Therefore, only three terms are important with $m = n$ and $m = n \pm 1$ in Eq. (5.9.10). We let

$$\int u_a^*(x - nc)\mathcal{H}_{1n}u_a(x - nc) \, d\tau = \alpha \int |u_a(x - nc)|^2 \, d\tau \tag{5.9.11}$$

$$\int u_a^*(x - nc \pm c)\mathcal{H}_{1n}u_a(x - nc) \, d\tau = \beta \int |u_a(x - nc)|^2 d\tau \tag{5.9.12}$$

Since any atom has an identical situation as any other atom, there are N such terms in the numerator and N terms of $\int |u_a|^2 \, d\tau$ in the denominator of Eq. (5.9.9) for a linear chain of N atoms. Thus if we neglect the contributions from the cross products $u_a(x - nc)u_a(x - mc)$ in the denominator we obtain from Eq. (5.9.9),

$$\langle\mathcal{E}\rangle = \mathcal{E}_a - \alpha - 2\beta \cos (kc) \tag{5.9.13}$$

The analysis above can be extended to a three-dimensional lattice. For an atomic s state with $\mathcal{E}_a = \mathcal{E}_1$,

$$\langle\mathcal{E}\rangle = \mathcal{E}_1 - \alpha_1 - 2\beta_1[\cos (k_x a) + \cos (k_y a) + \cos (k_z a)] \tag{5.9.14}$$

For an atomic p state (specifically for a p_x state),[†]

[†]See Problem 5.22 for an explanation of the sign change in front of the $2\gamma_2 \cos (k_x a)$ term.

$$\langle \mathscr{E} \rangle = \mathscr{E}_2 - \alpha_2 + 2\gamma_2 \cos (k_x a) - 2\beta_2 [\cos (k_y a) + \cos (k_z a)] \qquad (5.9.15)$$

Because the atomic orbitals p_x, p_y, and p_z are directed along the x, y, and z axes, respectively, the value of α_2 in Eq. (5.9.15) is larger than that of α_1 in Eq. (5.9.14). Therefore, the p band is expected to lie below the s band in a solid if $|\mathscr{E}_2 - \alpha_2| > |\mathscr{E}_1 - \alpha_1|$. The schematic energy-band diagram shown in Fig. 4.10 for NaCl is an example. There are two similar expressions for p_y and p_z states which can be obtained from Eq. (5.9.15) by permuting k_x, k_y, and k_z. Both Eqs. (5.9.14) and (5.9.15) are derived for simple cubic crystals. For a further discussion of these equations, the reader is referred to the books by Wilson and by Mott and Jones (A. H. Wilson, *Theory of Metals*, Cambridge University Press, Cambridge, 1953, pp. 39–41; N. F. Mott and H. Jones, *Theory of Metals and Alloys*, Dover Publications, Inc., New York, 1958, p. 70).

In Fig. 5.15 the energy $\langle \mathscr{E} \rangle$ curves for an s state and a p_x state are given as functions of k_x with constant k_y and k_z (Fig. 5.15a) and as functions of k_y or k_z with constant k_x (Fig. 5.15b). The curves are based on Eqs. (5.9.14) and (5.9.15). Near $k = 0$, expansion of cosine functions in Eq. (5.9.14) for the s band gives

$$\langle \mathscr{E} \rangle = \langle \mathscr{E} \rangle_1 - \alpha_1 - 6\beta_1 + \beta_1 a^2 (k_x^2 + k_y^2 + k_z^2) \qquad (5.9.16)$$

In the nearly free electron approximation, the energy of an electron is given by

$$\langle \mathscr{E} \rangle = \mathscr{E}_0 + \frac{\hbar^2}{2m^*} (k_x^2 + k_y^2 + k_z^2) \qquad (5.9.17)$$

where $\mathscr{E}_0$ is the potential energy. Even in the tight-binding approximation, the electrons behave as if they were free electrons with an effective mass

$$m^* = \frac{\hbar^2}{2\beta_1 a^2} \qquad (5.9.18)$$

One important function about which we like to get at least some qualitative information is the density-of-state function. According to our discussion in Section 4.4,

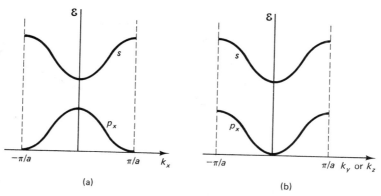

(a)   (b)

Figure 5.15 Plot of energy versus wave vector in the tight-binding approximation for s and p_x states in a cubic crystal.

the number of energy states per unit volume lying in the energy range between $\mathscr{E}$ and $\mathscr{E} + d\mathscr{E}$ is

$$\frac{dN}{V} = D(\mathscr{E})\, d\mathscr{E} = \frac{k^2}{\pi^2}\, dk \qquad (5.9.19)$$

For the energy function of Eq. (5.9.16), we have

$$D(\mathscr{E}) = 4\pi \left(\frac{2m^*}{h^2}\right)^{3/2} \mathscr{E}^{1/2} \qquad (5.9.20)$$

where $\mathscr{E}$ represents the kinetic energy of Eq. (5.9.16). The function $D(\mathscr{E})$ is shown as the dashed curve in Fig. 5.16b, where the point A corresponds to the center of the Brillouin zone. Equation (5.9.20) is the same as Eq. (4.4.15) except that the free-electron mass is substituted for by the effective mass.

Figure 5.16a is a two-dimensional plot of constant-energy contours in k space. As the value of k increases, the contours deviate considerably from the spherical shape. The value of $D(\mathscr{E}) = (k^2/\pi^2)\,(dk/d\mathscr{E})$ increases more rapidly than $\mathscr{E}^{1/2}$ as $\mathbf{k}$ approaches the zone boundary and reaches a maximum when $\mathbf{k}$ touches the zone boundary because $\partial\mathscr{E}/\partial k_x = 0$ and $\partial\mathscr{E}/\partial k_y = 0$ but $k \neq 0$ there. This point is denoted by B in Fig. 5.16. After this stage is reached, $D(\mathscr{E})$ drops very rapidly toward zero. At zone corners which are located at $(\pm\pi/a, \pm\pi/a, \pm\pi/a)$ in k space, we can again expand the cosine functions by letting $k_i' = \pm\pi/a - k_i$, where $i = x$, y, or z. These corners are represented by the point C in Fig. 5.16b. The reader is asked to show that near C the energy can be approximated by

$$\mathscr{E} = \mathscr{E}_{cn} - \frac{\hbar^2}{2m^*}\,(k_x'^2 + k_y'^2 + k_z'^2) \qquad (5.9.21)$$

where $\mathscr{E}_{cn} = \mathscr{E}_1 - \alpha_1 + 6\beta_1$.

Another important feature of the zone theory is that the energy $\mathscr{E}$ depends not only on the magnitude of k but its direction also. In Fig. 5.17 we show a two-dimensional plot of constant-energy contours. Two lines are drawn: one along the k_x axis and the

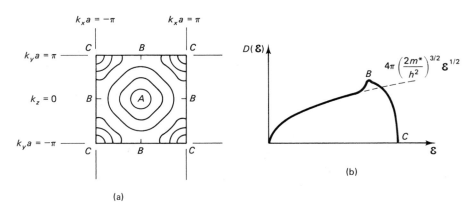

(a)

(b)

Figure 5.16 (a) Contours of constant energy in the plane $k_z = 0$ in k space. The point A is the center of the Brillouin zone. The lines $k_x = \pm\pi/a$ and $k_y = \pm\pi/a$ represent the zone boundaries. (b) Plot of the density-of-state function $D(\mathscr{E})$ versus energy.

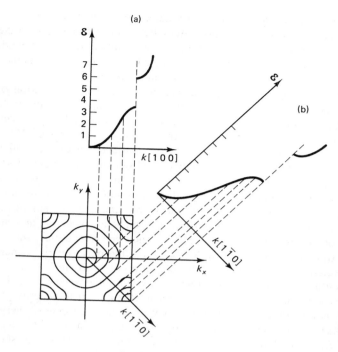

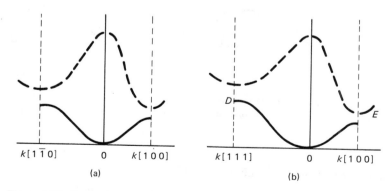

Figure 5.17 Diagram illustrating how to obtain the plot of energy versus wave vector from the plot of constant energy contours along (a) the $k[100]$ direction and (b) the $k[1\bar{1}0]$ direction. For a given contour, the energy is known. The intercept of the constant energy contour with the k_x axis gives the value of k_x. The contour and its intercept yield a set of values for $\mathscr{E}$ and k.

other along the diagonal direction. The intercepts of these two lines with the constant-energy contours produce two energy versus k diagrams: one along the [100] direction (a) and the other along the [1$\bar{1}$0] direction (b). As we can see, the two diagrams are quite different. A composite diagram is drawn in Fig. 5.18 to show the structure of the energy curve in k space. Such a diagram is called the *band-structure diagram*. As we shall see in Chapters 6, 10, and 12, knowing band structure is important in our discussion of the optical and high-field electrical properties of semiconductors.

In summary, we have used two approximate methods for the calculation of energy-band properties: the free-electron approximation in the discussion of the origin

Figure 5.18 Band-structure diagrams showing the energy as a function of the magnitude of k (a) along the $k[100]$ and $k[1\bar{1}0]$ directions and (b) along the $k[100]$ and $k[111]$ directions.

of energy gap in Sections 5.7 and 5.8 and the tight-binding approximation in the discussion of energy dependence on **k** in this section. Actually, both methods lead to the same qualitative conclusions: the existence of an energy discontinuity at the zone boundary and a dependence of electron energy on the wave vector **k**. We choose one method over the other in the discussion for reasons of mathematical simplicity.

Both methods are highly approximate. For a reasonably accurate calculation of the energy gap and the energy dependence on k, more elaborate methods have been proposed and used (for a general review of the subject, see, for example, J. M. Ziman, *Principles of the Theory of Solids*, Cambridge University Press, Cambridge, 1965, pp. 85–97). However, it is not our intent to give an extensive account of the mathematical methods. Our purpose is to illustrate the underlying physics in the simplest form possible. We also should point out that most of our discussion in this section is based on Eq. (5.9.14) for an s state in a simple cubic crystal. It is important that we separate general features from specific details. The features that there exist energy gaps at the zone boundaries and that there is a dependence of energy on **k** are general, and are most important. On the other hand, questions concerning the zone boundaries, the location of the energy maximum and minimum in **k** space, and the exact energy dependence on **k**, can be answered only if specific details about a given solid are known.

Finally, we mention a real possibility that one energy zone may overlap another in energy. Use the energy-band diagram of Fig. 5.18b as an example. If the energy gap at the extreme of $k[100]$ is not sufficiently large, we may find that the topmost energy in zone 1, which is represented by point D, exceeds the lowest energy in zone 2, which is represented by point E. This situation occurs in most metals. The density-of-state function $D(\mathscr{E})$ for overlapping bands is illustrated in Fig. 5.19a. The states in zone 2 begin to become occupied before the states in zone 1 are completely filled. On the other hand, if the energy gap is sufficiently large, no overlapping of two bands is possible. The situation that occurs in insulators and semiconductors is illustrated in Fig. 5.19b.

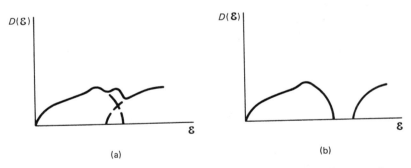

(a)

(b)

Figure 5.19 Density-of-state function for (a) overlapping bands and (b) nonoverlapping bands. The solid curve in (a) is a sum of the two dashed curves in the overlap region.

5.10 ELECTRON AND HOLE CONDUCTION

Consider the situation shown in Fig. 5.18b where the two zones overlap. Imagine that we are filling the zones with electrons. Initially, all the electrons go to the first zone because it is the lower-energy zone. When the filling reaches a certain stage, the electrons may find a situation in which the lowest part of the second zone lies below the highest part of the first zone. Under the circumstance, the electrons spill over from the first zone into the second zone. It is conceivable, therefore, that simultaneously we may have some empty states in an almost filled first zone and some occupied states in an almost empty second zone. It is the purpose of our present discussion to analyze the contributions of the two zones to electronic conduction.

Under an applied electric field E, the equation of motion is given by Eq. (5.5.10). For simplicity, we assume the constant-energy surface to be spherical. Thus Eq. (5.5.10) becomes

$$\frac{d\mathbf{v}_e}{dt} = \frac{\mathbf{F}}{m_e^*} = \frac{-e}{m_e^*}\mathbf{E} \qquad (5.10.1)$$

where m_e^* is the effective mass. Depending on whether m_e^* is positive or negative, two basically different situations arise. These two situations can best be illustrated by considering electrons near the bottom of a band (Fig. 5.20a) and electrons near the top of a band (Fig. 5.20b).

In Fig. 5.20a, the point A is at an energy minimum. By definition, the curvature of the $\mathscr{E}$ versus k curve must be positive (i.e., pointing upward), meaning a positive m_e^*. For positive m_e^* in Eq. (5.10.1), we see that electrons move in a direction opposite to $\mathbf{E}$. If an electron is initially at A, it moves to A_1, A_3, A_5 and so on, in succession in k space until collision occurs. Using $-e\mathbf{E}$ for $\mathbf{F}$ in Eq. (5.5.5), we see that $d\mathbf{k}/dt$ is negative and hence $\mathbf{k}$ of the electron should decrease with time. Therefore, Eqs. (5.5.5) and (5.10.1) lead to the same conclusion.

Next we consider an occupied state B_4 in Fig. 5.20b. The negative curvature of the $\mathscr{E}$ versus k curve near B_4 indicates a negative m_e^*. Note that the velocity $\mathbf{v}$ in Eq. (5.10.1) is the group velocity v_g defined in Eq. (5.5.1). A negative effective mass in Eq. (5.10.1) simply means that the value of $\partial\mathscr{E}/\partial k$ should increase with time. For $k_x >$

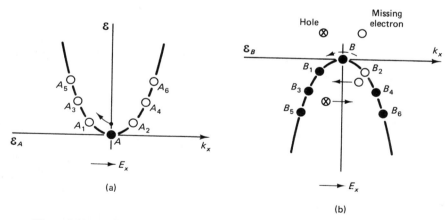

(a)

(b)

Figure 5.20 Motion of an electron in the $\mathscr{E}-k$ diagram for electrons (a) having energy near the bottom of a band which is almost empty and (b) having energy near the top of a band which is almost completely occupied. The empty states are indicated by open circles, whereas the occupied states are denoted by dots. The hole, in contrast to an empty state, is indicated by $\otimes$.

0 in Fig. 5.20b, $\partial\mathscr{E}/\partial k$ is negative. Therefore, to have $\partial\mathscr{E}/\partial k$ increasing with time, the electron must move from one position to another on the $\mathscr{E}$ versus k diagram with decreasing slope. In other words, the electron must move from B_6 to B_4 and then to B_2. This analysis leads to the conclusion that the electron near the top of a band still moves in a direction opposite to **E** even though it has a negative effective mass.

To account for the effects produced by a few vacant states near the top of an otherwise filled band, we introduce the concept of *hole*. The holes have physical properties that represent the physical properties of the totality of electrons in the band. For simplicity, we consider the situation in which there is a single empty state (B_2 in Fig. 5.20b). For a full band, there are as many states with $-\mathbf{k}$ as those with $+\mathbf{k}$; hence the sum $\Sigma\,\mathbf{k}$ is identically zero. Furthermore, the current density J is zero because a full band cannot conduct any electric current (Section 4.2). If there is a single vacant state (or a few vacant states), both $\Sigma\,\mathbf{k}$ and J are no longer zero. Our task is to make sure that the hole has proper physical properties such that the effects produced by a single hole are equivalent to the combined effects produced by electrons in a full band minus a single electron.

The equation of motion for a hole is given by

$$\frac{d\mathbf{v}_h}{dt} = \frac{e}{m_h^*}\,\mathbf{E} \tag{5.10.2}$$

Furthermore, a hole has the following properties:

$$\mathbf{k}_h = -\mathbf{k}_{me} \tag{5.10.3}$$

$$q_h = -q_{me} = +e \tag{5.10.4}$$

$$\mathscr{E}_h = -\mathscr{E}_e \tag{5.10.5}$$

$$m_h^* = -m_e^* = \text{positive} \tag{5.10.6}$$

$$\mathbf{v}_h = \mathbf{v}_{me} \tag{5.10.7}$$

Equations (5.10.3) to (5.10.7) relate the wave vector **k**, the charge q, the energy $\mathscr{E}$, the effective mass m^*, and the group velocity **v** of a hole (denoted by the subscript h) to the corresponding quantities associated with a missing electron or an empty state, had it been occupied by an electron (denoted, respectively, by the subscript me or e). For the situation shown in Fig. 5.20b, the sum of the wave vectors is $\Sigma_i\,\mathbf{k}_i - \mathbf{k}_{me}$. We equate this vector sum to $\mathbf{k}_h$. Since $\Sigma_i\,\mathbf{k}_i$ is zero over a full band, $\mathbf{k}_h = -\mathbf{k}_{me}$, which is the result expressed in Eq. (5.10.3). In Fig. 5.20b the locations of a missing electron and the corresponding hole along the k_x axis are shown.

Next, we consider the directions of motion for the missing electron and the hole. Equation (5.5.5) predicts that electrons will move with a negative dk_x/dt under an electric field E_x applied in the positive x direction. This means that electrons will move toward the left in Fig. 5.20b (the motion being indicated by the dashed arrow). A simultaneous motion of all the electrons toward the left will have the previously vacant state B_2 become occupied and the previously occupied state B become vacant. Therefore, the missing electron will also move toward the left in k space. Since $\mathbf{k}_h = -\mathbf{k}_{me}$, the hole will move in the direction of **E**. The directions of changing k for the missing electron and the corresponding hole are indicated by solid arrows in Fig. 5.20b.

The Band Theory of Electronic Conduction   Chap. 5

If the concept of hole developed thus far is a correct one, we expect that Eq. (5.5.5) should apply equally well to the hole. Thus we have

$$\frac{d\mathbf{k}_h}{dt} = -\frac{d\mathbf{k}_{me}}{dt} = \frac{-1}{\hbar}(q_{me}\mathbf{E}) = \frac{e\mathbf{E}}{\hbar} \qquad (5.10.8)$$

The condition $\mathbf{k}_h = -\mathbf{k}_{me}$ demands that the charge associated with the hole be opposite to that associated with a missing electron or $q_h = +e$. This property is stated in Eq. (5.10.4). Equation (5.10.8) also predicts a positive dk_{hx}/dt for positive E_x, which is consistent with our earlier prediction of the hole movement toward the right (Fig. 5.20b).

In our discussion thus far, the energy used in defining the $\mathscr{E}$–k relation represents the energy of an electron. From energy considerations, we know that electrons tend to stay in the lower-energy states which constitute the lower part of the band shown in Fig. 5.20b. If there are vacant states in the band, these vacant states will appear near the top of the band. For holes, the states near the top of a band are the lower-energy states. This reasoning leads to an assignment of a hole energy $\mathscr{E}_h$ which is opposite in sign to the energy of the state where the missing electron is located. Since the hole and the missing electron carry opposite electric charges, the assignment $\mathscr{E}_h = -\mathscr{E}_e$, which is stated in Eq. (5.10.5), also seems reasonable from charge considerations.

The combination of $\mathscr{E}_h = -\mathscr{E}_e$ and $\mathbf{k}_h = -\mathbf{k}_{me}$ further leads to other relations. Applying Eqs. (5.5.9) and (5.5.3) to the hole and the missing electron, we have

$$\frac{1}{m_h^*} = \frac{1}{\hbar^2}\frac{\partial^2\mathscr{E}_h}{\partial k_h^2} = \frac{-1}{\hbar^2}\frac{\partial^2\mathscr{E}_e}{\partial k_{me}^2} = -\frac{1}{m_e^*} \qquad (5.10.9)$$

$$v_h = \frac{1}{\hbar}\frac{\partial\mathscr{E}_h}{\partial k_h} = \frac{-1}{-\hbar}\frac{\partial\mathscr{E}_e}{\partial k_{me}} = v_{me} \qquad (5.10.10)$$

The results are the same as those stated in Eqs. (5.10.6) and (5.10.7). For simplicity, we have assumed a one-dimensional case in the derivation of Eq. (5.10.10).

Now we analyze the problem of current conduction in a partly filled band. For the situation shown in Fig. 5.20b, applying Eq. (A4.17) to a one-dimensional case yields

$$J = -e\sum(v_e - v_{me}) \qquad (5.10.11)$$

where the summation represents counting of electrons in a unit volume. The subscript e refers to all band states, whereas the subscript me refers to the vacant states (missing electrons). For a full band, $J = -e\sum v_e = 0$. Using the condition $\sum v_e = 0$ in Eq. (5.10.11), we obtain

$$J = e\sum v_{me} = e\sum v_h \qquad (5.10.12)$$

The summation represents counting of holes in a unit volume. The transformation from Eq. (5.10.11) to Eq. (5.10.12) is significant physically in two respects. In Eq. (5.10.11) the summation is over all occupied band states, while in Eq. (5.10.12) the summation is over all empty band states. Moreover, the sign associated with the charge changes from $-e$ to $+e$, meaning that a hole has the equivalent effect of a positive charge $+e$ insofar as electric current is concerned.

Let us now apply the results of our analysis to electronic conduction in solids. Using Eqs. (5.10.1) and (5.10.2) in place of Eq. (4.8.8) and following a similar pro-

TABLE 5.1 COMPARISON OF THE ESSENTIAL PROPERTIES OF ELECTRONS AND HOLES[a]

Electrons occupied band states	Physical state	Holes vacant band states
$-e$	Charge	$+e$
$m_e^* = $ positive	Effective mass	$m_h^* = $ positive
$\mathbf{v}_d = -\dfrac{e\tau_e}{m_e^*}\mathbf{E}$	Drift velocity	$\mathbf{v}_d = \dfrac{e\tau_h}{m_h^*}\mathbf{E}$
$m_e^* = \dfrac{1}{\hbar^2}\dfrac{\partial^2 \mathscr{E}}{\partial k^2}$	Effective mass	$m_h^* = -\dfrac{1}{\hbar^2}\dfrac{\partial^2 \mathscr{E}}{\partial k^2}$
$\dfrac{1}{\hbar^2}\dfrac{\partial^2 \mathscr{E}}{\partial k^2} = $ positive	Where applicable	$\dfrac{1}{\hbar^2}\dfrac{\partial^2 \mathscr{E}}{\partial k^2} = $ negative

[a]The energy $\mathscr{E}$ is the electron energy.

cedure as used in deriving Eq. (4.8.10), we find the drift velocities of electrons and holes to be

$$\mathbf{v}_{de} = -\frac{e\tau_e}{m_e^*}\mathbf{E}, \qquad \mathbf{v}_{dh} = \frac{e\tau_h}{m_h^*}\mathbf{E} \qquad (5.10.13)$$

where τ_e and τ_h are the respective mean free times. According to Eq. (4.8.11), we define two mobilities: μ_e for electrons and μ_h for holes,

$$\mu_e = \frac{e\tau_e}{m_e^*}, \qquad \mu_h = \frac{e\tau_h}{m_h^*} \qquad (5.10.14)$$

The conductivity of a solid with simultaneous electron and hole conduction is

$$\sigma = e\mu_e n + e\mu_h p \qquad (5.10.15)$$

where p is the hole concentration (number of empty states per unit volume). Equation (5.10.15) replaces Eq. (4.2.3).

In concluding this section it may be useful to summarize the important information obtained from the discussion above. In Table 5.1 we list the essential properties of electrons and holes.

5.11 HALL MEASUREMENT AND CYCLOTRON RESONANCE EXPERIMENT

In Section 5.10 we discussed two important situations in the zone theory of electronic conduction: the electron state with a negative electronic charge and the hole state with a positive electronic charge. In this section we describe experiments by which we can determine whether the charge carriers in a solid are electron-like or hole-like. Consider the experimental arrangement of Fig. 5.21 in which a dc magnetic field is applied to a rectangular specimen in a direction perpendicular to the applied dc electric field. Under the circumstance, the equation of motion for electrons is

$$m_e^* \frac{d\mathbf{v}}{dt} = -e(\mathbf{E} + \mathbf{v} \times \mathbf{B}) \qquad (5.11.1)$$

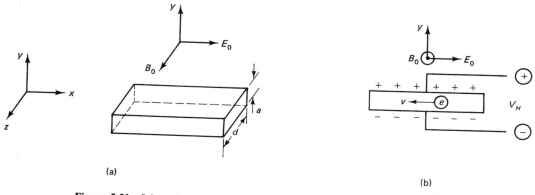

(a)

(b)

Figure 5.21 Schematic diagram showing (a) the arrangement of the electric field E_0 and the magnetic flux density B_0 and (b) the direction of the measured Hall voltage V_H with respect to E_0 and B_0. The Hall voltage is measured in a direction perpendicular to the plane containing $\mathbf{E}_0$ and $\mathbf{B}_0$.

The equation for holes is similar to Eq. (5.11.1) except that m_e^* and $-e$ should be replaced by m_h^* and $+e$, respectively.

Let us assume that the charge carriers are electrons. For an electric field $\mathbf{E}_0$ applied in the $+x$ direction, the electrons move in the $-x$ direction. With $\mathbf{B}_0$ pointing in the $+z$ direction, the Lorentz force exerted on electrons is

$$\mathbf{F} = q\mathbf{v} \times \mathbf{B} = -e\mathbf{v} \times \mathbf{B} \qquad (5.11.2)$$

This force drives electrons toward the bottom surface of the specimen, making the bottom surface negatively charged and leaving the top surface positively charged with uncompensated positive ions. Because of the space-charge buildup on the two surfaces, an electric field E_y develops in the $-y$ direction. Under an open-circuit condition (i.e., no current flow in the y direction), the electrostatic force $-eE_y$ must balance out the Lorentz force. In other words,

$$-eE_y + ev_x B_0 = 0 \qquad (5.11.3)$$

The drift velocity v_x is related to the current density J_x by $J_x = -env_x$. Using this information in Eq. (5.11.3), we find that

$$E_y = -\frac{J_x B_0}{ne} \qquad (5.11.4)$$

The ratio, defined as

$$R_H = \frac{E_y}{J_x B_0} = -\frac{1}{ne} \qquad (5.11.5)$$

is called the *Hall coefficient*. Experimentally, we measure a voltage V_H, called the Hall voltage, across the sample in the direction perpendicular to both $\mathbf{B}_0$ and $\mathbf{J}$. The voltage V_H is measured under open-circuit conditions. In terms of V_H, Eq. (5.11.5) becomes

$$R_H = \frac{V_H d}{B_0 I} = -\frac{1}{ne} \qquad (5.11.6)$$

Sec. 5.11   Hall Measurement and Cyclotron Resonance Experiment   **181** where I is the total current passing through the sample and d is the width of the sample in the direction of $\mathbf{B}_0$ (Fig. 5.21).

According to the free-electron theory of metals, the number of free electrons per unit volume is equal to $n_a N$ where n_a is the number of valence electrons per atom and N is the number of atoms per unit volume. Therefore, the product

$$eNR_H = \frac{1}{n_a} \tag{5.11.7}$$

is a measure of n_a. In Table 5.2 the values of eNR_H are given for a number of metals selected from the data compiled by Ziman (J. M. Ziman, *Electrons and Phonons*, Oxford University Press, Oxford, 1960, p. 488). Note that the polarity of V_H changes in Fig. 5.21 if the charge carriers are holes. The sign in Table 5.2 indicates the conventional sign of the charge of carriers ($-$ for electrons and $+$ for holes).

As we can see, a good agreement with the free-electron theory is obtained in alkali and noble metals listed in column I. For divalent (column II) and trivalent (column III) metals, the value of eNR_H is expected to be -0.50 and -0.33, respectively. The measured value is more or less consistent with the free-electron model in Mg, Ca, Hg, Al, and Ga. The positive sign found in Be, Zn, and Cd indicates that electric conduction is conducted by holes. The small value found in indium means that the positive contribution from hole-like carriers nearly matches the negative contribution from electron-like carriers. The application of a two-band model, electron and hole, to the analysis of the Hall effect is discussed in Chapter 6.

Next we discuss the cyclotron resonance experiment. Consider the motion of a free electron in the presence of a dc magnetic field B_0. Since the Lorentz force $e\mathbf{v} \times \mathbf{B}$ is always perpendicular to $\mathbf{v}$, only the direction of $\mathbf{v}$ changes but not its magnitude. As a result, the electron performs a circular motion (Fig. 5.22a). The acceleration in the radial direction is $a = evB_0/m$. The increment of the velocity in the radial direction is

$$v\,\Delta\theta = a\,\Delta t = \frac{evB_0}{m}\,\Delta t \tag{5.11.8}$$

From Eq. (5.11.8) the angular frequency of the circular motion is

$$\omega_c = \frac{\Delta\theta}{\Delta t} = \frac{eB_0}{m} \tag{5.11.9}$$

which is known as the *cyclotron resonance frequency*.

TABLE 5.2 THE PRODUCT OF eNR_H IN
A SELECTED NUMBER OF METALS

I		II		III	
Li	-1.3	Be	$+5.0$	Al	-0.4
Na	-0.9	Mg	-0.7	Ga	-0.5
K	-0.9	Ca	-0.7	In	-0.04
Rb	-1.0				
Cu	-0.8	Zn	$+0.4$		
Ag	-0.8	Cd	$+0.5$		
Au	-0.7	Hg	-0.6		

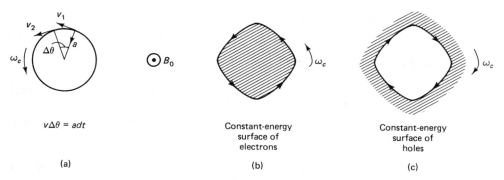

$v\Delta\theta = adt$

(a)

Constant-energy
surface of
electrons

(b)

Constant-energy
surface of
holes

(c)

Figure 5.22 (a) Motion of an electron in a magnetic field B in the free-electron model. Since the Lorentz force is always perpendicular to the velocity $\mathbf{v}$, the electron describes a circular motion with the magnitude of v being constant. (b) Electron orbit and (c) hole orbit in a magnetic field in the zone theory. The orbits follow the contour of constant-energy surface in k space in a plane normal to the magnetic field. The shaded areas represent the occupied states in the zone. The arrows indicate the direction of electron (or hole) orbital motion.

If an ac electric field is applied in a plane (assumed to be the xy plane) perpendicular to the direction (assumed to be the z axis) of the dc magnetic field B_0, resonant absorption of energy by the electron occurs when the frequency of the ac electric field coincides with the cyclotron frequency. In the presence of E_x and E_y, the equations of motion for a free electron become

$$m\frac{dv_x}{dt} + \frac{mv_x}{\tau} = -eE_x - ev_yB_0 \tag{5.11.10}$$

$$m\frac{dv_y}{dt} + \frac{mv_y}{\tau} = -eE_y + ev_xB_0 \tag{5.11.11}$$

where τ is the mean free time. The term mv/τ accounts for the retarding force due to collisions. Equations (5.11.10) and (5.11.11) are obtained from Eq. (4.8.15).

Note that the sense of the electron motion in Fig. 5.22a is counterclockwise. To supply energy continuously to the electron, the electric field must be rotating in the same sense. Therefore, we let

$$E_x = E_1 \cos(\omega t + \theta) \qquad E_y = E_1 \sin(\omega t + \theta) \tag{5.11.12}$$

The solution of Eqs. (5.11.10) and (5.11.11) can easily be found through the use of the phasor notation (complex notation). The task is left to the reader as a problem. If the use of real quantities is followed, the solution can be found by assuming that

$$v_x = v_1 \cos \omega t, \qquad v_y = v_1 \sin \omega t \tag{5.11.13}$$

Substituting Eqs. (5.11.12) and (5.11.13) in Eq. (5.11.10) and equating the respective coefficient of $\sin \omega t$, and $\cos \omega t$, we find that

$$m(\omega_c - \omega)v_1 = eE_1 \sin \theta \tag{5.11.14}$$

$$\frac{m}{\tau} v_1 = -eE_1 \cos \theta \tag{5.11.15}$$

The power delivered to the electron is

$$P = Fv = -e(v_x E_x + v_y E_y) = -eE_1 v_1 \cos \theta \qquad (5.11.16)$$

Substituting Eq. (5.11.15) in Eq. (5.11.16) for $\cos \theta$, we find that

$$P = \frac{m}{\tau} v_1^2 \qquad (5.11.17)$$

The quantity v_1^2 can be found from Eqs. (5.11.14) and (5.11.15). Use of v_1^2 in Eq. (5.11.17) yields

$$P = \frac{\tau(eE_1)^2}{m[1 + (\omega_c - \omega)^2 \tau^2]} \qquad (5.11.18)$$

Maximum absorption occurs when $\omega_c = \omega$. The reader is asked to show that if a clockwise field with

$$E_x = E_1 \cos (\omega t + \theta), \qquad E_y = -E_1 \sin (\omega t + \theta) \qquad (5.11.19)$$

is used and a corresponding change in the sign of v_y is made, a similar expression for P is obtained, with the term $(\omega_c - \omega)^2 \tau^2$ changed to $(\omega_c + \omega)^2 \tau^2$. Therefore, no appreciable absorption is expected from a clockwise field.

The foregoing discussion can easily be applied to cyclotron-resonance absorption due to electrons and holes in a solid. For electrons, the analysis is the same except that the effective mass m_e^* is used in Eq. (5.11.9). For holes, the sense of rotation is clockwise (Fig. 5.22c) because of the change in the sign of the charge. To observe resonant absorption due to holes, a field rotating in the clockwise direction must be used. This field is given in Eq. (5.11.19). In contrast, the field of Eq. (5.11.12) is suitable for electrons. The electron orbit and the hole orbit in the presence of a dc magnetic field are illustrated in Fig. 5.22 for comparison.

Cyclotron resonance experiments have contributed very significantly toward understanding the band structure of semiconductors. Figure 5.23 shows the resonance ab-

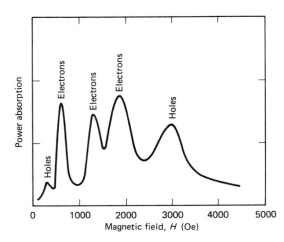

Figure 5.23 Cyclotron-resonance absorption observed in germanium at 24 GHz and 4 K. The magnetic field is in a (110) plane at 60° from a [001] axis. The five resonance peaks are due to the complicated band structure of germanium. An analysis of the data may be found in Section 6.10. (G. Dresselhaus, A. F. Kip and C. Kittel, *Phys. Rev.*, Vol. 98, p. 368, 1955.)

sorption in Ge measured at 24 GHz and 4 K reported by Dresselhaus, Kip, and Kittel (G. Dresselhaus, A. F. Kip, and C. Kittel, *Phys. Rev.*, Vol. 98, p. 368, 1955). In their experiment, a linearly polarized ac electric field was used. Since a plane-polarized wave can be resolved into two circularly polarized components, one of the form shown in Eq. (5.11.12) and the other shown in Eq. (5.11.19), both electron and hole resonant absorptions were observed. Imagine that in Fig. 5.20b there are two valence bands,

$$\mathscr{E}_1 = \mathscr{E}_B - \frac{\hbar^2 k^2}{2m_{h1}^*} \quad \text{and} \quad \mathscr{E}_2 = \mathscr{E}_B - \frac{\hbar^2 k^2}{2m_{h2}^*} \tag{5.11.20}$$

with two different effective masses. In Ge, the two observed resonances correspond to $m_{h1}^* = 0.04m_0$ and $m_{h2}^* = 0.32m_0$ where m_0 is the free-electron mass. The conduction-band structure in Ge is rather complicated and will be discussed in Chapter 6. The reader is referred to Section 6.10 for further discussion of cyclotron-resonance data.

Cyclotron-resonance experiments have also been exceedingly useful in yielding information about the Fermi surfaces in metals. Note that in Fig. 4.22 only those electrons near $\mathscr{E}_f$ have available vacant states for them to move into. The constant-energy surfaces in Fig. 5.22 are Fermi surfaces. In a dc magnetic field, electrons and holes can neither gain nor lose energy, and hence their orbits follow the Fermi surface in a plane perpendicular to $\mathbf{B}_0$. If T is the period of time required for an electron or a hole to complete a revolution, then $\omega_c = 2\pi/T$. This general definition applies to complicated Fermi surfaces other than spherical Fermi surfaces (see, for example, J. M. Ziman, *Principles of the Theory of Solids*, Cambridge University Press, Cambridge, 1965, p. 250).

One difficulty which arises in performing a cyclotron-resonance experiment in a metal is the skin-depth problem. To observe a reasonably sharp resonance [Eq. (5.11.18)], we must make $\omega_c \tau > 1$. At 4 K, τ is roughly 10^{-10} s. Let us take $\omega_c = 5 \times 10^{10}$ rad/s. For a typical metal such as copper $\sigma = 5.8 \times 10^7$ S/m, and the skin depth is

$$\delta = \left(\frac{2}{\omega\mu\sigma}\right)^{1/2} = 7.3 \times 10^{-7} \text{ m} \tag{5.11.21}$$

This value is to be compared with the mean free path l of 1.5×10^{-4} m calculated from Eq. (4.8.12). The small skin depth ($\delta \ll l$) that we encounter in metals not only makes the absorption weak to detect but also makes the experimental data difficult to interpret.

Despite the experimental difficulties, cyclotron-resonance experiments for metals have been performed successfully. Galt et al. used bismuth as part of the microwave cavity wall at frequencies of 24 and 72 GHz. They made a thorough analysis of the anomalous skin effect in relation to the experiment (J. K. Galt et al., *Phys. Rev.*, Vol. 114, p. 1396, 1959). In their experiment, a circularly polarized wave of the form shown in Eqs. (5.11.12) and (5.11.19) was used. Therefore, separate hole absorption and electron absorption were observed by changing the direction of $\mathbf{B}_0$. Bismuth is a semimetal in which electrons and holes exist in equal numbers. From the experimental data, they found the energy surface of Bi to be ellipsoidal [Eq. (5.5.11)] with

$$\frac{m_{11}}{m_0} = 0.0088, \quad \frac{m_{22}}{m_0} = 180, \quad \frac{m_{33}}{m_0} = 0.023, \quad \frac{m_{23}}{m_0} = \pm 0.16 \tag{5.11.22}$$

for electrons and

$$\frac{m_{11}}{m_0} = \frac{m_{22}}{m_0} = 0.068, \qquad \frac{m_{33}}{m_0} = 0.92 \qquad (5.11.23)$$

for holes.

PROBLEMS

5.1. Verify that Eqs. (5.2.2) and (5.2.3) are solutions of Eq. (5.2.1) for $0 < x < a$ and $-b < x < 0$ (Fig. 5.2b), respectively. Also show that both $d\psi/dx$ and ψ must be continuous at $x = a$ and $x = -b$ if $V(x)$ is finite. [*Hint:* Integrate Eq. (5.2.1) over a very short distance from $x = a - \delta$ to $a + \delta$, for example.]

5.2. In the simplified Kronig–Penney model, the lattice potential is assumed to take the form of periodic potential spikes of infinite amplitude V_0 and zero width b. Also, the product $V_0 b$ is taken to be finite and set equal to $(2P/a)\,(\hbar^2/2m)$. By integrating the one-dimensional Schrödinger equation over the span of the spike, show that the discontinuity in the first derivative of the wave function across the spike is given by

$$\left.\frac{d\psi}{dx}\right|_{0+} - \left.\frac{d\psi}{dx}\right|_{0-} = \frac{2P}{a}\,\psi\bigg|_0$$

Using the information above and the Floquet–Bloch theorem,

$$\psi(x + c) = \psi(x)\,\exp\,(ikc)$$

derive the following characteristic equation:

$$\frac{P}{\beta a}\,\sin\,\beta a + \cos\,\beta a = \cos\,ka$$

5.3. Consider a one-dimensional periodic lattice (Fig. 5.14) which has a cell wave function $u(x, k)$ given by ψ_1 of Fig. 4.2. Draw a schematic diagram to show the behavior of $\psi(x, k)$ of Eq. (5.3.1) in Fig. 5.14 over a distance of $10c$ for two cases: k being real and k being imaginary. Explain why the case for imaginary k corresponds to the nonpropagating mode of the electron-wave motion.

5.4. Using ψ_1 of Fig. 4.2 as the cell wave function, draw a schematic diagram to show the spatial variation of $\psi(x, k)$ of Eq. (5.3.1) in Fig. 5.14 over a distance of $12c$ for three cases: $kc = \pi$, $\pi/4$, and $\pi/6$. [*Hint:* Take either the real or the imaginary part of $\psi(x, k)$.]

5.5. The proof of Bloch's theorem presented in Section 5.3 can be put mathematically in a simpler but more abstract form. A periodic lattice possesses translational symmetries represented by translation operators T_j's such that $T_j F(\mathbf{r}) = F(\mathbf{r} + \mathbf{R}_j) = F(\mathbf{r})$, where $\mathbf{R}_j$ is a translation vector. Show that if $\mathcal{H}(\mathbf{r} + \mathbf{R}_j) = \mathcal{H}(\mathbf{r})$ where $\mathcal{H}$ is the Hamiltonian, then

$$\mathcal{H}(\mathbf{r} + \mathbf{R}_j)\psi(\mathbf{r} + \mathbf{R}_j) = \mathcal{E}\psi(\mathbf{r} + \mathbf{R}_j)$$
$$\mathcal{H}(\mathbf{r} + \mathbf{R}_j)\psi(\mathbf{r}) = \mathcal{E}\psi(\mathbf{r})$$

and

$$(\mathcal{H}T - T\mathcal{H})\psi = 0$$

Since $\mathcal{H}$ and T operators commute, eigenfunctions of $\mathcal{H}$ are also eigenfunctions of T, or

$$T\psi = \lambda\psi$$

which is identical to Eq. (5.3.6).

5.6. For the pulse shown in Fig. 5.5,

$$V(t) = V_0 \quad \text{for } -\tau < t < \tau \text{ and zero elsewhere}$$

show that the pulse has a frequency distribution

$$G(\omega) = V_0 \sqrt{\frac{2}{\pi}} \frac{\sin \omega\tau}{\omega}$$

Plot $G(\omega)$ as a function of $\omega\tau$. We define the bandwidth $\Delta\omega$ such that $G(\omega)$ drops to $2/\pi$ of its maximum value at $\omega = \Delta\omega/2$. Show that the product $\Delta\nu\,\Delta t = 1$, Δt being the duration of the pulse and $\nu = \omega/2\pi$.

5.7. Sketch the spatial variation of the function

$$\psi(x,\,t) = 2A \cos (\Delta k\,x - \Delta\omega\,t) \cos (kx - \omega t)$$

at two instants $t = 0$ and $t = 16\pi/\Delta\omega$. Only the part of $\psi(x,\,t)$ with $\phi = \Delta kx - \Delta\omega\,t$ lying between $= -\pi/2$ and $+\pi/2$ is to be drawn and the remaining part is to be omitted. Show clearly the movement of the wave packet in space. For definiteness, the following values for $k = 8\,\Delta k$ and $\omega = 8\,\Delta\omega$ are assumed.

5.8. Using Eq. (5.2.20), show that

$$\frac{d^2\beta}{dk^2} = \frac{-a \cos ka}{P(\cos \beta a/\beta a - \sin \beta a/\beta^2 a^2) - \sin \beta a}$$

at the band edges A, B, C, and so on. In obtaining the relation above, the condition $d\beta/dk = 0$ is used. Is this condition generally valid? (*Hint: $d\beta/dk$ is related to the group velocity of an electron wave.*)

5.9. (a) Show that for $P = 3\pi/2$ in Eq. (5.2.20), the band edge at B has a value of $\beta a = \pi$.
(b) Using the expression $d^2\beta/dk^2$ given in Problem 5.8, find the effective mass of electrons at the top of the band AB in Fig. 5.4.

5.10. (a) Show that for $P = 3\pi/2$ in Eq. (5.2.20), the band edge at C has a value of $\beta a = 3\pi/2$.
(b) Using the expression $d^2\beta/dk^2$ given in Problem 5.8, find the effective mass of electrons at the bottom of the band CD in Fig. 5.4.

5.11. It is stated in Fig. 5.1 that at k_{diff}, Bragg reflection of electron wave takes place. For total reflection, we expect the group velocity v_g of electrons to be zero at k_{diff} where an energy gap exists. Use the result of the one-dimensional Kronig–Penney model as an example to show that v_g is indeed zero at $ka = \pm\pi$, $\pm2\pi$, and so on, in Fig. 5.4. Also explain why the linear terms, such as the term proportional to $k_i - k_{i0}$, are not present in Eq. (5.5.11).

5.12. Show that the Bragg condition stated in Eq. (5.6.8) is satisfied if the k vector terminates either on the first-zone boundary or on the second-zone boundary in Fig. 5.8b. Also verify that Eq. (5.6.6) is satisfied on the zone boundaries.

5.13. Consider a two-dimensional square lattice (Fig. 5.8a). Construct its first Brillouin zone and indicate each allowed k value by a dot in the two-dimensional k space. Find (a) the number of primitive cells in an area L^2 in real space, (b) the number of independent k states in the first Brillouin zone, (c) the area of the first Brillouin zone in k space, and (d) the number of quantum states dN in an area $dk_x\,dk_y$ in k space. Check the result from part (b) by counting the number of dots in the first Brillouin zone.

5.14. Repeat Problem 5.13 for a two-dimensional rectangular lattice with lattice spacing $b = 2a$.

5.15. (a) As discussed in Sections 2.2 and 5.6, the three basis vectors for the primitive cell of a fcc structure (Fig. 2.4b) are the face diagonals given by Eq. (5.6.14). Calculate the volume of the unit cell.

(b) Using the basis vectors above, show that the three basis vectors $\mathbf{a}^*$, $\mathbf{b}^*$, and $\mathbf{c}^*$ for the reciprocal lattice for a fcc structure are given by Eq. (5.6.15). Thus a vector $\mathbf{r}^*$ in the reciprocal lattice can be expressed as

$$\mathbf{r}^* = h\mathbf{a}^* + k\mathbf{b}^* + l\mathbf{c}^*$$

(c) In Fig. 5.9a, there are eight hexagons along the body diagonals and six squares touching the faces of a cube. First, find the zone boundaries along the principal axes and along the body diagonals. Then show that the Bragg condition is satisfied at the zone boundaries.

5.16. Show that for a bcc structure, the vectors

$$\mathbf{a}^* = \frac{\hat{\mathbf{a}}_y + \hat{\mathbf{a}}_z}{a}, \qquad \mathbf{b}^* = \frac{\hat{\mathbf{a}}_x - \hat{\mathbf{a}}_y}{a}, \qquad \mathbf{c}^* = \frac{\hat{\mathbf{a}}_x - \hat{\mathbf{a}}_z}{a}$$

form a set of three basis vectors for the reciprocal lattice where $\hat{\mathbf{a}}_x$, $\hat{\mathbf{a}}_y$, and $\hat{\mathbf{a}}_z$ are three orthogonal unit vectors. The intersections of a [100] plane with the first Brillouin zone of Fig. 5.9b passing through its center form a square. Find the zone boundaries from $\mathbf{G}_\alpha$, $\mathbf{G}_\beta$, and $\mathbf{G}_\gamma$. Describe the shape of the first zone.

5.17. We know from experience that the properties of a laboratory-size crystal, such as electrical and thermal conductivity, is independent of the size of the crystal. Therefore, neglecting surface effects, we may treat a finite-size crystal (Fig. 5.10a) as if the crystal repeated itself at the boundaries. The one-dimensional mathematical model used to describe the situation can be represented by a ring (Fig. 5.10b).

Suppose that the wave functions at neighboring cells are related to one another by the relation

$$\psi(x + c) = C\psi(x)$$

where C is a constant. Employing the relation above, establish the relation between $\psi(x + Nc)$ and $\psi(x)$. Show that for $\psi(x + Nc) = \psi(x)$, we must have

$$\psi(x + c) = \psi(x) \exp\ (ikc)$$

with $k = q2\pi/Nc$, where q is an integer.

5.18. Starting with Eqs. (5.2.1) and (5.7.13), obtain the results expressed in Eqs. (5.7.18), (5.7.19), (5.7.21a), and (5.7.21b).

5.19. (a) Explain the physical origin of energy gap in the nearly free electron model.

(b) In our earlier discussions (Figs. 5.1b and 5.7), we have stated that reflection of electron wave occurs when the Bragg condition is satisfied. Is this statement consistent with the results expressed in Eqs. (5.7.21a) and (5.7.21b)?

5.20. Near a zone boundary (i.e., $k \cong n\pi/a$) both the incident wave $\exp\ (ikx)$ and the Bragg-reflected wave $\exp\ [i(k - nG)x]$ must be included in the wave function in finding $\mathscr{E}$. Starting from Eqs. (5.8.1) and (5.8.2), show that the characteristic equation for $\mathscr{E}$ is given by Eq. (5.8.14) and the energy $\mathscr{E}$ is given by Eq. (5.8.15). From Eq. (5.8.15) find the energy gap at $k = q\pi/a$.

5.21. Let $\psi_s = f(r)$ be the atomic wave function representing an s electron. The wave function is plotted as a function of x in Fig. P5.21, and is symmetric with respect to x, y, and z.

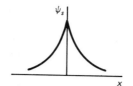

Figure P5.21

Consider the generalization of Eq. (5.9.13) so that it is applicable to a three-dimensional cubic lattice. One of the integrals used in the equation is

$$\int f(x \pm a, y, z)\mathcal{H}_1 f(x, y, z)\, d\tau = \beta_x \int |\psi^2|d\tau$$

which is used for the coefficient of the cos $k_x a$ term.

(a) Find the expressions for the coefficients of the cos $k_y a$ and cos $k_z a$ terms. Draw a diagram similar to Fig. P5.21 showing the two wave functions involved in the three integrals. Also use qualitative arguments to show that $\beta_x = \beta_y = \beta_z$ and they are positive.

(b) Find the width of the band for s electrons. Show that $\partial\mathcal{E}/\partial k_i = 0$ at the zone boundaries where $i = x, y,$ or z.

5.22. Let $\psi_x = xg(r)$ be the atomic wave function representing one of the three p states. The wave function is plotted as functions of $x, y,$ and z in Fig. P5.22. It is antisymmetric with

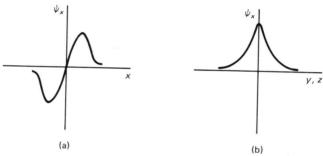

(a)   (b)

Figure P5.22

respect to x and symmetric with respect to y and z. For a three-dimensional cubic lattice, one of the integrals used in Eq. (5.9.14) is

$$\int (x \pm a)g(x \pm a, y, z)\mathcal{H}_1 xg(x, y, z)\, d\tau = \beta_x \int |\psi|^2\, d\tau$$

which is used for the coefficient for the cos $k_x a$ term.

(a) Find the expressions for the coefficients of the cos $k_y a$ and cos $k_z a$ terms. Draw diagrams similar to Fig. P5.22 showing the two wave functions involved in the integrals. Use qualitative arguments to show that $\beta_y = \beta_z = \beta_2 =$ positive and $\beta_x = -\gamma_2 =$ negative in Eq. (5.9.15).

(b) Find the width of the p_x band. Show that $\partial\mathcal{E}/\partial k_i = 0$ at the zone boundaries where $i = x, y,$ or z.

5.23. The mathematical treatment presented in Section 5.8 is based on the physical consideration discussed in Section 4.1. Using the wave functions ψ_s and ψ_x shown in Figs. P5.21 and P5.22a, sketch the composite electron wave function ψ of Eq. (4.1.4) for $ka = 0$ and

$ka = \pi$. Show by qualitative arguments that the energy is lower at $ka = 0$ than at $ka = \pi$ for the ψ_s state and that the energy is higher at $k_xa = 0$ than at $k_xa = \pi$ for the ψ_x state. Check this conclusion with the results presented in Eqs. (5.9.14) and (5.9.15). What are the physical origins of the terms α_1 in Eq. (5.9.14) and α_2 in Eq. (5.9.15)? [*Hint:* Compare $V(x)$ with $V_0 (x - nc)$.]

5.24. (a) From Eq. (5.9.15),. find the width of the p_x band. From Fig. 4.6, we see that the $2p$ band is much narrower than the $3p$ band. Explain why this is expected from Eq. (5.9.15).

(b) For symmetry reasons, it is expected in a cubic crystal that the wave function at $k[000]$ should have the same x, y, and z dependence. Consider the following wave function for a p electron at $k[000]$:

$$\psi_p = \frac{1}{\sqrt{3}} (\psi_x + \psi_y + \psi_z)$$

where $\psi_x = xg(r)$ and so on are the three atomic p orbitals. Modify Eq. (5.9.15) so that it can be used for the p electron. Use qualitative arguments to show that the modified equation is of the correct form.

5.25. The energy-band structure depends not only on the electron wave function (Problem 5.24) but also on the crystal structure. We illustrate this with a two-dimensional face-centered square structure shown in Fig. P5.25. Show that for an s electron the energy $\langle \mathscr{E} \rangle$ is of the form

$$\langle \mathscr{E} \rangle = \mathscr{E}_0 - \alpha - 4\beta \cos k_xa \cos k_ya$$

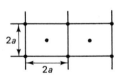

Figure P5.25

5.26. Based on the result obtained in Problem 5.25, deduce the form for $\langle \mathscr{E} \rangle$ for s electrons in (a) a faced-centered cubic structure, and (b) a body-centered cubic structure.

5.27. From Eq. (5.10.1) it might appear that an electron near the top of a band (Fig. 5.20b) would move in the direction of **E** because the electron has a negative effective mass. Is this conclusion correct? If not, give a correct interpretation of Eq. (5.10.1). Also show that your interpretation is consistent with the prediction based on Eq. (5.5.5). (*Hint:* Based on the $\mathscr{E}$–k diagram sketch the variation of $\partial \mathscr{E}/\partial k_x$ as a function of k_x.)

5.28. (a) From Fig. 5.20b, draw an $\mathscr{E}_h$ versus k_h plot for the hole associated with the empty state B_2. Sketch the variation of the group velocity as a function of wave vector k for both the missing electron and the hole. On each diagram, locate the position of the hole associated with B_2.

(b) Use the $\mathscr{E}$–k and v_g–k diagrams to show that by imposing the relation $\Sigma \, \mathbf{k} = 0$ for a full band and making $\mathscr{E}_h = -\mathscr{E}_e$, the other four relations expressed in Eqs. (5.10.3) to (5.10.7) are indeed observed.

5.29. (a) In Si, the energy of a conduction-band electron in the valley with minimum located at $\mathbf{k} = 1.4\pi/a$ [100] in Fig. 6.22 can be expressed as

$$\mathscr{E} = \mathscr{E}_c + 6.25 \times 10^{-39} \left(k_x - \frac{1.4\pi}{a} \right)^2 + 3.16 \times 10^{-38} (k_y^2 + k_z^2) \quad \text{(MKS)}$$

Find the effective mass or masses for Si. Also express m^* in terms of free-electron mass m_0.

(b) Calculate the momentum of the electron at $\mathbf{k} = (1.4\pi/a)$ [100]. Explain any discrepancy between the calculated value and the one predicted from Eq. (5.5.6).

5.30. Apply the equation of motion to holes under cross electric and magnetic fields of Fig. 5.21. Find the Hall coefficient R_H and indicate the polarity of the Hall voltage V_H.

5.31. Verify Eqs. (5.11.14) and (5.11.15). Further, show that the velocities v_x and v_y are given by

$$v_x = -\frac{e\tau}{m}\frac{1}{1+a^2}(E_1 \cos \omega t' + aE_1 \sin \omega t')$$

$$v_y = -\frac{e\tau}{m}\frac{1}{1+a^2}(E_1 \sin \omega t' - aE_1 \cos \omega t')$$

where $\omega t' = \omega t + \theta$ and $a = (\omega - \omega_c)\tau$.

5.32. (a) In the complex notation, the field in Eq. (5.11.12) may be expressed as

$$E_x = \text{Re} [E_1 \exp (i\omega t')], \qquad E_y = \text{Re} [- iE_1 \exp (i\omega t')]$$

where Re stands for the real part and $\omega t' = \omega t + \theta$. Show that by eliminating v_y, Eqs. (5.11.10) and (5.11.11) have a solution

$$v_x = -\frac{e\tau}{m}\frac{E_1 \exp (i\omega t')}{1 + i(\omega - \omega_c)\tau}$$

which is identical to the result obtained in Problem 5.31.

(b) Verify that for the electric field of Eq. (5.11.19), the solution becomes

$$v_x = -\frac{e\tau}{m}\frac{E_1 \exp (i\omega t')}{1 + i(\omega + \omega_c)\tau}$$

Which field, the field of Eq. (5.11.12) or the field of Eq. (5.11.19), should be used for resonance absorption by electrons and by holes? Give physical reasoning for your choice of the field in each case. (*Hint:* Draw diagrams to show the sense of rotation of the field.)

6

Semiconductor Fundamentals

6.1 INTRODUCTION

Semiconductors are an important class of material both for industrial use and for scientific study. Over the last two decades, semiconductors have come to be used in a wide range of electronic devices, such as transistors, switching devices, voltage regulators, photocells, and photodetectors. The rapid growth of the semiconductor industry has stimulated the demands for better material understanding and material quality. The technology of crystal growth developed during this period has made it possible to produce crystals of exceedingly high purity and crystal perfection. In the meantime, the list of semiconductor materials has expanded steadily. The availability of new and good-quality materials and the demand for a better understanding of material properties have prompted scientists to make concerted efforts in many directions. The interplay between technology and science in this area has been one of the most fruitful and rewarding experiences in human endeavor.

The wealth of knowledge accumulated on the properties of semiconductors has made possible a systematic presentation of subject matters. In this chapter we discuss those properties pertaining to fundamental properties of semiconductors. Most of our discussion is based on the free-electron theory presented in Chapter 4. Certain detailed analyses, however, require an understanding of the band structure. When such occasions arise, we shall refer to our discussion presented in Chapter 5. Semiconductors of the III–V compound family have gained in recent years increasing importance in many applications because they possess certain unique properties and capabilities which elemental semiconductors do not have. Advanced topics such as heterojunctions, two-dimensional electron gas, and quantum wells are discussed in Chapter 11. Semiconductor devices are discussed in Chapters 8, 9, 13, and 14, and are arranged in pedagogical order according to the degree of sophistication.

6.2 INTRINSIC AND EXTRINSIC SEMICONDUCTORS

As discussed in Section 4.2, the two bands that play a role in electronic conduction in semiconductors are the conduction band (the upper band) and the valence band (the lower band). Separating the two allowed energy bands, there exists an energy gap (Fig. 6.1). Refer to the energy-band diagram of diamond shown in Fig. 4.8 as an example. Electronic conduction is possible only when both electrons and vacant states are made available in the same band. At absolute zero temperature, there are electrons but no vacant states in the valence band, and there are vacant states but no electrons in the conduction band. Consequently, electron conduction is blocked in both bands.

At a finite temperature, electrons are excited thermally from the valence band into the conduction band, making electrons available in the conduction band and creating vacant states in the valence band. This process is indicated as process I in Fig. 6.1. Electronic conduction is now possible because there are electrons in an almost empty conduction band and there are holes in an almost-filled valence band. Since electrons and holes exist on account of thermal creation of electron–hole pairs, we have

$$p = n \qquad (6.2.1)$$

where n and p are the electron and hole concentration, respectively. Semiconductors in which free carriers exist through thermal excitation are called *intrinsic semiconductors*.

A semiconductor can also become conductive if impurity atoms are introduced. In Section 4.2 we mentioned several materials that are semiconductors. Among those mentioned, we are most familiar with element semiconductors Ge and Si. Later in this chapter, we shall have a general discussion on semiconductor materials. For our present discussion we use mostly Ge and Si as examples to illustrate the point in question. The elements on the right side of silicon and germanium in the periodic table (Table 1.1), such as phosphorus, arsenic, and antimony, have five valence electrons with an $s^2 p^3$ electron configuration. If atoms of these elements are used to substitute a silicon or germanium atom, only four of the five valence electrons are needed to form the covalent bond with neighboring Si or Ge atoms, leaving one extra electron relatively free to move about in the crystal. The situation can be summarized as follows:

$$\text{Sb}\binom{\text{neutral with five}}{\text{valence electrons}} \rightleftharpoons \text{Sb}^+\binom{\text{with four valence electrons}}{\text{to complete the covalent bond}}$$
$$+ \ e \ (\text{loosely attached to Sb}^+ \text{ atom}) \qquad (6.2.2)$$

Figure 6.2 shows (a) the covalent structure of Si with substitutional Sb impurity atoms and (b) the energy-band diagram of Si together with the energy level of Sb in Si.

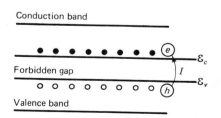

Figure 6.1 Energy bands in a semiconductor. Insofar as electric conduction is concerned, there are two bands of importance. The valence band is the lower band and the conduction band is the upper band. Intrinsic conductivity arises in a semiconductor as a result of thermal excitation of electrons from the valence band to the conduction band (process I).

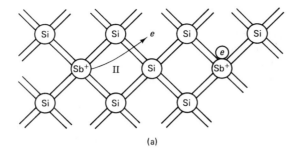

(a)

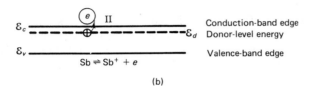

(b)

Conduction-band edge

$\mathcal{E}_d$ Donor-level energy

Valence-band edge

Figure 6.2 (a) Substitution of silicon atoms by antimony atoms which act as a donor impurity and (b) energy levels created in the forbidden gap by antimony atoms. Thermal excitation of an electron from an antimony atom to the conduction band (process II) may take place in Si doped with Sb.

The various energies $\mathcal{E}_c$, $\mathcal{E}_v$ and $\mathcal{E}_d$ are the energy of the conduction-band edge, the energy of the valence-band edge, and the energy of the donor level, respectively. The word "donor" refers to those impurities that are capable of donating an excess electron to the conduction band. The donor ionizaton energy is $\mathcal{E}_c - \mathcal{E}_d$, which is the amount of energy needed for the excess electron to break away from the Sb atom. When the thermal energy kT is sufficiently large compared to the donor ionization energy, practically all donors become ionized. Upon ionization, the donors deliver their electrons to the conduction band. The process is indicated as process II in Fig. 6.2.

Playing a role opposite to that of a donor are acceptors which have a valence of 3 with an s^2p electron configuration and hence must accept an electron from somewhere in order to complete the covalent bond with neighboring Si atoms. A missing electron in the covalent bond constitutes a hole. The elements on the left side of silicon and germanium in the periodic table, such as boron, aluminum, gallium, and indium, are acceptor impurities. The role of an acceptor, using Si as an example, can be stated in a way similar to Eq. (6.2.2) as follows:

$$\text{Ga} \begin{pmatrix} \text{neutral with three} \\ \text{valence electrons} \end{pmatrix} \rightleftharpoons \text{Ga}^- \begin{pmatrix} \text{with four valence electrons to} \\ \text{complete the covalent bond} \end{pmatrix}$$

$$+ \, h \begin{pmatrix} \text{a missing electron or a hole} \\ \text{loosely attached to a Ga}^- \text{ atom} \end{pmatrix} \tag{6.2.3}$$

Figure 6.3 is a pictorial illustration of Eq. (6.2.3), where the energy difference $\mathcal{E}_a - \mathcal{E}_v$ is the energy required for the missing electron to break away from the Ga$^-$ ion.

In summary, the donors possess an excess electron which upon thermal excitation jumps into the conduction band, becoming a free carrier. Since there are many vacant states in the conduction band, all the electrons that are ionized from the donor states contribute to electric conduction. The acceptors, on the other hand, need an electron to complete the covalent bond. They may capture the electron from the valence band. Once having given the electron to the acceptor, the valence band has a vacant state which is

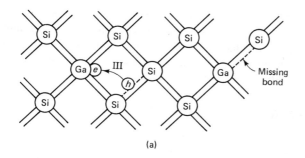

(a)

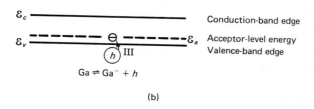

$\mathcal{E}_c$ ———————————— Conduction-band edge

$\mathcal{E}_v$ ——————————— $\mathcal{E}_a$ — Acceptor-level energy
Valence-band edge

$Ga \rightleftharpoons Ga^- + h$

(b)

Figure 6.3 (a) Two-dimensional schematic representation of covalent bonds in silicon with substitutional gallium atoms and (b) energy-band diagram of silicon together with energy levels of gallium. Thermal excitation of an electron from the valence band to the gallium atom (process III) may take place in Si doped with Ga. The gallium atom acts as an acceptor impurity.

now available to other electrons in the valence band. Since there are a large number of electrons in the valence band, all the states made vacant by giving electrons to the acceptors participate in the conduction process. The vacant states in the valence band are called *holes*.[†] Therefore, electric conduction in the conduction band is carried by electrons, whereas in the valence band it is carried by vacant states or holes.

In the presence of both electrons and holes, the drift current has a density equal to

$$J = e(\mu_n n + \mu_p p)E = \sigma E \quad A/m^2 \tag{6.2.4}$$

which replaces Eq. (4.2.3). In Eq. (6.2.4) we assume that the charges trapped by the donors and acceptors are immobile. Such is the case for impurity concentrations below 10^{17} cm^{-3}. The quantities n and p are the concentration of electrons in the conduction band and that of holes in the valence band, respectively, and μ_n and μ_p are the respective mobility for electrons and holes. A semiconductor is called *n type* if the current carriers are predominantly electrons and it is called *p type* if the current carriers are predominantly holes. The terms *n type* and *p type* refer to the charge type of majority carriers, with *negative type* meaning electrons and *positive type* meaning holes. Since the dominance of one type of carriers over the other is a result of the introduction of foreign atoms, such semiconductors are called *extrinsic semiconductors*. Without impurities, $p = n$ and a semiconductor is said to be *intrinsic*.

The principal quantities that determine the conductivity of a semiconductor are the concentrations of free carriers n and p in Eq. (6.2.4). These quantities depend not only on the concentrations N_d and N_a of donor and acceptor impurities but also on the temperature T, which determines the degree of thermal excitation of carriers from covalent bonds and from impurities: processes I, II, and III in Figs. 6.1, 6.2, and 6.3, respectively. The value of σ can change many orders of magnitude by changing either N_a, N_d,

[†]Here we used the terms *vacant states* and *holes* interchangeably. The subtle differences between them are discussed in Section 5.10.

Sec. 6.2   Intrinsic and Extrinsic Semiconductors

195

or T. The functional dependences of n and p upon N_a, N_d, and T are discussed in Sections 6.3 to 6.5. The mobilities μ_n and μ_p may not contribute as large a numerical factor as n and p in determining the value of σ, but they are of fundamental importance. A high mobility is desirable in many device applications. Various scattering mechanisms that limit the free-carrier mobility are discussed in Sections 6.7 and 6.9.

6.3 FREE-CARRIER CONCENTRATION AND FERMI LEVEL

The concentrations of electrons and holes in a semiconductor, n and p, are determined by the distribution of electrons among the various energy states, the valence and conduction band states, and the acceptor and donor impurity states. Since electrons in a semiconductor interact very strongly with other electrons, they are mutually exclusive insofar as occupying a given energy state is concerned. The mutual exclusiveness commonly known as the Pauli exclusion principle leads naturally to the use of Fermi–Dirac distribution presented in Section 4.5. Furthermore, it is assumed that the various energy states are in thermodynamic equilibrium, hence a common Fermi level is used for all energy states. Nonequilibrium situations in which different Fermi levels are used for different energy states are discussed in Chapter 7.

Figure 6.4 shows the relationship of the Fermi function $f(\mathscr{E})$ and the density-of-state function $D(\mathscr{E})$ with respect to the band edges. The function $f(\mathscr{E})$ which is defined

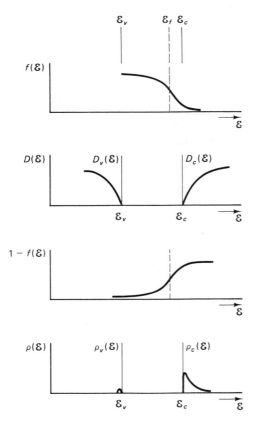

Figure 6.4 Plot of the Fermi function $f(\mathscr{E})$ with respect to the band edges. To calculate the electrons (occupied states) in the conduction band, $f(\mathscr{E})$ is used; to calculate the holes (vacant states) in the valence band, $1 - f(\mathscr{E})$ is used. That is, $\rho_c(\mathscr{E}) = D(\mathscr{E})f(\mathscr{E})$ and $\rho_v(\mathscr{E}) = D(\mathscr{E})[1 - f(\mathscr{E})]$ where $D(\mathscr{E})$ is the density-of-states function.

in Eq. (4.5.1) indicates the occupancy of a given energy state, whereas the function $D(\mathscr{E})$, which is defined in Eq. (4.4.15), represents the density of available energy states per unit energy interval. Depending on where the Fermi level $\mathscr{E}_f$ lies, semiconductors can be classified into two categories: degenerate semiconductor and nondegenerate semiconductor. If $\mathscr{E}_f$ is either in the conduction band or in the valence band, the two situations are similar to those in metals, the former corresponding to a metal where we have electron conduction and the latter to a metal where we have hole conduction. Such semiconductors are called *degenerate semiconductors*. Since degenerate semiconductors are discussed in Section 6.11, our present discussion is limited to nondegenerate semiconductors.

For *nondegenerate semiconductors*, the Fermi level $\mathscr{E}_f$ lies somewhere in the forbidden gap. We shall assume that semiconductors under discussion are nondegenerate semiconductors unless we state specifically otherwise. Note that in Eq. (4.5.1) $kT = 0.026$ eV at room temperature. In comparison, $\mathscr{E}_g$ is of the order of 1 eV in most semiconductors. As long as $\mathscr{E}_f$ is several kT below the conduction-band edge $\mathscr{E}_c$, the Fermi function for the conduction band electrons can be approximated by

$$f_c(\mathscr{E}) \cong \exp\left(-\frac{\mathscr{E}_c - \mathscr{E}_f + \mathscr{E}_1}{kT}\right) \tag{6.3.1}$$

where the energy $\mathscr{E}_1$ is the kinetic energy of the electron with

$$\mathscr{E}_1 = \mathscr{E} - \mathscr{E}_c = \frac{\hbar^2 k^2}{2m_e^*} = \frac{1}{2m_e^*}(p_x^2 + p_y^2 + p_z^2) \tag{6.3.2}$$

and m_e^* is the electron effective mass.

As pointed out in Section 5.5, the energy dependence of an electron can be expanded in a Taylor's series in the form of Eq. (5.5.11). Near the conduction-band minimum where most electrons are located, the $\mathscr{E}$ versus k diagram has the shape shown in Fig. 5.20a. Therefore, Eq. (6.3.2) is of correct form. For electrons near the top of the valence band, the energy decreases with increasing k (Fig. 5.20b) because of the negative effective mass. In terms of hole effective mass m_h^*, the energy is

$$\mathscr{E} = \mathscr{E}_v - \frac{\hbar^2 k^2}{2m_h^*} = \mathscr{E}_v - \mathscr{E}_2 \tag{6.3.3}$$

where $\mathscr{E}_2 = \hbar^2 k^2/2m_h^*$ represents the kinetic energy of a hole and is a positive quantity. In both Eqs. (6.3.2) and (6.3.3), we assume that the constant-energy surfaces are spherical and hence the effective masses are isotropic. Furthermore, the value of k in both Eqs. (6.3.2) and (6.3.3) refers to the respective increment from k of the band extremum.

For the valence band, we count the number of missing electrons. Therefore, the function $1 - f_v(\mathscr{E})$ is used instead. Further, $\mathscr{E}_f$ lies above $\mathscr{E}_v$ whereas $\mathscr{E}$ in Eq. (6.3.3) lies below $\mathscr{E}_v$. The condition $\exp[(\mathscr{E}_f - \mathscr{E})/kT] \gg 1$ is expected to be valid. Thus we have

$$1 - f_v(\mathscr{E}) = \frac{1}{1 + \exp[(\mathscr{E}_f - \mathscr{E})/kT]} \cong \exp\left(-\frac{\mathscr{E}_f - \mathscr{E}_v + \mathscr{E}_2}{kT}\right) \tag{6.3.4}$$

where $\mathscr{E}_2$ is related to $\mathscr{E}$ through Eq. (6.3.3). The function $1 - f_v(\mathscr{E})$ represents the probability that a valence-band state with energy $\mathscr{E}$ is vacant. Let us define two quantities $\rho_v(\mathscr{E})$ and $\rho_c(\mathscr{E})$ such that $\rho_v(\mathscr{E})\,d\mathscr{E}$ represents the density of hole (missing electrons) having energy between $\mathscr{E}$ and $\mathscr{E} + d\mathscr{E}$ in the valence band and $\rho_c(\mathscr{E})\, d\mathscr{E}$ represents the density of electrons having energy between $\mathscr{E}$ and $\mathscr{E} + d\mathscr{E}$ in the conduction band. By definition,

$$\rho_v(\mathscr{E})\, d\mathscr{E} = [1 - f_v(\mathscr{E})]D_v(\mathscr{E})\, d\mathscr{E} \tag{6.3.5}$$

$$\rho_c(\mathscr{E})\, d\mathscr{E} = f_c(\mathscr{E})D_c(\mathscr{E})\, d\mathscr{E} \tag{6.3.6}$$

where $D_v(\mathscr{E})$ and $D_c(\mathscr{E})$ are the density-of-state functions for the valence band and for the conduction band, respectively.

Using dN/L^3 from Eq. (4.4.16) for $D_c(\mathscr{E})\, d\mathscr{E}$ and $f_c(\mathscr{E})$ of Eq. (6.3.1) in Eq. (6.3.6), we find the electron density to be

$$n = \int_{\mathscr{E}_c}^{\infty} \rho_c(\mathscr{E})\, d\mathscr{E} = \frac{2}{h^3} \exp\left(-\frac{\mathscr{E}_c - \mathscr{E}_f}{kT}\right) \int\!\!\!\int\!\!\!\int_{-\infty}^{\infty} \exp\left(-\frac{\mathscr{E}_1}{kT}\right) dp_x\, dp_y\, dp_z \tag{6.3.7}$$

Strictly speaking, the integration limits in Eq. (6.3.7) should run from the bottom to the top of the conduction band. Since $f_c(\mathscr{E})$ drops very sharply due to the factor $\exp(-\mathscr{E}_1/kT)$ as $\mathscr{E}_1$ increases, very little error is introduced by making infinity as the upper integration limit. Substituting Eq. (6.3.2) in Eq. (6.3.7) and using Eq. (A3.1) of Appendix A3 for the three definite integrals, we obtain

$$n = N_c \exp\left(-\frac{\mathscr{E}_c - \mathscr{E}_f}{kT}\right) \qquad \text{electrons/m}^3 \tag{6.3.8}$$

where the quantity

$$N_c = 2\left(\frac{2\pi m_e^* kT}{h^2}\right)^{3/2} = 2.5 \times 10^{25} \left(\frac{m_e^*}{m_0}\frac{T}{300}\right)^{3/2} \qquad \text{m}^{-3} \tag{6.3.9}$$

is called the effective density of conduction-band states. By using Eqs. (6.3.3) and (6.3.4) in Eq. (6.3.5), the reader can easily show that the hole density is

$$p = \int_{-\infty}^{\mathscr{E}_v} \rho_v(\mathscr{E})\, d\mathscr{E} = N_v \exp\left(-\frac{\mathscr{E}_f - \mathscr{E}_v}{kT}\right) \qquad \text{holes/m}^3 \tag{6.3.10}$$

where the effective density of valence-band states is given by

$$N_v = 2\left(\frac{2\pi m_h^* kT}{h^2}\right)^{3/2} = 2.5 \times 10^{25} \left(\frac{m_h^*}{m_0}\frac{T}{300}\right)^{3/2} \qquad \text{m}^{-3} \tag{6.3.11}$$

Note that the product

$$pn = N_c N_v \exp\left(-\frac{\mathscr{E}_c - \mathscr{E}_v}{kT}\right) \tag{6.3.12}$$

is a constant independent of the Fermi level $\mathscr{E}_f$. In a pure (intrinsic) semiconductor where $N_a = N_d = 0$, electrons and holes exist on account of thermal excitation of electron–hole pairs. Use of Eq. (6.2.1) in Eq. (6.3.12) leads to

$$p = n = n_i = (N_c N_v)^{1/2} \exp\left(-\frac{\mathscr{E}_g}{2kT}\right) \qquad \text{m}^{-3} \tag{6.3.13}$$

where $\mathcal{E}_g = \mathcal{E}_c - \mathcal{E}_v$ is the gap energy. The quantity n_i is called the *intrinsic carrier concentration* because it is the carrier concentration in an intrinsic semiconductor. In an extrinsic semiconductor, p is no longer equal to n, and their values depend on the impurity concentration. Evaluation of p and n in extrinsic semiconductors is discussed in Section 6.5.

We should point out that Eqs. (6.3.8) and (6.3.10) are obtained through the use of Eqs. (6.3.1) and (6.3.4). Equation (6.3.8) is valid only for $\mathcal{E}_f < \mathcal{E}_c$, whereas Eq. (6.3.10) is valid only for $\mathcal{E}_f > \mathcal{E}_v$. This means that Eq. (6.3.12) is true only if $\mathcal{E}_v < \mathcal{E}_f < \mathcal{E}_c$. This is the condition for a nondegenerate semiconductor. In a degenerate semiconductor, the relation $pn = n_i^2$ no longer holds. Therefore, Eq. (6.3.12) is not a universal expression and is applicable only in nondegenerate semiconductors. Carrier concentrations in degenerate semiconductors will be presented in Section 6.11. Modifications of Eqs. (6.3.9) and (6.3.11) needed for nonspherical energy surfaces are discussed in Sections 6.9 and 6.10.

6.4 DONOR AND ACCEPTOR STATES

We may think of a donor atom as consisting of a positively charged core D^+ surrounded by a charge cloud with a charge $-e$. One physical quantity of importance is the energy required to excite this loosely bound electron into the conduction band. For ordinary donor impurities such as P, As, and Sb in Ge or Si, the hydrogenic model works quite well. According to Eq. (1.2.11) we may write

$$\mathcal{E}_d = -13.6 \left(\frac{m_e^*}{m_0} \right) \left(\frac{\epsilon_0}{\epsilon} \right)^2 \frac{1}{n_q^2} \quad \text{eV} \tag{6.4.1}$$

where $n_q = 1, 2, 3, \ldots$ is the principal quantum number and $\mathcal{E}_c$ is taken to be zero.

Because of the difference in the dielectric constant and the effective electron mass, the electron orbit in a semiconductor is much larger than that in a free atom and the ionization energy of a donor state is considerably smaller than that of a hydrogen atom. In Table 6.1 the ionization energies of donor and acceptor impurities in Ge and Si are

TABLE 6.1 MEASURED IONIZATION ENERGIES (eV) FOR DONORS AND ACCEPTORS IN Ge AND Si

Impurity	Ge	Si
Donors		
P	0.012	0.044
As	0.013	0.049
Sb	0.0096	0.039
Bi	—	0.067
Acceptors		
B	0.010	0.045
Al	0.010	0.057
Ga	0.011	0.065
In	0.011	0.16
Tl	0.01	0.26

given. Taking $\epsilon = 16\epsilon_0$ and $m_e^* = 0.26m_0$ in Eq. (6.4.1), we find the energy of the donor ground state ($n_q = 1$) to be 1.3×10^{-2} eV, which is of the same order of magnitude as the ionization energies shown in Table 6.1. In Eq. (6.4.1) the donor-state energy refers to the conduction-band edge as the zero energy. An equation similar to Eq. (6.4.1) applies to acceptor states with m_h^* substituting for m_e^* and with $\mathscr{E}_a$ lying above and referring to the valence-band edge. The acceptor states have a higher ionization energy than the donor states because of a higher m_h^*. The ionization energies of donors and acceptors are generally higher in Si than in Ge on account of a smaller ϵ and a larger effective mass.

We should point out that Eq. (6.4.1) is useful only for an order-of-magnitude estimate of the ionization energy of donors and acceptors. For an accurate calculation of the energy, the reader is referred to an excellent review article by Kohn (W. Kohn, "Shallow Impurity States in Si and Ge," in H. Ehrenreich, F. Seitz, and D. Turnbull, eds., *Solid State Physics*, Vol. 5, Academic Press, Inc., New York, 1957, p. 258). The main problems in an exact quantum-mechanical calculation are the following. First, Eq. (1.4.1) for a hydrogen atom is isotropic. In contrast, the effective masses are anisotropic for a nonspherical energy band (Sections 6.9 and 6.10). This means that an appropriate average of the masses must be used in Eq. (6.4.1) for the value of m_e^*. Second, there is a question regarding the position of $\mathscr{E}_d$ relative to $\mathscr{E}_c$. In our discussion we choose $\mathscr{E}_d(n_q = \infty)$ to coincide with $\mathscr{E}_c$. The assumption is not strictly true but is a reasonably good approximation for an estimate of the ionizaton energy. Finally, we notice that the ionization energy varies considerably from one element to another (Table 6.1). In an actual calculation, we must know reasonably accurately the ion-core potential and the electron wave function near the donor (or acceptor) ion core. The difference in ionization energy from one element to another is a result of the difference in the ion-core potential.

Next we discuss the distribution of electrons among the donor and the acceptor states. The Fermi functions for these states are slightly different from those for the conduction and valence bands. Take the donor state as an example. Since four out of five valence electrons are tied in the covalent bond, we focus our attention on the extra valence electron. We use the word *occupied* or *unoccupied* to indicate whether the extra electron is with the donor atom or free to move about in the crystal. The state that a donor impurity is in can be defined by the following equation:

Donor (electrically neutral, occupied by the extra electron) $\rightleftharpoons$
$\qquad$ Donor$^+$ (positively charged, unoccupied) + a free electron $\qquad$ (6.4.2)

Equation (6.4.2) is similar to Eq. (6.2.2) except that now the emphasis is on the extra electron. Unlike the four electrons that form the covalent bond, the extra electron has a choice of two possible spin orientations.

The two spin states (spin up and spin down) have the same energy, and the donor level has a twofold degeneracy in energy. For the derivation of the Fermi–Dirac distribution function for an impurity state, we refer to the four-level system of Fig. 4.18, and let the level 2 be the donor level. We also move the energy levels 1 and 3 up relative to the level 2 so that levels 1, 3, and 4 represent conduction-band states. When the donor level is occupied by an electron, the four-level system actually is made of two subsystems: one with the electron spin up and the other with the electron spin down. The situation is illustrated in Fig. 6.5a and b. When the donor level is empty, the situation is uniquely represented by Fig. 6.5c. The level 2 is simply empty. We do not

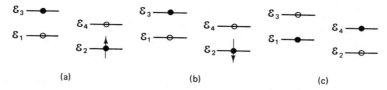

Figure 6.5 Interaction processes involving a donor state and three other energy states. Occupied states are denoted by dots and vacant states by open circles. The state 2 is assumed to be the donor state. When the donor state is occupied by an electron, two different situations arise with (a) spin up and (b) spin down. The situation in which the donor state is vacant is illustrated in (c).

have two empty states, one empty spin-up state and the other empty spin-down state, because a spin state must be associated with an electron and there is no electron at the level 2 in Fig. 6.5c.

Under thermal equilibrium, the three situations shown in Fig. 6.5a–c must be in balance. Let N_2' and N_2'' be the density of the occupied spin-up state and that of the occupied spin-down state, respectively, and N_2^e be the density of empty state. Since all the levels obey the Pauli exclusion principle, the reaction rate is proportional to the number of vacant final states as well as the number of occupied initial states. For rate balance between the two situations shown in Fig. 6.5a and c, we have

$$N_1^o N_3^e N_4^o N_2^e = N_3^o N_1^e N_4^e N_2'$$
(6.4.3)

where the superscripts o and e refer to occupied and empty states, respectively. A similar equation,

$$N_1^o N_3^e N_4^o N_2^e = N_3^o N_1^e N_4^e N_2''$$
(6.4.4)

applies to the spin-down state.

Rearranging terms in Eq. (6.4.3), we have

$$\frac{N_2'}{N_2^e} = \frac{N_1^o}{N_3^o} \frac{N_4^o}{N_4^e} \frac{N_3^e}{N_1^e}$$
(6.4.5)

From Eq. (4.5.1) and an energy-conservation relation, we find that

$$\frac{N_3^e}{N_1^e} \frac{N_1^o}{N_3^o} = \exp\left(\frac{\mathscr{E}_3 - \mathscr{E}_1}{kT}\right) = \exp\left(\frac{\mathscr{E}_4 - \mathscr{E}_2}{kT}\right)$$
(6.4.6)

$$\frac{N_4^o}{N_4^e} = \frac{f(\mathscr{E}_4)}{1 - f(\mathscr{E}_4)} = \exp\left(-\frac{\mathscr{E}_4 - \mathscr{E}_f}{kT}\right)$$
(6.4.7)

Using Eqs. (6.4.6) and (6.4.1) in Eq. (6.4.5), we obtain

$$\frac{N_2'}{N_2^e} = \exp\left(-\frac{\mathscr{E}_2 - \mathscr{E}_f}{kT}\right)$$
(6.4.8)

Note that the total density of occupied level-2 states is $N_2^o = N_2' + N_2''$. We define a Fermi distribution function f_d for the donor states such that

$$n_d = N_2^o = N_2' + N_2'' = N_d f_d(\mathscr{E}_d)$$
(6.4.9)

where n_d and N_d represent, respectively, the density of occupied donor states and that of the total donor states. In Eq. (6.4.9) the notations are changed from level 2 to donor states with $\mathscr{E}_2 = \mathscr{E}_d$. If N_d^+ represents the density of unoccupied donor states, then

$$N_d^+ = N_2^e = N_d[1 - f_d(\mathscr{E}_d)] \quad \text{and} \quad N_d^+ + n_d = N_d \qquad (6.4.10)$$

The notation N_d^+ is used to indicate the fact that a vacant donor state is positively charged.

Since the Fermi function depends on energy only and the two spin states are assumed to have the same energy, we expect that

$$N_2' = N_2'' = \frac{n_d}{2} = \frac{N_d}{2} f_d(\mathscr{E}_d) \qquad (6.4.11)$$

Substituting Eqs. (6.4.10) and (6.4.11) in Eq. (6.4.8), we find that

$$\frac{f_d(\mathscr{E}_d)}{1 - f_d(\mathscr{E}_d)} = 2 \exp\left(-\frac{\mathscr{E}_2 - \mathscr{E}_f}{kT}\right) \qquad (6.4.12)$$

Solving for $f_d(\mathscr{E}_d)$, we obtain

$$\frac{n_d}{N_d} = \frac{N_d - N_d^+}{N_d} = f_d(\mathscr{E}_d) = \frac{1}{1 + \frac{1}{2} \exp\left[(\mathscr{E}_d - \mathscr{E}_f)/kT\right]} \qquad (6.4.13)$$

We should point out that for a g-fold degenerate donor state,

$$f_d(\mathscr{E}) = \frac{1}{1 + g^{-1} \exp\left[(\mathscr{E}_d - \mathscr{E}_f)/kT\right]} \qquad (6.4.14)$$

where g is the degeneracy factor and $g = 2$ in Eq. (6.4.13). The derivation of Eq. (6.4.14) can be found in books on statistical mechanics.

The foregoing discussion can be extended to acceptor states. For the acceptor states, an additional electron is needed to complete the covalent bond. We again use the word *occupied* or *unoccupied* to indicate whether the added electron is with the acceptor atom or not, as illustrated by the following equation:

Acceptor (electrically neutral, unoccupied) + an added electron $\rightleftharpoons$
    Acceptor⁻ (negatively charged, occupied by the added electron) $\qquad (6.4.15)$

The physical situations representing the occupied and the unoccupied acceptor states are illustrated schematically in Fig. 6.6. Note that an occupied acceptor state (Fig. 6.6c) means a complete bond and hence corresponds to an unoccupied donor state. Since the

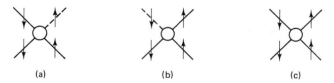

(a)   (b)   (c)

Figure 6.6 Diagram showing the covalent bond in diamond structure. The four electrons which form the covalent bond in the structure must have paired spins. The situation is illustrated in (c). When one bond is broken, the missing electron of the bond can have the electron spin up or down. The two situations are illustrated in (a) and (b).

four electrons in forming the covalent bond must have a certain spin arrangement, the situation shown in Fig. 6.6c is unique.

The situation with an unoccupied acceptor state is quite different. As shown in Fig. 6.6a and b, the missing bond must have a specific spin orientation in order to complete the bond. Depending on the spin orientations of the three existing electrons, the spin orientation of the missing electron can be either up (Fig. 6.7a) or down (Fig. 6.7b). Therefore, we have a spin degeneracy of 2 for the unoccupied acceptor state. We again refer to the four-level system of Fig. 4.18, and this time let the acceptor state be represented by level 2. Since there are two situations for the unoccupied (empty) state and there is only one situation for the occupied state, Eq. (6.4.8) is replaced by

$$\frac{N_2^o}{\frac{1}{2} N_2^e} = \exp\left(-\frac{\mathscr{E}_2 - \mathscr{E}_f}{kT}\right) \tag{6.4.16}$$

We use N_a^- to denote the density of occupied acceptor states because an occupied acceptor state is negatively charged. By definition,

$$N_a^- = N_2^o = N_a f_a(\mathscr{E}_a) \tag{6.4.17}$$

$$p_a = N_a - N_a^- = N_2^e = N_a[1 - f_a(\mathscr{E}_a)] \tag{6.4.18}$$

where f_a is the Fermi function for the acceptor states and p_a is the density of vacant acceptor states. Using Eq. (6.4.17) and (6.4.18) in Eq. (6.4.16), we obtain

$$\frac{N_a^-}{N_a} = f_a(\mathscr{E}_a) = \frac{1}{1 + 2 \exp\left[(\mathscr{E}_a - \mathscr{E}_f)/kT\right]} \tag{6.4.19}$$

A general expression of the Fermi function for the acceptor states is

$$f_a(\mathscr{E}_a) = \frac{1}{1 + g \exp\left[(\mathscr{E}_a - \mathscr{E}_f)/kT\right]} \tag{6.4.20}$$

if the state is g-fold degenerate.

The derivation of Eqs. (6.4.13) and (6.4.19) is based on the spin degeneracy alone, which is twofold. We have not considered the degeneracy of the orbital wave function. In Eqs. (6.4.14) and (6.4.20), the degeneracy factor g includes both the spin and the orbital degeneracy. In group IV element semiconductors (Ge, Si) and group III–V, II–VI compound semiconductors (GaAs, InSb, CdS, ZnS), the conduction band is known to have s-like and the valence band to have p-like wave functions at $k(000)$. The

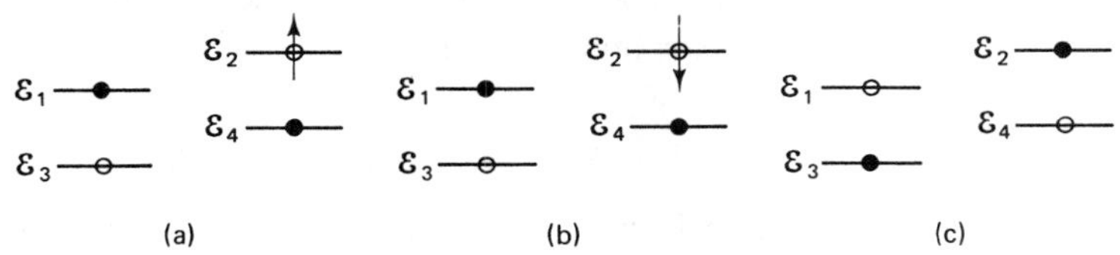

(a)   (b)   (c)

Figure 6.7 Diagrams used in the derivation of Eq. (6.4.16). The level 2 represents the acceptor state. For a vacant acceptor state (Fig. 6.6a and b), the spin of the missing electron can be either (a) up or (b) down. For an occupied acceptor state (Fig. 6.6c), all the bonds are complete and the situation is unique. This unique situation is shown in (c).

s state is nondegenerate, and hence $g = 2$. The p state is triply degenerate. However, the orbital degeneracy may be partly removed by the crystal field and the spin-orbit coupling. Therefore, the degeneracy factor can be 2, or 4 in Eq. (6.4.20), depending on the degeneracy of the valence band.

6.5 EVALUATION OF FERMI LEVEL AND CARRIER CONCENTRATIONS

In Section 6.3 we derived the equations relating the free-carrier concentrations n and p to the Fermi level $\mathscr{E}_f$, and in Section 6.4 we obtained the expressions for the densities of ionized donors and acceptors N_d^+ and N_a^- in terms of $\mathscr{E}_f$. The equation that ties n, p, N_d^+, and N_a^+ together is the charge-neutrality equation. Since electrons are merely redistributed among the various energy states in a semiconductor, but not taken out of or put into the semiconductor, the crystal remains electrically neutral. The equation that states this charge-neutrality condition reads

$$n + N_a^- = p + N_d^+ \qquad (6.5.1)$$

On the left side of Eq. (6.5.1), n represents the negative charge density due to conduction-band electrons and N_a^- the negative charge density due to charged (occupied) acceptor states. On the right side, p is the positive charge density due to valence-band holes and N_d^+ the positive charge density due to ionized (unoccupied) donor states.

Substituting Eqs. (6.3.8), (6.3.10), (6.4.13), and (6.4.19) in Eq. (6.5.1) we obtain

$$N_c \exp\left(-\frac{\mathscr{E}_c - \mathscr{E}_f}{kT}\right) + \frac{N_a}{1 + 2 \exp[(\mathscr{E}_a - \mathscr{E}_f)/kT]}$$

$$= N_v \exp\left(-\frac{\mathscr{E}_f - \mathscr{E}_v}{kT}\right) + \frac{N_d}{1 + 2 \exp[(\mathscr{E}_f - \mathscr{E}_d)/kT]} \qquad (6.5.2)$$

The only unknown in Eq. (6.5.2) is the Fermi level $\mathscr{E}_f$. In principle we can solve for $\mathscr{E}_f$. Once $\mathscr{E}_f$ is known, the values of n and p can be obtained from Eqs. (6.3.8) and (6.3.10). This process is exact but tedious. In the following discussion, we attempt to find the approximate values for n and p whenever approximations can be justified through physical reasoning.

The activation energies $\mathscr{E}_c - \mathscr{E}_d$ and $\mathscr{E}_a - \mathscr{E}_v$ for donor and acceptor impurities are around 0.01 eV in Ge and around 0.04 eV in Si (Table 6.1). These impurities with energies close to the respective band edge are known as *shallow impurities*. For comparison, the thermal energy kT is 0.026 eV at 300 K. At sufficiently high temperatures, all the donor and acceptor impurities are expected to be almost completely ionized. In other words,

$$N_a^- \simeq N_a \quad \text{and} \quad N_d^+ \simeq N_d \qquad (6.5.3)$$

The temperature region in which Eq. (6.5.3) is a good approximation is to be determined later. Using Eq. (6.5.3) in Eq. (6.5.1), we have

$$n - p = N_d - N_a \qquad (6.5.4)$$

Elimination of p (or n) from Eqs. (6.5.4) and (6.3.12) yields a quadratic equation in n (or p) which reads

$$n^2 - (N_d - N_a)n - n_i^2 = 0 \qquad (6.5.5)$$

The solution of Eq. (6.5.5) for n-type semiconductors is

$$n = \frac{N_d - N_a}{2} + \sqrt{\left(\frac{N_d - N_a}{2}\right)^2 + n_i^2} \qquad (6.5.6)$$

$$p = -\frac{N_d - N_a}{2} + \sqrt{\left(\frac{N_d - N_a}{2}\right)^2 + n_i^2} \qquad (6.5.7)$$

In ordinary semiconductors with shallow impurities, n is expected to be larger than p if $N_d > N_a$. Only the solution with a plus sign in front of the square root meets this condition, and this solution is given in Eqs. (6.5.6) and (6.5.7). Two different approximations can be taken, depending on the magnitude of $N_d - N_a$ relative to n_i. Note that the intrinsic carrier concentration n_i defined in Eq. (6.3.13) is a function of temperature. It is natural that we divide the temperature into regions in our discussion of the behavior of n and p.

There are three temperature regions: low, intermediate, and high. In the low-temperature region, the impurities may not be completely ionized. Because one of the relations expressed in Eq. (6.5.3) is no longer true, Eq. (6.5.1) should be used instead of Eq. (6.5.4). In the intermediate- and high-temperature regions, Eq. (6.5.4) is a good approximation. For $n_i \gg |N_d - N_a|/2$, Eqs. (6.5.6) and (6.5.7) can be approximated by

$$n = n_i + \frac{N_d - N_a}{2} \simeq n_i \qquad (6.5.8)$$

$$p = n_i - \frac{N_d - N_a}{2} \simeq n_i \qquad (6.5.9)$$

which are the solutions for the high-temperature region. On the other hand, if $n_i \ll (N_d - N_a)/2$, which is true in the intermediate-temperature region, n and p are approximately equal to

$$n = N_d - N_a - \frac{n_i^2}{N_d - N_a} \cong N_d - N_a \qquad (6.5.10)$$

$$p = \frac{n_i^2}{N_d - N_a} \qquad (6.5.11)$$

Equations (6.5.10) and (6.5.11) are for an n-type semiconductor where $N_d > N_a$. The expressions for a p-type semiconductor can be obtained from Eqs. (6.5.10) and (6.5.11) by interchanging p with n and interchanging N_d with N_a.

Before we discuss the low-temperature case, it is instructive to examine the variation of $\mathscr{E}_f$ with temperature. First we define an intrinsic Fermi level $\mathscr{E}_{fi}$ such that $n = p = n_i$ if $\mathscr{E}_f = \mathscr{E}_{fi}$. That means

$$N_c \exp\left(-\frac{\mathscr{E}_c - \mathscr{E}_{fi}}{kT}\right) = N_v \exp\left(-\frac{\mathscr{E}_{fi} - \mathscr{E}_v}{kT}\right)$$

$$= \sqrt{N_c N_v} \exp\left(-\frac{\mathscr{E}_c - \mathscr{E}_v}{2kT}\right) \qquad (6.5.12)$$

From Eq. (6.5.12), $\mathcal{E}_{fi}$ is defined as

$$\mathcal{E}_{fi} = \frac{\mathcal{E}_c + \mathcal{E}_v}{2} + \frac{kT}{2} \ln \frac{N_v}{N_c} \qquad (6.5.13)$$

If $N_c = N_v$, $\mathcal{E}_{fi}$ is exactly in the middle of the gap. In terms of $\mathcal{E}_{fi}$, the carrier concentrations are

$$n = n_i \exp\left(\frac{\mathcal{E}_f - \mathcal{E}_{fi}}{kT}\right) \qquad (6.5.14)$$

$$p = n_i \exp\left(\frac{\mathcal{E}_{fi} - \mathcal{E}_f}{kT}\right) \qquad (6.5.15)$$

In the high-temperature region, Eqs. (6.5.8) and (6.5.9) can be further approximated by $p \cong n \cong n_i$. Use of this information in Eqs. (6.5.14) and (6.5.15) shows that $\mathcal{E}_f \cong \mathcal{E}_{fi}$. In this region, the process of thermal excitation of carriers across the band gap (process I of Fig. 6.1) dominates, and even an extrinsic semiconductor behaves like an intrinsic semiconductor. In the intermediate temperature region, $\mathcal{E}_f$ can be obtained by equating n of Eq. (6.5.10) to n of Eq. (6.3.8). Thus we find that

$$\mathcal{E}_f = \mathcal{E}_c - kT \ln \frac{N_c}{N_d - N_a} \qquad (6.5.16)$$

Equation (6.5.16) applies to N-type semiconductors where $N_d > N_a$. The reader is asked to find a similar expression for p-type semiconductors.

Figure 6.8 shows the temperature variation of $\mathcal{E}_f$ in an n-type semiconductor. In the high-temperature region ($T > T_1$), the semiconductor is essentially intrinsic, with $\mathcal{E}_f \cong \mathcal{E}_{fi}$. As the temperature T decreases, the Fermi level moves toward the conduction-band edge. In the intermediate temperature region ($T_2 < T < T_1$), $\mathcal{E}_f$ varies almost linearly with T so as to maintain a constant $n = N_d - N_a$. The temperature T_1 can be obtained from the condition used in obtaining Eqs. (6.5.8) and (6.5.9). At T_1, $n_i =$

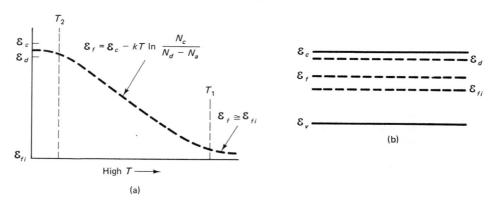

Figure 6.8 Schematic diagram showing the variation of the Fermi level as a function of temperature in an n-type semiconductor. For $N_d > N_a$, $\mathcal{E}_f$ stays above $\mathcal{E}_{fi}$. At a temperature $T > T_1$, the intrinsic carrier concentration n_i is larger than $N_d - N_a$ and $\mathcal{E}_f$ can be approximated by $\mathcal{E}_{fi}$. As the temperature decreases, $\mathcal{E}_f$ moves toward the conduction-band edge $\mathcal{E}_c$.

$(N_d - N_a)/2$. Using Eq. (6.3.13) for n_i, we have

$$T_1 = \frac{\mathcal{E}_g}{k \ln [4N_cN_v/(N_d - N_a)^2]} \tag{6.5.17}$$

Now we discuss the low-temperature case. As $\mathcal{E}_f$ approaches $\mathcal{E}_c$ in Fig. 6.8, the approximation $N_d^+ \cong N_d$ becomes poorer and poorer, the value of p becomes smaller and smaller, but the approximation $N_a^- \cong N_a$ remains good. Thus Eq. (6.5.2) becomes

$$N_c \exp \left(-\frac{\mathcal{E}_c - \mathcal{E}_f}{kT} \right) + N_a = N_d \left[1 + 2 \exp \left(\frac{\mathcal{E}_f - \mathcal{E}_d}{kT} \right) \right]^{-1} \tag{6.5.18}$$

by neglecting p. The reader is referred to elsewhere for a discussion of the general case with $N_a \neq 0$. (See, for example, S. Wang, *Solid-State Electronics,* McGraw-Hill Book Company, New York, 1966, pp. 150–151.) Here we consider only the case with $N_a = 0$. If we let

$$\lambda = \exp \left(\frac{\mathcal{E}_f}{kT} \right), \qquad \alpha = \exp \left(-\frac{\mathcal{E}_d}{kT} \right), \qquad \beta = \exp \left(-\frac{\mathcal{E}_c}{kT} \right), \qquad b = \frac{N_d}{N_c} \tag{6.5.19}$$

then Eq. (6.5.18) becomes

$$2\alpha\beta\lambda^2 + \beta\lambda - b = 0 \tag{6.5.20}$$

Solving for λ, we obtain

$$\lambda = \frac{1}{4\alpha} \left(-1 + \sqrt{1 + \frac{8b\alpha}{\beta}} \right) \tag{6.5.21}$$

Since λ is always positive, only the plus sign in front of the square root is admitted in Eq. (6.5.21). For $\beta > 8b\alpha$, Eq. (6.5.21) can be approximated by

$$\lambda \cong \frac{b}{\beta} \quad \text{or} \quad n = N_c\beta\lambda \cong N_d \tag{6.5.22}$$

which is the same as Eq. (6.5.10) for $N_a = 0$. Therefore, this solution corresponds to the intermediate-temperature case. For $\beta < 8b\alpha$, Eq. (6.5.21) becomes

$$\lambda \cong \left(\frac{b}{2\alpha\beta} \right)^{1/2} \tag{6.5.23}$$

From Eq. (6.5.23), the value of n can be found and is equal to

$$n = N_c\beta\lambda = \left(\frac{N_cN_d}{2} \right)^{1/2} \exp \left(-\frac{\mathcal{E}_c - \mathcal{E}_d}{2kT} \right) \tag{6.5.24}$$

Equation (6.5.24) corresponds to the low-temperature case. The temperature T_2 can be found by setting $\beta = 8b\alpha$. Using β, b, and α defined in Eq. (6.5.19), we obtain

$$T_2 = \frac{\mathcal{E}_c - \mathcal{E}_d}{k \ln (N_c/8N_d)} \tag{6.5.25}$$

Table 6.2 summarizes all the relevant quantities for a nondegenerate n-type semiconductor with $N_d < N_c$ and $N_a = 0$ in the three temperature regions. Note that for

TABLE 6.2 CARRIER CONCENTRATIONS AND FERMI LEVEL IN THE THREE TEMPERATURE REGIONS OF FIG. 6.9 FOR AN n-TYPE SEMICONDUCTOR WITH $N_a = 0$

1. In the high-temperature region, $T > T_1$ of Eq. (6.5.17),

$$\mathscr{E}_f = \mathscr{E}_{fi} \quad \text{and} \quad n = p = n_i$$

2. In the intermediate-temperature region, $T_2 < T < T_1$,

$$\mathscr{E}_f = \mathscr{E}_c - kT \ln \frac{N_c}{N_d}$$

$$n = N_d \quad \text{and} \quad p = \frac{n_i^2}{N_d}$$

3. In the low-temperature region, $T < T_2$ of Eq. (6.5.25),

$$\mathscr{E}_f = \frac{\mathscr{E}_c + \mathscr{E}_d}{2} - \frac{kT}{2} \ln \frac{2N_c}{N_d}$$

$$n = \sqrt{\frac{N_c N_d}{2}} \exp \left(-\frac{\mathscr{E}_c - \mathscr{E}_d}{2kT} \right)$$

$$p = N_v \sqrt{\frac{2N_c}{N_d}} \exp \left(-\frac{\mathscr{E}_c + \mathscr{E}_d - 2\mathscr{E}_v}{2kT} \right)$$

$N_d < N_c$, n is always smaller than N_c and hence $\mathscr{E}_f$ always lies below $\mathscr{E}_c$. The condition $\mathscr{E}_c > \mathscr{E}_f > \mathscr{E}_v$ is used previously to define a nondegenerate semiconductor. Figure 6.9 shows the majority-carrier concentration n as a function of reciprocal temperature. In the high-temperature region, n is equal to n_i of Eq. (6.3.13) and varies as $\exp(-\mathscr{E}_g/2kT)$ with temperature. Since intrinsic carriers due to thermal agitation dominate over carriers released from impurities, the region is referred to as the *intrinsic region*. In the intermediate- and low-temperature regions, majority carriers come from ionization of impurities, and hence these regions are called *extrinsic regions*. Furthermore, in the intermediate-temperature region, practically all impurities are ionized, and thus the region is called the *exhaustion region*.

Finally, we should mention that caution must be exercised in treating a compensated semiconductor. By *compensated semiconductor* we mean a semiconductor in which both types of impurities exist, so that the effect of impurities of one kind is partially canceled by that of the opposite kind. The reader might think that the ionization process of impurities in such a semiconductor would be a combination of processes II and III in Figs. 6.2 and 6.3 as illustrated in Fig. 6.10a. Hence the reader might conclude that $n = N_d^+$ and $p = N_a^-$. This result is incorrect. We must emphasize that bifurcation of Eq. (6.5.1) into two separate equations is not permissible.

The situation in a compensated semiconductor is illustrated correctly in Fig. 6.10b. Since p is by far the smallest quantity in Eq. (6.5.1), we ignore the valence band in finding the majority-carrier concentration n as we did in Eq. (6.5.18). The extra electrons released from the ionized donors are shared by the acceptors and conduction-band states. Since the acceptor states are the lower-energy states, they are filled first. If

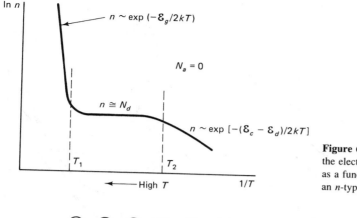

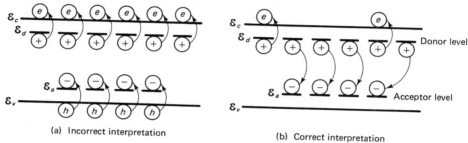

Figure 6.9 Schematic diagram showing the electron (majority-carrier) concentration as a function of reciprocal temperature in an *n*-type semiconductor.

(a) Incorrect interpretation

(b) Correct interpretation

Figure 6.10 Schematic diagrams showing the distribution of electrons among the available energy states in a compensated *n*-type semiconductor (a) if Eq. (6.5.1) is interpreted incorrectly and (b) if Eq. (6.5.1) is interpreted correctly. The arrows start from the places where the electrons originate.

the N_d donor impurities are completely ionized, of the N_d electrons released from donors, N_a electrons go to acceptor states, leaving $N_d - N_a$ electrons for the conduction band. This is exactly what Eq. (6.5.10) says.

6.6 CARRIER CONCENTRATION AND MOBILITY MEASUREMENTS

As discussed in Section 5.10, the Hall measurement yields information about carrier concentration. According to Eq. (5.11.5), the Hall coefficient R_H is equal to $(ne)^{-1}$ in materials where electron conduction is dominant. Figures 6.11 and 6.12 show the measured values of Hall coefficient and resistivity in several germanium samples reported by Debye and Conwell (P. P. Debye and E. M. Conwell, *Phys. Rev.*, Vol. 93, p. 693, 1954). In these samples, the acceptor concentration is much smaller than N_d; hence the equations in Table 6.2 apply. Take sample 51 as an example. The curve *ABCDE* is divided into three regions: (1) the high-temperature region *AB*, (2) the intermediate-temperature region *BC*, and (3) the low-temperature region *DE*. Because the low-temperature region is expanded relative to the high- and intermediate-temperature regions on the $1/T$ scale, we can see the gradual transition region *CD* from the intermediate- to the low-temperature case.

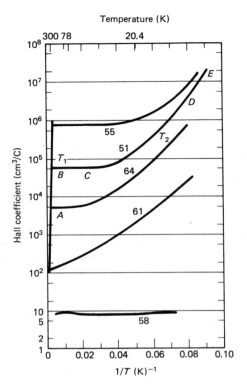

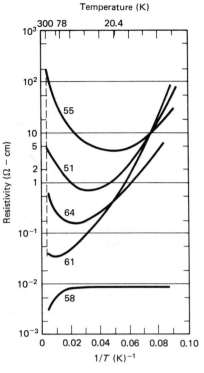

Figure 6.11 Plot of the measured Hall coefficient as a function of reciprocal temperature in several germanium samples. The curves from top to bottom are for samples with increasing impurity concentration. (P. P. Debye and E. M. Conwell, *Phys. Rev.*, Vol. 93, p. 693, 1954.)

Figure 6.12 Plot of the measured resistivity as a function of reciprocal temperature in several germanium samples. The same samples were used in the measurements of resistivity and Hall coefficients (Fig. 6.11). (P. P. Debye and E. M. Conwell, *Phys. Rev.*, Vol. 93, p. 693, 1954.)

The middle region BC in Fig. 6.11 is the exhaustion region, where n is equal to N_d. The Hall coefficient in this region gives a value of $N_d = 1.4 \times 10^{14}$ cm^{-3}. In the high-temperature region $n = n_i$ and n varies as $T^{3/2} \exp(-\mathscr{E}_g/2kT)$. The slope of the portion AB of the curve gives a value of $\mathscr{E}_g = 0.665$ eV for the gap energy. The value of $\mathscr{E}_g$ obtained this way is generally referred to as the *thermal gap energy*. In the low-temperature region, n varies as $T^{3/4} \exp[(\mathscr{E}_c - \mathscr{E}_d)/2kT]$. The slope of the portion ED of the curve gives the donor activation energy. In Fig. 6.11 the samples were doped with arsenic, and its activation energy was found to be 0.0125 eV. Using Eq. (6.5.25) and a value of $N_c = 10^{19}$ cm^{-3} for Ge, we find $T_2 = 16$ K, which is the temperature of transition from the intermediate- to the low-temperature case and is marked on the curve $ABCDE$.

Now we make some brief comments about the other curves of Fig. 6.11. The curves from top to bottom are given with increasing impurity concentration. As N_d increases, the transition temperature T_2 moves toward higher temperature, in agreement

Semiconductor Fundamentals   Chap. 6

with Eq. (6.5.25). A more accurate determination of the activation energy $\mathscr{E}_c - \mathscr{E}_d$ is obtained by fitting curves of Fig. 6.11 to the more exact equation (6.5.18) instead of the approximate equation (6.5.24). In terms of n, Eq. (6.5.18) becomes

$$\frac{2(n + N_a)n}{N_d - N_a - n} = N_c \exp\left(-\frac{\mathscr{E}_c - \mathscr{E}_d}{kT}\right) \tag{6.6.1}$$

Figure 6.13 shows the values of $\mathscr{E}_c - \mathscr{E}_d$, obtained from a fit of Fig. 6.11 to Eq. (6.6.1), as a function of the impurity concentration of the various Ge samples. As the impurity concentration increases beyond 10^{14} cm^{-3}, the activation energy begins to drop. The activation energy finally goes to zero for $N_d > 2 \times 10^{17}$ cm^{-3}. This occurs because the donor energy level is no longer discrete but merges into the conduction band. The phenomenon is attributed to the formation of an impurity band. When the donor activation energy becomes zero, n in Eq. (6.6.1) and hence R_H should be independent of temperature. This happens in sample 58.

In the high-temperature region where both types of carriers exist in comparable concentrations, the expression for the Hall coefficient must be generalized. The equations of motion under a cross electric and magnetic field (E_x and B_z) of Fig. 5.21 are:

$$\frac{dv_{ey}}{dt} + \frac{v_{ey}}{\tau_e} = -\frac{e}{m_e^*}(E_y - v_{ex}B_0) \qquad \text{for electrons} \tag{6.6.2}$$

$$\frac{dv_{hy}}{dt} + \frac{v_{hy}}{\tau_h} = \frac{e}{m_h^*}(E_y - v_{hx}B_0) \qquad \text{for holes} \tag{6.6.3}$$

where E_y is the Hall field. For dc measurements, dv/dt is zero; hence we have

$$v_{ey} = -\mu_n(E_y + \mu_n E_0 B_0) \tag{6.6.4}$$

$$v_{hy} = \mu_p(E_y - \mu_p E_0 B_0) \tag{6.6.5}$$

In obtaining Eqs. (6.6.4) and (6.6.5), we note that $\mu = e\tau/m$ and $v_x = \pm\mu E_0$, with plus sign for holes and minus sign for electrons.

Since the Hall voltage is measured under open-circuit conditions, $J_y = 0$, requiring that

$$ep v_{hy} - en v_{ey} = 0 \tag{6.6.6}$$

Substituting Eqs. (6.6.4) and (6.6.5) in Eq. (6.6.6) and solving for E_y, we obtain

$$E_y = \frac{\mu_p^2 p - \mu_n^2 n}{\mu_p p + \mu_n n} E_0 B_0 \tag{6.6.7}$$

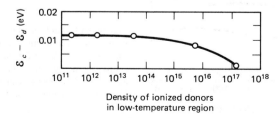

Figure 6.13 Activation energy of donor impurities as a function of donor concentration. (P. P. Debye and E. M. Conwell, *Phys. Rev.*, Vol. 93, p. 693, 1954.)

Thus the Hall coefficient is

$$R_{\mathrm{H}} = \frac{E_y}{J_x B_0} = \frac{\mu_p^2 p - \mu_n^2 n}{e(\mu_p p + \mu_n n)^2} \qquad (6.6.8)$$

Because the mobilities for electrons and holes are different, it is possible that within a certain temperature range R_{H} is negative and yet $p > n$. This happens in semiconductors with $\mu_n > \mu_p$ when the sample is about to leave the intermediate- and enter the high-temperature region. For an intrinsic semiconductor, Eq. (6.6.8) becomes

$$R_{\mathrm{H}} = \frac{1}{en_i} \frac{\mu_p - \mu_n}{\mu_p + \mu_n} \qquad (6.6.9)$$

The product of R_{H} and σ gives the mobility of the majority (dominant) carriers, known as the *Hall mobility,*

$$R_{\mathrm{H}}\sigma = \begin{cases} \mu_p & \text{for } p\text{-type semiconductors} \\ \mu_n & \text{for } n\text{-type semiconductors} \end{cases} \qquad (6.6.10)$$

Figure 6.14 shows the mobility of electrons in the various Ge samples. The curves are deduced from the curves shown in Figs. 6.11 and 6.12. Since the theory concerning carrier mobility will be presented in Section 6.7, only the general features of Fig. 6.14 are discussed here. For the purest sample, number 55, the mobility varies as $T^{-1.66}$ over quite a wide temperature range. The theoretical temperature dependence of mobility is $T^{-3/2}$ if the scattering of electron motion is due to lattice vibration. It is generally believed that in pure samples, lattice scattering dominates.

As impurity concentration increases, however, not only does the mobility drop but it deviates further from the $T^{-3/2}$ dependence. The effect of impurity on mobility can be seen more clearly from the curve of Fig. 6.15, which is taken from Fig. 6.14 at 300 K. Since the resistivity is inversely proportional to N_d in the exhaustion region, Fig. 6.15 can be converted into a plot of μ_n versus N_d. The dependence of mobility upon

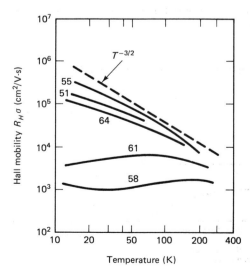

Figure 6.14 Hall mobility of electrons in germanium as a function of temperature. The curves are deduced from the Hall-coefficient and resistivity data of Figs. 6.11 and 6.12. (P. P. Debye and E. M. Conwell, *Phys. Rev.,* Vol. 93, p. 693, 1954.)

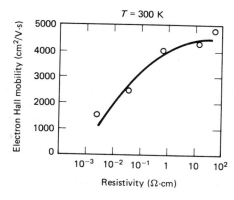

Figure 6.15 Plot of Hall mobility of electrons in germanium at 300 K as a function of resistivity. Since resistivity is inversely proportional to impurity concentration, the curve shows implicitly the dependence of mobility on impurity concentration. (P. P. Debye and E. M. Conwell, *Phys. Rev.*, Vol. 93, p. 693, 1954.)

impurity concentration is even more pronounced if data are selected at a lower temperature. We can summarize the picture as follows. Lattice scattering is the dominant mechanism in limiting the mobility of free carriers at high temperatures and low impurity concentrations. At low temperatures and high impurity concentrations, however, scattering by impurities may take over as the limiting mechanism. The same conclusion applies to the hole mobility.

We should point out that the product $R_H\sigma$ actually differs from the drift mobility μ. In Eqs. (4.8.11) and (5.11.5) we assumed that all carriers had the same mean free time τ. The same assumption is implied in Eq. (6.6.10). In semiconductors, a proper average of τ must be made. A formal analysis of the Boltzmann transport equation yields

$$\mu = \frac{e}{m^*}\frac{\langle v^2\tau\rangle}{\langle v^2\rangle} \quad \text{and} \quad R_H = -\frac{1}{ne}\frac{\langle v^2\tau^2\rangle\langle v^2\rangle}{\langle v^2\tau\rangle^2} \qquad (6.6.11)$$

The reader is referred to Appendices A5 and A6 [Eqs. (A5.16) and (A6.16)] or the book by Shockley for a derivation of Eq. (6.6.11) (W. Shockley, *Electrons and Holes in Semiconductors*, D. Van Nostrand Company, Inc., New York, 1950, pp. 270–277). From Eq. (6.6.11), the ratio of $R_H\sigma$ to μ is

$$\frac{\mu_H}{\mu} = \frac{R_H\sigma}{\mu} = \frac{\langle v^2\tau^2\rangle\langle v^2\rangle}{\langle v^2\tau\rangle^2} \qquad (6.6.12)$$

where the symbol $\langle\ \rangle$ stands for an averaging process over the Boltzmann distribution of carriers. The mobility μ is called the *drift mobility* and the product $R_H\sigma$ is called the *Hall mobility*. Measurement of the drift mobility is discussed in Section 7.8 in conjunction with the Haynes–Shockley experiment. The ratio μ_H/μ is generally not far from 1.

Finally, we should mention that precautions should be taken in the Hall and resistivity measurements. To avoid spurious results due to nonuniform distribution of current and nonlinear behavior of the end contacts, the sample used in such measurement preferably has a shape similar to that shown in Fig. 6.16. The broad areas on both ends are used for the current leads, the center arms for the Hall measurement and either pair of the outside arms on the same side for the resistivity measurement. Spurious results also may be caused by effects other than the one we want to measure. For example, an additional voltage difference results between different parts of the sample if the temperature is not uniform over the sample. The effect is known as the *thermoelectric effect*.

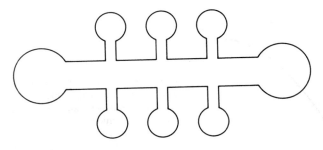

Figure 6.16 Dumbbell-shaped slab of semiconductor sample used for accurate measurements of Hall coefficient and resistivity.

Note that a reversal of the magnetic field changes the polarity of the Hall voltage. By averaging the two Hall voltages taken with $\pm B_0$, we can eliminate errors introduced by effects that are independent of B.

6.7 CARRIER MOBILITY AND SCATTERING MECHANISMS IN Si AND Ge

Impurity Scattering

The two most important mechanisms that limit the mobility of free carriers in a semiconductor are scattering by ionized impurities and scattering by lattice vibrations. Figure 6.17 illustrates the deflection of electron trajectories in the Coulomb field of charged particles. The electron trajectory can be analyzed through classical mechanics and the scattering angle θ is found to be

$$\tan \frac{\theta}{2} = \frac{e^2}{4\pi\epsilon am^*v^2} \qquad (6.7.1)$$

where v is the electron velocity and the distance a is defined in Fig. 6.17. For electrons whose trajectories are within an annular ring of area $dA = 2\pi a \, da$, the deflections will

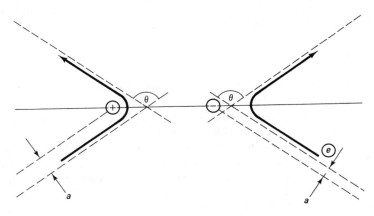

Figure 6.17 Deflection of the trajectory of an electron by (a) an ionized donor and (b) an ionized acceptor. The trajectory takes the asymptotic form of a straight line if the electron is far away from the ionized impurity. The parameter a is the distance between the asymptotic line (or lines) and a parallel line passing through the ionized impurity.

be within a solid angle $d\Omega = 2\pi \sin \theta \, d\theta$. Thus a differential cross section $\sigma(\theta)$ is defined as the ratio $dA/d\Omega$:

$$\sigma(\theta) = \frac{2\pi a \, da}{2\pi \sin \theta \, d\theta} = \left(\frac{e^2}{8\pi\epsilon m^* v^2}\right)^2 \csc^4 \frac{\theta}{2} \tag{6.7.2}$$

which is known as the *Rutherford law* (E. Rutherford, *Philos. Mag.*, Vol. 21, p. 669, 1911; see also for example, R. B. Leighton, *Principles of Modern Physics*, McGraw-Hill Book Company, New York, 1959, p. 485).

Imagine that each impurity atom has a region of influence that is taken to be cubic. If there are N_i impurity atoms per cubic centimeter, the edge b of the cube is of course given by $b^3 = N_i^{-1}$. Note that an electron suffers one deflection in each cube and that the distance between two adjacent cube centers is b. Therefore, the average time t between successive deflections is $t = b/v$. However, not all deflections are equally effective in changing the direction of v. For small θ, the direction of v is hardly changed. We define the free time τ_d between successive deflections as

$$\tau_d = \frac{b}{v(1 - \cos \theta)} \tag{6.7.3}$$

The factor $(1 - \cos \theta)$ measures the relative change in the component of the electron velocity along the initial direction of motion. If a carrier moves in a certain direction, its velocity component in the initial direction will not be lost after the first deflection but will be dissipated upon successive deflections. The effectiveness of each collision depends on the deflection angle θ. The factor $(1 - \cos \theta)$ accounts for this fact. For example, if $\theta = 0$, the electron trajectory is effectively not deflected and the free time between two successive deflections should be infinite. This is indeed the case with Eq. (6.7.3).

Refer again to Fig. 6.17. The deflection angle θ depends on the distance a through Eq. (6.7.2). For all trajectories within an annular ring of area $dA = 2\pi a \, da$, the deflection angle is the same. The reciprocal of the mean free time τ is equal to the average of τ_d^{-1} over dA. Thus we have

$$\frac{1}{\tau} = \frac{v}{b} \frac{1}{A} \int_0^{b/2} (1 - \cos \theta) 2\pi a \, da \tag{6.7.4}$$

For simplicity, the integration is taken to be over a circle of radius $b/2$ instead of a square. In Eq. (6.7.4), $A = \pi b^2/4$. Converting Eq. (6.7.4) into an integration over θ through the use of Eq. (6.7.2), we obtain[†]

$$\frac{1}{\tau(\mathscr{E})} = \frac{1}{b} \left(\frac{2\mathscr{E}_1}{m^*}\right)^{1/2} \left(\frac{\mathscr{E}_1}{\mathscr{E}}\right)^{3/2} \ln \left[1 + \left(\frac{2\mathscr{E}}{\mathscr{E}_1}\right)^2\right] \tag{6.7.5}$$

where $\mathscr{E}_1 = e^2/2\pi\epsilon b$ and $\mathscr{E} = m^* v^2/2$.

Note that the free time τ is a function of electron energy $\mathscr{E}$. Averaging $\tau(\mathscr{E})$ over the velocity distribution of electrons, we find that

$$\mu_i = \frac{e}{m^*} \frac{\langle v^2 \tau \rangle}{\langle v^2 \rangle} = \frac{2^{7/2}(kT)^{3/2}(4\pi\epsilon)^2}{\pi^{3/2} e^3 N_i(m^*)^{1/2}} \frac{1}{\ln[1 + (12\pi\epsilon kT/e^2 N_i^{1/3})^2]} \tag{6.7.6}$$

[†]Note that $\int (1 - \cos \theta) \csc^4 (\theta/2) \sin \theta \, d\theta = -8 \ln [\sin (\theta/2)]$ by expressing all trigonometric functions in terms of the argument $\theta/2$.

In obtaining Eq. (6.7.6), the following steps are taken. First the following relations exist:

$$\langle v^m \rangle = \frac{\int_0^\infty v^m \exp\left(-m^*v^2/2kT\right)4\pi v^2\, dv}{\int_0^\infty \exp\left(-m^*v^2/2kT\right)4\pi v^2\, dv} = v_T^m \frac{2}{\sqrt{\pi}} \left(\frac{m+1}{2}\right)! \qquad (6.7.7)$$

for any odd integer m and

$$\langle v^2 \rangle = \tfrac{3}{2} v_T^2 \qquad (6.7.8)$$

where $v_T^2 = 2kT/m^*$. Second, since the logarithm term is a slowly varying function of $\mathscr{E}$, it is taken outside of the integral in the averaging process and is evaluated at $\mathscr{E} = 3kT$. Third, the substitution $b = N_i^{-1/3}$ is made.

Substituting numerical constants in Eq. (6.7.6) gives

$$\mu_i = \frac{1.65 \times 10^{19}}{N_i \ln\left[1 + (3 \times 10^{11}/N_i^{2/3})(T\epsilon/300\epsilon_0)^2\right]} \left(\frac{T}{300}\right)^{3/2} \left(\frac{\epsilon}{\epsilon_0}\right)^2 \left(\frac{m_0}{m^*}\right)^{1/2} \text{cm}^2/\text{V-s}$$

$$\qquad (6.7.9)$$

where N_i is ionized-impurity concentration expressed in cm^{-3}, T is expressed in kelvin, ϵ is the dielectric constant, and m_0 is the free electron mass. The important characteristic of impurity scattering is the $T^{3/2}$ dependence. At higher temperatures, electrons on the average move faster and hence are less deflected. Taking $N_i = 10^{18}$ cm^{-3}, $T = 300$ K, $\epsilon = 16\epsilon_0$, and $m^* = 0.13m_0$ in Ge, we find μ_i to be 1.5×10^5 $\text{cm}^2/\text{V-s}$. Since the drift mobility μ for electrons in semiconductor-grade Ge is 3900 $\text{cm}^2/\text{V-s}$ at 300 K, scattering by charged impurities apparently is not the mobility-limiting process at 300 K. We use the subscript i to indicate that μ_i would be the true mobility if impurity scattering were the only mobility-limiting mechanism. The actual mobility is represented by μ without the subscript.

Lattice Scattering

Another important mobility-limiting mechanism is from lattice vibrations. The scattering of the motion of an electron by lattice vibrations is a wave phenomenon; therefore, it is conceptually difficult to comprehend from a classical viewpoint. The following presentation is based on a physical model originally proposed by Shockley and Bardeen (W. Shockley and J. Bardeen, *Phys. Rev.*, Vol. 77, p. 407, 1950; *Phys. Rev.*, Vol. 80, p. 72, 1950). Figure 6.18a, which is a reproduction of Fig. 4.8a in part, shows the variation of the conduction- and valence-band edges as a function of the lattice constant. As lattice atoms vibrate about their equilibrium positions, their vibrational motion causes a local variation in the energies of the conduction- and valence-band edges, as shown in Fig. 6.18b. Since the energy of an electron is conserved, the gain in potential energy must be at the expense of the kinetic energy. Thus, from $\mathscr{E} = \mathscr{E}_c + \hbar^2 k^2/2m_e^*$, we have

$$-\delta\mathscr{E}_c = \frac{\hbar^2 k}{m_e^*} \delta k \qquad (6.7.10)$$

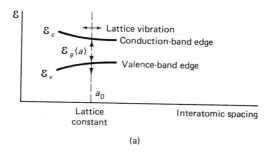

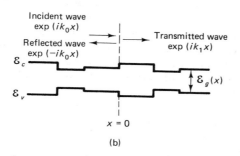

Figure 6.18 A physical model illustrating the effect of lattice vibrations on the wave motion of electrons. The effect is caused by modulation of lattice constant by lattice vibration. (a) Variation of energy-band edges as functions of lattice constant. (b) As atoms vibrate, the vibrations cause local variations of the lattice constant which in turn modulate the positions of the band edges. This local variation of band positions changes the propagation property of an electron, causing the scattering of electron wave motion.

and a similar expression for valence-band holes.

For ease of discussion, we assume that the electron motion is one-dimensional and that the change in $\mathscr{E}_c$ is abrupt. The word *abrupt* is used in the sense that the change in $\mathscr{E}_c$ occurs over a distance short compared with the wavelength associated with the wave motion of electrons. At the boundary where $\delta\mathscr{E}_c$ occurs, the propagation constant of the wave changes, resulting in a reflected wave. Referring to Fig. 6.18b, we have

$$\psi_1 = A \exp(ik_0x) + AR \exp(-ik_0x) \qquad \text{for} \quad x < 0 \qquad (6.7.11)$$

$$\psi_2 = AT \exp(ik_1x) \qquad \text{for} \quad x > 0 \qquad (6.7.12)$$

where ψ is the electron wave function and R and T represent, respectively, the reflection and transmission coefficient.

The boundary conditions at $x = 0$ require that ψ and $d\psi/dx$ be continuous, meaning that

$$1 + R = T \quad \text{and} \quad k_0(1 - R) = k_1T \qquad (6.7.13)$$

Solving for R and using Eq. (6.7.10), we find

$$|R|^2 = \left|\frac{k_0 - k_1}{k_0 + k_1}\right|^2 = \left(\frac{m_e^*}{2\hbar^2 k_0^2}\right)^2 (\delta\mathscr{E}_c)^2 \qquad (6.7.14)$$

The change in $\mathscr{E}_c$ can be expressed in terms of lattice dilation $\Delta = \delta V/V$ as follows:

$$\delta\mathscr{E}_c = \mathscr{E}_{1c}\Delta \qquad (6.7.15)$$

where Δ is the percentage volume change and $\mathscr{E}_{1c}$ is a proportionality constant commonly referred to as the *deformation potential*. Further, $\Delta = Kp$ under a pressure p, where K is the compressibility of a semiconductor. Since the stored mechanical energy is given by $p \, \delta V/2$, which on the average is equal to $kT/2$ for a one-dimensional model, we have

$$kT = p \, \delta V = \frac{V}{K} \Delta^2 \tag{6.7.16}$$

Using Eqs. (6.7.15) and (6.7.16) in Eq. (6.7.14), we obtain

$$|R|^2 = \left(\frac{m_e^* \mathscr{E}_{1c}}{2\hbar^2 k_0^2}\right)^2 \frac{KkT}{V} \tag{6.7.17}$$

In applying Eq. (6.7.17) to scattering problems, we must know the meaning of V. Since thermal vibration can be analyzed into Fourier components as running waves, part of the lattice experiences a positive δV and part a negative δV. If we take the crystal as a whole, the effect averages out. Therefore, the volume V in Eq. (6.7.17) must be an effective volume in which the lattice and electron waves interact strongly. This region extends over $\lambda/4$ in each dimension, where $\lambda = 2\pi/k_0$ is the wavelength of the electron wave. The mean free path l of an electron depends on the number of reflections that the electron suffers within $\lambda/4$. We define

$$l = \frac{\lambda/4}{2|R|^2} = \left(\frac{h^2}{4m_e^* \mathscr{E}_{1c}}\right)^2 \frac{1}{8KkT} \tag{6.7.18}$$

The factor 2 in $2|R|^2$ accounts for the fact that electrons suffer reflections at both ends of the region. The exact three-dimensional result based on quantum-mechanical calculations is

$$l = \left(\frac{h^2}{4m_e^* \mathscr{E}_{1c}}\right)^2 \frac{c}{\pi^3 kT} \tag{6.7.19}$$

where c is the elastic constant and has a magnitude of the order of K^{-1}.

According to our previous discussion in conjunction with impurity scattering, the mean free time τ is given by

$$\frac{1}{\tau} = \frac{1}{4\pi} \int \frac{v}{l} (1 - \cos \theta) \, d\Omega = \frac{v}{l} \frac{1}{4\pi} \int (1 - \cos \theta) 2\pi \sin \theta \, d\theta \tag{6.7.20}$$

Since l of Eq. (6.7.19) is isotropic (i.e., independent of θ), Eq. (6.7.20) reduces to

$$\frac{1}{\tau} = \frac{v}{l} = \frac{\mathscr{E}_{1c}^2 (m_e^*)^2 kT}{\pi \hbar^4 \rho v_s^2} v \tag{6.7.21}$$

where ρ is the mass density and v_s is the sound velocity in the semiconductor with $v_s^2 = c/\rho$. Using Eq. (6.7.7) for $\langle v^2 \tau \rangle$ in Eq. (6.6.11), we find that

$$\mu_l = \frac{e}{3kT} \langle v^2 \tau \rangle = \frac{2\sqrt{2\pi}}{2} \frac{e\hbar^4 \rho v_s^2}{\mathscr{E}_{1c}^2} (m_e^*)^{-5/2} (kT)^{-3/2} \quad \text{m}^2/\text{V-s} \tag{6.7.22}$$

The important characteristic of lattice mobility is the $T^{-3/2}$ dependence. Electron and hole drift mobilities in Ge and Si in the temperature range where only lattice scattering should be important have been measured by several workers and the results reported by Morin and Maita are shown in Fig. 6.19 (F. J. Morin and J. P. Maita, *Phys. Rev.*, Vol. 94, p. 1526, 1954; *Phys. Rev.*, Vol. 96, p. 28, 1954). The accepted room-

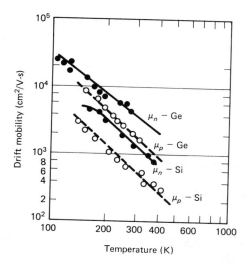

Figure 6.19 Measured values of drift mobility in germanium and silicon as functions of temperature. (F. J. Morin and J. P. Maita, *Phys. Rev.*, Vol. 94, p. 1526, 1954; Vol. 96, p. 28, 1954.)

temperature values of the drift mobilities and their temperature dependences are summarized in Table 6.3.

As we can see, the temperature dependences deviate quite considerably from the $T^{-3/2}$ law even in the region dominated by lattice scattering. As discussed in Section 2.10, there are two modes of lattice vibration in Ge and Si: the high-frequency (the optical) mode and the low-frequency (the acoustical) mode. In Eq. (6.7.22) only the long-wavelength acoustical mode is considered. In semiconductors such as Ge and Si, where there are several equivalent conduction-band minima (Section 6.10), electrons may be scattered from one band minimum to another as well as scattered within the same band minimum. These processes are known as *intervalley* and *intravalley scattering* processes, respectively. For intervalley scattering, a large change of electron momentum is involved, thus requiring the participation of short-wavelength phonons, both acoustical and optical.

When several scattering mechanisms exist simultaneously, Eq. (4.8.15) becomes

$$\frac{d\langle v \rangle}{dt} = -\langle v \rangle \left(\frac{1}{\tau_1} + \frac{1}{\tau_2} + \frac{1}{\tau_3} \right) + \frac{F}{m^*} \qquad (6.7.23)$$

where τ_1, τ_2, and τ_3 are the mean free time associated with each scattering mechanism. The resultant τ is equal to

$$\frac{1}{\tau} = \frac{1}{\tau_1} + \frac{1}{\tau_2} + \frac{1}{\tau_3} \qquad (6.7.24)$$

According to Eq. (4.8.11), the actual mobility μ is

$$\frac{1}{\mu} = \frac{1}{\mu_1} + \frac{1}{\mu_2} + \frac{1}{\mu_3} \qquad (6.7.25)$$

Strictly speaking, Eq. (6.6.11) should be used instead of Eq. (4.8.11). Therefore, Eq. (6.7.25) is true only if all the τ's have the same energy dependence. However, Eq. (6.7.25) is still useful for qualitative discussions. The optical-mode-scattering mobility,

	Ge		Si		GaAs	
	μ_n	μ_p	μ_n	μ_p	μ_n	μ_p
Value at 300 K	3900	1900	1400[a]	470	8000[b]	340
Temperature dependence	$T^{-1.66}$	$T^{-2.33}$	$T^{-2.5}$	$T^{-2.7}$	—	$T^{-2.3}$

[a]Note from Eq. (A6.16) that the effective-mass ratio must be taken into account in calculating μ from μ_H.

[b]Electron mobility is dominated by polar optical-phonon scattering.

according to Howarth and Sondheimer (D. J. Howarth and E. H. Sondheimer, *Proc. R. Soc. A,* Vol. 219, p. 53, 1953), is given by

$$\mu_{op} = \frac{\pi e \hbar^2 \rho (\hbar \omega_{op})^{3/2}}{2^{5/2} \mathscr{E}_{2c}^2 q^2} (m_e^*)^{-5/2} (kT)^{-1} \left[\exp \left(\frac{\hbar \omega_{op}}{kT} \right) - 1 \right] \qquad (6.7.26)$$

where q is the wave-number change involved in the intervalley-scattering process and $\hbar \omega_{op}$ is the energy of an optical phonon involved in the scattering process. The quantity ρ is the mass density and the parameter $\mathscr{E}_{2c}$ is a deformation potential, similar to $\mathscr{E}_{1c}$ defined in Eq. (6.7.15). Equation (6.7.26) is applicable to covalent semiconductors at a temperature $T < \hbar \omega_{op}/k$. We should point out that optical-mode scattering described in Eq. (6.7.26) is caused by the modulation of band edges by lattice vibration (Fig. 6.18a). The difference between Eqs. (6.7.26) and (6.7.22) is that short-wavelength phonons are involved in the former, whereas long-wavelength phonons are involved in the latter, resulting in different temperature dependence. Curves in Fig. 6.19 can be fitted quite well with a combination of μ_{op} and μ_l. Other scattering mechanisms are discussed in Section 6.8 (J. M. Ziman, *Electrons and Phonons,* Clarendon Press, Oxford, 1960).

We should point out that in both Eqs. (6.7.22) and (6.7.26) only longitudinal lattice vibrations are involved. For simplicity, we consider a cubic crystal with three orthogonal axes $\hat{x}$, $\hat{y}$, and $\hat{z}$. Under a small uniform deformation, the three axes are slightly distorted in orientation and in length, and the new axes become (C. Kittel, *Introduction to Solid-State Physics,* 4th ed., John Wiley & Sons, Inc., New York, 1971)

$$\hat{x}' = (1 + \delta_{xx})\hat{x} + \delta_{xy}\hat{y} + \delta_{xz}\hat{z}$$

$$\hat{y}' = \delta_{yx}\hat{x} + (1 + \delta_{yy})\hat{y} + \delta_{yz}\hat{z} \qquad (6.7.27)$$

$$\hat{z}' = \delta_{zx}\hat{x} + \delta_{zy}\hat{y} + (1 + \delta_{zz})\hat{z}$$

If the displacement of an atom is represented by $\mathbf{R} = u\hat{x} + v\hat{y} + w\hat{z}$, the various strain components are given by $\delta_{xx} = \partial u/\partial x$, $\delta_{xy} = \partial v/\partial x$, and so on. Obviously, the tensile strains are associated with a longitudinal lattice wave, and the shear strains with a transverse lattice wave. From Eq. (6.7.27) the volume V' of a distorted cell is related to the volume V of an unstrained unit cell by

$$V' = \hat{\mathbf{x}}' \times \hat{\mathbf{y}}' \cdot \hat{\mathbf{z}}' = V\left[1 + \sum_i \delta_{ii} + \sum_{ij} (\delta_{ii}\delta_{jj} - \delta_{ij}\delta_{ji}) + \sum \delta^3 \right] \qquad (6.7.28)$$

where $i\,j = x,\ y,$ or z. Therefore, the first-order dilation $\Delta = (V' - V)/V$ is due to δ_{ii} and caused by the longitudinal lattice wave only. The second-order terms such as $\delta_{ij}\delta_{ji}$ caused by a transverse lattice wave are negligible as compared to the first-order terms. Therefore, the effect of transverse lattice waves is not considered in the expressions μ_l and μ_{op}.

6.8 SEMICONDUCTOR MATERIALS

The word *semiconductor* is customarily used to refer to a class of material in which electric conduction is carried by occupied states in the conduction band and empty states in the valence band. Unlike metals, the concentrations of free electrons and holes can be controlled by the amount of impurities which are purposely introduced into a semiconductor. Common examples include (1) element semiconductors such as silicon and germanium; (2) compound semiconductors formed by elements belonging to the third and fifth columns of the periodic table, of the type IIIb–Vb, such as GaAs, InSb, AlP, and GaP; (3) compound semiconductors formed by elements belonging to the second and sixth columns of the periodic table, of the type IIb–VIb, such as CdS, ZnS, CdTe, and ZnO; and (4) compound semiconductors of the type IVb–VIb, such as PbS and PbTe.

Silicon is the most important semiconductor commercially because it is most extensively used in microelectronics (integrated circuits). The existence of a stable native oxide is contributory to the dominance of silicon in IC technology. Accurate control of device property is made possible through impurity diffusion and ion implantation. A vast amount of knowledge about silicon technology has been accumulated over the years specifically for making large-scale circuit integration. Gallium arsenide and the related III–V compounds are gaining in importance because they are used both as materials for injection lasers and as materials for ultrafast electronic devices. As our discussion in Chapters 8, 9, 13, and 14 will illustrate, each device application requires a certain specific property of a semiconductor. The purpose of this section is to present a general discussion of some common properties of the various kinds of semiconductors. Table 6.4 shows some of the common properties for a selected number of semiconductors.

Chemical Bond

The reader may have noticed that the gap energy has a tendency to be progressively larger in going from element to II–VI compound semiconductor. This tendency is a result of increasing degree of ionicity in the chemical bond. The difference between covalent and ionic bonding is that charge transfer is involved in forming an ionic bond. Germanium and silicon are known to be covalent semiconductors, and lead salts are considered to be polar semiconductors. Information concerning charge transfer and hence the degree of ionicity can be obtained by comparing the static and optical dielectric constant.

As will be discussed in Chapter 12, the displacement of positive and negative charges in an ionic crystal gives rise to an electric polarization that contributes to the dielectric constant. According to an analysis by Szigeti (B. Szigeti, *Trans. Faraday*

	Crystal[a] structure	Energy gap (eV)	Mobility[b] (cm²/V-s)		Static ϵ/ϵ_0	Refractive index,[c] n^2
			μ_n	μ_p		
Si	D	1.11	1,400	470	11.7	11.7
Ge	D	0.67	3,900	1,900	16	16.0
GaAs	ZB	1.42	8,000	340	13.2	10.9
GaSb	ZB	0.67	4,000	1,400	15	14.0
InSb	ZB	0.16	77,000	750	17	15.7
CdTe	ZB	1.45	700	~60	11	—
CdS	W, ZB	2.45(W)	200	—	11.6	5.9
ZnO	W	3.30	200	—	8.5	3.73
PbS	NaCl	0.41	600	700	174.4	18.5
PbTe	NaCl	0.32	1,800	900	310	42.2

[a]D, diamond; ZB, zinc blende; W, wurtzite; NaCl, NaCl structure.

[b]These values represent the highest values of drift mobilities reported thus far.

[c]For optical frequencies below the gap energy.

The table is compiled from values quoted in the following references:

1. E. M. Conwell, "Properties of Silicon and Germanium: II," *Proc. IRE,* Vol. 46, p. 1281, 1958.
2. T. S. Moss, *Optical Properties of Semiconductors,* Academic Press, Inc., New York, 1959.
3. C. Hilsum and A. C. Rose-Innes, *Semiconducting III–V Compounds,* Pergamon Press, Inc., New York, 1961.
4. N. B. Hannay, *Semiconductors,* American Chemical Society Monograph Series, Reinhold Publishing Corporaton, New York, 1959.
5. P. Airgrain and M. Balkanski, eds., *Selected Constants Relative to Semiconductors,* Pergamon Press, Inc., New York, 1961.
6. H. C. Casey, Jr., and M. B. Panish, *Heterostructure Lasers,* Academic Press, Inc., New York, 1978.
7. J. S. Blakemore, *J. Appl. Phys.,* Vol. 10, p. R123, October 1982.

Soc., Vol. 45, p. 155, 1949), the difference between the static and optical dielectric constant in a cubic crystal is given by

$$\frac{\epsilon}{\epsilon_0} - n^2 = \left(\frac{n^2 + 2}{3}\right)^2 S^2 \frac{(Ze)^2}{\epsilon_0} \frac{N}{\omega_t^2}\left(\frac{1}{M_1} + \frac{1}{M_2}\right) \qquad (6.8.1)^{\dagger}$$

where N is the number of ion pairs per unit volume, M_1 and M_2 the masses of the ions, ω_t the angular frequency of the transverse optical mode of lattice vibration, Z the valence of the compound, and n the optical index of refraction. In deriving Eq. (6.8.1), it is assumed that the effective charge of an ion is equal to SZe; hence the parameter S obtained in fitting Eq. (6.8.1) to the difference between the experimental values (Table 6.4) of ϵ/ϵ_0 and n^2 should be of value in estimating the ionic character of a compound.

In III–V compounds, the value of S is found to vary from 0.10 in GaAs to 0.20 in GaP, taking $Z = 3$ in Eq. (6.8.1) (G. Picus et al., *J. Phys. Chem. Solids,* Vol. 8,

†Except for the factor $[(n^2 + 2)/3]^2$, this equation can be obtained from Eq. (12.3.5) by noting that $\Delta\epsilon = N\alpha_i$.

p. 282, 1959; D. A. Kleinman and W. G. Spitzer, *Phys. Rev.*, Vol. 118, p. 110, 1960). In II–VI compounds, the value of S is much higher, as it is evident from Table 6.4 that a large discrepancy exists between static ϵ/ϵ_0 and n^2. The following values are quoted by Hutson in Hannay's book: $S = 1$ for MgO, 0.63 for ZnO, and 0.48 for ZnS (N. B. Hannay, ed., *Semiconductors,* American Chemical Society Monograph Series, Reinhold Publishing Corporation, New York, 1959). Therefore, it is fair to say that the degree of ionicity increases markedly from III–V to II–VI compounds. This is consistent with the tendency of having a larger gap energy in II–VI compounds. We list in Table 6.5 the gap energy in various element and compound semiconductors. Since the gap energy is sensitive to interatomic distances, we choose for comparison materials with closest interatomic distances possible in the table.

Another measure of ionicity in a compound is the difference in the electronegativity of its constituent elements. By *electronegativity* we mean the power of an atom to attract an electron to itself in a compound. In Table 6.6 we list the electronegativity values for the elements used in III–V and II–VI compounds. In comparing the static and optical dielectric constants in Eq. (6.8.1), only the magnitude of the charge transfer is determined, leaving the polarity of the charge unspecified. It is clear from the electronegativity value listed in Table 6.6 that electrons will be transferred from the IIb element to the VIb element in a II–VI compound and from the IIIb element to the Vb element in a III–V compound, making II and III atoms positively charged. The values of percentage of ionic bonding, known as *ionicity,* based on an extensive theoretical analysis of experimental data by Phillips (J. C. Phillips, *Rev. Mod. Phys.,* Vol. 42, p. 317, 1970), are also given in Table 6.6, for example, $f_i = 0.31 = 31\%$ in GaAs.

Other comments concerning Table 6.6 are summarized as follows. The elements are so arranged that the degree of ionicity generally decreases from the top toward the bottom, from the left toward the right, and from the upper left corner toward the lower right corner. According to Pauling, the percentage of ionic bonding in a compound increases from $f_i = 0.22$ to 0.63 to 0.89 as the difference between the electronegativity values of its elements increases from 1 to 2 to 3. Where the value of electronegativity is not sufficiently accurate in comparing the relative degree of ionicity in different compounds, data from the difference in the static and optical dielectric constant are used in

TABLE 6.5 COMPARISON OF GAP ENERGY IN A SELECTED NUMBER OF ELEMENT AND COMPOUND SEMICONDUCTORS[a]

Semiconductors	Si	AlP	ZnS	Ge	GaAs	InP	CdS
Interatomic distance (Å)	2.34	2.36	2.34	2.44	2.44	2.54	2.52
Gap energy (eV)	2.5 (1.10)	3.0	3.7	0.80 (0.665)	1.42	1.29	2.4
Semiconductors	InAs	AlSb	ZnTe	Gray-Sn	InSb	CdTe	—
Interatomic distance (Å)	2.615	2.655	2.635	2.80	2.80	2.805	—
Gap energy (eV)	0.36	1.62	2.0	0.17	0.36	1.6	—

[a]The values for Si and Ge are direct-gap energies measured across energy bands at $k[000]$, with the values of indirect-gap energies quoted in parentheses.

TABLE 6.6 ELECTRO NEGATIVITY VALUES (ACCORDING TO THE PAULING SCALE) AND IONICITY VALUES (ACCORDING TO PHILLIPS) FOR ELEMENTS III–V AND II–VI COMPOUNDS[a]

III V	Al 1.61	In 1.78	Ga 1.81	II VI	Mg 1.31	Cd 1.69	Zn 1.65	Be 1.57
N 3.04	0.45 W	0.58 W	0.50 W	O 3.44	0.84 NaCl	0.79 NaCl	0.62 W	0.60 W
P 2.19	0.31 ZB	0.42 ZB	0.37 ZB	S 2.58	0.79 NaCl	0.69 ZB, W	0.62 ZB, W	0.31 ZB
As 2.18	0.27 ZB	0.36 ZB	0.31 ZB	Se 2.55	0.79 NaCl	0.70 W, ZB	0.68 ZB, W	0.30 ZB
Sb 2.05	0.43 ZB	0.32 ZB	0.26 ZB	Te 2.1	0.55 W	0.68 ZB	0.55 ZB	0.17 ZB

[a] W, wurtzite structure; ZB, zinc-blende structure.

Sources: L. Pauling, *The Nature of the Chemical Bond,* Cornell University Press, Ithaca, N.Y., 1948; J. C. Phillips, "Ionicity of the Chemical Bonds in Crystals," *Rev. Mod. Phys.,* Vol. 42, p. 317, 1970.

the arrangement of Table 6.6. The percentage of ionic bonding, based on Pauling's criterion, is estimated to be 22% in CdS and ZnS and 9% in InAs. The values of ionicity quoted by Phillips and given in Table 6.6 are more accurate in comparing the relative degree of ionicity in II–VI and III–V compounds. Theoretical analyses by Birman and by Browne have led to a value close to 65% ionic bonding in ZnS, which agrees well with the value of 67% concluded from an analysis of the piezoelectric constant by Saksena (J. L. Birman, *Phys. Rev.,* Vol. 109, p. 810, 1958; P. F. Browne, *J. Electron.,* Vol. 2, p. 154, 1956; B. D. Saksena, *Phys. Rev.,* Vol. 81, p. 1012, 1951).

The crystal structure of the various compounds are also given in Table 6.6. The zinc-blende structure is the same as the diamond structure (Figs. 2.11 and 2.12) except that the two interpenetrating cubic close-packed structures are occupied by different atoms. The wurtzite structure (Fig. 6.20) may be considered as two interpenetrating hexagonal close-packed structures displaced with respect to each other by a distance $3c/8$ along the c axis. In the two structures, the first nearest neighbors form identical units (tetrahedral) as shown in Figs. 2.12c and 6.20b. Significant differences come only in the third nearest neighbors. This is the reason why some compounds exist in both structures. However, a close examination shows that the wurtzite structure is more favorable for crystals with large charge differences of electronegativity between the two kinds of atoms. In other words, the general tendency is such that the wurtzite structure is more prone than the zinc-blende structure to having a higher degree of ionicity. Of course, the NaCl structure is the structure most favorable for ionic bonding as most ionic crystals have this structure. Lead salts PbS and PbTe crystallize in the NaCl structure.

Polar and Piezoelectric Scattering and Mobility Comparison

The difference between element and compound semiconductors is further reflected in the magnitude as well as the temperature dependence of their mobility. As discussed in Section 6.7, the mobility in Ge and Si (purely covalent semiconductors) is limited by

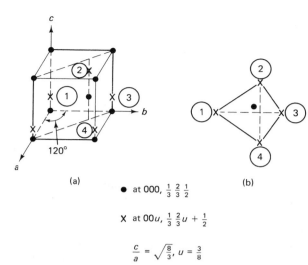

(a)

● at 000, $\frac{1}{3}\frac{2}{3}\frac{1}{2}$

✕ at 00u, $\frac{1}{3}\frac{2}{3}u + \frac{1}{2}$

$\frac{c}{a} = \sqrt{\frac{8}{3}}$, $u = \frac{3}{8}$

(b)

Figure 6.20 (a) The ideal wurtzite structure. The structure consists of two interpenetrating hcp lattices with atoms of one lattice denoted by ● and atoms of the other lattice by ✕. (b) The basic tetrahedral unit of the wurtzite structure.

three scattering mechanisms: impurity scattering, acoustical-mode lattice scattering, and optical-mode lattice scattering. According to Eqs. (6.7.9), (6.7.22), and (6.7.26), the various scattering-limited mobilities have the following principal dependences:

$$\mu_i = AN_i^{-1}m^{*-1/2}\epsilon^2 T^{3/2} \tag{6.8.2}$$

$$\mu_l = B\mathcal{E}_1^{-2}m^{*-5/2}T^{-3/2} \tag{6.8.3}$$

$$\mu_{op} = Cm^{*-5/2}T^{-1}\exp\left(\frac{\hbar\omega_{op}}{kT}\right) \tag{6.8.4}$$

where A, B, and C are three proportionality constants.

In compound semiconductors, scattering of carriers by optical modes of lattice vibration plays a very important role. In the optical modes, the two neighboring atoms move in opposite directions relative to each other (Section 2.11). Because of the ionic charge carried by each atom, their relative movement constitutes an electric polarization, which in turn produces an electric field. Deflection of the motion of free carriers by the field limits the mobility of carriers. According to an analysis by Howarth and Sondheimer (D. J. Howarth and E. H. Sondheimer, *Proc. R. Soc. London*, Vol. 70, p. 124, 1957), the mobility limited by polar scattering is

$$\mu_{po} = \frac{1}{2\alpha\omega_l}\frac{e}{m^*}\frac{8}{3\sqrt{\pi}}\frac{F(\theta_l/T)[\exp(\theta_l/T) - 1]}{(\theta_l/T)^{1/2}} \tag{6.8.5}$$

where $\theta_l = \hbar\omega_l/k$ is the equivalent temperature and $\hbar\omega_l$ is the energy associated with longitudinal optical phonons. In Eq. (6.8.5), α is a dimensionless interaction parameter known as the *Fröhlich coupling constant* and defined as

$$\alpha = \frac{e^2}{4\pi\epsilon_0}\left(\frac{m^*}{2\hbar\omega_l}\right)^{1/2}\frac{1}{\hbar}\left(\frac{\epsilon_0}{\epsilon_\infty} - \frac{\epsilon_0}{\epsilon}\right) \tag{6.8.6}$$

where ϵ and $\epsilon_\infty = n^2\epsilon_0$ are the static and optical dielectric constant, respectively. A similar expression for polar-scattering-limited mobility derived by Conwell (E. M. Conwell, *High Field Transport in Semiconductors*, Academic Press, Inc., New York, 1967) reads as follows:

$$\mu_{po} = \frac{3\sqrt{\pi}}{4E_0} \left(\frac{2k\theta_l}{m^*}\right)^{1/2} \frac{K_1^{-1}(\theta_l/2T)}{(\theta_l/T)^{3/2} \exp (\theta_l/2T)} [\exp (\theta_l/T) - 1] \qquad (6.8.7)$$

where K_1 is the modified Bessel function of the second kind. The quantity E_0, which is related to α and given by

$$E_0 = \frac{m^* e\omega_l}{4\pi\epsilon_0\hbar} \left(\frac{\epsilon_0}{\epsilon_\infty} - \frac{\epsilon_0}{\epsilon}\right) \qquad (6.8.8)$$

represents the strength of the electric polarization field associated with longitudinal optical modes of lattice vibration. Both Eqs. (6.8.6) and (6.8.8) are in MKS system of units.

The polarization created by relative displacement of ions not only is responsible for the interaction between the wave motion of electrons and the optical mode of lattice vibration but also contributes to the dielectric constant. Hence it is expected that Eqs. (6.8.8) and (6.8.1) should be related. The function $F(\theta_l/T)$ is a slowly varying function which decreases from unity at high temperatures to a minimum value 0.6 when $T = \theta_l$, then increases steadily with θ_l/T, and is equal to $(3/8)(\pi\theta_l/T)^{1/2}$ at low temperatures. Thus Eq. (6.8.5) has the following principal dependence

$$\mu_{po} = D\left(\frac{\epsilon}{\epsilon_0} - n^2\right)^{-1} m^{*-3/2} \left(\frac{T}{\theta_l}\right)^{-1/2} \qquad \text{for} \quad T > \theta_l \qquad (6.8.9)$$

$$D\frac{3\sqrt{\pi}}{8} \left(\frac{\epsilon}{\epsilon_0} - n_2\right)^{-1} m^{*-3/2} \left[\exp \left(\frac{\theta_l}{T}\right) - 1\right] \qquad \text{for} \quad T < \theta_l \qquad (6.8.10)$$

where D is a proportionality constant. The reader is referred to the book by Ziman for a quantum-mechanical derivation of Eq. (6.8.5) and also to an excellent review article by Hutson for a discussion of the role of longitudinal and transverse optical modes of lattice vibration as well as for a physical picture of the electron–phonon interaction. (J. M. Ziman, *Electrons and Phonons*, Oxford University Press, Oxford, 1960; A. R. Hutson, "Semiconducting Properties of Some Oxides and Sulfides," in N. B. Hannay, ed., *Semiconductors*, Reinhold Publishing Corporation, New York, 1959.) Using the limiting forms for small arguments ($z << 1$), $K_0(z) \cong -\ln z$ and $K_v(z) \cong \frac{\Gamma(v)}{2} (z/2)^{-v}$, and

the asymptotic expression for large arguments ($z >> 1$), $K_v(z) \cong \sqrt{\frac{\pi}{2z}} \exp (-z)$ for the

modified Bessel functions (M. Abramowitz and A. Stegun, *Handbook of Mathematical Functions*, National Bureau of Standards, U.S. Government Printing Office, Washington, D.C., 1964, pp. 375 and 378), we can show that Eq. (6.8.7) reduces to Eq. (6.8.9) for $T > \theta_l$ except that D is replaced by $D(3\sqrt{\pi}/4)$. At low temperatures, $T < \theta_l$, Eq. (6.8.7) differs from Eq. (6.8.10) by the factor $3T/2\theta_l$.

The difference between μ_l of Eq. (6.8.3) due to acoustic-phonon scattering, μ_{op} of Eq. (6.8.4) due to optical-phonon scattering and μ_{po} of Eq. (6.8.5) or (6.8.7) due to polar optical scattering is best understood with reference to the energy-band structure discussed in Section 6.9 and shown in Figs. 6.22 and 6.23. The conduction band in element semiconductors Si and Ge (Fig. 6.22) has several equivalent minima. Scattering within the same conduction-band minimum is called *intravalley scattering* and that between two equivalent conduction-band minima is called *intervalley scattering*. While Eq. (6.8.3) applies to intravalley scattering, Eq. (6.8.4) applies to intervalley scattering, where a large change in the electron wave vector **k** is involved. Therefore, for element semiconductor Si and Ge, both Eqs. (6.8.3) and (6.8.4) are applicable even though μ_I of Eq. (6.8.3) is more important than μ_{op} of Eq. (6.8.4). Whereas Eq. (6.8.4) is caused by modulation of band edges by lattice vibration, Eq. (6.8.5) is caused by opposite movements of positive and negative ions in lattice vibration. Therefore, Eq. (6.8.5) or (6.8.7) does not apply to Si and Ge, which are nonpolar. Since most electrons in GaAs are located in the conduction-band minimum at $k(000)$, Eq. (6.8.5) or (6.8.7) concerns only intravalley scattering, for which the change in electron wave vector **k** is small. Therefore, Eqs. (6.8.3) to (6.8.5) all apply to direct-gap compound semiconductors such as GaAs. The reader is referred to Section 11.10 and Fig. 11.31 for further discussions on scattering mechanisms in GaAs.

We should again point out that in the mobility expression for polar optical-phonon scattering, Eq. (6.8.5) or (6.8.7), the energy $k\theta_l = \hbar\omega_l$ is that of longitudinal optical phonons. The polarization **P** in an ionic crystal can be expressed in terms of its ionic and electronic contributions as

$$\mathbf{P} = N_0(e^*\mathbf{s} + \alpha\mathbf{E}_{\text{eff}}) \qquad (6.8.11)$$

where N_0 is the density of polarizable ion pairs per unit volume, e^* the effective charge associated with each ion, **s** the relative displacement of ions, and α the electronic polarizability of the ion pair. In a strongly ionic crystal, the effective electric field $\mathbf{E}_{\text{eff}}$ is equal to $\mathbf{E} + \mathbf{P}/3\epsilon_0$, where **E** is the applied field and $\mathbf{P}/3\epsilon_0$ is an internal field produced by local polarization **P**. Incorporating the effect of the local field, Eq. (6.8.11) becomes

$$\mathbf{P} = N_0 e^* \frac{\epsilon_\infty + 2\epsilon_0}{3\epsilon_0} \mathbf{s} + (\epsilon_\infty - \epsilon_0)\mathbf{E} \qquad (6.8.12)$$

where $N\alpha_0/3\epsilon_0 = (\epsilon_\infty - \epsilon_0)/(\epsilon_\infty + 2\epsilon_0)$.

The displacement vector **s** can be decomposed into the longitudinal component $\mathbf{s}_l$ and the transverse component $\mathbf{s}_t$. If $\mathbf{s} = \hat{\mathbf{a}}s \exp i\,(\mathbf{q} \cdot \mathbf{r} - \Omega t)$, where $\hat{\mathbf{a}}$ is a unit vector representing the direction of ionic movement, then $\hat{\mathbf{a}} \times \mathbf{q} = 0$ for $\mathbf{s}_l$ and $\hat{\mathbf{a}} \cdot \mathbf{q} = 0$ for $\mathbf{s}_t$. Therefore, we have

$$\text{curl } \mathbf{s}_l = 0 \quad \text{and} \quad \text{div } \mathbf{s}_t = 0 \qquad (6.8.13)$$

In a crystal free of space charges,

$$\text{div } \mathbf{D} = \epsilon_0 \text{ div } \mathbf{E} + \text{div } \mathbf{P} = 0 \qquad (6.8.14)$$

Applying the div operator to Eq. (6.8.12) and using Eq. (6.8.14), we find that

$$\epsilon_\infty \text{ div } \mathbf{E} = -Ne^* \frac{\epsilon_\infty + 2\epsilon_0}{3\epsilon_0} \text{ div } \mathbf{s}_l \qquad (6.8.15)$$

Associated with longitudinal lattice waves are an electric field $\mathbf{E} = -\text{grad } \phi$ and an electric potential ϕ. This potential ϕ acts as a scattering potential on the motion of free carriers in a manner similar to Coulomb potential of ionized impurities. Because div $s_t = 0$, the transverse lattice wave does not generate a scattering potential. Furthermore, the polar scattering process as represented by μ_{po} of Eq. (6.8.5) or Eq. (6.8.7) applies to intravalley scattering for which the change of the wave vector of the electron is small. Therefore, the value of $\hbar\omega_l$ is taken at a very small value of the lattice-vibrational wave number $q \cong 0$, or $k = 0$ in Fig. 2.38d.

Unlike element semiconductors, the electron mobility in III–V compound semiconductors seems to be limited by polar scattering instead of acoustical-mode scattering. To estimate the value of μ_l in III–V compounds, we choose InAs and Ge for comparison. The effective mass m^* for Ge in Eq. (6.8.3) is the average of transverse and longitudinal masses $1/m^* = (2/m_t + 1/m_l)/3$ or $m^*/m_0 = 0.12$, while that for InAs is equal to $m^*/m_0 = 0.024$. Therefore, based on the ratio of the effective masses, Eq. (6.8.3) predicts a value of $\mu_l = 220,000$ cm²/V-s in InAs. This value is to be compared with the observed mobility of 33,000 cm²/V-s and the calculated polar mobility of $\mu_{po} = 40,000$ cm²/V-s from Eq. (6.8.5). Similar comparisons can be made for other III–V compounds.

The large difference in electron and hole mobilities in III–V compounds seems to be explainable by Eq. (6.8.9) or a combination of equations from Eqs. (6.8.2) to (6.8.4) as being due to the difference in the effective masses. For example, the ratio of hole to electron effective mass in InAs is equal to 17, yielding a mobility ratio of 1/69 from Eq. (6.8.9). The observed ratio is 460:33,000, or 1:72. On the other hand, from Eq. (6.8.3) for acoustical-mode lattice scattering, the ratio would be 1:1200, assuming the same deformation potential $\mathscr{E}_1$ for conduction- and valence-band edges. However, from Eq. (6.8.2), the ratio would be only 1:4.2. A combination of the two would give a mobility ratio somewhere between the two extremes. Therefore, it is hazardous to draw any conclusion about the scattering mechanism based on the mass ratio without knowing the impurity concentration.

Now let us discuss the scattering process which limits the mobilities in II–VI compounds. Among the II–VI compounds ZnO and CdS have been studied most extensively. It is known that both crystals are strongly piezoelectric. By *piezoelectric effect* we mean that a polarization field (electric) is induced in a crystal by the application of a mechanical stress, or vice versa. For our present discussion, the piezoelectric effect means that lattice vibrations can now interact with electron motion through the electric field that accompanies the lattice wave. According to an analysis by Hutson (A. R. Hutson, *J. Appl. Phys.*, Vol. 32, p. 2287, 1961), the piezoelectric-scattering mobility is

$$\mu_{pe} = 1.44 \left(\frac{m_0}{m^*}\right)^{3/2} \left(\frac{300}{T}\right)^{1/2} \frac{\epsilon}{K^2 \epsilon_0} \qquad \text{cm}^2/\text{V-s} \qquad (6.8.16)$$

where K is a dimensionless quantity known as the *electromechanical coupling constant*. The lack of exponential temperature dependence indicates that only low-frequency acoustical-mode vibrations are considered in Eq. (6.8.16). Using $m^* = 0.38m_0$, $\epsilon/\epsilon_0 = 8.5$, and $K^2 = 0.074$ in ZnO, we find that $\mu_{pe} = 640$ cm²/V-s at $T = 300$ K from Eq. (6.8.16), as compared to the measured value of 200 cm²/V-s.

In summary, lattice vibrations can deflect the motion of free carriers in many different ways: (1) by modulating the band edges, (2) by inducing a polarization field caused by the relative movement of ionic charges, and (3) by inducing a polarization field through piezoelectric effect. In element semiconductors, only the first mechanism exists. In compound semiconductors with zinc-blende and wurtzite structures, all three mechanisms exist simultaneously. From the measured values of mobilities in III–V compounds, it seems that the second mechanism is the dominant mechanism around room temperature for electrons. For holes, acoustic-phonon scattering process of Eq. (6.8.3) and nonpolar optical-phonon scattering process of Eq. (6.8.4) appear to be of comparable importance. The reader is cautioned against using the temperature dependence of mobility as the sole judge in determining the relative importance of different scattering processes. Ehrenreich has shown that the temperature dependence in III–V compounds can be explained by a combination of polar scattering and other scattering processes, for example, electron–hole scattering in InSb and ionized impurity scattering in GaAs (H. Ehrenreich, *Phys. Rev.*, Vol. 120, p. 1951, 1960; *J. Phys. Chem. Solids*, Vol. 9, p. 129, 1959). The conclusion about polar scattering being the dominant mechanism for electrons was reached after a careful analysis of the experimental data.

The most serious difficulty in comparing the theoretically calculated and experimentally measured values of mobility is the lack of controlled purity in III–V compound semiconductors. A high-resistivity sample does not necessarily mean a high-purity sample because the sample may be highly compensated with donors and acceptors. The situation is even worse with II–VI compounds. Until recently, most measurements of the electrical properties of II–VI compounds were made on powder or polycrystalline samples and *p*-type conductivity was not generally observed. Therefore, we still wait for a thorough analysis of the mobility data in II–VI compounds. However, judging from the experimental data on ZnO, it seems most probable that piezoelectric scattering is dominant over acoustic and polar scattering. The statement is based on the following observations. Since $\theta_l = 920$ K is relatively high in ZnO, polar scattering alone would predict around room temperature a much steeper temperature dependence of mobility than the temperature dependence observed experimentally. Furthermore, to explain the magnitude of the thermoelectric voltage in ZnO, an unrealistically high deformation potential $\mathscr{E}_1$ in Eq. (6.7.15) would be required if acoustical-mode scattering were the dominant mechanism.

In all three lead compounds, the $T^{-5/2}$ dependence of mobility for both electrons and holes is now well established. A study of the mobility data shows that acoustical-mode lattice scattering is the dominant scattering mechanism. Several theories have been proposed to explain the observed $T^{-5/2}$ dependence instead of the theoretical $T^{-3/2}$ dependence given in Eq. (6.8.3). The role of impurity scattering is reduced in lead salts because of the large value of the dielectric constant (Table 6.4). Even in impure samples, the mobility remains high at 4 K. However, in samples with carrier concentrations ranging from 10^{17} to 10^{21} cm^{-3}, the mobility shows a progressive change to a $T^{-1.3}$ dependence as well as a lowering of its value. These results indicate that the role of impurity scattering is emerging in highly impure samples. For a review of as well as references to mobility studies in lead salts, the reader is referred to an excellent review article by Putley (E. H. Putley, "Lead Sulfide, Selenide, and Telluride," in C. A. Hogarth, ed., *Materials Used in Semiconductor Devices*, Interscience Publishers, New York, 1965).

Impurity States

In Section 6.4 we discussed the role of donor and acceptor impurities in Ge and Si, and used Eq. (6.4.1) for the ionization energy of the impurities. Equation (6.4.1) should apply equally well to donor and acceptor states in III–V compounds. The group IIb elements, such as Be, Zn, Cd, and Hg, and the group VIb elements, such as Se and Te, enter the lattice substitutionally, with the group II atoms replacing III atoms as acceptors and group VI atoms replacing V atoms as donors. Since electrons in III–V compounds have a smaller effective mass than holes, the ionization energy of donors is expected to be much smaller than that of acceptors. This is indeed found to be the case.

Because purification is still a problem, most III–V compounds either have a high degree of compensation or have a high impurity concentration. As a consequence, impurity states are no longer discrete but form an impurity band that merges into either the conduction or the valence band. Because of the high degree of compensation and the formation of impurity bands, the measurement of Hall coefficients at low temperatures may not yield as much useful information as we expect from a sample of controlled purity, as in the case of Ge and Si. Among III–V compounds, extensive data about impurity states are available only in GaAs.

The situation with II–VI compounds is even more complicated than that with III–V compounds. According to the covalent bond model, we expect that the group I elements will serve as acceptors replacing the group II element, while the group VII elements will act as donors replacing the group VI element. To some extent, this covalent bond picture still applies. However, we also find that excess group II elements, such as excess Zn in ZnO and excess Cd in CdS and CdTe, act as donors. As a matter of fact, most II–VI insulating compounds can be made n-type by exposing the crystal to Zn or Cd vapor at high temperatures. The ionization energies for the donor level seem to be in fair agreement with the value predicted from the hydrogenic model, for example, $\mathscr{E}_d = -0.05$ eV in ZnO and -0.02 eV in CdTe.

The p-type conduction has been observed only in CdTe, with an ionization energy that may vary from 0.2 eV to 0.4 eV. The level is tentatively assigned as being associated with copper. The dopants commonly used in CdS, CdTe, and CdSe are Ag and Cu as acceptors and Ga and Cl as donors. However, a definite identification of the donor and acceptor species with the energy levels has not been achieved. Even in high-resistivity materials, the concentrations of donors and acceptors are still high but they are nearly compensated. This makes a precise analysis of the Hall measurement very difficult.

Figure 6.21 summarizes the information on energy levels created in the forbidden gap in Ge, Si, and GaAs. There are two distinct groups of impurity atoms. The regular donor and acceptor impurities form one group. These impurity atoms have energy levels relatively close to the respective band edge and they are called *shallow states*. In this group we have (1) Li, P, As, Sb, and Bi behaving as donors in Ge and Si; (2) B, Al, Ga, In, and Tl acting as acceptors in Ge and Si; (3) Se, Te, Ge, Si, and Sn behaving as donors in GaAs; and (4) Cd, Zn, Ge, and Si acting as acceptors in GaAs. Since Ge, Si, and Sn are situated between Ga and As in the periodic table, they can be either donors or acceptors, depending on whether they go into the lattice to substitute a Ga or As atom. Such impurities are called *amphoteric*. It is found experimentally that Si behaves predominantly as donors and Ge predominantly as acceptors in GaAs.

The nature of the impurity states belonging to the other group is more complex. These states are generally referred to as the *deep-lying* impurity states. They not only have energy levels deep in the forbidden gap but may have several levels, depending on their state of ionization. Take Zn in Ge as an example. Since Zn has a $4s^2$ configuration outside closed shells, it can accept a maximum of two electrons in Ge and Si. The two energy states (0.03 eV) and (0.09 eV) in Ge correspond to a singly ionized

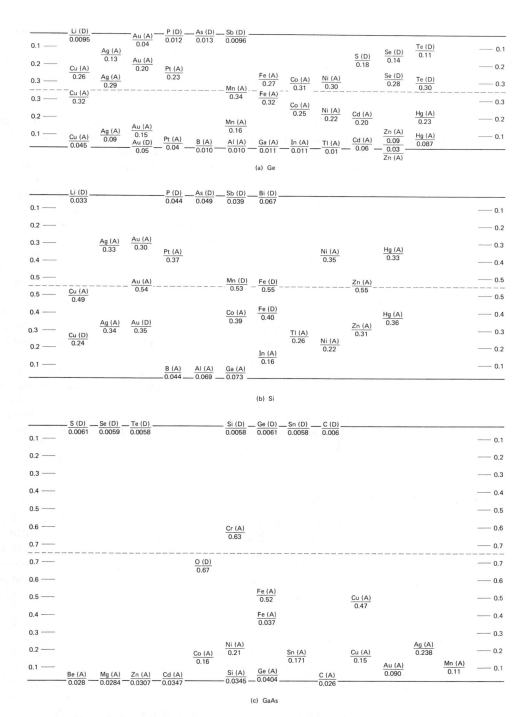

Figure 6.21 Energy levels of impurities in (a) Ge, (b) Si, and (c) GaAs (E. M. Conwell, ''Properties of Silicon and Germanium,'' *Proc. IRE*, vol. 40, p. 1281, 1958 © 1958 IEEE. Reprinted with permission from *Solid-State Electron.*, Vol. 11, p. 599, 1968; S. M. Sze and J. C. Irwin, ''Resistivity, Mobility, and Impurity Levels in GaAs at 300 K,'' © 1968, Pergamon Press Publication.)

zinc atom Zn^- and a doubly ionized zinc atom Zn^{2-}, respectively. In the same way, the transition elements Mn, Fe, Co, and Ni with two $4s$ electrons in the outer shell act as double acceptors in Ge. Electrons in the incomplete $3d$ shell apparently are unaffected. On the other hand, both Mn and Fe act as donors in Si. This unexpected result indicates that Mn and Fe may occupy interstitial positions in silicon.

The other situation worth mentioning is that of gold in germanium. Since Au belongs to the column Ib of the periodic table, it can accept a maximum of three electrons. The three states from top to bottom in Fig. 6.21a correspond to Au^{3-}, Au^{2-}, and Au^-. In addition to these three levels, Au can also behave as a donor, indicating that it may lose its single valence electron to assume a noble gas configuration. The reader is referred to the review article by Conwell for further discussion of impurity states in Ge and Si as well as for references to the data presented in Fig. 6.21a and b (E. M. Conwell, *Proc. IRE*, Vol. 46, p. 1281, 1958). For information concerning impurity states in GaAs, the article by Sze and Irvin is useful (S. M. Sze and J. C. Irvin, *Solid-State Electron.*, Vol. 11, p. 599, 1968).

Like II–VI compounds, the lead salts can be made extrinsic by deviation from stoichiometric composition. An excess of Pb atoms makes a sample n type, whereas an excess of the group VI elements makes a sample p type. Bloem, Kroger, and Vink have shown that each excess Pb atom contributes one free electron to the conduction band and each excess group VI atom contributes one free hole to the valence band (J. Bloem, F. A. Kroger, and H. J. Vink, *Defects in Crystalline Solids*, Physical Society, London, 1955, p. 273). The Hall coefficient in lead salts is found to be practically independent of temperature in the temperature region where Eq. (6.5.24) should apply. This finding suggests that the impurity levels associated with foreign atoms lie within either the conduction or the valence band. There are reports of experimental evidence that monovalent metals such as Na, Ag, or Cu substituting for Pb atoms act as acceptors, and that the halogens (group VII elements) replacing group VI atoms and the trivalent metal Bi replacing Pb atoms behave as donors (J. Bloem, *Philips Res. Rep.*, Vol. 11, p. 272, 1956; T. L. Koval'chik and Y. P. Maslakovets, *Sov. Phys.-Tech. Phys.*, Vol. 1, p. 2337, 1960). However, their energy levels have not been identified.

6.9 ENERGY-BAND STRUCTURE

It is of practical importance as well as of theoretical interest to know the energy-band structure of semiconductors. For example, laser action has been achieved in III–V, II–VI, and IV–VI compounds but not in Ge and Si. What makes the difference is the band structure. The Gunn effect (Sections 10.3 and 10.9), which can be used for microwave generation, has been observed only in materials such as GaAs with specific band properties. In this section we review the essential features of the energy-band structure of semiconductors.

Figure 6.22 shows a two-dimensional plot of $\mathscr{E}$ versus k curve for Ge and Si. Since the wave vector **k** consists of three components, k_x, k_y, and k_z, a two-dimensional plot requires either holding two components constant (e.g., k_y and k_z) or maintaining the direction of **k** constant. Because of the crystal symmetry, the three axes k_x, k_y, and k_z are identical; therefore, it is far more meaningful to express $\mathscr{E}$ along a certain direction in the k space (or the reciprocal lattice space). In Fig. 6.22, the two principal directions are chosen as [100] and [111], and the variations of $\mathscr{E}$ along those two direc-

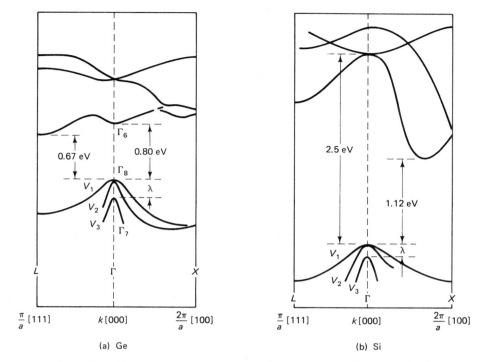

Figure 6.22 Diagrams showing the variations of electron energy with wave number in (a) Ge and (b) Si along the [100] and [111] directions in k space. Electrons are located near the minimum (shown with heavy lines) of the conduction band, whereas holes are located near the maximum (shown with heavy lines) of the valence band. The band structures of Ge and Si are examples of indirect-gap semiconductors.

tions in the k space are plotted, respectively, in the right half and the left half of the diagram.

We focus our attention on the minimum of the conduction band and the maximum of the valence band because most electrons and holes are located there. These places are shown with heavy lines in Fig. 6.22. We notice that the band extrema occur at different values of **k**. These semiconductors are called *indirect-gap semiconductors*. Besides Ge and Si, examples include AlSb, GaP, and BP, which all have Si-like band structures with conduction-band minima lying along the $k\langle 100\rangle$ axes (i.e., along the $k[100]$ and its equivalent axes).

In contrast, the band structure of GaAs, hexagonal CdS, and PbS are shown in Fig. 6.23. One common feature is that the conduction-band minimum and the valence-band maximum occur at the same value of **k**. These semiconductors are examples of *direct-gap semiconductors*. For II–VI and most III–V compounds, the band extrema are located at $k[000]$, the center of the Brillouin zone. For lead chalcogenides, PbS, PbSe, and PbTe, the band extrema have been identified to be at $\pi/a[111]$, the edge of the Brillouin zone along the $\langle 111\rangle$ axes. We should point out that the lattice constant a for the diamond and zinc-blende structures refers to that of a face-centered cubic crystal.

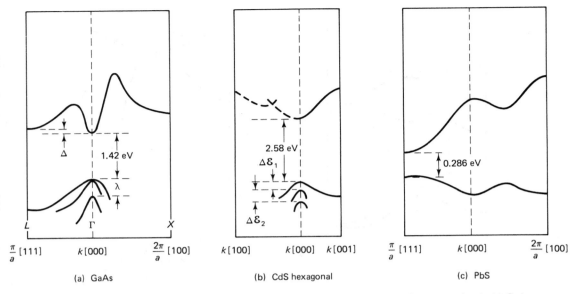

(a) GaAs   (b) CdS hexagonal   (c) PbS

Figure 6.23 Diagrams showing the variations of electron energy with wave number in (a) GaAs, (b) hexagonal CdS, and (c) PbS. These band structures are examples of direct-gap semiconductors in which the conduction-band minimum and the valence-band maximum occur at the same point in k space.

TABLE 6.7 BAND STRUCTURES AND EFFECTIVE MASSES OF A SELECTED NUMBER OF ELEMENT AND COMPOUND SEMICONDUCTORS AT 300 K

	Band structure	Effective masses, m^*/m_0	
		Conduction band	Valence band
Si	Indirect gap $\mathcal{E}_g = 1.11$ eV $\mathcal{E}_{g0} = 2.50$ eV $\lambda = 0.28$ eV	m_l : 0.98 m_t : 0.19 $m(k = 0)$ —	m_1 : 0.50 m_2 : 0.16 m_3 : 0.245
Ge	Indirect gap $\mathcal{E}_g = 0.67$ eV $\mathcal{E}_{g0} = 0.80$ eV $\lambda = 0.28$ eV	m_l : 1.58 m_t : 0.082 $m(k = 0)$: 0.037	0.32 0.04 0.077
GaAs	Direct gap $\mathcal{E}_g = 1.42$ eV $\lambda = 0.34$ eV	0.063	0.50 0.076 0.145
GaSb	Direct gap $\mathcal{E}_g = 0.72$ eV (0.87 eV) $\lambda = 0.86$ eV	0.044	0.23 0.06 —

| Band structure | Effective masses, $m*/m_0$ | |
	Conduction band	Valence band
InAs Direct gap $\mathscr{E}_g = 0.36$ eV $\lambda = 0.43$ eV	0.022	0.41 0.025 0.083
InSb Direct gap $\mathscr{E}_g = 0.18$ eV (0.236 eV) $\lambda = 0.90$ eV	0.014	0.40 0.015 —
CdS Direct gap $\mathscr{E}_g = 2.58$ eV $\Delta\mathscr{E}_1 = 0.016$ eV $\Delta\mathscr{E}_2 = 0.057$ eV	0.204	$m_\perp : 0.7, m_{//} : 5$ ($\perp$ and $//$ refer to the c axis)
ZnO Direct gap $\mathscr{E}_g = 3.35$ eV $\Delta\mathscr{E}_1 = 0.007$ eV $\Delta\mathscr{E}_2 = 0.038$ eV	0.38	1.8 (estimated)
PbS Direct gap $\mathscr{E}_g = 0.41$ eV (0.286 eV)	$m_l : 0.105$ $m_t : 0.080$	$m_l : 0.105$ $m_t : 0.075$
PbTe Direct gap $\mathscr{E}_g = 0.32$ eV (0.19 eV)	$m_l : 0.24$ $m_t : 0.024$	$m_l : 0.31$ $m_t : 0.022$

Gap energies in parenthesis refer to values at 0 K.

References

Group IV semiconductor

G. Dresselhaus, A. F. Kip, and C. Kittel, *Phys. Rev.,* Vol. 98, p. 368, 1955.

B. Lax, H. J. Zeiger, R. N. Dexter, and E. S. Rosenblum, *Phys. Rev.,* Vol. 93, p. 1418, 1954.

S. Zwerdling, B. Lax, L. M. Roth, and K. J. Button, *Phys. Rev.,* Vol. 114, p. 80, 1959.

III–V compound

H. Ehrenreich, *J. Appl. Phys.,* Vol. 32, p. 2155, 1961.

B. Lax et al., *Phys. Rev.,* Vol. 122, p. 33, 1961.

A discussion of the details may be found in C. Hilsum and A. C. Rose-Innes, *Semiconducting III-V Compounds,* Pergamon Press, Inc., New York, 1961.

J. S. Blakemore, "Major Properties of GaAs," *J. Appl. Phys.,* Vol. 53, p. R123, October 1982.

R. A. Smith, *Semiconductors,* Cambridge University Press, Cambridge, 1978.

H. C. Casey, Jr. and M. B. Parrish, *Heterostructure Lasers,* Academic Press, Inc., New York, 1978.

II–VI compound

J. J. Hopfield and D. G. Thomas, *Phys. Rev.,* Vol. 122, p. 35, 1961.

D. G. Thomas and J. J. Hopfield, *Phys. Rev.,* Vol. 116, p. 573, 1959.

R. E. Dietz, J. J. Hopfield, and D. G. Thomas, *J. Appl. Phys.,* Vol. 32, p. 2282, 1961.

A. R. Hutson, *J. Appl. Phys.,* Vol. 32, p. 2287, 1961.

IV–VI compound

The following articles, which appeared in M. Hulin, ed., *Physics of Semiconductors,* Dunod, Paris, 1964:

L. Kleinman and P. J. Lin, p. 64; G. W. Pratt, Jr. and L. G. Ferreira, p. 69; D. L. Mitchell, E. D. Palik, and J. N. Zemel, p. 325; R. S. Allgaier, B. B. Houston, Jr., J. Babiskin, and P. G. Siebenmann, p. 659; K. F. Cuff, M. R. Ellett, C. D. Kuglin, and L. R. Williams, p. 677.

A summary of the results may be found in C. A. Hogarth, ed., *Materials Used in Semiconductor Devices,* Interscience Publishers, New York, 1965.

The distance between nearest neighbors is $\sqrt{3}\,a/4$. Therefore, in a unit cell of volume a^3, there are eight germanium or silicon atoms and four gallium arsenide molecules. In the reciprocal lattice for a face-centered cubic solid, the first Brillouin zone comprises a truncated octahedron (Fig. 5.9a). The distance between the zone boundaries along the $k\langle 100\rangle$ directions is $4\pi/a$, and the zone boundaries are located at $k_x = \pm 2\pi/a$, $k_y = 0$, $k_z = 0$ and its equivalents. Along the $k\langle 111\rangle$ directions, the zone boundaries are located at $k_x = \pm\pi/a$, $k_y = \pm\pi/a$, and $k_z = \pm\pi/a$.

In Table 6.7 we list the gap energy and effective masses of a selected number of element and compound semiconductors. For indirect-gap semiconductors, two gap energies are given: the *direct-gap energy*, which measures the energy difference across the bands at $k[000]$, and the *indirect-gap energy*, which is the energy difference between the conduction-band minimum and valence-band maximum. The indirect-gap energy $\mathscr{E}_g$ is usually determined by a measurement of the temperature dependence of the Hall coefficient in the high-temperature (or intrinsic) region, while the direct-gap energy $\mathscr{E}_{g0}$ is directly measurable only in an optical absorption experiment in which a semiconductor absorbs a photon energy $\hbar\omega = \mathscr{E}_{g0}$ by exciting an electron from the valence band into the conduction band at $k[000]$. If a light beam of intensity $I(0)$ enters into a semiconductor at $x = 0$, the intensity of the beam decreases with distance according to

$$I(x) = I(0)\,\exp\left(-\alpha x\right) \tag{6.9.1}$$

inside the semiconductor. The constant α (in cm^{-1}) is called the *absorption coefficient*. Figure 6.24 shows the absorption data in Ge and GaAs (W. C. Dash and R.

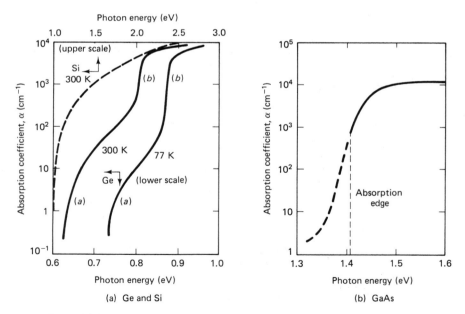

Figure 6.24 Absorption coefficient α as a function of photon energy in (a) Ge (solid curves) and Si (dashed curve) and (b) GaAs. (W. C. Dash and R. Newman, *Phys. Rev.*, Vol. 99, p. 1151, 1955; T. S. Moss, *J. Appl. Phys.*, Vol. 32, p. 2136, 1961.)

Newman, *Phys. Rev.*, Vol. 99, p. 1151, 1955; T. S. Moss, *J. Appl. Phys.*, Vol. 32, p. 2136, 1961). In an optical transition in a semiconductor, the total momenta should be conserved. If $\mathbf{k}_i$ and $\mathbf{k}_f$ are the initial and final wave vectors of the electron involved in the transition, then

$$\hbar \mathbf{k}_i + \hbar \boldsymbol{\eta} = \hbar \mathbf{k}_f \qquad (6.9.2)$$

where $\boldsymbol{\eta}$ is the wave vector of the electromagnetic radiation. The magnitude of k is of the order of π/a, a being the dimension of the unit lattice cell in angstrom units, whereas the magnitude of η is equal to $2\pi/\lambda$, where λ is in the neighborhood of 1 μm in the infrared and near-visible regions. Therefore, the magnitude of η is very much smaller than that of k, and the quantum-mechanical selection rule simply reduces to

$$\mathbf{k}_i \cong \mathbf{k}_i \qquad (6.9.3)$$

The transition process for which Eq. (6.9.3) is obeyed is called the *direct optical transition*. In Fig. 6.25 the optical processes associated with the absorption data of Fig. 6.24 are shown, with the direct transition being indicated as process b. The direct transition begins for $\hbar\omega > 0.80$ eV in Ge and for $\hbar\omega > 1.42$ eV in GaAs. The other absorption process, indicated as process a in Figs. 6.24 and 6.25, is often referred to as the *indirect transition*. Note that the electrons in the conduction-band minimum of Ge (Fig. 6.25a) have momentum (or $\mathbf{k}$ value) different from the momentum at the top of the valence band. To make an optical transition from the top of the valence band to the conduction-band minimum, an electron must interact with lattice imperfection, such as phonons or impurities, to conserve momentum. In other words, an electron in the valence band must be scattered by phonons or impurities from an intermediate state in making an optical transition to the conduction band (process a of Fig. 6.25a).

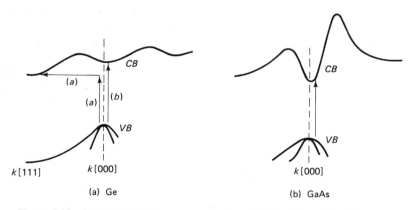

Figure 6.25 Optical absorption processes in Ge and GaAs. The part b of the curves in Fig. 6.24 is associated with a vertical (or direct) transition of an electron from the valence band to the conduction band (the process being indicated by arrow b). The part a of the curves in Fig. 6.24 is associated with an indirect transition in which electron momentum is not conserved.

Since the indirect transition is a two-step process, the transition probability of such a process is much smaller than that of the direct process. This explains why the absorption coefficient in the indirect transition region is much smaller than that in the direct transition region. Only in indirect-gap semiconductors such as Ge can both the direct transition and the indirect transition be seen, because the latter occurs first at a lower photon energy. The gap energy across the bands at $k[000]$ is called the direct-gap energy and is denoted by $\mathscr{E}_{g0}$ in Table 6.7. The direct-gap energy enters into the picture when we discuss optical processes, whereas the indirect-gap energy is the gap energy that determines the intrinsic carrier concentration in Eq. (6.3.13). In direct-gap semiconductors, the direct-gap energy is the one and only gap energy; it appears as the absorption edge in an optical measurement as well as the activation energy for the intrinsic carrier concentration.

Now we return to Figs. 6.22 and 6.23. For element, III–V compound, and zinc-blende II–VI compound semiconductors, the electrons in the valence band near $k[000]$ have p-like wave functions. Since p wave functions are triply degenerate, so is the valence band at $k[000]$. Spin-orbit interaction removes part of the orbital degeneracy. Therefore, the valence band in diamond and zinc-blende structures actually consists of three bands: the heavy-hole band V_1, the light-hole band V_2, and the split-off band V_3. Bands V_1 and V_2 are degenerate at $k[000]$ and are separated from the split-off band. The quantity λ in Table 6.7 is the energy separation between them. In the wurtzite structure (Fig. 6.23b), the combination of crystal field and spin-orbit interaction completely removes the orbital degeneracy. The energies $\Delta\mathscr{E}_1$ and $\Delta\mathscr{E}_2$ are the energy separation between V_1 and V_2 and that between V_2 and V_3, respectively.

6.10 EXPERIMENTAL STUDIES OF ENERGY-BAND STRUCTURE AND EFFECTIVE MASSES

In Section 6.9 we presented the band structures of some typical semiconductors. The band structure of germanium was first determined theoretically by Herman (F. Herman, *Phys. Rev.*, Vol. 93, p. 1214, 1954; *Proc. IRE*, Vol. 43, p. 1703, 1955). He showed that whereas the maximum of the valence band in Ge was located at $k[000]$, the minimum of the conduction band was not. Since his early work on group IV semiconductors, theoretical work on the calculation of energy-band structure has been extended by many workers to semiconductors having other crystal structures: the III–V compounds (zinc-blende structure), the II–VI compounds (zinc-blende and wurtzite structures), and the lead chalcogenides (NaCl structure) (see, for example, F. Herman, *J. Phys. Chem. Solids*, Vol. 8, p. 380, 1959; E. O. Kane, *J. Phys. Chem. Solids*, Vol. 1, p. 249, 1957; M. L. Cohen and T. K. Bergstresser, *Phys. Rev.*, Vol. 141, p. 789, 1966; D. G. Thomas, ed., *II–VI Compounds*, W. A. Benjamin, Inc., New York, 1967, pp. 462–551; P. J. Lin and L. Kleinman, *Phys. Rev.*, Vol. 142A, p. 478, 1966). Used in the theoretical calculation, there are interaction parameters whose values must be obtained from experimental data such as the gap energy and the effective masses. Therefore, an accurate determination of the band structure involves a careful analysis of the results of many experiments, as well as a quantum-mechanical calculation. Since a presentation of the calculation is outside the scope of this book, our discussion is limited to experimental studies of energy-band structure and effective masses.

Cyclotron-Resonance Experiment

The cyclotron-resonance experiment discussed in Section 5.11 provides a direct way of determining the effective masses in semiconductors. For an ellipsoidal energy surface (Fig. 6.26a) defined by

$$\mathscr{E} = \frac{\hbar^2 k_1^2}{2m_1} + \frac{\hbar^2 k_2^2}{2m_2} + \frac{\hbar^2 k_3^2}{2m_3} \tag{6.10.1}$$

the cyclotron-resonance frequency is no longer isotropic but depends on the orientation of the dc magnetic field with respect to the axes of the ellipsoid. If α, β, and γ are the direction cosines of $\mathbf{B}_0$ (Fig. 6.26b), the cyclotron resonance frequency ω_c is

$$\omega_c^2 = (eB_0)^2 \frac{m_1\alpha^2 + m_2\beta^2 + m_3\gamma^2}{m_1 m_2 m_3} \tag{6.10.2}$$

For $m_1 = m_2 = m_t$ and $m_3 = m_l$, we have

$$\omega_c = eB_0 \left(\frac{1 - \gamma^2}{m_l m_t} + \frac{\gamma^2}{m_t^2} \right)^{1/2} = \frac{eB_0}{m_c^*} \tag{6.10.3}$$

where m_c^* is defined as the cyclotron-resonance effective mass.

Let us now apply Eq. (6.10.3) to conduction-band electrons in Ge and Si. Figure 6.27 shows the constant-energy surfaces in these semiconductors near their conduction-band minimum. In Ge, there are eight constant-energy ellipsoids (Fig. 6.27a) along the principal diagonals, $k[111]$ and its equivalent directions. In Si, there are six ellipsoids along the $k\langle 100\rangle$ directions. The two masses m_l and m_t in Eq. (6.10.3) are called the *longitudinal mass* and *transverse mass,* respectively, because m_l is defined along the $k\langle 111\rangle$ axis in Ge and along the $k\langle 100\rangle$ axis in Si, whereas m_t is defined in a plane transverse to the said axis. The situation in lead salts is similar to that in Ge.

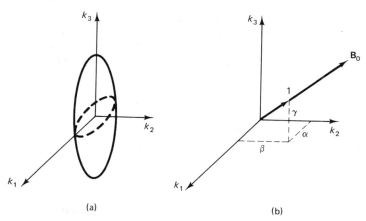

(a)   (b)

Figure 6.26 Diagrams used in the calculation of cyclotron-resonance frequency: (a) constant-energy ellipsoid in k space and (b) the orientation of an applied dc magnetic induction $\mathbf{B}_0$. The direction cosines of the vector $\mathbf{B}_0$ are given by the projections α, β, and γ of the unit vector $\mathbf{I}$ upon the three axes.

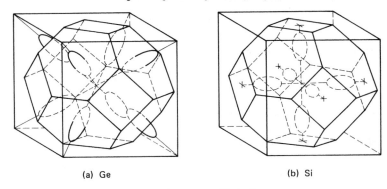

Eight hexagons facing the body diagonals

(a) Ge   (b) Si

Six squares in contact with the faces

Figure 6.27 Constant-energy surfaces in k space for electrons (a) in Ge and (b) in Si. The surfaces are ellipsoids described by Eq. (6.10.1), and the center of each ellipsoid is located at the point in k space where the conduction band is a minimum.

Figure 6.28 shows the ratio of cyclotron-resonance effective mass to free electron mass in Ge and Si as functions of the orientation of B_0. According to Eq. (6.10.3), $m_c^* = eB_0/\omega_c$. The following values are used to fit the curves:

$$m_l = 1.58m_0, \quad m_t = 0.082m_0 \qquad \text{for Ge} \qquad (6.10.4)$$

$$m_l = 0.98m_0, \quad m_t = 0.19m_0 \qquad \text{for Si} \qquad (6.10.5)$$

These values are listed in Table 6.7. In the experiment, $\mathbf{B}_0$ was in the (110) plane and the angle θ was the angle between $\mathbf{B}_0$ and the [001] direction. For cubic crystals, the reciprocal-lattice basis vectors, $\mathbf{a}^*$, $\mathbf{b}^*$, and $\mathbf{c}^*$ defined in Eq. (5.6.3) point in the same direction as the basis vectors, $\mathbf{a}$, $\mathbf{b}$, and $\mathbf{c}$ of the cubic crystals. Once this is understood, the relation between γ in Eq. (6.10.3) and θ in Fig. 6.28 can easily be found.

Take the ellipsoid along the $k[111]$ axis in Ge as an example. Its k_3 axis lies in the [111] direction, and hence the direction of $\mathbf{k}_3$ is represented by a unit vector.

$$\hat{\mathbf{k}}_3 = \frac{1}{\sqrt{3}}(1, 1, 1) \qquad (6.10.6)$$

The unit vector $\hat{\mathbf{B}}_0$ representing the direction of $\mathbf{B}_0$ is

$$\hat{\mathbf{B}}_0 = \left(\frac{\sin\theta}{\sqrt{2}}, -\frac{\sin\theta}{\sqrt{2}}, \cos\theta\right) \qquad (6.10.7)$$

Thus, for the $k(111)$ ellipsoid, the direction cosine is

$$\gamma = \hat{\mathbf{B}}_0 \cdot \hat{\mathbf{k}}_3 = \frac{\cos\theta}{\sqrt{3}} \qquad (6.10.8)$$

For $\theta = 90°$, $\gamma = 0$ and Eq. (6.10.3) reduces to $m_c^* = (m_l m_t)^{1/2}$. The quantity m_c^* has a value of $0.36m_0$ based on Eq. (6.10.4), and this calculated value is in agreement with the observed value. The same value of m_c^* is found for the ellipsoids along the $k[1\bar{1}\bar{1}]$,

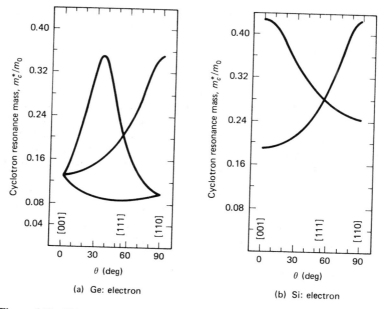

(a) Ge: electron   (b) Si: electron

Figure 6.28 Plot of the cyclotron-resonance effective masses as a function of θ for conduction-band electrons in (a) Ge and (b) Si measured at 4K, where θ is the angle which $\mathbf{B}_0$ makes with the [001] axis. The magnetic induction $\mathbf{B}_0$ is in the (110) plane. Therefore, $\mathbf{B}_0$ is in the [001] direction for θ = 0°, in the [111] direction for θ = $\cos^{-1}$ ($\sqrt{3}/3$) = 55°, and in the [110] direction for θ = 90°. (G. Dresselhaus, A. F. Kip, and C. Kittel, *Phys. Rev.*, Vol. 98, p. 368, 1955.)

$k[\bar{1}11]$, and $k[\bar{1}\bar{1}1]$ directions. The other values of m_c^* shown in Fig. 6.28a are for the other ellipsoids.

In most III–V compounds (GaAs, GaSb, InAs, InSb, and InP), the conduction band minimum is located at $k(000)$. For this energy valley, the constant-energy surface is almost spherical, and hence there is only one effective mass m_e^* in Table 6.7. This conclusion also applies to the $k(000)$ valley in Ge, whose presence can be ascertained only in an optical experiment. The value $m(k = 0) = 0.037m_0$ given in Table 6.7 is deduced from optical experiments. The conduction-band minimum and the valence-band maximum in lead compounds are located along the $k\langle 111\rangle$ directions; therefore, Eq. (6.10.3) applies to both electrons and holes. Further, their analyses are similar to that for conduction-band electrons in Ge.

The valence band in element semiconductors, III–V compounds, and zinc-blende II–VI compounds actually consists of three bands, as illustrated in Figs. 6.22 and 6.23a. In a cyclotron-resonance experiment, only the two top bands, V_1 and V_2, which are degenerate at $k[000]$ will participate because almost all the holes are in these two bands. Based on theoretical considerations, the kinetic energy of the valence-band holes in these materials can be expressed in terms of $\mathbf{k}$ as

$$\mathscr{E}\,(\mathbf{k}) = \frac{\hbar^2}{2m_0}\{Ak^2 \pm [B^2k^4 + C(k_x^2k_y^2 + k_y^2k_z^2 + k_z^2k_x^2)]^{1/2}\} \qquad (6.10.9)$$

For II–VI compounds having the wurtzite structure, the valence band splits into three bands and is no longer degenerate even at $k[000]$, as illustrated in Fig. 6.23b. Symmetry considerations show that the top valence-band (V_1) holes have an energy dependence of the form

$$\mathscr{E}(\mathbf{k}) = \frac{\hbar^2}{2m_0} [A(k_x^2 + k_y^2) + Bk_z^2] \qquad (6.10.10)$$

Figure 6.29 shows the cyclotron-resonance data for valence-band holes in Ge. Because of the $\Sigma k_x^2 k_y^2$ terms in parentheses in Eq. (6.10.9), the constant-energy surface deviates from a spherical surface, and hence the two cyclotron-resonance effective masses are slightly anisotropic. However, for all practical purposes, we may average the anisotropic effective masses and approximate them by two constant effective masses. The two masses are $m_1 = 0.32m_0$ and $m_2 = 0.04m_0$. The valence band V_1 with effective mass m_1 is called the heavy-hole band, whereas the valence band V_2 with m_2 is called the light-hole band. For the wurtzite CdS (Table 6.7), the anisotropy is large in Eq. (6.10.10). Therefore, two masses $m_\parallel$ and $m_\perp$ are given for the top band V_1, with $m_\parallel$ defined along the c axis and $m_\perp$ in a plane perpendicular to the c axis.

Density-of-State and Mobility Effective Masses

In our previous discussions, the effective mass m^* enters into the density-of-state expressions, N_c of Eq. (6.3.9) and N_v of Eq. (6.3.11), and into the mobility expressions of Eq. (6.7.22), for example. Although the same symbol m^* is used, the two effective masses are actually different in the density-of-state and mobility expressions. If there are two parabolic bands with spherical energy surfaces, the density of states N is simply a sum of two densities N_1 and N_2. From $N = N_1 + N_2$, we have

$$(m_N^*)^{3/2} = (m_1)^{3/2} + (m_2)^{3/2} \qquad (6.10.11)$$

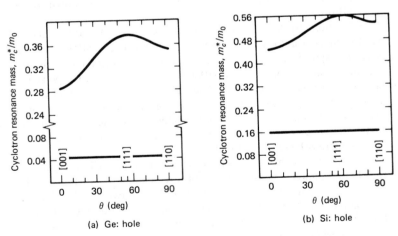

(a) Ge: hole

(b) Si: hole

Figure 6.29 Plot of the cyclotron-resonance effective masses as functions of θ for valence-band holes in (a) Ge and (b) Si measured at 4K. The angle θ is defined in the caption of Fig. 6.28. (G. Dresselhaus, A. F. Kip, and C. Kittel, *Phys. Rev.*, Vol. 98, p. 368, 1955.)

Semiconductor Fundamentals   Chap. 6

The average mobility of the two bands is

$$\mu = \frac{\mu_1 p_1 + \mu_2 p_2}{p_1 + p_2} = \frac{e\tau}{m_\mu^*} \tag{6.10.12}$$

Since the two carrier concentrations p_1 and p_2 are proportional to N_1 and N_2, the effective mass used in the mobility expression should be

$$\frac{1}{m_\mu^*} = \frac{m_1^{1/2} + m_2^{1/2}}{m_1^{3/2} + m_2^{3/2}} \tag{6.10.13}$$

The two masses m_N^* and m_μ^* are called the *density-of-state* and *mobility effective masses* and are distinguished from each other by the subscripts N and μ, respectively. Equations (6.10.11) and (6.10.13) apply to the valence-band holes in element, III–V compound, and II–VI (zinc-blende) compound semiconductors.

Now we turn our attention to nonspherical bands. Note that the term $m_e^{*3/2}$ in Eq. (6.3.9) actually comes from $dp_x\, dp_y\, dp_z$ in Eq. (6.3.7). For ellipsoidal bands,

$$(m_e^*)^3 = m_l m_t^2 \tag{6.10.14}$$

If there are q equivalent valleys, the density-of-state effective mass is

$$(m_N^*)^{3/2} = q(m_l m_t^2)^{1/2} \tag{6.10.15}$$

In the mobility expression, the effective mass is given by

$$\frac{1}{m_\mu^*} = \frac{1}{3}\left(\frac{2}{m_t} + \frac{1}{m_l}\right) \tag{6.10.16}$$

Equations (6.10.15) and (6.10.16) apply to conduction-band electrons in element semiconductors and lead salts, and to valence-band holes in wurtzite CdS and lead salts. In Si, $q = 6$, as is obvious from Fig. 6.27b. In Ge and lead salts, $q = 4$ because of the fact that the two ellipsoids along any principal diagonal in Fig. 6.27a are merely a continuation of each other across the zone boundary in the reduced-zone scheme. For wurtzite CdS, $q = 1$ because the top of the valence band is at $k[000]$. In Table 6.8 we list the values of N_c and N_v calculated from the effective masses and the value of n_i computed from Eq. (6.3.13) for a selected number of semiconductors.

TABLE 6.8 VALUES OF N_c AND N_v (cm^{-3}) AT 300 K FOR A SELECTED NUMBER OF SEMICONDUCTORS

	Ge	Si	GaAs	PbS	Cds (wurtzite)
N_c	1.04×10^{19}	2.9×10^{19}	4.7×10^{17}	2.6×10^{18}	2.3×10^{18}
N_v	4.8×10^{18}	1.1×10^{19}	1.5×10^{19}	2.4×10^{18}	3.9×10^{19}
n_i	1.9×10^{13}	0.9×10^{10}	3.3×10^{6}	9.1×10^{14}	2.1×10^{-3}
	(2.3×10^{13})	(1.4×10^{10})	(1.1×10^{7})		

a) The calculation is based on values of effective masses measured at 4K.

b) The measured values of n_i are in parentheses.

Interband Magnetooptic Effect

Another effect important to the study of energy-band structures is the interband magnetooptic effect. In the presence of a strong magnetic field, the motion of carriers in a plane perpendicular to $\mathbf{B}_0$ is quantized and the energy of the carriers is given by

$$\mathscr{E}_e = \mathscr{E}_c + \hbar\omega_{ce}\left(l_e + \frac{1}{2}\right) + \frac{\hbar^2 k_z^2}{2m_e^*} \qquad \text{for electrons} \qquad (6.10.17)$$

$$\mathscr{E}_h = \mathscr{E}_v - \hbar\omega_{ch}\left(l_h + \frac{1}{2}\right) - \frac{\hbar^2 k_z^2}{2m_h^*} \qquad \text{for holes} \qquad (6.10.18)$$

where ω_{ce} and ω_{ch} are the cyclotron-resonance frequencies and l_e and l_h are the orbital quantum numbers for electrons and holes, respectively. The energy levels with different intergers of l are called *Landau levels*. Figure 6.30 shows the energy of electrons and holes without and with B_0.

Whereas cyclotron resonance involves transitions between Landau levels in the same band ($\Delta l_e = 1$ for electron cyclotron resonance and $\Delta l_h = 1$ for hole cyclotron resonance), the interband magnetooptic effect is a result of transition between Landau levels across the band with $l_e = l_h$ (i.e., $\Delta l = 0$). From Eqs. (6.10.17) and (6.10.18), the energy for the transition is given by

$$\mathscr{E}_{m0} = \mathscr{E}_e - \mathscr{E}_h = \mathscr{E}_g + \frac{\hbar^2 k_z^2}{2m_r} + \left(l + \frac{1}{2}\right)\hbar\omega_{ceh} \qquad (6.10.19)$$

where m_r is the reduced mass,

$$\frac{1}{m_r} = \frac{1}{m_e^*} + \frac{1}{m_h^*} \qquad (6.10.20)$$

and ω_{ceh} is the combined electron and hole cyclotron frequency,

$$\omega_{ceh} = \omega_{ce} + \omega_{eh} = \frac{eB_0}{m_r} \qquad (6.10.21)$$

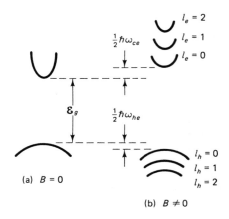

(a) $B = 0$

(b) $B \neq 0$

Figure 6.30 Energy bands for a simple semiconductor (a) in the absence of magnetic induction and (b) in the presence of magnetic induction B_0. In (b), the two bands are split into a series of one-dimensional subbands.

Semiconductor Fundamentals   Chap. 6

In an ordinary optical-absorption experiment, a light beam is incident on a semiconductor. The percentage of light transmitted is a measure of absorption in the semiconductor. When the absorption experiment is performed in a magnetic field (magnetooptic effect), optical absorptiion is modified. Theoretical analysis shows that in the presence of B_0, the density-of-state function is no longer a smooth function but has peaks at regular energy intervals with $\mathscr{E}_e = \mathscr{E}_c + (l_e + \frac{1}{2})\hbar\omega_{ce}$ for electrons and $\mathscr{E}_h = \mathscr{E}_v - (l_h + \frac{1}{2})\hbar\omega_{ch}$ for holes. As a consequence, the magnetooptic absorption shows peaks around the photon energy

$$h\nu = \mathscr{E}_e - \mathscr{E}_h = \mathscr{E}_g + (l + \tfrac{1}{2})\hbar\omega_{ceh} \qquad (6.10.22)$$

Figure 6.31 shows the ratio of the transmitted radiation, observed in Ge, with and without a magnetic field. For direct magnetooptic transition across the band, $\mathscr{E}_g$ in Eq. (6.10.22) is the direct-gap energy and m_e^* is the effective mass at $k[000]$, which is equal to $0.037m_0$, shown in Table 6.7. The first peak occurs when $l = 0$ in Eq. (6.10.22), and it is followed by additional peaks at regular energy intervals of $\hbar\omega_{ceh}$. Note that the locations of the peaks are shifted slightly with the orientation of the magnetic field. As pointed out earlier in this section, the hole mass based on Eq. (6.10.9) is slightly anisotropic. This anisotropy accounts for the orientation dependence. For a detailed discussion of the magnetooptic experiment, the reader is referred to a review article by Lax and Zwerdling (B. Lax and S. Zwerdling, in A. F. Gibson, ed., *Progress in Semiconductors,* Vol. 5, John Wiley & Sons, Inc., New York, 1961, p. 226).

Reflection Experiment

When a light beam is incident on a semiconductor, part of the beam is transmitted and part is reflected on account of the change in the dielectric constant from air to semiconductor. Therefore, in an optical experiment, we can measure either the intensity of the

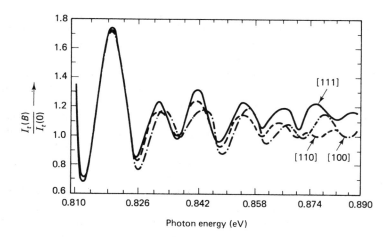

Figure 6.31 Oscillatory magnetooptic effect observed in Ge at room temperature. The ratio of the transmitted radiation with a magnetic field to that without a magnetic field is plotted as the ordinate. The experiment was performed at a field $H_0 = 35.7\text{kG}$. At such a high field, ω_{ceh} of Eq. (6.10.22) is in the optical region. The directions indicated are the direction of $\mathbf{H}_0$. (S. Zwerdling, B. Lax, and L. M. Roth, *Phys. Rev.*, Vol. 108, p. 1402, 1957.)

transmitted beam or that of the reflected light to detect the change in the dielectric constant. There is one serious limitation with the transmission experiment. As Fig. 6.24 shows, the absorption coefficient increases very rapidly for $h\nu > \mathscr{E}_g$. For $\alpha = 10^3$ cm^{-1} and $x = 10^{-2}$ cm in Eq. (6.9.1), the ratio $I(x)/I(o)$ is equal to exp (-10) or 4.5×10^{-5}. This ratio drops very rapidly with increasing photon energy. As a result, the transmission experiment is limited practically to the region near $h\nu \cong \mathscr{E}_{g0}$.

The reflection measurement, on the other hand, does not have this limitation and can be used to probe the energy-band structure, not only near $\mathscr{E}_{g0}$ but also deeply in energy. Figure 12.10 shows the variation of the real and imaginary parts of the dielectric constant in Ge deduced form the reflection measurement. Note that the density of state is proportional to $m^{*3/2}$ in Eq. (4.4.15) and that $m^{*-1} = \hbar^{-2}\partial^2\mathscr{E}/\partial k^2$. For those points in the $\mathscr{E}$ versus k diagram where $\partial^2\mathscr{E}/\partial k^2$ is zero, the density of state is infinitely large. These points are called *critical points*. Significant changes in the dielectric constant occur when $h\nu$ is near one of the critical points. Valuable information concerning the band structures of many III–V compounds has been obtained from the reflection measurement. For a detailed discussion of the subject, the reader is referred to several articles by Phillips (J. C. Phillips, *Phys. Rev.*, Vol. 125, p. 1931, 1962; *Phys. Rev.*, Vol. 133, p. A452, 1964; "The Fundamental Optical Spectra of Solids," in F. Seitz and D. Turnbull, eds., *Solid State Physics*, Vol. 18, Academic Press, Inc., New York, 1968, p. 55). Remarkable details of the band structure are revealed in the reflection experiment when an electric field is applied to a semiconductor surface. The experiment with the applied electric field is called the *electroreflectance experiment* (see, for example, B. O. Seraphin and N. Bottka, *Phys. Rev.*, Vol. 139, p. A560, 1965; Vol. 140, p. A1716, 1965; D. E. Aspnes, *Phys. Rev.*, Vol. 147, p. 554, 1966; M. Cardona, K. L. Shaklee, and F. H. Pollak, *Phys. Rev.*, Vol. 154, p. 696, 1967).

Magnetoresistance and Piezoresistance Measurements

We should also mention that useful information concerning the energy-band structure near band extrema can be deduced from magnetoresistance and piezoresistance measurements. For spherical energy surfaces, the resistance of a semiconductor should remain unchanged if the applied magnetic field is parallel to the direction of the current. This is no longer true, however, for ellipsoidal energy surfaces (Fig. 6.32). Furthermore, as the analysis in Appendix A6 shows, the value of J and hence the resistivity change $\Delta\rho$ are sensitive to the relative orientation of $\mathbf{B}$ and $\mathbf{E}$ [Eq. (A6.23)], in agreement with the results shown in Fig. 6.32. Since a detailed analysis of the magnetoresistance data is exceedingly involved, the reader is referred to the literature for the analysis (for reviews of the magnetoresistance effect, see A. C. Beer, *Galvanomagnetic Effects in Semiconductors*, Academic Press, Inc., New York, 1963; M. Glicksman, "Magnetoresistivity of Germanium and Silicon," in A. F. Gibson, ed., *Progress in Semiconductors*, Vol. 3, John Wiley & Sons, Inc., New York, 1958, p. 1).

For cubic crystals, the conductivity should be isotropic. If a uniaxial stress is applied, the cubic symmetry is destroyed. Take Si as an example. Under a stress applied in the z direction, the two conduction band minima along [001] and [00$\bar{1}$] axes (Fig. 6.27b) are shifted in energy relative to the other four minima. This shift causes an unequal distribution of electrons among the six ellipsoids, which in turn results in anisotropy of the conduction current. The location of the band extrema in k space can be deduced from a plot of the piezoresistance as a function of the orientation of the applied

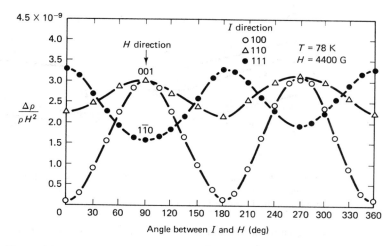

Figure 6.32 Magnetoresistive effect ($\Delta\rho/\rho B^2$) as functions of orientation of magnetic induction **B** with respect to the current **I** observed in n-type Si. One gauss = 10^{-4} Wb/m^2. (G. L. Pearson and C. Herring, *Physica*, Vol. 20, p. 975, 1954).

stress. For further discussions of the piezoresistance effect, the reader is referred to two review articles by Keyes and by Paul and Brooks (R. W. Keyes, "The Effects of Elastic Deformation on the Electrical Conductivity of Semiconductors," in F. Seitz and D. Turnbull, eds., *Solid State Physics*, Vol. 11, 1960, p. 149; W. Paul and H. Brooks, "Effect of Pressure on the Properties of Germanium and Silicon," in A. F. Gibson, ed., *Progress in Semiconductors*, Vol. 7, John Wiley & Sons, Inc., New York, 1963, p. 135).

Thermoelectric Effects

Since electrons in a solid are carriers of energy as well as electric charges, thermal effects naturally can produce electrical effects, and vice versa. When a current flows across a junction between two different materials, heat is generated or absorbed at the junction, depending on the direction of the current. This effect is known as the *Peltier effect*. Another commonly known thermoelectric phenomenon is the *Seebeck effect*. When two contacts to a metal or semiconductor are kept at different temperatures, a potential difference develops in the circuit connecting the two contacts. The third effect is known as the *Thomson effect*. When a temperature gradient is maintained along a conductor or semiconductor carrying a current, heat is either generated or absorbed, depending on the direction of the current flow with respect to the thermal gradient (D. K. C. MacDonald, *Thermoelectricity: An Introduction to the Principles*, John Wiley & Sons, Inc., New York, 1962).

Of the three thermoelectric effects, the Seebeck effect is the best known partly because it underlies the operation of thermocouples and partly because the Seebeck voltage is readily measurable. To see the physical origin of the Seebeck voltage, we refer to the energy diagram of Fig. 6.33. Since the Seebeck effect is much larger in semiconductors than in metals, only the effect in the semiconductor needs to be consid-

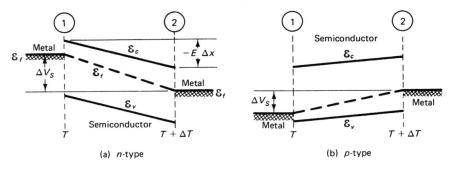

Figure 6.33 Schematic diagrams showing the variation of the Fermi level in (a) *n*-type and (b) *p*-type semiconductors where a temperature difference ΔT exists. As a result, a Seebeck voltage V_s arises.

ered. As discussed in Section 6.5, the Fermi level is a function of temperature. From Eq. (6.3.8), we have

$$\mathscr{E}_c - \mathscr{E}_f = kT(\ln N_c - \ln n) \tag{6.10.23}$$

Differentiating Eq. (6.10.23) with respect to T, we obtain

$$\frac{d(\mathscr{E}_c - \mathscr{E}_f)}{dT} = \frac{\mathscr{E}_c - \mathscr{E}_f}{T} + \frac{3K}{2} - kT\frac{d(\ln n)}{dT} \tag{6.10.24}$$

The shift in the Fermi level produces a difference in the values of $\mathscr{E}_c - \mathscr{E}_f$ at the two junctions.

The temperature gradient also produces an internal electric field. For the *n*-type semiconductor, the electron current density is given by Eq. (7.1.5), or

$$J_e = e\mu_n nE + e\frac{d}{dx}(D_n n) \tag{6.10.25}$$

Under an open-circuit condition, J_e is zero. Since electrons diffuse at a much faster rate near the hot junction than near the cold junction, an electric field is established to counterbalance the diffusion tendency. From Eq. (6.10.25),

$$E = \frac{-1}{\mu_n n}\frac{d}{dx}(D_n n) = \frac{-kT}{e}\left[\frac{d(\ln n)}{dx} + \frac{d(\ln D_n)}{dx}\right] \tag{6.10.26}$$

Using $\mu = e\tau/m^*$ and assuming that $\tau = A\mathscr{E}^q$, we find

$$D_n = \frac{kT}{e}\mu_n = \frac{kT}{m^*}A\mathscr{E}^q = \frac{AB^q}{m^*}(kT)^{q+1} \tag{6.10.27}$$

The last step which we take by letting $\mathscr{E} = BkT$ is justified because the average kinetic energy of electrons is proportional to kT. Substituting Eq. (6.10.27) in Eq. (6.10.26) yields

$$E = -\frac{kT}{e}\left[\frac{d(\ln n)}{dx} + \frac{q+1}{T}\frac{dT}{dx}\right] \tag{6.10.28}$$

From Fig. 6.33, the magnitude of the Seebeck voltage is

$$\Delta V_S = -E \, \Delta x + \frac{1}{e} [(\mathscr{E}_c - \mathscr{E}_f)_2 - (\mathscr{E}_c - \mathscr{E}_f)_1] \qquad (6.10.29)$$

Substituting Eqs. (6.10.24) and (6.10.28) in Eq. (6.10.29), we find

$$\Delta V_S = \frac{-1}{eT} \left[(\mathscr{E}_c - \mathscr{E}_f) + kT \left(\frac{5}{2} + q \right) \right] \Delta T \qquad (6.10.30)$$

The value of q changes from $-\frac{3}{2}$ for lattice scattering to $\frac{1}{2}$ for impurity scattering. In general, the value of q can be inferred from the temperature dependence of carrier mobility. It is expected that μ has a T^{q-1} dependence if τ varies as $\mathscr{E}^q$. Therefore, a study of the temperature variations of thermoelectric power and mobility yields information concerning the relative importance of different scattering mechanisms in different temperature regions. For p-type semiconductors, $\mathscr{E}_c - \mathscr{E}_f$ is replaced by $\mathscr{E}_f - \mathscr{E}_v$ and the polarity of ΔV_S is changed. The ratio $\Delta V_S / \Delta T = S$ is known as the Seebeck coefficient or the thermoelectric power and is given by

$$S = \frac{\mu_p p S_p + \mu_n n S_n}{\mu_p p + \mu_n n} \qquad (6.10.31)$$

A similar analysis for metals yields

$$S_M = -\frac{\pi^2}{3} \frac{k^2 T}{e \mathscr{E}_f} \qquad (6.10.32)$$

Equation (6.10.32) also applies to degenerate semiconductors with $\mathscr{E}_f$ replaced by $\mathscr{E}_f - \mathscr{E}_c$ for n type and by $\mathscr{E}_v - \mathscr{E}_f$ for p type.

Figure 6.34 shows the measured value (dashed curve) and the calculated value (solid curve) of the thermoelectric power in several germanium samples of different impurity concentration (V. A. Johnson and K. Lark-Horowitz, *Phys. Rev.*, Vol. 92, p. 226, 1953). Take the sample with $N_a = 5.7 \times 10^{15}$ cm^{-3}. With $N_v = 5.1 \times$

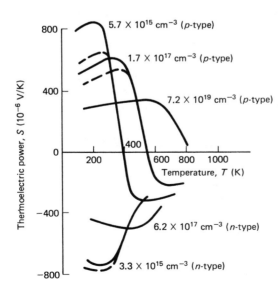

Figure 6.34 Values of the thermoelectric power (Seebeck coefficient) observed (dashed curves) as a function of temperature in n-type and p-type Ge samples. The solid curves are calculated from Eq. (6.10.30). (V. A. Johnson and K. Lark–Horowitz, *Phys. Rev.*, Vol. 92, p. 226, 1953.)

10^{18} cm^{-3} for Ge, we find a value of $7kT$ for $\mathscr{E}_f - \mathscr{E}_v$. Thus at 300 K, $S_p = 6.5 \times 10^{-4}$ V/K from Eq. (6.10.30) modified for p type and with $q = -\frac{3}{2}$. The value of S_p in metals and degenerate semiconductors is much smaller because the contributions from the E and $\mathscr{E}_f$ terms almost cancel out. The observed value of S_M/T in sodium is -2.8×10^{-8} V/K^2 and the calculated value is -0.8×10^{-8} V/K^2 from Eq. (6.10.32) with $\mathscr{E}_f = 3.12$ eV. Since the value of S is of the order of several hundred μV/K in typical semiconductors as compared to a value of less than 10 μV/K in good metallic conductors, the contribution to the Seebeck voltage from the metal can be neglected in a metal–semiconductor junction. The Seebeck voltage is used mainly to ascertain the dominant scattering mechanism through q. It is also very useful in determining the conductivity type of a semiconductor by its polarity.

6.11 DEGENERATE SEMICONDUCTORS

In Section 6.3 we used Boltzmann–Maxwell distribution to calculate the carrier density. This approximation is valid for nondegenerate semiconductors where the Fermi level lies in the forbidden gap. However, some applications may require a heavy doping concentration, causing the Fermi level to move into either the conduction or the valence band. In this section we discuss the consequences of heavy doping, which is indicated by n^+ or p^+. First, let us establish a procedure by which the carrier concentration can be found with reasonable accuracy. For a nondegenerate semiconductor, the Fermi function $f(y) = y/(1 + y)$ of Eq. (4.5.1) can be expanded into a power series

$$f(y) = y(1 - y + y^2 - y^3 + \cdots) \tag{6.11.1}$$

where $y = \exp - (\mathscr{E} - \mathscr{E}_f)/kT$. Therefore, for $x = (\mathscr{E}_f - \mathscr{E}_c)/kT < 0$, the electron concentration is

$$n = n_{\text{ND}}\left[1 - \frac{\exp x}{2\sqrt{2}} + \frac{\exp (2x)}{3\sqrt{3}} - \frac{\exp (3x)}{4\sqrt{4}} + \cdots \right] \tag{6.11.2}$$

where n_{ND} is the approximate expression for nondegenerate semiconductors given by

$$n_{\text{ND}} = 2\left(\frac{2\pi m_e^* kT}{h^2}\right)^{3/2} \exp (x) = A \exp (x) \tag{6.11.3}$$

In Fig. 6.35, Eq. (6.11.3) is plotted as $A \exp (x)$ with $A = 2.54 \times 10^{19}$ cm^{-3} calculated for $m_e^* = m_0$ and $T = 300$ K. For a particular semiconductor at a given temperature T, n_{ND} can be written as

$$n_{\text{ND}} = A\left(\frac{m_e^*}{m_0}\frac{T}{300}\right)^{3/2} \exp (x) = A' \exp (x) \tag{6.11.4}$$

which is the same as Eq. (6.3.8).

For a highly degenerate semiconductor with $x \geq 5$, the calculation proceeds in two steps. First, we let $\eta = \mathscr{E} - \mathscr{E}_f$ and $\xi = \mathscr{E}_f - \mathscr{E}_c$, and then expand $(\mathscr{E} - \mathscr{E}_c)^q$ as

$$(\mathscr{E} - \mathscr{E}_c)^q = (\xi + \eta)^q = \xi^q\left[1 + q\frac{\eta}{\xi} + \frac{q(q - 1)}{2}\left(\frac{\eta}{\xi}\right)^2 + \cdots \right] \tag{6.11.5}$$

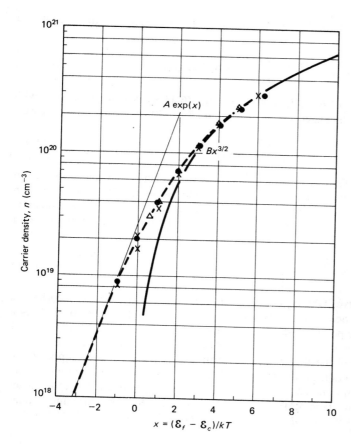

Carrier density, n (cm^{-3})

$A \exp(x)$

$Bx^{3/2}$

$x = (\mathcal{E}_f - \mathcal{E}_c)/kT$

Figure 6.35 Carrier density n computed from Eq. (6.11.11) by Blakemore as a function of the position of the Fermi level $\mathcal{E}_f$ relative to $\mathcal{E}_c$ (both normalized to kT). The two solid curves $A \exp(x)$ and $Bx^{3/2}$ represent expressions commonly used for nondegenerate and degenerate semiconductors, respectively. The points marked by x represent approximate formulas Eq. (6.11.2) for $x < 0$ and Eq. (6.11.8) for $x > 1$. The dots represent approximate formula, Eq. (6.11.12) used by Joyce and Dixon. The curves are computed for $m^* = m_0$ and $T = 300$ K.

Next we apply integration by parts to the integral to obtain

$$\int_{-\xi}^{\infty} (\xi + \eta)^{q-1} f(\eta)\, d\eta = -\frac{1}{q} \int_{-\xi}^{\infty} (\xi + \eta)^q \left(-\frac{\partial f}{\partial \eta} \right) d\eta \qquad (6.11.6)$$

In the integral expression for n, $q = \frac{3}{2}$. Furthermore, $\partial f/\partial \eta = \exp(y)/\{kT[1 + \exp(y)]^2\}$ and is an even function of η. Therefore, the second term in Eq. (6.11.5), which is odd in η, will not make a significant contribution to Eq. (6.11.6). Furthermore, for $x \geq 5$, the lower integration limit $-\xi = xkT$ can be replaced by $-\infty$ because $\partial f/\partial \eta$ drops to insignificant values. Using the definite integral

$$\int_{-\infty}^{\infty} \frac{y^2 \exp(y)\, dy}{[1 + \exp(y)]^2} = \frac{\pi^2}{3} \qquad (6.11.7)$$

after substitution of Eq. (6.11.5) in Eq. (6.11.6), we obtain

$$n = n_{D0}\left(1 + \frac{\pi^2}{12x^2} \right) \qquad (6.11.8)$$

Sec. 6.11   Degenerate Semiconductors

where the quantity n_{D0} is the electron concentration in a degenerate semiconductor at 0 K or

$$n_{D0} = \frac{8\pi}{3} \frac{(2m_e^*)^{3/2}}{h^3} (kTx)^{3/2} = Bx^{3/2} \tag{6.11.9}$$

which is the same as Eq. (4.6.3). In Fig. 6.35, Eq. (6.11.9) is plotted as $Bx^{3/2}$ with $B = 1.91 \times 10^{19}$ cm^{-3} calculated for $m_e^* = m_0$ and $T = 300$ K. For a general case,

$$n_{D0} = B \left(\frac{m_e^*}{m_0} \frac{T}{300} \right)^{3/2} x^{3/2} \tag{6.11.10}$$

The results from different computations of the relation between n and x are shown in Fig. 6.35. The two relations n_{ND} of Eq. (6.11.3) and n_{D0} of Eq. (6.11.9) are drawn as solid curves while the numerical result (J. S. Blakemore, *Semiconductor Statistics*, Pergamon Press, Inc., New York, 1962, Appendices B and C) calculated from the exact expression

$$n = \frac{8\pi}{h^3} (2m_0)^{3/2} \int_{\mathscr{E}_c}^{\infty} (\mathscr{E} - \mathscr{E}_c)^{1/2} f(\mathscr{E}) \, d\mathscr{E} \tag{6.11.11}$$

is shown as a dashed curve. Obviously, for the transition region ($-1 < x < 5$) from a nondegenerate to a degenerate semiconductor, neither Eq. (6.11.3) nor (6.11.9) is a good approximation. The points marked by $\times$ are obtained from Eq. (6.11.8) for $x \geq 1$ and from Eq. (6.11.2) for $x < $ or $= 0$. For the latter case, there are two x points, with the lower point using only the first correction term $(2)^{-3/2} \exp(x)$ and the upper point using the first two correction terms in Eq. (6.11.2). We see that an accuracy within 10% is obtained even with one correction term. To obtain the Fermi level from carrier concentration, we need to transform Eqs. (6.11.2) and (6.11.8). An approximate relation is (W. B. Joyce and R. W. Dixon, *Appl. Phys. Lett.*, Vol. 31, p. 354, 1977)

$$x = \ln \frac{n}{A} + 2^{-3/2} \left(\frac{n}{A} \right) \tag{6.11.12}$$

which is represented by dots in Fig. 6.35.

In summary, all three approximations give reasonably accurate results in appropriate regions. For $x < $ or $= 0$, Eq. (6.11.2) converges and it gives exact values if many terms are included. For $x > 3$, Eq. (6.11.8) yields an accuracy within 5%, and accuracy improves with increasing x. The inaccuracy of Eq. (6.11.8) mainly comes from the exclusion of contribution from the second term $q(\eta/\xi)$ in Eq. (6.11.5), which will have a dependence Dx. We find

$$n = n_{D0} \left(1 + \frac{0.20}{x} + \frac{\pi^2}{12x^2} \right) \tag{6.11.13}$$

to have an accuracy better than 3% for $x > $ or $= 0.5$. The points calculated from Eq. (6.11.13) are shown as open triangles. The constant 0.20 is chosen empirically by fitting the dashed curve. The accuracy of Eq. (6.11.12) is better than 5% for $x < 4$ and becomes poor for $x > 6$. Therefore, Eq. (6.11.12) is suitable for lightly degenerate semiconductors at room temperature ($x < 3$) and Eq. (6.11.8) is suitable for heavily degenerate semiconductors at low temperatures ($x > 3$). We should point out that for

Semiconductor Fundamentals   Chap. 6

degenerate p-type semiconductors, we simply change m_e^* to m_h^* and $\mathscr{E}_f - \mathscr{E}_c$ to $\mathscr{E}_v - \mathscr{E}_f$ in the expressions given above and in Fig. 6.35.

Even though we have analytical and graphical relations between carrier concentration and Fermi level, the situation in a heavily doped semiconductor is not as simple as that in a lightly or moderately doped semiconductor with regards to defining the band edge, $\mathscr{E}_c$ or $\mathscr{E}_v$, and to determining the degree of ionization for impurities. In extrinsic semiconductors with low impurity concentrations, it is generally assumed that the conduction- and valence-band edges as well as the donor- and acceptor-state energies are sharply defined. These energies coincide with their respective positions in the intrinsic material. In a degenerate semiconductor, however, we must take into account modifications introduced into the band structure by heavy impurity concentrations. The two effects of heavy doping can be described as follows. First, the hydrogen-like impurity-atom wave functions begin to overlap. As a result, the energy of the impurities broadens to form a band, known as the *impurity band*. Second, an impurity atom introduces a local variation in the potential energy of an electron because of the difference in the nuclear potentials of the impurity and host atom. Such a local, random variation in the potential energy modifies the position of the band edges. As a result, the bands extend beyond their respective positions in the intrinsic material. The extended part of a band is called the *band tailing*.

In Fig. 6.36, the band structure of a heavily doped semiconductor (solid curves) is drawn for comparison with the band structure of the intrinsic material (dashed curves). If we recall our discussion in Section 6.6, the Hall coefficient in heavily doped Ge (curve 58 in Fig. 6.11) was nearly independent of temperature and this behavior was

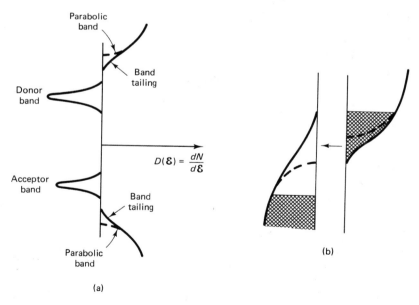

(a)

(b)

Figure 6.36 (a) Density-of-state function $D(\mathscr{E}) = dN/d\mathscr{E}$ for a heavily doped semiconductor (solid curves) as compared to that for a very lightly doped semiconductor (dashed curves) and (b) energy-band diagram showing possible tunneling between band-tailing states in a forward-biased degenerate junction.

Sec. 6.11   Degenerate Semiconductors

attributed to the formation of an impurity band. Further, from the value of donor energy (Fig. 6.13), we inferred that the donor-impurity band actually merged into the conduction band for $N_d > 10^{17}$ cm^{-3}. Several attempts have been made to calculate the impurity band structure. The models used include: a lattice of hydrogen-like impurities (W. Baltensperger, *Philos. Mag.*, Vol. 44, p. 1355, 1953), a one-dimensional lattice of random impurities (H. M. James and A. S. Ginzberg, *J. Phys. Chem.*, Vol. 57, p. 840, 1953), and a one-dimensional lattice of random delta function (M. Lax and J. C. Phillips, *Phys. Rev.*, Vol. 110, p. 41, 1958). For a random spatial distribution of impurities, the impurity band is expected to have a Gaussian shape. However, the models are not accurate enough to give quantitative estimates of the mean band width. Such estimates can be made by measuring the current in forward-biased, degenerate diodes (Section 13.3), caused by tunneling between the band and impurity states and between the impurity states (Fig. 6.36b).

The tunnel current in degenerate Ge diodes has been carefully studied by Meyerhofer, Brown, and Sommers (D. Meyerhofer, G. A. Brown, and H. S. Sommers, Jr., *Phys. Rev.*, Vol. 126, p. 1329, 1962). Information concerning the band properties of a degenerate semiconductor can be obtained from the tunneling experiment. Figure 6.37 shows the density-of-states function deduced from the dV/dI curve for *p*-type GaAs. The curve clearly shows the band tailing states. For details of the experiment and its analy-

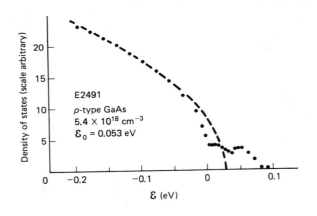

E2491

p-type GaAs

5.4×10^{18} cm^{-3}

$\mathcal{E}_0 = 0.053$ eV

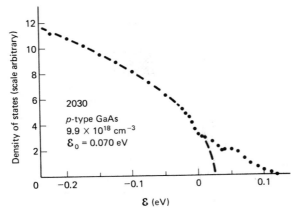

2030

p-type GaAs

9.9×10^{18} cm^{-3}

$\mathcal{E}_0 = 0.070$ eV

Figure 6.37 The density-of-state function for heavily doped *p*-type GaAs ($N_a = 5.4 \times 10^{18}$ cm^{-3} for the upper curve and $N_a = 9.9 \times 10^{18}$ cm^{-3} for the lower curve) deduced from the dV/dI curve of Schottky-barrier diodes. (G. D. Mahan and J. W. Conley, *Appl. Phys. Lett.*, Vol. 11, p. 29, 1967.)

sis, the reader is referred to a series of papers by Conley and coworkers (J. W. Conley, C. B. Duke, G. D. Mahan, and J. J. Tiemann, *Phys. Rev.*, Vol. 150, p. 466, 1966; J. W. Conley and J. J. Tiemann, *J. Appl. Phys.*, Vol. 38, p. 2880, 1967; G. D. Mahan and J. W. Conley, *Appl. Phys. Lett.*, Vol. 11, p. 29, 1967). The reader is also asked to consult two papers by Kane and by Halperin and Lax for theoretical discussions of the effects of randomly located impurity ions on band tailing (E. O. Kane, *Phys. Rev.*, Vol. 131, p. 79, 1963; B. I. Halperin and M. Lax, *Phys. Rev.*, Vol. 148, p. 722, 1966).

We note from Fig. 6.37 that extrapolation of the density-of-state curve $D(\mathcal{E})$ to zero indicates a shift of the band edge to a lower value. This effect is known as *band shrinkage*. In the presence of both band shrinkage and band tailing, the electron concentration is given by

$$n = n_{bt} + n_{D0}\left(1 + \frac{\pi^2}{12x^2}\right) \qquad (6.11.14)$$

where n_{bt} is the electron concentration in the band tail; the other two terms represent the electron concentration in the regular band with $x = (\mathcal{E}_f - \mathcal{E}_c')/kT$, and $\mathcal{E}_c'$ is the new extrapolated band edge. The density of ionized donors N_d^+ for a band of $\mathcal{E}_d$ is given by

$$N_d^+ = \int \frac{D(\mathcal{E}_d)\, d\mathcal{E}_d}{1 + 2\exp\,(\mathcal{E}_f - \mathcal{E}_d)/kT} \qquad (6.11.15)$$

where $D(\mathcal{E}_d)\, d\mathcal{E}_d$ represents the density of donors having energy between $\mathcal{E}_d$ and $\mathcal{E}_d + d\mathcal{E}_d$, and the integration is over the donor band. Thus the total donor density is $N_d = \int D(\mathcal{E}_d)\, d\mathcal{E}_d$. If $N_a << N_d$, setting $n = N_d^+$ should yield x or $\mathcal{E}_f - \mathcal{E}_c'$. However, the situation is complicated by the fact that the band shift $\mathcal{E}_c - \mathcal{E}_c'$ depends on n. Therefore, a more direct way of finding n is to determine it experimentally from the Hall measurement. From Fig. 6.37, the value of n_{bt} can be found. Knowing n and n_{bt}, the Fermi level can be determined from Eq. (6.11.12) or (6.11.14). The total donor density N_d is known from doping; therefore, information about $D(\mathcal{E}_d)$ can be deduced from Eq. (6.11.15). It is obvious that in a degenerate semiconductor we must have $N_d > N_d^+$ in order to keep $\mathcal{E}_f$ in the conduction band.

PROBLEMS

6.1. Verify Eq. (6.3.10). Is Eq. (6.3.12) valid in a degenerate semiconductor?

6.2. Given $N_c = 1.04 \times 10^{19}$ cm^{-3} and $N_v = 0.48 \times 10^{19}$ cm^{-3} in germanium at 300 K, find m_e^* and m_h^*. Also calculate N_c and N_v at 77 K.

6.3. Given $N_c = 2.8 \times 10^{19}$ cm^{-3} and $N_v = 1.04 \times 10^{19}$ cm^{-3} in silicon at 300 K. Using these values and the corresponding values given in Problem 6.2 for Ge, calculate the intrinsic carrier concentrations in Ge and Si. It is known that $\mathcal{E}_g = 0.665$ eV in Ge and 1.11 eV in Si at 300 K.

6.4. Applying the Sommerfeld–Wilson quantization rule to a donor atom, show that the Bohr radius of the donor electron is

$$a_n = \frac{4\pi\epsilon\hbar^2}{m^*e^2}\, n^2$$

and the quantized energy of the donor state is

$$\mathscr{E}_n = \frac{m^* e^4}{2(4\pi\epsilon\hbar)^2} \frac{1}{n^2}$$

where n is the principal quantum number. Find the values for a_1 and $\mathscr{E}_1$ in Ge and Si, given $\epsilon = 16\epsilon_0$ and $m^* = 0.12m_0$ in Ge and $\epsilon = 11.7\epsilon_0$, and $m^* = 0.26m_0$ in Si.

6.5. Write down a set of equations, corresponding to Eqs. (6.4.3) and (6.4.4), for the occupied and vacant acceptor states. From these equations, derive Eq. (6.4.19).

6.6. (a) Find an expression relating the intrinsic carrier concentrations at two different temperatures.
   (b) Given $n_i = 1.9 \times 10^{13}$ cm^{-3} in Ge and $n_i = 0.9 \times 10^{10}$ cm^{-3} in Si at 300 K, find the intrinsic carrier concentrations in Ge and Si at 400 K.
   (c) We have two n-type semiconductors, one Ge and the other Si, with impurity concentrations $N_d = 2 \times 10^{14}$ cm^{-3} and $N_a = 0$. Comment on whether or not the electron concentration in the two semiconductors is sensitive to temperature changes at 300 K and 400 K.

6.7. Draw diagrams showing the distribution of electrons among available energy states in **(a)** an intrinsic semiconductor, **(b)** a compensated n-type semiconductor, and **(c)** a compensated p-type semiconductor. In cases (b) and (c), the temperature is assumed to be in the intermediate-temperature region. In other words, the semiconductor is in the exhaustion region. Write down the appropriate expressions for the electron and hole concentrations in each case.

6.8. Consider a nondegenerate semiconductor doped with both donors and acceptors with the donor concentration N_d larger than the acceptor concentration N_a. Find the approximate expressions for the electron and hole concentrations for $n_i > N_d - N_a$ and for $n_i < N_d - N_a$. The impurities are assumed fully ionized. Discuss the temperature dependence of the carrier concentrations for each case.

6.9. (a) Find the intrinsic carrier concentration in Ge and Si at 300 K, and also the resistivity of intrinsic Ge and Si.

	N_c (cm^{-3})	N_v (cm^{-3})	$\mathscr{E}_g$ (eV)	μ_n (cm^2/V-s)	μ_p
Ge	1.04×10^{19}	0.48×10^{19}	0.665	3900	1900
Si	2.9×10^{19}	1.1×10^{19}	1.11	1400	470

   (b) Suppose that the temperature is raised from 300 K to 300 K + ΔT. Calculate the temperature rise ΔT required to double the intrinsic carrier concentration in each case.

6.10. A Si sample at room temperature contains 10^{16} cm^{-3} of phosphorus and 10^{15} cm^{-3} of boron.
   (a) Assume that all donors and acceptors are ionized. Find the electron and hole concentrations and the location of the Fermi level. Use the values given in Problem 6.9 for Si.
   (b) From the location of the Fermi level, find the fractions of donors and acceptors that are not ionized.

6.11. Draw a diagram, similar to Fig. 6.8, showing the variation of the Fermi level as a function of temperature in a p-type semiconductor. From the hole concentration, find an expression that relates the Fermi level to the temperature in the exhaustion region (i.e., the intermediate-temperature region). Find the location of the Fermi level in a germanium sample with $N_a = 3 \times 10^{15} \text{ cm}^{-3}$ and $N_d = 2 \times 10^{15} \text{ cm}^{-3}$. The calculation is made for $T = 300$ K and 75 K, and the value of $N_c = 0.48 \times 10^{19} \text{ cm}^{-3}$ at 300 K. Is the approximation of complete ionization of impurities still valid at 75 K?

6.12. Starting from Eq. (6.5.2), find an expression for the hole concentration in a p-type semiconductor ($N_d = 0$) in the low-temperature region. Whenever approximations are made, give reasons to show that the steps taken are justified.

6.13. Find the concentrations of electrons and holes at 300 K and 450 K in a silicon sample with $N_d = 5 \times 10^{15} \text{ cm}^{-3}$ and $N_a = 3 \times 10^{15} \text{ cm}^{-3}$. Given $n_i = 0.9 \times 10^{10} \text{ cm}^{-3}$ at 300 K and $\mathscr{E}_g = 1.11$ eV in Si.

6.14. The following data are obtained from sample 55 in Figs. 6.11 and 6.12:

T (K)	20	33	50	100	200
R_H (10^5 cm^3/C)	10.2	9.2	8.8	8.6	8.6
ρ (Ω – cm)	4	6	10	28	80

Find the electron concentrations at 20 K, 50 K, and 100 K. Assuming that $N_a = 2.5 \times 10^{12} \text{ cm}^{-3}$, calculate N_d and the percentage of ionized donors at 20 K. Find the activation energy $\mathscr{E}_c - \mathscr{E}_d$ of donors from the values of $\mathscr{E}_d - \mathscr{E}_f$ and $\mathscr{E}_c - \mathscr{E}_f$. Given at 300 K, $N_c = 1.04 \times 10^{19} \text{ cm}^{-3}$ and $kT = 0.0259$ eV.

6.15. **(a)** From the data given in Problem 6.14, calculate and then plot the values of the Hall mobility of electrons as a function of temperature on a logarithmic scale and compare it with the $T^{-3/2}$ dependence.
(b) It is known that the mean free path l of electrons is inversely proportional to T for lattice scattering. Use heuristic arguments to find the temperature dependences of the relaxation time τ and the mobility μ of electrons.

6.16. The following data are obtained from Figs. 6.11 and 6.12:

R_H (cm^3/C)	8.8×10^5	6×10^4	6×10^3	3.5×10^2	8
ρ (Ω – cm)	10	0.9	0.16	0.05	0.008

for various germanium samples at 50 K. Calculate the Hall mobility of electrons in different samples. Examine the tendency of the variation of μ_H with respect to ρ and cite the possible cause for the variation. Compare your result with the theoretical plot at 300 K (E. M. Conwell, *Proc. IRE,* Vol. 40, p. 1327, 1952).

6.17. Derive Eq. (6.6.1) from Eq. (6.5.18) by expressing $\mathscr{E}_f$ in terms of n. The values of R_H (cm^3/C) for sample 51 in Fig. 6.11 are 5.5×10^4 at 50 K, 1.8×10^5 at 20 K, and 4.5×10^6 at 12.5 K. Find the activation energy of the donor state by averaging the values obtained at the three temperatures. The reason for averaging is that there is a slight dependence of $\mathscr{E}_c$ on temperature. Given $N_c = 7.1 \times 10^{17} \text{ cm}^{-3}$ at 50 K and $N_a = 4 \times 10^{12} \text{ cm}^{-3}$.

6.18. The drift mobilities (cm^2/V-s) in Ge and Si are given in Problem 6.9. The values for the intrinsic carrier concentration are given in Problem 6.6. Find the resistivities and Hall coefficients in intrinsic Ge and intrinsic Si.

6.19. A gold (Au) atom in Si can be neutral (denoted by Au^0), accept an electron (to become Au^-), or denote an electron (to become Au^+). As shown in Fig. 6.21, the acceptor level Au^- is $\mathscr{E}_a = \mathscr{E}_c - 0.54$ eV, while the donor level Au^+ is $\mathscr{E}_d = \mathscr{E}_v + 0.35$ eV.

Apply the neutral-donor and the neutral-acceptor reactions

$$Au^0 \rightleftharpoons Au^+ + e, \qquad Au^0 + e \rightleftharpoons Au^-$$

to Eqs. (6.4.13) and (6.4.19), respectively, to obtain the ratios N_{Au}^+/N_{Au}^0 and N_{Au}^-/N_{Au}^0. Show that for $g = 1$,

$$N_{Au}^+ : N_{Au}^0 : N_{Au}^- = \exp\left(\frac{\mathscr{E}_d - \mathscr{E}_f}{kT}\right) : 1 : \exp\left(\frac{\mathscr{E}_f - \mathscr{E}_a}{kT}\right)$$

6.20. Refer to the result of Problem 6.19 and the data given in Problem 6.9. Using the charge-neutrality equation, discuss the movement of $\mathscr{E}_f$ as a function of temperature from, say, 600 K to 2 K. A value for $N_{Au} = 10^{16}$ cm^{-3} in Si is assumed.

6.21. A GaAs sample is doped with Zn (a shallow acceptor) and O (a deep donor). Use the values given in Table 6.8 for N_c, N_v, and n_i, and a degeneracy factor $g = 2$ for both donors and acceptors. Find the electron and hole concentrations n and p at 300 K for the following doping concentrations:
(a) N_a (acceptor) $= 4 \times 10^{16}$ cm^{-3}, N_d (donor) $= 3 \times 10^{16}$ cm^{-3}
(b) $N_a = 4 \times 10^{16}$ cm^{-3}, $N_d = 8 \times 10^{16}$ cm^{-3}
Draw diagrams indicating the general behavior of the Fermi level as a function of temperature T for the two cases.

6.22. A GaAs sample is doped with Cr (a deep acceptor) and Te (a shallow donor). Given: $\mathscr{E}_c - \mathscr{E}_d = 0.005$ eV and $\mathscr{E}_a - \mathscr{E}_v = 0.79$ eV; also at 300 K, $kT = 0.0259$ eV. Use the values given in Table 6.8 for N_c, N_v, and n_i, and a degeneracy factor $g = 2$ for both donors and acceptors. Find the electron and hole concentrations at 300 K for the following doping concentrations:
(a) N_d (donor) $= 5 \times 10^{16}$ cm^{-3}, N_a (acceptor) $= 1 \times 10^{16}$ cm^{-3}
(b) $N_d = 1 \times 10^{16}$ cm^{-3}, $N_a = 5 \times 10^{16}$ cm^{-3}

6.23. Consider a GaAs sample doped with a deep acceptor and a shallow donor with energies as given in Problem 6.22. Given $N_a = 2 \times 10^{17}$ cm^{-3}, find the range of N_d that can be tolerated in order to keep $n < 10^{10}$ cm^{-3} and $p < 10^{10}$ cm^{-3}.

6.24. The mobility and Hall coefficient in the Boltzmann formulation of the transport problem are given by

$$\mu = \frac{e}{3kT}\langle v^2\tau \rangle, \qquad R_H = \frac{3kT}{em^*}\frac{\langle v^2\tau^2\rangle}{\langle v^2\tau\rangle^2}\frac{1}{n}$$

where the symbol $\langle\ \rangle$ indicates an average over the electron (or hole) distribution. Find μ and R_H for two cases: (a) τ independent of v of carriers and (b) $l = v\tau$ independent of v of carriers. Also find the ratio of Hall to drift mobility $\mu_H/\mu = R_H\sigma/\mu$ for the two cases.

6.25. Start from the following expression of relaxation time τ for impurity scattering:

$$\frac{1}{\tau(\mathscr{E})} = \frac{v}{d}\frac{1}{d^2}\int_0^{d/2}(1 - \cos\theta)2\pi a\,da = A\ln\left(\sin^2\frac{\theta}{2}\right)$$

The integration limits are $a = 0$ to $a = d/2$, corresponding to $\theta = 180°$ to θ_1. Show that

$$A = \frac{-2\pi N_i e^4}{(4\pi\epsilon)^2} (2\mathscr{E})^{-3/2}(m^*)^{-1/2}$$

Also from $\mu = \dfrac{e}{3kT}\langle v^2\tau\rangle$, obtain the expression for μ_i.

6.26. In the quantum-mechanical treatment of impurity scattering, the scattering probability $S(k, k')$ is given by

$$S(k, k') = \frac{2\pi}{\hbar}\left| \int_0^\infty \exp(-i\mathbf{k}\cdot\mathbf{r}) \frac{e^2}{4\pi\epsilon r} e^{-qr} \exp(i\mathbf{k}'\cdot\mathbf{r})4\pi r^2\, dr \right|^2$$

$$= \frac{2\pi}{\hbar}\left\{ \frac{e^2}{\epsilon(|\mathbf{k}-\mathbf{k}'|^2 + q^2)} \right\}^2$$

where $1/q = (\epsilon kT/e^2 n)^{1/2}$ is the Debye screening length. The density of available quantum states is given by

$$N(\mathscr{E})d\mathscr{E} = \frac{2}{h^3} dp_x\, dp_y\, dp_z = \frac{2m^*}{h^3}(2m^*\mathscr{E})^{1/2}\, d\mathscr{E}d\Omega$$

where

$$d\Omega = 2\pi \sin\theta\, d\theta.$$

Show that for $q = 0$ (no screening),

$$S(k, k')N(\mathscr{E}) = \sigma(\theta)\, d\Omega$$

where $\sigma(\theta)$ is the Rutherford scattering cross section. In impurity scattering, the value of k is unchanged; therefore,

$$|\mathbf{k} - \mathbf{k}'| = k(1 - \cos\theta)$$

where θ is the scattering angle.

6.27. **(a)** The two dominant scattering mechanisms in Si and Ge are lattice scattering and ionized-impurity scattering. Given:

$$\mu_l = A(m^*)^{-5/2}(kT)^{-3/2} = A'\left(\frac{T}{200}\right)^{-3/2}$$

$$\mu_i = BN_i^{-1}(m^*)^{-1/2}(kT)^{3/2} = B'\left(\frac{N_i}{10^{13}}\right)^{-1}\left(\frac{T}{200}\right)^{3/2}$$

Refer to the data taken by Debye and Conwell and given in Figs. 6.11 and 6.14. Assume that μ_l dominates μ in sample 55 between 30 and 200 K. Find the constant A'.

(b) Now we try to fit the data for sample 64, which has $R_H = 4 \times 10^3$ cm^3/C at 300 K. Given

T	200 K	100 K	50 K	30 K	20 K
μ	6×10^3	1.4×10^4	3.5×10^4	6×10^4	7×10^4

find the average value of B' obtained from μ. From the values of A' and B' thus found, calculate the mobility μ for a sample with $N_i = 10^{16}$ cm^{-3} at $T = 100$ K.

6.28. Using Eq. (6.7.6), verify the numerical coefficients 1.65×10^{19} and 3×10^{11} in Eq. (6.7.9) and determine the proper units for these coefficients. Assuming that sample 58 in Figs. 6.11 and 6.14 is uncompensated (i.e., $N_a = 0$) and using $\epsilon = 16 \, \epsilon_0$ and $m_e^* = 0.13m_0$ in Ge, find the values of the ionized impurity concentration N_i, the impurity-scattering-limited mobility μ_i, and the mean free path l at 20 K.

6.29. The mean free time τ can be related to the quantum-mechanical transition probability W. The transition probability for an electron to be scattered from a $\mathbf{k}$ to $\mathbf{k}'$ state is

$$W_i = \frac{2\pi}{\hbar} \left| \frac{e^2}{\epsilon(|\mathbf{k} - \mathbf{k}'|^2 + q^2)} \right|^2 N_i \delta(\mathscr{E} - \mathscr{E}')$$

for impurity scattering and

$$W_l = \frac{2\pi}{\hbar} \frac{kT}{2\rho} \left(\frac{q}{\omega} \right)^2 \mathscr{E}_{1c}^2 \, \delta(\mathscr{E} - \mathscr{E}')$$

for lattice scattering where the Dirac delta function $\delta(\mathscr{E} - \mathscr{E}')$ is a mathematical statement of energy conservation. The quantity q in W_i is screening parameter used in the Coulomb potential $(e^2/4\pi\epsilon r) \exp(-qr)$, and the quantity q in W_l is the wave number for acoustic phonons. Check the dimensionality of W_i and W_l. We define a quantity called the scattering cross section $\sigma(\theta)$ (in m^2) such that $\sigma(\theta)v$ represents the transition (i.e., scattering event) per unit time and

$$\frac{1}{\tau} = \frac{1}{V} \int \sigma(\theta)v(1 - \cos\theta) \, d\Omega$$

Find the relation between $\sigma(\theta)$ and the transition probability. (*Hint:* To obtain the total transition rate, we must count all possible final states for electrons to be scattered into.) Show that if we let $q = 0$ in W_i, $\sigma(\theta)$ thus found is identical to $\sigma(\theta)$ of Eq. (6.7.2) and that $1/\tau$ thus found from W_l is identical to Eq. (6.7.21).

6.30. (a) The stress σ_{ij} in a solid can be expressed as a linear function of strain δ_{kl} in the form of Hooke's law as follows:

$$\sigma_{xx} = c_{11}\delta_{xx} + c_{12}\delta_{yy} + c_{13}\delta_{zz} + c_{14}\delta_{yz} + c_{15}\delta_{zx} + c_{16}\delta_{xy}$$
$$\sigma_{yz} = c_{41}\delta_{xx} + c_{42}\delta_{yy} + c_{43}\delta_{zz} + c_{44}\delta_{yz} + c_{42}\delta_{zx} + c_{46}\delta_{xy}$$

where σ_{ij} is the force acting on a unit area in the i direction. If the coordinate of a lattice point is at (u, v, w), the strain δ_{kl} has six components, given by

$$\delta_{xx} = \frac{\partial u}{\partial x}, \qquad \delta_{yy} = \frac{\partial v}{\partial y}, \qquad \delta_{zz} = \frac{\partial w}{\partial z}$$

$$\delta_{yz} = \frac{\partial w}{\partial y} + \frac{\partial v}{\partial z}, \qquad \delta_{zx} = \frac{\partial u}{\partial z} + \frac{\partial w}{\partial x}, \qquad \delta_{xy} = \frac{\partial v}{\partial x} + \frac{\partial u}{\partial y}$$

Write down similar expressions for tensile stress components, σ_{yy} and σ_{zz}, and shear stress components, σ_{zx} and σ_{xy}. Show that for cubic crystals

$$c_{11} = c_{22} = c_{33}, \qquad c_{44} = c_{55} = c_{66}$$
$$c_{12} = c_{21} = c_{23} = c_{32} = c_{13} = c_{31}$$

and other $c_{ij} = 0$ by applying the symmetry elements to the crystal, mainly 120° rotations about the body diagonal (threefold axis), and 90° and 180° rotations about the x, y, and z axes.

(b) In terms of stress components, the equation of motion in the x direction is given by

$$\rho \frac{\partial^2 u}{\partial t^2} = \frac{\partial \sigma_{xx}}{\partial x} + \frac{\partial \sigma_{xy}}{\partial y} + \frac{\partial \sigma_{xz}}{\partial z}$$

Show that for cubic crystals

$$\rho \frac{\partial^2 u}{\partial t^2} = c_{11} \frac{\partial^2 u}{\partial x^2} + c_{44}\left(\frac{\partial^2 u}{\partial y^2} + \frac{\partial^2 v}{\partial z^2}\right) + (c_{12} + c_{44})\left(\frac{\partial^2 v}{\partial x\, \partial y} + \frac{\partial^2 w}{\partial x\, \partial z}\right)$$

where ρ is the mass density.

6.31. (a) A crystal is said to be *centro-symmetrical* if it possesses an inversion center such that $I(x, y, z) = (-x, -y, -z)$, where I represents the inversion operation about the center. Show that

$$I\delta_{ij} = \delta_{ij} \quad \text{and hence} \quad I\sigma_{ij} = \sigma_{ij}$$

for strain and stress in a centro-symmetrical crystal.

(b) *Piezoelectric effect* refers to the electric displacement D_i induced by stress σ_{kl} described by the relation

$$D_i = d_{ixx}\sigma_{xx} + d_{iyy}\sigma_{yy} + d_{izz}\sigma_{zz} + d_{iyz}\sigma_{yz} + d_{izx}\sigma_{zx} + d_{ixy}\sigma_{xy}$$

where d_{ikl} is the piezoelectric constant. Apply the inversion operation to D_i to show that in a centro-symmetrical crystal, $d_{ikl} \equiv 0$.

(c) Is silicon or germanium piezoelectric? What is the value of μ_{pe} of Eq. (6.8.16) for Si and Ge?

(d) Using the equation of motion derived in Problem 6.30, find expressions for the longitudinal and transverse acoustic velocities, v_l and v_t. The values of elastic stiffness constants for Si and GaAs are

Si (10^{12} dyn/cm^2)	$c_{11} = 1.656$	$c_{12} = 0.639$	$c_{44} = 0.796$
GaAs	1.19	0.538	0.595

(e) The mass density (g/cm^3) is $\rho = 2.328$ for Si and 5.316 for GaAs. Calculate the values of v_l and v_t for Si and GaAs. Which value v_l or v_t should be used for v_s in Eq. (6.7.22)?

6.32. (a) The following values are given for GaAs and Si:

	v_l (cm/s)	ρ (g/cm^3)	m_e^*/m_0	$\mathscr{E}_{1c}$ (eV)
GaAs	4.74×10^4	5.32	0.067	6.7
Si	8.47×10^5	2.32	0.26	12

where v_l is the longitudinal acoustic velocity. Using the values above, calculate the value of lattice-scattering-limited mobility μ_l. Explain possible sources for the discrepancy between the calculated values of μ_l and the observed mobility values.

(b) One parameter the value of which is somewhat uncertain is the deformation potential $\mathcal{E}_{1c}$. However, two quantities, pressure coefficient of energy gap $\Delta\mathcal{E}_g/\Delta\sigma$ and Young's modulus Y, can be determined experimentally and are found to be 0.0126 eV/kbar and 0.85×10^{12} dyn/cm^2 for GaAs and 0.015 eV/kbar and 1.30×10^{12} dyn/cm^2 for Si. Assuming that $\Delta\mathcal{E}_g$ is split equally between $\mathcal{E}_c$ and $\mathcal{E}_v$, estimate the value of $\mathcal{E}_{1c}$ for GaAs and Si.

6.33. The mobility μ in relatively pure p-type germanium has a value $\mu = 1950$ cm^2/V-s at 300 K and 2.75 m^2/V-s at 100 K. The μ versus temperature curve can be fit by using (D. M. Brown and R. Bray, *Phys. Rev.*, Vol. 127, p. 1598, 1962)

$$\mu_l = 7000\left(\frac{T}{300}\right)^{3/2}$$

$$\mu_{op} = 900\left(\frac{300}{T}\right)\left[\exp\left(\frac{\hbar\omega_{op}}{kT}\right) - 1\right]$$

in Eq. (6.7.25) with both expressed in cm^2/V-s. Because both optical-mode deformation potential $\mathcal{E}_{v2}$ and magnitude of reciprocal-lattice wave vector q in Eq. (6.7.26) are not known with certainty, the factor 900 in μ_{op} is just a fitting parameter. Using the values of μ given above, find the value of $\hbar\omega_{op}$ and check the calculated value with Fig. 2.38 to see whether it is reasonable. Also, using $m_\mu^* = 0.25m_0$ for holes, $\rho = 5.32$ g/cm^3, and $v_l = 5.4 \times 10^5$ cm/s in Ge, find the value of the acoustic deformation potential $\mathcal{E}_{1c}$ for Ge and compare the calculated value with the values given in Problem 6.32 for GaAs and Si.

6.34. (a) Show that Eq. (6.8.7) can be expressed in terms of α of Eq. (6.8.6) as

$$\mu_{po} = \frac{3\sqrt{\pi}}{4\alpha}\frac{e}{m^*\omega_l}\frac{K_1^{-1}(x_l/2)}{(x_l)^{3/2}\exp(x_l/2)}[\exp(x_l) - 1]$$

Using $\epsilon/\epsilon_0 = 13.2$, $n^2 = 10.9$, and $\omega_{l0} = 52.8 \times 10^{12}$ rad/s in GaAs, find the values of μ_{po} at 300 and 100 K. The values of the modified Bessel function $K_1(z)$ are

z	0.5	0.6	0.7	0.8	1.0	2.0	3.0
$K_1(z)$	1.66	1.30	1.05	0.862	0.602	0.140	0.0347

(b) The electromechanical constant in Eq. (6.8.16) has a maximum value for interaction with transverse acoustic waves along the [110] direction, and K^2 is given by $d_{14}^2/\epsilon c_{44}$, where $d_{14} = d_{ikl}$ (with $k \neq l$) is the piezoelectric constant defined in Problem 6.31. Given $d_{14} = -0.16$ C/m^2 and $c_{44} = 0.595 \times 10^{12}$ dyn/cm^2, calculate the values of μ_{pe} at 300 K and 100 K. Compare the calculated values of μ_{po} and μ_{pe} with those given in Fig. 11.31.

6.35. Starting from a set of equations of motion similar to Eq. (5.11.10) for carriers in a magnetic field, derive Eq. (6.10.2), which is a general expression for cyclotron resonance.

6.36. The cyclotron-resonance effective mass is, according to Eq. (6.10.2), given by

$$m_c^* = \frac{eB}{\omega_c} = \left(\frac{m_1 m_2 m_3}{m_1\alpha^2 + m_2\beta^2 + m_3\gamma^2}\right)^{1/2}$$

where α, β, and γ are directional cosines of magnetic flux with respect to the three axes of the constant-energy ellipsoid. Suppose that the applied B field is in the (110) plane and makes an angle θ with the [001] axis. Find the expression for m_c^* in terms of m_0 for Si. Given the following values for m_c^*/m_0:

θ	$0°$	$90°$
m_c^*/m_0	0.18, 0.42	0.42, 0.26

find m_t and m_l in Si.

6.37. The conduction band minima in Ge are located along the k [111] and equivalent directions. Find the directional cosines of the constant-energy ellipsoids with respect to the applied B field, which is in the (110) plane and makes an angle θ with the [001] axis. From the experimental curves, it is obvious that you should get three different sets of directional cosines. By fitting the experimental curves, find m_t and m_l in Ge.

6.38. (a) Starting from Eq. (6.3.7), show that in a semiconductor, such as Ge and Si, with multiple conduction-band minima, the density-of-state effective mass m_N^* is given by

$$m_N^* = q^{2/3} (m_l m_t^2)^{1/3}$$

where q is the number of equivalent conduction-band minima.
    Given $m_t = 0.19 m_0$, $m_l = 0.98 m_0$ in Si and $m_t = 0.082 m_0$, $m_l = 1.58 m_0$ in Ge, find the values of m_N^*.

6.39. (a) Apply the equation of motion to electrons in Si in one of the six ellipsoids along the $k[100]$ direction. Find the components v_x, v_y, and v_z of the drift velocity under an applied electric field having components $E\alpha$, $E\beta$, and $E\gamma$ along the x, y, and z directions, where α, β, and γ are the directional cosines.
    (b) Find the average velocity over the six ellipsoids along the direction of the electric field and along the direction perpendicular to the field. Calculate the value of m_μ^* in Si.

6.40. The valence band in Si and Ge actually consists of two bands, one heavy-hole band with mass m_{hh}^* and one light-hole band with mass m_{hl}^*. Show that for $\mathscr{E}_f > \mathscr{E}_v$, the hole concentration is given by

$$p = 2\left(\frac{2\pi m_N^* kT}{h^2}\right)^{3/2} \exp\left[-(\mathscr{E}_f - \mathscr{E}_v)kT\right]$$

where $m_N^{*3/2} = m_{hh}^{3/2} + m_{hl}^{3/2}$.

6.41. Given $N_c = 1.04 \times 10^{19}$ cm^{-3} and $N_v = 4.8 \times 10^{18}$ cm^{-3} in Ge, calculate the Seebeck voltage at 200 K, 300 K, and 400 K for two Ge samples, one with $N_a = 5.7 \times 10^{15}$ cm^{-3} and the other with $N_d = 3.3 \times 10^{15}$ cm^{-3}. Compare the calculated values with the measured values shown in Fig. 6.34.

6.42. From the Boltzmann transport equation, show that

$$J_x = e^2 E_x \int v_x^2 \tau \frac{df_0}{d\mathscr{E}} D(\mathscr{E}) \, d\mathscr{E} + e \frac{d}{dx} \int v_x^2 \tau f_0 D(\mathscr{E}) \, d\mathscr{E}$$

Also show that in nondegenerate semiconductors,

$$D = \frac{kT}{e} \mu$$

which is known as the *Einstein relation*.

Chap. 6   Problems

6.43. Consider a degenerate n-type semiconductor where $\mathcal{E}_f > \mathcal{E}_c$. Show that

$$D = \frac{2}{5}\left(\frac{\tau}{m^*}\right)(\mathcal{E}_f - \mathcal{E}_c) \quad \text{and} \quad \mu = \frac{e\tau}{m^*}$$

In the derivation above, assume that $T = 0$ K, so that f_0 is represented by a step function.

6.44. (a) Derive Eq. (6.11.2). Show that Eq. (6.11.12) can be obtained from Eq. (6.11.2) by keeping only the first two terms and by approximating $\ln(1 + z) = z$ for $z < 1$. Compare the values of n obtained from Eqs. (6.11.2) and (6.11.12) for $x = -2$ and 0.

  (b) Because of the approximation used, Eq. (6.11.12) is not accurate for $x > 0$ (degenerate semiconductors). Compare the values of n obtained from Eqs. (6.11.12) and (6.11.13) for $x = 4$ and $x = 8$.

6.45. For this problem we approximate the Fermi function by a step function. The semiconductor is heavily doped.

  (a) Donors form a band extending from $\mathcal{E}_{d1}$ to $\mathcal{E}_{d2}$ (Fig. P6.45) and have a density

Figure P6.45

$dN_d/d\mathcal{E} = 10^{22}/cm^3/eV$. Find the density of ionized donors N_d^+ for three cases: assuming that (1) $\mathcal{E}_f < \mathcal{E}_{d1}$; (2) $\mathcal{E}_f = (\mathcal{E}_{d1} + \mathcal{E}_{d2})/2$; and (3) $\mathcal{E}_f > \mathcal{E}_{d2}$.

  (b) Apply the charge-neutrality condition to set up an equation from which the Fermi level can be found. Find the values of $\mathcal{E}_c - \mathcal{E}_f$ or $\mathcal{E}_f - \mathcal{E}_c$ for Si.

6.46. Assume the following form for the tunneling current:

$$I = A \int_{\mathcal{E}_{fs} - eV_a}^{\mathcal{E}_{fs}} TD(\mathcal{E}) \, d\mathcal{E}.$$

from a degenerate p-type semiconductor to a metal (see Fig. P6.46), where T is the tun-

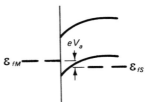

Figure P6.46

neling probability and A is a proportionality constant. Since T is a slowly varying function of V_a, we let

$$\frac{dT}{dV_a} = \frac{T}{V_0}$$

Show that the density-of-state function $D(\mathcal{E})$ can be found from

$$D(\mathcal{E}) = (AeT)^{-1}\left(\frac{dI}{dV_a} - \frac{I}{V_0}\right)$$

7

Transport and Recombination of Excess Carriers

7.1 DIFFUSION CURRENT

In Chapter 6 we discussed the fundamental properties of semiconductors, and in this chapter we concentrate our discussion on those properties that are important in device analyses. The operation of many semiconductor devices, among which the transistor is a notable example, depends on the existence of excess carriers. By *excess carriers* we mean carriers whose concentration exceeds the thermal-equilibrium concentration. For example, the carrier concentration near a *p-n* junction under a forward-bias condition is greater than the equilibrium concentration given in Section 6.3. It is the motion of excess carriers that causes current flow in the junction.

In the derivation of the expression for electric conductivity, Eq. (5.10.15), we assumed a homogeneous semiconductor in which the carrier concentrations n and p are uniform throughout. In cases where there are excess carriers, the carrier concentrations will no longer remain uniform throughout the semiconductor. A nonuniform distribution of carriers gives rise to a new phenomenon known as *diffusion*. When a concentration gradient exists, there is a net flow of carriers from the region of high concentration to the region of low concentration. Therefore, in an inhomogeneous semiconductor, a new current must be added to the drift current of Eq. (5.10.15). This new current is called *diffusion current*.

To analyze the diffusion phenomenon, we refer to Fig. 4.23 and consider the thermal motion of particles (electrons or holes) in semiconductors where a gradient exists in carrier concentration instead of temperature. If v_x is the x component of thermal velocity of particles, the number of particles moving across the boundary line *BB* from the left to the right per unit time per unit area is $n_x v_x/2$, while the corresponding number from the right to the left is $n_1 v_x/2$, where n_2 and n_1 are the concentrations of particles to the left and to the right of *BB*, respectively. The net rate of flow of particles per unit time per unit area is

$$S_x = \frac{v_x(n_2 - n_1)}{2} \tag{7.1.1}$$

The factor $\frac{1}{2}$ accounts for the fact that the particles can move either toward or away from the boundary. Only those moving toward the boundary are counted. Since only particles within one mean free path l on each side of the boundary are able to move across the line BB, the difference or $n_2 - n_1$ in Eq. (7.1.1) is evaluated within such a distance, or $n_1 = n_0 + (\partial n/\partial x)l$ and $n_2 = n_0 - (\partial n/\partial x)l$. Thus Eq. (7.1.1) becomes

$$S_x = -v_x \frac{\partial n}{\partial x} l \qquad (7.1.2)$$

If the particles under consideration are charged, the thermal motion of these particles constitutes a current flow, the density of which is given by

$$J_x = -eS_x = ev_x l \frac{\partial n}{\partial x} = \frac{ekT\tau}{m^*} \frac{\partial n}{\partial x} = eD \frac{\partial n}{\partial x} \qquad (7.1.3)$$

The proportionality constant D in the relation between current density and concentration gradient is called the *diffusion constant*. In nondegenerate semiconductors, a relation called the *Einstein relation* exists:

$$D = \frac{kT}{m^*} \tau = \frac{kT}{e} \mu \qquad (7.1.4)$$

between the diffusion constant D and the mobility μ of free carriers. In deriving Eq. (7.1.3), we have assumed a one-dimensional case and used the relations $l = v_x \tau$ and $\langle v_x^2 \rangle = v_{th}^2 = kT/m^*$, where v_{th} is the thermal velocity of carriers and the symbol $< \ >$ indicates a statistical average. For a three-dimensional case, the factor $\frac{1}{2}$ in Eq. (7.1.1) is replaced by $\frac{1}{6}$ because of the six possible directions. For nondegenerate semiconductors, $m^* v_{th}^2/2 = 3kT/2$ by equipartition theorem. Therefore, the Einstein relation applies only to carrier distributions that either obey or can be approximated by the Boltzmann distribution.

Combining the drift (conduction) and diffusion currents, we find the total electron current density to be

$$J_e = e\mu_n nE + eD_n \frac{\partial n}{\partial x} \qquad (7.1.5)$$

A similar equation is obtained for the total hole current density by substituting μ_p for μ_n and $-D_p$ for D_n. In the base region of a transistor, minority carriers are injected at the emitter junction and drained at the collector junction. The flow of the injected minority carriers from the emitter junction to the collector junction is governed by the diffusion process. To have a complete description of the dynamic behavior of a semiconductor device, we need to know not only the spatial variation but also the time variation of free carrier concentrations. In the following sections we discuss the physical processes that influence the time and space variations of free carriers. This discussion is followed by the development of the time-dependent diffusion equation that governs the motion of excess carriers.

7.2 THE THERMAL EQUILIBRIUM CONDITION

Many electronic devices are made of two different types of materials. For example, a simple semiconductor-junction diode has n-type material on one side and p-type material on the other. Semiconductors can also be used as thermoelectric elements to convert thermal energy into electric energy. In such applications, what happens at the semiconductor–metal contact is important. The first step toward understanding a junction made of two different materials is to study the junction under thermal equilibrium.

First, let us examine what we mean by thermal equilibrium. Consider a homogeneous semiconductor in which the impurity concentration is uniform. Under the condition of no electric field and no temperature gradient [i.e., $E_x = 0$ and $\partial T/\partial x = 0$ (and hence $\partial n/\partial x = 0$)], electrons can migrate freely from one part to another part of a semiconductor without gaining or losing their energy. This state of free migration is known as *thermal equilibrium*. Under thermal equilibrium, there are as many electrons migrating away from one region as migrating to the same region; hence J_e should be identically zero. For $E = 0$ and $\partial n/\partial x = 0$, the current density J_e is indeed zero, as predicted from Eq. (7.1.5). This condition $J = 0$ may be considered as the macroscopic definition of thermal equilibrium.

Let us apply this macroscopic condition to a *p-n* junction. Figure 7.1a shows the joining of two uniform semiconductors of the same material but of different types. Clearly, the electron concentration n_n on the n side is higher than that n_p on the p side; similarly, $p_p > p_n$. Since there must be a transition region in which $\partial n/\partial x$ and $\partial p/\partial x$ are nonzero, an electric field E must also exist in the transition region so that both J_e and J_h are identically zero. This transition region is indicated in Fig. 7.1a by a gradual change of band edges. However, what is shown in Fig. 7.1a may not represent exactly the thermal-equilibrium situation.

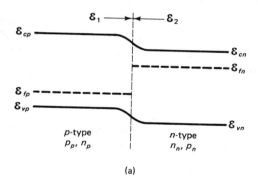

(a)

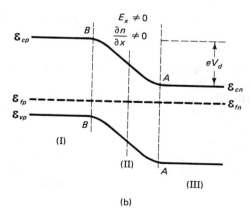

(b)

Figure 7.1 (a) Energy-band diagram illustrating the situation which arises from joining *n*-type and *p*-type semiconductors together. What happens near the boundary region is a matter of special interest. (b) Under thermal equilibrium, the Fermi levels on the two sides are equal. This condition requires an adjustment of the potential energies so that a potential difference is established. The electrostatic potential must change in such a manner that the drift and diffusion currents exactly balance out each other in Eq. (7.1.5).

To find the thermal-equilibrium situation in a composite system of which *p-n* junctions and semiconductor–metal contacts are examples, we must consider the Fermi levels. A statement is often made that in a system under thermal equilibrium, the Fermi level $\mathscr{E}_f$ stays constant throughout the entire system. The same applies to a composite system, or $\mathscr{E}_{fn} = \mathscr{E}_{fp}$. This condition can be obtained by counting electron flow across the junction. Consider electrons having energies $\mathscr{E}_2$ and $\mathscr{E}_1$ and bordering, respectively, on the *n* side and *p* side of the junction, as shown in Fig. 7.1a. For simplicity, a spherical energy surface with an effective mass m^* is assumed. The number of electrons dN moving in a time interval dt from the *p* side toward the *n* side in an area $dy\,dz$ is equal to $nv_{x1}\,dy\,dz\,dt$, and hence is given by

$$dN = \left[2\left(\frac{m^*}{h}\right)^3 \frac{dv_{x1}\,dv_{y1}\,dv_{z1}}{1 + \exp\,(\mathscr{E}_1 - \mathscr{E}_{fp})/kT} \right] v_{x1}\,dt\,dy\,dz \qquad (7.2.1)$$

In Eq. (7.2.1), v_{x1} is the x component of the velocity of the electrons and the term in brackets represents the density of occupied electrons in the conduction band. A similar expression is obtained for the corresponding number of electrons moving from the *n* side toward the *p* side.

Detailed balancing requires that the number of electrons from $\mathscr{E}_1$ to $\mathscr{E}_2$ must be equal to that from $\mathscr{E}_2$ to $\mathscr{E}_1$ under thermal equilibrium. This means that

$$[1 - f(\mathscr{E}_2)] \frac{v_{x1}\,dv_{x1}\,dv_{y1}\,dv_{z1}}{1 + \exp\,(\mathscr{E}_1 - \mathscr{E}_{fp})/kT} = \frac{v_{x2}\,dv_{x2}\,dv_{y2}\,dv_{z2}}{1 + \exp\,(\mathscr{E}_2 - \mathscr{E}_{fn})/kT} [1 - f(\mathscr{E}_1)] \qquad (7.2.2)$$

The factors $[1 - f(\mathscr{E}_2)]$ and $[1 - f(\mathscr{E}_1)]$ are added on account of the exclusion principle. Out of dN electrons expressed in Eq. (7.2.1), only $[1 - f(\mathscr{E}_2)]\,dN$ electrons are accepted by the *n* side. To simplify Eq. (7.2.2), we make the following observations. The energy of an electron must be conserved in crossing the boundary (i.e., $\mathscr{E}_1 = \mathscr{E}_2$), or

$$\mathscr{E}_{cp} + \frac{m^*}{2}\,(v_{x1}^2 + v_{y1}^2 + v_{z1}^2) = \mathscr{E}_{cn} + \frac{m^*}{2}\,(v_{x2}^2 + v_{y2}^2 + v_{z2}^2) \qquad (7.2.3)$$

The potential energies $\mathscr{E}_{cn}$ and $\mathscr{E}_{cp}$ are equal at the boundary. Also, the velocity of an electron remains the same in the y and z directions $v_{y1} = v_{y2}$ and $v_{z1} = v_{z2}$. Thus we have

$$v_{x1}\,dv_{x1}\,dv_{y1}\,dv_{z1} = v_{x2}\,dv_{x2}\,dv_{y2}\,dv_{z2} \qquad (7.2.4)$$

In view of $\mathscr{E}_1 = \mathscr{E}_2$ stated in Eq. (7.2.3), use of Eq. (7.2.4) in Eq. (7.2.2) yields

$$\mathscr{E}_{fp} = \mathscr{E}_{fn} \qquad (7.2.5)$$

Equation (7.2.5) is the same as Eq. (4.6.6) applied to a *p-n* junction, and results in a band-edge alignment, shown in Fig. 7.1b.

Let us examine whether Eq. (7.2.5) is consistent with the macroscopic condition $J_e = 0$ for thermal equilibrium. It is obvious from Fig. 7.1b that in order to make $\mathscr{E}_{fp} = \mathscr{E}_{fn}$, a difference in the potential energy results such that $eV_d = \mathscr{E}_{cp} - \mathscr{E}_{cn}$. This potential difference is established across a transition region in which there exists an internal electric field E_x. This transition is shown as region II in Fig. 7.1b. In the transition region, both E and dn/dx in Eq. (7.1.5) are nonzero. Under thermal equilibrium, the drift and diffusion currents should cancel out, yielding

$$\frac{-eE_x\,dx}{kT} = \frac{dn}{n} \qquad (7.2.6)$$

For ease of discussion, it is assumed that the transition region ends abruptly at points A and B in Fig. 7.1b, or in other words, $E_x = 0$ for $x < x_B$ and $x > x_A$. Justification for the use of such an idealized model will be given in Chapter 8, where a more detailed treatment of a p-n junction will be made. Integrating Eq. (7.2.6) from A to B gives

$$\frac{n_{p0}}{n_{n0}} = \exp \int_A^B \frac{dn}{n} = \exp\left(\frac{-eV_d}{kT}\right) \qquad (7.2.7)$$

where n_{p0} and n_{n0} are the equilibrium concentrations of electrons in the bulk p and n regions, respectively.

The potential difference $V_d = (\mathcal{E}_{cp} - \mathcal{E}_{cn})/e$ in Eq. (7.2.7) is generally referred to as the *built-in* or *diffusion potential* in a p-n junction. To prove that the macroscopic condition $J_e = 0$ is satisfied in the junction of Fig. 7.1b, we need to show that Eq. (7.2.7) is indeed true. Substituting Eq. (6.3.8) into Eq. (7.2.7) gives

$$(\mathcal{E}_{cp} - \mathcal{E}_{fp}) - (\mathcal{E}_{cn} - \mathcal{E}_{fn}) = eV_d \qquad (7.2.8)$$

or $\mathcal{E}_{fp} = \mathcal{E}_{fn}$, which is the same as Eq. (7.2.5). Therefore, the thermal equilibrium condition $\mathcal{E}_{fp} = \mathcal{E}_{fn}$ is the same as demanding zero macroscopic current. The reader is asked to show that an equation similar to Eq. (7.2.5) and leading to the same result of Eq. (7.2.8) is obtained for valence-band holes.

7.3 EXCESS CARRIERS AND QUASI-FERMI LEVELS

From discussions in the preceding section it is clear that for a nonvanishing J_e in a p-n junction, the Fermi levels $\mathcal{E}_{fn}$ and $\mathcal{E}_{fp}$ must be different. This situation happens when an external voltage V_a is applied to the junction as shown in Fig. 7.2. Since the difference in the potential energy $\mathcal{E}_{cp} - \mathcal{E}_{cn}$ now becomes $e(V_d - V_a)$, this new potential difference should replace eV_d in Eq. (7.2.5). Thus the minority carrier concentration on the p-side under the applied voltage is given by

$$n_p = n_n \exp\left[-\frac{e(V_d - V_a)}{kT}\right] = (n_p)_0 \exp\left(\frac{eV_a}{kT}\right) \qquad (7.3.1)$$

In Eq. (7.3.1) the majority-carrier concentration is assumed to remain the same, or, in other words, $n_n = n_{n0}$. We shall justify the use of Eq. (7.3.1) or rather the assumptions leading to Eq. (7.3.1) in Chapter 8 when we discuss the theory concerning a p-n junc-

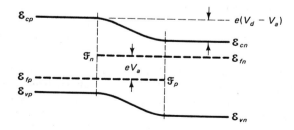

Figure 7.2 A p-n junction under a forward bias. When a voltage V_a is applied across a semiconductor junction, the two Fermi levels are separated by eV_a. The quasi-Fermi levels $\mathcal{F}_n$ and $\mathcal{F}_p$ are used in place of the respective Fermi level for semiconductors under a nonequilibrium situation.

tion. Accepting Eq. (7.3.1), we find that an excess minority carrier concentration $n_1 = n_p - n_{p0}$ is created near the junction such that

$$n_1 = n_p - n_{p0} = n_{p0} \left[\exp \left(\frac{eV_a}{kT} \right) - 1 \right] \qquad (7.3.2)$$

This phenomenon is called *minority-carrier injection*.

There are other means by which a nonequilibrium situation can be created so that an excess carrier concentration exists. These include the generation of electron–hole pairs by light (photons) and by energetic charged particles. Many semiconductor devices, among which transistors and photodetectors are good examples, have the useful information carried by excess carriers. In other words, an incoming signal creates an excess carrier concentration, which in turn causes an output voltage or current. In this chapter we develop tools for analyzing the behavior of excess carriers. The two questions immediately facing us are: How are the excess carriers distributed among available states in the bands? Do the excess carriers affect the charge neutrality inside a semiconductor? These two questions are analyzed separately in the remainder of this section and in Section 7.4.

In Section 4.5 we discussed the Fermi–Dirac distribution function, and in Section 6.3 we used the distribution function to derive the concentrations of free electrons and holes. The important point relevant to our present discussion is that we used a single Fermi level in both concentrations. We can do this only if the valence band and the conduction band are in thermal equilibrium. When excess carriers are present, the conduction and valence bands are no longer in thermal equilibrium; therefore, two different Fermi levels must be used, one for the conduction band and another for the valence band. However, a state of equilibrium still prevails among energy states of the same kind. If there are excess electrons in the conduction band, we cannot distinguish these electrons from the electrons that already existed in the conduction band under thermal equilibrium conditions. This means that the occupancy of conduction-band states can still be described by a single Fermi–Dirac distribution with the same Fermi level for all conduction-band electrons.

In summary, we treat the nonequilibrium situation due to the existence of excess carriers by proposing several Fermi levels, one for the conduction-band states, one for the donor states, and so on. These Fermi levels are called the *quasi-Fermi level*. In terms of the quasi-Fermi level $\mathscr{F}_n$ for conduction-band electrons, the distribution of electrons in the conduction band can be written as

$$f_c(\mathscr{E}) = \frac{1}{1 + \exp (\mathscr{E} - \mathscr{F}_n)/kT} \qquad (7.3.3)$$

and the concentration of conduction-band electrons as

$$n = n_0 + n_1 = \int_{\mathscr{E}_c}^{\infty} D_c(\mathscr{E}) f_c(\mathscr{E}) \, d\mathscr{E} = N_c \exp \left(-\frac{\mathscr{E}_c - \mathscr{F}_n}{kT} \right) \qquad (7.3.4)$$

where $D_c(\mathscr{E}) \, d\mathscr{E}$ is the density of conduction-band states in an energy interval $d\mathscr{E}$ and $N_c = 2(2\pi m^* kT/h^2)^{3/2}$. Similar expressions are obtained for valence-band states. We again emphasize that Eq. (7.3.4) is physically meaningful only if a state of equilibrium prevails among electrons in the conduction band. Otherwise, it would not be possible to describe the distribution of electrons by a single quasi-Fermi level $\mathscr{F}_n$ in Eq. (7.3.3).

Let us illustrate the use of quasi-Fermi level in the analysis of a nonequilibrium situation. In a p-n junction, if it is assumed that $\mathscr{F}_n$ runs from $\mathscr{E}_{fn}$ on the n side straight to the p side as shown in Fig. 7.2, then

$$n_p = N_c \exp \left(-\frac{\mathscr{E}_c - \mathscr{F}_n}{kT} \right) = n_{p0} \exp \left(\frac{eV_a}{kT} \right) \tag{7.3.5}$$

yielding the same result as Eq. (7.3.1). As will be shown in Chapter 8, the current density J_e of Eq. (7.1.5) can also be expressed in terms of $\mathscr{F}_n$.

The concept of quasi-Fermi level can also be applied to discrete states. Suppose that there exists in the forbidden gap a certain kind of energy state called an *electron trap* which can be in either of two possible states, the occupied and the unoccupied. When occupied, the electron-trap state traps an electron and is negatively charged. The ratio of occupied trap states to the total trap states can be written as

$$\frac{n_t}{N_t} = \frac{N_t(\text{occupied})}{N_t} = f_t(\mathscr{E}_t) = \frac{1}{1 + \exp [(\mathscr{E}_t - \mathscr{F}_t)/kT]} \tag{7.3.6}$$

where $\mathscr{E}_t$ is the trap energy and $\mathscr{F}_t$ the quasi-Fermi level for traps. The fact that two different quasi-Fermi levels $\mathscr{F}_n$ and $\mathscr{F}_t$ are used in Eqs. (7.3.5) and (7.3.6) simply means that the distribution of excess electrons between the conduction-band and trap states will not be the same as the thermal equilibrium distributions of electrons.

In the following example we show that under a nonequilibrium situation quasi-Fermi levels are different for different energy states. The electron and hole concentrations under thermal equilibrium are, respectively, given by

$$n_0 = N_c \exp \left(-\frac{\mathscr{E}_c - \mathscr{E}_f}{kT} \right) \tag{7.3.7a}$$

$$p_0 = N_v \exp \left(-\frac{\mathscr{E}_f - \mathscr{E}_v}{kT} \right) \tag{7.3.7b}$$

Suppose that a light beam generates n_1 electron–hole pairs per unit volume by lifting electrons from the valence band into the conduction band. As a result, the new electron and hole concentrations, respectively, become

$$n = n_0 + n_1 = N_c \exp \left(-\frac{\mathscr{E}_c - \mathscr{F}_n}{kT} \right) \tag{7.3.8a}$$

$$p = p_0 + p_1 = N_v \exp \left(-\frac{\mathscr{F}_p - \mathscr{E}_v}{kT} \right) \tag{7.3.8b}$$

with $p_1 = n_1$. Since $n > n_0$ and $p > p_0$, $\mathscr{E}_c - \mathscr{F}_n < \mathscr{E}_c - \mathscr{E}_f$ and $\mathscr{F}_p - \mathscr{E}_v < \mathscr{E}_f - \mathscr{E}_v$. In other words, the quasi-Fermi level for electrons and that for holes must move in opposite directions, as shown in Fig. 7.3.

In conclusion, the various expressions derived for thermal equilibrium conditions such as Eq. (7.3.7) are still useful under a nonequilibrium situation if the respective quasi-Fermi level is used in the place of the Fermi level. The fact that a single quasi-Fermi level is assumed for all electrons in Eq. (7.3.8) presupposes that electrons in the conduction band are in thermodynamic equilibrium among themselves. Therefore, the use of Eqs. (7.3.6) and (7.3.8) implies that a state of thermodynamic equilibrium exists

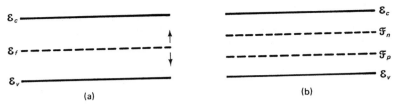

Figure 7.3 (a) A single Fermi level for a semiconductor under thermal equilibrium. (b) Two different quasi-Fermi levels, one for electrons and the other for holes, in a nonequilibrium situation.

among particles in each individual species (electrons in the traps, for example). On the other hand, the use of different quasi-Fermi levels ($\mathcal{F}_t$ for traps, $\mathcal{F}_n$ for conduction-band electrons, and $\mathcal{F}_p$ for valence-band holes) in Eqs. (7.3.6) and (7.3.8) is necessary because a state of nonequilibrium exists for particles of different species.

7.4 THE CHARGE-NEUTRALITY CONDITION

Associated with a charge density ρ is an electric field $\mathbf{E}$. The relation between $\mathbf{E}$ and ρ is governed by *Poisson's equation*,

$$\nabla \cdot \mathbf{E} = \frac{\rho}{\epsilon} \qquad (7.4.1)$$

To appreciate the consequence of a nonzero ρ, we examine the hypothetical situation shown in Fig. 7.4. Suppose that in region AB there existed a nonzero negative ρ_0, and in region CD there existed a nonzero positive ρ_0. The electric field created by this charge distribution can be calculated from Eq. (7.4.1) and is also plotted in Fig. 7.4.

If d is the distance between A and B and between C and D, the electric field in region BC is

$$E = -\frac{\rho_0 d}{\epsilon} \qquad (7.4.2)$$

For $\rho_0 = 1.6 \times 10^{-9}$ C/cm^3, $d = 10^{-1}$ cm, and $\epsilon = 1.04 \times 10^{-12}$ F/cm (a value taken to be the dielectric constant in Si), we find a value of $E = 154$ V/cm. In a material of resistivity, 1 Ω-cm, this electric field would create a current of density $J = 154$ A/cm^2, which is a fairly large number. What would happen is that the field E of Eq. (7.4.2) would push carriers of opposite polarity to move toward each other. Ultimately, the two charge distributions would land on top of each other, resulting in

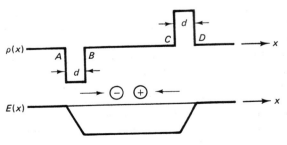

Figure 7.4 A hypothetical situation showing a negative space-charge region AB and a positive space-charge region CD. The electric field $E(x)$ is the field produced by the space charges. Because of $E(x)$, charges of opposite polarities are attracted toward each other (as indicated by arrows), and the space charges collapse very rapidly.

$\rho = \rho_0 - \rho_0 = 0$. Since the total amount of charge involved in $\rho_0 d = 1.6 \times 10^{-10}$ C/cm^2, it would only take a time $t = \rho_0 d/J = 1.04 \times 10^{-12}$ to reach the ultimate situation. This means that with the values chosen for the case, the situation shown in Fig. 7.4 could have lasted only for a time of the order of 10^{-12} s.

To analyze the situation properly, we must use the continuity equation

$$\nabla \cdot \mathbf{J} = -\frac{\partial \rho}{\partial t} \tag{7.4.3}$$

Using $\mathbf{J} = \sigma \mathbf{E}$ in Eq. (7.4.3) and eliminating $\nabla \cdot \mathbf{E}$ from Eq. (7.4.1), we obtain

$$\frac{\partial \rho}{\partial t} = -\frac{\sigma}{\epsilon} \rho \tag{7.4.4}$$

where σ is the conductivity of the medium. Equation (7.4.4) says that if there existed a charge density ρ, the charge would be swept away by the field E due to the charge's own presence with a time constant ϵ/σ. This time constant ϵ/σ is called the *dielectric relaxation time*.

We should point out that here we want to consider only the effect produced by E; hence only the drift term $\sigma \mathbf{E}$, but not the diffusion term, is included in the expression of J. The transient behavior governed by the diffusion process of excess carriers is discussed in Section 7.7. As the calculations (given below and in Section 7.7) show, the characteristic time involved in a diffusion process is much longer than the dielectric relaxation time associated with the drift process. In other words, the drift current density due to the electric field E created by a space-charge density ρ is normally much larger, by several orders of magnitude, than the diffusion current density. This justifies the use of $J = \sigma \mathbf{E}$ in analyzing the effect produced by space charges.

Now let us apply the analysis above to the nonequilibrium situation involving excess carriers. Equation (7.4.4) shows that any excess carrier density, n_1 or p_1, if it exists alone, can only last for a time on the order of the dielectric relaxation time ϵ/σ. Take $\epsilon = 11.7\epsilon_0 = 1.04 \times 10^{-12}$ F/cm, $\sigma = 1$ (Ω-cm)$^{-1}$, $\epsilon/\sigma = 1.04 \times 10^{-12}$ s. The values of ϵ and σ are chosen because most semiconductors used for device applications have values in the same range as these values. Therefore, unless we are dealing with transient phenomena of time duration shorter than 10^{-12} s or treating insulators with σ many orders of magnitude smaller than 1 (Ω-cm)$^{-1}$, the charge neutrality condition $\rho = 0$ can be assumed. The charge-neutrality condition does not apply, of course, to the space-charge region of a semiconductor junction such as the one shown as region AB in Fig. 7.1b. In semi-insulating materials such as high-resistivity CdS, the charge-neutrality condition can also be violated for a time short compared with the dielectric relaxation time. Later, in Chapter 13, we discuss a novel class of semiconductor devices known as Gunn oscillators. It is the setting up of space-charge waves that makes these devices capable of generating coherent electromagnetic waves in the microwave region.

Summarizing the discussion above, we conclude that the charge-neutrality condition is expected to be observed in the bulk semiconductor, but not in the junction region, when we deal with low-frequency phenomena whose period is longer than the dielectric relaxation time. In this chapter we deal exclusively with such low-frequency phenomena. Consider a semiconductor that has electron traps besides donors and acceptors. Applying the charge-neutrality condition to such a case, we have

$$n_0 + N_a^- + N_t^- = p_0 + N_d^+ \tag{7.4.5}$$

Sec. 7.4   The Charge-Neutrality Condition

where N_t^- represents the thermal-equilibrium density of negatively charged trap states. In the presence of excess carriers, the excess electrons will be shared by the conduction-band and electron-trap states. Thus Eq. (7.4.5) becomes

$$n_0 + n_1 + N_a^- + N_t^- + n_t = p_0 + p_1 + N_d^+ \qquad (7.4.6)$$

where n_1 and p_1 are the excess conduction-electron and valence-hole concentrations, respectively, and n_t is the concentration of excess electrons captured by trap states. In Eqs. (7.4.5) and (7.4.6) it is assumed that the donor and acceptor states are almost completely ionized and hence N_a^- and N_d^+ are not affected by excess carriers. Subtracting Eq. (7.4.5) from Eq. (7.4.6) gives

$$n_1 + n_t = p_1 \qquad (7.4.7)$$

A similar expression $n_1 = p_1 + p_t$ applies to semiconductors having hole traps.

In using Eq. (7.4.7), we must distinguish two types of traps. A temporary storage trap captures only one type of carrier, and hence either n_t or p_t is nonzero. A recombination trap acts as an intermediary for the recombination of excess carriers by simultaneously capturing an electron from the conduction band and a hole from the valence band. Hence for materials possessing recombination centers only, $n_1 + n_t = p_1 + p_t$ and $n_t = p_t$, yielding

$$n_1 = p_1 \qquad (7.4.8)$$

Of course, Eq. (7.4.8) also applies to a situation in which no traps whatsoever are involved. Equation (7.4.8) applies to germanium and to silicon at room temperature. We should point out that the definition adopted here for temporary-storage and recombination traps is highly idealized. In materials such as CdS and other II–VI compounds where traps are many and are very important, the demarcation line between the two species may not be so clear-cut. For such cases, we must use the general formula

$$n_1 + n_t = p_1 + p_t \qquad (7.4.9)$$

because Eq. (7.4.8) is no longer valid. An analysis of such cases is analytically involved and rather tedious and hence will not be given here.

7.5 CARRIER RECOMBINATION PROCESSES

In a semiconductor, electrons and holes are constantly generated by thermal excitation of electrons from the valence band to the conduction band. This generation process is counterbalanced, under thermal equilibrium, by a recombination process in which electrons and holes annihilate each other. When excess carriers are present, the recombination process outweighs the generation process. As a result, more free carriers are lost through recombination than are gained through generation. It is the purpose of this section to analyze the recombination processes and thus to obtain the time rate of change of the excess-carrier concentrations.

There are two basic processes by which electrons and holes may recombine with each other. In the first process, electrons in the conduction band make direct transitions to vacant states in the valence band. In the second process, electrons and holes recombine through intermediary states known as *recombination centers*. The recombination centers are usually impurities and lattice imperfections of some sort. These two processes are illustrated in Fig. 7.5. It is clear that the first process is an inherent property

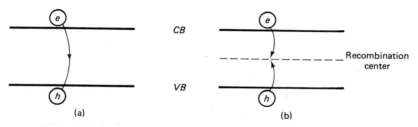

Figure 7.5 Schematic illustration showing two basic recombination processes: (a) the intrinsic process by which electrons make a direct band-to-band transition, and (b) the extrinsic process by which electrons and holes recombine through an intermediary state known as a recombination center. The intrinsic process is an inherent property of a given semiconductor, whereas the extrinsic process depends on the nature of the recombination center.

of the semiconductor itself, while the second process depends very much on the nature of the impurity or the lattice imperfection. Therefore, we refer to the first and second recombination processes, respectively, as the *intrinsic* and *extrinsic recombination processes.*

The Intrinsic Recombination Process (Band-to-Band Transition)

A proper treatment of the recombination process must be based on quantum-mechanical transitions between two different band states. Here, however, the process will be treated on a phenomenological basis only. Since the recombination process of Fig. 7.5a involves both the occupied states in the conduction band and the vacant states in the valence band, the total rate of recombination is given by

$$R = rnp \qquad (7.5.1)$$

as for any bimolecular process. In Eq. (7.5.1), r is a constant of proportionality. Opposite to the recombination process, there is a constant generation of carriers. Under thermal equilibrium, the generation rate G must be equal to the recombination rate R, that is, $G = rn_0p_0$, where n_0 and p_0 are, respectively, the thermal-equilibrium electron and hole concentration. The net rate of recombination, U, is equal to

$$U = R - G = r(np - n_0p_0) \qquad (7.5.2)$$

If n_1 and p_1 are the excess carrier concentrations such that $n = n_0 + n_1$ and $p = p_0 + p_1$, the charge-neutrality condition requires that $n_1 = p_1$. Under the condition that $n_1, p_1 \ll (n_0 + p_0)$, Eq. (7.5.2) becomes

$$\frac{dn_1}{dt} = -U = -r(p_0 + n_0)n_1 \qquad (7.5.3)$$

In the macroscopic description of the recombination process, a lifetime τ_0 is often so defined that

$$\frac{dn_1}{dt} = -\frac{n_1}{\tau_0} \qquad (7.5.4)$$

A comparison of Eq. (7.5.3) with Eq. (7.5.4) gives

$$\tau_0 = \frac{1}{r(n_0 + p_0)} \qquad (7.5.5)$$

The value of r, and thus the value of τ_0 in Eq. (7.5.5), depends, to a great extent, on the energy-band structure of a semiconductor. The transition of an electron between bands is subject to the quantum-mechanical selection rule that the wave vector $\mathbf{k}$ of an electron must be conserved [Eq. (6.9.3)]. In a direct-gap semiconductor such as GaAs, conduction-band electrons can make direct quantum-mechanical transitions to vacant valence-band states. This direct transition that results in the emission of light is shown as process II in Fig. 7.6a. In an indirect-gap semiconductor, such as Ge, because the conduction-band electrons and valence-band holes have different values of k in the reciprocal lattice, electrons must be first scattered by lattice vibrations to an intermediate state A having a proper value of k (process I) and then make a transition from A to a vacant valence-band state (process II). The recombination processes presented in Fig. 7.6a and b are, respectively, called the *direct* and *indirect radiative recombination processes*.

The lifetimes of excess carriers associated with the direct and indirect band-to-band transitions are drastically different, by several orders of magnitude. For the direct process, the value of r, and hence the value of τ_0, can be estimated from the spontaneous lifetime which can be calculated quantum mechanically and will be defined shortly. Note that if n_1 or p_1 is comparable to or greater than $(n_0 + p_0)$, Eq. (7.5.5) must be replaced by

$$\tau_0 = \frac{1}{r(n_0 + p_0 + n_1)} \qquad (7.5.6)$$

Thus τ_0 decreases with increasing excess carrier concentration n_1. However, this decrease does not continue indefinitely. An ultimate situation is reached if for every occupied conduction-band state, there is a corresponding vacant valence-band state, and vice versa. This condition is fulfilled if n and p in Eq. (7.5.2) are so large that both the

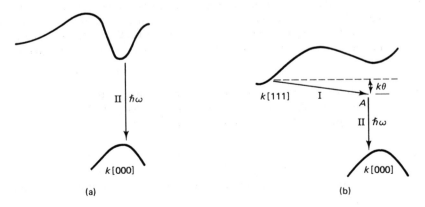

(a)   (b)

Figure 7.6 Diagram showing the intrinsic recombination process (a) in a direct-gap semiconductor and (b) in an indirect-gap semiconductor. The process shown in (b) requires phonon participation for momentum conservation.

conduction and valence bands become degenerate. Under such circumstances, the lifetime τ_0 reaches an ultimate lower limit. For the direct radiative recombination process of Fig. 7.6a, this lower limit is called the *spontaneous lifetime* and is of the order of 10^{-10} s in typical semiconductors such as GaAs. This value is shorter than a typical value of 10^{-8} s for atomic transitions because of the differences in dielectric constants and effective masses for electrons in the crystal lattice and in the isolated atom.

Now let us make some estimate of the intrinsic lifetime of excess carriers. In direct-gap semiconductors, if we take 10^{-10} s as the value of the spontaneous lifetime at a degenerate carrier concentration of $n_0 + p_0 + n_1 = 10^{18}$ cm^{-3}, then from Eq. (7.5.6) the value of r is found to be 10^{-8}. Using this value of r in Eq. (7.5.5), we find a value of $\tau_0 = 10^{-6}$ s at $(n_0 + p_0) = 10^{14}$ cm^{-3}. The situation in indirect gap semiconductors is quite different. Because the indirect (phonon-assisted) radiative recombination process of Fig. 7.6b is a two-step process, the probability of its occurrence is greatly reduced. Using optical absorption data, Van Roosbroeck and Shockley and later Dumke have concluded that the value of τ_0 is of the order of 0.5 s in intrinsic germanium (W. Van Roosbroeck and W. Shockley, *Phys. Rev.*, Vol. 94, p. 1558, 1954; W. P. Dumke, *Phys. Rev.*, Vol. 105, p. 139, 1957). Similar estimates made for silicon also give a value of lifetime of the order of seconds. Experimentally measured value of lifetime in Ge and Si ranges from 10^{-3} s in quality-controlled crystals to 10^{-7} s in crystals having an appreciable concentration of undesirable impurities. We conclude that the band-to-band transition does not play an important role for the recombinaton of excess carriers in indirect-gap semiconductors such as Ge and Si. The extrinsic recombination process that takes place via impurities is the dominant process in such materials. In direct-gap semiconductors such as GaAs, however, both the band-to-band transition and the recombination process via impurities should be considered.

The Extrinsic Recombination Process (Recombination via Impurities)

Next, we discuss the recombination process through lattice imperfections, mostly due to impurities having energy levels deep in the forbidden gap. This type of recombination process was originally proposed by Shockley and Read and independently suggested by Hall (W. Shockley and W. T. Read, *Phys. Rev.*, Vol. 87, p. 835, 1952; R. N. Hall, *Phys. Rev.*, Vol. 87, p. 387, 1952). Referring to Fig. 7.7, a recombination center (or trap) can be in either one of two possible states, the occupied and the unoccupied. In its unoccupied state, it is ready to receive an electron from either the conduction band (process 1) or the valence band (process 4). In its occupied state, it can give its electron to either the conduction band (process 2) or the valence band (process 3). For process 1, the rate of capturing an electron is proportional to the concentration of the unoccupied trap states and also to that of the occupied conduction-band states; hence the rate can be written as

$$U_1 = C_n(N_t - n_t)n \qquad (7.5.7)$$

where N_t is the concentration of recombination traps, n_t the concentration of occupied traps, and C_n the proportionality constant for the electron-capturing process. The rate of the reverse process (or releasing electrons to the conduction band by occupied traps) is equal to

$$\cdot U_1' = E_n n_t \qquad (7.5.8)$$

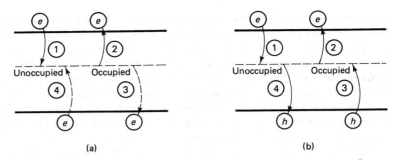

Figure 7.7 Extrinsic recombination process through recombination centers: (a) in terms of capturing electrons from conduction band (process 1) and valence band (process 4) by unoccupied centers, and releasing electrons to conduction band (process 2) and valence band (process 3) by occupied centers; and (b) in terms of capturing electrons from conduction band (process 1) and releasing holes to valence band (process 4) by unoccupied centers, and releasing electrons to conduction band (process 2) and capturing holes from valence band (process 3) by occupied centers.

where E_n is the proportionality constant for the electron-emission process. We should point out that the rate of electron emission actually should read $E_n'(N - n)n_t$, where N is the concentration of total conduction-band states. Since n is always insignificantly small compared to N, we can neglect n in $N - n$ and replace $E_n'N$ by E_n.

The net rate of capturing electrons from the conduction band is therefore equal to

$$U_{cn} = C_n(N_t - n_t)n - E_n n_t \qquad (7.5.9)$$

Under thermal equilibrium, detailed balancing requires $U_{cn} = 0$; thus Eq. (7.5.9) gives

$$E_n = C_n n_0 \left(\frac{N_t - n_t}{n_t} \right)_0 \qquad (7.5.10)$$

The subscript 0 refers to concentrations under thermal equilibrium. When excess carriers are present, U_{cn} of Eq. (7.5.9) is no longer zero. Substituting E_n from Eq. (7.5.10) into Eq. (7.5.9), we have

$$U_{cn} = C_n[(N_t - n_t)n - n_t n'] \qquad (7.5.11)$$

where the quantity n' is defined as

$$n' = n_0 \left(\frac{N_t - n_t}{n_t} \right)_0 = N_c \exp\left(-\frac{\mathscr{E}_c - \mathscr{E}_t}{kT} \right) \qquad (7.5.12)$$

The expression for the rate U_{cp} of hole capture by the traps (processes 3 and 4 of Fig. 7.7b) can be obtained in a similar manner, and is given by

$$U_{cp} = C_p[n_t p - (N_t - n_t)p'] \qquad (7.5.13)$$

where the quantity p' is defined as

$$p' = p_0 \left(\frac{n_t}{N_t - n_t} \right)_0 = N_v \exp\left(-\frac{\mathscr{E}_t - \mathscr{E}_v}{kT} \right) \qquad (7.5.14)$$

In the steady state, the net rate of capture of electrons must be equal to that of holes. Equating $U_{cn} = U_{cp}$ gives

$$\frac{n_t}{N_t} = \frac{C_n n + C_p p'}{(p + p')C_p + (n + n')C_n} \tag{7.5.15}$$

Substituting Eq. (7.5.15) back into either Eq. (7.5.11) or Eq. (7.5.13), we obtain

$$U = U_{cn} = \frac{N_t C_n C_p (pn - p'n')}{C_p(p + p') + C_n(n + n')} \tag{7.5.16}$$

Following the same procedure as used in obtaining Eq. (7.5.5) from Eqs. (7.5.2) and (7.5.3), and noting that $p'n' = p_0 n_0$, we find that

$$\tau_0 = \frac{\tau_{n0}(p_0 + p') + \tau_{p0}(n_0 + n')}{p_0 + n_0} \tag{7.5.17}$$

under the condition that the excess carrier concentrations n_1 and p_1 are much smaller than the equilibrium carrier concentrations n_0 and p_0. In Eq. (7.5.17), $\tau_{n0} = 1/C_n N_t$ and $\tau_{p0} = 1/C_p N_t$.

In the early development of transistor technology, it was found that the current amplification factor, that is, the ratio of the collector current to the emitter current, was seriously limited by a low lifetime. As shown in Chapter 9, because of carrier recombination in the base region of a transistor, part of the current injected by the emitter is diverted to the base terminal. The ratio of this current (lost to the base) to the emitter current is approximately equal to $W^2/(2D\tau_0)$, where W is the width of the base region and D is the diffusion constant. Before the modern technique for making diffused transistors was developed, the value of W was limited to around 1 mil or 2.5×10^{-3} cm. For $D = 50$ cm^2/V-s and $\tau_0 = 10^{-6}$ s, the loss would amount to 12.5%. It was technically important, therefore, to understand the recombination process so that the origins of the recombination centers could be traced.

To analyze the behavior of the carrier lifetime, we plot in Fig. 7.8 log τ_0 (solid curve) as a function of the Fermi level $\mathscr{E}_f$ for a given trap density N_t and trap energy $\mathscr{E}_t$. The trap energy is taken to be in the lower half of the forbidden gap and C_p is assumed to be much larger than C_n. Since p_0 and n_0 are functions of $\mathscr{E}_f$, as is the resistivity of a given sample, Fig. 7.8 is in effect also a plot of log τ_0 against resistivity. As we move from the left toward the right (i.e., $\mathscr{E}_f$ moves from $\mathscr{E}_v$ to $\mathscr{E}_c$ in Fig. 7.8), we are dealing with samples of different resistivities and different types, from low-resistivity p type, to high resistivity p type, to near-intrinsic, to high-resistivity n type, and finally, to low-resistivity n-type.

In view of Eqs. (6.5.14), (6.5.15), (7.5.12) and (7.5.14), the following inequalities hold:

$$\begin{aligned} p_0 > n_0 \quad &\text{for} \quad \mathscr{E}_f < \mathscr{E}_{fi} \\ p_0 < n_0 \quad &\text{for} \quad \mathscr{E}_f > \mathscr{E}_{fi} \end{aligned} \tag{7.5.18}$$

and

$$\begin{aligned} n_0 < n', \quad p_0 > p' \quad &\text{for} \quad \mathscr{E}_f < \mathscr{E}_t \\ n_0 > n', \quad p_0 < p' \quad &\text{for} \quad \mathscr{E}_f > \mathscr{E}_t \end{aligned} \tag{7.5.19}$$

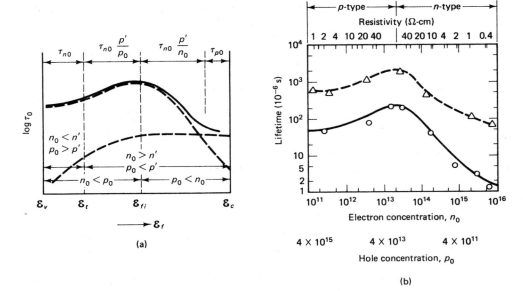

Figure 7.8 Variation of the lifetime of excess carriers as a function of (a) the Fermi level and (b) the carrier concentration. The curves in (a) and (b) are theoretical curves based on Eq. (7.5.17). In both cases, the trap energy $\mathscr{E}_t$ is taken to be in the lower half of the forbidden gap and C_p is assumed to be much larger than C_n. The experimentally measured values of lifetime in germanium are indicated by open triangles and open circles in (b). (G. Bemski, "Recombination in Semiconductors" *Proc. IEEE*, Vol. 46, p. 990, 1958 © 1958 IEEE.)

According to which are the dominating terms in Eq. (7.5.17), we divide Fig. 7.8 into four regions. In each region the lifetime can be approximated by the appropriate expression given in the figure. When $\mathscr{E}_f$ is very close to $\mathscr{E}_v$, the sample is almost degenerate *p*-type and p_0 is the dominating term in Eq. (7.5.17). Thus we have $\tau_0 = \tau_{n0} = 1/(C_n N_t)$. Physically, the sample is flooded with holes, so the lifetime is controlled by the ability of the trap to capture electrons from the conduction band. The opposite is true when $\mathscr{E}_f$ is close to $\mathscr{E}_c$. If $\mathscr{E}_f$ gradually moves away from $\mathscr{E}_v$, soon the term $\tau_{n0}p'$ catches up with $\tau_{n0}p_0$ and hence τ_0 can be approximated by $\tau_{n0}p'/p_0$. The rise of τ_0 in the center portion of Fig. 7.8 is due to scarcity of free carriers, n_0 and p_0. We should point out that Fig. 7.8 assumes that $C_p \gg C_n$. In other words, it is much harder for traps to capture electrons than to capture holes. Therefore, in the middle range, the lifetime is limited by the ability of traps to capture electrons as indicated by the $\tau_{n0}p'$ dependence of τ_0 in Fig. 7.8.

The reader may be confused by the use of two parameters τ_{n0} and τ_{p0} in Eq. (7.5.17) as meaning two separate lifetimes, one for excess electrons and the other for excess holes. In a given sample, there is only one lifetime—the one given by Eq. (7.5.17) for both excess electrons and holes. The value of the lifetime τ_0 may change, however, as we vary the resistivity of the sample even if we keep the trap density N_t constant. The two parameters τ_{n0} and τ_{p0} are used in Eq. (7.5.17) in order that we can describe the variation of τ_0 with n_0 and p_0 of the sample.

Figure 7.8b shows the measured values of lifetime in germanium as functions of carrier concentration (lower scale) and sample resistivity (upper scale). For the dashed curve that is obtained in highly purified Ge samples, we use the following values in Eq. (7.5.17): $\tau_{n0} = 500 \times 10^{-6}$ s, $\tau_{p0} = 50 \times 10^{-6}$ s, $p' = 1.6 \times 10^{14}$ cm^{-3}, and $n' = 3.2 \times 10^{12}$ cm^{-3}. Several values of τ_0 calculated for a selected number of resistivity values are as follows:

Resistivity (Ω − cm)	0.5 p type	5 p type	47 intrinsic	5 n type	0.5 n type
p_0 (cm^{-3})	6.5×10^{15}	6.5×10^{14}	2.3×10^{13}	1.7×10^{12}	1.7×10^{11}
n_0 (cm^{-3})	8.0×10^{10}	8.0×10^{11}	2.3×10^{13}	3.1×10^{14}	3.1×10^{15}
τ_0 (s)	510×10^{-6}	620×10^{-6}	2000×10^{-6}	310×10^{-6}	76×10^{-6}

These calculated values indeed agree well with the measured values. In the calculation, we have assumed the same trap density N_t for all samples since we used the same set of values for τ_{n0} and τ_{p0}. The trap is presumably due to residual impurities left in the sample during purification or due to lattice imperfections. Because all samples used in the experiment underwent the same crystal growing and purification process, it is reasonable to assume that they have the same N_t.

The solid curve of Fig. 7.8b is obtained in Ge samples doped with the same concentration of nickel. In these samples, N_t is known to be the same. Note that the curve also has the shape of the curve predicted from Eq. (7.5.17) and shown in Fig. 7.8a. The analysis of the dashed curve is left to the reader as a problem.

For comparison of the ratio of capturing electrons and holes by different recombination centers, the quantities τ_{n0} and τ_{p0} are often expressed in terms of the capture cross section σ such that

$$\tau_{n0} = \frac{1}{N_t \sigma_n v_{\text{th}}}, \qquad \tau_{p0} = \frac{1}{N_t \sigma_p v_{\text{th}}} \qquad (7.5.20)$$

where $v_{\text{th}} = (3kT/m^*)^{1/2}$ is the thermal velocity of free carriers and is about 10^7 cm/s at 300 K. The room-temperature capture cross sections of a selected number of impurities are given in Table 7.1 for both Ge and Si. Take nickel in Ge as an example. In a p-type sample, $\sigma_n = 8 \times 10^{-17}$ cm^2. For a Ni concentration of $N_t = 1.2 \times 10^{13}$ cm^{-3}, $\tau_{n0} = 100 \times 10^{-6}$ s. Since N_t is smaller than the intrinsic carrier concentration, the amount of nickel is not expected to change the resistivity of the sample. However, the

TABLE 7.1 CAPTURE CROSS SECTIONS OF A SELECTED NUMBER OF IMPURITIES IN SILICON AND GERMANIUM

Material:	Silicon		Germanium		
Impurity:	Fe	Au	Cu	Ni	Au
σ_n (cm^2)	$> 1.5 \times 10^{-15}$	3.5×10^{-15}	1×10^{-17}	8×10^{-17}	2×10^{-14}
σ_p (cm^2)	3×10^{-16}	5×10^{-15}	1×10^{-16}	$> 4 \times 10^{-15}$	

lifetime of the carriers is lowered from a value of 500×10^{-6} s (dashed curve for highly purified samples, Fig. 7.8b) to 100×10^{-6} s. For gold, which has a higher capture cross section, the amount needed to shorten the lifetime is even smaller than that for Ni.

Besides the dependence of carrier concentrations, Eq. (7.5.17) also predicts the temperature dependence of τ_0. Consider the case shown in Fig. 7.8a, where the trap density $\mathscr{E}_t$ lies in the lower half of the forbidden gap, and $\tau_{n0} \gg \tau_{p0}$. In a p-type sample, the term $\tau_{n0}(p_0 + p')$ is the dominant term in the numerator, whereas the term p_0 is the dominant term in the denominator of Eq. (7.5.17). Thus Eq. (7.5.17) can be approximated by

$$\tau_0 = \tau_{n0} + \tau_{n0} \frac{p'}{p_0} \qquad (7.5.21)$$

From room temperature (300 K) down to liquid nitrogen temperature (77 K), the majority-carrier concentration p_0 stays nearly constant in moderately doped materials. In this temperature range, the lifetime τ_0 varies as $\tau_{n0}p'/p_0$ or $(\tau_{n0}N_v/p_0) \exp{[-(\mathscr{E}_t - \mathscr{E}_v)/kT]}$. From the temperature variation of τ_0, the energy $\mathscr{E}_t$ of the recombination center can be found, and in turn the type of impurity can be identified. We discuss the experimental determination of lifetime in Sections 7.9 and 7.11.

Of the impurities investigated, copper and nickel are known to be lifetime shorteners in germanium, and gold in silicon. To be an effective recombination center, an impurity must have large values of C_n, C_p, and N_t. The impurities mentioned above meet these conditions. Not only do they readily capture electrons and holes (large C_n and C_p), but these metals also diffuse rapidly and have fairly large solid solubilities in germanium and silicon (large N_t). Besides impurities, lattice imperfections generated in germanium and silicon during bombardment by high-energy particles such as fast neutrons, gamma rays, and high-energy electrons, which create effective recombination centers having energy levels deep in the forbidden gap. Experimental evidences about these recombinaton centers are discussed in Section 7.11.

In summary, two different mechanisms for carrier recombination have been discussed: one due to band-to-band transition and the other through an intermediary state known as the *recombination center*. In indirect-gap semiconductors such as germanium and silicon, since the rate for band-to-band transition is slow, the lifetime of excess carriers is determined almost solely by the amount and nature of impurities and lattice imperfections present in the semiconductor. To be an effective recombination center, an impurity or a lattice imperfection must have its energy level deep in the forbidden gap, that is, away from either band edge. For same τ_{p0}, τ_{n0}, p_0, and n_0 in Eq. (7.5.17), τ_0 has a lower value if both the values of p' and n' are lower. In cases where the energy level is close to one band edge, either p' of Eq. (7.5.14) or n' of Eq. (7.5.12) will become very large because of the ratio $n_{t0}/(N_t - n_t)_0$. Physically, a recombination center plays an effective role only if it readily receives both types of carriers. This means that the ratio of the occupied traps (states ready to receive holes) to the unoccupied traps (states ready to receive electrons) should be neither too large nor too small. To minimize both n' and p', we must have $n' = p' = n_i$. This fact explains why most effective recombination centers have energy levels deep in the forbidden gap (i.e., close to the intrinsic Fermi level $\mathscr{E}_{fi}$).

In concluding this section, we discuss briefly the role of carrier recombination in affecting device performance. As mentioned earlier, the loss factor in a transistor is equal to $W^2/2D\tau_0$. By careful preparation and purification of Ge and Si crystals, a lifetime on the order of several milliseconds in Ge and on the order of 100 μs in Si can be obtained. Since the base width W can be made very small by modern techniques, carrier recombination does not seem to be a serious factor affecting the efficiency of a transistor. For semiconductors to be used as photoconductors, a longer lifetime means a higher photoconductance and hence a higher sensitivity (Section 7.11). However, in many applications, a fast response is required. For such applications, the lifetime must be made short even at the expense of sensitivity. The same applies to a switching transistor. Normally, we want a long lifetime in the collector junction in order to keep the reverse saturation current low. For fast switching, however, the carrier lifetime must be shortened. Finally, we should mention that a GaAs (direct-gap semiconductor) diode emits electromagnetic radiation very efficiently when it is biased in the forward direction. It is the recombination of carriers through band-to-band transition that causes the emission of light. We shall discuss more thoroughly the role of recombination processes when we treat these semiconductor devices in later chapters.

7.6 THE CONTINUITY EQUATION AND THE TIME-DEPENDENT DIFFUSION EQUATIONS

In many applications of semiconductor devices and in measurements of transport properties of semiconductors, electric pulses and time-varying signals are used. The purpose of our present discussion is to develop a set of general equations which are capable of describing the behavior of free carriers under time-varying as well as dc (constant) excitations. Since most of our past discussions started with electrons, this time we speak first of holes. The hole current density, similar to Eq. (7.1.5), is given by

$$J_h = -eD_p \frac{\partial p}{\partial x} + e\mu_p pE_x \tag{7.6.1}$$

The negative sign in front of the diffusion term accounts for the fact that carriers move in a direction toward lower concentration. Thus a negative hole concentration gradient constitutes a positive hole current. The opposite is true for electron diffusion. A comparison of the hole-diffusion current with the electron diffusion current is made in Fig. 7.9.

The equation that relates the time variation of charge density to current density is the continuity equation. Consider a one-dimensional flow of current as shown in Fig. 7.10. The number of holes entering, minus those leaving a box of length dx and cross-sectional area A in a time interval Δt, is given by

$$\frac{A}{e}[J_h(x) - J_h(x + dx)]\Delta t = -\frac{\partial J_h}{\partial x}dx\frac{A}{e}\Delta t \tag{7.6.2}$$

The net increase in the number of holes inside the box is equal to the number supplied by the current minus the number lost through recombination, or mathematically,

$$\Delta p\, A\, dx = -\frac{\partial J_h}{\partial x}\frac{A}{e}dx\,\Delta t - \frac{\Delta t}{\tau_0}p_1 A\, dx \tag{7.6.3}$$

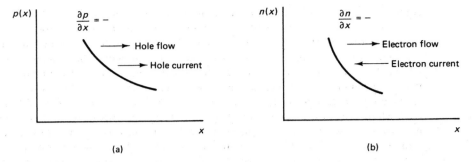

Figure 7.9 Diagrams showing that (a) a negative hole concentration gradient constitutes a positive hole diffusion current but (b) a negative electron concentration gradient constitutes a negative electron diffusion current.

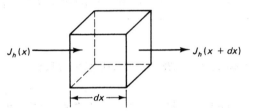

Figure 7.10 Diagram used in establishing the hole continuity equation by considering the time rate of change in the number of holes in an elementary box.

Canceling $A\,dx$ on both sides and letting $\Delta t \to 0$, we find that

$$\frac{\partial p}{\partial t} = -\frac{1}{e}\frac{\partial J_h}{\partial x} - \frac{p_1}{\tau_0} \qquad (7.6.4)$$

In Eq. (7.6.4), p_1 is the concentration of excess holes, and $p = p_0 + p_1$ where p_0 is the equilibrium hole concentration. Equation (7.6.4) is known as the *continuity equation*. In cases where no excess carriers are involved, the second term (recombination term) on the right side of Eq. (7.6.4) is zero. In the latter form, the continuity equation appears in many other places, for example in the electromagnetic theory. In semiconductors, however, the recombination terms must be included when we are dealing with excess carriers.

Substituting Eq. (7.6.1) into Eq. (7.6.4) gives

$$\frac{\partial p}{\partial t} = D_p \frac{\partial^2 p}{\partial x^2} - \frac{\partial}{\partial x}(\mu_p p E_x) - \frac{p_1}{\tau_0} \qquad (7.6.5)$$

The corresponding equation for electrons,

$$\frac{\partial n}{\partial t} = D_n \frac{\partial^2 n}{\partial x^2} + \frac{\partial}{\partial x}(\mu_n n E_x) - \frac{n_1}{\tau_0} \qquad (7.6.6)$$

can be obtained in a similar manner. Equations (7.6.5) and (7.6.6) are known as the time-dependent diffusion equations for holes and electrons, respectively. To help simplify the two equations above, the following observations are made. For most applications, the thermal-equilibrium concentrations n_0 and p_0 are independent of x and t; hence n and p can be replaced by n_1 and p_1, respectively, in terms involving differentiations with respect to x or t. Thus Eqs. (7.6.5) and (7.6.6), respectively, become

Transport and Recombination of Excess Carriers   Chap. 7

$$\frac{\partial p_1}{\partial t} = D_p \frac{\partial^2 p_1}{\partial x^2} - \mu_p p \frac{\partial E_x}{\partial x} - \mu_p E_x \frac{\partial p_1}{\partial x} - \frac{p_1}{\tau_0} \tag{7.6.7}$$

$$\frac{\partial n_1}{\partial t} = D_n \frac{\partial^2 n_1}{\partial x^2} + \mu_n n \frac{\partial E_x}{\partial x} + \mu_n E_x \frac{\partial n_1}{\partial x} - \frac{n_1}{\tau_0} \tag{7.6.8}$$

In treating Eqs. (7.6.7) and (7.6.8), we have to be careful about the terms involving the electric field E_x. In many electrical problems, we equate E_x to the field created by an applied voltage. We can do so only for media free of electric charges. In the presence of space charges, as in the case that we are presently treating, we must include the field created by the space charges. From Poisson's equation we obtain the relation

$$\frac{\partial E_x}{\partial x} = \frac{e(p_1 - n_1)}{\epsilon} \tag{7.6.9}$$

for the one-dimensional case we are discussing.

To help us visualize what can be done about Eqs. (7.6.7) to (7.6.9), we review our discussion in Section 7.4 on the charge-neutrality condition. We showed that if two charge distributions of opposite polarities were separated from each other in space, the two distributions would move toward each other and finally come together in a time of the order of the dielectric relaxation time. Any deviation from the charge-neutrality condition must be small and localized. If the deviation were large and extensive, the electric field E_x created by the space charge itself would be large enough to sweep the space charge out of existence. Therefore, in Eqs. (7.6.7) and (7.6.8), E_x must be largely due to externally applied voltage, but the significance of $\partial E_x/\partial x$ is yet to be determined.

Now let us concentrate on the terms involving $\partial E_x/\partial x$. We see that $\partial E_x/\partial x$ cannot be identically zero because if $p_1 = n_1$, the behavior of p_1 and n_1 should be described by a single equation, not by two different equations. The fact that Eqs. (7.6.7) and (7.6.8) are different makes it necessary to modify the charge-neutrality condition. We say that p_1 is approximately equal to n_1 but $p_1 - n_1$ is nonzero. We shall later show that the ratio $(p_1 - n_1)/n_1$ is indeed very small. Therefore, in Eqs. (7.6.7) and (7.6.8), we set $p_1 = n_1$ but retain the terms involving $\partial E_x/\partial x$. To eliminate the $\partial E_x/\partial x$ terms, we multiply Eq. (7.6.7) by $\mu_n n$ and Eq. (7.6.8) by $\mu_p p$, and add them. Thus we obtain

$$\frac{\partial n_1}{\partial t} = D' \frac{\partial^2 n_1}{\partial x^2} + \mu' E_x \frac{\partial n_1}{\partial x} - \frac{n_1}{\tau_0} \tag{7.6.10}$$

In Eq. (7.6.10), the quantities D' and μ' are called the *effective diffusion constant* and *effective mobility of excess carriers*, and are related to D_n, D_p, μ_n, and μ_p as follows:

$$D' = \frac{D_n D_p (n + p)}{D_p p + D_n n} \tag{7.6.11}$$

$$\mu' = \frac{\mu_n \mu_p (p - n)}{\mu_p p + \mu_n n} \tag{7.6.12}$$

Since D' and μ' are functions of p and n and hence are functions of n_1, Eq. (7.6.10) is nonlinear. However, for $p = p_0 + p_1 >> n = n_0 + n_1$,

$$D' = D_n, \qquad \mu' = \mu_n \tag{7.6.13}$$

and for $p = p_0 + p_1 \ll n = n_0 + n_1$,

$$D' = D_p, \qquad \mu' = -\mu_p \qquad (7.6.14)$$

Therefore, for extrinsic materials under low-level injection, Eq. (7.6.10) becomes linearized, and the coefficients D' and μ', respectively, take the values of the diffusion constant and the mobility of minority carriers. By *low-level injection* we mean injection under the condition that the injected excess-carrier concentration is much smaller than the equilibrium majority-carrier concentration. When the excess-carrier concentration is comparable to or exceeds the equilibrium majority-carrier concentration, Eq. (7.6.10) is nonlinear. Nonlinear effects as a result of high-level injection have been observed. We also should point out that the electric field E_x in Eq. (7.6.10) now stands for the externally applied field because we have already eliminated the term $\partial E_x/\partial x$ created by the space charge.

To estimate the value of $p_1 - n_1$, we subtract Eq. (7.6.8) from Eq. (7.6.7). Letting $p_1 - n_1 = n_1'$ and using Eq. (7.6.9) to eliminate $\partial E_x/\partial x$, we find that

$$\frac{\partial n_1'}{\partial t} + e(\mu_n n + \mu_p p)\frac{n_1'}{\epsilon} + \frac{n_1'}{\tau_0} = (D_p - D_n)\frac{\partial^2 n_1}{\partial x^2} - (\mu_n + \mu_p)E_x\frac{\partial n_1}{\partial x} \qquad (7.6.15)$$

In obtaining Eq. (7.6.15), we set $p_1 = n_1$ in Eq. (7.6.7) for terms involving $\partial^2 p_1/\partial x^2$ and $\partial p_1/\partial x$. Knowing the solution of n_1 from Eq. (7.6.10), the value of n_1' can be calculated from Eq. (7.6.15). However, we shall not do this. Instead, we estimate the value of n_1' by making the following observations. In Eq. (7.6.15), the relative magnitudes of the terms can be expressed in terms of three time constants: t_1 (the duration of a pulse or the period of a sine wave), t_2 ($= \epsilon/\sigma$, the dielectric relaxation time), and t_3 ($= \tau_0$, the lifetime).

On the left side of Eq. (7.6.15), the first term $\partial n_1'/\partial t$ is of the order n_1'/t_1, the second term is on the order of n_1'/t_2, and the third term is n_1'/t_3. Obviously, the second term n_1'/t_1 is the dominating term for pulses of nanosecond or longer duration. On the right side of Eq. (7.6.15), the value of the terms is on the order of $\partial n_1/\partial t$ or n_1/τ_0, whichever is larger, as can be seen from Eq. (7.6.10). Therefore, the ratio of n_1'/n_1 is on the order of t_2/t_1, where $t_2(= \epsilon/\sigma)$ is the dielectric relaxation time and t_1 is the pulse duration or the lifetime, whichever is smaller. Take $t_1 = 10^{-8}$ s for the pulse duration, $t_2 = 10^{-12}$ s for the dielectric relaxation time, and $t_3 = 10^{-5}$ s for the lifetime. Then the ratio of n_1'/n_1 is estimated to be the order of $10^{-12}/10^{-8} = 10^{-4}$.

In summary, we have shown that the deviation n_1' from charge neutrality is indeed much smaller, by several orders of magnitude, than the excess-carrier concentration n_1. Insofar as the movement of excess carriers is concerned, the motion is adequately described by the behavior of the charge-neutrality part n_1. Therefore, in the following discussion, we shall set $p_1 = n_1$ and use Eq. (7.6.10) to describe the motion of excess carriers. Note that Eq. (7.6.10) is different from Eqs. (7.6.7) and (7.6.8). The part n_1', however small it may be as compared to n_1, creates a large enough local variation in the field to account for a nonzero $\partial E_x/\partial x$ in Eqs. (7.6.7) and (7.6.8). It is the influence of the $\partial E_x/\partial x$ term which links p_1 and n_1 together so that their motion can be described by a single equation, Eq. (7.6.10). The difference between Eqs. (7.6.7) and (7.6.8), on the one hand, and Eq. (7.6.10), on the other, is that the term $\partial E_x/\partial x$ created by the space charge has been eliminated in Eq. (7.6.10). Therefore, the three equations may look different but do not contradict one another.

7.7 DISCUSSION OF THE SOLUTIONS OF THE TIME-DEPENDENT DIFFUSION EQUATION

In this section we discuss the solutions of the time-dependent diffusion equation, and actually find the specific solution for a number of simple physical systems. Consider an n-type sample under low-level injection. Use of Eq. (7.6.14) and the relation $p_1 = n_1$ in Eq. (7.6.10) gives

$$\frac{\partial p_1}{\partial t} = D_p \frac{\partial^2 p_1}{\partial x^2} - \mu_p E_x \frac{\partial p_1}{\partial x} - \frac{p_1}{\tau_0} \tag{7.7.1}$$

Figure 7.11 shows an experimental setup for measuring the lifetime, in which the excess carriers excited by a light beam are collected at a point C. The contact at C can be either a p-n junction or a metal–semiconductor contact. The carriers collected at C create a current through the resistor R. Thus the voltage V_R developed across R is directly proportional to the excess-carrier concentration at point C.

Let us comment briefly on the bias condition of the collector contact C. Since both excess electrons and holes exist at the point C (i.e., $n_1 = p_1 \neq 0$ at C, it appeared at first that the collector contact C could be biased either positive (to collect electrons) or negative (to collect holes) with respect to the sample. However, this is not true. In Fig. 7.12 the equilibrium carrier concentrations n_0 and p_0 in 0.5-Ω–cm n-type Ge and Si are shown ($n_0 = 3.1 \times 10^{15}$ cm^{-3} and $p_0 = 1.7 \times 10^{11}$ cm^{-3} in Ge, and $n_0 \doteq 9.2 \times 10^{15}$ cm^{-3} and $p_0 = 2.3 \times 10^4$ cm^{-3} in Si). Let us take the Ge sample, and suppose that an excess carrier concentration $n_1 = p_1 = 5 \times 10^{13}$ cm^{-3}. The addition of n_1 hardly changes the majority-carrier concentration (from 3.10×10^{15} to $3.15 \times$

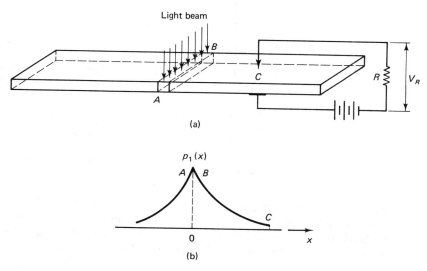

(a)

(b)

Figure 7.11 (a) Schematic diagram showing an experimental arrangement for measuring lifetime. (b) Under a steady illumination of the portion AB of the sample, excess carriers are generated and a steady-state distribution of excess carriers is established. This distribution is shown as the solid curve. At point C, the excess minority carriers are collected, and the current thus generated develops a voltage across R.

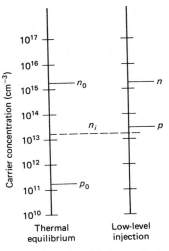

(a) Germanium: 0.5 Ω-cm, n-type,
$p_1 = n_1 = 5 \times 10^{13}$ cm^{-3}

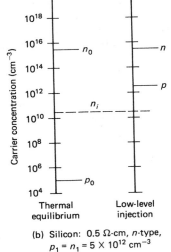

(b) Silicon: 0.5 Ω-cm, n-type,
$p_1 = n_1 = 5 \times 10^{12}$ cm^{-3}

Figure 7.12 A comparison of the carrier concentrations under low-level injection (scale on the right) and the equilibrium carrier concentrations (scale on the left) in (a) Ge and (b) Si. Existence of excess carriers hardly changes the majority-carrier (electron in the illustration) concentration, but increases drastically the minority-carrier concentration.

10^{15} cm^{-3}), whereas the addition of $p_1 (= n_1)$ increases drastically the minority-carrier concentration (from 1.7×10^{11} to 5.02×10^{13} cm^{-3}). Therefore, the collector contact C should be biased in such a way as to detect the change in the minority-carrier concentration. In an n-type sample, the point C should be biased negative, as shown in Fig. 7.11.

Now let us return to Eq. (7.7.1), and discuss its steady-state solution. Under a constant excitation (e.g., the excitation by a light beam of constant intensity), the term $\partial p_1/\partial t$ is zero when the steady state is reached. In cases where no external field E_x is applied, Eq. (7.7.1) becomes

$$D_p \frac{\partial^2 p_1}{\partial x^2} - \frac{p_1}{\tau_0} = 0 \tag{7.7.2}$$

The solution of Eq. (7.7.2) is given by

$$p_1 = A \exp\left(\frac{-x}{L_p}\right) + B \exp\left(\frac{x}{L_p}\right) \tag{7.7.3}$$

where the quantity L_p defined as

$$L_p = (D_p \tau_0)^{1/2} \tag{7.7.4}$$

is the *diffusion length* of holes.

First, we apply the solution of Eq. (7.7.3) to a sample of infinite length. At $x = \pm\infty$, the excess-carrier concentration must decay to zero because of the recombination process. Thus one of the coefficients, either A or B, must be zero. Let p_{10} be the excess-hole concentration at $x = 0$, where the light beam is shone. Then we have

$$p_1 = \begin{cases} p_{10} \exp\left(\frac{-x}{L_p}\right) & \text{for } x > 0 \tag{7.7.5a} \\[2em] p_{10} \exp\left(\frac{x}{L_p}\right) & \text{for } x < 0 \tag{7.7.5b} \end{cases}$$

where x is the distance between the point C and the plane AB. The variation of p_1 as a function of x is shown in Fig. 7.11b. Since the voltage V_R is proportional to p_1, a plot of $\ln V_R$ against x gives the value of L_p and in turn the value of the lifetime τ_0. For a sample of finite length, Eq. (7.7.5) is still valid if the length of the sample is longer than the diffusion length by a factor of 10 or more. As an example, we take $D_p = 12.3$ cm^2/V-s, which is the value in Si. For $\tau_0 = 50 \times 10^{-6}$ s, $L_p = 2.48 \times 10^{-2}$ cm. Therefore, the condition for the validity of Eq. (7.7.5) can easily be satisfied.

Next we consider another simple case in which a semiconductor slice is uniformly illuminated with a light beam so that there is a uniform rate G (per unit time per unit volume) of generation of electron–hole pairs throughout the slice. Since the generation of excess carriers is uniform, the terms $\partial^2 p_1/\partial x^2$ and $\partial p_1/\partial x$ are zero in Eq. (7.7.1). Thus Eq. (7.7.1) becomes

$$\frac{\partial p_1}{\partial t} = -\frac{p_1}{\tau_0} + G \tag{7.7.6}$$

where the term G accounts for the generation process. The solution of Eq. (7.7.6) is given by

$$p_1 = G\tau_0\left[1 - \exp\left(\frac{-t}{\tau_0}\right)\right] \tag{7.7.7}$$

If the light beam is turned off, p_1 will decay exponentially to zero with a time constant τ_0. The situation is very much like the charging and discharging of a capacitor in an RC circuit.

Now let us discuss the solution of Eq. (7.7.1) for a general case in which $p_1(x, t)$ is a function of both x and t. Let $f(x, t)$ be a source function that represents the injection of excess carriers. In the presence of a source, Eq. (7.7.1) reads

$$\frac{\partial p_1}{\partial t} = D_p \frac{\partial^2 p_1}{\partial x^2} - \mu_p E_x \frac{\partial p_1}{\partial x} - \frac{p_1}{\tau_0} + f(x, t) \qquad (7.7.8)$$

The term $f(x, t)$ corresponds to the term G in Eq. (7.7.6) except that the generation rate now is a function of both x and t.

Equation (7.7.8) may look quite formidable to those who have not been exposed to the partial differential equation. The first step in making our discussion simple enough and yet useful is to take a transformation of space and time coordinates so that Eq. (7.7.8) can be readily recognized. Using a moving-coordinate system such that

$$x_1 = x - \mu_p E_x t \quad \text{and} \quad t_1 = t \qquad (7.7.9)$$

and letting

$$p_1 = p' \exp\left(\frac{-t}{\tau_0}\right) \qquad (7.7.10)$$

we can rewrite Eq. (7.7.8) as

$$\frac{\partial p'}{\partial t_1} = D_p \frac{\partial^2 p'}{\partial x_1^2} + f(x_1, t_1) \qquad (7.7.11)$$

Partial differential equations of the form given by Eq. (7.7.11) are used in many engineering applications. Perhaps the best-known equation is the heat-conduction equation:

$$\frac{\partial T}{\partial t} = \frac{K}{\rho C} \frac{\partial^2 T}{\partial x^2} \qquad (7.7.12)$$

where $T(x, t)$ is the local temperature in a solid, K the thermal conductivity, ρ the density, and C the specific heat of the solid. The equation of heat conduction is exactly the same in mathematical form as the time-dependent diffusion equation except that the dependent physical variable is the temperature in heat conduction against the excess-carrier concentration in Eq. (7.7.11). In deriving the continuity equation, consideration is given to the conservation of heat-energy flux instead of the conservation of electric charges. Since the solutions of many problems were known in the theory of heat conduction long before semiconductor theory was developed, the reader is well advised to consult a book on heat conduction before tackling the time-dependent diffusion equation. Chances are good that an analogous problem has already been solved in the theory of heat conduction (see, for example, H. S. Carslaw and J. C. Yaeger, *Conduction of Heat in Solids,* Oxford University Press, Oxford, 1959).

An important solution to the time-dependent diffusion equation is the response to a unit pulse. Let us consider a simple case in which the unit pulse is injected into a semiconductor infinite in extent and having no initial excess carriers, that is, $p_1(t = 0) = 0$. For an infinite semiconductor, $p_1(x = \pm\infty)$ must be zero. The condition $p_1(t = 0) = 0$ is called the *initial condition* (specified in the time domain) and the condition $p_1(x = \pm\infty) = 0$ is called the *boundary condition* (specified in the space domain). The initial and boundary conditions are needed to define a physical problem uniquely. For a unit pulse injected at $x = 0$ at $t = 0$, $f(x_1, t_1) = \delta(x)\delta(t)$, where $\delta(x)$ and $\delta(t)$ are Dirac delta functions defined in Appendix A7. The solution of Eq. (7.7.11) under these conditions is discussed in Appendix A7 and is given by Eq. (A7.18):

$$p_1(x, t) = \frac{1}{(4\pi D_p t)^{1/2}} \exp\left(-\frac{x^2}{4D_p t}\right) \tag{7.7.13}$$

The function expressed in Eq. (7.7.13) is known as the *Gaussian distribution*.

Let us examine the behavior based on Eq. (7.7.13) of the excess-carrier concentration p_1. For simplicity, we assume that $E_x = 0$ and $\tau_0 = \infty$. Thus $x_1 = x$, $t_1 = t$, and $p_1(x, t) = p'(x_1, t_1)$. The case of a nonzero E_x and a finite τ_0 is discussed in Section 7.8. Figure 7.13 shows the injection of excess carriers (e.g., through excitation by photons from a light beam) at $x = 0$ and $t = 0$, as indicated by arrows. At $t = 0$, the excess-carrier concentration is zero everywhere except at $x = 0$ (Fig. 7.13a). In Eq. (7.7.13), because the exponential term goes to zero much faster than the $t^{-1/2}$ term goes to infinity, the product of these two terms is zero. The situation is different, however, at the point $x = 0$.

The definite integral

$$\int_{-\infty}^{\infty} \exp\left(-a\xi^2\right) d\xi = 2 \int_{0}^{\infty} \exp\left(-a\xi^2\right) d\xi = \sqrt{\frac{\pi}{a}} \tag{7.7.14}$$

is useful toward understanding the situation at $x = 0$. Applying Eq. (7.7.14) to Eq. (7.7.13), we find that

$$\int_{-\infty}^{\infty} p_1(x, t)\, dx = 1 \tag{7.7.15}$$

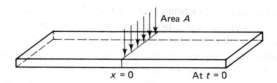

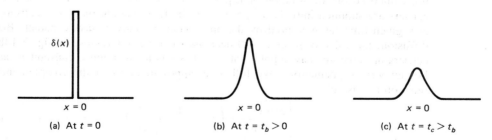

(a) At $t = 0$ (b) At $t = t_b > 0$ (c) At $t = t_c > t_b$

Figure 7.13 Diagrams showing the response to an impulse excitation. At $t = 0$, the semiconductor is subject to a pulse injected at $x = 0$. The pulse lasts only for an infinitely short duration, but injects a finite amount of excess carriers. The curves in (a), (b), and (c) show the time variation of the excess-carrier distribution inside the semiconductor. The excess-carrier distribution spreads as a result of diffusion.

Since $p_1(x, t)$ is zero everywhere except at $x = 0$, the function must go to infinity at $x = 0$. The function $p_1(x, t)$ is indeed a Dirac delta function $\delta(x)$ at $t = 0$ (Fig. 7.13). As time progresses, the distribution broadens on account of diffusion. Since the excess carriers disperse but do not disappear during the diffusion process, the amount of carriers injected at $t = 0$ is conserved. In other words, the area under the three curves in Fig. 7.13 must be the same as required by Eq. (7.7.15).

We should point out that Eq. (7.6.10) and hence Eq. (7.7.11) are based on a one-dimensional flow of excess carriers; therefore, Eq. (7.7.13) applies only to a long sample having a uniform distribution of excess carriers over the transverse plane. A factor N/A should appear in front on the right-hand side of Eq. (7.7.13) if the source injects N holes at $x = 0$ and $t = 0$ into the sample over a cross-sectional area A.

Another case of importance is one in which the excess-carrier concentration $p_1(x, t)$ is everywhere zero for $t < 0$ and a source is applied at $x = 0$ for $t > 0$ so as to maintain unity concentration at $x = 0$. Again we assume the electric field to be negligibly small and the lifetime to be infinitely long, so that the second and third terms on the right-hand side of Eq. (7.7.8) can be ignored. Under these conditions the solution of Eq. (7.7.11) is given by Eq. (A7.23):

$$p_1(x, t) = \operatorname{erfc}\left(\frac{x}{\sqrt{4D_p t}}\right) = \frac{2}{\sqrt{\pi}} \int_y^\infty \exp\left(-\xi^2\right) d\xi \qquad (7.7.16)$$

where $y = x/\sqrt{4D_p t}$. The function expressed in Eq. (7.7.16) is called the *complementary error function*. At $x = 0$, Eq. (7.7.16) reduces to a definite integral. Using Eq. (7.7.14), we find that $p_1(0, t) = 1$ for all $t > 0$. This is the condition imposed on the solution of Eq. (7.7.11). The reader is referred to Appendix A7 for the derivation of Eq. (7.7.16) and also for the solution of a modified case in which the lifetime τ_0 is finite.

In Fig. 7.14 we plot the Gaussian distribution and the complementary error function against space coordinate x for two selected values of time $\sqrt{4Dt}$. To give us the numerical value of the spread of the excess-carrier concentration, we use $D_p = 12.3$ cm^2/V-s, that is, the value in Si. Qualitatively, we see that the value of p_1 is appreciable, as compared to its value at $x = 0$, for a distance of $\sqrt{4D_p t}$. Thus at 5×10^{-7} s, the excess-carrier concentration p_1 spreads over a distance of about 5×10^{-3} cm on either side of $x = 0$. The two curves are different because the boundary conditions imposed on the solution of Eq. (7.7.11) are different. In Fig. 7.14a, the excess carriers are supplied initially at $t = 0$, and the total amount of excess carriers existing at a given time (as measured by the area under the curve) stays constant. Because of diffusion, the value of p_1 at $x = 0$ decreases as time progresses. In Fig. 7.14b, excess carriers are continuously supplied at $x = 0$ so as to maintain a constant p_1 at $x = 0$. For numerical calculations, the following approximations of the complementary error function are useful:

$$\operatorname{erfc} y = \begin{cases} \dfrac{\exp\left(-y^2\right)}{\sqrt{\pi} y}\left(1 - \dfrac{1}{2y^2} + \dfrac{3}{4y^4}\right) & \text{for } y > 1 \qquad (7.7.17a) \\[1.5em] 1 - \dfrac{2}{\sqrt{\pi}}\left(y - \dfrac{y^3}{3} + \dfrac{y^5}{10}\right) & \text{for } y < 1 \qquad (7.7.17b) \end{cases}$$

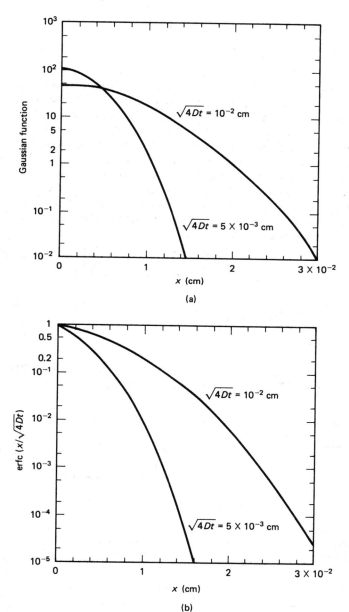

Figure 7.14 Plot of (a) the Gaussian function and (b) the complementary error function. The curves are plotted as a function of the space coordinate x for two selected values of $\sqrt{4Dt}$.

Also, a definite integral exists with

$$\int_0^\infty \text{erfc } y \, dy = \frac{1}{\sqrt{\pi}} \qquad (7.7.18)$$

Therefore, the total amount of excess carriers increases as $\sqrt{D_p t}$ as the time progresses.

Many problems that we encounter in practice may not have precisely the same initial and boundary conditions as the ones used in the discussion above. Now let us discuss methods by which we can extend the usefulness of the results of our analyses to cases slightly different from the cases presented. One useful method is the *method of superposition*. Consider the case where the concentration of excess carriers at $x = 0$ is an arbitrary function $\phi(t)$ of time for $t > 0$. To apply the method of superposition, we divide the time scale of the curve shown in Fig. 7.15a into small intervals of $\Delta\lambda$ each. The response to the function $\phi(\lambda)$ of Fig. 7.15b can be considered as the difference of the responses to the source functions of Fig. 7.15c and d. Therefore, the response to the source function of Fig. 7.15b is given by $\Delta u(x, t, \lambda)$:

$$\Delta u(x, t, \lambda) = \phi(\lambda) \left\{ \text{erfc}\left[\frac{x}{2\sqrt{D(t-\lambda)}}\right] - \text{erfc}\left[\frac{x}{2\sqrt{D(t-\lambda-\Delta\lambda)}}\right] \right\}$$

$$= \phi(\lambda) \, \Delta\lambda \, \frac{\partial}{\partial\lambda} \text{erfc}\left[\frac{x}{2\sqrt{D(t-\lambda)}}\right] \qquad (7.7.19)$$

The total response $p_1(x, t)$ is equal to

$$p_1(x, t) = \int_0^t \frac{\partial u}{\partial\lambda} \, d\lambda = \int_0^t \frac{x}{2\sqrt{\pi D(t-\lambda)^3}} \exp\left[-\frac{x^2}{4D(t-\lambda)}\right] \phi(\lambda) \, d\lambda \qquad (7.7.20)$$

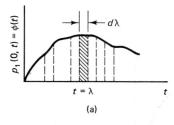

(a)

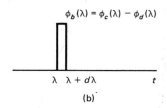

$$\phi_b(\lambda) = \phi_c(\lambda) - \phi_d(\lambda)$$

(b)

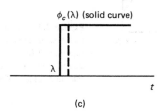

$\phi_c(\lambda)$ (solid curve)

(c)

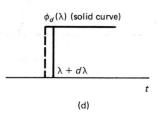

$\phi_d(\lambda)$ (solid curve)

(d)

Figure 7.15 Diagrams showing how the response to an arbitrary excitation can be obtained from the response to a step function by applying the method of superposition. (a) The excitation function is such that the excess-carrier concentration at $x = 0$ is an arbitrary function $\phi(t)$ of time. The boundary condition $p_1(o,\lambda)$ at $t = \lambda$ shown in diagram (b) is expressed as the difference of diagrams (c) and (d) because the erfc function is the solution for p_1 with the boundary condition being a step function of time.

Transport and Recombination of Excess Carriers   Chap. 7

Another useful method is the *method of image*. Let us consider the problem shown in Fig. 7.16a. A unit pulse is applied at x_0 to a semi-infinite semiconductor at time $t = 0$. The boundary condition is such that

$$p_1 = 0 \quad \text{at } x = 0 \quad \text{and} \quad x = \infty \tag{7.7.21}$$

To apply the solution given by Eq. (7.7.13) to the present problem, we first must change x to $x - x_0$ in Eq. (7.7.13). The condition $p_1 = 0$ at $x = 0$ can be satisfied if a fictitious negative pulse is assumed to exist at $x = -x_0$ (Fig. 7.16b). Therefore, the excess-carrier concentration p_1 is a superposition of two solutions (one with a positive source at $x = x_0$ and the other with a negative source at $x = -x_0$):

$$p_1(x, t) = \frac{1}{(4\pi D_p t)^{1/2}} \left\{ \exp\left[-\frac{(x - x_0)^2}{4D_p t} \right] - \exp\left[-\frac{(x + x_0)^2}{4D_p t} \right] \right\} \tag{7.7.22}$$

At $x = 0$, the two exponential terms cancel out. Thus p_1 is identically zero at $x = 0$, as required.

In summary, we have introduced the two basic solutions of the time-dependent diffusion equation, the Gaussian function and the complementary error function. In the transient analysis of semiconductor devices, the solution of the diffusion equation can be expressed in terms of these two basic functions and functions derived from them. For this reason we have also shown how we can extend the usefulness of these solutions by method of superposition. Strictly speaking, the one-dimensional treatment given here applies only to cases where the distribution of excess carriers is assumed uniform over the transverse plane. In cases where we must consider the recombination of excess carriers at the surface of a semiconductor, the one-dimensional analysis is no longer rigorous. In Section 7.10 we modify our one-dimensional analysis to include surface effects.

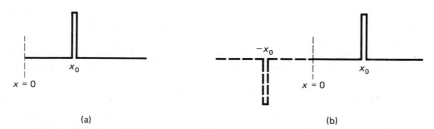

Figure 7.16 Diagrams illustrating the use of the method of image. (a) The problem is specified as follows. A unit impulse is applied at $x = x_0$ and the excess-carrier concentration at $x = 0$ is kept at zero. (b) To maintain the boundary condition $p_1 = 0$ at $x = 0$, a fictitious negative impulse of equal amplitude is placed at $x = -x_0$.

7.8 THE HAYNES–SHOCKLEY EXPERIMENT

In previous sections we discussed the mechanisms governing the flow as well as the decay of excess carriers and worked out mathematical expressions describing the motion of excess carriers. To demonstrate the actual operation of these mechanisms in accordance with theoretical predictions, we consider the experimental setup of Fig. 7.17a

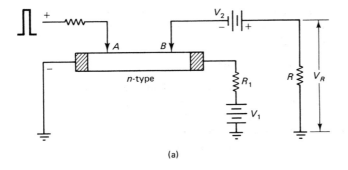

(a)

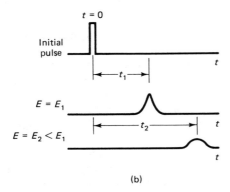

$t = 0$

Initial
pulse

t_1

$E = E_1$

t_2

$E = E_2 < E_1$

(b)

Figure 7.17 Modified version of the Haynes–Shockley experiment: (a) the experimental setup for an n-type semiconductor and (b) the experimental observation of the arrival of the injected holes at the collector. The detected pulse at B is delayed by a time t with respect to the initial pulse.

designed for an n-type semiconductor. The contacts A and B, respectively, serve to inject and collect the minority carriers (holes in the present case), and the voltages V_1 and V_2 are used, respectively, to create a uniform electric field along the sample and to bias the collector contact B. For a p-type sample, the polarity of V_1 and V_2 should be reversed, and a negative pulse should be used at contact A to inject electrons into the sample. In order that the expressions worked out in Section 7.7 can be applied directly to the experiment, the sample must have the shape of a long bar with a uniform cross section.

The experiment can be described as follows. At time $t = 0$, holes are injected into the bulk sample in the form of a pulse of very short duration (top trace of Fig. 7.17b). After hole injection occurs, two things happen. Within a time of the order of the dielectric relaxation time, an excess electron concentration neutralizes the injected hole concentration in the injection region. The increase in the carrier concentration immediately reduces the resistance of sample, thus creating a transient voltage across the resistor R_1. However, this voltage is not important to the experiment, for reasons that will become clear shortly. What is important is the observation of the arrival of the injected minority carriers at the collector contact B.

The movement of excess carriers, both minority and majority, is governed by Eq. (7.6.10). As pointed out in Section 7.6, the charge-neutrality condition demands that the excess majority carriers move with excess minority carriers. This conclusion might seem incorrect at first because majority and minority carriers carry opposite electric charges. The explanation lies in the fact that excess carriers behave differently from thermal equilibrium carriers. In the case of thermal equilibrium carriers, the movement of electrons in one direction and that of holes in the opposite direction do not create a space-charge region. When carriers move away from a certain region, they are constantly replaced by other carriers. A state of charge neutrality persists despite the opposite movements of the positive and negative carriers.

Whether a space charge will be built up or not can be seen from the continuity equations, Eqs. (7.6.5) and (7.6.6). For ease of discussion, we consider the drift terms only, and thus have

$$\frac{\partial p}{\partial t} = -\frac{\partial}{\partial x}(\mu_p p E_x) \tag{7.8.1}$$

$$\frac{\partial n}{\partial t} = +\frac{\partial}{\partial x}(\mu_n n E_x) \tag{7.8.2}$$

The density of space charge ρ is equal to $e(p - n)$. Therefore, the equation of importance is the following one:

$$\frac{1}{e}\frac{\partial \rho}{\partial t} = -\frac{\partial}{\partial x}(\mu_p p E_x + \mu_n n E_x) \tag{7.8.3}$$

Under thermal equilibrium, $E_x = E_a$, the applied field, $p = p_0$, and $n = n_0$. Since all these quantities are constants independent of x, the right-hand side of Eq. (7.8.3) is identically zero and $\partial \rho/\partial t = 0$. Even though electrons and holes move in opposite directions, no space charge results.

In the presence of injected carriers, there is local field E_{sc} due to local space charges in addition to the applied field E_a. Thus Eqs. (7.8.1) and (7.8.2) becomes

$$\frac{\partial p}{\partial t} = -\mu_p p \frac{\partial E_{sc}}{\partial x} - \mu_p(E_a + E_{sc})\frac{\partial p}{\partial x} \tag{7.8.4a}$$

$$\frac{\partial n}{\partial t} = +\mu_n n \frac{\partial E_{sc}}{\partial x} + \mu_n(E_a + E_{sc})\frac{\partial n}{\partial x} \tag{7.8.4b}$$

To prevent space-charge accumulation, that is, to make $\partial \rho/\partial t = 0$, we must have a local variation of the space-charge field of the following magnitude:

$$\frac{\partial E_{sc}}{\partial x} = -\frac{\mu_p + \mu_n}{\mu_n n + \mu_p p}(E_a + E_{sc})\frac{\partial p_1}{\partial x} \tag{7.8.5}$$

Substituting Eq. (7.8.5) into Eqs. (7.8.4a) and (7.8.4b) yields

$$\frac{\partial p}{\partial t} = \frac{\partial n}{\partial t} = \frac{\mu_n \mu_p(p - n)}{\mu_n n + \mu_p p} E_a \frac{\partial p_1}{\partial x} = \mu' E_a \frac{\partial p_1}{\partial x} \tag{7.8.6}$$

where we have set $\partial p/\partial x = \partial n/\partial x$ and $p_1 = n_1$, and neglected E_{sc} in comparison with E_a. Note that μ' in Eq. (7.8.6) is identical to that of Eq. (7.6.12).

The main difference between the behavior of thermal equilibrium carrier concentrations p_0 and n_0 and that of excess carrier concentrations p_1 and n_1 is due to the fact that p_1 and n_1 have a spatial variation as manifested by a nonzero $\partial p_1/\partial x$. Therefore, any relative movement of p_1 and n_1 will create a space-charge region and thus a local space-charge field E_{sc}. Imagine that the space-charge density $\rho(x)$ in Fig. 7.4 was caused by an instantaneous spatial separation of p_1 to the right and n_1 to the left. The space-charge field would immediately act on the carriers to neutralize the space charges, as indicated by the two arrows in Fig. 7.4. Since there are much more majority carriers than minority carriers, neutralizing excess minority carriers by majority carriers is much faster than neutralizing excess majority carriers by minority carriers. Therefore, the movement of excess carriers follows that of excess minority carriers. This reasoning is consistent with the result that μ' of Eq. (7.8.6) reduces to $-\mu_p$ if $n \gg p$ and to μ_n if $p \gg n$. Furthermore, in actual situations, p_1 and n_1 not only stay together but are nearly equal as well. This observation leads to the conclusion that E_{sc}, compared to E_a, can be neglected but $\partial E_{sc}/\partial x$ cannot be ignored.

The observations above serve to explain why the excess majority carriers can move and will move in the same direction as the excess minority carriers. In describing the physical situation in the experiment of Fig. 7.17a, therefore, we consider the excess carriers as a charge cloud consisting of both minority and majority carriers. This charge cloud moves as a whole. The motion of this charge cloud is dictated by Eq. (7.6.10), and in the presence of an electric field, the charge cloud drifts in the same direction as minority carriers. With the polarity of the voltage V_1 indicated in Fig. 7.17a, the charge cloud drifts toward the right.

The transient voltage developed across R_1 will last as long as there are excess carriers inside the sample; therefore, this voltage does not give much useful information except that it decays as excess carriers recombine. The situation is quite different, with the transient voltage developed across another resistor R. A voltage V_R appears across R only when excess holes drifting down the sample are collected at the contact B, thus creating a current flow through R. The lower two traces of Fig. 7.17b show V_R in relation to the initial pulse on the time scale for two different values of E_x. The time lapse between the initiation of the injection pulse and the arrival of excess carriers at the contact B is a measure of the drift velocity of excess carriers. Simultaneously with the drift motion, excess carriers are also dispersing because of diffusion. This explains why the pulse shown in the lower trace is not as sharp as the initial pulse shown in the top trace.

In Eq. (7.6.10), the three parameters that characterize the processes of drift, diffusion, and recombination are mobility, diffusion constant, and lifetime. In the following analysis we discuss how the values of the three parameters can be determined from such an experiment. The experiment just described is known as the *Haynes–Shockley experiment* because of their original work (J. R. Haynes and W. Shockley, *Phys. Rev.*, Vol. 81, p. 835, 1951). The setup shown in Fig. 7.17a using pulse techniques is an improved version of the original experiment.

The solution of Eq. (7.6.10) applicable to the analysis of the experiment is given by Eq. (A7.19):

$$p_1(x, t) = \frac{\exp(-t/\tau_0)}{(4\pi D_p t)^{1/2}} \exp\left[-\frac{(x - \mu_p E_x t)^2}{4 D_p t}\right] \tag{7.8.7}$$

To satisfy the condition on which Eq. (7.8.7) is based, the injection contact must be far away from both ends of the sample. At these ends, the excess-carrier concentration is zero because of the ohmic contacts. In an exact analysis, we must include the contributions from the images [Eq. (7.7.22)] to Eq. (7.8.7). If the injection contact is remote from both ends, however, the contributions from the images arrive at the contact B not only at a much later time but also with a much smaller amplitude because of recombination. In that case, the use of Eq. (7.8.7) is sufficient to describe the motion of excess carriers.

For a fixed distance $x = d = x_B - x_A$ in Eq. (7.8.7), the time at which the voltage across R is a maximum can be obtained from Eq. (7.8.7) by setting $\partial p_1 / \partial t = 0$ and solving for t. As a first-order approximation, we neglect the variation of p_1 due to $\exp(-t/\tau_0)$ and $1/t$. Thus the maximum of the hole density occurs at a time t_0 given by

$$d = x_B - x_A = \mu_p E_x t_0 \tag{7.8.8}$$

As will be shown shortly, Eq. (7.8.8) holds if p_1 remains sharply defined in x and t. For sharp pulses, the maximum moves along the sample with a drift velocity $\mu_p E_x = d/t_0$. Knowing t_0, d, and E_x, the value of μ_p can be calculated from Eq. (7.8.8).

If the time variation of the dependence $\exp(-t/\tau_0)$ and $1/t$ is taken into consideration, the maximum of p_1 occurs at a time t given by

$$t = t_0 - \frac{kT}{eE_x} \frac{1}{\mu_p E_x} \left(1 + \frac{2t_0}{\tau_0} \right) \tag{7.8.9}$$

Physically, Eq. (7.8.9) can be interpreted as follows. In order that the function p_1 remain sharply defined, the dispersion Δx caused by diffusion must be small compared to the drift distance, that is,

$$\Delta x = (D_p t_0)^{1/2} < \mu_p E_x t_0 \tag{7.8.10}$$

which leads to $t_0 \gg kT/(e\mu_p E_x^2)$ in Eq. (7.8.9). Furthermore, the time required for carriers to drift across Δx must be shorter than the lifetime, that is,

$$\frac{\Delta x}{\mu_p E_x} < \tau_0 \tag{7.8.11}$$

which together with Eq. (7.8.10) leads to $D_p t_0/(\mu_p^2 E_x^2 \tau_0) \ll t_0$ in Eq. (7.8.9).

The diffusion constant D_p can be measured from the width of the Gaussian distribution function. Neglecting again the time variation of $\exp(-t/\tau_0)$ and $1/t$, the signal V_R drops to $1/e$ ($e = 2.718$) of its peak value when

$$(d - \mu_p E_x t)^2 = 4D_p t \tag{7.8.12}$$

Let $t = t_0 \pm \Delta t/2$ be the two values of time that satisfy Eq. (7.8.12). For $\Delta t/t_0 \ll 1$, $4D_p t$ can be approximated by $4D_p t_0$, and hence the value of D_p is found to be

$$D_p = (\mu_p E_x)^2 \frac{(\Delta t)^2}{16t_0} \tag{7.8.13}$$

Again, the approximation is reasonably good only if the factor $\exp(-t/\tau_0)$ does not change appreciably over the range Δt. Another factor that may affect the measurement of D_p is the physical size of the collector contact. Theoretically, the condition stated in Eq. (7.8.12) applies only to a single point. For a collection of collector points, Eq. (7.8.7) should be summed over a range of x. In other words, the spread in distance between contacts A and B effectively increases the apparent spread Δt of the signal. To check whether the Haynes–Shockley experiment gives genuine results in the mobility and diffusion-constant measurements, the experiment should be repeated for different values of E_x.

It is also possible to obtain the value of the lifetime of excess carriers from the

Haynes–Shockley experiment. Note that the excess-hole concentration p_1 of Eq. (7.8.7) has the following property:

$$\int_{-\infty}^{\infty} p_1(x,\,t)\,dx = \exp\left(-\frac{t}{\tau_0}\right) \qquad (7.8.14)$$

In Eq. (7.8.7) we assume that the amount of carriers initially injected into the sample is unity. Therefore, Eq. (7.8.14) says that the amount of excess carriers injected into the sample decays as $\exp\,(-t/\tau_0)$. If η is the collector efficiency, defined as the ratio of holes collected by the contact B to the total number of holes passing through a transverse plane at B, then for N holes injected into the sample, $N\exp\,(-t_0/\tau_0)$ holes will survive the recombination process, and of these, $\eta N\exp\,(-t_0/\tau_0)$ holes will be collected by the contact B, where t_0 is the drift time between contacts A and B. The area S under the curves of Fig. 7.17b is of course directly proportional to the total holes collected, or

$$S = \int_{\text{pulse}} V_R\,dt = A\eta N \exp\left(\frac{-t_0}{\tau_0}\right) = A\eta N \exp\left(\frac{-d}{\mu_p E_x \tau_0}\right) \qquad (7.8.15)$$

where A is a proportionality constant. For different values of E_x, the area under the curve will be different. A plot of $\ln S$ against $d/\mu_p E_x$ gives a straight line of slope $1/\tau_0$. Note that the maximum of p_1 varies as $t_0^{-1}\exp\,(-t_0/\tau_0)$, and hence it does not decay as $\exp\,(-t_0/\tau_0)$. The factor t_0^{-1} should be taken into account if we use the maximum of p_1 to calculate τ_0.

In a single setup, the Haynes–Shockley experiment gives a vivid demonstration of the drift, diffusion, and recombination processes of excess carriers. Insofar as the actual determination of μ, D, and τ_0 is concerned, however, only the mobility measurement is simple and straightforward enough to yield accurate results. The thing to be careful about is that the electric field should be high enough so that the value of t_0 obtained from Eq. (7.8.8) is a good approximation. To avoid heating of the sample, we should also use a pulse generator instead of a dc source for the voltage V_1. For the measurement of D, extreme care must be taken to ensure that no erroneous effects exist to increase the apparent width of the pulse. The difficulty lies in the fact that it is hard to tell whether the pulse is genuinely Gaussian merely from looking at it. Insofar as the measurement of τ_0 is concerned, there are other methods that give a direct reading of τ_0 and hence may be preferred over the Haynes–Shockley experiment.

The values of drift mobility measured in germanium by Prince and in silicon by Ludwig and Watters are given in Table 7.2 (M. B. Prince, *Phys. Rev.*, Vol. 93, p. 1204, 1954; G. W. Ludwig and R. L. Watters, *Phys. Rev.*, Vol. 101, p. 1699, 1956). Since the measurements were made on high-purity samples, with resistivities

TABLE 7.2 DRIFT MOBILITIES (cm²/V-s) AND DIFFUSION CONSTANTS $D = \mu(kT/e)$ (cm²/s) in Ge AND Si AT 300 K

Ge	$\mu_n = 3900 \pm 100$	$\mu_p = 1900 \pm 50$	$D_n = 101$	$D_p = 49$
Si	$\mu_n = 1400 \pm 100^{\text{a}}$	$\mu_p = 470 \pm 15$	$D_n = 36$	$D_p = 12.2$

[a]A value of 1350 was originally reported and is still quoted in some textbooks. The value 1400 is in better agreement with Hall mobility $\mu_H = \sigma R_o$ through the use of Eq. (A6.18).

higher than 10 Ω-cm in germanium and ranging from 19 to 170 Ω-cm in silicon, the mobilities given in the table are lattice mobilities. In other words, the relaxation time τ in the expression of mobility $\mu = e\tau/m^*$ is dominated by scattering of free carriers by lattice vibrations. Another point worth mentioning is that since there are heavy-mass and light-mass holes, conceivably two separate pulses with different drift velocities would be observed. This is not found experimentally. Because holes make rapid transitions from the light- to heavy-mass band and vice versa, light-mass and heavy-mass holes propagate together as a single pulse.

7.9 SURFACE STATES AND FIELD-EFFECT EXPERIMENTS

In solving the time-dependent diffusion equation, we treated the flow of excess carriers as a one-dimensional problem by assuming that nothing extraordinary happens at the surface of a semiconductor. This oversimplification is a necessary step to the derivation of physically meaningful results without getting deeply involved in the mathematics. However, the assumption is not a realistic one. The manufacture of high-performance semiconductor devices with stable electrical characteristics requires not only a careful surface preparation but also a stable ambient condition for the device. To analyze surface effects, we first present a physical description of a semiconductor surface followed by a discussion of what may happen at the surface. Examples are given in Section 7.11 to illustrate how surface effects can be accounted for in the mathematical formulation of a physical problem.

It is generally agreed that chemically etched semiconductor surfaces such as those of germanium and silicon are covered with a thin layer of oxides. Because of the mismatch of crystallographic structures and lattice constants at the semiconductor–oxide interface boundary, imperfections in the structure are likely to exist and to create energy levels within the forbidden gap. The extra energy levels created at a semiconductor surface are commonly called *surface states*.

There are two different kinds of surface states. The energy states created at the semiconductor–oxide interface boundary are generally called the *fast surface states*. Since these states are in intimate electrical contact with the bulk semiconductor, they can reach a state of equilibrium with the bulk within a relatively short time (on the order of nanoseconds or less). There are other kinds of surface states known as *slow surface states*. The slow states may be attributed to either chemisorbed ambient ions or imperfections in the oxide. Carrier transfer from such states to the bulk semiconductor either has to overcome a high potential barrier presented by the large energy gap of the oxide or has to tunnel through the oxide layer; therefore, such a charge-transfer process involves a long time constant, typically on the order of seconds or longer.

The existence of two types of surface states, fast and slow, is illustrated schematically in Fig. 7.18. In this section we discuss the effects of the slow states on the electrical properties of a semiconductor. Like ionized donor and acceptor states, the slow states generally carry either positive or negative electric charges. These charges carried by slow states attract free carriers of opposite polarity toward the surface from the bulk of a semiconductor. As a result, the carrier concentration near the surface is different from that inside the bulk. As we shall see shortly, the semiconductor near the surface is no longer electrically neutral, but has a net charge. The slow-state charges and the semiconductor charges form a dipole layer, the opposite charges being separated by the oxide.

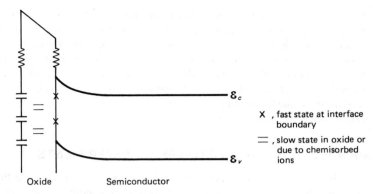

X , fast state at interface
boundary

≡ , slow state in oxide or
due to chemisorbed
ions

Oxide   Semiconductor

Figure 7.18 Surface states that exist near the surface of a semiconductor. Refer to
Fig. 7.19 for an explanation of the bending of the energy bands.

Figure 7.19 shows two possible situations that may exist at a semiconductor surface. In Fig. 7.19a, the slow-state charges are negative. To repel electrons from and
attract holes toward the surface, the potential energy of an electron must increase toward
the surface. Another way of stating the same phenomenon is that the dipoles at the
surface create an electric field in the negative x direction (toward the surface), and the
field in turn results in an increase in the potential energy of electrons toward the surface.
The situation shown in Fig. 7.19b is opposite to that shown in Fig. 7.19a. If the slow
states carry positive charges, the dipole field E_x is positive and the potential energy of
electrons decreases toward the surface. In either case, the band edges $\mathscr{E}_c$ and $\mathscr{E}_v$ will
either bend up (Fig. 7.19a) or bend down (Fig. 7.19b). We should emphasize, however,
that the Fermi level $\mathscr{E}_f$ stays constant throughout the semiconductor, which is the condition for thermal equilibrium.

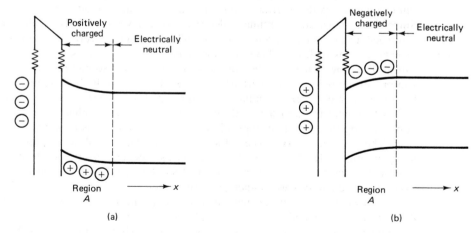

Figure 7.19 Potential barrier created near the surface of a semiconductor due to (a)
negatively charged, slow surface states, and (b) positively charged, slow surface states.
A surface space-charge region exists, and as a result the potential energy of free carriers
changes in the space-charge region.

Now let us examine whether the change in the band edges is consistent with the change in carrier concentrations. The net charge density in the bulk is $\rho = e(p_0 + N_d^+ - n_0 - N_a^-) = 0$. As we approach the surface in Fig. 7.19a, $\mathcal{E}_c - \mathcal{E}_f$ increases and $\mathcal{E}_f - \mathcal{E}_v$ decreases. Therefore, p_0 increases and n_0 decreases. For ease of discussion, we assume that the bending of the band edges is small, and hence N_d^+ and N_a^- are unaffected. Clearly, the combined result is that $\rho = e(p_s + N_d^+ - n_s - N_a^-)$ becomes nonzero and is positive, where the subscript s refers to carrier concentrations near the surface. The fact that p_s and n_s vary with distance near a semiconductor surface is an important surface phenomenon.

Not only can the carrier concentrations at the semiconductor–oxide interface change, but they may change so drastically that the surface conductivity may be of the type opposite to that in the bulk. In other words, if the bulk semiconductor is n-type with $n_0 > p_0$, the hole and electron concentrations, p_s and n_s, at the semiconductor–oxide interface may have the inequality sign reversed, that is, $p_s > n_s$. Thus an *inversion layer* is established with p-type conductivity near the surface and n-type conductivity in the bulk (Fig. 7.19a). Figure 7.19b shows a different situation in which $n_s > p_s$ and $n_s > n_0$. The transition region from the bulk to the surface, which retains the same conductivity type but has accumulated more free carriers, is given the name *accumulation layer*. The words *inversion* and *accumulation* indicate the state of surface carriers with reference to that of bulk carriers.

We now discuss a simple experiment, known as the *field-effect experiment*, which is a very useful tool in investigating the nature of the surface states and in determining the conductivity type of a semiconductor surface. In the experiment, a semiconductor slice is used as one plate of a parallel-plate condenser, the other plate being a metal plate in close proximity to the semiconductor. When a positive voltage is applied to the metal plate with respect to the semiconductor, negative charges are induced near the semiconductor surface opposite the metal plate. Part of the induced negative charges will be trapped by the surface states and thus become immobilized. The remainder will go into the surface space-charge region (region A of Fig. 7.19). It is the latter part that provides the useful information about the original surface.

Referring to the energy-band diagram near the surface of a semiconductor shown in Fig. 7.20, the quantity $e\phi$ is the difference (in electron volts) between the Fermi level $\mathcal{E}_f$ and the intrinsic Fermi level $\mathcal{E}_{\text{fi}}$. We should point out that $\mathcal{E}_{\text{fi}}$ defined in Eq. (6.5.13) is introduced only for operational convenience, and is unrelated to $\mathcal{E}_f$. In Fig. 7.20, $\mathcal{E}_f$ stays constant because the system is under thermal equilibrium. On the other hand, $\mathcal{E}_{\text{fi}}$ changes with distance near the surface in the same way as $\mathcal{E}_c$ and $\mathcal{E}_v$ so as to satisfy Eq. (6.5.13), in which $\mathcal{E}_{\text{fi}}$ is defined. The quantities $e\phi_s$ and $e\phi_b$ are equal to the difference $\mathcal{E}_f - \mathcal{E}_{\text{fi}}$ at the oxide–semiconductor boundary (taken as $x = 0$ in Fig. 7.20) and that inside the bulk, respectively. In terms of ϕ, the carrier concentrations at any point x inside the surface space-charge region are given by

$$n = n_i \exp\left(-\frac{\mathcal{E}_{\text{fi}} - \mathcal{E}_f}{kT}\right) = n_i \exp\left(\frac{e\phi}{kT}\right) \tag{7.9.1a}$$

$$p = p_i \exp\left(\frac{\mathcal{E}_{\text{fi}} - \mathcal{E}_f}{kT}\right) = n_i \exp\left(-\frac{e\phi}{kT}\right) \tag{7.9.1b}$$

where n_i is the intrinsic carrier concentration.

The variation of ϕ with respect to x can be obtained from Poisson's equation,

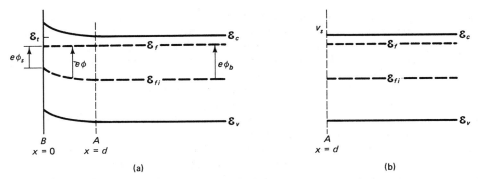

Figure 7.20 (a) Energy-level diagram used in defining the various energy levels for an analysis of the surface effects. (b) Schematic diagram indicating that the effect of fast states with trap energy $\mathcal{E}_t$ can be incorporated into the mathematical formulation by a parameter v_s called the surface recombination velocity. The arrows in (a) indicate the directions in which the quantities ϕ_s and ϕ_b- ϕ_s are measured as positive quantities.

$$\frac{d^2V}{dx^2} = \frac{d^2\phi}{dx^2} = -\frac{\rho}{\epsilon} \tag{7.9.2}$$

Note that $e\phi$ is a measure of the potential energy of a hole; hence $e\phi$ is equal to eV except for a constant where V is the electrostatic potential. In Eq. (7.9.2), ϵ is the dielectric constant and ρ is the net charge density given by

$$\rho = e(p + N_d^+ - N_a^- - n) \tag{7.9.3}$$

Let us consider a simple case in which the bending of energy bands at the surface is relatively small so that the donors and acceptors in the surface space-charge region remain completely ionized. This means that we can set $N_d^+ - N_a^-$ equal to $n_0 - p_0$ in Eq. (7.9.3), where n_0 and p_0 are the carrier concentrations in the bulk semiconductor. Thus, from Eq. (7.9.1),

$$N_d^+ - N_a^- = n_0 - p_0 = 2n_i \sinh \frac{e\phi_b}{kT} \tag{7.9.4}$$

Using Eqs. (7.9.1) and (7.9.4) in Eq. (7.9.3), we obtain

$$\rho = e(p - p_0 - n + n_0) = 2en_i \left(\sinh \frac{e\phi_b}{kT} - \sinh \frac{e\phi}{kT} \right) \tag{7.9.5}$$

Substituting Eq. (7.9.5) into Eq. (7.9.2), multiplying both sides of Eq. (7.9.2) by $(d\phi/dx)$, and then integrating with respect to x from A to B (Fig. 7.20a), we find that

$$\left(\frac{d\phi}{dx} \right)^2 = \frac{-4n_ikT}{\epsilon} [\cosh Y_b - \cosh Y - (Y_b - Y) \sinh Y_b] \tag{7.9.6}$$

where $Y = e\phi/kT$ and $Y_b = e\phi_b/kT$. The integration constant in Eq. (7.9.6) is so chosen as to make $d\phi/dx = 0$ for $\phi = \phi_b$.

In the field-effect experiment, two quantities of importance are the surface charge density Q_s and the surface conductance G_s. For a surface of unit area, the relative change in the number of negative carriers in the surface space-charge region of width d is equal to

$$Q_n = \int_0^d e(n - n_0)\, dx = \int_{\phi_s}^{\phi_b} e(n - n_0) \frac{dx}{d\phi}\, d\phi \qquad \text{C/m}^2 \qquad (7.9.7)$$

whereas the corresponding change in the number of positive carriers is equal to

$$Q_p = \int_0^d e(p - p_0)\, dx = \int_{\phi_s}^{\phi_b} e(p - p_0) \frac{dx}{d\phi}\, d\phi \qquad \text{C/m}^2 \qquad (7.9.8)$$

The reason for defining Q_n and Q_p in this manner is that the two quantities so defined can be used in both expressions of Q_s and G_s. In terms of Q_n and Q_p, the net surface charge per unit surface area is given by

$$Q_s = Q_p - Q_n \qquad \text{C/m}^2 \qquad (7.9.9)$$

The surface conductance G_s (per unit surface area) is defined as the change in the conductance of a semiconductor caused by the change in the carrier concentrations near its surface. In other words,

$$G_s = e\mu_n \int_0^d n\, dx + e\mu_p \int_0^d p\, dx - e\mu_n \int_0^d n_0\, dx - e\mu_p \int_0^d p_0\, dx \qquad (7.9.10)$$

The last two terms are subtracted from the first two terms in Eq. (7.9.10) because only the change in, not the absolute value of, sample conductance is of relevance. Combining $n - n_0$ and $p - p_0$ terms in Eq. (7.9.10), changing the variable from x to ϕ, and realizing that the sum of the integration is nonzero only in the surface space-charge region, we find that

$$G_s = \mu_n Q_n + \mu_p Q_p \qquad (7.9.11)$$

Although a semiconductor slice has two broad surfaces, only one surface is considered in Eq. (7.9.10) because only the surface opposite to the metal plate is affected in the field-effect experiment. In cases where both surfaces need to be considered, we should multiply the right side of Eq. (7.9.10) by a factor of 2.

Knowing $d\phi/dx$ as a function of ϕ from Eq. (7.9.6), the values of Q_n and Q_p can be calculated from Eqs. (7.9.7) and (7.9.8). Substituting the calculated values of Q_n and Q_p in Eqs. (7.9.9) and (7.9.11) gives a set of values of Q_s and G_s for a given ϕ_b and ϕ_s. Figure 7.21 shows the general behavior of Q_s and G_s as functions of ϕ_s in n- and p-type samples. Since ϕ_b depends on resistivity, each set of curves is calculated for a given resistivity of the sample. Only for an intrinsic sample where $Y_b = 0$ can integrations involved in Eqs. (7.9.7) and (7.9.8) be carried out analytically to yield explicit functions of ϕ_s. For nonzero Y_b, however, numerical methods must be employed. We shall actually evaluate the values of Q_s and G_s for silicon surfaces when we discuss the property of a metal-oxide-semiconductor (MOS) structure in Chapter 9 in connection with field-effect transistors.

The purpose of our present discussion is to give a qualitative picture of a semiconductor surface. Let us illustrate how we can use the information contained in Fig. 7.21 in the field-effect experiment. Suppose that we have an n-type sample and that the total charge in the surface states is $+Q_{s0}$. To neutralize this charge, $+Q_{s0}$, the total

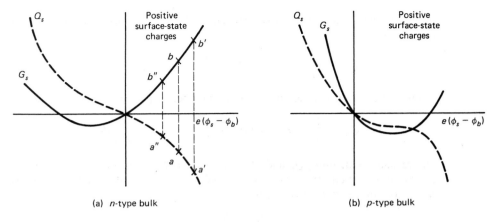

(a) n-type bulk   (b) p-type bulk

Figure 7.21 Plots of surface charge Q_s and surface conductance G_s as functions of surface potential $e(\phi_s - \phi_b)$ in cases (a) where the bulk semiconductor is n-type, and (b) where the bulk semiconductor is p-type. For negative Q_s in the semiconductor, the charges carried by slow surface states are positive.

charge in the surface space-charge region must be $-Q_{s0}$. The condition $Q_s = -Q_{s0}$ establishes point a on the Q_s curve (Fig. 7.21a), which in turn determines the value of ϕ_s and hence point b on the G_s curve.

Figure 7.22 shows the measuring circuit for the field-effect experiment. Besides the field-effect signal, there exists at the output a voltage due to charging currents (charging the semiconductor–metal condenser). The circuit in Fig. 7.22 is arranged in bridge form so that effects due to charging currents can be balanced out by adjusting R_1 and R_2 until a zero v_{out} is observed on the scope with zero dc bias voltage. The effect of the applied voltage (i.e., the probing voltage v_1) is to cause a change in Q_s, and the output voltage v_{out} is directly proportional to the consequent change in G_s brought about by the change in Q_s.

One immediate application of the field-effect experiment is to determine whether the surface conductivity of a semiconductor is n-type or p-type. Suppose that the surface

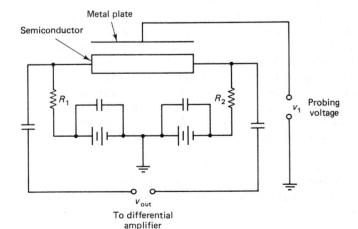

Figure 7.22 Schematic diagram showing a circuit arrangement for the field-effect experiment.

condition of an *n*-type semiconductor is initially represented by point *a* in Fig. 7.21a. Referring to the energy-band diagram of Fig. 7.20a, we see that for positive $\phi_s - \phi_b$, the band edges must bend down as shown in Fig. 7.19b. Thus the free carriers near the surface are predominantly electrons. If a positive voltage is applied to the metal plate with respect to the semiconductor, the negative charges induced near the semiconductor surface add further to the free-electron concentration and thus cause an increase in G_s. In Fig. 7.21 we can visualize the consequent action of the applied voltage as being equivalent to moving points *a* and *b*, respectively, to *a'* and *b'*. If v_1 is negative, positive charges are added to the surface space-charge region, which is equivalent to taking negative carriers away from it. As a result, both Q_s and G_s decrease in magnitude, and points *a* and *b*, respectively, move to *a"* and *b"*. An oscilloscope display of v_{out} as a function of v_1 is illustrated in Fig. 7.23a.

The polarities of v_1 and v_{out} in Fig. 7.23 are so chosen that a positive v_1 means the metal being biased positive with respect to the semiconductor as in a MOS field-effect transistor, whereas a positive v_{out} represents an increase in surface conductance. Three general types of oscilloscope traces can be expected, depending on the surface conductivity type. The case just discussed (Fig. 7.23a) represents an *n*-type surface, where the surface carriers are predominantly electrons. For a *p*-type surface, where the surface carriers are predominantly holes, adding positive charges to the surface space-charge region further increases the hole concentration and thus raises the surface conductance (Fig. 7.23c). If ϕ_s in Fig. 7.20 is close to zero, both *n* and *p* in Eq. (7.9.1) are close to n_i and hence the surface is close to being intrinsic. Any change in Q_s for a near-intrinsic surface must be accomplished by increasing *n* (or *p*) and reducing *p* (or *n*). Take $n_i = 1.4 \times 10^{10}$ cm^{-3} in Eq. (7.9.1). For $e\phi_s/kT = 2$, *n* increases from 1.4×10^{10} to 1.04×10^{11} cm^{-3} and *p* decreases from 1.4×10^{10} to 1.9×10^9 cm^{-3}. Thus the increase in *n* is larger than the decrease in *p* in Eq. (7.9.11). For $e\phi_s/kT = -2$ the increase in *p* overcomes the decrease in *n*. In either case, G_s increases with v_1 (Fig. 7.23b) irrespective of whether v_1 is positive or negative.

Another interesting aspect of the field-effect experiment concerns the variation of v_{out} with time. As pointed out earlier, the charges induced by the applied voltage v_1 are shared by the surface states and the semiconductor. However, any charge flow from or to the surface states must go through the semiconductor. Initially, all the induced charges are in the surface space-charge region of the semiconductor. The interface fast

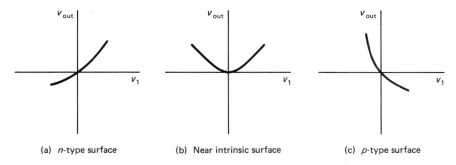

(a) *n*-type surface   (b) Near intrinsic surface   (c) *p*-type surface

Figure 7.23 Schematic diagrams showing the field-effect signal v_{out} (Fig. 7.22) as functions of the probing voltage v_1 for (a) an *n*-type surface, (b) a near-intrinsic surface, and (c) a *p*-type surface. The polarities of the voltages are so chosen that a positive v_{out} represents an increase in G_s and a positive v_1 means a negative ΔQ_s.

states, which are in intimate contact with the semiconductor space-charge region, immediately (within a very short time of the order of nanosecond or so) trap part of the field-induced charges. As time progresses, most field-induced charges leak slowly through the oxides and become immobilized in the slow states.

Figure 7.24a shows the variation of v_{out} with time for an n-type surface. In the experiment, the applied voltage v_1 is a constant voltage turned on at $t = 0$ and with negative polarity connected to the metal side. For $t < 0$, the original surface condition is specified by point o on the Q_s and G_s curves (Fig. 7.24b). If the values of v_i and v_f are converted into changes in G_s, two new points i and f are located on the G_s curve and, in turn, on the Q_s curve. The total amount of induced charge is equal to Cv_1, where C is the capacitance of the metal–semiconductor condenser. Out of Cv_1 charges, the amount that goes to the surface space-charge region is initially equal to $A(Q_{s0} - Q_{si})$ and finally equal to $A(Q_{s0} - Q_{sf})$, where A is the area of the semiconductor surface facing the metal plate.

It is found from experiments on germanium surfaces that the part v_f in Fig. 7.24a is negligibly small compared to v_i. This means that in the steady state of a field-effect experiment, the slow states absorb practically all the field-induced charges. Based on this fact, we conclude that the density of slow states is overwhelmingly large and that the surface condition of a semiconductor is determined by the amount and the type of electric charges stored in the slow states. As we discuss in Section 7.10, the fast states are responsible for the recombination of excess carriers at a semiconductor surface. Surface states are classified into two types, fast and slow, not only because fast and slow states have different physical origins but because they affect different electrical properties of a semiconductor.

One important observation in the early study of semiconductor surfaces is that the surface property of a semiconductor changes as the ambient gas changes through a cycle

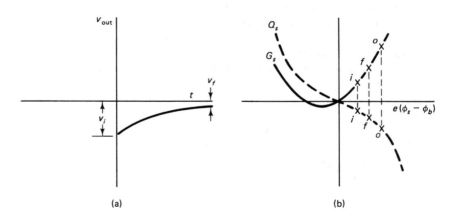

(a)   (b)

Figure 7.24 (a) Schematic diagram showing the time variation of the field-effect signal v_{out} (Fig. 7.22), and (b) the Q_s and G_s curves which are used to represent the surface condition of a semiconductor. The point 0 on the curves represents the original surface condition before a dc probing voltage v_1 is applied. The point i represents the initial condition at $t = 0$ when v_1 is just turned on and the point f represents the final condition after v_1 is turned on for a long time.

of ozone, dry oxygen, and dry and wet nitrogen (known as the Brattain–Bardeen cycle). In ozone the field-effect experiment shows a characteristic pattern of Fig. 7.23c, whereas in wet nitrogen, that of Fig. 7.23a. From this experimental observation we infer that ozone makes a germanium surface p type, and hence it must carry negative charges. On the other hand, water (carried in wet nitrogen), being responsible for an n-type surface, must carry positive charges.

During the early development of transistor technology, a great deal of effort has been devoted to stabilize and passivate surface conditions. One step taken by semiconductor manufacturers is to keep a semiconductor device in a hermetically sealed container in which a stable nonreactive atmosphere can be kept. Another step is to produce a relatively thick oxide layer on a semiconductor surface so that the underlying semiconductor is made less sensitive to ambient changes. The fact that a stable layer of silicon oxide can grow on top of silicon through oxidation is one reason why silicon has replaced germanium in many semiconductor devices.

7.10 FAST STATES AND SURFACE-RECOMBINATION VELOCITY

In this section we discuss the role of the fast states. The reader who has some experience with semiconductor surfaces will have observed that a semiconductor with a mechanically roughened surface such as a sandblasted surface has a much lower value for the effective lifetime of excess carriers. Undoubtedly, the interface fast surface states play an important role in carrier-recombination processes. The analysis of the surface-recombination process is based on the Shockley–Read–Hall model of recombination through traps. For the following discussion, we again refer to Fig. 7.20a for the energy bands of a typical semiconductor surface and the relevant energy levels. The energy $\mathscr{E}_t$ is the energy of the fast states participating in the surface-recombination process.

In applying the result obtained in Section 7.5, or specifically Eq. (7.5.17), to the surface-recombination process, the following points must be made clear. First, the bulk carrier concentrations p and n in Eq. (7.5.17) must be replaced by the carrier concentrations p_s and n_s defined at the interface boundary. Second, a quasi-equilibrium situation prevails between electrons at the surface and electrons in the bulk and also between holes at the surface and holes in the bulk. If we recall our discussion in Section 7.3, the concentration of electrons in the conduction band in the presence of excess carriers can be described by Eq. (7.3.4) with the Fermi level $\mathscr{E}_f$ replaced by the quasi-Fermi level $\mathscr{F}_n$. A similar expression with $\mathscr{E}_f$ replaced by $\mathscr{F}_p$ applies to valence-band holes.

Applying Eq. (7.5.16) to the surface-recombination process, we find the net rate of recombination to be

$$U = \frac{N_t C_n C_p (p_s n_s - p' n')}{C_p (p_s + p') + C_n (n_s + n')} \tag{7.10.1}$$

According to Eqs. (7.5.12) and (7.5.14), n' and p' in Eq. (7.10.1) are given by

$$n' = N_c \exp\left(-\frac{\mathscr{E}_c - \mathscr{E}_t}{kT}\right) = n_i \exp\left(\frac{\mathscr{E}_t - \mathscr{E}_{fi}}{kT}\right) \tag{7.10.2a}$$

$$p' = N_v \exp\left(-\frac{\mathscr{E}_t - \mathscr{E}_v}{kT}\right) = n_i \exp\left(-\frac{\mathscr{E}_t - \mathscr{E}_{fi}}{kT}\right) \tag{7.10.2b}$$

where n_i is the intrinsic carrier concentration. Letting

$$\frac{C_p}{C_n} = \exp\left(\frac{2e\phi_0}{kT}\right) \quad \text{and} \quad C_p C_n = C^2 \tag{7.10.3}$$

we can rewrite $C_p p' + C_n n'$ as

$$C_p p' + C_n n' = 2Cn_i \cosh \frac{\mathscr{E}_t - \mathscr{E}_{fi} - e\phi_0}{kT} \tag{7.10.4}$$

The quantity ϕ_0 is simply a parameter defining the ratio C_p/C_n, and does not have any other physical meaning.

Note that under nonequilibrium conditions, Eq. (7.9.1) should be written as

$$n = n_i \exp\left(-\frac{\mathscr{E}_{fi} - \mathscr{F}_n}{kT}\right) \tag{7.10.5}$$

and a similar expression is obtained for valence-band holes. Equation (7.10.5) also applies to conduction-band electrons in the surface space-charge region if the quantity $\mathscr{E}_{fi}$ is taken to be the intrinsic Fermi in the surface space-charge region. Since the difference between $\mathscr{E}_{fi}$ at the semiconductor–oxide interface and $\mathscr{E}_{fi}$ in the bulk is $e(\phi_s - \phi_b)$, n_s can be expressed in terms of the electron concentration n in the bulk as

$$n_s = n \exp\left[-\frac{e(\phi_b - \phi_s)}{kT}\right] \tag{7.10.6}$$

Similarly, the hole concentration p_s at the semiconductor–oxide interface is related to that of p in the bulk by

$$p_s = p \exp\left[\frac{e(\phi_b - \phi_s)}{kT}\right] \tag{7.10.7}$$

Using Eqs. (7.10.2), (7.10.6), and (7.10.7), we find that

$$p_s n_s - p'n' = np - n_i^2 = (n_0 + p_0 + n_1)n_1 \tag{7.10.8}$$

where $p_1 = n_1$ is the excess carrier concentration and n_0 and p_0 are, respectively, the equilibrium electron and hole concentration. Substituting Eq. (7.10.8) into Eq. (7.10.1) gives

$$U = \frac{N_t C_n C_p (n_0 + p_0 + n_1)n_1}{C_p(p_s + p') + C_n(n_s + n')} \tag{7.10.9}$$

Before we simplify Eq. (7.10.9) further, we should discuss the physical significance of the surface recombination process.

Because carriers recombine at the surface of a semiconductor, the distribution of excess carriers is no longer uniform over a given cross section of the semiconductor. Hence the flow of excess carriers along a bar sample discussed in Section 7.8 is no longer a one-dimensional problem. The boundary conditions at the surfaces must be incorporated into the mathematical formulation. To do this we remove the surface space-charge region AB in Fig. 7.20a as if the semiconductor ended at A. The situation is illustrated in Fig. 7.20b. To account for the carrier recombination at the surface, we introduce a parameter v_s defined as

$$U = v_s n_1 = v_s p_1 \qquad (7.10.10)$$

where $n_1 = p_1$ is the concentration of excess carriers in the bulk. In other words, the effect of the region AB on the excess carriers is taken into account by demanding that the rate of carrier recombination at the boundary A of Fig. 7.20b be given by U of Eq. (7.10.10).

Note that N_t in Eq. (7.10.10) is the density of surface states and hence it has a dimension of cm^{-2}, while N_t in Eq. (7.5.16) is of cm^{-3}. This means that the quantity U in Eq. (7.10.10) has a dimension of $cm^{-2} s^{-1}$ instead of $cm^{-3} s^{-1}$ in Eq. (7.5.16). Therefore, the parameter v_s has a dimension of velocity (cm/s), and for this reason it is called the *surface-recombination velocity*.

Now let us examine the dependence of v_s on the surface condition of a semiconductor. Under low-level injection, we may use in Eq. (7.10.9) the values of p_s and n_s under thermal equilibrium. Realizing that in Eqs. (7.10.6) and (7.10.7),

$$n_0 = n_i \exp\left(\frac{\mathscr{E}_f - \mathscr{E}_{\mathrm{fi}}}{kT}\right) = n_i \exp\left(\frac{e\phi_b}{kT}\right) \qquad (7.10.11a)$$

$$p_0 = n_i \exp\left(\frac{\mathscr{E}_{\mathrm{fi}} - \mathscr{E}_f}{kT}\right) = n_i \exp\left(-\frac{e\phi_b}{kT}\right) \qquad (7.10.11b)$$

we have

$$n_{s0} = n_i \exp\left(\frac{e\phi_s}{kT}\right) \qquad (7.10.12a)$$

$$p_{s0} = n_i \exp\left(-\frac{e\phi_s}{kT}\right) \qquad (7.10.12b)$$

Using Eqs. (7.10.3) and (7.10.12), we may write $C_p p_s + C_n n_s$ as

$$C_p p_s + C_n n_s = 2C n_i \cosh\left[\frac{e(\phi_s - \phi_0)}{kT}\right] \qquad (7.10.13)$$

Substituting Eqs. (7.10.4) and (7.10.13) into Eq. (7.10.9), using the definition of v_s expressed in Eq. (7.10.10), and neglecting n_1 in comparison with $n_0 + p_0$, we find that

$$v_s = \frac{U}{n_1} = \frac{N_t C(p_0 + n_0)/2n_i}{\cosh\left[(\mathscr{E}_t - \mathscr{E}_{\mathrm{fi}} - e\phi_0)/kT\right] + \cosh\left[e(\phi_s - \phi_0)/kT\right]} \qquad (7.10.14)$$

The important feature of Eq. (7.10.14) is the dependence of v_s on ϕ_s. We should point out that the first term in the denominator of Eq. (7.10.14) is a known constant and hence is independent of changes of surface conditions. As the ambient condition changes, the quantity $e\phi_s$ changes. Since the current amplification factor of a transistor will be affected by recombination of excess carriers at the surface, the performance of the device is sensitive to ambient changes. This is one reason why a stable ambient condition is essential for the stability of a semiconductor device.

Now let us illustrate the use of Eq. (7.10.14). Experimentally, it is found that a

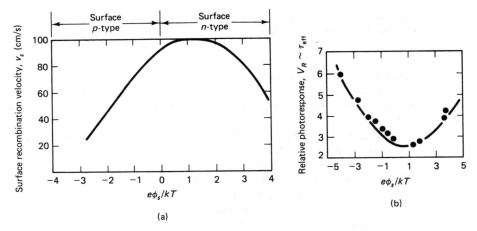

Figure 7.25 (a) Plot of the surface recombination velocity v_s versus $e\phi_s/kT$ for intrinsic germanium sample with $p_0 = n_0 = n_i$ and $N_tC = 900$ cm/s. The curve is based on Eq. (7.10.15). (b) Experimentally measured values (dots) of v_s as deduced from the relative photo-response V_R discussed in Section 7.11. (S. Wang and G. Wallis, *Phys. Rev.*, Vol. 105, p. 1459, 1957.)

chemically etched germanium surface has $\mathcal{E}_t - \mathcal{E}_{fi} = 0.10$ eV and $C_p/C_n = 9$. From Eq. (7.10.3), $e\phi_0 = 0.029$ eV. Using these values and the value of $kT = 0.026$ eV in Eq. (7.10.14), we obtain

$$v_s = \frac{N_tC(p_0 + n_0)/2n_i}{8 + \cosh\left[e(\phi_s - \phi_0)/kT\right]} \tag{7.10.15}$$

Figure 7.25 shows a plot of v_s versus $e\phi_s/kT$ for an intrinsic germanium sample with $p_0 = n_0 = n_i$ and $N_tC = 900$ cm/s. The maximum of v_s occurs at $e\phi_s/kT = e\phi_0/kT = 1.1$ eV. The region near $e\phi_s/kT = 0$ represents the surface condition of negligible surface charge. As mentioned in Section 7.9, the surface condition can be changed drastically if a semiconductor is subjected to ambient changes through the Brattain–Bardeen cycle. In Fig. 7.25, the right side (*n*-type surface) corresponds to surface conditions in a wet atmosphere, whereas the left side (*p*-type surface) corresponds to conditions in an oxidizing atmosphere. The value of $N_tC = 900$ cm/s is typical of a chemically etched germanium surface, although a value as low as 400 cm/s can be obtained. We also should mention that v_s increases with $n_0 + p_0$. Thus it is expected that the value of v_s can be appreciably higher than 100 cm/s in low-resistivity materials. In silicon, the value of v_s is of the order of 1000 cm/s.

7.11 EFFECTS OF SURFACE CONDITION, PARTICLE BOMBARDMENT, AND IMPURITY INCORPORATION ON CARRIER LIFETIME

In this section we first discuss the effect of the surface-recombination process on the distribution and the decay of excess carriers. This discussion is followed by the presentation of experimental data showing the effect of surface condition on the surface-recombination velocity and the effects of impurity and bombardment on the bulk lifetime. Consider a semiconductor bar (Fig. 7.26) of uniform, rectangular cross section. After the bar sample has been exposed to illumination for some time, we turn off the light beam. Let us refer to the case discussed in Section 7.7 in connection with Eq. (7.7.7). For the previous case we did not consider the surface-recombination process; hence the distribution of excess carriers decayed with a time constant equal to the bulk lifetime after the light beam was turned off.

Figure 7.26 shows the distribution of excess carriers in a real situation. Because of the surface-recombination process, there is a constant flow of excess carriers toward the surface. This flow is caused by the diffusion of carriers. The rate of diffusion toward the surface is given by $D_p \, \partial p_1 / \partial y$ and this rate must be equal to the rate of recombination at the surface. Thus, we have

$$D_p \frac{\partial p_1}{\partial y} = v_s p_1 \qquad \text{at} \quad y = 0$$

$$-D_p \frac{\partial p_1}{\partial y} = v_s p_1 \qquad \text{at} \quad y = b$$

(7.11.1)

In Eq. (7.11.1) it is implied that holes are the minority carriers. Similar expressions are also obtained at $z = 0$ and $z = c$. To deal with surface effects, we must first generalize the one-dimensional time-dependent equation, Eq. (7.7.1), and then impose on the solution the appropriate boundary conditions similar to those presented in Eq. (7.11.1).

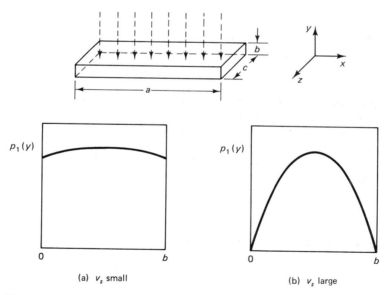

(a) v_s small   (b) v_s large

Figure 7.26 Excess-carrier distribution for two extreme cases: (a) where the rate of surface recombination is slow and (b) where the rate of surface recombination is fast. Because of recombination of excess carriers at the surfaces, the excess-carrier concentration decreases toward the surfaces. The distribution is taken along the direction of the smaller dimension in a semiconductor slab of rectangular shape.

The general time-dependent diffusion equation for a three-dimensional case reads

$$\frac{\partial p_1}{\partial t} = D_p \left(\frac{\partial^2 p_1}{\partial x^2} + \frac{\partial^2 p_1}{\partial y^2} + \frac{\partial^2 p_1}{\partial z^2} \right) + \mu_p E_x \frac{\partial p_1}{\partial x} - \frac{p_1}{\tau_0} \qquad (7.11.2)$$

It is understood from Eqs. (7.11.1) and (7.11.2) that the semiconductor under discussion is n-type and the applied electric field is in the x direction. To simplify the solution of Eq. (7.11.2), we make the following assumptions. The excess carriers are uniformly distributed in the xz plane so that the $\partial^2 p_1/\partial x^2$ and $\partial^2 p_1/\partial z^2$ terms in Eq. (7.11.2) can be taken as zero. Furthermore, the applied electric field is so small that the term $\mu_p E_x \partial p_1/\partial x$ is negligible in comparison with the other terms. Thus Eq. (7.11.2) becomes

$$\frac{\partial p_1}{\partial t} = D_p \frac{\partial^2 p_1}{\partial y^2} - \frac{p_1}{\tau_0} \qquad (7.11.3)$$

As we shall see shortly, the step that we take in neglecting the x and z dependence of p_1 is indeed justified if the sample dimension is such that $b << a$ and c.

Let us find the solution of Eq. (7.11.3) after the light beam is turned off. Since no generation of excess carriers is involved, the situation is completely described by Eqs. (7.11.1) and (7.11.3). We propose a solution of the following form:

$$p_1(y, t) = A \exp(-\alpha t) \cos \zeta \left(y - \frac{b}{2} \right) \qquad (7.11.4)$$

Equation (7.11.4) is a solution of Eq. (7.11.3) if

$$\alpha = D_p \zeta^2 + \tau_0^{-1} \qquad (7.11.5)$$

The boundary conditions at $y = 0$ and $y = b$ require that

$$D_p \zeta \sin \frac{\zeta b}{2} = v_s \cos \frac{\zeta b}{2} \qquad (7.11.6)$$

Equation (7.11.6) determines the value of ζ, which in turn determines the decay constant α of Eq. (7.11.5) and the y dependence of the excess-carrier concentration.

Equation (7.11.6) has a simple solution under either one of two extreme conditions. If v_s is very small, the value of ζ is expected to be small. In this extreme case we approximate $\sin(\zeta b/2)$ by $\zeta b/2$ and $\cos(\zeta b/2)$ by 1. Thus, from Eq. (7.11.6), we find that

$$\zeta^2 = \frac{2v_s}{bD_p} \qquad (7.11.7)$$

and from Eq. (7.11.5), we obtain

$$\frac{1}{\tau_{\text{eff}}} = \alpha = \frac{2v_s}{b} + \frac{1}{\tau_0} \qquad (7.11.8)$$

The other extreme case is one in which v_s is very large. This means that $\zeta b/2$ must be very close to $\pi/2$ to satisfy Eq. (7.11.6). As a first-order approximation, we set

$$\zeta = \frac{\pi}{b} \qquad (7.11.9)$$

Thus, from Eq. (7.11.5), we obtain

$$\frac{1}{\tau_{\text{eff}}} = \alpha = \frac{D_p \pi^2}{b^2} + \frac{1}{\tau_0} \qquad (7.11.10)$$

If we consider, in addition to the y dependence, the x and z variation of p_1, two more terms, $2v_s/a$ and $2v_s/c$, should be added to Eq. (7.11.8). However, for $b \ll a$ and c, the term $2v_s/b$ is the dominant term among the three. Under such circumstances, the step that we took earlier in neglecting the x and z dependence in Eq. (7.11.2) is justified. Similarly, Eq. (7.11.10) is valid if $b \ll a$ and c.

The quantity τ_{eff} is called the *effective lifetime of excess carriers,* which takes into account the recombination processes both at the surface and in the bulk. Figure 7.26 shows the distribution of excess carriers along the y direction. Had it not been for surface recombination, the distribution of excess carriers would be uniform. For small v_s, p_1 drops slightly toward the surface, and the spatial variation of p_1 is represented by a small portion of a cosine curve (Fig. 7.26a). For large v_s, on the other hand, Eq. (7.11.1) demands that p_1 at the surface be small so as to keep $D_p \, \partial p_1/\partial y$ finite. The distribution of excess carriers in the latter case is represented by a half-cosine curve with p_1 practically zero at the surface (Fig. 7.26b).

Now let us give a physical interpretation of the quantity τ_{eff}. Due to surface recombination, a current of density $eD_p \, \partial p_1/\partial y$ is drawn toward the surface. We define an average diffusion velocity $\bar{v}_d$ such that

$$e\bar{p}_1\bar{v}_d = eD_p \frac{\partial p_1}{\partial y} \qquad (7.11.11)$$

represents the current density drawn toward the surface, where $\bar{p}_1$ is the spatial average of p_1. For the case shown in Fig. 7.26a,

$$eD_p \frac{\partial p_1}{\partial y} = eD_p \zeta A \sin \frac{\zeta b}{2} \cong \frac{eD_p A \zeta^2 b}{2} \qquad (7.11.12)$$

and $\bar{p}_1$ can be approximated by A, the maximum value of p_1. Thus we have

$$\bar{v}_d = \frac{D_p \zeta^2 b}{2} = v_s \qquad (7.11.13)$$

The average transit time $\bar{t}$ needed for excess carriers to reach the surface is

$$\bar{t} = \frac{b}{2\bar{v}_d} = \frac{b}{2v_s} \qquad (7.11.14)$$

Comparing Eq. (7.11.14) with Eq. (7.11.8), we see that the effective lifetime is the combined time constant of two decay processes in parallel operation. For the bulk-recombination process, the decay constant is clearly τ_0. For the surface-recombination process, carriers must first diffuse toward the surface and then recombine at the surface. Therefore, the decay rate for the latter process is controlled by either the rate of surface recombination (proportional v_s) or the rate of diffusion (proportional to D_p), whichever is smaller. For small v_s, the decay rate is controlled by the rate of surface recombination, and hence $\bar{t}$ in the expression of the effective lifetime

$$\tau_{\text{eff}}^{-1} = \bar{t}^{-1} + \tau_0^{-1} \qquad (7.11.15)$$

Sec. 7.11   Effects on Carrier Lifetime

315

is given by $\bar{t}$ of Eq. (7.11.14). For large v_s, the average diffusion velocity $\bar{v}_d$ is controlled by the rate of diffusion and is equal to

$$\bar{v}_d = \frac{D_p \zeta^2 b}{2} = \frac{D_p \pi^2}{2b} \tag{7.11.16}$$

Thus we find that

$$\bar{t} = \frac{b}{2\bar{v}_d} = \frac{D_p \pi^2}{b^2} \tag{7.11.17}$$

If we use this value of $\bar{t}$ in Eq. (7.11.15), we find that Eq. (7.11.15) is indeed identical to Eq. (7.11.10). The reader is asked to verify Eq. (7.11.16).

Next we consider a photoconduction experiment (Fig. 7.27a) and discuss the steady-state solution. With the light beam shining on the sample, we must add a term G in Eq. (7.11.3) to account for the generation process. In the steady state, this new equation becomes

$$\frac{\partial p_1}{\partial t} = D_p \frac{\partial^2 p_1}{\partial y^2} - \frac{p_1}{\tau_0} + G = 0 \tag{7.11.18}$$

Equation (7.11.18) is similar to Eq. (7.7.6). Because of surface recombination, the distribution of excess carriers is no longer uniform, and hence the term $D_p \, \partial^2 p_1/\partial y^2$ can no longer be neglected in Eq. (7.11.18). In Section 7.7, we showed from Eq. (7.7.7) that the steady-state solution was given by $p_1 = G\tau_0$. In the following discussion, we see how the process of surface recombination affects the steady-state solution.

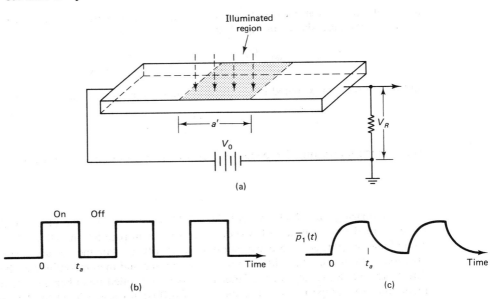

Figure 7.27 (a) Experimental arrangement for measuring photoconductance of a semiconductor, (b) illumination as a function of time, and (c) average excess-carrier concentration $\bar{p}_1$ as a function of time. The voltage V_R developed across resistance R is proportional to $\bar{p}_1(t)$.

In Eq. (7.11.18), we again assume that G is uniform throughout the sample. The solution of Eq. (7.11.18) is given by

$$p_1(y) = A \cosh \frac{2y - b}{2L_p} + G\tau_0 \qquad (7.11.19)$$

where $L_p = (D_p\tau_0)^{1/2}$ and A is a constant to be determined. As pointed out in Appendix A7, the complete solution of a differential equation consists of the complementary solution and the particular solution. In Eq. (7.11.19) the first term is the complementary solution, whereas the second term is the particular solution. Since the problem is symmetrical with respect to $y = b/2$, only the cosh term is kept, which is an even function with respect to $y = b/2$.

Applying the boundary conditions of Eq. (7.11.1) to $p_1(y)$ of Eq. (7.11.19), we find the constant A to be

$$A = \frac{-v_sG\tau_0}{(D_p/L_p) \sinh (b/2L_p) + v_s \cosh (b/2L_p)} \qquad (7.11.20)$$

Let $\bar{p}_1$ be the value of p_1 averaged over the sample thickness b. Averaging $p_1(y)$ of Eq. (7.11.19) over y and using the value of A in Eq. (7.11.19), we obtain

$$\bar{p}_1 = G\tau_0 \left[1 - \frac{\sinh w}{(b/2v_s\tau_0) \sinh w + w \cosh w} \right] \qquad (7.11.21)$$

where $w = b/2L_p$. For small v_s, Eq. (7.11.21) reduces to $\bar{p}_1 = G\tau_{\text{eff}}$.

We should point out that the steady-state distribution of Eq. (7.11.19) is exact, whereas the transient distribution of Eq. (7.11.4) is approximate. The exact transient solution is given by

$$p_1(y, t) = \sum_i A_i \exp (-\alpha_i t) \cos \zeta_i \left(y - \frac{b}{2} \right) \qquad (7.11.22)$$

Besides the value of ζ given by either Eq. (7.11.7) or (7.11.9), there are other values of ζ that satisfy Eq. (7.11.6). In Eq. (7.11.22), the summation is over all possible values of ζ. However, under certain conditions, we need only consider one value of ζ, the value of ζ given by Eq. (7.11.7), because the term with this value of ζ decays most slowly and hence is the dominant term in Eq. (7.11.22). This approximation simplifies greatly our interpretation of the experimental result. With a periodic light source (Fig. 7.27b), the on-portion of Fig. 7.27c can be represented by the equation

$$\bar{p}_1 = G\tau_{\text{eff}} \left[1 - \exp \left(-\frac{t}{\tau_{\text{eff}}} \right) \right] \qquad (7.11.23)$$

and the off-portion of Fig. 7.27c by

$$\bar{p}_1 = G\tau_{\text{eff}} \exp \left(-\frac{t - t_a}{\tau_{\text{eff}}} \right) \qquad (7.11.24)$$

after $t > \tau_{\text{eff}}$ and $t > t_a + \tau_{\text{eff}}$, respectively.

The photoconductive decay method shown in Fig. 7.27a offers a convenient way of measuring the lifetime. To provide a periodic light source, we simply cut off the light beam periodically by a rotating mechanical chopper. We also can use a flash lamp, the firing of which is controlled electronically by an *RC* circuit. However, the use of light beam as an excitation source for generating electron–hole pairs is limited to the range of lifetime above 1 μs because turning on and off a light beam usually requires a time in the neighborhood of 1 μs. For studying a short lifetime, say of the order of 10^{-8} s, we must use energetic electrons as the source of excitation. A short pulse of electron beam can be obtained either by sweeping a dc beam across an aperture or by gating the electron source.

Generally speaking, we can classify excitation sources into two categories, penetrating and nonpentrating. Refer to Eq. (6.9.1) which describes the decay in the intensity of a light beam inside a semiconductor. The absorption coefficient α depends on the photon energy of a light beam (Fig. 6.24). We list in Table 7.3 the values of α at selected values of photon energy measured in germanium and silicon at room temperature. The assumption of uniform generation, on which the foregoing analysis is based, is a good one only if the value of α*d* in Eq. (6.9.1) is on the order of 1 or less. For $d = 0.2$ cm, α must be smaller than 5 cm^{-1}. This limits the use of light beams to those having photon energies close to the gap energy, that is, 0.665 eV in Ge and 1.12 eV in Si. These energies lie in the infrared region of the light spectrum. Nearly uniform generation of electron–hole pairs is also possible with a beam of energetic electrons having electron energy in excess of 700 keV. Wertheim and Augustyniak (G. K. Wertheim and W. M. Augustyniak, *Rev. Sci. Instrum.*, Vol. 27, p. 1062, 1956) used a van de Graaff accelerator to produce a beam of high-energy electrons in lifetime measurements.

Table 7.3 shows that light beams in the visible region will be strongly absorbed in Ge and Si. For $\alpha = 10^4$ cm^{-1}, the intensity of a light beam is attenuated by a factor of 2.7 in a distance $d = 10^{-4}$ cm. In such cases, the generation of excess carriers is confined to a very shallow region near the surface of a semiconductor. Two modifications must be made in our calculation of the excess-carrier concentrations. First, $G = 0$ in Eq. (7.11.28) because no appreciable generation exists in the bulk. Therefore, the solution of Eq. (7.11.18) is given by

$$p_1(y) = A \sinh \frac{y}{L_p} + B \cosh \frac{y}{L_p} \qquad (7.11.25)$$

The second modification concerns the boundary condition. If the rate of generating excess carriers is *G*, the boundary conditions at the illuminated and unilluminated surfaces are, respectively, given by

$$D_p \frac{\partial p_1}{\partial y} = G - v_s p_1 \qquad \text{at} \quad y = b \qquad (7.11.26a)$$

$$= v_s p_1 \qquad \text{at} \quad y = 0 \qquad (7.11.26b)$$

Equation (7.11.26) replaces Eq. (7.11.1). Because of the term *G* in Eq. (7.11.26a), the problem is no longer symmetrical with respect to $y = b/2$; hence both the sinh and cosh terms must be kept in Eq. (7.11.25). The steady-state solution of such a problem is discussed in Problem 7.33. For the transient solution, we must add sine terms in the summation of Eq. (7.11.22). A complete analysis again shows that after a time $t > \tau_{\text{eff}}$, the distribution of excess carriers is adequately described by $p_1(y, t)$ of Eq. (7.11.4), with τ_{eff} given by either Eq. (7.11.8) or (7.11.10).

Now let us return to Fig. 27a. The ac voltage V_R developed across *R* is propor-

TABLE 7.3 ABSORPTION COEFFICIENT α (AT ROOM TEMPERATURE) IN Ge AND Si

	Material							
	Germanium				Silicon			
Photon energy (eV)	0.665	0.70	0.80	0.90	1.12	1.15	1.30	1.70
α (cm^{-1})	5	20	400	7000	4	10	300	2000

tional to the change in the conductance of the sample, which in turn is proportional to the average excess-carrier concentration $\bar{p}_1$. Using the definition of conductivity σ and the condition $p_1 = n_1$, the reader can easily show that

$$V_R = V_0 \frac{R(e\mu_n + e\mu_p)}{(1 + R\sigma_0 bc/a)^2} \frac{bc}{a'} \bar{p}_1 \qquad (7.11.27)$$

where σ_0 is sample conductivity without illumination and a' is the length of the illuminated portion of the sample. The value of $\bar{p}_1$ in Eq. (7.11.27) is given by either Eq. (7.11.23) or (7.11.24), if the generation of electron–hole pairs is uniform.

Either the steady-state or the decay method may be used to measure the effective lifetime. In Eq. (7.11.26) the steady-state value of $\bar{p}_1$ is $G\tau_{\text{eff}}$. Therefore, the steady-state value of V_R is directly proportional to τ_{eff}. Once the value of τ_{eff} is determined for one specific value of V_R, the constant of proportionality is known. The steady-state method is very convenient for observation of the variation of τ_{eff} in surface studies.

Figure 7.25b shows the variation of V_R as a function of $e\phi_s/kT$ observed experimentally by Wang and Wallis. According to Eq. (7.10.14), v_s has a maximum at $\phi_s = \phi_0$. Since V_R is proportional to τ_{eff} and thus inversely proportional to v_s, the position of minimum V_R in Fig. 7.25b determines the value of ϕ_0 in Eq. (7.10.14). The experiment was performed on a germanium slice 0.04 cm thick, with a resistivity of 35 Ω-cm, n-type, and having a bulk lifetime of 800 μs. The value of ϕ_s was varied by changing the ambient conditions through the Brattain–Bardeen cycle. The following values were used in fitting the theoretical curve, except for a proportionality constant, of τ_{eff} [Eqs. (7.11.8) and (7.10.14)] to the experimental points: $(v_s)_{\text{max}} = 55$ cm/s, $\mathscr{E}_t - \mathscr{E}_{\text{fi}} = 0.10$ eV, and $e\phi_0 = 0.029$ eV (S. Wang and G. Wallis, *Phys. Rev.*, Vol. 105, p. 1459, 1957). We see that the agreement between the theory and the experiment is good.

Whereas the steady-state method is useful in measuring the relative change in v_s, the decay method is convenient for measuring the absolute value of the bulk lifetime τ_0. For a semiconductor sample with well-etched surfaces, $v_s = 30$ cm/s, for example, and of fair thickness, say 0.2 cm, $b/2v_s = 3.3 \times 10^{-3}$ s in Eq. (7.11.8). In such cases, the decay method provides a direct measurement of τ_0 up to a value of about 500×10^{-6} s. For the excitation source, we can use ordinary light beams or pulse injections (as in the Haynes–Shockley experiment) for $\tau_0 > 10^{-6}$ s, and high-energy electrons or pulsed lasers for $\tau < 10^{-6}$ s.

To conclude our discussion, we present observations of the effects of bombardment and impurity on the bulk lifetime. Let τ_{eff} and τ'_{eff} be the effective lifetime, respectively, before and after the bombardment took place or the impurity was introduced. To make sure that the surface term in the expression of τ_{eff} did not change during the experiment, we should sandblast the surfaces. In other words, Eq. (7.11.10) should be used so that the term $D_p\pi^2/b^2$ remained the same in the expressions of τ_{eff} and τ'_{eff}. Thus the change in the effective lifetime is purely a bulk effect, that is,

$$\frac{1}{\tau'_{eff}} - \frac{1}{\tau_{eff}} = \frac{1}{\tau'_0} - \frac{1}{\tau_0} = BN_t \qquad (7.11.28)$$

where N_t is the density of recombination centers introduced during the bombardment or by the impurity, and B is a constant of proportionality which can be found through the use of Eq. (7.5.17).

Figure 7.28 shows the results of lifetime measurements in Ge reported by Curtis et al. It is believed that irradiation by fast neutrons and gamma rays produce interstitials with energy levels at 0.20 eV below the conduction band. These energy levels act as recombination centers. It can be seen from Fig. 7.28 that the density of such recombination centers increases and hence the lifetime decreases almost linearly with the amount of exposure (O. L. Curtis, Jr., J. W. Cleland, J. H. Crawford, Jr., and J. C. Pigg, *J. Appl. Phys.*, Vol. 28, p. 1161, 1957).

High-energy electron bombardment also produces recombination centers in a semiconductor. Work by Wertheim shows that lattice imperfections are produced at a rate of 0.18 center/cm³ in a silicon sample 5×10^{-2} cm thick for an electron beam of density n_e electron/cm² at 700 keV (G. K. Wertheim, *Phys. Rev.*, Vol. 105, p. 1730, 1957). In a lifetime measurement (for $\tau_0 < 10^{-6}$ s), the bombarding current density must be kept low. For $J = 10$ μA/cm², the number of bombarding electrons amounts to only 6×10^7 electrons/cm² during a period of 10^{-6} s. In other words, if the electron pulse is 10^{-6} s long, these high-energy electrons produce, on the average, a density N_t of 1.3×10^7 cm^{-3} of recombination centers per pulse in the lifetime measurement. This value is still insignificant even for repeated measurements.

However, the situation changes drastically if a semiconductor is under constant bombardment. For $t = 1$ min, an electron-beam current of density 10 μA/cm² and 700 keV will produce a density $N_t = 7 \times 10^{14}$ of recombination centers. This density

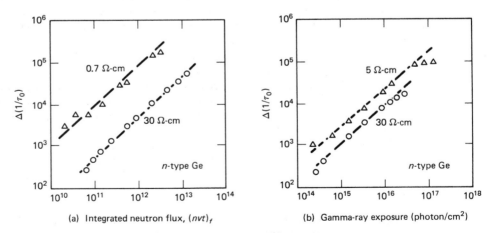

(a) Integrated neutron flux, $(nvt)_f$

(b) Gamma-ray exposure (photon/cm²)

Figure 7.28 Experimental curves showing the effects of bombardment (a) by fast neutrons and (b) by gamma rays on the lifetime of excess carriers in Ge. (O. L. Curtis, Jr., J. W. Cleland, J. H. Crawford, Jr., and J. C. Pigg, *J. Appl. Phys.*, Vol. 28, p. 1161, 1957.)

is already large enough to have a noticeable effect on the bulk lifetime. Figure 7.29 shows the measurement of lifetime as a function of temperature in 7-Ω-cm n-type bombarded silicon. As reported by Wertheim, the curves in Fig. 7.29 can be fitted by the expression

$$\tau_0 = \frac{1}{n_e}\left[1.44 \times 10^7 + 8.6 \times 10^{13} \exp\left(-\frac{0.31}{kT}\right)\right] \qquad (7.11.29)$$

where n_e is the bombarding-electron density. Equation (7.11.29) is obtained by using $N_t = 0.18 n_e$, $C_p = 4.0 \times 10^{-7}$ cm³-s, and $C_n = 1.3 \times 10^{-9}$ cm³-s in Eq. (7.5.17). According to our discussion in Section 7.5, the lifetime in an n-type semiconductor varies as

$$\tau_0 = \tau_{p0} + \tau_{p0}\frac{n'}{n_0} \quad \text{or} \quad \tau_0 = \tau_{p0} + \tau_{n0}\frac{p'}{n_0} \qquad (7.11.30)$$

depending on whether $n' >> p'$ or $p' >> n'$. Therefore, if the trap energy is near the conduction band, τ_0 should vary as $\exp\left[-(\mathscr{E}_c - \mathscr{E}_t)/kT\right]$, and if it is near the valence band, as $\exp\left[-(\mathscr{E}_t - \mathscr{E}_v)/kT\right]$. In either case the exponential variation with temperature is borne out by the experiment. In Table 7.4 we present the data on energy levels and capture cross sections of irradiation-related defects as compiled by Wertheim and by Curtis.

Besides vacancies and interstitials, impurities can also act as recombination centers. Figure 7.30 shows the temperature dependence of lifetime reported by Bemski (G. Bemski, *Phys. Rev.*, Vol. 111, p. 1515, 1958) in p-type silicon having a gold concentration of $N_t = 5 \times 10^{14}$ cm⁻³. As we discuss in Chapter 8, a relatively heavy concentration of gold (say, $N_t > 10^{16}$ cm⁻³) is needed in silicon diodes for fast switching. Of the two energy levels of gold, the donor level at 0.35 eV from the valence band is active in p-type silicon, whereas the acceptor level at 0.54 eV from the conduction band is active in n-type silicon. Therefore, the lifetime in Fig. 7.30 can be expressed as

$$\tau = \tau_{n0}\left[1 + \frac{N_v}{p_0}\exp\left(-\frac{0.35}{kT}\right)\right] \qquad (7.11.31)$$

since the term involving τ_{p0} in Eq. (7.5.17) can be neglected. To fit the experimental

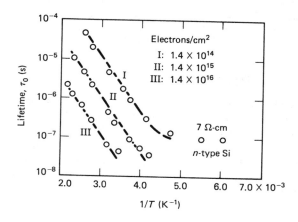

Electrons/cm²
I: 1.4×10^{14}
II: 1.4×10^{15}
III: 1.4×10^{16}

7 Ω-cm
o   o
n-type Si

Figure 7.29 Experimental curves showing the effects of electron bombardment on the lifetime of excess carriers in Si. The different curves are for different intensities of electron bombardment (electrons/cm²). (G. K. Wertheim, *Phys. Rev.*, Vol. 105, p. 1730, 1957.)

TABLE 7.4 ENERGY LEVELS AND CAPTURE CROSS SECTIONS OF IRRADIATION-INTRODUCED DEFECTS IN Si AND Ge

Material	Irradiation source	Defect level (eV)	Cross section[a] ($\times 10^{16}$ cm^2)
Si, n type	Neutrons	Midgap	$\tau = 3.9 \times 10^5 \, \Phi^{-1}$
Si, p type	Neutrons	Midgap	$\tau = 4.3 \times 10^5 \, \Phi^{-1}$
Si, n type	0.7-MeV electrons	$\mathcal{E}_t - \mathcal{E}_v = 0.27$	$800(\sigma_p), 95(\sigma_n)$
Si, p type	0.7-MeV electrons	$\mathcal{E}_c - \mathcal{E}_t = 0.16$	$18(\sigma_p), 19(\sigma_n)$
Ge, n type	Neutrons	$\mathcal{E}_c - \mathcal{E}_t = 0.20$	$9(\sigma_p), 45(\sigma_n)$
Ge, n type	^{60}Co gamma rays	$\mathcal{E}_c - \mathcal{E}_t = 0.13$	$0.6(\sigma_p), 15(\sigma_n)$
Ge, n type	2-MeV electrons	$\mathcal{E}_c - \mathcal{E}_t = 0.18$	$16(\sigma_p), 1.6(\sigma_n)$

[a]The total neutron flux density Φ is given by $\Phi = nvt$ and has the dimension cm^{-2} where n is the neutron density (cm^{-3}), v the velocity (cm/s), and t the time of exposure (s).
Source: Data from W. K. Wertheim, *J. Appl. Phys.*, Vol. 30, p. 1166, 1959, and O. L. Curtis, Jr., *J. Appl. Phys.*, Vol. 30, p. 1174, 1959.

curve to Eq. (7.11.31), however, it is necessary to assume a T^2 temperature dependence of τ_{n0}. We should point out that in samples with smaller p_0, the second term in Eq. (7.11.31) becomes relatively more important. Hence it is expected that the lifetime will decrease faster with decreasing temperature than the curve shown in Fig. 7.30. In n-type silicon, the active level is very close to the intrinsic Fermi level. This means that we can neglect the n' and p' terms in Eq. (7.5.17). Therefore, the temperature dependence of τ_0 comes almost entirely from either τ_{n0} or τ_{p0}.

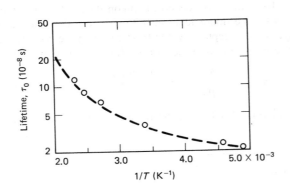

Figure 7.30 Temperature dependence of lifetime of excess carriers in gold-doped, p-type silicon. (G. Bemski, *Phys. Rev.*, Vol. 111, p. 1515, 1958.)

PROBLEMS

7.1. Derive the thermal-equilibrium condition expressed in Eq. (7.2.5) by applying the method of detailed balancing to valence-band holes. Also find the ratio of the hole concentrations p_{n0}/p_{p0}, following the same approach as that leading to Eq. (7.2.7).

7.2. A germanium (silicon or gallium-arsenide) diode is made of materials with resistivities $\rho = 1$ Ω-cm p-type and $\rho = 0.1$ Ω-cm n-type. Use the values given in Tables 6.4 and 6.8 for N_c, N_v, μ_n, and μ_p. Find the Fermi level in n- and p-type materials and also the value of V_d in the three diodes.

7.3. (a) From the Boltzmann transport equation, show that

$$J_x = - e^2 E_x \int v_x^2 \tau \frac{df_0}{d\mathscr{E}} D(\mathscr{E}) d\mathscr{E} + e \frac{d}{dx} \int v_x^2 \tau f_0 D(\mathscr{E}) \, d\mathscr{E}$$

(b) Then derive the current equation for electrons, Eq. (7.1.5).

(c) Also show that in nondegenerate semiconductors,

$$D = \frac{kT}{e} \mu$$

and under thermal equilibrium,

$$J_n = 0$$

7.4. (a) Repeat Problem 7.3 for a degenerate n-type semiconductor where $\mathscr{E}_f > \mathscr{E}_c$. Show that

$$D = \frac{2}{5} \frac{\tau}{m^*} (\mathscr{E}_f - \mathscr{E}_c) \quad \text{and} \quad \mu = \frac{e\tau}{m^*}$$

In the derivation above, assume that $T = 0$ K, so that f_0 is represented by a step function.

(b) Using the result in part (a), verify that

$$J_n = J_{\text{drift}} + J_{\text{diffusion}} = 0$$

in a degenerate junction under thermal equilibrium (i.e., without applied bias). In this case, what is the proper form for $J_{\text{diffusion}}$? Also explain why the form is proper.

7.5. (a) Apply J_n of Eq. (7.1.5) and

$$J_p = e\mu_p pE - eD_p \frac{\partial p}{\partial x}$$

to a p-n junction under thermal equilibrium. By setting $J_n = 0$ and $J_p = 0$, show that

$$eV_d = -e \int_{x_A}^{x_B} E \, dx = kT \ln \left(\frac{n_n}{n_p}\right)_0 = kT \ln \left(\frac{p_p}{p_n}\right)_0$$

where the subscripts n and p refer to the n and p regions, respectively, and the symbol 0 indicates the thermal equilibrium condition. In the derivation, make clear the assumptions used.

(b) Discuss the validity of the expression for eV_d in a degenerate junction.

7.6. According to Poisson's equation, we see that should a spatial separation of excess carrier concentrations n_1 and p_1 develop, a local space-charge field E_{spch} would result, [$E(x)$ of Fig. 7.4]. Because of E_{spch} there would exist a difference in J inside and outside the space-charge region. This current difference would bring back the excess carriers together. Refer to Fig. P7.6 and use the continuity equation

$$e \frac{\partial p_1 w(x_n)}{\partial t} = -(J_{\text{sp}})_p$$

to define an effective velocity $(v_p)_{\text{eff}}$ with which ep_1 moves:

$$(v_p)_{\text{eff}} = \frac{-(J_{\text{sp}})_p}{ep_1}$$

A similar expression is defined for electrons. Under an applied field E_{app}, the quantities x_n and x_p in Fig. P7.6a are

$$x_n = \mu_n E_{app} t, \qquad x_p = \mu_p E_{app} t$$

Show that because of the difference in $(v_p)_{eff}$ and $(v_n)_{eff}$, the final position x_f in Fig. P7.6b is given by

$$x_f = \frac{\mu_n \mu_p (p - n)}{\mu_n n + \mu_p p} E_{app} t$$

within a time given by the dielectric relaxation time.

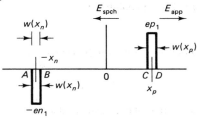

(a)

(b)

Figure P7.6

7.7. Suppose that the charge distribution of Fig. 7.4 existed in a germanium sample ($\epsilon = 16\epsilon_0$) with $n_0 = 10^{16}$ cm^{-3} and $p_0 = 5.3 \times 10^{10}$ cm^{-3} and that the distribution was a result of separation of excess carriers with $n_1 = 1 \times 10^{13}$ cm^{-3} on the left and $p_1 = 1 \times 10^{13}$ cm^{-3} on the right for a distance $AB = CD = 10^{-3}$ cm.
   (a) Calculate the maximum space-charge field created by the polarization of opposite charges.
   (b) Find the rate dQ/dt of charge depletion in regions AB and CD by taking the difference of current densities at points A and B and at points C and D.
   (c) Show by simple semiquantitative arguments that the position of n_1 moves within a time ϵ/σ toward that of p_1.

7.8. Derive Eq. (7.5.13) for the net rate of hole capture. Show that p' in Eq. (7.5.13) is given by Eq. (7.5.14).

7.9. Refer to our discussion in Section 6.5 and specifically to Fig. 6.8 about the change of the Fermi level as a function of temperature in an n-type semiconductor. If it is assumed that the trap energy $\mathcal{E}_t$ lies halfway between $\mathcal{E}_d$ and $\mathcal{E}_{fi}$, mark on Fig. 6.8 the temperature T_3 at which $n_0 = n'$ in Eq. (7.5.17). Discuss the temperature dependence of lifetime in the intermediate temperature region $T_2 < T < T_1$.

7.10. Repeat Problem 7.9 for the case in which the trap energy lies halfway between $\mathcal{E}_{fi}$ and $\mathcal{E}_v$. Assuming that $N_c = N_v$, mark on Fig. 6.8 the temperature T_3 at which $n_0 = p'$ in Eq.

(7.5.17). Discuss the temperature dependence of lifetime for $T_2 < T < T_1$, assuming that τ_{n0} and τ_{p0} are of the same order of magnitude. The temperatures T_1 and T_2 are defined in Section 6.5.

7.11. Refer to Fig. 7.8b and the values of τ_{p0}, τ_{n0}, and n' associated with the dashed curve. Decide which term is the dominant term in Eq. (7.5.17) if the resistivity of the sample is **(a)** 5×10^{-3} Ω-cm, n type; **(b)** 5 Ω-cm, n type; **(c)** 5 Ω-cm, p type; or **(d)** 5×10^{-2} Ω-cm, p type. Discuss the temperature dependence of the lifetime around room temperature in each case.

7.12. Given $\tau_n = 50 \times 10^{-6}$ s and $\tau_p = 1 \times 10^{-6}$ s in a group of germanium samples with the same type and concentration of recombination centers but with different resistivities, find the values of p' and n' by fitting the following experimental values of τ_0 reported by R. N. Hall (*Phys. Rev.*, Vol. 83, p. 228, 1951; Vol. 87, p. 387, 1952):

	Resistivity (Ω-cm)					
	1.5	30	48	10	3	0.5
p_0 (cm^{-3})	$p_0 = n_i^2/n_0$, where $n_i = 2 \times 10^{13}$ cm^{-3}					
n_0 (cm^{-3})	3×10^{11}	5×10^{12}	2×10^{13}	2×10^{14}	8×10^{14}	8×10^{15}
τ_0 (10^{-6} s)	50	120	200	35	8	1.5

From either p' or n', determine the trap energy $\mathcal{E}_t$.

7.13. In Eqs. (7.6.7), (7.6.8), and (7.6.9), there are three unknowns p_1, n_1, and $\partial E_x/\partial x$. Therefore, we can eliminate two of these and obtain a partial differential equation in either p_1 or n_1. Show that under dc steady state, the equation is given by

$$L_4^4 \frac{\partial^2 p_1}{\partial x^4} - L_2^2 \frac{\partial^2 p_1}{\partial x^2} + L_1 \frac{\partial p_1}{\partial x} + p_1 = G$$

where G is the generation rate and the various coefficients are

$$L_4^4 = D_p D_n \tau \tau_r \qquad L_2^2 = D'\tau + \bar{D}\tau_r + \mu_n \mu_p \tau \tau_r E_0^2 \qquad L_1 = (\mu'\tau + \bar{\mu}\tau_r)E_0$$

The quantity $\tau_r = \epsilon/\sigma (= \epsilon/e(\mu_n n + \mu_p p)$ is the dielectric relaxation time; D' and μ' are given by (7.6.11) and (7.6.12) respectively. The above equation can be obtained by letting $n_1 = p_1 - n_2$ in Eqs. (7.6.8) and (7.6.9) and then by eliminating $\partial E_x/\partial x$ and n_2. Find the expressions for $\bar{D}$ and $\bar{\mu}$. Discuss the conditions under which the equation above reduces to Eq. (7.6.10). (F. Stockmann, *Photoconductivity Conference,* John Wiley & Sons, Inc., New York, 1956, p. 269; *Z. Phys.*, Vol. 147, p. 544, 1957.)

7.14. In spherical coordinates, the steady-state diffusion equation reads

$$\frac{D_p}{r^2} \frac{d}{dr}\left(r^2 \frac{dp_1}{dr}\right) - \frac{p_1}{\tau_0} = 0$$

Show that the equation above has the following solution:

$$p_1(r, t) = Ar^{-1} \exp\left(-\frac{r}{L_p}\right)$$

Describe an experiment in which the lifetime τ_0 can be determined, and discuss the experimental condition under which the solution is valid.

7.15. Show that the dc steady-state solution of Eq. (7.7.1) is given by

$$p_1(x) = A \exp\left(\frac{x}{L_1}\right) + B \exp\left(\frac{x}{L_2}\right)$$

Express L_1 and L_2 in terms of D_p, μ_p, and E. One case of practical importance corresponds to the situation prevailing in the base region of a transistor. At the emitter ($x = 0$), $p_1 = p_1(0)$; and at the collector ($x = W$), $p_1 = 0$. Assuming that $E = 0$ in the base region, show that

$$p_1(x) = p_1(0) \frac{\sinh\left[(W - x)/L\right]}{\sinh\left(W/L\right)}$$

where $L = \sqrt{D_p\tau_0}$.

7.16. Verify by substitution that Eq. (7.7.13) satisfies Eq. (7.7.8) for $t > 0$ if $E = 0$ and $\tau = \infty$. Show that at $t = 0$, $p_1(x, t)$ of Eq. (7.7.13) behaves like a delta function. In other words, at the instant $t = 0$, the function $p_1(x, t) = \infty$ at $x = 0$ and zero elsewhere, and

$$\int_{-\infty}^{\infty} p_1(x, t)\, dx = 1.$$

7.17. Given $\tau_0 = 10^{-5}$ s, $\mu_p = 1900$ cm^2/V-s, and $D_p = 50$ cm^2/s in a typical germanium sample. Plot $p_1(x, t)$ of Eq. (7.8.7) as a function of x with $E = 5$ V/cm for $t = 10^{-6}$, 2×10^{-6}, 5×10^{-6}, and 10^{-5} s. Compare the area under the curve at $t = 10^{-6}$ and 10^{-5} s, and verify that they are in the ratio of $\exp\left[-(t_1 - t_2)/\tau_0\right]$.

7.18. Verify by substitution that Eq. (7.7.16) satisfies Eq. (7.7.8) for $t > 0$ if $E = 0$ and $\tau = \infty$. Find the values of $p_1(x, t)$ in Eq. (7.7.16) at $x = 0$ and $x = 10^{-2}$ cm for $t = 0$, 10^{-7}, and 5×10^{-7} s. The value of D is 50 cm^2/s.

7.19. (a) Sketch the behavior of Eq. (7.8.7) as a function of x at two instants t_1 and t_2 for the simple case in which $E = 0$. Explain the differences of the two curves in terms of known physical processes.

(b) Next consider the more complicated case for which there is an electric field E. Repeat part (a).

7.20. Equation (7.7.13) is the solution of the time-dependent diffusion equation for a unit impulse input at $x = 0$. Suppose that in Eq. (A7.2), we have

$$p_1(t = 0) = \frac{1}{\sqrt{\pi}\,\Delta} \exp\left[-\frac{(x - x_0)^2}{\Delta^2}\right]$$

and $f(x_1, t_1) = 0$. Use physical reasoning to show that the solution $p_1(x, t)$ to the diffusion equation is also a Gaussian function. Express $p_1(x, t)$ of the new solution in terms of τ_0, μ_p, D_p, E, x_0, and Δ.

7.21. (a) In the Haynes–Shockley experiment, we observe the arrival and subsequent collection of minority carriers at a fixed point. In other words, the pulse observed at point B in Fig. 7.17 is proportional to $p_1(x, t)$ of Eq. (7.8.7) with $x = AB = d$ being fixed. The magnitude of the pulse changes because t in Eq. (7.8.7) varies. Show that the maximum of $p_1(d, t)$ occurs when

$$-\frac{1}{\tau_0} - \frac{1}{2t} + \mu_p E_x \frac{d - \mu_p E_x t}{2D_p t} + \frac{(d - \mu_p E_x t)^2}{4D_p t^2} = 0$$

(b) From the equation above, obtain the result expressed in Eq. (7.8.9) by expressing $t = t_0 + t_1$ and neglecting terms quadratic in t_1. Show that Eq. (7.8.8) is a good approximation only if the spread Δx caused by diffusion is much smaller than the drift distance, and the time required for excess carriers to drift across Δx is much smaller than the lifetime τ_0.

   (c) Using the numerical values given in Problem 7.17, calculate the value of E_x for $d = 0.5$ cm such that Eq. (7.8.8) is a valid approximation.

7.22. (a) Sketch the energy-band diagram near the surface of a p-type semiconductor. Consider two different situations: (1) with the surface states carrying positive charges, and (2) with the surface states carrying negative charges.

   (b) Discuss the behavior of free-carrier concentrations near the surface and also the resultant change of the conductivity of the semiconductor as the amount of Q_s increases (again for both $+Q_s$ and $-Q_s$).

7.23. Find analytical expressions for the surface charge density Q_s and surface conductance G_s as functions of ϕ_s for an intrinsic sample, that is, $Y_b = 0$ in Eq. (7.9.6).

7.24. Show that near a surface of strong inversion or accumulation, the carrier concentration varies as

$$ n = \frac{n_s}{(1 + x/\sqrt{2}\, L_D)^2} $$

where $L_D = \sqrt{\epsilon kT/e^2 n_s}$ is the Debye length. [Hint: Keep only the dominant term in Eq. (7.9.5).]

7.25. (a) Show that Eqs. (7.9.5) and (7.9.6) can be rewritten as

$$ \rho = -2n_i kT \frac{dz}{d\phi} $$

$$ \frac{d\phi}{dx} = \left(\frac{4n_i kT}{\epsilon}\right)^{1/2} z^{1/2} $$

where

$$ z = \cosh Y - \cosh Y_b - (Y - Y_b)\sinh Y_b $$

Further, using Eqs. (7.9.7) and (7.9.8) in Eq. (7.9.9), obtain the expression

$$ Q_s = \int_{\phi_s}^{\phi_b} 2n_i kT \frac{dz}{d\phi}\left(\frac{d\phi}{dx}\right)^{-1} d\phi = (4\epsilon n_i kT z_s)^{1/2} $$

for the surface charge per unit area.

   (b) Check the result in part (a) by applying Gauss's law directly to the relation between the surface charge density and the electric field.

7.26. Following the procedure outlined in Problem 7.25, show that for an intrinsic semiconductor

$$ \frac{d\phi}{dx} = -\sqrt{\frac{8n_i kT}{\epsilon}}\sinh\frac{Y}{2}, \qquad Q_s = -\sqrt{8n_i \epsilon kT}\sinh\frac{Y_s}{2} $$

Given $n_i = 1.9 \times 10^{13}$ cm^{-3} and $\epsilon = 16\epsilon_0$ in Ge, $n_i = 0.9 \times 10^{10}$ cm^{-3} and $\epsilon = 11.7\epsilon_0$ in Si. The surface-charge density of a germanium or silicon sample can reach a value of -1×10^{-8} C/cm^2 in an ozone atmosphere and a value of $+1 \times 10^{-8}$ C/cm^2 in a wet (H$_2$O) atmosphere. Calculate the quantity ϕ_s and the maximum electric field in the surface space-charge region in intrinsic Ge and Si.

7.27. (a) Find expressions in integral form relating the surface free-carrier densities (per unit surface area) $\int p(x)\, dx$ and $\int n(x)\, dx$ to the quantity $e\phi_s/kT$ by converting the variable of integration from x to ϕ. Unlike the expression for the total surface-charge density

Q_s of Eq. (7.9.9), the two integral expressions in general cannot be written in explicit form and must be evaluated by numerical computation (e.g., through the use of a digital computer).

   (b) Consider a special case in which the energy band at the surface is bent upward (or downward) sufficiently so that the quantity p (or n) is the dominant term in Eq. (7.9.5) in the space-charge region near the surface. Find the explicit expressions for $\int p(x)\,dx$ and $\int n(x)\,dx$. Also find the expression for surface conductance G_s.

7.28. **(a)** Describe the ac field-effect experiment for measuring differential surface conductance (Fig. 7.22). Explain how and why information about a semiconductor surface can be ascertained from the measurement.

   (b) Suppose that we know the surface-charge density Q_s induced in the semiconductor and measure the surface conductance G_s. Discuss the circumstances under which the following expression may be true:

$$G_s = -\mu_n Q_s \quad \text{or} \quad G_s = \mu_p Q_s$$

7.29. Take $C_p/C_n = 9$ and $\mathscr{E}_t - \mathscr{E}_{fi} - e\phi_0 = 0.068$ eV in Eq. (7.10.14). The other pertinent data for a germanium sample are $(v_s)_{\max} = 55$ cm/s, $\tau_0 = 8 \times 10^{-4}$ s, and $b = 0.04$ cm. Plot τ_{eff} as a function of ϕ_s between $\pm 6kT$.

7.30. Discuss qualitatively how surface conductance G_s and surface recombination velocity v_s change as a function of $e\phi_s/kT$. Draw diagrams similar to Figs. 7.21 and 7.25 to illustrate the respective change, and indicate the dominant dependence of G_s and v_s for $e\phi_s/kT > 4$ and $e\phi_s/kT < -4$. The semiconductor under discussion is intrinsic. We assume further that $\mu_n = \mu_p$, $C_n = C_p$, and $\mathscr{E}_t - \mathscr{E}_{fi} = 3kT$.

7.31. Consider the distributions of excess carriers shown in Fig. 7.26. Calculate the average diffusion velocity $\bar{v}_d$ of excess carriers from Eq. (7.11.11). Also find the effective lifetime of excess carriers in a semiconductor. The sample is assumed to be of a rectangular bar (Fig. 7.26) and to have dimensions b, $c \ll a$. In other words, the effects of recombination at four surfaces perpendicular to either y axis or z axis should be considered.

7.32. Consider the series solution of Eq. (7.11.3) given by Eq. (7.11.22). Show that the other roots of ξ (referred to as *higher modes*) in Eq. (7.11.6) can be approximated by

$$\xi_n = \frac{n-1}{b}\,2\pi \quad \text{where} \quad n = 2, 3, 4$$

This result can easily be seen from a graphical solution of Eq. (7.11.6) by letting $\eta = v_s/D_p\xi$ and $\eta = \tan(\xi b/2)$. Find the effective decay constant for the higher modes. Also show that the higher modes decay much faster than the lowest mode. Thus, after a time $t > \tau_{\text{eff}}$, where τ_{eff} is given by either Eq. (7.11.8) or Eq. (7.11.10), the dominant term in Eq. (7.11.22) is $p_1(y, t)$ of Eq. (7.11.4).

7.33. Show that $p_1(y)$ of Eq. (7.11.25) is

$$p_1(y) = \frac{G[\sinh(y/L_p) + \beta \cosh(y/L_p)]}{v_s[(1 + \beta^2)\sinh(b/L_p) + 2\beta \cosh(b/L_p)]}$$

where $\beta = D_p/L_p v_s$ and $L_p = (D_p\tau)^{1/2}$.

7.34. From the result obtained in Problem 7.33, find the average value $\bar{p}_1$ (the value of p_1 averaged over the sample thickness b). Show that for $b/L_p > 1$,

$$b\bar{p}_1 = G\tau_{\text{eff}} = G\left(\frac{1}{\tau_0} + \frac{v_s}{L_p}\right)^{-1}$$

and for $b/L_p < 1$,

$$b\bar{p}_1 = G\tau_{\text{eff}} = G\left(\frac{1}{\tau_0} + \frac{2v_s}{b}\right)^{-1}$$

8

Simple Semiconductor Junction Devices: Theory of p-n Homojunction and Methods of Junction Formation

8.1 *p-n* JUNCTION DIODES

Many semiconductor devices are made of materials of inhomogeneous composition. Figure 8.1a shows the structure of a semiconductor diode that consists of two regions: one p type and the other n type. Since the theory of junction diodes which is about to be developed will be found to underlie many other semiconductor devices, such as bipolar and field-effect transistors, we must carefully examine the physical mechanisms that cause the electrical properties of a p-n junction to be different from those of a homogeneous semiconductor. We start our discussion with an idealized model of a p-n junction diode. Then we discuss the methods of fabricating junction devices. After we are familiar with the actual situation in a real diode, we reexamine the simple diode theory and make proper modifications of the theory so that the modified theory can be applied adequately to practical use.

If we recall our discussion in Section 7.1, we consider a p-n junction diode as a thermodynamic system consisting of two subsystems, one being the n region and the other the p region. Under thermal equilibrium, the Fermi level $\mathscr{E}_f$ remains the same throughout the entire system. This equalization of $\mathscr{E}_f$ makes necessary the establishment of a potential barrier in the junction region, as is evident from Fig. 8.1b. Because holes and electrons carry opposite charges, the potential-energy diagram for holes is the

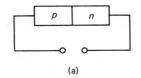

(a)

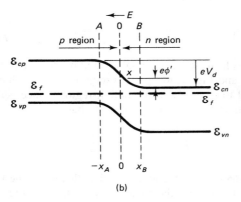

(b)

Figure 8.1 (a) A semiconductor junction made of n-type material on one side and p-type material on the other side, and (b) the energy-band diagram representing the junction. Under thermal equilibrium (no bias condition), the two Fermi levels on the two sides are equal. The potential difference $e\phi'$ is the gain in potential energy for a hole to go from a point x to the point x_B, or the gain in potential energy for an electron to go from the point x_B to a point x.

one for electrons turned upside down. Therefore, holes will experience a potential barrier (or an uphill climb in colloquial terms) in going from the p region to the n region. For the following discussion it is assumed that the barrier region is bounded by $-x_A < x < x_B$.

Outside the barrier region, the physical situation is not different from that normally existing in a homogeneous semiconductor. The electron and hole concentrations at $x = -x_A$ are given, respectively, by

$$n_{p0} = N_c \exp\left(-\frac{\mathscr{E}_{cp} - \mathscr{E}_f}{kT}\right) \tag{8.1.1a}$$

$$p_{p0} = N_v \exp\left(-\frac{\mathscr{E}_f - \mathscr{E}_{vp}}{kT}\right) \tag{8.1.1b}$$

and their corresponding concentrations at $x = x_B$ by

$$n_{n0} = N_c \exp\left(-\frac{\mathscr{E}_{cn} - \mathscr{E}_f}{kT}\right) \tag{8.1.2a}$$

$$p_{n0} = N_v \exp\left(-\frac{\mathscr{E}_f - \mathscr{E}_{vn}}{kT}\right) \tag{8.1.2b}$$

where the subscript 0 refers to equilibrium carrier concentrations, and the subscripts p and n refer to carrier concentrations in the p and n regions, respectively. Taking the ratio of the equations above, we find that

$$\frac{eV_d}{kT} = \frac{1}{kT}(\mathscr{E}_{cp} - \mathscr{E}_{cn}) = \frac{1}{kT}(\mathscr{E}_{vp} - \mathscr{E}_{vn}) = \ln\frac{n_{n0}}{n_{p0}} = \ln\frac{p_{p0}}{p_{n0}} \tag{8.1.3}$$

From Eq. (8.1.3) we see that if no external voltage is applied across the junction, the barrier potential is adjusted such that the carrier concentrations on the p and n sides are in the ratio of the Boltzmann factor. Since we are dealing with nondegenerate semiconductors, the two concentrations in the two subsystems should be in this ratio under thermal equilibrium.

Now we turn our attention to the barrier region bounded by $-x_A < x < x_B$. To be specific, we take point x in the n region shown in Fig. 8.1b. Because of the potential barrier, the electron concentration at x has a value given by

$$n = N_c \exp \left(-\frac{\mathscr{E}_{cn} + e\phi' - \mathscr{E}_f}{kT} \right) = n_{n0} \exp \left(-\frac{e\phi'}{kT} \right) \qquad (8.1.4)$$

The quantity $-e\phi'$ is the electrostatic potential created by the potential barrier. It differs from $e\phi$ in Section 7.10 and Fig. 7.20 by a constant and a sign change such that their sum $(e\phi + e\phi')$ is equal to $\mathscr{E}_f - \mathscr{E}_{fi}$ in the bulk n-region. Since $e\phi'$ varies with x, an electrostatic field E_x exists in the barrier region such that

$$E_x = -\frac{\partial V}{\partial X} = \frac{\partial \phi'}{\partial x} \qquad (8.1.5)$$

According to Fig. 8.1b, $\partial\phi'/\partial x$ is a negative quantity; therefore, the motion of electrons caused by E_x is from the left toward the right (in the positive x direction). On the other hand, the motion of electrons caused by diffusion is from the right toward the left (in negative x direction) because of a much higher electron concentration on the right. Under thermal equilibrium, no electron current flows in the junction. In other words, the two motions mentioned above should exactly cancel out. This is indeed the case, as the reader can easily show that

$$J_n = e\mu_n n E_x + eD_n \frac{dn}{dx} = 0 \qquad (8.1.6)$$

through the use of Eqs. (8.1.4) and (8.1.5) together with the Einstein relation $\mu_n = eD_n/kT$. The same applies to the hole-current density.

The existence of a potential barrier across a p-n junction is the key factor that makes p-n junctions have different electrical characteristics from those of homogeneous semiconductors. First, as ϕ' varies in Eq. (8.1.4), so does n. As a consequence of the variation of n and p, a space-charge region AB (Fig. 8.1b) is formed. Similar to the surface space-charge region discussed in Section 7.9, the region AB stores electric charges. The amount of charges stored in this region changes when a voltage is applied across the junction. Therefore, the current though a p-n junction has a capacitive component in addition to the conductive component. This behavior is quite different from that of a homogeneous semiconductor.

The potential barrier also makes the junction possess a very useful current–voltage characteristic. On an intuitive basis, we see that since holes proceeding from the left to the right in Fig. 8.1b (i.e., from the p side to the n side) have to climb a potential hill, their chances of reaching the other side (i.e., the n side) are greatly reduced. Referring to our discussion of thermionic emission in Section 4.7, we find that the thermionic current depends on the work function $\mathscr{E}_w$ (the surface-barrier potential) as $\exp(-\mathscr{E}_w/kT)$. Thus we expect the current $I(\rightarrow)$ as a result of hole flow from the left to the right to be

$$I(\rightarrow) = A \exp\left(-\frac{eV_d}{kT}\right) \tag{8.1.7}$$

and the net current to be

$$I(\text{net}) = I(\rightarrow) - I(\leftarrow) = A \exp\left(-\frac{eV_d}{kT}\right) - B \tag{8.1.8}$$

In Eqs. (8.1.7) and (8.1.8), A and B are two constants of proportionality.

If no external voltage is applied, $I(\text{net})$ must be zero. This condition for zero current relates the constant A to B. When an external bias voltage is applied to the junction so that the barrier potential is lowered from V_d to $V_d - V_a$, a nonzero current results. Replacing V_d by $V_d - V_a$ in eq. (8.1.8), we obtain

$$I(\text{net}) = A \exp\left[-\frac{e(V_d - V_a)}{kT}\right] - B = B\left[\exp\left(\frac{eV_a}{kT}\right) - 1\right] \tag{8.1.9}$$

after the elimination of A.

Figure 8.2 shows the electrical characteristic of an ideal junction diode, based on Eq. (8.1.9). From the energy-band diagram of Fig. 8.1b, we see that the potential barrier is lowered if the positive polarity of the voltage is applied to the p side. The diode in this situation is said to be biased in the forward direction. For positive V_a in Eq. (8.1.9), the current increases very rapidly with increasing V_a. On the other hand, the potential barrier is raised if the positive polarity is applied to the n side. In the reverse direction, V_a in Eq. (8.1.9) is negative, and I approaches a constant value I_s known as the *reverse saturation current*.

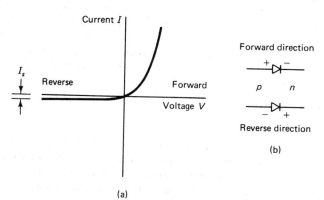

Forward direction

p n

Reverse direction

(b)

(a)

Figure 8.2 Current-voltage characteristic of an ideal p-n junction diode.

8.2 SPACE-CHARGE REGION AND JUNCTION CAPACITANCE

In Section 8.1 we discussed the reason for the existence of a potential barrier across a p-n junction but we did not explain the physical process through which the potential barrier is established. Let us concentrate our attention on the barrier region (region AB of Fig. 8.1b). As the potential barrier rises, the thermal-equilibrium electron concentration in region AB is given by Eq. (8.1.4) and not by n_{n0}. A similar situation applies to holes in the barrier region. The equilibrium concentrations are reached through carrier recombination and generation processes described in Chapter 7. Note that $n < n_{n0}$ for $0 < x < x_B$; however, electrons in the region $x > x_B$ are held back from replenishing the barrier region by the barrier potential $e\phi'$. Thus a space-charge region is created. We define a net charge density $\rho(x)$ such that

$$\rho(x) = e\,(p + N_d^+ - n - N_a^-) \qquad (8.2.1)$$

For simplicity, we assume that the donor and acceptor impurities are completely ionized and that $N_a^- - N_d^+ = N_1$ in the p region and $N_d^+ - N_a^- = N_2$ in the n region. Outside the barrier region AB, the charge neutrality condition prevails or $\rho(x) = 0$ for $x > x_B$ and for $x < -x_A$. In the barrier region $x_B > x > -x_A$, however, the charge-neutrality condition is violated. At a point x in the n region, the electron (majority carrier) concentration is given by Eq. (8.1.4), in which the quantity n_{n0} is equal to N_2. Note that electrons are still the majority carriers. Hence in this region we can neglect the hole (minority carrier) concentration p. Thus Eq. (8.1.2) for $x > 0$ becomes

$$\rho(x) = eN_2 \left[1 - \exp\left(-\frac{e\phi'}{kT} \right) \right] \qquad (8.2.2)$$

The profile of the electrostatic potential V is easily obtainable through the use of one-dimensional Poisson's equation

$$\frac{d^2V}{dx^2} = -\frac{d^2\phi'}{dx^2} = -\frac{\rho(x)}{\epsilon} \qquad (8.2.3)$$

in conjunction with Eq. (8.2.2), ϵ being the dielectric constant of a given semiconductor. To facilitate solving Eq. (8.2.3), we make the following observation. The value of n given by Eq. (8.1.4) drops very sharply as soon as $e\phi'$ reaches a value of the order of kT. Since the value of kT at room temperature is around 0.0259 eV, which is at least an order of magnitude smaller than eV_d, V_d being the difference in the electrostatic potential between the n and p regions, the net charge density $\rho(x)$ takes the value of eN_2 practically for the whole length of the transition region OB. As a first-order approximation, Poisson's equation becomes

$$\frac{d^2V}{dx^2} = -\frac{eN_2}{\epsilon} \qquad \text{for} \quad 0 < x < x_B \qquad (8.2.4)$$

As we discuss in Section 8.6, two principal techniques used in junction-device fabrication are the alloying or epitaxial technique and the diffusion technique. In alloyed diodes, the impurity concentration changes abruptly from N_d in the n region to N_a in the p region. In separate n and p regions, the doping concentration is uniform, or in other words, the value of N_a or N_d is independent of x in Eq. (8.2.4). Such a junction is often referred to as a *step* (or *abrupt*) *junction*. On the other hand, the doping concentration in a junction grown during crystal pulling changes gradually from a value $N_2 = N_d - N_a$ in the n region, to zero at the junction boundary, and then to a different value $N_1 = N_a - N_d$ in the p region. Therefore, a grown junction is also known as a *graded junction*. The situation in a diffused junction, which is more complicated, is discussed in Section 8.7.

In this section we consider only step junctions. For a uniform impurity concentration, that is, an N_2 independent of x in the n region, integrating Eq. (8.2.4) gives the following results:

$$\frac{dV}{dx} = -\frac{eN_2}{\epsilon}(x - x_B) \tag{8.2.5a}$$

$$V = -\frac{eN_2}{2\epsilon}x(x - 2x_B) \qquad \text{for} \quad 0 < x < x_B \tag{8.2.5b}$$

Applying similar arguments to the p region, we have

$$\frac{dV}{dx} = \frac{eN_1}{\epsilon}(x + x_A) \tag{8.2.6a}$$

$$V = \frac{eN_1}{2\epsilon}x(x + 2x_A) \qquad \text{for} \quad -x_A < x < 0 \tag{8.2.6b}$$

For further clarification of the steps taken in arriving at Eqs. (8.2.5) and (8.2.6), we refer to Fig. 8.3. The space-charge region is assumed to end abruptly at $x = x_B$ and $x = -x_A$. Therefore, the value of the electric field $E_x = -dV/dx$ is taken to be zero at $x = x_B$ and $x = -x_A$. Furthermore, the electrostatic potential at the boundary between the n and p regions is used as the preference potential (i.e., $V = 0$ at $x = 0$). Following these steps, the reader should be able to verify Eqs. (8.2.5) and (8.2.6).

Before we proceed further with Eqs. (8.2.5) and (8.2.6), it may be instructive to make the following general observations. According to electrostatics, the electric flux starts from a positive charge and ends at a negative charge. Furthermore, accompanied by the flux, there is an electric field. In the space-charge region of a p-n junction, the electric flux starts from the positively charged donors and ends at the negatively charged acceptors. Since the path of the flux is complete inside the space-charge region, the total positive charges stored on the n side must be equal to the total negative charges stored on the p side, and the electric field at the boundary of the space-charge region is zero. The situation is illustrated in Fig. 8.4a. If we define a differential transition-region capacitance C_T as the ratio of the change in the stored charge Q to the change in the applied voltage V_a, then

$$C_T = \left| \frac{dQ}{dV_a} \right| = \frac{A\epsilon}{W} \tag{8.2.7}$$

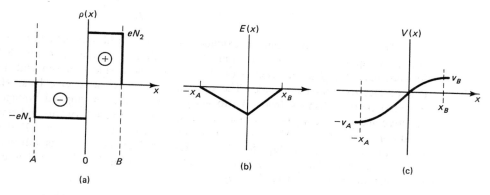

Figure 8.3 Diagrams showing the variations of (a) the space-charge density $\rho(x)$, (b) the electric field E_x, and (c) the electrostatic potential $V(x)$ in a step junction.

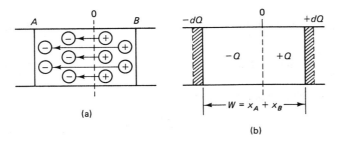

Figure 8.4 Diagrams used in analyzing the effects of stored charges in the space-charge region. (a) The electric flux starts from a positively charged donor on the n side and ends at a negatively charged acceptor on the p side. (b) There are equal amounts of positive and negative charges stored, respectively, on the n side and p side of a p-n junction. Any change in the stored charge dQ must also be equal for the two sides.

where A is the cross-sectional area of the junction and $W = x_A + x_B$ is the width of the space-charge region. As shown in Fig. 8.4b, the charges $\pm dQ$ are separated apart at a distance W; therefore, Eq. (8.2.7) is true for any arbitrary impurity distribution.

Equating the values of E_x at $x = 0$ from Eqs. (8.2.5) and (8.2.6) gives

$$eN_2 x_B = eN_1 x_A \qquad (8.2.8)$$

Equation (8.2.8) simply says that the total positive charges stored in the space-charge region on the n side must equal the corresponding negative charges stored on the p side. Setting $V = -V_A$ at $x = -x_A$, $V = V_B$ at $x = x_B$ and $V_d = V_A + V_B$, we find that

$$V_d = \frac{e}{2\epsilon} (N_1 x_A^2 + N_2 x_B^2) \qquad (8.2.9)$$

Eliminating x_A or x_B from Eqs. (8.2.8) and (8.2.9) gives

$$x_A = \left(\frac{2\epsilon V_d N_2}{eN_1} \frac{1}{N_1 + N_2} \right)^{1/2} \qquad (8.2.10a)$$

$$x_B = \left(\frac{2\epsilon V_d N_1}{eN_2} \frac{1}{N_1 + N_2} \right)^{1/2} \qquad (8.2.10b)$$

and the total width of the transition region W:

$$W = x_A + x_B = \left(\frac{2\epsilon V_d}{e} \frac{N_1 + N_2}{N_1 N_2} \right)^{1/2} \qquad (8.2.11)$$

The analysis above can be extended to a nonequilibrium situation in which a voltage is applied to the p-n junction. The situation is illustrated in Fig. 8.5. For a forward bias, the potential energy of electrons on the n side is raised with respect to that of electrons on the p side. The opposite is true for a reverse bias. Hence the potential barrier changes from eV_d to $e(V_d \pm V_a)$ under an applied voltage V_a, the negative and positive signs corresponding, respectively, to the forward- and reverse-bias conditions. As long as $e(V_d \pm V_a)$ is an order of magnitude larger than kT, the approximation used in obtaining Eq. (8.2.9) from Eqs. (8.2.2) and (8.2.3) is still valid.

Let us show that Eq. (8.2.7) is indeed true. From Eq. (8.2.8), the charge Q stored in a junction of an area A is given by

$$\frac{Q}{A} = eN_2 x_B = eN_1 x_A = \left(\frac{2e\epsilon N_1 N_2}{N_1 + N_2} \right)^{1/2} (V_d \pm V_a)^{1/2} = W \frac{eN_1 N_2}{N_1 + N_2} \qquad (8.2.12)$$

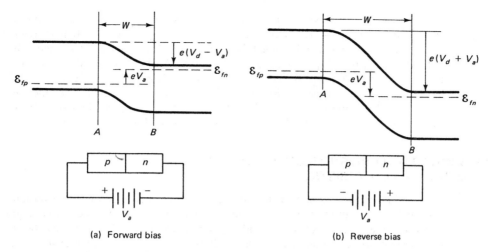

(a) Forward bias   (b) Reverse bias

Figure 8.5 Energy-band diagrams for a *p-n* junction biased (a) in the forward direction and (b) in the reverse direction. In (a), the polarity of the bias voltage is such that the potential energy of electrons on the *n* side is raised with respect to that on the *p* side. As a result, the potential barrier is lowered. In (b), the potential barrier is raised by the bias voltage.

Therefore, the capacitance C_T associated with the stored charge dQ in the transition region is equal to

$$C_T = \left|\frac{dQ}{dV_a}\right| = \frac{A}{2}\left(\frac{2e\epsilon N_1 N_2}{N_1 + N_2}\right)^{1/2}(V_d \pm V_a)^{-1/2} = \frac{A\epsilon}{W} \qquad (8.2.13)$$

We should point out that the average separation between the stored charges in Fig. 8.4 is not equal to W. Hence only the ac (or differential) capacitance C_T, not the dc capacitance Q/V_a, is equal to $A\epsilon/W$. In practical applications to be discussed later, it is the ac capacitance C_T that enters into the circuit equation. Therefore, we shall not be concerned with the dc capacitance Q/V_a.

In many applications, Eq. (8.2.13) can be simplified. For an alloyed diode, the alloyed side usually has an impurity concentration far greater than the base side; therefore, the expression $N_1 N_2/(N_1 + N_2)$ in Eq. (8.2.13) reduces to $N = N_1$ or N_2, whichever is smaller. It is also obvious from Eq. (8.2.10) and Fig. 8.3 that the space-charge region extends farther into the region with lower impurity concentration. Thus, for alloyed diodes,

$$C_T = A\left(\frac{\epsilon}{2\mu\rho}\right)^{1/2}(V_d \pm V_a)^{-1/2} \qquad (8.2.14)$$

$$W = (2\epsilon\rho\mu)^{1/2}(V_d \pm V_a)^{1/2} \qquad (8.2.15)$$

where ρ is the resistivity of the base region and μ is the mobility of the majority carriers in the base region. As we shall see in Section 8.7, Eqs. (8.2.14) and (8.2.15) are also applicable to diffused junctions under a large reverse bias.

To estimate the width of the space-charge region and the magnitude of the junction capacitance, we take as an example an alloyed germanium diode with a base resistivity of 1.6 Ω-cm, n type. From the value of the resistivity, the carrier concentrations are found to be $n_{n0} = 10^{15}$ cm^{-3} and $p_{n0} = 5.3 \times 10^{11}$ cm^{-3}. If it is assumed that the alloyed side is so heavily doped that the hole concentration p_{p0} takes the value $N_v = 6 \times 10^{18}$ cm^3, then Eq. (8.1.3) gives $V_d = 0.42$ V. For a forward bias V_a of 0.10 V, W is found to be 7.5×10^{-5} cm and C_T/A to be 1.86×10^{-8} F/cm^2. For a Si diode having the same base resistivity and p-side doping concentration, we find that $W = 5.7 \times 10^{-5}$ cm and $C_T/A = 1.82 \times 10^{-8}$ F/cm^2 at a forward bias of 0.10 V. In the calculation, we use $\epsilon = 1.42 \times 10^{-12}$ F/cm in Ge and $\epsilon = 1.04 \times 10^{-12}$ F/cm in Si.

8.3 MINORITY-CARRIER INJECTION AND IDEAL DIODE CHARACTERISTICS

When a bias voltage is applied to a p-n junction, the Fermi level $\mathscr{E}_f$ can no longer stay constant throughout the junction, as can be seen from Fig. 8.5. Under a nonequilibrium situation, quasi-Fermi levels $\mathscr{F}_n$ and $\mathscr{F}_p$ introduced in Section 7.3 must be used. To facilitate our using the quasi-Fermi levels in the present analysis, we examine the variation of $\mathscr{F}_n$ and $\mathscr{F}_p$ in relation to current flow.

Using Eq. (7.3.4) for n and realizing that the junction field E is given by $eE_x = -e \, \partial V/\partial x = \partial \mathscr{E}_c/\partial x$, we have

$$eD_n \frac{dn}{dx} = eD_n \frac{n}{kT} \left(\frac{\partial \mathscr{F}_n}{\partial x} - \frac{\partial \mathscr{E}_c}{\partial x} \right) = \mu_n n \left(\frac{\partial \mathscr{F}_n}{\partial x} - eE_x \right) \tag{8.3.1}$$

Thus the current density due to conduction-band electrons can be expressed in terms of the quasi-Fermi level $\mathscr{F}_n$ for electrons as

$$J_n = e\mu_n n E_x + eD_n \frac{\partial n}{\partial x} = \mu_n n \frac{\partial \mathscr{F}_n}{\partial x} \tag{8.3.2}$$

The spatial variation of J_n is, in turn, related to the temporal variation of the electron concentration n. Under steady-state conditions, it is required by the continuity equation

$$\frac{\partial n}{\partial t} = \frac{1}{e} \frac{\partial J_n}{\partial x} - \frac{n_1}{\tau_0} \tag{8.3.3}$$

that if recombination of excess carriers in the space-charge region is negligible, J_n should be a constant independent of x. From the constancy of J_n, we can obtain from Eq. (8.3.2), the ratio of $\partial \mathscr{F}_n/\partial x$ inside and outside the space-charge region.

In Section 7.3 we assumed a constant $\mathscr{F}_n$ across the space-charge region to arrive at Eq. (7.3.1), but we have not shown whether this assumption is valid. The purpose of this preliminary discussion is to confirm that under certain conditions the assumption is indeed valid. The variation of the quasi-Fermi levels $\mathscr{F}_n$ and $\mathscr{F}_p$ is illustrated in Fig. 8.6a.

Let us follow the path of electrons. At the far right (region beyond D), where excess carriers are practically zero, we have $\mathscr{F}_n = \mathscr{E}_{fn}$ because the system is in thermal equilibrium. As we approach the junction boundary B, excess carrier concentration $(p_1 = n_1)$ is no longer zero. However, electrons are still the majority carriers. Under low-level injection $n_1 \ll n_0$, that is, $n \cong n_0$; hence $\mathscr{F}_n$ remains practically equal to $\mathscr{E}_{fn}$, the thermal-equilibrium Fermi level. At the far left (region beyond C), where excess carriers again become practically zero, the quasi-Fermi level $\mathscr{F}_n$ finally must be equal

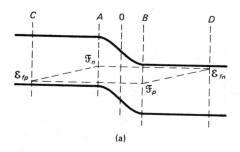

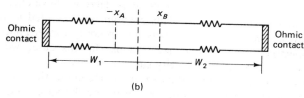

Figure 8.6 Diagrams showing (a) the variation of the quasi-Fermi levels $\mathcal{F}_n$ for electrons and $\mathcal{F}_p$ for holes, and (b) the geometry of a *p-n* junction.

to $\mathscr{E}_{fp}$. Therefore, there is a region BC in which the value of $\mathcal{F}_n$ changes from $\mathscr{E}_{fn}$ to $\mathscr{E}_{fp}$. Similarly, in the region AD, the value of $\mathcal{F}_p$ changes from $\mathscr{E}_{fp}$ to $\mathscr{E}_{fn}$.

The validity of the assumption that $\mathcal{F}_n$ and $\mathcal{F}_p$ are constant in the space-charge region now hinges on the relative change of $\mathcal{F}_n$ and $\mathcal{F}_p$ inside and outside the space-charge region. As we shall see later, the excess minority-carrier concentration eventually drops to zero either through the recombination process or through the action of the base contact. In other words, $\mathcal{F}_n$ moves gradually toward $\mathscr{E}_{fp}$ in the p region and $\mathcal{F}_n$ practically equals $\mathscr{E}_{fp}$ within a distance d determined by the diffusion length $\sqrt{D_n\tau_0}$ or the proximity of the base contact, whichever is smaller. Since the current density J_n must be continuous at point A, the change in the quasi-Fermi level in the space-charge region can be estimated from Eq. (8.3.2) as follows:

$$\Delta\mathcal{F}_n\Big|_B^A \cong \Delta F_n\Big|_A^C \frac{\Delta x_{AB}}{\Delta x_{AC}} \frac{n_{AC}}{n_{AB}} \tag{8.3.4}$$

Since the ratio of the average electron concentration in region AC to that in region AB, n_{AC}/n_{AB}, is much less than 1,

$$\frac{\Delta\mathcal{F}_n\Big|_B^A}{\Delta\mathcal{F}_n\Big|_A^C} \leq \frac{\Delta x_{AB}}{\Delta x_{AC}} \tag{8.3.5}$$

The width of the space-charge region was estimated previously to be of the order of 10^{-4} cm, while $\sqrt{D_n\tau_0} = 5.9 \times 10^{-3}$ cm with $D_n = 35$ cm²/s and $\tau_0 = 10^{-6}$ s. It is obvious that almost all of the drop in $\mathcal{F}_n$ occurs outside the space-charge region.

In summary, if recombination of excess carriers in the space-charge region is negligible and if the width of the space-charge region is very thin compared with the diffusion length, the assumption is valid that the quasi-Fermi levels remain approximately constant in the space-charge region. Under such circumstances, the quasi-Fermi levels are related to the Fermi levels by

$$\mathcal{F}_n = \mathscr{E}_{fp} + eV_a \qquad \text{at} \quad x = -x_A$$
$$\mathcal{F}_p = \mathscr{E}_{fn} - eV_a \qquad \text{at} \quad x = x_B \tag{8.3.6}$$

Thus the minority-carrier concentrations are given by

$$n_p = N_c \exp\left(-\frac{\mathscr{E}_c - \mathscr{F}_n}{kT}\right) = n_{p0} \exp\left(\frac{eV_a}{kT}\right) \qquad \text{at} \quad x = -x_A \qquad (8.3.7a)$$

$$p_n = N_v \exp\left(-\frac{\mathscr{F}_p - \mathscr{E}_v}{kT}\right) = p_{n0} \exp\left(\frac{eV_a}{kT}\right) \qquad \text{at} \quad x = x_B \qquad (8.3.7b)$$

where n_{p0} and p_{n0} represent the equilibrium minority-carrier concentrations in the p and n regions, respectively. We again refer to Fig. 8.5. For a forward bias, V_a is positive in Eq. (8.3.7) and hence the minority-carrier concentrations given by Eq. (8.3.7) exceed the corresponding equilibrium minority-carrier concentrations. Here we have excess minority carriers in the respective p and n regions. This phenomenon is called *minority-carrier injection*. On the other hand, for a reverse bias, the minority-carrier concentrations fall below their respective equilibrium concentrations. In this case, we have minority-carrier extraction or a deficiency of minority carriers.

The foregoing analysis leads to one important result. If recombination of excess carriers is insignificant in the space-charge region, we can ignore the space-charge region entirely insofar as the current flow across a junction is concerned. Because of the continuity of the current, we can evaluate the current anywhere in a junction diode. The analysis is simplest in the region where the current flow is controlled entirely by the diffusion of minority carriers. Therefore, in the following discussion, we consider in Fig. 8.6a only region AC for electron diffusion and only region BD for hole diffusion.

To find the current flow through a junction diode, we refer to Eq. (7.7.1). Since almost all the applied voltage V_a appears across the space-charge region, the drift term is unimportant outside the space-charge region. Thus, under steady-state dc conditions, the diffusion equation applicable to Fig. 8.6 reads

$$D_p \frac{\partial^2 p_1}{\partial x^2} - \frac{p_1}{\tau_0} = 0 \qquad \text{for} \quad x > x_B \qquad (8.3.8)$$

A similar expression is obtained for excess electrons in the region $x < -x_A$. The general solution of Eq. (8.3.8) is given by

$$p_1 = A \sinh \frac{W_2 - x}{L} + B \cosh \frac{W_2 - x}{L} \qquad (8.3.9)$$

We should point out that the solution of Eq. (8.3.9) can be expressed in terms of either $\exp(-x/L)$ and $\exp(x/L)$ functions or sinh and cosh functions. The present form given in Eq. (8.3.9) is chosen for easy application of the boundary conditions. In Eq. (8.3.9), W_2 is length of the n region (Fig. 8.6b) and $L = (D_p\tau_0)^{1/2}$. Equation (8.3.9) is subject to the boundary condition that at the base contact $x = W_2$, $p_1 = 0$, and that at the boundary of the space-charge region $x = x_B$, $p_{10} = p_n - p_{n0}$ where p_n is given by Eq. (8.3.7). Thus Eq. (8.3.9) becomes

$$p_1(x) = p_{n0}\left[\exp\left(\frac{eV_a}{kT}\right) - 1\right] \frac{\sinh\left[(W_2 - x)/L\right]}{\sinh\left[(W_2 - x_B)/L\right]} \qquad (8.3.10)$$

Two different approximations can be taken, depending on whether W_2' is greater or smaller than L, where $W_2' = W_2 - x_B$. The case $W_2' > L$ is referred to as the *wide-base diode* and the case $W_2' < L$ as the *narrow-base diode*. For a wide-base diode, the sinh function can be approximated by the exponential function; thus Eq. (8.3.10) becomes

$$p_1(x) = p_{n0} \left[\exp\left(\frac{eV_a}{kT}\right) - 1 \right] \exp\left(-\frac{x - x_B}{L}\right) \qquad (8.3.11)$$

For a narrow-base diode, the sinh function can be replaced by its argument; thus Eq. (8.3.10) becomes

$$p_1(x) = p_{n0} \left[\exp\left(\frac{eV_a}{kT}\right) - 1 \right] \frac{W_2 - x}{W_2 - x_B} \qquad (8.3.12)$$

Note that Eq. (8.3.12) satisfies Eq. (8.3.8) if the p_1/τ_0 term is neglected. By a narrow-base diode, we therefore mean a diode in which recombination of excess carriers is negligible in the base region. From Eq. (8.3.11) or (8.3.12), the current density due to diffusion of holes is equal to

$$J_p = \left| -eD_p \frac{\partial p_1}{\partial x} \right|_{x=x_B} = \left(\frac{eD_p}{d}\right) p_{n0} \left[\exp\left(\frac{eV_a}{kT}\right) - 1 \right] \qquad (8.3.13)$$

In the equation above, the distance d is given by either L or $W_2 - x_B$, whichever is smaller.

The voltage dependence of the current given by Eq. (8.3.13) is indeed the same as that given by Eq. (8.1.9). Refer to the current–voltage characteristic of an ideal diode shown in Fig. 8.2. In the forward direction, the current increases very rapidly with V_a as $\exp(eV_a/kT)$. In the reverse direction, the current approaches a constant value I_s. Since the reverse saturation current I_s is an important factor determining the heat dissipation in a junction device, we discuss briefly the choice of semiconductor material in this regard. In Table 8.1 we present the data for Ge, Si, and GaAs, pertinent to our

TABLE 8.1 ENERGY GAP $\mathscr{E}_g$ AND INTRINSIC CARRIER CONCENTRATION, n_i IN Ge, Si, AND GaAs[a]

	Material		
	Ge	Si	GaAs
$\mathscr{E}_g$ (eV)	0.665	1.11	1.42
n_{n0} (cm^{-3})	1.6×10^{15}	4.6×10^{15}	7.3×10^{14}
n_i (cm^{-3})[b]			
At 300 K	2.3×10^{13}	1.4×10^{10}	1×10^7
At 350 K	1.7×10^{14}	3.7×10^{11}	6×10^8
p_{n0} (cm^{-3})			
At 300 K	3.3×10^{11}	4.3×10^4	1.4×10^{-1}
At 350 K	1.8×10^{13}	3.0×10^7	4.9×10^2
Ratio of n_{p0}	56	700	3600

[a]The thermal-equilibrium carrier concentrations, n_{n0} and n_{p0}, are calculated for a 1-Ω-cm, n-type material.

[b]The values of n_i quoted here are the measured values given in Table 6.8.

discussion. For comparison of numerical values, we choose all the three materials to be n-type and to have a resistivity of 1 Ω-cm.

Everything except p_{n0} in Eq. (8.3.13) being equal, the reverse saturation current I_s is directly proportional to p_{n0}. At 300 K, we expect I_s in Si to be smaller than I_s in Ge by 10^7 and I_s in GaAs to be smaller than I_s in Si by 10^6. Of course, in Eq. (8.3.13) we did not consider the current generated in the space-charge region. Consideration of this current will lower the two figures by a factor of 10^2 to 10^3, depending on the temperature range and on the relative importance of the two currents. Even with the lower figure, it is advantageous to use a semiconductor having a high energy gap in order to achieve a low reverse saturation current. However, a p-n junction made of a larger gap material requires a larger forward bias to reach the same current as a p-n junction made of a smaller gap material.

The other point worth mentioning is the temperature dependence of the minority-carrier concentration p_{n0}. From 300 to 350 K, the value of p_{n0} increases 56 times in Ge, 700 times in Si, and 3500 times in GaAs. Neglecting the temperature dependence of N_c and N_v, we find that

$$\frac{\Delta p_{n0}}{p_{n0}} = \frac{\mathscr{E}_g}{kT} \frac{\Delta T}{T} \qquad (8.3.14)$$

An increase of $\Delta T = 11$ K at 300 K in Ge (7.0 K in Si, 5.5 K in GaAs) doubles the value of p_{n0}. If heat dissipation in a diode is not taken care of properly, any increase in the reverse current increases the heat dissipation and hence raises the temperature, which further increases the reverse current. If not properly controlled, this regenerative cycle can cause a thermal runaway.

Now we clarify the steps taken in securing the boundary condition at the base contact and in obtaining the current expression J_p of Eq. (8.3.13). Theoretically, a rectifying junction results between two materials in contact if excess (or deficit) carriers exist at the junction. This usually happens at a metal–semiconductor contact. However, the degree of rectification is relative. A contact is said to be *ohmic* if no appreciable excess (or deficit) carriers may exist at the contact, that is, $n_i = p_1 = 0$. This ohmic boundary condition is used in evaluating the coefficient B in Eq. (8.3.9).

The other point concerns the composition of the diode current. Actually, the diode current consists of two components: one component, J_1, due to the injection of electrons into the p region and the other component, J_2, due to the injection of holes into the n region. The current density given by Eq. (8.3.13) is the component J_2 due to hole injection. A similar expression can be obtained for the component J_1 due to electron injection. For each component, say J_2, the current density can be further decomposed into $J_{p2} + J_{n2}$, one carried by holes and the other carried by electrons. It is instructive to examine the nature of the current in detail, based on and derived from Eqs. (8.3.11) and (8.3.12).

First we consider Eq. (8.3.11). As the injected holes diffuse from B toward D in Fig. 8.6, they recombine with electrons. This explains why p_1 decreases with x in Eq. (8.3.11), but it is not the whole story. The n region must supply electrons during the recombination process, creating a movement of electrons toward the left and thus an electron current. Refer to the following set of continuity equations:

$$\frac{\partial p_1}{\partial t} = -\frac{1}{e} \frac{\partial J_p}{\partial x} - \frac{p_1}{\tau_0} \qquad (8.3.15a)$$

$$\frac{\partial n_1}{\partial t} = \frac{1}{e}\frac{\partial J_n}{\partial x} - \frac{n_1}{\tau_0} \tag{8.3.15b}$$

Since the term p_1/τ_0 is nonzero, so is n_1/τ_0. Accompanied with a nonzero n_1/τ_0, there is a nonzero $\partial J_n/\partial x$. Subtracting Eq. (8.3.15b) from Eq. (8.3.15a) and realizing that $p_1 = n_1$, we find that the total current density $J_2 = J_p + J_n$ is independent of x. In other words, it is immaterial where we evaluate J_2. The reason that we evaluated J_p at $x = x_B$ is that at $x = x_B$, the current is constituted by hole diffusion alone, or in other words, $J_2 = J_p$ at $x = x_B$.

As we move from B toward D, the electron current increases and the hole current decreases. After the excess-carrier concentration drops to zero through recombination, the current is carried entirely by electrons, the majority carriers. Figure 8.7a shows the composition of the current density J_2 created by diffusion of holes as a function of x for cases where recombination processes play an important role, that is, for cases where Eq. (8.3.11) applies. The situation is quite different for cases where recombination of excess carriers is relatively insignificant. Note that Eq. (8.3.12) predicts a constant hole current. The situation is shown in Fig. 8.7b, in contrast to that shown in Fig. 8.7a. The current density J_2 in the p region is, of course, carried entirely by holes (the majority carriers) in both cases. We should also point out that an expression J_1 similar to Eq. (8.3.13) is obtained for electron injection into the p region, and the total current density J is a sum of J_1 and J_2. Figure 8.8 shows the composition of the total current density as a function of distance when both electron and hole injections are considered.

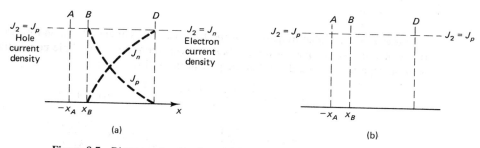

(a)   (b)

Figure 8.7 Diagrams showing the spatial variation of the current composition in (a) wide-base diodes and (b) narrow-base diodes. The current density J_2 is due to injection of holes into the n side. In case (a), recombination of injected holes with electrons draws a flow of electrons. As p_1 of Eq. (8.3.11) decreases, the electron component J_n of the current density increases, whereas the hole component J_p decreases so as to maintain a constant J_2. In case (b) carrier recombination is neglected. Hence J_2 is entirely due to J_p. The lines A and B indicate the boundaries of the space-charge region.

8.4 CARRIER STORAGE AND DIFFUSION CAPACITANCE

In this section we work out the ac diode-current expression so that we can represent a diode by equivalent circuit elements in small-signal, ac circuit analysis. To find the steady-state response of a junction diode to a sinusoidally time-varying signal, we can write the time-dependent diffusion equation as

$$\frac{\partial p_1}{\partial t} = i\omega p_1 = D_p\frac{\partial^2 p_1}{\partial x^2} - \frac{p_1}{\tau_0} \tag{8.4.1}$$

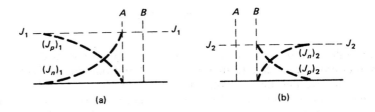

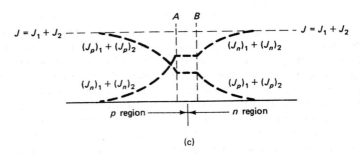

Figure 8.8 Diagrams showing the spatial variations of the current composition (a) for the component J_1 due to electron injection into the p side, (b) for the component J_2 due to hole injection into the n side, and (c) for the total diode current density $J = J_1 + J_2$. The illustrations are for wide-base diodes.

where ω is the angular frequency of the time-varying voltage. If V_0 and v_1 are, respectively, the dc and ac part of the applied voltage, then the excess-hole concentration can also be separated into a dc and an ac part in the following manner:

$$
\begin{aligned}
p_1 &= p_{n0} \left\{ \exp \left[\frac{e(V_0 + v_1)}{kT} \right] - 1 \right\} = (p_1)_{dc} + (p_1)_{ac} \\
&= p_{n0} \left[\exp \left(\frac{eV_0}{kT} \right) - 1 \right] + \frac{ev_1}{kT} p_{n0} \exp \left(\frac{eV_0}{kT} \right)
\end{aligned}
\tag{8.4.2}
$$

In Eq. (8.4.2), the ac applied voltage is assumed to be small such that $ev_1/kT < 1$; hence $\exp (ev_1/kT)$ can be approximated by $1 + ev_1/kT$. The solution of Eq. (8.4.1) is similar to that of Eq. (8.3.8), the dc solution, except that τ_0 should be replaced by $\tau_0/(1 + i\omega\tau_0)$ and at $x = x_B$, $p_1 = (p_1)_{ac}$ of Eq. (8.4.2). Thus the ac part of the excess-hole concentration obeys the following equation:

$$
(p_1)_{ac} = \frac{ev_1}{kT} p_{n0} \exp \left(\frac{eV_0}{kT} \right) \frac{\sinh (W_2 - x)/L'}{\sinh (W_2 - x_B)/L'}
\tag{8.4.3}
$$

where $L' = \sqrt{D_p \tau_0/(1 + i\omega\tau_0)}$.

Now we work out the current expression for the short-base case $(W_2 - x_B) \ll L$. It is further assumed that the p region is much more heavily doped than the n region, so that the current due to diffusion of holes predominates. Differentiating Eq. (8.4.3) with respect to x and evaluating the derivative at $x = x_B$, we find the ac current i_1 in a junction of area A to be

$$i_1 = \frac{ev_1}{kT} p_{n0} \exp\left(\frac{eV_0}{kT}\right) \frac{eD_p}{W_2 - x_B}\left[1 + \frac{i\omega}{3D_p}(W_2 - x_B)^2\right] A \qquad (8.4.4)$$

In obtaining Eq. (8.4.4), we expand coth y into $y^{-1}(1 + y^2/3)$ and neglect the term $(W_2 - x_B)^2/D_p\tau_0$ in comparison to 1 after the expansion.

Obviously, the ac current has a conductive and a capacitive component. A comparison of Eq. (8.4.4) with Eq. (8.3.13) shows that if the dc bias voltage is relatively large so that $eV_0/kT > 1$, the equivalent conductance and capacitance, G and C_D, of a junction diode can be approximated, respectively, as follows:

$$G = \frac{eI_0}{kT} \quad \text{and} \quad C_D = \frac{eI_0}{kT}\frac{(W_2 - x_B)^2}{3D_p} \qquad (8.4.5)$$

where I_0 is the dc current flowing through the diode. The ac conductance G given in Eq. (8.4.5) is valid only in the forward direction, and is equal to the slope of the dc current–voltage curve shown in Fig. 8.2. The ac capacitance C_D, however, is new and is unrelated to the transition-region capacitance C_T discussed in Section 8.2.

Figure 8.9 shows the small-signal ac equivalent circuit for a junction diode. Note that in Fig. 8.8a, the current J_1 in the n region is constituted by majority carriers and hence must be supported by an electric field. The same is true for the current J_2 in the p region. The voltage necessary to support J_1 and J_2 outside the diffusion region is represented by the voltage drop developed across R in Fig. 8.9, R being generally called the *spreading resistance* of a junction diode. The value of R is in the neighborhood of few ohms in diffused diodes. The capacitance C_T in Fig. 8.9 is the transition-region capacitance due to charges stored in the space-charge region. The capacitance C_D given by Eq. (8.4.5) is generally referred to as the *diffusion capacitance* because it is associated with diffusion of excess carriers. As the voltage across a junction changes, so do the width of the space-charge region and the charges stored in the region. The variation in the stored charges is accounted for by C_T. At the same time, the distribution $p_1(x, t)$ of excess carriers also varies as v_1 changes. Any redistribution of excess carriers takes place through the process of diffusion and thus requires a finite time. The phase difference between the applied voltage and the excess-carrier distribution makes the current have a capacitive component which is accounted for by C_D in Fig. 8.9.

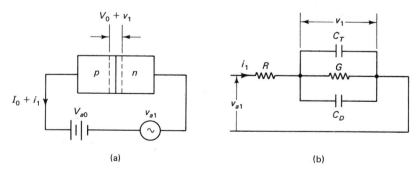

(a)   (b)

Figure 8.9 (a) A *p-n* junction diode operated under simultaneous dc and ac bias conditions, and (b) small-signal ($ev_1/kT \ll 1$) ac equivalent circuit for the diode.

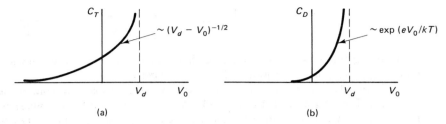

(a)

(b)

Figure 8.10 Diagrams showing the dependences of (a) the transition-region capacitance C_T and (b) the diffusion capacitance C_D upon the dc bias voltage V_0 of a junction diode. The maximum voltage which can be applied across a p-n junction is V_d.

Let us compare the magnitude of C_T and C_D. For the example for Si discussed in Section 8.2, we calculated C_T/A to be 1.82×10^{-8} F/cm^2. For a junction area A of 4×10^{-4} cm^2, C_T is equal to 7.3×10^{-12} F. Take $I_0 = 6.5 \times 10^{-3}$ A, $kT = 0.0259$ eV, $W_2 - x_B = 2 \times 10^{-3}$ cm, and $D_p = 12.4$ cm^2/V-s in Eq. (8.4.5), we obtain $C_D = 8.1 \times 10^{-8}$ F. The reason that $C_D \gg C_T$ is that in the calculation the diode is assumed to be biased in the forward direction. Figure 8.10 shows the variation of C_T and C_D as function of the dc bias voltage V_0. In the forward direction, C_D increases much faster with increasing forward bias than C_T. In the reverse direction, C_T decreases much more slowly with increasing reverse bias than C_D. Therefore, C_D usually dominates over C_T for forward-biased diodes and C_T dominates over C_D for reverse-biased diodes.

8.5 SWITCHING RESPONSE AND RECOVERY TIME

The equivalent circuit of Fig. 8.9 is useful only for small variations of the applied voltage. As shown in Fig. 8.10, the value of C_D changes drastically with the junction voltage V_0. For switching diodes the applied voltage swings from a positive to a negative value in order to make a diode switch from the on mode to the off mode of operation. Obviously, for such a large swing of the applied voltage, an entirely different approach other than the small-signal, equivalent-circuit analysis must be undertaken. Consider the arrangement of Fig. 8.11, in which the applied voltage in a diode circuit is suddenly switched from a positive to a negative direction at time $t = 0$. The question is: How does the current respond to this voltage change? To answer this question we must know the minority-carrier distribution in the base region. For ease of discussion,

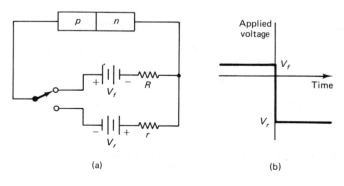

(a)

(b)

Figure 8.11 (a) Circuit arrangement for switching diode operation from a positive to a negative applied voltage at $t = 0$ and (b) diagram showing the time variation of the applied voltage.

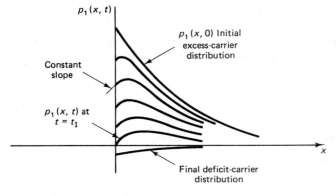

$p_1(x, 0)$ Initial excess-carrier distribution

Constant slope

$p_1(x, t)$ at $t = t_I$

Final deficit-carrier distribution

x

Figure 8.12 Excess-hole distribution in the base region of a wide-base diode. The different curves are for different times after switching. The distribution changes progressively from the initial (excess-carrier) distribution to the final (deficit-carrier) distribution. At $t = t_I$, $p_1 = 0$ at $x = x_B$ and the diode switches from a forward to a reverse-bias condition.

we again assume that the p side of the junction is much more heavily doped than the n side so that only the injection of holes into the n region need be considered.

First, we realize that any change of hole distribution in the n region must be accomplished through the process of diffusion, and hence any redistribution of holes is a time-consuming process. In changing from the steady-state distribution in the forward direction to that in the reverse direction, the excess-hole concentration right at the junction must pass through zero. The situation is illustrated in Fig. 8.12. Depending on whether the excess-hole concentration p_1 at the edge of the space-charge region is greater or smaller than zero, the transient behavior of a junction diode can be divided into two phases, I and II, as shown in Fig. 8.13. In both Figs. 8.12 and 8.13, the time t_I marks the transition from the phase I to the phase II operation.

During phase I operation, the hole concentration p_n at the junction is larger than the equilibrium hole concentration p_{n0}. According to Eq. (8.3.7), the junction voltage is still biased in the forward direction, although the applied voltage is already in the reverse direction. Since the voltage drop across the junction cannot be large in the forward direction, the current is mainly determined by the external resistance r or $I_r = V_r/r$. During phase II operation, the hole concentration at the junction goes down below the equilibrium hole concentration and hence the junction becomes reversely biased. The junction voltage builds up gradually from zero to the full reverse applied voltage

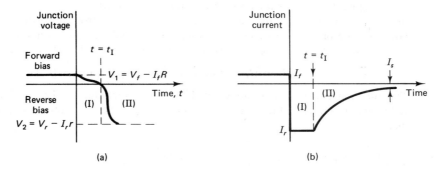

Figure 8.13 Plots showing the variation of (a) junction voltage and (b) junction current with time. The time t_I at which the junction voltage is zero marks the transition from a forward-bias condition (phase I operation) to a reverse-bias condition (phase II operation).

V_r except for the small drop across resistance r and the current decreases gradually from V_r/r to I_s, the reverse saturation current of the diode.

Let us discuss the situation in a wide-base diode first. The mathematical analysis of such a problem is presented in Appendix A8. During the phase I operation, the excess-hole concentration $p_1(x, t)$ is given by Eq. (A8.7), or

$$
p_1(x, t) = \frac{L}{2eAD_p} \left\{ 2I_f \exp(-v) - (I_r + I_f) \right.
$$
$$
\left. \left[\exp(-v) \, \text{erfc} \left(\frac{v}{2u} - u \right) - \exp(v) \, \text{erfc} \left(\frac{v}{2u} + u \right) \right] \right\}
$$
(8.5.1)

where $v = x/L$, $u = \sqrt{t/\tau_0}$, A is the junction area, and erfc is the complementary error function. In Eq. (8.5.1) we move the origin to x_B. At $t = 0$, $v/2u = \infty$; hence both erfc functions are zero because the lower limit of integration in Eq. (7.7.16) becomes infinite in both cases. Using the relation between current and excess carrier concentration similar to that between Eqs. (8.3.13) and (8.3.11), the reader can easily show that the first term in Eq. (8.5.1), that is, the nonzero term at $t = 0$, indeed is equal to the initial excess-hole distribution $p'(0)$ of Eq. (A8.3). Therefore, Eq. (8.5.1) satisfies the initial condition specified in the problem.

Next we evaluate the derivative $\partial p_1/\partial x$ at $x = 0$. Noting that

$$
\frac{d}{dv} \, \text{erfc} \left(\frac{v}{2u} \pm u \right) \bigg|_{v=0} = \frac{1}{\sqrt{\pi} u} \exp \left[\left(\frac{v}{2u} \pm u \right)^2 \right] \bigg|_{v=0} = \frac{\exp(-u^2)}{\sqrt{\pi} u}
$$
(8.5.2)

we find that

$$
eAD_p \frac{\partial p_1}{\partial x} \bigg|_{x=0} = I_f - \frac{I_r + I_f}{2} [\text{erfc}(-u) + \text{erfc}(u)] = -I_r
$$
(8.5.3)

As shown in Fig. 8.13b, the current during the phase I operation is equal to I_r. Through Eq. (8.5.3) we prove that the excess-hole concentration of Eq. (8.5.1) indeed satisfies the boundary condition required by the current I_r. Since we have already proved in the preceding paragraph that Eq. (8.5.1) satisfies the initial condition, $p_1(x, t)$ of Eq. (8.5.1) is indeed the proper solution for the problem.

Figure 8.12 shows the progressive change with time of the excess-hole distribution in the base region after the applied voltage is switched from a positive to a negative polarity. The phase I operation ends when the excess-hole concentration at the junction drops to zero. Thus, letting $p_1 = 0$ at $t = t_1$ and $x = 0$ in Eq. (8.5.1), we obtain

$$
2I_f - (I_r + I_f)[\text{erfc}(-u_1)] - \text{erfc}(u_1)] = 0
$$
(8.5.4)

The function $\exp(-\xi^2)$ is plotted in Fig. 8.14. According to Eq. (7.7.16), the erfc functions are equal to the shaded areas under the curve. It is obvious from Fig. 8.14 that

$$
\frac{2}{\sqrt{\pi}} \left[\int_{-u}^{\infty} \exp(-\xi^2) \, d\xi - \int_{u}^{\infty} \exp(-\xi^2) \, d\xi \right]
$$
(8.5.5)

$$
= \frac{4}{\sqrt{\pi}} \int_{0}^{u} \exp(-\xi^2) \, d\xi = 2 \, \text{erf} \, u
$$

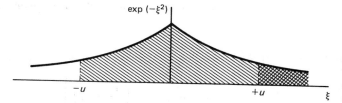

Figure 8.14 Plot of the function exp $(-\xi^2)$. The erfc y function is equal to the area under the curve from $\xi = y$ to ∞. Therefore, the difference between erfc $(-u)$ and erfc (u) is equal to the shaded area under the curve from $-u$ to u.

Thus Eq. (8.5.4) becomes

$$\text{erf } u_1 = \text{erf}\left(\sqrt{\frac{t_1}{\tau_0}}\right) = \frac{I_f}{I_r + I_f} \tag{8.5.6}$$

The error function erf (u) is related to the erfc function through

$$\text{erf }(u) + \text{erfc }(u) = \frac{2}{\sqrt{\pi}} \int_0^\infty \exp(-\xi^2)\, d\xi = 1 \tag{8.5.7}$$

Thus we can use Fig. 7.14b to find the value of the error function. For $u < 1$, Eq. (7.7.17) gives

$$\text{erf }(u) = 1 - \text{erfc }(u) \cong \frac{2u}{\sqrt{\pi}} \tag{8.5.8}$$

In practical operation, we want the switching time from the on mode to the off mode as short as possible. The value of t_1 can be made small if we make $I_r > I_f$ in Eq. (8.5.6). Using the approximation of Eq. (8.5.8) for the error function, we can write Eq. (8.5.6) as

$$t_1 \cong \frac{\pi}{4}\left(\frac{I_f}{I_r + I_f}\right)^2 \tau_0 \tag{8.5.9}$$

After $t > t_1$, phase II operation begins. During this phase, the junction voltage is biased in the reverse direction and the junction current decreases toward the reverse saturation current (Fig. 8.13b). Because the current is no longer a constant, Eq. (8.5.1) does not apply to phase II operation. As pointed out in Appendix A8, because of a rather complicated initial condition, it would be very difficult to find the exact solution for phase II operation. Here we follow the approach taken by Kingston in his analysis of the switching response of a junction diode (R. H. Kingston, *Proc. IRE*, Vol. 42, p. 829, 1954).

The initial excess-hole distribution for phase II operation is chosen to be the same as that used for phase I operation. However, the boundary condition is that for a reverse-biased diode, that is,

$$p_1(x, t) = p_{n0}\left[1 - \exp\left(-\frac{eV_r}{kT}\right)\right] \cong p_{n0} \cong 0 \qquad \text{at} \quad x = 0 \tag{8.5.10}$$

The solution of the time-dependent diffusion equation under such conditions is given by Eq. (A8.10), or

$$p_1(x, t) = \frac{LI_f}{2eAD_p}\left[2 \exp{(-v)} - \exp{(-v)} \operatorname{erfc}\left(\frac{v}{2u} - u\right)\right.$$

$$\left. - \exp{(v)} \operatorname{erfc}\left(\frac{v}{2u} + u\right)\right] \tag{8.5.11}$$

At $t = 0$, $v/2u = \infty$; thus the two erfc functions become zero. We note that the first term in Eq. (8.5.11) is identical to the first term in Eq. (8.5.1), and hence represents the initial excess-hole distribution. At the boundary of the space-charge region $x = 0$, $v = 0$, and $v/2u = 0$. Thus the last two terms in Eq. (8.5.11) add to 2 as

$$\frac{2}{\sqrt{\pi}}\left[\int_u^\infty \exp{(-\xi^2)}\,d\xi + \int_{-u}^\infty \exp{(-\xi^2)}\,d\xi\right] \tag{8.5.12}$$

$$= \frac{2}{\sqrt{\pi}}\left[\int_{-\infty}^{-u} \exp{(-\xi^2)}\,d\xi + \int_{-u}^\infty \exp{(-\xi^2)}\,d\xi\right] = 2$$

which is evident from Fig. 8.14 and Eq. (7.7.14). Use of Eq. (8.5.12) in Eq. (8.5.11) gives $p_1(x, t) = 0$ at $x = 0$. Therefore, Eq. (8.5.11) indeed satisfies both the initial and boundary conditions.

Figure 8.15 shows the transient behavior of $p_1(x, t)$ based on Eq. (8.5.11). For $u = \infty$, erfc $(u) = 0$ and erfc $(-u) = 2$. Thus at $t = \infty$, $p_1(x, t)$ is zero everywhere. This means that the excess-hole distribution decreases gradually toward zero in phase II operation, as illustrated in Fig. 8.15. However, our main interest here is not the time variation of p_1 but rather the behavior of the diode current as a function of time. Differentiating Eq. (8.5.11) with respect to x gives

$$I = -eAD_p\left.\frac{\partial p_1}{\partial x}\right|_{v=0} = -I_f\left[\operatorname{erfc}{(u)} + \frac{\exp{(-u^2)}}{\sqrt{\pi}u} - 1\right] \tag{8.5.13}$$

We have used Eq. (8.5.2) in obtaining Eq. (8.5.13).

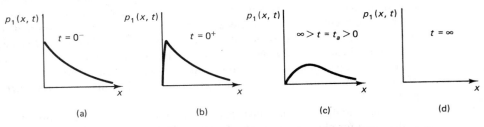

$p_1(x, t)$ $t = 0^-$ $p_1(x, t)$ $t = 0^+$ $p_1(x, t)$ $\infty > t = t_a > 0$ $p_1(x, t)$ $t = \infty$

          (a)   (b)   (c)   (d)

Figure 8.15 Schematic diagrams showing the progressive changes in the excess-hole distribution. The curves are for (a) the instant $t = 0^-$ before switching of the applied voltage, (b) the instant $t = 0^+$ right after switching of the applied voltage, (c) the instant when $p_1(x, t) = 0$ at $x = 0$, the boundary of the space-charge region, and (d) $t = \infty$. The curves are based on $p_1(x, t)$ of Eq. (8.5.11). Note that $\partial p/\partial x$ is infinite at $x = 0$ in (b).

Figure 8.16 shows the general shape of the diode current I in the second phase. The current goes to infinity at $t = 0$ because of the artificial boundary condition imposed on the solution. The condition that p_1 at $x = 0$ is nonzero for $t < 0$ and zero for $t > 0$ requires an infinite slope for p_1 at $x = 0$ and hence an infinite current. This is illustrated in Fig. 8.15a and b. Apparently, the solution p_1 of Eq. (8.5.11) near $t = 0$ is far from being the actual excess-hole distribution. Note, however, that at a time $t = t_a$ in Fig. 8.16a, the current I has the same value I_r as the current in the phase I operation. This means that the excess-hole distribution given by Eq. (8.5.11) at $t = t_a$ has the same value and the same slope at $x = 0$ as the actual excess-hole distribution given by Eq. (8.5.1) at $t = t_1$. Therefore, we can reasonably except that for $t \geq t_a$, the approximate solution will be close to the actual solution. It has been shown by Kingston and by Lax and Neustadter that this is indeed the case (B. Lax and S. F. Neustadter, *J. Appl. Phys.*, Vol. 25, p. 1148, 1954).

To obtain a complete solution of the current, we match the current I (phase I) $= I$ (phase II) at $t = t_1$ by shifting I (phase II) in time scale from t_a to t_1 as illustrated in Fig. 8.16b. Equation (8.5.13) would represent the true current had the resistance r in Fig. 8.11a been zero. In practice, however, the current I_r must be limited to a finite value by a nonzero r. Because the value of I_r is finite, the diode takes a finite time to reach a reverse-bias condition. The shift in time scale of the current curve in Fig. 8.16 is to account for the finite delay introduced by a finite I_r. In Fig. 8.16, the current I approaches zero instead of I_s, the reverse saturation current of a diode, as t goes to infinity in Eq. (8.5.13). In our calculation, the reverse saturation current is taken to be negligibly small because of the approximation made in Eq. (8.5.10).

The recovery time t_r of a diode can be defined as the time required for the current to drop to a certain percentage of I_r, say to ϵI_r, where $\epsilon < 1$ is a preset value. From Fig. 8.16b, $t_r = t_1 + t_{\text{II}}$. In Fig. 8.17, the behavior of I/I_f is plotted as a function of u. The time lapse for I/I_f to drop from I_r/I_f to $\epsilon I_r/I_f$ is the value for t_{II}. As an example, we suppose that the external circuit is designed such that $I_r = 5I_f$ in Fig. 8.13b. Using Eq.

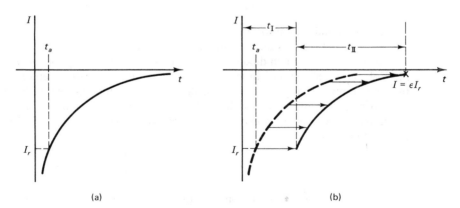

(a)   (b)

Figure 8.16 (a) Transient behavior of the diode current based on Eq. (8.5.13). At $t = 0$, the current is infinite on account of an infinite slope $\partial p / \partial x$ at $x = 0$ in Fig. 8.15b. (b) Diagram illustrating the procedure used in obtaining the transient behavior of the diode current for a real case. For $t < t_1$, the current I is set equal to $I_r = V_r/r$ in the phase I operation. For $t > t_1$, the current is set equal to I of Eq. (8.5.13) by shifting I in time scale from t_a to t_1 so that $I(t = t_a) = I_r$.

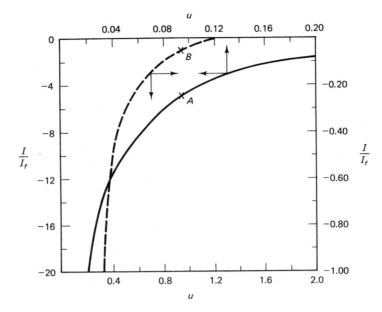

Figure 8.17 A numerical plot of I/I_f based on Eq. (8.5.13) where $u = \sqrt{t/\tau_0}$. For the dashed curve, use the right and bottom scales. For the solid curve, use the left and top scales.

(8.5.9), we obtain $t_1 = 2.2 \times 10^{-2}\tau_0$. The condition $I_r = 5I_f$ determines point A in Fig. 8.17. If we let $\epsilon = 0.01$ as a measure of recovery, the condition $I = 0.05I_f$ specifies point B. From Fig. 8.17 we find that

$$t_{\mathrm{II}} = (u_A^2 - u_B^2)\tau_0 = 0.89\tau_0 \tag{8.5.14}$$

Combining t_1 and t_{II}, we have $t_r = 0.91\tau_0$.

Figure 8.18 shows the measurement of recovery time reported by Bakanowski and Forster in diffused silicon diodes (A. E. Bakanowski and J. H. Forster, *Bell Syst. Tech. J.*, Vol. 39, p. 87, 1960). In the experiment, the recombination centers were provided by gold atoms. During the process of diffusion, the gold concentration in the junction was so high that its value was limited only by the solid solubility of gold in silicon. Using the known values of capture cross sections for gold in Eq. (7.5.17), the recovery time is found to be

$$t_r = 0.9\tau_0 = \frac{2.53 \times 10^7}{N} \tag{8.5.15}$$

where N is the gold concentration. In Fig. 8.18, the circles represent the average t_r for a group of diodes and the solid bars indicate the range of measured t_r within each group. The dashed lines represent the calculated value of t_r, using Eq. (8.5.15) and the values of N taken from the solid-solubility data. As can be seen from Fig. 8.18, the agreement between the measured and calculated values is indeed good. Note that the lifetime of excess carriers in these diodes is very short. Therefore, the foregoing analysis for wide-base diodes applies to the data shown in Fig. 8.18. We again emphasize that the defi-

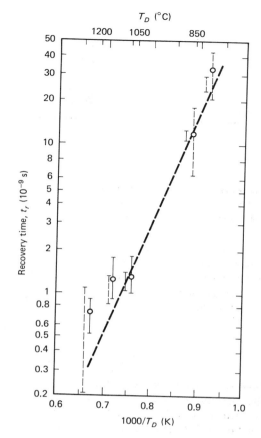

Figure 8.18 Recovery time t_r for gold-doped silicon diodes plotted as a function of the gold-diffusion temperature. (A. E. Bakanowski and J. H. Forster, "Electrical Properties of Gold-doped Diffused Silicon Computer Diodes" *Bell Syst. Tech. J.*, Vol. 39, p. 87, 1960, reprinted with permission from the *Bell System Technical Journal* copyright 1960 AT&T.)

nition of wide-base and narrow-base diodes does not depend on the absolute value of the basewidth, but depends on its value relative to the diffusion length.

The analysis above can easily be extended to narrow-base diodes where $W' < L$. The only modification is that in the expression of u, t is normalized by a different time constant instead of the lifetime τ_0 which is used in wide-base diodes ($u = \sqrt{t/\tau_0}$). Therefore, the procedure above may still be followed in estimating the recovery time of a narrow-base junction diode.

To find the new time constant, we compare the excess-hole distribution in a wide-base diode (Fig. 8.19a) with that in a narrow-base diode (Fig. 8.19b). In the wide-base diode, our attention should be concentrated on the region within a diffusion length of the junction because it is the diffusion of minority carriers but not the drift of majority carriers which causes the time lag between the applied voltage and the diode current. The diffusion capacitance calculated earlier and the recovery time discussed above originate from the same physical process—the diffusion of minority carriers across the region where minority carriers are stored.

To calculate the new time constant, we define a diffusion velocity v_d such that

$$v_d = \frac{I}{Aep_1} = \frac{D_p}{p_1}\frac{\partial p_1}{\partial x} \tag{8.5.16}$$

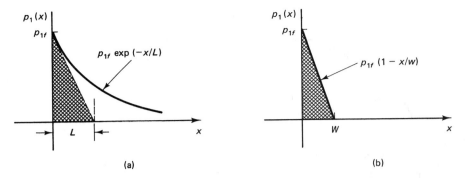

Figure 8.19 A comparison of the excess-carrier distributions in (a) wide-base diodes and (b) narrow-base diodes. In (b), the base contact is at $x = W_2$. Cross hatched areas represent the region where most excess carriers are stored.

For the distribution of Fig. 8.19a, $v_d = D_p/L$. Hence the time required for diffusion across the region where excess holes are stored is given by $t = L/v_d = \tau_0$, which is used as the time basis for normalization in the case of wide-base diodes. For the distribution of Fig. 8.19b, $v_d = 2D_p/W$ if p_1 in Eq. (8.5.16) is taken as the average value of the excess-hole concentration in the base region. The transit time in this case is equal to

$$t = \frac{W}{v_d} = \frac{W^2}{2D_p} \tag{8.5.17}$$

where we have approximated W' by W by neglecting the width of the space-charge region. The lifetime and the transit time, τ_0 and $W^2/2D_p$, can be considered as the characteristic response times of wide-base and narrow-base diodes, respectively. Therefore, we expect that Eq. (8.5.9) and Fig. 8.17 are still useful in calculating the recovery time of a narrow-base diode, provided that the new u^2 is made equal to $2D_p t/W^2$. A complete and more accurate analysis of the narrow-base diode problem will be left for the reader as a problem. The reader is also referred to the aforementioned article by Lax and Neustadter and an article by Gossick for further discussions of the transient behavior of junction diodes (B. R. Gossick, *J. Appl. Phys.*, Vol. 27, p. 950, 1956).

8.6 JUNCTION-FORMATION AND FILM-GROWTH TECHNIQUES

Fabrication is an important and integral part of device technology. In this section we introduce the basic principles used in device fabrication. The four principal methods of making junction structures are alloying, diffusion, implantation, and epitaxy. The alloying process was important in the early development of semiconductor industry but has been replaced by the diffusion and implantation techniques. The epitaxy method is a relatively recent development, now widely used for depositing thin layers of controlled doping profile and in making special diodes such as IMPATT and laser diodes. The evolution of device-fabrication technology is necessitated not only by the need to reduce production cost but also by the stringent demand on quality control.

To understand the alloying and liquid-phase epitaxial processes, we first study the phase diagram of a two-component system. As an example, we show the germanium–

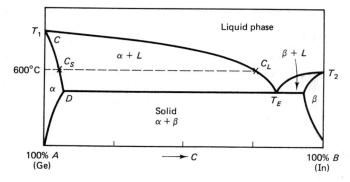

Figure 8.20 Phase diagram for a binary system completely soluble in each other. The diagram shows the relations between the different phases or structures that are encountered when the two principal variables, temperature and composition, are varied. For example, the curve DC for a Ge–In system shows that when a liquid Ge–In solution is in contact with solid Ge, the saturation concentration of In in Ge is about 20% at 600°C.

indium phase diagram in Fig. 8.20. The abscissa is the composition of the system, and the ordinate is the temperature. The melting point of Ge and that of In are, respectively, marked T_1 and T_2. For a given composition and at a given temperature, the system is represented by a point in the diagram. Therefore, a phase diagram is a graphical way of presenting information about a system.

Figure 8.20 is typical for a binary system whose two elements are partially soluble in each other in the solid state. Donor and acceptor impurities such as Sb, P, Ga, and In form such a binary system with either Ge or Si. In Fig. 8.20 there are two solid phases, α and β, and there is only one liquid phase, L. If the percentage of In and the temperature are such that the point defining the system falls in the region marked α, the two elements form a single solid phase with a small amount of In completely soluble in Ge. However, if the point happens to fall in the region marked $\alpha + L$, the two elements form a two-phase system, that is, a mixture of the solid phase α and the liquid phase L. In the region marked β, the two elements form a single solid phase β with a small amount of Ge dissolved in In. The lowest temperature at which a solution remains completely liquid, called the *eutectic point*, is marked T_E in Fig. 8.20.

Alloying Process

In the alloying process, a small pellet of metal dopant, say indium, is placed on top of a semiconductor slab (Fig. 8.21a), say n-type germanium doped with Sb, and the combination is then placed in an oven heated to a temperature $T_2 < T < T_1$ (Fig. 8.20). The molten indium rapidly attacks the germanium. A molten zone of In–Ge mixture forms (Fig. 8.21b), of which the composition is determined by the Ge–In phase diagram. At 600°C, for example, the saturated solution of In–Ge contains about 20% In. If the temperature is lowered toward the eutectic point, part of the molten zone begins

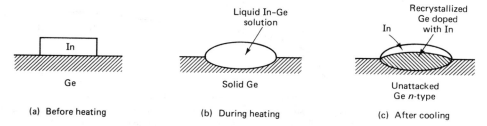

(a) Before heating   (b) During heating   (c) After cooling

Figure 8.21 Schematic diagrams illustrating the alloying process.

to solidify. The recrystallized region, which contains mostly Ge, is deposited on the parent Ge slab, which acts as a seed for recrystallization (Fig. 8.21c). Since In has a finite solid solubility in Ge, the recrystallized germanium is converted into *p* type, creating an abrupt *p-n* junction. The impurity distribution in the base region is not disturbed during the alloying process and hence can be assumed uniform in alloyed diodes. Furthermore, at the temperature of alloying, little diffusion of impurities takes place; thus the impurity concentration changes very abruptly across the junction. Therefore, the assumptions made in Section 8.2 in our analysis of an alloyed *p-n* junction are indeed valid.

Liquid-Phase Epitaxy

A growth technique whose principle is similar to that of the alloying process is the liquid-phase epitaxy (LPE). The word *epitaxy* means an oriented overgrowth of one crystalline material upon another crystalline material of a similar crystal structure. Figure 8.22 shows an example of the boat assembly used in LPE. It consists of several bins which contain melts of different chemical elements. The substrate is housed in the depression of a slide. When the boat is pushed relative to the slide, the substrate can be made in contact with the melt in a specific bin. Figure 8.23 shows the atomic fraction *x* of P, As, or Sb required to saturate liquid Ga or In. For example, a Ga plus As liquid

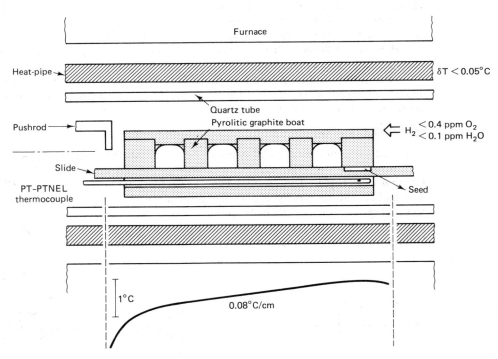

Figure 8.22 A typical boat assembly used in liquid-phase epitaxy. Relative motion of the boat with respect to the slide brings the substrate into contact with a melt of specific composition. Temperature uniformity is required for uniform film quality.

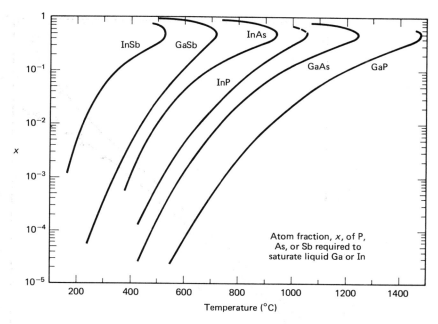

Figure 8.23 Solubility curves for the arsenides, phosphides, and antimonides in a Ga or In melt, for example, GaAs in Ga or InP in In. (R. N. Hall, "Solubility of III–V Compound Semiconductors in Column III Liquids," *J. Electrochem. Soc.*, Vol. 110, p. 385, 1963, reprinted with permission of the publisher, The Electrochemical Society.)

solution is saturated at 810°C for 2.5 at % of As. Precipitation of GaAs onto the substrate occurs upon lowering the temperature below 810°C.

Dopants can be incorporated into the LPE-grown layer by putting the dopant elements into the melt. For example, germanium is commonly used as an acceptor in LPE-grown GaAs films. Figure 8.24 shows the germanium concentration incorporated into GaAs crystals as a function of Ge concentration in the liquid. For each impurity there exists a definite ratio C_S/C_L, called the *distribution coefficient* $K = C_S/C_L$, where C_S and C_L are, respectively, the solubility of the given impurity in the liquid phase and that in the solid phase. The Ge concentration in GaAs shown in Fig. 8.24 is given by $C_S = KC_L$. A multiple-layered structure is grown on the substrate upon successive contacts with and recrystallization from different melts of Fig. 8.22. Also shown in Fig. 8.24 is the hole concentration determined from the Hall measurement. The reason the incorporated Ge atoms are not fully activated as acceptors is not completely understood. Suggested possible mechanisms include the existence of dopant species other than the shallow (singly ionized) acceptor, and the formation of electrically inactive complexes.

Whereas Ge is generally used as a *p* dopant, Te is generally used as a *n* dopant in LPE-grown GaAs. If the impurity incorporation is controlled by the equilibrium between the liquid solution and the solid surface, a linear relation between C_S and C_L is expected. Such is the case with Ge, Te, and Sn in GaAs. A notable exception is Zn, which shows a square-root dependence of C_S on C_L in GaAs and GaP. Another interesting behavior is that of Si in GaAs. At high growth temperatures T and for a moderate

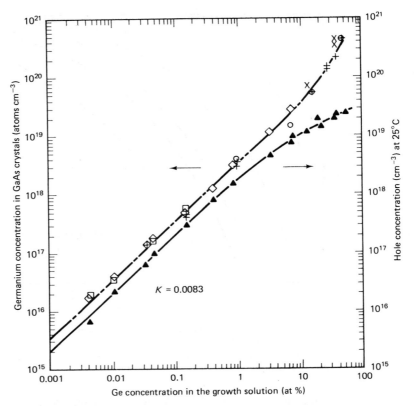

Figure 8.24 Measured Ge and hole concentrations in crystalline GaAs versus Ge concentration in GaAs-Ga melt. (F. E. Rosztoczy and K. B. Wolfstirn, *J. Appl. Phys.,* Vol. 42, p. 426, 1971.)

atomic fraction x_{si} in the liquid, such as $T > 800°C$ and $X_{si} > 4 \times 10^{-4}$, silicon-doped GaAs is n type. However, at low T and for a high X_{si}, such as $T < 800°C$ and $x_{si} > 2 \times 10^{-3}$, silicon-doped GaAs is p type. There is evidence that both Si_{Ga}(donor) and Si_{As}(acceptor) are present in Si-doped GaAs. The reader is referred to the book by Casey and Panish for further discussions on impurity incorporation in solution-grown GaAs (H. C. Casey, Jr. and M. B. Panish, *Heterostructure Lasers,* Part B, Academic Press, Inc., New York, 1978, Chap. 6). We should point out that the choice of dopants depends on the growth process used. For molecular-beam epitaxy (MBE), the common dopants are Be as acceptors and Si as donors, and for diffusion, Zn is commonly used as acceptors in both GaAs and InP.

The values of the distribution coefficients K in elemental semiconductors Ge and Si are given in Table 8.2 for a selected number of shallow and deep impurities. Except for B, most impurities have a distribution coefficient considerably smaller than unity. If a melting zone is formed and made to move through a Ge or Si crystal, the recrystallized solid will have a new impurity concentration C_S' given by $C_S' = KC_L$. Since the liquid zone is melted from the solid, C_L is equal to the old C_S. Therefore, after each passage, the impurity concentrations in the recrystallized Ge or Si are reduced by a factor K.

TABLE 8.2 DISTRIBUTION COEFFICIENTS K FOR A SELECTED NUMBER OF IMPURITIES[a]

Solute element	$K = C_S/C_L$	
	In Ge (at 937°C)	In Si (at 1410°C)
Group III		
Al	0.073	0.002
B	17.0	0.8
Ga	0.087	0.008
In	0.001	0.0004
Group V		
As	0.02	0.3
P	0.08	0.35
Sb	0.003	0.023
Group IV		
Pb	1.7×10^{-4}	—
Sn	0.02	0.016
Group I		
Au	1.3×10^{-5}	2.5×10^{-5}
Cu	1.5×10^{-5}	4×10^{-4}
Transition elements		
Fe	3×10^{-5}	8×10^{-6}
Mn	$\sim 10^{-6}$	$\sim 10^{-5}$
Co	$\sim 10^{-6}$	8×10^{-6}

[a]The data are taken at the melting points of Ge and Si.

This method, known as *zone purification,* is used to produce highly pure Ge and Si crystals.

Diffusion

One of the most important tasks in the fabrication of transistors is accurate control of the base width W_0. For high-frequency transistors, say with a cutoff frequency $\omega_\alpha = 10^8$ rad/s, we require that $W_0 = 10^{-3}$ cm for $D = 50$ cm^2/s. Obviously, accurate control of such a thin base width with the alloying process is out of the question. It is made possible, however, through the diffusion technique. If a semiconductor slab originally doped with an acceptor concentration N_a is placed in an oven in an inert atmosphere containing a partial pressure of a donor impurity, part of the semiconductor near the surface is converted into n type through the diffusion of donor impurities into the solid. The transport of the diffusant within the solid is determined by the time-dependent diffusion equation

$$\frac{\partial C(x, t)}{\partial t} = D \frac{\partial^2 C(x, t)}{\partial x^2} \tag{8.6.1}$$

where $C(x, t)$ is the concentration of diffusant and D is the diffusion coefficient.

Figure 8.25 shows a typical arrangement for the diffusion process. The furnace is heated by a resistance heater to a temperature in the vicinity of 1100°C which is below

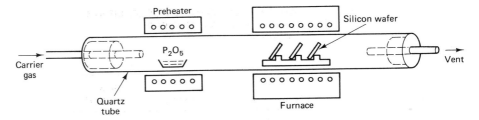

Figure 8.25 Experimental arrangement for diffusion of phosphorus into silicon.

the melting point (1410°C) of silicon. The impurities to be introduced into silicon come from a source material, usually in the form of oxides (e.g., P_2O_5 for phosphorus). The source material evaporates under the heating of the preheater. A carrier gas carries the source material to the semiconductor wafers. The temperature in the preheater region determines the partial pressure of the source material. At the semiconductor surface, the following reaction takes place:

$$P_2O_5 \ (g) \ + \ Si \ (s) \rightarrow P \ + \ \text{phosphosilicate glass} \tag{8.6.2}$$

The phosphorus atoms, thus reduced, diffuse into the silicon wafer.

Before phosphorus atoms can diffuse, they must be incorporated into the silicon lattice. At a given temperature, there is a definite maximum solubility of the diffusant in the host lattice. The solid solubilities of a number of important impurities in Si and Ge are plotted as a function of temperature in Fig. 8.26 (F. A. Trumbore, *Bell Syst. Tech. J.*, Vol. 39, p. 205, 1960). In the diffusion process, if the partial vapor pressure of the source material is sufficiently high, the impurity concentration C_0 at the surface is limited by the solid solubility. Thus we can assume a constant $C_0 = C(x = 0, t)$ in solving Eq. (8.6.1). Note the similarity between our present problem and the problem discussed in Section 7.7 in connection with Eq. (7.7.16). At a distance x below the surface, the impurity concentration is given by

$$C(x, t) = C_0 \ \text{efrc} \left(\frac{x}{2\sqrt{Dt}} \right) \tag{8.6.3}$$

where t is the duration of the diffusion process.

In Fig. 8.27a we show the impurity distribution as a result of diffusion of donor impurities into an originally p-type substrate. The point at which the $C(x, t)$ versus x curve intercepts the N_a line defines the boundary of a p-n junction. The transistor structure shown in Fig. 8.27b can be obtained (1) through separate diffusion of donors and acceptors at two different temperatures, or (2) through a postalloy diffusion process in which both donors and acceptors are first alloyed into the semiconductor and then diffuse at different rates inside the semiconductor. The penetration depth d and hence the base width of the structures in Fig. 8.27 are determined by D and t in Eq. (8.6.3). The diffusion of impurities, unlike the diffusion of electrons and holes, usually requires that impurity atoms in changing positions must overcome a potential barrier $\mathscr{E}$. Therefore, D in Eq. (8.6.3) varies exponentially with temperature T as

$$D = D_0 \exp \left(-\frac{\mathscr{E}}{kT} \right) \tag{8.6.4}$$

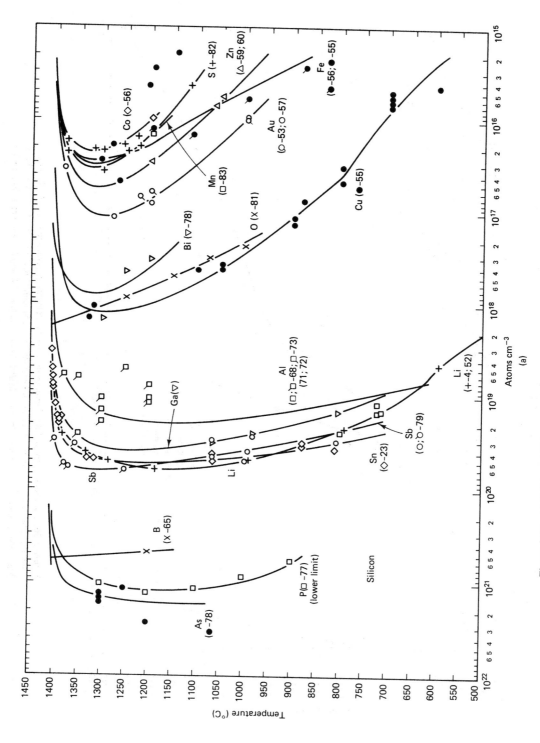

Figure 8.26 (a) Solid solubilities of impurity elements in silicon. (*Figure continues on next page.*)

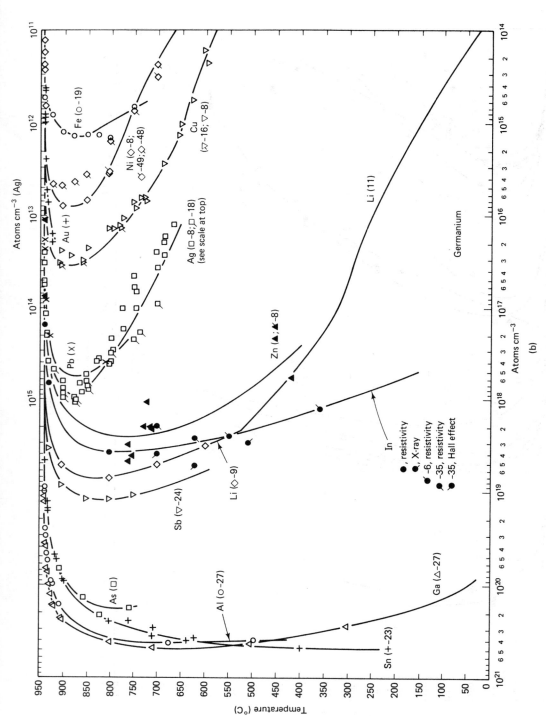

Figure 8.26, continued. (b) Solid solubilities of impurity elements in germanium. (F. A. Trumbore, "Solid Solubilities of Impurity Elements in Germanium and Silicon," *Bell Syst. Tech. J.*, Vol. 39, p. 205, 1960, reprinted with permission from the *Bell System Technical Journal*, copyright 1960 AT&T.)

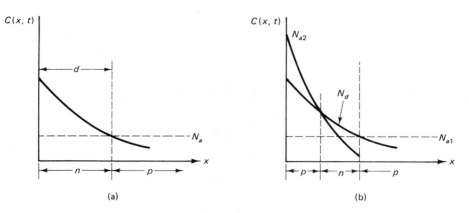

Figure 8.27 (a) Formation of a *p-n* junction by diffusing donor impurities into a *p*-type substrate. The depth of the junction is controlled by the duration (t) and temperature (T and hence the value of D) of the diffusion process. (b) Formation of a *p-n-p* transistor by a double-diffusion process. The original acceptor concentration in the substrate is indicated by the dashed line whereas the diffused impurity concentrations are represented by the solid curves.

By accurately controlling the duration and temperature of diffusion, the penetration depth d in Fig. 8.27a can be held to a value within reasonable limits.

The values of the diffusion coefficient at several temperatures are given in Table 8.3 for a number of important impurities in Ge and Si. We further plot the variation of D with $1/T$ in Si in Fig. 8.28. We see that the exponential dependence of Eq. (8.6.4) is indeed observed. Let us apply the discussion to the diffusion of phosphorus into silicon. Suppose that the diffusion process takes place at $T = 1200°C$ and that the Si wafer has $N_a = 10^{14}$ cm^{-3}. If the vapor pressure of phosphorus is sufficiently high so that C_0 in Eq. (8.6.3) is limited by the solid-solubility value, then from Fig. 8.26a we find that $C_0 = 1.4 \times 10^{21}$ cm^{-3}. From Eq. (8.6.3) we obtain

$$\text{erfc}\left(\frac{x}{2\sqrt{Dt}}\right) = \frac{10^{14}}{1.4 \times 10^{21}} \quad \text{or} \quad \frac{x}{2\sqrt{Dt}} = 3.82 \qquad (8.6.5)$$

through the use of Eq. (7.7.17). At $T = 1200°C$, $D = 3.5 \times 10^{-12}$ cm^2/s. Thus for $t = 10$ h, the penetration depth d (Fig. 8.27a) is found from Eq. (8.6.5) to be $d = 7.64 \sqrt{Dt} = 27$ μm.

We note from Fig. 8.28 that at $T = 1100°C$, $D = 4.2 \times 10^{-13}$ cm^2/s. For the same penetration depth $d = 27$ μm, $t = 83$ h. Because of the exponential dependence of D on $1/T$, the time duration of diffusion required to achieve a given penetration depth decreases rapidly with increasing temperature. This aspect points out the importance of accurate temperature control during the diffusion process. In practice, we choose the temperature such that the time t in Eq. (8.6.5) is in the neighborhood of several hours. In Fig. 8.28, the right-hand scale for D is expressed in μm^2/h. For $t = 1$ h, a value of $D = 1$ gives $\sqrt{Dt} = 1$ μm. Therefore, the convenient range of D lies in the vicinity of the line $D = 3.6 \times 10^{-1}$ (right-hand scale).

TABLE 8.3 DIFFUSION COEFFICIENT OF IMPURITIES IN GERMANIUM AND SILICON

D (cm^2/s)	Ge Temperature (°C)						Si Temperature (°C)						
	As	Sb	P	B	Ga	In	As	Sb	P	Al	B	Ga	In
10^{-10}	850	860	925	910	900	875	1270	1280	1170	1300	1180	1140	1200
10^{-12}	680	690	725	810	730	700	1060	1070	970	1070	980	950	1020
10^{-14}	550	530	575	725	610	575							
10^{-16}			475	650									

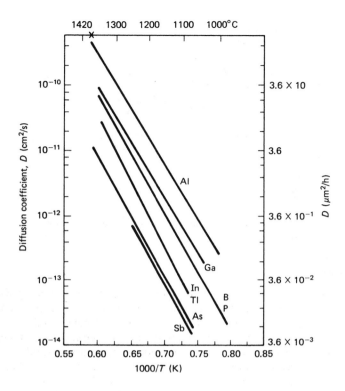

Figure 8.28 Temperature variation of the diffusion constants in silicon for a selected number of impurities. (C. S. Fuller and J. A. Ditzenberger, *J. Appl. Phys.*, Vol. 27, p. 544, 1956.)

Chloride Vapor-Phase Epitaxy

Figure 8.29 shows a typical arrangement used for vapor-phase epitaxy-growing process. The growth of an epitaxial layer is achieved through the following chemical reaction:

$$SiCl_4 + 2H_2 \rightarrow 4HCl + Si(s) \qquad (8.6.6)$$

A similar reaction takes place if $GeCl_4$ is used. Both $SiCl_4$ and $GeCl_4$ are liquid at room temperature and have sufficiently high vapor pressure that the rate of their delivery can be regulated by the flow of the hydrogen gas. A quartz tube forms the reactor chamber, which contains a graphite susceptor. Induction heating is used so that only the graphite susceptor and the silicon wafers supported by it are heated. This arrangement keeps the wall of the quartz tube cool, thus preventing the reaction from taking place there. Doping is achieved either by addition of a volatile impurity to the tetrachloride or by the separate introduction of a gaseous impurity as shown in Fig. 8.29. Cave and Czorny have reported that 1120 to 1350°C is the suitable range for the deposition temperature and that the corresponding growth rate in this temperature range is 0.1 to 3 μm/min, with a hydrogen flow velocity of 2000 cm/min (E. F. Cave and D. Czorny, *RCA Rev.*, Vol. 24, p. 523, December 1963). Because the epitaxial growth takes place within minutes, little diffusion of impurities takes place. Without subsequent diffusion, the impurity distribution in an epitaxial-grown junction can be taken as that of an abrupt junction.

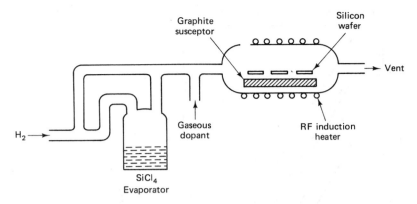

Graphite
susceptor

Silicon
wafer

Vent

H₂ →

Gaseous
dopant

RF induction
heater

SiCl₄
Evaporator

Figure 8.29 Experimental arrangement for the vapor-phase deposition of a doped semiconductor layer on a substrate.

Molecular-Beam Epitaxy

While the chloride vapor-phase system is used extensively for epitaxial growth of Si films, molecular-beam epitaxy (MBE) and metalorganic chemical-vapor deposition (MOCVD) are used almost exclusively for epitaxial growth of films of compound semiconductors for devices that require accurate control of the doping concentrations and the layer thicknesses. In the MBE growth technique, constituents of the resultant solid film are simultaneously evaporated onto the substrate surface (A. Y. Cho and J. R. Arthur, *Prog. Solid-State Chem.*, Vol. 10, p. 157, 1975; A. Y. Cho, *Thin Solid Films*, Vol. 100, p. 191, 1983). Figure 8.30 shows schematically a typical MBE growth chamber equipped with effusion cells containing elemental sources. Each cell consists of a pyrolytic boron nitride (PBN) crucible housed in an oven assembly, and has a shutter in front. The cells are shielded from each other by cryopanels (liquid nitrogen shrouds) to prevent cross-contamination.

The attributes of the MBE growth technique are many and can be summarized as follows. While Ga has almost a unity sticking coefficient, As has a very small sticking coefficient. Therefore, the growth rate of a GaAs film is controlled by the partial pressure of Ga, and has a value of the order of 1 μm/h. The LPE growth process, on the other hand, is controlled by the transport of the solute to the substrate surface, which is limited by diffusion. If the melt is supercooled by lowering the temperature by an amount ΔT in the so-called *step-cooling method,* the thickness d of the grown film increases as $t^{1/2}$ with the growth time (J. J. Hsieh, *J. Crys. Growth*, Vol. 27, p. 49, 1974). For $\Delta T = 8°C$ at $T = 825°C$, it takes less than 1 min to grow a 1-μm-thick film. Therefore, the rate of MBE growth is smaller than the rate of LPE growth by almost two orders of magnitude. The slow growth rate makes possible precise control of the doping concentration and the layer thickness by MBE. The MBE growth technique has been used extensively for fabricating novel devices, such as modulation-doped field-effect transistors (MODFET), which, to have an abrupt interface, require such accurate control.

Unlike the LPE growth process, the MBE growth process is not an equilibrium process. Therefore, the incorporation of impurities is not limited by their solid solubilities. Maximum carrier concentrations of $n \sim 5 \times 10^{18}$ cm^{-3} and $p > 10^{19}$ cm^{-3} have

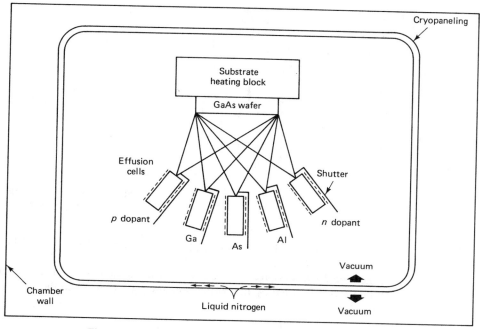

Figure 8.30 Schematic representation of an MBE growth chamber.

been reported in GaAs, using Si and Be, respectively, as dopants (A. Y. Cho and I. Hayashi, *Metall. Trans.*, Vol. 2, p. 777, 1971; M. Illegems, *J. Appl. Phys.*, Vol. 48, p. 1278, 1977). Both elements have a unity sticking coefficient on GaAs surface. Another important attribute of the MBE growth process is the uniformity of the grown layers. An improved MBE growth chamber provides a rotatable sample holder. Also, the source-to-substrate distance and the crucible diameter are increased. All these steps are taken to smooth the angular dependences of the impinging fluxes from the various sources. With these provisions, a uniformity within $\pm 5\%$ for both thickness and doping has been achieved over a 3-in. wafer. The high-vacuum condition allows use of instruments inside the chamber for *in situ* film characterization.

Metalorganic Chemical-Vapor Deposition

The basic reaction in a MOCVD process is illustrated in the following example:

$$(CH_3)_3Ga + AsH_3 \rightarrow GaAs + 3CH_4 \tag{8.6.7}$$

Figure 8.31 shows a schematic representation of a MOCVD reaction system for the growth of (GaAl)As films. The liquids trimethylgallium (TMGa) and trimethylaluminum (TMAl) are contained in stainless-steel bubblers, and the vapor pressures of the source materials are carefully controlled. These species are transported by passing purified H_2 gas through the bubblers at controlled rates. The hydride, which in the present case is arsine AsH_3, is in gaseous form (H. M. Manasevit, *Appl. Phys. Lett.*, Vol. 12, p. 156, 1968; *J. Electrochem. Soc.*, Vol. 118, p. 647, 1971). The sources, hydrogen selenide (H_2Se) and dimethylzinc (DMZn), are found to be most easily controlled for *n*-type and *p*-type doping, respectively, over a wide range of doping concentrations (J. J. Yang,

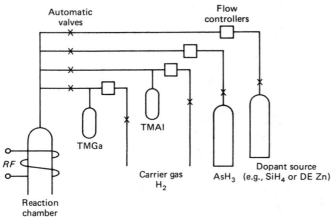

Figure 8.31 Schematic representation of a MOCVD reactor system.

(labels in figure:)
Automatic valves
Flow controllers
TMAl
TMGa
RF
Reaction chamber
Carrier gas H₂
AsH₃
Dopant source (e.g., SiH₄ or DE Zn)

L. A. Moudy, and W. I. Simpson, *Appl. Phys. Lett.*, Vol. 40, p. 244, 1982; M. Sakamoto and T. Okada, *J. Appl. Phys.*, Vol. 55, p. 3613, 1984).

Since the MOCVD process is carried out under conditions of excess arsenic, the growth rate g is controlled by the partial pressure p of TMGa and the flow rate F of the carrier gas H_2 or

$$g = KpF \qquad (8.6.8)$$

where K is a proportionality constant dependent on the specific design of a reactor. Generally, the growth rate for the MOCVD process lies between those of the LPE and MBE growth processes. The MOCVD process has been used to grow (GaAl)As lasers, which are of comparable quality to those grown by MBE (R. D. Dupuis and P. D. Dapkus, *IEEE J. Quantum Electron.*, Vol. QE-15, p. 128, 1979). It is especially valuable in growing quaternary materials (GaIn)(AsP), for which the high vapor pressure of P makes the MBE system very complicated and its operation cumbersome (M. Razeghi, M. A. Poisson, J. P. Lavirain, and J. P. Duchemin, *J. Electron. Mater.*, Vol. 12, p. 371, 1983).

Finally, we should mention a hybrid technology, the chemical or gas-source MBE, which combines the advantages of MBE and MOCVD techniques (W. T. Tsang, *Appl. Phys. Lett.*, Vol. 45, p. 1234, 1984; M. B. Panish and S. Sumski, *J. Appl. Phys.*, Vol. 55, p. 3571, 1984). In the hybrid technique, trimethylgallium and trimethylarsine (TMGa and TMAs) are used as the sources. The TMAs is cracked at 950 to 1200°C after mixing with palladium-diffused H_2, while TMGa is completely dissociated above 550°C. The use of separate ovens for the two gases prevents the possible formation of polymer products, which causes reduced and irreproducible growth rates in conventional atmospheric MOCVD. The chemicals in the chemical beam epitaxy (CBE) come out from the ovens as molecular beams which impinge directly along the line of sight at the heated substrate surface. In contrast, the chemicals in MOCVD has to diffuse through a stagnant boundary layer of carrier gas before they reach the heated substrate surface. For the growth of InP films, triethylphosphine (TEP) and trimethylindium (TMIn) are used instead. The growth rate of GaAs increases with temperature until it saturates above $T > 550°C$. The saturated rate is 2 μm/h for GaAs at a TEGa flow rate $F = 3cm^3/min$ and 3.5 μm/h for InP at a TMIn flow rate of 0.75 cm^3/min. Both GaAs and

InP films of excellent device quality have been obtained. The ease of controlling growth parameters, the low growth temperature (versus MBE), and the possible use of *in situ* RHEED (reflection high-energy electron diffraction) apparatus for film characterization (possible for MBE but not for MOCVD) make CBE a very attractive technique for growing films of compound semiconductors.

Ion Implantation

Besides diffusion, another controlled means of introducing dopants into a semiconductor is ion implantation. An ionized beam of dopants is accelerated by an electric field to an energy typically around 200 keV. When a high-energy ion beam strikes a semiconductor, it penetrates the semiconductor to an average depth dependent on the orientation of the beam with respect to the crystal plane, the ion energy, and the ion mass. In principle, scattering of host atoms by an ion beam depends on the number of atoms encountered by the beam along its path, and hence it is sensitive to the beam orientation if the beam is close to being in one of the principal planes. This effect is known as *channeling*. The channeling effect can be greatly reduced by implanting the semiconductor through a thin amorphous layer such as SiO_2 or by slightly misorienting the beam by 6 to 8°. Since the ion beam, after passing through the amorphous layer, is no longer precisely directed along a certain direction, the process of ion implantation into the semiconductor can be treated as if the semiconductor were amorphous. This procedure is generally followed in the use of ion implantation for processing integrated-circuit devices.

For ion implantation into an amorphous solid, the distribution $C(x)$ of implanted ions can be approximately described by a Gaussian function

$$C(x) = \frac{N}{\sqrt{2\pi}\, \sigma_p} \exp\left[-\frac{(x - R_p)^2}{2\sigma_p^2} \right] \qquad (8.6.9)$$

where R_p is the *projected range* indicating the position of the distribution maximum and $\sqrt{2}\, \sigma_p$ is the standard deviation representing the distribution spread, commonly known as *straggling* (J. F. Gibbons, *Proc. IEEE*, Vol. 56, p. 295, 1968; D. H. Lee and J. W. Mayer, *Proc. IEEE*, Vol. 62, p. 1241, 1974). As discussed in Section 7.7 and Appendix A7, integrating a normalized Gaussian over x yields a value of unity. Therefore, the quantity N represents the dose or total implanted ions per unit area. For a current density J, the ion flux density (or dose) rate is J/e (ions/cm^2-s). The total dose N is given by

$$N = \frac{Jt}{e} \qquad (8.6.10)$$

where t is the duration (seconds) of implantation. To reach a dose of 5×10^{16} ions/cm^2, for example, a time $t = 8 \times 10^3$ s or approximately 2 h is required at a dose rate of $J = 1$ μA/cm^2.

Figure 8.32 shows the values of projected range R_p and standard deviation σ_p reported by Lee and Mayer for implantation of boron, phosphorus, and arsenic into silicon, materials commonly used in integrated-circuit fabrication, as functions of the ion-beam energy (D. H. Lee and J. W. Mayer, *Proc. IEEE*, Vol. 62, p. 1241, 1974). Different ions have different masses, resulting in different values of R_p and σ_p. Because our knowledge of the scattering process is somewhat limited, it is useful to compare the simple theory presented in Eq. (8.6.9) and Fig. 8.32 with the experimental results and

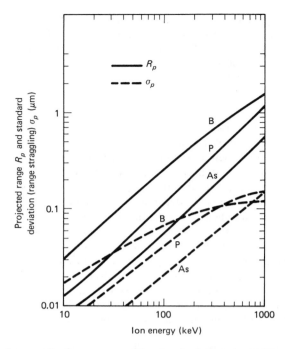

Figure 8.32 Projected range R_P (solid curves) and standard deviation σ_P (dashed curves) for implantation of boron, phosphorus, and arsenic into Si as functions of ion-beam energy. (D. H. Lee and J. W. Mayer, "Ion-Implanted Semiconductor Devices," *Proc. IEEE*, Vol. 62, p. 1241, 1974 © 1974 IEEE.)

the results from more refined calculations, which are shown in Fig. 8.33 for boron and arsenic (L. A. Christel, J. F. Gibbons, and S. Mylroie, *J. Appl. Phys.*, Vol. 51, p. 6176, 1980). The points represent the experimental values of $C(x)$ determined from SIMS measurements (secondary-ion mass spectroscopy) reported by Hofker et al. for B and by Hirao et al. for As (W. K. Hofker, D. P. Oosthoek, N. J. Koeman, and H. A. M. DeGrefte, *Radiat. Eff.*, Vol. 24, p. 223, 1975; T. Hirao, K. Inoue, S. Takayanagi, and Y. Yaegashi, *J. Appl. Phys.*, Vol. 50, p. 193, 1979). The various curves represent the calculated distributions based on the LSS (Lindhard–Schaff–Schiott) stopping theory and on the Boltzmann transport equation, using different sets of fitting parameters, as reported by Christel et al.

From Fig. 8.32 we find $R_p \sim 0.30$ μm and $\sigma_p \sim 0.065$ μm for a 100-keV boron ion beam and $R_p \sim 0.19$ μm and $\sigma_p \sim 0.06$ μm for a 355-keV arsenic ion beam. The experimental values are $R_p \sim 0.33$ μm and $\sqrt{2}\,\sigma_p \sim 0.08$ μm for B, and $R \sim 0.18$ μm and $\sqrt{2}\,\sigma_p \sim 0.08$ μm for As, from Fig. 8.33. Note that the value of maximum doping concentration C_{max} can be computed from

$$C_{max} = \frac{N}{\sqrt{2\pi}\,\sigma_p} \qquad (8.6.11)$$

Equation (8.6.11) gives a value of $C_{max} \sim 7 \times 2^{19}$ cm^{-3} for B and $C_{max} \sim 6 \times 10^{20}$ cm^3 for As from the given values of dose N and the values of σ_p obtained from Fig. 8.32. We see that the values of R_p, σ_p, and C_{max} obtained from Fig. 8.32 are in good agreement with the experimental values. Therefore, the curves shown in Fig. 8.32 are useful in planning an ion-implantation experiment. We should point out, however, that the distribution profiles $C(x)$ of Fig. 8.33 deviate considerably from a Gaussian curve for low values of C. The curve for boron is not even symmetrical. Because ion

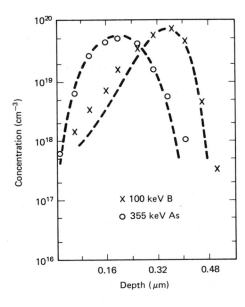

Figure 8.33 Comparison of theoretical and experimental results of the range profile $C(x)$ for 100-keV boron to a dose of 10^{15} cm^{-2} and 355-keV arsenic to a dose of 10^{16}cm^{-2} into silicon. (L. A. Christel, J. F. Gibbons, and S. Mylroie, *J. Appl. Phys.*, Vol. 51, p. 6176, 1980.)

implantation introduces a substantial amount of lattice damage, thermal annealing is needed after implantation. After annealing, the distribution profile changes considerably from that before annealing. The curves shown in Fig. 8.32 are useful in obtaining an approximate distribution profile after annealing. Finally we should cite two excellent references, a treatise edited by Ziegler for a detailed discussion of the procedure used in the calculation of the projected range R_p and range straggling σ_p (J. F. Ziegler, J. P. Biersack, and U. Littmark, "The Stopping and Range of Ions in Solids," Pergamon Press, New York, 1985) and a paper by Gibbons for an informative review of implantation-produced damage and its annealing characteristics (J. F. Gibbons, *Proc. IEEE*, Vol. 60, p. 1062, 1972).

8.7 FABRICATION TECHNIQUES FOR PLANAR STRUCTURES

In Section 8.6 we discussed the four principal processes by which *p-n* junctions can be made. However, not all the processes are suitable for mass production of high-quality semiconductor devices. By *high-quality devices* we mean devices whose electrical characteristics can be made according to preset specifications. To achieve a high uniformity of device characteristics, we must be able to control with a high degree of precision both the doping concentrations in and the geometry of a junction. The fact that we can control the depth of a *p-n* junction in the diffusion process has led to a rapid development of device-fabrication techniques for the diffused diodes and transistors. In this section we concentrate our discussion on the diffusion process and the masking technique for the fabrication of planar structures which are widely used in integrated circuits.

As discussed in Section 8.6, the penetration depth during a diffusion process can be determined accurately by controlling the temperature and duration of diffusion. However, we still need techniques to confine the area of a junction. This is done through masking and photoetching procedures. It was found by Frosch and Derick (C. J. Frosch and L. Derick, *J. Electrochem. Soc.*, Vol. 104, p. 547, 1957) that a silicon dioxide layer acts effectively as a mask against the diffusion of most important donor and acceptor impurities such as As, B, P, and Sb. Silicon-dioxide layers can be grown by the thermal oxidation of silicon in either H_2O or O_2 atmosphere through the following chemical reaction:

$$Si + 2H_2O \rightarrow SiO_2 + 2H_2 \qquad (8.7.1a)$$

$$Si + O_2 \rightarrow SiO_2 \qquad (8.7.1b)$$

Next, we expose the areas selected for diffusion through the use of photoetching technique. The oxidized wafers are coated with a photoresist material (KPR). A stencil with transparent and dark regions is placed on top of the photoresist material. Under an ultraviolet radiation, the photoresist material becomes polymerized. Thus, corresponding to the transparent region on the stencil, there will be a region on the SiO_2 layer protected by the polymerized photoresist material. Subsequent chemical etching, say by HF, of the Si wafer removes the unprotected regions of SiO_2 and hence opens up windows for diffusion.

Figure 8.34 shows an early version of a planar transistor structure. We start with an n^+ (heavily doped) Si substrate and grow an epitaxial layer of lightly doped n silicon to a thickness around 15 microns. After successive diffusions, we have an n^+-p-n-n^+ structure. The collector capacitance C_T can be reduced if the charges stored in the space-charge region are separated farther apart. This can be done by inserting a near-intrinsic region between the n^+ and p regions of the collector junction. The lightly doped n region serves this purpose. In Fig. 8.34 the emitter junction is represented by the n^+p junction, whereas the collector junction is constituted by the p-n-n^+ structure. The boron diffusion is performed with silicon wafer partly masked by SiO_2. In other words, the p region in Fig. 8.34 is restricted by SiO_2 masking. We also show in Fig. 8.34 that the regions where junctions intersect the surface are protected by an oxide layer. This layer minimizes the effect of changing ambient conditions on the junction and thus stabilizes the electrical characteristics of the junction.

So far we have discussed the general principles as to how commercial diodes and transistors are made. Associated with each process, there are technical details that we must master. In the following discussion, we call to the reader's attention some of the more important technological aspects of the problem. First, we consider the gas-phase epitaxial growth. Besides the reaction given by Eq. (8.6.6), there is a competing reaction of the following form:

$$SiCl_4 + Si \rightleftharpoons 2SiCl_2(g) \qquad (8.7.2)$$

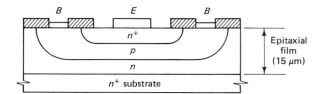

B, base contact
E, emitter contact

Figure 8.34 A typical planar n^+-p-n-n^+ transistor. The epitaxial layer is lightly doped. The purpose of the layer (i.e., the middle n region) is to reduce the transition-region capacitance C_T of the junction.

Therefore, the concentration of $SiCl_4$ relative to that of H_2 is important. If the mole fraction (i.e., the ratio of the number of $SiCl_4$ molecules to the total gas molecules) of $SiCl_4$ is too high, Eq. (8.7.2) dominates over Eq. (8.6.6). Under the circumstance, Si will be etched away rather than grown epitaxially. Theuerer (H. C. Theuerer, *J. Electrochem. Soc.*, Vol. 108, p. 649, 1961) has reported a turning point around 0.1 mole fraction of $SiCl_4$. The commonly used figure is around 0.02 mole fraction. At this value, the epitaxial layer grows at a rate of about 1 μm/min. The reader is also referred to the article by Shepherd (W. H. Shepherd, *J. Electrochem. Soc.*, Vol. 112, p. 988, 1965) for a discussion of the kinetics of the epitaxial-growing and etching processes. As expected, the rates of such processes are sensitive functions of the substrate temperature.

Now let us turn to the diffusion process. We note that the solid solubility of impurities shown in Fig. 8.25a is around 10^{21} cm^{-3} for As, B, and P, and around 4×10^{19} cm^{-3} for Al, Ga, and Sb. If we allow impurities of such a high concentration to diffuse for a long time, the region of Si wafer within a distance $\sqrt{Dt}$ from the surface will become very heavily doped. This is not desirable. We can avoid this difficulty by performing the diffusion process in two steps: the predeposition step and the drive-in step. During the predeposition step, the surface concentration indeed reaches the solid solubility value. However, the duration t for this step is relatively short, so that the total amount of impurities incorporated into silicon lattice is not too large. The drive-in step takes place in an atmosphere that does not contain any impurities. During the drive-in phase of the diffusion process, impurities redistribute themselves but their amount remains the same. Therefore, the analysis of the problem is similar to that associated with Eq. (7.7.13), and the impurity distribution is given by

$$C(x,\ t) = \frac{Q}{A\sqrt{4\pi Dt}} \exp\left(-\frac{x^2}{4Dt}\right) = C_0(t) \exp\left(-\frac{x^2}{4Dt}\right) \qquad (8.7.3)$$

where Q is the total amount of impurities incorporated during the predeposition phase and A is the area of the wafer exposed during the diffusion process.

The foregoing discussion appears relatively simple. However, our analysis becomes much more complicated when other effects are considered. If the semiconductor wafer is protected by an oxide layer during the drive-in diffusion, we must include the effect of the oxide on the distribution of impurities in our analysis. On the other hand, if the surface is unprotected, we must consider the process of outdiffusion. The entire problem is further complicated by the fact that both the process of outdiffusion and that of diffusion through an oxide layer depend on the ambient gas. The reader is well advised to consult the literature for information concerning these processes. A useful review can be found in a book by Grove (A. S. Grove, *Physics and Technology of Semiconductor Devices*, John Wiley & Sons, Inc., New York, 1967).

Finally, we discuss briefly the techniques used in evaluating the results of the diffusion process. The junction depth d (Fig. 8.35) can be determined by following the procedure outlined below. The specimen is first lapped. To increase the length of the lapped region, the lapping is done at a small angle θ to the plane of the junction. The lapped region is then polished. A solution, such as aqueous solution of HNO_3 (0.1 mol, 70%) and HF (100 mol, 48%), is then applied to the polished surface. Because the solution reacts selectively with the two sides of the junction, the two sides will show a color variation under illumination. The *p* region appears to be stained dark relative to the *n* region.

Sec. 8.7   Fabrication Techniques for Planar Structures

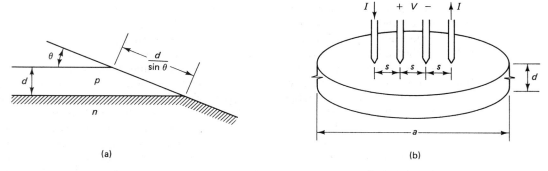

(a)

(b)

Figure 8.35 (a) Determination of penetration depth d by exposing the junction through lapping at an angle to the junction plane, and (b) measurement of average resistivity of the diffused layer by a four-point probe.

Figure 8.35b illustrates the measurement of sheet resistivity of the diffused layer by a four-point probe. First, we define an average conductivity $\bar{\sigma}$ such that

$$\bar{\sigma} = \frac{e}{d} \int_0^d \mu[C(x) - C(d)] \, dx \qquad (8.7.4)$$

where $C(x)$ is the concentration of diffusant. In Eq. (8.7.4) we have assumed that impurities are completely ionized. Referring to Fig. 8.27a, we see that N_d is represented by $C(x)$ and N_a is represented by $C(d)$. Therefore, the difference $C(x) - C(d)$ is equal to the majority-carrier concentration. Two important impurity distributions are the Gaussian distribution of Eq. (8.7.3) and the complementary error function of Eq. (8.6.3). Substituting these distributions in Eq. (8.7.4), we obtain the functional dependence of $\bar{\sigma}$ upon C_0 and C_d. Such curves have been calculated by Bachenstoss (G. Bachenstoss, *Bell Syst. Tech. J.,* Vol. 37, p. 699, 1958) and by Irvin (J. C. Irvin, *Bell Syst. Tech. J.,* Vol 41, p. 387, 1962). Since the values of C_0 and C_d are known, the calculated value of $\bar{\sigma}$ from Eq. (8.7.4) should agree with the measured value. A comparison of these two values can be used as a check to see whether the impurity distribution indeed follows one of the two functions.

To obtain the relation between σ and the measured value of V and I (Fig. 8.35b), we must solve an electrostatic problem that takes into account the nonuniform spreading of the current I. Smits has shown (F. M. Smits, *Bell Syst. Tech. J.,* Vol. 37, p. 711, 1958) that the following relation is useful in the present analysis:

$$\rho = 4.53 \frac{V}{I} \, dF\left(\frac{d}{s}\right) G\left(\frac{a}{s}\right) \qquad (8.7.5)$$

where s is the spacing between adjacent probes and a is the diameter of the wafer, which is assumed to be of circular shape. In Eq. (8.7.5) F and G are two correction factors whose values are given in Table 8.4. The spacing s is about 10^{-1} cm and the penetration depth d is on the order of 100 μm $= 10^{-2}$ cm or less; therefore, the factor F is almost unity. The other condition $a \gg s$ for G to be unity usually is not met. However, the value of G changes only by a factor of 2 between the two extreme values of $a/s = 3$ (minimum) and $a/s = \infty$ (maximum). Therefore, the sheet resistivity ρ can be accurately evaluated from Eq. (8.7.5).

TABLE 8.4 CORRECTION
FACTORS F AND G FOR USE IN
EQ. (8.7.5)

$\dfrac{d}{s}$	F	$\dfrac{a}{s}$	G
0	1.000	3.0	0.501
0.5	0.997	5.0	0.742
1.0	0.921	20.0	0.980
2.0	0.634	∞	1.000

8.8 TRANSITION-REGION CAPACITANCE OF DIFFUSED DIODES

In Section 8.2 we derived an expression for the transition-region capacitance in alloyed diodes. In this section we apply a similar treatment to diffused diodes. For the case shown in Fig. 8.27a (a diffused n region on a p substrate), the space-charge density in the junction region is given by

$$\rho(x) = e(N_d^+ - N_a^-) = e[C(x) - C(d)] \tag{8.8.1}$$

where $C(x)$ is either the erfc function of Eq. (8.6.3) or the Gaussian function of Eq. (8.7.3). In either case we see that it would be very difficult to solve Poisson's equation. Therefore, reasonable approximations must be made.

Figure 8.36a is a redrawing of Fig. 8.27a in which we show the impurity distribution. In a practical diode, the ratio C_0/C_d is very large, say of the order of 10^3 or larger. This means that the value of $x/\sqrt{Dt}$ in Eq. (8.6.3) or (8.7.3) is bigger than 1. Therefore, near point O, the function $C(x)$ changes very rapidly because of the $\exp(-y^2)$ dependence, where $y^2 = x^2/4\,Dt$. Here we use Eq. (7.7.17a) for $y > 1$ to approximate the erfc function. Since the $\exp(-y^2)$ factor is the predominating dependence of $C(x)$ upon x, we can approximate both Eqs. (8.6.3) and (8.7.3) around the point O as

$$C(x) = C(d) \exp\left(-\frac{x^2}{4Dt}\right) \exp\left(\frac{d^2}{4Dt}\right) \tag{8.8.2}$$

Shifting the origin to d and letting $x = d + x'$, we find that

$$C(x) = C(d) \exp\left(-\frac{2dx'}{4Dt}\right) \cong C(d)\left(1 - \frac{d}{2Dt}x'\right) \tag{8.8.3}$$

Noting that $C(d) = N_a$ and $C(x) = N_d$, we obtain

$$\rho(x) = e(N_d - N_a) = -\frac{eN_a d}{2Dt}x' \tag{8.8.4}$$

Figure 8.36b shows the actual space-charge density $\rho(x)$ given by Eq. (8.8.1). In the region AB, the function $\rho(x)$ is approximated by a linear function (Fig. 8.36c):

$$\rho(x) = -eax' \qquad \text{for} \quad |x'| < \frac{2Dt}{d} \tag{8.8.5}$$

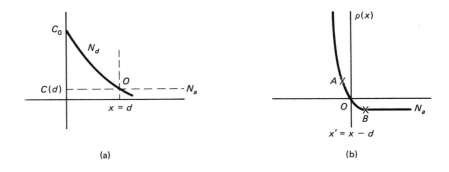

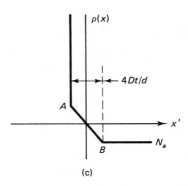

Figure 8.36 Diagrams used in evaluating the transition-region capacitance of a diffused diode: (a) donor and acceptor distributions, (b) variation of the space-charge density, (c) the approximate charge density used in the calculation. In (a), C_0 is the surface concentration of donor and $C(d)$ is the donor concentration at the junction boundary $x = d$. In (b) and (c), the origin is shifted from the surface to the junction boundary.

where the proportionality constant

$$a = \frac{N_a d}{2Dt} \tag{8.8.6}$$

has a dimension of cm^{-4}. Following a procedure similar to that outlined in Section 8.2, we find the width W of the space-charge region to be

$$W = \left[\frac{12\epsilon}{ea} (V_d \pm V_a) \right]^{1/3} \quad \text{for} \quad W < \frac{4Dt}{d} \tag{8.8.7}$$

The transition-region capacitance C_T can be calculated from the relation $C_T = \epsilon A/W$ of Eq. (8.2.7), where A is the junction area. Note that the impurity concentration on the n side increases very rapidly as x' goes beyond the point A. As pointed out in Section 8.2, in a junction with two vastly different impurity concentrations, the space-charge region extends much farther into the less heavily doped side. This means that if $W > 4Dt/d$, we can replace Fig. 8.36c by a step junction with $N_a \ll N_d$. Under the circumstances, the width W is given by Eq. (8.2.11), or

$$W = \left[\frac{2\epsilon}{eN_a} (V_d + V_a) \right]^{1/2} \quad \text{for} \quad W > \frac{4Dt}{d} \tag{8.8.8}$$

The $+$ sign and $-$ sign in Eqs. (8.8.7) and (8.8.8) refer to a reverse-biased junction and a forward-biased junction, respectively. The quantities V_d and V_a are the built-in voltage and the applied voltage, respectively.

The result of the discussion above can be summarized as follows. At low junction voltages such that $W < 4Dt/d$, a diffused junction can be treated as a graded junction with a linearly varying space-charge density. At high junction voltages, a diffused junction can be considered as a one-side step junction because the base side has a much lower impurity concentration than the diffused side. Our analysis is based on a junction having a p-type base and an n-type diffused region. For a p-type diffused region on an n-type substrate, we should substitute N_d for N_a in Eqs. (8.8.6) and (8.8.8).

Using numerical methods, Lawrence and Warner (H. Lawrence and R. M. Warner, *Bell Syst. Tech. J.*, Vol. 39, p. 389, 1960) have computed a family of curves relating W to C_0, $C(d)$, and d for both the Gaussian distribution and the erfc distribution. Take a specific case with $C_0 = 10^{20}$ cm^{-3}, $C(d) = 10^{15}$ cm^{-3}, and $d = 10^{-3}$ cm. For a Gaussian distribution, the value of $d^2/4Dt$ is calculated from Eq. (8.7.3) to be 11.5, and thus the value of Dt is 2.2×10^{-8} cm^{-2}. At $V_d - V_a = 0.1$ V, the width W is calculated from Eq. (8.8.7) to be 7.0×10^{-5} cm, the same as the exact value from numerical integration. Note that the value of $4Dt/d$ is 8.8×10^{-5} cm. Thus the condition $W < 4Dt/d$ is satisfied. At $V_d + V_a = 10$ V, we expect that the other condition $W > 4Dt/d$ is valid. Using Eq. (8.8.8), we obtain $W = 3.5 \times 10^{-4}$ cm, in agreement with the exact value from numerical integration. In the region where $W \sim 4Dt/d$, the exact value from numerical integration lies between the values calculated from Eqs. (8.8.7) and (8.8.8).

8.9 GENERATION AND RECOMBINATION CURRENT

In Section 8.3 we derived the expression of junction current, which is based on the diffusion of carriers outside the space-charge region. We assumed that generation and recombination of carriers were insignificant inside the space-charge region. This assumption, however, is not always a valid one. In the following discussion we compare the current arising from generation and recombination of carriers inside the space-charge region with the diffusion current.

First, we consider a diode biased in the reverse direction, and refer to Fig. 8.5b. The electron and hole concentrations in the space-charge region are given by

$$n = N_c \exp \left(-\frac{\mathscr{E}_c - \mathscr{F}_n}{kT} \right) \tag{8.9.1a}$$

$$p = N_v \exp \left(-\frac{\mathscr{F}_p - \mathscr{E}_v}{kT} \right) \tag{8.9.1b}$$

When the value of $\mathscr{E}_c - \mathscr{F}_n$ exceeds that of $\mathscr{E}_c - \mathscr{E}_{fi}$, the value of n becomes less than n_i. A similar situation applies to the hole concentration p. The condition in which both n and p are smaller than n_i exists for the greater part of the transition region in a reverse-biased junction. Note that $p < n_i$ and $n < n_i$ can happen simultaneously only inside the space-charge region, where $\mathcal{F}_n$ and $\mathcal{F}_p$ are no longer equal.

For $p < n_i$ and $n < n_i$, the rate of generation exceeds that of recombination in Eq. (7.5.16). Neglecting p and n in comparison with other terms, we obtain from Eq. (7.5.16) the net rate of generation:

$$G = \frac{N_t C_n C_p n_i^2}{C_p p' + C_n n'} \tag{8.9.2}$$

We assume that the trap energy lies in the upper half of the energy gap such that $C_p p' \ll C_n n'$. Thus Eq. (8.9.2) becomes

$$G = N_t C_p \frac{n_i^2}{n'} = \frac{1}{\tau_{p0}} \frac{n_i^2}{n'} \tag{8.9.3}$$

The electron–hold pair thus generated creates a current called the *generation current*. The quantity G has a dimension of s^{-1} cm^{-3}. For a space-charge region of width W, the generation current density is

$$J_{gen} = \frac{eW}{\tau_{p0}} \frac{n_i^2}{n'} \tag{8.9.4}$$

For cases where $C_n n' \ll C_p p'$, we replace $\tau_{p0} n'$ by $\tau_{n0} p'$ in Eq. (8.9.4).

For comparison, the current density due to diffusion in a reverse-biased diode is

$$J_{diff} = \frac{eD_p}{d} p_{n0} = \frac{eD_p}{d} \frac{n_i^2}{N_d} = \frac{e}{\tau_{p0}} \frac{L_p^2}{d} \frac{n_i^2}{N_d} \tag{8.9.5}$$

In Eq. (8.9.5) we assume that the base region is n type and hence hole diffusion dominates. For simplicity, we take a wide-base diode. According to our discussion in Section 8.3, $d = L_p$. Thus the two currents are in the ratio

$$J_{gen} : J_{diff} = \frac{W}{n'} : \frac{L_p}{N_d} = WN_d : L_p n' \tag{8.9.6}$$

The ratio of L_p/W is of the order of 100 in a reverse-biased diode having a lifetime τ_{p0} around 10^{-4} s. Therefore, the diffusion current dominates if n' is about equal to or greater than N_d. This seems to be the prevailing situation in germanium diodes at room temperature. Note that n' decreases with decreasing temperature according to Eq. (7.5.12). At low temperatures, the generation current becomes more important than the diffusion current. In silicon and gallium-arsenide diodes, the generation current is bigger than the diffusion current even at room temperature. Only when the temperature is considerably above 300 K is the reverse saturation current controlled by the diffusion of minority carriers.

The generation current can easily be identified by observing both its voltage dependence and its temperature dependence. The diffusion current saturates in the reverse direction. The generation current, on the other hand, increases as the width W of the space-charge region increases. There is no true saturation of the reverse current in diodes where space-charge generation dominates the current. The temperature dependence of J_{gen} is also quite different from that of J_{diff}. According to Eqs. (8.9.4) and (7.5.12), J_{gen} varies as $\exp[-(\mathcal{E}_t - \mathcal{E}_v)/kT]$ with temperature. The temperature dependence of J_{diff}, on the other hand, is determined by n_i^2 in a narrow-base diode and by $n_i^2 \tau_0^{-1/2}$ in a wide-base diode.

The situation in a forward-biased diode is more complicated than that in a reverse-biased diode. As pointed out in Section 8.1, the voltage dependence of junction current in an ideal diode is given by exp $(eV_a/kT) - 1$. If the generation or recombination current is significant, a general voltage dependence of the form

$$I = I_s \left[\exp\left(\frac{eV_a}{mkT}\right) - 1 \right]$$

(8.9.7)

is observed, where m is called the *ideality factor* and has a value between 1 and 2. In the reverse direction, V_a is negative. As long as the magnitude of eV_a/kT is on the order of 10 or larger, we can approximate Eq. (8.9.7) by letting $I \approx I_s$ without a precise knowledge of m. For a diode biased in the forward direction, however, the exp (eV_a/mkT) dependence is important. To explain why m may be different from 1, we need a careful examination of the quasi-Fermi levels $\mathscr{F}_n$ and $\mathscr{F}_p$ in relation to the Fermi levels $\mathscr{E}_{fn}$ and $\mathscr{E}_{fp}$.

As discussed in Section 8.3, the quasi-Fermi level is related to the current density through Eq. (8.3.2). Since

$$n = n_i \exp\left(\frac{\mathscr{F}_n - \mathscr{E}_{\text{fi}}}{kT}\right)$$

(8.9.8)

we obtain from Eq. (8.3.2) the following equation:

$$kT \exp\left(\frac{\mathscr{F}_n}{kT}\right) = \int \exp\left(\frac{\mathscr{E}_{\text{fi}}}{kT}\right) \frac{J_n}{\mu_n n_i} \, dx$$

(8.9.9)

where the integration limits run across the space-charge region. In Section 8.3 we showed that $\Delta\mathscr{F}_n$ across the space-charge region is indeed very small if recombination of carriers is assumed negligible in the space-charge region. By using suitable, approximate dependencies of J_n and $\mathscr{E}_{\text{fi}}$ upon x, Sah, Noyce, and Shockley have carried out the integration in Eq. (8.9.9) (C. T. Sah, R. N. Noyce, and W. Shockley, *Proc. IRE*, Vol. 45, p. 1228, 1957) and concluded that the assumption of constant $\mathscr{F}_n$ and $\mathscr{F}_p$ in a forward-biased junction is indeed valid. Referring to Fig. 8.6 and using Eq. (8.9.1), we obtain

$$np = n_i^2 \exp\left(\frac{\mathscr{F}_n - \mathscr{F}_p}{kT}\right) = n_i^2 \exp\left(\frac{eV_a}{kT}\right)$$

(8.9.10)

Let us apply Eq. (8.9.10) to the recombination process in the space-charge region. As discussed in Section 7.5, most recombination centers have energies close to the middle of the energy gap. In Eq. (7.5.16), $n' \approx p' \approx n_i$; hence either p or n is the dominating term in the denominator, and $pn \gg p_0 n_0$ in the numerator. Furthermore, the total rate is a maximum when the denominator is a minimum. This occurs when $p = n$ due to the restriction imposed by Eq. (8.9.10). Assuming that $C_n = C_p$, we obtain the maximum rate of recombination in the space-charge region:

$$R_{\text{max}} = \frac{N_t C_p p}{2} = \frac{N_t C_n n}{2}$$

(8.9.11)

Sec. 8.9   Generation and Recombination Current

For simplicity, we further assume that this maximum rate occurs for the greater part of the space-charge region of width W. Thus the recombination of carriers creates a current of density

$$J_{rec} = \frac{eW}{2\tau_{p0}} p = \frac{eW}{2\tau_{p0}} n_i \exp\left(\frac{eV_a}{2kT}\right) \qquad (8.9.12)$$

Equation (8.9.12) is obtained through the use of Eq. (8.9.10) with $n = p$.

For comparison, the diffusion current in the forward direction in a wide-base diode is given by

$$J_{diff} = \frac{e}{\tau_{p0}} L_p p_n = \frac{e}{\tau_{p0}} L_p \frac{n_i^2}{N_d} \exp\left(\frac{eV_a}{kT}\right) \qquad (8.9.13)$$

We should point out that outside the space-charge region, $n = N_d$ in Eq. (8.9.10). This explains why Eq. (8.9.12) has an $\exp\left(eV_a/2kT\right)$ dependence, whereas Eq. (8.9.13) has an $\exp\left(eV_a/kT\right)$ dependence. The two currents are in the ratio

$$J_{rec} : J_{diff} = WN_d : 2L_p n_i \exp\left(\frac{eV_a}{2kT}\right) \qquad (8.9.14)$$

For numerical values, we refer to Table 8.1. The ratio of N_d/n_i for a 1-Ω-cm base material is 70 in Ge, 4.2×10^5 in Si, and 1.8×10^8 in GaAs. However, the factors $(L/W) \exp\left(eV_a/2kT\right)$ are in favor of J_{diff}. In germanium diodes, the factor L/W alone can counterbalance the ratio N_d/n_i; therefore, the diffusion current J_{diff} dominates over the recombination current J_{rec}.

In Fig. 8.37a, the solid curve (curve I) shows the voltage dependence of the diffusion current, whereas the dashed curve illustrates the voltage dependence of the

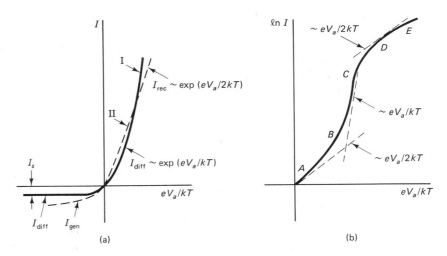

Figure 8.37 (a) Composition of diode current: I_{diff} caused by diffusion of minority carriers outside the space-charge region, and I_{gen} or I_{rec} due to generation or recombination of free carriers in the space-charge region, and (b) the voltage dependence of current in forward-biased Si and GaAs diodes. In (b), the current is plotted on a logarithmic scale.

recombination or generation current. The characteristics of germanium diodes at room temperature follow curve I. In silicon diodes, the factor $\exp(eV_a/2kT)$ at low bias voltage is not large enough to overcome the factor N_d/n_i in Eq. (8.9.14). Therefore, the characteristics of silicon diodes follow curve II all the way from a reverse bias to a low forward bias. At high forward bias, the diffusion current eventually takes over because of the $\exp(eV_a/2kT)$ factor in Eq. (8.9.14).

In Fig. 8.37b, we plot $\ln I$ as a function of eV_a/kT for Si and GaAs diodes. In the low-bias region AB, the recombination current dominates. In the high-bias region BCD, the diffusion current takes over. Near point C, the injection of minority carriers into the base region becomes so high that $n_1 = p_1 > n_0 + p_0$. When this happens, p becomes nearly equal to n in Eq. (8.9.10). Thus both the diffusion current and the recombination current vary as $\exp(eV_a/2kT)$. In region DE the applied voltage V_a is practically equal to the built-in voltage V_d, and the junction almost disappears. In this region the current is limited by the spreading resistance of the diode. Most silicon diodes at room temperature show a characteristic like the one illustrated in Fig. 8.37b. In gallium-arsenide diodes, the factor N_d/n_i is much larger, so that the $\exp(eV_a/2kT)$ dependence extends over a much larger range of currents than in silicon diodes.

8.10 JUNCTION BREAKDOWN

Before we discuss the phenomenon of junction breakdown, it may be useful to recapitulate the results of our analysis thus far regarding the $I–V$ characteristic of a p-n junction. The simple diode theory presented in Section 8.3 is based on diffusion of minority carriers outside the space-charge region. This simple theory predicts that the current in the reverse direction approaches a constant value known as the *reverse saturation current*. In a practical diode, however, we usually find that the diode current rises slowly as bias voltage is increased in the reverse direction until the voltage approaches a certain critical voltage near which the current increases very rapidly with increasing voltage. In Fig. 8.38 the $I–V$ characteristics of ideal and practical diodes are drawn for comparison.

The slow rise of current in region I of Fig. 8.38b may occur for several reasons. As discussed in Section 8.9, the current due to generation and recombination of carriers in the space-charge region has a different voltage dependence from that of the diffusion

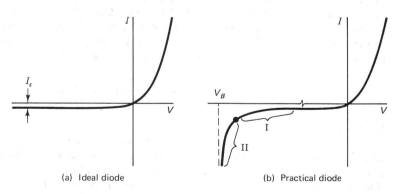

(a) Ideal diode

(b) Practical diode

Figure 8.38 Comparison of the current–voltage characteristics of (a) ideal diodes and (b) practical diodes.

current. In silicon diodes, the generation of carriers in the space-charge region is the dominant mechanism for the slow current rise in Fig. 8.38b. In germanium diodes, the generation current is not as important as the diffusion current. Another factor that may cause the slow rise of current is the reduction of effective base width. In a narrow-base diode, the space-charge region extends a fair distance into the base region. As the voltage increases, the width of the space-charge region increases, reducing the effective base width and hence raising the reverse saturation current. Surface effects can also cause a slow rise of reverse current. As pointed out in Section 7.9, an inversion layer may exist on a semiconductor surface, creating a surface p-n junction in parallel with the bulk p-n junction. As the voltage increases, the effective area of the inversion layer increases, thereby raising the reverse saturation current.

The rapid rise of current in region II of Fig. 8.38b is different in nature from the slow rise in region I, and the phenomenon is generally known as *breakdown* in a junction diode. There are two physical processes that can cause a sudden rise of current: direct tunneling of electrons across the forbidden gap (Fig. 10.7b) and creation of electron–hole pairs through impact ionization by energetic carriers (Fig. 10.7a). The theory governing the physics of these two processes is presented in Chapter 10; therefore, only the simple and practical aspects of junction breakdown are discussed here.

Refer to Eq. (10.6.18) for the tunneling current density in a reverse-biased diode. The exponential factor $\exp(-E_0/E)$ represents the probability that an electron may penetrate the barrier to appear in the conduction band at B. The quantity E_0 is given by Eq. (10.6.19) and has the dimension V/m. The quantity m_r^* is the reduced mass. In germanium with $\mathscr{E}_g = 0.67$ eV and $m_r^* = 0.032m_0$, we find that $E_0 = 4 \times 10^8$ V/m.

In the presence of a tunneling process, the current in a junction diode biased in the reverse direction can be written as

$$I = I_s + I_1 \exp\left(-\frac{E_0}{E}\right) \qquad (8.10.1)$$

where I_s is the reverse saturation current due to diffusion mechanism and $I_1 \exp(-E_0/E)$ is the current due to tunneling mechanism. Since I_1 is generally several orders of magnitude greater than I_s, tunneling current becomes important long before E approaches the value of E_0. A practical limit for breakdown in the reverse direction may be chosen at $E = 0.1E_0$. Thus, in germanium, we find the critical field E_{cr} for tunnel breakdown to be of the order of 5×10^5 V/cm. Once the second term becomes comparable to the first term in Eq. (8.10.1), it increases very rapidly with further increase of the reverse bias V_a. The maximum electric field existing in a reverse-biased junction can be calculated from Eq. (8.2.5) or (8.2.6) and is given by

$$E_{\max} = \left[\frac{2e}{\epsilon} \frac{N_1 N_2}{N_1 + N_2}(V_d + V_a)\right]^{1/2} \qquad (8.10.2)$$

for an abrupt junction.

Now we discuss the impact ionization process. The process is similar to that of electrical discharge of a gas. In a gas discharge, gas molecules become ionized into electrons and positive ions when they make inelastic collisions with energetic electrons. Similarly, the covalent bond of an atom can be broken if sufficient energy is given to the atom to break the bond. During such inelastic collisions, electron–hole pairs are created. We define a quantity α called the *ionization coefficient* such that the number of electron–hole pairs generated per unit distance is equal to αn. In other words,

$$\frac{dn}{dx} = \alpha n \quad \text{or} \quad \alpha p \tag{8.10.3}$$

where n (or p) is the electron (or hole) concentration.

If we apply Eq. (8.10.3) to the space-charge region AB of a junction (Fig. 8.5b), we find that the electron concentration n in the space-charge region goes to infinity when

$$\int_A^B \alpha \, dx = 1 \tag{8.10.4}$$

Equation (8.10.4) simply says that an electron in going through the space-charge region from A to B creates one electron–hole pair, and the hole thus generated in turn creates another electron–hole pair on its way from B to A. As the process goes on, the carriers inside the region AB multiply themselves and the current increases indefinitely. This type of breakdown is called *avalanche breakdown*. To make an estimate of the field required for avalanche breakdown in junctions, we refer to the plot of α versus E in Fig. 10.9. The width of the space-charge region is on the order of 10^{-4} cm in a reverse-biased diode. From Eq. (8.10.4) we find that α should be of the order of 10^4 cm^{-1}. This value of α is reached for a field of the order of 3×10^5 V/cm.

As we can see, both the tunnel breakdown and the avalanche breakdown require a very high electric field to exist in the space-charge region of a reverse-biased diode. However, there is a subtle difference between the two processes. In Eq. (8.10.1), the electric field alone determines the magnitude of the tunnel current. In Eq. (8.10.4), the junction width and the electric field together determine the condition for breakdown. In a heavily doped junction, the electric field may reach a value exceeding 5×10^5 V/cm for tunnel breakdown to occur, yet at such a field strength, the integral $\int \alpha \, dx$ may be below the value 1 required for avalanche breakdown to occur, owing to a small value for the junction width. It is expected, therefore, that tunnel breakdown occurs first and hence dominates in a heavily doped junction. On the other hand, in a lightly doped junction, the field required to meet the condition stated in Eq. (8.10.4) is considerably lowered because of a relatively large value of the junction width. In such a diode, avalanche breakdown occurs first. A rough demarcation line may be drawn for the two processes. In diodes with $N_I > 10^{18}$ cm^{-3}, the tunneling process dominates; and in diodes with $N_I < 10^{17}$ cm^{-3}, the avalanche process dominates.

Because of impact ionization process, the current in a reverse-biased diode exceeds the reverse saturation current I_s by a factor M; that is, $I = MI_s$. The factor M is known as the *multiplication factor,* and it obeys an empirical relation

$$M = \frac{1}{1 - (V/V_B)^r} \tag{8.10.5}$$

where V_B is the breakdown voltage at which the current increases indefinitely. The value of r is found experimentally to be around 3 in *p-n-p* and around 6 in *n-p-n* Ge transistors (S. L. Miller, *Phys. Rev.*, Vol. 99, p. 1234, 1955). The value of V_B can be calculated by using the experimental value of α (Fig. 10.9) in Eq. (8.10.4). The result of such a calculation is plotted in Fig. 8.39 for Ge, Si, and GaAS step junctions (S. M. Sze and

Sec. 8.10   Junction Breakdown

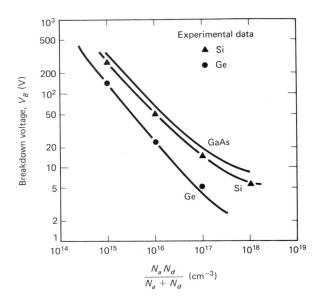

Figure 8.39 Plots of breakdown voltage versus impurity concentration for Ge, Si, and GaAs diodes. The values of V_B are for a one-sided step junction. (S. M. Sze and G. Gibbons, *Appl. Phys. Lett.*, Vol. 8, p. 111, 1966.)

G. Gibbons, *Appl. Phys. Lett.*, Vol. 8, p. 111, 1966). Breakdown voltages for graded and diffused diodes are also available in the literature (S. M. Sze and G. Gibbons, *Appl. Phys. Lett.*, Vol. 8, p. 111, 1966; D. P. Kennedy and R. R. O'Brien, *IRE Trans. Electron Devices*, Vol. ED-9, p. 478, 1962). The calculated values of V_B are in good agreement with the measured values except for $V_B < 6$ V, where the tunneling process becomes the dominant process.

PROBLEMS

Refer to Tables 6.3 and 6.8 for the values of μ_n, μ_p, n_i, N_c, and N_v in Ge, Si, and GaAs at 300 K.

8.1. Calculate the energies $\mathscr{E}_c - \mathscr{E}_f$ and $\mathscr{E}_f - \mathscr{E}_v$ in Ge, Si, and GaAs for **(a)** an n-type sample of 1 Ω-cm resistivity and **(b)** a p-type sample of 1 Ω-cm resistivity. Also find the built-in voltage in Ge, Si, and GaAs diodes made of these materials.

8.2. A p-n germanium junction is made of an alloyed p region with a hole concentration of 1×10^{18} cm^{-3} and a base n region of 1 Ω-cm resistivity. Calculate the built-in voltage V_d. Also compute the width W of the space-charge region and the junction capacitance per unit area (C_T/A) at zero applied bias voltage. What is the ratio of the extension of the space-charge region into the n and p regions?

8.3. **(a)** A silicon abrupt junction has a hole concentration of 5×10^{18} cm^{-3} in the p region and an electron concentration of 1×10^{15} cm^{-3} in the n region. Calculate the junction capacitance (C_T/A) of the diode for (1) a forward bias of 0.2 V, (2) no applied bias voltage, and (3) a reverse bias of 10 V.

   (b) Also find the value and the location of maximum electric field in the junction for (1)–(3) in part (a).

8.4. In a graded p-n junction, the impurity concentrations in the transition region vary linearly with the distance away from the junction. In other words, $N_1 = ax$ for $x < 0$ and $N_2 =$

bx for $x > 0$ in Eq. (8.2.4). Sketch $\rho(x)$ and $E(x)$ as a function of x for both abrupt and graded junctions. Show that the maximum electric field is

$$E_{max} = \frac{3}{2}\left[\frac{e}{3\epsilon}\frac{ab}{(\sqrt{a}+\sqrt{b})^2}(V_d \mp V_a)^2\right]^{1/3}$$

in a graded junction and

$$E_{max} = \left[\frac{2e}{\epsilon}\frac{N_1N_2}{N_1+N_2}(V_d \mp V_a)\right]^{1/2}$$

in an abrupt junction.

8.5. (a) Refer to the description of a graded p-n junction in Problem 8.4. Starting from Poisson's equation, show that $ax_A^2 = bx_B^2$. State the physical implication of this equation.

   (b) Also verify that the width of the space-charge region is

$$W = \left[\frac{3\epsilon}{e}\frac{(\sqrt{a}+\sqrt{b})^2}{ab}(V_d \mp V_a)\right]^{1/3}$$

and the junction capacitance is

$$C_T = \frac{dQ}{dV_a} = \frac{\epsilon A}{W}$$

where A is the junction area. Check the dimensions of the two expressions.

8.6. (a) Consider a p-i-n diode that has a middle region made of intrinsic material sandwiched between p and n materials. Draw diagrams (similar to Fig. 8.3) showing $\rho(x)$, $E(x)$, and $V(x)$ as a function of x in the transition region. Verify the expression

$$V_d \mp V_a = \frac{Qd}{\epsilon} + \frac{(N_1 + N_2)Q^2}{2e\epsilon N_1N_2}$$

where Q is the charge stored per unit area of the junction and d is the width of the intrinsic region. The other symbols have the same meaning as the corresponding ones in Eq. (8.2.12).

   (b) Comparing the equation above with Eq. (8.2.12), show that a p-i-n junction can be considered as two capacitors in series with

$$\frac{1}{C_T} = \frac{d + W}{\epsilon A}$$

where W is combined depletion width of p and n regions.

8.7. Write down the expression corresponding to Eq. (8.3.13) due to diffusion of electrons in the n region. Apply these two equations to an alloyed p-n junction in which the p side is the much more heavily doped side. Show that the current is predominantly determined by hole diffusion into the n region. Discuss the temperature dependence of the reverse saturation current (current under large reverse bias) for the case $L > W_2 - x_B$ and for the case $L < W_2 - x_B$.

8.8. (a) Show that the ratio of the hole current to the electron-current density in a wide-base diode is

$$\frac{J_p}{J_n} = \frac{L_n}{L_p}\frac{\sigma_p}{\sigma_n}$$

where the symbols L and σ represent, respectively, the diffusion length and conductivity. Thus, for $\sigma_p \gg \sigma_n$, the total current density is given approximately by J_p, or

$$J_{total} \cong \frac{eD_p}{L_p} p_{n0} \left[\exp\left(\frac{eV_a}{kT}\right) - 1 \right]$$

(b) It is generally observed that it takes a much larger applied bias to reach a certain forward current in GaAs diodes than in Ge diodes as shown in Fig. P8.8. Explain why this behavior is expected from the theory.

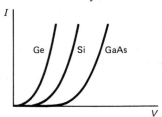

Figure P8.8

(c) Suppose that we want to reduce the reverse-saturation current in a diode made of materials with $\sigma_p \gg \sigma_n$. Comment on the effect of (1) raising σ_n, (2) lowering σ_n, (3) raising σ_p, and (4) lowering σ_p on the current.

8.9. **(a)** Consider the use of Ge, Si, and GaAs in making junction diodes. Suppose that the starting material is n type and has a resistivity of 2 Ω-cm. The p side of the junction is heavily doped through alloying. Calculate the hole concentration in the base region of Ge, Si, and GaAs diodes.

  (b) Assuming equal lifetime, base width, and junction area, find the ratio of the reverse saturation current in the three diodes. It is assumed that the minority carrier concentration has to be raised to a value of 10^{15} cm^{-3} for a forward current of 10^{-4} A. Calculate the forward bias needed to give such a current in the three diodes. The calculations are to be made at 300 K and 350 K.

8.10. **(a)** Verify Eq. (8.3.14). Assuming D_p and d to be temperature independent, find the percentage change of the reverse saturation current for $\Delta T = 1°C$ at 300 K in Ge, Si, and GaAs diodes.

  (b) Calculate the percentage change of the forward current for $\Delta T = 1°C$ at $eV_a = 0.20$ eV.

8.11. **(a)** Using Eqs. (8.3.11) and (8.3.12), find J_p as a function of x. Explain why J_p depends on x in one case and is independent of x in the other.

  (b) Analyze the composition of current in a narrow-base diode with the aid of diagrams similar to Figs. 8.7 and 8.8.

8.12. For a diode biased in the forward direction, it is more convenient to express junction and diffusion capacitances as functions of diode current than as functions of junction voltage. Show that in a wide-base diode,

$$C_D = \frac{eI_0}{kT} \frac{\tau_0}{2}$$

$$C_T = A \left(\frac{\epsilon e N_d}{2} \frac{e}{kT} \bigg/ \ln \frac{eAN_aD_p}{I_0L_p}\right)^{1/2}$$

where I_0 is the dc forward current and A is the junction area. In obtaining the expression for C_T, we assume an abrupt junction with a heavily doped p region. Which quantity increases faster, C_D or C_T?

8.13. For a narrow-base diode, the excess-hole distribution during phase I operation can be written as

$$p_1(x, t) = -\frac{I_r}{eAD_p}(W - x) + \sum_n \frac{2(I_r + I_f)W}{eAD_p}\frac{(-1)^n}{(\theta_n)^2}\sin\left[\theta_n\left(1 - \frac{x}{W}\right)\right]\exp(-\alpha_n t)$$

where $\theta_n = (2n + 1)\pi/2$, $\alpha_n = D_p(\theta_n/W)^2$ and n is an integer. Verify that the distribution above indeed satisfies the time-dependent diffusion equation and the boundary condition at $x = 0$. Also show that at $t = 0$, the distribution reduces to

$$p_1(x, t) = \frac{I_f}{eAD_p}(W - x)$$

which is the initial hole distribution. (*Hint:* Expand the part $(I_r + I_f)(W - x)/eAD_p$ into a Fourier series as $\Sigma C_n \sin[\theta_n(1 - x/W)]$.)

8.14. **(a)** For a narrow-base diode, the excess-hole distribution during phase II operation can be written as

$$p_1(x, t) = \sum_n B_n \sin\frac{n\pi x}{W_2}\exp(-\alpha_n t)$$

This equation corresponds to Eq. (8.5.11) for a wide-base diode. Find the suitable value of α_n such that the time-dependent diffusion equation is satisfied.

(b) By expanding the initial excess-hole distribution of phase II operation into a Fourier series, show that

$$B_n = \frac{2I_f W_2}{eAD_p}\frac{1}{n\pi}, \qquad n = 1, 2, 3, \ldots$$

8.15. Using the result of Problem 8.14, show that the current in phase II operation of a narrow-base diode is

$$I = 2I_f \sum \exp(-\alpha_n t), \qquad n = 1, 2, 3$$

Also find the percentage of charge removed from the base region during phase II operation.

8.16. The recovery time of a diode can be estimated from the time required to move the stored charges out of the base region. Show that the total charge Q initially stored in the base region of a wide-base diode is

$$Q = I_f \tau_0$$

where I_f is the forward current and τ_0 is the lifetime. Assume that $I_r = I_f$ in Fig. 8.13b. Find the amount of charge removed during phase I operation. Also calculate from Eq. (8.5.13) the amount removed during $t = 0$ to τ_0 of phase II operation. Use Eq. (8.5.8) and approximate $\exp(-u^2)$ by 1 for $u < 1$.

8.17. **(a)** Describe the alloying and liquid-phase-epitaxy processes in making p-n junctions.

(b) Redraw curve CD in Fig. 8.20 so that it shows 20% In in Ge at 600°C.

(c) Note the linear relation between Ge concentrations in GaAs crystal and Ga plus GaAs liquid solution. Given $a = 5.65$ Å for the unit cell of GaAs, estimate the distribution coefficient K of Ge from Fig. 8.24.

8.18. Impurities with distribution coefficients $K < 1$ can be purified from a host crystal by running a melting zone over the crystal repeatedly. Given $a = 5.431$ Å for the unit cell of Si, find the number of runs needed to reduce impurity concentration below 10^{14} cm^{-3} if the starting Si crystal is 99.9% pure. Make calculations for B, Ga, P, and Sb.

8.19. Consider the diffusion of phosphorus into a *p*-type silicon wafer with $N_a = 10^{15}$ cm^{-3}. Suppose that the surface concentration C_0 of phosphorus is limited by the solid solubility of P in Si. Find the location of the *p*-*n* junction if the diffusion process is held at a temperature $T = 1150°$C and for a duration $t = 10$ h. Repeat the problem for $T = 1000°$C.

8.20. Repeat Problem 8.19 for diffusion of antimony into the silicon wafer under the same experimental conditions.

8.21. From the data given in Table 8.3, find the activation energy in Eq. (8.6.4) for phosphorus diffusion in silicon. The diffusion process shown in the example given in the chapter was carried out at $T = 1200°$C and for $t = 10$ h so as to obtain the desired penetration depth d of 27 μm. Calculate the value of d and the percentage error introduced **(a)** for $T = 1200°$C and $t = 9$ h and **(b)** for $T = 1150°$C and $t = 10$ h. For an accurate determination of the penetration depth, which quantity needs a more precise control, the temperature T or the duration t?

8.22. **(a)** Explain why it is often desirable to carry out the diffusion process in two steps. Using Eq. (7.7.18), find an expression for Q/A, the amount of impurities incorporated per unit area during the predeposition phase.

   (b) In a diffusion experiment, the following steps are taken: predeposition with BBr$_3$ at 1100°C for 10 min, and drive-in diffusion at 1200°C for 5 h. Find the value of Q/A and the maximum impurity concentration in Eq. (8.7.3) after the diffusion process. Comment on the dependences of the maximum impurity concentration on (1) the predeposition temperature, (2) the predeposition duration, (3) the drive-in temperature, and (4) the drive-in duration.

8.23. The fabrication process of a *p*-*n* junction can be described as follows: (1) the starting material, *n*-type silicon with 5×10^{15} phosphorus atoms/cm^3; (2) the predeposition diffusion of boron at 1150°C for 5 min; and (3) the drive-in diffusion of boron at 1250°C for 10 h.

   (a) Find the location of the junction, expressed in terms of distance from the surface.

   (b) Sketch the progressive change in the boron concentration as a function of distance at two different instants during the predeposition phase and at two different instants during the drive-in phase.
   [*Hint:* Use Eq. (7.7.18) to find the quantity Q/A in Eq. (8.7.3).]

8.24. Describe qualitatively the behavior of the transition-region capacitance of a diffused junction as a function of the bias voltage as the applied voltage changes from a large reverse bias to a forward bias V_a close to the built-in voltage V_d. Given $C_0 = 10^{20}$ cm^{-3}, $C(d) = 10^{15}$ cm^{-3}, and $d = 10^{-3}$ cm, find the bias voltage at which $W = 4Dt/d$ for Si. Comment on the applicability of Eqs. (8.8.7) and (8.8.8) in the forward- and reverse-bias regions.

8.25. Figure P8.25a shows the reverse currents in typical Ge, Si, and GaAs diodes and Fig. P8.25b shows the reverse current–voltage characteristics of a Si diode. Notice that there is a change in the slope of the current versus temperature curve at a certain temperature in Fig. P8.25a. We assume that the recombination centers have a trap energy in the middle of the gap and $C_p = C_n$ in Eq. (8.9.2). Find the dominant temperature dependences of J_{gen} and J_{diff}, and associate the dashed and solid portions of the curves in Fig. P8.25a with J_{gen} and J_{diff}. Explain the essential difference in the reverse characteristics shown in Fig. P8.25b at 25°C and 225°C, and show that your explanation is consistent with your assignment of J_{gen} and J_{diff} in Fig. P8.25a.

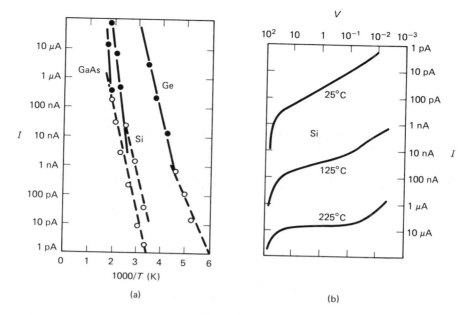

Figure P8.25

8.26. Check the dimension of E_0 in Eq. (10.6.19). Given $m_r^* = 0.10m_0$ and $\epsilon = 12\epsilon_0$ in silicon, find the value of E_0. Setting $E = 0.1E_0$ as the condition for tunnel breakdown, calculate the value of V_B in a step junction with $N_a = N_d = 5 \times 10^{18}$ cm^{-3}.

8.27. **(a)** Find the value of avalanche breakdown voltage for a one-sided step junction made of silicon and having an impurity concentration $N_I = 10^{16}$ cm^{-3}. Also calculate the maximum field E_{max} in the junction and the width of the junction.

(b) Refer to Fig. 10.9 for the ionization coefficients α_e and α_h in Si to obtain $\alpha = (\alpha_e\alpha_h)^{1/2}$. From Figs. 10.9 and 8.3b, sketch the variation of α in the space-charge region. Estimate the effective width (where $\alpha = 0.2\alpha_{max}$) of the multiplication region, and give an approximate calculation to show that $\int \alpha \, dx$ has a magnitude of the order of unity at a bias voltage $V = V_B$.

9

Multijunction and Interface Devices

9.1 BIPOLAR TRANSISTORS WITH ABRUPT DOPING PROFILES: SMALL-SIGNAL EQUIVALENT CIRCUITS

Of all semiconductor devices, the bipolar transistor is by far the most important device from historical perspective. The invention of the transistor has indeed revolutionized the entire electronic industry. It is fair to say that without transistors many of our achievements in science and technology, such as space exploration and computer development, could not have been made at such a rapid pace. Transistors have many advantages over vacuum tubes. Perhaps the most important are speed, reliability, size, and costs. The fact that transistor-fabrication technique is adaptable to automation for mass production has made many complex electronic systems feasible not only in terms of the engineering capability but also in terms of the production cost.

The first semiconductor device that showed the capability of a vacuum-tube triode for amplification of an electric signal was the point-contact transistor discovered by Bardeen and Brattain (J. Bardeen and W. H. Brattain, *Phys. Rev.,* Vol. 74, p. 230, 1948; *Phys. Rev.,* Vol. 75, p. 1208, 1949). The junction transistor action predicted by Shockley was first observed by Shockley, Sparks, and Teal (W. Shockley, *Bell Syst. Tech. J.,* Vol. 28, p. 435, 1949; W. Shockley, M. Sparks, and G. K. Teal, *Phys. Rev.,* Vol. 83, p. 151, 1951). Bardeen, Brattain, and Shockley were awarded the Nobel Prize in physics in 1953 for their pioneering work on transistors. The essential element in the point-contact transistor is the metal–semiconductor contact. In terms of transistor materials, a point-contact transistor is a metal–semiconductor transistor, whereas a junction transistor is a semiconductor–semiconductor transistor. During the early development of the semiconductor industry, junction transistors occupied practically a monopolistic position. However, since the work on circuit integration was started, metal–semiconductor interface devices have become increasingly more important. These interface devices will be discussed later because the mechanism of current conduction in a metal–semiconductor contact is more complex than that in a semiconductor–semiconductor junction.

Figure 9.1 shows the physical structures of *p-n-p* and *n-p-n* transistors. As we can see, a transistor can be described as two diodes connected back to back, one being

390

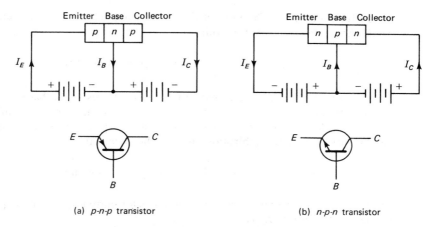

(a) *p-n-p* transistor

(b) *n-p-n* transistor

Figure 9.1 Schematic diagrams showing the physical structures and symbolic circuit representations of (a) a *p-n-p* transistor and (b) a *n-p-n* transistor. The arrows indicate the actual directions of current flow.

operated in the forward direction and the other in the reverse direction. Take the *p-n-p* transistor (Fig. 9.1a) as an example and refer to the energy-band diagram shown in Fig. 9.2. Holes and electrons are injected, respectively, into the *n* and *p* regions of the forward-biased diode on the left. The focus of our attention is on holes for a *p-n-p* transistor. After being injected into the middle *n* region, holes will continue to move across the *n* region into the *p* region on the right which, because of the reverse bias, represents the lower-energy region for holes.

In Fig. 9.1 the middle region is called the *base* region, whereas the forward-biased diode is called the *emitter* and the reverse-biased diode the *collector*. Since the emitter current consists of two components, I_{Eh} due to hole injection and I_{Ee} due to electron injection,

$$I_E = I_{Eh} + I_{Ee} \tag{9.1.1}$$

Of the two components of emitter current, only one component will appear in the collector circuit and hence become useful. In a *p-n-p* transistor, the hole component I_{Eh} is

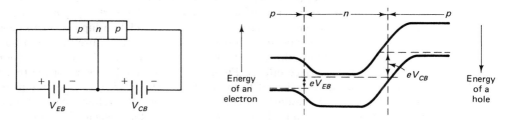

Figure 9.2 Energy-band diagram for a *p-n-p* transistor under normal operation with a forward-biased emitter junction and a reverse-biased collector junction.

the useful component. The ratio of the useful component to the total emitter current is called the *emitter efficiency* γ, with

$$\gamma = \frac{I_{Eh}}{I_E} \quad \text{for a } p\text{-}n\text{-}p \text{ transistor} \tag{9.1.2}$$

Owing to the presence of recombination processes in the base region, part of the injected minority carriers will be lost during their journey from the emitter to the collector. We define a *transport efficiency* α_T as the percentage of the injected carriers having survived the recombination process; this means that

$$\alpha_T = \frac{I_{Ch}}{I_{Eh}} \quad \text{in a } p\text{-}n\text{-}p \text{ transistor} \tag{9.1.3}$$

In view of the foregoing discussion, the collector current I_C can be expressed in terms of the emitter current I_E as

$$I_C = I_{Ch} + I_{C0} = \alpha I_E + I_{C0} \tag{9.1.4}$$

Since the collector is a reverse-biased diode, an additional current I_{C0} appears, where I_{C0} is the reverse saturation current of the collector junction.

In Eq. (9.1.4) the quantity α is called the *current gain*, with

$$\alpha \text{ (current gain)} = \gamma \text{ (emitter efficiency)} \times \alpha_T \text{ (transport efficiency)} \tag{9.1.5}$$

Obviously, the current gain is less than 1. The reader may wonder how transistors can be used as amplifiers. The answer lies in the fact that the emitter circuit is a low-impedance circuit, while the collector circuit is a high-impedance circuit. Even though the current gain is less than unity, power amplification is possible because of the large ratio of output and input voltages. The base current I_B is the difference between I_E and I_C, that is,

$$I_B = I_E - I_C = (1 - \alpha)I_E - I_{C0} \tag{9.1.6}$$

The configuration shown in Fig. 9.1 with the base region connected to both the emitter and the collector is called the *common-base* configuration. Instead, if the emitter is common to both the base and collector circuits, the *common-emitter* current gain β is

$$\beta = \frac{\Delta I_C}{\Delta I_B} = \frac{\alpha}{1 - \alpha} \tag{9.1.7}$$

For a quantitative analysis of the amplifying action of a transistor, we must treat the diffusion of minority carriers in the base region under ac operating condition, and express γ, α_T, and α in terms of the semiconductor properties of the transistor. Since I_{C0} is a dc current, it will not appear in the dynamic relationship between I_E and I_C, and hence will be ignored. For the following analysis, we again use the p-n-p transistor as an example. To simplify the analysis, we treat the transistor as a one-dimensional problem by neglecting surface effects. Figure 9.3 shows the geometry of the transistor and the physical quantities relevant to the analysis. The shaded areas represent the space-charge regions of the emitter and collector junctions. Therefore, the base region AB is from the edge of the emitter junction to the edge of the collector junction.

Under simultaneous dc and ac bias conditions (Fig. 9.3a), the injected hole con-

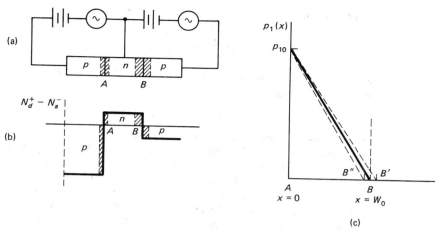

Figure 9.3 Schematic diagrams showing (a) a *p-n-p* transistor under combined dc and ac bias voltages, (b) the doping profile, and (c) the hole concentration in the base region. The shaded areas in (a) and (b) represent the space-charge regions of the emitter and collector junctions. Under a time-varying bias, the boundary of the collector junction is modulated as indicated by points B' and B''.

centration p_1 at the emitter junction can be separated into a dc and an ac part. Following the same procedure as outlined in Section 8.4, we have for the ac part,

$$(p_1)_{a-c} = \frac{ev_E}{kT} p_{10} \quad \text{at} \quad x = 0 \tag{9.1.8}$$

According to Eq. (8.4.2), v_E is the ac bias of the emitter junction, and the quantity p_{10} is

$$p_{10} = p_{n0} \exp\left(\frac{eV_E}{kT}\right) \tag{9.1.9}$$

where p_{n0} is the equilibrium hole concentration at the point A and V_E is the dc emitter bias. Hence p_{10} is the hole concentration at the emitter junction under a dc bias V_E.

As pointed out in Section 8.4, the time-dependent diffusion equation for the steady-state response to a sinusoidally time-varying function is given by Eq. (8.4.1), or

$$i\omega p_1 = D_p \frac{d^2 p_1}{dx^2} - \frac{p_1}{\tau_0} \tag{9.1.10}$$

Since we are interested only in the ac component of the hole concentration, the subscript ac of p_1 is dropped in Eq. (9.1.10). The solution of Eq. (9.1.10) is

$$p_1 = A \sinh \frac{W - x}{L'_p} + B \cosh \frac{W - x}{L'_p} \tag{9.1.11}$$

where

$$L'_p = \left(\frac{D_p \tau_0}{1 + i\omega\tau_0}\right)^{1/2} \tag{9.1.12}$$

At the collector junction ($x = W$), the hole concentration is

$$p_1 = p_{n0}\left[\exp\left(\frac{eV_C}{kT}\right) - 1\right] \cong -p_{n0} \qquad (9.1.13)$$

because of the reverse bias, that is, V_C = negative. Since p_{n0} is much smaller than p_{10}, we can approximate Eq. (9.1.13) by

$$p_1 \cong 0 \qquad \text{at} \quad x = W \qquad (9.1.14)$$

The transistor problem now appears to have been reduced to the same form as the diode problem discussed in Section 8.4. However, there is one important consideration. We must consider the effect of the ac voltage on the boundary condition. Since the collector is reverse biased, the ac voltage will not affect significantly the condition $p_1 = 0$ at $x = W$, but it will modulate the width of the space-charge region and thereby the width of the base region. The effect is called the base-width modulation effect or the Early effect (J. M. Early, *Proc. IRE*, Vol. 40, p. 1401, 1952) and is due to the collector-junction bias alone. The width of the emitter space-charge region is much smaller than that of the collector junction because of the forward-bias condition; hence the effect of the emitter bias on the base width can be neglected.

The fact that W in Eq. (9.1.11) varies with time makes the proposed form of the solution unsuitable. We proceed as follows. First, we let

$$W = W_0 + w(t) \qquad (9.1.15)$$

where W_0 and $w(t)$ denote, respectively, the dc and time-varying part of W. Since $p_1(W) = 0$, $p_1(W_0)$ is now a function of time. The situation is illustrated in Fig. 9.3c, with dashed lines indicating the moving boundary of the collector junction. As a first-order approximation,

$$p_1(W_0) = p_1(W) - \frac{dp_1}{dx} w(t) = -\frac{dp_1}{dx} w(t) \qquad (9.1.16)$$

Second, we use a solution of the form

$$p_1(x) = A \sinh\frac{W_0 - x}{L_p'} + B \sinh\frac{x}{L_p'} \qquad (9.1.17)$$

as the proposed solution instead of Eq. (9.1.11).

To find the ac value of p_1 at $x = W_0$, we use the dc distribution of p_1 in Eq. (9.1.16). Further, according to our discussion of p_1 in Section 8.3, the distribution for $W < L$ can be approximated by a straight-line distribution. Since the condition $W < L$ applies to all commercial-grade transistors, we may write Eq. (9.1.16) as

$$p_1(W_0) = +\frac{p_{10}}{W_0} w(t) \qquad (9.1.18)$$

Applying the boundary conditions of Eqs. (9.1.9) and (9.1.18) to Eq. (9.1.17), we have

$$p_1 = p_{10}\left\{\frac{ev_E}{kT}\frac{\sinh\left[(W_0 - x)/L_p'\right]}{\sinh(W_0/L_p')} + \frac{w(t)}{W_0}\frac{\sinh(x/L_p')}{\sinh(W_0/L_p')}\right\} \qquad (9.1.19)$$

From Eq. (9.1.19), the ac emitter and collector current can be evaluated, respectively, as follows:

$$i_{Eh} = -AeD_p \frac{dp_1}{dx}\Big|_{x=0} = \frac{AeD_p}{L'_p} p_{10} \left[\frac{ev_E}{kT} \coth \frac{W_0}{L'_p} - \frac{w(t)}{W_0} \operatorname{csch} \frac{W_0}{L'_p} \right] \qquad (9.1.20)$$

$$i_{Ch} = -AeD_p \frac{dp_1}{dx}\Big|_{x=W_0} = \frac{AeD_p}{L'_p} p_{10} \left[\frac{ev_E}{kT} \operatorname{csch} \frac{W_0}{L'_p} - \frac{w(t)}{W_0} \coth \frac{W_0}{L'_p} \right] \qquad (9.1.21)$$

where A is the cross-sectional area of the junction. Let us temporarily ignore the base-width modulation effect, and proceed to find γ and α_T. We can do so because the terms involving $w(t)$ in Eqs. (9.1.20) and (9.1.21) are comparatively small. The transport efficiency is

$$\alpha_T = \frac{i_{Ch}}{i_{Eh}} \cong \frac{1}{\cosh (W_0/L'_p)} \qquad (9.1.22)$$

Note that electrons are injected from the base into the emitter region. Therefore, insofar as the electron component of the emitter current is concerned, the situation is the same as that in a diode. Using a corresponding equation for electrons similar to Eq. (8.4.4), we find that

$$i_{Ee} = \frac{ev_E}{kT} \frac{eAD_n}{L'_n} n_{p0} \exp \left(\frac{eV_E}{kT} \right) \qquad (9.1.23)$$

Since the emitter is heavily doped, the lifetime of minority carriers is likely to be very short in the emitter region. Therefore, the condition $W > L'$ and hence L' instead of $W_2 - x_B$ are used in obtaining Eq. (9.1.23). From Eqs. (9.1.20) and (9.1.23), we have for the emitter efficiency

$$\gamma = \frac{i_{Eh}}{i_{Eh} + i_{Ee}} = \frac{1}{1 + (n_{p0}/p_{n0})(D_n/D_p) (L'_p/L'_n) \tanh (W_0/L'_p)} \qquad (9.1.24)$$

The hole component of the dc emitter current can be found through a similar procedure and is equal to

$$I_{Eh} = \frac{AeD_p}{L_p} p_{n0} \left[\exp \left(\frac{eV_E}{kT} \right) - 1 \right] \coth \frac{W_0}{L_p} \cong \frac{AeD_p}{L_p} p_{10} \coth \frac{W_0}{L_p} \qquad (9.1.25)$$

The ac variation of the base width can be calculated as follows:

$$w(t) = \left| \frac{\partial W_{CB}}{\partial V_C} \right| v_C \qquad (9.1.26)$$

where W_{CB} is the width of the collector-junction space-charge region on the base side and v_C is the ac collector bias. Next we define

$$Y_E = \frac{eI_{Eh}}{kT}, \qquad Y_C = \frac{I_{Eh}}{W_0} \left| \frac{\partial W_{CB}}{\partial V_C} \right| \qquad (9.1.27)$$

In terms of Y_E and Y_C, Eqs. (9.1.20) and (9.1.21) become

$$i_{Eh} = AY_E v_E - ABY_C v_C \qquad (9.1.28)$$

$$i_{Ch} = ABY_E v_E - AY_C v_C \qquad (9.1.29)$$

where the two parameters A and B are given by

$$A = \frac{L_p \coth (W_0/L_p')}{L_p' \coth (W_0/L_p)}, \qquad B = \frac{1}{\cosh (W_0/L_p')} \qquad (9.1.30)$$

For $W_0 < L_p'$ we made the following approximations:

$$\cosh \frac{W_0}{L_p'} \cong 1 + \frac{1}{2} \left(\frac{W_0}{L_p'}\right)^2 \cong \left[1 + \frac{1}{2}\left(\frac{W_0}{L_p}\right)^2\right]\left(1 + \frac{i\omega}{\omega_\alpha}\right) \qquad (9.1.31)$$

$$\frac{W_0}{L_p'} \coth \frac{W_0}{L_p'} = \frac{\cosh (W_0/L_p')}{(L_p'/W_0) \sinh (W_0/L_p')} \cong \frac{[1 + \frac{1}{2}(W_0/L_p)^2](1 + i\omega/\omega_\alpha)}{[1 + \frac{1}{6}(W_0/L_p)^2](1 + i\omega/3\omega_\alpha)} \qquad (9.1.32)$$

We define two quantities, ω_α and α_{T0}, useful in later discussions:

$$\omega_\alpha = \frac{2D_p}{W_0^2} \qquad (9.1.33)$$

$$\alpha_{T0} = \left[1 + \frac{1}{2}\left(\frac{W_0}{L_p}\right)^2\right]^{-1} \qquad (9.1.34)$$

In terms of ω_α and α_{T0}, Eqs. (9.1.22) and (9.1.30) become

$$\alpha_T = \frac{\alpha_{T0}}{1 + i\omega/\omega_\alpha} \qquad (9.1.35)$$

$$A = \frac{1 + i\omega/\omega_\alpha}{1 + i\omega/3\omega_\alpha}, \qquad B = \frac{\alpha_{T0}}{1 + i\omega/\omega_\alpha} \qquad (9.1.36)$$

In view of Eqs. (9.1.35) and (9.1.36), the current relations given in Eq. (9.1.28) and (9.1.29) can be expressed in terms of input, output, and transfer admittances as

$$i_1 = i_E = Y_{11}v_1 + Y_{12}v_2 \qquad (9.1.37)$$

$$i_2 = -i_C = Y_{21}v_1 + Y_{22}v_2 \qquad (9.1.38)$$

The equivalent circuit representing Eqs. (9.1.37) and (9.1.38) is shown in Fig. 9.4, and the values of the admittances are

$$Y_{11} = Y_E \frac{1 + i\omega/\omega_\alpha}{1 + i\omega/3\omega_\alpha}, \qquad Y_{21} = -Y_E \frac{\alpha_0}{1 + i\omega/3\omega_\alpha} \qquad (9.1.39)$$

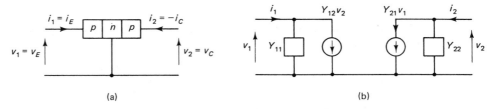

(a)   (b)

Figure 9.4 Small-signal circuit model representing carrier injection in a p-n-p transistor: (a) definition of ac terminal voltages and currents indicated by i and v with subscript 1 referring to the input and subscript 2 referring to the output, and (b) circuit representation with two admittances and two current sources.

$$Y_{12} = -Y_C \frac{\alpha_0}{1 + i\omega/3\omega_\alpha}, \qquad Y_{22} = Y_C \frac{1 + i\omega/\omega_\alpha}{1 + i\omega/3\omega_\alpha} \qquad (9.1.40)$$

In Eqs. (9.1.39) and (9.1.40), α_0 is the dc current gain with

$$\alpha_0 = \alpha_{T0}\gamma = \frac{1}{1 + \frac{1}{2}(W_0/L_B)^2} \frac{1}{1 + (\sigma_B/\sigma_E)(L_B/L_E)\tanh(W_0/L_B)} \qquad (9.1.41)$$

where σ_B and σ_E are the conductivity of the base and emitter regions, respectively, and L_B and L_E are the diffusion length of the minority carriers in the base and emitter regions, respectively. In Eq. (9.1.38), we have added the electron component [i_{Ee} of Eq. (9.1.23)] to the hole component [i_{Eh} of Eq. (9.1.28)] to get the total emitter current. Since $i_{Ee} \ll i_{Eh}$, this addition does not significantly change Eqs. (9.1.28) and (9.1.29). The only change is the inclusion of the emitter efficiency, which accounts for the fact that only γi_E of the total emitter current goes to the collector.

The following comments may be instructive in clarifying the physical meaning of the terms in Eqs. (9.1.37) and (9.1.38). First, the dependence of the emitter and collector currents on the collector voltage can be explained as follows. A positive v_2 makes the total collector bias (ac plus dc bias) less negative, resulting in a smaller extension of the collector-junction space-charge region into the base region and hence a larger base width. As can be seen from Fig. 9.3c, a larger base width means a smaller dp_1/dx, resulting in smaller emitter and collector currents. The decrease of i_E with v_C is accounted for by the minus sign in Y_{12}. Note that i_2 is defined as $-i_C$. Therefore, i_2 should increase with v_2. This is indeed the case in Eq. (9.1.38), for Y_{22} is a positive quantity.

As pointed out in Section 8.5, the excess minority-carrier distribution in the base region can be approximated by a straight line for $W_0 < L_B$. Further, for a straight-line distribution, the time required for minority carriers to diffuse across the base region is

$$t_B = \frac{W_0^2}{2D} \qquad (9.1.42)$$

as t_B can be calculated from Eq. (8.5.17). We call t_B the *base-transit time*. The various phase factors $(1 + i\omega/\omega_\alpha)$ and $(1 + i\omega/3\omega_\alpha)$ in Eqs. (9.1.39) and (9.1.40) account for the delay introduced by the transit time. Next, we turn our attention to the transport efficiency α_{T0}. The excess carriers decay according to

$$\frac{p_1(t)}{p_1(0)} = \exp\left(-\frac{t}{\tau_0}\right) \cong \frac{1}{1 + t/\tau_0} \qquad \text{for} \quad t < \tau_0 \qquad (9.1.43)$$

Since the excess carriers spend, on the average, a time t_B in the base region, the transport efficiency is

$$\alpha_{T0} = \frac{p_1(t_B)}{p_1(0)} \cong \frac{1}{1 + t_B/\tau_0} = \left[1 + \frac{1}{2}\left(\frac{W_0}{L_B}\right)^2\right]^{-1} \qquad (9.1.44)$$

9.2 DIFFUSED TRANSISTORS AND DESIGN CONSIDERATIONS

We know that the current gain of a transistor decreases at high frequencies. One of the factors contributing to this decrease is the transport efficiency. Substituting Eq. (9.1.31) in Eq. (9.1.22) and using Eq. (9.1.34), we find that the ac transport efficiency is related to the dc transport efficiency by

$$\alpha_T = \frac{\alpha_{T0}}{1 + i\omega/\omega_\alpha} \qquad (9.2.1)$$

As ω approaches ω_α, α_T begins to decrease. For $D_n = 36$ cm^2/s in silicon, a value of $\omega_\alpha = 6.3 \times 10^9$ rad/s would require a base width $W_0 < 1.1 \times 10^{-4}$ cm. This means that we must be able to control the base width to better than 1 μm in order to produce high-frequency transistors of uniform quality. Such an accurate control is made possible with the diffusion process.

In Section 9.1 the analysis of transistor performance was based on a uniform impurity distribution in the base region. Such an analysis applies to transistors made by alloying and rate-grown processes. Modern transistors are made by the diffusion process, for which the impurity distribution in the base region is no longer uniform, as illustrated in Fig. 8.27b. As a matter of fact, the impurity concentration varies as

$$C(x, t) = C_0 \text{ erfc} \left(\frac{x}{2\sqrt{Dt}}\right) \qquad (9.2.2)$$

according to Eq. (8.6.3). In Eq. (9.2.2), D is the diffusion constant of the diffusant and t is the duration of diffusion. For ease of discussion, we approximate the complementary error function by an exponential function. For a p-n-p transistor illustrated in Fig. 8.27b, we let

$$N = N_d(x) = N_0 \exp \left(\frac{-x}{l}\right) \qquad (9.2.3)$$

where l is a characteristic distance and $N = C(x, t)$ is the donor concentration in the base region.

One consequence of a nonuniform impurity distibution is the existence of a built-in electric field. The situation is illustrated in Fig. 9.5. In the intermediate-temperature (exhaustion) region, donors are fully ionized; thus we have

$$n = N_c \exp \left(-\frac{\mathscr{E}_c - \mathscr{E}_f}{kT}\right) \cong N_d = N_0 \exp \left(-\frac{x}{l}\right) = N \qquad (9.2.4)$$

Since under thermal equilibrium $\mathscr{E}_f$ stays constant throughout the base region, $\mathscr{E}_c$ must change with x (Fig. 9.5b), creating an electric field. If V represents the electrostatic potential experienced by electrons, the potential energy of electrons is equal to $-eV$. Since $\mathscr{E}_c$ is the potential energy, we have

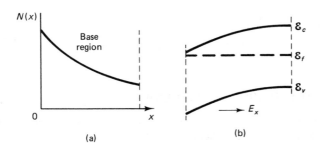

Figure 9.5 Schematic diagrams showing (a) the profile of doping concentration in the base region, and (b) the variation of conduction-band and valence-band energies resulting from nonuniform doping.

Multijunctions and Interface Devices   Chap. 9

$$E_x = -\frac{\partial V}{\partial x} = \frac{1}{e}\frac{\partial \mathscr{E}_c}{\partial x} = -\frac{kT}{en}\frac{dn}{dx} = -\frac{kT}{eN}\frac{dN}{dx} \qquad (9.2.5)$$

For negative dN/dx (Fig. 9.5a), there are more electrons on the left. This means that the diffusion force is toward the right; hence the electric force must be toward the left to counterbalance the diffusion force.

In Section 9.1 the excess-hole density p_1 was approximated by a straight-line distribution (Fig. 9.3c). This distribution was used to calculate the diffusion time t_B of Eq. (9.1.42). The excess-hole distribution under a built-in electric field can be found by solving the time-dependent diffusion equation. However, we do not choose to take this approach. We prefer an alternative approach which is mathematically simpler and yet physically more instructive. The hole current density under a built-in field is

$$J_p = e\mu_p p_1 E_x - eD_p\frac{dp_1}{dx} = -eD_p\left(\frac{p_1}{N}\frac{dN}{dx} + \frac{dp_1}{dx}\right) \qquad (9.2.6)$$

where p_1 is the excess-hole density in the base region. If it is assumed that recombination of excess carriers is insignificant as in Fig. 9.3c, then J_p must be independent of x, as demanded by the continuity equation. Thus we have

$$\frac{dp_1}{dx} + \frac{p_1}{N}\frac{dN}{dx} = -\frac{J_p}{eD_p} \qquad (9.2.7)$$

The solution of Eq. (9.2.7) for constant J_p is

$$p_1(x) = \frac{J_p}{eD_p N(x)}\int_x^W N(x)\,dx \qquad (9.2.8)$$

The upper limit of integration is chosen such that at the collector junction $x = W$, $p_1(W) = 0$. Note that for constant N, Eq. (9.2.8) reduces to

$$p_1(x) = \frac{J_p}{eD_p}(W - x) = p_1(0)\frac{W - x}{W} \qquad (9.2.9)$$

which is the straight-line distribution shown in Fig. 8.19b. The charge stored in the base region due to minority carriers is equal to

$$Q = A\int_0^W ep_1(x)\,dx \qquad (9.2.10)$$

The time required to move Q carriers across the base region is

$$t_B = \frac{Q}{I_p} = \frac{Q}{AJ_p} = \frac{e}{J_p}\int_0^W p_1(x)\,dx \qquad (9.2.11)$$

which is the general expression for the transit time. The reader is asked to show that Eq. (9.2.11) yields the result given in Eq. (9.1.42) for a straight-line distribution of $p_1(x)$.

Substituting Eq. (9.2.3) into Eqs. (9.2.8) and (9.2.11), we find that

$$p_1(x) = \frac{lJ_p}{eD_p}\left[1 - \exp\left(-\frac{W - x}{l}\right)\right] \qquad (9.2.12)$$

$$t_B = \frac{l}{D_p} \left[W - l + l \exp\left(-\frac{W}{l}\right) \right] \qquad (9.2.13)$$

The three possible cases are: case A for a constant N, case B for N decreasing toward the collector junction, and case C for N increasing toward the collector junction. The three situations are illustrated in Fig. 9.6a. In comparing the three cases, we make the following common constraints: (1) the same doping concentration N_0 at the emitter junction, (2) the same emitter current density J_p, and (3) the same base width. In Eqs. (9.2.12) and (9.2.13), $l = \infty$ for case A, l = positive for case B, and l = negative for case C.

Using Eq. (9.2.3) in Eq. (9.2.5), we see that $E_x = 0$ for case A, E_x = positive and hence the drift term is aiding the diffusion term in Eq. (9.2.6) for case B, and E_x = negative and hence the drift term is opposing the diffusion term for case C. This information is summarized in Fig. 9.6b. With Fig. 9.6b in mind, we can turn our attention to the excess-hole distribution shown in Fig. 9.6c. At the collector junction $x = W$, $p_1(W) = 0$ in Eq. (9.2.6). Therefore, for the same current density J_p, the slope dp_1/dx at the collector junction must be the same for the three cases. At the emitter junction, the electric field is aiding the hole flow for case B. For the same J_p, case B requires a smaller dp_1/dx at the emitter junction. The opposite is true for case C.

The charge Q stored in the base region is proportional to the area under the curve shown in Fig. 9.6c. Since $Q_B < Q_A < Q_C$, case B has the shortest transit time according to Eq. (9.2.11). The same conclusion is reached from the quantitative analysis expressed in Eq. (9.2.13). For illustration, we assume that $W < l$. Expansion of the exp $(-W/l)$ term into a Taylor's series gives

$$t_B \cong \frac{W^2}{2D} \left(1 - \frac{W}{3l} \right) \qquad (9.2.14)$$

It is obvious that case B with positive l has the shortest t_B. We should emphasize that the analysis above is based on the assumption that all three cases have the same base width. The situation can be entirely different if a different constraint is assumed.

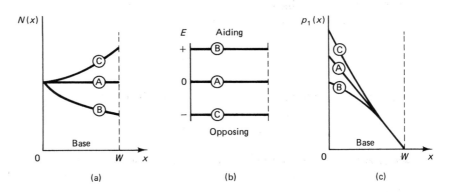

(a) (b) (c)

Figure 9.6 Schematic diagrams showing the effects of nonuniform doping in the base region: (a) the doping profiles for three cases labeled A, B, and C, (b) the resulting fields E_x and (c) the profiles of injected hole concentration $p_1(x)$ for the three cases. Refer to the text for the common constraints imposed on the three cases.

One of the important parameters in a transistor is the base resistance. The equivalent circuit shown in Fig. 9.4 is incomplete in two respects. First, since the base current is nonzero, there must be a potential drop to support this current. In Fig. 9.7a we add the base resistance r_B in the base circuit to account for the potential drop. Second, the space-charge-region capacitances C_{TC} and C_{TE} of the form given in Eq. (8.2.13) must be attached to the collector and emitter junctions, respectively. In Fig. 9.7a, the two capacitances are added. For $\omega < \omega_\alpha$, all the admittances become real with $r_E = 1/Y_E$. Therefore, Fig. 9.7a represents the complete equivalent circuit of a transistor operated at low frequencies in the common-base connection.

It can be shown from circuit analysis that the feedback current generator $\alpha v_C/r_C$ can be incorporated into the base–emitter circuit by changing r_B to r_B'. Further, since $r_C > r_E$ and $C_{TC} > C_{TE}$, the frequency response of the transistor is determined by the time constant $r_C C_{TC}$, but not by $r_E C_{TE}$. The equivalent circuit shown in Fig. 9.7b is a simplified version of the circuit shown in Fig. 9.7a. Suppose that the transistor is connected on the input side to a voltage generator with an input resistance R_G and on the output side to a load with load resistance R_L. If η is defined as the ratio of the output power to the maximum available power, then under matched input and output conditions,

$$\eta_{\max} = \left. \frac{i_C^2 R_L}{v_G^2/4R_G} \right|_{\max} = \frac{\alpha_0^2 r_C}{r_B'} \qquad (9.2.15)$$

Equation (9.2.15) is valid in the low-frequency region where $\omega < \omega_\alpha$ and $\omega < 1/r_C C_{TC}$.

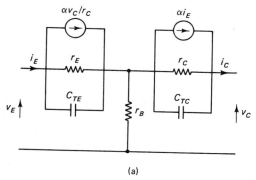

(a)

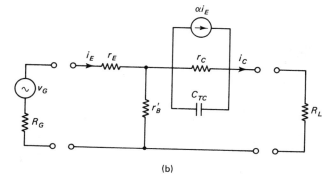

(b)

Figure 9.7 Diagrams showing (a) the small-signal equivalent circuit for a bipolar transistor and (b) a simplified version incorporating the feedback current generator into r_b' and neglecting C_{TE}. Figure 9.7 differs from Fig. 9.4 in that the junction capacitances and the base resistance are added to complete the circuit representation of a bipolar transistor. The base-width modulation effect is represented by r_C.

At high frequencies, not only does the transit-time effect come into play, but the shunting effect of C_{TC} also becomes noticeable. The bandwidth of the collector circuit is

$$\omega_C = \frac{1}{r_C C_{TC}} \qquad (9.2.16)$$

Therefore, the gain–bandwidth product of a transistor is

$$GB = \eta_{max}\omega_C = \frac{\alpha_0^2}{r_B' C_{TC}} \qquad (9.2.17)$$

For high gain, both r_B' and C_{TC} must be small. It can be argued that we should use the same base resistance as a common constraint instead of the same base width. The case is briefly discussed below.

The base resistance, except for a geometric factor, is equal to R, which has a dimension of resistance and is defined as

$$\frac{1}{R} = e\mu_n \int_0^W N(x)\,dx = e\mu_n N_0 l\left[1 - \exp\left(-\frac{W}{l}\right)\right] \qquad (9.2.18)$$

for a p-n-p transistor. To maintain the same base resistance, we must have

$$W_A = l\left[1 - \exp\left(-\frac{W_B}{l}\right)\right] = l\left[-1 + \exp\left(\frac{W_C}{l}\right)\right] \qquad (9.2.19)$$

The situation is shown in Fig. 9.8a. For the same area under the curve, the condition is $W_C < W_A < W_B$. From Eq. (9.2.8), we see further that $p_1(0)$ must be the same for all three cases because $\int_0^W N(x)\,dx$ and $N(0)$ are the same. Therefore, the excess-hole concentration $p_1(x)$ in the base region takes the general shape shown in Fig. 9.8. From the area under the various curves, it is clear that t_B is lowest for case C.

From the foregoing discussion concerning the transit time and the gain–bandwidth product, it appears that we could perform the diffusion process either from the emitter side or from the collector side, the former corresponding to case B and the latter to case C. Figure 9.9 shows the actual structure of a planar n^+-p-n^+ transistor. The fabrication procedure for such a structure was discussed in Section 8.7 in connection with Fig. 8.34. Another important consideration in the design of a transistor is the collector-junction capacitance C_{TC}. As Eq. (9.2.17) indicates, a wide bandwidth requires a low C_{TC}. The collector-junction capacitance can be lowered by inserting a near-intrinsic region between the n^+ and p regions of the collector junction. The lightly doped n

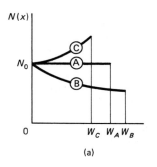

(a)

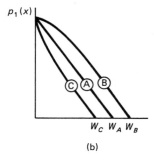

(b)

Figure 9.8 Schematic diagrams showing (a) the doping profiles and (b) the profiles of injected hole concentration $p_1(x)$ for the same three cases A, B, and C in Fig. 9.6 except that the same base resistance replaces the same base width as one common constraint.

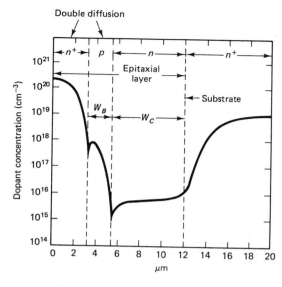

Figure 9.9 Doping profiles in a double-diffused n^+-p-n-n^+ transistor.

region grown through the epitaxial process serves this purpose. The base region and the emitter junction are formed through a double-diffusion process. Therefore, practical transistors are made through diffusion from the emitter side, and the actual situation is similar to case B. The n^+-p-n-n^+ structure is preferred over the p^+-n-p-p^+ structure for high-frequency response because the diffusion of minority carriers is faster and hence the transit time is shorter in the base region.

Design and fabrication of bipolar transistors for high-speed operations are presented in Section 13.15. Especially interesting is the heterostructure bipolar transistor, which makes possible the use of heavy doping in the base region without lowering the emitter efficiency. The factors limiting the high-frequency response of bipolar and field-effect transistors are discussed generally in Section 13.10.

9.3 DESIGN CONSIDERATIONS FOR POWER TRANSISTOR, MICROWAVE TRANSISTOR, AND SWITCHING TRANSISTOR

In Sections 9.1 and 9.2 we discussed several aspects concerning the performance of transistors: emitter efficiency, transport efficiency, and transit time. In this section we expand the scope of the earlier discussion to include other aspects of transistor operation and to associate them with actual problems that may arise in specific applications. We realize that transistors used for different purposes may have different requirements on transistor parameters. We divide special-purpose transistors into three functional groups: (1) power transistor, which is designed for handling a large amount of power; (2) microwave transistor, which is used for high-frequency analog applications; and (3) switching transistor, which is operated as an on–off switch for digital applications.

Power Transistor

The problem of distributing the currents I_E and I_B uniformly across the emitter function and in the base region becomes important at high current densities. Figure 9.10 shows two typical transistor geometries, one circular geometry and the other linear geometry.

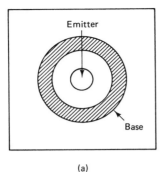

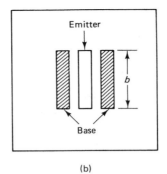

(a)

(b)

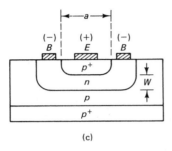

(c)

Figure 9.10 Schematic diagrams showing (a) the circular geometry (top view), (b) the linear geometry (top view), and (c) the side view of a power transistor. The potential drop due to the base current causes the emitter current to crowd near the periphery.

The linear geometry is used in power transistors for the following reason. The base current in a transistor flows in directions parallel to the plane containing the emitter junction, and establishes a potential drop that decreases as we move from the center toward the periphery of the emitter junction. The situation is illustrated in Fig. 9.10c for a *p-n-p* transistor. Since the emitter side is heavily doped, it can be considered as an equipotential surface. The potential difference created by the base current makes the center region of the emitter junction less forwardly biased than the region near the periphery. This means that the excess carriers, and hence the emitter current, will be heavily concentrated near the periphery. This crowding of the emitter current reduces the effective area of the emitter junction.

We could do two things to minimize the effect. First, we realize that because of the base-current bias effect there is a practical limitation to the emitter-junction size. When the base-current bias effect becomes important, the central portion of the emitter junction is practically nonfunctional. A structural approach to improve the situation is to use several small emitter junctions connected in parallel instead of a large emitter junction. This arrangement is possible with the linear geometry shown in Fig. 9.10b. We simply use more stripes arranged in alternate and side-by-side fashion for emitter and base regions to handle the large current and to distribute the current uniformly.

The second approach is preventive. The base current in a transistor has four components:

$$I_B = \left[(1 - \gamma) + \frac{W^2}{2D\tau_0} + \frac{2v_s W^2}{Da} \right] I_E + I_{RE} \qquad (9.3.1)$$

The term $I_{B1} = (1 - \gamma)I_E$ accounts for the fact that of the two emitter-current components, the useful component γI_E goes to the collector, whereas the other component $(1 - \gamma)I_E$ circulates in the base–emitter circuit. The second term, $I_{B2} = I_E(W^2/2D\tau_0)$, which is equal to $I_{Eh} - I_{Ch}$ in a p-n-p transistor, is caused by recombination of excess carriers in the base region. The third term, $I_{B3} = I_E(2v_sW^2/Da)$, which has not been discussed before, is due to recombination of excess carriers at the surface of the base region. The fourth term, $I_{B4} = I_{RE}$, represents the recombination current in the emitter junction. The origin of I_{RE} was discussed in Section 8.9. According to Eq. (8.9.14), the ratio J_{rec}/J_{diff} becomes smaller as the forward bias increases; therefore, I_{RE} is important only at low emitter currents.

The most effective region for surface recombination is near the emitter, where the concentration of injected carriers is high. However, injected carriers are drawn toward the collector as well as toward the surface, the former process being the result of the collector bias, and the latter being that of recombination. Since these two processes are competing directly, the effective region of surface recombination is limited to an area that is within a distance W from the emitter junction, where W is the base width. If v_s is the surface-recombination velocity and p_{10} is the excess-hole concentration at the emitter junction, the rate of recombination is $v_s p_{10} A$, where A is the area of the effective recombination region. As a result, a current of $ev_s p_{10} A$ is generated. Like the bulk-recombination current I_{B2}, the surface-recombination current contributes to the base current. For linear transistor structure shown in Fig. 9.10b with $b > a$, $A = 2Wb$. Thus

$$I_{B3} = ev_s p_{10} A = \frac{2v_s W^2}{Da} I_E \qquad (9.3.2)$$

after the substitution $I_E = eDp_{10}ab/W$.

Now we return to the $(1 - \gamma)I_E$ term because it is the most important term in Eq. (9.3.1). According to Eq. (9.1.24), the quantity $(1 - \gamma)$ can be approximated by

$$1 - \gamma \cong \frac{1 - \gamma}{\gamma} = \frac{n_{p0}}{p_{n0}} \frac{D_n}{D_p} \frac{L_p}{L_n} \tanh \frac{W}{L_P} \cong \frac{n_{p0}}{p_{n0}} \frac{D_n}{D_p} \frac{W}{L_n} \qquad (9.3.3)$$

for a p-n-p transistor with $W < L_p$. Note that I_{B1} is proportional to W, whereas I_{B2} and I_{B3} are proportional to W^2. For transistors with very narrow base width, the component I_{B1} of the emitter current due to nonunity emitter efficiency is much larger than I_{B2} and I_{B3}. For $W = 3 \times 10^{-4}$ cm, $D = 39$ cm²/s, and $\tau_0 = 10^{-6}$ s, $W^2/2D\tau_0 = 1.2 \times 10^{-3}$. The observed value of $\beta = \Delta I_C/\Delta I_B$ is typically around 50, corresponding to $I_B \cong 0.02 I_E$. The fourth component, $I_{B4} = I_{RE}$, is unimportant at high emitter currents. Therefore, we concentrate on the I_{B1} term in the following discussion.

Equation (9.3.3) is valid only at low emitter currents. When the injected hole (minority-carrier) concentration becomes comparable to, or greater than, the equilibrium electron (majority carrier) concentration in the base region of a p-n-p transistor, the electron concentration n_n in the base region increases with the excess-hole concentration. This means that $n_n = n_{n0} + p_1$. The electron concentrations n_p and n_n on the emitter and base side, respectively, are related by

$$n_p = n_n \exp \left[-\frac{e(V_d - V_E)}{kT} \right] \qquad (9.3.4)$$

where V_d and V_E are the diffusion voltage and applied bias of the emitter junction. Therefore, under high emitter currents, n_{p0} should be replaced by $n_p = n_{p0}(1 + p_1/n_{n0})$, and Eq. (9.3.3) becomes

$$1 - \gamma \cong \frac{n_{p0}}{p_{n0}} \frac{D_n}{D_p} \frac{W}{L_n} \left(1 + \frac{p_1}{n_{n0}}\right) = \frac{n_{p0}}{p_{n0}} \frac{D_n}{D_p} \frac{W}{L_n} \left(1 + \frac{WI_E}{eDn_{n0}ab}\right) \qquad (9.3.5)$$

To reduce the base current, we must make $n_{p0} << p_{n0}$. In other words, the emitter side of the emitter junction must be heavily doped as compared to the base side. Boron and phosphorus are almost universally employed as the *p*-type impurity in *p-n-p* transistors and as the *n*-type impurity in *n-p-n* transistors, respectively. These impurities not only have the desirable property of slow diffusion in SiO_2 and rapid diffusion in Si for device fabricatoin (Section 8.7), but also have very high solid solubilities in silicon (Fig. 8.26a). The latter property permits a high doping concentration on the emitter side.

Besides increasing the emitter doping concentration, we should increase the length of the emitter stripe (Fig. 9.10b) to minimize the base current. At high emitter currents, $(1 - \gamma)$ increases with I_E according to Eq. (9.3.5). Since $(1 - \alpha)$ can be approximated by $(1 - \gamma)$, the common-emitter current-gain β defined in Eq. (9.1.7) begins to fall off when $I_E > eDn_{n0}ab/W$. The effect that has been observed is called the β falloff. In Eq. (9.3.5), the width a of the emitter stripe is limited in size because of the base-current bias effect. Therefore, higher current operations can be obtained by making the emitter stripe longer. If we observe the β falloff at $I_E = 20 \times 100^{-3}$A in a transistor with a 1- by 1.5-mil ($a \times b$) emitter stripe, we expect the β falloff to occur at $I_E = 80 \times 10^{-3}$ A in a transistor with a 1- by 6-mil emitter stripe.

In summary, two important effects which occur at high emitter currents are the base-current bias effect and the β falloff effect. For a given emitter junction area $a \times b$ and a given base width W, the amount of emitter current that can pass through the junction without causing a deterioration in performance is limited by the two effects. The crowding of emitter current can cause a high concentration of injected minority carriers near the periphery of the emitter junction, thus raising both the I_{B1} and I_{B3} terms. To minimize the two effects, we should use a heavily doped emitter so that $n_{p0} >> p_{n0}$ in Eq. (9.3.5), and we should make the dimension of the emitter stripe such that $a < b$, a being the dimension longitudinal to the base-current flow. Many emitter stripes are sometimes needed to handle the large current.

Another important consideration is heat dissipation. First, an adequate heat sink must be provided for power transistors. The power-handling capacity of a transistor is ultimately limited by the temperature rise at the junctions. For the same doping concentration, a semiconductor with a larger gap energy remains extrinsic over a wider temperature range. Further, the temperature rise is smaller in a semiconductor with a higher thermal conductivity for the same amount of heat dissipation. Of the three common semiconductors Ge, Si, and GaAs, gallium arsenide has the largest gap and silicon has the highest thermal conductivity. Since gallium-arsenide technology has not been as extensively developed as silicon technology, most power transistors are made from silicon.

Microwave Transistors

Every transistor has a cutoff or break frequency beyond which the gain drops below unity. The frequency at which the common-emitter short-circuit current gain drops to unity is called the *cutoff frequency*. The frequency can be obtained from a calculation of the gain of a transistor based on the equivalent-circuit model. However, since the calculation is tedious but straightforward, it is omitted in the following discussion. Physically, the gain of a transistor drops because of the various time delays introduced between the input and output circuits. Referring to our discussion in Section 9.2 and specifically to Figs. 9.7 and 9.9, we see that there are four possible places where a delay (or a phase shift) may be introduced. The four delays are (1) the time t_E to charge the emitter junction, (2) the transit time t_B across the base region, (3) the time t_C to charge the collector junction, and (4) the transit time t_{DC} across the epitaxial layer.

In the common-emitter configuration with a short-circuit load ($R_L = 0$), the input circuit sees an equivalent capacitance which is equal to $C_{TC} + C_{TE}$. Thus the emitter-junction charging time is

$$t_E = r_E(C_{TC} + C_{TE}) = \frac{kT}{eI_E}(C_{TC} + C_{TE}) \tag{9.3.6}$$

For a diffused transistor, the base transit time is

$$t_B = \frac{\eta W_B^2}{2D} \tag{9.3.7}$$

According to Eq. (9.2.13), the transit time can be considerably shortened by an aiding built-in field. The factor η in Eq. (9.3.7) accounts for this fact. In the epitaxial-layer region, the carriers drift under an electric field. If v_{ds} is the drift velocity of carriers, then

$$t_{DC} = \frac{W_C}{v_{ds}} \tag{9.3.8}$$

where W_C is the thickness of the epitaxial layer. The collector-junction charging time is given by $t_C = r_C C_{TC}$. In transistors with an epitaxial layer in the collector junction, r_C is substantially reduced. In the common-emitter short-circuit configuration, t_C can be ignored.

Summarizing the foregoing discussion, we have

$$t_{EC} = \frac{kT}{eI_E}(C_{TC} + C_{TE}) + \frac{\eta W_B^2}{2D} + \frac{W_C}{v_{ds}} \tag{9.3.9}$$

for the time delay between the input and output circuits. The result expressed in Eq. (9.3.9) is different from that presented in Section 9.2 because the two situations are different. First we have, at present, a common-emitter configuration with a short-circuit load. Second, we have considered the effect of the epitaxial layer. The common-emitter configuration is used extensively for high-frequency transistor oscillators.

Now let us estimate the value of t_{EC}. First, we note that $C_{TE} > C_{TC}$ for the emitter junction is forwardly biased. The value of C_{TE}/A is on the order of 4×10^{-8} F/cm^2, where A is the emitter-junction area. For an emitter-current density $I_E/A = 10^2$ A/cm^2, we have $t_E = 10^{-11}$ s. The value of $W_B^2/2D$ is 3×10^{-11} s for $W_B = 5 \times 10^{-5}$ cm and $D = 39$ cm^2/s. The value of t_B is actually much less than the calculated value because of the factor η in Eq. (9.3.7). The drift velocity v_{ds} in the epitaxial layer is on the order of 10^7 cm/s. For $W_c = 3 \times 10^{-4}$ cm, the value of t_{DC} is 3×10^{-11} s. Adding t_E and t_{DC}, we find $t_{EC} = 4 \times 10^{-11}$ s, or a cutoff frequency of $v_T = (2\pi t_{EC})^{-1} = 4$ GHz.

In the estimate above, we use a rather low value of I_E/A. The value probably could be raised to 10^3 A/cm^2. Beyond that we would encounter the crowding effect. As a matter of fact, microwave transistors are made of many separate emitters and base contacts connected together with metal overlays to increase periphery-to-area ratio and thus minimize current crowding. The base width could also be further reduced by a factor of 2 using the present state of the art. Because η is a small number, the delay time t_B does not appear to be the limiting factor. This leaves only the term $t_{DC} = W_C/v_{ds}$ for further improvement. The reader may ask why we need the epitaxial layer in the first place. The obvious reason for introducing the layer is to reduce C_{TC} and r_C. What is even more important is that the epitaxial layer provide an active region for the microwave signal to interact with the drifting carriers.

The operation of a microwave transistor may be compared to that of a vacuum tube. The emitter junction serves the function of the cathode as a source to provide charge carriers. It is in the drift region where the carriers acquire energy from the dc electric field and convert it into microwave energy. Since the cutoff frequency is related to the time delay through $v_T = (2\pi t_{EC})^{-1}$, the drift region must be made shorter for higher cutoff frequencies. The shorter drift region means a shorter interaction length, which inevitably leads to less power. There exists a fundamental relation between available power and frequency for devices utilizing the interaction of drift carriers with the electric field. The microwave transistor is one of such devices. There are many versions of the power–frequency relation applicable to high-frequency transistors. The following relation is due to Johnson (E. O. Johnson, *RCA Rev.*, Vol. 26, p. 163, 1965):

$$P_m X_C (2\pi v_T)^2 = E_m v_{ds} \qquad (9.3.10)$$

where P_m is the maximum power that can be delivered to the drifting carriers, v_T the cutoff frequency due to transit-time limitations, and X_C the reactance of the drift region at v_T.

On the right side of Eq. (9.3.10), E_m and v_{ds} are the material parameters, with E_m representing the maximum electric field that can be sustained in the epitaxial layer without having avalanche breakdown and v_{ds} being the saturation value of the drift velocity. As will be discussed in Section 10.4, the drift velocity of carriers does not increase indefinitely with the field but reaches a saturation value at high electric fields. Since the transistor must be matched in impedance to the microwave circuitry for maximum power transfer, the value of X_C is not arbitrary. For example, we cannot make the junction area arbitrarily large without suffering mismatch. Therefore, we may conclude that the output power of microwave transistors is expected to drop at high frequencies, following a $1/v^2$ relation. A survey of the state of the art shows that a power as high as 40 W is available at 200 MHz and a frequency as high as 5 GHz is possible with a 0.02-W output-power.

Switching Transistor

A switching transistor simply acts as an on–off switch. The on–off operation requires a large swing of the operation conditions. Consider a *n-p-n* transistor operated in the common-emitter configuration (Fig. 9.11a). Under normal operation conditions, the emitter junction is forward biased and the collector junction is reverse biased. Elimination of I_E from Eq. (9.1.4) and (9.1.6) yields

$$I_C = \frac{\alpha}{1 - \alpha} I_B + \frac{1}{1 - \alpha} I_{C0} \qquad (9.3.11)$$

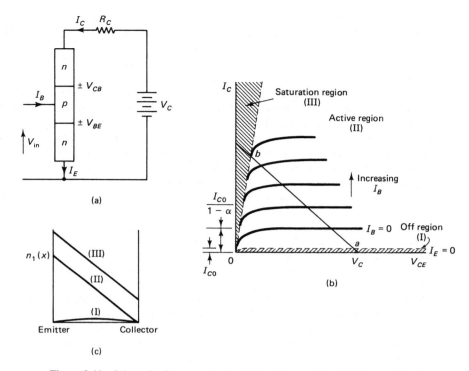

Figure 9.11 Schematic diagrams showing (a) the bias condition, (b) the current–voltage characteristics, and (c) the injected electron concentration in the base region of a *n-p-n* switching transistor operated in the common-emitter configuration. The load is represented by line *ab* in diagram (b).

The family of I_C versus V_{CE} curves for various I_B is shown in Fig. 9.11b. For $V_{BE} = 0$, a small current approximately equal to I_{C0} flows into the base. For $I_B = 0$, $I_C = I_{C0}/(1 - \alpha)$. Since $I_C > I_{C0}$, a small input voltage V_{in} must be present and the emitter junction is slightly forward biased. For $V_{in} \leq 0$ the emitter–base junction is reverse biased, as is the collector–base junction, and the transistor is said to be in the *off state*.

As we increase V_{in}, I_B increases and so does I_C. As long as the collector–base junction remains reverse biased, Eq. (9.3.11) holds and the transistor is in the active region (i.e., the normal operational mode). Note that there is a resistance R_C in the collector–emitter circuit. This load resistance is represented in Fig. 9.11b by the load line *ab*. The operating point moves up the load line as I_B increases. Upon further increase of I_B, the operating point finally reaches point *b*, where $V_{CB} = 0$. From Fig. 9.11a,

$$V_{CB} = V_C - I_C R_C - V_{BE} \tag{9.3.12}$$

According to Eq. (9.1.25), the emitter current has a voltage dependence given by

$$I_E = I_{E0} \exp\left(\frac{eV_{BE}}{kT} - 1\right) \tag{9.3.13}$$

where I_{E0} is a proportionality constant. Since I_C is related to I_E through Eq. (9.1.4), the point at which V_{CB} becomes zero can be obtained from Eqs. (9.1.4), (9.3.12), and (9.3.13) by solving for V_{BE}. The line *ob* represents the locus of $V_{CB} = 0$.

Beyond point *b*, the collector–base junction becomes forward biased, and the transistor is said to be in the *saturation region*. The minority-carrier concentration in the base region is shown in Fig. 9.11c for the three phases of operation to illustrate the physical condition of the transistor. In the off region (curve I), both junctions are reverse biased and hence the minority-carrier concentration is very low throughout the base region. In the active region (curve II), the electron (minority carrier) concentration is raised at the emitter junction by the forward emitter bias but remains zero at the collector junction. In the saturation region (curve III), the electron concentration at the collector junction is also raised because both junctions are now forward biased.

When both junctions are forward biased, the sum $V_{CB} + V_{BE}$ is a small quantity and hence Eq. (9.3.12) can be approximated by $I_C = V_C/R_C$, which is the maximum current allowed in the collector circuit. Once this situation is reached, any further increase of I_B will not affect I_C. The term "saturation" is used to indicate that the value of I_C remains almost constant in this region. In Fig. 9.11c the slope of curve III remains unchanged even though the level of the electron concentration is raised upon further increase of I_B. The reader may ask what happens to I_B if it does not raise I_C further. Note that since the level of the electron concentration is raised, there are more excess carriers in the base region and hence there is a corresponding increase in the rate of recombination. Any increase in the base current after the saturation region is reached is to supply more carriers to the base region for recombination. Further discussions of the I_C versus V_{CE} characteristics are given in Section 9.4.

The foregoing analysis is for the dc operation of the switching transistor. If a pulse is applied to the base to change the state of the switch, the collector circuit cannot respond instantaneously to the change in the base circuit. As discussed in Section 8.5, stored charges must be established or removed before a junction can change its state from the reverse-biased to forward-biased condition, or vice versa. The same applies to a switching transistor. The situation is illustrated in Fig. 9.12. To change the transistor state from *a* to *b* (i.e., I_C from zero to V_C/R_C), we need only a base current of amplitude $V_C(1 - \alpha)/\alpha R_C$. Initially, the transistor is in the off state. At point *a*, the carrier distri-

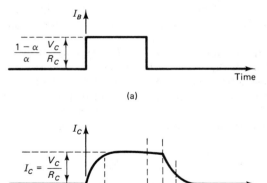

(a)

(b)

Figure 9.12 Time variations of (a) the base current I_B, which is the input signal in the form of a pulse, and (b) the collector current I_C, which appears in the output circuit. The various time delays are the times required for the injected minority carriers to reach the respective steady states.

bution in the base region is represented by curve I in Fig. 9.11c. The turn-on time t_0 in Fig. 9.12 is the time required to establish a carrier distribution represented by curve II in Fig. 9.11c. Since the collector current remains saturated at V_C/R_C during the time when the carrier distribution changes from curve II to curve III, the time t_0 is the only time delay insofar as the output current pulse is concerned.

The situation with the turn-off time is very much similar to that discussed in Section 8.5. At the time when the control pulse I_B is turned off, the collector and emitter junctions are forwardly biased. It takes a finite time for excess carriers at the collector junction to drop to zero. This finite time causes the time delay t_I shown in Fig. 9.12. After t_I, the collector junction becomes reverse biased. However, because there are still minority carriers in the base region, I_C does not return to zero suddenly, but decays gradually to zero. As in a switching diode, the recovery phase in a switching transistor can be divided into two phases. The two times t_I and t_{II} can be calculated in a manner similar to that outlined in Section 8.5. One important difference is that the carrier concentration in the base region and the boundary condition at the collector junction are different.

Two most important parameters for a switching transistor are the current gain and the switching time. The current gain determines the amplitude of the control current needed for switching. The switching time limits the speed of switching operations. To reduce the switching time, the transistor may be doped with gold. As mentioned in Section 8.5, gold creates recombination centers and thus reduces the recovery time of diodes (Fig. 8.18). However, lowering lifetime τ also means reducing α_T. To compensate for a lower α_T, we may improve γ by using a lower doping concentration in the base region. A switching time of the order of few nanoseconds is now practical in transistors with a current gain β of 50 and a maximum current of 0.02 A. These figures are expected to improve with advanced device technology. Transistors with much higher current capacity (of the order of amperes) but with a much slower switching speed (about 10^{-6} s) are also available.

9.4 LARGE-SIGNAL ANALYTICAL MODELS

As shown in Figs. 9.11 and 9.12, the operation of a switching transistor involves a large swing of the output current. Therefore, the small-signal equivalent circuit derived in Section 9.1 is no longer useful for analyzing the operation of a switching transistor. In this section we present two models, the Ebers–Moll model and the charge-control model. The formulation of the functional relations between the currents and the applied bias voltages is based on phenomenological considerations. The values of the parameters used in the formulation are determined by experiments rather than by theoretical calculations. These models are especially useful for numerical simulation study of the behavior of a switching transistor.

In Section 9.1 we considered only transistors operating in the active region (Fig. 9.11b), with the emitter junction biased in the forward direction and the collector junction biased in the reverse direction. To establish a general analytical tool applicable to switching transistors, we use the method of superposition. Consider a n-p-n transistor (Fig. 9.11a). When it operates in the saturation region, injection of minority carriers (electrons) takes place at both the emitter and the collector junctions. The excess-electron concentration $n_1(x)$ in the base region is shown as curve III in Fig. 9.11c. In the *Ebers–Moll model,* the concentration $n_1(x)$ is decomposed into two components: $n_{1F}(x)$, due to forward injection from the emitter junction to the collector junction, and $n_{1R}(x)$, due to reverse injection from the collector junction to the emitter junction (J. J. Ebers and J. L. Moll, *Proc. IRE*, Vol. 42, p. 1761, 1954). In Fig. 9.13a, the excess-electron concentration $n_1(x)$ is redrawn as the solid curve, and the two component concentrations are drawn as dashed curves. Therefore, we have

$$n_{1F}(C) = 0 \quad \text{and} \quad n_{1R}(E) = 0 \tag{9.4.1}$$

for the forward injection and the reverse injection, respectively. The symbols C and E refer to the boundaries of the collector and emitter junctions, respectively.

For transistors with a reasonably high transport efficiency, the excess-carrier concentration $n_1(x)$ can be approximated by a straight line. Therefore, it is a simple matter to decompose $n_1(x)$ as the sum of two straight lines, $n_{1F}(x)$ and $n_{1R}(x)$. Applying the definition of current gain as given in Eq. (9.1.4) to both the forward and reverse injections, we have

$$I_E = I_{ES}\left[\exp\left(\frac{eV_{BE}}{kT}\right) - 1\right] - \alpha_R I_{CS}\left[\exp\left(\frac{eV_{BC}}{kT}\right) - 1\right] \tag{9.4.2}$$

$$I_C = -I_{CS}\left[\exp\left(\frac{eV_{BC}}{kT}\right) - 1\right] + \alpha_F I_{ES}\left[\exp\left(\frac{eV_{BE}}{kT}\right) - 1\right] \tag{9.4.3}$$

Note that the direction of I_E defined in Fig. 9.11a is opposite to assigning current flowing into a junction as the conventional direction. The two proportionality constants I_{ES} and I_{CS} as well as the two current gains α_F and α_R are to be determined experimentally.

If the collector junction is reverse biased and the emitter circuit is open, Eqs. (9.4.2) and (9.4.3) become

$$0 = I_{ES}\left[\exp\left(\frac{eV_{BE}}{kT}\right) - 1\right] + \alpha_R I_{CS} \tag{9.4.4}$$

$$I_C = I_{CS} + \alpha_F I_{ES}\left[\exp\left(\frac{eV_{BE}}{kT}\right) - 1\right] \tag{9.4.5}$$

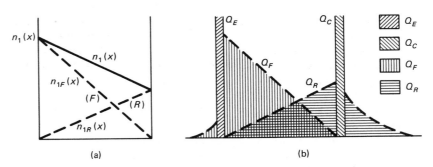

(a)   (b)

Figure 9.13 Schematic diagrams showing (a) the decomposition of minority-carrier (electron) concentration in the base region into two components $n_1 = n_{1F} + n_{1R}$, and (b) the charges Q_F and Q_R injected by the emitter–base junction and by the collector–base junction and the charges Q_E and Q_C stored in the space-charge regions of the two junctions. All the quantities represent the absolute values of the charges, and hence are positive, regardless of the polarities of the charges.

where V_{BE} is the floating potential of the emitter–base junction. Eliminating the exp (eV_{BE}/kT) terms from Eqs. (9.4.4) and (9.4.5), we obtain

$$I_C = I_{C0} = I_{CS} (1 - \alpha_F \alpha_R) \qquad (9.4.6)$$

Similarly, if the emitter junction is reverse biased and the collector circuit is open, then the emitter current is given by

$$I_{E0} = I_{ES}(1 - \alpha_F \alpha_R) \qquad (9.4.7)$$

Both I_{C0} and I_{E0} can be experimentally measured. The reader is asked to show that the current ($I_C = I_E$) under negative bias V_{BC} and open-base circuit conditions is given by

$$I_C = I'_{C0} = \frac{I_{C0}}{1 - \alpha_F} = \frac{1 - \alpha_F \alpha_R}{1 - \alpha_F} I_{CS} \qquad (9.4.8)$$

and the current ($I_E = I_C$) under negative bias V_{BE} and open-base circuit conditions is given by

$$I_E = I'_{E0} = \frac{I_{E0}}{1 - \alpha_R} = \frac{1 - \alpha_F \alpha_R}{1 - \alpha_R} I_{ES} \qquad (9.4.9)$$

Note that I_{C0} of Eq. (9.4.6) and I_{C0} of Eq. (9.1.4) are the same and that Eq. (9.4.8) agrees with Eq. (9.3.11) with $I_B = 0$. From measurements of the four currents, the values of α_R, α_F, I_{ES}, and I_{CS} can be determined. For example, the currents I_{C0} and I'_{C0} are shown in Fig. 9.11b. We should also point out that the transistor of Fig. 9.11 is connected in the common-emitter configuration, with the emitter junction always being forward biased, and that the collector junction is reverse biased for both curves $I_B = 0$ and $I_E = 0$. Therefore, the current gain α shown in Fig. 9.11b is the forward current gain α_F. The values of I_{E0} and I'_{E0} can be obtained similarly in the common-collector configuration.

Another quantity of interest concerning Fig. 9.11b is the value of V_{CE} in the saturation region. For both junctions forwardly biased, it is a good approximation to write Eqs. (9.4.2) and (9.4.3), respectively, as

$$I_E = x - \alpha_R y \qquad (9.4.10)$$

$$I_C = -y + \alpha_F x \qquad (9.4.11)$$

where $x = I_{ES} \exp(eV_{BE}/kT)$ and $y = I_{CS} \exp(eV_{BC}/kT)$. Solving for x and y, expressing $I_E = I_C + I_B$, and realizing that $V_{CE} = V_{BE} - V_{BC}$, we find that

$$(V_{CE})_{\text{sat}} = \frac{kT}{e} \ln \left[\frac{I_{CS}}{I_{ES}} \frac{I_B + (1 - \alpha_R)I_C}{\alpha_F I_B - (1 - \alpha_F)I_C} \right] \qquad (9.4.12)$$

Since the parameters α_F, α_R, I_{ES}, and I_{CS} are known, the values of $(V_{CE})_{\text{sat}}$ and I_C can be found from Eqs. (9.4.12) and (9.3.12) for a given set of values for I_B, V_C, and R.

While the Ebers–Moll model is useful in obtaining the steady-state values of I_B and I_C in the saturation region, the charge-control model is used extensively in describing the transient behavior of a switching transistor. Before we introduce the charge-control model, it may be instructive to make the following observations. First, the current–voltage dependences shown in Eqs. (9.4.2) and (9.4.3) are highly nonlinear. Second, in dealing with transient behaviors, we need to consider the charging and discharging of the stored charges, including the charges associated with ionized impurities in the space-charge regions as well as the charges due to injected minority carriers. The dependence of these charges are also very nonlinear. Therefore, differential equations relating the currents to the time variations of the voltages will be very nonlinear and hence very difficult to solve. In the small-signal analysis presented in Section 9.1, the various current–voltage and charge–voltage relations are linearized. However, the linearized model is no longer applicable to problems involving a large swing of the current.

In the *charge-control model,* the various currents are expressed in terms of the charges involved and their time derivatives. In Fig. 9.13b, we show the relevant charges with Q_E and Q_C representing the space charges associated with the emitter and collector junctions, respectively, and with Q_F and Q_R representing the injected minority carriers due to the forward-biased emitter and collector junctions, respectively. In terms of these charges, the various currents can be expressed as

$$I_C = \frac{Q_F}{t_F} - \frac{dQ_R}{dt} - Q_R\left(\frac{1}{t_R} + \frac{1}{t_{BR}}\right) - \frac{dQ_C}{dt} \qquad (9.4.13)$$

$$I_E = -\frac{Q_R}{t_R} + \frac{dQ_F}{dt} + Q_F\left(\frac{1}{t_F} + \frac{1}{t_{BF}}\right) + \frac{dQ_E}{dt} \qquad (9.4.14)$$

$$I_B = I_E - I_C \qquad (9.4.15)$$

The directions of the currents are defined in Fig. 9.11a for a *n-p-n* transistor. Again we note that $I_E' = -I_E$ is the conventional assignment, that is, the current into the emitter as being positive.

Refer to Eq. (9.2.11) where we defined a transit time t_B across the base region such that the collector current I_C is related to the stored charge Q in the base region by $Q = I_C t_B$. The time constant t_F in Eq. (9.4.13) is similar to t_B except that Q_F in Fig. 9.13b includes both Q_{Fe} of injected electrons into the base region and Q_{Fh} of injected holes into the emitter region. For transistors with reasonable emitter efficiency, Q_{Fh} is much smaller than Q_{Fe}. Therefore, as a first-order approximation, we have

$$t_F \cong \eta_F \frac{W_B^2}{2D} \quad \text{and} \quad t_R = \eta_R \frac{W_B^2}{2D} \qquad (9.4.16)$$

or the same as t_B of Eq. (9.3.7). The factor η_F accounts for the effect of a built-in field in the base region of a diffused transistor. According to Eq. (9.2.14), η_F is smaller than unity for the doping profile given in Fig. 9.9. For current injection from the collector side, the field is opposing and $\eta_R > 1$ is expected. Therefore, two different time constants are used: t_F for Q_F and t_R for Q_R.

The situation with t_R, however, is subject to further examination. We see from Fig. 9.9 that the doping concentration in the collector is kept low to minimize the junction capacitance C_{TC}. A lower collector doping means a smaller ratio of injected carrier concentrations n_p/p_n, resulting in a lower injection efficiency. Furthermore, the thickness of the epitaxial low-doping layer is either comparable to or larger than the base width (Fig. 9.9). The combined effects of the two factors are a substantial contribution from stored charge Q_{Rh} on the collector side to Q_R and a long time for injected holes to diffuse out of the epitaxial n region. Therefore, the value of η_R in Eq. (9.4.16) is expected to be much larger than that based on the effect of an opposing built-in field alone.

Based on the preceding discussions, we see that the terms Q_F/t_F and Q_R/t_R represent, respectively, the collector current due to injection from the emitter junction and the emitter current due to injection from the collector. Since the currents due to hole injection and electron injection add together to constitute the total current, Q_F and Q_R are the numerical sums of the magnitudes of the charges due to injected holes and electrons. In other words, both Q_F and Q_R are always positive. The signs in front of Q_F and Q_R are so assigned that they give the correct directions of current flow, with reference to the directions assigned in Fig. 9.11a. The two terms $Q_F(t_F^{-1} + t_{BF}^{-1})$ and $Q_R(t_R^{-1} + t_{BR}^{-1})$ represent the emitter-injection and collector-injection currents, respectively. Therefore, those two currents and the other two currents should be in the following ratios:

$$\frac{t_F^{-1}}{t_F^{-1} + t_{BF}^{-1}} = \alpha_F \quad \text{and} \quad \frac{t_R^{-1}}{t_R^{-1} + t_{BF}^{-1}} = \alpha_R \qquad (9.4.17)$$

Besides the four components given above, any changes in the stored charges will draw currents. For example, in Fig. 9.11, increases in the injected electron concentration in the base region require the supply of electrons from the emitter, resulting in a current I_E away from the emitter. Since Q_F is the absolute value of the stored charges, a positive dQ_F/dt means a positive I_E according to direction of I_E specified in Fig. 9.11a. Similarly, the quantity $-dQ_R/dt$ represents the component of the collector current caused by increases in Q_R as a result of increased injection from the collector junction. The negative sign accounts for the fact that injections from the collector and emitter junctions are in opposite directions. The two other currents dQ_E/dt and $-dQ_C/dt$ represent the currents caused by increased space charges. Again both Q_E and Q_C are positive quantities because they are the absolute values of space charges.

To illustrate the use of the charge-control model, we refer to Fig. 9.11a. Suppose that the transistor is operated in the forward active region and that the base current is switched instantaneously from I_{B1} to I_{B2}. We want to find the time variation of the collector current. We assume that the change in I_B is small and hence the collector reverse bias is not significantly changed. Therefore, we can set $Q_R = 0$ and neglect dQ_C/dt in Eqs. (9.4.13) and (9.4.14). Furthermore, because the emitter–base junction is forward biased, dQ_E/dt is considered negligible as compared with dQ_F/dt. Applying these conditions to Eq. (9.4.15), we find that

$$I_B = \frac{dQ_F}{dt} + \frac{Q_F}{t_{BF}} \qquad (9.4.18)$$

Solving for Q_F from Eq. (9.4.18) and using $I_C = Q_F/t_F$ from Eq. (9.4.13), we obtain

$$I_C = \frac{t_{BF}}{t_F} \left[I_{B2} + (I_{B1} - I_{B2}) \exp\left(\frac{-t}{t_{BF}}\right) \right] \qquad (9.4.19)$$

According to Eq. (9.4.17), the ratio t_{BF}/t_F is simply β_F.

From the example above, we make the following observations. Because the charges Q_E and Q_C stored in the space-charge region are unrelated to the charged Q_F and Q_R, the approximations of ignoring dQ_E/dt and dQ_C/dt become necessary. The assumption that I_B makes an abrupt change at $t = 0$ is also artificial. Therefore, the purpose of the charge-control model is not to give an accurate description of the time variation of I_C. Intuitively, we expect that any change in Q_F in the base region would take a time of the order of the carrier lifetime. Equation (9.4.19) serves to support this qualitative argument. Application of Eqs. (9.4.13) and (9.4.15) to a switching transistor will require piecewise approximation of constant I_B and I_C in order to get a second-order differential equation in Q_F or Q_R. For further discussions on the subject, the reader is referred to the book by Gray and Searle (P. E. Gray and C. L. Searle, *Electronic Principles: Physics, Models, and Circuits,* John Wiley & Sons, Inc., New York, 1969).

9.5 THE *p-n-p-n* STRUCTURE

Another useful multijunction semiconductor device is the *p-n-p-n* structure shown in Fig. 9.14. This structure can be operated as a two-, three-, or four-terminal device. As a two-terminal device (Fig. 9.14a), the structure has the property of a switch. It possesses a low-impedance on state and a high-impedance off state, and it can change from either one of the states to the other. If a third terminal is attached to the middle *p* region (Fig. 9.14b), the switching of the structure can be controlled by a current passing through the third terminal. Because of low power dissipation, the *p-n-p-n* structure has been extensively used as static switches, power inverters, and dc choppers. The three-terminal device is commercially called the semiconductor controlled rectifier or *thyristor*. It is available with current ratings ranging from few milliamperes to hundreds of amperes and voltage ratings exceeding 1000 V.

Consider the *p-n-p-n* structure under a bias voltage V with the positive polarity applied to the outer *p* region (Fig. 9.14c). At low voltages, the *p-n-p-n* structure can be considered as two transistors, one *p-n-p* and the other *n-p-n* with a common middle region. Note that the middle *n-p* junction is reverse biased, while the two outside *p-n* junctions are forward biased. Figure 9.14c shows an artificial cut of the *p-n-p-n* structure so that the currents associated with either the *p-n-p* or the *n-p-n* section of the structure can be treated in the same manner as they are treated in a transistor. For example, I_{B1} is the collector current of the *p-n-p* transistor, and hence it is equal to

$$I_{B1} = \alpha_1 I_{E1} + I_{C01} \qquad (9.5.1)$$

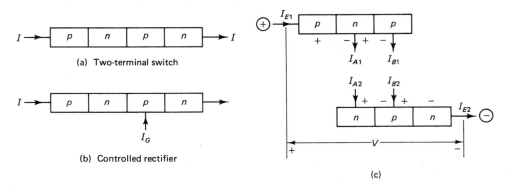

Figure 9.14 Schematic diagrams showing the operations of a *p-n-p-n* diode with (a) the connection as a two-terminal switch, (b) the connection as a controlled rectifier having the middle *p* region used as a control terminal, and (c) the model used in the analysis.

On the other hand, I_{B2} represents the base current in the *n-p-n* transistor,

$$I_{B2} = (1 - \alpha_2)I_{E2} - I_{C02} \qquad (9.5.2)$$

In Eqs. (9.5.1) and (9.5.2), α_1 and α_2 are the respective current gain.

Since no electric connection is made to the middle *p* and *n* regions, $I_{B1} = I_{B2}$ and $I_{E1} = I_{E2} = I$. Equating I_{B1} with I_{B2} gives

$$I = \frac{I_{C0}}{1 - (\alpha_1 + \alpha_2)} \qquad (9.5.3)$$

where $I_{C0} = I_{C01} + I_{C02}$ is the reverse saturation current of the middle *p-n* junction. In Eq. (9.5.3) we have ignored two effects that must be taken into account. First, under high electric fields, there is appreciable carrier multiplication and Eq. (9.5.3) should be replaced by

$$I = \frac{MI_{C0}}{1 - M(\alpha_1 + \alpha_2)} \qquad (9.5.4)$$

where M is the multiplication factor defined in Eq. (8.10.5). Figure 9.15 shows the *I–V* characteristic of a *p-n-p-n* structure. In the low-voltage region 1, the current is essentially that for a transistor operated as a two-terminal device (in the common-emitter configuration with open base). As the bias voltage increases, the multiplication factor increases. When the quantity $M(\alpha_1 + \alpha_2)$ approaches unity, the current increases very rapidly. This corresponds to region 2 in Fig. 9.15.

As the current increases, the current gain $\alpha_1 + \alpha_2$ rises. The rise in α is due to the fact that the ratio I_{rec}/I_{diff} in Fig. 8.37a decreases and has the same physical origin as the initial rise in β with the emitter current in transistors. (For a calculation of the dependence of α on I, see E. S. Yang and N. C. Voulgaris, *Solid-State Electron.*, Vol. 10, p. 641, 1967.) Note that the *I–V* characteristic shows a negative resistance in region 3 of Fig. 9.15. A larger α means a smaller M needed for the same I, resulting in the negative resistance. At a certain current, the value of $(\alpha_1 + \alpha_2)$ may become sufficiently large so that $\alpha_1 + \alpha_2$ may be greater than unity even without multiplication in the middle *n-p* junction. When this happens, the middle junction suddenly switches from a reverse- to a forward-biased condition. This switch corresponds to a transition from region 3 to region 4 of the curve in Fig. 9.15. We should point out that in region 4 of

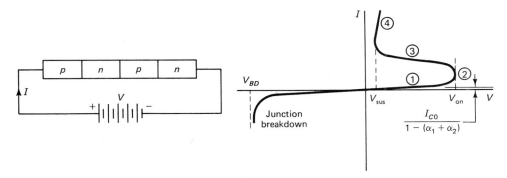

Figure 9.15 The current–voltage characteristic of a *p-n-p-n* diode.

the curve, the voltage across the structure is low and hence multiplication is insignificant. The switching action is entirely due to rise in the value of $\alpha_1 + \alpha_2$.

The physical process during switching can be explained as follows. If I is the current flowing through the p-n-p-n structure, the rate of hole injection from the left into the n region and that of electron injection from the right into the p region of the middle junction are, respectively, $\alpha_1 I/e$ and $\alpha_2 I/e$. Since the current going through the middle junction is also I, only I/e injected carriers out of a total of $(\alpha_1 + \alpha_2)I/e$ carriers will be able to pass through the middle junction per second. The rest of the injected carriers must pile up near the boundary of the middle junction, holes in the n region, and electrons in the p region. The situation is illustrated in Fig. 9.16 for the middle n-p junction. When $\alpha_1 + \alpha_2 < 1$, the minority-carrier concentrations at the junction boundaries A and B are zero (Fig. 9.16a). When $\alpha_1 + \alpha_2 > 1$, $p(x)$ at A and $n(x)$ at B are no longer zero (Fig. 9.16b).

The piling up of the minority carriers at the middle junction produces two effects. First, it reduces the rate of hole diffusion in the n region by reducing dp/dx and the rate of electron diffusion in the p region by reducing dn/dx. Second, it may change the bias condition at the junction. As discussed in Section 8.5 in connection with the recovery phase of a switching diode, the bias of a junction is determined by the concentrations of minority carriers at the junction. The piling up of carriers at the middle junction causes the minority-carrier concentrations to exceed their respective equilibrium concentrations, thus changing the bias condition of the middle junction. The change occurs as soon as $(\alpha_1 + \alpha_2)$ becomes greater than unity. This explains the sharp drop in voltage in region 3 of the I–V curve in Fig. 9.15.

In region 4 of Fig. 9.15, all three junctions are forward biased. If V_1, V_m, and V_2 are, respectively, the bias voltages on the left, middle, and right junctions, the current flowing in the middle junction is

$$I_m = \alpha_1 I_{S1}[\exp(x) - 1] - I_{Sm}[\exp(y) - 1] + \alpha_2 I_{S2}[\exp(z) - 1] \qquad (9.5.5)$$

where $x = eV_1/kT$, $y = eV_m/kT$, and $z = eV_2/kT$. In Eq. (9.5.5), I_{S1}, I_{Sm}, and I_{S2} are the reverse saturation currents of the left, middle, and right junctions. The first term in Eq. (9.5.5) is obtained by considering the action of V_1 alone with $V_2 = V_m = 0$, the second term is due to V_m alone with $V_1 = V_2 = 0$, and the last term is caused by V_2

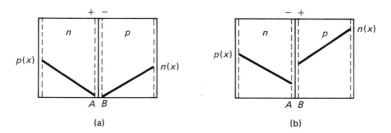

(a)

(b)

Figure 9.16 Schematic diagrams showing (a) the injected hole concentration $p(x)$ in the middle n region, the base of the p-n-p transistor, and the injected electron concentration $n(x)$ in the middle p region, the base of the n-p-n transistor, when the p-n-p-n diode is operated in region 1 in Fig. 9.15, and (b) the corresponding injected hole and electron concentrations when the diode is operated in region 4 of Fig. 9.15, which is similar to the saturation region of a bipolar transistor.

alone with $V_1 = V_m = 0$. In other words, the first term is the result of a forward-biased left junction with the middle junction acting as a collector having a zero bias voltage.

Expressions for the currents I_1 and I_2 through the left and right junctions can be obtained in the same way. From these expressions, the voltage V in region 4 of Fig. 9.15 can be solved in terms of the current I, realizing that $I = I_1 = I_m = I_2$ and $V = V_1 - V_m + V_2$. If it is assumed that all the α's are equal and the two components (hole and electron) of I_{Sm} are equal, we find that

$$V = \frac{kT}{e} \ln \left[\frac{I_{Sm}}{I_{S1} I_{S2}} \frac{(IA_1 + I_{S1})(IA_2 + I_{S2})}{IA_m + I_{Sm}} \right] \tag{9.5.6}$$

where $A_1 = A_2 = (1 - \alpha/2)/(1 - \alpha^2)$ and $A_m = (2\alpha - 1)/(1 - \alpha^2)$. Region 4 of Fig. 9.15, whose I–V characteristic is described by Eq. (9.5.6), represents the on mode of the p-n-p-n structure. Although the avalanche breakdown does not play a role in region 4, it is responsible for bringing about the transition from the off mode into the on mode of operation. Impact ionization is instrumental in increasing the current, which in turn raises the value of $\alpha_1 + \alpha_2$.

The I–V characteristic in regions 1, 2, and 3 can be computed if the values of M as a function of V and the value of α as a function of I are known. The reader is asked to show that the equation

$$\frac{1}{M} = \alpha_1 + \alpha_2 \frac{I + I_G}{I} + \frac{I_{C0}}{I} = f(I, I_G) \tag{9.5.7}$$

applies to the semiconductor controlled rectifier shown in Fig. 9.14b, where I_G is the current flowing into the control gate. In Fig. 9.17 we illustrate the procedure followed by Gibbons in obtaining the graphical solution of the I–V characteristics of the rectifier (J. F. Gibbons, *Proc. IEEE*, Vol. 55, p. 1366, 1967). The left side of Eq. (9.5.7) is a function of V only, whereas the right side of Eq. (9.5.7) is a function of I only. The I–V characteristic is obtained by selecting values of I and V at which the values on both sides of Eq. (9.5.7) are equal.

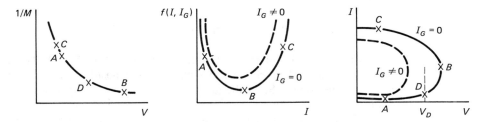

Figure 9.17 Schematic diagrams used to explain the action of the gate current I_G in switching a controlled rectifier from the off state to the on state. A stable operating point exists when the value of M in the diagram on the left and the value of $f(I, I_G)$ in the diagram in the middle are equal. The points A, B, and C are stable. The values of V and I at these points define the I–V characteristic of a p-n-p-n diode. The solid and dashed curves represent situations without and with the gate current, respectively. With $I_G \neq 0$, $V = V_D$ can no longer intersect the I–V curve, and the rectifier switches to the on state.

The curve $f(I, I_G)$ falls as I increases in the region where the term I_{C0}/I is dominant in Eq. (9.5.7). The curve rises beyond the point B because the current gain $\alpha_1 + \alpha_2$ increases with I. The effect of I_G is to increase the value of the function $f(I, I_G)$, thus raising the value of $1/M$ and reducing the value of V at which switching takes place. Note that the point D with applied voltage V_D represents a stable point on the curve with $I_G = 0$. If a control-current pulse I_G of sufficient magnitude is sent to the gate, the device is no longer stable at $V = V_D$ and switching from the off mode to the on mode of operation occurs. The function of a semiconductor controlled rectifier is similar to that of a vacuum-tube thyratron. Further discussion of the subject may be found in the book by Gentry et al. (F. E. Gentry, F. W. Gutzwiler, N. H. Holonyak, and E. E. Van Zastrow, *Semiconductor Controlled Rectifiers,* Prentice-Hall, Inc., Englewood Cliffs, N.J., 1964).

9.6 JUNCTION FIELD-EFFECT (UNIPOLAR) TRANSISTOR

A field-effect transistor is in essence a voltage-controlled variable resistor. Unlike the bipolar transistor, the field-effect (unipolar) transistor operates on electric conduction carried by majority carriers. The structure of a double-diffused field-effect transistor is shown in Fig. 9.18a. The middle p region serves as a conductive channel. Two ohmic contacts are made to the channel, one acting as the source to supply carriers and the other as the drain to collect carriers. A voltage V_D is applied between the source and the drain. For a channel made of p-type semiconductor, the drain is negative biased with respect to the source. For ease of discussion, we concentrate our attention on the region AB, where the action of resistance modulation takes place, and represent the field-effect transistor by an idealized structure (Fig. 9.18b).

The basic idea behind the field-effect transistor is quite simple. The middle p region that forms a conductive channel is only moderately doped as compared to the two n regions; therefore, the depletion (space-charge) region extends almost entirely into the channel. As we increase either V_G or V_D or both, the depletion region widens, reducing the channel width. The progressive change in the effective width of the channel is illustrated in Fig. 9.19. For simplicity, we assume a uniform doping concentration N_a in the p region. Let $-V(x)$ be the voltage at a point x in the channel. The depletion-region depth h is given by Eq. (8.2.15), or

$$h(x) = \left(\frac{2\epsilon}{eN_a}\right)^{1/2} [V_d + V_G + V(x)]^{1/2} \tag{9.6.1}$$

where V_d is the diffusion (or built-in) voltage defined in Eq. (8.1.3). The polarities of V_G and $V(x)$ are defined in Fig. 9.18b with V_G being positive and V_D being negative with respect to the source. Since the junction is biased in the reverse direction, all three voltages are additive (Section 8.2).

The width of the conductive channel is $2(a - h)$, and the hole concentration in the p region is N_a. Thus the current through the channel is

$$I_D = e\mu_p N_a E_x Z[2a - 2h(x)] \tag{9.6.2}$$

where Z is the lateral dimension of the transistor structure. Realizing that $E_x = -d[-V(x)]/dx$, we have

$$\int_0^L I_D \, dx = \int_0^{V_D} 2e\mu_p N_a Z[a - h(x)] \, dV \tag{9.6.3}$$

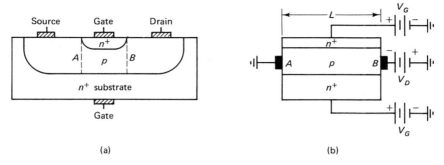

(a)   (b)

Figure 9.18 Diagrams showing (a) the structure of a junction field-effect transistor, and (b) the idealized model used in the analysis. The gate voltage is V_G and the drain voltage is V_D.

Substituting Eq. (9.6.1) into Eq. (9.6.3), we obtain

$$I_D = G_m \left\{ V_D - \frac{2}{3a} \left(\frac{2\epsilon}{eN_a} \right)^{1/2} \left[\left(V_d + V_G + V_D \right)^{3/2} - \left(V_d + V_G \right)^{3/2} \right] \right\} \quad (9.6.4)$$

where $G_m = 2e\mu_p N_a a Z/L$. Equation (9.6.4) holds as long as $h(x) < a$ for $0 < x < L$. We define a *pinch-off voltage* V_p' such that when $V_G = V_p'$ at $x = 0$, the value of $h(0) = a$ in Eq. (9.6.1). Thus

$$V_p' = \frac{eN_a}{2\epsilon} a^2 - V_d = V_p - V_d \quad (9.6.5)$$

The situation shown in Fig. 9.19b corresponds to pinch-off at the drain with $V_G + V_D = V_p'$. In terms of V_p' Eq. (9.6.4) becomes

$$I_D = G_m \left\{ V_D - \frac{2}{3(V_d + V_p')^{1/2}} \right.$$
$$\left. \left[\left(V_d + V_G + V_D \right)^{3/2} - \left(V_d + V_G \right)^{3/2} \right] \right\} \quad (9.6.6)$$

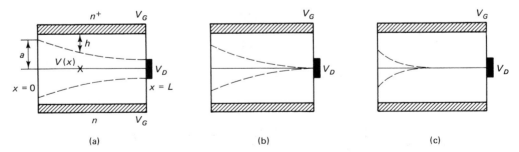

(a)   (b)   (c)

Figure 9.19 Schematic diagrams illustrating the effect of the gate voltage V_G on the width of the conduction channel. The conduction channel is bounded by the dashed lines (Fig. 9.19a). The channel is pinched off at the drain (Fig. 9.19b) and the pinch-off point is moved toward the source end (Fig. 9.19c) as V_G is progressively increased.

Figure 9.20 shows the I_D–V_D characteristics based on Eq. (9.6.6). The numbers for I_D, V_G, and V_D are put in to give the reader some idea of their magnitudes in typical field-effect transistors. The characteristics are similar to those of a vacuum-tube pentode. For low V_D, expansion of the $(V_d + V_G + V_D)^{3/2}$ term in Eq. (9.6.6) gives

$$I_D = G_m \left[1 - \left(\frac{V_d + V_G}{V_d + V_p'} \right)^{1/2} \right] V_D \qquad (9.6.7)$$

In this region, I_D increases linearly with V_D. At $V_D = V_p' - V_G$, the current I_D becomes saturated. We denote the values of V_D and I_D at this point by V_{DS} and I_{DS} in Fig. 9.20. Differentiating I_D with respect to V_D, we find that

$$g_D = \frac{\partial I_D}{\partial V_D} = G_m \left[1 - \left(\frac{V_d + V_G + V_D}{V_d + V_p'} \right)^{1/2} \right] \qquad (9.6.8)$$

As we can see, g_D becomes zero when $V_D = V_p' - V_G$. In other words, the current remains constant at V_{DS} for any incremental change in V_D. The quantity g_D is called the *drain conductance*.

We further define a *transconductance*

$$g_m = \frac{\partial I_D}{\partial V_G} = \frac{-G_m}{(V_d + V_p')^{1/2}} \left[(V_d + V_G + V_D)^{1/2} - (V_d + V_G)^{1/2} \right] \qquad (9.6.9)$$

In terms of g_D and g_m, the ac component of the drain current is

$$i_D = g_m v_G + g_D v_D \qquad (9.6.10)$$

where v_G and v_D are the ac gate and drain voltages. Amplification is possible if a load of resistance R_L is connected to the drain circuit. The unipolar transistor differs from the bipolar transistor in one important aspect. The gate–channel junction is reverse biased under normal operating conditions; hence it offers a very high input impedance which is desirable for certain low-frequency applications.

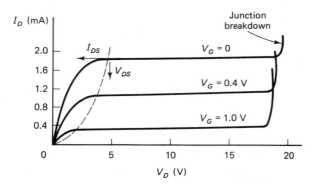

Figure 9.20 The drain-current I_D versus drain-voltage V_D characteristics for several values of the gate voltage V_G.

9.7 METAL–SEMICONDUCTOR CONTACT AND MEASUREMENT OF BARRIER HEIGHT

In Section 4.6 we showed that when two different metals were connected electrically, a potential difference was established between the vacuum levels of the two metals. This potential difference is known as the contact potential. A contact potential also exists between the vacuum levels of a metal and a semiconductor. The situation is illustrated in Fig. 9.21a. The energy difference between the vacuum level and the bottom of the conduction band is called the *electron affinity* and is denoted by $e\chi$. Thus the contact potential ϕ_{MS} between a metal and a semiconductor is given by

$$e\phi_{MS} = e\phi_M - e\chi - (\mathscr{E}_c - \mathscr{E}_f) \qquad (9.7.1)$$

where $\mathscr{E}_W = e\phi_M$ is the work function of the metal and $\mathscr{E}_f$ is the Fermi level.

When the metal and the semiconductor are brought into intimate contact, the energies of the two sides adjust themselves so that the two vacuum levels as well as the two Fermi levels become equal. This equalization establishes a space-charge region in the semiconductor near the metal–semiconductor interface boundary. The situation is illustrated in Fig. 9.21b. Electrons in going from the metal to the semiconductor must overcome a potential barrier. The barrier height is equal to

$$e\phi_B = e\phi_M - e\chi \qquad (9.7.2)$$

Equation (9.7.2) applies to an ideal metal–semiconductor contact.

As pointed out in Section 7.9, a real semiconductor surface is covered with an oxide layer. Referring to Fig. 7.19, we see that a space-charge region may have existed near the semiconductor–oxide interface even before the metal is brought into contact. If a metal–semiconductor contact is indeed made of metal–oxide–semiconductor, a potential drop eV_{ox} exists across the oxide. Because of this drop, Eq. (9.7.2) should be replaced by

$$e\phi_B = e\phi_M - eV_{ox} - e\chi \qquad (9.7.3)$$

The situation is illustrated in Fig. 9.22.

Since the quantity eV_{ox} in Eq. (9.7.3) is an unknown quantity, being dependent on semiconductor-surface conditions, an experimental determination of the barrier height becomes necessary. The most direct and accurate method of determining the barrier height is the photoelectric measurement (Fig. 9.23a). When a monochromatic light of photon energy $h\nu$ is incident on a metal, a photocurrent may result if $h\nu > e\phi_B$,

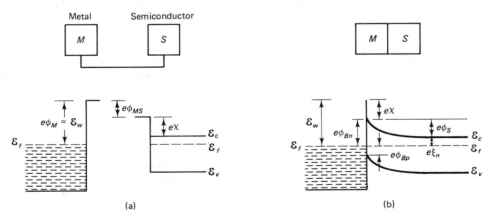

Figure 9.21 Energy-band diagrams for (a) a metal and a semiconductor connected electrically but separated physically, and (b) a metal–semiconductor contact. The potential barrier is ϕ_{Bn} for electrons and ϕ_{Bp} for holes.

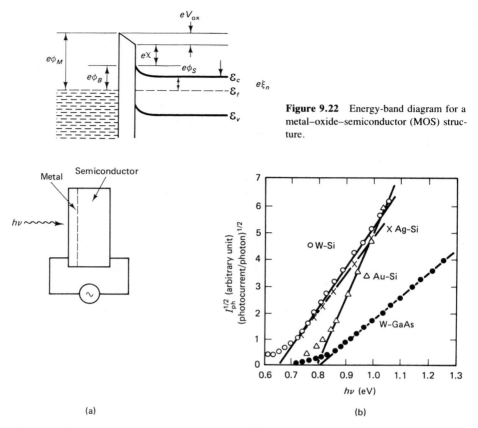

Figure 9.22 Energy-band diagram for a metal–oxide–semiconductor (MOS) structure.

(a)

(b)

Figure 9.23 Photoelectric measurement to determine the barrier height at a metal–semiconductor contact with diagram (a) showing the experimental arrangement and diagram (b) showing the experimental results. (C. R. Crowell, W. G. Spitzer, L. E. Howarth, and E. E. LaBate, *Phys. Rev.*, Vol. 127, p. 2006, 1962.)

so that electrons in the metal are given sufficient energy to overcome the potential barrier. The photoresponse I_{ph} is defined as the photocurrent generated per absorbed photon. For $hv - e\phi_B > 3kT$, the quantity I_{ph} is found to be proportional to $(hv - e\phi_B)^2$, or

$$I_{ph} = A(hv - e\phi_B)^2 \tag{9.7.4}$$

where A is a proportionality constant. (For a derivation of I_{ph}, see R. H. Fowler, *Phys. Rev.*, Vol. 38, p. 45, 1931. Also refer to C. R. Crowell, W. G. Spitzer, L. E. Howarth, and E. E. LaBate, *Phys. Rev.*, Vol. 127, p. 2006, 1962 for experimental details.) The experimental results reported by Crowell et al. are shown in Fig. 9.23b. The two extrapolated values are the barrier heights.

The barrier height can also be determined from measurements of the interface capacitance as a function of the applied voltage. If a voltage V is applied across a metal–semiconductor junction with the positive polarity connected to the semiconductor, the

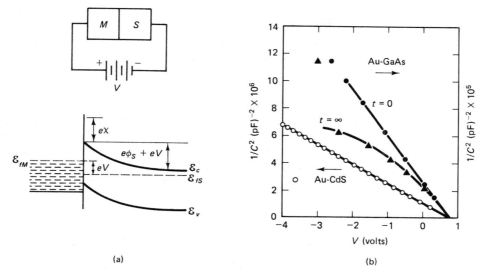

Figure 9.24 Capacitance-voltage measurement to determine the barrier height at a metal–semiconductor interface with diagram (a) showing the energy bands near the interface and diagram (b) showing the experimental results. (A. M. Goodman, *J. Appl. Phys.*, Vol. 34, p. 329, 1963; R. R. Senechal and J. Basinski, *J. Appl. Phys.*, Vol. 39, p. 4851, 1968.)

TABLE 9.1 MEASURED VALUES OF THE SCHOTTKY BARRIER HEIGHT

Semiconductor	Metal	Barrier height, $\phi_B{}^a$(V)		
		I	II	III
n-CdS (vac-cleaved)	Au	0.78	0.80	
	Cu	0.36	0.35	
n-CdS (chem-treated)	Au	0.68	0.66	0.68
	Cu	0.50	0.41	0.47
n-GaAs (vac-cleaved)	Au	0.90	0.95	
	Cu	0.82	0.87	
p-GaAs (vac-cleaved)	Au	0.42	0.48	
	Pt		0.48	
n-Ge (vac-cleaved)	Au		0.45	
	Al		0.48	
n-Si (chem-treated)	Au	0.78	0.80	0.79
	W	0.65	0.65	0.67
p-Si (chem-treated)	Au			0.25

[a]I, photoelectric measurement; II, capacitance measurement; III, measurement of *I–V* characteristic.

potential energy of electrons in the semiconductor is lowered with respect to that in the metal. As a result, the surface barrier (Fig. 9.24a) becomes $e\phi_s + eV$. Note that the quantity ϕ_s corresponds to the built-in potential V_d in a p-n junction. Thus, according to Eq. (8.2.13), the metal–semiconductor interface capacitance is

$$\frac{C}{A} = \left[\frac{e\epsilon N_{\mathrm{I}}}{2(\phi_S + V)}\right]^{1/2} \quad \mathrm{F/m^2} \tag{9.7.5}$$

Figure 9.24b shows the experimental results reported by Goodman and by Senechal and Basinski (A. M. Goodman, *J. Appl. Phys.*, Vol. 34, p. 329, 1963; R. R. Senechal and J. Basinski, *J. Appl. Phys.*, Vol. 39, p. 4581, 1968). The intercept for infinite C gives the value of ϕ_S in Eq. (9.7.5). A compilation of the measured values of barrier heights may be found in articles by Mead and Spitzer (C. A. Mead and W. G. Spitzer, *Phys. Rev.*, Vol. 134, p. A713, 1964; W. G. Spitzer and C. A. Mead, *J. Appl. Phys.*, Vol. 34, p. 3061, 1963). Selected values are listed in Table 9.1.

9.8 CURRENT CONDUCTION IN METAL–SEMICONDUCTOR BARRIER

The electric current in a metal–semiconductor barrier is conducted by majority carriers. The nature of the current transport is similar to that of thermionic emission. We note that the barrier height $e\phi_B$ is around 0.6 eV (Table 9.1). The value of $e\phi_s$ is even less by an amount $\mathscr{E}_c - \mathscr{E}_f$. In contrast, the work function $\mathscr{E}_w$ is about 2 eV in alkali metals and about 4.5 eV in other metals (Table 4.2). Because of the low barrier, the thermionic-emission current in a metal–semiconductor barrier is appreciable even at room temperature. Another significant difference is that the barrier height on the semiconductor side can be controlled by an applied voltage. As a result of electron emission from both the metal and the semiconductor, the emission current density is given by

$$J = A \exp\left(-\frac{e\phi_B}{kT}\right)\left[\exp\left(\frac{eV}{kT}\right) - 1\right] \tag{9.8.1}$$

where A is a proportionality constant and V is the voltage applied across the metal–semiconductor junction. The values of ϕ_B listed in Table 9.1 under (III) are obtained from an analysis of the I–V characteristics (C. R. Crowell, J. C. Sarace, and S. M. Sze, *Trans. Metall. Soc. AIME*, Vol. 233, p. 478, 1965).

The derivation of Eq. (9.8.1) can be carried out in a manner similar to that outlined in Section 4.7 for the derivation of Eq. (4.7.9). First, we treat the electron emission from the semiconductor. The density of electrons in the conduction band is, according to Eq. (6.3.6), given by

$$\rho(\mathscr{E}) = 4\pi\left(\frac{2m^*}{h^2}\right)^{3/2}\mathscr{E}_1^{1/2}\exp\left(-\frac{\mathscr{E}_c - \mathscr{E}_f}{kT}\right)\exp\left(-\frac{\mathscr{E}_1}{kT}\right) \tag{9.8.2}$$

where $\mathscr{E}_1$ is the kinetic energy with

$$\mathscr{E}_1 = \frac{1}{2m^*}(p_x^2 + p_y^2 + p_z^2) \tag{9.8.3}$$

Note that under the bias condition shown in Fig. 9.24, the barrier height for conduction-band electrons is $e\phi_s' = e\phi_s + eV$. Therefore, the emission current density from the semiconductor to the metal is

$$J_{S \to M} = \int_{e\phi'_s}^{\infty} e v_x \rho(\mathscr{E}) \, d\mathscr{E} \tag{9.8.4}$$

which replaces Eq. (4.7.2). In Eq. (9.8.4), $v_x = \partial\mathscr{E}/\partial p_x$.

Again, the key point in the derivation is recognition of the fact that since the potential barrier exists in only one direction, which is chosen to be the x direction, p_y and p_z remain unchanged. Letting $\mathscr{E}_\perp = p_x^2/2m^*$, we may write Eq. (9.8.4) as

$$J_{S \to M} = \frac{2e}{h^3} \exp\left(-\frac{\mathscr{E}_c - \mathscr{E}_f}{kT} \right)$$

$$\int_{e\phi'_s}^{\infty} \exp\left(-\frac{\mathscr{E}_\perp}{kT} \right) d\mathscr{E}_\perp \int\int_{-\infty}^{\infty} \exp\left(-\frac{p_y^2 + p_z^2}{2m^*kT} \right) dp_y \, dp_z \tag{9.8.5}$$

Equation (9.8.5) replaces Eq. (4.7.8) on account of the difference in the barrier height and the difference in the $\rho(\mathscr{E})$. Carrying out the integration in Eq. (9.8.5), we find that

$$J_{S \to M} = \frac{4\pi e m^*}{h^3} (kT)^2 \exp\left(-\frac{e\phi_B}{kT} \right) \exp\left(\frac{-eV}{kT} \right) \tag{9.8.6}$$

in view of the fact that $e\phi'_s = e\phi_s + eV$ and $e\phi_B = e\phi_s + \mathscr{E}_c - \mathscr{E}_f$. Counting electron emission from the metal, we have

$$J = J_{M \to S} - J_{S \to M} \tag{9.8.7}$$

Since both emission currents are carried by electrons and they are in opposite directions, the total current density J is a difference of the two densities.

Note that the barrier height ϕ_B on the metal side remains unchanged under the applied voltage V. Therefore, the current density $J_{M \to S}$ should be a constant independent of V. This constant can be evaluated at $V = 0$. For $V = 0$ in Eq. (9.8.7), J should be zero. Otherwise, we would have perpetual current flow, which is a thermodynamic impossibility. Applying the condition $J = 0$ at $V = 0$ to Eq. (9.8,7), we obtain

$$J = \frac{4\pi e m^*}{h^3} (kT)^2 \exp\left(-\frac{e\phi_B}{kT} \right) \left[1 - \exp\left(-\frac{eV}{nkT} \right) \right] \tag{9.8.8}$$

The bias condition shown in Fig. 9.24a with the positive polarity connected to the semiconductor represents a reverse-biased condition. Therefore, Eq. (9.8.8) applies to a reverse-biased metal–semiconductor junction. On the other hand, Eq. (9.8.1) is for a forward-biased junction. The two equations are the same except for the bias condition. Also, a factor n, called the *ideality factor,* is added to Eq. (9.8.8) to account for possible derivations from an ideal Schottky-barrier diode.

Several comments about Eq. (9.8.8) are in order. First, it is assumed that the oxide layer is thin, so that electrons can tunnel through the layer. Second, electron collision in the surface-space-charge region is ignored. The condition for the validity of the latter assumption is that $E_{max} > kT/el$, where E_{max} is the maximum electric field in the depletion region and l is the electron mean free path (H. A. Bethe, ''Theory of the Boundary Layer of Crystal Rectifiers,'' M.I.T. Radiation Laboratory, *Report 43-12,* 1942). On the other hand, if electron collision, instead of electron emission, is the dominant mechanism that limits the current flow, then an analysis of the current flow must be based on the diffusion of carriers in the depletion region (W. Schottky, *Naturwissenschaften*, Vol. 26, p. 843, 1938). An attempt has been made to incorporate the essential features of the emission and diffusion theories (C. R. Crowell and S. M. Sze, *Solid-State Electron.*, Vol. 9, p. 1035, 1966). The result of such an analysis shows that the current expression shown in Eq. (9.8.1) is of the correct form but the constant A is modified.

In concluding this section we should point out that Eq. (9.8.1) applies only to an ordinary metal–semiconductor contact. There are other special types of metal-semiconductor contacts to which Eq. (9.8.1) does not apply. A point-contact diode is made by pressing a small metal wire with a sharp point against a semiconductor. The electric contact is usually formed by passing a large current through the diode by electric discharge processes. During the discharge, a small *p-n* junction actually may be formed. Such a metal–semiconductor contact behaves like a *p-n* junction. In contrast to rectifying contacts, metal and semiconductor can also form ohmic contacts in the form of metal-n^+-n or metal-p^+-p arrangements. Ohmic contacts to Ge and Si are made in practice by first evaporating Sn-Sb alloy onto the semiconductor and then alloying Sn into Ge or Si at the respective eutectic temperature.

A common variation of the ordinary metal–semiconductor contact is the addition of an epitaxial layer in between the metal and the semiconductor arranged in either metal-p-p^+ or metal-n-n^+ form. The middle *p* or *n* region is very lightly doped. Two extreme cases are of practical interest. One extreme case is the one in which the epitaxial layer is so thin that the depletion layer actually reaches the heavily doped p^+ or n^+ region. Such a contact is known as the *Mott barrier*. Only slight modification of Eq. (9.8.1) is needed for a Mott barrier. For further discussions, the reader is referred to the book on rectifying contacts by Henisch (H. K. Henisch, *Rectifying Semiconductor Contacts*, Clarendon Press, Oxford, 1957). The other extreme case is the space-charge-limited diode in which the epitaxial layer is comparatively thick. The current flow in such a diode is limited by the drift of carriers across the epitaxial layer but not by the emission of carriers from the semiconductor (or the metal). Further, since the impurity concentration in the epitaxial layer is very low, carriers flowing through the layer affect the field distribution in the layer. The situation is very much like that in a space-charge-limited vacuum tube. For a plane-parallel geometry, the current density is proportional to V^2. The reader is referred to the book by Lampert, and the article by Buget and Wright for an analysis of the space-charge-limited solid-state diode (M. A. Lampert *Injection Currents in Solids*, Academic Press, Inc., New York, 1965; U. Buget and G. T. Wright, *Solid-State Electron.*, Vol. 10, p. 199, 1967).

9.9 METAL–INSULATOR–SEMICONDUCTOR STRUCTURE

In our discussion of a metal–semiconductor contact in Section 9.7, we ignored the oxide layer entirely. The treatment is valid if the layer is very thin so that electrons can tunnel through it. In certain applications to be presented here, we need a relatively thick oxide layer to block the conduction current. The oxide layer is usually grown through the thermal oxidation process (Section 8.7). The structure with an oxide layer separating a metal and a semiconductor is known as the MOS (metal–oxide–semiconductor) structure. The term "MIS structure" is often used interchangeably with the term "MOS structure" because the oxide layer serves as an insulator in the structure.

The MIS structure was first used in the study of surface states (L. M. Terman, *Solid-State Electron.*, Vol. 5, p. 285, 1962; K. Lehovec and A. Slobodskoy, *Phys. Status Solidi*, Vol. 3, p. 447, 1963). Such a study has been useful toward understanding certain aspects of device performance. (For a comprehensive review of the subject, the reader is referred to A. S. Grove, *Physics and Technology of Semiconductor Devices,* John Wiley & Sons, Inc., New York, 1967; and to S. M. Sze, *Physics of Semiconductor Devices,* John Wiley & Sons, Inc., New York, 1981.) The basic physics concerning the MIS structure was actually expounded in Section 7.9. To refresh our memory, we redraw Fig. 7.19 as Fig. 9.25, in which the relevant energy levels of the metal are added. The surface condition of a semiconductor can be changed from the one shown in Fig. 7.19a to that shown in Fig. 7.19b by exposing the semiconductor to different ambient gases. The reason for the change is that different chemisorbed ions may have different sticking power and may carry opposite charges.

As we recall our discussion of the field-effect experiment, the surface condition of a semiconductor can also be changed by induced charges. When a positive voltage is applied to the semiconductor with respect to the metal (Fig. 9.25a), positive charges are induced in the surface space-charge region. If the semiconductor is *n* type, a surface inversion layer with *n*-type bulk and *p*-type surface may be formed at a sufficiently high bias voltage. On the other hand, an accumulation of majority carriers (electrons) occurs near the surface with *n*-type bulk and n^+ surface (Fig. 9.25b) if a negative voltage is applied to the semiconductor.

One of the possible applications of an MIS structure is in integrated circuits as a voltage-controlled variable capacitor. In Fig. 9.26 we draw the equivalent circuit for an

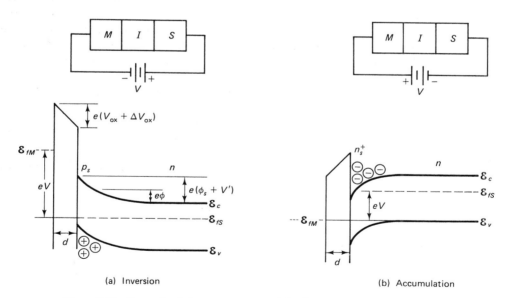

(a) Inversion
(b) Accumulation

Figure 9.25 Energy-band diagrams near a metal–insulator-(*n*-type)–semiconductor interface showing the formation of (a) an inversion layer and (b) an accumulation layer through the application of a gate voltage of proper polarity. The applied voltage *V* is equal to $\Delta V_{ox} + V'$, where ΔV_{ox} is the change in voltage drop across the oxide and V' is the added voltage drop across the space-charge region.

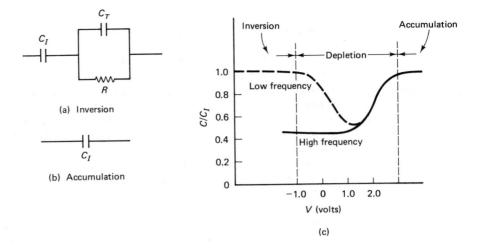

Figure 9.26 Diagrams showing the capacitance associated with a MI(n-type)S interface with (a) an inversion layer and (b) an accumulation layer, and (c) the expected behavior of capacitance as a function of the gate voltage V. The solid and dashed curves are for measurements at high frequency (>1 MHz) and at low frequency (<100 Hz), respectively.

MIS structure. The oxide layer, which serves simply as a dielectric, is represented by the capacitance C_I. The inversion layer of Fig. 9.25a behaves like a one-sided p-n junction. To transport carriers to the semiconductor–oxide interface, a current must flow through the inversion layer. Like a p-n junction, a surface inversion layer can be represented by a junction capacitance C_T and an effective resistance R in parallel (Fig. 9.26a). On the other hand, for the MIS structure having an accumulation layer (Fig. 9.25b), the only capacitance associated with the structure is C_I (Fig. 9.26b).

The reader is asked to show that the equivalent capacitance of the parallel combination of C_T and R is

$$C'_T = \frac{1 + \omega^2 C_T^2 R^2}{\omega^2 C_T^2 R^2} C_T \qquad (9.9.1)$$

At low frequencies such that $\omega C_T R \ll 1$, $C'_T \to \infty$; thus the total capacitance in Fig. 9.26a is

$$C = \frac{C_I C'_T}{C_I + C'_T} \cong C_I \qquad (9.9.2)$$

At high frequencies with $\omega C_T R \gg 1$, $C'_T \cong C_T$; hence we have

$$C = \frac{C_I C'_T}{C_I + C'_T} \cong \frac{C_I C_T}{C_I + C_T} \qquad (9.9.3)$$

The variation of C with the applied voltage is illustrated in Fig. 9.26c for an n-type semiconductor. The polarity of V is so chosen that it is positive if the positive side is connected to the metal gate. Therefore, for an appreciable $V(+)$, an accumulation layer forms. In between the accumulation and inversion regions in Fig. 9.26c lies the depletion region for which both the electron and hole concentrations near the surface are considerably less than the majority-carrier concentration in the bulk. Equations (9.9.2) and (9.9.3) apply to the depletion region as well as the inversion region, whereas $C = C_I$ is for the accumulation region.

Let A be the cross-sectional area of the MIS structure, d be the thickness of the oxide layer, and W be the width of the inversion layer. According to Eq. (8.2.13), $C_T = \epsilon A / W$. Hence at high frequencies,

$$C = \frac{C_I C_T}{C_I + C_T} = \frac{\epsilon' \epsilon A}{\epsilon' W + \epsilon d} \tag{9.9.4}$$

where ϵ' is the dielectric constant of the oxide layer. The quantity W reaches a maximum value when the hole concentration p_s at the surface is equal to the electron concentration n in the bulk. Once this situation is reached, any further increase in V will not appreciably raise the value of W. Referring to Fig. 9.25a, we see that an increase in V results in a further increase in the value of $E = -dV/dx$ at the interface. If the value of dV/dx is already high, the resultant change in Δx is comparatively small for the same change in ΔV.

The charge density in the surface-space-charge region shown in Fig. 9.25a is

$$\rho = e(N_d + p - n) \tag{9.9.5}$$

which has a maximum value

$$\rho_{\max} = e(N_d + p_s) \tag{9.9.6}$$

at the semiconductor–metal interface. As pointed out earlier, the maximum value of W is reached when $p_s = n$ (bulk) $= N_d$. This means that the value of ρ in Eq. (9.9.5) starts with a value of zero and reaches a maximum value of $2N_d$. We further note that the quantities p and n in Eq. (9.9.5) change very rapidly with distance through their dependence on ϕ, (Fig. 9.25a), but the quantity N_d remains constant. Because ρ changes from 0 to eN_d and from eN_d to $2eN_d$ within a very short distance but ρ takes the value eN_d for practically the whole length of the space-charge region, we may approximate Eq. (9.9.5) by

$$\rho = eN_d \tag{9.9.7}$$

With a constant ρ, Poisson's equation can easily be solved. The result is

$$\phi_s + V' = \frac{eN_d}{2\epsilon} W^2 \tag{9.9.8}$$

where $\phi_s + V'$ is the potential barrier across the space-charge region and V' is the added voltage drop across the space-charge region. Note that the carrier concentrations on the two sides of a potential barrier are given by Eq. (8.1.3), which for surface inversion takes a value ϕ_{SI} given by

$$\phi_{SI} = \phi_s + V' = \frac{kT}{e} \ln \frac{n}{n_s} = \frac{kT}{e} \ln \frac{N_d p_s}{n_i^2} \tag{9.9.9}$$

We set $p_s = n$ (bulk) $= N_d$ in Eq. (9.9.9). From Eqs. (9.9.9) and (9.9.8), we obtain

$$W_{\max} = \left(\frac{2\epsilon}{eN_d} \frac{2kT}{e} \ln \frac{N_d}{n_i} \right)^{1/2} \tag{9.9.10}$$

Thus the ratio of the maximum to minimum value of C is

$$\frac{C_{max}}{C_{min}} = 1 + \frac{\epsilon'}{\epsilon} \frac{W_{max}}{d} \qquad (9.9.11)$$

Taking $N_d = 5 \times 10^{15}$ cm^{-3}, $n_i = 1.5 \times 10^{10}$ cm^{-3} (Si), $\epsilon = 12\epsilon_0$ (Si), $\epsilon' = 3.8\epsilon_0$ (SiO$_2$), and $d = 10^{-7}$ m in Eqs. (9.9.10) and (9.9.11), we find $W = 5.7 \times 10^{-7}$ m and $C_{max}/C_{min} = 1.8$. Curves similar to those shown in Fig. 9.26c have been calculated theoretically and observed experimentally (A. Goetzberger, *Bell Syst. Tech. J.*, Vol. 45, p. 1097, 1966; A. S. Grove et al., *J. Appl. Phys.*, Vol. 33, p. 2458, 1964; *Solid-State Electron.*, Vol. 8, p. 145, 1965).

9.10 METAL–OXIDE–SEMICONDUCTOR FIELD-EFFECT TRANSISTOR

One device most widely used in silicon-based integrated circuits is the metal–oxide–semiconductor field-effect transistor (MOSFET). The success of the device is attributable to several technologically important factors. First, silicon can be thermally oxidized to produce a stable oxide which is an excellent insulator. Second, the surface-state density at the silicon–oxide interface is sufficiently low to ensure reproducibility. Third, being planar, the structure is amenable to large-scale integration. An example of a *p*-channel MOSFET is shown in Fig. 9.27a, using an MIS structure as the gate. The device is also known as an insulated-gate field-effect transistor (IGFET). The word "insulated" is used to indicate the fact that an ideal MIS structure does not allow any conduction current. If a sufficient gate voltage V_G with a proper polarity is applied, an inversion layer is formed. The formation of the inversion layer (a *p*-type surface layer on an *n*-type substrate) establishes a conductive channel between the source and the drain (the two p^+ regions in Fig. 9.27b). Let $V(x)$ be the voltage at a point x in the channel. The voltage drop at the point x across the oxide layer is $V_G - V(x)$. The value of $V(x)$ changes from 0 at the source to $-V_D$ at the drain.

From Gauss's theorem, the field E across the oxide layer is related to the surface charge density ρ_s (C/m^2) by $E = \rho_s/\epsilon'$, where ϵ' is the dielectric constant of the oxide layer. The total induced charge for an MIS gate of cross-sectional area A is equal to

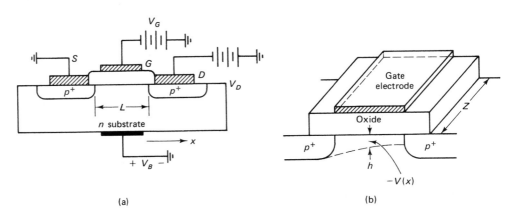

(a)   (b)

Figure 9.27 Schematic diagrams showing (a) the structure and the bias condition, and (b) the geometry and the channel formation of a MOSFET.

$Q = \rho_s A$. The induced charge Q actually consists of two parts: Q_1 stored in the surface p-n junction region and Q_2 in the p^+ channel. Only the part Q_2 participates in the conduction process. Similarly, we separate ρ_s into two parts ρ_{s1} and ρ_{s2}. The part ρ_{s1} is given by

$$\rho_{s1} = eN_d W \tag{9.10.1}$$

where W is the width of the inversion layer. Therefore, the part ρ_{s2} contributing to conduction is

$$\rho_c = \rho_{s2} = \rho_s - \rho_{s1} = \epsilon' \frac{-V_G + V(x)}{d} - eN_d W \tag{9.10.2}$$

The dimensions of the channel are defined by three quantities: the length L (Fig. 9.27a), the depth h (Fig. 9.27b), and the lateral dimension Z (Fig. 9.27b). The current I_D going through the p channel is equal to

$$I_D = \int_0^h JZ \, dy = \int_0^h e p \mu_p E_x Z \, dy \tag{9.10.3}$$

Noting that the hole concentration p (m^{-3}) at the surface is related to ρ_{s2} (C/m^2) through

$$\rho_c = \rho_{s2} = \int_0^h ep \, dy \tag{9.10.4}$$

Using Eqs. (9.10.4) and (9.10.2) in Eq. (9.10.3), we find that

$$I_D = Z\mu_p E_x \left[\epsilon' \frac{-V_G + V(x)}{d} - eN_d W \right] \tag{9.10.5}$$

Since $E_x = d[-V(x)]/dx$, Eq. (9.10.5) becomes

$$I_D \, dx = \frac{\epsilon' Z \mu_p}{d} [V_G - V(x) + V_0] \, dV(x) \tag{9.10.6}$$

where $V_0 = eN_d W d/\epsilon'$. The physical meaning of V_0 becomes clear in section 9.11.

Integrating Eq. (9.10.6) and using the limits of integration $V(x) = 0$ at $x = 0$ and $V(x) = V_D$ at $x = L$, we find that

$$I_D = \frac{\epsilon' Z \mu_p}{2Ld} (2V_G + 2V_0 - V_D)V_D \tag{9.10.7}$$

Like the junction field-effect transistor, the insulated-gate field-effect transistor is represented in operation by two conductances. The channel conductance is

$$g_D = \frac{\partial I_D}{\partial V_D} = \frac{\epsilon' Z \mu_p}{Ld} (V_G + V_0 - V_D) \tag{9.10.8}$$

and the transconductance is

$$g_m = \frac{\partial I_D}{\partial V_G} = \frac{\epsilon' Z \mu_p}{Ld} V_D \tag{9.10.9}$$

The I_D-V_D characteristics of an IGFET structure are similar to those of an FET structure shown in Fig. 9.20 except that the polarity and the scale of the bias voltages may be different. The saturation region is defined by V_{DS} such that $g_D = 0$ at $V_D = V_{DS}$ with

$$V_{DS} = V_G + V_0 \qquad (9.10.10)$$

The current I_D at $V_D = V_{DS}$ is equal to

$$I_{DS} = \frac{\epsilon' Z \mu_p}{2Ld} V_{DS}^2 \qquad (9.10.11)$$

Both quantities V_{DS} and I_{DS} are indicated in Fig. 9.20 except that both V_G and V_D are negative for the present case.

In concluding our discussion on the MOSFET structure, we should like to add the following comments. The mobility of carriers in the inversion layer may be substantially lower than that of carriers in the bulk semiconductor because carriers near the surface suffer additional scattering due to the surface electric field. The effect of surface scattering was first treated by Schrieffer and has been reviewed by Frankl (J. R. Schrieffer, *Phys. Rev.*, Vol. 97, p. 641, 1955; D. R. Frankl, *Electrical Properties of Semiconductor Surfaces*, Pergamon Press Ltd., Oxford, p. 93, 1967). During the early development of the MIS structure, one major difficulty was that the capacitance–voltage curve (Fig. 9.26c) was subject to drift under constant bias at elevated temperature. The drift was found to be caused by sodium contamination of the oxide layer (E. Yon, W. H. Ko, and A. B. Kuper, *IEEE Trans. Electron Devices*, Vol. ED-13, p. 276, 1966). When appropriate precautions in device fabrication are undertaken to eliminate possible impurity-ion contamination, the MIS structure can be made stable at elevated temperatures (P. Lamond, J. Kelley, and M. Papkoff, *Electro-Technology*, p. 40, 1965). The effect of ion drift on the MIS characteristics was discussed by Snow et al. (E. H. Snow, A. S. Grove, B. E. Deal, and C. T. Sah, *J. Appl. Phys.*, Vol. 36, p. 1664, 1965). Finally, we should mention that the equivalent circuit of an MIS structure was discussed by Ihantola and Moll (H. K. Ihantola and J. L. Moll, *Solid-State Electron.*, Vol. 7, p. 423, 1964). The use of Schottky barrier contacts for the source and drain was investigated by Lepselter and Sze, and the operation of the IGFET in the common-gate configuration was discussed by Lukes (M. P. Lepselter and S. M. Sze, *Proc. IEEE*, Vol. 56, p. 1400, 1968; Z. Lukes, *Solid-State Electron.*, Vol. 9, p. 21, 1966).

9.11 REFINED MOSFET ANALYSES: IMPORTANT DEVICE PARAMETERS

The analysis of a MOSFET given in Section 9.10 is based on an idealized model on several accounts. Let us examine the situation carefully by referring to the energy-band diagram of a MIS structure shown in Fig. 9.28a. Equating the energies, we find that

$$e\phi_M - eV_G = eV_{ox} + e\chi + e\phi_s + e\xi_n \qquad (9.11.1)$$

We define a quantity called the *flat-band voltage* V_{FB} such that $\phi_s = 0$ and hence $V_{ox} = 0$ for $V_G = V_{FB}$. From Eq. (9.11.1), we have

$$e\phi_M - eV_{FB} = e\chi + e\xi_n \qquad (9.11.2)$$

Subtracting Eq. (9.11.2) from Eq. (9.11.1), we find that

$$V_{ox} = V_{FB} - V_G - \phi_s \qquad (9.11.3)$$

Therefore, the total surface-charge density ρ_s on the semiconductor side is

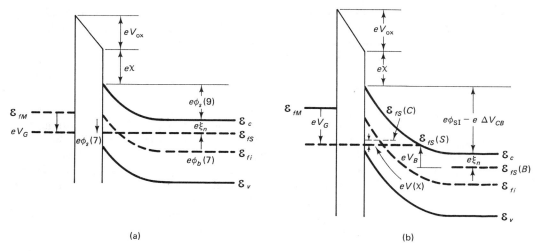

Figure 9.28 Energy-band diagram of a MIS structure under a negative gate voltage ($V_G = -$) with (a) $V_B = 0$ and $V_D = 0$, and (b) $V_B = +$ and $V_D = -$. The potential ϕ is defined as $e\phi = \mathscr{E}_c - \mathscr{E}_c$ (bulk), which is the same as in Figs. 9.22, 9.24, and 9.25, but is different from that defined in Fig. 7.20. The two quantities are related by $e\phi(9) = e\phi_b(7) - e\phi(7)$, where the number in parentheses refers to the chapter in which ϕ is used. A quantity defined in Fig. 9.28a is positive if the arrow is pointed upward. In Fig. 9.28b, the letters S, C, and B in parentheses refer to $\mathscr{E}_{fs}$ at the source, in the channel, and of the substrate, respectively. Also $eV(x) = \mathscr{E}_{fs}(S) - \mathscr{E}_{fs}(C)$.

$$\rho_s \ (C/m^2) = \frac{\epsilon'(V_{FB} - V_G - \phi_s)}{d_{ox}} \tag{9.11.4}$$

where d_{ox} is the oxide thickness. In Eq. (9.11.4), V_G is positive if the positive polarity is applied to the gate.

The charge density associated with the ionized donors is eN_dW, where W is the width of the surface space-charge region. Therefore, the charge density ρ_c associated with the conduction channel is

$$\rho_c \ (C/m^2) = \frac{\epsilon'(V_{FB} - V_G - \phi_s)}{d_{ox}} - eN_dW \tag{9.11.5}$$

Equation (9.11.5) applies to a channel with substrate (bulk) bias $V_B = 0$ and with drain bias $V_D = 0$ in Fig. 9.27. If V_G is made more and more negative, a situation is reached when the hole concentration ρ_s at the interface is equal to the electron concentration n_b in the bulk. At this value of V_G, ϕ_s becomes ϕ_{SI} of Eq. (9.9.9) and W becomes W_{max} of Eq. (9.9.10).

The value of ρ_c is small at the beginning of surface inversion. As a first order of approximation, we set it equal to zero. The value of V_G with $\rho_c \cong 0$ is called the *threshold voltage* V_T, which according to Eq. (9.11.5) is given by

$$V_T = V_{FB} - \phi_{SI} - \frac{d_{ox}}{\epsilon'} \sqrt{2e\epsilon N_d\phi_{SI}} \tag{9.11.6}$$

Once a surface inversion is reached, the value of ϕ_s and hence that of W increase only very slowly and remain nearly equal to ϕ_{SI} and W_{max} upon further increase in the negative gate bias. Therefore, in terms of V_T, ρ_c can be expressed as

$$\rho_c = \frac{\epsilon'(V_T - V_G)}{d_{ox}} \tag{9.11.7}$$

for a *p*-channel MOSFET.

Now we turn our attention to an active MOSFET with drain and substrate bias voltages (Fig. 9.27). Note that the Fermi level in Fig. 9.28a refers to that at the source or that in the substrate (bulk) because $V_B = 0$. If a potential difference ΔV_{CB} exists between the channel and the bulk, the potential barrier across the surface space-charge region is increased from ϕ_{SI} to $\phi_{SI} - \Delta V_{CB}$ as shown in Fig. 9.28b. The voltage drop across the oxide is changed by an amount $V(x)$, the channel voltage. In Fig. 9.28b, the Fermi level at the source, $\mathscr{E}_{fS}(S)$, is used as the reference energy. In view of these changes, the charge density in the surface channel becomes

$$\rho_c = \frac{\epsilon'[V_{FB} - V_G + V(x) - \phi_{SI}]}{d_{ox}} - \sqrt{2e\epsilon N_d[\phi_{SI} - V(x) + V_B]} \tag{9.11.8}$$

where $V(x)$ is the potential along the channel. Equation (9.11.8), unlike Eq. (9.10.2), takes into account both the interface condition and the variation of W along the channel.

Since $V(x)$ is function of x, a threshold gate voltage V_T can only be defined at a given x. The chosen point is the source where $V(x) = 0$. Setting $\rho_c = 0$ and $V(x) = 0$ in Eq. (9.11.8), we obtain

$$V_T = V_{FB} - \phi_{SI} - \frac{d_{ox}}{\epsilon'} \sqrt{2e\epsilon N_d(\phi_{SI} + V_B)} \tag{9.11.9}$$

If we ignore the dependence of W on $V(x)$, we can express ρ_c in terms of V_T as

$$\rho_c = \frac{\epsilon'[V_T - V_G + V(x)]}{d_{ox}} \tag{9.11.10}$$

This step is taken so that we may have an expression for ρ_c similar to that given in Eq. (9.10.2).

Several comments regarding Eqs. (9.11.8) to (9.11.10) are in order. First, the polarity of the voltages is in reference to the source which is grounded. For a *p*-channel MOSFET, V_G is negative, as indicated by the polarity of the gate voltage in Fig. 9.27a. Since V_D is negative, $V(x)$ is also negative. The sign of V_B is positive if the polarity of the substrate (bulk) voltage is positive with respect to the source terminal. We further note that $e\phi_s$ is the potential energy for an electron. Since $e\phi_s$ increases toward the interface in Fig. 9.28a, it is positive. However, the convention of the voltage refers to the potential energy of a positive charge. Therefore, the potential energy of an electron is $-eV$. This difference in sign explains why ϕ_s is substituted by $\phi_s - \Delta V_{CB}$ in Eq. (9.11.8), where $\Delta V_{CB} = V(x) - V_B$.

Following a procedure similar to that used in obtaining Eq. (9.10.6) from ρ_c of Eq. (9.10.4), we obtain

$$I_D \, dx = \frac{\epsilon' Z \mu_p'}{d_{\text{ox}}} [V_G - V_T - V(x)] \, dV(x) \qquad (9.11.11)$$

from Eq. (9.11.10). Obviously, V_0 in Eq. (9.10.6) is equal to $-V_T$ and V_T is a negative quantity. Integrating Eq. (9.11.11), we find that

$$I_D = \frac{\epsilon' Z \mu_p'}{2 L d_{\text{ox}}} (2V_G - 2V_T - V_D)V_D \qquad (9.11.12)$$

where ϵ' is the dielectric constant and d_{ox} the thickness of the oxide, and L and Z are the length and width (lateral dimension) of the gate. In Eqs. (9.11.11) and (9.11.12), we use μ_p' to account for the fact that the channel mobility may be different from the bulk mobility. Equation (9.11.12) is the same as Eq. (9.10.7) except that V_T is used instead of $-V_0$. However, this substitution is important because the I_D–V_D characteristic of a MOSFET is now related to the property of the MOS gate through V_T.

Depending on the value of the metal work function, a semiconductor can have an interface layer ranging from accumulation to depletion and then to inversion. Two situations for n-type semiconductors with surface accumulation and surface inversion are illustrated in Fig. 9.29. It is obvious that a negative gate voltage is needed to raise the hole concentration p_s at the interface in Fig. 9.29a and hence $V_T = -$. On the other hand, if the metal work function is large relative to the electron affinity of the semiconductor, a strong inversion (Fig. 9.29b) exists even at $V_G = 0$. In this case, a positive V_G is needed to deplete holes in the surface channel; therefore, $V_T = +$. A p-channel MOSFET with $V_T = -$ is operated in the *enhancement mode* because a negative V_G is needed to turn on the conduction channel. On the other hand, a n-channel MOSFET with $V_T = -$ is operated in the *depletion mode* because a conduction channel already exists at $V_G = 0$ and needs a $V_G = -$ to turn it off.

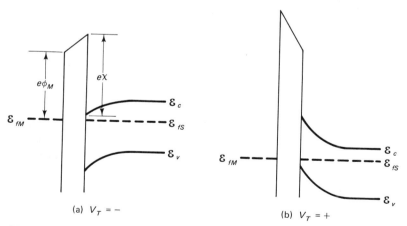

Figure 9.29 Energy-band diagrams at a MOS interface showing the existence of (a) an accumulation layer and (b) an inversion layer in an n-type semiconductor. Situation (a) happens when the metal work function is small, and situation (b) happens when the work function is large relative to the electron affinity of the semiconductor.

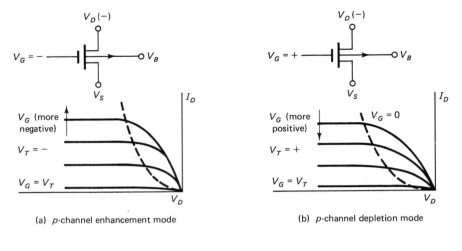

(a) *p*-channel enhancement mode

(b) *p*-channel depletion mode

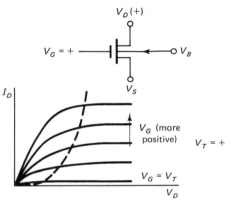

(c) *n*-channel enhancement mode

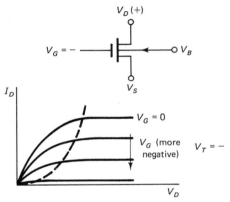

(d) *n*-channel depletion mode

Figure 9.30 Schematic diagrams showing the I_D–V_D characteristics and the circuit symbols for (a) *p*-channel enhancement-mode, (b) *p*-channel depletion-mode, (c) *n*-channel enhancement-mode, and (d) *n*-channel depletion-mode MOSFETs.

The I_D–V_D characteristics together with the circuit symbols for *p*-channel and *n*-channel MOSFETs are shown in Fig. 9.30. One important quantity describing the operation of a MOSFET is the threshold gate voltage V_T defined in Eq. (9.11.9). It determines whether a MOSFET is normally on or off with $V_G = 0$. Another quantity of importance is the saturation drain voltage V_{DS}, above which the drain current saturates, or $\partial I_D/\partial V_D = 0$. According to Eq. (9.10.10),

$$V_{DS} = V_G - V_T \qquad (9.11.13)$$

and the saturation drain current is

$$I_{DS} = \frac{C'_{ox} Z \mu'_p}{2L} V_{DS}^2 \qquad (9.11.14)$$

where $C'_{ox} = \epsilon'/d_{ox}$ is the oxide capacitance per unit area. The quadratic dependence of I_{DS} on V_{DS} is shown in Fig. 9.30 by the dashed curve. In terms of circuit performance, the quantity of importance is the transconductance

$$g_m = \frac{C'_{ox} Z \mu'_p}{L} V_D \qquad (9.11.15)$$

Using $C'_{ox} = 7 \times 10^{-8}$ F/cm^2, $L = 2$ μm, and $V_D = V_{DS} = 2$ V, we find that $g_m/Z = 17$ mS/mm for a p-channel MOSFET with $\mu'_p = 240$ cm^2/V-s and $g_m/Z = 47$ mS/mm for a n-channel MOSFET with $\mu'_n = 675$ cm^2/V-s. Here we assume that the channel mobility is only half of the bulk mobility.

We should point out that Eq. (9.11.12) is not exact because of the various assumptions made in its derivation. First, the variation of W and $V(x)$ in Eq. (9.11.8) is ignored. However, its effect on ρ_c is quite significant. The derivation of I_D taking this effect into account is left to the reader as a problem. Second, according to Eq. (9.11.7), the channel charge density ρ_c is identically zero for $V_G = V_T$. This oversimplification is rather artificial. At the point $p_s = n$ (bulk), the channel conductivity is small but not zero. This finite conductivity results in a small drain current called *subthreshold current*. Third, for a gate length $L = 2$ μm, the average channel field reaches a value of 2.5×10^4 V/cm at $V_D = 5$ V. In the high-field region, the carrier mobility is much smaller than the low-field mobility. The effect of high field on the carrier velocity will be discussed in Sections 10.2 and 10.8.

Despite the shortcomings cited above, Eq. (9.11.12) and its associated equations, such as Eq. (9.11.14), are commonly used in the literature. A comparison with experimental results generally yields useful information about the mechanisms controlling current flow in the channel. These equations are also simple enough to use in device modeling for circuit analysis. The equivalent circuit and the frequency response of MOSFETs are discussed in Section 13.10. Considerations on how the dimensions of MOSFETs should be scaled down for very large scale integration are presented in Section 13.13.

9.12 GATE STRUCTURES, OHMIC CONTACTS, AND ELECTRODE INTERCONNECTS

For field–field transistors, the choice of a gate structure is of the utmost importance. In this chapter so far we have presented two different types of gates: the junction gate in Section 9.6 and the MOS gate in Sections 9.10 and 9.11. The MOS gate is preferred over the junction gate because of the simplicity and the ease of control in the fabrication process. MOSFETs are exclusively used in silicon integrated circuits. However, the MOS technology may not be suitable for other semiconductors, for reasons which will become clear shortly. Another gate structure, especially suited for III–V compound semiconductors, is the metal–semiconductor gate in the form of a Schottky diode. This type of FET is called MESFET.

For MOS or MIS gates, the overriding consideration is the density of surface states. As discussed in Section 7.9, extra energy states may exist either inside an oxide (or an insulator) or at a semiconductor-oxide (or insulator) interface. The density of these states must be sufficiently low so that they will not significantly affect the electrical characteristics of a MOSFET or MISFET. Take a MOS structure as an example. For an oxide of thickness $d_{ox} = 500$ Å with a dielectric constant $\epsilon' = 3.45 \times 10^{-13}$

F/cm, the induced surface-charge density is 2.76×10^{-7} C/cm^2 at a voltage of 4 V. The surface state density N_{ss}(cm^{-2}) should be such that

$$eN_{ss} << \frac{\epsilon' V}{d_{ox}} \tag{9.12.1}$$

The Si–SiO$_2$ interface has a surface-state density between 2 and 5×10^{10} cm^{-2}, corresponding to a value of eN_{ss} between 3 and 8×10^{-9} C/cm^2. Therefore, the inequality in Eq. (9.12.1) is well satisfied.

In contrast, an In$_{0.53}$Ga$_{0.47}$As n-channel MOSFET was reported to have a surface-state density $\Delta N_{ss}/\Delta \mathscr{E}$ about 5×10^{12} (cm^2-eV)$^{-1}$. The native oxide was grown using the plasma oxidation technique (A. S. H. Liao, B. Tell, R. F. Leheny, and T. Y. Chang, *Appl. Phys. Lett.*, Vol. 41, p. 280, 1982). Therefore, a change in $e\phi_s$ by 0.2 eV will produce a shift of the Fermi level relative to the surface-state energy by the same amount, resulting in a net charge flow into the surface state by 1.6×10^{-7} C/cm^2. With such a large surface-state density, a significant, if not a major, part of the induced charge is expected to be absorbed in the surface states.

Figure 9.31 shows (a) the transient response of I_D to a gate-voltage pulse and (b) the I_D–V_D characteristics measured by using a box-car integrator with a time aperture of 0.5 µs and a time delay of 1 µs for V_D. Note from Fig. 9.31a that there is an initial rapid decay of I_D followed by a slower drift to lower drain current. This decrease is due to flow of induced carriers into the surface states. Second, the device has an oxide capacitance $C'_{ox} = 1.24 \times 10^{-7}$ F/cm^2, a gate width $Z = 100$ µm, a gate length $L = 3.5$ µm, and a threshold voltage $V_T = 1.1$ V. Using $I_D = 6$ mA at $V_D = 1$ V and $V_G = 5$ V shown in Fig. 9.31b, we obtain from Eq. (9.11.12) an effective channel mobility $\mu'_n = 500$ cm^2/V-s compared to a measured bulk electron mobility of 6000 cm^2/V-s. Such a large reduction in the apparent mobility cannot be entirely due to surface scattering, mentioned in Section 9.11.

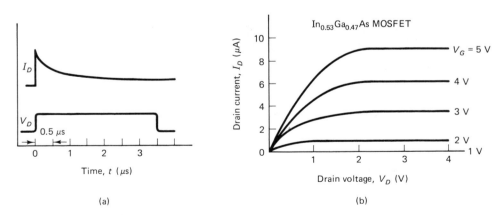

Figure 9.31 Experimental curves showing (a) the drain-current response (upper trace) to a gate-voltage pulse (lower trace) and (b) recorder traces of I_D versus V_D characteristics sampled by a boxcar integrator with a time delay of 1 µs (of the drain voltage with respect to the gate voltage) and a time aperture of 0.5 µs. (A. S. H. Liao, B. Tell, R. F. Leheny, and T. Y. Chang, *Appl. Phys. Lett.*, Vol. 41, p. 280, 1982.)

Let us now consider another effect, the high-field effect discussed in Sections 10.2 and 10.8. It will be shown that the drift velocity of carriers reaches a saturation value v_s at high fields. To approximately account for this effect, we replace $\mu'E$ by v_s and $V(x)$ by an average value $V_D/2$ in Eq. (9.11.11). Thus we have

$$I_D = ZC'_{ox}\left(V_G - V_T - \frac{V_D}{2}\right)v_s \qquad (9.12.2)$$

From Eq. (9.12.2), we obtain, in place of Eq. (9.11.15),

$$g_m = ZC'_{ox}v_s \qquad (9.12.3)$$

The measured value of $g_m/Z = 33$ mS/mm yields a value of 2.6×10^6 cm/s for v_s, which is again nearly an order of magnitude lower than the bulk value. From this observation we can conclude that the partition of induced charges between the channel and the surface states is the main cause for the reduced g_m and μ'_n as deduced from the experiment. In other words, the channel charge density ρ_c is substantially smaller than that given in Eq. (9.11.10).

There may be several reasons for the inferior quality of a MOS structure on III–V compounds. First, separate oxides of group III elements and group V elements exist. Second, there are two, possibly more, oxidation states for the group V elements, for example, As_2O_3 and As_2O_5. Third, oxides of III–V compounds have a relatively small bonding energy to the III–V compounds. For example, oxide of GaAs can be removed from GaAs surface around 525°C while SiO_2 can be removed from Si surface around 950°C, both under high vacuum. These characteristics of oxides of III–V compounds could be fundamental to preventing the MOS technology for III–V compounds from further development. The use of SiO_2 as the insulator in a MIS structure would appear to be a possible alternative. However, at the interface, bonds must be formed either between silicon atoms and III–V atoms or between oxygen atoms and III–V atoms. Referring to the discussion in Section 10.7 on the bonding requirement at a Si–SiO_2 interface, we see that breaking a bond means the creation of a surface state. Since a III–V–SiO_2 interface involves bonding between atoms of many different valencies, satisfying the bonding requirements at such an interface may also be a fundamental problem difficult to overcome.

There are two alternatives to a MOS or MIS gate structure. One involves the use of a heterostructure such as the one used in modulation-doped FETs. MODFETs are discussed in Section 13.12. Another approach uses a Schottky barrier in the form of a metal–semiconductor contact. MESFETs are discussed in Sections 9.13 and 13.12. A MES structure can be used either as a rectifying contact or as an ohmic contact, depending on the magnitudes of the barrier height ϕ_B and the substrate doping, that is, ξ_n in Fig. 9.21b. In this section we review the requirement of a MES structure to be used as a gate and that for an ohmic contact.

Referring to Eq. (9.8.8), we see that current density is a sensitive function of $e\phi_B/kT$. Using $m^* = 0.067m_0$ in n-type GaAs, we have

$$J = 7.4 \times 10^5 \exp\left(\frac{-e\phi_B}{kT}\right) \qquad \text{A/cm}^2 \qquad (9.12.4)$$

in the reverse direction. The value of J at 300 K is 7.4×10^{-8} A/cm^2 for $e\phi_B = 0.78$ eV. Thus for a gate with width $Z = 100$ μm and length $L = 3$ μm, we obtain a value of 2.2×10^{-13} A for the current. In this calculation we have not considered image force and have ignored tunneling. For moderately and lightly doped semiconductors ($N_d < 10^{18}$ cm^{-3}), thermionic emission is the dominant mechanism for current transport across a metal–semiconductor interface.

Let us consider the effect of image force. When an electron leaves the metal, a positive charge is induced on the metal surface. The lowering of the potential barrier by the image force is given by

$$\Delta\phi_B = -\left(\frac{eE}{4\pi\epsilon}\right)^{1/2} \tag{9.12.5}$$

At a field E of 5×10^7 V/m, $\Delta\phi_B$ is about 0.08 eV. This will raise the current to a value about 5×10^{-12} A. For p-type substrates, the current will be about 10 times larger because of the larger effective mass. Even so, the current through a metal–semiconductor contact is small enough for it to be useful as a FET gate. MESFET structures are discussed in Sections 9.13 and 13.12.

Next we turn our attention to the use of a MES structure as an ohmic contact. As indicated by Eq. (9.12.4), thermionic emission across a barrier is a sensitive function of temperature. Furthermore, there is always the question of how low a contact resistance or how high a current we can achieve. This depends on the availability of metal having a low ϕ_B with the semiconductor. A more acceptable and commonly used way of making an ohmic contact is to have a metal-n^+-n or metal-p^+-p contact. Tunneling is the dominant mechanism for current transport across the interface between a metal and a heavily doped semiconductor. Since tunneling is discussed in Sections 10.5 and 13.3, here we quote the results given in Section 10.5 without derivation. The results derived in Section 10.5 for a p-n junction will be modified properly so that they can be applied to a metal–semiconductor contact.

The tunneling-current density in a p-n junction is given by Eq. (10.6.18). For a metal–semiconductor contact, $\mathscr{E}_g$ should be replaced by $e\phi_B$. Therefore, we define a specific contact conductance G per unit area as J_t/V_a, or

$$G = \frac{\pi e^2 m^*}{h^3}(e\phi_B)\frac{E}{E_0}\exp\left(\frac{-E_0}{E}\right) \qquad \text{S/m}^2 \tag{9.12.6}$$

where E is the average electric field in the space-charge region and E_0 is a parameter given by

$$E_0 = \frac{\pi(2m^*)^{1/2}(e\phi_B)^{3/2}}{4e\hbar} \qquad \text{V/m} \tag{9.12.7}$$

Note that there is no temperature dependence in Eq. (9.12.6). To have a low contact resistance or high contact conductance, the barrier height ϕ_B must be low and the doping concentration must be high, so the ratio E/E_0 can be made as large as possible.

Using an effective mass $m^* = 0.19m_0$ for tunneling along $\langle 100 \rangle$ direction in silicon and $\phi_B = 0.25$ V, we find that

$$G = 1.86 \times 10^{12}\frac{E}{E_0}\exp\left(\frac{-E_0}{E}\right) \qquad \text{S/m}^2 \tag{9.12.8}$$

and

$$E_0 = 2.2 \times 10^8 \quad \text{V/m} \tag{9.12.9}$$

A metal–semiconductor contact can be considered as a one-sided junction; therefore, according to Eq. (8.2.11), the average field is

$$E_{\text{av}} = \frac{\phi_B}{W} = \left(\frac{e\phi_B N_I}{2\epsilon}\right)^{1/2} \quad \text{V/m} \tag{9.12.10}$$

Using an impurity concentration $N_I = 10^{26} \text{m}^{-3}$ and $\epsilon = 11.7\epsilon_0$ in silicon, we have $E_{\text{ave}} = 2 \times 10^8$ V/m for $\phi_B = 0.25$ V. Using the values of E_{av} for E and E_0 from Eq. (9.12.9) in Eq. (9.12.8), we obtain $R = G^{-1} = 1.8 \times 10^{-12}$ $\Omega\text{-m}^2$ or 1.8×10^{-8} $\Omega\text{-cm}^2$ for the specific contact resistance.

We should point out that heavy doping is essential to achieving a low contact resistance. If we use $N_I = 10^{19}$ cm^{-3} instead of 10^{20} cm^{-3}, we find $E_0/E_{\text{av}} = 3.5$ and $R = 1 \times 10^{-6}$ $\Omega\text{-cm}^2$. Heavy doping can compensate a moderate ϕ_B to achieve a low E_0/E_{av}. In Table 9.2 we list the values of specific resistance obtained in heavily doped Si/W contacts and in Si/Al–Si alloy contact. We see that a specific contact resistance of the order of 10^{-6} $\Omega\text{-cm}^2$ or lower can be achieved if the silicon is heavily doped. With $R = 10^{-6}$ $\Omega\text{-cm}^2$, we have a contact resistance of 0.5 Ω for a contact area of 2 μm $\times$ 100 μm. At a current of 10 mA, the voltage drop across the contact is only 5 mV, which is much smaller than the thermal voltage of 26 mV at room temperature.

TABLE 9.2 SPECIFIC CONTACT RESISTANCE R ($\Omega\text{-cm}^2$) AND RESISTIVITY ρ ($\Omega\text{-cm}$) FOR A SELECTED NUMBER OF MATERIALS[a]

R ($\Omega\text{-cm}^2$)

W/Si[b]	Al–Si/Si[b]	Ag/TiN/Pt/Mg/GaAs[c]	Au/Ge/Ni/GaAs[d]
2.5×10^{-7} (As, 10^{20})			$<1 \times 10^{-6}$ (Ge, 10^{20})
9.8×10^{-7} (B, 10^{19})	1.8×10^{-6}	$2\pm1 \times 10^{-4}$ (Zn, 10^{18})	

ρ ($\mu\Omega\text{-cm}$)[e]

TiSi$_2$	TaSi$_2$	WSi$_2$	Ti	Ta	W	Al-Si	poly-Si
13-17	8-45	14-70	43-47	13-16	5.3	2.6-3.0	500

[a]The dopants and their concentrations are shown in parentheses.

[b]M. L. Green and R. A. Levy, in R. S. Blewer, ed., "Tungsten and Other Refractory Metals for VLSI Applications," *Mater. Res. Soc. Conf. Proc.*, MRS, Pittsburgh, 1985, p. 423.

[c]J. L. Tandon, K. D. Douglas, G. Vendura, E. Kolawa, F. C. T. So, and M.-A. Nicolet, in R. S. Blewer, ed., "Tungsten and Other Refractory Metals for VLSI Applications," *Mater. Res. Soc. Conf. Proc.*, MRS, Pittsburgh, 1985, p. 331.

[d]A. A. Ketterson, F. Ponse, T. Henderson, J. Klem, C.-K. Peng, and H. Morkoç, "Characterization of Extremely Low Resistance on Modulation-Doped FETs," *IEEE Trans. Electron Devices*, Vol. ED-32, p. 2257, 1985 © 1985 IEEE.

[e]R. S. Blewer, M. E. Tray, and V. A. Wells, in R. S. Blewer, ed., "Tungsten and Other Refractory Metals for VLSI Applications," *Mater. Res. Soc. Conf. Proc.*, MRS, Pittsburgh, 1985, p. 407.

Contact formation on GaAs is generally carried out in two steps: (1) evaporating Cr or Ti for improving adhesion and then Au or Pt for contacting on p^+ substrates, and evaporating Au–Ge alloy on n^+ substrate, and (2) sintering at a temperature around 450°C for a short time, typically of the order of 1 to 3 min. A specific contact resistance in the range 2 to 5×10^{-6} Ω-cm^2 can be achieved. However, the stability of the contact after subsequent high-temperature processing has not been extensively studied. The intermediate TiN layer shown in Table 9.2 is to prevent diffusion of Au or Ag into GaAs. Apparently, the use of a diffusion barrier has increased the specific contact resistance considerably.

Finally, we examine the problem related to interconnections. In Table 9.2, we also show the resistivity ρ in $\mu\Omega$-cm for a selected number of materials used for interconnects. We should point out that ρ in thin film is generally larger than that in the bulk. The values given in Table 9.2 are for thick films, typically exceeding 1 μm. For a $\rho = 10 \times 10^{-6}$ Ω-cm line of 200 μm in length, 2 μm in width, and 0.1 μm in thickness, we have a resistance 1 Ω/square, and a total resistance of 100 Ω. At a current of 10 mA, the voltage drop is 1 V and the power dissipation is 10 mW.

Besides power dissipation, another important consideration is propagation delay. The results reported by Yamamoto et al. have indicated an increase of the delay time (distributed RC time) from 1 ns for W and Mo, to 10 ns for their silicides, and to 100 ns for polysilicon, gate electrode and interconnect in 1-megabyte memory. Therefore, from both power-dissipation and delay-time points of view, refractory metals are better than their silicides which are better than polysilicon as interconnects and gate materials.

In summary, we have discussed possible gate structure for FETs. The MOS structure is most suitable for Si because of the low surface-state density. A low N_{ss} value of 2.5×10^{10} (cm^2/eV)$^{-1}$ has been reported even in the newly developed W–SiO$_2$–Si gate technology (N. Yamamoto, H. Kume, S. Iwata, K. Yagi, N. Kobayashi, N. Mori, and H. Miyazaki, in R. S. Blewer, ed., "Tungsten and Other Refractory Metals for VLSI Applications," *Mater. Res. Soc. Conf. Proc.*, MRS, Pittsburgh, 1985, p. 297). For III–V compounds, the gate structure may be made of either a heterostructure or a metal–semiconductor Schottky contact. MODFETs and MESFETs will be introduced in Sections 13.12 and 9.13. For a Schottky gate, the potential barrier must be reasonably large to limit the gate current due to thermionic emission. We have also considered the use of a MES structure for ohmic contact and have shown that a specific contact resistance of 10^{-6} Ω-cm^2 can be achieved. While polysilicon has been used widely in 64-kilobyte memories, silicides have been used in higher-density memories. Refractory metals are considered to be the materials for both gate electrode and interconnect in the future generation of MOS very-large-scale integration. However, in the development of a new technology, problems regarding reliability and reproducibility have to be studied.

9.13 METAL–SEMICONDUCTOR FIELD-EFFECT TRANSISTOR

Consider the structure shown in Fig. 9.32 for a metal–semiconductor field-effect transistor (MESFET) consisting of an epitaxial n-GaAs layer formed on a semi-insulating (SI) GaAs substrate either through ion implantation or by epitaxial growth. For ion-implanted samples, the dopant distribution $N(y)$ can be approximated by a Gaussian function

$$N(y) = N(\max)\exp\left[\frac{-(y - R_p)^2}{2\sigma_p^2}\right] \qquad (9.13.1)$$

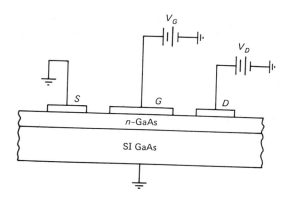

Figure 9.32 Schematic diagram showing the structure of a MESFET. The width of the conducton channel is controlled by the reverse-bias voltage of the Schottky-barrier diode.

according to Eq. (8.6.9), where R_p is the projected range, $\sqrt{2}\,\sigma_p$ is the standard deviation, and $N(\text{max})$ is the dopant concentration at $y = R_p$ where it is maximum. We note from Figs. 8.32 and 8.33 that the value of R_p is larger than that of $\sqrt{2}\,\sigma_p$ by a factor ranging from 2.5 to 5.4. Since N_d drops very rapidly as y moves away from R_p, the implanted n region actually consists of two subregions: a highly conductive region having a full width $2\sqrt{2}\,\sigma_p$ and a nearly SI region. Therefore, an ion-implanted MESFET with the nearly SI region serving as an insulator behaves very much like a MOSFET.

For ion-implanted channels, the I–V characteristics of a MESFET can be described by the MOSFET-like relation as follows:

$$I_D = 2K\left[(V_G - V_T)V_D - \frac{V_D^2}{2}\right] \tag{9.13.2}$$

where the parameter K is generally referred to as the K factor. Equation (9.13.2) is similar to Eq. (9.11.12) with two modifications. First, the gate capacitance per unit area ϵ'/d_{ox} is replaced by ϵ/a_{eff}, where ϵ is the dielectric constant of GaAs and a_{eff} is the effective depth of the implanted region given by

$$a_{\text{eff}} = R_p + 2\sigma_p\sqrt{\frac{2}{\pi}} \tag{9.13.3}$$

In terms of a_{eff}, the K factor is

$$K = \frac{\mu_n'\epsilon Z}{2La_{\text{eff}}} \tag{9.13.4}$$

where L is the length and Z the width of the channel. Again the channel mobility μ_n' is considerably lower than the bulk mobility μ_n because of velocity saturation and additional channel scattering (K. Lehovec and R. Zuleeg, *IEEE Trans. Electron Devices*, Vol. ED-27, p. 1074, 1980).

The drain current I_D expressed in Eq. (9.13.2) saturates when $dI_D/dV_D = 0$, or

$$V_D = V_{DS} = V_G - V_T \tag{9.13.5}$$

which is identical to Eq. (9.11.13). The saturation drain current is given by

$$I_{DS} = K(V_G - V_T)^2 \qquad \text{for} \quad V_D > V_{DS} \tag{9.13.6}$$

The threshold voltage V_T is the applied gate voltage at which the space-charge region pinches off the channel at the source. The total potential drop, denoted by V_p, across a fully extended space-charge region is

$$V_p = V_T + \phi_{bi} = \frac{e}{\epsilon} \int_0^\infty N(y)y \, dy \qquad (9.13.7)$$

where ϕ_{bi} is the built-in potential equal to ϕ_s' in Fig. 9.21b. For a well-localized Gaussian distribution, that is $\sqrt{2} \, \sigma_p \ll R_p$ in Eq. (9.13.1), y can be approximated by R_p. Thus we have

$$V_p = V_T + \phi_s' = \frac{e}{\epsilon} N(\max) R_p \sigma_p \sqrt{\frac{\pi}{2}} \qquad (9.13.8)$$

The localized Gaussian distribution upon which the analysis above is based may not be strictly true. As discussed in Section 8.6, the incorporation of impurities may take place in two steps: (1) a shallow, high-dose implantation, and (2) a subsequent diffusion. In this case the impurity distribution can be approximated by a uniform concentration of N_D extending a distance a, and hence Eq. (9.13.8) should be replaced by

$$V_p = V_T + \phi_{bi} = \frac{eNa^2}{2\epsilon} \qquad (9.13.8a)$$

The quantity ϕ_{bi} corresponds to V_d in Eq. (9.6.4) for a JFET. We note further that V_G and V_D are assigned different polarities in Fig. 9.18b for a JFET and the same polarity in Fig. 9.32 for a MESFET. Therefore, Eq. (9.6.4) should be replaced by

$$I_D = G_m' \left\{ \frac{2}{3V_p^{1/2}} [(V_p + V_G - V_T)^{3/2} - (V_p + V_G - V_T - V_D)^{3/2}] - V_D \right\} \qquad (9.13.9)$$

where $G_m' = e\mu_n' NaZ/L$ because of the fact that the active layer thickness is a instead of $2a$.

In the nonsaturation region of I_D, $V_D < V_{DS}$ and the term V_p is much larger than $V = V_G - V_T$, or $V_G - V_T - V_D$. Therefore, we can expand $(V_p + V)^{3/2}$ in a Taylor's series of V/V_p by keeping only terms up to $(V/V_p)^2$. Thus we find that

$$I_D \cong \frac{\epsilon\mu_n' Z}{aL} \left[(V_G - V_T)V_D - \frac{V_D^2}{2} \right] \qquad \text{for} \quad V_D < V_{DS} \qquad (9.13.10)$$

which is the same as Eq. (9.13.2). For $V_D > V_{DS}$, I_D is given by the saturation value

$$I_{DS} = KV_{DS}^2 = K(V_G - V_T)^2 \qquad (9.13.11)$$

and the transconductance is given by

$$g_m = 2KV_{DS} = 2K(V_G - V_T) \qquad (9.13.12)$$

where $K = \epsilon\mu_n' Z/2La$. Since I_{DS} and g_m are evaluated at $V_D = V_{DS}$, the approximation is still valid. Equations (9.13.10) to (9.13.12) are generally preferred over Eqs. (9.6.4) and (9.6.9) because they are simple to use in device modeling and because the physical meanings of K and V_T are easily recognized.

In the analysis above, we have focused our attention on the conduction of current in channel directly below the gate in Fig. 9.32. However, there is a physical separation between the source and gate electrode. Therefore, in obtaining the terminal I_D–V_D characteristics of a MESFET, we must add the voltage drops across the metal–semiconductor contact and in the region between the source and the gate. Two important considerations in the design of a MESFET are (1) to minimize the contact resistance, and (2) to place the source as close to the gate as possible. These two objectives can be met through a two-stage ion implantation. The first stage is an n implantation with moderate doping ($N_d \sim 10^{17}$ cm^{-3}) to establish a n channel. The second stage is an n^+ implantation with heavy doping ($N_d > 10^{19}$ cm^{-3}) in the source and drain regions for the formation of low-resistance ohmic contacts.

The procedure is illustrated in Fig. 9.33a with the gate metal TiW serving as the mask for the second-stage implantation (N. Yokoyama, T. Ohnishi, K. Odani, H. Onodera, and M. Abe, *IEEE Trans. Electron Devices*, Vol. ED-29, p. 1541, 1982). After implantation, the surface was covered by SiO$_2$, and the sample was annealed at 800°C for 10 min. The TiW silicide that formed during heavy Si implantation remained

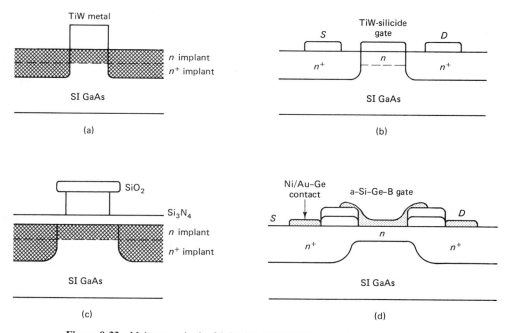

Figure 9.33 Major steps in the fabrication of MESFETs: (a) a two-step ion implantation with TiW metal to block n^+ implant, (b) the structure for a TiW-gate MESFET, (c) a two-step ion implantation with the SiO$_2$–polyimide composite mask to block n^+ implant, and (d) the structure for a a–Si–Ge–B gate MESFET. (Y. Yokoyama, T. Ohnishi, K. Odani, and M. Abe, "TiW Silicide Gate Self-Aligned Technology for Ultra-High Speed GaAs MESFET LSI/VLSIs," *IEEE Trans. Electron Devices*, Vol. ED-29, p. 1541, 1982 © 1982 IEEE; M. Suzuki, K. Murase, N. Kato, M. Togashi, and M. Shirayama, "Advantages of Metallic-Amorphous-Silicon-Gate Applications," *IEEE Trans. Electron Devices*, Vol. ED-33, p. 919, 1986 © 1986 IEEE.)

stable during annealing. The final structure of a Ti-silicide MESFET is shown in Fig. 9.33b. The self-aligned gate technology greatly reduces the source–gate resistance. The self-aligned TiW-silicide-gate MESFET with a $L = 1$ μm gate length was reported to have a transconductance $g_m/Z = 100$ mS/mm as compared to a value about 30 mS/mm for conventional MESFETs.

A third consideration, especially important for large-scale integration of MESFETs, is the choice of gate material. Because of the existence of surface states of sufficiently high density at a metal–semiconductor interface, the Fermi level at the interface is pinned. As a result, the potential barrier for electrons, ϕ_{Bn} in Fig. 9.21b, is about 0.80 eV nearly independent of the gate metal (e.g., Al, Ti–Pt–Au, or Cr–Pt–Au) (C. A. Mead, *Solid-State Electron.*, Vol. 9, p. 1023, 1966). The value of ϕ_{Bn} limits the logical swing of V_G to less than 0.7 eV. Furthermore, because of the nonuniform property of GaAs substrates, the value of V_T has a considerable spread with an about 50-meV standard deviation. Therefore, a Schottky gate with a barrier potential larger than 0.8 eV is highly desirable.

A high Schottky barrier with $\phi_{Bn} = 1$ V with a ternary amorphous–silicon–germanium–boron (a–Si–Ge–B) film as the gate material was reported by Suzuki et al. (M. Suzuki, K. Murase, N. Kato, M. Togashi, and M. Shirayama, *IEEE Trans. Electron Devices*, Vol. ED-33, p. 919, 1986). Figure 9.33c shows the two-step ion implantation using a SiO_2–polyimide composite mask to block the second-stage n^+ implantation. After the implanted sample was annealed at 800°C for 20 min and a window was opened in the SiN mask, the a–Si–Ge–B film was evaporated and the Ni/Au–Ge ohmic contacts were formed, resulting in the final MAS (metal–amorphous–semiconductor) FET structure of Fig. 9.33d. The MASFET was reported to have a transconductance $g_m/Z = 175$ mS/mm for a $L = 1$ μm gate length, compared to a value of 115 mS/mm observed in a MESFET with a similar structure but a Au/Pt/Ti gate. The logical swing in E/D (enhancement/depletion) DCFL (direct-coupled FET logic) for the MASFET inverters was about 1.5 times larger than that for the MESFET inverters. The increased values of g_m and logical swing are due to the fact the MASFET can be driven to a higher V_G than the MESFET.

Finally, we briefly comment on the origin and the adverse effects of crystal nonuniformity. In semi-insulating GaAs ingots grown by the LEC (liquid-encapsulated-Czochralski) method, large dislocation densities are observed near the periphery and, to a lesser extent, near the center of the crystal, resulting in a W-shaped distribution in the [011] direction. The region surrounding a dislocation generally has a strong variation in resistivity, photoluminescence, and cathodoluminescence, possibly due to the gettering of impurities (A. K. Chin, R. Caruso, M. S. S. Young, and A. R. von Neida, *Appl. Phys. Lett.*, Vol. 45, p. 552, 1984; K. Watanabe, H. Nakanishi, Y. Yamada, and K. Koshikawa, *Appl. Phys. Lett.*, Vol. 45, p. 643, 1984). Therefore the gettered impurities must be removed from a SI GaAS crystal while the dislocations, especially the dislocation networks, are reduced by proper heat treatment. Using a combination of backside damage to getter Cr and Au and heat treatment at 400 to 600°C, a substantial improvement in g_m, from 120 mS/mm to 180 mS/mm, was reported (F. C. Wang and M. Bujatti, *IEEE Trans. Electron Devices*, Vol. ED-32, p. 2839, 1985). Reducing dislocation density and controlling unwanted defects and impurities are important problems that need to be overcome if large-scale-integrated (LSI) GaAs circuits are to become commercially competitive with LSI Si circuits.

9.14 CIRCUIT INTEGRATION AND THIN-FILM WORK

By *integrated circuits* we mean assemblies of small but otherwise conventional electronic components which are arranged in some orderly fashion to perform certain electronic functions. Figure 9.34 shows the schematic of a three-input current-mode logic gate and the actual functional block of the logic gate. The contacts indicated by numerals are used for electric connections to other functional blocks. Electronic systems for modern industrial application and scientific exploration are becoming increasingly com-

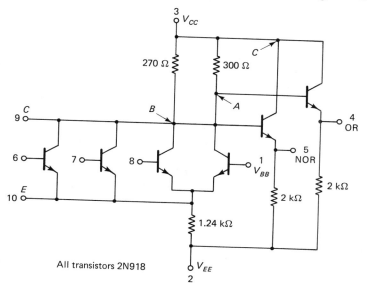

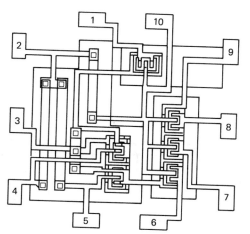

Figure 9.34 Schematic diagrams illustrating integration of devices (bipolar transistors) into a functional (logic gate) block.

plex in nature. As the complexity of a system increases, the number of electronic components used in the system increases accordingly and the requirements for high reliability, low cost, and small size become paramount. Circuit integration into functional blocks is our answer to meet the requirements.

With the terminals marked with numerals, the reader should be able to identify the transistors and resistors in the functional block with the corresponding ones in the schematic diagram (Fig. 9.34). The basic processes involved in making the resistors and the transistors are illustrated in Fig. 9.35. In integrated circuits, we use a common substrate for all circuit components belonging to the same functional block. The substrate may be an insulating material, such as sapphire, or a semiconductor. In Fig. 9.35, the substrate is a p-type semiconductor. Isolation between circuit components is provided by a reverse-biased p-n junction. For the resistors the middle n region provides the isolation, whereas for the transistors the p substrate, or if necessary a diffused p^{+} region, provides the isolation.

One of the outstanding features of semiconductor functional blocks is that many dissimilar structures can be processed simultaneously. The resistor and the transistor shown in Fig. 9.35 are different only in the final stage of fabrication. The transistor needs an n^{+} diffusion to form the emitter junction. Of course, the dimensions of the various n and p regions as well as the arrangements for electric contacts are different for the resistor and transistor structures. The control of the dimension and the etching of the oxide to provide electric contacts are done through the use of photoetching techniques (Section 8.7). With a proper modification of the fabrication process, either the junction or the metal–oxide–gate FET structure can also be made. Another distinct feature is that devices with matching characteristics are readily obtainable in a functional block because they are fabricated under identical conditions. For discussions of engineering design and circuit considerations concerning functional blocks, the reader is

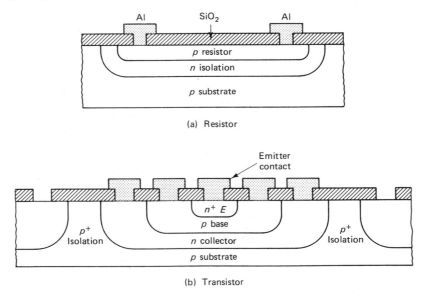

Figure 9.35 Schematic diagrams illustrating the structures of (a) a resistor and (b) a bipolar transistor.

referred to books on integrated circuits (see, for example, H. C. Lin, *Integrated Electronics,* Holden-Day, Inc., San Francisco, 1967; and R. M. Warner, Jr. and J. N. Fordemwalt, *Integrated Circuits,* McGraw-Hill Book Company, New York, 1965).

The IC (integrated-circuit) technology has undergone many significant changes as it evolves from medium-scale to large-scale device integration. First, MOSFETs have become the dominant device in digital applications. Second, the demand of low sheet resistance and low contact resistance for devices with reduced sizes has necessitated the constant search for new materials for gate electrodes and source–drain contacts. Figure 9.36 shows the self-aligned structure of (a) MOSFETs using silicides $TiSi_2$ or CoSi as the electrode material, and (b) MOSFETs using refractory metal W as the electrode material. The choice of electrode material has changed from heavily doped polycrystalline silicon to silicides and then to refractory metal as the device density is increased from medium-scale to very-large-scale integration. Refractory metals, such as Mo, Ta, and W, have low resistivity ($\sim$10 $\mu\Omega$-cm), as shown in Table 9.2. Tungsten has also

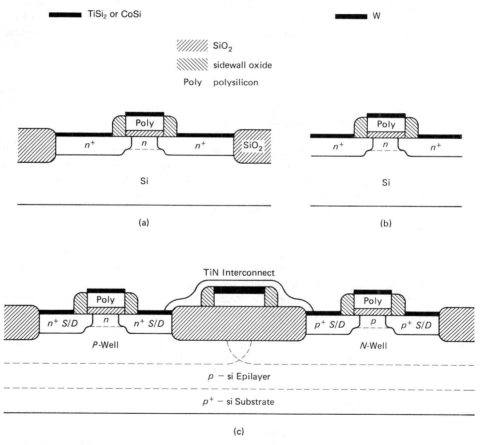

Figure 9.36 Schematic diagrams illustrating the use of silicides and refractory metals as the gate electrode and as the source and drain contacts. The structure is under study for future very-large-scale integration (VLSI).

been shown to have low contact resistance ($<10^{-6}$ Ω-cm^2). Another desirable property of W is its ability for conformal coating. Tungsten is highly resistant to electromigration. Because of these advantages, tungsten is considered and studied extensively as a primary candidate for electrode material in VLSI applications.

Figure 9.36c shows schematically the structure of a CMOS inverter. The diagram is used to illustrate the complexity in the fabrication process. As the dimension of the device is reduced, for example the gate length to micrometer or submicrometer size, precise control of the doping profile and the gate location becomes essential. Ion implantation and self-aligned gate technology have become indispensable. Furthermore, the structure has to be almost free of defects. Even though refractory metals have a thermal expansion coefficient $\sim 5 \times 10^{-6}/°C$ closer to that of Si ($\sim 2.5 \times 10^{-6}/°C$) than that of silicides ($\sim 8 \times 10^{-6}/°C$), stress-caused failure in the gate oxide is still a concern. In addition, tungsten is susceptible to oxidation above 400°C and to silicidation (forming WSi$_2$) above 600°C. All these technological problems must be overcome before a decision can be made in the choice of the electrode material in future VLSI circuits. The reader is referred to the literature for further discussions on the subject (see, for example, R. S. Blewer, ed., "Tungsten and Other Refractory Metals for VLSI Applications," *Mater. Res. Soc. Conf. Proc.*, MRS, Pittsburgh, 1985; Special Issue on "Interconnections for Contacts for VLSI," *IEEE Trans. Electron Devices*, Vol. ED-34, March 1987).

The integrated circuits illustrated above are built on a semiconductor substrate. Such circuit integration is called the *monolithic type*. Another approach is the use of an insulating material as the substrate. Thin-film layers are deposited on the substrate to form circuit components. Materials proven satisfactory as the dielectric layer in thin-film capacitors include SiO$_2$, Ta$_2$O$_5$, and Al$_2$O$_3$. For thin-film resistors, SnO$_2$, Ta, and nichrome (an alloy of nickel and chromium) are commonly used. These thin-film capacitors and resistors can also be used in monolithic integrated circuits together with junction capacitors and diffused resistors. The situation with active thin-film circuit components is not as far advanced. Two thin-film devices that have been investigated extensively are the insulated-gate thin-film transistor and the hot-electron transistor (P. K. Weimer, *Proc. IRE*, Vol. 50, p. 1462, 1962; C. A. Mead, *Proc. IRE*, Vol. 48, p. 359, 1960).

The thin-film FET structure is in principle identical to the IGFET structure. The only difference is that the semiconductor layer that forms the conductive channel is deposited on the substrate. Among semiconductors that have been studied are Si, GaAs, CdS, CdTe, PbTe, InAs, and Te. Because the art of thin-film deposition has not reached such an advanced technological stage as the diffusion or epitaxial-grown process, semiconductor films formed by vacuum deposition generally have many crystalline defects, which give rise to trapping centers and surface states. Trapping effects and nonohmic contacts (drain and source) are thought to be responsible for poor performance and surface states responsible for poor reproducibility of such devices. Recent advances in Si growth on SiO$_2$ and on Al$_2$O$_3$ have made semiconductor-on-insulator (SOI) technology useful in applications requiring radiation hardness.

The hot-electron transistor is made of the S-M-S, M-I-M-I-M, or M-I-M-S structure, with S, M, and I representing semiconductor, metal, and insulator, respectively. The first S-M barrier acts as the emitter, the middle metal region as the base, and the last M-S barrier as the collector. Many combinations, including the Au–Al$_2$O$_3$–Al–Ge, the Si–Au–Ge, and the In–Cd–Au–Ge systems, have been investigated. However, none of the hot-electron transistors can compete with junction transistors (J. L. Moll, *IEEE Trans. Electron Devices,* Vol. ED-10, p. 299, 1963). One of the reasons for poor performance is that the lifetime of hot electrons is exceedingly short in the metal (base) region. The reader is referred to the article on hot-electron transport by Crowell and Sze for further discussion (C. R. Crowell and S. M. Sze, *Physics of Thin Films,* Vol. 4, Academic Press, Inc., New York, 1967). Recent advances in growth technology have revived interests in hot-electron transistors using semiconductor as the base. Such hot-electron devices are discussed in Section 11.8.

PROBLEMS

9.1. (a) The dc characteristics of a typical *p-n-p* transistor in the common-base and the common-emitter configuration are shown in Fig. P9.1. Estimate the values of α and I_{CO} from Fig. P9.1a.

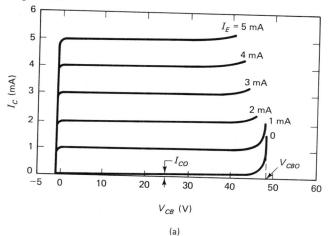

(a)

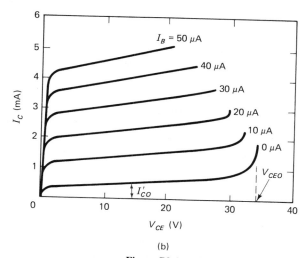

(b)

Figure P9.1

(b) Find the expression relating I_C to I_B. From Fig. P9.1b, obtain the values of β and I'_{C0}, and more accurate values of α and I_{C0}.

(c) Give reasons why I_C may increase with the collector voltage. Note that the variation of I_C is more pronounced in Fig. P9.1b than in Fig. P9.1a. Deduce from this fact the main cause for the lack of saturation in I_C.

9.2. **(a)** Show that γ of Eq. (9.1.24) can be expressed as

$$\gamma^{-1} = 1 + \frac{\sigma_B}{\sigma_E} \frac{L'_B}{L'_E} \tanh \frac{W_0}{L'_B}$$

(b) Calculate the values of the emitter efficiency and the transport efficiency for a silicon n-p-n transistor with (1) $\rho = 5 \times 10^{-2}$ Ω-cm, $\tau = 10^{-6}$ s in the base region; (2) $\rho = 8 \times 10^{-4}$ Ω-cm, $\tau = 10^{-8}$ s in the emitter region; (3) $W_0 = 2$ μm of base width, and (4) $D_n = 11$ cm^2/s and $D_p = 3$ cm^2/s due to heavy doping (Fig. 13.53b). Also calculate the common-emitter current gain β.

9.3. **(a)** Show that the electron components of the dc emitter and the dc collector currents in an n-p-n transistor are, respectively, given by

$$I_{Ee} = \frac{AeD_n}{L_n} n_{10} \coth \frac{W}{L_n}$$

$$I_{Ce} = \frac{AeD_n}{L_n} n_{10} \operatorname{csch} \frac{W}{L_n}$$

where $n_{10} = n_{po} [\exp(eV_E/kT) - 1]$.

(b) Find the expression for the transport efficiency α_T. Also write down the expression for the emitter efficiency γ. Comment on whether I_{Ee} and I_{Eh} will increase or decrease (1) if W is reduced, (2) if the doping on the emitter n side is increased, and (3) if the doping in the base region is increased, and then consider their effects on γ and α_T.

(c) Are the values of γ and α_T independent of the dc collector voltage? Explain.

9.4. **(a)** Show that Y_{11} of Eq. (9.1.39) for the forward-biased emitter–base junction can be represented by the equivalent junction conductance G and diffusion capacitance C_D of Eq. (8.4.5).

(b) Referring to Fig. 9.3, explain the physical origins of the real and imaginary parts of Y_{22} of Eq. (9.1.40).

(c) Given $N_a = 10^{17}$ cm^{-3}, $\mu_p = 320$ cm^2/V-s, $D_n = 19$ cm^2/s, and $W_0 = 2.5$ μm for the base of a Si p-n-p transistor, calculate the base resistance per square $R_\square = (e\mu_p pW_0)^{-1}$ and the base transit time. Using Fig. 13.61b, find the base width W_0 needed to keep $R_\square$ unchanged if N_a is raised to 10^{18} cm^{-3}.

9.5. **(a)** Modify Fig. 9.7a properly so that it represents the ac, small-signal equivalent circuit of a transistor by changing r to Z and using two different α's. Express the transistor parameters α_1, α_2, Z_E, and Z_C in terms of ω_α, Y_E, and Y_C.

(b) The ratio of the emitter resistance to the collector resistance is called the *base width-modulation factor* η and the ratio $I_c/|\partial I_c/\partial V_c| = V_{EA}$ is known as the *Early voltage*. Show that

$$\eta = \frac{kT}{eW_0} \frac{\partial W_{CB}}{\partial V_C} \quad \text{and} \quad V_{EA} = \frac{kT}{e\eta}$$

Calculate the values of η and V_{EA} at $V_C = 10$ V for a silicon transistor in which the collector side is heavily doped, the base side of the collector junction has an impurity concentration $N_I = 10^{15}$ cm^{-3}, and the base width is 3×10^{-4} cm.

9.6. (a) Consider a symmetrical silicon transistor in which the collector and emitter sides are heavily doped with equal impurity concentrations and the base region is lightly doped. It is also assumed that the areas of the collector and emitter junctions are equal. Find the expression for the ratio of the time constants $(r_E C_{TE}/r_C C_{TC})$ of the emitter and collector junctions.

(b) Given $V_d = 0.75$ eV, $N_I = 10^{15}$ cm^{-3} in the base region, and $W_0 = 3 \times 10^{-4}$ cm, find the value of this ratio if the transistor is operated at a forward emitter bias of 0.55 V and a reverse collector bias of 10 V.

9.7. Refer to the equivalent circuit shown in Fig. 9.7b. Consider the operation of the circuit at very low frequencies such that the effect of C_{TC} can be ignored. Show that matched conditions at the input and output terminals require

$$R_g = r_1 \left(1 - \frac{r_3 r_4}{r_1 r_2}\right)^{1/2}, \qquad R_L = r_2 \left(1 - \frac{r_3 r_4}{r_1 r_2}\right)^{1/2}$$

where $r_1 = r_B' + r_E$, $r_2 = r_B' + r_C$, $r_3 = r_B' + \alpha_0 r_C$, and $r_4 = r_B'$. Verify that under matched conditions

$$\eta = \left[\frac{r_3}{(r_1 r_2)^{1/2} + (r_1 r_2 - r_3 r_4)^{1/2}}\right]^2$$

Since $r_C \gg r_B' \gg r_E$ and $\alpha_0 \cong 1$, the equation above reduces to $\eta_{max} \cong \alpha_0^2 r_C/r_B'$ of Eq. (9.2.15).

9.8. Using Eqs. (9.2.4) and (9.2.5), find the built-in field in the base region if N_a changes from 8×10^{17} cm^{-3} to 10^{15} cm^{-3} in 2 µm as in Fig. 9.9. Also calculate the ratio of the drift velocity to the diffusion velocity $(2D/W)$. Show that for such a diffused base, the base transit time is given by

$$t_B = \frac{lW}{D}$$

9.9. (a) The choice of a suitable doping concentration in the base region of a transistor is based on many factors. Comment on the effects of raising the base-region doping concentration on the following transistor properties: (1) the base resistance r_B, (2) the collector resistance r_C (due to base-width modulation effect), (3) the emitter efficiency γ, (4) the collector capacitance C_{TC}, and (5) the collector-junction breakdown voltage V_{BD}. The transistor is assumed to have a plain p-n-p or n-p-n structure.

(b) Comment on the benefits expected from using the doping profiles shown in Fig. 9.9, in particular, a decreasing N_a in the base toward the collector and an epitaxial n layer for the collector.

9.10. (a) An ideal planar transistor (Fig. 8.34) has a p^+-n-i-p^+ or n^+-p-i-n^+ structure. The main purpose of the intrinsic region i sandwiched between the collector and the base regions is to reduce the collector capacitance C_{TC}. In practice, however, the sandwiched-between region is not truly intrinsic, but is slightly extrinsic. Find the location of the space-charge region and the value of C_{TC} (per unit area) for collector-bias voltages $V_{CB} = 5$, 10, 15, and 20 V in a silicon junction where the middle near-intrinsic region has an impurity concentration $N_I = 5 \times 10^{14}$ cm^{-3} and has a width $W_I = 6 \times 10^{-4}$ cm. The following data are also known: $N_a = 10^{17}$ cm^{-3} (acceptors) on the base side and $N_d = 10^{19}$ cm^{-3} (donors) on the collector side. Given: $N_c = 2.9 \times 10^{19}$ cm^{-3}, $N_v = 1.1 \times 10^{19}$ cm^{-3}, $\epsilon = 12\epsilon_0$, and $\mathcal{E}_g = 1.11$ eV in Si.

(b) Since the middle region is extrinsic, we must decide whether it should be n type or p type. What is your choice for the case under consideration? Explain.

(c) Calculate the value of V_{CB} needed for the space-charge region to extend all way across the middle region. Would it be preferable to bias the collector junction above or below this voltage? Explain.

9.11. (a) Plot $p_1(x)$ of Eq. (9.2.12) as a function of x, assuming the same base width and the same current density for the three cases. Compare the excess-hole concentrations at the emitter and then the transit times.

(b) Comment on the effects of the three distributions upon E_{max} in the space-charge region, the emitter efficiency, the collector-junction capacitance, and the base-width modulation effect.

9.12. Repeat Problem 9.11 for the case in which the base resistance r_B and the current density are assumed to be the same for the three distributions.

9.13. (a) Consider a p-n-p transistor with a uniformly doped base region. Verify Eq. (9.3.5).

(b) In the analysis presented in this chapter, we have assumed a zero electric field in the base region. In reality, an electric field exists. The value of the field can be found by equating the majority current (electron current in a p-n-p transistor) to zero. Show that the effect of the electric field can be accounted for by letting

$$D = \frac{D_p(1 + 2Z)}{1 + Z}$$

be the effective diffusion constant for holes. The parameter Z is defined as the ratio of the excess minority-carrier concentration to the equilibrium majority-carrier concentration.

(c) Show that for a narrow-base transistor, the parameter Z is related to the emitter current I_E as follows:

$$Z \frac{1 + 2Z}{1 + Z} = \frac{\mu_n W I_E}{D_p \pi a^2 \sigma_n}$$

where W is the base width, σ_n the base conductivity, and a the emitter radius.

(d) Given $W = 2 \times 10^{-3}$ cm, $\mu_n = 3900$ cm^2/V-s, $D_p = 50$ cm^2/s, $\sigma_n = 1$ (Ω-cm)$^{-1}$, and $a = 0.1$ cm for a Ge transistor, find the value of I_E at which $Z = 1$.

9.14. (a) Find the relative magnitudes of the terms in Eq. (9.3.1) for a Ge p-n-p transistor operated in the low injection region. In addition to the data given in Problem 9.13, the following data are known: $D_n = 100$ cm^2/s, $\tau_0 = 10^{-7}$ s and $\sigma_p = 500$ (Ω-cm)$^{-1}$ in the emitter region, $\tau_0 = 10^{-4}$ s in the base region, and $v_s = 50$ cm/s.

(b) Modify Eq. (9.3.1) so that it can be used under high injection conditions ($Z > 1$). The variations of τ_0 and D with injection level should be considered.

9.15. Refer to the transistor structure shown in Fig. 9.9. Approximate the impurity distribution in the base region by an exponential function. If it is assumed that the value of N_a changes from a value of 10^{18} cm^{-3} at the emitter junction to a value of 10^{15} cm^{-3} at the collector junction, find the value of η in Eq. (9.3.7). The base width is chosen to be 2×10^{-4} cm.

9.16. (a) For switching operations, the following set of equations, first introduced by Ebers and Moll (J. J. Ebers and J. L. Moll, *Proc. IRE*, Vol. 42, p. 1791, 1954), is often useful.

$$I_E = \frac{I_{E0}}{1 - \alpha_F \alpha_R} x - \frac{\alpha_R I_{C0}}{1 - \alpha_F \alpha_R} y$$

$$I_C = \frac{\alpha_F I_{E0}}{1 - \alpha_F \alpha_R} x - \frac{I_{C0}}{1 - \alpha_F \alpha_R} y$$

where $x = \exp (eV_{BE}/kT) - 1$ and $y = \exp (eV_{BC}/kT) - 1$. Show that they are identical to Eqs. (9.4.2) and (9.4.3).

(b) Derive the Ebers–Moll equations by following the procedure described below. First, we express the currents as

$$I_E = a_{11}x + a_{12}y$$
$$I_C = a_{21}x + a_{22}y$$

Then apply the following conditions to the equations to find the coefficients a's: (1) With collector open circuited, I_E should reduce to $I_{E0}x$, where I_{E0} is the reverse saturation current of the emitter junction. (2) With emitter open circuited, I_C should reduce to $I_{C0}y$, where I_{C0} is the reverse saturation current of the collector junction. (3) Under normal operation as a transistor, $I_C = \alpha_F I_E + I_{C0}$, where α_F is the normal current gain. (4) Under inverted operating condition (with normal emitter as collector and normal collector as emitter), $I_E = \alpha_R I_C + I_{E0}$, where α_R is the inverse current gain.

9.17. (a) Verify Eq. (9.4.17) and the relations

$$t_{BF} = \beta_F t_F \quad \text{and} \quad t_{BR} = \beta_R t_R$$

(b) For the prototype transistor of Fig. 9.3b with constant doping in the base region and equal emitter and collector area, show that $t_{BF} = t_{BR} = \tau_0$. Given $D = 20$ cm²/s, $W = 2$ μm, and $\beta = 100$ in a Si n-p-n transistor with unity emitter efficiency, calculate t_F, t_R, t_{BF}, and t_{BR}.

9.18. (a) Consider the n-p-n transistor of Problem 9.17 arranged in the common-emitter configuration. Set up differential equations for I_B and I_C. Neglect the space-charge terms dQ_C/dt and dQ_E/dt.

(b) In view of $t_{BF} \gg t_F$ and $t_{BR} \gg t_R$, show that the transient response of Q_F and Q_R has two time constants, one fast, t_{fast}, and one slow, t_{slow}, approximately given by

$$t_{\text{fast}}^{-1} = t_{BF}^{-1} + t_{BR}^{-1} + t_F^{-1} + t_R^{-1}$$
$$t_{\text{slow}}^{-1} = t_{\text{fast}} [t_{BF}^{-1} (t_{BR}^{-1} + t_R^{-1}) + t_{BR}^{-1} t_F^{-1}]$$

(c) Calculate the values of t_{fast} and t_{slow} for the n-p-n transistor.

9.19. (a) Consider a narrow-base p-n-p transistor. Using Eq. (8.9.12) for the recombination current in the emitter junction, show that the ratio of the recombination current to the diffusion current can be expressed in terms of the diffusion current I_{diff} as

$$\frac{I_{\text{rec}}}{I_{\text{diff}}} = \frac{W_{SC}W_B}{2L_p^2} \sqrt{\frac{N_d e D_p A_E}{I_{\text{diff}}W_B}}$$

where W_{SC} is the width of the emitter-space-charge region, W_B the base width, N_d the donor concentration in the base region, and A_E the emitter-junction area.

(b) Given: $N_d = 10^{17}$ cm^{-3}, $W_B = 5 \times 10^{-4}$ cm, $A_E = 0.05$ cm², $V_d = 0.95$ V, $D_p = 12$ cm²/s, $L_p = 10^{-3}$ cm, $\epsilon = 12\epsilon_0$, and $n_i = 10^{10}$ cm^{-3}. Find the ratio $J_{\text{rec}}/J_{\text{diff}}$ and the value of the current-amplification factor α at $I_{\text{diff}} = 10^{-6}$, 10^{-4}, and 10^{-2} A. It is assumed that $I_{B1} + I_{B2} + I_{B3} = 0.20I_{\text{diff}}$ in Eq. (9.3.1).

9.20. (a) A silicon thyristor (p-n-p-n device) is generally used to convert ac into dc power at sufficiently large voltage and high current. The impurity concentration N_I in the middle junction must be reasonably low and the width of the two middle regions must be reasonably large, so that a large voltage may be maintained in the reverse direction without breakdown and punch-through. It is known that the breakdown voltage in an abrupt Si junction obeys the following empirical relation:

$$V_{BD} = 55 \left(\frac{10^{16}}{N_I} \right)^{0.70} \quad \text{volts}$$

where N_I is the impurity concentration in cm^{-3}. Find the values of N_I and W needed for $V_{BD} = 400$ V in Fig. 9.15. It is assumed that the values of N_I are the same in the middle n and p regions.

(b) The multiplication factor M in a Si junction is found experimentally to obey

$$M = \frac{1}{1 - (V/V_{BD})^3}$$

Find the value of the turn-on voltage V_{on} in a thyristor where $\alpha_1 = \alpha_2 = 0.45$ at switching.

(c) After switching, the α of the thyristor is limited by the transport efficiency. Find the sustaining voltage V_{sus} needed in a thyristor where $L_n = L_p = 4 \times 10^{-3}$ cm in the middle junction, $I = 1$ A and $I_S = I_{S1} = I_{S2} = 10^{-6}$ A. Explain why we have to use two different values for α in evaluating V_{on} and V_{sus}.

9.21. **(a)** Explain the physical processes taking place inside a p-n-p-n structure during switching.

(b) Obtain expressions I_1 and I_2, similar to Eq. (9.5.5) for the currents flowing through the outer two junctions. Make sure that you count the right amount of contributions to I_{Sm} from hole and electron injection in these expressions.

(c) Verify Eq. (9.5.6) under the assumptions stated in the chapter.

9.22. **(a)** Expressing V_d in terms of V_p and expressing $(V + V_p)^{3/2}$ terms in a Taylor's series in V/V_p, show that Eq. (9.6.6) and the drain-saturation current I_{DS}, respectively, can be approximated by

$$I_D = \frac{G_m(2V_T - 2V_G - V_D)V_D}{4V_p}$$

$$I_{DS} = \frac{\epsilon \mu_n Z}{aL} (V_G - V_T)^2$$

where $V_T = V_p'$ is the threshold voltage.

(b) Given: $\epsilon = 11.7\epsilon_0$, $\mu_n = 1200$ cm^2/V-s, $N_d = 2 \times 10^{15}$ cm^{-3}, $L = 150$ μm, $a = 1$ μm, and $L = 2$ μm, find I_{DS} at $V_G = 0$ V.

9.23. **(a)** The drain-saturation current in a JFET reaches a maximum value $(I_{DS})_m$ when $V_G = 0$ and $V_D = V_T$. Referring to Problem 9.22 for I_{DS}, show that

$$(I_{DS})_m = \frac{e\mu_n N_d \, aZV_D}{2L}$$

(b) One important consideration concerning field-effect and bipolar transistors is the temperature sensitivity of the currents. Verify the following expressions:

$$\frac{1}{(I_{DS})_m} \frac{d(I_{DS})_m}{dT} = -\frac{3}{2T}, \qquad \frac{1}{I_C} \frac{dI_C}{dT} = \frac{5}{2T} + \frac{\mathscr{E}_g}{kT^2} - \frac{eV_{BE}}{kT^2}$$

for the transistors if mobility is controlled by lattice scattering.

(c) Given: $\mathscr{E}_g = 1.12$ eV and $V_{BE} = 0.9$ V, calculate the percentage change in currents if T changes from 300 K to 320 K.

9.24. The work functions of Al and Au are known:

$$\mathscr{E}_w(\text{Al}) = 4.2 \text{ eV} \quad \text{and} \quad \mathscr{E}_w(\text{Au}) = 4.8 \text{ eV}$$

The electron affinity $e\chi$ (from the conduction band) in silicon is 4.15 eV. Draw energy-band diagrams to show the Schottky barrier in Al–Si and Au–Si contacts. Find the potential barriers ϕ_B on the metal side and ϕ_S on the silicon side (n type with $N_d = 10^{15} \text{ cm}^{-3}$) in the two metal–semiconductor contacts. Find the value of the bias voltage needed to maintain a flat band in silicon for each case. Also indicate the polarity of the applied bias.

9.25. Draw an energy-band diagram showing the potential barrier at a metal contact with a p-type semiconductor. Find the potential barriers ϕ_B on the metal side and ϕ_S on the semiconductor side for hole conduction across the Schottky barrier. (*Hint:* Compare the situation with a metal contact on n-type semiconductor except that the attention is now on the vacant states.) Show that the two potential barriers, ϕ_{Bp} for p-type semiconductor and ϕ_{Bn} for n-type semiconductor, are related by

$$e(\phi_{Bn} + \phi_{Bp}) = \mathscr{E}_g$$

9.26. In Section 9.8 we derived an expression, Eq. (9.8.8), for current conduction through a metal contact with an n-type semiconductor. Derive an analogous expression for current conduction through a metal–p-type semiconductor contact. Indicate the polarity for a forward-biased Schottky diode.

9.27. The following data are given for a silicon MOS device: $N_d = 5 \times 10^{15} \text{ cm}^{-3}$ (Si), $n_i = 1.4 \times 10^{10} \text{ cm}^{-3}$ (Si), $\epsilon = 11.7\epsilon_0$(Si), $\epsilon' = 3.8\epsilon_0$(SiO$_2$), and $d = 100$ nm (SiO$_2$). Find the ratio of the capacitance change C_{max}/C_{min} possible in the MOS device.

9.28. **(a)** Derive analytic expressions for the surface charge density Q_s and surface conductance G_s (per unit area) as functions of ϕ_s (Fig. 7.20a) for an intrinsic sample, that is, $Y_b = 0$ in Eq. (7.9.6).

(b) Given $\epsilon = 11.7\epsilon_0$ and $n_i = 1.4 \times 10^{10} \text{ cm}^{-3}$ in Si and $\epsilon' = 3.8\epsilon_0$ in SiO$_2$, find the values of ϕ_s and G_s for $Q_s = 3.2 \times 10^{-8}$ C/cm^2, that is, an areal carrier density of $2 \times 10^{11} \text{ cm}^{-2}$. Also calculate the electric field E in the oxide and in Si at the interface.

9.29. **(a)** Verify the following relation for a MOS structure:

$$\mathscr{E}_w - \frac{eQ_s d_{ox}}{\epsilon_{ox}} = e\chi + e(\phi_b + \xi_b - \phi_s)$$

where $\xi = (\mathscr{E}_c - \mathscr{E}_f)/e$ and the subscript b refers to the bulk. For a semiconductor remaining nondegenerate at the interface, show that

$$\frac{Q_s}{\epsilon} = \sqrt{\frac{4n_i kT}{\epsilon}} [\cosh Y_s - \cosh Y_b - (Y_s - Y_b) \sinh Y_b]^{1/2}$$

where $Y = e\phi/kT$. Note that ϕ_s and ϕ_b are defined in Fig. 7.20. Refer to Fig. 9.28 for the difference in the definition of ϕ in Sections 7.9 and 9.11. Discuss how the sign of Q_s is determined.

(b) For a given set of values for d_{ox} and ξ, the values of Q_s and ϕ_s can be found from solving the two equations. Given $\mathscr{E}_w(\text{Al}) = 4.10$ eV, $\chi(\text{Si}) = 4.15$ V, and $N_d = 5 \times 10^{15} \text{ cm}^{-3}$, find the value of ϕ_s for two extreme cases of an Al–SiO$_2$–Si MOS structure: (1) $d_{ox} = 0$ and (2) $d_{ox} \gg W_{sc}$, the width of space-charge region near the interface.

(c) Obviously, for a given ξ (i.e., substrate doping concentration), ϕ_s depends on the oxide thickness d_{ox}. Find the values of d_{ox} in order to have $\phi_s = 0.50$ V and $\phi_s = 0.40$ V.

9.30. (a) Consider the establishment of a p-channel in an Al–SiO$_2$–Si MOSFET built on a n-Si substrate with $N_d = 5 \times 10^{15}$ cm^{-3}. Show that for a nondegenerate semiconductor, the surface-charge density needed to make $p_s = n_b$ is given by

$$Q_s = \sqrt{2\epsilon e N_d(2|\phi_b|)}$$

and that the result above can be obtained from Eq. (7.9.6) by setting $\phi_s = -\phi_b$.

(b) Given: $\mathcal{E}_w(\text{Al}) = 4.10$ eV, $\chi(\text{Si}) = 4.15$ V, $\epsilon(\text{Si}) = 11.7\epsilon_0$, $\epsilon_{\text{ox}} = 3.8\epsilon_0$, and $d_{\text{ox}} = 20$ nm, find the threshold voltage V_T for the MOSFET. Refer to Fig. 9.28 for the difference between ϕ of Section 9.11 and ϕ of Section 7.9.

9.31. Repeat Problem 9.30 for the establishment of a n-channel in an Al–SiO$_2$–Si MOSFET built on a p-Si substrate with $N_a = 5 \times 10^{15}$ cm^{-3}.

9.32. (a) If we set the condition $\mathcal{E}_c - \mathcal{E}_f > 2kT$ at the interface as the condition for a nondegenerate surface, use Eq. (7.9.6) to calculate the maximum surface charge density Q_s achievable in a nondegenerate n channel on a p substrate with $N_a = 5 \times 10^{15}$ cm^{-3}. For a given n-MOSFET with $\epsilon_{\text{ox}} = 3.8\epsilon_0$ and $d_{\text{ox}} = 500$ Å, does $\mathcal{E}_f$ at the interface stay below $\mathcal{E}_c$ at a gate voltage $V_G - V_T = 2$ V?

(b) Also calculate the change in $\mathcal{E}_c - \mathcal{E}_f$ at the interface from a flat-band situation to $\mathcal{E}_c - \mathcal{E}_f = 2kT$ at the interface. From this change, find the maximum value for the surface-state density N'_{ss} (per eV) allowable if the induced charge density $\Delta e\, N_{ss}$ is kept below $0.05Q_s$, where N_{ss} is assumed to be uniformly distributed in the forbidden gap, or in other words, $N'_{ss} = N_{ss}/1.12$ for Si.

9.33. (a) Consider Au/n-GaAs MESFETs fabrication by Se implantation into semi-insulating GaAs substrates. Depending on the implantation dose and range, two operational modes are possible: one being the normally-on or depletion and the other being the quasi-normally-off or enhancement mode. Given $\phi_{Bn} = 0.95$ V and $N = 10^{17}$ cm^{-3}, find the maximum value of a in Eq. (9.13.8a) in order for the MESFET to be in the quasi-normally-off mode. For an E-MESFET with $L = 1$ μm and $Z = 50$ μm, find the maximum gate voltage V_G allowed in order to keep the gate current $I_G < 10$nA. Also calculate the value of g_m at the maximum gate voltage if $\mu'_n = 4000$ cm^2/V-s.

(b) Suppose that we now use $a = 0.18$ μm. Find the threshold voltage V_T and the value of g_m at $V_G = 0$. Is the MESFET a D-MESFET or E-MESFET?

9.34. A quasi-normally-off (E-mode) MESFET consumes much less power than a normally-on (D-mode) MESFET; therefore, the former is more preferable than the latter. Consider the choice of the doping concentration N in Eq. (9.13.8a) for an E-MESFET with $\phi_{Bn} = 0.95$ V.

(a) Find the maximum value of a in Eq. (9.13.8a) for three E-MESFETs with $N = 10^{16}$ cm^{-3}, 10^{17} cm^{-3}, and 10^{18} cm^{-3}. Given: $L = 1$ μm and $Z = 50$ μm, find the maximum gate voltage V_G allowed to keep the gate current $I_G < 10$nA. Also calculate the value of g_m at the maximum gate voltage if $\mu'_n = 7000$ cm^2/V-s for $N = 10^{16}$ cm^{-3}, 4000 cm^2/V-s for $N = 10^{17}$ cm^{-3}, and 3000 cm^2/V-s for $N = 10^{18}$ cm^{-3}.

(b) Based on considerations of ease of control in implantation process and attainment of a reasonable g_m, choose the value of N for an E-MESFET.

9.35. (a) The values of work function $\mathcal{E}_w = e\phi_M$ are given below for a number of metals for possible use as gate contact with GaAs. The measured values of the barrier potential ϕ_{Bn} (Fig. 9.21) are also given.

	Au	Ag	Pt	Al	W	Cu
ϕ_M (V)	4.8	4.3	5.3	4.2	4.5	4.4
ϕ_{Bn} (V)	0.95	0.90	0.90	0.80	0.78	0.85

From the electron affinity $e\chi = 4.09$ eV for GaAs, find the expected values of ϕ_{Bn} and compare these values with the measured values. Obviously, the two values do not agree.

(b) Refer to the energy-band diagram of Fig. 9.28a in which the midregion represents an interfacial region of thickness d_{if} between the metal and the semiconductor. We suppose that surface states exist at the interface–semiconductor boundary with a density D_{ss} per cm^2 per eV, starting from $\mathscr{E}_v + e\phi_0$, and that they are negatively charged when occupied. By counting the charge density Q_{ss} (C/cm^2) stored in the surface states and neglecting the charge density stored in the semiconductor, show that the barrier potential ϕ_{Bn} is given by

$$\phi_{Bn} = \frac{b}{1+b}\left(\frac{\mathscr{E}_g}{e} - \phi_0\right) - \frac{1}{1+b}(\phi_M - \chi)$$

where $b = e^2 D_{ss} d_{if}/\epsilon_{if}$ and ϵ_{if} is the permittivity of the interfacial region. For $b \gg 1$, ϕ_{Bn} becomes independent of $\phi_M - \chi$, and the Fermi level is pinned at $\mathscr{E}_v + e\phi_0$. Given $d_{if} = 5$ Å and $\epsilon_{if} = 3\epsilon_0$, calculate the value of D_{ss} needed to make $b = 10$.

9.36. Draw an energy-band diagram similar to Fig. 9.21b for a degenerate, n-type semiconductor. Suppose that the conduction band edge crosses the Fermi level at a distance d from the interface. In the region $0 < x < d$, the space charge is mainly due to ionized donors. Express the average field $E_{av} = \phi_{Bn}/d$ in terms of N_I and show that the ratio E_0/E_{av} is given by

$$\frac{E_0}{E_{av}} = \left(\frac{\pi\phi_{Bn}}{2\hbar}\right)\left(\frac{\epsilon m^*}{N}\right)^{1/2}$$

Given $\phi_{Bn} = 0.65$ V for W, 0.63 V for Al, 0.80 V for Au, and 0.86 V for PtSi, and $m^* = 0.19\, m_0$ for Si, find the values of E_0/E_{av} and thus the value of the specific contact resistance (Ω/cm^2) for the metals on Si with $N = 10^{20}$ cm^{-3}.

10

High-Field Phenomena and Hot-Electron Effects

10.1 INTRODUCTION

The continually growing field of communications has pushed information carriers toward higher and higher frequencies in order to accommodate more and more communication channels and to provide a broader band for each channel needed for transmission of information of high resolution at a fast rate. The inventions of the magnetron, the klystron, and the traveling-wave tube have made operation of communication systems possible and reliable in the microwave and millimeter wave region (for the X band, for example, frequency $\nu = 10$ GHz and wavelength $\lambda = 3$ cm). These devices belong to the general category of vacuum tubes, in which the current is derived from electron emission from a heated cathode. With the rapid advance of the semiconductor technology, two new microwave devices using semiconductors were invented and have been developed. One device, generally referred to as the *Gunn-effect device*, named after the discoverer (J. B. Gunn, "Microwave Oscillations of Current in III–V Semiconductors," *IBM J. Res. Dev.*, Vol. 8, pp. 141–159, 1964; *Solid State Commun.*, Vol. 1, p. 88, 1963), utilizes the effect of electron transfer in the conduction band from low-energy, high-mobility states to high-energy, low-mobility states (Fig. 6.23a) in GaAs under a high electric field. The second device, generally referred to as the *IMPATT diode*, which was evolved from the Read diode proposed by Read (W. T. Read, "A Proposed High-Frequency Negative Resistance Diode," *Bell Syst. Tech. J.*, Vol. 37, pp. 401–406, 1958), utilizes the avalanche process in a reverse-biased *p-n* junction for generation of energetic carriers. The two semiconductor microwave devices promise to provide simple and reliable means of generating and amplifying microwaves and millimeter waves.

Although coherent sources are used to generate signals to carry information, fast electronic devices are needed to process the signal, without which a high-frequency or high-speed information system will not function. The rapid development of the compound–semiconductor technology has opened a new frontier for semiconductor devices, namely a new class called *heterostructure devices,* capable of operation at ultrahigh speed. By *heterostructure* we mean device structures made of two different semiconductors. With heterostructures, it is possible to invent new device concepts and to

462

achieve device performances, such as mobility enhancement, which are unattainable in homostructure devices. These fast electronic devices, which are based on high electron mobility or improved RC constants or a combination of both, are important to systems using analog circuits as well as systems using digital circuits. It is also hoped that the greatly expanded and still expanding horizon of heterostructure physics may lead to discoveries of devices drastically different from devices based on homostructure technology (G. H. Heilmeier, *Tech. Dig.*, International Electron Devices Meeting (IEDM), p. 2, 1984). We should also point out that carriers of information need not be limited to microwaves and millimeter waves. The advent of low-loss optical fibers has made possible the use of guided optical waves as information carriers. Although heterostructure devices are still in their infancy, it is fair to say that their impact has already been, and will increasingly be, felt in microelectronics as well as optoelectronics.

Before we discuss the operation of the semiconductor high-frequency and high-speed devices, some preliminary considerations are useful. Here we touch on one aspect important to the operation of a high-frequency device, namely the transit-time effect. In a bipolar transistor, the transit time τ_t required for injected minority carriers to diffuse across the base region is given by

$$\tau_{t1} = \frac{W^2}{2D} \qquad (10.1.1)$$

where W is the base width and D is the diffusion constant. It is the transit-time effect that results in a capacitive component in the base current (represented by the diffusion capacitance in the equivalent circuit). In semiconductor devices employing majority carriers, such as the field-effect transistor, the transit time τ_t required for majority carriers to move a distance d is

$$\tau_{t2} = \frac{d}{\mu E} = \frac{d^2}{\mu V} \qquad (10.1.2)$$

where μ is the drift mobility and V is the voltage drop across a distance d, for example, the channel length. Using the Einstein relation $eD = \mu kT$, we find that

$$\frac{\tau_{t1}}{\tau_{t2}} = \left(\frac{W}{d}\right)^2 \frac{eV}{2kT} \qquad (10.1.3)$$

At room temperature, kT/e is only 0.026 eV; hence the factor $eV/2kT$ has a value on the order of 100 with $V = 5$ V. For a base width W of 0.1 μm and a channel length d of 1 μm, the two transit times are comparable. The choice between bipolar and field-effect transistor can be determined, therefore, only after a careful and detailed examination of all the factors involved, including the transit-time effect. Even though high-frequency and high-speed operation of a semiconductor device would certainly be affected, and in some cases even be controlled by RC constants, the transit-time effect would set an upper limit for the operating frequency.

Because of the transit time, the conduction current flowing through a device possesses a reactive component in phase quadrature with the applied voltage as well as the resistive component in phase with the applied voltage. The phase shift caused by the transit time is equal to $\theta = \omega \tau_{t2}$. An appreciable reactive component is expected if θ becomes comparable to $\pi/4$. Here we set $\theta = \pi/4$ as the condition for the transit-time effect to become important. Using Eq. (10.1.2), we can express this condition in terms of the drift velocity v_d as

$$v_d \geq 8 \, dv \qquad (10.1.4)$$

where $v = \omega/2\pi$ is the operating frequency. For $v = 10$ GHz we find that $v_d = 4 \times 10^7$ cm/s if $d = 5$ μm and $v_d = 8 \times 10^6$ cm/s if $d = 1$ μm. If the mass of the carrier is assumed equal to the free-electron mass, the energy associated with the drift motion is calculated to be 0.45 eV and 0.018 eV for the two cases, which is considerably larger than or at least comparable to the thermal energy 0.026 eV. We refer to these electrons as hot electrons. Assuming that $\mu = 2 \times 10^3$ cm²/V-s, the field E required to attain the high drift velocity is calculated to be 2×10^4 V/cm and 4×10^3 V/cm, respectively.

Although the analysis above is crude in the sense that the values of v_d and E thus obtained are not accurate, it serves to bring to the fore two salient points regarding the operation of a high-frequency device: the need for carriers to attain a high drift velocity and the likelihood for the device to operate under a high electric field. The purpose of our discussion in this chapter is twofold: to acquaint ourselves with new phenomena which may lead to devices suitable for high-frequency operation and to prepare ourselves with the background needed for an analysis of the high-frequency devices.

There are many legitimate and interesting questions concerning the behavior of energetic carriers and carriers under a high electric field. One important question concerns the validity of the linear relation $v_d = \mu E$. In Section 10.2 we discuss the behavior of v_d at high fields in elemental semiconductors, where the low-field mobility is dominated by acoustic-phonon scattering. It is shown that the whole electron or hole distribution can be heated to a temperature considerably higher than the lattice temperature. Once carriers become energetic, there exists the interesting possibility of creating new situations. In Section 10.3 we discuss the effect of electron transfer in GaAs from low-energy, high-mobility states to high-energy, low-mobility states in the conduction band. If the energy of carriers is further increased to a level sufficient to break the covalent bond, electron–hole pairs may be generated through collisions of carriers with atoms. In Section 10.4 we discuss the avalanche process of carrier multiplication. We all know that a dielectric breaks down under sufficiently high fields. One of the breakdown mechanisms is through tunneling of electrons directly from the valence band into the conduction band. In Section 10.5 we discuss the tunneling process in semiconductors. In Section 10.6 we discuss the breakdown in a reverse-biased p-n junction in terms of the avalanche and tunneling processes. Field-effect transistors have become the workhorse in silicon integrated-circuit technology. In Section 10.7 we discuss hot electron effects in Si MOSFETs. In device analysis, knowing the drift velocity versus field relation is indispensible. Transport equations are presented in Section 10.8 for Si and in Section 10.9 for GaAs. We reserve a discussion of high-frequency and high-speed devices for subsequent chapters.

10.2 HIGH-FIELD DRIFT VELOCITY OF CARRIERS

The behavior of drift velocity of carriers at high electric fields deviates substantially from the linear relation $v_d = \mu E$ observed at low fields. Figure 10.1 shows the experimentally determined values of drift velocity as a function of the applied electric field in Ge, Si, and GaAs. Apart from the appearance of a negative differential mobility in

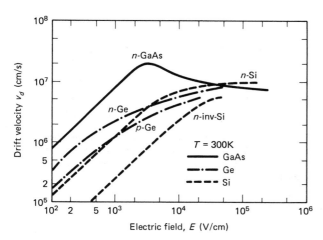

Figure 10.1 Measured drift velocity of carriers as functions of the applied field in Ge, Si, and GaAs. The curve labeled *n*-inv-Si was taken in a Si *n*-inversion channel. (E. J. Ryder, *Phys. Rev.*, Vol. 90, p. 766, 1953; R. D. Larrabee, *J. Appl. Phys.*, Vol. 30, p. 857, 1959; C. B. Norris and J. F. Gibbons, "Measurements of High-Field Carrier Drift Velocities by a Time-of-Flight Technique," *IEEE Trans. Electron Devices*, Vol. ED-14, p. 38, 1967 © 1967 IEEE; F. F. Fang and A. B. Fowler, *J. Appl. Phys.*, Vol. 30, p. 857, 1959; J. G. Ruch and G. S. Kino, *Appl. Phys. Lett.*, Vol. 10, p.40, 1967; P.M. Smith, M. Inoue, and J. Frey, *Appl. Phys. Lett.*, Vol. 37, p. 797, 1980.

GaAs (to be discussed in Section 10.3), the drift velocities in the three semiconductors all tend to saturate at high fields. To understand the high-field behavior, let us review our discussion in Chapter 4. The linear relation $v_d = \mu E$ is based on the assumption that the average velocity acquired by a free carrier in the direction of the applied field during the time between collisions is given by $v_d = (e\tau/m^*)E$, where m^* is the effective mass and τ is the mean free time. This linear relation breaks down when the energy of carriers becomes considerably higher than the thermal energy. In this section we treat the field dependence of mobility for cases where the low-field mobility is dominated by acoustic-phonon scattering.

The behavior of carrier mobility at high fields was first analyzed by Shockley (W. Shockley, *Bell Syst. Tech. J.*, Vol. 30, p. 990, 1951). Here we present a much simplified version of his treatment. We follow the approach used by Shockley to gain physical insight into the problem without getting deeply involved with the mathematics even though the treatment is a first-order approximation. Advanced treatments of hot carriers based on the Boltzmann transport equation are discussed in Section 10.8. We know that the electrons (from here on we use electrons as an example of free carriers) acquire energy from the applied field and dissipate it to the host lattice. Since energy flow is from the electron heat reservoir to the lattice heat reservoir, the electron temperature must exceed the lattice temperature. At low fields, the temperature difference is nonzero but insignificant. The dependence of drift velocity at high fields is quite different from that at low fields because the electron temperature is significantly above the lattice temperature. We can find the electron temperature by analyzing the rate of energy transfer between electrons and the lattice. For this purpose we use the model presented in Fig. 10.2, where the electron is represented by a particle of mass m^* and the lattice vibration is represented by a particle (a quantum particle called a *phonon*) of mass M. The interaction of an electron with the host lattice can be thought of, in the classical approximation, as a collision between two hard spheres.

Before we can analyze the electron–phonon interaction, we must know the phonon property. Figure 10.3 shows the phonon spectra in Ge, Si, and GaAs. Refer to our discussion in Section 2.11. In a lattice wave, the movement of atoms can be either in

(a) Process 1   (b) Process 2

Before collision

After collision

Figure 10.2 Schematic representation of electron–phonon interaction as collisions between two particles. The model is useful only as a first-order approximation of the interaction. In process 1, the phonon is assumed at rest initially. In process 2, the electron is assumed at rest initially.

$$\mathcal{E} \ (eV) = \frac{h\nu}{e} = \frac{6.625 \times 10^{-34}}{1.6 \times 10^{-19}} \nu = 4.14 \times 10^{-15} \nu$$

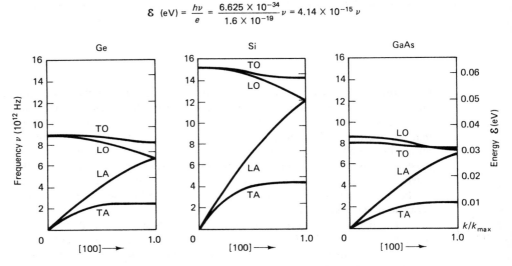

Figure 10.3 Phonon spectra in Ge, Si, and GaAs. (B. N. Brockhouse, *Phys. Rev. Lett.*, Vol. 2, p. 256, 1959; J. L. T. Waugh and G. Dolling. *Phys. Rev.*, Vol. 132, p. 2410, 1963).

the same direction as the direction of wave propagation or in a direction transverse to it. The former is referred to as the *longitudinal mode* and the latter as the *transverse mode*. Furthermore, in both the diamond (Ge and Si) and zinc-blende (GaAs) structures, there are two atoms per unit cell. The two atoms can move either in phase or out of phase, resulting in different energy of lattice vibration. Obviously, the mode with out-of-phase movements involves more mechanical energy than the mode with in-phase movements. The high-energy branch of the phonon spectra designated by LO and TO is called the *optical-phonon branch* and the low-energy branch designated by LA and TA is called the *acoustic-phonon branch*.

As discussed in Sections 6.7 and 6.8, the dominant scattering mechanism in GaAs is the polar scattering by optical phonons. On the other hand, the temperature dependence of mobility in Si indicates that carrier scattering at room temperature is dominantly by acoustic phonons even though there is a significant contribution from optical phonons. As a first-order approximation, the situation in Si can be stated as follows. At low and moderate field strengths, only acoustic phonons are important because not many electrons in the electron distribution have sufficient energy to interact with optical phonons. (In the quantum-mechanical picture, electrons give away their energy by emitting a phonon.) On the other hand, optical phonons become important once electrons have sufficient energy to interact with them, as they are much more effective than acoustic phonons in removing energy from electrons. Therefore, we consider only the interaction with acoustic phonons at low and moderate fields and only the interaction with optical phonons at high fields. This greatly simplified treatment correctly predicts the qualitative behavior of the drift velocity. A phenomenological treatment of hot electrons in GaAs is given in Section 10.3 and more formal treatment is given in Section 10.9.

For acoustic phonons, the two quantities that enter into the calculation are the lattice temperature T_l and the acoustic velocity u. The equivalent mass M of the acoustic phonons can be found from the equipartition theorem $Mu^2 = kT_l$. For simplicity, only a one-dimensional model is considered in Fig. 10.2 and treated here. The acoustic velocity u is given by the slope of the phonon dispersion curve or $d\omega/dk$ at $k = 0$ in Fig. 10.3. In most solids, the acoustic velocity is in the range 10^5 to 10^6 cm/s. If we take $u = 5 \times 10^5$ cm/s, we find $M = 1.7 \times 10^{-28}$ kg, which is larger than the effective mass of free carriers by several orders of magnitude. Therefore, for acoustic phonons, the inequality $M >> m^*$ holds. It becomes clear shortly that this large difference in masses makes the energy transfer per collision very small from an energetic electron to the acoustic phonon.

To analyze the interaction of electrons with acoustic phonons, we first consider the collision process involving an electron with an initial momentum p_1 and a phonon of mass M initially at rest as illustrated in Fig. 10.2a. For simplicity, we treat here only collinear interactions, that is, collisions in a one-dimensional space. From classical mechanics we know that the momentum p_2 of the electron reverses its direction after collision in cases where $m^* << M$. Conservation of energy and momentum requires that

$$\frac{p_1^2}{2m^*} - \frac{p_2^2}{2m^*} = \frac{p_r^2}{2M} \tag{10.2.1}$$

$$p_1 + p_2 = p_r \tag{10.2.2}$$

Dividing Eq. (10.2.1) by Eq. (10.2.2), we find that

$$p_1 - p_2 = \frac{m^*}{M}(p_1 + p_2) \tag{10.2.3}$$

Since $M^* << M$, we can approximate p_2 by p_1, which is $p_1 \cong p_2$. Thus the amount of energy transfer per collision can be approximated by

$$\langle \delta \mathscr{E} \rangle_1 = \frac{p_r^2}{2M} \cong \frac{(2p_2)^2}{2M} \cong \frac{2u^2 p_1^2}{kT_l} \tag{10.2.4}$$

where T_l is the lattice temperature such that $kT_l = Mu^2$. The reverse process in which a phonon with initial momentum p_r collides with an electron initially at rest is illustrated in Fig. 10.2b. For $m^* << M$, the scattered electron moves in the same direction as the phonon. Thus we have

$$\frac{p_r^2}{2M} - \frac{p_r'^2}{2M} = \frac{p^2}{2m^*} \tag{10.2.5}$$

$$p_r - p'_r = p \qquad (10.2.6)$$

From Eqs. (10.2.5) and (10.2.6), we obtain

$$p_r + p'_r = \frac{M}{m^*} p \gg p_r - p'_r \qquad (10.2.7)$$

The energy transferred from phonon to electron per collision is

$$\langle \delta \mathscr{E} \rangle_2 = \frac{p^2}{2m^*} = \frac{m^*}{2} \left(\frac{p_r + p'_r}{M} \right)^2 \cong 2m^* u^2 \qquad (10.2.8)$$

We note from Eqs. (10.2.3) and (10.2.7) that the change of momentum during collision is a very small fraction (of the order of m^*/M) of the initial momentum. Hence the fraction of energy transferred during collision is also exceedingly small. This makes the interaction of electrons with acoustic phonons ineffective in removing energy from electrons. In the meantime, the direction of the electron motion becomes randomized after each collision, which cannot be seen from a one-dimensional analysis but can be seen from a three-dimensional analysis. As a result, the thermal energy of electrons increases. Thermodynamically, we may think of electrons and host lattice as two heat reservoirs. If the rate of energy transfer is slow, the electrons acquire an average random velocity v_e exceeding the normal thermal velocity $v = (kT_l/m^*)^{1/2}$. This situation happens at moderately high fields. We define an electron temperature T_e such that

$$\tfrac{1}{2} m^* v_e^2 = \tfrac{1}{2} kT_e \qquad (10.2.9)$$

For constant mean free path l, the mobility of electrons can be expressed as

$$\mu = \frac{e}{m^*} \tau = \frac{e}{m^*} \frac{l}{v_e} = \mu_0 \left(\frac{T_l}{T_e} \right)^{1/2} \qquad (10.2.10)$$

where μ_0 is the low-field mobility. At high fields, T_e becomes larger than T_l, resulting in a lower μ and hence a lower v_d than the value $\mu_0 E$, where μ_0 is the low-field mobility.

To find the dependence of μ on E, we must obtain the power-balance equation, from which the electron temperature can be found. Since electrons on the average suffer one collision per mean free time τ, the time rate of net energy transfer (from electron to lattice) is

$$\left\langle \frac{d\mathscr{E}}{dt} \right\rangle = \frac{1}{\tau} (\langle \delta \mathscr{E} \rangle_1 - \langle \delta \mathscr{E} \rangle_2) = 2u^2 \left(\frac{p_1^2}{kT_l} - m^* \right) \frac{v_e}{l} \qquad (10.2.11)$$

from Eqs. (10.2.4) and (10.2.8). Equation (10.2.11) is based on a one-dimensional model and holds for a specific electron momentum p_1. To make it applicable to a real situation, certain modifications are needed. First, in a three-dimensional treatment, the constants in front of the two terms are different. Second, the value of p_1^2 must be averaged over the velocity distribution of electrons and thus expressed in terms of the average velocity v_e. The treatment of electron-lattice collision by Shockley yields

$$\left\langle \frac{d\mathscr{E}}{dt} \right\rangle = \frac{8u^2}{\sqrt{\pi}} \left(\frac{m^{*2} v_e^2}{3kT_l} - m^* \right) \frac{v_e}{l} \qquad (10.2.12)$$

The rate with which electrons acquire energy from the field is

$$\left\langle \frac{d\mathcal{E}}{dt} \right\rangle = eEv_d = e\mu E^2 = e\mu_0 E^2 \frac{v}{v_e} \qquad (10.2.13)$$

Under steady-state conditions, the two rates must be equal, yielding an expression relating v_e to E.

Equating Eq. (10.2.12) to Eq. (10.2.13) and using $m^* v_e^2 = 3kT_e$ for a three-dimensional case, we obtain

$$\left(\frac{T_e}{T_l} - 1 \right) \frac{T_e}{T_l} = \frac{3\pi}{32} \left(\frac{\mu_0 E}{u} \right)^2 \qquad (10.2.14)$$

The mean free path l was eliminated from Eq. (10.2.12) by using the relation

$$\mu_0 = \frac{4}{3\sqrt{\pi}} \frac{el}{m^* v} \qquad (10.2.15)$$

The factor $4/3\sqrt{\pi}$ results from averaging τ over the electron velocity distribution. From Eq. (10.2.14), T_e/T_l can be solved in terms of $\mu_0 E/u$ and from Eq. (10.2.10), the field-dependent mobility can be found. At low fields such that $\mu_0 E < u$, the solution can be approximated by

$$\mu = \mu_0 \left(\frac{T_l}{T_e} \right)^{1/2} = \mu_0 \left[1 - \frac{3\pi}{64} \left(\frac{\mu_0 E}{u} \right)^2 \right] \qquad (10.2.16)$$

At moderately high electric fields when $\mu_0 E > u$, $T_e \gg T_l$. Neglecting the linear term T_e/T_l in Eq. (10.2.14), we obtain: for the mobility,

$$\mu = \mu_0 \left(\frac{T_l}{T_e} \right)^{1/2} = 1.36 \mu_0 \left(\frac{u}{\mu_0 E} \right)^{1/2} \qquad (10.2.17)$$

and for the drift velocity,

$$v_d = \mu E = 1.36 (u\mu_0 E)^{1/2} \qquad (10.2.18)$$

So far we have limited our discussion to acoustic (low-energy) phonons. The role of optical (high-energy) phonons was first discussed by Shockley, and the importance of their role was further stressed in the analyses of Conwell and Stratton (E. M. Conwell, *Phys. Chem. Solids,* Vol. 8, p. 234, 1959; R. Stratton, *J. Electron. Control,* Vol. 5, p. 157, 1958). Referring to Fig. 10.3, we see that the optical phonons for intravalley scattering in Ge and Si are those at $k = 0$ and hence have an energy corresponding to a frequency v of 9×10^{12} and 15×10^{12} Hz or an equivalent temperature of 430 K and 710 K, respectively. The optical phonon energies involved in intervalley scattering is somewhat lower. Therefore, if μ drops to $0.5\mu_0$, we expect from Eq. (10.2.14) that the average electron energy will have an equivalent temperature of $4T_l = 1200$ K sufficient to excite optical phonons. This should occur at $\mu_0 E = 11u$ from Eq. (10.2.14). A reference to Fig. 10.1 indicates that optical phonons should become important for $E > 10^4$ V/cm, as compared to a value 3×10^3 V/cm calculated from $\mu_0 E = 11u$. The discrepancy is due to the simplifying step we took in our calculation by ignoring the optical phonons.

In the high-field region, electrons dissipate their energy to the lattice predominantly through emission of optical phonons. Let τ_{op} be the mean free time of electrons for optical phonon scattering and $\mathcal{E}_{op}$ be the optical phonon energy. For each optical-phonon scattering process, an electron gives up an energy $\mathscr{E}_{op}$ to create an optical phonon. If it is assumed that the optical-phonon scattering process dominates, energy balance requires that

$$\left\langle \frac{d\mathscr{E}}{dt} \right\rangle_{field} = \frac{e^2 E^2 \tau_{op}}{m^*} = \left\langle \frac{d\mathscr{E}}{dt} \right\rangle_{phonon} = \frac{\mathscr{E}_{op}}{\tau_{op}} \qquad (10.2.19)$$

From Eq. (10.2.19), we obtain

$$v_d = \mu E = \frac{eE}{m^*}\tau_{op} = \left(\frac{\mathscr{E}_{op}}{m^*} \right)^{1/2} \qquad (10.2.20)$$

which is independent of the applied electric field. Using $\mathscr{E}_{op} = 0.050$ eV for the LO phonon at $k[100]$ and $m^* = 0.27m_0$ for Si, we find a saturation value for v_d to be 1.7×10^7 cm/s, in general agreement with the experimentally observed value of 1×10^7 cm/s.

In summary, the general behavior of the drift velocity can be qualitatively explained by the theoretical analysis presented by Shockley and outlined above. In the moderate field region, the drift velocity deviates from a linear dependence on the applied field as a result of rising electron temperature due to ineffectiveness of acoustic phonons to remove energy acquired by electrons from the applied field. In the high-field region, the optical-phonon scattering process dominates and leads to saturation of the drift velocity. Equation (10.2.20) predicts a value of the saturation velocity in the neighborhood of 10^7 cm/s. The experimentally measured values are in the same range. Simple calculations based on Eqs. (10.2.16) and (10.2.17), on the other hand, show that the calculated departure from the linear dependence is too large as compared to the experimental result. Improved calculations by Conwell and by Stratton, which include a proper amount of optical phonon scattering at low and moderate fields, show a reasonably good fit to the experimentally measured field dependence of the drift mobility. Therefore, the theoretical treatments give a correct explanation of the physical mechanisms affecting the high-field mobility. However, because high-frequency field-effect transistors operate in the nonlinear velocity region, a more accurate v_d versus E relation is needed and is discussed in Section 10.8.

10.3 THE ELECTRON-TRANSFER EFFECT

As mentioned in the introduction (Section 10.1) the Gunn-effect device is based on the effect of electron transfer from low-energy, high-mobility states to high-energy, low-mobility states in the conduction band. In 1963, Gunn discovered current oscillations in n-type GaAs when he applied electric fields larger than 3000 V/cm. However, he did not give a definitive explanation for the effect. Earlier Ridley and Watkins (B. K. Ridley and T. B. Watkins, *Proc. Phys. Soc. London,* Vol. 78, pp. 293–304, 1961) and Hilsum (C. Hilsum, *Proc. IRE,* Vol. 50, pp. 185–189, 1962) had independently suggested on theoretical grounds that current instabilities could take place in certain many-valley semiconductors. It is apparent now that the drift-velocity characteristic of GaAs shown in Fig. 10.1 will result in a negative differential conductance which can lead to current instabilities. It took several years to gather sufficient experimental evidence to show conclusively that the electron-transfer mechanism is responsible for the Gunn effect. In this section we present a simple electron-transfer theory which shows the possible occurrence of a negative differential mobility in many-valley semiconductors. This simple treatment is not accurate but serves well as a starting point. A more refined treatment of the electron-transfer effect and other important aspects of the Gunn-effect device is discussed in Section 10.9.

The electron-transfer effect as manifested by a negative differential mobility is intimately connected with the band structure of a semiconductor. In general, the energy of conduction-band electrons is not a monotonic function of the wave number k but instead consists of several minima in the k space. Figure 10.4 shows an energy versus k diagram of a semiconductor capable of exhibiting the electron-transfer effect. In the calculation of such quantities as the electron concentration and the drift velocity, ordinarily only the properties, such as the effective mass and the mobility, of the lowest-energy minimum are involved because almost all electrons are located there. The situation can be changed at high fields because electrons may have sufficient energy to populate the next-to-lowest energy minimum. Since only two minima are involved in the electron-transfer effect, we refer to the lowest-energy minimum as the low-energy valley (or minimum) and the next-to-lowest energy minimum as the high-energy valley. For a semiconductor to show a negative differential mobility, the high-energy valley must have a lower mobility than the low-energy valley. Imagine a situation in which most electrons are located at the low-energy, high-mobility valley at a field strength E_1 and most electrons are transferred to the high-energy, low-mobility valley at a field strength E_2. It is conceivable that even though $E_2 > E_1$, the overall drift velocity for the whole electron distribution is lower at E_2 than at E_1 if the electron-transfer process takes place within a sufficiently small range of E.

The average drift velocity in a two-valley semiconductor is given by

$$v_d = \frac{n_1 \mu_1 + n_2 \mu_2}{n_1 + n_2} E \qquad (10.3.1)$$

For simplicity, we assume that $\mu_2 \ll \mu_1$ and thus can approximate Eq. (10.3.1) by

$$v_d = \mu_1 E \left(1 + \frac{n_2}{n_1} \right)^{-1} \qquad (10.3.2)$$

If N_{c1} and N_{c2} are the effective density of states for the low-energy and high-energy valleys, respectively, then

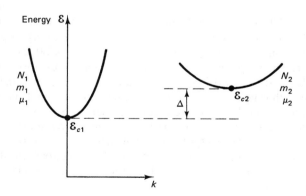

Figure 10.4 Schematic representation of a two-valley semiconductor.

$$n_2 = N_{c2} \exp \left(-\frac{\mathscr{E}_{c2} - \mathscr{E}_f}{kT_{e2}} \right) \tag{10.3.3}$$

$$n_1 = N_{c1} \exp \left(-\frac{\mathscr{E}_{c1} - \mathscr{E}_f}{kT_{e1}} \right) \tag{10.3.4}$$

We further assume that the electron temperatures are the same for the two valleys. Thus we have

$$\frac{n_2}{n_1} = B \exp \left(\frac{-\Delta}{kT_e} \right) \tag{10.3.5}$$

where $B = N_{c2}/N_{c1}$ and $\Delta = \mathscr{E}_{c2} - \mathscr{E}_{c1}$.

The other equation we need is the energy-balance equation. Notice that the average kinetic energy of electrons obeying the Boltzmann distribution is $3kT_e/2$. Furthermore, there is no net energy exchange between electrons and the lattice if $T_e = T_l$. Therefore, the net rate of energy exchange can be expressed in terms of the thermal energies (three-dimensional) of electrons and the lattice as

$$\left\langle \frac{d\mathscr{E}}{dt} \right\rangle = \frac{3}{2} (kT_e - kT_l) \frac{1}{\tau_{\mathscr{E}}} \tag{10.3.6}$$

where $\tau_{\mathscr{E}}$ is a phenomenological energy-relaxation time. The interaction with the applied field is still given by Eq. (10.2.13). Equating Eq. (10.3.6) to Eq. (10.2.13) and using Eqs. (10.3.2) and (10.3.5), we obtain

$$\theta_e = \theta_l + \left(\frac{E}{E_0} \right)^2 \left[1 + B \exp \left(-\frac{1}{\theta_e} \right) \right]^{-1} \tag{10.3.7}$$

where $E_0^2 = 3\Delta/(2e\mu_1\tau_{\mathscr{E}})$, $\theta_e = kT_e/\Delta$, and $\theta_l = kT_l/\Delta$. Equation (10.3.7) is expressed in a normalized form so that for a given B, the value of θ_e can be found from Eq. (10.3.7) as a function of θ_l and E/E_0. Once the value of θ_e is known, the value of v_d can be found from Eq. (10.3.2) as

$$\frac{v_d}{v_0} = \frac{E}{E_0} \left[1 + B \exp \left(-\frac{1}{\theta_e} \right) \right]^{-1} \tag{10.3.8}$$

where $v_0 = \mu_1 E_0$.

The computed results are shown in Fig. 10.5, where the values of v_d/v_0 are plotted as a function of E/E_0 for $B = 50$ and for several values of θ_l. The fact that Eqs. (10.3.7) and (10.3.8) are expressed in terms of normalized quantities makes the family of curves almost universal. The only quantity that is specified is $B = N_{c2}/N_{c1} = 50$, and information about the other relevant properties of a specific material is contained in the normalization factors Δ, E_0, and v_0. Two important features of the curves stand out. First, there is a well-defined threshold for the onset of negative differential mobility. Second, the negative mobility diminishes as θ_l increases, and ceases to exist if θ_l is too large. This happens because the ratio n_2/n_1 in Eq. (10.3.2) does not increase fast enough to overcome the increase in E. The other factor besides θ_l which controls the ratio n_2/n_1 is the quantity B. A large value of B means a significant portion of electrons transferred to the high-energy valley within a short range of E. Therefore, a large B value has an effect equivalent to a small θ_l value.

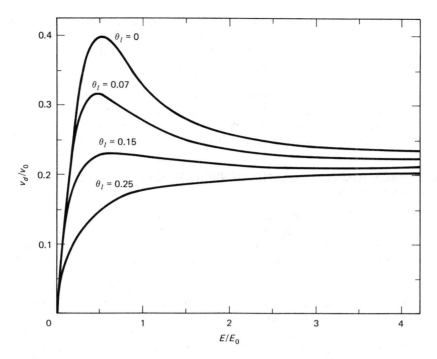

Figure 10.5 A plot of v_d/v_0 versus E/E_0. The curves are computed from Eqs. (10.3.8) and (10.3.7) for $B = 50$ and several values of θ_l, and are used to show the possibility of exhibiting a negative differential mobility in a two-valley semiconductor.

To show that the conclusions drawn above are qualitatively correct, we refer to the velocity-field characteristics shown in Fig. 10.1. In Ge and Si, the low-energy valley has a much heavier mass than the high-energy valley. As a result, $\mu_2 > \mu_1$ and a negative differential mobility is not expected. Furthermore, $B << 1$ and no significant amount of electron transfer in the field range of 10^4 V/cm is expected. The situation in GaAs is just opposite to those in Ge and Si. The two conditions $\mu_2 < \mu_1$ and $B >> 1$ are both satisfied in GaAs. Furthermore, the energy separation between the two valleys is $\Delta = 0.3$ eV, which gives a value of $\theta_l = 0.08$. From Fig. 10.5, the curve for $\theta_l = 0.07$ indeed shows a negative differential mobility.

Although the simple electron-transfer theory predicts qualitatively the appearance of negative differential mobility in GaAs, it does not give accurate quantitative results. A case in point is the threshold for the onset of negative differential mobility, which according to Fig. 10.5 is about equal to $0.5E_0$. To find E_0, some discussion on the relation between $\tau_\mathscr{E}$ and τ is in order. If acoustic-phonon scattering dominates, Eq. (10.2.12) leads to

$$\tau_\mathscr{E} = \tau\left(\frac{\sqrt{\pi M}}{16m^*}\right) \tag{10.3.9}$$

by comparing it with Eq. (10.3.6). The factor M/m^* enters into the expression because for each collision an electron loses only a fraction m^*/M of its energy to the lattice according to Eq. (10.2.4). As discussed in Section 6.8, scattering by transverse (polar) optical phonons is the dominant process in GaAs. For electrons with energy comparable to optical phonon energy, the energy loss is almost total; therefore, we expect $\tau_{\mathscr{E}} = \tau$ and $E_0^2 \cong (\Delta/\mu_l^2 m_l^*)$. Using $\Delta = 0.3$ eV, $\mu_l = 8000$ cm^2/V-s, and $m_l^* = 0.068m_0$, we find E_0 to be of the order of 1×10^4 V/cm. The observed threshold field for the onset of negative differential mobility in GaAs is only 3500 V/cm. The discrepancy is an indication that even optical phonons become rather ineffective in removing energy from electrons at sufficiently high fields. A more refined electron-transfer theory which produces reasonably accurate quantitative results will be presented in Section 10.9.

10.4 IMPACT IONIZATION AND CARRIER MULTIPLICATION PHENOMENA

The ideal diode equation predicts saturation of the current under a reverse-bias voltage. What we find in real diodes is that the current I in a reverse-biased diode remains more or less constant until the reverse-bias voltage V approaches the breakdown voltage V_B. Figure 10.6 shows typical current–voltage characteristics of practical junction diodes. The current I increases very rapidly as V approaches V_B, and becomes indefinitely large at $V = V_B$. The rapid increase of current is due to production of electron–hole pairs by energetic carriers in the space-charge region of a p-n junction. As discussed in Sections 10.2 and 10.3, a free carrier can gain from an applied electric field a kinetic energy well above the thermal energy. In Fig. 10.1 this field is only in the range of 10^4 V/cm. If the electric field is increased further, the kinetic energy of free carriers is expected to increase accordingly. At sufficiently high electric fields, typically in the range 2×10^5 V/cm or larger, an electron (or hole) can have kinetic energy in excess of the binding energy of a valence electron to its parent atom. Thus, in a collision with an atom, such an electron can break the covalent bond and produce an electron–hole pair. This process is very similar to the impact-ionization process in gaseous discharge. We use the term *impact ionization* here also to distinguish this mechanism of junction breakdown from tunneling of electrons, which is important in highly doped p-n junctions and is discussed in Sections 10.5 and 10.6.

One important quantity characterizing an impact-ionization process is the ionization energy $\mathscr{E}_i$, which is the minimum energy required by an electron or hole to ionize

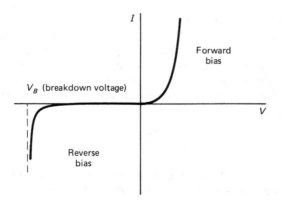

Figure 10.6 Current–voltage (I–V) characteristic of a p-n junction diode. For semiconductors with $\mathscr{E}_g > 1$ eV, the reverse current before breakdown is negligibly small as compared to the forward current if they are plotted on the same linear scale.

an atom (i.e., to break the covalent bond to produce an electron–hole pair). In the classical approximation, we may think of the process as an inelastic collision between an energetic electron (or hole) and a stationary atom. The energy lost by the electron is used to produce an electron–hole pair. The quantity $\mathscr{E}_i$ is generally bigger than the gap energy $\mathscr{E}_g$ because of the conditions imposed by conservation of energy and momentum in the collision.

If we let v_0 and v_1 be, respectively, the initial and final velocity of the incident electron and v_e and v_h be the velocity of the newly produced electron and hole, respectively, conservation of energy requires that

$$\tfrac{1}{2} m_e^* v_0^2 = \mathscr{E}_g + \tfrac{1}{2} m_e^* v_1^2 + \tfrac{1}{2} m_e^* v_e^2 + \tfrac{1}{2} m_h^* v_h^2 \qquad (10.4.1)$$

where $\mathscr{E}_g$ is the potential energy required to break the covalent bond, which is assumed to be the gap energy. A general treatment of momentum conservation requires formulation in a three-dimensional space. Here we are interested in finding the minimum value of $m_e^* v_0^2/2$. Obviously, from Eq. (10.4.1), a minimum is reached when all the momenta are collinear because any nonvanishing components of v_1, v_e, and v_h in the perpendicular direction would raise the value of $m_e^* v_0^2/2$. Thus the condition for momentum conservation at the minimum (that is, in the collinear case) becomes

$$m_e^* v_0 = m_e^* v_1 + m_e^* v_e + m_h^* v_h \qquad (10.4.2)$$

It can be shown (in Problem 10.10) that the kinetic energy of the incident electron is further minimized when $v_1 = v_2 = v_h$. This minimum energy is designated as $\mathscr{E}_i$ and is given by

$$\mathscr{E}_i = \left(\frac{1}{2} m_e^* v_0^2 \right)_{\min} = \frac{2m_e^* + m_h^*}{m_e^* + m_h^*} \mathscr{E}_g \qquad (10.4.3)$$

A similar expression with m_e^* and m_h^* interchanged applies to energetic holes. However, at present, it is not possible to distinguish meaningfully between the values of $\mathscr{E}_i$ for electrons and holes. Thus we take an average of the two values, which yields $\mathscr{E}_i = 1.5\mathscr{E}_g$.

In Fig. 10.7, we illustrate schematically the processes of (a) impact ionization and (b) tunneling. At very high electric fields, the band edges change significantly with distance. Two situations may arise as a result of the high field. As illustrated in Fig. 10.7a, an electron may gain substantial kinetic energy over a short distance. The large kinetic energy gain is then used to produce an electron–hole pair. The other situation is illustrated in Fig. 10.7b, where the distance AB that an electron wave packet in the valence band (here quantum-mechanical language must be used) must penetrate in order to reach the conduction band is greatly shortened by the field. Because tunneling involves a single electron, the energy for that electron is conserved. This is not true for the incident electron in the impact ionization process. For the tunneling process, the important quantity is the probability with which the electron wave packet can penetrate the distance AB. For the impact-ionization process, the important quantity is the rate of electron–hole production. Returning to Fig. 10.7a, we see that the hole generated can also gain sufficient kinetic energy to produce an additional electron–hole pair. Since electrons and holes travel in opposite directions, the carriers can multiply few times in the space-charge region before they reach the exit electrode.

The number of electron–hole pairs produced by an energetic carrier per unit distance traveled is called the *ionization coefficient*. For ease of discussion, we consider an

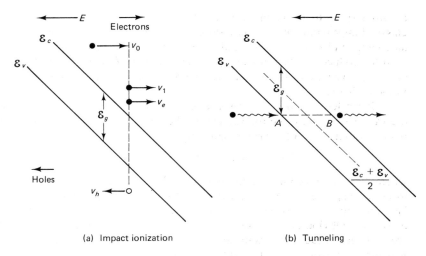

(a) Impact ionization   (b) Tunneling

Figure 10.7 Schematic diagrams comparing (a) the impact ionization process and (b) the tunneling process.

idealized case of two parallel-plate electrodes of distance W apart (Fig. 10.8) between which a high electric field exists. A *p-i-n* diode with heavily doped p and n regions is a practical example of the situation. We assume that electrode A emits a flux of N_0 electrons per second per cm^2, whereas electrode B emits a flux of P_0 holes per second per cm^2. In an elementary distance dx, the number of electron–hole pairs produced per second per cm^2 by energetic electrons is, by definition, given by

$$dN = dP = \alpha_e N \, dx \qquad (10.4.4)$$

where α_e is the ionization coefficient for energetic electrons. A similar equation

$$dN = dP = \alpha_h P \, dx \qquad (10.4.5)$$

represents the number of electron–hole pairs created per second per cm^2 by energetic holes. In general, the two ionization processes have different ionization coefficients α_e and α_h. We postpone a discussion of the nature of α_e and α_h until we have completed our analysis of the carrier multiplication phenomenon. We should point out that the

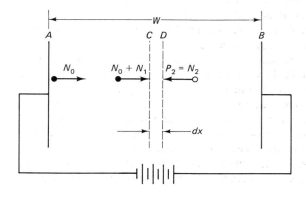

Figure 10.8 Schematic diagram used to find the number of electron–hole pairs generated within an elementary volume $A \, dx$, where A is the cross-sectional area.

carrier fluxes N and P (with a dimension $\text{cm}^{-2}\text{s}^{-1}$) are related, respectively, to the electron and hole concentration (with a dimension cm^{-3}) by a velocity. Since the carrier velocity is a constant (the saturation velocity v_s) at very high fields, $N = nv_s$ and $P = pv_s$. Thus Eqs. (10.4.4) and (10.4.5) also hold for carrier concentrations n and p.

To simplify our discussion, we assume that only electron emission takes place from electrode A but no hole emission from electrode B. The situation corresponds to that in a $p\text{-}i\text{-}n^+$ diode, where the reverse current is controlled by extraction of electrons from the less heavily doped p region. For analyzing carrier multiplication, we focus our attention on what is going on within the elementary distance dx. Due to the impact ionization process the number of electrons arriving per second per cm^2 from the left is equal to $N_0 + N_1$, where N_1 is the number of electrons generated per second per cm^2 in the space between A and C of Fig. 10.8. Even though no holes are emitted from electrode B, electron–hole pairs are generated in the space between B and D. The holes thus generated will move toward D. If P_2 is the number of holes created per second per cm^2 in the space between B and D, the total number of electrons generated per second per cm^2 within the elementary distance dx is the sum of Eqs. (10.4.4) and (10.4.5). Thus

$$dN = \alpha_e(N_0 + N_1)\,dx + \alpha_h P_2\,dx \qquad (10.4.6)$$

To proceed from Eq. (10.4.6), we make the following observations. First, since electrons and holes are created in pairs, the number of electrons created in the region BD is $N_2 = P_2$. Under steady-state conditions, electrons must arrive at electrode B at a rate equal to the total rate by which electrons have left electrode A and electrons are produced within the space $ACDB$. In other words, if N_f is the number of electrons arriving at B per second per cm^2, then $N_f = N_0 + N_1 + N_2$ with $N_2 = P_2$. Second, since electrons are moving from the left toward the right, dN created within dx adds to N_1. In other words, $dN = dN_1$. Using these two conditions and eliminating N_2 from Eq. (10.4.6), we obtain

$$\frac{dN_1}{dx} = (\alpha_e - \alpha_h)(N_0 + N_1) + \alpha_h N_f \qquad (10.4.7)$$

In applying Eq. (10.4.7) to a junction diode, it is obvious that the electrodes A and B of Fig. 10.8 correspond, respectively, to the boundaries of the space-charge region on the p and n sides of a $p\text{-}i\text{-}n$ diode. The fluxes N_0 and P_0 ($N_0 = n_p v_s$ and $P_0 = p_n v_s$) multiplied by the electronic charge e and the area A of the junction give the component of the reverse saturation current I_s in a junction diode due to electrons and holes, respectively. Because of the impact-ionization processes, however, the electron current will not be proportional to N_0 but to N_f. We define a *multiplication factor* M_e for electrons as

$$M_e = \frac{N_f}{N_0} = \frac{I_r}{I_s} \qquad (10.4.8)$$

In diodes with a heavily doped n region ($n_n \gg p_p$ or $n_p \gg p_n$), the electron current dominates. If the hole current is neglected, the ratio of the reverse current I_r to the reverse saturation current I_s yields M_e. Similarly, the ratio of I_r/I_s in diodes with a heavily doped p region gives the multiplication factor M_h for holes.

Since a general treatment of Eq. (10.4.7) is mathematically involved and does not contribute toward gaining physical insight, only the special case with $\alpha_e =$

$\alpha_h = \alpha$ is treated. Integrating Eq. (10.4.7) from A to B and realizing that $N_1 = 0$ at A and $N_f = N_0 + N_1$ at B, we obtain

$$N_1|_A^B = N_f - N_0 = N_f \int_0^W \alpha \, dx \qquad (10.4.9)$$

Using Eq. (10.4.8), we find that

$$M_e = \frac{I_r}{I_s} = \frac{1}{1 - \int_0^W \alpha \, dx} \qquad (10.4.10)$$

Equation (10.4.10) says that M_e goes to infinity if

$$\int_0^W \alpha \, dx = 1 \qquad (10.4.11)$$

When this state is reached, the diode current increases indefinitely and breakdown of the junction occurs. The phenomenon is known as *avalanche breakdown*. The carrier multiplication process is like avalanches triggering snow. In cases where $\alpha_e = \alpha_h$ the same breakdown condition as stated in Eq. (10.4.11) applies to M_h in diodes with a heavily doped p region ($p_p >> n_n$ or $p_n >> n_p$).

General expressions for M_e and M_h can be obtained from Eq. (10.4.7) and a similar equation for dP_1/dx. They are given by

$$M_e^{-1} = 1 - \int_0^W \alpha_e \left\{ \exp \left[-\int_0^x (\alpha_e - \alpha_h) \, dx' \right] \right\} dx \qquad (10.4.12)$$

$$M_h^{-1} = 1 - \int_0^W \alpha_h \left\{ \exp \left[-\int_x^W (\alpha_x - \alpha_e) \, dx' \right] \right\} dx \qquad (10.4.13)$$

In building up a chain reaction, a feedback loop is needed. The two impact-ionization processes, one by energetic electrons and the other by energetic holes, form such a loop. Even in diodes where one current component, say the electron current, dominates the reverse current before breakdown, the breakdown condition $M_e = \infty$ involves both α_e and α_h. Therefore, the ionization coefficients α_e and α_h must be determined from the combined results of two separate measurements: one for M_e in diodes where the electron current dominates and the other for M_e in diodes where the hole current dominates (S. L. Miller, *Phys. Rev.* Vol. 99, p. 1234, 1955; A. G. Chynoweth, *Phys. Rev.*, Vol. 109, p. 1537, 1958). Information about α_e and α_h can also be obtained from measuring the photo-excited current and avalanche noise in reverse-biased p-n junctions and Schottky diodes. However, there are considerable discrepancies in the reported values from different methods and in different device structures. For a discussion of precautions in interpreting the results and for a comparison of different methods, the reader is referred to extensive works by Stillman and coworkers. (G. E. Bulman, V. M. Robbins, and G. E. Stillman, *IEEE Trans. Electron Devices,* Vol. ED-32, p. 2454, 1985; G. E. Stillman, V. M. Robbins, and N. Tabatabaie, *IEEE Trans. Electron Devices,* Vol. ED-31, p. 1643, 1984).

The ionization coefficients for Ge, Si, GaAs, and GaP and InP as deduced from measurements of M_e and M_h are shown in Fig. 10.9. To understand the field dependence

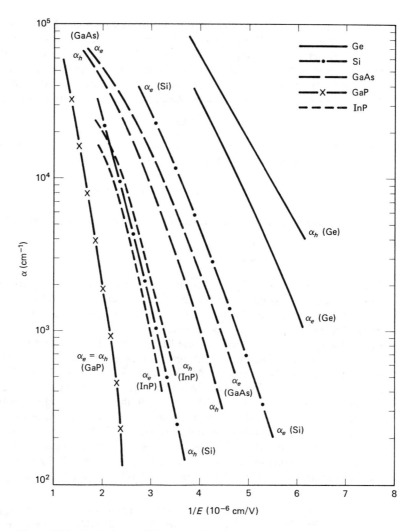

Figure 10.9 Ionization coefficients α_e and α_h in Ge, Si, GaAs, GaP, and InP as functions of the reciprocal electric field. The values are deduced from measured multiplication factors M_e and M_h. (S. L. Miller, *Phys. Rev.*, Vol. 99, 1234, 1955; C. A. Lee, R. A. Logan, R. L. Batdorf, J. J. Kleimack, and W. Wiegmann, *Phys. Rev.*, Vol. 134, p. A761, 1964; G. E. Bulman, V. M. Robbins, and G. E. Stillman, "The Determination of Impact Ionization Coefficient in (100) Gallium Arsenide Using Avalanche Noise and Photocurrent Multiplication Measurements" *IEEE Trans. Electron Devices*, Vol. ED-32, p. 2454, 1985 © 1985 IEEE; R. A. Logan and H. G. White, *J. Appl. Phys.*, Vol. 36, p. 3945, 1965; L. W. Cook, G. E. Bulman, and G. E. Stillman, *Appl. Phys. Lett.*, Vol. 40, p. 589, 1982.)

of α, we present here a rather simple argument which can be stated as follows. Since the electrons obey the Boltzmann distribution, the percentage of electrons having kinetic energy greater than $\mathcal{E}_i$ is equal to $\exp(-\mathcal{E}_i/kT_e)$. The electron temperature can be found from the energy-balance equation by equating Eq. (10.3.6) to Eq. (10.2.13). Both equations are applicable to the present case. We further note that at very high fields, $T_e > T_l$ and $v_d = v_s$. Thus electron temperature T_e is approximately given by

$$kT_e \cong \tfrac{2}{3} eEv_s\tau_e \tag{10.4.14}$$

Since dN/N in Eq. (10.4.4) is proportional to the percentage of electrons having $\mathcal{E} > \mathcal{E}_i$, the ionization coefficient is expected to vary as

$$\alpha = A \exp\left(-\frac{B}{E}\right) \tag{10.4.15}$$

where A is a proportionality constant having a dimension of inverse distance and

$$B = \frac{3\mathcal{E}_i}{2ev_s\tau_e} = \frac{\mathcal{E}_i}{el_r} \tag{10.4.16}$$

The quantity $l_r = 2v_s\tau_e/3$ is the distance an electron travels to gain energy $\mathcal{E}_i$ and thus can be interpreted as the mean free path of an energetic electron.

In Table 10.1 we list the values of the two parameters A and B obtained from fitting Eq. (10.4.15) to the experimental curves of Fig. 10.9. We should point out that Eq. (10.4.15) holds only for a limited range of α because of the simplistic model on which it is based. Therefore, different values of A and B are obtained from fitting different parts of the curve (S. M. Sze and G. Gibbons, *Appl. Phys. Lett.*, Vol. 8, pp. 111–113, 1966). The values in Table 10.1 are obtained by fitting Eq. (10.4.15) to α values in the range 10^3 to 10^5 cm^{-1} for this is the range mostly encountered in practical devices. From the value of B, we can estimate the value of l_r. In the high-field region, scattering by optical phonons dominates. In Si, $\mathcal{E}_i = 1.6$ eV. Thus we obtain $l_r = 80$ Å from Eq. (10.4.16). So far the best theory of impact ionization has been the calculation by Baraff involving computer solutions of the Boltzmann transport equation (G. A. Baraff, *Phys. Rev.*, Vol. 128, pp. 2507–2517, 1962). Fitting the experimental

TABLE 10.1 PARAMETERS A AND B USED IN THE IONIZATION COEFFICIENT $\alpha = A \exp(-B/E)$ WITH E EXPRESSED IN V/cm

Semiconductors	Electrons		Holes	
	A (cm^{-1})	B (V/cm)	A (cm^{-1})	B (V/cm^{-1})
Ge	1.61×10^7	1.60×10^6	1.04×10^7	1.28×10^6
Si	2.24×10^6	1.61×10^6	1.01×10^6	2.11×10^6
GaAs	1.05×10^6	1.52×10^6	1.70×10^6	1.85×10^6
GaP	2.26×10^6	3.33×10^6	2.26×10^6	3.33×10^6
InP	3.48×10^6	2.76×10^6	2.64×10^6	2.40×10^6

curves to the computed curves of α gives the following values of l_r in angstrom: (1) 65 for electrons and 44 for holes in Si, and (2) 65 for both electrons and holes in Ge. Therefore, Eq. (10.4.15), although based on a rather simplistic model, is qualitatively correct in predicting the general behavior of α.

A very important point which is not considered in the simple model is multiple phonon scattering. Even if an electron attains the threshold energy $\mathscr{E}_i$, it may not immediately have an ionizing collision. The electron may have a succession of collisions with optical phonons before an ionizing collision takes place. However, because $\mathscr{E}_{op} << \mathscr{E}_i$, the electron still possesses an energy close to $\mathscr{E}_i$ after a phonon collision. Multiple phonon scattering was considered in the analyses by Moll and by Carroll (J. L. Moll, *Physics of Semiconductors*, McGraw-Hill Book Company, New York, 1964, pp. 215–222; J. E. Carroll, *Hot Electron Microwave Generators*, Edward Arnold (Publishers) Ltd., London, 1970, pp. 277–280). Reasonable agreement between their results and the computer result has been obtained.

The situation as regards to how multiple phonon scattering affects the value of A can be qualitatively stated as follows. Notice that A has a dimension of reciprocal distance. If the factor $\exp(-B/E)$ represents the percentage of energetic carriers, A^{-1} is obviously the average distance d for an electron to gain the necessary energy. At high fields (say, in region $\alpha > 10^4$ cm^{-1}), because an energetic electron has an average energy close to $\mathscr{E}_i$, it needs only an energy $\mathscr{E}_{op}$ to make up the energy lost through phonon collision. Thus $eEd = \mathscr{E}_{op}$, where $\mathscr{E}_{op}$ is the optical phonon energy. At low fields (say, in region $\alpha < 10^2$ cm^{-1}), the electrons must all start from practically zero energy. Hence $eEd = \mathscr{E}_i$. Therefore, we expect the value of A to lie between $eE/\mathscr{E}_{op}$ and $eE/\mathscr{E}_i$. Using $\mathscr{E}_{op} = 0.063$ eV, $\mathscr{E}_i = 1.6$ eV, and $E = 4 \times 10^5$ V/cm^{-1} for Si, we find the range for A to be between 6.3×10^6 cm^{-1} (at high fields) and 2.5×10^5 cm^{-1} (at low fields). The value of A given in Table 10.1 for Si indeed falls in this range. Furthermore, the apparent change of slope of the experimental curves (Fig. 10.9) can be attributed to a decreasing value for A with decreasing field.

10.5 TUNNELING

One problem commonly used in quantum mechanics to illustrate the use of the Schrödinger equation

$$-\frac{\hbar^2}{2m_0}\frac{d^2\psi}{dx^2} + U\psi = \mathscr{E}\psi \qquad (10.5.1)$$

is the potential-barrier problem (Fig. 10.10). Classically, an electron of energy $\mathscr{E}$ cannot enter into a region with a potential energy $U_0 > \mathscr{E}$ because it would have a negative kinetic energy. The potential-barrier region, region AB of Fig. 10.10, is a region forbidden by classical mechanics. In quantum mechanics, a negative kinetic energy simply means a wave with attenuated amplitude. If we propose

$$\psi = A \exp(\pm ikx) \qquad (10.5.2)$$

as a solution to Eq. (10.5.1), k becomes imaginary, or $ik = \alpha$, with α being real in the potential-barrier region. The important point is that although the wave functions ψ attenuates as it propagates, it is not zero in the barrier region. An electron with energy $\mathscr{E} < U_0$ incident on the barrier will have a finite probability of getting through the

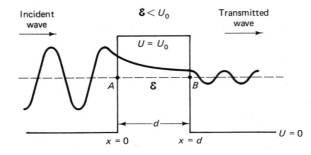

Incident wave
$\mathscr{E} < U_0$
Transmitted wave

$U = U_0$

A | $\mathscr{E}$ | B

$U = 0$

d

$x = 0$ $x = d$

Figure 10.10 Schematic diagram used to calculate the probability of tunneling through a potential barrier. For simplicity, a rectangular barrier is assumed. Here the potential energy is represented by U instead of V, which is reserved to represent voltage, a quantity that appears often in this chapter.

barrier region. The transmission coefficient T (ratio of the transmitted to incident wave amplitude) is given by

$$|T|^2 = \frac{(2k/\alpha)^2}{(1 - k^2/\alpha^2)^2 \sinh^2\alpha d + (2k/\alpha)^2 \cosh^2\alpha d} \qquad (10.5.3)$$

where $\hbar k = \sqrt{2m_0\mathscr{E}}$, $\hbar\alpha = \sqrt{2m_0(U_0 - \mathscr{E})}$, and d is the width of the barrier region. Equation (10.5.3) is obtained by applying the boundary conditions (ψ and $d\psi/dx$ being continuous) to the solutions at $x = 0$ and $x = d$ in Fig. 10.10 (see, for example, J. L. Powell and B. Crasemann, *Quantum Mechanics*, Addison-Wesley Publishing Company, Reading, Mass., 1961, pp. 107–109).

To get a feeling about the magnitude of the transmission coefficient, we set $U_0 - \mathscr{E} = 1$ eV and thus obtain $\alpha = 5.1 \times 10^7$ cm^{-1}. For a barrier region of width $d = 1$ μm $= 10^{-4}$ cm, the value of $|T|^2$ is on the order of exp (-10^4). Therefore, the probability of an electron tunneling through the barrier region is extremely small. To enhance the tunneling probability (or the quantity $|T|^2$), we must reduce the product αd. This is the reason why tunneling in semiconductor junctions is important only at very high fields. The energy-band diagram under a high electric field is shown in Fig. 10.7b. Consider the tunneling process from point A in the valence band to point B in the conduction band. If it is assumed that a tunneling electron enters the conduction band immediately after it leaves the valence band, a potential barrier of height $\mathscr{E}_g$ exists for the electron at A. The situation is illustrated in Fig. 10.11a. On account of the electric field E, however, the barrier U decreases with distance as

$$U = \mathscr{E}_g - eEx \qquad (10.5.4)$$

At point B, $U = 0$; therefore, the barrier region is limited to a distance $d = \mathscr{E}_g/eE$, which decreases with increasing E.

If the potential varies slowly with distance, we can use the WKB (Wentzel–Kramers–Brillouin) approximation to find the solution for the electron wave function. First, we propose a solution of the form

$$\psi = A \exp[\phi(x)] \qquad (10.5.5)$$

Substituting Eq. (10.5.5) into Eq. (10.5.1) gives

$$\frac{d^2\phi}{dx^2} + \left(\frac{d\phi}{dx}\right)^2 = -\frac{2m_0}{\hbar^2}(\mathscr{E} - U) \qquad (10.5.6)$$

As a first-order approximation, we neglect the second derivative term and thus obtain

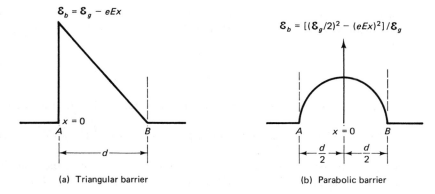

$\mathcal{E}_b = \mathcal{E}_g - eEx$

$\mathcal{E}_b = [(\mathcal{E}_g/2)^2 - (eEx)^2]/\mathcal{E}_g$

$x = 0$

A

B

$\leftarrow\!\!\!-\!\!\!-\!\!\!- d -\!\!\!-\!\!\!-\!\!\!\rightarrow$

(a) Triangular barrier

A $x = 0$ B

$\leftarrow\!\dfrac{d}{2}\!\rightarrow\!\leftarrow\!\dfrac{d}{2}\!\rightarrow$

(b) Parabolic barrier

Figure 10.11 Representation of tunneling process in a semiconductor under a high electric field. In calculating the tunneling probability $|T|^2$, the region AB in Fig. 10.7b is represented by (a) a triangular barrier and by (b) a parabolic barrier.

$$\left(\frac{d\phi}{dx}\right)^2 = -\frac{2m_0}{\hbar^2}(\mathcal{E} - U) = -k^2 \tag{10.5.7}$$

Using Eq. (10.5.7), we can find $d^2\phi/dx^2$. With this correction term counted in Eq. (10.5.6), we find the equation

$$\left(\frac{d\phi}{dx}\right)^2 = -k^2 + \frac{m_0}{i\hbar^2 k}\frac{dU}{dx} \tag{10.5.8}$$

as the improved solution. The spatial variation of the potential, dU/dx, is simply eE. If the condition $|\mathcal{E} - U| > eE/|k|$ is satisfied, Eq. (10.5.7) is a good approximation and we can write Eq. (10.5.5) as

$$\psi = A \exp\int ik(x)\,dx \tag{10.5.9}$$

The solution for ψ in the form stated in Eq. (10.5.9) is known as the *WKB approximation*. The condition for the validity of the WKB approximation is that the change in the potential energy over the decay length as represented by eE/α be smaller than the magnitude of the kinetic energy as represented by $|\mathcal{E} - U|$. For a barrier height of 1 eV, the value of $|k| = |\alpha|$ has previously been calculated as 5.1×10^7 cm^{-1}. Therefore, even at a field of 10^6 V/cm, we find $eE/|k| = 0.02$ eV, which is much smaller than the barrier height of 1 eV.

The value of k can be found from energy considerations. The energy of a tunneling electron is equal to $\mathcal{E}_v - \hbar^2 k_v^2/2m_v^*$ when it is in the valence band and equal to $\mathcal{E}_c + \hbar^2 k_c^2/2m_c^*$ when it is in the conduction band. The energy acquired from the electric field is eEx. Therefore, conservation of energy requires

$$\mathcal{E}_v - \frac{\hbar^2 k^2}{2m_v^*} + eEx = \mathcal{E}_c + \frac{\hbar^2 k^2}{2m_c^*} \tag{10.5.10}$$

In obtaining Eq. (10.5.10), we have assumed momentum conservation or $k_c = k_v = k$. The imposition of energy and momentum conservation is justified only in tunneling processes, which do not require the participation of phonons. Such a tunneling process is generally referred to as *direct tunneling*. Solving for k from Eq. (10.5.10), we find that

$$k = \frac{i}{\hbar} \sqrt{2m_r(\mathscr{E}_g - eEx)} \qquad (10.5.11)$$

where $m_r = m_c^* m_v^*/(m_c^* + m_v^*)$ is the reduced mass. The electron wave functions at A and B are related to each other through $\psi_B = \psi_A \exp\left(i \int_A^B k\, dx\right)$. Since $|\psi|^2 d\tau$ represents the probability of finding the electron within a volume element $d\tau = dx\, dy\, dz$, the tunneling probability $P = |T|^2$ is equal to $|\psi_B/\psi_A|^2$, or

$$P = |T|^2 = \exp\left(2i \int_A^B k\, dx\right) = \exp\left(-\frac{4\sqrt{2}}{3} \frac{m_r^{1/2}\, \mathscr{E}_g^{3/2}}{e\hbar E}\right) \qquad (10.5.12)$$

The integration in Eq. (10.5.12) can easily be carried out by changing the variable from x to θ with $x = (\mathscr{E}_g/eE)\sin^2\theta$.

The quantity $\mathscr{E}_g - eEx$ in Eq. (10.5.11) represents an effective barrier $\mathscr{E}_b$ which an electron experiences in tunneling. The triangular barrier is based on the assumption that a tunneling electron suddenly changes its potential energy at the starting point of tunneling. This assumption is rather artificial. To find a more realistic barrier, we review the energy-band structure. Refer to our discussion in Section 5.2. The characteristic equation for the simplified Kronig–Penney model is given by Eq. (5.2.20), or

$$Q\frac{\sin \beta a}{\beta a} + \cos \beta a = \cos ka \qquad (10.5.13)$$

where $\beta = \sqrt{2m_0\mathscr{E}}/\hbar$, a is the lattice period, and Q is a parameter same as P in Eq. (5.2.20). The symbol Q is used here because P is used in Eq. (10.5.12) to represent the tunneling probability. The function $F(\beta a)$ is sketched as a function of βa in Fig. 5.3. Knowing the value of $F(\beta a)$ we can find the value of ka from Eq (10.5.13). The energy versus ka diagram along the real k axis is shown in Fig. 5.4 and redrawn in Fig. 10.12a with different labels for the band edges. In the range AB, $F(\beta a) > 1$ and the values of ka become complex with $ka = \pi + i\alpha a$ so that $|\cos ka| = \cosh \alpha a > 1$. This range of energy corresponds to the forbidden gap. The energy $\mathscr{E}$ versus ka diagram along the imaginary k axis is shown in Fig. 10.12b. It is the behavior of k along the imaginary axis, which we are concerned with during the tunneling process.

Qualitatively, we see that as $\mathscr{E}$ moves away from either band edge A or B and further into the forbidden gap, the value of $F(\beta a)$ and hence the value of α increase. Therefore, the value of α is zero at the band edges and reaches a maximum somewhere in the forbidden gap. The exact form of $\mathscr{E} - U = \mathscr{E}_b$ (the effective barrier) is not known. The simplest algebraic function, which has the correct behavior ($\mathscr{E}_b = 0$ and hence $\alpha = \text{Im } k = 0$ at the band edges, and $\mathscr{E}_b$ reaches a maximum for $\mathscr{E}_v < \mathscr{E} < \mathscr{E}_c$), is

$$\mathscr{E}_b = \frac{(\mathscr{E}_c - \mathscr{E})(\mathscr{E} - \mathscr{E}_v)}{\mathscr{E}_g} \qquad (10.5.14)$$

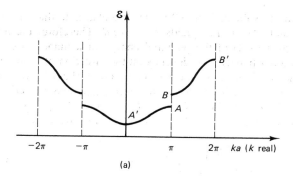

(a)

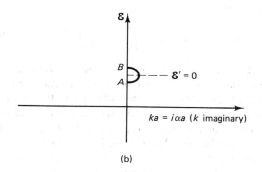

$\mathscr{E}' = 0$

$ka = i\alpha a$ (k imaginary)

(b)

Figure 10.12 Schematic diagrams used in gaining insight into the behavior of electrons in tunneling. Diagram (a) is based on the Kronig–Penney model discussed in Sec. 5.2, and redrawn from Fig. 5.4. Note that the letterings for the band edges have been changed. Here points A and B represent, respectively, the top of the valence band and the bottom of the conduction band. Diagram (b) shows the expected behavior of k on the $\mathscr{E}$-k diagram as electrons move from point A to point B.

The factor $\mathscr{E}_g$ is arbitrary and makes Eq. (10.5.14) dimensionally correct. Letting $\mathscr{E}_g = \mathscr{E}_c - \mathscr{E}_v$ and $\mathscr{E}' = \mathscr{E} - (\mathscr{E}_c + \mathscr{E}_v)/2$, we have

$$\mathscr{E}_b = \frac{(\mathscr{E}_g/2)^2 - (\mathscr{E}')^2}{\mathscr{E}_g} \tag{10.5.15}$$

For a semiconductor not under an applied field as in Fig. 10.12b, both $\mathscr{E}_c$ and $\mathscr{E}_v$ are constant.

In the presence of a high electric field, the two points A and B in Fig. 10.12b become the starting and end points (the classical turning points at which $\mathscr{E}_b = 0$) in a tunneling process as shown in Fig. 10.11b. We refer to the energy-band diagram of Fig. 10.7b. Under an applied field E, both $\mathscr{E}_c(x)$ and $\mathscr{E}_v(x)$ depend linearly on x. The quantity $(\mathscr{E}_c + \mathscr{E}_v)/2$ is represented by the dashed line while the energy $\mathscr{E}$ of the electron is represented by the horizontal line joining A and B. Obviously, the quantity $\mathscr{E}'$ equal to the difference of the two energies is given by $\mathscr{E}' = eEx$. Thus Eq. (10.5.15) can be written as

$$\mathscr{E}_b = \frac{(\mathscr{E}_g/2)^2 - (eEx)^2}{\mathscr{E}_g} \tag{10.5.16}$$

The origin of x is chosen at the middle between A and B so that $\mathscr{E}_b$ is a maximum there. In other words, $\mathscr{E} = (\mathscr{E}_{c0} + \mathscr{E}_{v0})/2$ or $\mathscr{E}' = 0$ in Eq. (10.5.15) at the midpoint.

The potential barrier of Fig. 10.11b, unlike the triangular barrier, is symmetrical with respect to electron tunneling from the valence to the conduction band (A to B) and that from the conduction to the valence band (B to A). We also note that the condition $\mathscr{E}_b = 0$ at $A(x = -d/2)$ and $B(x = d/2)$ yields $\mathscr{E}_g = eEd$. Therefore, the total gain (from A to B) or loss (from B to A) of the potential energy of a tunneling electron is $\mathscr{E}_g$, the same as in the triangular barrier, and depends on the direction of tunneling (gain if it is opposite to E). The parabolic potential barrier of Eq. (10.5.16) was first suggested by Kane and used in his analysis of tunneling in p-n junctions (E. O. Kane, *J. Phys. Chem. Solids*, Vol. 12, p. 181, 1960; *J. Appl. Phys.*, Vol. 32, p. 83, 1961).

Using Eq. (10.5.16) to replace $\mathscr{E}_g - eEx$ in Eq. (10.5.11), we have

$$k = \frac{i}{\hbar} \sqrt{\left(\frac{2m_r}{\mathscr{E}_g}\right)\left[\left(\frac{\mathscr{E}_g}{2}\right)^2 - (eEx)^2\right]} \qquad (10.5.17)$$

The tunneling probability is thus given by

$$P = \exp\left(2i \int_A^B k \, dx\right) = \exp\left(-\frac{\pi}{2\sqrt{2}} \frac{m_r^{1/2} \mathscr{E}_g^{3/2}}{e\hbar E}\right) \qquad (10.5.18)$$

Equation (10.5.18) agrees with the expression obtained by Kane from a quantum-mechanical treatment of the tunneling process except for a numerical factor $\pi^2/9$ in front. A comparison of Eqs. (10.5.12) and (10.5.18) shows the same functional dependence of the tunneling probability on E, m_r, and $\mathscr{E}_g$ for the triangular and parabolic potential barriers except for the numerical constant in the exponent ($4\sqrt{2}/3 = 1.88$ for the former and $\pi/2\sqrt{2} = 1.11$ for the latter). Had we used $s\mathscr{E}_g$ instead of $\mathscr{E}_g$ in Eq. (10.5.14) we would have a numerical constant of $1.11/\sqrt{s}$. The numerical constant s is a parameter that can be used to adjust the barrier height in Fig. 10.11b.

In summary, we have derived the tunneling probability for both the triangular and parabolic barriers. Since the two probabilities have the same functional dependence but two different numerical constants, the field E required to have appreciable tunneling should not be very sensitive to the shape of the barrier. Letting $m_r = m_0$ and $\mathscr{E}_g = 1$ eV, we find from Eq. (10.5.18) that $P \cong 5 \times 10^{-5}$ at $E = 4 \times 10^6$ V/cm. Because of the exponential dependence on $1/E$, the value of P is expected to be vanishingly small for $E < 4 \times 10^5$ V/cm. With this estimate of the order of magnitude for E, we conclude our introductory discussion of tunneling. Discussions on tunneling in p-n junctions, including the derivation of current expressions, are given in Section 10.6. Esaki diodes and resonant tunneling devices including the effect of energy-band structure on tunneling are presented in Section 13.3.

10.6 ANALYSIS OF JUNCTION BREAKDOWN

When a sufficiently high field is applied to a p-n junction in the reverse direction, the junction conducts a very large current. From the ideal diode equation, we expect a very small current in a reverse-biased junction. The phenomenon of an excessively large reverse current is generally known as *junction breakdown*. Two important mechanisms for junction breakdown are carrier multiplication (discussed in Section 10.4) and tunneling (discussed in Section 10.5). The breakdown due to the former is generally known as *avalanche breakdown* while the breakdown due to tunneling is generally known as *Zener breakdown*. Direct tunneling or excitation of electrons from a lower to an upper band was first suggested by Zener as a possible mechanism for breakdown in insulators

(C. Zener, *Proc. R. Soc.*, London, Vol. 145, p. 523, 1934). In this section we apply the results obtained in Sections 10.4 and 10.5 to analyzing breakdown in *p-n* junctions.

First we treat the avalanche breakdown. To make the integral in Eq. (10.4.10) readily integrable, we replace the dependence of α on E of Eq. (10.4.15) by

$$\alpha = aE^m \tag{10.6.1}$$

where a and m are two constants to be determined. Since the contribution to the integral in Eq. (10.4.10) comes mostly from the field region, where α is large (say, $\alpha > 3 \times 10^3$ cm^{-1} in Fig. 10.9), the approximate dependence of Eq. (10.6.1) has to be well fitted only over a very small range of E. Of the two parameters, the exponent m is the important one and can be found from a two-point fitting with $m = [\ln (\alpha_1/\alpha_2)]/\ln (E_1/E_2)$. For example, the α_e curve of silicon in Fig. 10.9 can be approximated with reasonable accuracy by

$$\alpha_e = 5 \times 10^4 \left(\frac{E}{4 \times 10^5} \right)^{5.3} \text{cm}^{-1} \tag{10.6.2}$$

where E is in V/cm. This approximate expression fits exactly the values at $E_1 = 4 \times 10^5$ V/cm and at $E_2 = 2.5 \times 10^5$ V/cm, but underestimates the value of α_e for E within the range $E_1 > E > E_2$ and overestimates the value of α_e outside this range.

To make the mathematical steps tractable, we limit our discussion to the case $\alpha_e = \alpha_h = \alpha$ and use Eq. (10.4.10) instead of Eqs. (10.4.12) and (10.4.13). For an abrupt or step junction, the field in the space-charge region is given by

$$E = \frac{E_{max}(x + x_A)}{x_A} \tag{10.6.3}$$

where $E_{max} = eN_1 x_A/\epsilon$ is the field, a maximum field, at the metallurgical boundary (chosen at $x = 0$) for the *p-n* junction, $N_1 = N_D - N_A$ or $N_A - N_D$ is the net impurity concentration on one side, and x_A is the extent of the space-charge region on one side of the junction. Using Eqs. (10.6.1) and (10.6.3) in Eq. (10.4.10), we find that

$$\int_{-x_A}^{x_B} \alpha \, dx = \frac{\alpha_{max} \, W}{m + 1} = 1 - M^{-1} \tag{10.6.4}$$

where $\alpha_{max} = a(E_{max})^m$, $W = x_A + x_B$ is the width of the space-charge region, and the integration is carried over both sides of the junction from $x = -x_A$ to 0 and from 0 to $x = x_B$.

One quantity important to device applications is the breakdown voltage V_B. Breakdown of a junction occurs when $M = \infty$, meaning that

$$aE_{max}^m \, W = m + 1 \tag{10.6.5}$$

Using the following expressions for E_{max} and W,

$$E_{max} = \left(\frac{2eVN}{\epsilon} \right)^{1/2}, \qquad W = \left(\frac{2\epsilon V}{eN} \right)^{1/2} \tag{10.6.6}$$

in Eq. (10.6.5) and setting $V = V_B$, we obtain

$$V_B = \frac{1}{2} \left[\frac{2}{(1 - q)a} \right]^{1-q} \left(\frac{eN}{\epsilon} \right)^{-q} \tag{10.6.7}$$

where $q = (m - 1)/(m + 1)$ and $N = N_1N_2/(N_1 + N_2)$. For $N_1 >> N_2$, N reduces to N_2, the impurity concentration in the less heavily doped region. In germanium, the following empirical relation is found by Miller (S. L. Miller, *Phys. Rev.*, Vol. 99, p. 1234, 1955)

$$V_B = 26(N \times 10^{-16})^{-0.725} \qquad (10.6.8)$$

where V_B is expressed in volts and N in cm^{-3}. If we use the geometric mean $\sqrt{\alpha_e\alpha_h} = \alpha$, that is, a curve midway between α_e and α_h in Fig. 10.9, we find for germanium: $\alpha = 4.35 \times 10^4 \ cm^{-1}$ at $E = 2.5 \times 10^5$ V/cm and $\alpha = 5.25 \times 10^3 \ cm^{-1}$ at $E = 1.8 \times 10^5$ V/cm. These values yield $m = 6.44$ and $q = 0.731$, in good agreement with the experimentally found value of 0.725. In terms of V_B, the multiplication factor M is given by

$$M^{-1} = 1 - \left(\frac{V}{V_B}\right)^r \qquad (10.6.9)$$

where $r = 1/(1 - q) = (m + 1)/2$ and $r = 3.7$ in Ge.

We should point out that the dependence of M on V as expressed in Eq. (10.6.9) and the dependence of V_B on N as expressed in Eq. (10.6.7) are based on the dependence of α on E as expressed in Eq. (10.6.1). Since α_e and α_h have very similar dependence on E (Fig. 10.9), the value of r in Eq. (10.6.9) and the value of q in Eq. (10.6.7) can be readily obtained from the value of m by fitting Eq. (10.6.1) to the experimental curves. On the other hand, the value of V_B depends on how we determine the constant a in Eq. (10.6.1), which appears in Eq. (10.6.7). Since there is no unique way of making an appropriate average of α_e and α_h to get α, only an approximate value for V_B can be expected. If we again choose $\alpha = \sqrt{\alpha_e\alpha_h}$, we obtain for Ge $a = 2.7 \times 10^{-29}$ with both E and α expressed in MKS units in Eq. (10.6.1). Using this value in Eq. (10.6.7), we find $V_B = 29$ V at $N = 10^{22} \ m^{-3}$, in reasonable agreement with the experimental value of 26 V from Eq. (10.6.8).

To obtain an accurate value for V_B, we must use the experimental data of α_e and α_h in either Eq. (10.4.12) or (10.4.13) and obtain the condition $M_e = 1$ or $M_h = 1$ numerically. It is left as an exercise for the reader to show that the two conditions are satisfied simultaneously. Therefore, there is only one value for V_B. Figure 10.13 shows the value of V_B computed by Sze and Gibbons (S. M. Sze and G. Gibbons, *Appl. Phys. Lett.*, Vol. 8, p. 111, 1966) for abrupt junctions in Ge, Si, GaAs, and GaP. As indicated in Eq. (10.6.7), the dependence of V_B on N is slightly less than inversely linear. In germanium, every 10-fold decrease in N results in an increase of V_B by a factor of 5.3 according to Eq. (10.6.8), in good agreement with the computed curve of Fig. 10.13. Figure 10.14 shows a similar calculation also by Sze and Gibbons for linearly graded junctions in which the impurity concentration varies as $N = Ax$, where A is the impurity-concentration gradient. The reader is asked to show that in this case, the breakdown voltage V_B is expected to vary as A^{-q} with $q = (m - 1)/(2m + 1)$. In germanium, $m = 6.44$ or $q = 0.39$ meaning an increase of V_B by a factor of 2.5 for every 10-fold decrease in A.

The dashed lines in Figs. 10.13 and 10.14 roughly indicate a demarcation for N and A beyond which the tunneling process becomes important. When the avalanche process dominates, the reverse current increases very rapidly as V approaches V_B and becomes infinitely large at $V = V_B$. This sharp increase in current clearly defines the breakdown voltage. The tunneling process, unlike the avalanche process, is not a regen-

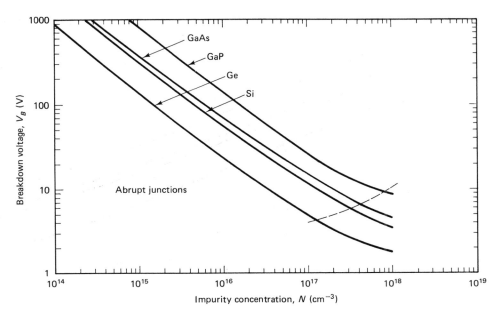

Figure 10.13 Computed values of the avalanche breakdown voltage V_B as functions of the doping concentration N in abrupt junctions of Ge, Si, GaAs, and GaP. The quantity N is related to the impurity concentrations on the two sides by $N = N_1N_2/(N_1 + N_2)$. (S. M. Sze and G. Gibbons, *Appl. Phys. Lett.*, Vol. 8, p. 111, 1966.)

erative process. Once an electron is tunneled from the valence band into the conduction band (Fig. 10.7b), the process is completed. Because of lack of any feedback mechanism, the increase in tunneling current with applied voltage is rather gradual as compared to the sharp dependence of M in Eq. (10.6.9) for the avalanche current $I = MI_s$. Under the circumstance, we can determine the importance of the tunneling process only from the magnitude of the tunneling current.

Referring to Fig. 10.7b, we see that in a reverse-biased junction, the current is dominated by tunneling of electrons from the valence band to the conduction band. The number of electrons striking the junction per unit area of the junction is equal to the product of the density of occupied states and the velocity of these electrons. Thus the incident current density is given by

$$dJ_i = ev_x \, dn = ev_x \frac{2}{h^3} f(\mathscr{E}) \, dp_x \, dp_y \, dp_z \qquad (10.6.10)$$

Of the incident electrons, however, only a few can tunnel across the junction. Let P_1 be the probability with which an incident electron can tunnel through the junction. Thus the tunneling current is

$$J_t = e \iiint \frac{2}{h^3} P_1(v_x \, dp_x)(dp_y \, dp_z) \qquad (10.6.11)$$

Here we have assumed that the valence band is completely occupied and the conduction band is completely empty. Therefore, we have neglected electron tunneling from the conduction band to the valence band and have set $f(\mathscr{E}) = 1$ in Eq. (10.6.10).

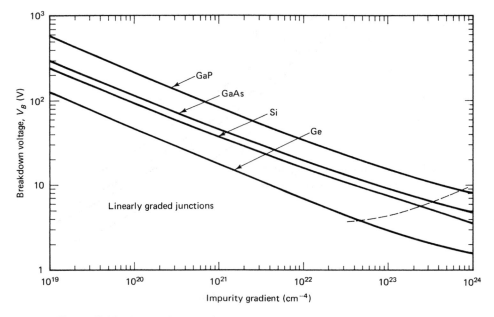

Figure 10.14 Computed values of the avalanche breakdown voltage V_B as a function of the impurity concentration gradient dN/dx in linearly graded junctions of Ge, Si, GaAs, and GaP. It is assumed that the gradients $dN_1/dx = dN_2/dx$ are equal. (S. M. Sze and G. Gibbons, *Appl. Phys. Lett.*, Vol. 8, p. 111, 1966.)

To carry out the integration in Eq. (10.6.11), we let

$$\mathscr{E}_x = \frac{p_x^2}{2m_r}, \qquad \mathscr{E}_\perp = \frac{p_y^2 + p_z^2}{2m_r} = \frac{p_\perp^2}{2m_r} \qquad (10.6.12)$$

and change $dp_y\, dp_z$ to $2\pi p_\perp\, dp_\perp$ in cylindrical coordinates. Further, we note that the tunneling probability of Eq. (10.5.18) is based on a one-dimensional analysis. For a three-dimensional case, Eq. (10.5.17) can be rewritten as

$$\left(\frac{\hbar^2 k_x^2}{2m_r}\right) + \mathscr{E}_\perp = -\mathscr{E}_b = -\frac{(E_g/2)^2 - (eEx)^2}{\mathscr{E}_g} \qquad (10.6.13)$$

Since tunneling is along the x direction, it is k_x, not k, which should enter into the integral $\int k_x\, dx$ in Eq. (10.5.18). As can be seen from Eq. (10.6.13), the result of this change is to substitute $(\mathscr{E}_g/2)^2$ in Eq. (10.5.17) by $(\mathscr{E}_g/2)^2 + \mathscr{E}_\perp\mathscr{E}_g$ to get the x dependence for k_x. However, $\mathscr{E}_g$ in the denominator remains. Using this new expression in the integral, we find that

$$P_1 = P \exp\left(-\frac{\mathscr{E}_\perp}{\overline{\mathscr{E}}}\right) \qquad (10.6.14)$$

where P is the same as P of Eq. (10.5.18) for zero transverse momentum and $\overline{\mathscr{E}}$ is given by

$$\overline{\mathscr{E}} = \frac{e\hbar E}{\pi(2m_r\mathscr{E}_g)^{1/2}} \qquad (10.6.15)$$

Equation (10.6.14) can be obtained from Eq. (10.5.18) by noting $\mathscr{E}_g^{3/2} = \mathscr{E}_g^2/\mathscr{E}_g^{1/2}$ and substituting $\mathscr{E}_g^2 + 4\mathscr{E}_\perp \mathscr{E}_g$ for $\mathscr{E}_g^2$ in the numerator.

Substituting Eq. (10.6.14) in Eq. (10.6.11) and using $\mathscr{E}_x$ and $\mathscr{E}_\perp$ as variables, we have

$$J_t = \frac{4\pi e m_r}{h^3} \int\int P \exp\left(-\frac{\mathscr{E}_\perp}{\overline{\mathscr{E}}}\right) d\mathscr{E}_\perp \, d\mathscr{E}_x \qquad (10.6.16)$$

The integration over $\mathscr{E}_\perp$ yields $\overline{\mathscr{E}}$. Figure 10.15 shows the energy-band diagram of a heavily doped p-n junction under a reverse bias. Since the energy is conserved in a direct tunneling process, only those electrons in the valence band having energies between the two band edges participate in the tunneling process. Therefore, the integration over $\mathscr{E}_x$ gives $eV_a - \xi_p - \xi_n$, where V_a is the reverse applied voltage, $\xi_p = \mathscr{E}_f - \mathscr{E}_v$ on the p side, and $\xi_n = \mathscr{E}_c - \mathscr{E}_f$ on the n side. In a heavily doped junction, the values of ξ_p and ξ_n are small and hence negligible compared to eV_a. Therefore, we can simply use eV_a and thus obtain

$$J_t = \frac{4\pi e m_r}{h^3} (eV_a)(\overline{\mathscr{E}}) \exp\left(-\frac{\mathscr{E}_g}{4\overline{\mathscr{E}}}\right) \qquad (10.6.17)$$

In obtaining Eq. (10.6.17), we have assumed a constant field in the junction and consequently taken P outside the integral in Eq. (10.6.16). In subsequent calculations, we use an average field $E_{\max}/2$ for E.

For computation purposes, we express Eq. (10.6.17) in terms of the field E as

$$J_t = \frac{\pi e m_r}{h^3} eV_a \mathscr{E}_g \frac{E}{E_0} \exp\left(-\frac{E_0}{E}\right) \qquad (10.6.18)$$

where the parameter E_0 is a constant given by

$$E_0 = \frac{\pi (2m_r)^{1/2} \mathscr{E}_g^{3/2}}{4e\hbar} \qquad (10.6.19)$$

From $m_r = 0.068 m_0$, $\mathscr{E}_g = 1.4$ eV (in GaAs) and $V_a = 6$ V, we have

$$J_t = 2.3 \times 10^{13} \left(\frac{E}{E_0}\right) \exp\left(-\frac{E_0}{E}\right) \qquad \text{A/m}^2 \qquad (10.6.20)$$

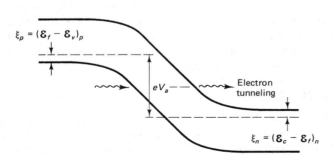

Current direction

$\xi_p = (\mathscr{E}_f - \mathscr{E}_v)_p$

eV_a

Electron tunneling

$\xi_n = (\mathscr{E}_c - \mathscr{E}_f)_n$

Figure 10.15 Energy-band diagram of a reverse-biased junction. Only the states lying within an energy range $eV_a - \xi_p - \xi_n$ satisfy the condition that the initial (valence-band) state is occupied and the final (conduction-band) state is empty. Therefore, only those states are counted in finding the tunneling current.

$$E_0 = 1.7 \times 10^9 \qquad \text{V/m} \tag{10.6.21}$$

To get an order-of-magnitude estimate of the tunneling current, we use $E_{max}/2$ from Eq. (10.6.6) for E and compute the value of J_t from Eq. (10.6.20) at two very different values of impurity concentration. Letting $\epsilon = 13\epsilon_0$ (again in GaAs) and $V_a = 6$ V, we find (1) $E = 6.5 \times 10^7$ V/m and $J_t = 10^2$ A/m^2 at $N = 10^{24}$ m^{-3}, and (2) $E = 6.5 \times 10^6$ V/m and $J_t = 10^{-99}$ A/m^2 at $N = 10^{22}$ m^{-3}. For comparison, the reverse-saturation current density due to minority-carrier diffusion is

$$J_s = \frac{eD}{L} \frac{n_i^2}{N} \tag{10.6.22}$$

where D is the diffusion constant, L the diffusion length, and n_i the intrinsic carrier concentration. For $D = 210$ cm^2/s, $L = 10^{-6}$ m, and $N = 10^{22}$ m^{-3}, $n_i = 10^{13}$ m^{-3} (in GaAs), we find that $J_s = 2.2 \times 10^{-7}$ A/m^2. Obviously, the tunneling current density J_t is negligible at $N = 10^{22}$ m^{-3}, but J_t becomes dominant at $N = 10^{24}$ m^{-3}.

The significant difference between Eqs. (10.6.18) and (10.6.22) can be stated as follows. Equation (10.6.22) is basically a product of three quantities: the electronic charge e, the minority-carrier concentration n_i^2/N, and the diffusion velocity $v_{dif} = D/L$. Similarly, we can think of Eq. (10.6.18) as being the product of the electronic charge, an effective velocity $v_{eff} = \sqrt{\mathscr{E}_g/m_r}$, an effective density of states $n_{eff} = (\pi e V_a \mathscr{E}_g^{1/2} m_r^{3/2}/h^3)$, and the tunneling probability $P = (E/E_0) \exp(-E_0/E)$. Because a tremendous number of quantum states, those lying within the energy range eV_a in Fig. 10.15, participate in the tunneling process and because the velocity associated with these states is very large, the ratio $n_{eff} v_{eff}/(n_i^2 v_{dif}/N)$ is a tremendously large number, on the order of 10^{20} for GaAs in the calculation above. The tunneling process can be important if P has a reasonable value, say $P > 10^{-15}$.

Since the value of J_s for GaAs is extremely small because of a small n_i, we choose a current density $J \sim 10^{-2}$ A/m^2 (or a current of 1 nA for a junction area of 0.1 mm^2) as a practical value for comparison with J_t. Based on this value of J, we expect the tunneling process to dominate if $P > 4 \times 10^{-16}$ or $E/E_0 > 3 \times 10^{-2}$. This condition requires a maximum field strength $E_{max} = 2E > 6 \times 10^{-2} E_0$ of Eq. (10.6.19), or

$$E_{max} > \left(\frac{m_r}{m_0}\right)^{1/2} (\mathscr{E}_g)^{3/2} \times 2 \times 10^6 \text{ V/cm} \tag{10.6.23}$$

for the tunneling process to become dominant in a reverse-biased junction where $\mathscr{E}_g$ is expressed in units of eV. The dashed line in Figs. 10.13 and 10.14 is an approximate location where E_{max} exceeds the value in Eq. (10.6.23). For an impurity concentration N or gradient A to the left of the dashed line, the avalanche breakdown occurs first before the tunneling current becomes appreciable and the breakdown voltage is given by the numerically computed value of V_B shown in the two figures. For N or A to the right of the dashed line, the tunneling current is already very large before the avalanche multiplication factor becomes significant, and the value of V_B should be calculated by using the value of E_{max} from Eq. (10.6.23) in Eq. (10.6.6)

10.7 HOT-ELECTRON EFFECTS IN MOSFET

It is generally recognized that adverse effects caused by hot electrons can pose a limit to the scaling down of gate length of a MOSFET (T. H. Ning, P. W. Cook, R. H. Dennard, C. M. Osburn, S. E. Schuster, and H. N. Yu, *IEEE Trans. Electron Devices*,

Vol. ED-26, p. 346, 1979). The device-performance degradation is manifested in a rapid increase in the gate current with increasing gate bias and a substantial substrate current. The various currents and voltages are defined in Fig. 10.16, where the structure of a MOSFET is shown. The increase in gate current can be understood on the basis of electron emission from silicon into the SiO_2 layer. Using an experimental arrangement in which electron–hole pairs were generated inside Si by incident photons from a tungsten light bulb, Ning et al. measured the substrate and drain currents as functions of gate and substrate voltages under the condition that both source and gate were grounded (T. H. Ning, C. M. Osburn, and H. N. Yu, *J. Appl. Phys.*, Vol. 48, p. 286, 1977). From the experimental data, the emission probability P was shown to have a dependence $P = A \exp(-d/\lambda)$, where λ is the mean free path for optical-phonon scattering, d the distance from the interface of the point in Si at which the potential energy is equal to the barrier potential, and A a proportionality constant. The mean free path λ is related to λ_0 at $T = 0$ K by

$$\lambda = \lambda_0 \tanh \frac{\mathscr{E}_{op}}{2kT} \tag{10.7.1}$$

where $\mathscr{E}_{op}$ is the optical-phonon energy. In Si, $\lambda_0 = 108$ Å and $\mathscr{E}_{op} = 0.063$ eV (C. R. Crowell and S. M. Sze, *Appl. Phys. Lett.*, Vol. 9, p. 242, 1966).

Since the early work of Ning et al., a number of investigations have been reported. An excellent discussion and review of hot-electron effects in MOSFET can be found in a paper by Hu et al. (C. Hu, S. C. Tam, F. C. Hsu, P. K. Ko, T. Y. Chan, and K. W. Terrill, *IEEE Trans. Electron Devices*, Vol. ED-32 p. 375, 1985). The analysis of Hu et al. is based on a phenomenological model of impact ionization and redirection scattering. As illustrated in Fig. 10.17a, electrons gain energy as they move from source to drain. In the low-field region, the acquired energy is lost through collisions, but in the high-field region, impact ionization can take place if the field is sufficiently high for electrons to acquire within a mean free path λ the minimum energy for impact ionization. The line AB in Fig. 10.16 indicates the beginning of the impact-ionization region. After impact ionization, there are three particles: the primary electron and the generated electron–hole pair. An interesting question arises as to the direction in which the three particles will move because the movement of electrons toward the drain constitutes I_D, that of electrons through SiO_2 constitutes I_G, and that of holes toward the p-substrate constitutes I_{SUB}.

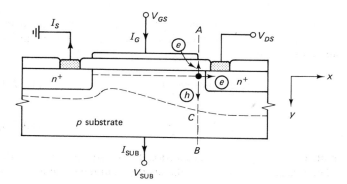

Figure 10.16 Schematic diagram showing hot-electrons effects in a Si MOSFET. The gate current is denoted by I_G and the substrate current by I_{SUB}.

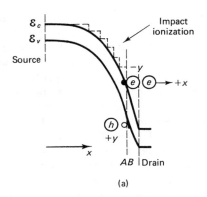

(a)

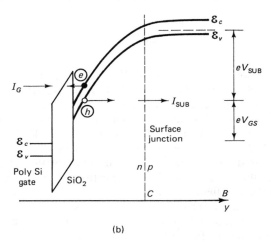

(b)

Figure 10.17 Energy-band diagrams showing (a) impact ionization in the channel and (b) injection of hot electrons into the gate oxide. The drain–source current is indicated by I_D, the gate current by I_G and the substrate current by I_{SUB}.

Figure 10.17b shows the energy-band diagram of the MOSFET along the line AB. It may appear that electrons would move toward the gate and holes would move toward the substrate if we ignore the x component (i.e., the component along the channel) of the carrier velocity. We note, however, that based on conservations of energy [Eq. (10.4.1)] and momentum [Eq. (10.4.2)] the carriers after impact ionization have a substantial velocity (compared to the initial velocity of the primary electron) along the channel. Therefore, the transport of electrons in the channel depends on both E_x and E_y, the former determined by the drain voltage and the latter by the gate voltage. In the absence of a quantitative analysis, Hu and coworkers have proposed the following dependences for the substrate and gate currents:

$$I_{SUB} = C_1 I_D \exp\left(-\frac{\mathscr{E}_i}{eE\lambda}\right) \qquad (10.7.2)$$

$$I_G = C_2 I_D \exp\left(-\frac{\mathscr{E}_b}{eE\lambda}\right) \qquad (10.7.3)$$

where I_D is the drain current, C_1 and C_2 are two proportionality constants, and E is the unspecified channel field. The ionization energy $\mathscr{E}_i$ is the energy required of an energetic

High-Field Phenomena and Hot-Electron Effects   Chap. 10

electron to generate an electron–hole pair, and λ is the same as l_r in Eq. (10.4.16). Therefore, the factor exp $(-\mathscr{E}_i/eE\lambda)$ represents the fraction of electrons in I_D which have sufficient energy to generate electron–hole pairs. The quantity $\mathscr{E}_b$ in Eq. (10.7.3) is the barrier height 3.2 eV at Si–SiO$_2$ interface minus barrier lowering due to image force. The mechanism of converting a predominantly x-directed carrier motion into a motion having a significant y-directed component of velocity, however, was not specified.

If it is assumed that some unspecified scattering is responsible for the redirection of carrier motion, the two constants C_1 and C_2 should be more or less independent of the gate and substrate voltages. Based on this phenomenological model, the two currents are thus related to each other by

$$\frac{I_G}{I_D} = C_2 \left(\frac{I_{\text{SUB}}}{I_D C_1}\right)^p \tag{10.7.4}$$

where $p = \mathscr{E}_b/\mathscr{E}_i$. Figure 10.18 shows the correlation between the experimentally measured values of I_G/I_D and I_{SUB}/I_D as reported by Hu et al. Indeed, the relation shown in Eq. (10.7.4) is observed. However, the value of $\mathscr{E}_i$ is found to be 1.24 eV versus a value of 1.65 eV from a fit to the ionization-coefficient curve of Fig. 10.9. In Fig. 10.18, each curve was obtained for a given V_{GD} to keep a constant E in Eqs. (10.7.2) and (10.7.3). In a tunneling or emission process, the momentum of a carrier is con-

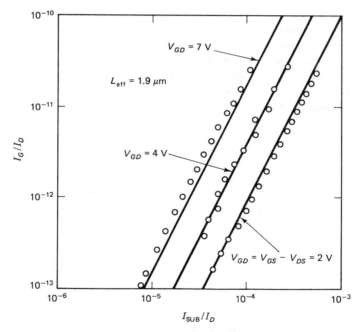

Figure 10.18 Plots showing the correlation between measured values of I_G/I_D and I_{SUB}/I_D. The data were taken at various V_{DS} but the same V_{GD}. (C. Hu, S. C. Tam, F. C. Hsu, P. K. Ko, T. Y. Chan, and K. W. Terril, "Hot-Electron-Induced MOSFET Degradation—Model, Monitor and Improvements," *IEEE Trans. Electron Devices*, Vol. ED-32, p. 375, 1985 © 1985 IEEE.)

served in the transverse plane. Therefore, the part of the electron energy associated with the electron momentum in the y direction needs to be high enough to overcome the barrier height $\mathscr{E}_b$ at the Si–SiO$_2$ interface. Unless a satisfactory mechanism can be found in redirecting the carrier momentum, the validity of Eq. (10.7.4) may be open to question even though a good fit to the experimental data has been obtained.

The substrate current, if unchecked, can overload the substrate-bias generator. However, a more serious problem is associated with the gate current, although it is much smaller than the substrate current. A number of studies have provided strong evidence that MOSFET degradation with normally good oxides may be caused mainly by generation of interface traps as a result of electron capture in the oxide. Figure 10.19 shows schematically the bonding requirement of a silicon atom on the surface of a silicon crystal and that on the surface of a SiO$_2$ film. To satisfy the covalent-bond requirement in a crystal, a Si atom must have four nearest neighbors. A surface Si atom, however, has only three nearest neighbors. Therefore, one electron of the Si atom has an unpaired spin. The term *dangling bond* (DB) refers to the electron with unpaired spin. A dangling bond acts as an electron trap because it is ready to capture an electron to complete the covalent bond. The symbols Si$_{SS}$ and Si$_S$ used in Fig. 10.19a, respectively, refer to a Si atom on the silicon surface with a dangling bond and that with the covalent bond completed. The bonding requirement in SiO$_2$ is different. Electrons are transferred from silicon to oxygen. Inside a SiO$_2$ film, a Si atom has four oxygen neighbors through which the bonds are extended to other silicon atoms. Each silicon atom transfers a total of four electrons with one electron to each oxygen neighbor, making Si^{4+} and O^{2-} in SiO$_2$. A silicon atom on the surface of a SiO$_2$ film, however, has only three neighboring oxygen atoms. Therefore, it has an extra electron not being transferred. The symbols Si$_{OS}$ and Si$_O$ used in Fig. 10.19b, respectively, refer to a Si atom on the oxide surface with the extra electron and that without the extra electron.

Since a Si$_{SS}$ atom needs an electron to complete the covalent bond and a Si$_{OS}$ atom has an extra electron, it is natural for an electron transfer from the Si$_{OS}$ atom to the Si$_{SS}$ atom converting them to Si$_O^+$ and Si$_S^-$, respectively. The completed Si–SiO$_2$

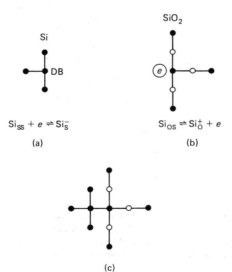

Si$_{SS}$ + e ⇌ Si$_S^-$

(a)

SiO$_2$

Si$_{OS}$ ⇌ Si$_O^+$ + e

(b)

(c)

Figure 10.19 Schematic diagrams showing (a) a silicon atom at silicon surface, denoted by Si$_{SS}$, having a dangling bond, (b) a silicon atom at oxide surface denoted by Si$_{OS}$, having an extra electron left over from forming ionic bonds with only three oxygen neighbors instead of four oxygen neighbors in the interior of SiO$_2$, and (c) completion of bond requirements by joining Si$_{SS}$ and Si$_{OS}$ so that the electron from Si$_{OS}$ can pair with the electron from Si$_{SS}$ to satisfy the covalent-bond requirement.

bond is also illustrated in Fig. 10.19c, and a completed Si–SiO₂ bond is electrically neutral. However, the interatomic distances between Si atoms in silicon and between Si atoms in silicon oxide are different. Even though a SiO₂ film grown on Si substrate is amorphous and hence has a certain degree of flexibility in the bond length, a perfect atomic arrangement at a Si–SiO₂ interface is not expected. We refer to Fig. 10.20 for a schematic two-dimensional representation of the Si–SiO₂ interface as proposed by Jeppson and Svensson in their model of degradation of MOS devices under negative-bias stress (NBS) (K. O. Jeppson and C. M. Svensson, *J. Appl. Phys.*, Vol. 48, p. 2004, 1977). It is well known from the work on amorphous silicon (a-Si) that hydrogenation is an effective way of reducing dangling bonds in a-Si by contributing an electron to a Si_{SS} atom in forming a Si_{SS}–H bond. Since infrared measurements have shown that a large number of Si–H groups exist in the SiO₂ interior, it is reasonable to propose that Si_S–H groups also exist at a Si–SiO₂ interface. The situation is illustrated in Fig. 10.20a.

Figure 10.21 shows the results of capacitance measurements, C_{HF} measured at 1 MHz and C_{LF} being the quasi-static or low-frequency capacitance, in a MNOS (metal–nitride–oxide–silicon) device as reported by Jeppson and Svensson. The different sets of curves were taken with increasing accumulated stress (NBS) time as a parameter. Experiments were performed at fields (4 to 7 MV/cm). The surface-trap density N_{ST} can be calculated from

$$N_{ST} = \frac{C_{ox}}{eA} \left(\frac{C_{LF}/C_{ox}}{1 - C_{LF}/C_{ox}} - \frac{C_{HF}/C_{ox}}{1 - C_{HF}/C_{ox}} \right) \qquad (10.7.5)$$

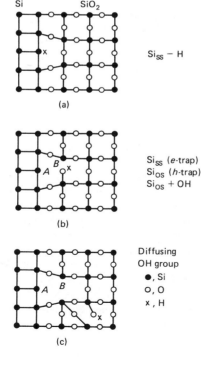

Figure 10.20 A proposed two-dimensional representation of a Si-SiO₂ interface. Diagram (a) shows the formation of Si_S–H bond at the interface. Diagram (b) shows the formation of a Si_{OS}–OH bond, thus leaving behind an electron trap at Si_{SS} (denoted by A) and a hole trap at Si_{OS} (indicated by B). Diagram (c) shows diffusion of hydrogen from the interface into oxide. (K. O. Jeppson and C. M. Svenson, *J. Appl. Phys.*, Vol. 48, p. 2004, 1977.)

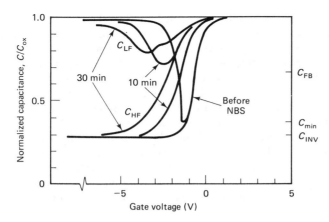

Figure 10.21 Measured low-frequency and high-frequency capacitances, C_{LF} and C_{HF} of a metal–nitride–oxide–Si device as a function of the gate voltage. The set of C_{LF} and C_{HF} curves before NBS are typical C–V curves similar to those shown in Fig. 9.26c. The other two sets of capacitance curves were taken after NBS for 10 min and 30 min at fields 4 to 7 MV/cm. The change in C_{LF} and C_{HF} is used to calculate the increase in surface trap density. (K. O. Jeppson and C. M. Svensson, *J. Appl. Phys.*, Vol. 48, p. 2004, 1977.)

where A is the capacitor area. With an oxide thickness of 950 Å, $C_{ox}/eA = 2.2 \times 10^{11}$ cm^{-2}(eV)$^{-1}$. Therefore, a change in N_{ST} on the order of 10^{10} cm^{-2}(eV)$^{-1}$ can be detected (M. Kuhn, *Solid-State Electron.*, Vol. 13, p. 873, 1970). The following mechanism has been proposed by Jeppson and Svensson as being responsible for the degradation of MOS devices. When a surface defect (Si$_{SS}$–H) is electrically activated, the dissociation of the Si$_{SS}$–H bond is followed by a reassociation of the H atom with an oxygen atom by breaking an existing Si–O bond and forming a silanol Si–OH bond, thus leaving behind a surface electron trap Si$_{SS}$ and a surface hole trap Si$_{OS}$, indicated, respectively, by A and B in Fig. 10.20b. The process is field dependent since a transfer of charge from a Si$_{SS}$ atom to a Si$_{OS}$ atom takes place. It is shown further that as the H atom diffuses away from the interface into the oxide (Fig. 10.20c), the surface trap density increases with the accumulated stress time t as

$$N_{ST} = R_1 t^{1/4} \qquad (10.7.6)$$

where R_1 is a proportionality constant related to the rate of surface-trap formation. This relation is in agreement with the observed change in the measured capacitance. Furthermore, experiments carried out at different temperatures (25 to 125°C) yield an activation energy $e\phi_0 = 0.3$ eV for the process.

The increase in the interface trap density causes degradation of MOSFET performance, which is manifested in a number of device characteristics, such as a shift in the threshold voltage ΔV_{th}, a shift in the subthreshold current swing $\Delta S(= \partial V_G/\partial \log I_D)$, and a reduction in the transconductance in the saturation region Δg_m. A systematic study to correlate these changes has been reported by Hu and coworkers. Figure 10.22 shows (a) the change ΔV_{th} as a function of stress time t and (b) the one-to-one correlation between $\Delta g_m/g_m$ and ΔV_{th}. The data for a MOSFET with an oxide thickness of 35.8 nm and under the applied bias conditions $V_{DS} = 6.5$ V, $V_G = 3$ V, and $V_{SUB} = 0$ V are used for presentation in Fig. 10.22. Based on correlation curves like the one in Fig. 10.22b, Hu and coworkers have concluded that the same degradation mechanism is responsible for the observed changes. They also have proposed breaking the Si$_{SS}$–H bond by hot

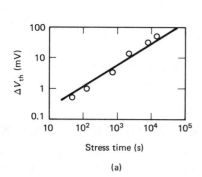
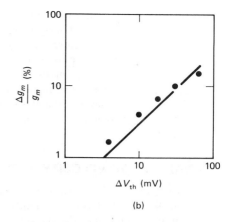

(a)

(b)

Figure 10.22 Experimental results showing (a) the threshold-voltage shift, ΔV_{th} as a function of stress time and (b) correlation between the measured transconductance reduction $\Delta g_m/g_m$ and the threshold-voltage shift. (C. Hu, S. C. Tam, F. C. Hsu, P. K. Ko, T. Y. Chan, and K. W. Terril, "Hot-Electron-Induced MOSFET Degradation—Model, Monitor and Improvements," *IEEE Trans. Electron Devices*, Vol. ED-32, p. 375, 1985 © 1985.)

electrons as the main mechanism responsible for generating interface traps and arrived at the relation

$$N_{ST} = C_3 \left[t \frac{I_D}{Z} \exp\left(-\frac{\mathscr{E}_T}{eE\lambda} \right) \right]^n \tag{10.7.7}$$

for the interface-trap density where Z is the cannel width and $\mathscr{E}_T = (3.2 + 0.3)$ eV is the activation energy with 3.2 eV for the interface barrier and 0.3 eV for the dissociation of Si_{ss}–H bond. The quantity n has a value between 0.5 and 0.75, and C_3 is a proportionality constant. The operating life τ of a MOSFET is defined as the time $t = \tau$ at which N_{ST} reaches a preset value. From Eq. (10.7.7) it is obvious that $\tau = C_4(Z/I_d)$ exp $(\mathscr{E}_T/eE\lambda)$, from which a useful expression

$$\tau = H(I_{SUB})^{-2.9}(I_D)^{1.9}(\Delta V_{th})^{1.5}Z \tag{10.7.8}$$

is derived for the prediction of τ through the use of Eq. (10.7.2). The exponent of I_{SUB} should be equal to $-\mathscr{E}_T/\mathscr{E}_i$, which has a value of -2.2 if we use $\mathscr{E}_i = 3\mathscr{E}_g/2$. The exponent -2.9 is the empirical value determined from experiments. The value of H, however, spans a range over 100, depending on the state of oxide and nitride technology of Si. For MOSFETs fabricated by using the same technology, a measurement of I_{SUB} under NBS should yield useful information about the operating life of the device.

   In summary, we have presented in this section hot-electron effects on the MOSFET performance. In the channel near the drain, electrons have acquired sufficient energy to generate electron–hole pairs through impact ionization. The direct consequences of energetic carriers of significant density are a nonnegligible gate current and an appreciable substrate current. A more serious consequence is the generation of interface traps. The experimental evidences are strong, supporting the view that the increase in interface traps is the dominant mechanism for MOSFET degradation. A MOSFET under a high field and at elevated temperature shows progressive shifts in the threshold voltage and in the subthreshold current swing and continuing reduction in the transconductance. If the proportionality constant H in Eq. (10.7.8) does not improve with new dielectric (oxide and nitride) technology, the hot-electron problem may pose a limit to scaling the gate length down to submicrometer dimensions. The hot-electron problem is discussed further in Sections 13.10 and 13.11 where we present general considerations for devices of ultrahigh speed and ultrasmall dimension and compare Si MOSFET and GaAs MESFET.

10.8 ANALYSIS OF VELOCITY SATURATION BY TRANSPORT EQUATIONS

In Sections 10.2 and 10.3, we presented analyses of the velocity-saturation and electron-transfer effects based on greatly simplifed models. These analyses are very useful in gaining insight into the physical mechanisms underlying the observed phenomena, but they are not accurate. Furthermore, the two treatments appear very different even though the underlying physical principals are very similar. To unify the theoretical analyses and to arrive at reasonably accurate predictions, we have established a set of four equations in Appendix A9. The four equations, derived by Hänsch and Miura-Mattausch from the Boltzmann equation of Eq. (A5.7) for a one-valley semiconductor, are the continuity equation of Eq. (A9.8), the momentum (or current) equation of Eq. (A9.14), the energy equation of Eq. (A9.22), and the energy-flow equation of Eq. (A9.27). Under the dc steady-state conditions, they are, respectively, given by

$$\nabla \cdot \mathbf{J}_e = 0 \tag{10.8.1}$$

$$\mathbf{J}_e = e\mu n\mathbf{E} + \frac{2}{3}\mu\nabla\mathscr{E} \tag{10.8.2}$$

$$\frac{\mathscr{E} - \mathscr{E}_0}{\tau_\mathscr{E}} = \mathbf{E} \cdot \mathbf{J}_e - \nabla \cdot \mathbf{C}_\mathscr{E} \tag{10.8.3}$$

$$\mathbf{C}_\mathscr{E} = -\frac{5}{3}\mu\mathbf{E}\mathscr{E} = -\frac{5}{3en}\mathbf{J}_e\mathscr{E} + \frac{9}{10en}\mu\mathscr{E}\nabla\mathscr{E} \tag{10.8.4}$$

In the equations above, $\mathscr{E}$ is the energy density (J/m^3), $\mathbf{C}_\mathscr{E}$ is the power density (W/m^2), which is a vector, n is the electron concentration, and μ is the mobility given by

$$\mu = \frac{e\tau_m}{m^*} \tag{10.8.5}$$

The quantity $\tau_\mathscr{E}$ is the energy relaxation time given by Eq. (A9.21). The quantity $(\mathscr{E} - \mathscr{E}_0)/\tau_\mathscr{E}$ represents the rate at which the electrons dissipate their energy to the lattice, where

$$\mathscr{E} = \tfrac{3}{2}kT_e n \tag{10.8.6}$$

is the energy density at an electron temperature T_e and $\mathscr{E}_0$ is the energy at the lattice temperature T_l. The quantity τ_m is the momentum relaxation time given by Eq. (A9.11), or

$$\tau_m^{-1} = \frac{e}{J_e^2} \mathbf{J}_e \cdot \int\int_{-\infty}^{\infty} (\mathbf{v} - \mathbf{v}')S(\mathbf{v}, \mathbf{v}')f(\mathbf{r}, \mathbf{v}')dv\, dv' \qquad (10.8.7)$$

where $dv = dv_x\, dv_y\, dv_z$ and $dv' = dv_x'\, dv_y'\, dv_z'$. Because electrons lose their momentum in a given direction at a rate much faster than the rate with which they lose their excess energy, τ_m is in general much faster than $\tau_\mathscr{E}$.

Let us now apply Eqs. (10.8.2), (10.8.3), and (10.8.6) to an analysis of the velocity–saturation phenomenon. The drift velocity v_d versus the field E curve shown in Fig. 10.1 is measured in homogeneous samples under a constant E. Therefore, Eqs. (10.8.2) and (10.8.3) reduce, respectively, to

$$\mathbf{J}_e = e\mu n\mathbf{E} \qquad (10.8.8)$$

$$\mathscr{E} - \mathscr{E}_0 = (\mathbf{E} \cdot \mathbf{J}_e)\tau_\mathscr{E} = e\mu nE^2\tau_\mathscr{E} \qquad (10.8.9)$$

Equation (10.8.9) simply states that the power gained from the electric field, represented by the $e\mu nE^2$ term, must be equal to the power dissipated to the lattice, represented by the $(\mathscr{E} - \mathscr{E}_0)/\tau_\mathscr{E}$ term. It is, in spirit, the same as Eq. (10.2.13). The difference is in the approach. The two important quantities in both approaches are the energy-relaxation time and the mobility. The approach taken in Section 10.2 involves the derivation of μ and $\tau_\mathscr{E}$ from the scattering theory. The approach taken by Hänsch and Miura-Mattausch, on the other hand, emphasizes an accurate formulation of the principal dependences of μ and $\tau_\mathscr{E}$, so that the results thus obtained can be fitted to the terminal v_d versus E characteristic.

As we recall, the derivation of the v_d versus E relation in Section 10.2 was based on the separate treatments of the acoustic-phonon scattering in the intermediate field region and the optical-phonon scattering in the high-field region. Ignoring the optical-phonon scattering in the immediate field region, however, makes the prediction of the v_d versus E relation inaccurate in the region. Hänsch and Miura-Mattausch took a very different approach by deriving a set of macroscopic equations based on the first principles without inquiring into the details of the scattering processes. The results thus obtained are general. The parameters introduced in the formulation of the problem can be obtained from the experimentally measured v_d versus E curve. The emphasis of this approach is on the correct energy dependence of the parameters.

One important parameter in the formulation is the momentum relaxation time, which is denoted by τ_m here but by τ in Section 10.2. The energy dependence of τ_m is implicit in Eq. (10.8.7). To derive an explicit dependence, we make the following observation. The Boltzmann equation, Eq. (A5.7), for a uniform semiconductor under a dc field can be written as

$$\frac{\partial f}{\partial t} = -\mathbf{a} \cdot \mathbf{\nabla}_v f + \left.\frac{\partial f}{\partial t}\right|_{collision} \qquad (10.8.10)$$

If we heuristically set the collision term as

$$\left.\frac{\partial f}{\partial t}\right|_{collision} = -\frac{f - f_0}{\tau_c} \qquad (10.8.11)$$

where τ_c is a characteristic time, the dc steady-state solution must satisfy the equation

$$f - f_0 = -\tau_c\mathbf{a} \cdot \mathbf{\nabla}_v f \qquad (10.8.12)$$

Sec. 10.8   Analysis of Velocity Saturation by Transport Equations   **501**

The solution of Eq. (10.8.12) can be obtained by iteration and is given by

$$f = f_0(1 - A + A^2 - A^3 + \cdots) \tag{10.8.13}$$

where $A = (\mathbf{a} \cdot \mathbf{v})(m^*\tau_c/kT)$. Therefore, in general, the solution of the Boltzmann equation can be expressed as

$$
\begin{aligned}
f(\mathbf{r}, \mathbf{v}) = \\
f_0(\mathbf{r}, \mathbf{v}) + (\mathbf{E} \cdot \mathbf{v})g_1(\mathbf{r}, \mathbf{v}) + (\mathbf{E} \cdot \mathbf{v})^2 g_2(\mathbf{r}, \mathbf{v}) + (\mathbf{E} \cdot \mathbf{v})^3 g_3(\mathbf{r}, \mathbf{v}) + \cdots \tag{10.8.14}
\end{aligned}
$$

in view of the fact that $m^*\mathbf{a} = -e\mathbf{E}$.

The functions $g_i(\mathbf{r}, \mathbf{v})$ are all even functions of v. From our discussion on lattice and impurity scatterings in Section 6.7, we see that the reflection coefficient expressed in Eq. (6.7.14) and the scattering angle θ expressed in Eq. (6.7.1) depend only on the magnitude of carrier velocity, meaning that the scattering matrix element $S(\mathbf{v}', \mathbf{v})$ is an even function in v. Since $(\mathbf{v} - \mathbf{v}')$ in Eq. (10.8.7) is an odd function, nonzero contributions to Eq. (10.8.7) must come from g_1 and g_3 of Eq. (10.8.14). We further note that the contributions from g_3 and g_1 are in the ratio of E^2 according to Eq. (10.8.14). However, E^2 is proportional to $(\mathscr{E} - \mathscr{E}_0)/n$ according to Eq. (10.8.9). Therefore, we can express τ_m^{-1} of Eq. (10.8.7) as

$$\tau_m^{-1} = \tau_{m0}^{-1} \left(1 + \eta \frac{\mathscr{E} - \mathscr{E}_0}{n}\right) \tag{10.8.15}$$

where η has a dimension of reciprocal energy or J^{-1} and τ_{m0}^{-1} is related to the low field mobility μ_0 by $\tau_{m0}^{-1} = e/m^*\mu_0$.

From Eqs. (10.8.9) and (10.8.15) in conjunction with Eq. (10.8.5), we obtain an expression relating the excess electron energy $U = \mathscr{E} - \mathscr{E}_0$ to the applied field E as follows:

$$U^2 + \frac{n}{\eta} U = e\mu_0 n^2 E^2 \frac{\tau_\mathscr{E}}{\eta} \tag{10.8.16}$$

From Eq. (10.8.16), U can be expressed in terms of E and used in Eq. (10.8.15). Thus we find that

$$U = -\frac{n}{2\eta} + \frac{n}{2\eta}(1 + 4e\mu_0\eta E^2\tau_\mathscr{E})^{1/2} \tag{10.8.17}$$

At high fields, the electron approaches the saturation velocity v_s. Taking the high-field limit, we have, from Eq. (10.8.17), $\eta U/n \cong \sqrt{e\mu_0\eta\tau_\mathscr{E}}\, E$, and from Eq. (10.8.15), $\mu(\eta U/n) \cong \mu_0$. Using the relation $\mu E = v_s$, we have $\eta\tau_\mathscr{E} = \mu_0/ev_s^2$. In terms of v_s and $\tau_\mathscr{E}$, the electron energy density is

$$\frac{\mathscr{E}}{n} = \frac{3}{2}kT_l - \frac{e\Delta}{2}\left[1 - \left(1 + 4\frac{\mu_0^2 E^2}{v_s^2}\right)^{1/2}\right] \tag{10.8.18}$$

and the electron velocity is

$$v_d = 2\mu_0 E\left[1 + \left(1 + 4\frac{\mu_0^2 E^2}{v_s^2}\right)^{1/2}\right]^{-1} \tag{10.8.19}$$

The quantity $\Delta = \tau_\mathscr{E} v_s^2/\mu_0$ has a dimension of volts.

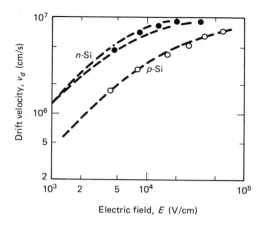

Figure 10.23 Comparison of the computed drift velocity v_d versus field E curves (dashed) with experimental values for electrons (dots) and holes (open circles) measured in silicon at $T = 300$ K.

In Fig. 10.23 we compare the v_d versus $\mathscr{E}$ curves computed from Eq. (10.8.19) for the electrons and holes in silicon (dashed curves) with the experimental value (dots for electrons and open circles for holes). For electrons, the values $\mu_0 = 1420$ cm^2/V-s and $v_g = 1 \times 10^7$ cm/s are used for the upper curve, and $\mu_0 = 1350$ cm^2/V-s and $v_s = 9 \times 10^6$ cm/s are use for the lower curve. For holes, the values $\mu_0 = 480$ cm^2/V-s and $v_s = 8 \times 10^6$ cm/s are used. A good agreement between the theory and the experiment is obtained. We should point out that an empirical formula of the form

$$\mu = 2\mu_0 \left\{ 1 + \left[1 + \left(\frac{2\mu_0 E}{v_s} \right)^\beta \right]^{1/\beta} \right\}^{-1} \tag{10.8.20}$$

with β being a fitting parameter has been used by several authors (see, for example, R. Jaggi, *Helv. Phys. Acta,* Vol. 42, p. 941, 1969; Y. Ohno, *Jpn. J. Phys. Soc.,* Vol. 55, p. 590, 1986). The analysis by Hänsch and Miura-Mattausch gives Eq. (10.8.20) a theoretical basis for its field dependence. However, Eq. (10.8.19) applies only to carrier velocity in bulk semiconductors. In devices such as MOSFET and MODFET, a high electric field exists also in the direction normal to the oxide–semiconductor and semiconductor–semiconductor interface. For device simulation, Eq. (10.8.19) needs to be modified in order to describe channel mobility with reasonable accuracy. Suitable expressions for carrier mobility in FET channels are presented in Section 13.13.

10.9 ELECTRON TRANSFER AND VELOCITY-FIELD CHARACTERISTICS IN TWO-VALLEY SEMICONDUCTORS

In 1963, Gunn discovered a new bulk phenomenon in an n-type GaAs sample (Section 10.1). When a high electric field was applied, the sample showed a strong oscillation of current with frequency in the microwave region. Both sinusoidal and nonsinusoidal waveforms were observed when the field exceeded a threshold value around 3500 V/cm. Figure 10.1 shows that v_d in n-GaAs decreases with increasing E for $E > 3500$ V/cm. For a semiconductor to exhibit negative differential mobility the high-mobility valley must lie below the low-mobility valley in energy (Fig. 10.4). Such is the case in GaAs. Any semiconductor that possesses a low-energy, high-mobility valley and high-energy, low-mobility valley (or valleys), such as GaAs and InP, is a potential candidate for the Gunn-effect oscillator and related devices.

In Section 10.3 we used Eq. (10.3.5) for the ratio n_2/n_1 of electron concentrations in the satellite and central conduction-band minima (Γ and L valley in Fig. 6.23a) and Eq. (10.3.6) for the rate $\langle d\mathscr{E}/dt \rangle$ of energy dissipation to compute the drift velocity versus electric field curve (Fig. 10.5) at several normalized lattice temperatures ($\theta = kT_l/\Delta$), where Δ is the energy separation between the central and satellite valley. However, the computation is based on the assumption that the electron temperature T_e is the same for central and satellite valley electrons. Although this assumption leads to simplified results, it is not realistic. In this section we take into consideration that the two types of carriers can have drastically different thermal velocities. A general treatment of the electron-transfer problem requires (1) the assignment of two electron temperatures, T_{e1} for central-valley electrons of concentration n_1 and T_{e2} for satellite-valley electrons of concentration n_2; and (2) the consideration of intervalley scattering as well as intravalley scattering of electrons. Such a general treatment was carried out by Butcher and Fawcett (P. N. Butcher and W. Fawcett, *Proc. Phys. Soc. London*, Vol. 86, p. 1205, 1965; *Phys. Lett.*, Vol. 1, p. 489, 1966). Figure 10.24 shows their calculated curves of $n_1/(n_1 + n_2)$ and T_e as a function of the applied field after several refinements of the early calculation (W. Fawcett and I. B. Bott, "The Gunn Effect in Gallium Arsenide," in L. Young, ed., *Advances in Microwaves*, Vol. 3, Academic Press, Inc., New York, 1968). Figure 10.25 shows the theoretical curve (dashed) of drift velocity v_d calculated by Butcher and Fawcett as compared to the experimental curve (solid) measured by Ruch and Kino (J. G. Ruch and G. S. Kino, *Appl. Phys.*

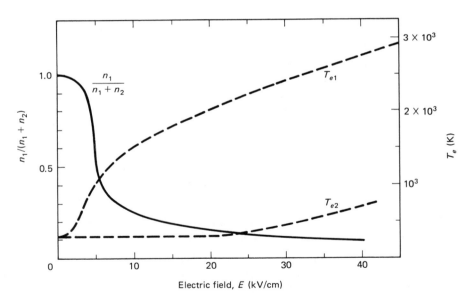

Figure 10.24 The electron temperatures, T_{e1} and T_{e2}, and the fraction of electrons remaining in the central valley, n_1/n, computed by Fawcett and Butcher as a function of applied electric field. (W. Fawcett and I. B. Bott, "The Gunn Effect in Gallium Arsenide," *Advances in Microwaves*, L. Young, ed., Vol. 3, p. 223, Academic Press, 1968.)

High-Field Phenomena and Hot-Electron Effects   Chap. 10

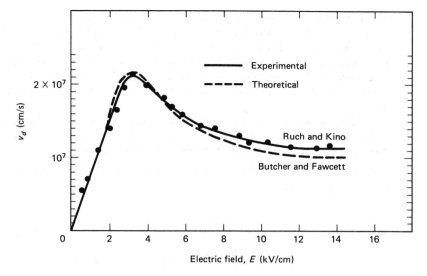

Figure 10.25 Comparison of measured (solid) and calculated (dashed) curves of drift velocity v_d as a function of applied electric field. (J. G. Ruch and G. S. Kino, *Appl. Phys. Lett.*, Vol. 10, p. 40, 1967.)

Lett., Vol. 10, p. 40, 1967; *Phys. Rev.*, Vol. 174, p. 921, 1968). We see good agreement between theory and experiment. For details of the theoretical calculations of the v_d versus E curve, the reader is referred to the articles by Butcher and Fawcett and by Conwell and Vassel (E. M. Conwell and M. O. Vassel, *Phys. Rev.*, Vol. 166, p. 797, 1966; *IEEE Trans. Electron Devices*, Vol. ED-13, p. 22, 1966). In the following discussion, we concentrate on the important aspects of the electron-transfer process so that we may learn the basic physics from analyzing the process.

The first point concerns the ratio n_1/n_2 of electron distribution. To transfer an electron from the central valley to a satellite valley, an energy $\Delta\mathscr{E}$ is required. Out of the n_1 electrons in a unit volume, only $n_1 \exp(-\Delta\mathscr{E}/kT_{e1})$ electrons have the sufficient energy. Therefore, the rate r_{12} of upward transfer is given by

$$r_{12} = R_{12}n_1 \exp\left(\frac{-\Delta\mathscr{E}}{kT_{e1}}\right) \tag{10.9.1}$$

where R_{12} is a proportionality constant. The rate for the downward transfer is simply

$$r_{21} = R_{21}n_2 = \frac{n_2}{\tau_{21}} \tag{10.9.2}$$

because no minimum energy is required. In the steady state, the two rates must be equal or $r_{21} = r_{12}$, yielding

$$\frac{n_1}{n_2} = \frac{R_{21}}{R_{12}} \exp\left(\frac{\Delta\mathscr{E}}{kT_{e1}}\right) = B' \exp\left(\frac{\Delta\mathscr{E}}{kT_{e1}}\right) \tag{10.9.3}$$

At low fields, $T_{e1} = T_l$ and Eq. (10.9.3) must reduce to Eq. (10.3.5); therefore, R_{12}/R_{21} must be equal to B in Eq. (10.3.5). At $E = 6.5$ kV/cm, we find from Fig.

10.24 a value of 1160 K for T_{e1} or $kT_{e1} = 0.10$ eV. For $\Delta\mathscr{E} = 0.36$ eV and $B' = 2.4 \times 10^{-2}$, we obtain from Eq. (10.9.3) $n_1 = 0.84n_2$ and $n_1/(n_1 + n_2) = 0.46$ in reasonable agreement with the value $n_1/(n_1 + n_2) = 0.48$ given in Fig. 10.24. We should point out that the old value 0.36 eV used here was for energy separation between Γ and X conduction band minima. Recently, it is found that the L minima lie below the X minima and the energy separation between Γ and L conduction-band minima is 0.30 eV.

The second point concerns the energy balance. The electrons in transferring from the central valley to a satellite valley remove, on the average, an energy $\Delta\mathscr{E} + \frac{3}{2} kT_{e1}$ from the central valley, while the electrons in a downward transfer bring, on the average, an energy $\Delta\mathscr{E} + \frac{3}{2} kT_{e2}$. Therefore, the net power transfer from the central valley to the satellite valleys is

$$P_{12} = r_{12}\left(\Delta\mathscr{E} + \tfrac{3}{2} kT_{e1}\right) - r_{21}\left(\Delta\mathscr{E} + \tfrac{3}{2} kT_{e2}\right)$$
$$= \tfrac{3}{2} k(T_{e1} - T_l)r_{21} = \tfrac{3}{2}\frac{k(T_{e1} - T_l)n_2}{\tau_{21}} \tag{10.9.4}$$

where $r_{21} = r_{12}$ and expressed in units of $cm^{-3}s^{-1}$. Since the mobility μ_2 of the satellite valleys is very low (estimated to be about $0.02\mu_1$), T_{e2} remains at the lattice temperature T_l even though T_{e1} is much larger than T_l. We set $T_{e2} = T_l$ in Eq. (10.9.4). Energy balance requires that

$$\left\langle\frac{d\mathscr{E}}{dt}\right\rangle = eEv_{d1} - \frac{P_{12}}{n_1} - \tfrac{3}{2}\frac{kT_{e1} - kT_l}{\tau} \tag{10.9.5}$$

where the first term represents the power acquired from the applied field and the last term is the power given to the lattice, which is the same as the term in Eq. (10.3.6). Under the steady state, setting $\langle d\mathscr{E}/dt \rangle = 0$ yields

$$kT_{e1} = kT_l + \tfrac{2}{3} eEv_{d1}\tau\left(1 + \frac{\tau n_2}{n_1\tau_{21}}\right)^{-1} \tag{10.9.6}$$

Equation (10.9.6) in effect says that τ is now replaced by $\tau_{\text{eff}} = \tau[1 + (\tau n_2/n_1\tau_{21})]^{-1}$ due to scattering of electrons between central and satellite valleys. Thus the drift velocity v_{d1} in Eq. (10.9.6) should be given by

$$v_{d1} = \frac{e}{m_1}\tau_{\text{eff}}E = \mu_1 E\left(1 + \frac{\tau n_2}{n_1\tau_{21}}\right)^{-1} \tag{10.9.7}$$

The quantity τ is the intravalley scattering time (i.e., the time constant for polar optical-phonon scattering within the central valley) and can be obtained from the mobility at low fields by $\tau = m^*\mu/e$. The quantity $\tau_{21} = R_{21}^{-1}$ is the intervalley scattering time (i.e., the time constant for electrons scattering from the satellite valleys to the central valley) and is estimated to be about 2.0×10^{-12} s in GaAs. Thus $\tau/\tau_{21} = 0.15$. At $E = 6.5$ kV/cm, we find from Eq. (10.9.6) a value of 0.076 eV for kT_{e1}, using a value of $n_2/n_1 = 1.1$ from Fig. 10.24. This value of kT_{e1} is smaller than the value $kT_{e1} = 1160$ K given in Fig. 10.24.

The discrepancy between the value of T_{e1} calculated from Eq. (10.9.6) and that shown in Fig. 10.24 is due to the fact that Eq. (10.9.6) does not take into account the details of the intravalley scattering processes. As the electric field increases, the range of **k** vectors involved in polar scattering becomes larger and consequently polar scattering becomes less effective. Then intravalley acoustic scattering must be considered. The consequence of this consideration is that we must use two relaxation times, $\tau_{\mathscr{E}}$ for energy transfer and τ_m for momentum transfer as we did in Section 10.8. The value of $\tau_{\mathscr{E}}$ which determines the electron temperature in Eq. (10.9.6) increases with increasing E. The value of τ_m, on the other hand, decreases with increasing E because of the increasing importance of acoustic scattering. Theoretical calculations show that a μ value for the central valley drops to 6200 cm^2/V-s at $E = 5$ kV/cm and to 4000 cm^2/V-s at $E = 10$ kV/cm. The dashed curve of Fig. 10.25 takes into account the decrease in μ with E. Therefore, two different mobilities are used for the central-valley electrons. Since the electron temperature is determined mainly by polar scattering, the value of $\mu_1 = 8000$ cm^2/V-s is still used in the energy equation. The ineffectiveness of polar scattering is taken into account phenomenologically by increasing the factor $\frac{2}{3}$ to 1 in Eq. (10.9.6). Thus we have

$$kT_{e1} = kT_l + e\mu_1 E^2 \tau \left(1 + \frac{n_2}{n_1 \tau_{21}}\right)^{-2} \qquad (10.9.8)$$

The mobility in the momentum equation, on the other hand, is significantly affected by acoustic phonon scattering. Thus we use

$$v_d = \frac{(\mu_1' n_1 + u_2 n_2)E}{n_1 + n_2} \qquad (10.9.9)$$

The use of two different μ's in Eqs. (10.9.8) and (10.9.9) is another way of stating the fact that the energy-relaxation time $\tau_{\mathscr{E}}$ and the momentum-relaxation time τ_m are different.

In summary, we have shown that the ratio n_1/n_2 is in general given by Eq. (10.9.3). At high fields, the temperature T in Eq. (10.3.5) should be the electron temperature T_{e1} of the central valley. Furthermore, there is a significant transfer of energy at high fields associated with the electron-transfer process. Since electrons in the satellite valleys acquire less kinetic energy from the field than electrons in the central valley on account of a lower mobility, the satellite valley has a cooling effect on the central valley. At high fields where n_2/n_1 becomes appreciable, Eq. (10.9.6) should be used instead of Eq. (10.3.6). In principle, we can solve for T_{e1} and n_1/n_2 from Eqs. (10.9.3) and (10.9.6) as a function of E. This is best left for numerical computations. The important point is that the v_d versus E curve (dashed curve in Fig. 10.25) based on the computed values of n_1/n_2 indeed shows a differential negative conductance and agrees reasonably well with the measured values (solid curve in Fig. 10.25). At high fields, acoustic-phonon scattering must be included in μ because of increased electron k value. The dashed curve of Fig. 10.25 takes the decrease in μ_1 with E into account. Again the high-field transport theory requires two relaxation times: one for momentum and the other for energy.

For practical applications it is very inconvenient to use the information given in Fig. 10.24. Even if we can represent the dependence of n_2/n_1 on E by a simple E^q relation, we still have to account for the dependence of τ_{eff} on n_2/n_1 according to Eq.

(10.9.7). Since our main interest is in computing the macroscopic current, it is useful to find a simple power dependence of v_d on E as follows:

$$v_d = \frac{\mu_1 + \mu_2 x}{1 + x} E \tag{10.9.10}$$

where $x = AE^q$. In Eq. (10.9.10) we assume that the two mobilities μ_1 and μ_2 are independent of E. Thus the parameter x no longer represents the ratio n_2/n_1 alone but incorporates in it the variation of mobilities with E as well. One set of parameters that seem to fit the experimental curve of Ruch and Kino (solid curve in Fig. 10.25) reasonably well can be represented by

$$x = \frac{n_2}{n_1} = 1.35 \left(\frac{E}{5}\right)^3 \tag{10.9.11}$$

together with $\mu_1 = 8000$ cm^2/V-s and $\mu_2 = 500$ cm^2/V-s. In Eq. (10.9.11), E is expressed in kV/cm. The values of mobility are different from the actual low-field values ($\mu_1 = 8500$ and $\mu_2 \sim 160$ cm^2/V-s) for the reasons stated above.

We should point out that when the Gunn effect was discovered, the Ridley–Watkins–Hilsum mechanism had not yet been confirmed. Even though connection between the two was suggested (H. Kroemer, *Proc. IEEE,* Vol. 52, p. 1736, 1964), positive identification of the transferred-electron effect as being the mechanism responsible for the Gunn effect was made only after many theoretical and experimental investigations. The calculation of the velocity-field characteristics by Butcher and Fawcett and the experimental measurement by Ruch and Kino indeed confirmed the existence of a negative differential conductance region after the applied field exceeded a certain threshold value. In a definitive experiment that verified electron transfer as being the cause of the Gunn effect, Hutson et al. used hydrostatic pressure to change the energy separation $\Delta\mathscr{E}$ between the two valleys (A. R. Hutson, A. Jayaraman, A. G. Chynoweth, A. S. Coriell, and W. L. Feldman, *Phys. Rev. Lett.,* Vol. 14, p. 639, 1967; A. R. Hutson, A. Jayaraman, and A. S. Coriell, *Phys. Rev.,* Vol. 155, p. 786,, 1967). Since $\Delta\mathscr{E}$ decreases with hydrostatic pressure, we expect that the Gunn oscillation would cease above a certain threshold pressure. This behavior was indeed confirmed at a pressure of 26 kbar. Further confirmation of the electron-transfer mechanism was obtained by Allen et al., using GaAs$_{1-x}$P$_x$ to change $\Delta\mathscr{E}$. As the phosphorus content x is increased, the energy separation $\Delta\mathscr{E}$ is decreased. For $x \geqslant 0.5$, the GaAs$_{1-x}$P$_x$ compound becomes an indirect-gap semiconductor with the satellite valleys having the lower energy as in Si. In the experiment of Allen et al., the Gunn oscillation ceased when $x \geqslant 0.35$ (J. W. Allen, M. Shyam, Y. S. Chen, and G. L. Pearson, *Appl. Phys. Lett.,* Vol. 7, p. 78, 1965).

Finally, we should mention that the presence of a negative differential mobility is intimately connected with the polar-scattering mechanism and the existence of a critical field. According to the analyses by Fröhlich and Parajape (H. Fröhlich and B. V. Parajape, *Proc. Phys. Soc. London,* Vol. B69, p. 21, 1956) and by Stratton (R. Stratton, *Proc. R. Soc. London,* Vol. A246, p. 406, 1958), the energy-transfer process through polar-mode scattering becomes rather ineffective as the electric field increases. When the field approaches a critical field E_{cr}, the electron temperature increases rapidly. This critical field is

$$E_{\text{cr}} = 0.6 \frac{m^* e \omega_0}{4 \pi \hbar} \left(\frac{1}{\epsilon_\infty} - \frac{1}{\epsilon} \right) \qquad (10.9.12)$$

where ϵ_∞ and ϵ are the optical and low-frequency dielectric constant and $\hbar \omega_0$ is the optical phonon energy. For GaAs, $m^* = 0.067 m_0$, $\epsilon_\infty = 10.82 \epsilon_0$, $\epsilon = 13.23 \epsilon_0$, and $\omega_0 = 5.4 \times 10^{13}$ rad/s (or $\hbar \omega_0 = 0.035$ eV). Using these values in Eq. (10.9.12), we find $E_{\text{cr}} = 3.5$ kV/cm. Table 10.2 shows the relevant data on the Gunn effect compiled by Foyt and McWhorter for a number of compound semiconductors (A. G. Foyt and A. L. McWhorter, *IEEE Trans. Electron Devices*, Vol. ED-13, p. 79, 1966). The drift velocity v_d and hence the thermal energy $\frac{3}{2} k T_e$ of electrons are expected to rise rapidly as the applied field approaches E_{cr}. The threshold field given in Table 10.2 is the value of the field experimentally observed at the onset of the Gunn oscillation. No Gunn effect was observed in InAs and InSb. Notice that the value of $\Delta \mathscr{E}$ is larger than that of $\mathscr{E}_g$ in these materials. Therefore, it is more likely for energetic electrons to participate in an impact ionization process than in an electron-transfer process.

In summary, we have shown that the negative differential conductance responsible for the Gunn effect is a direct consequence of the transferred-electron effect. The material requirements for Gunn-effect devices can be stated as follows. (1) The effective mass in the satellite (higher energy) valleys must be appreciably larger than that in the central (lower energy) valley. (2) The electron mobility in the satellite valleys must be much smaller than that in the central valley. (3) The transfer process must be efficient so that n_2 becomes appreciably larger than n_1 within a small range of E to ensure a negative differential mobility. (4) The energy separation $\Delta \mathscr{E}$ between the central and satellite valleys must be several times larger than $k T_l$ so that $n_2 = 0$ at low fields. (5) The energy gap $\mathscr{E}_g$ must be larger than $\Delta \mathscr{E}$ to prevent impact ionization of electrons from occurring before intervalley transfer of electrons. For further discussions of the transferred-electron effect, the reader is referred to the book by Bulman et al. (P. J. Bulman, G. S. Hobson, and B. C. Taylor, *Transferred Electron Devices*, Academic Press, Inc., New York, 1972). Special reference should be made to the work by Fawcett et al. on the development of the Monte Carlo numerical technique to overcome theoretical difficulties in the computation of the v_d versus E curve and to the work by Ruch and Fawcett on the temperature dependence of and the effect of impurity scattering on the v_d versus E curve (W. Fawcett, A. D. Bourdman, and S. Swain, *J. Phys. Chem. Solids*, Vol. 31, p. 1963, 1970; J. G. Ruch and W. Fawcett, *J. Appl. Phys.*, Vol. 41, p. 3843, 1970).

TABLE 10.2 PREDICTED CRITICAL FIELD AND OBSERVED THRESHOLD FIELD

	T(K)	$\mathscr{E}_g$(eV)	$\Delta \mathscr{E}$(eV)	Gunn effect	Predicted E_{cr}(kV/cm)	Threshold field (kV/cm)
GaAs	300	1.42	0.36	Yes	3.5	2.5–4.0
InP	300	1.29	0.4	Yes	7.8	7.2
CdTe	300	1.45	—	Yes	12.9	13.0
InAs	300	0.33	1.1	No	—	—
InSb	77	0.22	0.6	No	—	—

PROBLEMS

10.1. Consider a silicon MOSFET with a n-channel of 1 μm length and a silicon n-p-n transistor with doping concentrations $N_a = 10^{18}$ cm^{-3} in the base and $N_d = 10^{16}$ cm^{-3} in the collector. The drain–source voltage and the base–collector voltage are both 5 V. Calculate the average drift velocity in the channel and in the collector depletion region. Compare the velocity thus calculated with the thermal velocity.

10.2. (a) Consider the collision between an electron with an initial momentum p_1 and a phonon initially at rest. Show that applying the laws of momentum and energy conservation yields

$$(1 + r)^2 p_r^2 = 2p_1^2(1 + r\sin^2\theta) - 2p_1^2 \cos\theta(1 - r^2 \sin^2\theta)^{1/2}$$

where $r = m^*/M$ is the mass ratio, p_r the phonon momentum after collision, and θ the angle between the initial and final momenta of the electron.

(b) For $m^* \ll M$, the equation above reduces to

$$p_r^2 = 2p_1^2(1 - \cos\theta)$$

By averaging p_r^2 over $d\Omega = 2\pi \sin\theta\, d\theta$, show that the average energy transferred to the phonon is equal to

$$\langle \delta\mathcal{E} \rangle_l = \frac{T_e}{T_l} m^* u^2$$

which is the same as Eq. (10.2.4).

10.3. (a) Consider the collision between a phonon with an initial momentum p_r and an electron initially at rest. Show that applying the laws of momentum and energy conservation yields

$$(1 + r^{-1})^2 p^2 = 2p_r^2(1 + r^{-1}\sin^2\theta) + 2p_r^2 \cos\theta(1 - r^{-2} \sin^2\theta)^{1/2}$$

where $r^{-1} = M/m^*$ is the mass ratio, p the electron momentum after collision, and θ the angle between the initial and final momenta of the phonon.

(b) Show that for $M \gg m^*$, the collision must be collinear and that the average energy transferred to the electron is given by Eq. (10.2.8).

10.4. The velocity of longitudinal and transverse acoustic waves in a cubic crystal is equal to $(c_{11}/\rho)^{1/2}$ and $(c_{44}/\rho)^{1/2}$, where c_{11} and c_{44} are the elastic constants and ρ is the mass density. Given $c_{11} = 1.66$ and $c_{44} = 0.79$, both in 10^{12} dyn/cm^2 and $\rho = 2.328$ g/cm^3 in Si, find the two acoustic velocities. Using the appropriate acoustic velocity, calculate the drift velocity and the temperature of electrons in Si at $E = 5 \times 10^3$ V/cm. Compare the calculated v_d with the value given in Fig. 10.1.

10.5. (a) Show that for electrons obeying Boltzmann distribution, the average velocity $\langle v \rangle$ is given by

$$\langle v \rangle = 2\left(\frac{2kT}{\pi m^*}\right)^{1/2}$$

Thus, for lattice scattering, the relation given by Eq. (10.2.15) can be rewritten as

$$2el = 3\mu_0\left(\frac{\pi m^* kT}{2}\right)^{1/2}$$

(b) Given $\mu = 1400$ cm^2/V-s in Si at 300 K, find the mean free path and the energy gained by an electron within the mean free path at $E = 5 \times 10^3$ V/cm.

10.6. (a) Show that Eq. (10.2.12) can be written in the form of Eq. (10.3.6) with the energy relaxation time defined as

$$\tau_{\mathcal{E}} = \frac{\sqrt{\pi M}}{16 m^*} \tau$$

(b) The electron mobility in Si can be fit by a combination of lattice scattering and optical-phonon scattering by letting

$$\mu^{-1} = \mu_l^{-1} + \mu_{op}^{-1}$$

Given $\mu_l = 1500$ cm²/V-s, $\mu_{op} = 15{,}000$ cm²/V-s, $\mathcal{E}_{op} = 0.047$ eV, $m^* = 0.26 m_0$, and $u = 8.44 \times 10^5$ cm/s in Si, find the rates of energy dissipation to host crystal by acoustic-phonon and optical phonon scattering for $T_e = 1.25 T_l$. From the energy balance, calculate the drift velocity of electrons.

10.7. The instantaneous power acquired by an electron from the applied electric field E is

$$P = \frac{d\mathcal{E}}{dt} = Fv = q\mu E^2$$

Therefore, during the mean free time τ, the average energy $\mathcal{E}_{av}$ gained by an electron is

$$\mathcal{E}_{av} = \tfrac{1}{2} P\tau = \tfrac{1}{2} m^*(\mu E)^2$$

The mobility effective mass $m_\mu^* = 3m_t m_l/(2m_l + m_t)$ is given by $m_\mu^* = 0.26 m_0$ in Si and $0.063 m_0$ in GaAs.

(a) Assuming μ to remain unchanged ($\mu_e = 1400$ cm²/V-s in Si and 8000 cm²/V-s in GaAs), estimate the electric field E required for electrons to acquire an energy to excite optical phonons in Si and GaAs.

(b) The minimum optical phonon energies $\hbar\omega_{op}$ are also known: $\hbar\omega_{op} = 0.047$ eV in Si and 0.029 eV in GaAs. Estimate the saturation velocity v_s in Si and GaAs.

10.8. (a) Consider the transferred-electron effect. Plot $v_d/\mu_1 E_0$ as a function of E/E_0 (at values 0.2, 0.5, 1, 2, 5, 10) for $kT_l = 0.052$ eV, 0.026 eV, and 0.0065 eV.

(b) In GaAs, $\Delta = 0.30$ eV and $B = 26$. Find the approximate lattice temperature above which the negative differential conductance caused by the transferred-electron effect is expected to disappear.

10.9. The v_d versus E curve for GaAs can be fitted empirically by the following expression:

$$v_d = \frac{\mu_1 n_1 + \mu_2 n_2}{n_1 + n_2} E$$

where $\mu_1 = 8000$ cm²/V-s, $\mu_2 = 500$ cm²/V-s, and

$$\frac{n_2}{n_1} = 1.35 \left(\frac{E}{5000}\right)^3$$

with E in V/cm. The ac mobility μ_{ac} is defined as $\mu_{ac} = dv_d/dE$. Find the value of μ_{ac} at $E = 1$ kV/cm and 5 kV/cm. Comment on whether a space charge will decay or grow in its intensity at the two field strengths.

10.10. The ionization energy $\mathcal{E}_i$ can be found by eliminating one of three quantities $p_1 = m_e^* v_1$, $p_e = m_e^* v_e$, or $p_h = m_h^* v_h$ from Eqs. (10.4.1) and (10.4.2) and then minimizing $p_0 = m_e^* v_e$ with respect to the other two quantities. Show that setting $\partial p_0/\partial p_e = 0$ and $\partial p_0/\partial p_h = 0$ yields

$$2p_e^2 + 4p_e p_h + \left(1 - \frac{m_e^*}{m_h^*}\right)p_h^2 = 2\mathscr{E}_g m_e^*$$

$$2\left(1 + \frac{m_e^*}{m_h^*}\right)p_e p_h + \left(1 + \frac{m_e^*}{m_h^*}\right)p_h^2 = 2\mathscr{E}_g m_e^*$$

From the equations above, find the condition on p_e and p_h and then obtain Eq. (10.4.3).

10.11. Show that

$$(N_0 + N_1)f(x) = N_f \int_0^x \alpha_h f(x') \, dx' + A$$

satisfies Eq. (10.4.7) if

$$f(x) = \exp\left[-\int_0^x (\alpha_e - \alpha_h) \, dx'\right]$$

Then apply the boundary conditions to find the integration constant A, and to derive Eq. (10.4.12).

10.12. From Eqs. (10.4.12) and (10.4.13), show that

$$M_e^{-1} = f(W)M_h^{-1}$$

where $f(x)$ is given in Problem 10.11. Therefore, if one multiplication factor goes to infinity, the other also goes to infinity.

10.13. Consider an abrupt p^+-i-n^+ diode in which the middle i region is made of intrinsic material. The doping concentrations of two outer p^+ and n^+ regions are so high that the total width of the two space-charge regions $W_n + W_p$ is negligibly small compared to the width of the intrinsic region W.

(a) Draw diagrams showing the field and potential variations in the diode under a reverse bias.

(b) Given $W = 4 \times 10^{-4}$ cm and $\alpha(\text{cm}^{-1}) =$ ionization coefficient $= 10^5 \, (E/5 \times 10^5)^5$, where E is in V/cm, find the breakdown voltage V_B of the diode.

10.14. (a) Expressing Eq. (10.5.12) as $T = \exp(-B/E)$, find the dimension of B. Given $m_r^* = 0.062m_0$ and $\mathscr{E}_g = 1.42$ eV in GaAs, calculate the value of B.

(b) If we set $E_{\max} = B/20$ as the condition for tunneling breakdown, calculate the breakdown voltage V_B in an abrupt junction with $N_a = N_d = 10^{18}$ cm^{-3}.

10.15. (a) The ionization coefficient $\alpha = \sqrt{\alpha_e \alpha_h}$ in GaAs can be approximated by

$$\alpha = 6 \times 10^3 (E/3 \times 10^5)^5$$

and the breakdown voltage in an abrupt junction by

$$V_B = 70\left(\frac{N_I}{10^{16}}\right)^{-q}$$

where N_I is the joint impurity concentration. Find the maximum electric field for $N_I = 10^{15}, 10^{16}, 10^{17},$ and 10^{18} cm^{-3} at the respective breakdown voltage.

(b) Use Eq. (10.6.18) with $E = E_{\max}/2$ to calculate the tunneling-current density J_t at $0.9V_B$ for each given impurity concentration. Also use Eq. (8.9.3) with $\tau_{p0} = 10^{-7}$ s and $n' = n_i$ to calculate the corresponding generation-current density J_{gen}. Find the approximate demarcation line between tunneling-dominated ($J_t > J_{\text{gen}}$) and avalanche-dominated ($J_{\text{gen}} > J_t$) region in terms of N_I.

10.16. (a) Use Eq. (10.5.13) with $Q = 3\pi/2$ to calculate the value of $ka = i\alpha a$ in the forbidden gap from $\beta a = \pi$ to $\beta a = 1.5\pi$. From the plot, find the energy gap and the maximum value of αa.

(b) Use Eq. (10.5.14) with $m_r = 0.063m_0$, $\mathscr{E}_g = 1.4$ eV, and $a = 5$ Å to find the value of αa as a function of energy from $\mathscr{E} = \mathscr{E}_v$ to $\mathscr{E} = \mathscr{E}_c$.

(c) Compare the plot of αa based on Eq. (10.5.13) with that based on Eq. (10.5.14), and explain why the parabolic barrier of Eq. (10.5.14) is physically more reasonable than the triangular barrier of Eq. (10.5.4).

10.17. (a) Show that for a linearly graded junction, the breakdown voltage V_B follows the relation

$$V_B = BA^{-(m-1)/(2m+1)}$$

where A is the impurity gradient and B is a proportionality constant. Also derive the following dependence for the multiplication factor

$$M^{-1} = 1 - \left(\frac{V}{V_B}\right)^{(2m+1)/3}$$

(b) For Si graded junctions, we have, with $m = 5$

$$V_B = 38\left(\frac{A}{10^{21}}\right)^{-0.364}$$

Calculate V_B for $A = 10^{23}$ cm^{-4} and 10^{19} cm^{-4}, and compare the calculated values with the values given in Fig. 10.14.

10.18. (a) Calculate the value of λ in Si from Eq. (10.7.1) at 300 K. Using the values of $C_1 = 2$, $C_2 = 2 \times 10^{-3}$, $\mathscr{E}_b = 3.2$ eV, and $\mathscr{E}_i = 3\mathscr{E}_g/2 = 1.68$ eV quoted in the paper by Hu et al., calculate the values of E separately from Eqs. (10.7.2) and (10.7.3) for the following sets of data points: (1) $x = I_{SUB}/I_D = 4 \times 10^{-4}$, $y = I_G/I_D = 1.3 \times 10^{-11}$, (2) $x = 10^{-4}$, $y = 7 \times 10^{-13}$, (3) $x = 10^{-4}$, $y = 1.9 \times 10^{-11}$, and (4) $x = 2 \times 10^{-5}$, $y = 7 \times 10^{-13}$ taken from the $V_{GD} = 2$ V curve for (1) and (2) and from the $V_{GD} = 7$ V curve for (3) and (4) in Fig. 10.18. The various data points with increasing values of x and y on each curve were taken with increasing value of V_D above V_{Dsat}. Explain qualitatively why the calculated behavior of E is expected.

(b) A quantitative evaluation of E requires a two-dimensional solution of Poisson's equation by numerical simulation. Such a calculation shows a general increase of E toward the drain and then a sharp increase of E near the drain. The field E in Eqs. (10.7.2), (10.7.3), and (10.7.7) represents the maximum E close to the drain, which is about four times the average field V_D/L. Check whether the calculated E falls within the expected range for $V_D = 6$ V and $L = 1$ to 1.5 μm.

10.19. (a) Indicate the electron-pair bond in Si by two bars as illustrated in Fig. 6.2, and explain why a Si atom on the surface of a Si crystal (Fig. 10.19a) acts as an electron trap. Show the ionic bond in SiO$_2$ by charges associated with Si and O ions, and explain why a Si atom on the SiO$_2$ surface acts as a hole trap by counting the charges it carries.

(b) Indicate whether or not the bond requirement is satisfied for a Si atom at site A, a Si atom at site B, at the Si-H site, and at the Si-OH site in Fig. 10.20. If not, assign the site either as an electron trap or as a hole trap.

10.20. If a charge density $\rho(x)$ exists in the gate oxide of a MOS structure, show that the flat-band voltage is shifted by an amount

$$\Delta V_{FB} = -\frac{1}{\epsilon_{\text{ox}}} \int_0^{d_{\text{ox}}} x\rho(x)\, dx$$

In Fig. 10.21, the movement of both C_{LIF} and C_{HF} curves after negative-bias stress (NBS) is a result of increased oxide charges. Assuming a uniform distribution of oxide charges, estimate the oxide trap density (in cm^{-2}) generated after 30 min of NBS.

10.21. (a) The change in the charge ΔQ stored on the semiconductor side of a MOS structure is given by

$$\Delta Q = eAN_{\text{ST}}\, \Delta V + C_d\, \Delta V$$

where $\Delta V = \Delta \phi_s$ is the change in the surface potential, C_d is the depletion region capacitance, and A is the MOS area. Assuming that the surface traps are slow surface states, derive Eq. (10.7.5).

(b) As discussed in Problem 10.20, the shift of the capacitance curves toward negative V_G is caused by oxide traps. The difference between C_{LF} and C_{HF} at minimum C_{LF} gives a good estimate of N_{ST}. Find the values of N_{ST} before NBS, after 10 min of NBS, and after 30 min of NBS.

10.22. (a) The dependence of N_{ST} on $t^{1/4}$ as given in Eq. (10.7.6) is based on diffusion model of oxide defects. Assuming a linear relation between ΔV_{th} and N_{ST}, find the N_{ST} dependence on stress time t from Fig. 10.22a.

(b) Next, starting from Eqs. (10.7.7) and (10.7.2), derive the following expression for the operating life τ of a MOSFET:

$$\tau = H(\Delta V_{\text{th}})^x I_D^{y-1}\, I_{\text{SUB}}^{-y}\, Z$$

Determine the value of x from Fig. 10.22a and y from given values of $\mathscr{E}_i = 1.68$ eV and $\mathscr{E}_T = 3.5$ eV. A MOSFET with $I_{\text{SUB}} = 0.1$ mA has an operating life of 2×10^4 s. After NBS, I_{SUB} increases to 1 mA. What is the projected operating life of the MOSFET?

10.23. To impart further physical meanings to Eqs. (10.8.1) to (10.8.4), we make the following observations.

(a) Show that for a uniform, nondegenrate semiconductor, Eq. (10.8.2) reduces to Eq. (7.1.5).

(b) Note that Eq. (7.4.6) for charge conservation can be written as

$$\frac{\partial \rho}{\partial t} = -\nabla \cdot \mathbf{J} - \frac{\rho - \rho_0}{\tau_0}$$

where $\rho - \rho_0$ represents the excess charge density and τ_0 represents the time constant for ρ to return to ρ_0. Show that under thermal equilibrium, the equation above reduces to Eq. (10.8.1).

(c) Another fundamental law of physics is energy conservation. Obtain the equation for energy conservation from the particle-conservation equation by replacing charge density by energy density and current density by power-flow density. Compare the equation thus obtained with Eq. (A9.22). Explain the physical meanings of the terms $(\mathscr{E} - \mathscr{E}_0)/\tau_{\mathscr{E}}$ and $\mathbf{E} \cdot \mathbf{J}_e$ in Eq. (10.8.3).

(d) Equation (A9.14), from which Eq. (10.8.2) is obtained, can be considered as a momentum equation. Using proper substitutions, obtain a corresponding equation for power-flow density $\mathbf{C}_{\mathscr{E}}$. Compare the equation thus obtained with Eq. (A9.27). Explain their differences. Using Eq. (A5.11) for Eq. (A9.24), derive the expression for the missing term in Eq. (A9.27) or (10.8.4) arising from the contribution from the fourth moment.

10.24. (a) Derive Eq. (10.8.17) from Eqs. (10.8.9) and (10.8.15). Further obtain Eq. (10.8.19) from Eq. (10.8.17) through the use of Eq. (10.8.9) and the definition $v_d = \mu E$.

(b) Using the values of μ_0 and v_s given in the text, calculate the drift velocity of electrons and holes in Si at $E = 10^4$ V/cm.

10.25. Derive Eq. (10.9.6).

10.26. Given $\mu_1 = 8000$ cm^2/V-s, $\tau = 3 \times 10^{-13}$ s, $\tau_{21} = 2 \times 10^{-12}$ s, $\Delta\mathscr{E} = 0.36$ eV, and $B' = 2.4 \times 10^{-2}$, find the values of T_{e1} and n_1/n_2 at $E = 6.5$ kV/cm from Eqs. (10.9.3) and (10.9.8). Also calculate the value of v_d from $v_d = (\mu'_1 n_1 + \mu_2 n_2)/(n_1 + n_2)$ using $\mu_2 = 175$ cm^2/V-s and the value μ'_1 quoted in the text.

10.27. Repeat Problem 12.26 for $E = 10$ kV/cm.

10.28. Use Eqs. (10.9.10) and (10.9.11) to plot the v_d versus E curve from $E = 1$ to 13 kV/cm, and compare the computed curve with the measured curve of Fig. 10.25.

10.29. The interatomic distance between Ga and As atoms in GaAs is 2.44 Å. Using Eq. (6.8.1) and Table 6.4, find the amount of ionic charge carried by a Ga atom. The following values are also given: $\omega_t = 5.4 \times 10^{13}$ rad/s, $M_{Ga} = 69.7M$, $M_{As} = 74.9M$, and $M = 1.67 \times 10^{-27}$ kg.

10.30. The following values are known in InSb: $\epsilon = 17\epsilon_0$, $\epsilon_\infty = 15.7\epsilon_0$, and $m_e^* = 0.013m_0$. Assuming that $\omega_0 = 5.4 \times 10^{13}$ rad/s, calculate the critical field. Experimentally, it is observed that the current in InSb increases very rapidly but does not show any sign of instability (i.e., Gunn oscillation) as the electric field approaches 250 V/cm. Explain why no Gunn oscillation takes place.

11

Properties of Heterostructures

11.1 MATERIAL REQUIREMENTS

Shortly after transistor action was demonstrated in germanium with point-contact metal–semiconductor junctions, several techniques were investigated and developed to make metallurgical *p-n* junctions, including the alloying process, the zone-melting rate-grown process, and the diffusion process (J. S. Shay and W. C. Dunlap, Jr., *Phys. Rev.*, Vol. 90, p. 630, 1953; R. N. Hall, *Phys. Rev.*, Vol. 88, p. 139, 1952; W. C. Dunlap, Jr., *Phys. Rev.*, Vol. 94, p. 1531, 1954). At the same time, the idea of using two different semiconductors to form *p-n* junctions (heterojunctions) was considered (A. J. Gubanov, *Zh. Eksp. Teor. Fiz.*, Vol. 21, p. 721, 1951; H. Kroemer, *Proc. IRE*, Vol. 45, p. 1535, 1957). Attempts were made experimentally to grow Ge–Si junctions by the alloying process and Ge–GaAs junctions by the vapor-deposition method (J. Shewchun and L. Y. Wei, *J. Electrochem. Soc.*, Vol. 111, p. 1145, 1964; R. L. Anderson, *Solid-State Electron.*, Vol. 5, p. 341, 1962). However, the junctions thus made had high dislocation density at the interface, and activities in heterojunction research waned. The interests in heterostructures revived after the successful fabrication and CW room-temperature operation of double heterostructure (DH) GaAs–(GaAl)As lasers in 1971. The GaAs–(GaAl)As system is unique in the sense that the lattice constant of (GaAl)As closely matches that of GaAs, resulting in a high-quality interface. However, alloys of other compound semiconductors, except the indirect-gap GaP and AlP, do not have this desirable property. Therefore, in order to grow, for example (GaIn)As or Ga(AsSb) on GaAs substrates, buffer layers were grown first to grade the composition either smoothly or in many small abrupt steps to make a gradual transition from the lattice of GaAs to that of (GaIn)As or Ga(AsSb). Using this approach, DH (GaIn)As/(GaIn)P and Ga(AsSb)/(AlGa)(AsSb) injection lasers were made to operate continuously at room temperature (C. Nuese, G. H. Olsen, M. Ettenberg, J. J. Gannon, and T. J. Zamerowski, *Appl. Phys. Lett.*, Vol. 29, p. 807, 1976; R. E. Nahory, M. A. Pollack, E. D. Beebe, J. C. DeWinter, and R. W. Dixon, *Appl. Phys. Lett.*, Vol. 28, p. 19, 1976). Simultaneous with the graded buffer approach is the approach taken by Hsieh to find an alloy with a composition that can be lattice matched to the substrate. Using the composition-lattice-match approach, first pulsed and then CW room-temperature operation of the (GaIn)(AsP)/InP injection laser was achieved (J. J. Hsieh, *Appl. Phys. Lett.*, Vol. 28, p. 283, 1976). The latter approach has been widely used because its fabrication process is simpler to control and hence the lasers thus made are more reliable.

The brief review given above of the development of DH lasers serves to underscore the importance of lattice match to producing a heterojunction interface with low dislocation density. Figure 11.1 shows the band-gap energy and lattice constant of a selected number of III–V compound, II–VI compound, and elemental semiconductors (H. C. Casey, Jr., and M. B. Panish, *Heterostructure Lasers*, Academic Press, Inc., New York, 1978, Chap. 5; D. Botez and G. J. Herskowitz, *Proc. IEEE*, Vol. 68, p. 689, 1980; S. Wang, *Solid State Electronics*, McGraw-Hill Book Company, New York, 1966, p. 180). For our present discussion we concentrate on III–V compounds. One desirable attribute of compound semiconductors is their miscibility either over the entire or over a part of the composition range. In Fig. 11.1 the lines or curves connecting two compounds indicate the range of gap energies and lattice constants covered, with the heavy ones showing the miscible range and the light ones showing the miscibility gap. The dashed lines indicate the range in which the alloy is an indirect-gap semiconductor. Referring to the band structure of GaAs shown in Fig. 11.2, we see that the direct-gap energy is equal to the difference in energy between the Γ_6 (conduction band minimum) and the Γ_8 (valence band maxima) points which are both located at $k[000]$ in the Brillouin zone. The indirect-gap energy commonly refers to the smallest energy difference among band extrema located at different k. Since the energy separation between conduction-band extrema is smaller than that between valence-band extrema, the indirect-gap energy is given by the energy difference between the X_6 and Γ_8 points or between the L_6 and Γ_8 points, which are denoted by $\mathscr{E}_X$ and $\mathscr{E}_L$, respectively. The symbols Γ_n, X_n, and L_n are theoretic-group notations used to indicate the symmetry of the Brillouin zone at a given reciprocal lattice point (M. Timken; *Group Theory and Quantum Mechanics*, McGraw-Hill Book Company, New York, 1964, D. Long, *Energy Bands in Semiconductors*, Wiley-Interscience, New York, 1968; W. A. Harrison, *Solid State Theory*, McGraw-Hill Book Company, New York, 1970). In some alloys the indirect-gap energy $\mathscr{E}_X$ changes very little with composition, causing a sudden change in the slope of the solid and dashed lines, such as the lines for the (InGa)P and (InAl)Sb alloys.

The development of the heterostructure science and technology is largely influenced by the availability of suitable substrates and the need from industrial applications. Among III–V compounds, the ternary (GaAl)As system lattice-matched to GaAs substrate and the quaternary (GaIn)(AsP) system lattice-matched to InP substrate have been extensively studied, the former for high-speed integrated-circuit applications and the latter for optical fiber communication applications. At room temperature, the lattice constant a for the unit cell of GaAs is 5.65325 Å, which is $4\sqrt{3}$ of the distance between nearest neighbors in a zinc-blende structure, and that of AlAs is 5.6612 Å. Therefore, the lattice mismatch $\Delta a/a$ between GaAs and AlAs is only 0.14%, and even less between GaAs and (GaAl)As. For the quaternary system, the procedure to find the lattice-matched composition can be decomposed into two steps: The first step is to lattice-match the ternary $Ga_{1-z}In_zAs$ to InP. Assuming a linear dependence of lattice constant a on composition, the lattice-matching condition is

$$a_1 + (a_2 - a_1)z = a_3 \qquad (11.1.1)$$

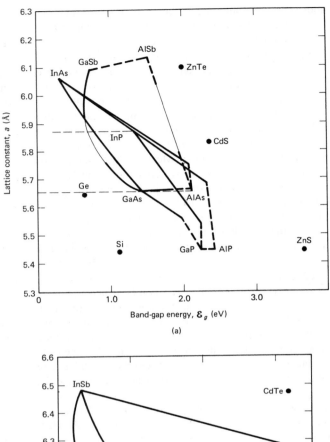

(a)

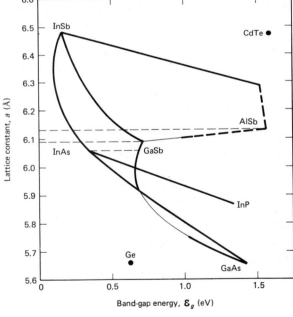

(b)

Figure 11.1 Lattice constant a and energy gap $\mathscr{E}_g$ for a selected number of III–V and II–VI compounds. The lines joining two compounds indicate the range of a and $\mathscr{E}_g$ covered by the compound alloy with the heavy lines showing the miscibility range, the light lines the miscibility gap, and the dashed lines the indirect-gap region. (H. C. Casey, Jr. and M. B. Panish, "Heterostructure Lasers," Part B, Academic Press, Inc., New York, 1978; L. Esaki, "A Bird's Eye View on the Evolution of Semiconductor Superlattices and Quantum Wells," *IEEE J. Quantum Electron.*, Vol. QE-22, p. 1611, 1986 © 1986 IEEE.)

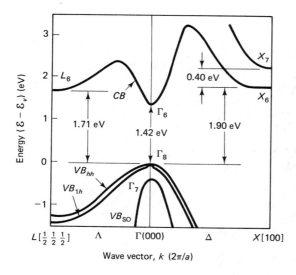

Figure 11.2 Energy-band structure of GaAs along the $k[100]$ and $k[111]$ directions of **k** space. The Γ_n, X_n, Δ, L_n, and Λ are theoretic group notations indicating the symmetry of conduction and valence bands (CB and VB) at different positions in the Brillouin zone with Γ at the zone center, X and L at the zone boundary, and Δ and Λ on the axis along the $\langle 100 \rangle$ and $\langle 111 \rangle$ directions, respectively. The values of **k** are expressed in units of $2\pi/a$.

where a_1 is a for GaAs, a_2 is a for InAs, and a_3 is a for InP. Using $a_3 = 5.86875$ Å, $a_2 = 6.0584$ Å, and $a_1 = 5.65325$ Å, we find $z = 0.532$. Experimentally, the ternary compound $Ga_{0.47}In_{0.53}As$ is found to lattice-match InP (S. Akiba, K. Sakai, and T. Yamamoto, *Jpn J. Appl. Phys.*, Vol. 17, p. 1899, 1978). The quaternary compound $Ga_xIn_{1-x}As_{1-y}P_y$ can be considered as an alloy of $(InP)_y$ and $[Ga_{1-z}In_zAs]_{1-y}$ with $x = (1 - y)(1 - z)$. Thus if we again assume a linear relation between the energy gap, the energy gap of the quaternary is given by

$$\mathscr{E}_g = \mathscr{E}_{g1}x + \mathscr{E}_{g2}(1 - x - y) + \mathscr{E}_{g3}y \qquad (11.1.2)$$

where $\mathscr{E}_{g1}$, $\mathscr{E}_{g2}$, and $\mathscr{E}_{g3}$ are the band-gap energies for GaAs, InAs, and InP, respectively.

We should point out that Eqs. (11.1.1) and (11.1.2) are also applicable to other ternary and quaternary alloys with appropriate substitution of the corresponding quantities as long as both lattice constant and band-gap energy are linear functions of the composition. Figure 11.3 shows the constant lattice-constant contours (dashed lines) and constant gap-energy contours (solid lines) in the composition x-y plane for the $Ga_xIn_{1-x}P_yAs_{1-y}$ compound (R. L. Moon, G. A. Antypas, and L. W. James, *J. Electron. Mater.*, Vol. 3, p. 635, 1974). The intercept of the InP lattice-constant line with the $y = 0$ axis gives the value of z in Eq. (11.1.1), which is 0.54. Another example is the quaternary with $x = 0.11$ and $y = 0.76$, which is also on the InP line with a value for $z = (1 - x - y)/(1 - y) = 0.54$. From Eq. (11.1.2), $\mathscr{E}_g = 1.20$ eV, as compared to $\mathscr{E}_g = 1.17$ eV from Fig. 11.3. The discrepancy is due to a slightly nonlinear dependence of $\mathscr{E}_g$ on the composition. The reader is asked to show that the ternary alloy $Ga_{1-z}In_zP$ with $z = 0.51$ and the quaternary alloy $Ga_xIn_{1-x}P_yAs_{1-y}$ with $x = 0.61$ and $y = 0.80$ are lattice-matched to GaAs (Zh. I. Alferov, I. N. Arcentev, D. Z. Gakbuzov, and V. D. Rumyantsev, *Pisma Zh. Tekh. Fiz.*, Vol. 1, p. 406, 1975). In Table 11.1 we list the values of lattice constant and band-gap energy compiled by Casey and Panish (H. C. Casey, Jr., and M. B. Panish, *Heterostructure Lasers*, Part B, Academic Press, Inc., New York, 1978, pp. 8–9).

Sec. 11.1　　Material Requirements

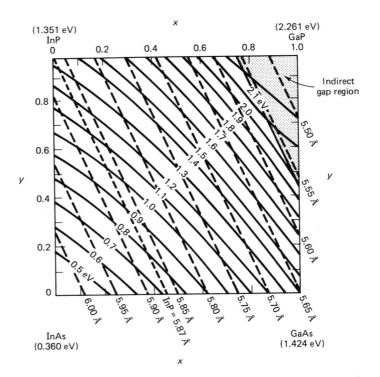

Figure 11.3 Contours of constant lattice constant (dashed lines) and constant direct-energy-gap (solid lines) in the compositional plane x and y for $Ga_xIn_{1-x}P_yAs_{1-y}$. (R. L. Moon, G. A. Antypas, and L. W. James, *J. Electron. Mater.*, Vol. 3, p. 635, 1974.)

Referring again to Fig. 11.1, we see that besides the (GaAl)As/GaAs system, the (GaIn)(AsP)/InP system and the (GaIn)(AsP)/GaAs system being discussed, three other quaternary alloys capable of lattice-match to a binary substrate are the (GaInAl)Sb/AlSb system, the (GaIn)(AsSb)/GaSb system, and the (GaIn)(AsSb)/InAs system. The range of direct-gap energies for the six systems above are arranged in the order mentioned above: (1) 1.42 to 1.96 eV, (2) 0.74 to 1.35 eV, (3) 1.42 to 1.75 eV, (4) 0.6 to 1.0 eV, (5) 0.3 to 0.7 eV, and (6) 0.35 to 0.69 eV. The (GaIn)(AsSb)/GaSb system is interesting because it is capable of emitting in the long-wavelength region with $\lambda >$ 2μm. For a comprehensive and extensive discussion of the above and other lattice-matchable systems of III–V compound alloys, the reader is referred to books devoted to heterojunction lasers (H. C. Casey, Jr., and M. B. Panish, *Heterostructure Lasers*, Academic Press, Inc., New York, 1978; H. Kressel and J. K. Butler, *Semiconductor Lasers and Heterojunction LEDs*, Academic Press, Inc., New York, 1977). Finally, we should mention that the advent of advanced technology of thin-film growth, such as molecular-beam epitaxy (MBE) and metal–organic chemical–vapor deposition (MOCVD), has prompted investigations of thin-film growth involving group IV elements and II–VI compound alloys, in conjunction with III–V compounds. The condition for growth of these films will be discussed when appropriate. The remainder of this chapter is focused on the electronic properties of heterojunctions. The optical properties of heterojunctions are discussed in Chapter 14.

Compound	Lattice constant (Å)	Gap	Energy at 300 K(eV)	Electron affinity (eV)
AlP	5.451	2.45	$(\mathscr{E}_X)$	
AlAs	5.6605	2.163	$(\mathscr{E}_X)$	$(3.50)^a$
AlSb	6.1355	1.58	$(\mathscr{E}_X)$	3.64
GaP	5.45117	2.261	$(\mathscr{E}_X)$	4.0
GaAs	5.65325	1.424	$(\mathscr{E}_\Gamma)$	$4.05\ (4.07)^b$
GaSb	6.09593	0.726	$(\mathscr{E}_\Gamma)$	$4.03\ (4.06)^b$
InP	5.86875	1.351	$(\mathscr{E}_\Gamma)$	4.4
InAs	6.0585	0.360	$(\mathscr{E}_\Gamma)$	$4.54\ (4.90)^b$
InSb	6.47937	0.172	$(\mathscr{E}_\Gamma)$	$4.59\ (4.59)^b$

[a]S. Adachi, *J. Appl. Phys.,* Vol. 58, pp. R1–R29, August 1, 1985.
[b]G. W. Gobelli and F. G. Allen, *Phys. Rev.,* Vol 137, p. A245, 1965.

11.2 THE (GaAl)As/GaAs and (GaIn)(AsP)/InP SYSTEM

One of the most extensively studied heterostructure systems is the (GaAl)As/GaAs system. In this section we use this system as an example to illustrate some important considerations regarding the physical properties of a heterostructure system. Figure 11.4a shows the variation of the lattice constants of GaAs and AlAs as a function of temperature. The two lattice constants are matched at around 900°C. Due to a slight difference in the thermal expansion coefficients, $\Delta a/a\ \Delta T = 6.4 \times 10^{-6}/°C$ in GaAs and $5.2 \times 10^{-6}/°C$ in AlAs, there is a slight lattice mismatch

$$\frac{\Delta a}{a} = 13x \times 10^{-4} \qquad \text{at} \quad 300 \text{ K} \qquad (11.2.1)$$

between GaAs and $Ga_{1-x}Al_xAs$. For example, for $Ga_{0.7}Al_{0.3}As$ grown on GaAs, the lattice mismatch is about 0.04%, producing an average stress of 6×10^8 dyn/cm². We note from Fig. 11.1 that GaP has a larger lattice constant than GaAs. Therefore, addition of GaP to (GaAl)As by a small amount can make (GaAl)(AsP) lattice matched to GaAs at 300 K, as indicated by the curve for $Ga_{0.7}Al_{0.3}As_{0.99}P_{0.01}$ in Fig. 11.4a (H. C. Casey, Jr. and M. B. Panish, *Heterostructure Lasers,* Part B, Academic Press, Inc., New York, 1978, pp. 34–36). Figure 11.4b shows stress compensation in the grown layer as a function of phosphorus content in $Ga_{0.66}Al_{0.34}As_{1-y}P_y$ grown on GaAs by liquid-phase epitaxy (LPE). The curves indicate a perfect lattice match with $y \sim 0.013$. We should point out that if compound semiconductors and their alloys have different thermal expansion coefficients, perfect lattice match can be achieved only at one temperature. Therefore, we have to choose between lattice match near growth temperature or at room temperature. So far, there is no conclusive evidence to favor one choice over the other. This is a common problem we face in dealing with alloys of compound semiconductors.

While crystal perfection in epitaxial films grown on a substrate requires a nearly perfect match of the lattice constants, many novel devices utilize the fact that an abrupt

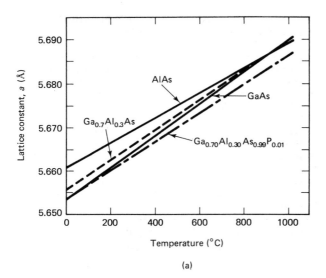

(a)

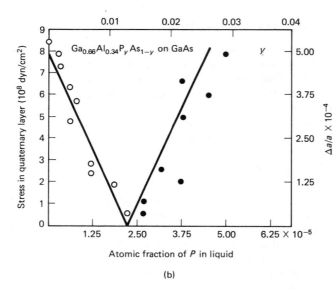

(b)

Figure 11.4 (a) Lattice constants of GaAs, AlAs, $Ga_{0.7}Al_{0.3}As$ and $Ga_{0.7}Al_{0.3}As_{0.99}P_{0.01}$ as functions of temperature, and (b) stress in $Ga_{0.66}Al_{0.34}P_yAs_{1-y}$ as a function of phosphorus content. (M. Ettenberg and R. J. Paff, *J. Appl. Phys.*, Vol. 41, p. 3926, 1970; G. A. Rozgonyi and M. B. Panish, *Appl. Phys. Lett.*, Vol. 23, p. 533, 1973.)

change exists in other physical properties at a heterostructure junction. One such property is the gap energy. While GaAs is a direct-gap semiconductor, AlAs is an indirect-gap semiconductor. In a mixed compound $Ga_{1-x}Al_xAs$ the transition from direct to indirect gap occurs at $x = 0.45$ with $\mathscr{E}_g = \mathscr{E}(\Gamma_6) - \mathscr{E}(\Gamma_8)$ varying as

$$\mathscr{E}_g = 1.424 + 1.247x \qquad (11.2.2)$$

in the direct-gap region and with $\mathscr{E}_g = \mathscr{E}(X_6) - \mathscr{E}(\Gamma_8)$ varying as

$$\mathscr{E}_g = 1.985 + 1.147(x - 0.45)^2 \qquad (11.2.3)$$

in the indirect-gap region. Therefore, a gap discontinuity exists at the (GaAl)As–GaAs interface. Another property is the dielectric constant ϵ which for $Ga_{1-x}Al_xAs$ is given by

$$\epsilon = (13.1 - 3.0x)\epsilon_0 \qquad (11.2.4)$$

using a linear interpolation between $13.1\epsilon_0$ for GaAs and $10.1\epsilon_0$ for AlAs (I. Strazalkowski, S. Joshi, and C. R. Crowell, *Appl. Phys. Lett.*, Vol. 28, p. 350, 1976; R. E. Fern and A. Onton, *J. Appl. Phys.*, Vol. 42, p. 3499, 1971). The gap discontinuity establishes a potential well for carriers on the side of smaller-gap material. As discussed later in this chapter, this property is used in many novel electronic devices. The refractive-index ($n = \sqrt{\epsilon/\epsilon_0}$) discontinuity makes total reflection possible for optical beams incident from the higher-index (i.e., smaller-gap) side. As discussed in Chapter 14, the latter property is used in double-heterostructure injection lasers to guide a laser beam.

In Table 11.2 we list a number of material and band parameters for GaAs, AlAs, InAs and InP selected from data compiled by Adachi. The property P_{ABC}

TABLE 11.2 SELECTED MATERIAL AND BAND PARAMETERS FOR GaAs, AlAs, InAs AND InP[a]

Parameter		GaAs	AlAs	InAs	InP
Lattice constant a (Å)		5.6533	5.661	6.0584	5.8688
Density (g/cm³)		5.360	3.760	5.667	4.784
Thermal expansion coefficient ($\Delta L/L$ $\Delta T, 10^{-6}/°C$)		6.4	5.2	5.16	4.36
Elastic stiffness constant (10^{11} dyn/cm²)					
c_{11}		11.88	12.08	8.33	10.22
c_{44}		5.94	5.89	3.96	4.60
Young's modulus (10^{11} dyn/cm²)		8.53	8.35	5.14	6.07
Compressibility (10^{-12} cm²/dyn)		1.33	1.28	1.73	1.38
LO phonon energy (meV)		36.25	50.09	(not given)	
TO phonon energy (meV)		33.29	44.88	(not given)	
Debye temperature (K)		370	446	(not given)	
Thermal resistivity (deg-cm/W)		2.27	1.10	(not given)	
Band-gap energy (eV)		1.424	2.168	0.36	1.35
Effective masses, m^*/m_0					
Γ-valley electrons		0.067 (0.063)	0.150	0.023	0.08
X-valley electrons m_t^*		0.23	0.19		
m_l^*		1.3	1.1		
L-valley electrons m_t^*		0.0754	0.0964		
m_l^*		1.9	1.9		
Light hole	m_{lh}^*	0.087 (0.076)	0.150	0.027	0.089
Heavy hole	(m_{hh}^*)	0.62 (0.50)	0.76	0.60	0.85
Split-off band	(m_{so}^*)	0.15 (0.145)	0.24	0.089	0.17
Electron affinity (eV)		4.07	3.50	4.90	4.37
Dielectric constants					
ϵ_s/ϵ_o		13.18	10.06	14.6	12.4
$\epsilon_\infty/\epsilon_0$		10.88	8.16	12.25	9.55

[a]Selected from data compiled by S. Adachi, "GaAs, AlAs, and $Al_xGa_{1-x}As$: Material Parameters for Use in Research and Device Applications, *J. Appl. Phys.*, Vol. 58, pp. R1–R29, August 1, 1985; "Material Parameters of $In_{r-x}Ga_xAs_yP_{i-y}$ and Related Binaries," *J. Appl. Phys.*, Vol. 53, p. 8775, 1982. The values in parentheses are quoted by J. S. Blakemore, *J. Appl. Phys.*, Vol. 53, p. R123, October, 1982.

of a ternary compound $A_xB_{1-x}C$ can be expressed in terms of the corresponding properties P of binary compounds as

$$P_{ABC} = xP_{AC} + (1 - x)P_{BC} + x(1 - x)P_{AB} = a + bx + cx^2 \qquad (11.2.5)$$

where $a = P_{BC}$, $b = P_{AC} - P_{BC} + P_{AB}$, and $c = -P_{AB}$. The parameter c, generally known as the *bowing* or *nonlinear parameter*, arises from contributions to P from the lattice disorder generated by intermixing of atoms A and B on the lattice sites normally occupied by one kind of atom in a binary compound. A linear interpolation of P_{AC} and P_{BC} is a good approximation for P_{ABC} only if the bowing parameter is small. For physical quantities related to lattice vibration such as phonon energy and Debye temperature, linear interpolation is no longer adequate, and the quadratic term needs to be included. This is also the case for band-gap energy, which is sensitive to interatomic spacing.

One physical quantity commonly used in constructing the energy-band diagram at a semiconductor–metal interface is electron affinity χ_e. For $Ga_{1-x}Al_xAs$, the quantity varies with composition as

$$\chi_e = 4.07 - 1.06x \qquad (11.2.6)$$

in the direct-gap region $x < 0.45$ and as

$$\chi_e = 3.64 - 0.14x \qquad (11.2.7)$$

in the indirect-gap region $0.45 < x < 1.0$. Figure 11.5 shows the conduction-band edge $\mathscr{E}_c$ in GaAs and (GaAl)As relative to the vacuum level. Thus we expect that a conduction-band discontinuity $\Delta\mathscr{E}_c$ would exist at the interface with

$$\Delta\mathscr{E}_c = \chi_2 - \chi_1 = 1.06x \qquad (11.2.8)$$

based on Eq. (11.2.6). Since $\Delta\mathscr{E}_g = \Delta\mathscr{E}_c + \Delta\mathscr{E}_v$, the valence-band discontinuity would be

$$\Delta\mathscr{E}_v = 0.153x \qquad (11.2.9)$$

Until recently, it was generally assumed that the division of the band-gap discontinuity is such that $\Delta\mathscr{E}_c = 0.85\,\Delta\mathscr{E}_g$ and $\Delta\mathscr{E}_v = 0.15\,\Delta\mathscr{E}_g$, based on the measured values of χ_e (R. S. Bauer, P. Zurcher, and H. W. Wang, Jr., *Appl. Phys. Lett.*, Vol. 43, p. 663, 1983; R. Dingle, W. Wiegmann, and C. H. Henry, *Phys. Rev. Lett.*, Vol. 33, p. 827,

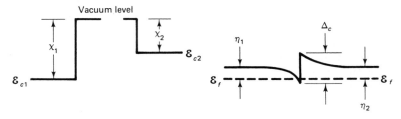

Figure 11.5 Energy-band diagrams showing the existence of a conduction-band-edge discontinuity at the interface between two semiconductors with different values of electron affinity.

1974). Recent results on capacitor–voltage C–V and current density–voltage J–V measurements in heterojunctions, however, yield the following empirical relation:

$$\Delta\mathscr{E}_c \cong 0.60 \; \Delta\mathscr{E}_g \tag{11.2.10}$$

(M. O. Watanabe, J. Yoshida, M. Mashita, T. Nakanisi, and A. Hojo, *J. Appl. Phys.*, Vol. 57, p. 5340, 1985; T. W. Hickmott, P. M. Solomon, R. Fischer, and H. Morkoç, *J. Appl. Phys.*, Vol. 57, p. 2844, 1985; J. Batey and S. L. Wright, *J. Appl. Phys.*, Vol. 59, p. 200, 1986). Since the values of $\Delta\mathscr{E}_c$ and $\Delta\mathscr{E}_v$ are important parameters used in characterizing heterojunction devices, a discussion of the experiments is given in Section 11.3.

The discrepancy between the experimentally measured band-edge discontinuity and that predicted from Eq. (11.2.8) could indicate that the linear interpolation of electron affinity in $Ga_{1-x}Al_xAs$ may not be a good approximation. Among the physical quantities listed in Table 11.2, the phonon energies, the Debye temperature, and the thermal resistivity are known to have an appreciable bowing parameter. The following expressions have been proposed by Adachi as the effective phonon energies in $Ga_{1-x}Al_xAs$:

$$h\nu_{LO} = 36.25 + 1.83x + 1.12x^2 - 5.11x^3 \quad \text{meV} \tag{11.2.11}$$

$$h\nu_{TO} = 33.29 + 10.70x + 0.03x^2 + 0.86x^3 \quad \text{meV} \tag{11.2.12}$$

In the direct-gap region, linear interpolation for $\mathscr{E}_g$ [Eq. (11.2.2)] works well. Variation of X- and L-valley energies with x is presented in Section 11.10. Linear interpolation is also a good approximation for the effective masses, the thermal expansion coefficient, and the elastic constants.

Besides the (GaAl)As/GaAs system, another system of technological importance is the (GaIn)(AsP)/InP system. If linear interpolation is a good approximation, Eq. (11.1.2) can be used for physical quantities other than gap energy. The reader is referred to an informative article by Adachi (S. Adachi, *J. Appl. Phys.*, Vol. 53, p. 8775, 1982). The following expressions for direct-gap energies are useful:

$$\mathscr{E}_g^\Gamma(InAs_yP_{1-y}) = 1.35 - 1.083y + 0.091y^2 \quad \text{eV} \tag{11.2.13}$$

$$\mathscr{E}_g^\Gamma(In_{1-x}Ga_xP) = 1.35 + 0.643x + 0.786x^2 \quad \text{eV} \tag{11.2.14}$$

With $x = 1$, Eq. (11.2.14) yields $\mathscr{E}_g^\Gamma = 2.78$ eV as compared to $\epsilon_g = 2.26$ eV for the indirect gap. Obviously, GaP is an indirect-gap semiconductor. For $Ga_xIn_{1-x}As_yP_{1-y}$ lattice matched to InP, the expressions for the gap energy, the dielectric constants, and the effective masses are given, respectively, as follows:

$$\mathscr{E}_g^\Gamma = 1.35 - 0.72y + 0.12y^2 \quad \text{eV} \tag{11.2.15}$$

$$\epsilon_s = (12.40 + 1.5y)\epsilon_0 \tag{11.2.16}$$

$$\epsilon_\infty = (9.55 + 2.2y)\epsilon_0 \tag{11.2.17}$$

$$m_l^* = (0.080 - 0.039y)m_0 \tag{11.2.18}$$

$$m_{hh}^* = (0.85 - 0.23y)m_0 \tag{11.2.19}$$

$$m_{lh}^* = (0.075 - 0.025y)m_0 \qquad (11.2.20)$$

$$m_{s0}^* = (0.174 - 0.057y)m_0 \qquad (11.2.21)$$

Since $Ga_{0.47}In_{0.53}As$ is lattice matched to InP, the value of x is given by $0.47y$ in the quaternary. Among the family of III–V compounds, the (GaIn)(AsP)/InP system is studied most extensively for applications in optical communications and the (GaAl)As/GaAs system for applications in ultrafast electronics.

11.3 BAND-EDGE DISCONTINUITY: CAPACITANCE–VOLTAGE AND CURRENT–VOLTAGE MEASUREMENTS IN SIS STRUCTURES

The band-edge discontinuities $\Delta\mathscr{E}_c$ and $\Delta\mathscr{E}_v$ at a GaAs–Ga$_{1-x}$Al$_x$As interface have been determined from capacitance–voltage (C–V) and current density–voltage (J–V) measurements in the same sample. Although both experiments are relatively straightforward in themselves, the interpretation of the measurement results need careful consideration. The main problem comes from uncertainty about the amount of charges which unknowingly exist either in the insulating Ga$_{1-x}$Al$_x$As layer or at the interface. However, before we discuss the complicated case involving unknown charges, we start with an ideal semiconductor–insulator–semiconductor (SIS) structure consisting of a thin undoped Ga$_{1-x}$Al$_x$As layer sandwiched between two p GaAs layers which are contacted by p^+ GaAs serving as electrodes. The ideal SIS structure and its energy-band diagram at zero bias are shown in Fig. 11.6 (J. Batey and S. L. Wright, *J. Appl. Phys.*, Vol. 59, p. 201, 1986). The reason for choosing p-type structure in our discussion is twofold. There are no satellite valence-band maxima, so the measurements can cover the whole composition range $0 < x \leq 1$. The effective density of states for valence band N_v is relatively large, so Fermi–Dirac distribution can be approximated by Boltzmann distribution in the calculations to follow.

In the following discussion, we further assume the SIS structure to be symmetric. Therefore, at zero bias, a flat-band situation exists. Figure 11.7 shows the energy-diagram of the SIS capacitor (a) under a large gate bias and (b) under a small gate bias.

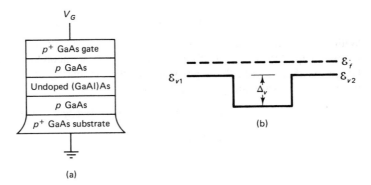

Figure 11.6 (a) A semiconductor–insulator–semiconductor structure with a symmetrical doping profile and an undoped (GaAl)As region serving as the insulating layer and (b) its valence-band diagram showing discontinuities of the valence band edges at the two (GaAl)As–GaAs interfaces. The band-edge discontinuity is denoted by Δ.

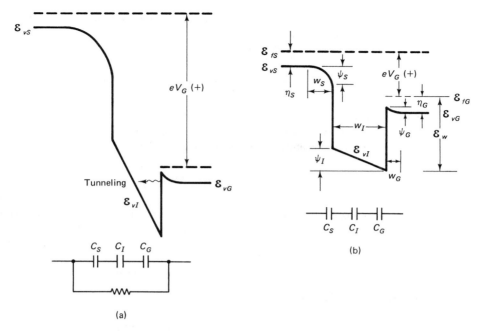

Figure 11.7 Energy-band diagrams and circuit representations of the SIS structure of Fig. 11.6a under (a) a large gate bias and (b) a small gate bias.

For case (a), the gate interface is in the accumulation region while the substrate interface is in the inversion region. As the gate voltage is reduced, the substrate interface undergoes a transition from inversion to depletion and then to flat band at zero bias. Therefore, the SIS capacitance has a V-shaped curve similar to the one shown in Fig. 9.26c for a MIS capacitor except that the curve stops at the flat-band position. Since the structure is symmetric, the C–V curve has a shape of W. The two capacitance minima are located on opposite sides of the flat-band situation, with the minimum on the positive V_G side signifying depletion on the substrate side and that on the negative V_G side signifying depletion on the gate side.

We note from Fig. 11.7a that under a large V_G, positive or negative, a high field exists in the insulator which results in appreciable current flow across the insulator through tunneling. The tunneling current not only renders impossible an accurate capacitance measurement but also masks the current component, the thermionic current, to be used in the determination of the band-edge discontinuity Δ. Therefore, in both C–V and J–V measurements, the gate bias V_G is limited within a small range, so the energy-band diagram does not deviate too much from that of the flat-band situation shown in Fig. 11.6b. The energy-band diagram under a small gate bias is shown in Fig. 11.7b. In this limited range of V_G, the C–V curve has the shape of a hump, that is, the middle portion of a W-shaped curve, and the J–V curve is in the form of the Dushman–Richardson equation [Eq. (4.7.10)]. The various symbols in Fig. 11.7b have the following physical meaning: $\mathcal{E}_f$ being the Fermi level, $\mathcal{E}_v$ the valence-band edge, $\psi = e\phi$ the potential energy, $\eta = \mathcal{E}_f - \mathcal{E}_v$, and w the width of the space-charge region. The subscripts I, G, and S refer to the insulator, and the two GaAs layers on the gate and substrate side, respectively.

From Poisson's equation and Gauss's law, we have

$$\psi_S = \frac{e^2 N_S w_S^2}{2\epsilon_S} \quad \text{and} \quad \psi_I = \frac{e^2 N_S w_S w_I}{\epsilon_I} \tag{11.3.1}$$

in the substrate and insulator, respectively, where $N_S = N_G$ is the acceptor concentration and ϵ is the dielectric constant. For small positive V_G, the gate side is slightly in the accumulation region, and the hole concentration is given by

$$p_G = N_G \exp\left(\frac{-\psi}{kT}\right) \tag{11.3.2}$$

where ψ is taken to be zero in the gate bulk and ψ is negative in the gate-interface region. For $\psi < kT$, expansion of Eq. (11.3.2) yields

$$\frac{d^2\psi}{dx^2} = \frac{d^2 eV}{dx^2} = -\frac{e\rho}{\epsilon_G} = \frac{e^2 N_G}{\epsilon_G kT}\psi \tag{11.3.3}$$

The solution of Eq. (11.3.3) is

$$\psi = -\psi_G \exp\left(-\frac{\sqrt{2}x}{L_D}\right) \tag{11.3.4}$$

where x is taken to be zero at the gate interface and $x = \infty$ in the gate bulk. The quantity L_D is the Debye screening length given by

$$L_D = \frac{\sqrt{2\epsilon_G kT}}{e^2 N_G} \tag{11.3.5}$$

From the continuity of $\epsilon \, d\psi/dx$ at the interfaces, we obtain

$$\psi_G = \frac{e^2 N_S w_S L_D}{\sqrt{2}\,\epsilon_G} \tag{11.3.6}$$

From Eqs. (11.3.1) and (11.3.6) we have

$$eV_G = (\psi_S + \psi_I + \psi_G) = e^2 N_S\left(\frac{w_S^2}{2\epsilon_S} + \frac{w_S w_I}{\epsilon_I} + \frac{w_S L_D}{\sqrt{2}\,\epsilon_S}\right) \tag{11.3.7}$$

Therefore, the capacitance per unit area of the ideal SIS capacitor is

$$\frac{C}{A} = \frac{dQ}{dV_G} = \frac{\epsilon_S}{w_S + w_e} \tag{11.3.8}$$

where the space-charge width w_S on the substrate side is given by

$$w_S = \left(\frac{2\epsilon_S V_G}{e N_S} + w_e^2\right)^{1/2} - w_e \tag{11.3.9}$$

and the equivalent width w_e of the other two regions is

$$w_e = \frac{\epsilon_S}{\epsilon_I} w_I + \frac{L_D}{\sqrt{2}} \tag{11.3.10}$$

In the low-bias region, the current flow across an ideal SIS structure is a result of thermionic emission of energetic holes across the potential barrier on the gate side in Fig. 11.7b. The Dushman–Richardson equation for the thermionic emission–current density J is given by Eq. (4.7.10), or

$$J = A^* T^2 \exp\left(\frac{-\mathscr{E}_w}{kT}\right) \tag{11.3.11}$$

where the work function $\mathscr{E}_w$ is

$$\mathscr{E}_w = \Delta_v - \psi_G + \eta_G \tag{11.3.12}$$

and ψ_G can be calculated from Eq. (11.3.6) and $\eta_G = kT \ln (N_v/N_G)$. An Arrhenius plot of $\ln (J/T^2)$ as a function of $1/kT$ gives

$$\ln \frac{J}{T^2} + \frac{\eta_G - \psi_G}{kT} = \ln A^* - \frac{\Delta_v}{kT} \tag{11.3.13}$$

or the valence-band-edge discontinuity $\Delta_v = \Delta\mathscr{E}_v$ if the gate bias V_G is held constant. To compare the theoretical result with the experimental results of Batey and Wright, we take the sample V352 quoted in their paper with $x = 1.0$, $w_I = 50$ nm, and $N_S = N_G = 10^{22}$ m^{-3}. For $\epsilon_I = 10.1\epsilon_0$ and $\epsilon_G = 13.1\epsilon_0$, Eqs. (11.3.5) and (11.3.10) yield at 77 K values of $L_D = 30.5$ nm and $w_e = 86.2$ nm. From Eq. (11.3.9) we find $w_S = 195$ nm at $V_G = 0.5$ V. These values give a value of 2.60 for the ratio $C(V_G = 0)/C(V_G = 0.5\ \text{V})$ versus an experimental value of 2.42. The good agreement justifies the assumption of $N_I = 0$. From Eq. (11.3.7) we find $\psi_S = 60$ meV with $V_G = 0.5$ V at 77 K. For comparison, $\eta_G = 29$ meV at 77 K. In the paper by Batey and Wright, the effect of ψ_G was neglected. This is true only for very small V_G.

Figure 11.8 shows the values of $\Delta\mathscr{E}_v$ and $\Delta\mathscr{E}_c$ deduced from the Arrhenius plots of Eq. (11.3.13) in various samples with different x. The band-edge discontinuity can be fit by a simple linear relation

$$\Delta\mathscr{E}_v \cong 0.55 x_{\text{Al}} \qquad \text{eV} \tag{11.3.14}$$

for the valence band over the entire x range. For the conduction band, the corresponding relation is

$$\Delta\mathscr{E}_c \cong 0.75 x_{\text{Al}} \quad \text{for } 0 < x_{\text{Al}} < 0.45 \tag{11.3.15}$$

in the direct-gap region, yielding a $\Delta\mathscr{E}_c = 0.60 \, \Delta\mathscr{E}_g$ which is given as Eq. (11.2.5). Previous C–V and J–T measurements yield the following values for the ratio $\Delta\mathscr{E}_c/\Delta\mathscr{E}_g$ 0.62 (M. O. Watanabe, J. Yoshida, M. Mashita, T. Nakanisi, and A. Hojo, *J. Appl. Phys.*, Vol. 57, p. 5340, 1985), 0.63 (T. W. Hickmott, P. M. Solomon, R. Fischer, and H. Morkoc, *J. Appl. Phys.*, Vol. 57, p. 2844, 1985), and 0.65 (D. Arnold, A. Kettersen, T. Henderson, J. Klem, and H. Morkoç, *J. Appl. Phys.*, Vol. 57, p. 2880, 1985). To fit the experimental data to the theoretical curves, either an interface state density ($\sigma \sim -1.3$ to $+5.8 \times 10^{10}$ cm^{-2}) or an insulator impurity density ($N_I \sim 3$ to

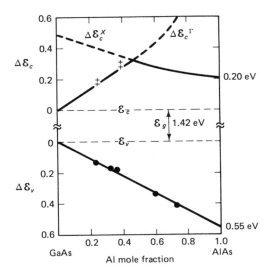

Figure 11.8 Values of band-edge discontinuities $\Delta\mathscr{E}_c$ and $\Delta\mathscr{E}_v$ as functions of Al mole fraction (J. Batey and S. L. Wright, *J. Appl. Phys.*, Vol. 59, p. 201, 1986.)

6×10^{16} cm^{-3}) was assumed. One suspected unintentional dopant is carbon, which is an acceptor with a moderately deep energy level. The existence of these charges makes interpretation of the experimental results rather complicated. The reader is referred to the above-mentioned papers for a detailed analysis of the experiments. Despite these complications, a general agreement on the band-edge discontinuity evolves. However, the origin of the discrepancy between the value of $\Delta\mathscr{E}_c/\Delta\mathscr{E}_g$ of 0.6 deduced from C–V and J–T measurements and that of 0.85 concluded from the electron affinity data is still unresolved.

11.4 TWO-DIMENSIONAL ELECTRON (HOLE) GAS

In Section 9.9 we showed that the carrier density in a semiconductor near the interface of a MIS structure can be changed by applying a bias voltage across the structure. From the energy-band diagram of Fig. 9.25, it is easy to see that a sufficiently large negative (positive) bias can make the surface conductivity degenerate n type (p type). A similar situation exists near a GaAs–(GaAl)As interface. Figure 11.9a shows the energy-band diagram for a nGaAs–n(GaAl)As–nGaAs heterostructure. Because the donors in (GaAl) are higher in energy than the conduction band edge in GaAs, electrons ionized from the donors aggregate near the interface on the two GaAs sides. As a result, potential wells form which attract electrons. The potential variation (or band bending) as well as the electron concentration in the interface (space-charge) region can be found in a self-consistent manner from Poisson's equation and Gauss's law similar to the manner used for a MIS structure. It is well known from quantum mechanics that the energy of electrons (or holes) in a potential well is quantized. The word "two-dimensional" refers to the situation in which electrons or holes have quantized energy levels in one spatial dimension (i.e., in the dimension of the potential well), but are free to move in the other two spatial dimensions.

The MIS structure shown in Fig. 9.25 and the GaAs–(GaAl)As heterostructure shown in Fig. 11.9a have one feature in common. There exists a band-edge disconti-

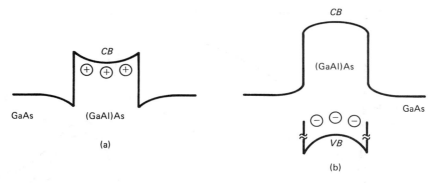

Figure 11.9 Energy-band diagrams of a GaAs–(GaAl)As–GaAs structure with (GaAl)As being doped (a) with donors and (b) with acceptors.

nuity at the interface. However, two potential barriers are needed to form a potential well. In the case of the MIS structure, whether a potential well forms or not depends on the work function of the metal. In the case of the heterostructure, the doping type in the large-gap semiconductor is an important factor in determining the potential near the interface. Figure 11.9b shows the energy-band diagram of a n-GaAs-p-(GaAl)As-n-GaAs heterostructure. The acceptors in (GaAl)As take away electrons from GaAs, resulting in a potential drop, because they are lower in energy than the conduction band edge of GaAs. Therefore, while the metal work function is important to the performance of a MIS-based device, the dopant type and the dopant distribution are important to the performance of a heterostructure-based device. By *heterostructure* we mean specifically structures made of two different semiconductors. The effect of dopants on such devices is discussed in Section 11.7.

The electronic properties of two-dimensional systems have been studied by many authors. For a detailed discussion of these properties, the reader is referred to an excellent review article by Ando, Fowler, and Stern (T. Ando, A. B. Fowler, and F. Stern, *Rev. Mod. Phys.*, Vol. 54, p. 437, 1982). Even though the emphasis of the article is on the properties of Si MIS structures, the general physics of two-dimensional electronic systems applies equally well to the III–V compound heterostructures. One fundamental property of a two-dimensional system is the quantization of energy. Figure 11.10 shows schematically the potential well together with the electron wave function on an expanded scale in both energy and spatial coordinates. If we approximate the variation of electron potential energy $\psi(z)$ by a triangular well with

$$\psi(z) = eEz \qquad \text{for} \quad z > 0 \tag{11.4.1}$$

and $\psi(z) = \infty$ for $z < 0$, the solutions of the Schrödinger equation are Airy functions (M. Abramowitz and I. A. Stegun, eds., *Handbook of Mathematical Functions*, National Bureau of Standards Applied Mathematics Series, No. 55, U.S. Government Printing Office, Washington, D.C., 1964).

$$\zeta_i(z) = \text{Ai}\left[\left(\frac{2m_z eE}{\hbar^2}\right)^{1/3}\left(z - \frac{\mathscr{E}_i}{eE}\right)\right] \tag{11.4.2}$$

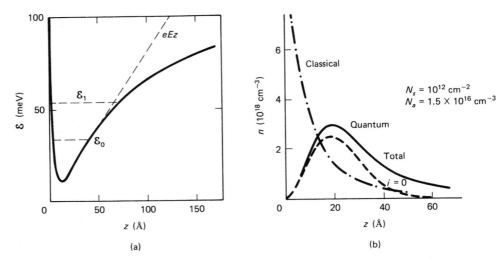

Figure 11.10 (a) Energy and (b) electron concentration n near a Si MIS interface with a surface electron density $N_s = 10^{12}$ cm^{-2} and a background (uniform) hole concentration $N_a = 1.5 \times 10^{16}$ cm^{-3}. (T. Ando, A. B. Fowler, and F. Stern, *Rev. Mod. Phys.*, Vol. 54, p. 437, 1982.)

where the eigenvalues are given by

$$\mathscr{E}_i \cong \left(\frac{\hbar^2}{2m_z}\right)^{1/3}\left[\frac{3\pi eE}{2}\left(i + \frac{3}{4}\right)\right]^{2/3} \quad \text{with} \quad i = 0, 1, 2, \dots \quad (11.4.3)$$

The quantity E in Eq. (11.4.1) is an average electric field. Before we can proceed to find E, we must know how to calculate the carrier density for a two-dimensional electron gas. In the x and y directions, the motion of electrons is not restricted. Therefore, we can define a two-dimensional density of state $D_2(\mathscr{E})$ such that the available quantum states dN/A per unit area is

$$\frac{dN}{A} = 2 \times \frac{2\pi k_\perp \, dk_\perp}{(2\pi)^2} = D_2(\mathscr{E}_\perp) \, d\mathscr{E}_\perp \quad (11.4.4)$$

where $k_\perp = (k_x^2 + k_y^2)^{1/2}$. The energy $\mathscr{E}_\perp$ is associated with the motion of electrons in xy plane. The total energy is

$$\mathscr{E} = \mathscr{E}_\perp + \mathscr{E}_i = \mathscr{E}_i + \frac{\hbar^2 k_\perp^2}{2m_\perp} \quad (11.4.5)$$

where $\mathscr{E}_i$ is the quantized energy and $m_\perp$, assumed isotropic, is the effective mass in xy plane. From Eq. (11.4.5) we see that the energy levels in a two-dimensional electron gas form many subbands, with each band having a minimum energy $\mathscr{E}_i$. Using Eq. (11.4.5) in Eq. (11.4.4) and adding contributions from equivalent valleys, we obtain

$$D_2(\mathscr{E}_\perp) = \frac{gm_\perp}{\pi\hbar^2} \quad \text{m}^{-2} \, \text{eV}^{-1} \quad (11.4.6)$$

Properties of Heterostructures   Chap. 11

where g is the valley-degeneracy factor. If $\mathscr{E}_f$ is the Fermi level, then within the energy interval $\Delta\mathscr{E}_\perp = \mathscr{E}_f - \mathscr{E}_i$ there are N_{si} electrons per unit area equal to

$$N_{si} = \frac{gm_\perp}{\pi\hbar^2}(\mathscr{E}_f - \mathscr{E}_i) \qquad m^{-2} \qquad (11.4.7)$$

The subscript s signifies surface density and the subscript i stands for ith subband. In deriving Eq. (11.4.7), we approximate the Fermi function by a step function which is true only at 0 K.

The charge density $\rho_i(m^{-3})$ associated with electrons in the subband i can be found by letting

$$\rho_i(z) = B e \zeta_i^2(z) \qquad (11.4.8)$$

where B is a constant of proportionality to be determined. Since $\int \rho(z)\, dz = -eN_s$, we have

$$\rho_i(z) = -eN_{si}\zeta_i^2(z) \qquad (11.4.9)$$

Once we know $\rho_i(z)$, we can proceed to find the electric field E. There are two contributions to E: (1) E_c due to accumulated carriers, and (2) E_i due to ionized impurities. Both can be found from Poisson's equation. The part E_c is

$$E_c = e \sum \frac{N_{si}f_i}{\epsilon} \qquad (11.4.10)$$

where the factor f_i defined as

$$f_i(z) = \int_z^{z_{sc}} \zeta_i^2(z)\, dz \qquad (11.4.11)$$

has a value of 1 at the interface and a value of 0 in the bulk ($z = z_{sc}$). Similarly, the part E_i is

$$E_i = -\frac{eN_I}{\epsilon}(z - z_{sc}) \qquad (11.4.12)$$

where N_I is the impurity concentration with $N_I = N_d$ for donors $N_I = -N_a$ for acceptors. The difference in sign accounts for the different polarity of ionized donors and acceptors. The integration constants after integrating Poisson's equation are chosen such that $E_c = 0$ and $E_i = 0$ in the bulk, that is, at the edge of the space-charge region $z = z_{sc}$. In the potential well, $z \ll z_{sc}$ in Eq. (11.4.12). Furthermore, we take $f_i = 1/2$ for the average field in the well. Thus the total field in the well is

$$E = \frac{e}{\epsilon}\left(N_I z_{sc} + \frac{N_s}{2}\right) \qquad (11.4.13)$$

where $N_s = \sum_i N_{si}$ is the total interface carrier density (per unit area).

The energy-subband structure and the carrier-transport properties have been extensively studied for inversion-layer electrons in Si MOSFETs. The triangular potential used in finding the subband-energy minima $\mathscr{E}_i$ and its associated electron wave function $\zeta_i(z)$ is an approximation. The reader is referred to the review article by Ando, Fowler, and Stern for a discussion of improved calculations for the subband energy and wave function as well as for a discussion of carrier-transport properties. In the following discussion we present a brief summary of transport properties of carriers again in Si surface-inversion layers. Figure 11.11 shows the measured channel mobility as functions of (a) channel-carrier density N_S at several temperatures and (b) temperature for several values of N_s. The charge concentration in the oxide has a reported value of $N_{ox} = 6 \times 10^{10}$ cm^{-2} for the solid curves and $N_{ox} = 13 \times 10^{10}$ cm^{-2} for the dashed curves in Fig. 11.11a and has a value of 2×10^{11} cm^{-2} in Fig. 11.11b. The channel-carrier density N_s in Fig. 11.11b can be calculated from $N_s = 2 \times 10^{11} V'_G$.

The important scattering mechanisms are (1) phonon scattering, (2) Coulomb scattering from both ionized impurities in the inversion layer and unwanted ions in the oxide, (3) scattering by surface potential, and (4) Si-SiO$_2$ interface-roughness scattering. Theoretical calculations for acoustic-phonon scattering show $\mu \propto T^{-1}N_s^{-1/3}$ for a two-dimensional electron gas (2DEG). From Fig. 11.11 it is obvious that the agreement between theory and experiment around room temperature is worse for the N_s dependence than for the T dependence. Intervalley phonon scattering and polar optical phonon scattering have been suggested to play a part in the discrepancy. At low temperatures, both Coulomb scattering and surface-potential scattering become important, as evidenced by the decrease in μ with increased N_{ox} and with decreased T. The theoretical temperature dependence of μ for Coulomb scattering is $T^{1.5}$ for bulk carriers and T for 2DEG. Theoretical behavior of μ considering the combined effects of the two mechanisms indeed shows rising μ and then falling μ as N_s is increased. This behavior in Si 2DEG is discussed further in Section 13.13. The charge distribution and the transport properties of GaAs 2DEG are discussed in Sections 11.6 and 11.7.

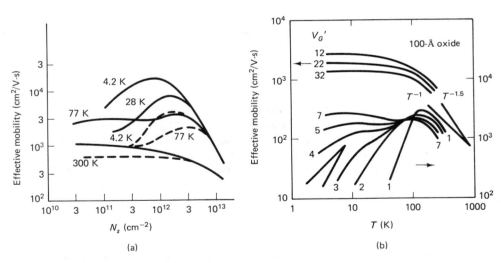

Figure 11.11 Measured channel mobility in Si MIS structures as (a) functions of areal carrier density N_s(cm^{-2}) at several temperatures and (b) functions of temperature at various gate voltages V'_G. (T. Ando, A. B. Fowler, and F. Stern, *Rev. Mod. Phys.*, Vol. 54, p. 437, 1982.)

11.5 MAGNETORESISTANCE AND QUANTUM HALL EFFECT

In Section 11.4 we presented quantization of energy levels in a potential well. In this section we discuss an experiment that not only confirms the two-dimensionality of such a system but also shows the quantity e^2/h as the fundamental unit of two-dimensional conductivity (K. v. Klitzing, G. Dorda, and M. Pepper, *Phys. Rev. Lett.*, Vol. 45, p. 494, 1980). The experiment involves the measurements of Hall effects at very high magnetic fields. To prepare for the background for the experiment we refer to Appendix A6 for an analysis of the galvanomagnetic effect based on the classical Boltzmann transport equation.

In the presence of a magnetic flux density B applied in the z direction, the current densities J_x and J_y are given by Eqs. (A6.8) and (A6.9),

$$J_x = \sigma_{xx}E_x - \sigma_{xy}E_y \tag{11.5.1}$$

$$J_y = \sigma_{yx}E_x + \sigma_{yy}E_y \tag{11.5.2}$$

The coefficients σ_{xx} and σ_{yy} and $\sigma_{xy} = \sigma_{yx}$ are given by Eqs. (A6.10) and (A6.11)

$$\sigma_{xx} = \int -\frac{\partial f_0}{\partial \mathscr{E}} \frac{e^2\tau}{1+s^2} v_x^2 \, dk_x \, dk_y \, dk_z \tag{11.5.3}$$

$$\sigma_{xy} = \int -\frac{\partial f_0}{\partial \mathscr{E}} \frac{e^2\tau s}{1+s^2} v_x^2 \, dk_x \, dk_y \, dk_z \tag{11.5.4}$$

where $s = e\tau B/m^*$. Note that the terms involving $v_x v_y$ are odd functions and $\partial f_0/\partial \mathscr{E}$ is an even function of k_x and k_y. Therefore, any integral involving $(\partial f_0/\partial \mathscr{E}) v_x v_y$ is zero. The Hall coefficient can be found from Eqs. (11.5.1) and (11.5.2) by setting $J_y = 0$. Thus we find that

$$J_x = \frac{\sigma_{xx}^2 + \sigma_{xy}^2}{\sigma_{xx}} E_x \tag{11.5.5}$$

$$R_H = \frac{E_y}{J_x B} = \frac{-\sigma_{xy}}{\sigma_{xx}^2 + \sigma_{xy}^2} \frac{1}{B} \tag{11.5.6}$$

The change of longitudinal resistivity E_x/J_x with B under $J_y = 0$ is known as magnetoresistance (R. A. Smith, *Semiconductors*, Cambridge University Press, Cambridge, 1978; S. Wang, *Solid State Electronics*, McGraw-Hill Book Company, New York, 1966).

Under a strong B such that $s \gg 1$, we have $\sigma_{xy} \gg \sigma_{xx}$. Thus Eq. (11.5.6) becomes

$$R_H = -\frac{1}{ne} \tag{11.5.7}$$

The reader can easily show that for $s \gg 1$,

$$\sigma_{xy} = \int -\frac{\partial f_0}{\partial \mathscr{E}} \, e \frac{m^* v_x^2}{B} 4\pi k^2 \, dk = \frac{e}{B} \int f_0 4\pi k^2 \, dk \tag{11.5.8}$$

through integration by parts. Equation (11.5.8) is valid for both degenerate and nondegenerate electron concentrations. The important point is that under a very strong B, the Hall coefficient R_H is independent of scattering mechanism because σ_{xy} becomes independent of τ. The condition $s \gg 1$ can be translated into

$$s \cong \mu B \gg 1 \qquad (11.5.9)$$

for R_H to be exactly equal to $-1/ne$.

Equation (11.5.7) can also be made applicable to a two-dimensional electron gas. Referring to Fig. 11.10 and Eq. (11.4.9), the current density J_x in the conduction channel is $J_x = \rho v_x$. Thus for a surface-conduction channel of width W in the inset of Fig. 11.12 (chosen to be in the y direction), the total current I_x along the channel is

$$I_x = \int J_x(z)W\,dz = W \int efv_x\,dk_x\,dk_y \int \zeta^2(z)\,dz \qquad (11.5.10)$$

In Eq. (11.5.10), the z axis is in the direction of depth of the surface channel (i.e., in the direction normal to the surface) along which the magnetic field is applied. The Hall

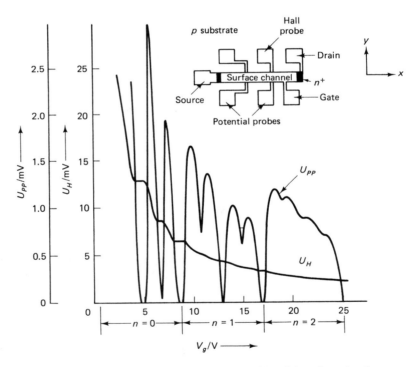

Figure 11.12 Recordings of the Hall voltage $U_H = V_H$ and the voltage drop between the potential probes $U_{pp} = V_{xx}$ as functions of the gate voltage V_G at a temperature $T = 1.5$ K. The recordings were taken at a constant magnetic field $B = 18$ T (Wb/m^2) $= 1.8 \times 10^5$ G and a source-drain current $I_D = 1$ μA. The top view of the Si MOSFET used in the experiment and shown in the inset had a gate length $L = 400$ μm and a gate width $W = 50$ μm. The distance between the potential probes is 130 μm. The symbol U was used for voltage in the paper by von Klitzing et al. (K. v. Klitzing, G. Dorda, and M. Pepper, *Phys. Rev. Lett.*, Vol. 45, p. 494, 1980.)

voltage V_H measured across the channel in the y direction is $V_H = E_y W$. For the two-dimensional electron gas, the quantity σ_{xy}^{2D} corresponding to Eq. (11.5.4) is

$$\sigma_{xy}^{2D} = -\int \frac{\partial f_0}{\partial \mathscr{E}} \frac{e^2 \tau s}{1 + s^2} v_x^2 \, dk_x \, dk_y \qquad (11.5.11)$$

which under $s \gg 1$ reduces to

$$\sigma_{xy}^{2D} = -\int e \frac{\partial f_0}{\partial \mathscr{E}} \frac{m^* v_x^2}{B} 2\pi k \, dk = \frac{e}{B} N_s \qquad (11.5.12)$$

Using the information above, we obtain the Hall coefficient R_H^{2D} for a two-dimensional electron gas

$$R_H^{2D} = \frac{V_H}{I_x B} = \frac{-1}{e N_s} \qquad (11.5.13)$$

Therefore, a measurement of the ratio of Hall voltage to the product of the channel current and the applied magnetic flux density yields the surface charge density in the channel $e N_s$ (C/m^2).

Figure 11.12 shows the experimental results on the Hall voltage V_H (U_H in the paper) and the voltage drop V_{xx} (U_{pp} in the paper) across the potential probes as a function of the gate voltage V_G reported by Klitzing and coworkers (K. v. Klitzing, G. Dorda, and M. Pepper, *Phys. Rev. Lett.*, Vol. 45, p. 494, 1980). A typical MOSFET device (the inset) built on a p-Si substrate was used in the experiment. Under a sufficiently large positive gate voltage, a surface inversion layer is established to form a conduction channel. The gate field which is normal to the surface produces subbands (Fig. 11.10) in the direction normal to the Si–oxide interface. A strong magnetic field applied normal to the surface, that is, in the z direction, produces quantized Landau levels for electron motion parallel to the interface, that is, in the xy plane. Referring to the discussion of cyclotron resonance in Section 6.10, we see that under an applied $\mathbf{B}$, electrons are in circular motion about the $\mathbf{B}$ axis. Balancing the magnetic force with the centrifugal force, we have $evB = mv^2/r$, yielding an angular frequency

$$\omega_c = \frac{v}{r} = \frac{eB}{m^*} \qquad (11.5.14)$$

which is known as the *cyclotron frequency*. Quantum-mechanical calculations show that the quantized energy level, called the *Landau level*, is given by

$$\mathscr{E}_L = (n + \tfrac{1}{2})\hbar\omega_c, \qquad n = 1, 2, 3, \ldots \qquad (11.5.15)$$

Therefore, the total quantized energy of the electrons in an inversion layer under a strong B is

$$\mathscr{E} = \mathscr{E}_{ij} + (n + \tfrac{1}{2})\hbar\omega_c \qquad (11.5.16)$$

where $\mathscr{E}_{ij}$ is given by Eq. (11.4.3), with j indicating possible spin and valley degeneracy.

To find the value of N_s in Eq. (11.5.13), we express $dN_s = dk_x\, dk_y/(2\pi)^2$ instead of $dN_s = D_2(\mathscr{E})\, d\mathscr{E}$ as in Eq. (11.4.4). Not counting the spin and valley degeneracy, the electron surface density is

$$N_s = \frac{k_\perp^2}{4\pi} \qquad (11.5.17)$$

where $k_\perp^2 = k_x^2 + k_y^2$ in cylindrical coordinate. The maximum value of $k_\perp^2$ is determined by the separation between two Landau levels with $\Delta n = 1$, that is,

$$\frac{\hbar^2 k_\perp^2}{2m^*} = \hbar\omega_c = \frac{eB\hbar}{m^*} \qquad (11.5.18)$$

yielding $k_\perp^2 = 2eB/\hbar$. Using this value in Eq. (11.5.17) and taking spin and valley degeneracy into account, we have

$$N_s = \frac{jeB}{h} \qquad (11.5.19)$$

where $j = 2g_v$ is the degeneracy factor and g_v is the valley degeneracy. Substituting Eq. (11.5.19) in Eq. (11.5.13) and defining a Hall conductance $\sigma_{xy} = I_x/V_H$, we obtain

$$\sigma_{xy}^{-1} = -R_H^{2D}B = \frac{h}{e^2 j} \qquad (11.5.20)$$

Table 11.3 lists the effective masses and valley degeneracies for Si and Ge MOSFET with different surface orientations. For Landau levels, the effective mass m^* in Eq. (11.5.14) is the conductivity effective mass m_{con}^* with

$$m_{con}^* = \frac{2m_x m_y}{m_x + m_y} \qquad (11.5.21)$$

TABLE 11.3 EFFECTIVE MASSES AND VALLEY DEGENERACIES FOR ELECTRONS IN SURFACE INVERSION LAYERS IN Si AND Ge MOSFETS WITH DIFFERENT SURFACE ORIENTATIONS[a]

Surface orientation	Si				Ge			
	m_x	m_y	m_z	g_v	m_x	m_y	m_z	g_v
{100}	m_t	m_t	m_l	$2(l)$	m_t	$\dfrac{m_t + 2m_l}{3}$	$\dfrac{3m_t m_l}{m_t + 2m_l}$	4
	m_t	m_l	m_t	$4(h)$				
{110}	m_t	$\dfrac{m_t + m_l}{2}$	$\dfrac{2m_t m_l}{m_t + m_l}$	$4(l)$	m_t	$\dfrac{m_l + 2m_t}{3}$	$\dfrac{3m_t m_l}{m_l + 2m_t}$	$2(l)$
	m_t	m_l	m_t	$2(h)$	m_t	m_l	m_t	$2(h)$
{111}	m_t	$\dfrac{m_t + 2m_l}{3}$	$\dfrac{3m_t m_l}{m_t + 2m_l}$	6	m_t	m_t	m_l	$1(l)$
					m_t	$\dfrac{m_t + 8m_l}{9}$	$\dfrac{9m_t m_l}{m_t + 8m_l}$	$3(h)$

[a]The letters l and h in parentheses indicate the lower- and higher-energy subband, respectively.
Source: After F. Stern and W. E. Howard, *Phys. Rev.,* Vol. 163, p. 816, 1967.

For density of states, $m_\perp$ in Eq. (11.4.6) is given by m_d^* with

$$m_d^* = (m_x m_y)^{1/2} \qquad (11.5.22)$$

For a {100} surface orientation, $m_{\text{con}}^* = m_d^*$ in Si, resulting in the equivalence of two masses in Eq. (11.5.18). For the subband energy $\mathscr{E}_i$, the effective mass m_z in Eq. (11.4.2) is the m_z given in Table 11.3. Therefore, the valley with a larger m_z produces the subband with a lower $\mathscr{E}_i$.

In the experiment of von Klitzing et al., the MOSFET had a length $L = 400$ μm, a width $W = 50$ μm, and a distance d between potential probes $d = 130$ μm. The device was subject to a magnetic flux density $B = 18$ T applied normal to the surface (in the z direction) and the source–drain current was maintained at $I_D = 1$ μA. The measurement was taken at 1.5 K with $kT = 1.3 \times 10^{-4}$ eV. For comparison, the energy $\hbar\omega_c = 11 \times 10^{-3}$ eV with $m_{\text{con}}^* = m_t = 0.190m_0$ in Eq. (11.5.14). From Table 11.3, we see that for {100} oriented surface, the lowest subband has a valley degeneracy of 2 which combined with spin degeneracy gives a total degeneracy factor of 4. The degeneracy is lifted by the applied B, giving rise to four separate values of j in Eq. (11.5.20) from $j = 1, 2, 3,$ and 4 when the states are successively occupied. As the gate voltage V_G is raised, the Fermi level is swept successively across $\mathscr{E}_{ij}$ with $i = 0$ and $j = 1, 2, 3,$ and then 4. The first three plateaus in the measured V_H correspond to $j = 2, 3,$ and 4. For $j = 4$, the measured V_H is 6.4532 mV, which in turn gives a value of $\sigma_{xy}^{-1} = 6453.2$ Ω for Hall resistance V_H/I_x. This measured value is in excellent agreement with the theoretical value of $h/4e^2 = 6453.2$ Ω. The plateau state in V_H corresponds to a situation where the Fermi level lies between two $\mathscr{E}_{0j}$ and $\mathscr{E}_{0j'}$ with $j' = j + 1$. Since the carriers have to be thermally activated in order to participate in the conduction process, σ_{xx} undergoes a minimum. From Eq. (11.5.5) we see that for constant J_x, E_x also goes to a minimum in the plateau region since $\sigma_{xy} > \sigma_{xx}$ in this region.

Von Klitzing was awarded the Nobel Prize in Physics in 1985 for his discovery of the quantized Hall effect. This work allows a high-precision measurement of e^2/h to 1 part in 10 million as the ratio I_x/V_H is an integral multiple of e^2/h. Not only was the result of integral multiple of e^2/h duplicated in GaAs/Ga$_{1-x}$Al$_x$As structures (D. C. Tsui and A. C. Gossard, *Appl. Phys. Lett,* Vol. 38, p. 550, 1981) but also fractional quantum Hall effect was discovered (D. C. Tsui, H. Störmer, and A. Gossard, *Phys. Rev. Lett.*, Vol. 48, p. 1559, 1982). The observation of a Hall-voltage plateau and that of a fractional e^2/h have generated a great deal of theoretical interests in their origins (R. Laughlin, *Phys. Rev.*, Vol. B23, p. 5652, 1981; R. Laughlin, *Phys. Rev. Lett.*, Vol. 50, p. 1395, 1983; S. Luryi and R. Kazarinov, *Phys. Rev.*, Vol. B27, p. 1386, 1983). For a general review of von Klitzing's work and a background discussion of quantum Hall effect, the reader is referred to a news article by Schwarzschild (B. Schwarzschild, *Phys. Today,* Vol. 38, No. 12, p. 17, December 1985). The properties of GaAs/Ga$_{1-x}$Al$_x$As heterostructures are discussed in Sections 11.6 and 11.7.

11.6 SEMICONDUCTOR-HETEROSTRUCTURE INTERFACE

One important physical quantity used to characterize the electronic property of a semiconductor-heterostructure interface is the areal (or sheet) free-carrier density N_s (m^{-2}). Referring to Fig. 11.10 and Eq. (11.4.5), we see that the electrons confined in a one-dimensional potential well can still move freely in the other two dimensions. Such a two-dimensional electron gas forms a surface-conduction channel with a channel conductance between the source and drain given by

$$g_{SD} = \frac{eN_s\mu W}{L} \qquad (11.6.1)$$

where μ is the electron mobility, W the channel width, and L the channel length. Many semiconductor heterostructure devices, such as the modulation-doped (MOD)FET, utilize 2DEG to form a conduction channel. As we discuss in Section 11.7, the areal charge density N_s at a semiconductor-heterostructure interface, unlike that in the inversion layer of a Si MOSFET, is limited in its value. In this section we use N_s as a parameter to characterize other properties, such as the Fermi level and quantized energy, of a semiconductor-heterostructure interface.

In Section 11.4 we obtained the expressions for the quantized energy levels, Eq. (11.4.3), and for the average electric field, Eq. (11.4.13), in a triangular well. Even though the triangular well is an approximation of the actual well, it provides a useful base on which the physics of a heterostructure interface can be characterized. One important device parameter is the areal carrier density N_s in Eq. (11.6.1). If we ignore the contribution of ionized impurities in Eq. (11.4.13), we can express Eq. (11.4.3) in terms of N_s as

$$\mathcal{E}_i \cong \left(\frac{\hbar^2}{2m_l}\frac{9\pi^2 e^4}{4\epsilon^2}\right)^{1/3}\left(\frac{N_s}{2}\right)^{2/3}\left(i + \frac{3}{4}\right)^{2/3} \qquad (11.6.2)$$

In many device applications the impurity concentration in the semiconductor where the two-dimensional electron gas is situated is kept low to keep the carrier mobility high by minimizing impurity scattering. The areal charge density $N_I z_{sc}$ associated with ionized impurities is much smaller than N_s for $N_I < 10^{20}$ m^{-3}, which is achievable in unintentionally doped semiconductors.

Applied to GaAs, Eq. (11.6.2) becomes

$$\mathcal{E}_0(\text{eV}) = 1.76 \times 10^{-12}(N_s)^{2/3} \qquad (11.6.3)$$

$$\mathcal{E}_1(\text{eV}) = 2.95 \times 10^{-12}(N_s)^{2/3} \qquad (11.6.4)$$

respectively, for the ground state $i = 0$ and the first excited state $i = 1$. Experimental results as gathered by Delagebaudeuf and Linh indicate a numerical constant 2.5×10^{-12} for $\mathcal{E}_0$ and 3.2×10^{-12} for $\mathcal{E}_1$ (D. Delagebaudeuf and N. T. Linh, *IEEE Trans. Electron Devices*, Vol. ED-29, p. 955, 1982). The discrepancy can be due to an impurity concentration $N_I > 10^{21}$ m^{-3} or 10^{15} cm^{-3}. We should also point out that the wave function ζ_1 extends farther than ζ_0 into the region where f_i in Eq. (11.4.10) is small. Therefore, $|\zeta_1|^2$ should experience a smaller average E than $|\zeta_0|^2$. This consideration can explain why the ratio $\mathcal{E}_1/\mathcal{E}_0 = 3.2/2.5$ deduced from experiment is lower than the theoretical ratio 2.95/1.76, which assumes an equal average field E for $|\zeta_0|^2$ and $|\zeta_1|^2$.

The areal carrier density N_{si} occupying each subband i can be calculated from the two-dimensional density of state, D_2 of Eq. (11.4.6) and the Fermi function $f(\mathcal{E})$, as follows:

$$N_{si} = \frac{m^*}{\pi\hbar^2}\int f(\mathcal{E})\,d\mathcal{E}_\perp = \frac{m^* kT}{\pi\hbar^2}\ln(1 + x_i) \qquad (11.6.5)$$

where $\mathcal{E} = \mathcal{E}_i + \mathcal{E}_\perp$ for the ith subband and

$$x_i = \exp \left(\frac{\mathcal{E}_f - \mathcal{E}_i}{kT} \right) \tag{11.6.6}$$

Summing the areal carrier densities over two lowest-energy levels, we obtain

$$N_s = \frac{m^* kT}{\pi \hbar^2} \ln \left[(1 + x_0)(1 + x_1) \right] \tag{11.6.7}$$

The fraction of N_s in the lowest subband is equal to $N_{s0}/N_s = \ln (1 + x_0)/[\ln (1 + x_0) + \ln (1 + x_1)]$. We also note that at low temperatures, $x_0 \gg 1$ and Eq. (11.6.5) reduces to Eq. (11.4.7). For GaAs the value of D_2 is

$$D_2 = \frac{m^*}{\pi \hbar^2} = 2.86 \times 10^{17} \text{ m}^{-2} (\text{eV})^{-1} \tag{11.6.8}$$

using the bulk value $m^* = 0.067 m_0$ at 0 K.

Figure 11.13 shows (a) the energy levels at $T = 0$ K and (b) the Fermi level at $T = 0$ K, 77 K, and 300 K computed by Stern and Das Sarma for a GaAs/(GaAl)As interface with $\Delta \mathcal{E}_c = 0.3$ eV by solving the Schrödinger equation under a varying electric field $E(z) = E_c + E_i$ of Eqs. (11.4.10) and (11.4.12) (F. Stern and S. Das

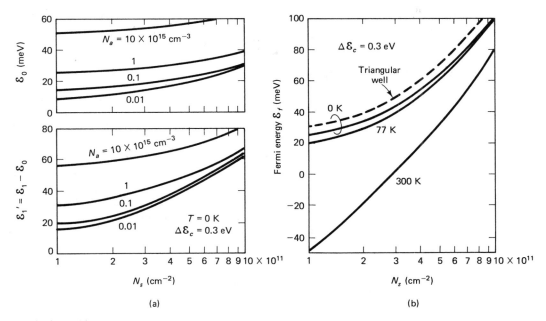

Figure 11.13 (a) The energies of the lowest sub-bands, $\mathcal{E}_0$ and $\mathcal{E}_1$ and (b) the Fermi level $\mathcal{E}_f$ computed as functions of the areal carrier density N_s (in cm^{-2} instead of the m^{-2} used in the text). Curves in (a) were computed for several values of background acceptor concentration N_a. (F. Stern and S. Das Sarma, *Phys. Rev,* Vol. B30, p. 840, 1984.)

Sarma, *Phys. Rev.*, Vol. B30, p. 840, 1984). The $\mathcal{E}_0$ curve for $N_a = 1 \times 10^{14}$ cm^{-3} can be fit reasonably well by

$$\mathcal{E}_0 = 1.31 \times 10^{-12}(N_s + N_{SI})^{2/3} \quad \text{eV} \tag{11.6.9}$$

with $N_{SI} = 8.2 \times 10^{14}$ m^{-2} and the $\mathcal{E}_1' = \mathcal{E}_1 - \mathcal{E}_0$ curve by

$$\mathcal{E}_1' = 0.60 \times 10^{-12}(N_s + N_{SI}')^{2/3} \quad \text{eV} \tag{11.6.10}$$

with $N_{SI}' = 1.1 \times 10^{15}$ m^{-2}. The quantity $N_{SI} = N_I z_{sc}$ represents the areal impurity density associated with ionized impurities. Therefore, the width of the interface space-charge region z_{sc} is about 8.2×10^{-6} m as compared to a value on the order of 8×10^{-9} m at $T < 77$ K and 16×10^{-9} m at $T = 300$ K for the average electron distance in GaAs from the heterostructure interface.

At low temperatures ($T < 77$ K), Eq. (11.4.7) is a good approximation and counting the areal carrier densities in the two lowest subbands is sufficient. Therefore, the Fermi level can be calculated from

$$\mathcal{E}_f = \frac{(\mathcal{E}_0 + \mathcal{E}_1) + N_s/D_2}{2} \tag{11.6.11}$$

At $N_s = 3 \times 10^{15}$ m^{-2}, use of $\mathcal{E}_0$ and $\mathcal{E}_1$ from Fig. 11.13a yields $\mathcal{E}_f = 43$ meV, in agreement with the value given in Fig. 11.13b. At 300 K, however, Eq. (11.6.7) must be used. For $N_s = 3 \times 10^{15}$ m^{-2}, we find $\mathcal{E}_f = 0$ meV, $\mathcal{E}_0 = 33$ meV, and $\mathcal{E}_1 = 48$ meV from Fig. 11.13. Using these values and a value 6.68×10^{15} for D_2 (based on $m^* = 0.063m_0$ at 300 K) in Eq. (11.6.5), we obtain $N_{s0} = 1.9 \times 10^{15}$ m^{-2} and $N_{s1} = 0.9 \times 10^{15}$ m^{-2}. Therefore, there is an appreciable population in the upper subbands.

11.7 MODULATION-DOPED HETEROSTRUCTURE

The idea of modulation doping is to separate carriers from ionized impurities so that carriers can attain a mobility not affected by impurity scattering (R. Dingle, H. L. Störmer, A. C. Gossard, and W. Wiegmann, *Appl. Phys. Lett.*, Vol. 33, p. 665, 1978). This idea can be realized in a semiconductor heterostructure in which only part of the higher-gap semiconductor is doped. Figure 11.14 shows the energy-band diagram of a (GaAl)As/GaAs heterostructure. The GaAs is not intentionally doped. However, since the conduction-band edge of GaAs lies below the donor states in (GaAl)As, electrons move from the donor states into GaAs near the heterostructure interface, forming a two-dimensional electron gas. The first experimental observation of a dramatic increase of electron mobility at low temperatures in modulation-doped (MD) heterostructures over that in uniformly doped (UD) heterostructures is reported by Dingle et al. and is illustrated in Fig. 11.15. Since then, electron mobility in excess of 10^5 cm^2/V-s at 77 K has been reported, which corresponds to the value measured in bulk samples with an estimated donor concentration of $N_I = 10^{14}$ cm^{-3} (G. E. Stillman, C. M. Wolfe, and J. O. Dimmock, *J. Phys. Chem. Solids*, Vol. 31, p. 1199, 1970). At room temperature, there is only a slight improvement in electron mobility because of the predominance of polar scattering over impurity scattering.

There are three major considerations in the design of a modulation-doped (MOD) heterostructure: (1) the channel (or areal) carrier density N_s obtainable, (2) the channel

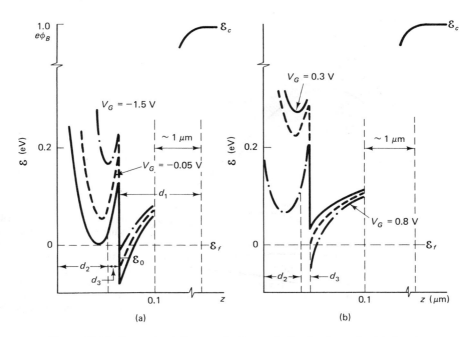

Figure 11.14 Energy-band diagrams for the conduction band near the interface between (GaAl)As on the left and GaAs on the right in two samples: (a) sample 3468 and (b) sample 3469 under different bias conditions. Under zero gate bias (solid curves), sample 3468 has an appreciable N_s while sample 3468 does not. Whether a MOD sample is normally on or off depends on the thickness d_3 of the undoped and doping concentration N_d of the doped (GaAl)As layers. (B. Vinter, *Appl. Phys. Lett.*, Vol. 44, p. 307, 1984.)

(or gate) capacitance C_c associated with the charges in the heterostructure, and (3) the effect of doping on carrier mobility for low-temperature operation (B. Vinter, *Appl. Phys. Lett.*, Vol. 44, p. 307, 1984; F. Stern, *Appl. Phys. Lett.*, Vol. 43, p. 974, 1983). The first two subjects were discussed by Vinter and the third by Stern. Referring again to Fig. 11.14, we anticipate that for a given barrier height $e\phi_B$ at the metal–(GaAl)As interface, the positions of the quantized energy levels $\mathscr{E}_0$ and $\mathscr{E}_1$ will depend on the donor concentration N_d and the width d_2 of the doped (GaAl)As, the conduction-band-edge discontinuity $\Delta\mathscr{E}_c$, and the width d_3 of the undoped (GaAl)As, generally referred to as the *spacer region*. In Table 11.4 the material and physical parameters used by Vinter in his calculation of N_s and C_c are given for two samples, numbers 3468 and 3469. In sample 3468, $\mathscr{E}_0$ lies below $\mathscr{E}_f$ and hence the device is normally on. In sample 3469, $\mathscr{E}_0$ lies above $\mathscr{E}_f$ and the device is normally off. The two situations are illustrated, respectively, in (a) and (b) of Fig. 11.14.

Figure 11.16 shows (a) the computed areal carrier density N_s in the channel, (b) the computed areal carrier density N_s' in (GaAl)As, (c) a sum of $N_s + N_s'$, and (d) the areal density $N_d^0 d_2$ of neutral donors as function of the gate voltage V_G, V_G being positive if the polarity of the metal electrode is positive. Take as an example sample 3468, which is normally on. As V_G becomes more negative (< -1.5 V), more positive charges are added to the semiconductors, resulting in a less areal electron density $N_s + N_s'$ and a

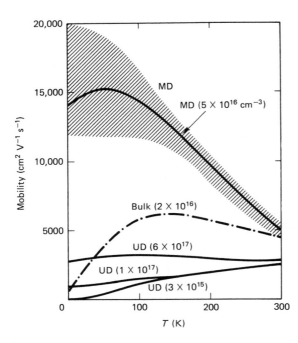

Figure 11.15 Electron mobility versus temperature for bulk GaAs, uniformly doped (UD) GaAs and modulation-doped (MD) (GaAl)As–GaAs samples. The shaded region indicates the range of mobility observed in early MD samples. (R. Dingle, H. L. Störmer, A. C. Gossard, and W. Wiegmann, *Appl. Phys. Lett.*, Vol. 33, p. 665, 1978.)

more areal hole concentration (which is not shown in the figure), and the device is turned off. As V_G is less negative (> -1 V), both conduction band edges of GaAs and (GaAl)As move lower, resulting in a larger $N_s + N_s'$ and at the same time a larger percentage of neutral donors. The reason that curve (a) increases only moderately with increasing V_G is due to the fact that the negative charges induced on the semiconductor side by the gate voltage are shared by the channel, the (GaAl)As, and the donors and that the donors receive an increasing share of the induced negative charges as evidenced by the rapid rise of curve (d). The same physical reasoning applies to the behavior of sample 3469 except that the voltage scale is shifted. Figure 11.17 shows the differential capacitance per unit area $C_c = edN_T/dV_G$ as functions of V_G where N_T is the areal charge density with $N_T = N_s + N_s' - N_d d_2$ for curve (a), $N_T = N_s + N_s'$ for curve (b), and $N_T = N_s$ for curve (c). The rise of curve (a) for $V_G > -1.5$ V in sample 3468 and for $V_G > 0.5$ V in sample 3469 is again a manifestation of the fact that an increasing amount of induced negative charges are to neutralize donors and to a less extent to populate the well region in (GaAl)As.

From the discussion above, we see that the design as well as the operating range of V_G for a MODFET is limited by a rapid increase in C_c on the one side and a rapid

TABLE 11.4 PARAMETERS USED BY VINTER

Sample	Al content	N_d (10^{18} cm^{-3})	d_2 (Å)	d_3 (Å)	d_1 (μm)	$\Delta\mathscr{E}_c$ (eV)	$e\phi_B$ (eV)
3468	0.26	1.3	550	75	1.0	0.26	1
3469	0.28	0.6	400	65	0.8	0.28	1

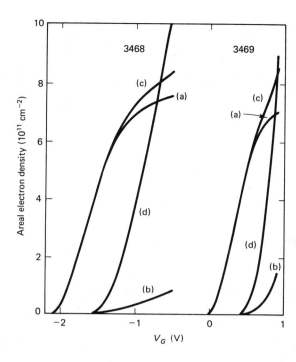

Figure 11.16 Computed areal electron density as a function of the gate voltage V_G for the two samples of Fig. 11.14 with curve (a) for N_s in the GaAs channel, curve (b) for N'_s in (GaAl)As, curve (c) for the sum $N_s + N'_s$, and curve (d) for areal density of unionized donors in (GaAl)As. (B. Vinter, *Appl. Phys. Lett.*, Vol. 44, p. 307, 1984.)

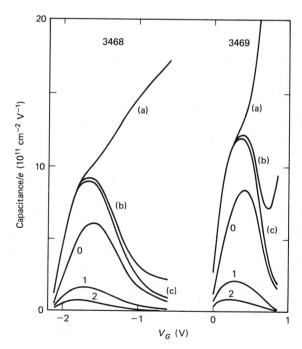

Figure 11.17 Computed differential capacitance expressed as $C_c/e = dN_T/dV_G$ as functions of the gate voltage V_G with $N_T = N_s + N'_s - N_d d_2$ counting all charges in curve (a), $N_T = N_s + N'_s$ counting all electrons in curve (b), and $N_T = N_s$ counting only channel electrons in curve (c). The curves designated 0, 1, and 2 are for the lowest three subbands, that is $N_T = N_{s0}$, N_{s1}, and N_{s2}, respectively. (B. Vinter, *Appl. Phys. Lett.*, Vol. 44, p. 307, 1984.)

Sec. 11.7   Modulation-Doped Heterostructure

decrease in N_s on the other side of the useful range of V_G. One major difference between a MODFET and a MISFET is that (GaAl)As, which plays the role of a gate insulator, can store charges. The change in the stored charges is the combined results of changing d_2 in $N_d d_2$ and changing N'_s in (GaAl)As. Finally, we should mention that the curves shown in Figs. 11.16 and 11.17 are obtained from a self-consistent calculation based on Poisson's equation,

$$\frac{d^2\psi}{dz^2} = \frac{e^2}{\epsilon}\,[N_d^+(z) - n(z)] \qquad (11.7.1)$$

and the effective-mass Schrödinger equation,

$$-\frac{\hbar^2}{2m^*}\frac{d^2\zeta_i}{dz^2} + \psi(z)\zeta_i(z) = \mathscr{E}_i\zeta_i(z) \qquad (11.7.2)$$

where $\psi(z) = e\phi(z)$ is the potential energy for electrons,

$$n(z) = \sum N_{si}|\zeta_i(z)|^2 \qquad (11.7.3)$$

and N_{si} is given by Eq. (11.6.5). The two boundary conditions are

$$\psi(z = 0) = \psi_B - eV_G \quad \text{and} \quad \psi(\text{GaAlAs}) = \psi(\text{GaAs}) + \Delta\mathscr{E}_c \qquad (11.7.4)$$

at the metal–(GaAl)As interface and at the (GaAl)As–GaAs interface, respectively. In the calculation by Vinter, the Fermi level in the two semiconductors is used as a reference energy; therefore, the conduction-band edge in (GaAl)As at the metal–(GaAl)As interface is shifted by an amount $-eV_G$ by the gate voltage.

The mechanism that limits the electron mobility at low temperatures ($\sim$ 4 K) in a MOD heterostructure is of considerable interest as the measured value in GaAs/(GaAl)As heterojunctions with reduced background (unintentionally doped) impurity concentration approaches 10^6 cm^2/V-s (H. L. Störmer, A. Chang, D. C. Tsui, J. C. M. Hwang, A. C. Gossard, and W. Wiegmann, *Phys. Rev. Lett.*, Vol. 50, p. 1953, 1983). In the calculation by Stern on ungated MODFET, the ionized donors in the (GaAl)As barrier are replaced by a sheet of charge located at d_b, the center of the space-charge region away from the heterojunction interface. It is found that at low N_s ($< 10^{11}$ cm^{-2}), the mobility is limited by scattering from background impurities in the channel (F. Stern, *Appl. Phys. Lett.*, Vol. 43, p. 974, 1983). At high N_s ($> 3 \times 10^{11}$ cm^{-2}), the mobility is controlled by scattering from the impurities in the barrier space-charge region. A maximum value of $\mu \sim 10^6$ cm^2/V-s is obtained for heterostructures with a background impurity concentration of 10^{14} cm^{-3} in the channel and spacer regions and a donor concentration of 7×10^{17} cm^{-3} in the barrier. This value is relatively insensitive to the barrier donor concentration. However, the calculation is made for an ungated MODFET. It is obvious from Figs. 11.14 and 11.16 that the extent of the barrier space-charge region and hence the value of d_b vary considerably with the gate voltage. Therefore, the value of 10^6 cm^2/V-s is useful as a point of reference for low-temperature ($\sim$ 4 K) electron mobility, but a comparison between the theoretical and experimental values still awaits a detailed analysis of low-temperature mobility under specific gate-bias conditions.

11.8 HOT ELECTRONS AND BALLISTIC ELECTRONS IN HETEROSTRUCTURES

In Chapter 10 we discussed effects manifested by hot carriers, that is, carriers with an average velocity exceeding the thermal velocity. Examples of hot-carriers effects include carrier trapping in oxides, discussed in Section 10.7, and negative differential conductance, discussed in Section 10.3. The motivation for our interests in hot electrons in heterostructures is also twofold. One is preventive. We know that electrons in interface channels like those in bulk materials and in MOSFETs will become hot if the applied field is sufficiently high. The hot-electron effects are expected to occur at lower fields in higher mobility structures such as the MODFET. Therefore, we need to study the effects of hot electrons on device performance and to find ways to minimize undesirable effects. The other motivation is exploratory. Hot electrons, by definition, have an energy distribution higher than that for carriers under thermal equilibrium. Therefore, they can either go over or tunnel through a potential barrier with greater probability than low-energy carriers. Besides, in heterostructures, hot carriers can be injected from one semiconductor to another by tunneling through a barrier layer. Both possibilities may lead to new device concepts. In this section we present experiments on hot electrons to illustrate their possible role in heterostructure devices. We concentrate our discussion on electrons at low temperatures because of the high mobility achievable in MOD heterostructures.

One interesting and direct experiment on hot electrons is the measurement of luminescence spectra in MOD heterostructures reported by Shah et al. (J. Shah, W. Pinczuk, H. L. Störmer, A. C. Gossard, and W. Wiegmann, *Appl. Phys. Lett.*, Vol. 44, p. 322, 1984). The sample has a low-field mobility of 7.9×10^4 cm^2/V-s at low temperatures, and an areal electron concentration of 3.9×10^{11} cm^{-2}. The experiment was performed at $T_l = 2$ K, and an infrared dye laser was used to generate holes. The luminescence is caused by vertical transitions of electrons in the subband states with photo-excited holes. Figure 11.18 shows the luminescence spectra $I(\nu)$ observed under various applied electric field E. At the high-energy side, the electron distribution can be approximated by a Boltzmann distribution; hence $I(\nu)$ is expected to vary as

$$I(\nu) = A \exp\left(-\frac{h\nu}{kT_e}\right) \qquad (11.8.1)$$

where A is proportionality constant, $h\nu$ the photon energy, and $T_e = T_c$ is the electron or carrier temperature. The application of an electric field raises the average velocity of electrons and hence the electron temperature. The values of T_e for different values of E are obtained by fitting Eq. (11.8.1) to the various spectra curves.

Figure 11.19 shows the value of T_e deduced from measurement of $I(\nu)$ at $T_l = 2$ K. The bars with a dot represent the results deduced from the experiment. An analysis of the data has been reported by Lee et al., based on the treatment of hot-electron transport by Conwell (E. M. Conwell, "High Field Transport in Semiconductors," *Solid State Physics Suppl.*, Vol. 9, p. 156, 1967; H. P. Lee, D. Vakhshoori, Y. H. Lo, and S. Wang, *J. Appl. Phys.*, Vol. 57, p. 4814, 1985). In the analysis, two basic equations are used: one for momentum balance and the other for energy balance. The equation for momentum balance is actually an equation for mobility μ given as

$$\mu^{-1} = \mu_{lf}^{-1} + 1.75\, x_e \exp\left(-x_e\right) \qquad (11.8.2)$$

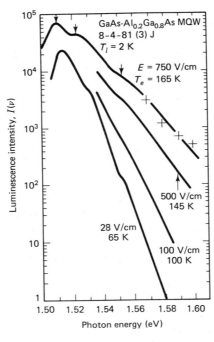

Figure 11.18 Luminescence spectra observed at $T_l = 2$ K in a modulation-doped GaAs–Ga$_{0.8}$Al$_{0.2}$As sample under four different electric fields. The arrows indicate energy of transition between conduction and valence subbands. The spectra can be fit to Eq. (11.8.1), using different values for the electron temperature T_e. (J. Shah, A. Pinczuk, H. L. Störmer, A. C. Gossard, and W. Wiegmann, *Appl. Phys. Lett.*, Vol. 44, p. 322, 1984.)

where μ_{lf} is the low-field mobility, which in the experiment is due to some unknown scattering mechanism and assumed to be given by $\mu_{lf} = 7.9$ m^2/V-s. The quantity $x_e = h\nu/kT_e$ is the photon energy $h\nu$ normalized to the average electron energy kT_e. The energy-balance equation reads

$$\mu E^2 \frac{1}{e}\left\langle \frac{d\mathscr{E}}{dt}\right\rangle_{po} = C\,2.4 \times 10^{11}\,x_e^{1/2}\exp\left(-x_e\right) \tag{11.8.3}$$

where the term $e\mu E^2$ represents the rate of energy increase acquired from the field and $\langle d\mathscr{E}/dt\rangle_{po}$ represents the rate of energy dissipation through optical-phonon excitation. The mobility for acoustic-phonon scattering and that for piezoelectric scattering are, respectively, given by

$$\mu_{ac} = 1.7 \times 10^5\,T_l^{-1}T_e^{-1/2} \tag{11.8.4}$$

$$\mu_{pe} = 7 \times 10^2\,T_l^{-1}T_e^{1/2} \tag{11.8.5}$$

All the equations are in MKS units, and the constants in the equations are based on the following values: 3×10^{11} N/m for longitudinal elastic constant, 8.4×10^{12} Hz for optical-phonon frequency, 7 eV for deformation potential, 0.067 for polar constant, 0.052 for electromechanical constant, and 5.3×10^3 kg/m^3 for GaAs density. The reader is referred to Conwell's article or Seeger's book for the analytical expressions (K. Seeger, *Semiconductor Physics*, Springer-Verlag, New York, 1973). At low temperatures, the product $K_{1,0}(x_e/2)\exp(x_e/2)$ varies very slowly with x_e and a value of 0.4 is used, where $K_{1,0}$ represents the two modified Bessel functions appearing in the mobility and energy expressions (Problem 11.31) for polar optical-phonon scattering. In the analysis by Lee et al. it is found that the mobility term μ_{po} due to polar optical-phonon scattering dominates because of its rapid exp $(-x_e)$ dependence. Hence only polar optical-phonon scattering is considered in Eqs. (11.8.2) and (11.8.3).

The value of x_e can be found from Eqs. (11.8.2) and (11.8.3) as a function of E. The computed values of T_e are also shown in Fig. 11.19 for two cases: curve 2a with an adjustable parameter $C = \frac{1}{9}$ in Eq. (11.8.3) and curve 2b according to the theory without adjustment, that is, $C = 1$. A reasonable agreement is obtained. Also measured by Shah et al. and shown in Fig. 11.19 as dots is the electron mobility normalized to μ_{lf}. The computed normalized mobility from Eqs. (11.8.2) and (11.8.3) is plotted as dashed curves for comparison with experiments for two cases: curve 3a with $C = \frac{1}{9}$ and curve 3b with $C = 1$. In view of the fact that the model has only one adjustable parameter C, the agreement between theory and experiment is considered good. To test the validity of the theoretical model over an extended temperature range, the computed electron mobility is compared in Fig. 11.20 with the experimental mobility measured by Drummond et al. (T. J. Drummond, M. Keever, W. Kopp, H. Morkoç, K. Hess, and B. G. Streetman, *Electron. Lett.*, Vol. 17, p. 545, 1981). Good fit to the experimental results is obtained from $T_l = 10$ to 100 K by using a value $C = \frac{1}{4}$ for the adjustable parameter. Note that for both Figs. 11.19 and 11.20, a value for $C < 1$ is used. The wave vector **k** of electrons in a two-dimensional electron gas is quantized in one direction, so the number of optical phonons available for scattering the electrons is expected to be smaller for 2DEG than for bulk GaAs. From Fig. 11.20 we also see that hot-electron effects become prominent only at low temperatures.

Another interesting experiment on hot electrons is the measurement of substrate current reported by Luryi et al. (S. Luryi, A. Kastalsky, A. C. Gossard, and R. H. Hendel, *IEEE Trans. Electron Devices,* Vol. ED-31, p. 832, 1984). Consider a MOD heterostructure grown on n^+-GaAs substrate and consisting of two GaAs/(GaAl)As interfaces. The top, doped (GaAl)As layer is used to supply electrons to the undoped GaAs channel, while the bottom undoped (GaAl)As layer is used to block electrons into

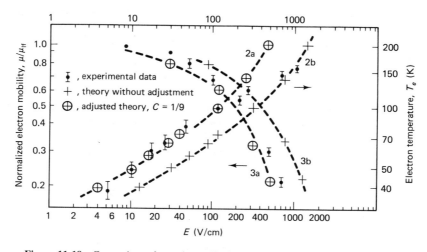

Figure 11.19 Comparison of experimentally determined values of electron temperature T_e and normalized mobility μ/μ_{lf} from Fig. 11.18 and computed values of T_e and μ/μ_{lf} at $T_l = 2$ K from Eqs. (11.8.2) and (11.8.3) as a function of the applied field. (H. P. Lee, D. Vakhshoori, Y. H. Lo, and S. Wang, *J. Appl. Phys.,* Vol. 57, p. 4814, 1985.)

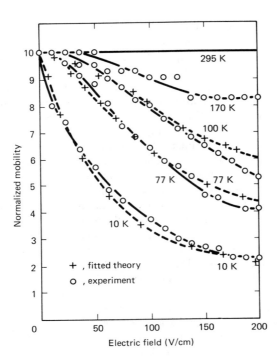

Figure 11.20 Comparison of measured and calculated values of normalized electron mobility μ/μ_{lf} as a function of the applied field at several temperatures. (H. P. Lee, D. Vakhshoori, Y. H. Lo, and S. Wang, *J. Appl. Phys.*, Vol. 57, p. 4814, 1985.)

the substrate. The drain is grounded and the gate is unbiased. Figure 11.21 shows the substrate current I_{SUB} measured at $T_l = 77$ K as a function of the source–drain voltage $V_{SD} = V_S - V_D$ for several substrate biases V_{SUB}. Note that at $V_{SD} = 0.4$ V, a field $E = 10^3$ V/cm exists in a channel of length 4 μm. Based on the mobility data shown in Fig. 11.20, it is expected that T_e far exceeds T_l. Since the potential barrier ψ at the bottom heterostructure interface is finite, there is a significant substrate current I_{SUB} caused by thermionic emission over the barrier, with

$$I_{SUB} = I_0 \exp\left(\frac{-\psi}{kT_e}\right) \qquad (11.8.6)$$

according to Eq. (4.7.9). Here we assume I_0 to be a constant by ignoring the weak T_e^2 dependence of the Richardson constant. The rapid rise of I_{SUB} shown in Fig. 11.21 can be attributed to an increase in T_e. Therefore, information about T_e as a function of V_{SD} can be deduced from the measured dependence of I_{SUB} on V_{SD}.

If a dependence of T_e on V_{SD} of the form

$$T_e = T_l(1 + \gamma V_{SD}^m) \qquad (11.8.7)$$

is assumed, with γ being a constant, then from Eq. (11.8.6) we obtain

$$f = \left(V_{SD}\frac{d \ln I_{SUB}}{dV_{SD}}\right)^{-2} = \frac{kT_e}{m\psi}\frac{T_e}{T_e - T_l} \qquad (11.8.8)$$

where $T_l = 77$ K is the lattice temperature. Figure 11.22 shows the value of T_e as a function of V_{SD}^2 deduced from the data shown in Fig. 11.21. For $V_{SUB} < 0.5$ V, the T_e

Properties of Heterostructures   Chap. 11

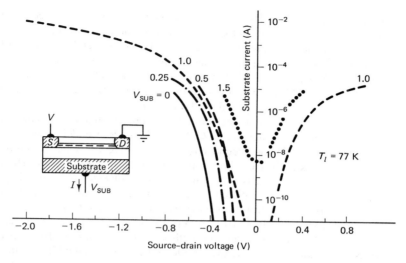

Figure 11.21 Substrate current I_{SUB} measured at 77 K as a function of the source–drain voltage $V_{SD} = V_S - V_D$ in a MOD structure (shown in the inset). The sharp rise of I_{SUB} with V_{SD} is due to an increased T_e. (S. Luryi, A. Kastalsky, A. C. Gossard, and R. H. Hendel, "Charge Injection Transistor Based on Real-Space Hot-Electron Transfer," *IEEE Trans. Electron Devices,* Vol. ED-31, p. 832, 1984 © 1984 IEEE.)

values can be fitted to Eqs. (11.8.7) and (11.8.8) with $m = 2$ and $\psi = 0.3$ eV, but the value of γ increases with V_{SUB}. Furthermore, the simple power dependence of Eq. (11.8.7) no longer holds for the $V_{SUB} = 1$ V and $V_{SUB} = 1.5$ V curves. Therefore, the value of T_e shown in Fig. 11.22 is subject to uncertainties. For a more accurate determination of T_e, we need to establish a more realistic model which incorporates the spatial variation of E along the channel when a substrate bias is applied.

One basic reason for building MODFET is the possibility of attaining high electron mobility, especially at low temperatures. However, hot-electron effects will neu-

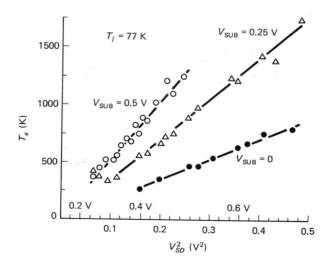

Figure 11.22 Electron temperature T_e deduced from Fig. 11.21 as a function of the source–drain voltage V_{SD} for several values of substrate bias. (S. Luryi, A. Kastalsky, A. C. Gossard, and R. H. Hendel, "Charge Injection Transistor Based on Real-Space Hot-Electron Transfer," *IEEE Trans. Electron Devices* Vol. ED-31, p. 832, 1984 © 1984 IEEE.)

tralize this advantage. Therefore, a great deal of attention has been directed toward two objectives: one to utilize hot electrons and the other to reduce scattering events so that electrons can acquire a high drift velocity. To the first end, a number of possibilities for novel devices have been investigated. These include the charge injection transistor (S. Luryi, A. Kastalsky, A. C. Gossard, and R. Hendel, *IEEE Trans. Electron Devices*, Vol. ED-31, p. 832, 1984), the negative-resistance field-effect transistor (A. Kastalsky, S. Luryi, A. C. Gossard, and R. Hendel, *IEEE Electron Device Lett.*, Vol. EDL-5, p. 57, 1984), and the hot-electron erasable programmable random-access memory element (S. Luryi, A. Kastalsky, A. C. Gossard, and R. Hendel, *Appl. Phys. Lett.*, Vol. 45, p. 1294, 1984). To the second end, the objective has been to study the transport of electrons across a distance short compared to the mean free path. The idea is simple. If electrons do not suffer any scattering within the distance traveled, they can attain a high drift velocity without its direction being randomized. Such electrons are called *ballistic electrons*, as distinct from *hot electrons*, whose velocity is randomized by scattering. Based on a value of 1×10^5 cm^2/V-s for μ_n at 77 K, we find a relaxation time τ of 3.8×10^{-12} s and a mean free path $l = v_{th}\tau$ of 8.4×10^{-7} m or 0.84 μm. For collisionless transit at 77 K, the gate length L_G of a MODFET and the base width W of a vertical heterostructure n^+-n-n^+ transistor must be made much smaller than 0.8 μm. Obviously, this condition is more easily met in a vertical transistor than a horizontal MODFET.

Figure 11.23 shows the energy diagram of an experimental structure used by Heiblum et al. to study ballistic transport in GaAs (M. Heiblum, M. I. Nathan, D. C. Thomas, and C. M. Knoedler, *Phys. Rev. Lett.*, Vol. 55, p. 2200, 1985). Applying a positive bias to the base with respect to the emitter causes electrons to tunnel to the base. The base is made very thin (310 Å thick) to reduce the chance for collisions. The collector is separated from the base by a relatively thick (GaAl)As layer (1000 Å thick,

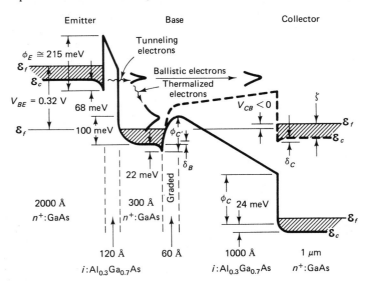

Figure 11.23 Experimental structure used to study ballistic transport across an n^+-GaAs base. (M. Heiblum, M. I. Nathan, D. C. Thomas, and C. M. Knoedler, *Phys. Rev. Lett.*, Vol. 55, p. 2200, 1985.)

undoped). The potential barrier ψ_C' at the collector junction not only prevents thermalized electrons from reaching the collector but also serves as an electron-energy spectrometer to determine the energy distribution of electrons injected into the base and collected by the collector. If $V_{CB} = V_C - V_B$ is sufficiently negative, the collector barrier prevents electrons, both ballistic and thermalized, from reaching the collector. As V_{CB} is increased or ψ_C' is lowered, energetic electrons having sufficient energy to overcome the barrier will be able to reach the collector. Therefore, the differential conductance $G_C = dI_C/dV_{CB}$ is a direct measure of the energy distribution of collected electrons.

Figure 11.24 shows the value of G_C deduced from I_C measured at 4.2 K as a function of V_{CB}. From the emitter side, the electron distribution has a maximum at an energy $eV_{BE} - \Delta$ above the Fermi level in the base, where Δ is the position of the distribution peak with respect to the quasi-Fermi level in the emitter. From the collector side, the measured peak position (V_p) of G_C has an energy at $eV_p + \psi_C - \zeta$ with respect to the Fermi level in the base, where ψ_C is the collector barrier and ζ is the Fermi energy in the collector as shown in Fig. 11.23. If the collected electrons are ballistic electrons, they should have the same energy as the electrons injected from the emitter, requiring

$$eV_p + \psi_C - \zeta = eV_{BE} - \Delta \qquad (11.8.9)$$

This relation is indeed observed with a constant $\psi_C = 213 \pm 8$ meV, a calculated $\zeta = 54$ meV based on a value of 1×10^{18} cm^{-3} for n and a calculated value for $\Delta = 15$ meV. As I_E increases, the value of V_p shifts toward higher values because of the higher value of V_{BE} in Eq. (11.8.9). The experimentally observed 60 meV width of the G_C peak also is in good agreement with the value of ζ. If optical phonon scattering were prevalent, Eq. (11.8.9) would yield a value for ψ_C several $h\nu_{op}$ below the value 213 meV based on conduction-band discontinuity. Although transport of hot electrons has been reported (J. R. Hayes, A. F. J. Levi, and W. Wiegmann, *Phys. Rev. Lett.*, Vol. 54, p. 1570, 1985; S. Muto, K. Imamura, N. Yokoyama, S. Hiyamizu, and H. Nishi, *Electron. Lett.*, Vol. 21, p. 555, 1985), the data shown in Fig. 11.24 are the

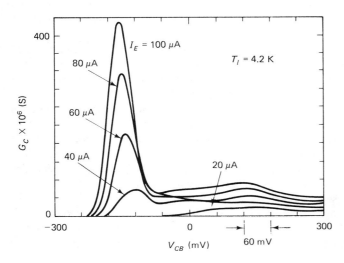

Figure 11.24 The values of differential conductance $G_c = dI_C/dV_{CB}$ in Siemens measured at 4.2 K as a function of the collector–base bias V_{CB}. (M. Heilblum, M. I. Nathan, D. C. Thomas, and C. M. Knoedler, *Phys. Rev. Lett.*, Vol. 55, p. 2200, 1985.)

first evidence that a significant fraction of injected electrons are ballistic in the sense that they reach the collector without measurable loss of energy.

In order for ballistic n^+-n-n^+ transistors to become practical, the transport efficiency of ballistic electrons must be greater than 0.5 at the least, so a current gain $\beta > 1$ in the common-emitter configuration can be achieved. Therefore, it is important to examine possible mechanisms for keeping injected electrons in or scattering electrons back to the base. The three possible mechanisms are (1) back scattering of electrons by optical phonons, during transit across the collector (GaAl)As layer, (2) quantum-mechanical reflection at the base–collector–barrier interface, and (3) scattering of electrons from Γ valley into L valley in the base. The first mechanism has been analyzed by Vakhshoori and Wang (D. Vakhshoori and S. Wang, *IEEE Electron Device Lett.*, Vol. EDL-7, p. 262, 1986). Assuming that the electron energy is much larger than the phonon energy, the scattering angle θ is given by

$$q = 2k \sin \frac{\theta}{2} \qquad (11.8.10)$$

where q and k are the phonon and electron wave number. However, after back scattering, an electron must have a minimum energy $\hbar^2 k_z^2/2m^*$ to climb the potential hill eEz, where z is the distance from the base–collector–barrier interface. Therefore, the minimum angle θ_m for back scattering obeys

$$\cos^2 \theta_m = \left(\frac{k_z'}{k}\right)^2 = \frac{eEz}{\mathscr{E}_0 + eEz \pm h\nu_{op}} \qquad (11.8.11)$$

where $\mathscr{E}_0$ is the initial energy of the electron at $z = 0$.

The probability for scattering can be found by following Conwell's treatment [Eqs. (3.1.8) and (3.6.21) in E. M. Conwell, "High Field Transport in Semiconductors," *Solid State Physics Suppl.*, Vol. 9, p. 156, 1967]. The important point here is that the matrix element is proportional to q^{-2}. For forward scattering, θ in Eq. (11.8.10) is small, and hence q is small. For back scattering, on the other hand, θ has a minimum value θ_m given by Eq. (11.8.11), and the value for q is large. Therefore, this process strongly favors scattering in the forward direction. Figure 11.25 shows the calculated collector efficiency α as a function of the structure parameter $a = eEZ_C/\mathscr{E}_0$ and as a function of E, where Z_C is the width of the collector barrier region, E is the field in the region, and $\mathscr{E}_0$ is initial electron energy. The calculated results show that high collector efficiency ($> 97\%$) can be achieved even at room temperature. Therefore, low β values observed in experimental structures must be due to mechanisms other than back scattering (M. Heiblum, D. C. Thomas, C. M. Knoedler, and M. I. Nathan, *Appl. Phys. Lett.*, Vol. 47, p. 1105, 1985). The second mechanism, quantum-mechanical reflection, can be minimized by using a barrier region with graded composition. The third mechanism, scattering into an upper valley, deserves special consideration. We note that if the electron energy is reduced practically to zero after intervalley scattering, the value of q connecting the initial and final electron states will be equal to k and independent of direction of scattering. Numerical calculation, including intervalley scattering, shows that for $\mathscr{E}_0 > \Delta\mathscr{E}_{\Gamma L}$, the separation of Γ- and L-valley minima, intervalley scattering becomes dominant, resulting in a very short mean free path (M. A. Littlejohn, R. W. Trew, J. R. Hauser, and J. M. Golio, *J. Vac. Sci. Technol.*, Vol. B1(2), p. 449, 1983). Therefore, intervalley scattering may be a fundamental obstacle for realizing novel de-

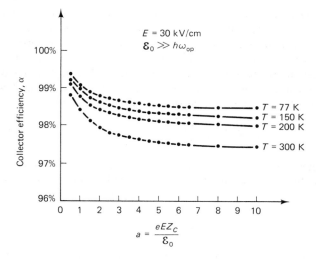

Figure 11.25 Computed collector efficiency as a function of the structure parameter $eEZ_C/\mathscr{E}_0$, where $\mathscr{E}_0$ is initial energy of electrons and Z_C is the width of the collector barrier region, having a value of 1000 Å for the structure of Fig. 11.23. In the computation, only the effect of back scattering was considered. (D. Vakhshoori and S. Wang, "Calculation of Base-Collector Efficiency for Hot-Electron Devices," *IEEE Electron Device Lett.*, Vol. EDL-7, p. 262, 1986 © 1986 IEEE.)

vices based on transport of ballistic electrons (C. R. Crowell and S. M. Sze, *Solid-State Electron.*, Vol. 8, p. 979, 1965).

11.9 QUANTUM WELLS AND SUPERLATTICES

In the discussion presented so far in this chapter, our attention has focused on the heterostructure interface. The property of a two-dimensional electron (or hole) gas is important to devices, such as MODFET or HEMT (high electron mobility transistor), in which electric current is conducted in a conduction channel parallel to the interface. Such a two-dimensional conduction channel can also be used in the base region of a HBT (heterojunction bipolar transistor) to lower the base resistance. In this section we discuss properties of heterostructures important to carrier transport in the direction perpendicular to the heterojunction. Figure 11.26 shows schematically two novel structures made of heterojunctions: (a) an isolated quantum well (QW), and (b) a one-dimensional periodic potential structure known as superlattice (SL). The term *superlattice* generally refers to structures where the barrier layer, the (GaAl)As layer in Fig. 11.26b, is thin, so there is appreciable interaction between the wave functions in neighboring

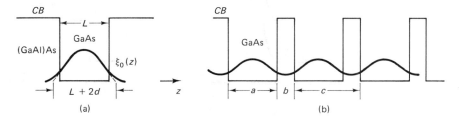

Figure 11.26 Structures for (a) a GaAs quantum well sandwiched between two (GaAl)As barrier layers and (b) a superlattice made of alternate GaAs well layers and (GaAl)As barrier layers.

wells. The term *multiple quantum wells* (MQW), on the other hand, refers to structures where the wells are isolated from each other by a relatively thick barrier region.

First let us start with quantum wells. For carriers confined in one dimension, the fundamental dispersion ($\mathscr{E}$ versus k) relation must be modified. One such relation applicable to the potential well in Si-MOSFET is given by Eq. (11.4.5). For a GaAs–(GaAl)As quantum well, the electron wave function spreads partly to the barrier layers. Therefore, the dispersion relation for conduction-band electrons is given by

$$\mathscr{E} = \mathscr{E}_i + \frac{\hbar^2 k_\perp^2}{2m_e^*} \tag{11.9.1}$$

where $\mathscr{E}_i$ is the confinement energy of the ith subband of a flat well. If ζ_i is the normalized electron wave function, the probability P_b of finding the electron in the barrier is

$$P_b = \int_b |\zeta_i(z)|^2 \, dz \tag{11.9.2}$$

where the subscript b means integration over the barrier layers. In terms of P_b, the effective mass m_e^* is

$$\frac{1}{m_e^*} = \frac{1}{m_{ew}^*}(1 - P_b) + \frac{1}{m_{eb}^*} P_b \tag{11.9.3}$$

where the subscripts w and b refer to the well and barrier layer, respectively. Complication may arise if the composition of the (GaAl)As is in the indirect-gap region. In that case, the value of m_{eb}^* in Eq. (11.9.3), similar to $m_\perp$ in Eq. (11.4.5), is equal to $(m_x m_y)^{1/2}$ with m_x and m_y given in Table 11.3 for the Si-like X minima. Figure 11.27 shows schematically (a) the dispersion relation of Eq. (11.9.1) as a function of $k_\perp = \sqrt{k_x^2 + k_y^2}$, and (b) the two-dimensional density of states $D_2(\mathscr{E}_\perp)$ as a function of $\mathscr{E}_\perp$. For $\mathscr{E} < \mathscr{E}_1$, there is only one subband and $D_2 = m_e^*/\pi\hbar^2$. In the energy region $\mathscr{E}_2 > \mathscr{E} > \mathscr{E}_1$, there are two subbands and $D_2 = 2m_e^*/\pi\hbar^2$. Therefore, $D_2(\mathscr{E}_\perp)$ takes the form of a staircase step function. However, for higher-energy subbands, the electron wave function spreads farther into the barrier region, resulting in a different m_e^* for each step.

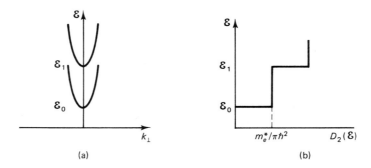

(a)   (b)

Figure 11.27 (a) The energy of an electron in a quantum well as a function of $k_\perp$ in the transverse plane and (b) the density of state function $D_2(\mathscr{E})$ as a function of electron energy for a two-dimensional electron gas.

The case for the valence band is complicated by spin-orbit coupling and by the need to include terms such as $k_x^2 k_y^2$ in a second-order perturbation calculation. In a bulk GaAs, the heavy-hole and light-hole bands at $\mathbf{k} = 0$ are characterized by a total angular momentum quantum number $j = \frac{3}{2}$ with the total magnetic quantum number $m_j = \pm\frac{3}{2}$ for the heavy-hole band and $m_j = \pm\frac{1}{2}$ for the light-hole band. In a one-dimensional confinement structure, however, there is considerable mixing of the four eigenstates for $k_\perp \neq 0$. Figure 11.28 shows the dispersion relation as a function of $k_\perp$ near $\mathbf{k} = 0$ in a GaAs–Ga$_{0.7}$Al$_{0.3}$As quantum well for two well sizes: (a) $L_z = 100$ Å and (b) $L_z = 150$ Å, as reported by Bastard and Brum (G. Bastard and J. A. Brum, *IEEE J. Quantum Electron.*, Vol. QE-22, p. 1625, 1986). The dashed curves are obtained when only diagonal elements are included in the perturbation calculation while the solid curves are obtained when off-diagonal elements are included. Calculation of such in-plane (k_x and k_y) dispersion diagram requires information about band parameter for both the well and barrier materials. Therefore, the dispersion diagram depends not only on the Al content x in the barrier layer but also on the well size L. The reader is referred to the article by Bastard and Brum for details of the dispersion-relation calculations as well as for references to such calculations, including the work by Altarelli and coworkers, by Ando, and by Broido and Sham (U. Ekenberg and M. Altarelli, *Phys. Rev.*, Vol. B30, p. 3569, 1984; T. Ando, *J. Phys. Soc. Jpn.*, Vol. 54, p. 1528, 1985; D. A. Broido and L. J. Sham, *Phys. Rev.*, Vol. B31, p. 888, 1985).

Before we leave the subject of quantum well, the following observations regarding Fig. 11.28 are worth noting. First, the barrier ψ_B at the GaAs–(GaAl)As interface in Fig. 11.26a is 170 meV calculated from Eq. (11.3.14) for $x = 0.3$. The penetration depth d of the electron wave function into (GaAl)As as estimated from $\hbar/(2m_{hh}^* \psi_B)^{1/2}$ using m_{hh}^* from Table 11.2 is $d = 6$ Å. Therefore, the confinement ener-

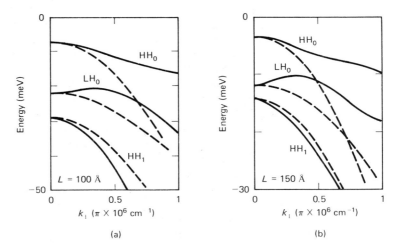

(a)   (b)

Figure 11.28 Dispersion relation, energy versus $k_\perp$ relation, computed for valence-band holes in a GaAs–Ga$_{0.7}$Al$_{0.3}$As quantum well for two well dimensions: (a) $L_z = 100$ Å and (b) $L_z = 150$ Å. (G. Bastard and J. A. Brum, "Electronic States in Semiconductor Heterostructures," *IEEE J. Quantum Electron.*, Vol. QE-22, p. 1625, 1986 © 1986 IEEE.) The symbols HH and LH refer to heavy hole and light hole, respectively, with the subscript 0 or 1 indicating the subband.

gies $\mathscr{E}_{0a}$ and $\mathscr{E}_{0b}$ for HH_1 are expected to be in the ratio $(L_b + 2d)^2/(L_a + 2d)^2$ for cases a and b, respectively. This ratio is about $2:1$, as compared to the ratio 7.5 meV/3.5 meV from Fig. 11.28. Second, neglecting the dependence of d on subband energy, the energies $\mathscr{E}_1$ for HH_1 and $\mathscr{E}_0$ for HH_0 are expected to be in the ratio $2^2:1$. The corresponding ratios are $29/7.5$ from Fig. 11.28a and $14/3.5$ from Fig. 11.28b. As long as $L > 3d$, the expression

$$\mathscr{E}_i = \frac{\hbar^2}{2m_{hz}^*}\left(\frac{\pi}{L + 2d}\right)^2 (i + 1)^2 \qquad \text{with} \quad i = 0, 1, 2 \qquad (11.9.4)$$

is a good approximation for the confinement energy $\mathscr{E}_i$. The values of m^* are given in Table 11.2. Using $m_{hh}^* = 0.45m_0$ for heavy holes (J. S. Blakemore, *J. Appl. Phys.*, Vol. 53, p. R165, October 1982), we find $\mathscr{E}_0 = 7.0$ meV for $L = 100$ Å as compared to a value of 7.5 meV from Fig. 11.28a. For light holes with $m_{lh}^* = 0.093m_0$ in (GaAl)As with $x = 0.3$, the penetration depth is estimated to be about 16 Å. Thus, using $m_{lh}^* = 0.087m_0$ in Eq. (11.9.4), we find $\mathscr{E}_0 = 25$ meV as compared to 22 meV from Fig. 11.28a. Third, analogous to Eq. (11.9.1), the dispersion relation for valence-band holes is

$$\mathscr{E} = \mathscr{E}_i + \frac{\hbar^2 k_\perp^2}{2m_{h\perp}^*} \qquad (11.9.5)$$

Different effective masses m_{hz}^* and $m_{h\perp}^*$ are used in Eqs. (11.9.4) and (11.9.5), respectively. For m_{hz}^*, the values of m_{hh}^* and m_{lh}^* given in Table 11.2 appear to yield reasonable values for $\mathscr{E}_i$. However, the values of $m_{h\perp}^*$ do not agree at all with those given in Table 11.2. Therefore, additional comments are in order.

Note from Fig. 11.28 that even for the diagonal approximation (dashed curves), the heavy-hole subbands having a larger curvature exhibit lighter effective masses, whereas the light-hole subbands having a smaller curvature exhibit heavier effective masses. This phenomenon is known as mass reversal effect. When off-diagonal elements are included, the hole wave function for $k_\perp \neq 0$ contains an admixture of heavy- and light-hole wave functions at $k_\perp = 0$. As a result, the dispersion relation (the solid curves in Fig. 11.28) becomes very sensitive to k_z and hence to well thickness L and barrier height ψ_B. This means that the dispersion curve must be computed for each specific quantum-well structure. More important, the dispersion relation can no longer be approximated by a parabolic subband as given in Eq. (11.9.5). Consequently, the mass $m_\perp$ in Eq. (11.4.6) for the two-dimensional density of states and the mass m^* in Eq. (11.6.5) for the areal carrier density become ill defined. Therefore, an alternative way must be found in determining the hole population in a quantum well instead of Eq. (11.6.5).

Let us now turn to superlattices. Consider a one-dimension periodic structure shown in Fig. 11.26b. This dispersion relation in the z direction, that is, the direction of superlattice, is given by a Kronig–Penney-like relation,

$$\tfrac{1}{2}(\gamma - \gamma^{-1})\sin \beta a \sinh \alpha b + \cos \beta a \cosh \alpha b = \cos qc \qquad (11.9.6)$$

where $c = a + b$ is the period, q the superlattice wave vector ($-\pi/c < q < \pi/c$),

$$\beta = \frac{(2m_a^* \mathscr{E})^{1/2}}{\hbar}, \qquad \alpha = \frac{[2m_b^*(\psi_B - \mathscr{E})]^{1/2}}{\hbar} \qquad (11.9.7)$$

Properties of Heterostructures   Chap. 11

ψ_B the barrier height, and m_a^* and m_b^* the effective mass in the well and barrier region, respectively. Since only the effective mass in the z direction is involved in Eq. (11.9.6), the values given in Table 11.2 can be used for both electrons and holes. One important difference between Eq. (11.9.6) and Eq. (5.2.19) is the expression for γ, which is equal to α/β in the Kronig–Penney relation. Because of different effective masses in the well and barrier region, charge and flux (velocity times charge density) conservation at the boundary requires continuity of ψ and $(m^*)^{-1} d\psi/dz$. Thus the quantity γ is given by

$$\gamma = \frac{\alpha m_a^*}{\beta m_b^*} \qquad (11.9.8)$$

When the product αb is much greater than 1, the wells become isolated from each other. Mathematically, letting $\sinh \alpha b$ and $\cosh \alpha b$ approach infinity, we can approximate Eq. (11.9.6) by $\tan \beta a = 2/(\gamma^{-1} - \gamma)$, or

$$\tan \frac{\beta a}{2} = \gamma \quad \text{and} \quad \cot \frac{\beta a}{2} = -\gamma \qquad (11.9.9)$$

for the even and odd solutions, respectively. Equation (11.9.4) is obtained if we let $\gamma = \infty$ and $a = L + 2d$ in Eq. (11.9.9). For moderate αb, coupling between neighboring wells broadens the allowed value for $\mathscr{E}_i$ and a band for $\mathscr{E}_i$ forms. For a superlattice of $2p$ periods, there are $2p$ discrete values for $\mathscr{E}_i$, with each level having the following value for q:

$$q = \frac{j}{p} \frac{\pi}{c}, \qquad j = -p, -(p-1), \ldots, p \qquad (11.9.10)$$

Figure 11.29 shows the energy-band structure of a one-dimensional InAs-GaAs superlattice, with E, LH, and HH denoting electron, light hole, and heavy hole, respectively. Because of a smaller effective mass, interaction between electrons or light holes in neighboring wells is significant and hence $\mathscr{E}_i$ broadens at a larger period for E and LH than for HH (G. Bastard and J. A. Brum, *IEEE J. Quantum Electron.*, Vol. QE-22, p. 1625, 1986).

Research in semiconductor superlattices was initiated in 1969 in a paper by Esaki and Tsu (L. Esaki and R. Tsu, *IBM J. Res. Dev.*, p. 61, January 1970). Although the original proposal was to achieve a negative differential conductivity in semiconductors, the subject of superlattice has since generated a great deal of activities scientifically to gain a better understanding of these artificial semiconductor structures as well as technologically to invent new devices to meet engineering needs. The list of quantum-well and superlattice devices is too extensive to give an adequate coverage of each device here. Instead, selected devices will be presented when specific applications are discussed. For example, resonant tunneling device will be presented in Section 13.3 and superlattice avalanche device in Section 14.12. Superlattices of II–VI compounds are discussed in Section 11.11. To conclude this section, we describe briefly the doping superlattices as originally proposed by Döhler (G. H. Döhler, *Phys. Status Solidi*, Vol. B52, p. 79, 1972; *IEEE J. Quantum Electron.*, Vol. QE-22, p. 1682, 1986). Figure 11.30 shows the energy-band diagram of a *nipi* superlattice (a) under thermal equilibrium and (b) under an applied field. The structure consists of alternating thin p and n layers separated regularly by thin i layers. The periodic p-n-p structures establish quan-

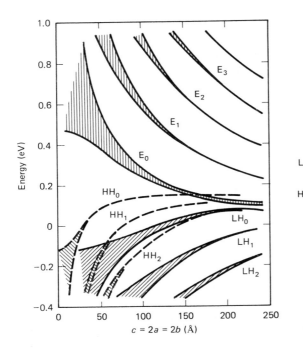

Figure 11.29 Broadening of energy levels $\mathcal{E}_i$ in an isolated quantum well into bands in a superlattice with a period of $c = a + b$ and equal well and layer thickness $a = b$. The computed results are for the conduction-band electrons designated by E and the valence-band light and heavy holes designed by LH and HH in a InAs–GaAs superlattice. (G. Bastard and J. A. Brum, "Electronic States in Semiconductor Heterostructures," *IEEE J. Quantum Electron.*, Vol. QE-22, p. 1625, 1986 © 1986 IEEE.)

tum wells in the *n* layers while the periodic *n-p-n* structures establish quantum wells in the *p* layers. The quantized level is indicated by $\mathcal{E}_{0e}$ for conduction-band electrons and by $\mathcal{E}_{0h}$ for valence-band holes. Note that under an applied electric field, the energy separation $\mathcal{E}_{0e} - \mathcal{E}_{0h}$ is blue-shifted for $\mathcal{E}^+$ on the left and red-shifted for $\mathcal{E}^-$ on the right. This effect can be used for the realization of tunable light sources and detectors.

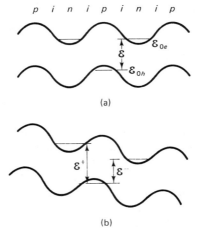

Figure 11.30 Energy-band diagrams of a dopant superlattice under (a) thermal equilibrium and (b) an applied bias. The periodic variations in $\mathcal{E}_c$ and $\mathcal{E}_v$ are due to the field created by alternate regions of ionized donors and acceptors.

Properties of Heterostructures   Chap. 11

11.10 SELECTED PROPERTIES OF GaAs AND (GaAl)As

Among III–V compounds, GaAs is most extensively studied for material properties and for device potentials. In this section we present some selected properties of GaAs and (GaAl)As. As our discussion in Section 11.9 indicates, one important band parameter is the effective mass. The wave functions at $\mathbf{k} = 0$ are known to be s-like for the conduction band and to consist of an admixture of p_x, p_y, and p_z orbitals for the valence band. Using the $s\alpha$, $p_+\beta$, $p_z\alpha$, $p_-\beta$, $s\beta$, $p_-\alpha$, $p_z\beta$, $p_+\alpha$, and $s\alpha$ wave functions as the base vectors and applying $\mathbf{k} \cdot \mathbf{p}$ perturbation calculation to the Schrödinger equation, (Problem 12.21), we can reduce the 8×8 matrix into two identical 4×4 matrices which lead to the Kane dispersion relations:

$$(\mathscr{E}' - \mathscr{E}_s)\left(\mathscr{E}' - \mathscr{E}_p - \frac{\Delta}{3}\right)\left(\mathscr{E}' - \mathscr{E}_p + \frac{2\Delta}{3}\right)$$

$$= \frac{\hbar^2 k^2 P^2}{m_0^2}\left(\mathscr{E}' - \mathscr{E}_p + \frac{\Delta}{3}\right) \tag{11.10.1}$$

$$\mathscr{E}' = \mathscr{E}_p + \frac{\Delta}{3} \tag{11.10.2}$$

where α and β represent the two spin wave functions with $s_z = +\frac{1}{2}$ and $s_z = -\frac{1}{2}$, and $p_\pm = (p_x \pm ip_y)/\sqrt{2}$ (E. O. Kane, *J. Phys. Chem. Solids*, Vol. 1, p. 249, 1957). The quantity P defined as

$$P = -\frac{\hbar}{i}\int p_j \frac{\partial}{\partial j} s \, d\tau = \frac{\hbar}{i}\int s \frac{\partial}{\partial j} p_j \, d\tau \tag{11.10.3}$$

is the momentum matrix element connecting the s and p wave functions with $j = x$, y, or z and $d\tau = dx \, dy \, dz$. The energy $\mathscr{E}' = \mathscr{E} - \hbar^2 k^2/2m_0$ is the eigenvalue for conduction- and valence-band electrons.

For conduction-band electrons, we can rearrange Eq. (11.10.1) in the form of $\mathscr{E}' - \mathscr{E}_s = k^2 P^2 f(\mathscr{E}')$ and treat the term $k^2 P^2 f(\mathscr{E}')$ as perturbation by setting $f(\mathscr{E}') = f(\mathscr{E}_s)$. Thus we obtain

$$\mathscr{E} = \mathscr{E}_c + \frac{\hbar^2 k^2}{2m_0} + \frac{\hbar^2 k^2 P^2}{3m_0^2}\left(\frac{2}{\mathscr{E}_g} + \frac{1}{\mathscr{E}_g + \Delta}\right) \tag{11.10.4}$$

where we let $\mathscr{E}_s = \mathscr{E}_c$, $\mathscr{E}_p = \mathscr{E}_v - \Delta/3$, and $\mathscr{E}_s = \mathscr{E}_c - \mathscr{E}_v$. For valence-band holes, Eq. (11.10.2) leads to

$$\mathscr{E}_{hh} = \mathscr{E}_v + \frac{\hbar^2 k^2}{2m_0} \tag{11.10.5}$$

with hh indicating heavy holes. Had $k^2 P^2$ been zero, the middle factor in Eq. (11.10.1) would give the energy expression for light holes. Therefore, we treat this factor in a manner similar to that applied to the factor $\mathscr{E}' - \mathscr{E}_s$. Thus we find

$$\mathscr{E}_{lh} = \mathscr{E}_v + \frac{\hbar^2 k^2}{2m_0} - \frac{2\hbar^2 k^2 P^2}{3\mathscr{E}_g m_0^2} \tag{11.10.6}$$

for light holes. Similarly, we have

$$\mathscr{E}_{so} = \mathscr{E}_v - \Delta + \frac{\hbar^2 k^2}{2m_0} - \frac{\hbar^2 k^2 P^2}{3(\mathscr{E}_g + \Delta)m_0^2} \qquad (11.10.7)$$

for holes in the split-off valence band. The quantity Δ introduced in Eqs. (11.10.1) and (11.10.2) represents the energy split between valence bands caused by spin-orbit coupling, and $\Delta = 0.34$ eV in GaAs.

From the energy expressions, we note that the effective masses for electrons, light holes, and split-off holes are all related through $k^2 P^2$ by the relation

$$\frac{1}{m_e^*} - \frac{1}{m_0} = \frac{1}{m_{lh}^*} + \frac{1}{m_{so}^*} - \frac{2}{m_0} \qquad (11.10.8)$$

where m_0 is the free-electron mass. Use of effective masses from Table 11.2 in Eq. (11.10.8) shows that the relation is indeed satisfied to the first order. The discrepancy is due to second-order perturbation terms. As discussed in Section 12.11, the matrix element P^2 is also related to the probability of optical transition across the energy gap and hence to the optical absorption coefficient through Eqs. (12.11.23) and (12.11.25). Effective masses deduced from cyclotron-resonance and optical absorption measurements are consistent with the values shown in Table 11.2. For heavy holes, however, there is no direct relation of its effective mass to those of other carriers in the first-order perturbation calculation. A second-order perturbation calculation by Dresselhaus et al. for valence bands in Si and Ge yields the following energy expressions (G. Dresselhaus, A. F. Kip, and C. Kittel, *Phys. Rev.*, Vol. 98, p. 368, 1955):

$$\mathscr{E}_{so} = \mathscr{E}_v - \Delta - \frac{\hbar^2}{2m_0} A k^2 \qquad (11.10.9)$$

for split-off band and

$$\mathscr{E} = \mathscr{E}_v - \frac{\hbar^2}{2m_0} [A k^2 \pm (B k^4 + C \sum k_x^2 k_y^2)^{1/2}] \qquad (11.10.10)$$

with + sign for heavy holes and − sign for light holes. Since GaAs has valence-band structures similar to those of Si and Ge, Eqs. (11.10.9) and (11.10.10) are also applicable to GaAs. From these equations we have

$$\frac{1}{m_{so}^*} = \frac{1}{2}\left(\frac{1}{m_{lh}^*} + \frac{1}{m_{hh}^*}\right) \qquad (11.10.11)$$

The values of effective masses given in Table 11.2 are in general agreement with this relation. The reader is referred to an excellent review article by Blakemore for discussions of the band structure near zone center and the experimental determination of effective masses in GaAs. The values of effective masses compiled by Blakemore are listed in Table 11.2 in parentheses (J. S. Blakemore, *J. Appl. Phys.*, Vol. 53, p. R123, October 1982). The value of $m_{hh}^* = 0.45 m_0$ used in connection with Eq. (11.9.4) was deduced by Vrehen from interband magnetoabsorption measurements (Q. H. F. Vrehen, *J. Phys. Chem. Solids*, Vol. 29, p. 129, 1968).

Another important property is carrier mobility. As discussed in Section 6.8, the important scattering mechanisms are polar optical scattering, lattice or acoustic-phonon scattering, impurity scattering, and piezoelectric scattering with the associated mobility indicated by μ_{po}, μ_I, μ_i, and μ_{pe}, respectively. Figure 11.31 shows the measured electron mobility (solid curves) as a function of temperature for three GaAs samples with different donor concentrations, $N_d = 5 \times 10^{13}$ cm^{-3}, 10^{15} cm^{-3}, and 5×10^{15} cm^{-3} from top to bottom, as reported by Stillman et al. (G. E. Stillman, C. M. Wolfe, and J. O. Dimmock, *J. Phys. Chem. Solids*, Vol. 31, p. 1199, 1970). Using Eq. (6.8.16) with $K = 0.0617$, $m_c^* = 0.067m_0$, and $\epsilon = 13\epsilon_0$, we find $\mu_{pe} = 3.1 \times 10^5$ cm^2/V-s at 300 K. The temperature variation of μ_{pe} is represented by the dashed curve labeled μ_{pe}. Using Eq. (6.7.22) with $\rho = 5.36 \times 10^3$ kg/m^3, average acoustic velocity $v_s = 5 \times 10^3$ m/s, and an acoustic deformation potential $\mathcal{E}_{c1} = 3.6$ eV, we find $\mu_I = 16$ m^2/V-s at 300 K. The temperature dependence of μ_I is represented by the dashed line labeled μ_I. Using Eq. (6.7.9) with an ionized impurity concentration $N_i = N_a + N_d = 1.4 \times 10^{15}$ cm^{-3}, we find $\mu_i = 8 \times 10^5$ cm^2/V-s at 300 K. Obviously, the three mobilities are much higher than the measured mobility at 300 K. In Table 11.5 we list some useful physical quantities in GaAs and AlAs selected from two review articles by Adachi and by Blakemore.

From Fig. 11.31 it is obvious that the electron mobility at room temperature is dominated by polar optical scattering. However, in impure samples, impurity scattering also plays a part even at 300 K. Figure 11.32 shows the variation of mobility as a function of carrier concentration for (a) *p*-GaAs and (b) *n*-GaAs based on measured data compiled by Blakemore. Because of the high degree of dopant compensation in GaAs, a family of curves are shown in Fig. 11.32 with different $(N_d + N_a)/(N_d - N_a)$ ratios. Impurity scattering becomes significant for doping concentration exceeding 10^{17} cm^{-3} for holes and 10^{16} cm^{-3} for electrons. Referring to Figs. 11.14 and 11.16, we see that the maximum areal carrier density that can be achieved in practice is about 8×10^{11} cm^{-2} and is confined in a region less than 100 Å in width. These values give an equivalent volume density of 8×10^{17} cm^{-3}. If the ionized donors were in the same region as the carriers, impurity scattering would reduce the carrier mobility by a substantial factor even at 300 K. The use of modulation doping to spatially separate carriers

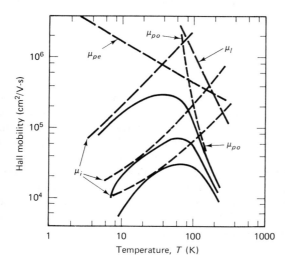

Figure 11.31 Comparison of measured Hall mobilities (solid curves) in GaAs samples having $N_d = 5 \times 10^{13}$ cm^{-3}, 10^{15} cm^{-3}, and 5×10^{15} cm^{-3} from top to bottom and the theoretical values of μ_{pe}, μ_{po}, μ_I, and μ_i as functions of temperature. (J. S. Blakemore, *J. Appl. Phys.*, Vol. 53, p. R123, October 1982.)

Sec. 11.10   Selected Properties of GaAs and (GaAl)As   **563**

TABLE 11.5 USEFUL QUANTITIES FOR GaAs AND AlAs (IN PARENTHESES)

	GaAs	AlAs
Static dielectric constant	$\epsilon_s = 13.18\epsilon_0$	(10.06)
Index of refraction squared	$n^2 = 10.89$	(8.16)
Electromechanical coupling constant	$K = 0.0617$	(0.094)
Acoustic deformation potential (eV)	$\mathcal{E}_{c1} = 3.6$ eV	(2.9)
Optical deformation potential (eV)	$\mathcal{E}_{c2} = 5.9$ eV	(5.9)
Intravalley deformation potential (eV)	$\mathcal{E}_{g1} = 6.8$ eV	(6.3)
Intervalley deformation field (eV/cm)	$E_{\Gamma x} = 1 \times 10^9$	a
Density (10^3 kg/cm^3)	$\sigma = 5.36$	(3.76)
Longitudinal acoustic velocity	$v_{100} = 4.73$	(5.6)
(10^5 cm/s)	$v_{110} = 5.24$	(6.2)
	$v_{111} = 5.40$	a
Phonon energy (meV) at selected	LO(Γ) = 36.1	LO(X) = (50.1)
reciprocal lattice points	TO(Γ) = 33.2	TO(X) = (44.9)
	LO(L) = 29.6	
	TO(L) = 32.4	
	LA(L) = 25.9	
	TA(L) = 7.7	

[a]Value not given.
Source: Data from S. Adachi, *J. Appl. Phys.*, Vol. 58, p. R1, August 1985, and J. S. Blakemore, *J. Appl. Phys.*, Vol. 53, p. R123, October 1982.

from ionized dopants as discussed in Section 11.7 will have the benefit for keeping μ high while achieving a high areal carrier density N_s.

The use of doped (GaAl)As to supply electrons to GaAs, however, is not without problems. Some of the problems have been discussed in Section 11.7 in connection with Figs. 11.16 and 11.17. Another important problem that has not been discussed concerns the behavior of donors in (GaAl)As. The three relevant energies are

$$\mathcal{E}_g^\Gamma (x) = 1.425 + 1.247x \qquad (0 < x < 0.45) \qquad (11.10.12)$$

$$\mathcal{E}_g^X (x) = 1.900 + 0.125x + 0.143x^2 \qquad (11.10.13)$$

$$\mathcal{E}_g^L (x) = 1.708 + 0.642x \qquad (11.10.14)$$

representing the energy difference between conduction-band minima at Γ, X, and L points and valence-band maximum at Γ point in reciprocal lattice space. In $Ga_{0.75}Al_{0.25}As$, the L minimum is about 1.868 eV above the valence band maximum, compared to a value of 1.737 eV for the Γ minimum. Using the conductivity effective mass, the donor ionization energy $\mathcal{E}_d$ is estimated from Eq. (6.4.1) to be 7 meV for Γ-like donors and 12 meV for L-like donors. Even though the calculation above shows that the L-like donors have energy above the Γ-like donors, DLTS (deep-level transient spectroscopy) measurements show deep traps whose density closely follows the Si doping concentration (S. Dhar, W. P. Hong, P. K. Bhattacharya, Y. Nishimoto, and F. Y. Juang, *IEEE Trans. Electron Devices*, Vol. ED-33, p. 698, 1986).

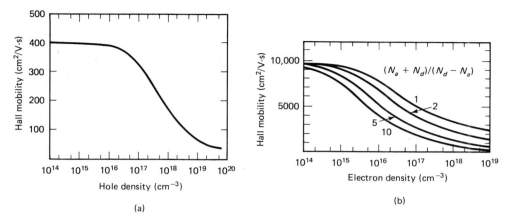

Figure 11.32 Variation of Hall mobility as a function of doping concentrations in (a) p-GaAs and (b) n-GaAs, as compiled by Blakemore. (J. S. Blakemore, *J. Appl. Phys.*, Vol. 53, p. R123, October 1982.)

Early observation of persistent photoconductivity in (GaAl)As has also been attributed to donor-related traps, known as D-X centers (D. V. Lang and R. A. Logan, *Phys. Rev. Lett.*, Vol. 39, p. 635, 1977). It has been suggested that the D-X center is made of a donor coupled to an As vacancy. However, recent pressure-dependent DLTS data indicate that the deep trap is simply a substitutional donor (M. Tachikawa, M. Mizuta, H. Kukimoto, and S. Minomura, *Jpn. J. Appl. Phys.*, Vol. 24, p. L821, 1985). Earlier pressure experiment also shows that a shallow donor evolves into a deep-level impurity as the pressure is increased to change a direct-gap into an indirect-gap (GaAl)As. The activation energy increases from a value about 100 meV at the L–Γ crossover point to a maximum value of about 170 meV at the X–Γ crossover point and then decreases again to about 90 meV in the X-minimum indirect-gap region ($x = 1$). If we use an activation energy of 140 meV, we find the L-like donor level now lying below the Γ-like donor level. Since the L minimum has a much larger density of state than the Γ minimum, it can be expected that the L-like donors can be a very effective electron trap. Even though the model above seems highly plausible, it is unclear why a Si-substitutional donor at the L minimum in GaAs has an activation energy much larger than an ordinary donor such as P in Ge, which has an L band minimum. This problem is not only theoretically interesting but also practically important because the deep-level behavior of donors will have an adverse effect on the performance of electronic and optical devices.

11.11 HETEROSTRUCTURES AND SUPERLATTICES OF DIFFERENT KINDS

In the discussions presented so far in this section we have concentrated on the (GaAl)As/GaAs system because it is a relatively simple system and because it is studied most extensively. However, the (GaAl)As/GaAs interface may not be representative of all heterointerfaces. Depending on the band-edge offsets and the band-gap energies, heterointerfaces can be divided into four different kinds: type I, type II–misaligned, type II–staggered, and type III, as illustrated in Fig. 11.33 (L. Esaki, *IEEE J. Quantum Electron.*, Vol. QE-22, p. 1611, 1986.) At a type I interface, the energy gap of one semiconductor, *A,* is completely within that of the other semiconductor, *B* (Fig. 11.33a). Therefore, potential barriers exist for both electrons and holes in going from the smaller gap semiconductor, *A,* to the larger gap semiconductor, *B.* Interfaces formed by III–V compounds either with different group III elements such as GaAs–AlAs or with different group V elements such as GaAs–GaP belong to type I. Obviously, only one type of two-dimensional carrier gas, either electron gas (Fig. 11.33a) or hole gas (not shown), can form at a type I interface—which carrier type depends on the doping profile in the two semiconductors.

Interfaces formed by III–V compounds with both different group III elements and different group V elements belong to type II. For example, the difference in the electron affinities of GaSb (4.03 eV) and InAs (4.90 eV) is so large that the two forbidden gaps are completely misaligned in energy (Fig. 11.33b), resulting in a type II–misaligned interface (L. Esaki, ''InAs-GaSb Superlattices—Synthesized Narrow-Gate Semiconductors and Semimetals,'' in W. Zawadzki, ed., *Narrow-Gap Semiconductors—Physics and Applications,* Lecture Notes in Physics, Vol. 133, Springer-Verlag, Berlin, 1980; L. L. Chang, ''Semiconductor-Semimetal Transitions in InAs-GaSb Superlattices,'' *Proc. 15th Int. Conf. Phys. Semiconduct.,* Kyoto, 1980; *J. Phys. Soc. Jpn.,* Vol. 49, Suppl. A., p. 997, 1980). Because the conduction bandedge $\mathcal{E}_{cA}$ of InAs lies below the valence bandedge $\mathcal{E}_{vB}$ of GaSb, electrons are attracted to InAs and holes to GaSb at the interface, thus forming a two-dimensional electron gas on the InAs side and a two-dimensional hole gas on the GaSb side. The term ''semimetal'' is sometimes used even though the two regions of higher $\mathcal{E}_v$ and lower $\mathcal{E}_c$ refer to the valence band and the conduction band of two different materials. In other words, the electron gas and the

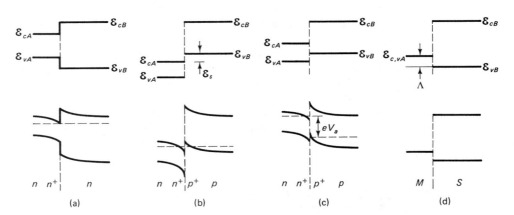

Figure 11.33 Schematic diagrams showing four representative types of heterostructure interfaces: (a) type I with $\mathcal{E}_{gA}$ completely within $\mathcal{E}_{gB}$, (b) type II–misaligned with the two energy gaps completely misaligned, (c) type II–staggered with overlapping energy gaps, and (d) type III with the overlapping bands in a semimetal lying within the energy gap of a semiconductor. (L. Esaki, ''A Bird's Eye View on the Evolution of Semiconductor Superlattices and Quantum Wells,'' *IEEE J. Quantum Electron.,* Vol. QE-22, p. 1611, 1986 © 1986 IEEE.)

hole gas are spatially separated. Therefore, the situation represented in Fig. 11.33b is somewhat different from that represented in Fig. 4.11d.

The type II–staggered interface results if we reduce the electron-affinity difference and raise the gap energy of the smaller-gap semiconductor by mixing group III elements or group V elements or both. Examples of type II–staggered interface (Fig. 11.33c) include the (InGa)As–Ga(SbAs) and (AlIn)As–InP systems. Referring to Fig. 11.1 and Table 11.1, we see that the lattice constant changes from 6.06 Å (for $x = 0$) to 5.65 Å (for $x = 1$) in $In_{1-x}Ga_xAs$ and from 6.10 Å (for $y = 0$) to 5.65 (for $y = 1$) in $GaSb_{1-y}As_y$. Therefore, if x and y are nearly equal, the two lattice constants can be matched. Also, $Al_xIn_{1-x}As$ with $x = 0.48$ is lattice-matched to InP. Experiments on n-(InGa)As–p-Ga(SbAs) junctions show a transition from p-n junction characteristic to ohmic characteristic, signaling the formation of a semimetallic interface, for $x < 0.16$ and $y < 0.10$ (H. Sakaki, L. L. Chang, R. Ludeke, C. A. Chang, G. A. Sai-Halasz, and L. Esaki, *Appl. Phys. Lett.*, Vol. 31, p. 211, 1977). Based on the electron-affinity and gap-energy values, the values of $\mathscr{E}_g = \mathscr{E}_{cA} - \mathscr{E}_{vB}$ are found to be 0.47, 0.33, 0.14, and -0.09 eV at the following alloy compositions (x, y): (0.62, 0.64), (0.52, 0.56), (0.50, 0.28), and (0.16, 0.10). Therefore, the observation of a semimetallic interface is in general agreement with the theoretical prediction. Note also from Fig. 11.33c that the energy separation between electrons in smaller-gap semiconductor A and holes in larger-gap semiconductor B is smaller than the gap energy of A or B. Therefore, in a type II–staggered p-n junction, the emission spectrum is expected to be below the gap of either semiconductor. Such is indeed the case for a p-$Al_{0.48}In_{0.52}As$ ($\mathscr{E}_g \sim 1.55$ eV) n-InP ($\mathscr{E}_g = 1.45$ eV) junction with emission around 0.96 eV at $T = 1.4$ K (E. J. Caine, S. Subbanna, H. Kroemer, J. L. Merz, and A. Y. Cho, *Appl. Phys. Lett.*, Vol. 45, p. 1123, 1984).

The II–VI compound semiconductors constitute an important class of material for applications in the infrared (Hg compounds with S, Se, or Te) and visible (Cd or Zn compounds with S, Se, or Te) regions. The properties of high-gap compounds are discussed in Chapter 12, especially in Section 12.14. In this section we concentrate our discussion on infrared II–VI compounds. Figure 11.34 shows the energy-band structure of bulk HgTe and CdTe. The symbols E, HH, and LH refer to electrons (conduction band), heavy holes and light holes, respectively (Y. Guldner, G. Bastard, and M. Voos, *J. Appl. Phys.*, Vol. 57, p. 1403, 1985). Referring to the band structure of GaAs shown in Fig. 11.2, we see that the band which has the Γ_6 wave function is the topmost band and thus becomes the conduction band, while the two degenerate bands with Γ_8 wave functions are heavy-hole and light-hole valence bands. This situation applies to all the direct-gap III–V compounds and the high-gap II–VI compounds. In the low-gap II–VI compounds, however, the Γ_6 and one of the Γ_8 bands are inverted not only in curvature but also in energy. As a result, the Γ_8 light-hole band in CdTe becomes the Γ_8 conduction band in HgTe (J. P. Faurie, *IEEE J. Quantum Electron.*, Vol. QE-22, p. 1656, 1986). Since the conduction Γ_8^E band and the heavy-hole Γ_8^{HH} are degenerate, HgTe behaves like a semimetal. A type III interface (Fig. 11.33d) forms when one constituent is semimetallic with the degenerate bands lying in the forbidden gap of the normal semiconductor.

Superlattices with type III interfaces offer interesting optical and transport properties (Y. C. Chang, J. N. Schulman, G. Bastard, Y. Guldner, and M. Voos, *Phys. Rev.*, Vol. B31, p. 2557, 1985). As discussed in Section 11.9, information about subbands can be found from a Kronig–Penney-like dispersion relation described by Eq.

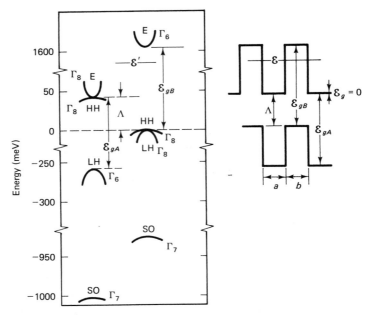

Figure 11.34 Energy-band structure of HgTe (on the left) and CdTe (on the right). Note that the Γ_6 conduction band (E) and the Γ_8 light-hole valence band (LH) in large-gap CdTe are inverted to become LH and E band, respectively, in small-gap HgTe. (J. P. Faurie, "Growth and Properties of HgTe-CdTe and Other Hg-Based Superlattices," *IEEE J. Quantum Electron.*, Vol. QE-22, p. 1656, 1986 © 1986 IEEE.) The inset shows the energy-band diagram of HgTe–CdTe superlattice.

(11.9.6) if the effective masses m_a^* in the well and m_b^* in the barrier are known. However, the practical situation as represented in Fig. 11.34 is more complicated than the idealized Kronig–Penney model. Since there are three bands, it is not clear how the proper effective mass can be found. In the treatment by Guldner et al. (Y. Guldner, G. Bastard, and M. Voos, *J. Appl. Phys.*, Vol. 57, p. 1403, 1985), the heavy-hole bands are first disregarded. Further, in following the Kane $\mathbf{k} \cdot \mathbf{p}$ calculation, the split-off bands (with $p_+\beta$ and $p_-\alpha$ wave functions) are ignored. Thus the dispersion relation of Eq. (11.10.1) is replaced by (see Problem 11.33)

$$(\mathscr{E}' - \mathscr{E}_s)\left(\mathscr{E}' - \mathscr{E}_p - \frac{\Delta}{3}\right) = \frac{2}{3}\frac{\hbar^2 k^2 P^2}{m_0^2} \tag{11.11.1}$$

Using the top of the Γ_8^{HH} band in CdTe as reference energy and applying Eq. (11.11.1) separately to HgTe and CdTe, we find that

$$(\mathscr{E}' - \Lambda)(\mathscr{E}' - \Lambda + \mathscr{E}_{gA}) = \frac{2}{3}\frac{\hbar^2 k_A^2 P_A^2}{m_0^2} \tag{11.11.2}$$

for HgTe and

$$\mathscr{E}'(\mathscr{E}' - \mathscr{E}_{gB}) = \frac{2}{3}\frac{\hbar^2 k_B^2 P_B^2}{m_0^2} \tag{11.11.3}$$

for CdTe where $\mathscr{E}_{gA}$ and $\mathscr{E}_{gB}$ are the interaction gap energies $|\mathscr{E}(\Gamma_6) - \mathscr{E}(\Gamma_8)|$ in bulk HgTe and CdTe, respectively, and $\Lambda = \mathscr{E}_A(\Gamma_8) - \mathscr{E}_B(\Gamma_8)$ is the valence-band offset as defined in Fig. 11.34. Furthermore, as in Eq. (11.10.1), we let $\mathscr{E}_c = \mathscr{E}_s$ and $\mathscr{E}_v = \mathscr{E}_p + \Delta/3$.

The energy $\mathscr{E}'$ in Eq. (11.11.1) is defined as

$$\mathscr{E}' = \mathscr{E}_A(\Gamma_6) + \frac{\hbar^2 k_A^2}{2m_a^*} = \mathscr{E}_B(\Gamma_6) + \frac{\hbar^2 k_B^2}{2m_b^*} \qquad (11.11.4)$$

Eliminating m_a^* and m_b^* from Eq. (11.9.8) and realizing that $\mathscr{E}_A(\Gamma_6) = \Lambda - \mathscr{E}_{gA}$ and $\mathscr{E}_B(\Gamma_6) = \mathscr{E}_{gB}$ from Fig. 11.34, we obtain

$$\gamma = \frac{-\beta}{\alpha} \frac{\mathscr{E}' - \mathscr{E}_{gB}}{\mathscr{E}' + \mathscr{E}_{gA} - \Lambda} \qquad (11.11.5)$$

where $k_A = \beta$ and $k_B = i\alpha$. If the matrix elements P_A and P_B defined by Eq. (11.10.3) are known in HgTe and CdTe, then for a given energy $\mathscr{E}'$ the values of k_A and k_B and hence those of α and β can be calculated from Eqs. (11.11.2) and (11.11.3). Substituting the values of α and β thus found in Eq. (11.9.6), we can calculate from Eq. (11.9.6) the allowed values of $\mathscr{E}'$ for which the value of $|\cos\ qc|$ is smaller than or equal to unity. Figure 11.35 shows the position and width of the various subbands (conduction, heavy-hole, and light-hole subbands indicated by E, HH, and LH, respectively) at $T = 4$ K as a function of layer thickness for HgTe–CdTe superlattices with $a = b$ in Eq. (11.9.6) as reported by Guldner et al. The matrix element P is calculated by using $m_c^* = 0.03m_0$ in Eq. (11.10.4) for both HgTe and CdTe. The heavy-hole subband is obtained by assuming equal mass m_{hh}^* in HgTe and CdTe. Hence $\gamma = \alpha/\beta$ is used in Eq. (11.9.8).

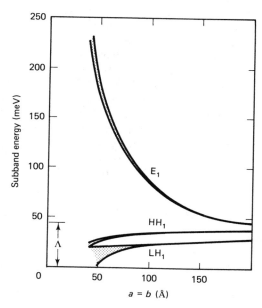

Figure 11.35 Computed energy subbands (E for electrons, LH for light holes, and HH for heavy holes) in a HgTe–CdTe superlattice. (Y. Guldner, G. Bastard, and M. Voos, *J. Appl. Phys.*, Vol. 57, p. 1403, 1985.)

The HgTe/CdTe superlattice was originally proposed by Schulman and McGill as a new infrared material to substitute for the $Hg_{1-x}Cd_xTe$ random alloy (J. N. Schulman and T. C. McGill, *Appl. Phys. Lett.*, Vol. 34, p. 663, 1979). The ability to control the subband energy gap in a superlattice by the well width a is considered an advantage of the superlattice over the random alloy. Experimental measurements of infrared absorption and photoluminescence spectra indicate subband energy gaps in general agreement with calculated values. The electron mobility reported to be in the range 10^4 to 10^5 cm^2/V-s is comparable to the value obtained in good-quality (HgCd)Te alloys. For an analysis of optical and transport properties and for additional references, the reader is referred to an excellent review article by Faurie (J. P. Faurie, *IEEE J. Quantum Electron.*, Vol. QE-22, p. 1656, 1986). Since research on the type III superlattices is still in its initial stage, new information about the subbands is expected from both experiments and calculations. Even so, some general remarks about Fig. 11.34 can be made. As $a = b$ approaches zero, the energy separation between E_1 and HH_1 tends toward the band gap of $Hg_{0.5}Cd_{0.5}Te$, which is 564 meV. As $a = b$ becomes infinite, an interface state forms from evanescent waves in both HgTe and CdTe. This interface E_1 state has an energy $\Lambda - 6$ meV at 4 K, while the energy of the HH_1 state tends to $\Lambda = 40$ meV.

In summary, we have described here three additional kinds of superlattices different from the type I superlattice of which (GaAl)As/GaAs is a representative. The type II–misaligned superlattice has two-dimensional electron and hole gas at the opposite sides of the interface. Therefore, it may serve as an ohmic contact from a small-gap semiconductor to a large-gap semiconductor. The type II–staggered superlattice offers, in principle, the possibility of photon emission with energy smaller than the gap energy of either material. However, the scheme may prove inefficient on two counts. First, the potential barrier is relatively small for electron and hole diffusion away from the interface. Therefore, under a forward bias V_a (Fig. 11.33c), there is an appreciable diffusion current in addition to the recombination current. Second, the overlap between the electron and hole wave functions is much reduced by their spatial separation. Both problems are common to type II–staggered and doping (*nipi*) superlattices, which may limit their potential as a tunable light source. The type III superlattice of which HgTe/CdTe is an example looks promising as an infrared material. However, much research is needed not only on device physics but also on material problems. A potential problem is the possible interdiffusion at the interface. A superlattice of 2 μm thickness requires approximately a growth run of 2 h by molecular beam epitaxy. With a diffusion coefficient of 3×10^{-18} cm^2/s at the growth temperature $T = 185°C$, a 2-h growth run could result in an interdiffusion region of about 10 Å on each side of the interface. For further discussions of (HgCd)Te heterostructures, especially its epitaxial growth and its diode property, the reader is referred to an excellent review by Herman and Pessa (M. A. Herman and M. Pessa, *J. Appl. Phys.*, Vol. 57, p. 2671, 1985).

PROBLEMS

11.1. Find the relation between x and y in the quaternary compound $Ga_xIn_{1-x}As_{1-y}P_y$ so that it is lattice-matched to InP. Then verify Eq. (11.1.2) if a linear dependence of energy gap on composition is assumed. Finally, find the composition parameters x and y for a quaternary compound that emits $\lambda = 1.30μm$ radiation. Check the values of x and y thus found against those obtained from Fig. 11.3. Explain the origin of the discrepancy.

11.2. Using the data given in Table 11.1, show that the ternary compound $Ga_{1-z}In_zP$ with $z = 0.484$ is lattice-matched to GaAs. For the quaternary compound $Ga_xIn_{1-x}P_yAs_{1-y}$, find the relation between x and y so that it is lattice-matched to GaAs. Decomposing the quaternary compound into $(GaAs)_{1-y}(InP)_{1-x}(GaP)_z$, express the lattice constant and energy gap of the quaternary compound in terms of the lattice constants and energy gaps of the constituent binary compounds. Show that the quaternary compound with $x = 0.613$ and $y = 0.80$ is lattice-matched to GaAs and find the value of the energy gap for the compound. Check the values thus calculated against those obtained from Fig. 11.3. Explain the origin of any discrepancy.

11.3. **(a)** The direct energy gap $\mathscr{E}_{g\Gamma}$ in $Al_xGa_{1-x}As$ varies as

$$\mathscr{E}_{g\Gamma} = 1.424 + 1.247x \qquad \text{for} \quad x < 0.45$$

and the two indirect energy gaps vary as

$$\mathscr{E}_{gX} = 1.900 + 0.125x + 0.143x^2$$
$$\mathscr{E}_{gL} = 1.708 + 0.642x$$

where the indices Γ, X, and L refer to the Γ, X, and L points in k space (Fig. 11.2). Plot the variation of the various energy gaps as functions of x, and compare their values at $x = 0.35$.

   (b) Using linear interpolation, find the density-of-state effective masses for electrons in the various conduction-band minima, and the percentages of electrons in these minima in a nondegenerate $Al_{0.35}Ga_{0.65}As$ sample at 300 K.

11.4. Repeat Problem 11.3 for $x = 0.25$. Also calculate the electron concentrations in the X and L minima for a degenerate sample with $n_\Gamma = 2 \times 10^{18}$ cm^{-3} in the central Γ valley.

11.5. Starting from Poisson's equation and using Gauss's law, derive Eq. (11.3.7) and verify Eq. (11.3.8). Make proper modification of Eq. (11.3.10) so that Eq. (11.3.8) is applicable to the case $V_G = 0$.

11.6. The measured capacitance C/A (per unit area) of an AlAs/GaAs symmetrical MOD structure (Fig. 11.6) with $N_a = 1 \times 10^{16}$ cm^{-3} and $w_l = 500$ Å has the following values at 77 K: (1) 1.04×10^{-7} F/cm^2 at $V_G = 0$, (2) 6.20×10^{-8} F/cm^2 at $V_G = 0.2$ V and (3) 4.17×10^{-8} F/cm^2 at $V_G = 0.5$ V. The capacitance measurement is necessary to verify that impurity concentration is negligible in the undoped AlAs region. Given $\epsilon_s(GaAs) = 13.1\epsilon_0$ and $\epsilon_l(AlAs) = 10.1\epsilon_0$, calculate the values of C/A, and compare them with the measured values. Explain why Eq. (11.3.10) needs to be modified for the $V_G = 0$ case and estimate the upper limit for N_l in the AlAs layer.

11.7. The current density of an $Al_{0.36}Ga_{0.64}As/GaAs$ symmetrical MOD structure (Fig. 11.6) with $N_a = 1.8 \times 10^{16}$ cm^{-3} measured at two bias voltages has the following values at various temperatures:

T (K)		79.6	98.1	121.4	155
J (A/cm^2)	($V_G = 17$ mV)	3×10^{-9}	4×10^{-7}	4.5×10^{-5}	2.5×10^{-3}
	($V_G = 150$ mV)	4×10^{-7}	4×10^{-5}	2×10^{-3}	6×10^{-2}

Calculate the value of $\Delta\mathscr{E}_v$ from an Arrhenius plot of the data obtained with $V_G = 17$ mV. Comment on the contribution from ψ_G in Eq. (11.3.13).

11.8. Repeat Problem 11.7 for the $V_G = 150$ mV case. Explain possible sources of error for measurements at high V_G.

11.9. The airy function is given by an integral

$$Ai(\zeta) = \frac{1}{\pi} \int_0^\infty \cos\left(\zeta x + \frac{x^3}{3}\right) dx \qquad \text{for } \zeta < 0$$

and can be approximated by

$$Ai(\zeta) = \frac{1}{2\sqrt{\pi}} \zeta^{-1/4} \exp\left(-\frac{2}{3}\zeta^{3/2}\right) \qquad \text{for } \zeta > 1$$

Letting $\zeta = B(eEz - \mathscr{E})$, show that both expressions satisfy (or approximately satisfy) the Schrödinger equation if

$$\zeta = \left(\frac{2m_z eE}{\hbar^2}\right)^{1/3}\left(z - \frac{\mathscr{E}_i}{eE}\right)$$

11.10. Draw energy-band diagrams illustrating possible formation of **(a)** a two-dimensional electron gas in an MOS (Si) structure, **(b)** a two-dimensional hole gas in a MOS (Si) structure, **(c)** a two-dimensional electron gas at a n-AlGaAs/GaAs interface, and **(d)** a two-dimensional hole gas at a p-AlGaAs/GaAs interface. Derive the condition for establishing the two-dimensional system in terms of the work function, the electron affinity, and the energy gap for cases (a) and (b), and in terms of the doping concentration and the band-edge discontinuity for cases (c) and (d) if both Si and GaAs are assumed intrinsic.

11.11. Calculate the average electric field E of Eq. (11.4.13) and the first eigenenergy $\mathscr{E}_0$ from Eq. (11.4.3) for the sample of Fig. 11.10 using $z_{sc} = 200$ Å. Compare the calculated values with those shown in Fig. 11.10a. The calculation is for two masses, $m_z = 0.98m_0$ and $m_z = 0.19m_0$.

11.12. Starting from Eqs. (A 6.1) and (A 6.2), derive Eqs. (A 6.6) and (A 6.7), and then Eqs. (11.5.3) and (11.5.4).

11.13. Show that if τ is independent of v and if $1 + s^2$ is approximated by 1, then Eq. (11.5.2) derived for electrons reduces to Eq. (6.6.4). Also show that for $s < 1$, Eq. (11.5.5) leads to μ and Eq. (11.5.6) leads to R_H of Eq. (6.6.11).

11.14. At room temperature, the mobility in Si is dominated by lattice scattering. Therefore, we can express $\tau = l/v$, where l is the mean free path, which according to Eq. (6.7.19) is independent of v. Show that for $s^2 \ll 1$, expansion of $(1 + s^2)^{-1}$ as $1 - s^2$ in Eqs. (11.5.3) and (11.5.4) for a nondegenerate semiconductor leads to

$$\sigma_{xx} = e\mu_0 n\left(1 - \frac{9\pi}{16}B^2\mu_0^2\right)$$

$$\sigma_{xy} = \frac{3\pi}{8}e\mu_0 n(B\mu_0)\left(1 - \frac{9\pi}{8}B^2\mu_0^2\right)$$

where n is the carrier concentration and $\mu_0 = 4el/(3\sqrt{2\pi m^* kT})$ is the carrier mobility at zero magnetic field.

11.15. Show that for $s \gg 1$, Eq. (11.5.3) reduces to

$$\sigma_{xx} = \frac{4}{3}\left(\frac{m^* n\langle v \rangle}{B^2 l}\right)$$

if l is independent of v, and Eq. (11.5.4) reduces to

$$\sigma_{xy} = \frac{en}{B}$$

irrespective of the scattering mechanism. The quantity $\langle v \rangle$ is defined as

$$\langle v \rangle = \frac{\int f_0 v \, dk_x \, dk_y \, dk_z}{\int f_0 \, dk_x \, dk_y \, dk_z}$$

Both σ_{xx} and σ_{xy} are applicable to degenerate as well as nondegenerate semiconductors.

11.16. **(a)** Using σ_{xx} and σ_{yy} from Problem 11.14 and keeping only terms to the first order in B^2 in Eqs. (11.5.5) and (11.5.6), show that

$$\sigma = e\mu_0 n \left[1 - \left(\frac{1}{2} - \frac{\pi}{8} \right) \frac{9\pi}{8} B^2 \mu_0^2 \right]$$

$$R_H = -\frac{3\pi}{8en} \left[1 - \left(1 + \frac{\pi}{8} \right) \frac{9\pi}{8} B^2 \mu_0^2 \right]$$

for a nondegenerate semiconductor.

(b) Apply the results in Problem 11.15 to a nondegenerate semiconductor. Show that

$$\langle v \rangle = \left(\frac{8kT}{\pi m^*} \right)^{1/2}$$

$$\sigma_{xx} = \frac{32}{9\pi} \frac{en}{B^2 \mu_0}$$

and that

$$\frac{\sigma(B = \infty)}{\sigma(B = 0)} = 0.884, \qquad \frac{R_H(B = \infty)}{R_H(B = 0)} = 0.849$$

(c) Plot $(\sigma_0 - \sigma)/\sigma_0$ and R_H/R_{H0} as functions of $B\mu$ in the regions $B\mu < 0.5$ and $B\mu > 2$.

11.17. According to Eqs. (11.4.4) and (11.4.9), the electron concentration (per unit volume) in a two-dimensional electron gas is given by

$$dn = \frac{2F_0 \, dk_x \, dk_y \, \zeta^2(z)}{(2\pi)^2}$$

where F_0 is the Fermi–Dirac distribution function. In other words f_0 in Eq. (11.5.11) is equal to $f_0 = 2F_0/(2\pi)^2$. Using this f_0 and replacing dk_z by $\zeta^2(z)$ in Eq. (11.5.4), show that for $s > 1$,

$$\sigma_{xy} = \sigma_{xy}^{2D} \zeta^2(z) = -\frac{eN_s}{B} \zeta^2(z)$$

Finally derive Eq. (11.5.13) from Eq. (11.5.6) by noting that $I_x = W \int J_x \, dz$.

11.18. **(a)** Derive Eq. (11.5.19) and find the quantized value of N_s. The values of V_H in Fig. 11.12 are 12.9 mV for $V_G = 4.9$ V, 8.65 mV for $V_G = 6.8$ V, and 6.45 mV for $V_G = 9$ V. Find the value of j in Eq. (11.5.19) for each measured V_H.

(b) Calculate for (100) Si the energy separation between two Landau levels $n = 0$ and $n = 1$ with $m_z = m_l$ and between two Landau levels $m_z = m_l$ and $m_z = m_t$ with $n = 0$. Show that the energy separation in both cases is much smaller than $e \, \Delta V_G$. Draw energy-band diagrams for a MOS structure on a p-Si substrate with increasing gate voltage V_G. Explain why once a surface inversion is established further increase in gate voltage is applied across the oxide. Assuming a value of $V_G = 0.9$ V to establish surface inversion, estimate the surface charge density eN_s induced at each V_G, for $d_{ox} = 0.1\mu m$, and compare the estimated value with the value obtained from V_H.

11.19. Starting from Eqs. (11.4.4) and (11.4.5), derive Eq. (11.6.5). From Fig. 11.13a, find the values of $\mathscr{E}_0$ and $\mathscr{E}_1$ at $N_s = 5 \times 10^{11}$ cm^{-3}. Assuming these values to be insensitive to temperature and using the value of $\mathscr{E}_f$ from Fig. 11.13b, calculate the value of N_s at 300 K from Eq. (11.6.5) and compare it with the initial value of 5×10^{11} cm^{-3}.

11.20. (a) Given $m^* = 0.063m_0$ and $\epsilon = 13.1\epsilon_0$, express the values of $\mathscr{E}_0$ and $\mathscr{E}_1$ in terms of N_s in Eq. (11.6.2) and check your expressions against Eqs. (11.6.3) and (11.6.4).

(b) The discrepancy between Eqs. (11.6.3) and (11.6.4), on the one hand, and Eqs. (11.6.9) and (11.6.10), on the other, can be accounted for by the fact that f_i is set equal to $\frac{1}{2}$ in Eqs. (11.4.13) and (11.6.2). Find the value of f_i for the first two subbands $i = 0$ and $i = 1$ to make the two sets of equations consistent. Explain why $f_1 < f_0$ is expected.

11.21. (a) Show that for $x > 1$, Eq. (11.6.7) leads to Eq. (11.6.11). Given $\mathscr{E}_0 = 43$ meV and $\mathscr{E}_1 = 65$ meV for the $N_s = 5 \times 10^{11}$ cm^{-2} and $N_a = 1 \times 10^{14}$ cm^{-3} case, find the value of $\mathscr{E}_f$ for $T < 77$ K and check the calculated value from Eq. (11.6.11) with the value from Fig. 11.13b.

(b) The value of $\mathscr{E}_f$ at 300 K from Fig. 11.13b is 31 meV. Calculate the values of N_{s0} and N_{s1} in the first two quantized states. From Eqs. (11.4.7) and (11.6.5), find the occupancy for the subbands $i = 0$ and $i = 1$ at 300 K relative to that at 77 K.

11.22. (a) In the following calculation we first assume that the donors in the AlGaAs layer are completely ionized. Find the values of V_G needed to obtain $N_s = 2 \times 10^{11}$ cm^{-2} and $N_s = 7 \times 10^{11}$ cm^{-2} for the sample 3468 of Table 11.4, first using the curves for $N_a = 10^{13}$ cm^{-3} in Fig. 11.13 and then using Eqs. (11.6.3), (11.6.4), and (11.6.7) to calculate $\mathscr{E}_0$, $\mathscr{E}_1$, and $\mathscr{E}_f$. Compare the values obtained from the two methods.

(b) In the calculation to obtain Fig. 11.16, a donor ionization energy of 50 meV is used. Calculate the energy separation $\mathscr{E}_d - \mathscr{E}_f$ for the two cases $N_s = 2 \times 10^{11}$ cm^{-2} and $N_s = 7 \times 10^{11}$ cm^{-2} to check whether the assumption of complete donor ionization is valid or not.

11.23. Repeat Problem 11.22 for sample 3469 of Table 11.4.

11.24. The photoluminescence spectra of Fig. 11.18 show that $I = 40$ at 1.56 eV and $I = 1$ at 1.58 eV for the $E = 28$ V/cm curve and that $I = 3 \times 10^3$ at 1.57 eV and $I = 3.5 \times 10^2$ at 1.60 eV for the $E = 750$ V/cm curve. Find the carrier temperature T_e at the two applied fields. Also, using Eq. (11.8.2), the value of T_e from curve 2a of Fig. 11.19, $h\nu_{op} = 34$ meV, and a value of $\mu_{lf} = 7.9$ m^2/V-s, calculate the value of μ/μ_{lf} for $E = 28$ V/cm, 100 V/cm, and 750 V/cm.

11.25. Given $h\nu_{op} = 34$ meV and $\mu_{lf} = 20$ m^2/V-s at 10 K for the MOD sample of Fig. 11.20, calculate the values of T_e and μ/μ_{lf} at $E = 100$ V/cm and 200 V/cm from Eqs. (11.8.2) and (11.8.3) with $C = 0.25$.

11.26. Derive Eq. (11.9.6) and show that γ is given by Eq. (11.9.8). Then verify that for quantum wells, Eq. (11.9.6) can be replaced by Eq. (11.9.9).

11.27. Finding the value of m_b^* in the $Al_{0.3}Ga_{0.7}As$ barrier and using α of Eq. (11.9.7), calculate the penetration depth for electrons, heavy holes, and light holes. Then using Eq. (11.9.4) and a similar equation for electrons, find the values of $\mathscr{E}_0$, $\mathscr{E}_1$, and $\mathscr{E}_2$ for electrons, heavy holes, and light holes for $L = 100$Å. Compare the value of $\mathscr{E}_0$ thus found for electrons with the value obtained from an exact solution of Eq. (11.9.9).

11.28. Starting from Eq. (11.10.1), derive Eqs. (11.10.4) to (11.10.7). Obviously, first-order perturbation calculation is inadequate for heavy holes because Eq. (11.10.5) shows an upward curvature or a negative effective hole mass.

11.29. (a) Using the values of $m_{hh}^* = m_1$ and $m_{lh}^* = m_2$ given in Table 6.7, calculate the value m_{s0}^* from Eq. (11.10.11) and compare the calculated value with the value of m_3 in Table 6.7 for Si, Ge, GaAs, and InAs.

   (b) Calculate the value of m_e^* from Eq. (11.10.8) and, if necessary, Eq. (11.10.11), using the values of hole masses given in Table 6.7. Compare the calculated value with the effective electron mass given in Table 6.7 for GaAs, GaSb, InAs, and InSb.

11.30. Using the physical parameters given in Table 11.5, calculate the values of μ_{pe} from Eq. (6.8.16) and the values of μ_l from Eq. (6.7.22) at 300 K and at 10 K.

11.31. (a) The mobility for polar optical-phonon scattering is given by (E. M. Conwell, in *Solid State Physics Suppl.*, Vol. 9, 1967)

$$\mu_{po} = \frac{3\sqrt{\pi}}{2\alpha} \frac{e}{m^*\omega_{l0}} \frac{\exp(x_l) - 1}{x_e^{3/2} \exp(x_e/2)} \times$$

$$\left\{ \left[\exp(x_l - x_e) - 1 \right] K_0\left(\frac{x_e}{2}\right) + \left[\exp(x_l - x_e) + 1 \right] K_1\left(\frac{x_e}{2}\right) \right\}^{-1}$$

where $x_l = \hbar\omega_{l0}/kT_l$, $x_e = \hbar\omega_{l0}/kT_e$, $\hbar\omega_{l0}$ is the longitudinal optical-phonon energy, T_l the lattice temperature, and T_e the carrier temperature. Show that μ_{po} reduces to the low-field μ_{po} of Eq. (6.8.7) for $T_e = T_l$.

   (b) At low temperatures such that $x_l - x_e > 3$, show that μ_{po} can be approximated by

$$\mu_{po} = \frac{3\sqrt{\pi}}{2\alpha} \frac{e}{m^*\omega_{l0}} \frac{\exp(x_e/2)}{x_e^{3/2}} \left[K_0\left(\frac{x_e}{2}\right) + K_1\left(\frac{x_e}{2}\right) \right]^{-1}$$

Using $\hbar\omega_{l0} = 34$ meV and $\alpha = 0.067$ in GaAs, and approximating the modified Bessel functions by

$$K_0\left(\frac{x_e}{2}\right) = K_1\left(\frac{x_2}{2}\right) = \sqrt{\frac{\pi}{x_e}} \exp\left(\frac{-x_e}{2}\right)$$

show that

$$\mu_{po} = 1.01 \sqrt{\frac{x_e}{\pi}} x_e^{-3/2} \exp(x_e)$$

which is the same as μ_{po} of Eq. (11.8.2).

11.32. Show that Eq. (6.8.7) can be expressed in terms of α of Eq. (6.8.6) as

$$\mu_{po} = \frac{3\sqrt{\pi}}{4\alpha} \frac{e}{m^*\omega_{l0}} \frac{K_1^{-1}(x_l/2)[\exp(x_l) - 1]}{x_l^{3/2} \exp(x_l/2)}$$

Given $\alpha = 0.067$, $\omega_{l0} = 5.28 \times 10^{13}$ rad/s and $m^* = 0.067m_0$ in GaAs, find μ_{po} at $T = 100$ K, 300 K, and 350 K. The modified Bessel function has the following values:

z	0.5	0.6	0.7	0.8	1.0	2.0	3.0
$K_1(z)$	1.565	1.303	1.050	0.862	0.602	0.140	0.0347

Next, we estimate the effect of impurity scattering. Assuming that $\mu_{po} = 8000 \text{ cm}^2/\text{V-s}$ and using Eq. (6.7.9) for μ_i, find the values of μ at 300 K for $N_I = 10^{16} \text{ cm}^{-3}$ and 10^{18} cm^{-3}. Compare the calculated values with the values shown in Fig. 11.32.

11.33. Letting $\mathscr{E}_c = \mathscr{E}_s$ and $\mathscr{E}_v = \mathscr{E}_p + \Delta/3$ and assuming Δ to be much larger than the energies involved, show that Eq. (11.10.1) reduces to

$$(\mathscr{E}' - \mathscr{E}_c)(\mathscr{E}' - \mathscr{E}_v) = \frac{2}{3} \frac{\hbar^2 k^2 P^2}{m_0^2}$$

Then verify Eqs. (11.11.2) and (11.11.3) by referring to Fig. 11.34. Finally, find expressions for the effective masses of the conduction band E and light-hole band LH of CdTe and HgTe.

12

Dielectric and Optical Properties

12.1 POLARIZATION AND DIPOLE MOMENT

In the macroscopic description of the dielectric properties of a substance, the principal parameter is ϵ, the dielectric constant. For example, this parameter enters in the calculation of the capacitance of a condenser, and it also appears in Maxwell's equations. However, to understand why different materials have different values of ϵ, we must start our discussion with a macroscopic quantity that is of more fundamental nature than ϵ. This macroscopic quantity is the *polarization*, denoted by $\mathbf{P}$, which appears in the relation between the electric flux density $\mathbf{D}$ and the electric field $\mathbf{E}$ as follows:

$$\mathbf{D} = \epsilon_0 \mathbf{E} + \mathbf{P} \qquad \text{C/m}^2 \qquad (12.1.1)$$

where $\epsilon_0 = 8.854 \times 10^{-12}$ F/m is the dielectric constant of free space (see, for example, J. D. Jackson, *Classical Electrodynamics*, John Wiley & Sons, Inc., New York, 1962, pp. 108 and 618).

The macroscopic polarization $\mathbf{P}$ has its origin in the charge distribution of atoms. A basic concept from an atomic viewpoint is the concept of electric dipole. For a system of charged particles, the dipole moment is defined as

$$\mathbf{p} = \sum_i q_i \mathbf{r}_i \qquad \text{C-m} \qquad (12.1.2)$$

where q_i is the charge associated with the ith particle and $\mathbf{r}_i$ is the displacement of the particle from the origin. Since a solid is made of initially neutral atoms, the dipole moment p must be the result of charge separations. We may think of a solid as consisting of many charged pairs. Figure 12.1 shows a representative pair with charges $\pm q$ separated at a distance d apart. From Eq. (12.1.2) we find that

$$\mathbf{p} = q_1 \mathbf{r}_1 + q_2 \mathbf{r}_2 = q(\mathbf{r}_1 - \mathbf{r}_2) = q\mathbf{d} \qquad (12.1.3)$$

where $\mathbf{d}$ is the displacement vector pointing from the $-q$ to the $+q$ charge. For an electrically neutral system, $\mathbf{p}$ as defined by Eq. (12.1.2) is independent of the origin of the coordinate system.

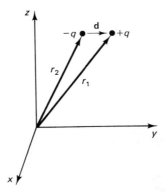

Figure 12.1 Definition of a dipole moment **p** = q**d**. The vector **d** points from −q toward +q.

The macroscopic polarization **P** is the density of atomic electric dipole per unit volume, or

$$\mathbf{P} = \frac{\sum \mathbf{p}}{V} = N\mathbf{p} \qquad (12.1.4)$$

where the summation is over a large volume V and N is the number of charged pairs per unit volume. A formal derivation of Eq. (12.1.4) can be found in texts on classical electromagnetic theory (see, for example, J. D. Jackson, *Classical Electrodynamics*, John Wiley & Sons, Inc., New York, 1962). Here we use a simple classical model to show that Eq. (12.1.4) is indeed valid.

In the classical model, we consider an atom as consisting of a stationary nucleus with charge $+q$ and some electrons with charge $-q$ moving around the nucleus. Without any applied electric field, the center of the electron distribution coincides with the position of the nucleus (Fig. 12.2a), and **p** is zero because of zero **d** in Eq. (12.1.3). Under an electric field, the center of the electron distribution shifts with respect to the nuclear position (Fig. 12.2b). Now we replace the electron distribution by a charge $-q$ at its center (Fig. 12.2c). The energy $\mathscr{E}_1$ stored in the atom is

$$\mathscr{E}_1 = \int \gamma x \, dx = \frac{\gamma}{2} x^2 \qquad (12.1.5)$$

where γx is the restoring force.

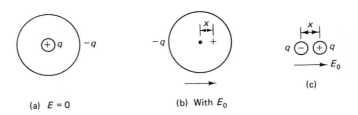

(a) $E = 0$ (b) With E_0 (c)

Figure 12.2 (a) A simple model for a hydrogen atom, (b) the displacement of the electron distribution relative to the position of the proton under an applied field E_0 and (c) the resultant dipole moment.

Applying force balance to the electronic charge in Fig. 12.2c, we have

$$qE_0 = \gamma x \qquad (12.1.6)$$

Realizing that $qx = p$, we find $\mathscr{E}_1 = pE_0/2$. For N atoms per unit volume, the stored energy density is

$$\frac{\mathscr{E}}{V} = N\mathscr{E}_1 = \frac{NpE_0}{2} \qquad (12.1.7)$$

Based on the macroscopic model, the energy density of a dielectric medium is $E_0 D/2$. However, according to Eq. (12.1.1), $\mathbf{D}$ consists of two parts: $\epsilon_0 E_0$ and P. The energy density stored in the atoms is represented by the $E_0 P/2$ term, or

$$\frac{\mathscr{E}}{V} = \frac{E_0 P}{2} \qquad (12.1.8)$$

A comparison of Eqs. (12.1.7) and (12.1.8) gives $P = Np$, which is consistent with Eq. (12.1.4).

The electric-dipole moment in the example considered above arises from the displacement of electrons in an atom relative to the nucleus in the presence of an electric field, and this contribution to the dipole moment is called the *electronic contribution*. There are other contributions to electric polarization. The *ionic contribution* comes from the displacement of positive ions relative to the negative ions in an ionic crystal. In substances possessing permanent electric dipoles, a contribution called *orientational polarization* arises as a result of reorientation of the dipoles in the presence of an electric field. The various polarizations will be treated in more detail in the following sections.

Finally, we should mention that in dielectric materials the polarization $\mathbf{P}$ is proportional to the applied field $\mathbf{E}$. A quantity defined as

$$\chi_e = \frac{P}{E} \qquad (12.1.9)$$

is called the *electric susceptibility*. For the simplest cases, the induced $\mathbf{P}$ is parallel to $\mathbf{E}$. Thus, using Eq. (12.1.9) in Eq. (12.1.1), we find that

$$\mathbf{D} = \epsilon_0 \left(1 + \frac{\chi_e}{\epsilon_0}\right) \mathbf{E} = \epsilon \mathbf{E} \qquad (12.1.10)$$

where ϵ is the dielectric constant of a medium. On the atomic scale, a corresponding ratio may be defined. However, the electric field E_{atom} experienced by an atom or an ion in a solid may be different from the applied field E. The ratio of the electric-dipole moment to E_{atom},

$$\alpha = \frac{p}{E_{\text{atom}}} \qquad (12.1.11)$$

is called the *polarizability*. The difference between E_{atom} and E is discussed in Section 12.5.

12.2 ELECTRONIC POLARIZABILITY

In Section 12.1 we used a simple model to show that an atom can become polarized in an electric field. Now we proceed to calculate the atomic polarizability. For simplicity, we consider the hydrogen atom as an example. To make calculations possible, we assume that the electron distribution is uniform and confined within the Bohr radius a (Fig. 12.3a). In the presence of $\mathbf{E}$, the entire electron distribution is shifted with respective to the position of the nucleus (Fig. 12.3b). Applying the Gauss theorem $\int D_n \, dS = Q$, that the surface integral of the normal component of $\mathbf{D}$ is equal to the total charge within the enclosure, we have

$$4\pi x^2 D_n = -e \frac{x^3}{a^3} \tag{12.2.1}$$

where x is the displacement of the electron distribution. In Eq. (12.2.1), the enclosure is taken to be a sphere of radius x drawn from the center of the electron distribution (Fig. 12.3c).

From Eq. (12.2.1) the Coulomb force acting on the nucleus is

$$F_x = eE_n = \frac{-e^2}{4\pi\epsilon_0 a^3} x \tag{12.2.2}$$

The force acting on the atom due to the field E_0 is $F_x = eE_0$. Force balance requires that

$$\alpha_h = \frac{ex}{E_0} = 4\pi\epsilon_0 a^3 \qquad \text{F-m}^2 \tag{12.2.3}$$

for a hydrogen atom. We should emphasize that the assumptions used in the derivation of Eq. (12.2.3) are rather simplistic, and Eq. (12.2.3) is useful only for an order-of-magnitude estimate.

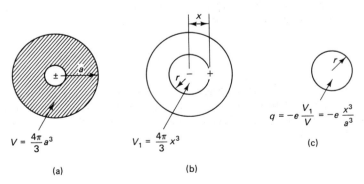

$$V = \frac{4\pi}{3} a^3$$

(a)

$$V_1 = \frac{4\pi}{3} x^3$$

(b)

$$q = -e \frac{V_1}{V} = -e \frac{x^3}{a^3}$$

(c)

Figure 12.3 A simple model used for the calculation of electronic polarizability of a hydrogen atom α_h, with diagram (a) assuming a uniform electron distribution within a sphere of Bohr radius a, diagram (b) showing the volume of a sphere of radius x, and diagram (c) showing the charge inside the sphere of radius x.

An accurate calculation of α_h must be based on quantum mechanics. Under an electric field $\mathbf{E}_0$, the Schrödinger equation becomes

$$\left(\frac{\hbar^2}{2m} \nabla^2 + \mathscr{E} + \frac{e^2}{4\pi\epsilon_0 r} - e\mathbf{E}_0 \cdot \mathbf{r} \right) \psi = 0 \qquad (12.2.4)$$

The solution of Eq. (12.2.4) is

$$\psi = \psi_0 \left[1 - \frac{4\pi\epsilon_0 a^2}{e} \frac{\mathbf{E}_0 \cdot \mathbf{r}}{r} \left(\frac{r}{a} + \frac{r^2}{2a^2} \right) \right] \qquad (12.2.5)$$

where $\psi_0 = (\pi a^3)^{-1/2} \exp(-r/a)$ is the wave function for an unperturbed hydrogen atom (see, for example, N. F. Mott and I. N. Sneddon, *Wave Mechanics and Its Applications,* Clarendon Press, Oxford, 1948). The quantum-mechanical counterpart of Eq. (12.1.3) is

$$\mathbf{p} = \int \psi^*(-e\mathbf{r})\psi \, d\tau \qquad (12.2.6)$$

where $d\tau = r^2 \, dr \sin\theta \, d\theta \, d\phi$ is the volume element. Substituting Eq. (12.2.5) in Eq. (12.2.6) yields

$$\alpha_h = \frac{p}{E_0} = 18\pi\epsilon_0 a^3 \qquad (12.2.7)$$

Now we turn to the situation in which the electric field varies sinusoidally with time. The dynamic equation of motion for the displacement x is

$$m \frac{d^2x}{dt^2} = -\beta \frac{dx}{dt} - \gamma x - eE_1 \exp(i\omega t) \qquad (12.2.8)$$

On the right-hand side of Eq. (12.2.8), the second term represents the restoring force and the first term, like resistive loss in an electric circuit, represents an equivalent retarding force.

The steady-state solution of Eq. (12.2.8) is

$$x = -\frac{eE_1/m}{\omega_0^2 - \omega^2 + i\omega\beta/m} \exp(i\omega t) \qquad (12.2.9)$$

where $\omega_0 = \sqrt{\gamma/m}$. Using $E_{\text{atom}} = E_1 \exp(i\omega t)$ in Eq. (12.1.11), we find that

$$\alpha_e = \frac{e^2/m}{\omega_0^2 - \omega^2 + i\omega\beta/m} = \alpha_e' - i\alpha_e'' \qquad (12.2.10)$$

The ac electronic polarizability α_e consists of real and imaginary parts with

$$\alpha_e' = \frac{e^2}{m} \frac{\omega_0^2 - \omega^2}{(\omega_0^2 - \omega^2)^2 + (\omega\beta/m)^2} \qquad (12.2.11)$$

$$\alpha_e'' = \frac{e^2}{m} \frac{\omega\beta/m}{(\omega_0^2 - \omega^2)^2 + (\omega\beta/m)^2} \qquad (12.2.12)$$

The current associated with the electronic motion is equal to

$$i = -e\dot{x} = (i\omega\alpha_e' + \omega\alpha_e'')E_1 \exp(i\omega t) \qquad (12.2.13)$$

The part of i associated with α_e'' is in phase with the field, and hence the imaginary part α_e'' contributes to loss in a dielectric medium.

Figure 12.4 shows the variation of α_e' and α_e'' with ω. Significant change in the electronic polarizability occurs in the region near ω_0, where the absorption is appreciable. Outside the absorption region, α_e' and hence the dielectric constant remain practically constant. We should point out that Eq. (12.2.10) and Fig. 12.4 are based on a single-oscillator model. A quantum-mechanical analysis shows that α_e is given by

$$\alpha_e = \frac{e^2}{m} \sum_j \frac{f_{j0}}{\omega_{j0}^2 - \omega^2} \qquad (12.2.14)$$

In the quantum-mechanical description, the resonant frequency ω_{j0} is associated with an atomic transition such that $\hbar\omega_{j0} = \mathscr{E}_j - \mathscr{E}_0$, where $\mathscr{E}_j$ and $\mathscr{E}_0$ are the energies of the excited and the ground state, respectively. An electron in the ground state is coupled to the various oscillators (various excited states j) through f_{j0}, which is a dimensionless quantity called the *oscillator strength*. Since we have as many ω_{j0}'s as the number of excited states, α_e' and α_e'' should be a superposition of many curves similar to the one presented in Fig. 12.4.

In concluding this section we estimate the magnitude of α_e. For $\omega < \omega_0$, Eq. (12.2.14) can be replaced by

$$\alpha_e = \frac{e^2}{m\omega_0^2} \qquad (12.2.15)$$

where ω_0 is a weighted average of all ω_{j0}'s. The reader is asked to show (in Problem 12.5) that except for a slightly different numerical constant, using the ionization energy of hydrogen atom for $\hbar\omega$ in Eq. (12.2.15) leads to α_h of Eq. (12.2.7). For most ionic and molecular crystals, $\mathscr{E}_j - \mathscr{E}_0$ falls in the optical region. Taking $\hbar\omega = 4$ eV or $\omega_0 = 6 \times 10^{15}$ rad/s, and m to be the free electron mass, we find that $\alpha = 7.9 \times 10^{-40}$ F-m^2. If we neglect the difference between E in Eqs. (12.1.11) and (12.1.9), $\chi_e = N\alpha_e$ in Eq. (12.1.10). Using $N = 2 \times 10^{28}$ atoms/m^3, we have $\epsilon/\epsilon_0 = 1 + N\alpha_e/\epsilon_0 = 2.8$, which is of correct order of magnitude in the optical region. We should point out that the calculation of ϵ/ϵ_0 just presented is for an order-of-magnitude estimate only because we have neglected the difference between E_{atom} and E.

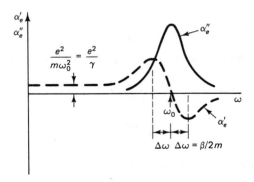

Figure 12.4 Schematic diagram showing the behavior of α_e' and α_e'', the real and imaginary part of the electronic polarizability as a function of angular frequency ω.

12.3 IONIC POLARIZABILITY

In an ionic crystal, such as NaCl, KBr, ZnO, and so on, relative displacement of positive ions with respect to the position of negative ions in the presence of an electric field gives rise to additional polarization in the solid. Since ionic polarization involves the motion of ions which are heavier than electrons by a factor of more than 1000, the resonant frequency of the ionic system is in the infrared region, as indicated by the mass ratio. Although the shapes of the absorption and dispersion curves due to ionic polarization are quite similar to those due to electronic polarization except for the frequency shift, it is instructive to derive the force equation governing the relative motion of ions.

In setting up Eq. (12.2.8), we consider an atom to be essentially isolated. The assumption is a valid one because the intra-atomic force is much stronger than the interatomic force. For ionic polarization, however, motion of neighboring ions is coupled together through interatomic forces. Consider a linear diatomic chain (Fig. 12.5) of negative ions having masses M_1 located at odd-numbered lattice points and positive ions having masses M_2 located at even-numbered lattice points. The equations of motion for M_1 and M_2 are given by Eq. (2.11.10). However, while we were interested in the natural response of Eq. (2.11.10), here we are concerned with the forced response of Eq. (2.11.10) to an applied field. Therefore, if we let

$$u_{2n} = v \exp [iq(2n)a] \tag{12.3.1a}$$

$$u_{2n+1} = u \exp [iq(2n + 1)a] \tag{12.3.1b}$$

the wave number $q = 2\pi/\lambda$ in Eq. (12.3.1) is that of the applied field E, and thus the symbol q is used instead of k, which is used in Eq. (2.11.11). In terms of u and v, the equations of motion are

$$M_1 \frac{d^2u}{dt^2} = 2\gamma(v \cos qa - u) - f_i e E_1 \exp (i\omega t) \tag{12.3.2}$$

$$M_2 \frac{d^2v}{dt^2} = 2\gamma(u \cos qa - v) + f_i e E_1 \exp (i\omega t) \tag{12.3.3}$$

where f_i is the ionicity value.

To study the effects of an electric field on the motion of ions, we make the following observation. Like electronic motion discussed in Section 12.2, ionic motion may also be considered as a driven resonant circuit under an applied field. This means that the phase constant q in Eq. (12.3.1) is the same as that of the field. For electromagnetic waves in the infrared region, the wavelength λ is of the order of 5 $\times$

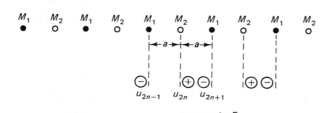

Figure 12.5 Schematic diagram showing relative displacements of positive and negative ions in an ionic crystal under an applied electric field E.

10^{-6} m, which is considerably larger than the interatomic distance $a \sim 3 \times 10^{-10}$ m. The value of $qa = 2\pi a/\lambda$ is thus very much smaller than 1, and cos qa can be approximated by 1 in Eqs. (12.3.2) and (12.3.3).

Letting cos $ka = 1$ and subtracting Eq. (12.3.3) from Eq. (12.3.2), we obtain an equation for the relative displacement $u - v$ of positive and negative ions, of which the steady-state solution is

$$u - v = \frac{f_i e/M}{\omega_0^2 - \omega^2} E_1 e^{i\omega t} \tag{12.3.4}$$

where $M = M_1 M_2/(M_1 + M_2)$ and $\omega_0 = \sqrt{2\gamma/M}$. If damping mechanisms are included in Eq. (12.3.4), the general expression for *ionic polarizability* is

$$\alpha_i = \frac{f_i e^2/M}{(\omega_0^2 - \omega^2) + i\omega\beta/M} \tag{12.3.5}$$

Equation (12.3.5) has the same form as Eq. (12.2.10) except for two important differences. First, the resonant frequency ω_0 is now in the infrared region because of a much larger mass M and a different force constant γ. The value of β is also different because different damping mechanisms are associated with the ionic polarization.

12.4 ORIENTATIONAL POLARIZABILITY

Molecules can be divided into two classes, polar and nonpolar, according to whether they do or do not possess an electric dipole moment. The most commonly known polar molecule is water. In water, the H—O—H bond is not straight but forms an angle of $105°$. Since hydrogen and oxygen atoms are slightly charged, water molecules have a permanent electric dipole pointing in the direction that bisects the bond angle. Without an electric field, the dipoles are randomly oriented in all directions with equal probability and hence the resultant polarization is zero. In the presence of an electric field, however, a realignment of dipoles takes place, giving rise to a nonvanishing polarization often referred to as *orientational polarization*.

Let N be the number of molecules whose dipole moment makes an angle θ with E_0. According to the Boltzmann distribution law,

$$N = A \exp\left(-\frac{\mathscr{E}}{kT}\right) \tag{12.4.1}$$

where $\mathscr{E} = -p_0 E_0 \cos\theta$ is the potential energy of the dipole. The proportionality constant A can be determined from the condition that $\int N d\Omega$ must be equal to the total number N_0 molecules, where $d\Omega = \sin\theta \, d\theta \, d\phi$ is the solid angle. The limits of integration are from 0 to π for θ and from 0 to 2π for ϕ. Of $N d\Omega$ dipoles, the component along the direction E_0 is

$$\bar{p}_0 = \frac{1}{N_0} \int_0^\pi \int_0^{2\pi} p_0(\cos\theta) N \, d\Omega = p_0\left(\coth\frac{p_0 E_0}{kT} - \frac{kT}{p_0 E_0}\right) \tag{12.4.2}$$

Because of the Boltzmann factor, there are more dipoles pointing in the direction of E_0 than dipoles pointing in opposite directions; and consequently, there is a net dipole moment along E_0. Equation (12.4.2) is often written as $\bar{p}_0 = p_0 L(x)$, where $L(x)$ is known as the *Langevin function* with $x = p_0E_0/kT$. Take q as the electronic charge and d as the interatomic distance 10^{-10} m; thus we have $p_0 \approx 10^{-29}$ C-m. For $E_0 = 10^5$ V/m, $p_0E_0 \approx 10^{-24}$ J and at room temperature $kT \approx 4 \times 10^{-21}$ J. For $x \ll 1$, the Langevin function can be approximated by $x/3$; hence the orientational polarizability is

$$\alpha_0 = \frac{\bar{p}_0}{E_0} = \frac{p_0^2}{3kT} \tag{12.4.3}$$

The phenomenological equation describing the dynamic behavior of $\bar{p}_0$ in the presence of changing electric field is given by

$$\tau \frac{d\bar{p}_0}{dt} = -\bar{p}_0 + \alpha_0 E_1 \exp{(i\omega t)} \tag{12.4.4}$$

Equation (12.4.4) was derived and used by Debye to correlate the frequency dependence with the relaxation phenomenon in polar substances (P. Debye, *Polar Molecules*, Dover Publications, Inc., New York, 1945, Chap. 5). The steady-state solution of Eq. (12.4.4) reads

$$\bar{p}_0 = \frac{\alpha_0 E_1}{1 + i\omega\tau} \exp{(i\omega t)} = (\alpha_d' - i\alpha_d'')E_1 \exp{(i\omega t)} \tag{12.4.5}$$

Thus the ac orientational polarizability is

$$\alpha_d' = \frac{\alpha_0}{1 + \omega^2\tau^2} \quad \text{and} \quad \alpha_d'' = \frac{\omega\tau\alpha_0}{1 + \omega^2\tau^2} \tag{12.4.6}$$

For $\omega > 1/\tau$ the dipoles no longer can follow the rapid field variations, and α_d drops rapidly with increasing ω. The value of τ can be obtained from a best fit to Eq. (12.4.6) of measured α_d as a function of ω.

The time constant τ governs the rate with which a dipole system approaches the steady state. In liquids, τ is related to the viscosity η through the following expression, originally due to Debye:

$$\tau = \frac{4\pi a^3 \eta}{kT} \tag{12.4.7}$$

where a is the radius of the polar molecule, which is assumed spherical. Experimental results on pure polar liquids give a value of τ ranging from 5×10^{-12} to 5×10^{-10} s. In solids a reorientation of dipoles involves the movement of an ion from one equilibrium position to another and must overcome a potential barrier, say of energy ψ. Hence τ is expected to have the following temperature dependence:

$$\tau = C \exp{\left(\frac{\psi}{kT}\right)} \tag{12.4.8}$$

where C is a constant in seconds. The experimental value of τ in ice can be best fitted by taking $C = 5 \times 10^{-16}$ s and $\psi = 9 \times 10^{-20}$ J. From Eq. (12.4.8) the values of τ are 6×10^{-5} s and 1.2×10^{-2} s at $-10°C$ and $-50°C$, respectively. Since the magnitude of τ is determined by the ease with which a polar molecule can reorientate inside a solid, polarization measurements can be used to detect any structural change of the solid which prohibits the reorientational freedom of polar molecules (see, for example, R. J. W. LeFèvre, *Dipole Moments*, Methuen and Company Ltd., London, 1948; C. P. Smyth, *Dielectric Behavior and Structure*, McGraw-Hill Book Company, New York, 1955).

12.5 CHEMICAL BOND, MOLECULAR STRUCTURE, AND DIELECTRIC PROPERTIES OF MATERIALS

In Sections 12.2, 12.3, and 12.4 we discussed three types of polarization: electronic, ionic, and orientational. The total polarizability α of a substance is

$$\alpha = \alpha_e + \alpha_i + \alpha_d \tag{12.5.1}$$

where α_e is given by Eq. (12.2.10), α_i by Eq. (12.3.5), and α_d by Eq. (12.4.6). Based on Eq. (12.5.1) the behavior of $\alpha = \alpha' - i\alpha''$ as a function of frequency has a general shape schematically illustrated in Fig. 12.6. Significant changes in α occur only when one component of polarization can no longer respond to fast changes in the applied field. For α_d, the change usually occurs in the radio or microwave region; for α_i, in the infrared region; and for α_e, in the optical (from near-infrared to ultraviolet) region. Since these changes happen at very different frequencies, the contribution of each component to α can easily be separated out by measuring the polarizability at three different frequencies, ω_1, ω_2, and ω_3, as shown in Fig. 12.6. We should emphasize, however, that in a real substance there may be several discrete resonances for α_i and several overlapping resonances for α_e. The single resonance shown in the figure is an oversimplification intended for illustration only.

According to the frequency variation of α, dielectric materials can be divided into three classes: (1) substances showing α_e only, (2) substances having both α_e and α_i and (3) substances having all three components. The first class contains all materials consisting of a single type of atom. Common examples are germanium, nitrogen (in solid, liquid, or gaseous state), and so on. By reason of symmetry, these materials cannot possibly either possess a permanent dipole moment or contain dipolar groups of atoms as in the case of ionic crystals. For materials in this group, the polarizability and hence the dielectric constant stay nearly constant up to the optical region, and then becomes frequency dependent as α_e changes near one of ω_{j0}'s. This can occur in

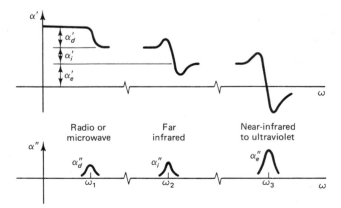

Radio or microwave   Far infrared   Near-infrared to ultraviolet

Figure 12.6 Schematic diagram showing the frequency dependence of the polarizability of a substance in which all three contributions α_e, α_i and α_d exist. The part for $\alpha_e = \alpha'_e - i\alpha''_e$ assumes a two-level system and hence has only one resonance frequency. Refer to Fig. 12.18 and Eq. (12.7.2) for the dielectric behavior of semiconductors.

semiconductors and insulators when the photon energy $\hbar\omega$ approaches or exceeds the gap energy, which may range from 0.18 eV ($\lambda = 6.9 \ \mu m$) in InSb to 9.4 eV ($\lambda = 1320 \ \text{Å}$) in KCl.

The second class contains materials whose molecular structure possesses certain degrees of symmetry so that the total dipole moment is zero even though the molecule may consist of dipolar groups of atoms. A simple example in CO_2. The three atoms are arranged in a collinear chain with the carbon atom $(+)$ halfway between the two oxygen atoms $(-)$, resulting in a cancellation of dipole moments. Another example is carbon tetrachloride, CCl_4, with the four chlorine atoms at the corners and the carbon atom at the center of a regular tetrahedron. Ionic crystals of alkali halides are also representatives of this group. In the face-centered cubic structure (Fig. 2.7), each Na^+ ion is surrounded symmetrically by six Cl^- ions. As a result of the cubic symmetry, there is no permanent dipole in NaCl crystal although the NaCl molecule is polar. The compound semiconductors, such as SiC (IV–IV compound), GaAs (III–V compound), and CdS (II–VI compound), have a moderate degree of ionicity; and hence they, like many other materials in this group, possess both α_e and α_i.

All materials made of dipolar molecules belong to the third group. Common examples are H_2O, CH_3Cl, NH_3, and so on. In contrast to CO_2, water has a bond angle of 105° and hence possesses a dipole moment. The replacement of a hydrogen atom in CH_4 by a chlorine atom destroys the symmetry of the tetrahedral structure, creating a net dipole moment. The dipole moment of a polar gas can be determined from a measurement of the temperature dependence of the static dielectric constant as α_0 of Eq. (12.4.3) is inversely proportional to T. The polarizability $\alpha_i + \alpha_e$ due to elastic displacement of charges, on the other hand, is independent of temperature. In Fig. 12.7 the dielectric constant of hydrogen sulfide measured at 5 kHz is plotted as a function of temperature. The sudden drop in ϵ at around 100 K is a direct consequence of a disorder–order transition. In the ordered state, the dipoles are all frozen in and hence the dielectric constant becomes temperature independent.

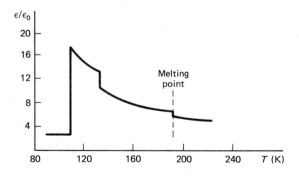

Figure 12.7 The dielectric constant of hydrogen sulfide measured at 5 kHz as a function of temperature.

12.6 INTERNAL FIELDS: DEPOLARIZATION FIELD AND LORENTZ FIELD

In the previous discussion of electric polarizability we purposely avoided the use of dielectric constant ϵ because the relationship between ϵ and α is by no means straightforward. As mentioned in Section 12.1, the electric polarizability α is defined as the ratio of the induced electric dipole moment p to the electric field E_{atom} experienced by an atom. In order that we may be able to use Eq. (12.1.11) in Eq. (12.1.4) to find the macroscopic polarization P, we must know the relation between E_{atom} and the applied field E_0.

For an ellipsoidal sample placed in a uniform field E_0, the field $\mathbf{E}$ inside the dielectric is given by

$$\mathbf{E} = \mathbf{E}_0 - N_x\mathbf{P}_x - N_y\mathbf{P}_y - N_z\mathbf{P}_z = \mathbf{E}_0 + \mathbf{E}_1 \qquad (12.6.1)$$

where the field $\mathbf{E}_1$ is called the *depolarizing field*, which is due to polarization charges induced on the surface (see, for example, J. D. Jackson, *Classical Electrodynamics*, John Wiley & Sons, Inc., New York, 1962; S. Ramo, J. R. Whinnery, and T. Van Duzer, *Fields and Waves in Communication Electronics*, John Wiley & Sons, Inc., New York, 1965). The three constants N_x, N_y, and N_z, called the *depolarizing factors*, are defined along the three principal axes of the ellipsoid. They obey the following general relation:

$$N_x + N_y + N_z = \frac{1}{\epsilon_0} \qquad (12.6.2)$$

The three geometrical shapes shown in Fig. 12.8 can be considered as extreme cases of an ellipsoidal shape. The three cases are important because these shapes are commonly used in dielectric and magnetic experiments.

Once the general relation expressed in Eq. (12.6.2) is known, the depolarizing factors can be found through physical arguments. For a spherical sample (Fig. 12.8b),

$$N_x = N_y = N_z = \frac{1}{3\epsilon_0} \qquad (12.6.3)$$

by reason of symmetry. For an infinitely long, circular cylinder (Fig. 12.8),

$$N_z = 0, \qquad N_x = N_y = \frac{1}{2\epsilon_0} \qquad (12.6.4)$$

if the axis of the cylinder is taken to be along the z direction. In Eq. (12.6.4), $N_z = 0$

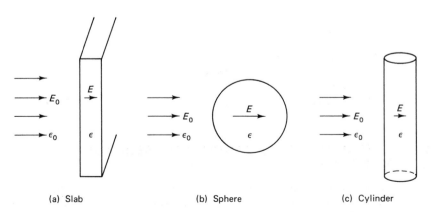

(a) Slab   (b) Sphere   (c) Cylinder

Figure 12.8 Three shapes of sample commonly used in dielectric and magnetic experiments: (a) slab, (b) sphere, and (c) cylinder.

in view of the fact that the polarization surface charges ρ_s are infinitely far away in the z direction and $N_x = N_y = \frac{1}{2}\epsilon_0$ by virtue of symmetry. For a thin slab infinite in extent in the xy plane (Fig. 12.8a),

$$N_x = N_y = 0, \qquad N_z = \frac{1}{\epsilon_0} \qquad (12.6.5)$$

The derivations of Eqs. (12.6.3) to (12.6.5) can be found in Problems 12.10 and 12.11. The requirement that a specimen is infinitely long or infinitely thin is met in practice if one dimension is larger or smaller by a factor of 5 or more than the other two dimensions. For a general calculation of the depolarizing factors in an ellipsoidal specimen, the reader is referred to similar calculations reported by Osborn and by Stoner for demagnetizing factors (J. A. Osborn, *Phys. Rev.*, Vol. 67, p. 351, 1945; E. C. Stoner, *Philos. Mag.*, Vol. 36, p. 803, 1945).

The field $\mathbf{E}$ inside a dielectric can be obtained by eliminating $\mathbf{P}$ from Eq. (12.6.2) and

$$\mathbf{D} = \epsilon_0 \mathbf{E} + \mathbf{P} = \epsilon \mathbf{E} \qquad (12.6.6)$$

which is a combination of Eq. (12.1.1) and (12.1.10). Thus the fields inside the dielectric samples of Fig. 12.8 are, respectively,

$$\mathbf{E} = \begin{cases} \dfrac{\epsilon_0}{\epsilon} \, \mathbf{E}_0 & \text{for a slab} \\[2ex] \dfrac{3\epsilon_0}{2\epsilon_0 + \epsilon} \, \mathbf{E}_0 & \text{for a sphere} \\[2ex] \dfrac{2\epsilon_0}{\epsilon_0 + \epsilon} \, \mathbf{E}_0 & \text{for a cylinder} \end{cases} \qquad (12.6.7)$$

However, the field $\mathbf{E}$ only combines the external field $\mathbf{E}_0$ and the depolarizing field $\mathbf{E}_1$, due to the surface charges of density $\rho_s = \hat{\mathbf{n}} \cdot \mathbf{P}$, where $\hat{\mathbf{n}}$ is a unit outward vector normal to the surface.

Besides the field $\mathbf{E}_1$ created by surface charges, we must consider the field produced by neighboring dipoles. The dipole field at atom site i caused by the dipole moment of all other atoms is given by (Problem 12.12)

$$\mathbf{E}_i = \sum_{j \neq i} \frac{3(\mathbf{p}_j \cdot \mathbf{r}_{ij})\mathbf{r}_{ij} - r_{ij}^2 \mathbf{p}_j}{4\pi\epsilon_0 r_{ij}^5} \qquad (12.6.8)$$

where $\mathbf{p}_j$ is the dipole moment of atom j and $\mathbf{r}_{ij}$ is a radial vector pointing from atom j toward i. Since the summation in Eq. (12.6.8) is over the whole crystal, it will be very difficult to carry out the summation. Special steps must be taken to evaluate $\mathbf{E}_i$.

Consider a dielectric material placed in an electric field $\mathbf{E}_0$ applied externally as shown in Fig. 12.9. Imagine that a microscopic spherical cavity is cut out of the specimen, the radius a of the sphere being very small compared with the physical dimensions of the specimen but very large compared with atomic dimensions. First, we separate the contribution to $\mathbf{E}_i$ due to dipoles inside the cavity and call it $\mathbf{E}_3$. For the remaining contribution, since the summation is over a large volume, we can replace the microscopic dipole moment $\mathbf{p}_j$ by a macroscopic polarization $\mathbf{P}$. As discussed earlier, the polarization $\mathbf{P}$ produces a surface polarization-charge of density $\rho_s = \hat{n} \cdot \mathbf{P}$. There are two surfaces to be considered: one being the specimen surface and the other being the cavity surface. The field $\mathbf{E}_1$ created by polarization charges at the specimen surface has been treated. The field $\mathbf{E}_2$ in Fig. 12.9 is due to polarization charges at the cavity surface.

In summary, the total field acting on an atom inside a dielectric is given by

$$\mathbf{E}_{\text{atom}} = \mathbf{E}_0 + \mathbf{E}_1 + \mathbf{E}_2 + \mathbf{E}_3 \qquad (12.6.9)$$

where $\mathbf{E}_0$ is the external field, $\mathbf{E}_1$ the field due to polarization charges on the surface of the specimen, $\mathbf{E}_2$ the field due to polarization charges on the surface of the cavity, and $\mathbf{E}_3$ the field due to dipoles inside the cavity. The field $\mathbf{E}_0$ and $\mathbf{E}_1$ can be combined as in Eq. (12.6.1), yielding

$$\mathbf{E}_{\text{atom}} = \mathbf{E} + \mathbf{E}_2 + \mathbf{E}_3 \qquad (12.6.10)$$

where $\mathbf{E}$ is the field existing inside the dielectric. Our next task is to find $\mathbf{E}_2$ and $\mathbf{E}_3$. Once $\mathbf{E}_2$ and $\mathbf{E}_3$ can be expressed in terms of either the microscopic dipole moment $\mathbf{p}$ or the macroscopic polarization $\mathbf{P}$, we can use Eqs. (12.1.4), (12.1.11), (12.6.6), and (12.6.10) to obtain the relation between the macroscopic dielectric constant ϵ and the microscopic polarizability α.

For the calculation of $\mathbf{E}_3$, we refer to Eq. (12.6.8). For simplicity, we assume $\mathbf{E}_0$ and hence $\mathbf{p}$ to be in the z direction. Thus we have, respectively, for the x, y, and z components in the numerator:

$$A_x = \sum 3(p_j z_{ij}) x_{ij} = 3p \sum z_{ij} x_{ij}$$

$$A_y = \sum 3(p_j z_{ij}) y_{ij} = 3p \sum z_{ij} y_{ij} \qquad (12.6.11)$$

$$A_z = \sum 3(p_j z_{ij}) z_{ij} - \sum p_j r_{ij}^2 = p \sum (3z_{ij}^2 - r_{ij}^2)$$

For crystals of cubic symmetry, $\sum z_{ij}^2 = \sum x_{ij}^2 = \sum y_{ij}^2 = r_{ij}^2/3$, and thus $A_z = 0$. The other two components A_z and A_y must also be zero on the following ground. The atom i sits at the center of a fictitious sphere. If the coordinates are rotated by $180°$ about the z axis, the surroundings of the atom i remain unchanged. Therefore, we expect A_x and A_y to remain the same. However, x_{ij} and y_{ij} now become $-x_{ij}$ and $-y_{ij}$ in Eq. (12.6.11) through the rotation. The only way to reconcile the two conflicting results is to require

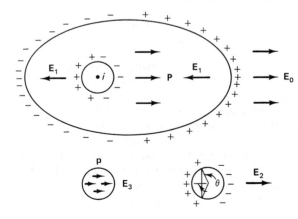

Figure 12.9 Schematic diagram showing the decomposition of the internal field into three components: the field $\mathbf{E}_1$ due to polarization charges induced on the surface of the specimen, the field $\mathbf{E}_2$ due to polarization charges on the surface of the fictitious cavity, and the field $\mathbf{E}_3$ due to dipoles inside the fictitious cavity.

Dielectric and Optical Properties   Chap. 12

that $A_x = A_y = 0$. The analysis shows that $\mathbf{E}_3$ is identically zero for crystals of cubic symmetry.

The field $\mathbf{E}_2$ was first introduced by Lorentz and hence is called the *Lorentz field*. Since the fictitious cavity is chosen to be spherical, the calculation of $\mathbf{E}_2$ is similar to that of $\mathbf{E}_1$ for a spherical sample. The only difference is that $\mathbf{E}_2$ is in the direction of $\mathbf{P}$. Thus we have

$$\mathbf{E}_2 = \frac{\mathbf{P}}{3\epsilon_0} \qquad (12.6.12)$$

Using Eq. (12.6.12) in Eq. (12.6.10) and setting $\mathbf{E}_3 = 0$, we obtain

$$\mathbf{E}_{\text{atom}} = \mathbf{E} + \frac{\mathbf{P}}{3\epsilon_0} \qquad (12.6.13)$$

Equation (12.6.13) holds for those ionic crystals whose atoms have an environment of cubic symmetry. For ionic crystals having other symmetry, we must compute $\mathbf{E}_3$ in Eq. (12.6.10) for $\mathbf{E}_3$ is in general nonzero. We should also point out that Eq. (12.6.13) does not hold for covalent semiconductors and polar substances. The reasons for its nonapplicability is discussed in Section 12.7.

12.7 POLARIZABILITY AND DIELECTRIC CONSTANT

To derive the relation between α and ϵ, we should distinguish three distinct types of materials: (1) metals and semiconductors, (2) ionic crystals, and (3) polar liquids and solids. In metals and semiconductors, Eq. (12.6.9) loses its meaning. Because such solids are highly polarizable, the effect of an external field on the electron distribution of an atom is not localized to the atom under consideration. In other words, the electric dipole moment $\mathbf{p}$ should be considered as a collective rather than a localized phenomenon. Theoretical analyses of Nozières and Pines have shown that no local field (the Lorentz field) correction is needed (P. Nozières and D. Pine, *Phys. Rev.*, Vol. 109, p. 762, 1958). With $\mathbf{E}_2 = \mathbf{E}_3 = 0$ in Eq. (12.6.10), we have $\mathbf{E}_{\text{atom}} = \mathbf{E}$. Substituting Eq. (12.1.11) in Eq. (12.1.4) and using Eq. (12.1.10), we find that

$$\epsilon = \epsilon_0 + N\alpha \qquad (12.7.1)$$

where N is the number of atoms per unit volume. Equation (12.7.1) also applies to a gaseous medium in which the value of P is small and hence the local field correction can be neglected.

For covalent semiconductors, such as Ge and Si, α is due to electronic polarizability alone. Using Eq. (12.2.14) in Eq. (12.7.1), we have

$$\epsilon = \epsilon_0 + \frac{Ne^2}{m} \sum_j \frac{f_{j0}}{\omega_{j0}^2 - \omega^2} = \epsilon_0 + \frac{e^2}{m} \int \frac{f_{cv}(k)}{\omega_{cv}^2(k) - \omega^2} \frac{k^2 \, dk}{\pi^2} \qquad (12.7.2)$$

Figure 12.10a shows the measurement of the index of refraction $n = \sqrt{\epsilon/\epsilon_0}$ in germanium reported by Philipp and Taft (H. R. Philipp and E. A. Taft, *Phys. Rev.*, Vol. 113, p. 1002, 1959). As we can see, the curve, in no way resembles the polarizability curve of Fig. 12.6 based on a single-oscillator model. Applying the band theory to the dielectric-constant expression, we first count the density of states in k space and replace N by $dN = k^2 \, dk/\pi^2$. Second, we replace the energy separation $\hbar\omega_{j0}$ by the energy gap $\hbar\omega_{cv}$,

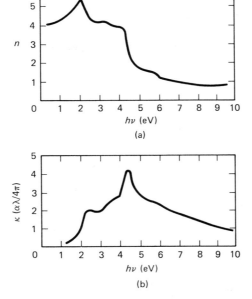

Figure 12.10 The spectral dependence of the real and imaginary part of the index of refraction, $n - i\kappa$, of Ge. (H. R. Philipp and E. A. Taft, *Phys. Rev.*, Vol. 113, p. 1002, 1959.)

which is a continuous function of k. The oscillator strength f_{cv} also depends on k. Following these steps, we obtain the integral expression for ϵ. However, the principal features of ϵ are more easily recognized when ϵ is considered as a superposition of several oscillators. In the many-oscillator model, we approximately represent the dielectric constant by a summation in Eq. (12.7.2), with each $\hbar\omega_{j0}$ representing the transition across the energy gap at a critical point where the effective density of state $D(\mathcal{E}) = D(k)\, dk/d\mathcal{E}$ is high.

For photon energies lower than the gap energy of a semiconductor, $\omega^2 << \omega_{j0}^2$ and Eq. (12.7.2) can be approximated by

$$\epsilon = \epsilon_0 + \frac{Ne^2}{m} \sum_j \frac{f_{j0}}{\omega_{j0}^2} = \epsilon_0 + \frac{N_e e^2}{m\omega_0^2} \qquad (12.7.3)$$

where ω_0 is a weighted average of all the ω_{j0}'s, and N_e is the number of electrons involved per unit volume. The value $\hbar\omega_0$ can be approximated by the energy at which $D(\mathcal{E})$ is the largest and hence the absorption curve shows a prominent maximum. In Ge, maximum absorption occurs at 4.3 eV (Fig. 12.10b), yielding $\omega_0 = 6.5 \times 10^{15}$ rad/s. The interatomic distance in Ge is 2.44 Å, corresponding to a value of $N = 4.5 \times 10^{28}$ atoms/m^3. Since there are four valence electrons per atom, $N_e = 1.8 \times 10^{29}$ electrons/m^3. Using the values of ω_0 and N_e in Eq. (12.7.3), we find that $\epsilon = 15\epsilon_0$, in fair agreement with the measured value of $16\epsilon_0$.

Figure 12.11 shows the dielectric constant (real part) of SrF$_2$ over a wide frequency range. Since SrF$_2$ is ionic, it possesses both α_e and α_i. The difference between the static dielectric constant ϵ_s (to which both α_e and α_i contribute) and the optical dielectric constant $\epsilon = n^2\epsilon_0$ (to which only α_e contributes) is due to ionic polarization. Ideal ionic crystals are electric insulators. In such crystals, the electron wave function

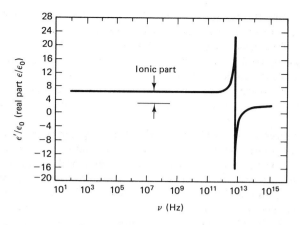

Figure 12.11 The spectral dependence of the real part of the dielectric constant ϵ'/ϵ_0 observed in SrF_2.

is highly localized, and the Lorentz field correction must be used. Using Eq. (12.1.11) in Eq. (12.1.4) and Eq. (12.6.13) for $\mathbf{E}_{\text{atom}}$, we have

$$\mathbf{P} = \frac{3\epsilon_0 N(\alpha_e + \alpha_i)\mathbf{E}}{3\epsilon_0 - N(\alpha_e + \alpha_i)}\,\mathbf{E} \tag{12.7.4}$$

Substituting Eq. (12.7.4) in Eq. (12.1.1) and rearranging terms, we obtain

$$\frac{\epsilon - \epsilon_0}{\epsilon + 2\epsilon_0} = \frac{1}{3\epsilon_0}\,N(\alpha_e + \alpha_i) \tag{12.7.5}$$

which is known as the *Clausius–Mossotti equation*.

We should point out that the presence of α_i is not limited to crystals whose bonding is purely ionic. In Table 12.1 the static and optical dielectric constant (ϵ_s and $\epsilon = n^2\epsilon_0$) for a selected number of insulators and semiconductors are given. As we can see from the difference between ϵ_s/ϵ_0 and n^2, the contribution from the ionic polarizability starts from zero in element semiconductors Ge and Si, and increases progressively from III–V compounds, to II–VI compounds, and then to I–VII compounds. This tendency is a direct result of the change in chemical bonding from a strongly covalent nature (III–V compounds) to a purely ionic nature (I–VII compounds). Theoretical

TABLE 12.1 STATIC AND OPTICAL DIELECTRIC CONSTANTS FOR A SELECTED NUMBER OF INSULATORS AND SEMICONDUCTORS

I–VII	ϵ_s/ϵ_0	$n^2 = \epsilon/\epsilon_0$	II–VI	ϵ_s/ϵ_0	$n^2 = \epsilon/\epsilon_0$
LiF	9.27	1.92	CdS	11.6	5.9
NaCl	5.62	2.25	ZnO	8.5	3.73
RbCl	5.0	2.9	CdTe	11	. . .

III–V	ϵ_s/ϵ_0	$n^2 = \epsilon/\epsilon_0$	IV	ϵ_s/ϵ_0	$n^2 = \epsilon/\epsilon_0$
GaAs	13.2	10.9	Ge	16	16
GaSb	15	14	Si	11.7	11.7
InSb	17	15.7			

analyses by Nozières and Pine (P. Nozières and D. Pine, *Phys. Rev.*, Vol. 109, p. 762, 1958) confirm the classical result of Darwin (C. G. Darwin, *Proc. R. Soc. London*, Vol. A146, p. 13, 1934). In essence, the Lorentz formula [Eq. (12.7.4)] should apply whenever the electrons are spatially well localized. This situation prevails in strongly ionic crystals, such as the alkali halides and the oxides (NaCl, MgO, Al_2O_3, for example). On the other hand, in strongly covalent semiconductors where electron distributions are highly polarizable and as a result $N\alpha \gg 1$, the Sellmeyer formula [Eq. (12.7.2)] is valid.

Let us turn our attention to the third group of materials, polar liquids and solids. In these materials, the total polarizability α consists of α_e, α_i, and α_d. For the electronic and ionic contributions, the treatment of the local-field correction is the same as outlined above. But for the dipolar contribution, the treatment is different. In the calculation of E_2 (the Lorentz field), we assumed $\mathbf{P}$ to be in the direction of $\mathbf{E}$. This assumption still holds for induced electronic and ionic dipoles. The permanent dipoles, however, are not all aligned in the direction of $\mathbf{E}$. It is true that the resultant macroscopic polarizaton $\mathbf{P}$ due to polar contribution is in the direction of $\mathbf{E}$; but insofar as the local field is concerned, it is the effect of the local dipole moment, not the effect of the macroscopic polarization, which is to be considered.

For a detailed analysis of the local-field effect in polar substances, the reader is referred to a paper by Onsager (L. Onsager, *J. Am. Chem. Soc.*, Vol. 58, p. 1486, 1936). The Onsager equation relating ϵ to α is

$$\epsilon - \epsilon_0 = \frac{3\epsilon C^2}{3\epsilon + \epsilon_0} N\alpha_d + \frac{\epsilon(2\epsilon_0 + \epsilon_\infty)}{\epsilon_0(2\epsilon + \epsilon_\infty)} N(\alpha_e + \alpha_i) \qquad (12.7.6)$$

where α_d is given by Eq. (12.4.6) and

$$C = \frac{(2\epsilon + \epsilon_0)(\epsilon_\infty + 2\epsilon_0)}{3\epsilon_0(2\epsilon + \epsilon_\infty)} \qquad (12.7.7)$$

The quantity ϵ_∞ which satisfies the relation

$$\frac{\epsilon_\infty - \epsilon_0}{\epsilon_\infty + 2\epsilon_0} = \frac{1}{3\epsilon_0} N(\alpha_e + \alpha_i) \qquad (12.7.8)$$

is the dielectric constant of the polar substance in the high-frequency region, where only α_e and α_i contribute to ϵ. Note that if there were no permanent dipoles, $\alpha_d = 0$ and Eq. (12.7.8) reduces to Eq. (12.7.5).

12.8 OPTICAL TRANSITIONS IN SOLIDS

In the preceding discussion, we used the classical harmonic-oscillator model to derive the electron polarizability. In this model the loss mechanism is represented phenomenologically by a damping term $\beta \, dx/dt$ in Eq. (12.2.8). From Eq. (12.2.12) and Fig. 12.4 we see that the lossy or imaginary part α_e'' of the electronic polarizability becomes appreciable when the angular frequency ω of an incident electromagnetic wave approaches one of the resonant frequencies ω_{j0} of an atomic system. Many optical devices, such as detectors, solar cells, luminescent devices, and lasers, utilize transitions between atomic levels either to absorb or to emit electromagnetic radiation to convert optical into electrical energy, or vice versa. In addition, measurements of optical absorption and emission spectra constitute an important experimental tool for material characterization. In this section and Section 12.9 we present a background discussion on the basic physics concerning the interaction of radiation with matter, which is followed by a more quantitative treatment of the subject in Sections 12.10 and 12.11.

In classical mechanics the Hamiltonian $\mathcal{H}_0$ is a sum of the kinetic and potential energies, or

$$\mathcal{H}_0 = \frac{p^2}{2m} + V(r) \tag{12.8.1}$$

In terms of $\mathcal{H}_0$ the equations of motion are

$$\frac{dp_x}{dt} = -\frac{\partial \mathcal{H}_0}{\partial x}, \quad \frac{dx}{dt} = \frac{\partial \mathcal{H}_0}{\partial p_x} \tag{12.8.2}$$

In the presence of an electromagnetic field, the motion of an electron in the x direction is given by

$$m \frac{d^2x}{dt^2} = -\frac{\partial V}{\partial x} - e(\mathbf{E} + \mathbf{v} \times \mathbf{B})_x \tag{12.8.3}$$

Because of $\mathbf{E}$ and $\mathbf{B}$, we must replace $\mathcal{H}_0$ of Eq. (12.8.1) by a new Hamiltonian,

$$\mathcal{H} = \frac{1}{2m} (\mathbf{p} + e\mathbf{A}) \cdot (\mathbf{p} + e\mathbf{A}) + V(r) \tag{12.8.4}$$

where $\mathbf{A}$ is the vector potential associated with the radiation field.

A formal treatment of the interaction of radiation with matter requires the use of quantum mechanics by replacing $\mathbf{p}$ by the momentum operator and expressing $\mathcal{H}$ as $\mathcal{H}_0 + \mathcal{H}_1$, or

$$\mathcal{H} = \mathcal{H}_0 + \frac{e}{2m} (\mathbf{p} \cdot \mathbf{A} + \mathbf{A} \cdot \mathbf{p}) + \frac{e^2 A^2}{2m} \tag{12.8.5}$$

where $\mathcal{H}_1$ is the perturbing Hamiltonian. Note that $\mathbf{A}$ contains a propagation factor and hence is dependent on spatial coordinates. A power series expansion of $\exp(i\mathbf{k}_r \cdot \mathbf{r})$ for the $\mathbf{p} \cdot \mathbf{A}$ and $\mathbf{A} \cdot \mathbf{p}$ terms in Eq. (12.8.5) gives rise to various interaction terms whose physical origin can be identified with that of its classical counterpart. In the following discussion, we obtain first the classical interaction terms and then use the correspondence principle to find their quantum-mechanical counterparts.

The electrical energy $\mathcal{E}_e$ of an electron distribution $\rho(x, y, z)$ in an electrical potential $\Phi(x, y, z)$ is

$$\mathcal{E}_e = - \iiint \rho(x, y, z) \; \Phi(x, y, z) \; dx \; dy \; dz \tag{12.8.6}$$

Expanding Φ around a chosen origin, for example, the position of the neucleus, we have

$$\Phi(x, y, z) = \Phi(0) + \sum x \frac{\partial \Phi}{\partial x} + \frac{1}{2} \sum xy \frac{\partial^2 \Phi}{\partial x \, \partial y} \tag{12.8.7}$$

where the summation signs are used to indicate a sum of all terms of similar form. For a nonpolar substance, the contribution of $\Phi(0)$ to $\mathscr{E}_e$ is zero. Realizing that the electric dipole and quadrupole moments are, respectfully, is given by

$$p_x = \int x\rho(x, y, z) \, dx \, dy \, dz \qquad (12.8.8)$$

$$Q_{xy} = \int xy\rho(x, y, z) \, dx \, dy \, dz \qquad (12.8.9)$$

we can rewrite Eq. (12.8.6) as

$$\mathscr{E}_e = -\mathbf{p} \cdot \mathbf{E} - \sum Q_{xy} \frac{\partial E_x}{\partial y} \qquad (12.8.10)$$

It is well known from electromagnetic theory that a circulating current constitutes a magnetic dipole (S. Ramo, J. R. Whinnery, and T. Van Duzer, *Fields and Waves in Communication Electronics,* John Wiley & Sons, Inc., New York, 1965). A relation exists between the magnetic moment $\mathbf{u}$ and the angular momentum $\mathbf{J}$ of an electron

$$\mathbf{u} = -g \frac{e}{2m} \mathbf{J} \qquad (12.8.11)$$

where g is known as the Landé splitting factor, and $g = 1$ for orbital motion and $g = 2$ for spinning motion of electrons. The magnetic dipole moment of Eq. (12.8.11) can be considered as the magnetic counterpart to the electric dipole moment represented by $-\mathbf{p} \cdot \mathbf{E}$ in Eq. (12.8.10). Therefore, in the presence of a radiation field, the magnetic energy associated with the orbital and spinning motion of electrons is

$$\mathscr{E}_m = -\mathbf{u} \cdot \mathbf{B} \qquad (12.8.12)$$

In quantum-mechanical treatment of the interaction of radiation with matter, the $\mathbf{p} \cdot \mathbf{A} + \mathbf{A} \cdot \mathbf{p}$ terms of the perturbation Hamiltonian $\mathscr{H}_1$ can be decomposed into its three principal components as

$$\mathscr{H}_1 = -e\mathbf{r} \cdot \mathbf{E} - \mathbf{u} \cdot \mathbf{B} - \sum exy \frac{\partial E_x}{\partial y} \qquad (12.8.13)$$

arranged in the order of descending importance. Although there are other terms in the series expansion due to higher-order multipoles, the three terms in Eq. (12.8.13) are sufficient to describe the interaction. Atomic transitions caused by the three terms are, respectively, called *electric-dipole, magnetic-dipole,* and *electric-quadrupole transitions.* Now let us estimate the relative orders of magnitude of the terms. Since electromagnetic energy is equally stored in E and H fields, $\epsilon E^2 = \mu H^2$. Thus we have

$$\frac{uB}{erE} \cong \left(\frac{e}{m} \hbar \right) \frac{me^2}{e4\pi\epsilon_0\hbar^2} (\mu\epsilon)^{1/2} = \frac{e^2}{4\pi\epsilon_0\hbar c} = \frac{1}{137} \qquad (12.8.14)$$

$$\frac{Q_{xy}}{erE} \frac{\partial E_x}{\partial y} \cong \frac{ea_1^2}{ea_1E} kE = ka_1 = 2\pi \frac{a_1}{\lambda} \qquad (12.8.15)$$

by taking (1) r to be the Bohr radius a_1, (2) gJ to be $2\hbar$ in Eq. (12.8.11), and (3) Q_{xy} to be ea_1^2. The Bohr radius a_1 is only 0.53 Å, while λ is on the order of 5000 Å in the visible region. Therefore, the three terms are approximately in the ratio $1:10^{-2}:10^{-4}$.

Examples of atomic transitions are given in Section 12.10 when we discuss transition probability and selection rules.

12.9 ABSORPTION AND EMISSION OF RADIATION

The correct description of absorption and emission processes in accordance with the quantum theory was first given by Einstein. If v is the frequency and λ the wavelength of an electromagnetic wave, only those atomic states of energy $\mathscr{E}_m$ and $\mathscr{E}_n$ that satisfy the energy relation

$$\Delta\mathscr{E} = \mathscr{E}_m - \mathscr{E}_n = hv \quad \text{or} \quad \lambda \text{ (in Å)} = \frac{1.24 \times 10^4}{\Delta\mathscr{E} \text{ (eV)}} \qquad (12.9.1)$$

will interact strongly with the light wave. We focus our attention on such states. According to the energy conservation relation expressed in Eq. (12.9.1), a transition from the higher-energy state m to the lower-energy state n results in the release of a photon of energy hv. Conversely, a transition from the lower-energy state to the higher-energy state requires the consumption of a photon of energy hv. Therefore, the emission and absorption of radiation is associated with a downward and upward transition, respectively.

According to the treatment by Einstein, the rate of downward transition is

$$R_{\text{down}} = [A + B_{mn}I(v)]N_m \qquad (12.9.2)$$

whereas the rate of upward transition is

$$R_{\text{up}} = B_{nm}I(v)N_n \qquad (12.9.3)$$

First, the two transition rates are proportional to the number of atoms in the initial atomic state as represented by N_m and N_n, respectively. Second, in the emission process, transitions from higher-energy states to lower-energy states either may take place without the help of electromagnetic radiation or may be induced by an optical beam of intensity $I(v)$. The former process is known as the *spontaneous emission process* and represented by the AN_m term, while the latter process is known as the *stimulated emission process* and represented by the $B_{mn}I(v)N_m$ term in Eq. (12.9.2). Third, in the absorption process, transitions from lower-energy states to higher-energy states are possible only when energy is provided by an optical beam. Therefore, only the stimulated absorption term appears in Eq. (12.9.3). Finally, since the spectral width Δv of an optical beam is finite, the beam intensity $I(v)$ has a dimension of W/m^2 per unit frequency interval.

To show validity of the two equations, we apply the results to the emission from an enclosed hot body known as *blackbody radiation*. Since by definition a blackbody absorbs electromagnetic radiation at all frequencies, the emitted radiation from a blackbody contains all frequency components. There exist in a blackbody so enormously many energy levels that their energy differences give a whole spectrum of v in Eq. (12.9.1), ranging from very low to very high frequencies. When a steady state is reached, the conversion of atomic energy into electromagnetic energy exactly balances the conversion of electromagnetic energy into atomic energy. The energy balance requires that $R_{\text{down}} = R_{\text{up}}$ in Eqs. (12.9.2) and (12.9.3), and this condition predicts a spectral distribution

$$I(\nu) = \frac{A}{B_{mn}} \frac{1}{(B_{nm}/B_{mn}) \exp{(h\nu/kT)} - 1} \tag{12.9.4}$$

Under thermal equilibrium, the populations of the states m and n are in the ratio of their respective Boltzmann factors. The relation $N_n/N_m = \exp{(\Delta\mathscr{E}/kT)}$ and Eq. (12.9.1) are used in obtaining Eq. (12.9.4).

The intensity distribution $I(\nu)$ of the blackbody radiation can be derived theoretically. The density of normal modes of electromagnetic waves in k space is given by

$$dN = \frac{2 \times 4\pi k^2 \, dk}{(2\pi)^3} \tag{12.9.5}$$

where $k = 2\pi/\lambda$ is the wave number of an electromagnetic wave. The factor 2 accounts for two independent orthogonal polarization directions. The average energy $\overline{\mathscr{E}}$ of each mode is given by the product of the photon energy $h\nu$ and the Bose–Einstein function $f(\nu)$, or

$$\overline{\mathscr{E}} = h\nu f(\nu) = \frac{h\nu}{\exp{(h\nu/kT)} - 1} \tag{12.9.6}$$

where $f(\nu)$ represents the occupancy of that mode. Multiplying Eq. (12.9.6) by Eq. (12.9.5) yields the energy density $U(\nu) \, d\nu$ of the electromagnetic radiation within a frequency interval between ν and $\nu + d\nu$. Since $I(\nu)$ is related to $U(\nu)$ by $I(\nu) = cU(\nu)$ and $k = 2\pi\nu/c$, where c is the velocity of light, we obtain

$$I(\nu) = \frac{8\pi h\nu^3}{c^2} \frac{1}{\exp{(h\nu/kT)} - 1} \tag{12.9.7}$$

which is known as the *Planck law of blackbody radiation*. A comparison of Eq. (12.9.4) with Eq. (12.9.7) yields

$$B_{mn} = B_{nm} = B \quad \text{and} \quad \frac{A}{B} = \frac{8\pi h\nu^3}{c^2} \tag{12.9.8}$$

The quantities $B_{mn}I(\nu)$ and $B_{nm}I(\nu)$ are known in quantum mechanics as the *transition probability* for the process from m to n and that from n to m, respectively. Using the time-dependent perturbation calculation, both quantities can be formulated in terms of the perturbation Hamiltonian $\mathscr{H}_1$ and the appropriate wave functions representing the states m and n. The result of the quantum-mechanical calculation verifies that the probability for downward transition and that for upward transition are indeed the same [Eq. (12.10.6)]. Since the quantum-mechanical transition probability $w = BI(\nu)$ represents transitions per unit time per state, the Einstein A coefficient has a dimension of s^{-1}. The reciprocal of A, that is A^{-1}, is generally known as the *spontaneous radiative lifetime*.

12.10 TRANSITION PROBABILITY AND SELECTION RULES

The quantum-mechanical transition probability w is derived in Appendix A10 and given by Eq. (A10.23):

$$w = \frac{2\pi}{\hbar} |\mathscr{H}_{mn}|^2 \delta(\mathscr{E}_m - \mathscr{E}_n - h\nu) \tag{12.10.1}$$

Dielectric and Optical Properties   Chap. 12

where $\delta(u)$ is a Dirac delta function with the following properties:

$$\int_{-\infty}^{\infty} \delta(u)\, du = 1 \quad \text{and} \quad \delta(u) = 0 \quad \text{for} \quad u \neq 0 \tag{12.10.2}$$

Letting $u = \mathscr{E}_m - \mathscr{E}_n - h\nu$, we see that the appearance of the delta function is a mathematical statement of the physical law of conservation of energy. The quantity $\mathscr{H}_{mn}$ is the matrix element [Eq. (A10.12)] connecting the initial state n with the final state m through the perturbation Hamiltonian $\mathscr{H}_1$. Mathematically, we have

$$\mathscr{H}_{mn} = \iiint \psi_m^* \mathscr{H}_1 \psi_n \, dx\, dy\, dz \tag{12.10.3}$$

where ψ_m and ψ_n are the wave functions of the two states involved in the transition.

In quantum mechanics, the expectation value of physical quantity L is given by

$$\langle L \rangle = \iiint \psi^* L \psi \, dx\, dy\, dz \tag{12.10.4}$$

Furthermore, operators that represent a physical quantity have the property

$$\iiint \psi_1^* L \psi_2 \, dx\, dy\, dz = \iiint \psi_2 L^* \psi_1^* \, dx\, dy\, dz \tag{12.10.5}$$

where ψ_1 and ψ_2 are any wave functions vanishing at infinity (J. L. Powell and B. Crasemann, *Quantum Mechanics,* Addision-Wesley Publishing Co., Inc., Reading, Mass., 1961, p. 295; N. F. Mott and I. N. Sneddon, *Quantum Mechanics and Its Applications,* Oxford University Press, Oxford, 1948, p. 37). Such operators are known as *Hermitian operators,* and the perturbation Hamiltonian is a Hermitian operator, as it can be expressed in terms of real functions of momentum p and spatial coordinate r. Using Eq. (12.10.5), we can easily show that

$$|\mathscr{H}_{mn}|^2 = |\mathscr{H}_{nm}|^2 \tag{12.10.6}$$

Equation (12.10.6) proves the statement that the probabilities for downward and upward transitions are equal or $B_{mn} = B_{nm}$ in Eqs. (12.9.2) and (12.9.3).

The conditions for nonzero matrix element are known as *selection rules.* For a hydrogen atom having a single electron, the relevant quantum numbers are the orbital quantum number l, the magnetic quantum number m, and the spin quantum number s. However, for a complex atom having several electrons in a solid, the atomic state is affected by various interaction energies. These include the electrostatic interaction between electrons having different orbitals and the interaction between the spin and the orbital motion of an electron. In an ionic crystal, the electrostatic potential caused by ions at neighboring lattice sites constitutes the third interaction term. The proper quantum numbers to use for complex atoms in a solid therefore depends on the relative strengths of the various interaction energies. Obviously, a thorough discussion of the selection rules is outside the scope of this book. Here we use simple examples to illustrate the basic principles.

The electric-dipole operator $e\mathbf{r}$ is an odd operator where $\mathbf{r}$ is the displacement vector of an electron with respect to the position of the nucleus. Under an inversion operation represented symbolically by I, the coordinates (x, y, z) change to $(-x, -y, -z)$. Therefore, $I(e\mathbf{r}) = -e\mathbf{r}$. Both the magnetic-dipole and electric-quadrupole operators are even operators as the sign of the orbital angular momentum $\mathbf{r} \times \mathbf{p}$ and that of the electric quadrupole moment exy remain unchanged after the inversion operation. Similarly, the electron wavefunctions can be classified into odd and even parity, with $I\psi = -\psi$ for the former and $I\psi = \psi$ for the latter. For nonzero $|\mathcal{H}_{mn}|^2$, the product $\psi_m^* \mathcal{H}_1 \psi_n$ must be an even function. Thus for electric-dipole transitions, the initial and final states must have opposite parity, that is, either an even–odd or an odd–even combination. These transitions are called *allowed electric-dipole transitions*. For magnetic-dipole and electric-quadrupole transitions, the two states m and n must have the same parity. Applying the parity selection rule to a hydrogen atom, we find that the allowed electric-dipole transitions involve $s \leftrightarrow p$ and $p \leftrightarrow d$ states, for example, whereas the $s \leftrightarrow d$ transitions are forbidden.

Ionic Crystals

Among the solid-state laser materials, Nd:YAG (yttrium aluminum garnet, $Y_3Al_5O_{12}$) and $Cr:Al_2O_3$ (ruby) are the most commonly known. In these materials, ionic crystals are used as the host, whereas either a rare-earth element such as Nd or a transition element of the iron group such as Cr is the active dopant. A Cr^{3+} ion has an electron configuration $3d^3$ outside the stable argon configuration, while a Nd^{3+} ion has an electron configuration $4f^3 5s^2 5p^6$ outside the stable xenon configuration. Since the $3d^3$ electrons occupy the outmost shell in Cr^{3+}, they feel strongly the influence of the crystalline potential. An analysis of the absorption and emission spectra would require knowledge of the group theory and hence will not be treated here. On the other hand, since the $4f^3$ electrons of Nd^{3+} are shielded by the $5s^2 5p^6$ electrons, they are not much affected by the crystalline potential. As a result, a Nd^{3+} ion in a solid can be treated as a free ion with a total orbital quantum number $L = 3 + 2 + 1 = 6$ and a total spin quantum number $s = \frac{3}{2}$. Since the $4f$ shell is less than half filled, the ground state of Nd^{3+} has a total angular momentum quantum number $J = L - S = \frac{9}{2}$, and in spectroscopic notation it is designated as $^4I_{9/2}$. The choice of $m_1 = 3$, $m_2 = 2$, and $m_3 = 1$ for the three electrons is to minimize the electrostatic interaction among electrons.

Figure 12.12 shows (a) the absorption spectrum observed in $Nd^{3+} : CaWO_4$ and (b) the relevant energy levels. There are four vectorial combinations of **L** and **S**, giving rise to a value for J ranging from $\frac{9}{2}$ to $\frac{15}{2}$. One level $^4I_{11/2}$ is the terminal state for the 1.06-μm laser transition. The superscript 4 describes the spin degeneracy $2S + 1$ of the state. The excited states represent the various combinations of the magnetic quantum numbers with (1) $m_1 = 2$, $m_2 = 1$, and $m_3 = 0$ for F; (2) $m_1 = 3$, $m_2 = 2$, and $m_1 = 0$ for H; (3) $m_3 = 3$, $m_2 = 1$, and $m_1 = 0$ for G; and (4) $m_3 = 3$, $m_2 = -2$, and $m_3 = -1$ for S. Since the ^4I, ^4F, ^4G, and ^4S levels of Fig. 12.12b all lie within the $4f$ shell, there is no parity change between them. In other words, the selection rule $\Delta l = \pm 1$ for electric-dipole transition is not observed. Therefore, electric-dipole transitions involving various ^4I, ^4F, ^4G, and ^4S states are forbidden in free Nd^{3+} ions. This conclusion also applies to levels of other rare-earth ions lying wholly within the $4f$ shell.

The situation is changed somewhat for ions in a crystalline environment. The crystal potential produces a small amount of admixture of orbitals of higher energy. Above the $4f$ orbital, there is the $4d$ orbital which is of opposite polarity to the $4f$ orbital. Therefore, for a rare-earth ion RE^{3+} having an electron configuration $4f^n 5s^2 5p^6$, the next group of energy levels will have an electron configuration $4f^{n-1} 5s^2 5p^6 5d$. The energy $\mathcal{E}_{fd}$ required to promote an electron from $4f$ to $5d$ is about 15,000 cm^{-1}, while the crystal-field splitting Δ is about 150 cm^{-1}. From the perturbation theory, the amount of admixture of the $4f^{n-1} 5d$ configuration into $4f^n$ configuration is expected to be about

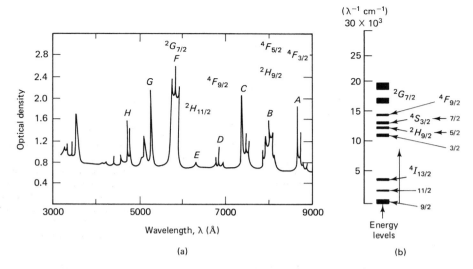

Figure 12.12 (a) The absorption spectrum of Nd^{3+} dopants in a host $CaWO_4$ crystal and (b) the relevant energy levels. Optical density is defined as $\log_{10}(I_{in}/I_{out}) = \alpha d/2.3$, where α is the absorption coefficient (cm^{-1}) and d is the sample thickness. For example, the values of $\alpha = 10\ cm^{-1}$ and $d = 0.4\ cm$ give a value of 1.73 for optical density. (L. F. Johnson, *J. Appl. Phys.*, Vol. 34, p. 897, 1963.)

$\Delta/\mathscr{E}_{fd}$. The oscillator strength for the $4f^n$-to-$4f^{n-1}5d$ transition is found to be 1.5×10^{-2} (Section 12.11). Therefore, the oscillator strength caused by the admixture should be on the order of $1.5 \times 10^{-2} \times (\Delta/\mathscr{E}_{fd}) = 1.5 \times 10^{-6}$, in general agreement with the observed value of 3×10^{-6} in Nd^{3+} : $CaWO_4$.

The transition in the Nd^{3+} : YAG and Cr^{3+} : Al_2O_3 lasers are all second-order electric-dipole transitions. Because in first-order theory, electric-dipole transitions are not allowed within the $4f^3$ configuration of Nd^{3+} and the $3d^3$ configuration of Cr^{3+}, these transitions are also known as *forbidden transitions*. In second-order theory, a small amount of interconfigurational mixing of states of opposite parity is introduced. The mixing of $4f$–$5d$ orbital in Nd^{3+} and the mixing of $3d$–$4p$ orbital in Cr^{3+} make the electric-dipole matrix element nonzero. Because the amount of admixture is small, the oscillator strength for the second-order transition is much smaller than that for the first-order transition.

Note that the ratio $\Delta/\mathscr{E}_{fd}$ is of the same order of magnitude as the ratio uB/erE in Eq. (12.8.14). Therefore, it is desirable to be able to experimentally distinguish between a second-order electric-dipole transition and a first-order magnetic-dipole transition. Refer to the experimental arrangement shown in Fig. 12.13 for a uniaxial crystal. The absorption spectrum can be observed in three different arrangements: (1) with propagation vector **k** along the optic axis (α spectrum), (2) with **k** and **E** perpendicular to the optic axis (σ spectrum), and (3) with **k** and **H** perpendicular to the optic axis (π spectrum). If the **E** field is active (as in electric-dipole transition), the σ spectrum should be identical to the α spectrum because the **E** field lies in the plane of symmetry. If the **H** field is active (as in magnetic-dipole transition), the π spectrum should be identical to the α spectrum. Examples of magnetic-dipole transitions are the $^5I_7$-to-$^5I_8$ transition in

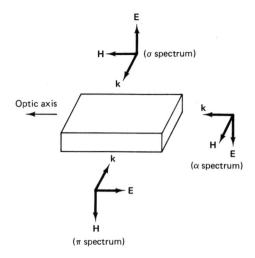

E

H ← (σ spectrum)

k

Optic axis ←

k

H

E
(α spectrum)

k

E

H
(π spectrum)

Figure 12.13 Three basic arrangements for measuring absorption spectra, illustrating different relationships between the radiation-field vectors and the crystal axis. The optic axis is the crystal axis along which the index of refraction is different from that in the plane perpendicular to the axis.

Dy^{2+} : CaF_2 (Z. J. Kiss, *Proc. IRE,* Vol. 50, p. 1531, 1962), the $^2F_{5/2}$-to-$^2F_{7/2}$ transition in Tm^{2+} : CaF_2 (Z. J. Kiss, *Phys. Rev.,* Vol. 127, p. 719, 1962), and the $^5D_0$-to-$^7F_1$ transition in Sm^{2+} : SrF_2 (D. L. Wood and W. Kaiser, *Phys. Rev.,* Vol. 126, p. 2079, 1962).

Semiconductors

In ion crystals the wave function ψ of the active dopant, for example Nd^{3+} in YAG, is highly localized and hence optical transitions take place between discrete levels. Now we turn our attention to optical transitions of a different kind involving continuum states. In semiconductors, transitions are possible among conduction- and valence-band states. According to our discussion in Section 5.3, the wave function ψ for a band state is in the Bloch form, or

$$\psi(r) = \frac{1}{\sqrt{N}} u(r) \exp(i\mathbf{k} \cdot \mathbf{r}) \qquad (12.10.7)$$

where $u(r)$ represents the cell wave function, $\exp(i\mathbf{k} \cdot \mathbf{r})$ is a phase factor, and N is a normalization constant representing the number of unit cells. Because $\psi(r)$ propagates throughout the crystal, the integration in Eq. (12.10.3) must be carried out over the whole crystal. In contrast, for ionic crystals, the integration in Eq. (12.10.3) is over an isolated laser-active atom.

To simplify the integration procedure, we take note of the two important properties of a Bloch wave:

$$u(\mathbf{r} + \mathbf{R}) = u(\mathbf{r}) \quad \text{and} \quad \psi(\mathbf{r} + \mathbf{R}) = \psi(\mathbf{r}) \exp(i\mathbf{k} \cdot \mathbf{R}) \qquad (12.10.8)$$

where $\mathbf{R}$ is a translation vector for the unit cell. In carrying out the integration in Eq. (12.10.3), we divide a crystal into unit cells, perform the integration over each cell, and then sum up all the contributions from individual cells. If the field $\mathbf{E}$ varies as $\mathbf{E}_1 \exp i\mathbf{k}_r \cdot \mathbf{r}$, the fields in different cells differ by a phase factor $\exp(i\mathbf{k}_r \cdot \mathbf{R})$, where $\mathbf{k}_r$ is the wave vector for the radiation field. Mathematically, we let

$$\mathcal{H}_{cv} = F_1 F_2 \qquad (12.10.9)$$

where the integral F_1 is over a unit cell

$$F_1 = \iiint_{\text{cell}} u_c^*(\mathbf{r}) e \mathbf{E}_1 \cdot \mathbf{r} u_v(\mathbf{r}) \, dx \, dy \, dz \qquad (12.10.10)$$

Because $u(\mathbf{r})$ is periodic, F_1 is the same for all cells. The subscripts c and v stand for conduction- and valence-band states, respectively.

The factor F_2 represents the sum of the phase factors from ψ_c^*, ψ_v and $\mathbf{E}$, and is given by

$$F_2 = \frac{1}{N} \sum_{\text{crystal}} \exp i(\mathbf{k}_v - \mathbf{k}_c + \mathbf{k}_r) \cdot \mathbf{R} \qquad (12.10.11)$$

Since $\mathbf{R}$ can be expressed in terms of the basis vectors of a unit cell, the sum is in the form of $\Sigma \exp (in\theta)$, with n being an integer. For nonzero θ, mutual cancellation of the terms reduces the sum to negligible magnitude. Therefore, the condition for nonzero F_2 is

$$\mathbf{k}_c = \mathbf{k}_v + \mathbf{k}_r \qquad (12.10.12)$$

When Eq. (12.10.12) is satisfied, $F_2 = 1$. Equation (12.10.12) can be considered as a statement of conservation of momentum. The electron and photon momenta are approximately in the ratio $\hbar k_e / \hbar k_r = \lambda / a$, where a is the lattice constant. Since k_e is much larger than k_r, Eq. (12.10.12) reduces to

$$\mathbf{k}_c = \mathbf{k}_v \qquad (12.10.13)$$

In addition to the parity selection rule, optical transitions between band states must also conserve momentum.

12.11 ABSORPTION COEFFICIENT, OSCILLATOR STRENGTH, AND SPONTANEOUS LIFETIME

When an electromagnetic wave propagates through a dielectric medium, it causes transitions between atomic states if the energy relation of Eq. (12.9.1) is satisfied. As a result of photon absorption and emission, the intensity of the wave changes as it propagates. Let $I(v)$ be the intensity of the beam. Since the rate of both stimulated processes is proportional to $I(v)$, the spatial variation of $I(v)$ is given by

$$\frac{dI}{dx} = -\alpha I \qquad (12.11.1)$$

the proportionality constant α is called the *absorption coefficient*. If we let

$$\epsilon = \epsilon' + \epsilon'' = \epsilon_0 (n + i\kappa)^2 \qquad (12.11.2)$$

the field varies as $\exp [i(n + i\kappa)k_0 x]$, where n is the index of refraction and κ is the extinction coefficient. Therefore, the absorption coefficient α is related to κ and ϵ'' by

$$\alpha = 2\kappa k_0 = \frac{\epsilon'' k_0}{\epsilon_0 n} \qquad (12.11.3)$$

where $k_0 = 2\pi/\lambda$ is the free-space wave number.

Note that if there are N dopant atoms in volume V, the contribution from the ground to excited state $(0 \rightarrow j)$ transition to the imaginary part of the electric constant, according to Eq. (12.2.10), can be approximated by

$$\epsilon'' = \frac{Ne^2}{2Vm\omega_{j0}} \frac{f_{j0}(\Delta\omega/2)}{(\omega_{j0} - \omega)^2 + (\Delta\omega/2)^2} \tag{12.11.4}$$

where $\Delta\omega = \beta/m$ is the full width of the absorption line. Substituting Eq. (12.11.4) in Eq. (12.11.3) and then integrating both sides with respect to ν, we obtain

$$f_{j0} = \left(\frac{V}{N}\right) \frac{4m\epsilon c}{e^2} \int \alpha(\nu) \, d\nu \tag{12.11.5}$$

As an example, we refer to the absorption measurement of the $4f^n$-to-$4f^{n-1}5d$ transitions in $Sm^{2+} : CaF_2$ reported by Wood and Kaiser (D. L. Wood and W. Kaiser, *Phys. Rev.*, Vol. 126, p. 2079, 1962). These transitions are responsible for the absorptions in the ultraviolet region, with λ^{-1} ranging from 33,000 to 41,841 cm^{-1}, with an average oscillator strength f of 1.5×10^{-2}.

The absorption coefficient and hence the oscillator strength can also be expressed in terms of the spontaneous lifetime τ, which can be measured experimentally. Note that the rate of stimulated process is $BI(\nu)N$ per unit time. Since for each upward transition, one photon of energy $h\nu$ is absorbed and for each downward transition one photon of energy $h\nu$ is emitted, the net change of energy per unit time is

$$\frac{d\mathcal{E}}{dt} = BI(\nu)(N_j - N_0)h\nu \tag{12.11.6}$$

Furthermore, we assume that the population difference distributes over a frequency spectrum

$$\rho(\nu) = \frac{\Delta\omega}{(\omega_{j0} - \omega)^2 + (\Delta\omega/2)^2} \tag{12.11.7}$$

such that $\int \rho(\nu) \, d\nu = 1$. Since $\mathcal{E}c/V \pm I(\nu) \, d\nu$ and $d\mathcal{E}/dt = c \, d\mathcal{E}/dx$, we have

$$\frac{dI(\nu)}{dx} = \frac{BI(\nu)(N_j - N_0)\rho(\nu)h\nu}{V} \tag{12.11.8}$$

In an absorption experiment, the upper energy state is not excited. Letting $N_j = 0$ and using Eq. (12.9.8), we find that

$$\alpha(\nu) = \frac{-1}{I} \frac{dI}{dx} = \frac{N_0}{V} B\rho(\nu) = \frac{c^2}{8\pi\nu^2} \frac{\rho(\nu)}{\tau} \frac{N_0}{V} \tag{12.11.9}$$

Integrating $\alpha(\nu)$ over $d\nu$, we obtain

$$\int \alpha(\nu) \, d\nu = \frac{N_0}{V} \frac{c^2}{8\pi\nu^2\tau} \tag{12.11.10}$$

where $\tau = A^{-1}$ is the spontaneous lifetime. Substituting Eq. (12.11.10) in Eq. (12.11.5) yields

$$f_{j0} = \frac{m\epsilon c^3}{2\pi e^2 \nu^2} \frac{1}{\tau} \tag{12.11.11}$$

Because part of the luminescent dopants may not occupy the desirable lattice site and hence may not participate in the absorption and emission process, it is difficult to know the value of N/V in Eqs. (12.11.5) and (12.11.10). On the other hand, the value of τ can be determined experimentally. Under the circumstances, Eq. (12.11.11) is more useful than Eq. (12.11.5) in finding the oscillator strength. Values of τ ranging from 2×10^{-6} to 1×10^{-2} were reported at 20 K for Sm^{2+} in CaF_2 for the $\lambda^{-1} = 14,114$ cm^{-1} line and in SrF_2 for the $\lambda^{-1} = 14,353$ cm^{-1} line. These values correspond to $f_{j0} = 10^{-2}$ and 2×10^{-6}, respectively. The value of f_{j0} indicates that the former line is due to an electric-dipole transition, while the latter line is due to either a first-order magnetic-dipole transition or a second-order electric-dipole transition.

The spontaneous lifetime τ can be calculated by quantum mechanics. Taking $E_x = E_1 \cos (k_r z - \omega t)$ and using Eq. (12.10.1), we find for the electric-dipole transition

$$w = \frac{2\pi}{\hbar} e^2 \langle x \rangle^2 \left(\frac{E_1}{2} \right)^2 \delta(\mathscr{E}_m - \mathscr{E}_n - h\nu) \qquad (12.11.12)$$

The factor $\frac{1}{2}$ comes from the fact that only the component $(E_1/2) \exp[i(\omega t - k_r z)]$ is involved in an absorption process; the other component $E_1/2 \exp[i(k_r z - \omega t)]$ is involved in an emission process. Note that the average density of an electromagnetic wave is given by the real part of $\mathbf{E} \times \mathbf{H}^*/2$, or $c\epsilon E_1^2/2$. Thus E_1^2 is related to $I(\nu)$ by

$$E_1^2 = \frac{2}{c\epsilon} \int I(\nu) \, d\nu = \frac{2}{c\epsilon h} \int I(\nu) \, dh\nu \qquad (12.11.13)$$

Substituting Eq. (12.11.13) in Eq. (12.11.12), we find

$$B = \frac{w}{I(\nu)} = \frac{e^2}{2\hbar^2} \sqrt{\frac{\mu}{\epsilon}} \langle x \rangle^2 \qquad (12.11.14)$$

From Eq. (12.9.8) we obtain

$$\frac{1}{\tau} = A = \frac{8\pi^2 e^2 \nu^3}{3\hbar c^2} \sqrt{\frac{\mu}{\epsilon}} (\langle x \rangle^2 + \langle y \rangle^2 + \langle z \rangle^2) \qquad (12.11.15)$$

For spontaneous emission, the polarization direction is unspecified; therefore, an average of $\Sigma \langle x \rangle^2$ is taken. If we define a dimensionless quantity f_{mn} called the *oscillator strength*,

$$\langle x \rangle^2 + \langle y \rangle^2 + \langle z \rangle^2 = \Sigma \left| \iiint \psi_m^* x \psi_n \, dx \, dy \, dz \right|^2 = \frac{3\hbar f_{mn}}{2m\omega} \qquad (12.11.16)$$

then Eqs. (12.11.15) and (12.11.11) are identical.

Let us now turn our attention to semiconductors. Because of the condition for momentum conservation expressed in Eq. (12.10.13) and because of the Pauli exclusion principle, the population difference $N_j - N_0$ is

$$N_j - N_0 = V \frac{8\pi k^2 \, dk}{(2\pi)^3} [f_c(1 - f_v) - f_v(1 - f_c)] \qquad (12.11.17)$$

Refer to the energy-band diagram of Fig. 12.14. In a vertical transition, the energies of the upper (conduction band) and lower (valence band) states are, respectively, given by

$$\mathscr{E}_m = \mathscr{E}_j = \mathscr{E}_c + \frac{\hbar^2 k^2}{2m_c} \tag{12.11.18}$$

$$\mathscr{E}_n = \mathscr{E}_0 = \mathscr{E}_v - \frac{\hbar^2 k^2}{2m_v} \tag{12.11.19}$$

Letting $m_r^{-1} = m_c^{-1} + m_v^{-1}$ and subtracting $\mathscr{E}_0$ from $\mathscr{E}_j$, we obtain

$$\mathscr{E}_m - \mathscr{E}_n = \mathscr{E}_g + \frac{\hbar^2 k^2}{2m_r} \tag{12.11.20}$$

Thus, using w of Eq. (12.11.12) for $BI(\nu)$ in Eq. (12.11.6), we find

$$\frac{d}{dt}\frac{\mathscr{E}}{V} = \frac{2\nu e^2 \langle x \rangle^2}{\hbar^3} \left(\frac{E_1}{2}\right)^2 (2m_r)^{3/2} (h\nu - \mathscr{E}_g)^{1/2}(f_c - f_v) \tag{12.11.21}$$

Thus, the absorption coefficient $\alpha = -\mathscr{E}^{-1} d\mathscr{E}/dx$ is given by

$$\alpha = \frac{\nu e^2}{\epsilon c \hbar^3} (2m_r)^{3/2}\langle x \rangle^2 (h\nu - \mathscr{E}_g)^{1/2} (f_v - f_c) \tag{12.11.22}$$

In direct-gap, III–V compound semiconductors, the cell function $u(r)$ in Eq. (12.10.10) at $k(000)$ is known to be s-like for conduction band states and p-like for valence band states. The fact that the valence band consists of the heavy-hole, the light-hole, an the split-off bands is a manifestation of the p-like wave function. The matrix element connecting the conduction- and valence-band states was calculated by Kane (E. O. Kane, *J. Phys. Chem. Solids*, Vol. 1, p. 249, 1957) using the $\mathbf{k} \cdot \mathbf{p}$ approximation, and is found to be related to the electron effective mass by (Problem 12.23)

$$\langle x \rangle^2 = \frac{3}{2m_{co}\omega^2} \frac{\mathscr{E}_{go}(\mathscr{E}_{go} + \Delta)}{3\mathscr{E}_{go} + 2\Delta}\left(1 - \frac{m_{co}}{m_0}\right) \tag{12.11.23}$$

where Δ is the energy separation of the split-off band from the top of the other two valence bands. The subscript 0 is added to refer to quantities at $k(000)$ so that Eq.

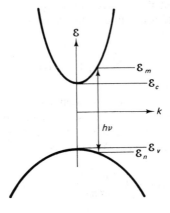

Figure 12.14 Relevant energies for a vertical transition between conduction-band and valence-band states in a direct-gap semiconductor.

(12.11.23) can also be applied to optical transitions at $k(000)$ in indirect-gap semiconductors.

The value of $\langle x \rangle^2$ can be determined either from Eq. (12.11.22) by measuring the absorption coefficient α or from Eq. (12.11.23) by using the known effective mass m_{co}. The absorption curve can be fitted to the following curve

$$\alpha = K(\hbar\omega - \mathcal{E}_g)^{1/2} \tag{12.11.24}$$

where the parameter K is given by

$$K = \frac{e^2 v}{\epsilon c \hbar^3} (2m_r)^{3/2} \langle x \rangle^2 \tag{12.11.25}$$

An ordinary absorption experiment is generally done in a pure sample. Therefore, we set $f_v = 1$ and $f_c = 0$ in Eq. (12.11.22). In GaAs, the value of K is found to be 5×10^4 cm^{-1}(eV)$^{-1/2}$. Also we have the values of $m_{co} = 0.068m$, $\epsilon = 11\epsilon_0$, $\mathcal{E}_{go} = 1.42$ eV and $\Delta = 0.33$ eV. With these values both Eqs. (12.11.23) and (12.11.25) yield a value of 1.5×10^{-18}m^2 for $\langle x \rangle^2$. Using this value of $\langle x \rangle^2$, we find a value of $\tau = 2.5 \times 10^{-9}$ s for the spontaneous lifetime. In semiconductors, the electron orbit is much larger than the corresponding orbit in atomic state due to a larger dielectric constant and a smaller effective mass. The large value of $\langle x \rangle^2$ results in a strong absorption and a fast lifetime.

12.12 LYDDANE–SACHS–TELLER AND KRAMERS–KRONIG RELATIONS

Some optical and dielectric properties of solids are difficult to measure experimentally. Therefore, it will be very useful to establish a theoretical relation between such a physical quantity to one that can readily be measured. One such relation concerns the longitudinal and transverse optical phonon frequencies v_{LO} and v_{TO} in Fig. 2.38. As discussed in Section 6.8 with reference to Eq. (6.8.13), the displacement vector $\mathbf{s}$ associated with lattice vibration can be decomposed into the longitudinal component $\mathbf{s}_l$ and the transverse component $\mathbf{s}_t$. The ionic contribution to the dielectric constant is discussed in Section 12.3. The vector $\mathbf{s}$ represents the relative displacement $\mathbf{u} - \mathbf{v}$ of positive and negative ions. Therefore, from Eqs. (12.3.2) and (12.3.3), we obtain an equation of motion for $\mathbf{s}$ as follows:

$$M_r \frac{d^2\mathbf{s}}{dt^2} + 2\gamma\mathbf{s} = e^*\mathbf{E}_{atom} \tag{12.12.1}$$

after letting $qa = 0$ (the Γ point). The mass $M_r = M_1 M_2/(M_1 + M_2)$ is known as reduced mass. The quantity $e^* = sze$ represents the effective charge carried by each ion and its value can be determined from the difference between ϵ_s/ϵ_0 and n^2 through the use of Eq. (6.8.1) and Table 12.1. The quantity $\mathbf{E}_{atom}$ is the effective field an ion experiences and is given by Eq. (12.6.13) in a cubic crystal.

Substituting Eq. (6.8.12) in Eq. (12.6.13) for $\mathbf{E}_{atom}$, we can express Eq. (12.12.1) in terms of the applied field $\mathbf{E}$ as

$$M_r \frac{d^2\mathbf{s}}{dt^2} + 2\gamma'\mathbf{s} = \frac{e^*(\epsilon_\infty + 2\epsilon_0)}{3\epsilon_0} \mathbf{E} \tag{12.12.2}$$

where γ' is modified by the coupling between polarization and displacement and is given by

$$2\gamma' = M_r\omega_0^2 = 2\gamma - \frac{N_0e^{*2}(\epsilon_\infty + 2\epsilon_0)}{(3\epsilon_0)^2} \qquad (12.12.3)$$

We should point out that even without an applied field, there is an electric field **E** associated with the LO lattice wave. According to Eq. (6.8.15), this field is given by

$$\mathbf{E} = \frac{-Ne^*(\epsilon_\infty + 2\epsilon_0)}{3\epsilon_0\epsilon_\infty}\,\mathbf{s}_l \qquad (12.12.4)$$

Substituting Eq. (12.12.4) in Eq. (12.12.2) and separating **s** into $\mathbf{s}_l$ and $\mathbf{s}_t$, we obtain

$$M_r\frac{d^2s_t}{dt^2} + 2\gamma's_t = 0 \qquad (12.12.5)$$

$$M_r\frac{d^2s_l}{dt^2} + 2\left[\gamma + (\gamma - \gamma')\frac{2\epsilon_0}{\epsilon_\infty}\right]s_l = 0 \qquad (12.12.6)$$

We note that at $\omega = 0$, Eq. (12.12.2) yields the relation between static displacement s_s and dc field E_0 as follows:

$$\mathbf{s}_s = \frac{e^*(\epsilon_\infty + 2\epsilon_0)}{3\epsilon_0\gamma'}\,\mathbf{E}_0 \qquad (12.12.7)$$

Use of Eq. (12.12.7) in Eq. (6.8.12) yields the static dielectric constant

$$\epsilon_s = \epsilon_\infty + \frac{(\epsilon_\infty + 2\epsilon_0)(\gamma - \gamma')}{\gamma'} \qquad (12.12.8)$$

Eliminating γ from Eq. (12.12.6), we obtain

$$M_r\frac{d^2s_l}{dt^2} + 2\gamma'\frac{\epsilon_s}{\epsilon_\infty}s_l = 0 \qquad (12.12.9)$$

Therefore, the LO and TO phonon frequencies near the zone center ($qa \ll 1$) are in the ratio

$$\frac{\nu_{\mathrm{LO}}}{\nu_{\mathrm{TO}}} = \sqrt{\frac{\epsilon_s}{\epsilon_\infty}} \qquad (12.12.10)$$

which is known as the *Lyddane–Sachs–Teller relation* (R. H. Lyddane, R. G. Sachs, and E. Teller, *Phys. Rev.*, Vol. 59, p. 673, 1941; A. Anselm, *Introduction to Semiconductor Theory* (English translation, Izdatelstvo Mir, Moscow), Prentice-Hall, Inc., Englewood Cliffs, N.J., 1981, p. 167). Since it is easier to measure ν_{TO} than ν_{LO}, Eq. (12.12.10) can be used as a check for the measured ν_{LO}. From the data given in Table 11.5 we find that this relation is indeed satisfied. We should point out, however, that the values of $\nu_{\mathrm{TO}}(\Gamma)$ and $\nu_{\mathrm{LO}}(\Gamma)$ are taken at the Γ point in the reciprocal lattice because of the condition $qa = 0$ used in obtaining Eq. (12.12.1).

In Sections 12.1 to 12.7 we derived expressions for the various contributions to the dielectric constant, and in Sections 12.10 and 12.11 we treated the problem of absorption. We note from Eqs. (12.7.2) and (12.11.4) that the real and imaginary parts of ϵ depend on a quantum-mechanical quantity f_{ij} called oscillator strength. Therefore, it is reasonable to expect that a general relation exists between ϵ' (real part) and ϵ'' (imaginary part). From Eq. (12.2.8) we see that a causality relation exists between displacement x and applied field E. We further note that the macroscopic polarization P is the volume average of the microscopic dipole moment p through Eq. (12.1.4). In describing the temporary (time-dependent) behavior of $\mathbf{D}$, Eq. (12.1.10) for the steady-state response must be replaced by a dynamic relation

$$\mathbf{D}(t) = \epsilon_0 \mathbf{E}(t) + \int_{-\infty}^{t} f(t - t') \, \mathbf{E}(t') \, dt' \qquad (12.12.11)$$

The integral represents the response of an atomic system at time t to the applied E at a prior time t', and corresponds to the contribution from $\mathbf{P}$ to $\mathbf{D}$ in Eq. (12.1.10). In other words, the contribution to polarization change $d\mathbf{P}(t)$ at time t due to an electric field $\mathbf{E}(t')$ applied between t' and $t' + dt'$ is given by

$$d\mathbf{P}(t) = f(t - t') \, \mathbf{E}(t') \, dt' \qquad (12.12.12)$$

The temporary behavior of a physical quantity is related to its spectral response through the Fourier transform. Therefore, we let

$$\mathbf{D}(t) = \frac{1}{2\pi} \int_{-\infty}^{\infty} \mathbf{D}(\omega) \exp(-i\omega t) \, d\omega \qquad (12.12.13)$$

$$\mathbf{E}(t) = \frac{1}{2\pi} \int_{-\infty}^{\infty} \mathbf{E}(\omega) \exp(-i\omega t) \, d\omega \qquad (12.12.14)$$

Substituting Eqs. (12.12.13) and (12.12.14) into Eq. (12.12.11) and performing the inverse transform, we obtain

$$\int_{-\infty}^{\infty} \{\mathbf{D}(\omega) - [\epsilon_0 + f(\omega)] \, \mathbf{E}(\omega)\} \exp(-i\omega t) \, d\omega = 0 \qquad (12.12.15)$$

where $f(\omega)$ is given by

$$f(\omega) = \int_{0}^{\infty} f(t) \exp(i\omega t) \, dt \qquad (12.12.16)$$

Since Eq. (12.12.15) must be satisfied for all t, a general relation exists between the Fourier components

$$\mathbf{D}(\omega) = [\epsilon_0 + f(\omega)] \, \mathbf{E}(\omega) \qquad (12.12.17)$$

From Eq. (12.12.17), we have

$$\epsilon(\omega) = \epsilon'(\omega) + i\epsilon''(\omega) = \epsilon_0 + \int_{0}^{\infty} f(t) \exp(i\omega t) \, dt \qquad (12.12.18)$$

In a passive system, the integral behaves properly in the upper half of the complex frequency $\omega = \omega_r + i\omega_i$ plane. In other words, the dielectric function $\epsilon(\omega)$ has no singularity in the upper $(\omega_r + i\omega_i)$ plane. Applying the Cauchy theorem to the function $[\epsilon(\omega) - \epsilon_0]/(\omega - \omega_0)$, we obtain (A. Anselm, p. 617; T. S. Moss, *Optical Properties of Semiconductors*, Academic Press, Inc., New York, 1959)

$$\epsilon'(\omega) = \epsilon_0 + \frac{2}{\pi} \int_0^\infty \frac{\omega' \epsilon''(\omega') \, d\omega'}{\omega'^2 - \omega^2} \qquad (12.12.19)$$

$$\epsilon''(\omega) = -\frac{2\omega}{\pi} \int_0^\infty \frac{[\epsilon'(\omega') - \epsilon_0] \, d\omega'}{\omega'^2 - \omega^2} \qquad (12.12.20)$$

The integrals are carried out along the real ω axis. These relations are known as the Kramer–Kronig relations (A. Anselm, *Introduction to Semiconductor Theory* (English translation, Izdatelstvo Mir, Moscow), Prentice-Hall, Inc., Englewood Cliffs, N.J., 1981, p. 401; J. D. Jackson, *Classical Electrodynamics,* 2nd ed., John Wiley & Sons, Inc., New York, 1975, p. 311; H. A. Kramers and R. de L. Kronig, *Z. Phys.,* Vol. 30, p. 521, 1929). These relations enable us to find the real part $\epsilon'(\omega)$ of the dielectric constant at a particular angular frequency ω from an integral over the whole spectrum of an expression containing the imaginary part $\epsilon''(\omega)$, and vice versa. These relations are especially useful for relating the experimentally observed change in one part, say $\epsilon''(\omega)$, to the expected change in the other part, say $\epsilon'(\omega)$. Suppose that the change $\delta\epsilon'$ occurs over a frequency range $\omega_0 \pm \Delta\omega_0$. Therefore, Eq. (12.12.19) reduces to

$$\delta\epsilon'(\omega) \cong \frac{1}{\pi} \int_{\omega_0 - \Delta\omega_0}^{\omega_0 + \Delta\omega_0} \frac{\delta\epsilon''(\omega') \, d\omega'}{\omega' - \omega} \qquad (12.12.21)$$

Further discussions of these relations can be found in Section 12.13 in connection with optical properties of semiconductors (G. H. B. Thompson, *Physics of Semiconductor Laser Devices,* John Wiley & Sons, Inc., New York, 1980, p. 537).

12.13 OPTICAL PROPERTIES OF GaAs

Among compound semiconductors, GaAs is most extensively studied in terms of both transport and optical properties. In this section we use GaAs as an example to illustrate the important physical processes determining the optical and dielectric properties of semiconductors, especially direct-gap semiconductors. The discussions are grouped into four subjects: (1) lattice absorption in the infrared region, (2) near-band-edge absorp-

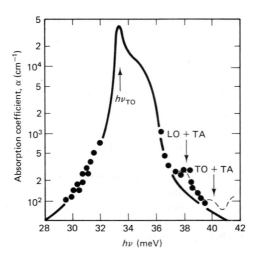

Figure 12.15 Lattice-absorption spectrum of GaAs (J. S. Blakemore, *J. Appl. Phys.,* Vol. 53, p. R142, October 1982).

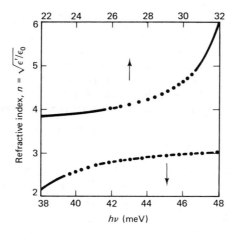

Figure 12.16 Variation of the refractive index of GaAs in the far-infrared region near $h\nu_{TO} = 33.2$ meV. (J. S. Blakemore, *J. Appl. Phys.*, Vol. 53, p. R142, October 1982.)

tion, (3) above-gap behavior, and (4) absorption in quantum wells. The effect of an applied electric field on the optical property is discussed in Section 12.14, and non-radiative Auger recombination is presented in Section 14.8.

Figure 12.15 shows the measured absorption coefficient $\alpha = 2\pi\kappa/\lambda$ compiled by Blakemore, and represented by data points (J. S. Blakemore, *J. Appl. Phys.*, Vol. 53, p. R142, October 1982). The absorption, called *restrahlen absorption*, is due to the excitation of Ga-cation and As-anion oscillations by the transverse electric field of an electromagnetic wave. The solid line represents a fit of the data points to a "single oscillator" similar to Eq. (12.3.5), or

$$\epsilon = \epsilon_\infty + \frac{\nu_{TO}^2(\epsilon_s - \epsilon_\infty)}{\nu_{TO}^2 - \nu^2 + i\gamma\nu} \tag{12.13.1}$$

with $\nu_{TO} = 33.25$ meV while the dashed line represents contributions due to multi-phonon processes. The damping constant γ has a value of 6×10^{10} s^{-1}. Figure 12.16 shows the measured refractive-index data $n = \sqrt{\epsilon'/\epsilon_0}$ also compiled by Blakemore in the restrahlen-absorption region. The solid curve again represents a fit to Eq. (12.13.1). We see that most of dispersion (variation of n with ν) can be accounted for by the interaction with TO phonons. Because an electromagnetic wave has transverse electric fields, it interacts only with transverse lattice waves in a first-order process. Therefore, information about LO phonons is relatively inaccessible to measurements with electromagnetic waves unless in second-order processes.

Figure 12.17 shows the absorption coefficient measured near the band edge of

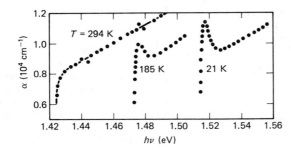

Figure 12.17 Absorption coefficient measured near the band edge of GaAs at $T = 294$ K, 185 K, and 21 K. The two absorption peaks at $h\nu$ slightly below the respective band gap at 185 K and 21 K are due to bound excitons. (M. D. Sturge, *Phys. Rev.*, Vol. 127, p. 768, 1962.)

GaAs at several temperatures as reported by Sturge (M. D. Sturge, *Phys. Rev.*, Vol. 127, p. 768, 1962). Plotted on a linear scale, the curves clearly show the absorption edge, that is, the gap energy. Except for the exciton peak at 185 K and 21 K, the curves are similar. The solid curve in Fig. 12.18 shows the refractive-index variation reported by Sell et al. as a function of photon energy (D. D. Sell, H. C. Casey, Jr., and K. W. Wecht, *J. Appl. Phys.*, Vol. 45, p. 2650, 1974). For $h\nu < \mathscr{E}_g$, the refractive index at 300 K can be approximated by an empirical Sellmeier type of equation,

$$n^2 = n_r^2 + \frac{(n_\infty^2 - n_r^2)(h\nu_0)^2}{(h\nu_0)^2 - (h\nu)^2} \tag{12.13.2}$$

with $h\nu$ and $h\nu_0$ expressed in eV, $h\nu_0 = 2.35$ eV, $n_\infty^2 = 10.88$, and $n_r^2 = 7.10$ (G. E. Stillman, C. M. Wolfe, and J. O. Dimmock, in R. K. Willardson and A. C. Beer, eds., *Semiconductors and Semimetals*, Vol. 12, Academic Press, Inc., New York, 1977, p. 169). The constant n_∞ is the high-frequency refractive index, corresponding to a value $n_\infty^2 = 10.88$ given in Table 11.5 and has a temperature dependence

$$n_\infty = 3.255(1 + 4.5 \times 10^{-5}T) \tag{12.13.3}$$

The absorption edge, on the other hand, varies with temperature as

$$\mathscr{E}_g = 1.519 - \frac{5.405 \times 10^{-4}T^2}{T + 204} \tag{12.13.4}$$

with T expressed in Kelvin (J. S. Blakemore, *J. Appl. Phys.*, Vol. 53, pp. R146 and R155, October 1982). According to Eq. (12.13.2), the refractive index n at 300 K has a value 3.31 at $h\nu = 0.3$ eV and a value 3.60 at $h\nu = 1.40$ eV. For $h\nu > \mathscr{E}_g$, the refractive index is strongly affected by the absorption caused by vertical band-to-band transition, and a simple empirical equation such as Eq. (12.13.2) is no longer adequate.

We should point out that Eq. (12.13.2) applies only to samples of high purity. In impure samples, high doping concentrations produce three effects. First, the impurity level broadens to form a band. Second, the band gap $\mathscr{E}_g$ is reduced because of lowering of potential energies of electrons and holes in the presence of opposite charges. Third, the band edges $\mathscr{E}_c$ and $\mathscr{E}_v$ are no longer sharply defined because of modulation by random distribution of impurities. These effects, known as impurity-band formation, band-gap shrinkage, and band tailing, respectively, are described in Section 6.11. The first two effects are discussed further in Section 14.6. The band-tail states can be rep-

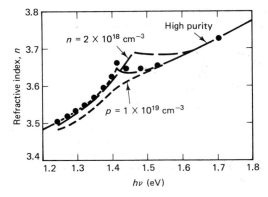

Figure 12.18 Variation of refractive index of GaAs near $h\nu = \mathscr{E}_g$ measured in samples with different dopant concentrations. (D. D. Sell, H. C. Casey, Jr., and K. W. Wecht, *J. Appl. Phys.*, Vol. 45, p. 2650, 1974.) For $h\nu < \mathscr{E}_g$, the variation can be described by an empirical equation, Eq. (12.13.2).

represented by an exponential energy dependence. Consequently, there is an additional contribution α_{bt} to absorption coefficient given by

$$\alpha_{bt}(h\nu) = A \exp\left[\frac{(h\nu - \mathcal{E}_g)}{\mathcal{E}_l}\right] \tag{12.13.5}$$

where A and $\mathcal{E}_l$ are two empirical constants.

Figure 12.19 shows the absorption coefficients measured in one n-type and one p-type GaAs sample. The curves are selected from data reported by Sell et al. (D. D. Sell, H. C. Casey, Jr., and K. W. Wecht, *J. Appl. Phys.*, Vol. 45, p. 2650, 1974). The variation of α for $h\nu < \mathcal{E}_g$ deviates considerably from Eq. (12.11.24) even in pure GaAs. Two possible causes for the deviation are band tailing if the samples are highly compensated and phonon-assisted optical transitions if the temperature is sufficiently high. The shift of the absorption curve towards lower photon energy observed in p-type samples is consistent with the energy-band shrinkage effect. The situation, however, is different for n-type samples. First, because of the closeness of the donor level to the conduction-band edge, the donor-impurity band merges into the conduction band. Second, because of a much smaller effective mass, n-type samples become degenerate at much lower doping concentrations than p-type samples. The filling of impurity-band and conduction-band states reduces absorption.

The dashed curves shown in Fig. 12.18 are the measured refractive index n at 300 K (D. D. Sell, H. C. Casey, Jr., and K. W. Wecht, *J. Appl. Phys.*, Vol. 45, p. 2650, 1974). An estimate of the contribution of the absorption-edge region to the dispersion of refractive index can be obtained from Eq. (12.12.19) as follows:

$$\Delta n(\omega) = \frac{c}{\pi} \int_{\omega_1}^{\omega_2} \frac{\Delta\alpha(\omega')\,d\omega'}{\omega'^2 - \omega^2} \tag{12.13.6}$$

where c is the velocity of light in free space and $\alpha(\omega')$ is the absorption coefficient near the absorption energy. Equation (12.13.6) was used by Stern and by Zoroofchi and Butler to compute the refractive index curves in doped GaAs samples from the measured absorption curves (F. Stern, *Phys. Rev.*, Vol. 133, p. A1653, 1964; J. Zoroofchi and J. K. Butler, *J. Appl. Phys.*, Vol. 44, p. 3697, 1973). The limits of integration ω_1 and ω_2 correspond to a range of photon energy from 0.80 to 1.44 eV. Reasonable agreements between the measured n and the computed n from α were obtained. However, as

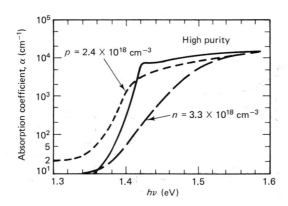

Figure 12.19 Absorption coefficient measured at 300 K near the band edge of GaAs in samples with different doping conditions. (H. C. Casey, Jr., D. D. Sell, and K. W. Wecht, *J. Appl. Phys.*, Vol. 46, p. 250, 1975.)

pointed by Sell et al., a good agreement between the two requires an accurate determination of α over an extended energy range. Even so, Eq. (12.13.6) is still very useful for estimating Δn in cases where Δn is inaccessible for experimental determination. Such is the case with semiconductor lasers.

While information about n and α near the absorption edge is important in practical applications, the behavior of n and α above the absorption edge provides valuable information about the structure of energy bands and the band states. One basic experiment in the infrared and ultraviolet spectral range is the reflectance R measurement with $R = [(n - 1)^2 + \kappa^2]/[(n + 1)^2 + \kappa^2]$. However, we also need the phase angle $\theta = \tan^{-1}[2\kappa/(n^2 + \kappa^2 - 1)]$ in order to determine n and κ. Information regarding θ can be obtained from one form of the Kramers–Kronig relations as follows:

$$\theta(\omega) = -\frac{1}{2\pi} \int_0^\infty \frac{d\ln R(\omega')}{d\omega'} \ln \left| \frac{\omega' + \omega}{\omega' - \omega} \right| d\omega' \qquad (12.13.7)$$

An accurate measurement of R over an extended spectral range enables us to compute $\theta(\omega)$. Using the measured values of $R(\omega)$ and the computed values of $\theta(\omega)$, the values of $\epsilon'(\omega)$ and $\epsilon''(\omega)$ in turn can be determined (H. Ehrenreich and H. R. Philipp, *Phys. Rev.*, Vol. 128, p. 1622, 1962). Figure 12.20 shows the results on ϵ' and ϵ'' reported by Philipp and Ehrenreich for Ge (light curves) and GaAs (heavy curves). Three peaks of ϵ'' in GaAs can be clearly seen near 3, 5, and 6.5 eV. In this energy range, the absorption coefficient attains a magnitude on the order of 10^6 cm^{-1}. The value of α increases from a value around 10^4 cm^{-1} at 1.5 eV to a value around 10^5 cm^{-1} at 2.7 eV and finally to a maximum value around 2×10^6 cm^{-1} at 5 eV (H. R. Philipp and H. Ehrenreich, *Phys. Rev.*, Vol. 129, p. 1550, 1963). Besides the simple reflectance measurement, there are various modulation experiments that measure the derivative of the reflectance by varying the wavelength or the strength of an applied electric

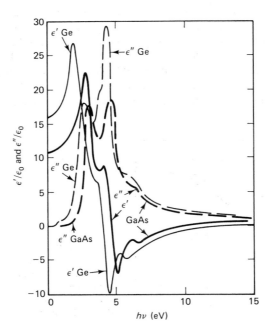

Figure 12.20 The values of the real part ϵ'/ϵ_0 and the imaginary part ϵ'/ϵ_0 of the relative dielectric constant in Ge and GaAs as deduced from the reflectance measurement. (H. R. Philipp, and H. Ehrenreich, *Phys. Rev.*, Vol. 129, p. 1550, 1963.)

Dielectric and Optical Properties   Chap. 12

field. These modulation experiments and the X-ray photoemission experiment provide valuable information for ascertaining the band structure of semiconductors (M. Cardona, *Modulation Spectroscopy,* Academic Press, Inc., New York, 1969; B. O. Seraphin, p. 1, and D. E. Aspnes and N. Bottka, p. 457, in R. K. Willardson and A. C. Beer, eds., *Semiconductors and Semimetals,* Vol. 10, Academic Press, Inc., New York, 1972; D. E. Eastman, W. D. Grobman, J. L. Freeouf, and M. Erbaduk, *Phys. Rev.,* Vol. B9, p. 600, 1974).

Let us now return to Fig. 12.17. The peaks observed at low temperatures are caused by exciton absorption. *Excitons* generally refer to bound states of an electron and a hole interacting with each other. In III–V compounds, the orbit of an exciton extends over a great many lattice sites. For such excitons, the electron–hole interaction can be treated as Coulomb interaction between two point charges. Therefore, according to Eq. (1.2.11), the quantized exciton energy is

$$\mathscr{E}_{xn} = \frac{-m_r e^4}{2(4\pi\epsilon\hbar)^2}\frac{1}{n_x^2} = \frac{-13.6}{n_x^2}\frac{m_r}{m_0}\left(\frac{\epsilon_0}{\epsilon}\right)^2 \quad \text{eV} \tag{12.13.8}$$

where $m_r = m_c^* m_v^*/(m_c^* + m_v^*)$ is the reduced mass and n_x an integer $(1, 2, 3, \ldots)$ representing the quantum number. Since it takes an energy $\mathscr{E}_g$ to excite an electron–hole pair, the exciton absorption occurs at

$$(h\nu)_x = \mathscr{E}_g - \mathscr{E}_x \tag{12.13.9}$$

Using the effective masses and dielectric constant for GaAs, we find the lowest energy $\mathscr{E}_{x1} = -4.4$ meV. Therefore, only at low temperatures, the exciton absorption peak will not be broadened by phonon scattering.

This situation, however, is changed in a quantum well (QW). Figure 12.21 shows the absorption spectrum of GaAs multiple quantum wells compared to that of a bulk GaAs at room temperature. Whereas the bulk spectrum shows only a slight hump, the MQW spectrum shows two distinct resonances, one at 1.463 eV from electron-heavy-hole exciton and the other at 1.474 eV from electron–light-hole exciton (D. A. B. Miller, D. S. Chemla, D. J. Eilenberger, P. W. Smith, A. C. Gossard, and W. T.

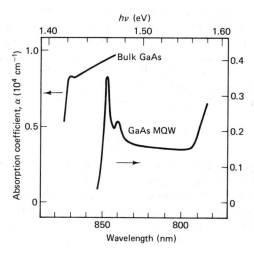

Figure 12.21 Absorption coefficient measured in a GaAs sample (right scale) containing multiple quantum wells as compared to that measured in an ordinary GaAs sample (left scale). Both curves are taken at 300 K. (D. A. B. Miller, D. S. Chemla, D. J. Eilenberger, P. W. Smith, A. C. Gossard, and W. T. Tsang, *Appl. Phys. Lett.,* Vol. 41, p. 679, 1982.) In a more refined experiment, a value of $\alpha = 1.15 \times 10^4$ cm^{-1} was found at the heavy-hole peak instead of $\alpha = 3.5 \times 10^3$ cm^{-1} shown here.

Tsang, *Appl. Phys. Lett.*, Vol. 41, p. 679, 1982). One important parameter is the layer thickness in comparison to the exciton dimension. In bulk GaAs, the exciton radius is given by

$$a_{xn} = \frac{4\pi\epsilon\hbar^2}{m_r e^2} n_x^2 = 0.53 \left(\frac{m_0}{m_r}\right) \frac{\epsilon}{\epsilon_0} n_x^2 \quad \mathring{A} \qquad (12.13.10)$$

which yields a value of $a_{x1} = 150$ Å for $n_x = 1$. In the experiment of Miller et al., a GaAs well thickness of 102 Å was used, which is much smaller than the exciton dimension (diameter) of 300 Å. For a purely two-dimensional (flattened) exciton, the binding energy is equal to $4\mathscr{E}_{xn}$ of Eq. (12.13.8) (M. Shinado and S. Sugano, *J. Phys. Soc. Jpn.*, Vol. 21, p. 1936, 1966). However, in a quantum well, the penetration of the exciton wave function into the large-gap (GaAl)As is not negligible. For very thin well layers, the large extension of the wave function makes excitons again behave like a three-dimensional exciton. Therefore, the binding energy of a QW exciton $\mathscr{E}_{x,QW}$ peaks between $2\mathscr{E}_{xn}$ and $3\mathscr{E}_{xn}$ for well thickness in the range $0.5a_{x1}$ to $3a_{x1}$.

The increase in the exciton binding energy would not make the exciton resonances discernible if it was accompanied by a corresponding broadening of the resonance line width by phonon scattering. The dominant broadening mechanism at 300 K in bulk GaAs is LO phonon absorption. Therefore, the line width Γ is expected to have a component proportional to the phonon population, suggesting the relation

$$\Gamma = \Gamma_0 + \frac{\Gamma_L}{\exp{(h\nu_{LO}/kT)} - 1} = \Gamma_0 + \Gamma_{BL} \qquad (12.13.11)$$

A good fit of Eq. (12.13.11) to the measured spectrum was obtained with $\Gamma_0 = 2.0$ meV and $\Gamma_L = 5.5$ meV. What is important is that Γ_L is actually less than 7.0 meV found in bulk GaAs. Using the value of $h\nu_{LO} = 36$ meV, we find $\Gamma_{BL} = 1.8$ meV and $\Gamma = 3.8$ meV. Since the thermal broadening of 3.8 meV is less than the ionization energy of about 9 meV, clear exciton resonances can be seen even at 300 K. The value of $\Gamma_{BL} = 1.8$ meV corresponds to a mean lifetime $\tau_{ex} = \hbar/\Delta\mathscr{E} = \hbar/\Gamma_{BL}$ of 0.4 ps. In other words, the broadening suggests a mean time of 0.4 ps for the ionization of excitons into electron–hole pairs.

Excitons posses interesting nonlinear optical properties. Figure 12.22 shows the saturation of absorption of the electron–heavy-hole peak in both bulk and MQW samples also reported by Miller et al. Later refinements of the experiment, reported by Chemla et al., show that the value of α in Fig. 12.21 was underestimated and a value of $\alpha = 1.15 \times 10^4$ cm^{-1} is measured at the heavy-hole peak (D. S. Chemla, D. A. B. Miller, P. W. Smith, A. C. Gossard, and W. Wiegmann, *IEEE J. Quantum Electron.*, Vol. QE-20, p. 265, 1984). Furthermore, the saturation behavior can be described by

$$\alpha(I) = \frac{\alpha_0}{1 + (I/I_s)} \qquad (12.13.12)$$

where I is the light intensity and I_s is a parameter called *saturation intensity*. The saturation phenomenon is attributed to screening of Coulomb interaction by free carriers. Let α_B be the absorption coefficient in bulk GaAs. Within a total well thickness L, the rate of electron–hole pair generation is $I\alpha_B L/(h\nu)$ per unit area per unit time. The recombination rate, on the other hand, is equal to $N_e/\tau = N_h/\tau$, where τ is the carrier lifetime and $N_e = N_h$ is the areal density of electrons or holes. Under the steady-state condition,

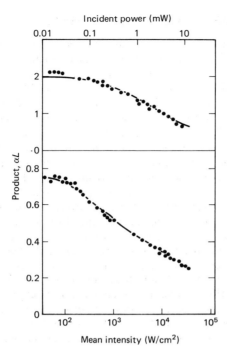

Figure 12.22 Dependence of optical absorption in GaAs bulk (top curve) and MQW (bottom curve) samples on the incident beam intensity. Optical absorption is expressed in terms of the product of absorption coefficient α and the sample and total well thickness, respectively. Both curves were measured at the heavy-hole peak wavelength in Fig. 12.21. (D. A. B. Miller, D. S. Chemla, D. J. Eilenberger, P. W. Smith, A. C. Gossard, and W. T. Tsang, *Appl. Phys. Lett.*, Vol. 41, p. 679, 1982.)

the two rates are equal, yielding $N_e = N_h = \tau\alpha_B LI/(h\nu)$. Saturation of exciton absorption occurs when the sum $N_e + N_h$ is equal to the areal exciton density N_x. Based on this simple physical model, the saturation intensity I_s is given by

$$I_s = \frac{h\nu}{2\tau\alpha_B LA_x} \tag{12.13.13}$$

where $A_x = \pi a_x^2$ is the exciton area.

As discussed in Section 12.12, the real and imaginary parts of ϵ are related by the Kramers–Kronig relations. Therefore, the observed saturation of absorption can be translated into a dependence of refractive index n on incident light intensity. If we write

$$n = n_1 + n_2 I, \quad \alpha = \alpha_1 - \alpha_2 I \quad \text{or} \quad P_l = 2\chi_3 E_i E_j E_k \tag{12.13.14}$$

the nonlinear terms can reach maximum values $n_2 \sim 2 \times 10^{-4}$ cm^2/W, $\alpha_2 \sim 39$ cm/W, and $\chi_3 \sim 6 \times 10^{-2}$ esu. These values are indeed very large. The value of χ_3 is about 10^6 times larger than a value of 8×10^{-8} esu measured at $\lambda = 1.06$ μm in bulk Si. A part of the enhancement is due to the fact that χ_3 in MQW structures is measured at a photon energy close to the resonant absorption peak of the exciton. The nonlinear properties of excitons in MQW structures have been utilized in a number of optical devices, such as optical bistability and four-wave mixing. For an extensive discussion of optical nonlinearity in MQW structures as well as for additional references, the reader is referred to the article by Chemla et al. cited in this section (D. S. Chemla, D. A. B. Miller, P. W. Smith, A. C. Gossard, and W. Wiegmann, *IEEE J. Quantum Electron.*, Vol. QE-20, p. 265, 1984).

12.14 FRANZ–KELDYSH, POCKELS, AND STARK EFFECTS

When a semiconductor is placed in an electrical field, the absorption coefficient changes with the applied field. The phenomenon observed independently by Franz and by Keldysh is known as the Franz–Keldysh effect (W. Franz, *Z. Naturforsch.*, Vol. 13a, p. 484, 1958; L. V. Keldysh, *Sov. Phys. JETP*, Vol. 7, p. 788, 1958). The physical origin of this change can be understood with the help of Fig. 12.23. The situation can be divided into two cases: (I) $hv < \mathscr{E}_g$ as illustrated in curves (a) and (b) with E in (b) $> E$ in (a), and (II) $hv > \mathscr{E}_g$ as illustrated in curve (c). The probability of lifting an electron from the valence band into the conduction band depends on the overlap of the two wave functions ψ_c and ψ_v. The points marked A and B represent the classical turning point for ψ_v and ψ_c, respectively, at which the wave function changes from oscillatory to decaying behavior. As the applied electric field is increased, the distance AB is decreased and as a result, the overlapping of the two wave functions is enhanced. Therefore, for photon energy $hv < \mathscr{E}_g$, the gap energy, the absorption coefficient is expected to increase rapidly with increasing E. The case for $hv > \mathscr{E}_g$ is more complicated. The two wave functions are made of an incident wave and a reflected wave. The relative phase of the two waves varies with the applied field E. The interference of the incident and reflected waves results in an oscillatory behavior of the absorption coefficient.

The Franz–Keldysh effect can be treated analytically in expressing the wave functions in the momentum representation [A. Anselm, *Introduction to Semiconductor Theory* (English translation, Izdatelstvo Mir, Moscow), Prentice-Hall, Inc., Englewood Cliffs, N.J., p. 447, 1981]. The absorption coefficient α_E under an applied E is found to be

$$\alpha_E = K(hv_E)^{1/2}\pi \int_\beta^\infty \text{Ai}^2(x)\,dx \qquad \text{m}^{-1} \qquad (12.14.1)$$

where K is given by Eq. (12.11.25), hv_E is given by

$$hv_E = \left(\frac{e^2E^2\hbar^2}{2m_r}\right)^{1/3} \qquad (12.14.2)$$

(a)   (b)   (c)

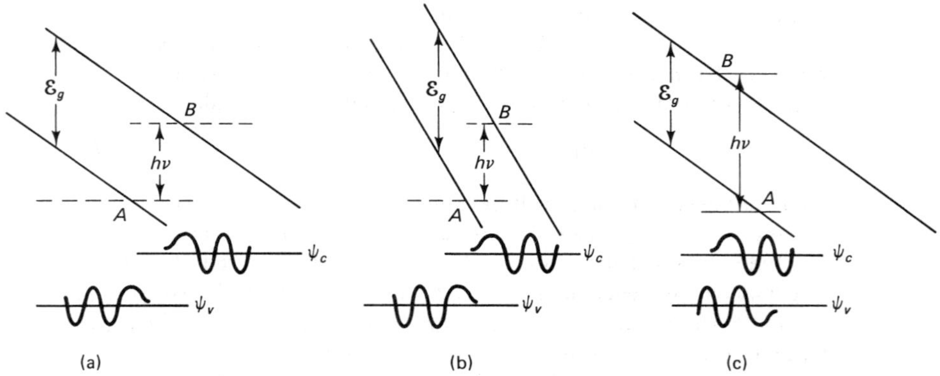

Figure 12.23 Energy-band diagrams used to illustrate the Franz–Keldysh effect. For $hv < \mathscr{E}_g$ (situation a), overlapping of the wave functions ψ_v and ψ_c increases with increased bias electric field E (situation b), resulting in a rapid increase in α with E. For $hv > \mathscr{E}_g$ (situation c), the relative phase of ψ_c and ψ_v changes with E, producing an oscillatory component in α.

E is expressed in V/m, and β is defined as

$$\beta = \frac{\mathscr{E}_g - h\nu}{h\nu_E} \tag{12.14.3}$$

Using a value $m_r^* = 0.060m_0$ for GaAs, we find $h\nu_E = 8.57$ meV at $E = 10^6$ V/m. Therefore, the absolute value of β is expected to be much greater than unity. The function Ai(x) in Eq. (12.14.1) is the Airy function, which has the following asymptotic forms:

$$\sqrt{\pi}\, \text{Ai}(x) = \frac{1}{2} x^{-1/4} \exp\left(-\frac{2}{3} x^{3/2}\right)\left(1 - \frac{5}{48}\frac{1}{x^{3/2}} + \cdots\right) \tag{12.14.4}$$

$$\sqrt{\pi}\, \text{Ai}(-y) = y^{-1/4}\left[\sin\left(\frac{2}{3} y^{3/2} + \frac{\pi}{4}\right) - \frac{5}{48} y^{-3/2} \cos\left(\frac{2}{3} y^{3/2} + \frac{\pi}{4}\right)\right] \tag{12.14.5}$$

for $x \gg 1$ and $y \gg 1$, respectively. The former approximation applies to curves (a) and (b) in Fig. 12.23 with $h\nu < \mathscr{E}_g$, while the latter applies to curve (c), with $h\nu > \mathscr{E}_g$ (M. Abramowitz and I. Stegun, *Handbook of Mathematical Functions,* Dover Publications, Inc., New York, p. 449, 1965).

Use of Eq. (12.14.4) in Eq. (12.14.1) yields

$$\alpha_E = K(h\nu_E)^{1/2}(8\beta)^{-1} \exp\left(-\frac{4}{3}\beta^{3/2}\right) \tag{12.14.6}$$

The quantity K has a value 5×10^4 cm^{-1}(eV)$^{-1/2}$ in GaAs. Thus for $\beta = 2$ and $h\nu_E = 8.57$ meV, the value of α_E is 4 cm^{-1} at a photon energy $h\nu = \mathscr{E}_g - 17$ meV with $E = 10^4$ V/cm. This value of α_E is much smaller than the value about 200 cm^{-1} shown at $h\nu = 1.4$ eV in Fig. 12.19. Therefore, the Franz–Keldysh effect is expected to be significant for $h\nu < \mathscr{E}_g$ only in p-n junctions where a large E can exist. The situation for $h\nu > \mathscr{E}_g$ is quite different because the exponential function is now replaced by sinusoidal functions. Even though an evaluation of α_E will require a numerical calculation of the integral in Eq. (12.14.1), a qualitative behavior of α_E can be seen from Eq. (12.14.5). As we vary β, that is, the photon energy $h\nu$, the value of Ai$^2(-y)$ and hence that of α_E contain an oscillatory component. This oscillatory behavior is a result of interference between the incident and reflected waves mentioned earlier. Furthermore, the amplitude of this oscillatory component is expected to be an appreciable part of the total α_E. Figure 12.24 shows the measured α_E as a function of $h\nu$, which indicates a modulation of the absorption coefficient greater than 20% at $E = 40$ kV/cm (E. G. S. Paige and H. D. Rees, *Phys. Rev. Lett.,* Vol. 16, p. 444, 1966). The period of $\Delta\alpha$ can be estimated from Eq. (12.14.5) by requiring y to change to $y + \Delta y$ such that the resultant phase angle is π or $\Delta y \sim \pi y^{-1/2}$. At $E = 40$ kV/cm used in the experiment, we find $h\nu_E = 21$ meV and $y \sim 4$ at $h\nu = 1.52$ eV. Therefore, the period in $\Delta h\nu$ is expected to be around 33 meV, in reasonable agreement with Fig. 12.24. For an analysis of the Franz–Keldysh effect at high fields, the reader is referred to the treatments by Callaway (J. Callaway, *Phys. Rev.,* Vol. 130, p. 549, 1963; Vol. 134, p. A998, 1964).

While the Franz–Keldysh effect relates to the change of absorption coefficient by an applied field, the Pockels effect concerns the change of refractive index by an applied field. The harmonic restoring force γx in Eq. (12.2.8) is only the first-order term in a Taylor series expansion of the energy in terms of displacement x. If we include anhar-

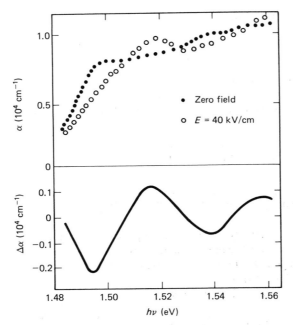

Figure 12.24 Absorption coefficient (top curves) and change in absorption coefficient $\Delta \alpha$ (bottom curve) measured in a GaAs sample under zero field and under $E = 4 \times 10^4$ V/cm. (E. G. S. Paige and H. D. Rees, *Phys. Rev. Lett.*, Vol. 16, p. 444, 1966.)

monic forces vx^2, wx^3, and so on, in the expansion, we obtain a general expression relating electric flux density D_i in the ith direction to applied electric fields E_k and E_j in the kth and jth direction as follows:

$$D_i = \epsilon_{ij}E_j + 2d_{ikj}E_kE_j + 4\chi_{ikjl}E_kE_jE_l \tag{12.14.7}$$

The second term is due to polarization generated by the anharmonic force vx^2, while the third term is due to polarization generated by wx^3 and identical to P of Eq. (12.13.14). The Pockels effect can be represented by the second term in the following sense. When one field E_k is a dc or low-frequency field and D_i and E_j are associated with a high-frequency electromagnetic wave, the term $2d_{ikj}$ can be considered as a perturbation to the dielectric constant ϵ_{ij}. As discussed in Section 2.9, crystal symmetry makes certain dielectric elements ϵ_{ij} zero. Similarly, symmetry considerations can eliminate many coefficients d_{ikj}. For a crystal possessing inversion symmetry, that is, the inversion operation ($x \rightarrow -x$, $y \rightarrow -y$, $z \rightarrow -z$), d_{ikj} is identically zero. Such is the case with silicon and germanium.

Even though III–V compounds do not have an inversion symmetry, the zinc-blende crystal possesses other symmetries, such as three- and fourfold rotational axes and mirror reflections. Using these symmetry elements, we can show that the only nonzero coefficients are those which relate the change in ϵ_{xy} to an applied field E_z, and so on. There are three such coefficients. Even though the form $2d_{ikj}E_kE_j$ can be used for describing the Pockels effect, it is customarily reserved for treating nonlinear optics problems where two high-frequency fields E_k and E_j are involved. For the Pockels effect, the following form for index change is used:

$$\Delta \frac{1}{n_{ij}^2} = r_{ijk}E_k^0 \tag{12.14.8}$$

In III–V compounds, only coefficients $r_{xyz} = r_{zxy} = r_{yzx} = r$ are nonzero. Therefore, in the presence of dc fields E_x^0, E_y^0, and E_z^0, the electrostatic energy can be written as

$$W_e = \frac{1}{2\epsilon_0} \left(\frac{D_x^2}{n^2} + \frac{D_y^2}{n^2} + \frac{D_z^2}{n^2} + 2rE_x^0 D_y D_z + 2rE_y^0 D_z D_x + 2rE_z^0 D_x D_y \right) \quad (12.14.9)$$

Let us consider an electromagnetic field propagating in the z direction. Setting $D_z = 0$ and performing the following coordinate transformation

$$x = \frac{x' - y'}{\sqrt{2}}, \qquad y = \frac{x' + y'}{\sqrt{2}} \quad (12.14.10)$$

through a $\pi/4$ rotation, we find that

$$W_e = \frac{1}{2\epsilon_0} \left[D_x'^2 \left(\frac{1}{n^2} + rE_z^0 \right) + D_y'^2 \left(\frac{1}{n^2} - rE_z^0 \right) \right] \quad (12.14.11)$$

In other words, the effective indices along the x' and y' axes are given, respectively, by

$$n_{x'} = n - \frac{n^3 r}{2} E_z^0 \quad (12.14.12)$$

$$n_{y'} = n + \frac{n^3 r}{2} E_z^0 \quad (12.14.13)$$

At $h\nu = 1.38$ eV, the value of $n^3 r$ is 51×10^{-12} m/V in GaAs. At a field $E_z^0 = 10^7$ V/m, a differential phase change of π requires a length $L = 10^3 \lambda$ or 900 μm (0.9 mm) at $\lambda = 0.9$ μm. The Pockels effect, which is linearly dependent on the applied field, is known as the *linear electrooptic effect*. It has been used not only in optical modulators (A. Carenco, L. Menigaux, F. Alexandre, M. Abadalla, and A. Brenac, *Appl. Phys. Lett.*, Vol. 34, p. 755, 1979) but also in electrooptic sampling of electric fields in GaAs devices (B. H. Kolner and D. M. Bloom, *IEEE J. Quantum Electron.*, Vol. QE-22, p. 69, 1986; Z. H. Zhu, J. P. Weber, S. Y. Wang, and S. Wang, *Appl. Phys. Lett.*, Vol. 49, p. 432, 1986).

A discussion of the electric field effects in semiconductors will be incomplete without the quantum-confined Stark effect recently observed in quantum wells. A great deal of interest has been generated in electrooptical properties of quantum wells because of the potential for practical applications. The *Stark effect* generally refers to a shift in the atomic energy upon the application of an electric field. It is well known in atomic physics that the energy of the 2s and 2p states in a hydrogen atom is split into $\mathscr{E}_2 + 3eaE$, $\mathscr{E}_2$, and $\mathscr{E}_2 - 3eaE$, with $\mathscr{E}_2$ given by Eq. (1.2.10) with $n = 2$ and a given by Eq. (1.2.9). This energy linear in E is known as the *linear Stark effect*. Here we are concerned with the relative shift or energy separation of the conduction-band and valence-band states in a quantum well. According to the second-order perturbation theory, the Stark shift $\Delta\mathscr{E}$ is given by

$$\Delta\mathscr{E} = \frac{-e^2 E^2 \; \Sigma \; |\langle c|x|v\rangle|^2}{\mathscr{E}_2 - \mathscr{E}_1} \quad (12.14.14)$$

where $\mathscr{E}_2$ and $\mathscr{E}_1$ are the energy of the conduction-band (excited) and valence-band (ground) states, respectively. The quadratic or second-order Stark effect has resulted in a number of interesting and potentially important phenomena.

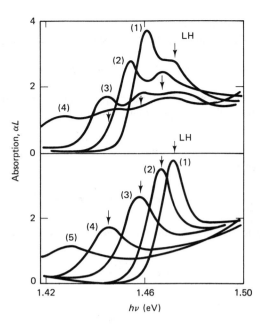

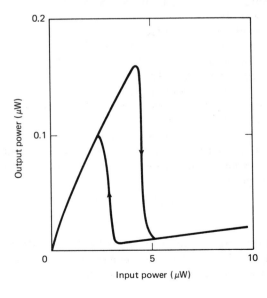

Figures 12.25 and 12.26 show, respectively, the absorption spectrum and the output–input characteristic observed in a QW structure as reported by Weiner et al. (J. S. Weiner, D. A. B. Miller, D. S. Chemla, T. C. Damen, C. A. Burrus, T. H. Wood, A. C. Gossard, and W. Wiegmann, *Appl. Phys. Lett.,* Vol. 47, p. 1148, 1985). The basic structure consists of two GaAs wells of width 90 Å separated by a superlattice barrier. The two wells are part of an optical waveguide, so the optical beam can be guided along the plane of the quantum wells. The sample is doped as a *p-i-n* diode, so an electric field can be applied perpendicular to the QW layers. The different curves in

Fig. 12.25 are for different field strengths of (1) 0.16, (2) 1, (3) 1.4, (4) 1.8, and (5) 2.2 in units of 10^5 V/cm. Theoretical calculations on the energy shift agree well with experiments (D. A. B. Miller, D. S. Chemla, T. C. Damen, A. C. Gossard, W. Wiegmann, T. H. Wood, and C. A. Burrus, *Phys. Rev.*, Vol. B32, p. 1043, 1985). For example, the calculated shift of -40 meV for a well width of 100 Å at a field $E = 1.5 \times 10^5$ V/cm compares well with the shift observed between curves (3) and (4) with respect to curve 1. For further discussions on excitonic states in a quantum well, the reader is referred to an analysis by Bastard and Brum (G. Bastard and J. A. Brum, *IEEE J. Quantum Electron.*, Vol. QE-22, p. 1625, 1986).

The output–input characteristic shown in Fig. 12.26 is observed when the wavelength of the input laser is set at the zero-field exciton wavelength, and a constant-current source is used to bias the device. When the photocurrent is below the bias current, all the voltage appears across the device. Since the exciton wavelength is longer than the laser wavelength as a result of the Stark shift, only a part of the incident light is absorbed and the output light increases with the input light. However, when the photocurrent exceeds the bias current, the voltage across the device begins to decrease, bringing the exciton wavelength to the zero-field position and resulting in strong absorption of the input light. The optical bistability shown in Fig. 12.26 is a direct consequence of Stark shifting of the exciton absorption spectrum toward and away from the input-laser wavelength. We should point out that optical bistable device belongs to a general class of devices called SEED, utilizing self-electrooptic effect by combining the function of detection and modulation in a single device. The reader is referred to an excellent review article by Miller et al. for references to various SEEDs and for extensive review of the quantum-confined Stark (QCS) effect (D. A. B. Miller, J. S. Weiner, and D. Chemla, *IEEE J. Quantum Electron.*, Vol. QE-22, p. 1816, 1986). In the following discussion, we present a summary of the important theoretical results relevant to Fig. 12.25.

The heavy-hole band is characterized by a set of quantum numbers ($J = \frac{3}{2}$, $M_J = \pm \frac{3}{2}$) and the light-hole band by ($\frac{3}{2}$, $\pm \frac{1}{2}$). The effect of the confinement lifts the degeneracy and imposes additional selection rules on interband transitions. For E parallel to the plane of layers (upper curves in Fig. 12.25), the transition probabilities and hence the absorption peaks should be in the ratio of $\frac{3}{4}$ to $\frac{1}{4}$ for the HH and the LH to conduction-band transitions. For E perpendicular to the layers (lower curves in Fig. 12.25), they should be in the ratio 0 : 1. Therefore, the absorption curves in the lower part of Fig. 12.25 are due to transitions from the LH subband to the conduction subband. Because there is a considerable overlap between the HH and LH absorption spectra, it is difficult to ascertain whether the areas under the two curves are in the ratio 3 : 1.

The effects of an applied field are summarized in Fig. 12.27. First, the electron and hole wave functions are pulled toward opposite sides of the well. The reduced overlap of the two wave functions results in a corresponding reduction in absorption and in luminescence. Second, the energy separation $\mathscr{E}_2 - \mathscr{E}_1$ is reduced by the field. A field E of 10^5 V/cm constitutes a potential drop of 100 meV across a well of width 100 Å. This drop is considerably larger than the combined effects of increased binding energy $\mathscr{E}_i$ in Eq. (11.4.3) and reduced exciton energy. Therefore, the observed absorption peak shifts toward lower photon energy at higher E. Third, the probability of tunneling especially for low-barrier holes increases rapidly with increasing E. Tunneling shortens carrier lifetime (nonradiative) and thus broadens the absorption spectrum at high electric fields. Theoretical calculation shows a drop of the tunneling lifetime from

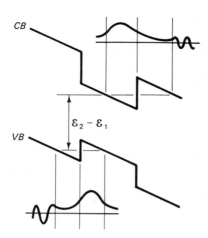

Figure 12.27 Energy-band diagram summarizing the effect of an electric field on the wave functions and the associated properties of a quantum well.

1000 ps at $E = 9 \times 10^4$ V/cm to 3 ps at $E = 1.5 \times 10^5$ V/cm in a well of width 70 Å (E. J. Austin and M. Jaros, *Appl. Phys. Lett.*, Vol. 47, p. 274, 1985). Even though the Coulomb interaction in the exciton state is relatively weak in affecting the energy separation $\mathscr{E}_2 - \mathscr{E}_1$, it is strong enough to keep electron and hole together in the plane of quantum wells. The relatively strong absorption, even for curve 4, is a strong indication that the exciton does not break up even at high fields.

12.15 II–VI COMPOUND AND DILUTED MAGNETIC SEMICONDUCTORS

Semiconductors of II–VI compounds offer some unique possibilities which complement those of III–V compound semiconductors. Figure 12.28 shows the band-gap energies and lattice constants of II–VI compounds with II = Zn, Cd, or Hg and VI = S, Se, or Te. In Section 11.11 we discussed the (HgCd)Te heterostructure and its potential use as material for long wavelength ($\lambda > 2$ μm) detectors. It is equally interesting to note that the mixed compounds of Zn and Cd have gap energies covering the visible regions of the electromagnetic spectrum. Most II–VI compounds crystallize in the zinc-blende structure, although some exist in both wurtzite and zinc-blende structures. Therefore, most II–VI and their mixed compounds can be grown on substrates of diamond or zinc-blende structure if the condition for lattice match can be met. Equally interesting to note is the possibility of incorporating the compound MnTe or MnSe into an appropriate II–VI compound. Although Mn compounds crystallize in the wurtzite structure, they can be grown into the zinc-blende structure if the atomic percent of Mn is not too large.

The development of II–VI compound materials has been hampered by poor crystal quality. The recent advent of advanced thin-film growth technology such as molecular beam epitaxy (MBE) offers hope that high-quality films can be grown if a set of optimal growth conditions to minimize lattice defects and to reduce background impurities can be found. The recent successful growth of (CdMn)Te thin films and (CdMn)Te-CdTe superlattices on GaAs substrates by MBE has generated a great deal of activities on the growth aspects and the optical properties of diluted magnetic semiconductors (L. A. Kolodziejski, T. Sakamoto, R. L. Gunshor, and S. Datta, *Appl. Phys. Lett.*, Vol. 44, p. 799, 1984; R. N. Bicknell, N. C. Giles-Taylor, J. F. Schetzina, N. G. Anderson, and W. D. Laidig, *Appl. Phys. Lett.*, Vol. 46, p. 238, 1985). Since then various epi-

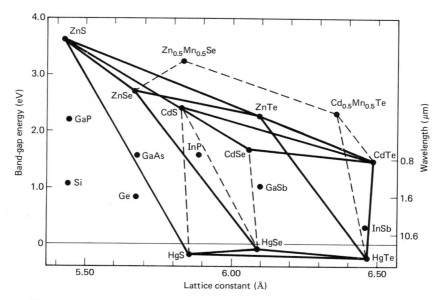

Figure 12.28 Band-gap energies and lattice constants of a selected number of II–VI compound semiconductors. (Courtesy of J. F. Schetzina.)

taxial layers of (ZnMn)Se and superlattices of (ZnMn)Se–ZnSe have also been grown successfully. Single zinc-blende phase is maintained up to an atomic percent $x = 0.60$ in the case of $Cd_{1-x}Mn_xTe$ and $x = 0.66$ in the case of $Zn_{1-x}Mn_xSe$. Because of a large lattice mismatch (14.6%) between CdTe and GaAs, the grown film normally has a (111) orientation on (100) substrates. A (100)-oriented film can be grown if a residual (10 Å) oxide is left on the GaAs surface. The (ZnMn)Se film, on the other hand, keeps the (100) orientation of the substrate. For further discussions of the various growth aspects, the reader is referred to an excellent review article by Kolodziejski et al. (L. A. Kolodziejski, R. L. Gunshor, N. Otsuka, S. Datta, W. M. Becker, and A. V. Nurmikko, *IEEE J. Quantum Electron.*, Vol. QE-22, p. 1666, 1986).

We note from Fig. 12.28 that a considerable mismatch between the lattice constants of ZnSe and MnSe (cubic) and those of CdTe and MnTe (cubic). Therefore, the superlattices presently under study have a considerable amount of strain, both in the plane of and perpendicular to the layers. As a result, the band structure and hence the optical properties will be different from a strain-free superlattice. With this reservation in mind, we can make some general observations based on the reported experiments. The band-edge offset is thought to occur almost entirely in the conduction band, as concluded from UV photoemission measurements on (CdMn)Te heterostructures (M. Pessa and O. Jylha, *Appl. Phys. Lett.*, Vol. 45, p. 646, 1984; M. Taniguchi, L. Ley, R. L. Johnson, J. Ghijsen, and M. Cardona, *Phys. Rev.*, Vol. B33, p. 1206, 1986). Studies of the splitting of exciton ground-state resonance under an applied magnetic field (B up to 4 T) also indicate a valence-band offset less than 20 meV and further confirm the theoretical prediction that the lower-energy valence band with $M_J = \pm\frac{1}{2}$ in the bulk becomes the higher-energy valence band in quantum wells and superlattices (Y. Hefetz, J. Nakahara, A. V. Nurmikko, L. A. Kolodziejski, R. L. Gunshor, and S. Datta, *Appl. Phys. Lett.*, Vol. 47, p 989, 1985).

Stimulated emission has been observed in both $Cd_{0.55}Mn_{0.45}Te/CdTe$ (at 1.59 eV) and $Zn_{0.67}Mn_{0.33}Se/ZnSe$ (at 2.77 eV) multiple quantum wells up to 80 K by optical pumping (R. N. Bicknell, N. C. Giles-Taylor, J. F. Schetzina, N. G. Anderson, and W. D. Laidig, *Appl. Phys. Lett.*, Vol. 46, p. 236, 1985; R. B. Bylsma, W. M. Becker, T. C. Bonsett, L. A. Kolodziejski, R. L. Gunshor, M. Yamanishi, and S. Datta, *Appl. Phys. Lett.*, Vol. 47, p. 1039, 1985). The results are encouraging in view of appreciable strain in these structures. However, the presence of strain complicates the interpretation of the luminescence data. In a strain-free quantum well, a blue shift of the emission spectrum is expected. Instead, a continual shift toward the red is observed as the Mn concentration is increased. This red shift is obviously caused by strain. A shift toward the blue is observed only in narrow quantum wells where the confinement energy is larger than the strain-caused energy shift. Therefore, an analysis of the emission spectrum requires a knowledge of the effect of strain on the band structure of both the well and the barrier material. Because the valence-band offset is small, the hole wave function spreads appreciably into the barrier region. Finally, we should mention one experimental observation of practical importance. In epitaxial layers containing Mn, the luminescence spectrum generally shows two competing optical transitions with the broad yellow luminescence caused by internal Mn-ion transitions dominating over the blue excitonic transition. In multiple quantum wells, however, the peak intensity of the near-band-gap blue (2.77 eV) emission is more intense by more than two orders of magnitude than the yellow (2.1 eV) emission as a result of carrier confinement in the ZnSe well.

The pseudobinary II–VI semiconductors, such as (CdMn)Te or (ZnMn)Se, in which the group II element sites are partly substituted by Mn are generally referred to as *diluted magnetic semiconductors*. A Mn^{2+} ion has a $3d^5$ electron configuration with spin parallel arrangement. Therefore, each Mn^{2+} ion possesses a magnetic moment $u = su_B$, where $s = \frac{5}{2}$ is the spin quantum number and

$$u_B = \frac{e\hbar}{2m_0} = 9.27 \times 10^{-24} \text{ A-m}^2 \qquad (12.15.1)$$

is the basic unit, called the *Bohr magneton,* for the magnetic moment. The spin-spin exchange interaction between localized Mn^{2+} ions and band electrons not only affects the band-edge energies (Fig. 12.28) but also gives rise to interesting magnetic properties. For a discussion of this interaction and references on the subject, the reader is referred to an excellent review article by Furdyna (J. K. Furdyna, *J. Appl. Phys.*, Vol. 53, p. 7637, 1982). One interesting magnetic property is a large Faraday rotation. In the presence of a magnetic flux density B, the index of refraction becomes different for two circularly polarized waves with n^+ for right-hand wave and n^- for left-hand wave. Thus a linearly polarized wave experiences a change in the polarization direction by an angle θ, called *Faraday rotation*. The rotation per unit distance is given by

$$\frac{\theta}{d} = \frac{\omega}{2c} (n^+ - n^-) \qquad (12.15.2)$$

where d is the distance traveled in the direction of B.

Figure 12.29 shows the value of the Verdet constant $\theta/(Bd)$ measured in $Cd_{0.85}Mn_{0.15}Te$ at $T = 293$ K (N. Kullendorff and B. Hök, *Appl. Phys. Lett.*, Vol. 46, p. 1016, 1985). It reaches a value of 120 deg/mm per tesla at $\lambda = 7550$ Å, which is close to the absorption edge λ_{abs} of 7500 Å. The value of $\theta/(Bd)$ increases to 200 deg/mm per tesla at $\lambda = 6150$ Å in $Cd_{0.55}Mn_{0.45}Te$ with $\lambda_{abs} = 6100$ Å (A. E. Turner,

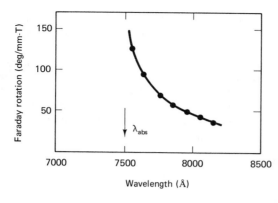

Figure 12.29 Faraday rotation $\theta/(Bd)$ measured at 293 K in $Cd_{0.85}Mn_{0.15}Te$ as a function of wavelength. (N. Kullendorff and B. Hök, *Appl. Phys. Lett.*, Vol. 46, p. 1016, 1985.)

R. L. Gunshor, and S. Datta, *Appl. Opt.*, Vol. 22, p. 3152, 1983). By comparison, the value of θ/d in an important magnetooptic material YIG (yttrium iron garnet, which is ferrimagnetic) is 23 deg/mm. In nonmagnetic semiconductors, such as CdTe and GaAs, Faraday rotation originates from the modulation of carrier orbit by B. Letting $v_y^+ = iv_x^+$ for right-hand polarized wave and $v_y^- = -iv_x^-$ for left-hand polarized wave in Eq. (5.11.10) or (5.11.11), the reader can readily show that

$$\frac{\theta}{d} = \frac{(n^+ - n^-)\omega}{2c} = \frac{Ne^3B}{2\epsilon_0 cn\omega^2(m^*)^2} \qquad \text{rad/m} \qquad (12.15.3)$$

where N is the carrier concentration, c the velocity of light, and n the refractive index without B. Equation (12.15.3) gives a value of $\theta/Bd = 7.5$ deg/mm per tesla with $N = 10^{18}$ cm^{-3}, $n = 3.6$, $v = 10^{14}$ Hz, and $m^* = 0.1m_0$. Not only is the Faraday rotation small but it is also accompanied by free-carrier absorption with an absorption coefficient

$$\alpha_{fc} = \frac{2\kappa\omega}{c} = \frac{Ne^2}{m^*\epsilon_0 cn\omega^2\tau} \qquad (12.15.4)$$

While α_{fc} always accompanies θ for Faraday rotation by free carriers, absorption can be minimized by choosing $\lambda > \lambda_{abs}$ for diluted magnetic semiconductors. Faraday rotator is a basic element in building an optical isolator, a nonreciprocal device allowing wave propagation in one direction only (e.g. from the input end to the output end) but not in the opposite direction.

PROBLEMS

12.1. Find the magnitude and direction of the electric dipole moments associated with the following charge distributions: **(a)** $+e$ at $(a, 0, 0)$ and $(0, a, 0)$ and $-e$ at $(-a, 0, 0)$ and $(0, -a, 0)$; **(b)** $+2e$ at $(0, 0, 0)$ and $-e$ at $(0, 0, a)$ and $(0, 0, -a)$; and **(c)** $+e$ at $(a, b, 0)$ and $(-a, -b, 0)$ and $-e$ at $(a, -b, 0)$ and $(-a, b, 0)$.

12.2. Using symmetry arguments, determine whether it is possible for molecules of the following types to possess a nonzero dipole moment: diatomic molecules AA (such as N_2, O_2); diatomic molecules AB (such as HCl, CO); triatomic molecules ABA in rectilinear arrangement (such as CO_2, CS_2) and in triangular arrangement (such as H_2O, SO_2); triatomic molecules ABC (such as HCN, KCN); tetratomic molecules AB_3 in a plane triangular arrangement with A at its center (such as BCl_3) and in a triangular pyramid with A at its apex (such as NH_3); and pentatomic molecules AB_4 in a tetrahedral arrangement (such as CH_4).

12.3. Consider an ionic molecule with positive and negative ions of equal charge occupying alternate corners of a regular hexagon. Determine the electric dipole moment for the molecule. Now the molecule is compressed along one body diagonal such that the bond length a remains the same but the bond angle θ at the compressed corners changes from $120°$ to $120° + \Delta\theta$. Find the electric dipole moment of the distorted molecule.

12.4. Letting $\Psi = \Psi_0 + \phi$ and $\mathbf{E}$ to be in the z direction, show that Eq. (12.2.4) becomes

$$\left(\frac{\hbar^2}{2m}\nabla^2 + \mathscr{E}_0 + \frac{e^2}{4\pi\epsilon_0 r}\right)\phi - eE_0 r\cos\theta\Psi_0 = 0$$

by neglecting terms in E_0^2. If we further let

$$\phi = \Psi_0 \frac{4\pi\epsilon_0 a^2}{e} E_0\cos\theta\, R(r)$$

verify that the differential equation for $R(r)$ is

$$\frac{d^2R}{dr^2} + 2\left(\frac{1}{r} - \frac{1}{a}\right)\frac{dR}{dr} - \frac{2R}{r^2} = -\frac{2r}{a^3}$$

Show that $R(r)$ of Eq. (12.2.5) indeed satisfies the equation above.

12.5. Using Eq. (12.2.5), verify the result given in Eq. (12.2.7). Further, using Eq. (1.2.11) for the ionization energy and Eq. (1.2.9) for the Bohr radius, show that Eq. (12.2.15) differs from Eq. (12.2.7) only by a numerical factor.

12.6. Although an exact calculation of the refractive index in a semiconductor from Eq. (12.7.2) requires a detailed knowledge of the band structure, an estimate of $n = \sqrt{\epsilon_\infty/\epsilon_0}$ can be made from Eq. (12.7.3) to test the correctness of the model. We note from Fig. 11.2 the conduction and valence bands at X and L points in k space have approximately the same curvature. As a result, many states participating in a direct optical transition across the gap will have approximately the same energy separation, producing peaks in the value of $n - i\kappa$ at photon energies $h\nu = \Delta\mathscr{E}_X$ and $\Delta\mathscr{E}_L$. An analysis of the reflectance measurements by Ehrenreich et al. (H. Ehrenreich, H. R. Philipp, and J. C. Phillips, *Phys. Rev. Lett.*, Vol. 8, p. 59, 1962) yields $\Delta\mathscr{E}_X = 4.3$ eV in Ge and 5.0 eV in GaAs for a major peak and $\Delta\mathscr{E}_{L1} = 2.2$ eV and $\Delta\mathscr{E}_{L2} = 6.0$ eV in Ge and 3.0 eV and 6.7 eV in GaAs for two minor peaks. Given $a = 5.66$ Å in Ge and 5.65 Å in GaAs, estimate the values of n^2 for Ge and GaAs from Eq. (12.7.3) using $\omega_0 = \Delta\mathscr{E}_X/\hbar$ and $m = m_0$.

12.7. Analysis of reflectance measurements in alkali halides by Tessman et al. (J. Tessman, A. H. Kahn, and W. Shockley, *Phys. Rev.*, Vol. 92, p. 800, 1953) yields the following values of electronic polarizability α_e (units in cgs $\times 10^{-24}$ cm^{-3}) for a selected number of alkali and halogen ions:

F^-	0.65	Cl^-	2.97	Br^-	4.17
Na^+	0.41	K^+	1.33	Rb^+	1.98

Given $a = 5.63$ Å in NaCl and $a = 3.74$ Å in RbCl (CsCl structure), find the refractive indices of NaCl and RbCl, and compare the calculated values with the values given in Table 12.1. Explain why the Clausius–Mossotti equation has to be used.

12.8. An exact calculation of ω_0 in Eq. (12.3.4) requires an analysis of lattice dynamics. However, an estimate of ω_0 can be made by using the compressibility data. Show that the force constant γ in Eq. (12.3.2) is related to compressibility K of Eq. (3.3.6) by $\gamma = a'/K$, where a' is the separation between positive and negative ions. Given $a = 5.63$ Å, $K = 3.3 \times 10^{-12}$ cm^2/dyn in NaCl, and the M_1 and M_2 values in Table 1.1, find the value of ω_0.

A strong lattice reflection is observed at $\lambda = 50$ μm in NaCl (C. Kittel, *Introduction to Solid State Physics*, 2nd ed., John Wiley & Sons, Inc., New York, 1956, p. 114). Check the calculated ω_0 value against the measured λ value. Then calculate the ionic contribution to the polarizability.

12.9. Consider the triangular arrangement of ions in a water molecule. Obtain an expression for the electrostatic energy containing only terms that depend on the bond angle θ. Show that for minimum energy

$$\sin^3 \frac{\theta}{2} = \frac{4\pi\epsilon_0 a^3}{8\alpha}$$

where a is the bond length and α is the polarizability of the oxygen ion. Using $\theta = 105°$ and $a = 1.02$ Å, find the value of α and the electric dipole moment of a water molecule.

12.10. A sphere of radius a with a dielectric constant ϵ is placed in a uniform field E_0. Show that the electrostatic potential Φ_0 outside the sphere and Φ inside the sphere as given by $\Phi_0 = -E_0 r \cos \theta + (A \cos \theta)/r^2$ and $\Phi = Br \cos \theta$ satisfy Laplace's equation in spherical coordinates. By applying boundary conditions to Φ_0 and Φ, verify that

$$A = \frac{\epsilon - \epsilon_0}{\epsilon + 2\epsilon_0} a^3 E_0, \qquad B = \frac{-3\epsilon_0}{\epsilon + 2\epsilon_0} E_0$$

Then find **P** and prove that Eq. (12.6.3) is indeed true.

12.11. Consider an infinitely long dielectric cylinder placed in a uniform electric field $\mathbf{E}_0$ normal to the axis of the cylinder (Fig. 12.8c). Show that the potential functions Φ_0 outside the cylinder and Φ inside the cylinder given by

$$\Phi_0 = -E_0 r \cos \theta + Ar^{-1} \cos \theta, \qquad \Phi = Br \cos \theta$$

satisfy Laplace's equation in cylindrical coordinates. Find **E** and **P** inside the cylinder by applying continuity conditions of normal **D** and tangential **E**. Verify that $N_x = N_y = 1/2\epsilon_0$ in Eq. (12.6.4).

12.12. Consider a dipole with $+q$ placed at $(r + d_x, +d_y)$ and $-q$ placed at $(r - d_x, -d_y)$. Find E_x and E_y at origin $(0, 0)$. Show that for $r >> \sqrt{d_x^2 + d_y^2}$,

$$E_x = \frac{2p_x}{4\pi\epsilon_0 r^3}, \qquad E_y = -\frac{p_y}{4\pi\epsilon_0 r^3}$$

and check the results with those from Eq. (12.6.8).

12.13. Show that using Eq. (12.8.4) in Eq. (12.8.2) results in Eq. (12.8.3) by noting that

$$\mathbf{E} = -\frac{\partial \mathbf{A}}{\partial t} \quad \text{and} \quad \mathbf{B} = \nabla \times \mathbf{A}$$

12.14. It can be shown quantum mechanically that Eqs. (12.9.2) and (12.9.3) for downward and upward transitions are replaced by

$$R_{\text{down}} = \Sigma \, r(N_p + 1)N_m, \qquad R_{\text{up}} = \Sigma \, r(N_p)N_n$$

if quantization of the radiation field is applied, where N_p is an integer representing number of photons, r is a proportionality constant, and the summation is over all possible radiation modes at frequency v. Identify the spontaneous and stimulated emission terms. Then verify that $B_{mn} = B_{nm}$ and relate N_p to $I(\nu)$ of Eqs. (12.9.2) and (12.9.3). Finally, count the number of radiation modes in a volume V within a spectral width $\Delta\nu$, and derive the ratio A/B from R_{down} and R_{up} given above.

12.15. The angular dependences of the atomic wave function are given in Eqs. (1.4.9) and (1.4.13). Refer to the associated Legendre polynomials $\Theta_{l,m}$ for the $s(l = 0)$, $p(l = 1)$, and $d(l = 2)$ states given in Eq. (1.4.20), and $\Theta_{3,m}$ for the $f(l = 3)$ state given below.

$$\Theta_{3,0} = 5\cos^3\theta - 3\cos\theta, \qquad \Theta_{3,\pm1} = \sin\theta(5\cos^2\theta - 1)$$
$$\Theta_{3,\pm2} = \sin^2\theta\cos\theta, \qquad \Theta_{3,\pm3} = \sin^3\theta$$

(a) Consider an inversion operation I that changes θ to $\pi - \theta$ and ϕ to $\pi + \phi$. Wave functions Ψ can be divided into classes of even or odd parity according to $I\Psi = \Psi$ or $I\Psi = -\Psi$, respectively. Perform $I\Psi$ for the s, p, d, and f states and assign the proper parity for the s, p, d, and f wave functions.

(b) Determine whether the electric-dipole operator $-e\mathbf{E} \cdot \mathbf{r}$ of Eq. (12.8.11) is an even or odd operator, and which of the following transitions $s \to p$, $p \to p$, $p \to d$, $p \to f$, $d \to f$, and $f \to f$ have a nonzero matrix element $\mathcal{H}_{mn}$ in Eq. (12.10.3).

12.16. **(a)** Convert the angular dependences $\Theta_{l,m}\Phi_m$ of the atomic wave function in Eq. (1.4.16) into dependences on x, y, and z for the s, p, and d states. Also refer to Problem 12.15 for the f state and show that

$$\Theta\Phi \sim (5z^3 - 3zr^2),\ (5z^2 - r^2)x,\ (5z^2 - r^2)y,\ (x^2 - y^2)z,\ xyz,\ 3x^2y - x^3,$$
$$\text{and}\quad 3xy^2 - y^3$$

(b) For $\mathbf{B}$ applied in the z direction, the magnetic dipole operator of Eq. (12.8.11) is given by $-u_z B_z$. Thus for orbital motion, $\mathbf{u}$ of Eq. (12.8.10) is given by

$$u_z = \frac{e}{2m}L_z = \frac{e\hbar}{2mi}\left(x\frac{\partial}{\partial y} - y\frac{\partial}{\partial z}\right)$$

according to Eq. (1.5.4). Determine whether u_z is an even or odd operator, and which of the following transitions $p \to p$, $p \to d$, $d \to f$, and $f \to f$ have a nonzero matrix element $\mathcal{H}_{mn}$ in Eq. (12.10.3).

12.17. An oscillating charged particle acts as an antenna, sending out electromagnetic radiation. The rate of energy radiated is given by

$$\frac{d\mathcal{E}}{dt} = -\frac{2}{3}\frac{e^2(\ddot{r})^2}{c^2}\frac{\eta}{4\pi}$$

(a) Consider a one-dimensional harmonic oscillator with $r = x = a\sin\omega t$. Verify that the result agrees with

$$\frac{d\mathcal{E}}{dt} = -\frac{\eta\pi I_0^2}{3}\left(\frac{a}{\lambda}\right)^2$$

from a dipole antenna where $\eta = \sqrt{\mu/\epsilon}$ and $I_0 = e\omega$ (S. Ramo, J. R. Whinnery, and T. Van Duzer, *Fields and Waves in Communication Electronics*, 2nd ed., John Wiley & Sons, Inc., New York, 1984). Further show that

$$\frac{d\mathcal{E}}{dt} = -\frac{\mathcal{E}}{\tau} = -\frac{2\pi e^2 \nu^2}{3m_0\epsilon c^3}\mathcal{E}$$

(b) Calculate τ for $\nu = 3 \times 10^{14}$ Hz. Compare the above τ with the τ of Eq. (12.11.11) and explain their differences.

12.18. The part (a) of the absorption curve shown in Fig. 6.24a for germanium is caused by phonon-assisted optical transitions (Fig. 6.25a). For indirect transitions, the momentum $\hbar\mathbf{k}$ for electrons is not conserved. Therefore, we need to count all possible initial and final states and expect α to be proportional to

$$\alpha \sim \int \frac{\delta(\mathscr{E}_m - \mathscr{E}_n - \mathscr{E}) \, k_m^2 \, dk_m k_n^2 dk_n}{\pi^4}$$

where $\mathscr{E} = \hbar\omega \pm k\theta$, $k\theta$ is the phonon energy, and the $\pm$ signs are for phonon absorption and emission, respectively.

Converting dk into $d\mathscr{E}$, show that

$$\alpha(\text{indirect}) \sim \int \sqrt{\mathscr{E}_1(A - \mathscr{E}_1)} \, d\mathscr{E}_1$$

where $\mathscr{E}_1$ is the kinetic energy of conduction-band electrons. Determine the integration limits. Consulting an integration table, show that

$$\alpha(\text{indirect}) \sim A_1(\hbar\omega + k\theta - \mathscr{E}_g)^2 + A_2(\hbar\omega - k\theta - \mathscr{E}_g)^2$$

Comment on the temperature dependence of A_1/A_2.

12.19. In rectangular coordinates, the wave functions of s and p electrons can be expressed in the form

$$u_s = f(r), \qquad u_x = xg(r), \qquad u_y = yg(r), \qquad u_z = zg(r)$$

where $r^2 = x^2 + y^2 + z^2$. The relations between u and $\psi_{l,m}$ of Eq. (1.4.16) are as follows:

$$\psi_{0,0} = u_s, \qquad \psi_{1,\pm 1} = u_\pm = \frac{\mp u_x - iu_y}{\sqrt{2}}, \qquad \psi_{1,0} = u_z$$

(a) Using the definition of angular momentum operator $\mathbf{L} = \mathbf{r} \times \mathbf{p}$ given in Eq. (1.5.3), verify the following relations:

$$L_z u_z = 0, \qquad L_z u_+ = \hbar u_+, \qquad (L_x + iL_y)u_z = \sqrt{2}\,\hbar u_+$$

and show that they agree with the general expressions

$$L_z \psi_{l,m} = m\hbar\psi_{l,m}, \qquad (L_x \pm iL_y)\psi_{l,m} = \hbar\sqrt{(l \mp m)(l \pm m + 1)}\,\psi_{l,m\pm 1}$$

(b) Also show that for $i, j = x, y,$ or z,

$$\iiint u_i p_{x,y \text{ or } z}\, u_j \, dx \, dy \, dz = 0$$

12.20. (a) In analyzing the band properties of III–V compound semiconductors, such as the one shown in Fig. 11.2, it is essential to include the spin-orbit interaction energy

$$\mathcal{H}_{SO} = \frac{\hbar^2}{2\hbar^2 m_0^2 c^2}\left(\frac{1}{r}\frac{\partial V}{\partial r}\right)\mathbf{L} \cdot \mathbf{S} = \frac{\xi(r)\mathbf{L} \cdot \mathbf{S}}{\hbar^2}$$

where $\xi(r)$ is generally referred to as the spin-orbit coupling constant. Show that

$$\mathcal{H}_{SO} = \frac{\xi}{\hbar^2}\left(L_z S_z + \frac{L_+ S_- + L_- S_+}{2}\right)$$

(b) The spin states $\alpha(\text{up})$ and $\beta(\text{down})$ obey the relations

$$S_z(\alpha, \beta) = \pm\frac{\hbar}{2}(\alpha, \beta), \qquad S_+\alpha = 0, \qquad S_+\beta = \hbar\alpha, \qquad S_-\alpha = \hbar\beta, \qquad S_-\beta = 0$$

Show that the relations above are consistent with the relations for $L_z\psi_{l,m}$ and $L_\pm\psi_{l,m}$ given in Problem 12.19 if we assign $l = \frac{1}{2}$ and $m = \pm\frac{1}{2}$ for α and β spin states, respectively.

(c) The wave functions for conduction-band electrons are given by $u_s\alpha$ and $u_s\beta$, while those for valence-band electrons are given by $u_+\alpha$, $u_+\beta$, $u_z\alpha$, $u_z\beta$, $u_-\alpha$, and $u_-\beta$. Refer to Problem 12.19 for the various orbital wave functions. Verify the following relations:

$$\mathcal{H}_{so}u_s\alpha = 0, \qquad \mathcal{H}_{so}u_z\alpha = \frac{\xi(u_+\beta)\sqrt{2}}{2}, \quad \mathcal{H}_{so}u_-\beta = \frac{\xi(u_-\beta)}{2}$$

$$\mathcal{H}_{so}u_+\beta = \frac{\xi(-u_+\beta + \sqrt{2}u_z\alpha)}{2}$$

12.21 (a) Using ψ of Eq. (12.10.7) for the Bloch wave, show that the cell wave function $u(r)$ satisfies the wave equation

$$\left(\mathcal{H}_0 - \frac{i\hbar^2}{m_0}\mathbf{k}\cdot\nabla + \mathcal{E}_k + \mathcal{H}_{so}\right)u(r) = \mathcal{E}u(r)$$

where $\mathcal{H}_0 = -(\hbar^2/2m_0)\nabla^2 + V(r)$, $\mathcal{E}_k = \hbar^2k^2/2m_0$, and $\mathcal{H}_{so}$ represents the spin-orbit interaction energy presented in Problem 12.20. The eight basis functions for conduction- and valence-band electrons are divided into two groups according to

$$v_1(r) = A_1u_s\alpha + B_1u_+\beta + C_1u_z\alpha + D_1u_-\beta$$
$$v_2(r) = A_2u_s\beta + B_2u_-\alpha + C_2u_z\beta + D_2u_+\alpha$$

(b) Substitute $v_1(r)$ into the wave equation, then multiply the resultant equation by $(u_s\alpha)^*$, $(u_+\beta)^*$, $(u_z\alpha)^*$, or $(u_-\beta)^*$ and integrate over $dx\,dy\,dz$, to obtain the following determinant:

$$
\begin{array}{c}
 \\
u_s\alpha \\
u_+\beta \\
u_z\alpha \\
u_-\beta
\end{array}
\begin{array}{|cccc|}
\overset{u_s\alpha}{} & \overset{u_+\beta}{} & & \\
\mathcal{E}_s - \mathcal{E}' & 0 & u_z\alpha & u_-\beta \\
0 & \mathcal{E}_p - \mathcal{E}' - \Delta/3 & i\hbar kP/m_0 & 0 \\
-i\hbar kP/ & \sqrt{2}\,\Delta/3 & \sqrt{2}\Delta/3 & 0 \\
m_0 & & \mathcal{E}_p - \mathcal{E}' & 0 \\
0 & 0 & 0 & \mathcal{E}_p - \mathcal{E}' + \Delta/3
\end{array} = 0
$$

where $\mathcal{E}_s$ and $\mathcal{E}_p$ are the eigenenergies of $\mathcal{H}_0 u_s$ and $\mathcal{H}_0 u_p$, respectively, and $\mathcal{E}' = \mathcal{E} - \mathcal{E}_k$. For simplicity, we take $\mathbf{k}$ to be in the z direction. In the equation above, P and Δ are defined as

$$P = \frac{\hbar}{i}\iiint u_s \frac{\partial}{\partial j} u_j dx\,dy\,dz = -\frac{\hbar}{i}\iiint u_j \frac{\partial}{\partial j} u_s\,dx\,dy\,dz$$

where $j = x, y,$ or z and

$$\Delta = (3/2)\iiint \xi(r)u^2_{x,y,z}\,dx\,dy\,dz$$

(c) Show that the matrix elements between the basis functions of two different sets are identically zero, for example,

$$\iiint (u_s\alpha)^*\,\mathcal{H}\,u_z\beta\,dx\,dy\,dz = 0$$

$$\iiint (u_s\alpha)^*\,\mathcal{H}\,u_-\alpha\,dx\,dy\,dz = 0$$

Dielectric and Optical Properties   Chap. 12

12.22. **(a)** From the determinant of Problem 12.21, obtain the following characteristic equations:

$$\mathscr{E}' = \mathscr{E}_p + \frac{\Delta}{3}$$

$$\left(\mathscr{E}' - \mathscr{E}_p + \frac{2\Delta}{3}\right)\left(\mathscr{E}' - \mathscr{E}_p - \frac{\Delta}{3}\right)(\mathscr{E}' - \mathscr{E}_s) = \frac{\hbar^2 k^2 P^2}{m_0^2}\left(\mathscr{E}' - \mathscr{E}_p - \frac{\Delta}{3}\right)$$

(b) Treating the term on the right-hand side as perturbation and letting $\mathscr{E}_s = \mathscr{E}_c$ and $\mathscr{E}_p = \mathscr{E}_v - \Delta/3$, show that

$$\mathscr{E}(\text{CB}) = \mathscr{E}_c + \frac{\hbar^2 k^2}{2m_0}\left[1 + \frac{2P^2}{3m_0}\left(\frac{2}{\mathscr{E}_g} + \frac{1}{\mathscr{E}_g + \Delta}\right)\right]$$

$$\mathscr{E}(\text{HH}) = \mathscr{E}_v + \frac{\hbar^2 k^2}{2m_0}$$

$$\mathscr{E}(\text{LH}) = \mathscr{E}_v + \frac{\hbar^2 k^2}{2m_0}\left(1 - \frac{4P^2}{3m_0}\frac{1}{\mathscr{E}_g}\right)$$

$$\mathscr{E}(\text{SO}) = \mathscr{E}_v - \Delta + \frac{\hbar^2 k^2}{2m_0}\left(1 - \frac{2P^2}{3m_0}\frac{1}{\mathscr{E}_g + \Delta}\right)$$

for the conduction band, the heavy-hole band, the light-hole band, and the split-off band, respectively.

12.23. **(a)** From $\mathscr{E}$ (CB) of Problem 12.22, show that

$$P^2 = \frac{3m_0(m_0 - m_c)}{2m_c}\frac{\mathscr{E}_g(\mathscr{E}_g + \Delta)}{3\mathscr{E}_g + 2\Delta}$$

(b) Using the quantum-mechanical equation $\langle p_x \rangle = m_0 d\langle x \rangle/dt$ for the expectation values, derive Eq. (12.11.23).

(c) Using the values of m_c, $\mathscr{E}_g$, and $\Delta = \lambda$ given in Table 6.7, find the values for $\langle x \rangle^2$ and τ (spontaneous lifetime) for direct optical transitions in Ge and GaAs. Then calculate the values of K for Ge and GaAs from Eq. (12.11.25) and compare the calculated values with the values obtained by fitting Eq. (12.11.24) to experimental absorption-coefficient curves of Fig. 6.24.

12.24. **(a)** Derive Eqs. (12.12.5) and (12.12.6) from Eq. (12.12.1). Referring to the discussion in Section 6.8, explain why the force constants for transverse and longitudinal lattice waves are different in ionic crystals.

(b) Use the values for GaAs given in Table 11.5 to check Eq. (12.12.10). From Fig. 2.38c, we see $\nu_{\text{TO}} = \nu_{\text{LO}}$ in Ge. Explain why this result is expected.

12.25. The Cauchy theorem states that if an analytic function $Z(z)$ in complex plane $z = x + iy$ has no singularities inside a closed contour C, then an integral around the contour yields

$$\oint_C Z(z)\, dz = 0$$

(a) Letting $z = \omega = \omega_1 + i\omega_2$ and $Z(z) = [\epsilon(\omega) - \epsilon_0]/(\omega - \omega_0)$, show that

$$\int_{-\infty}^{\infty} \frac{\epsilon(\omega_1) - \epsilon_0}{\omega_1 - \omega_0}\, d\omega_1 = i\pi[\epsilon(\omega_0) - \epsilon_0]$$

The contour follows the real ω axis which is divided into two segments by a semicircle of small radius ρ and joined at the two ends by a semicircle of large radius R. The Cauchy theorem yields the result above as we let ρ go to zero and R go to infinity. Comment on the behavior of $\epsilon(\omega)$ of Eq. (12.12.18) as $R \to \infty$.

    (b) Derive Eqs. (12.12.19) and (12.12.20) from the result above by separating $\epsilon(\omega)$ into real and imaginary parts and by converting the integral into one along the positive real ω axis.

12.26. Using $\epsilon_s = 13.2\epsilon_0$, $\epsilon_\infty = 10.9\epsilon_0$, and the values given for ν_{TO} and γ in Eq. (12.13.1), calculate the values of n and α at photon energies $h\nu = 33.25$ meV and 26 meV.

12.27. Show that for a narrow absorption line (i.e., $\gamma << \nu$), the imaginary part $\epsilon''(\omega)$ of Eq. (12.13.1) can be approximated by

$$\epsilon''(\omega) = \frac{\epsilon_s - \epsilon_\infty}{2} \frac{\gamma_1 \nu}{(\nu_{TO} - \nu)^2 + \gamma_1^2}$$

where $\gamma_1 = \gamma/2$. Using $\epsilon''(\omega)$ thus obtained in Eq. (12.12.19), verify that the contribution from the ionic part to $\epsilon'(\omega = 0)$ is given by $\epsilon_s - \epsilon_\infty$.

12.28. Show that Eqs. (12.13.6) and (12.12.21) can be transformed into each other. Also show that if $\Delta\alpha$ occurs within $\Delta\omega$ comparable to $\omega' - \omega$, then Δn is on the order of $\lambda \Delta\alpha/4\pi^2$. Discuss why an accurate measurement of $\Delta\alpha(\omega')$ over an extended range of photon energy is needed to predict $\Delta n(\omega)$ with reasonable accuracy.

12.29. For an estimate of $\epsilon'(\omega)$ at $h\nu < \mathscr{E}_g$, Eq. (12.12.19) is divided into segments as follows

$$\epsilon'(\omega) = \epsilon_0 + \frac{2}{\pi} \sum_i \frac{\langle \epsilon''(\omega')\rangle \, \Delta h\nu'}{\langle h\nu'\rangle_i}$$

where $\langle \epsilon''(\omega')\rangle \, \Delta h\nu'$ represents the area under the $\epsilon''(\omega')$ curve and $\langle h\nu'\rangle_i$ the average photon energy in segment i. Estimate the value of $\epsilon'(\omega')$ from Fig. 12.20 for two cases: **(a)** $\Sigma\Delta h\nu' = 5$ eV and **(b)** $\Sigma\Delta h\nu' = 15$ 1eV. Which value gives a better agreement with the value 10.9 in GaAs and 16 in Ge? Make $\Delta h\nu' = 1$eV.

12.30. **(a)** Using the value of $\alpha_B = 1 \times 10^4$ cm^{-1}, the value of I_s from Fig. 12.22, and other values quoted in the text, estimate the value of exciton lifetime τ from Eq. (12.13.13).

    (b) Show that for an order-of-magnitude estimate, Eq. (12.13.6) can be approximated by

$$\Delta n = -\frac{\lambda \, \Delta\alpha}{4\pi^2}$$

Check the value of n_2 calculated from α_2 in Eq. (12.13.14) with the value of n_2 quoted in the text. Are n_2 and α_2 functions of $h\nu$? If so, do they have the same dependence on $h\nu$?

12.31. **(a)** Referring to the behavior of ψ_c and ψ_v in Fig. 12.23, explain why the oscillation in $\Delta\alpha$ shown in Fig. 12.24 is expected. Calculate the value of $h\nu_E$ and hence the period of oscillation in $h\nu$ for $\Delta\alpha$ at $E = 4 \times 10^4$ V/cm.

    (b) Obtain Eq. (12.14.6) by using Eq. (12.14.4) in Eq. (12.14.1) and integrating the latter by parts. Show that Eq. (12.14.6) is a reasonable approximation only for $\beta^{3/2} >> 1$. For an estimate, however, Eq. (12.14.6) is useful even for $\beta \sim 1$. Estimate the value of α_E at $h\nu = \mathscr{E}_g - 17$ meV with $E = 20$ kV/cm.

12.32. Derive Eqs. (12.14.12) and (12.14.13) from Eq. (12.14.9). Consider an incident beam with x-directed polarization. Show that the polarization becomes y-directed after a distance L given by

$$L = \frac{\lambda}{2n^3 r E_z^0}$$

12.33. Read the papers by Miller et al. (D. A. B. Miller, J. S. Weiner, and D. S. Chemla, *IEEE J. Quantum Electron.*, Vol. QE-22, p. 1816, 1986) and by Bastard and Brum (G. Bastard and J. A. Brum, *IEEE J. Quantum Electron.*, Vol. QE-22, p. 1625, 1986). Present a theoretical discussion of the Stark shift and polarization dependence of αL shown in Fig. 12.25.

12.34. Using equations of motion for free carriers under a magnetic flux density B, derive Eq. (12.15.3) for Faraday rotation.

13

High-Frequency and High-Speed Devices

13.1 INTRODUCTION

In Chapters 8 and 9 we discussed the electrical characteristics of semiconductor junction and multijunction devices. These devices are intended for low-frequency applications, that is, for operating frequencies below 1 GHz. In the low-frequency region, a device can be adequately represented by the lumped-circuit model in analyzing the device performance. During recent years, a great deal of attention has been focused on the use of semiconductor junctions for device applications in the microwave and millimeter-wave regions and for ultrafast switching with a speed in the picosecond region. In this chapter we examine semiconductor devices for generation, amplification, and transmission of microwave and millimeter-wave signals. Such devices include varactor diodes, IMPATT diodes, tunnel diodes, and Gunn-effect devices, as discrete elements, and high-frequency bipolar and field-effect transistors in integrated-circuit applications. For devices operating in the high-frequency and high-speed regions, parasitic effects, such as lead inductance and stray capacitance, and propagation effects, such as transit-time delays, become important. Both parasitic and propagation effects are considered, wherever appropriate, in the device analyses presented in this chapter. For large-scale integrated circuits, important issues include uniformity and reproducibility of device performance in addition to high speed and small dimension. These issues are discussed in conjunction with performance of transistors.

13.2 VARACTOR DIODES AND PARAMETRIC INTERACTION

In many scientific endeavors as well as practical applications, much of the progress depends on our ability to detect very weak signals. In the microwave frequency range, very sensitive receivers are essential to the progresses being made in radio astronomy, satellite communications, and long-range radars. What limits our ability to detect weak signals is the noise generated in first-stage amplifiers. Some exciting developments in low-noise amplifiers include the invention of masers, the realization of parametric amplifiers, and the arrival of GaAs FETs. In this section low-noise amplifiers based on the principle of parametric interaction are treated.

636

By *parametric interaction* we mean the interaction of two or more electromagnetic waves caused by time-varying circuit elements (at low frequency) or time-varying material parameters (at high frequency). Through parametric interaction, we may amplify the energy of a signal at one frequency (parametric amplifier) or generate coherent oscillations at one frequency (parametric oscillator) by supplying energy to a nonlinear circuit element or medium at a higher frequency. To see that amplification of a signal is indeed possible with a time-varying circuit element, we consider an idealized tank circuit consisting of a lossless capacitor of capacitance C and a lossless inductor of inductance L. If there is initial energy stored in the tank circuit, the stored energy will oscillate from the inductor, where it is in the form of magnetic energy $LI^2/2$ to the capacitor, where it is in the form of electrostatic energy $Q^2/2C$, I and Q being the current through the inductor and the charge stored in the capacitor respectively. In Fig. 13.1a the charge Q is shown as a function of time, oscillating with an angular frequency $\omega = 1/\sqrt{LC}$. The amplitude of oscillation will remain constant as long as no energy is further added to or taken away from the tank circuit.

Suppose that the plates of the capacitor are pulled apart when Q reaches a maximum and are pushed back to the initial separation when Q is zero. Figure 13.1b shows the separation d as a function of time. Note that energy is pumped into the tank circuit at $t_1, t_3, t_5, \ldots$ (Fig. 13.1c) in the form of work done against coulombic attraction in pulling the plates apart but no energy is taken out from the tank circuit at $t_2, t_4, t_6,$ $\ldots$. The voltage V (Fig. 13.1d) developed across the capacitor jumps abruptly at t_1, $t_3, t_5, \ldots$, because of the change in d, and the amplitude of oscillation in Q (Fig. 13.1e) increases every half cycle because of the energy increase in the tank circuit. This simple example illustrates that amplification of a signal is indeed possible if a reactance element (capacitance or inductance) is made to vary properly with time.

As discussed in Section 8.2, the magnitude of transition-region capacitance of a *p-n* junction depends on the applied voltage. This voltage dependency makes junction diodes useful in many practical applications. For example, junction diodes can be used to achieve electronically tunable circuits. However, such an application is not the purpose of our present discussion. Semiconductor junctions are also natural candidates for the role required of a time-varying reactance element in a parametric amplifier. Junction diodes designed specifically for purposes of parametric interaction are called *varactor diodes*.

Consider a semiconductor junction biased under a dc voltage V_0 and an ac voltage v_1. According to Eq. (8.2.13) with $V_a = V_0 + v_1$, we can express the junction capacitance C_T as

$$C_T = C_0 + C_1 = K(V_d + V_0 + v_1)^{-m} = K(V_d + V_0)^{-m}\left(1 - \frac{mv_1}{V_d + V_0}\right) \quad (13.2.1)$$

for a reverse-biased diode such that $v_1 < V_d + V_0$. In Eq. (13.2.1) the capacitance C_0, which depends on V_0 only, is the steady-state part and the capacitance C_1, which depends on v_1, is the time-varying part. The two constants K and m can be identified by comparing Eqs. (13.2.1) and (8.2.13). Note that C_T as defined in Eq. (8.2.7) is a differential capacitance. Therefore, the ac current flowing through the diode is given by

$$i_1 = \frac{dQ}{dt} = \frac{dQ}{dV_a}\frac{dV_a}{dt} = C_0\left(1 - \frac{mv_1}{V_d + V_0}\right)\frac{dv_1}{dt} \quad (13.2.2)$$

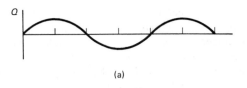

(a)

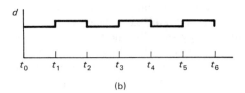

(b)

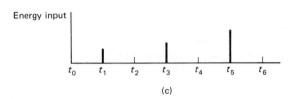

(c)

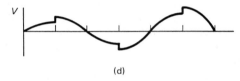

(d)

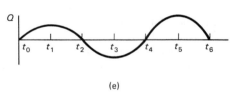

(e)

Figure 13.1 (a) Charge Q stored in the capacitor of a resonant circuit with constant inductance L and capacitance C and (c), (d), and (e) schematic diagrams showing, respectively, the time variation of the energy input to the resonant circuit, the voltage V across the capacitor and the charge Q stored in the capacitor for the case (b) where the separation d between the plates of the capacitor is varied periodically. The diagrams (b) to (e) are used to illustrate the principle of parametric interaction.

The condition for amplification is that the current i_1 must be 180° out of phase with v_1. This condition can be achieved at one frequency if energy is supplied to C_T at a higher frequency.

Suppose that the ac voltage v_1 consists of three components: one at ω_s, which is the signal to be amplified, another at ω_p, which supplies the energy and is generally referred to as the *pump*, and the third at ω_i, which is known as the *idler*. Thus $v_1 = e_s + e_p + e_i$, or

$$v_1 = v_s \sin(\omega_s t + \theta_s) + v_p \sin(\omega_p t + \theta_p) + v_i \sin(\omega_i t + \theta_i) \quad (13.2.3)$$

the subscripts s, p, and i being referred to the signal, the pump, and the idler voltage, respectively. The frequencies of the three components must be chosen such that

$$\omega_p = \omega_s + \omega_i \quad (13.2.4)$$

If this frequency condition is satisfied, mixing of the pump and signal voltages produces a term in Eq. (13.2.2) which varies at ω_i. This means that the idler voltage will be generated internally and hence we need not to supply v_i from an external source. For ease of discussion, however, we assume that an independent source for v_i exists.

Substituting Eq. (13.2.3) into Eq. (13.2.2), we find the current component at ω_s to be

$$i_1(\omega_s) = \omega_s C_0 v_s \cos(\omega_s t + \theta_s) + m v_p v_i \frac{1}{V_d + V_0} \frac{\omega_s C_0}{2} \sin(\omega_s t + \theta_p - \theta_i) \quad (13.2.5)$$

Next we express $\cos\theta$ as Re exp $(i\theta)$ and $\sin\theta$ as Re $[-i \exp(i\theta)]$, where Re stands for the real part, and write both v_1 and i_1 in phasor (complex) notation. The admittance Y which the input signal sees is simply the ratio of $i_1(\omega_s)/v_s(\omega_s)$ expressed in complex notation. Thus we have

$$Y(\omega_s) = i\omega_s C_0 + \frac{m v_p v_i}{(V_d + V_0) v_s} \frac{\omega_s C_0}{2} \exp[i(\theta_p - \theta_i - \theta_s)] \quad (13.2.6)$$

A similar expression is obtained for Y at the idler frequency ω_i with the exchange of subscripts i and s. Both the signal and idler voltages see a negative conductance if

$$\theta_p - \theta_s - \theta_i = \pi \quad (13.2.7)$$

Equation (13.2.7) is the condition for amplification. In actual operation of a parametric amplifier, the idler voltage is generated internally. Since maximum amplification occurs only when Eq. (13.2.7) is satisfied, the phase θ_i will adjust itself to this condition.

Figure 13.2a shows the structure of a mesa varactor to give the reader a feel of its physical dimensions. In addition to the transition-region capacitance C_T, a varactor has a series resistance R_s, which is a sum of the various resistances associated with the contact, the diffused p region, the mesa n region, and the n-type wafer. The values of R_c, R_d, and R_m may be kept less than 0.1 Ω by carefully preparing the contact, by minimizing the junction depth, and by carefully controlling the mesa structure through

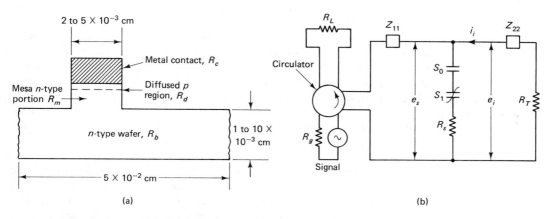

Figure 13.2 (a) Schematic diagram showing the physical dimension of a varactor diode operating in the 10-GHz region and (b) equivalent circuit of a parametric amplifier.

etching. Most contribution to R_s comes from R_b. Depending on the wafer thickness, the value of R_b may vary from 0.2 to 1 Ω. At microwave frequencies, consideration of skin-depth effect is important and may raise the value of R_b based on low-frequency calculations. For parametric amplifiers to be operated at low temperatures, special consideration should also be given to the freeze-out effect of impurities. Below 20 K, impurities become only partially ionized, and as a result, resistivity increases. Parametric amplifiers have been operated successfully at 4 K with varactors made of silicon wafer, which is heavily doped and has a carefully controlled doping profile.

Figure 13.2b shows the equivalent circuit of a cavity-type parametric amplifier. Since a diode is a two-terminal device, it is necessary to isolate the output circuit from the input circuit. The circulator provides for this isolation. The impedances Z_{11} and Z_{22} are the impedances associated with the signal and idler cavity, respectively. The resistance R_T is the termination resistance for the idler circuit, whereas R_L is the load resistance for the signal circuit. As to the varactor diode itself, we need to find the impedance matrix representing it. To do so, we use complex notation and let i and e with appropriate subscripts represent the respective current and voltage. Next, we define a time-varying capacitance (again in complex notation):

$$C_1' = C_0 \frac{m}{V_d + V_0} e_p = C_0 \frac{m}{V_d + V_0} \text{Re} \{ -iv_p \exp [i(\omega_p t + \theta_p)] \} \qquad (13.2.8)$$

In complex notation, Eq. (13.2.5) becomes

$$i_s = i\omega_s C_0 e_s + i\omega_s \frac{C_1'}{2} e_i^* \qquad (13.2.9)$$

and a similar equation for idler current reads

$$i_i^* = -i\omega_i \frac{C_1'^*}{2} e_s - i\omega_i C_0 e_i^* \qquad (13.2.10)$$

where the asterisk indicates the complex conjugate. Using Eqs. (13.2.9) and (13.2.10) in conjunction with Fig. 13.2, we can find the gain of a parametric amplifier by eliminating $e_i^* = i_i^* (Z_{22}^* + R_T)$.

As mentioned earlier, the important feature of a parametric amplifier is its low-noise figure. There are two noise sources: the shot noise associated with diode current and the thermal noise associated with diode resistance R_s. For a theoretical calculation of the noise figure, we add a noise-current source $2eI_0B$ in parallel with elastance S_0 and a noise-voltage source $4kTR_sB$ in series with resistance R_s, where B is the bandwidth, $S_0 = 1/C_0$, and $S_1 = C_1'/C_0^2$. Measured values of effective input noise temperature in a 4-GHz amplifier pumped at 36 GHz range from a value of 10 K for the amplifier operated at ambient temperature 4.2 K to a value of 85 K at room temperature. These measured values agree well with the values predicted theoretically. With a single-stage, double-tuned parametric amplifier, a gain of 15 dB and a bandwidth of 15% are reported to have been achieved. For experimental data and references to work on parametric amplifiers, the reader is referred to an excellent review article by Uenohara (M. Uenohara in L. Young, ed. *Advances in Microwaves,* Vol. 2, Academic Press, Inc., New York, 1967, p. 89).

13.3 NEGATIVE DIFFERENTIAL CONDUCTANCE: ESAKI DIODE AND RESONANT-TUNNELING BARRIER STRUCTURE

In Sections 8.10, 10.5, and 10.6 we presented discussions of the effect of tunneling process on junction characteristics. At that time our attention was concentrated on a reverse-biased diode. However, the tunneling process manifested in a forward-biased diode is far more interesting. Discovery of negative differential conductance was first reported by Esaki in a forward-biased, degenerate junction (L. Esaki, *Phys. Rev.,* Vol. 109, p. 603, 1958). The tunneling process may manifest itself in a S-S, M-S, or M-I-S junction, where S stands for semiconductor, M for metal, and I for insulator. Negative differential conductance was also predicted in superlattices (R. Tsu and L. Esaki, *Appl. Phys. Lett.,* Vol. 22, p. 562, 1973) and has been observed in quantum wells (T. C. L. G. Sollner, W. D. Goodhue, P. E. Tannenwald, C. D. Parker, and D. D. Peck, *Appl. Phys. Lett.,* Vol. 43, p. 588, 1983). For his work on tunneling in semiconductors L. Esaki shared the Nobel Prize in Physics in 1973 with I. Giaever and B. D. Josephson. The purpose of this section is to give an overview of the tunneling effect: the theory of tunneling, the practical use of a tunneling junction, and the use of the tunneling process for theoretical investigations.

We limit our discussion to the S-S junction. In order that a tunneling process may take place between any two states, the initial states must be occupied and the final state must be empty. Further, the two states must conserve energy. These two requirements are basic for an understanding of some of the device characteristics. Let us compare the tunneling process in a degenerate-semiconductor junction with that in a nondegenerate-semiconductor junction. By *degenerate semiconductor* we mean a semiconductor in which the Fermi level lies either in the conduction band (degenerate *n*-type) or in the valence band (degenerate *p*-type). Insofar as the tunneling process is concerned, the main difference between a degenerate and a nondegenerate semiconductor is the occupancy of the band states.

Figure 13.3 illustrates schematically the situations in a degenerate and a nondegenerate junction under forward-bias, and reverse-bias conditions. Tunneling is possible in both junctions biased in the reverse direction. During the process, an electron makes a quantum-mechanical transition from an occupied state in the valence band on the *p* side (the initial state) to an empty state in the conduction band on the *n* side (the final state). The situation in the forward direction is quite different. For the forward-bias case, the dominant process is the tunneling of electrons from the *n* side to the *p* side. In a nondegenerate junction (Fig. 13.3b), there are not many occupied states in the conduction band. What is even more important is that there is no corresponding allowed state on the *p* side. As Fig. 13.3b indicates, the final state would have to end up in the forbidden gap, where no state is allowed. Therefore, for a nondegenerate semiconductor junction, tunneling can occur only when the junction is biased in the reverse direction, not when the junction is biased in the forward direction.

The situation in a forward-biased degenerate junction is much more interesting and needs a more careful examination. When a small forward bias V_a is applied, electrons in the conduction band on the *n* side having energies between $\mathscr{E}_{fn}$ and $\mathscr{E}_{fn} - eV_a$ will be able to tunnel to vacant states in the valence band on the *p* side having energies between $\mathscr{E}_{fp}$ and $\mathscr{E}_{fp} + eV_a$. The situation is illustrated in Fig. 13.4a. As V_a further increases, the number of available states for tunneling increases and so does the tunneling current. However, a situation of maximum available states is reached when $\mathscr{E}_{fn}$ is

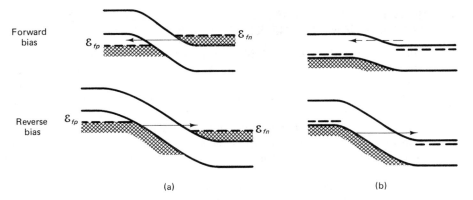

Forward
bias

$\mathcal{E}_{fp}$

$\mathcal{E}_{fn}$

Reverse
bias

$\mathcal{E}_{fp}$

$\mathcal{E}_{fn}$

(a)

(b)

Figure 13.3 Comparison of alignment of conduction-band states on n side and valence-band states on p side in (a) a degenerate junction (Esaki diode) and (b) a nondegenerate junction (conventional diode) under various bias conditions. Tunneling is possible (as indicated by a solid arrow) when occupied states (crosshatched) on one side are aligned in energy with empty states on the other side.

about to rise above $\mathcal{E}_v$ on the p side or $\mathcal{E}_{fp}$ to go below $\mathcal{E}_c$ on the n side, depending on whichever event comes first. This corresponds to the situation shown in Fig. 13.4b. Finally, upon still further increase of V_a, the conduction-band edge on the n side rises above the valence-band edge on the p side. When this happens, tunneling is no longer possible, as is apparent from Fig. 13.4c. Therefore, the I–V characteristic in the tunneling region follows the solid curve of Fig. 13.4d. Diodes having such characteristics are known as *Esaki diodes,* named after the discoverer of the phenomenon (L. Esaki, *Phys. Rev.,* Vol. 109, p. 63, 1958).

Now let us find the tunnel current. The number of electrons striking the junction per second per unit area of the junction is given by the product of the density of occupied states and the velocity of these electrons. Thus the incident current density associated with such a motion is equal to

$$dJ_i = -2ev_x h^{-3} f(\mathcal{E}_1) \, dp_x \, dp_y \, dp_z \qquad (13.3.1)$$

Of the incident electrons, however, only a few can tunnel across the junction. Let P_1 be the transmission coefficient for tunneling, defined as the probability with which an incident electron can actually tunnel through the junction if the states on the other side of the junction are completely empty. An expression for P_1 is given in Section 10.6 by Eq. (10.6.14). If the final states are partly occupied, the actual tunneling probability is only $P_1[1 - f(\mathcal{E}_2)]$. Therefore, the current density as a result of electron tunneling from the conduction band into the valence band is

$$J_t(c \rightarrow v) = \int -e \frac{4\pi}{h^3} P_1 f(\mathcal{E}_1)[1 - f(\mathcal{E}_2)](v_x \, dp_x)(p_\perp dp_\perp) \qquad (13.3.2)$$

A similar expression is obtained for the current tunneling in the opposite direction. Therefore, the net tunneling-current density is the difference of the two expressions, or

$$J_t = J_t(v \rightarrow c) - J_t(c \rightarrow v) = \int \frac{4\pi e}{h^3} P_1 [f(\mathcal{E}_2) - f(\mathcal{E}_1)] m^* \, d\mathcal{E}_x \, d\mathcal{E}_\perp \qquad (13.3.3)$$

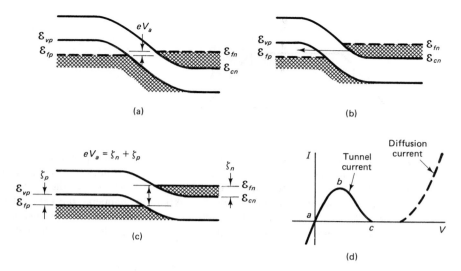

Figure 13.4 (a) Energy-band diagram for a degenerate junction under a slight forward bias, (b) forward-bias condition for maximum overlap of occupied conduction-band states on the n side and vacant valence-band states on the p side, (c) energy-band diagram at a forward bias where tunneling ceases, and (d) typical I–V characteristic of a tunnel (Esaki) diode.

In Eqs. (13.3.2) and (13.3.3), the x direction is chosen to be the direction of tunneling,

$$\mathscr{E}_x = \frac{p_x^2}{2m^*}, \qquad \mathscr{E}_\perp = \frac{p_y^2 + P_z^2}{2m^*} = \frac{p_\perp^2}{2m^*} \tag{13.3.4}$$

and $m^* = m_r$ is the reduced mass. Equation (13.3.3) can be evaluated numerically if the transmission coefficient P_1 is known. It is also possible to obtain an analytical expression from Eq. (13.3.3) if certain approximations are made. The reader is referred to an article by Kane for the details of the analytical calculation. Here we quote the results of his analysis (E. O. Kane, *J. Appl. Phys.*, Vol. 32, p. 83, 1961).

In the reverse direction, the tunneling current density is given as Eq. (10.6.17), or

$$J_t = \frac{4\pi em^*}{h^3} (\overline{\mathscr{E}})(eV_a) \exp\left(-\frac{\mathscr{E}_g}{4\overline{\mathscr{E}}}\right) \tag{13.3.5}$$

where the parameter $\overline{\mathscr{E}}$ is given by Eq. (10.6.15) or

$$\overline{\mathscr{E}} = \frac{e\hbar E}{\pi(2m^*\mathscr{E}_g)^{1/2}} \tag{13.3.6}$$

and E is the average field in the junction. In the forward direction,

$$J_t = \frac{4\pi em^*}{h^3} (D) \exp\left(-\frac{\mathscr{E}_g}{4\overline{\mathscr{E}}}\right) \tag{13.3.7}$$

where D is a quantity expressed in units of $(eV)^2$ and it represents, except for a constant, the number of states effectively participating in the tunneling process. Figure 13.5 shows the variation of D with the applied voltage. The quantity D increases with the

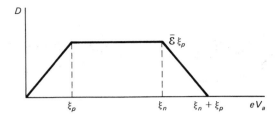

applied voltage V_a until the empty zone on the p side is completely within the occupied zone on the n side. Upon further increase of eV_a, the number of available states for tunneling is limited by the empty states on the p side. Therefore, the value of D remains constant until $\mathscr{E}_{fp}$ begins to fall below $\mathscr{E}_{cn}$. When this happens, the value of D decreases linearly with eV_a. Eventually, the empty zone on the p side and the occupied zone on the n side go completely out of alignment if $\mathscr{E}_{vp}$ falls below $\mathscr{E}_{cn}$. The value of D goes to zero at

$$eV_a = (\mathscr{E}_{fn} - \mathscr{E}_{cn}) + (\mathscr{E}_{vp} - \mathscr{E}_{fp}) = \xi_n + \xi_p \qquad (13.3.8)$$

The quantities ξ_n and ξ_p can be evaluated if the majority-carrier concentration is known. Note that the Fermi function in a degenerate semiconductor can be approximated in a manner similar to that applied to a metal. Following the procedure used in deriving Eq. (4.6.3) and making the appropriate changes in Eq. (4.6.3), we obtain

$$\xi_n = \mathscr{E}_{fn} - \mathscr{E}_{cn} = \frac{h^2}{2m_c^*}\left(\frac{3n}{8\pi}\right)^{2/3} \qquad J \qquad (13.3.9a)$$

$$\xi_p = \mathscr{E}_{vp} - \mathscr{E}_{fp} = \frac{h^2}{2m_c^*}\left(\frac{3p}{8\pi}\right)^{2/3} \qquad J \qquad (13.3.9b)$$

Take GaAs as an example with $m_c^* = 0.068m_0$ and $m_v^* = 0.64m_0$. At free-carrier concentrations $n = 1 \times 10^{19}$ cm^{-3} on the n side and $p = 5 \times 10^{19}$ cm^{-3} on the p side, we find that $\xi_n = 0.242$ eV and $\xi_p = 0.07$ eV. In this diode, the tunneling current is significant only for $V_a < 0.312$ V.

Note that the factor $\exp(-\mathscr{E}_g/4\overline{\mathscr{E}})$ is identical to P of Eq. (10.5.18) and can be expressed simply as $\exp(-E_0/E)$. For GaAs, $m_r^* = 0.064m_0$ and $\mathscr{E}_g = 1.40$ eV. Using these values in Eq. (10.6.19), we find that $E_0 = 1.6 \times 10^7$ V/cm. Noting that $\overline{\mathscr{E}} = \mathscr{E}_g E/4E_0$ and $D = \overline{\mathscr{E}}\xi_p$, and using $\xi_p = 0.070$ eV and $m^* = m_r = 0.064m_0$ in Eq. (13.3.7), we find that

$$j_t = 2.7 \times 10^{11} \frac{E}{E_0} \exp\left(-\frac{E_0}{E}\right) \qquad A/m^2 \qquad (13.3.10)$$

For a current density j_t of 3 A/cm^2, we need a field strength $E = 4.5 \times 10^{-2} E_0$ or 7×10^5 V/cm. Similar calculations can be made for Ge and Si tunnel diodes. For appreciable tunnel current in the forward direction, the field strength E should have a value somewhere between 5×10^5 and 10^6 V/cm. This value can easily be reached in a degenerate junction.

The important characteristic of a tunnel diode is that it possesses a negative differential conductance region separating two positive differential conductance regions. If

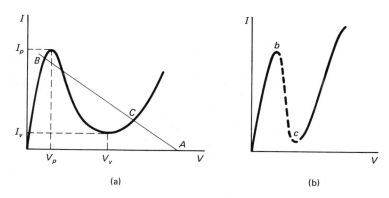

Figure 13.6 (a) *I–V* characteristic of a tunnel diode illustrating its use as a switch, and (b) *I–V* characteristic of a tunneling quantum-well structure.

properly biased and loaded with a resistance (line *AB* in Fig. 13.6a), a tunnel diode has two stable operating points (*B* and *C*), one corresponding to a high-current, low-voltage state (the on state *B*) and the other to a low-current, high-voltage state (the off state *C*). Such a device can be used as a switch. We should point out, however, that a larger current ratio and hence a more effective switching action can be achieved with other junction devices, such as the *p-n-p-n* structure, than with tunnel diodes. An ideal Esaki diode should have a negligible valley current I_v in the region (Fig. 13.4d) where the tunnel current has ceased to exist and the ordinary diffusion current has not become significant. However, in practical diodes, there may be deep-level states in the space-charge region of a tunnel diode. Electrons from the conduction band can now tunnel to these midgap states first and then make transitions from these states to the valence band. This indirect tunneling process results in an excess current which reduces the peak-to-valley current ratio of I_p/I_v.

Negative differential conductance has also been observed in quantum well structures. Figure 13.7a shows the experimental structure used by Inata et al. (F. Inata, S. Muto, T. Fujii, and S. Hiyamizu, *Extended Abstract of 18th Conference on Solid State Devices and Materials,* Tokyo, 1986, p. 767). The middle InGaAs layer (62 Å) and the two outer AlGaAS layers (41 Å) form a potential well that quantizes the energy state in the direction of confinement (Section 11.9). Referring to the energy-band diagram of Fig. 13.7b, we see that electrons originate near the Fermi level on the left of the first AlGaAs barrier, tunnel into the well, and then tunnel through the second InAlAs barrier on the right into the empty states in the conduction band of n^+ InGaAs. *Resonant tunneling* (or maximum current) occurs when the energy of the tunneling electron is approximately in line with the quantized energy level for electrons confined in the well (Problem 13.7). If the bias voltage *V* is raised, the energy levels become out of alignment (Fig. 13.7c) and the current decreases. Figure 13.6b shows qualitatively the *I–V* characteristic observed by Inata et al. at 77 K, with points *b* and *c* corresponding to the situations shown in Figs. 13.7b and c, respectively. A large peak-to-valley ratio, I_p/I_v, of about 12, and a large peak current density, J_p, of 2.2×10^4 A/cm^2, were achieved simultaneously at 77 K.

Previously, negative differential conductance was reported in GaAs–AlGaAs quantum wells with a I_p/I_v ratio of 8 and a peak current density J_p of 1.6×10^3 A/cm^2

Sec. 13.3   Negative Differential Conductance   **645**

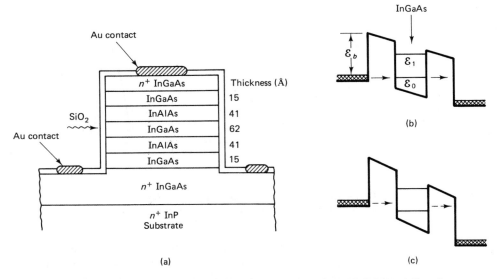

Figure 13.7 (a) Experimental quantum-well structure made of GaInAs well and AlInAs barriers, (b) its energy diagram for conduction-band electrons at a forward bias for resonant tunneling (solid arrow), and (c) its energy diagram at a higher forward bias. The misalignment of energy levels in (c) results in a much smaller tunneling current.

at 77 K (T. J. Shewchuk, P. C. Chapin, P. D. Coleman, W. Kopp, R. Fischer, and H. Morkoç, *Appl. Phys. Lett.*, Vol. 46, p. 508, 1985). There are several requirements for achieving a high I_p/I_v and a high J_p simultaneously. First, the total width ($L_W + 2L_B$) for the well and two barriers should be shorter than the electron mean path. Furthermore, the band-gap discontinuity and hence the barrier height $\mathscr{E}_b$ in relation to the barrier width L_B should be such that tunneling current dominates over the thermionic-emission current. These considerations point to the use of undoped well and barriers to minimize ionized impurity scattering and to better device performance because of reduced lattice scattering and reduced thermionic emission. The tunneling probability P across a flat barrier can be found from

$$P = \exp\left(-\frac{2\sqrt{2m^*\mathscr{E}_b}\,2L_B}{\hbar}\right) \qquad (13.3.11)$$

The much larger J_p observed in the InGaAs–InAlAs resonant-tunneling barrier (RTB) structure can be attributed to a smaller effective mass. The value of V_p is between 250 and 400 mV. Therefore, a field on the order of 2×10^5 V/cm exists in the well and barrier regions. In the presence of an electric field, both the tunneling probability P and the quantized energy level depend on E (Problem 13.8). Therefore, a quantitative analysis of the resonant tunneling behavior will require a knowledge of E as a function of the applied bias V. Finally, we should mention that upon further increase of V beyond point c in Fig. 13.6b, the current rises again and reaches a second peak when resonant tunneling to the first excited quantum state $\mathscr{E}_1$ occurs. For a discussion of design considerations for RTB diodes, the reader is referred to an article by Vakhshoori and Wang (D. Vakhshoori and S. Wang, *J. Appl. Phys.*, Vol. 62, p. 3474, 1987).

A device possessing a negative conductance can be used as an amplifier and as an oscillator if certain stability conditions are met. Tunnel-diode amplifiers have been built in waveguide, coaxial line, and strip transmission line, and operation has been achieved at frequencies as high as 85 GHz. Since a tunnel diode is a two-terminal device, a nonreciprocal device known as *circulator* is commonly used to isolate the output circuit from the input circuit. In principle, tunnel diodes, if operated as oscillators, can be used to provide microwave power. However, they do not appear as promising as IMPATT diodes and Gunn-effect devices for such use because the power level achieved in tunnel diodes is much lower than that in IMPATT diodes. The resonant-tunneling barrier (RTB) structure, on the other hand, offers not only a higher power capability because of the larger tunneling probability (due to smaller $\mathscr{E}_b$ and smaller L_B) but also a higher-frequency response because of elimination of junction capacitance. Experiments demonstrating the detecting and mixing capabilities of RTB structures at frequencies as high as 56 GHz have been reported (T. C. L. G. Sollner, E. R. Brown, W. D. Goodhue, and H. Q. Le, *Appl. Phys. Lett.*, Vol. 50, p. 332, 1987). Therefore, the RTB structure appears promising as a nonlinear element at millimeter and submillimeter wavelength.

For a discussion of the equivalent circuit and the performance of tunnel diodes in microwave applications, the reader is referred to other books (see, for example, W. F. Chow, *Principles of Tunnel Diode Circuits*, John Wiley & Sons, Inc., New York, 1964; L. S. Nergaard and M. Glicksman, eds., *Microwave Solid-State Engineering*, D. Van Nostrand Company, Inc., New York, 1964; F. Sterzer, ''Tunnel Diode Devices,'' in L. Young, ed., *Advances in Microwaves*, Vol. 2, Academic Press, Inc., New York, 1967, p. 1). We should mention, however, one important application of the tunnel diode as a backward diode. If the doping concentrations in a tunnel diode are decreased, the filled part of the conduction band ($\mathscr{E}_{fn} - \mathscr{E}_c$) and the empty part of the valence band ($\mathscr{E}_v - \mathscr{E}_{fp}$) are reduced (Fig. 13.4). As a result, there are fewer states available for tunneling in the forward direction. At certain doping concentrations N_a and N_d, tunneling becomes impossible in the forward direction. Since tunneling is still possible in the reverse direction, such a diode has a large current in the reverse direction and a low current in the forward direction.

The backward diode competes very favorably with the point-contact diode and the Schottky-barrier diode as microwave detector and frequency converter. Since the backward diode is based on tunneling of majority carriers, it has many advantages. First, the device has a fast frequency response because there is no minority-carrier storage effect. Second, the device is relatively insensitive to radiation effect. It is true that irradiation may have a degrading effect on the lifetime of excess carriers, but lifetime plays no part in majority-carrier devices. Third, the device is insensitive to temperature changes. Over a wide temperature range, the majority-carrier concentration stays practically constant. This is not true with the minority-carrier concentration. The backward diode also has a very low noise (S. T. Eng, *IRE Trans. Microwave Theory and Tech.*, Vol. MTT-8, p. 419, 1961). Insofar as detection and frequency conversion are concerned, the important parameter is nonlinearity, which is commonly measured by the curvature coefficient $\gamma = I''/I'$, where $I'' = d^2I/dV^2$ and $I' = dI/dV$. For a forward-biased junction, $\gamma = 40$ at room temperature. Values close to or higher than 40 can be obtained in a backward diode with proper doping concentrations (J. Karlovsky, *Solid-State Electron.*, Vol. 10, p. 1109, 1967).

A tunnel junction, besides being a useful device, can be employed as a tool for semiconductor research. First we consider the tunneling process in an indirect-gap semiconductor such as Ge. In Ge, there are two important conduction-band valleys (Fig. 6.22a): the conduction-band minima at $k\langle111\rangle$ and the central valley at $k[000]$. Tunneling between the valence band and the $k[000]$ valley conserves the electron momentum, whereas tunneling between the valence band and the $k[111]$ minimum does not. The former process is often called the *direct* tunneling process, and the latter the *indirect* tunneling process. In the indirect process, phonon participation is needed in order to conserve the total momentum. The effect of phonons has indeed been detected in tunneling experiments (A. G. Chynoweth, R. A. Logan, and D. E. Thomas, *Phys. Rev.,* Vol. 125, p. 877, 1962; H. Fritzsche and J. J. Tiemann, *Phys. Rev.,* Vol. 130, p. 617, 1963; J. J. Tiemann and H. Fritzsche, *Phys. Rev.,* Vol. 132, p. 2506, 1963).

Figure 13.8 shows the *I–V* characteristic of a Ge tunnel diode observed by Fritzsche and Tiemann at 4.2 K in the low-bias region. The current is practically zero for $-8 > V > 8$ mV. The current increases significantly when the bias voltage is large enough to permit the emission of a low-energy phonon needed for momentum conservation. This phonon must possess a wave vector **K** to make up the difference of the electron wave vectors $\mathbf{k}_{[000]}$ and $\mathbf{k}_{[111]}$, or $\mathbf{K} = \mathbf{k}_{[111]} - \mathbf{k}_{[000]}$, where $\mathbf{k}_{[111]}$ refers to electrons in the conduction-band minimum and $\mathbf{k}_{[000]}$ refers to electrons near the top of the valence band. The current increases again when a higher-energy phonon of the same wave number can be emitted. The energies 8 and 28 meV obtained from Fig. 13.8a have been identified as associated with transverse and longitudinal acoustic phonons.

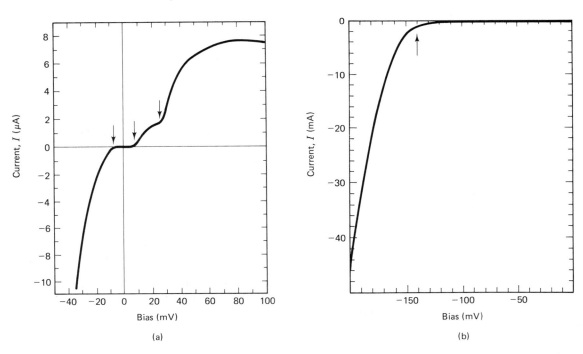

(a)   (b)

Figure 13.8 *I–V* characteristics of a Ge tunnel diode (a) showing phonon participation beyond certain applied bias voltages (indicated by arrows) and (b) showing the onset of direct tunneling. (H. Fritzsche and J. J. Tiemann, *Phys. Rev.,* Vol. 130, p. 617, 1963.)

Figure 13.8b shows the *I–V* characteristic of the diode in the -150 mV region. Note the vast difference in the scale of the two curves (Fig. 13.8). The contrast is caused by the difference in the magnitudes of the direct and indirect tunnel currents. Quantum mechanically, indirect tunneling is a second-order process because it is a result of second-order perturbation calculation. It is expected, therefore, that the direct tunnel current has a much larger amplitude than the indirect tunnel current. The sharp increase in current near -140 mV can be attributed to the onset of the direct tunneling process from the valence band to the conduction-band $k[000]$ valley. The diode used in the experiment had a doping concentration such that $\xi_n = 0.02$ eV. This value yields an energy separation of 0.16 eV between the conduction-band minimum and the $k[000]$ valley, which is in agreement with the separation obtained from optical measurements.

13.4 IMPATT AND TRAPATT DIODES

The development of semiconductor microwave generators has been one of the most significant advances in microwave engineering ($\lambda \sim 10$ cm to 1 mm, $\nu \sim 3$ to 300 GHz). The avalanche diodes and Gunn oscillators are perhaps the most important of solid-state microwave sources which have been developed. These devices are capable of generating microwave power at levels adequate for many applications in microwave systems and operating at wavelengths extending into the millimeter wavelength range. In this section and Sections 13.5 to 13.9, we devote our discussion on the underlying physics and performance characteristics of the avalanche and Gunn devices.

A device must exhibit a negative differential resistance or conductance to be capable of converting dc into ac power. The Gunn oscillator utilizes the negative conductance resulting from the electron-transfer effect. This mechanism depends on the band structure of a semiconductor and hence is a basic property of the semiconductor. Therefore, not all semiconductors possess the required property. The avalanche diode, on the other hand, utilizes the negative conductance, which is a property of a *p-n* junction in the avalanche breakdown region. Therefore, any semiconductor *p-n* junction can be used, and avalanche-diode oscillators have been constructed in Ge, Si, and GaAs. The rapid development of the avalanche-diode oscillator has been greatly aided by the vast amount of technology developed for Si in material preparation and device fabrication.

The idea of obtaining a negative resistance from a reverse-biased junction using the transit-time effect was first suggested by Shockley (W. Shockley, *Bell Syst. Tech. J.*, Vol. 33, pp. 799–826, 1954). He showed that by injecting an electron bunch into the space-charge region of a reverse-biased *p-n* junction, the transit time for the electrons to drift across the high-field region would introduce a phase difference between the terminal voltage and current, and could result in a negative resistance. He proposed the use of a forward-biased junction as the injecting electrode. This scheme, however, is impractical. Subsequently, Read suggested the use of impact ionization as an efficient means of injecting electrons and proposed a p^+-*n-i-n*$^+$ diode as the basic device structure (W. T. Read, *Bell Syst. Tech. J.*, Vol. 37, pp. 401–446, 1958). In this structure the p^+-*n* junction is the high-field region, where the avalanche process takes place, and the *i* region is the drift region, where the transit-time effect takes place. He showed that by combining properly the transit-time delay and the time delay required for the current to build up in the avalanche process, an efficient microwave oscillator could be constructed.

The Read diode, however, proved difficult to make. Although the Read diode was proposed in 1958, it was not until 1965 that the first experimental work on the Read oscillator was reported (C. A. Lee, R. L. Batdorf, W. Wiegmann, and G. Kaminsky, *Appl. Phys. Lett.*, Vol. 6, pp. 89–91, 1965). These diodes were low-frequency and low-power devices. At the same time, microwave oscillation was also observed from a simple silicon *p-n* junction (R. L. Johnston, B. C. DeLoach, and B. G. Cohen, *Bell Syst. Tech. J.*, Vol. 44, p. 369, 1965). Since then, significant advances have been made both in device technology, in the form of new structures and improved fabrication techniques, and in device physics, in terms of our understanding of the principles of the device operation. Continuous (CW) operation of Si avalanche diodes has been obtained up to 170 GHz with an output power of 28 mW (K. P. Weller, R. S. Ying, and D. H. Lee, *IEEE Trans. Microwave Theory Tech.*, Vol. MTT-24, p. 685, 1976) and with a stabilized power of 1 W at 51.9 GHz (Y. Hirachi, T. Nakagami, Y. Toyama, and Y. Fukukawa, *IEEE Trans. Microwave Theory Tech.*, Vol. MTT-24, p. 731, 1976).

There are two basic modes of operation for avalanche diodes. The mode of oscillation conceived by Read is known as the IMPATT mode. (The coined word "IMPATT" stands for *IMP*act-ionization *A*valanche *T*ransit *T*ime.) The IMPATT oscillator is best suited for low-power (a few watts in a diode array), high-frequency (up to over 100 GHz) CW applications. A different mode of operation is known as the TRAPATT mode, in which the plasma produced by the avalanche process is temporarily trapped in a low-field region and hence a large amount of power can be extracted at some subharmonic frequency (H. J. Prager, K. K. N. Chang, and J. Weisbrod, *Proc. IEEE*, Vol. 55, p. 586, 1967). (The coined word "TRAPATT" stands for *TRA*pped *P*lasma *A*valanche *T*riggered *T*ransit.) The TRAPATT oscillator is best suited for high-power (1 kW at 1 GHz and 10 W at 10 GHz), high-efficiency (over 50%) pulsed applications up to 10 GHz.

Since both IMPATT and TRAPATT diodes utilize impact ionization to generate carriers, we first present some background information on the electrical properties of a *p-n* junction under a high field near avalanche breakdown. For these applications, we need a steep change in the junction field in order to confine the impact ionization region. Therefore, an abrupt junction is used. The breakdown voltage V_B is given in Fig. 10.13 as a function of impurity concentration N. The maximum field E_{max} and the width of the space charge region W are related to V_B, respectively, by

$$E_{max} = \left(\frac{2eV_B N}{\epsilon} \right)^{1/2} \tag{13.4.1}$$

$$W = \left(\frac{2\epsilon V_B}{eN} \right)^{1/2} \tag{13.4.2}$$

where $N = N_a N_d/(N_a + N_d)$ is the joint impurity concentration. For one-sided step junctions, $N = N_a$ or N_d, whichever is smaller; for symmetrical step junctions, $N = N_a/2 = N_d/2$. The values of E_{max} and W in Ge, Si, GaAs and GaP are plotted as functions of N in Fig. 13.9. Therefore, with $N = 10^{16}$, E_{max} is around 4×10^5 V/cm and W is around 3 μm in Si and GaAs.

Figure 13.10 shows the electric field E, the drift velocity v_d, and the ionization coefficient α for (1) a Si one-sided junction with $N = 10^{16}$ cm^{-3} at $V_B = 60$ V, and (2) a Si symmetrical junction with $N = 5 \times 10^{15}$ cm^{-3} at $V_B = 96$ V. Referring to Fig. 10.1, we see that for $E > 2 \times 10^4$ V/cm, the drift velocity of electrons can be

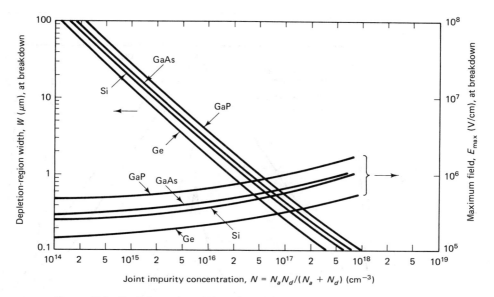

Figure 13.9 Depletion-region width W and maximum electric field E_{max} at the breakdown voltage for abrupt junctions, computed from Eqs. (8.2.11) and (8.2.12) by using V_B of Fig. 8.13.

approximated by a saturation velocity $v_s = 1 \times 10^7$ cm/s. Therefore, we can set $v_d = v_s$ for most part of the space-charge region. The curves for α are obtained by using the information given in Fig. 10.9. Because of the exp $(-B/E)$ dependence, α has an appreciable value only in a small portion of the space-charge region near the junction. The information on the behavior of v_d and α is useful in our analyses of IMPATT and TRAPATT diodes.

To explain the operational difference between an IMPATT and TRAPATT diode, we begin with their structural difference. An IMPATT diode is made of p^+-n-i-n^+ regions with the p^+-n junction serving as the avalanche region and the i region as the drift region. A small-signal analysis of the IMPATT diode is given in Section 13.5. A TRAPATT diode, on the other hand, is made of p^+-n-n^+ or p^+-p-n^+ regions, with part of the p^+-n or p-n^+ junction serving as the avalanche region. Here we give a qualitative discussion of the operation of a TRAPATT diode. Figure 13.11 shows the time sequence of the field distribution and the formation of a trapped plasma in a TRAPATT diode. At t_1, the field is below E_B, the field required for a significant avalanche multiplication. At t_2, the field exceeds E_B and carriers begin to multiply. If the point $E = E_B$ moves with a velocity v faster than that of generated carriers, the electron–hole pairs are trapped behind. The trapped plasma is indicated by the shaded area in Fig. 13.11. The velocity v with which the point $E = E_B$ moves can be found from the displacement current density J as follows:

$$v = \frac{\partial E}{\partial t} \bigg/ \frac{\partial E}{\partial x} = \frac{J}{\epsilon} \frac{\partial E}{\partial x} = \frac{J}{eN} \tag{13.4.3}$$

With a current density $J = 1 \times 10^4$ A/cm^2 and $N = 2 \times 10^{15}$ cm^{-3}, we find that $v = 3 \times 10^7$ cm/s, which is larger than the saturation velocity of carriers.

Sec. 13.4   IMPATT and TRAPATT Diodes

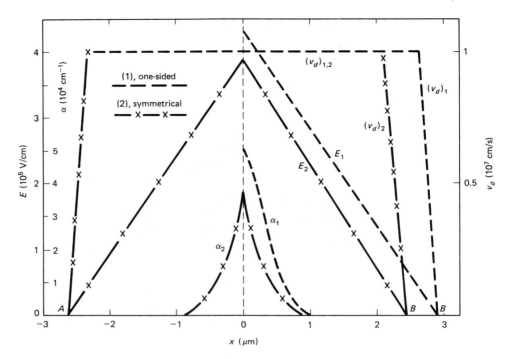

Figure 13.10 The electric field E, the drift velocity v_d, and the ionization coefficient in the space-charge region of abrupt Si junctions. The subscripts 1 and 2 refer to a one-side junction with $N = 10^{16}$ cm^{-3} and a symmetrical junction with $N = 5 \times 10^{15}$ cm^{-3}, respectively.

Once a plasma is formed, the field in the plasma region drops practically to zero with a very short dielectric relaxation time ϵ/σ because of its high conductivity σ. With E being very small, the plasma becomes really trapped. Upon further movement of the avalanche front toward the right, the plasma finally covers the entire p region (Fig. 13.11d) and the voltage across the diode drops almost to zero. Figure 13.11e shows schematically the voltage waveform from a TRAPATT diode with points a, b, c, and d corresponding to the field distributions shown in Fig. 13.11a, b, c, and d, respectively. The time interval from a to d represents the avalanche transient period during which the electron–hole plasma begins to form until it spreads over the entire p region in a n^+-p-p^+ (or entire n region in a p^+-n-n^+) TRAPATT diode. This transient phase lasts for a fraction of a nanosecond. During the recovery phase represented by the time interval d-f-g in Fig. 13.11e, electrons drift to the left and holes to the right, at first slowly under the influence of the residual electric field. This drift motion creates two charged zones, the left zone with a negative charge density and the right zone with a positive charge density. Between the two charged zones, there is the neutral plasma zone. While the field in the two zones gradually recovers, the field in the central neutral zone remains small. Therefore, this part of the recovery phase is slow and is represented by the time interval d-f in Fig. 13.11e. When the two charged zones meet, the charged carriers are swept away rapidly with a velocity equal to the saturation velocity as the field there already exceeds E_S. This rapid recovery phase is represented by the time

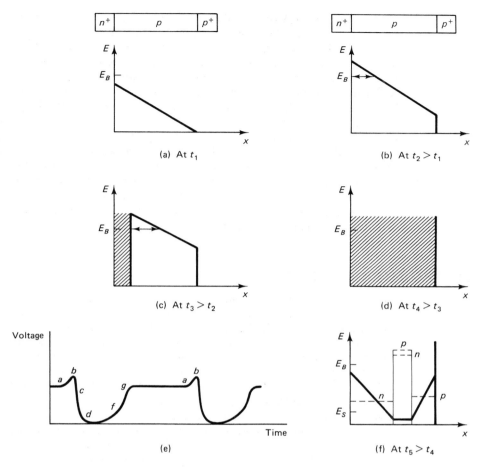

Figure 13.11 Time sequence of the electric field distribution in a TRAPATT diode during the formation phase (diagrams a to d) and the recovery phase (diagram f), and the terminal voltage across the diode (diagram e). The quantity E_B represents the minimum field required for significant impact ionization of electron–hole pairs, leading to the formation of an electron–hole plasma (shaded area) and thus a sudden drop of E to a very small value in the region. Diagram f also shows the nonuniform carrier concentrations (dashed lines) as a result of the movement of electrons and holes in opposite directions.

interval *f-g* in Fig. 13.11e (A. S. Clorfeine, R. J. Ikola, and L. S. Napoli, *RCA Rev.,* Vol. 30, p. 397, September 1969).

The voltage form shown in Fig. 13.11e is rich in harmonic content. For a discussion of the microwave circuit design, the reader is referred to articles by Swan and by Evans et al. (C. W. Swan, *Proc. IEEE,* Vol. 56, p. 1617, 1968; W. J. Evans, T. E. Siedel, and D. L. Scharfetter, *Proc. IEEE,* Vol. 58, p. 1284, 1970). The principal features are the use of a second harmonic choke and the placement of a capacitive load close to the TRAPATT diode to allow reflection of the high-frequency components of the voltage drop back to the TRAPATT diode. With a proper phase, for example tunable by a plunger, the reflected voltage adds to the fundamental component and the dc bias to raise the diode voltage above the breakdown voltage. This excess voltage then triggers a new traveling avalanche front and the cycle *abcdfg* repeats itself. Because the operation of a TRAPATT diode involves large swings of the voltage, numerical analyses are needed to simulate its behavior. Computer simulation and experimental results have been reported in a number of articles (R. L. Johnston, D. L. Scharfetter, and D. J. Bartelink, *Proc. IEEE,* Vol. 56, p. 1611, 1968; J. E. Carroll and R. H. Credé, *Int. J. Electron.,* Vol. 32, p. 273, 1972; N. A. Staymake and J. E. Carroll, *Electron. Lett.,* Vol. 6, p. 159, 1970; S. G. Liu, *Proc. IEEE,* p. 1216, 1971). The reader is also referred to the article by DeLoach and Scharfetter for a discussion of device physics and to the books by Carroll and by Gibbons for an extensive discussion on IMPATT and TRAPATT oscillators (B. C. DeLoach and D. L. Scharfetter, *IEEE Trans. Electron Devices,* Vol. ED-17, p. 9, 1970; J. E. Carroll, *Hot Electron Microwave Generators,* Edward Arnold (Publishers) Ltd., London, 1970; G. Gibbons, *Avalanche-Diode Microwave Oscillators,* Clarendon Press, Oxford, 1973).

13.5 TRANSIT-TIME EFFECTS

In Section 13.4 we presented a qualitative description of the operation of a TRAPATT diode. In this section we discuss the small-signal analysis of an IMPATT diode from which the device physics can be seen analytically. An electronic device capable of amplifying an electric signal must possess two essential elements: a source to supply free carriers and a physical mechanism to produce a negative resistance. In IMPATT diodes, free carriers are created through the impact-ionization process. As the free carriers drift across the space-charge region, a phase lag develops between the junction voltage and the diode current. By properly controlling the drift time, a negative resistance can be obtained within a certain frequency range.

Figure 13.12 shows the structure of a Read diode used by DeLoach and Johnston (B. C. DeLoach and R. L. Johnston, *IEEE Trans. Electron Devices,* Vol. ED-13, p. 186, 1966). The diode was made by first growing an epitaxial near-intrinsic layer on an n^+ silicon substrate and then converting part of the layer into p^+ and n regions through double diffusion. To visualize the electric-field distribution in the structure, we approximate the impurity profile of Fig. 13.12a by an idealized profile shown in Fig. 13.12b. According to Eqs. (8.2.5) and (8.2.6) (or Fig. 8.3), the electric field varies linearly with distance in the n region and remains practically constant in the near-intrinsic region. Because of the high impurity concentration, the electric field drops very sharply in the p^+ and n^+ regions. Therefore, the field distribution for the idealized impurity profile takes the shape shown in Fig. 13.13a. The actual field distribution calculated from the impurity profile of Fig. 13.12a is shown in Fig. 13.13b.

In analyzing the dynamic characteristics of a Read diode, we divide the intermediate region into two regions: the avalanche region and the drift region. As shown in Fig. 10.9, the ionization coefficients α_e and α_h increase rapidly with the field. Because of the strong field dependence, the highest value of α and hence most of the ionization process occur in a narrow region near the p^+n junction where the field is highest. The shaded region shown in Fig. 13.13c is the avalanche region. Since $w_a \ll W$, we neglect the transit-time (or drift) effect in this region. The remaining part of the intermediate region ($W - w_a$ in Fig. 13.13a and d) is the drift region. Since the drift

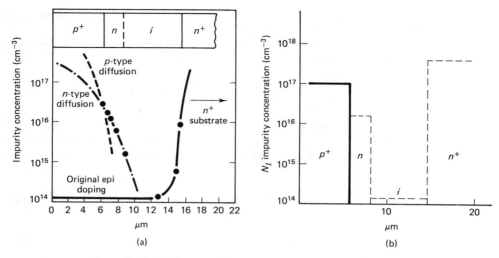

Figure 13.12 (a) Structure of a Read diode and its dopant profile and (b) the idealized dopant profile used in device analysis.

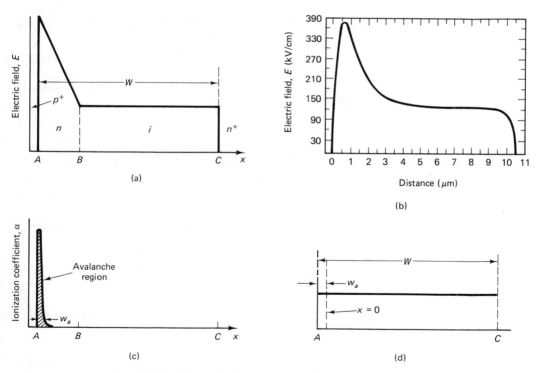

Figure 13.13 Electric field (a) used in the idealized model and (b) computed from the actual doping profile, (c) expected spatial variation of the ionization coefficient α from the field variation, and (d) division of a Read diode of length W into an avalanche region of length w_a and a drift region of length $W - w_a$.

velocity of carriers tends to saturate under high electric field (Fig. 10.1), we assume that carriers drift with a constant saturation velocity v_s.

We consider the avalanche region first. The rate of generation of excess carriers due to the impact-ionization process is

$$\left.\frac{\partial p}{\partial t}\right|_{\text{IMP}} = \left.\frac{\partial n}{\partial t}\right|_{\text{IMP}} = \frac{\partial n}{\partial x}\frac{dx}{dt} = \alpha(n + p)v_s \qquad (13.5.1)$$

Equation (13.5.1) is obtained by adding the contributions to $\partial n/\partial x$ from Eqs. (10.4.4) and (10.4.5). For simplicity, we assume that $\alpha_e = \alpha_h = \alpha$ and equal saturation velocity for electrons and holes. In the presence of carrier generation, the two continuity equations become

$$\frac{\partial p}{\partial t} = -\frac{1}{e}\frac{\partial J_h}{\partial x} + \alpha(n + p)v_s \qquad (13.5.2)$$

$$\frac{\partial n}{\partial t} = \frac{1}{e}\frac{\partial J_e}{\partial x} + \alpha(n + p)v_s \qquad (13.5.3)$$

The equations above are obtained by replacing the recombination term p_1/τ_0 in Eq. (7.6.4) by the generation term.

Our task is to find the relation between the field and the current density in the avalanche region. The current consists of two components: the displacement current and the conduction current. The displacement-current density is

$$J_d = i\omega\epsilon E_1(\omega) \qquad (13.5.4)$$

which is an ac current. In Eq. (13.5.4), $E_1(\omega)$ is the ac component of the field in the avalanche region. The conduction-current density is

$$J_c = J_h + J_e = e(p + n)v_s \qquad (13.5.5)$$

which contains both dc and ac components.

Adding Eqs. (13.5.2) and (13.5.3), we obtain a dynamic equation for J_c,

$$\frac{1}{v_s}\frac{\partial J_c}{\partial t} = \frac{\partial}{\partial x}(J_e - J_h) + 2\alpha J_c \qquad (13.5.6)$$

through the use of Eq. (13.5.5). Integrating Eq. (13.5.6) over the avalanche region, we find that

$$\frac{w_a}{v_s}\frac{dJ_c}{dt} = \left.(J_e - J_h)\right|_{-w_a}^{0} + 2J_c\int_{-w_a}^{0}\alpha\, dx \qquad (13.5.7)$$

Note that the dc component of the conduction-current density is, according to Eq. (10.4.10), given by

$$J_r = J_{c0} = J_0\left(1 - \int_{-w_a}^{0}\alpha\, dx\right)^{-1} \qquad (13.5.8)$$

where J_0 is the reverse-saturation-current density. In order that Eq. (13.5.7) may reduce to Eq. (13.5.8) for $dJ_c/dt = 0$, we must have

$$\left. \left(J_e - J_h \right) \right|_{-w_a}^{0} = 2(J_0 - J_c) \tag{13.5.9}$$

In other words, Eq. (13.5.7) must take the form

$$\frac{w_a}{v_s} \frac{dJ_c}{dt} = 2(J_0 - J_c) + 2J_c \int_{-w_a}^{0} \alpha \, dx \tag{13.5.10}$$

The reader is asked to show that Eq. (13.5.9) is indeed true by considering the boundary conditions at $x = -w_a$ and $x = 0$.

To simplify the notation, we let

$$\tau_a = \frac{w_a}{v_s} \quad \text{and} \quad \int_{-w_a}^{0} \alpha \, dx = \langle \alpha \rangle w_a \tag{13.5.11}$$

where $\langle \alpha \rangle$ is a spatial average of α. In terms of τ_a and $\langle \alpha \rangle$ Eq. (13.5.10) becomes

$$\frac{dJ_c}{dt} = \frac{2J_c}{\tau_a} (\langle \alpha \rangle w_a - 1) + \frac{2J_0}{\tau_a} \tag{13.5.12}$$

In the presence of both ac and dc fields, we can express

$$\langle \alpha \rangle = \langle \alpha_0 \rangle + \frac{d\langle \alpha \rangle}{dE} \Delta E = \langle \alpha_0 \rangle + \langle \alpha' \rangle E_1(\omega) \tag{13.5.13}$$

We use the indices 0 and 1 to denote the dc and ac components, respectively. The ac component of the conduction-current density is

$$J_{c1} = \frac{2J_r \langle \alpha' \rangle w_a}{i\omega\tau_a} E_1(\omega) \tag{13.5.14}$$

Since there is no dc component of the displacement current, the dc conduction-current density J_{c0} is the total dc current density J_r. Further, the term $J_{c1}(\langle \alpha_0 \rangle w_a - 1)$ can be neglected in Eq. (13.5.14) because the quantity $\langle \alpha_0 \rangle w_a$ is close to 1 under the avalanche condition.

Combining Eqs. (13.5.14) and (13.5.4), we obtain

$$J_1 = J_{c1} + J_d = \left[\frac{2J_r \langle \alpha' \rangle w_a}{i\omega\tau_a} + i\omega\epsilon \right] E_1(\omega) \tag{13.5.15}$$

Note that the total ac current flowing through the Read diode is $J_1 A$ and the ac voltage across the avalanche region is $E_1(\omega)w_a$, where A is the cross-sectional area. According to Eq. (13.5.15), the avalanche region can be represented by a parallel LC circuit as shown in Fig. 13.14a with

$$L = \frac{\tau_a}{2J_r \langle \alpha' \rangle A}, \quad C = \frac{\epsilon A}{w_a} \tag{13.5.16}$$

Furthermore, the ratio γ of the ac conduction-current density to the total ac current density is

$$\gamma = \frac{J_{c1}}{J_1} = \frac{1/i\omega L}{(1/i\omega L) + i\omega C} = \frac{1}{1 - (\omega/\omega_r)^2} \tag{13.5.17}$$

where $\omega_r^2 = 2J_r \langle \alpha' \rangle v_s / \epsilon$.

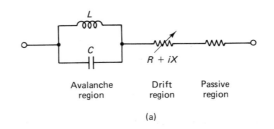

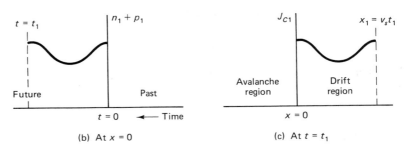

Figure 13.14 (a) Equivalent circuit representing an IMPATT diode, (b) time variation of total carrier concentrations $n + p$ by noting $n_1 + p_1 = J_{c1}/ev_s$, and (c) spatial variation of J_{c1}.

Now we analyze the situation in the drift region. Because of the ac electric field, the carrier concentrations $n + p$ also vary with time. This ac variation can be seen from Eqs. (13.5.14) and (13.5.15) and is illustrated in Fig. 13.14b. However, the carriers do not stay stationary but drift under the dc electric field. Suppose that $(n_1 + p_1)$ at $x = 0$ (the boundary of the avalanche-drift region) is given by

$$n_1 + p_1 = A \cos \omega t \tag{13.5.18}$$

At $t = t_1$ this carrier concentration moves to a new position $x = x_1 = v_s t_1$. In other words, an expression for $(n_1 + p_1)$, which takes drift into account, is

$$n_1 + p_1 = A \cos \left[\omega \left(t - \frac{x}{v_s} \right) \right] \tag{13.5.19}$$

Since the conduction current is related to the carrier concentration through Eq. (13.5.5), J_{c1} in the drift region also varies with distance. This spatial variation is illustrated in Fig. 13.14c. We write J_{c1} as

$$J_{c1} = B \exp \left(-\frac{i\omega x}{v_s} \right) \tag{13.5.20}$$

In Eq. (13.5.20), complex notation is used and the time dependence is omitted. However, Eq. (13.5.20) shows the same spatial dependence as Eq. (13.5.19). Besides J_{c1}, there is the ac displacement current-density J_d. The total current density in the drift region is

$$J_1 = i\omega\epsilon E_1(\omega) + B \exp \left(-\frac{i\omega x}{v_s} \right) \tag{13.5.21}$$

High-Frequency and High-Speed Devices   Chap. 13

Note that at $x = 0$ (the boundary of the avalanche-drift region) J_{c1} is related to J_1 through Eq. (13.5.17) or $B = \gamma J_1$. Using this information in Eq. (13.5.21) and solving for $E_1(\omega)$, we find that

$$E_1(\omega) = \frac{J_1}{i\omega\epsilon}\left[1 - \gamma \exp\left(-\frac{i\omega x}{v_s}\right)\right] \tag{13.5.22}$$

Before we proceed further, we should clarify several important points. We may consider the avalanche region as an equivalent cathode emitting free carriers. A space-charge wave is formed in the drift region as carriers drift through the region. This space-charge wave produces a spatially varying conduction current. However, the total current flowing through the diode must be a constant independent of position x. The electric field, therefore, must adjust itself so as to make J_1 a constant in Eq. (13.5.22). The situation in the avalanche region is different because the region is narrow. For small w_a such that $\omega w_a << v_s$, Eq. (13.5.22) reduces to Eq. (13.5.15). The constant electric field $E_1(\omega)$ in Eq. (13.5.15) is a result of an earlier assumption that the transit-time effect in the avalanche region is to be neglected.

Integrating Eq. (13.5.22) over the drift region, we obtain

$$V_d(\omega) = \frac{W - w_a}{i\omega\epsilon} J_1\left[1 - \gamma \frac{1 - \exp(-i\theta)}{i\theta}\right] \tag{13.5.23}$$

We can represent the drift region by an impedance (Fig. 13.14a):

$$Z_d = R + iX = \frac{V_d(\omega)}{J_1 A} = \frac{W - w_a}{i\omega\epsilon A}\left[1 - \frac{\gamma \sin \theta}{\theta} - \frac{\gamma(1 - \cos \theta)}{i\theta}\right] \tag{13.5.24}$$

In Eqs. (13.5.23) and (13.5.24), the quantity θ is the transit angle of the drift region defined as

$$\theta = \frac{\omega(W - w_a)}{v_s} \tag{13.5.25}$$

From Eq. (13.5.24), the equivalent resistance R is equal to

$$R = \frac{(W - w_a)^2}{\epsilon A v_s} \frac{\omega_r^2}{\omega_r^2 - \omega^2} \frac{1 - \cos \theta}{\theta^2} \tag{13.5.26}$$

The discussion presented above follows an analysis given by Gilden and Hines (M. Gilden and M. F. Hines, *IEEE Trans. Electron Devices*, Vol. ED-13, p. 169, 1966). The important result of the analysis is that a sinusoidally time-varying signal sees a negative resistance for $\omega > \omega_r$. This negative-resistance effect can be used for generation and amplification of microwave power. The value of $\alpha' = d\alpha/dE$ has been calculated by Misawa for Ge, Si, GaAs, and GaP (T. Misawa, *IEEE Trans. Electron Devices*, Vol. ED-14, p. 795, 1967). It lies between 4×10^{-2} to 1 (V)$^{-1}$ for a change of α from 10^3 to 10^5 cm^{-1}. In Si at the maximum field $E = 3.8 \times 10^5$ V/cm of Fig. 13.13b, α is around 4×10^4 cm^{-1}, and α' is about 1 (V)$^{-1}$. The value of v_s is around 8×10^4 m/s (Fig. 10.1). Thus for a current density $J_r = 1360$ A/cm^2, ω_r is found to be 1.4×10^{11} rad/s, which is in the microwave region.

The analysis of Gilden and Hines is based on the complete separation of the avalanche and drift regions. This approximation is good only for a Read diode. For diodes with other doping profiles, the electric field may not drop fast enough in the drift region and hence the impact ionization process must be considered there. The general case has been treated by Misawa and by Hoefflinger (T. Misawa, *IEEE Trans. Electron Devices*, Vol. ED-13, p. 137, 1966; B. Hoefflinger, *IEEE Trans. Electron Devices*, Vol. ED-14, p. 563, 1967). Note that the solution of Eqs. (13.5.2) and (13.5.3) in the drift region can be represented by Eq. (13.5.19) only for $\alpha = 0$. If ionization process is to be considered in the drift region, the spatial variation of n and p will be affected by the $\alpha(n + p)$ term in Eqs. (13.5.2) and (13.5.3).

The carrier concentrations p and n are related to electric field through Poisson's equation:

$$\frac{\partial E}{\partial x} = \frac{e}{\epsilon}(p + N_d^+ - n - N_a^-)$$ (13.5.27)

The ac component of Eq. (13.5.27) is

$$\frac{\partial E_1}{\partial x} = \frac{e}{\epsilon}(p_1 - n_1)$$ (13.5.28)

Similarly, we can expand the term $\alpha(n + p)v_s$ into dc and ac components and obtain from Eqs. (13.5.2) and (13.5.3) two differential equations for $\partial p_1/\partial x$ and $\partial n_1/\partial x$. Eliminating p_1 and n_1 from those two equations and Eq. (13.5.28), we find that

$$\frac{\partial^2 E_1}{\partial x^2} + k^2 E_1 = J_1\left(\frac{2\alpha_0}{\epsilon v_s} - \frac{i\omega}{\epsilon v_s^2}\right)$$ (13.5.29)

The reader is referred to Problem 13.14 for the details in the derivation of Eq. (13.5.29).

In Eq. (13.5.29), the quantity k is given by

$$k^2 = \frac{\omega^2}{v_s^2} - \frac{2\alpha' J_r}{\epsilon v_s} + \frac{i2\alpha_0\omega}{v_s}$$ (13.5.30)

and J_1 is the total ac current density

$$J_1 = ev_s(n_1 + p_1) + i\omega\epsilon E_1$$ (13.5.31)

The solution of Eq. (13.5.29) is

$$E_1 = C_1 \exp(ikx) + C_2 \exp(-ikx) + \frac{1}{k^2}\left(\frac{2\alpha_0}{\epsilon v_s} - \frac{i\omega}{\epsilon v_s^2}\right)J_1$$ (13.5.32)

which is proposed by Misawa as the general solution of an IMPATT diode. The two constants C_1 and C_2 can be found by applying boundary conditions to Eq. (13.5.32). Equations (13.5.30) and (13.5.32) reduce to the result expressed in Eq. (13.5.22) for $\alpha_0 = 0$.

The application of Eq. (13.5.32) to a practical case is complicated by the fact that α' and α_0 are functions of E_0. Since E_0 varies with distance, a complete knowledge of $E_0(x)$ is needed in the calculation. Further, Eq. (13.5.32) is a valid solution of Eq. (13.5.29) only if k is independent of x. One standard approach to get around this difficulty is to divide the space-charge region into several subregions, each with a constant E_0 and hence a constant k. The solutions are then matched at the boundaries of the subregions. Such a calculation is often done by a computer. The reader is referred to several articles by Misawa, by Hoefflinger, and by Gummel and Scharfetter for the computational result (H. K. Gummel and D. L. Scharfetter, *Bell Syst. Tech. J.*, Vol. 45, p. 1797, 1966). Again their results show that a negative conductance is obtained in the microwave region if the IMPATT diode is operated under avalanche conditions.

13.6 GROWTH AND PROPAGATION OF CARRIER WAVES IN TWO-VALLEY SEMICONDUCTORS

In treating ordinary semiconductor devices, it is generally stated that charge neutrality is maintained everywhere except in the space-charge region near an interface, such as in a *p-n* junction, near a semiconductor surface, or in a Schottky-barrier junction. The condition of zero charge density, is based on the continuity equation [Eq. (7.4.3)] and Poisson's equation [Eq. (7.4.1)]. The two equations result in a time-dependent equation or solution for the charge density ρ

$$\frac{\partial \rho}{\partial t} = -\frac{\sigma}{\epsilon} \rho \quad \text{or} \quad \rho = \rho_0 \exp\left(-\frac{t}{\tau_d}\right) \tag{13.6.1}$$

where $\tau_d = \epsilon/\sigma$ is called the *dielectric-relaxation time*. Equation (13.6.1) says that any charge imbalance in a bulk semiconductor will disappear with a time constant equal to τ_d. Thus, in a material with $\epsilon = 11\epsilon_0$ and $\sigma = 1$ $(\Omega\text{-cm})^{-1}$, we can set $\rho = 0$ for any operation with $\omega < \tau_d^{-1} \sim 10^{12}$ rad/s. What happens is that a charge imbalance sets up a large local field [Eq. (7.4.1)] which draws a large current to make the charges disperse [Eq. (7.4.3)].

The situation is quite different, however, for bulk devices operated in the negative differential conductance region. As discussed in Section 10.9, the drift velocity v_d in that region is a function of E. Using $J = env_d$ in Eq. (7.4.3), we have

$$\frac{\partial \rho}{\partial t} = -en \frac{dv_d}{dE} \frac{dE}{dx} \tag{13.6.2}$$

Eliminating dE/dx from Eqs. (7.4.1) and (13.6.2), we obtain

$$\frac{\partial \rho}{\partial t} = -\left(\frac{en}{\epsilon} \frac{dv_d}{dE}\right)\rho = -\frac{en\mu_{ac}}{\epsilon} \rho = \omega_{ac}\rho \tag{13.6.3}$$

where the quantity $\mu_d = dv_d/dE$ is the differential (or ac) mobility and the quantity $1/\omega_{ac} = \epsilon/en(-\mu_{ac})$ is the differential dielectric-relaxation time. When μ_{ac} is negative, ω_{ac} becomes positive. Therefore, a charge-carrier wave will grow in GaAs when biased for operation in the negative differential conductance region. However, we should not forget that the carrier wave also drifts with a velocity v_d along the sample. For significant growth in a sample of length l, the transit time $t = l/v_d$ must be longer than the differential dielectric-relaxation time ω_{ac}^{-1}, or

$$N_d l > \frac{\epsilon v_d}{e|\mu_{ac}|} \tag{13.6.4}$$

where $N_d = n$ is the donor concentration.

The treatment above is obviously too simplistic. In the following discussion, we first generalize the formulation of the problem by including the displacement and diffusion currents. Then we look for a small-signal solution for the propagation of a carrier wave and thus establish the stability condition for the wave. In the stable region, the negative differential conductance can be used to make traveling-wave amplifiers. We start with the following equation for the current density:

$$ J = -env + eD \frac{dn}{dx} + \epsilon \frac{dE}{dt} \qquad (13.6.5) $$

The quantities J, n, E, and v consist of a dc component denoted by subscript 0 and an ac component denoted by subscript 1. Separating out the ac components in Eq. (13.6.5) and using Eq. (7.4.1) or $\nabla \cdot \mathbf{E} = \rho/\epsilon$ to eliminate n_1, we have

$$ J_1 = en_0\mu_{ac}E_1 - \epsilon\mu_{dc}E_0 \frac{dE_1}{dx} - \epsilon D \frac{d^2E_1}{dx^2} + \epsilon \frac{dE_1}{dt} \qquad (13.6.6) $$

where the diffusion constant D is assumed to be independent of E. For small-signal analysis, we have neglected the product n_1v_1 and used $v_1 = -\mu_{ac}E_1$ in Eq. (13.6.6).

To find the solution of Eq. (13.6.6), we first examine the natural response (complementary solution) of the system by setting $J_1 = 0$ and propose a solution of the traveling-wave type:

$$ E_1 = A \exp(ikx - i\omega t) \qquad (13.6.7) $$

Using Eq. (13.6.7) in Eq. (13.6.6) and dividing it by $en_0\mu_{dc}$, we find that

$$ -iv_d\tau_d k - i\omega\tau_d + \mu_r + L_D^2 k^2 = 0 \qquad (13.6.8) $$

where $v_d = v_0 = \mu_{dc}E_0$ is the dc drift velocity, $\mu_r = \mu_{ac}/\mu_{dc}$ is the ac mobility relative to the dc mobility, $\tau_d = \epsilon/en_0\mu_{dc}$ is the dc dielectric-relaxation time, and $L_D = (D\epsilon/en_0\mu_{dc})^{1/2} = (\epsilon kT/n_0e^2)^{1/2}$ is the Debye length. If we neglect the effect of diffusion, we can rewrite Eq. (13.6.8) as

$$ k = \frac{\omega}{\mu_{dc}E_0} - i\frac{\mu_r}{\mu_{dc}E_0\tau_d} = \frac{\omega\tau_d - i\mu_r}{\mu_{dc}E_0\tau_d} \qquad (13.6.9) $$

For a dc field applied in the $-x$ direction, that is, a negative E_0, the electrons drift in the $+x$ direction with a drift velocity $|v_d| = |\mu_{dc}E_0|$, which is the phase velocity of the carrier wave. Furthermore, since both μ_r and E_0 are negative in Eq. (13.6.9), the wave has a gain coefficient

$$ g = \frac{\mu_r}{\mu_{dc}E_0\tau_d} = \frac{\omega_{ac}}{|v_d|} \qquad (13.6.10) $$

where $\omega_{ac} = en_0|\mu_{ac}|/\epsilon$ is the reciprocal of the differential dielectric relaxation time. Since n_1 is related to E_1 through $\nabla \cdot \mathbf{E} = \rho/\epsilon$, both E_1 and n_1 have the same spatial and temporal dependence. Thus we can write n_1 as

$$ n_1 = B \exp\left(i\frac{\omega x}{|v_d|} + \frac{\omega_{ac}x}{|v_d|} - i\omega t\right) \qquad (13.6.11) $$

A comparison of Eqs. (13.6.11) and (13.6.3) shows that ρ in Eq. (13.6.3) actually represents the amplitude $B \exp(\omega_{ac}x/|v_d|)$ of the carrier wave as $\partial\rho/\partial t = (\partial\rho/\partial x)(\partial x/\partial t)$ and $(\partial x/\partial t) = v_d$.

The general solution of Eq. (13.6.6) can be written as

$$E_1 = [C + A \exp{(ikx)}] \exp{(-i\omega t)} \qquad (13.6.12)$$

where the particular solution is found to be

$$C = \frac{J_1}{en_0\mu_{ac} - i\omega\epsilon} \qquad (13.6.13)$$

by substituting Eq. (13.6.12) into Eq. (13.6.6). Consider a situation in which a Gunn-effect device is fed from an ideal voltage source and current oscillation appears above a certain threshold dc bias. In this situation, the ac voltage V_1 is zero, yet the ac current I_1 is nonzero. In the language of circuit theory, we are dealing with short-circuit instability or looking for zeros in the impedance function $Z(\omega) = V_1/I_1$. Applying the boundary condition $E_1 = 0$ at the cathode $x = 0$ to Eq. (13.6.12) and integrating E_1 from $x = 0$ to the anode $x = l$, we find the condition for $V_1 = 0$ to be

$$ikl + 1 - \exp{(ikl)} = 0 \qquad (13.6.14)$$

Obviously, k is complex. Thus we let $ik = ik_1 + g$. The first solution is located at $ikl = 2.09 \pm i7.46$. From Eq. (13.6.9), the zero of $Z(\omega)$ in the ω plane occurs at

$$\omega = |\mu_{dc}E_0|k - \frac{i\mu_r}{\tau_d} \qquad (13.6.15)$$

The system is stable if the zero lies in the lower half of the ω plane, requiring that $2.09|\mu_{dc}E_0|/l > |\mu_r|/\tau_d$, or

$$N_d l < 2.09 \frac{\epsilon|v_d|}{e|\mu_{ac}|} = 1.4 \times 10^{11} \text{ cm}^{-2} \qquad (13.6.16)$$

In Eq. (13.6.16), the values $|\mu_{ac}| = 2 \times 10^3$ cm^2/V-s, $v_d = 2 \times 10^7$ cm/s, and $\epsilon = 13\epsilon_0$ are used. Equation (13.6.16) is the stability condition derived by McCumber and Chynoweth (D. E. McCumber and A. G. Chynoweth, *IEEE Trans. Electron Devices*, Vol. ED-13, p. 4, 1966).

   Now we consider the effect of diffusion on stability. A carrier wave as represented by Eq. (13.6.11) changes its sign for every half wavelength $\Lambda/2 = \pi|v_d|/\omega$. We may think of the wave as consisting of a train of dipoles alternating in polarity. The wave grows in amplitude because the spatial variation in E caused by local variation in n through $\nabla \cdot \mathbf{E} = \rho/\epsilon$ affects the drift velocity v_d (through dv_d/dE) in such a way as to populate carriers further in regions where $n_1 = +$ and to deplete carriers further in regions where $n_1 = -$. Since diffusion tends to smooth out any local variation in n, it has the effect of reducing the negative differential conductance. The Debye length L_D in Eq. (13.6.8) is a measure of the distance to which a local disturbance is limited by diffusion. A dipole domain will not grow at all when $L_D^2 k^2 > |\mu_{ac}|/\mu_{dc}$ in Eq. (13.6.8). Letting $k = 2\pi/\Lambda$ and $\Lambda = l$, the sample length, we obtain the following condition inhibiting domain growth

$$N_d l^2 < \frac{4\pi^2 D\epsilon}{e|\mu_{ac}|} = 2.4 \times 10^7 \text{ cm}^{-1} \qquad (13.6.17)$$

where the values $D = 200$ cm^2/s, $|\mu_{ac}| = 2 \times 10^3$ cm^2/V-s, and $\epsilon = 13\epsilon_0$ are used. Equation (13.6.17) is the condition for inhibition of domains derived by Ridley (B. K.

Ridley, *IEEE Trans. Electron Devices,* Vol. ED-13, p. 41, 1966). For a given donor concentration N_d, Eq. (13.6.17) sets a condition on the sample length l such that for $l < l_1$, no growth of carrier waves or domains will be possible. Equation (13.6.16), on the other hand, sets a limit on l such that for $l < l_2$, a carrier wave or domain will grow but will not go into spontaneous oscillation. Since in practical cases $l_2 > l_1$, we have three possible situations: (1) no growth region with $l < l_1$, (2) amplifying region with $l_1 < l < l_2$, and (3) oscillation or unstable region with $l > l_2$.

In the unstable region, current oscillation is accompanied by domain formation. Domain dynamics are discussed in Section 13.7. Additional comments on the device characteristics of and device operation in the amplifying region are in order. In the analysis outlined above, we have assumed a uniform dc field E_0. The presence of a carrier wave, however, distorts the field distribution. Therefore, a precise calculation of device parameters, such as impedance, requires the incorporation of the effect of field redistribution. The reader is referred to the book by Carroll [J. E. Carroll, *Hot Electron Microwave Generators,* Edward Arnold (Publishers) Ltd., London, 1970, pp. 83–89] and the paper by McWhorter and Foyt (A. L. McWhorter and A. G. Foyt, *Appl. Phys. Lett.,* Vol. 9, p. 300, 1966) for such a treatment. Not only a good agreement between computed and observed impedance curves is indeed obtained, which gives further support to the transferred-electron model and the calculation of Butcher and Fawcett, but also both curves show the existence of a negative conductance region useful for amplification. Work on two-port unidirectional growing-wave amplifiers operated in a stabilized negative-conductance medium (GaAs) has been reported by Robson, Kino, and Fay (P. N. Robson, G. S. Kino, and B. Fay, *IEEE Trans. Electron Devices,* Vol. ED-14, p. 612, 1967). Electromagnetic waves were coupled to the device near the cathode and amplified signals were collected by a pair of electrodes at the anode. It is also possible to construct a reflection amplifier using the negative conductance. If a transmission line with characteristic admittance Y_0 is terminated by a load with admittance Y, the reflection coefficient is given by $(Y_0 - Y)/(Y_0 + Y)$. When Y is negative, the reflected power is larger than the incident power. Amplification of carrier waves has been observed by Thim and Barber in the reflection-amplifier configuration (H. W. Thim and M. R. Barber, *IEEE Trans. Electron Devices,* Vol. ED-13, p. 110, 1966). However, both two-port and reflection amplifiers have poor noise figures, typically in the range 20 to 25 dB.

13.7 DOMAIN FORMATION

In Section 13.6 we have shown that a Gunn-effect device operated in the negative differential conductance region can become unstable if the product $N_d l$ exceeds a certain value [Eq. (13.6.16)]. When we speak of instability here, we mean the spontaneous appearance of a localized and propagating space-charge region, generally referred to as a *domain.* There are two types of domains: (1) the accumulation domain with an accumulation of electrons ($\Delta n = +$) and (2) the dipole domain made of an accumulation layer and a depletion layer ($\Delta n = -$) being next to each other, where $\Delta n = n - n_0 = n - N_d$ is the local electron concentration relative to the thermal-equilibrium value n_0 which is equal to N_d. Figure 13.15 shows (a) the structure of a Gunn-effect device, and (b) doping profile (N_d) across the cathode junction. Under thermal equilibrium, when no bias voltage is applied, the diffusion current exactly balances the drift current in the junction, giving rise to a space-charge region. Figure 13.15 also shows

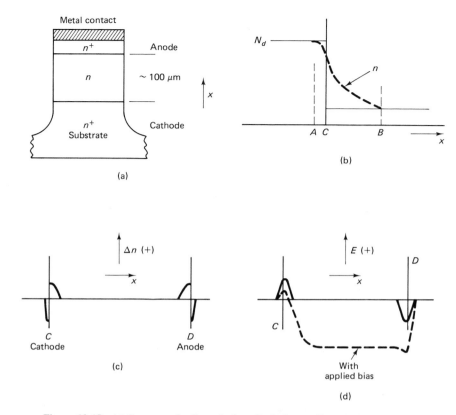

Figure 13.15 (a) Structure of a Gunn device, (b) doping profile and electron concentration (both in log scale), (c) electron accumulation regions ($\Delta n = +$), and (d) electric field distribution in the sample without bias (solid curves) and with bias (dashed curve).

(b) the electron-concentration profile, n, and (c) the quantity Δn near the electrodes. There is an accumulation of electrons (i.e., $\Delta n = +$) on the n side of a n^+-n junction. The situation at the anode junction is similar to that at the cathode except with the field direction reversed (solid lines in Fig. 13.15d).

We may think of an n^+-n-n^+ device as being analogous to an n-p-n device with the cathode serving as an emitter and the anode serving as a collector. When a bias voltage is applied to the Gunn-effect device with the negative polarity to the cathode, electrons are injected from the n^+ side to the n side (the drift region) at the cathode, thus forming an accumulation layer. However, the behavior of injected electrons in a Gunn-effect device is quite different from that of injected minority carriers in a transistor. In the base region of a transistor, the charge-neutrality condition $\rho = 0$ is still maintained even under injection because it is easy for injected minority carriers to draw majority carriers to neutralize them. In the drift region of a Gunn-effect device, on the other hand, the injected electrons will not be neutralized. The accumulation of charges sets up a space-charge field, which in turn modulates the drift velocity of electrons. As discussed in Section 13.6, the accumulated charges will grow if the field is biased in the negative differential mobility region.

Sec. 13.7   Domain Formation

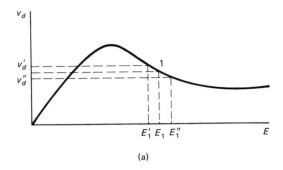

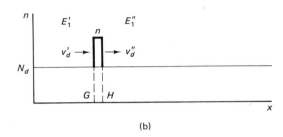

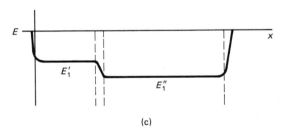

Figure 13.16 Diagrams showing (a) v_d versus E curve, (b) existence of an accumulation layer where $n > N_d$, and (c) spatial variation of E in the presence of an accumulation layer.

The situation in a Gunn-effect device is illustrated in Fig. 13.16 (B. K. Ridley, *Proc. Phys. Soc. London*, Vol. 82, p. 954, 1963; H. Kroemer, *IEEE Spectrum*, Vol. 5, p. 47, 1968). Imagine that an accumulation layer exists between *GH*. From Poisson's equation, the field change caused by Δn is given by

$$E_H - E_G = -\int_G^H \frac{e\,\Delta n}{\epsilon}\,dx \tag{13.7.1}$$

which makes $E_H < E_G$. Since E is negative, $|E_H| = E_1'' > E_1' = |E_G|$. From the drift velocity versus field diagram, we see that $v_d' > v_d''$. Therefore, there are more electrons moving into the accumulation layer at *G* than moving out of it at *H*. As a result, the accumulated charge Δn grows.

The accumulation domain is basically unstable. We see from Fig. 13.16 that the field E_1'' to the anode side is in the negative differential mobility region. Therefore, a small depletion caused by any nonuniformity of sample properties, such as a nonuniform doping concentration, will grow into a larger depletion. In Fig. 13.17 we consider two extreme situations for a dipole domain. At time $t = t_1$, the domain just begins to grow.

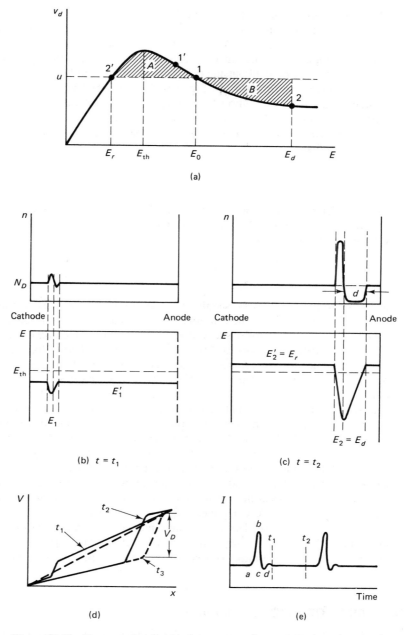

Figure 13.17 Diagrams showing (a) the v_d versus E curve, (b) the existence of a dipole domain and the resultant field distribution near the cathode, (c) the grown dipole and the resultant E as the dipole moves close to the anode, (d) the potential at various time, and (e) the current pulse caused by the dipole domain.

Due to the space charge, the field in the domain changes from a value E_1' outside the domain to a maximum value E_1 at the boundary of the accumulation and depletion layers (Fig. 13.17b). Since the field in the whole domain is in the negative differential mobility region, the space charge continues to grow. As the space charge becomes sufficiently large, the field outside the domain starts to fall below the threshold field E_{th}. A stable situation at time $t = t_2$ (Fig. 13.17c) is reached when the electrons inside and outside the domain move with the same average velocity. Let v_+ and v_- be the average drift velocity of electrons in the accumulation and depletion layer, respectively, and $u = v_2'$ be the drift velocity of electrons outside the domain. Since v_+ and v_- depend on E, we define

$$ v_+ = \int_{E_r}^{E_d} \frac{v(E)\ dE}{E_d - E_r} \qquad (13.7.2) $$

and a similar expression for v_-. In Eq. (13.7.2), $E_r = E_2'$ is the field outside the domain (the resistive region) and $E_d = E_2$ is the maximum field in the domain. The three velocities are equal when

$$ \int_{E_r}^{E_d} [v(E) - u]\ dE = 0 \qquad (13.7.3) $$

Referring to Fig. 13.17a, we see that the quantity $v - u$ changes sign at E_0. Therefore, Eq. (13.7.3) is satisfied when the two shaded areas A and B are equal. The condition stated in Eq. (13.7.3) for a stable domain is known as the *equal-area rule*, which is discussed further in Section 13.8 (P. N. Butcher, *Phys. Lett.*, Vol. 19, p. 546, 1965).

After the domain reaches a stable condition, the field pattern simply shifts in position as the domain propagates. When the domain reaches the anode, a large current pulse results. The voltage–distance and current–time variations for a dipole domain are illustrated in Fig. 13.17d and e. Computer simulation by McCumber and Chynoweth has indeed confirmed the predicted progression of the field distribution and current shown in Fig. 13.17 (D. E. McCumber and A. G. Chynoweth, *IEEE Trans. Electron Devices*, Vol. ED-13, p. 4, 1966). The points a and b in Fig. 13.17e correspond, respectively, the initiation of a dipole domain at the cathode and its arrival at the anode. Between the time interval after the domain grows to maturity and before it arrives at the anode, the electrical field at the anode is kept at E_r, and thus the current density is made entirely by the dc component, $J_{dc} = enu$. This corresponds to the flat portion d–a of the current curve. During the time interval when the domains grow to maturity, the field at the anode actually drops from E_1' to E_2'. The dip in current (portion c–d) below the dc value is a direct consequence of decreasing displacement current.

13.8 DYNAMICS OF DIPOLE DOMAINS

In Section 13.7 we have argued that for a mature dipole domain, the carriers outside and inside the domain must move with the same average velocity. This qualitative argument leads to the equal-area rule stated in Eq. (13.7.3). Here we present a formal derivation of the rule so that we may see more clearly the dynamics of a dipole domain during the maturing process. We start with Eq. (13.6.5), the current-density expression, and Eq. (7.4.1), Poisson's equation. Since a domain traveling with a velocity u' has a steady-state solution of the form $E = E(x - u't)$ we introduce a moving coordinate system (x', t') defined by

$$x' = x - \int_0^t u'(t)\, dt, \qquad t' = t \tag{13.8.1}$$

so that

$$\frac{\partial}{\partial t} = \frac{\partial}{\partial t'} - u' \frac{\partial}{\partial x'}, \qquad \frac{\partial}{\partial x} = \frac{\partial}{\partial x'} \tag{13.8.2}$$

In the moving-coordinate system, Eqs. (13.6.5) and (7.4.1), respectively, become

$$J = -env + eD \frac{\partial n}{\partial x'} - u'\epsilon \frac{\partial E}{\partial x'} + \epsilon \frac{\partial E}{\partial t'} \tag{13.8.3}$$

$$\epsilon \frac{\partial E}{\partial x'} = -e(n - N_d) \tag{13.8.4}$$

Next we change the variable to E instead of x inside the domain. From Eq. (13.8.4) we have

$$\frac{\partial}{\partial x'} = \frac{\partial}{\partial E} \frac{\partial E}{\partial x'} = -\frac{e(n - N_d)}{\epsilon} \frac{\partial}{\partial E} \tag{13.8.5}$$

In the moving-coordinate system, the steady-state value of E does not change with time. Setting $\partial E / \partial t' = 0$ and using Eq. (13.8.5) in Eq. (13.8.3), we find that

$$J = -env + e(n - N_d)u' - \frac{De^2}{\epsilon}(n - N_d)\frac{dn}{dE} \tag{13.8.6}$$

Dividing Eq. (13.8.6) by en and integrating it from E_r to E_d, we obtain

$$\int_{E_r}^{E_d} \frac{1}{en}(J + eN_d u')\, dE = -\int_{E_r}^{E_d}(v - u')\, dE - \frac{eD}{\epsilon}\int_{n_1}^{n_2}\left(1 - \frac{N_d}{n}\right) dn \tag{13.8.7}$$

where E_r is the field outside the domain and E_d is the maximum field inside the domain.

We note from Fig. 13.17 that the value of $n_2 = n(E_d)$ at the boundary of the accumulation and depletion layers is the same as the value of $n_1 = n(E_r)$ outside the domain because E_d must occur at the point where $\partial E / \partial x' = 0$ or $n_1 = n_2 = N_d$. Therefore, the last integral in Eq. (13.8.7) is identically zero, and we are left with two integrals. There are two possible integration paths: one over the accumulation layer and the other over the depletion layer. The first integral on the right-hand side of Eq. (13.8.7) is a function of E only and hence independent of which path we take for the integration. This is not true, however, for the other integral. Since J in Eq. (13.6.5) includes the displacement current, it must be continuous (that means independent of x). Therefore, the sum $J + eN_d u'$ is a constant in the integral on the left-hand side of Eq. (13.8.7). The term en in the denominator, on the other hand, is always smaller than eN_d in the depletion layer and greater than eN_d in the accumulation layer. The only way that this integral can be independent of path of integration is for the numerator to be identically zero, or $J = -eN_d u'$ inside the domain. Since the current density outside the domain is given by $J = -eN_d u$ and the two J's must be equal, we conclude that a mature domain moves with velocity u' equal to the velocity of electrons outside the domain or

$$u' = u = v(E_r) \tag{13.8.8}$$

where u is defined in Fig. 13.17a. Using Eq. (13.8.8) in Eq. (13.8.7), we obtain

$$\int_{E_r}^{E_d} [v(E) - u]dE = 0 \tag{13.8.9}$$

as the condition for a mature dipole domain, which is identical to Eq. (13.7.3).

Equation (13.8.9) is a statement of the steady-state condition of a mature dipole domain. To study the dynamic growth of a domain, we again refer to the current density inside the domain expressed in Eq. (13.6.5) and equate it to the current density

$$J = -eN_d u + \epsilon \frac{\partial E_r}{\partial t} \tag{13.8.10}$$

outside the domain. Rearranging terms and integrating the resultant equation over the dipole region, we find that

$$-\epsilon \frac{dV_D}{dt} = \int_{x_1}^{x_2} eN_d(v - u)dx + \int_{x_1}^{x_2} e(n - N_d)v \, dx \tag{13.8.11}$$

where x_1 and x_2 represent the extent of the dipole domain and V_D defined in Fig. 13.17d as

$$V_D = -\int_{x_1}^{x_2} (E - E_r) \, dx \tag{13.8.12}$$

is the excess voltage developed across the domain. In Eq. (13.8.11), the integral involving the diffusion current vanishes because it has identical values at x_1 and x_2, and hence it is dropped. We further note that we can change the variable to dE from dx by using $\nabla \cdot \mathbf{E} = -e(n - N_d)/\epsilon$. Thus the last integral in Eq. (13.8.11) is transformed into $\int_{E_r}^{E_r} v \, dE$, which also vanishes. The first integral on the right-hand side of Eq. (13.8.11), however, is nonzero. Figure 13.17c shows the physical situation in a fully grown dipole domain. The smallest value of n possible in the depletion layer is zero, but the largest possible value of n in the accumulation layer is almost unlimited. Thus the situation is very much like that in a one-sided step junction. After we convert dx to dE, we see that the contribution to the first integral mostly comes from the depletion layer because dx is in the ratio $(dx)_A/(dx)_D = |\rho_D/\rho_A| << 1$ for the accumulation (denoted by A) and depletion (denoted by D) layers. By neglecting the contribution from the accumulation layer and letting $\rho_D = eN_d$, we can rewrite Eq. (13.8.11) as

$$\frac{dV_D}{dt} = -\int_{E_d}^{E_r} [v(E) - u] \, dE = -\int_{E_r}^{E_d} (|v| - |u|) \, dE \tag{13.8.13}$$

We should point out that in the sign convention, an electron current in the $+x$ direction means a negative electron velocity. To avoid this confusion, we have transformed Eq. (13.8.13) into an integral over the magnitude of electron velocity. In Fig. 13.18 we present (a) the dynamic characteristic and (b) the operation characteristic for a dipole domain. For a given E_d we can always find a value for u such that the equal-area rule is satisfied. The dashed line in Fig. 13.18 is the locus of (u, E_d) satisfying this rule and is known as the *dynamic characteristic*. For $u^- < u$, the domain decays; for $u^+ > u$, the domain grows. To compute the domain voltage, we approximate the field

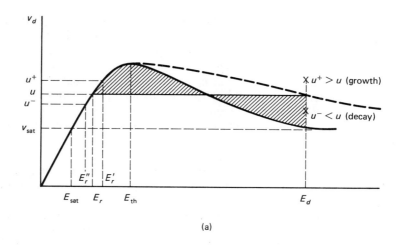

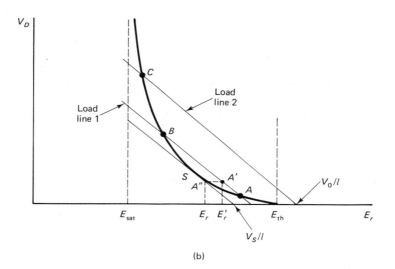

Figure 13.18 (a) Dynamic characteristic (domain velocity u versus E curve—dashed) and (b) operational characteristic (domain voltage V_D versus E_r curve—solid) for a dipole domain where E_r is the field outside the domain.

distribution in Fig. 13.17c by a rectangular triangle as in a one-sided step junction, and thus obtain

$$V_D = \frac{eN_d d^2}{2\epsilon} = \frac{\epsilon(E_d - E_r)^2}{2eN_d} \tag{13.8.14}$$

For a given E_r, the value of E_d for a mature domain can be found from Fig. 13.18a. Therefore, the value of V_D is a function of E_r which defines an operation characteristic ABC, as shown in Fig. 13.18b.

As Fig. 13.18a shows, the maximum value of E_r is E_{th}. At $E_r = E_{th}$, equal-area-rule requires $E_d = E_{th}$ for a mature domain, meaning $V_D = 0$ from Eq. (13.8.14). As

E_r decreases from E_{th}, the difference $E_d - E_r$ increases (as is evident from Fig. 13.18a) and so does V_D. However, there exists a lower limit for E_r. We note from Fig. 13.18a that v_d in GaAs approaches a constant value, the saturation velocity, at high fields. We denote this velocity by v_{sat} and the corresponding value of E_r for $v_d = v_{sat}$ by E_{sat}. As u approaches v_{sat}, the value of E_d becomes very large and goes asymptotically to infinity. Therefore, the operational value of E_r for a mature domain lies between E_{sat} and E_{th}. To find the actual operating point, we must fix the load. For a constant terminal voltage V_0, the domain voltage V_D and the field E_r must satisfy the relation

$$V_D + E_r l = V_0 \qquad (13.8.15)$$

where l is the length of the n region in Fig. 13.15a. Since the dipole width d is much smaller than l, we use l instead of $l - d$ in Eq. (13.8.15). This relation is represented by a load line in Fig. 13.18b. The intersection of the load line with the operation characteristic defines the operating point. In some cases there are two intersections, as represented by case 1. Point A, however, is an unstable point. Suppose that the domain voltage is increased by noise fluctuations from $V_D(A)$ to $V_D(A')$. Since the load-line condition must be satisfied at all times, the field E_r outside the domain takes a new value E_r'. Insofar as the stability of the domain is concerned, the new value of $V_D(A'')$ requires a value $E_r(A'')$ from the operation characteristic. Since $E_r' > E_r$, the domain grows (Fig. 13.18a). The domain continues to grow until it reaches the stable operating point B. The reader is asked to show that point B is stable. Case 2, on the other hand, has only one intersection point, C, which not only is stable but also produces a larger V_D than point B. The velocity with which a domain moves is given by $u \cong \mu E_r$. Therefore, the time required for the domain to traverse from the cathode to the anode is approximately $\tau = l/u$.

13.9 MODES OF OPERATION OF TRANSFERRED ELECTRON DEVICES

Figure 13.19 shows the classification of the different modes of operation of transferred electron devices based on a simulation analysis by Copeland (J. A. Copeland, *IEEE Trans. Electron Devices*, Vol. ED-14, p. 461, 1967). The three main classifications are stable amplification, Gunn oscillation, and limited-space-charge accumulation (LSA) oscillation. The vertical line $N_d l = 10^{12}$ cm^{-2} represents the stability condition stated in Eq. (13.6.16) except that a different value for $|\mu_{ac}|$ is used. For $N_d l < 10^{12}$ cm^{-2}, domain is prevented from forming; hence stable amplification can be achieved. In the region $N_d l > 10^{12}$ cm^{-2}, spontaneous oscillation takes place in three distinct modes: the transit-time mode, the delayed domain mode, and the quenched-domain mode. Depending on the product fl, where f is the cavity resonant frequency, the oscillation takes the form of one of the modes above. Note from Fig. 13.18 that maximizing V_D (or maximizing the current peak in Fig. 13.17e) requires operating point C close to $E_r = E_{sat}$. In other words, for maximum power, the load line should be so chosen that a domain moves with a velocity u equal to the saturation velocity v_s of electrons. Since the domain width d is much smaller than the cathode–anode separation l, the transit time τ is equal to $\tau = l/v_s$. The horizontal line $fl = v_s = 10^7$ cm/s defines the transit-time mode of operation for which the cavity resonant frequency is tuned nearly equal to the transit-time frequency τ^{-1}.

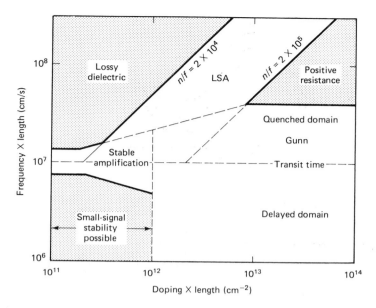

Figure 13.19 Classification of different modes of operation for a transferred-electron device in a two-dimensional plane with frequency × length as ordinate and doping × length as abscissa. (J. A. Copland, *J. Appl. Phys.*, Vol. 38, p. 3096, 1967.)

Figure 13.20 shows schematically the relations between the voltage waveform (which is assumed to consist of a dc bias and a sinusoidal voltage) and the current pulse for the three modes: (I) the transit-time mode, (II) the delayed-domain mode, and (III) the quenched domain mode. Because the time lapse between the arrival of a domain at the anode (point *b* of curve I) and the formation of a new domain at the cathode (point *d* of curve I) is very short, the output power is limited for the transit-time mode. The efficiency for such a mode is below 10%. If a delay is introduced in the formation of the new domain, the current pulse is stretched, resulting in an increased power. The delay τ_D is accomplished by allowing the voltage to swing below V_{th}, the voltage required to maintain a field equal to the threshold field E_{th}. During the time interval *B* to *D*, no domain exists and the current pulse simply follows the voltage variation. During the time interval *D* to *B*, a mature domain has formed and the current density is therefore given by *enu* (Fig. 13.18), which is smaller than the current density env_{th} at point *B* or *D*. In the delayed-domain mode, the time interval *DB* is equal to the transit time $\tau = l/v_s$. With the added delay τ_D, the period *T* of the voltage waveform is equal to $T = \tau + \tau_D$. Therefore, the delayed-domain mode (case II) operates in the region $fl < 10^7$ cm/s. In Fig. 13.20 point *D* (or *d*) represents the time at which a new domain forms at the cathode. Referring to Fig. 13.18, we see that for a given *l*, there is a minimum voltage required so that the load line and the V_D versus E_r curve have at least one intersection point. This voltage is called the *sustaining voltage* V_S, and the intersection point is represented by point *S*. If $V < V_S$, a domain collapses before reaching the anode. In the quenched-domain mode (case III), the voltage swing goes slightly below V_S at point *E* in Fig. 13.20, and a domain is quenched in transit from the cathode to the anode. Therefore, the value of τ is longer than *T*, and the mode operates in the region

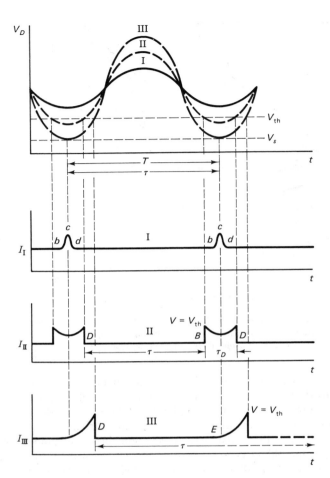

Figure 13.20 Voltage waveforms and current pulse forms for the three Gunn modes of operation: (I) transit-time mode, (II) delayed-domain mode, and (III) quenched-domain mode.

$fl > 10^7$ cm/s. In all three cases, a domain exists in the time interval d to b for (I), D to B for (II), and D to E for (III). The efficiency is about 20% for mode (II) and 13% for mode (III).

An extension of the quenched-domain mode of operation is the LSA mode of operation. If the voltage is allowed to swing into the positive μ_{ac} region and the period T is shortened to limit the time for space-charge growth, the size of the space charge can be controlled. Suppose that the field stays in the negative differential conductance region from $t = 0$ to $t = t_{th}$ and in the positive μ_{ac} region from $t = t_{th}$ to $t = T$, where t_{th} indicates the time when $E = E_{th}$. Further, we define a growth factor G and a decay factor D as follows:

$$G = \exp\left(\frac{n}{f}\frac{e}{\epsilon T}\int_0^{t_{th}} |\mu_{ac}|\, dt\right), \qquad D = \exp\left(-\frac{n}{f}\frac{e}{\epsilon T}\int_{t_{th}}^{T} |\mu_{ac}|\, dt\right) \qquad (13.9.1)$$

Setting the limits for $G < \exp 6$ and $D > \exp(-5)$, the condition for the LSA mode is found to be

$$2 \times 10^5 > \frac{n}{f} > 2 \times 10^4 \text{ s-cm}^{-3} \tag{13.9.2}$$

by using the v_d versus E curve of Fig. 10.25 in evaluating the two integrals (J. A. Copeland, *J. Appl. Phys.*, Vol. 38, p. 3096, 1967). The instantaneous power absorbed by an electron is $P_{\text{inst}} = ev_d E$. For $E = E_0 + E_1 \cos [\omega(t - t_s/2)]$, the dc (average) power absorbed is

$$P_0 = eE_0 T^{-1} \int_0^T v_d(t)\, dt \tag{13.9.3}$$

and the ac power delivered to the cavity is

$$P_1 = -eE_1 T^{-1} \int_0^T v_d(t) \cos \left[\omega \left(\frac{t - t_s}{2} \right) \right] dt \tag{13.9.4}$$

Because the ac voltage is negative when v_d is large and positive when v_d is small (i.e., in the negative differential mobility region), the integral in Eq. (13.9.4) has a negative value. In other words, the ac part of $v_d(t)$ and that of E are out of phase. This is the physical mechanism by which dc power is partly converted into ac power. The existence condition stated in Eq. (13.9.2) for the LSA mode is in agreement with experimental observations (J. A. Copeland, *Bell Syst. Tech. J.*, Vol. 46, p. 284, 1967). Because the large voltage swing and high frequency, the LSA mode is capable of delivering kilowatt pulsed power which is higher by more than an order of magnitude than that of the Gunn modes. For further discussions of the transferred-electron device, the reader is referred to the book by Carroll and the review article by Uenohara (J. E. Carroll, *Hot Electron Microwave Generators*, Edward Arnold (Publishers) Ltd., London, 1969; M. Uenohara,

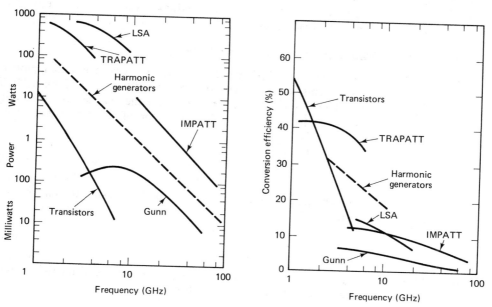

Figure 13.21 Power and efficiency as a function of frequency for various microwave generators. The curves are intended to indicate the general range and tendency only.

Sec. 13.9   Modes of Operation of Transferred Electron Devices

"Bulk Gallium Arsenide Devices," in H. A. Watson, ed., *Microwave Semiconductor Devices and Their Circuit Application*, McGraw-Hill Book Company, New York, 1969, Chap. 16).

The power and efficiency of various microwave generators are shown in Fig. 13.21. The maximum power achieved generally follows a f^{-2} relation. For devices whose maximum field is set by a certain physical phenomenon, such as impact ionization or electron transfer, the voltage V and the transit time τ are directly proportional to length l. Since power P increases as V^2 and the frequency f decreases as τ^{-1}, P is expected to decrease as f^{-2}. The dc-to-ac conversion efficiency of a microwave device also decreases with frequency, more drastically in transistors than in IMPATT and Gunn devices. Therefore, avalanche and Gunn devices are important solid-state sources for high-power moderate-efficiency operations in the range 10 to 100 GHz.

13.10 CONSIDERATIONS FOR ULTRAHIGH-SPEED OPERATION OF FIELD-EFFECT TRANSISTORS

The invention of the transistor and the subsequent realization of integrated circuits have brought about a revolution in the electronics industry. The use of integrated circuits has expanded rapidly into a wide range of industrial as well as consumer products: computers, communications, controls, instrumentations, and sensors. One primary push is for devices of higher speeds and smaller dimensions so as to enable us to analyze and solve problems of greater complexity in a shorter time. However, high speed and small dimensions alone are not sufficient criteria for a high-performance system. High performance also means reliability, which requires uniformity and reproducibility of device performance. In this and the following section we present a general discussion on three important aspects concerning integrated circuits: speed, dimension, and reliability. Specific devices are discussed in subsequent sections.

Characteristic Frequency f_t and Switching Delay τ_t

Since the FET is used extensively in integrated circuits, we use it as an example to illustrate the relevant elements affecting the speed. Referring to Fig. 9.27 for a MOSFET, we see that in order to modulate the conductance of the channel, we must vary the charge ΔQ in the conduction channel by the application of an ac gate voltage v_G. By definition,

$$g_m = \frac{\partial I_D}{\partial V_G} = \frac{\Delta I_D}{v_G} \qquad (13.10.1)$$

where ΔI_D is the variation in the drain current I_D. On the other hand, the gate capacitance C_{GS} is just the ratio of $\Delta Q/v_G$, or

$$C_{GS} = \frac{\partial Q}{\partial V_G} = \frac{\Delta Q}{v_G} \qquad (13.10.2)$$

Since ΔI depends on ΔQ, the response of FET is governed by a characteristic time $\tau_t = \Delta Q/\Delta I$ or a characteristic frequency f_τ such that

$$f_t = (2\pi\tau_t)^{-1} = \frac{g_m}{2\pi C_{GS}} \qquad (13.10.3)$$

The quantity τ_t is generally known as the *switching delay* and the quantity f_t as the *current-gain cutoff frequency*. Substituting Eqs. (9.9.4) and (9.10.9) in Eq. (13.10.3) gives

$$f_t = \frac{1}{2\pi} \frac{V_D \mu}{L_G^2} \left[\frac{\epsilon' W + \epsilon d}{\epsilon d} \right] \tag{13.10.4}$$

where $L_G = L$ is the gate length, d the thickness of the oxide, and μ the carrier mobility. For a MESFET, the mutual transconductance in the saturation region is given by Eq. (9.13.12) and the gate–source capacitance C_{GS} is equal to $C_{GS} = \epsilon Z L_G / a$, where a is the depth of the space-charge region. Using these expressions in Eq. (13.10.3), we obtain

$$f_t = \frac{1}{2\pi} \frac{\mu V_{DS}}{L_G^2} \tag{13.10.5}$$

in the pinch-off region. The factor in brackets in Eq. (13.10.4) takes into account the contribution to C_{GS}^{-1} in a MOSFET due to the extension of the space-charge region into the conduction channel.

Note that V_D/L_G in Eq. (13.10.4) or V_{DS}/L_G in Eq. (13.10.5) is related to the average field E existing in the conduction channel. For the MOSFET we ignore the contribution from $\epsilon' W$ in Eq. (13.10.4). Thus from Eq. (13.10.4) or (13.10.5) we have

$$f_t \cong \frac{1}{2\pi} \frac{1}{\tau_t} \cong \frac{1}{2\pi} \frac{1}{L_G/v} \tag{13.10.6}$$

where $\tau_t = L_G/v$ is the transit time across the channel and v is the carrier velocity. Two extreme cases are worth noting. In the low-field case, μ is a constant and hence f_t is expected to vary with L_G^{-2}, or

$$f_t = A\mu L_G^{-2} \tag{13.10.7}$$

In the high-field case, v is given by the saturation velocity and hence f_t is expected to follow the relation

$$f_t = B L_G^{-1} \tag{13.10.8}$$

In a practical device, the field E varies along the length of a channel. Therefore, it is expected that the dependence of f_t on L_G^n should have the value of n between -1 and -2. The quantity f_t sets an upper limit on the operating frequency of a FET above which the current gain of the transistor starts to decrease. It is transit-time limited.

Figure 13.22 shows the reported values of f_t and τ_t in Si-MOSFET, GaAs-MESFET, and GaAs-MODFET (or HEMT), mainly compiled by Drummond et al. and by Abe et al. plus additional references (T. J. Drummond, W. T. Masselink, and H. Morkoç, *Proc. IEEE*, Vol. 74, p. 773, 1986; M. Abe, T. Mimura, N. Yokoyama, and S. Suyama, in A. Goetzberger and M. Zerbst, eds., *Solid State Devices*, Verlag Chemie GmbH, Wienheim, West Germany, 1982, p. 25; MOSFET: K. Ohwada, Y. Omura, and E. Sans, *Tech. Dig.*, International Electron Devices Meeting (IEDM), p. 756, 1980; K. Nishiuchi, H. Shibayama, T. Nakamura, T. Hisatsugu, H. Ishikawa, and Y. Fukukawa, *ISSCC Dig. Tech. Pap.*, 1980, p. 60; I. Ito, H. Ishikawa, and Y. Fukukawa, *Proceedings 12th Conference on Solid State Devices*, 1980, p. 9; D. C. Shaver, *IEEE Electron Device Lett.*, Vol. EDL-6, p. 36, 1985; MESFET: B. Kim, H. Q. Tserng, and

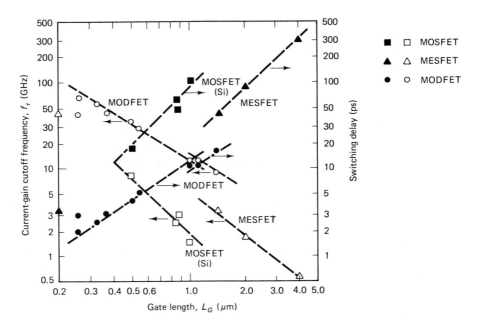

Figure 13.22 Current-gain cutoff frequency for GaAs-MODFET (open circles ○), GaAs-MESFET (open triangles △), and Si-MODFET (open squares □) and switching delay for GaAs-MODFET (closed dots ●), GaAs-MESFET (closed triangles ▲), and Si-MOSFET (closed squares ■) as a function of gate length L_G.

H. D. Shih, *IEEE Electron Device Lett.*, Vol. EDL-6, p. 1, 1985; MODFET: D. J. Arnold, R. Fischer, W. F. Kopp, T. Henderson, and H. Morkoç, *IEEE Trans. Electron Devices*, Vol. ED-32, p. 107, 1984; P. C. Chao, T. Hu, P. M. Smith, S. Wayna, J. C. M. Hwang, W. H. Perkins, H. Lee, and L. F. Eastman, *Electron. Lett.*, Vol. 19, p. 894, 1983; J. J. Berenz and K. P. Weller, *Tech. Dig.*, IEEE Microwave and Millimeter Wave Monolithic Circuits Symposium, 1984, p. 83). Two features of the curves stand out. First, the MODFETs have the highest f_t and the shortest τ_t, while the MOSFETs have the lowest f_t and the longest τ_t. Second, the exponent n in the dependence of f_t on L_G^n has a value close to -1.5 in the case of MODFETs and a value close to -2 in the case of MESFETs and MOSFETs. This observed dependence might imply a constant carrier mobility. However, this interpretation is not consistent with the mobility values. As mentioned earlier and shown in Fig. 11.15, the electron mobility in GaAs is larger than that in Si by a factor of 6 and that in GaAs-MOD samples is larger than that in bulk GaAs by about a factor of 2. For gate length $L_G = 0.5$ μm, the ratio of f_t values in GaAs MODFET and Si MOSFET is about 4 instead of 6. The curves would project a ratio of about 2 at $L_G = 0.25$ μm. At a drain voltage of 2 V in a $L_G = 1$ μm FET, the average field has a value of $E = 2 \times 10^4$ V/cm, which is already in the velocity saturation region (Fig. 10.1). It is likely that as L_G becomes smaller, a larger portion of the channel is in the velocity-saturation region and hence the f_t changes gradually from a L_G^{-2} dependence to a L_G^{-1} dependence. Since the drift velocity in GaAs deviates from the linear relation $v_d = \mu E$ at a smaller field ($E \sim 2 \times 10^3$ V/cm) than that in Si ($E \sim 6 \times 10^3$ V/cm), the transition from L_G^{-2} dependence to L_G^{-1} dependence is expected to occur at a longer L_G in GaAs than in Si. Therefore, we do not expect that a L_G^n dependence with a single n will hold. It is not surprising then to find that the reported values of f_t all fall below the projected curves for $L_G = 0.25$ μm FETs.

Power–Gain Cutoff Frequency f_{max}

Now we examine the factors affecting the high-frequency operation of the three common types of field-effect transistors, the MESFET, the MODFET, and the MOSFET, in a circuit environment and describe the steps taken toward reducing the dimensions of these devices. We start with the MESFET. Figure 13.23 shows (a) the structure and (b) the equivalent circuit of a MESFET (C. A. Liechti, *IEEE Trans. Microwave Theory Tech.*, Vol. MTT-24, p. 279, 1976). The terminals designated by S', G', and D' represent the source, gate, and drain of an ideal MESFET, while those designated by S, G, and D are the respective terminals for connection to an external circuit. Therefore,

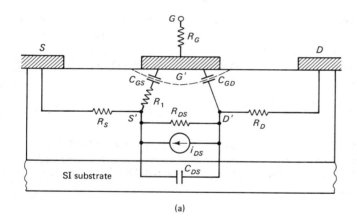

(a)

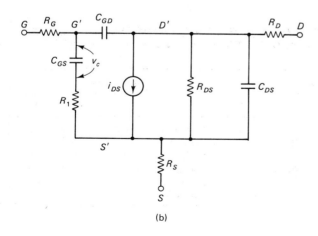

(b)

Figure 13.23 (a) Structure and (b) equivalent circuit of a MESFET.

the circuit elements connecting S', G', and D' in Fig. 13.23 are the intrinsic elements of the device, while those connecting the primed and unprimed terminals are the extrinsic or parasitic elements. The lumped capacitance C_{GS} and C_{GD} are the source part and the drain part of the Schottky-barrier capacitance. The capacitance C_{dc} associated with the charging and discharging of a dipole domain which exists near the drain end of the channel is not included here. The resistance $R_{DS} = (dI_{DS}/dV_D)^{-1}$ is the equivalent resistance due to nonsaturation of the drain current. The parasitic capacitance C_{DS} represents the capacitive coupling between the drain and the source through the semi-insulating (SI) substrate as a dielectric. The inductances L_S, L_G, and L_D associated with lead wire are not shown in the figure. The values for the various circuit elements for a selected number of GaAs FETs as compiled by Drummond et al. are given in Table 13.1 (T. J. Drummond, W. T. Masselink, and H. Morkoç, *Proc. IEEE*, Vol. 74, p. 773, 1986).

We should point out that the width of the space-charge region and hence the total mobile charges in the conduction channel are controlled by the ac voltage v_c developed across the lumped capacitance C_{GS} in the circuit model of Liechti. Therefore, the ac current source i_{DS} is related to v_c by

$$i_{DS} = v_c g_m \exp\left(-i\omega\tau_t\right) \tag{13.10.9}$$

Because of the phase delay of Eq. (13.10.9) and the shunting effect of the various capacitances, the gain of a MESFET or MODFET decreases with increasing frequency. The highest frequency at which a MESFET still possesses a power gain is known as f_{max}. This power–gain cutoff frequency is approximately given by (W. Fischer, *Solid State Electron.*, Vol. 9, p. 71, 1966)

$$f_{max} = \frac{f_t}{2}\left(\frac{R_1 + R_S + R_G}{R_{DS}} + 2\pi f_t R_G C_{GD}\right)^{-1/2} \tag{13.10.10}$$

Using the values given in Table 13.1, we find the two terms in parentheses to be, respectively, 0.044 and 0.012 for the 1 μm-gate MODFET, and 0.048 and 0.012 for the 1 μm-gate MESFET. These values yield f_{max} = 38 GHz for the former and f_{max} = 29 GHz for the latter.

TABLE 13.1 VALUES OF THE VARIOUS CIRCUIT ELEMENTS FOR A = MODFET WITH L_G = 0.5 μm, B = MODFET WITH L_G = 1 μm, AND B^* = MESFET WITH L_G = 1 μm

Intrinsic elements				Parasitic elements			
	A	B	B*		A	B	B*
g_m (mS/mm)	235	140	100	C_{DS} (pF/mm)	0.45	0.16	0.28
τ_t (ps)	2	2.4	4.4	R_G (Ω-mm)	1	1.2	1.2
C_{GS} (pF/mm)	1.3	1.3	1.1	R_S (Ω-mm)	0.8	1.5	0.93
C_{GD} (pF/mm)	0.05	0.09	0.11	R_D (Ω-mm)	1.0	0.84	0.15
R_1 (Ω-mm)	0.4	7.0	3.6	f_t (GHz)	29	18	14
R_{DS} (Ω-mm)	66	220	120	f_{max} (GHz)	96	38	30

Power–Delay Product $P\tau_d$

For analog applications the power–gain cutoff frequency f_{max} is an important quantity for comparing the performance of various FETs. For digital applications, one measure commonly used for device comparisons is the product $P\tau_d$ of power and switching delay time. This product represents the energy required to switch from one logic state to another. Figure 13.24 shows two types of logic circuits commonly used: (a) the direct-coupled FET logic (DCFL) inverter for its circuit simplicity, and (b) the low-power FET logic (LPFL) inverter for its tolerance to threshold-voltage nonuniformity. In the latter, the loads can be either depletion-mode FETs (DFETs) or ungated FETs used as saturated resistors. The logic delay τ_d per stage consists of two parts: the switching delay τ_t and the time to change the charges stored in the capacitances. The energy involved in a logic-voltage swing ΔV_L is given by $CV_{dd}\Delta V_L/2$, where V_{dd} is the power-supply voltage and C is the sum of the capacitances associated with a logic gate. The logic delay τ_d per stage is thus given by

$$\tau_d = \tau_t + \frac{CV_{dd}\,\Delta V_L}{2P} \tag{13.10.11}$$

and the power–delay product is given by

$$P\tau_d = P\tau_t + \frac{CV_{dd}\,\Delta V_L}{2} \tag{13.10.12}$$

Figure 13.25 shows the data on the speed–power relation for various logic technologies compiled by Drummond et al. The performance characteristics are based on ring-oscillator circuits, with the exception of the Josephson junction. As pointed out by Drummond et al., such characteristics may be overly optimistic in projecting the speed advantage of MODFET technology in large-scale integrated circuits, where a fan-out of greater than 2 is often required. As can be seen from Eq. (13.10.11), the logic delay τ_d is made of two parts. It is often argued that even though τ_t for GaAs logic is faster than τ_t for Si logic because of the higher electron mobility in GaAs, this advantage may disappear if the second term becomes comparable to or even larger than τ_t. The answer to this question lies in our ability to reduce V_{dd} and ΔV_L. As discussed in Sections 10.7 and 13.13, hot-electron injection into the oxide is a serious problem for MOSFETs. One

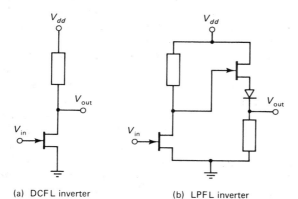

(a) DCFL inverter

(b) LPFL inverter

Figure 13.24 Two commonly used logic circuits: (a) direct-coupled FET logic inverter and (b) low-power FET logic inverter.

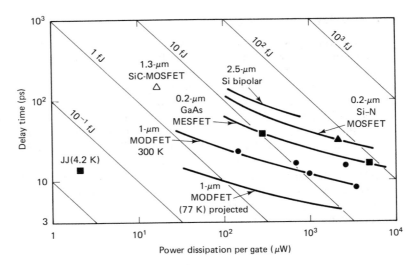

Figure 13.25 Gate delay τ_d versus power per gate P for MESFET-, MODFET-, MOS-FET-, bipolar-, and JJ (Josephson junction) logic in ring oscillator circuits. The straight lines represent the power–delay product $P\tau_d$. (After T. J. Drummond, W. T. Masselink, and H. Morkoç, "Modulation-Doped GaAs/(Al,Ga)As Heterojunction Transistors: MODFETs," *Proc. IEEE*, Vol. 74, p. 773, 1986, © 1986 IEEE.)

way to minimize the hot-electron effect is to reduce the power-supply voltage V_{dd} from 5 V to 2.5 V. A recent study of scaling issues in submicrometer MOSFETs has indeed suggested a reduction of power-supply voltages to 3 V (E. Sangiorgi, E. A. Hofstatter, R. K. Smith, P. F. Bechtold, and W. Fichtner, *IEEE Electron Device Lett.*, Vol. EDL-7, p. 115, 1986).

Besides reducing V_{dd}, a reduction in τ_d can also be accomplished by reducing the logic-voltage swing ΔV_L. Figure 13.26 shows the saturated drain current I_{DS} as a function of the gate voltage for a Si MOSFET, a GaAs MESFET, and an (AlGa)As/GaAs MODFET. The curves were shifted with respect to the voltage axis to simulate operation in a logic inverter with adequate margins. One important aspect as pointed out by Drummond et al. is the sharp turn-on characteristic of the MODFET and MESFET. A larger mutual transconductance g_m and a sharper turn-on characteristic mean a smaller ΔV_L required in a logic operation. The sharpness of the turn-on characteristic is strongly dependent on the source resistance R_S, which consists of two parts, the contact resistance R_{SC} and the semiconductor resistance R_{SS} between the contact and the source end of the gate. Given equal electron concentration, the part R_{SS} is much lower in GaAs than in Si because of a higher low-field mobility in GaAs. A specific contact resistivity as low as 1.3 $\mu\Omega$-cm^2 has been achieved through *in situ* metallization of a GaAs film grown by MBE and doped with Si to a concentration 1×10^{20} cm^{-3} (P. D. Kirchner, T. N. Jackson, G. D. Pettit, and J. M. Woodall, *Appl. Phys. Lett.*, Vol. 47, p. 26, 1985). For a 1-μm source, the value of R_{SC} is only 0.13 Ω-mm. The availability of advanced growth technology, such as MBE, to GaAs for *in situ* ohmic-contact formation certainly is a great advantage which may enable GaAs-based logic to preserve its speed advantage in large-scale integrated circuits.

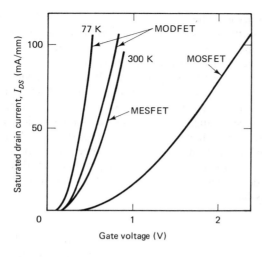

Figure 13.26 Comparison of saturated drain current in Si-MOSFET, GaAs-MES-FET, and (AlGa)As/GaAs MODFET as a function of gate voltage. (After T. J. Drummond, W. T. Masselink, and H. Morkoç, "Modulation-Doped GaAs/ (Al,Ga)As Heterojunction Transistors: MODFETs," *Proc, IEEE*, Vol. 74, p. 773, 1986, © 1986 IEEE.)

A small source resistance R_S is very desirable not only for achieving high speed in digital applications but also for attaining a low-noise figure in analog applications. The reported data on gain and noise as compiled by Drummond et al. have shown that a noise figure as low as 2.7 dB and a gain as high as 5.9 dB have been achieved at 34 GHz in discrete MODFETs with $L_G = 0.35$ μm. The corresponding data of 3.1 dB for noise figure and 7.5 dB for gain achieved at 20 GHz in a four-stage MODFET amplifier are equally impressive. For further discussions of the operation and performance of MODFETs, the reader is referred to the excellent review article by Drummond et al., from which many of the data given in this section are taken. Similar values of gain (somewhat lower) and noise (somewhat higher) have also been reported in MESFETs fabricated on epitaxial layers grown on SI substrates. The superior performance of MODFETs and MESFETs makes them the dominant devices used in microwave receivers. As shown in Fig. 13.22, the highest value of f_t is achieved in MODFETs. For digital applications, however, speed is not the only consideration, and performance is always weighed against cost. Processing of GaAs ICs is more costly than processing of Si ICs. It remains to be established that a processing technology can be developed for ultralarge-scale integration of GaAs devices with an acceptable yield. Much work is needed in materials research to achieve an integration level with GaAs devices comparable to that with Si devices.

13.11 ISSUES FOR INTEGRATION OF DEVICES OF ULTRASMALL DIMENSIONS

In Section 13.10 we considered various factors limiting the speed of field-effect transistors. From Eq. (13.10.6) we see that the intrinsic speed of a FET is limited by the transit time τ_t for carriers to traverse the length of a channel. For high f_t, a short gate length L_G is required. Therefore, ultrahigh speed operation of transistors necessitates reduction of device size to ultrasmall dimensions. Furthermore, output signal from one device is used as input signal to another device. A finite time τ_p is also required for signal propagation from one device to another. For a transmission line of length L, the propagation delay is $\tau_p = \sqrt{\mu\epsilon}\, L$, where μ and ϵ are the permeability and permittivity (dielectric constant) of the transmission medium. Assuming that $c = (\mu\epsilon)^{-1/2} = 1 \times 10^{10}$ cm/s and $L = 1$ mm, we find $\tau_p = 1 \times 10^{-11}$ s. Therefore, to achieve high speed in an operating subsystem, the overall length interconnecting devices must be short. The considerations for fast τ_t and τ_p require not only that the devices be made of small dimensions but also that they be densely packed. In order to integrate devices of ultrasmall dimensions on a very large scale, we must face such important issues as controllability of device dimension, uniformity of device characteristics, reliability of device performance, and proximity effects on device characteristics. In this section we describe approaches taken to overcome these problems. Some problems are common to both Si-based and GaAs-based devices. However, many problems are different because of their different stages of development. The latter problems are discussed separately for Si-based FETs and GaAs-based FETs.

Self-Aligned Gate

Self-aligned gate is used extensively for the fabrication of MESFET (Fig. 9.33) and MOSFET (Fig. 9.36). This technology is essential not only for accurate control of the gate length but also for minimization of parasitic source and drain resistances, R_S and R_D in Fig. 13.23. It generally involves the use of the gate metal as a mask for ion implantation to define the source and drain regions. Refer to Fig. 9.33 as an example of the self-aligned gate technology as applied to MESFET. The masked region defines the gate length. Furthermore, through ion implantation, the source and the drain make direct electric contact to the channel. However, ion implantation introduces damages and defects which must be removed by thermal annealing.

During an annealing process, generally at around 800°C for about 10 min, the gate must remain stable and must not form an alloy with GaAs. The TiW silicide fulfills these requirements even though the TiW metal alloy becomes somewhat unstable at the annealing temperature. Measurements of the I–V characteristics show that the barrier height ϕ_B and the ideality factor n in $\exp(-eV/nkT)$ of Eq. (9.8.8) remain comparatively constant with $\phi_B \sim 0.78$ V and $n = 1.1$ after annealing at temperatures up to 850°C. The adoption of a suitable self-alignment scheme is a necessary step in achieving reproducible results on short-gate FETs. As discussed in Section 9.14, the self-aligned gate technology based on ion implantation has also been applied to Si-MOSFET.

The case with MODFET, however, is different. Figure 13.27 shows schematically a self-alignment process used only to deposit a gate metal, such as Al for the fabrication of a MODFET. The sidewall profile of the photoresist film defines the length of the gate. Ion implantation is not suitable to define the gate length in MODFET. It is found that heavy silicon implantation needed to make n^+ source and drain regions results in

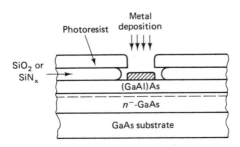

Figure 13.27 Example of self-alignment schemes for gate: deposition of gate metal through photoresist opening.

disordering of atoms in GaAs and (GaAl)As films. Subsequent thermal annealing further causes intermixing of Ga and Al atoms across the interface, thus converting the implanted GaAs and (GaAl)As regions into a single region having an average and uniform composition. This intermixing action makes it necessary to keep ion implantation away from the channel under the gate. As a result, the source and the drain regions cannot be made as close to the gate region in a MODFET as in a MESFET, resulting in an increased parasitic source and drain resistance.

Defects and Threshold-Voltage Variation

Uniformity and reproducibility of device performance are important factors determining the feasibility of large-scale integration. The self-alignment process is just one of the steps in the fabrication procedure which may introduce nonuniformity in the device performance. During high-temperature processing, the surface composition of a III–V compound semiconductor may change due to high vapor pressure of As or P and due to creation or annihilation of defects. The former effect can be greatly reduced by growing a cap layer on the surface or by maintaining a sufficiently high partial vapor pressure of As or P to prevent thermal dissociation. This effect is now well under control even though its prevention makes the processing of a III–V wafer more complicated than that of a Si wafer. The major problem is caused by inherent defects. One restraint on the rapid development of GaAs-based and InP-based technology is the relatively poor quality of GaAs and InP substrates. The growth of III–V crystals is compounded by several factors: difficulty in controlling stoichiometry due to the high vapor pressure of one binary component, and thermal stress due to poor thermal conductivity. Not only is the dislocation density in GaAs or InP much higher than that in Si, but the defect density in GaAs or InP has a much larger spatial variation than that in Si. The defects are manifested in two particularly noted effects on device performance: threshold-voltage variation and dependence of drain current on voltage of a proximity electrode, which is known as the *backgating effect*.

Figure 13.28 shows (a) the variation of the threshold voltage and (b) the variation of the sheet (areal) carrier concentration as functions of the distance to dislocation (etch pit) observed in MESFETs as reported by Miyazawa and Hyuga (S. Miyazawa and F. Hyuga, *IEEE Trans. Electron Devices,* Vol. ED-33, p. 227, 1986). The labels HD, T, and DF indicate region of high dislocation density, transition region, and region relatively free of dislocations, respectively. The observed variations can be explained on the following model proposed by Mizakawa, Hyuga, and coworkers (S. Miyazawa and K. Wada, *Appl. Phys. Lett.,* Vol. 48, p. 905, 1986; F. Hyuga, K. Watanabe, J. Osaka, and K. Hoshikawa, *Appl. Phys. Lett.,* Vol. 48, p. 1072, 1986). An increased concentration of Ga vacancies is expected near dislocations. The higher carrier density near a dislocation is a direct result of a higher activation of Si atoms as a donor because of a higher probability of occupying the Ga site. Since the threshold voltage V_T is a function of activated donor concentration, the higher is the electron concentration, the lower becomes V_T. Based on this model, Monte Carlo calculations of threshold-voltage variations have been made and a general agreement between the experimental and calculated results has been reported (R. Anholt and T. R. Sigmon, *IEEE Electron Device Lett.,* Vol. EDL-8, p. 16, 1987). Static random-access memories (SRAM) fabricated on In-doped dislocation-free LEC crystals showed a threshold-voltage variation of only 20 mV versus a variation of 200 mV in the HD region of Fig. 13.28 (M. Hirayama, M.

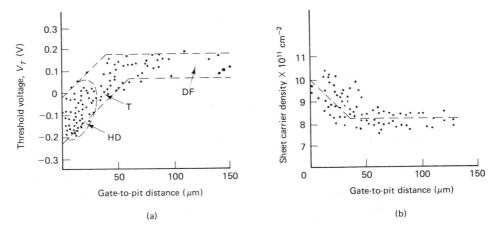

Figure 13.28 Variations of (a) threshold voltage and (b) sheet carrier density as functions of gate-to-dislocation (etch pit) distance. The symbols HD, DF, and T stand for high (dislocation) density, dislocation free, and transition (region), respectively. (After S. Miyazawa and F. Hyuga, "Proximity Effect of Dislocations on GaAs MESFET Threshold Voltage," *IEEE Trans. Electron Devices,* Vol. ED-33, p. 227, 1986, © 1986 IEEE.)

Togashi, N. Kato, M. Suzuki, Y. Matsuoka, and Y. Kawasaki, *IEEE Trans. Electron Devices,* Vol. ED-33, p. 104, 1986).

Backgating Effect

Besides threshold-voltage variation, another material-related problem is the backgating effect observed in devices fabricated on semi-insulating substrates. Referring to Fig. 6.21c and Eq. (6.5.2), we see that an interesting situation arises if we have a deep donor, such as oxygen (or a deep acceptor such as chromium), of a concentration considerably higher than that of a shallow acceptor (or a shallow donor). Because of the charge neutrality condition, the Fermi level is locked to the deep level within ± few kT, the exact amount depending on the ratio of deep to shallow impurity concentrations. As a result, the free-carrier concentrations are very small and the resistivity of the material is very high ($\sim 10^8$ Ω-cm). The semi-insulating property of GaAs or InP can be used to our advantage, such as providing electric isolation between two devices. Besides oxygen and chromium, undoped liquid-encapsulated Czochralski (LEC) GaAs is semi-insulating because of the presence of "EL2" centers which compensate the shallow acceptors caused by the background carbon doping (D. E. Holmes, R. T. Chen, K. R. Elliot, C. G. Kirkpatrick, and P. W. Yu, *IEEE Trans. Electron Devices,* Vol. ED-29, p. 1045, 1982). Figure 13.29 shows the variation of the drain current in the saturation region, I_{DS} reported by Birrittella et al. as a function of the voltage applied to a backgate electrode ($V_{BG} = V_B - V_G$) at a distance of L_{BG} from the source electrode. We see that I_{DS} decreases very rapidly as V_{BG} exceeds a certain threshold voltage V_T. The value of V_T corresponds to the voltage required to extend the acceptor space-charge region throughout the source–backgate region (M. S. Birrittella, W. C. Seelbach, and H. Goronkin, *IEEE Trans. Electron Devices,* Vol. ED-29, p. 1135, 1982).

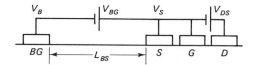

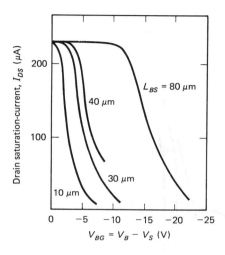

Figure 13.29 Drain saturation current as a function of backgate voltage for various distances between backgate and source. (After M. S. Birrittella, W. C. Seelbach, and H. Goronkin, "The Effect of Backgating on the Design and Performance of GaAs Digital Integrated Circuits," *IEEE Trans. Electron Devices,* Vol. ED-29, p. 1135, 1982, © 1982 IEEE.)

To gain a better understanding of the origin of the backgating effect, we present in Fig. 13.30 the variation of the substrate current through the backgate I_{BG} as a function of the backgate voltage V_{BG} (C. P. Lee, S. J. Lee, and B. M. Welch, *IEEE Electron Device Lett.,* Vol. EDL-3, p. 97, 1982). Note that the current flow between the source and backgate is similar to that in an *n-i-n* structure which has been analyzed by Horio et al. (K. Horio, T. Ikoma, and H. Yanai, *IEEE Trans. Electron. Devices,* Vol. ED-33, p. 1242, 1986). With $V_{BG} = 0$, an *n-i-n* structure consists of five regions with two end *n* regions separated from the middle near-intrinsic region by a space-charge region of width w_0. In the low-V_{BG} region marked as (I) in Fig. 13.30, the current flow is ohmic, and determined by the resistivity of the *i* region. As V_{BG} is raised, electrons are injected from the cathode (the backgate) into the *i* region, neutralizing donors and hence extending the acceptor space-charge region in part of the *i* region. This donor-trap filling part of the curve is marked as (II) in Fig. 13.30. When all the deep donors are filled, the curve goes into the space-charge-limited regime, marked as (III), in which the current increases quadratically as the applied voltage (M. A. Lampert and P. Mark, *Current Injection in Solids,* Academic Press, Inc., New York, 1970). Because of the low concentration of deep donors ($\sim 10^{15}$ cm^{-3}), the transition from (I) to (III) takes place within a small range of V_{BG} which appears very much like a threshold voltage V_T. The values of the transition voltage V_T are indicated by an arrow in Fig. 13.30. The backgating effect can be eliminated by electrically isolating the conduction channel of a FET from the semi-insulating substrate as demonstrated by Lo et al. in a buried-gate GaAs junction FET (Y. H. Lo, S. Wang, J. Miller, D. Mars, and S. Y. Wang, *IEEE Electron. Device Lett.,* Vol. EDL-8, p. 36, 1987). Figure 13.31 shows the device structure and the normalized drain–source saturation current as a function of V/L^2, where V is the

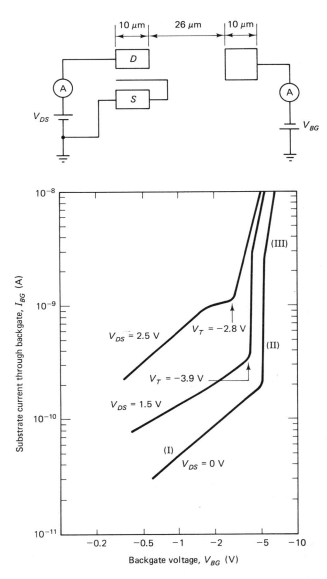

Figure 13.30 Variation of substrate current through backgate as a function of backgate voltage at several drain–source bias voltages. The behavior can be classified into three regions: (I) ohmic, (II) trap-filling, and (III) space-charge limited. (After C. P. Lee, S. J. Lee, and B. M. Welch, "Carrier Injection and Backgating in GaAs MESFETs," *IEEE Electron Device Lett.*, Vol. EDL-3, p. 97, 1982, © 1982 IEEE.)

backgate bias and L is the distance between the backgate and FET-gate electrodes for (a) the junction FET and (b) various MESFETs with different L. The drop in I_{DS} in MESFETs is due to an increased I_{BG} through the semi-insulating substrate (Fig. 13.30). No such effect is observed for the FET with junction isolation.

Intrinsic Defects in GaAs and Si

Many deep levels have been identified in GaAs in a number of experiments, including deep level transient spectroscopy (DLTS), photo-excited DLTS, and photoluminescence (H. Z. Zhu, Y. Adachi, and T. Ikoma, *J. Cryst. Growth*, Vol. 55, p. 154, 1981; M.

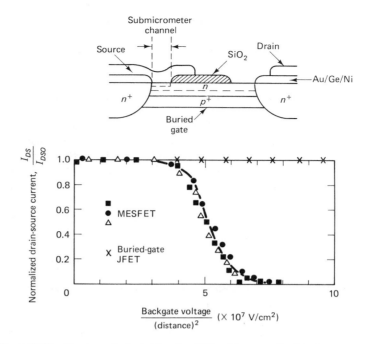

Figure 13.31 Structure of a buried-junction FET and the normalized drain–source saturation current as a function of V/L^2 (the ratio of backgate voltage to the backgate distance squared from the FET gate). (After Y. H. Lo, S. Wang, J. Miller, D. Mars, and S. Y. Wang, "A Self-Aligned Quarter-to-Half Micrometer Buried-Gate Junction FET," *IEEE Electron Device Lett.*, Vol. ED-8, p. 36, 1987, © 1987 IEEE.)

Taniguchi and T. Ikoma, *Inst. Phys. Conf. Ser. No. 65*, 1982, p. 65, paper presented at the International Symposium on GaAs and Related Compounds, Albuquerque). Of these, the EL2 center is the major defect associated with the anti-site defect As_{Ga}. By *anti-site defect,* we mean an As atom occupying a normal Ga site or a Ga atom occupying a normal As site denoted by As_{Ga} or Ga_{As} (J. Lagowski, H. C. Gatos, J. M. Parsey, K. Wada, M. Kaminska, and W. Walukiewicz, *Appl. Phys. Lett.*, Vol. 40, p. 342, 1982). Extensive investigations of EL2 and other deep levels indicate that many of these levels may be related to defect complexes or to impurities such as oxygen and iron. The many levels being detected may be caused by an impurity complex which is formed by an impurity atom attached to a lattice defect. The spatial distribution of the deep levels has been investigated in a number of experiments, including direct infrared imaging (M. S. Skolnick, M. R. Brozel, L. J. Reed, I. Grant, D. J. Stirland, and R. M. Ware, *J. Electron. Mater.*, Vol. 13, p. 107, 1984), cathodoluminescence (CL), secondary ion mass spectroscopy (SIMS) (T. Kamejima, F. Shimura, Y. Matsumoto, H. Watanabe, and J. Matsui, *Jpn. J. Appl. Phys.*, Vol. 21, p. L721, 1982), and electrooptic probing (Z. H. Zhu, Y. H. Lo, S. Y. Wang, and S. Wang, to be published in *Journal of the Electrochemical Society*). All the experiments show a macroscopic inhomogeneous distribution of deep levels in the form of a W shape (Y. Nanishi, S. Ishida, and S. Miyazawa, *Jpn. J. Appl. Phys.*, Vol. 22, p. L54, 1983; D. E. Holmes, R. T. Chen, and J. Yang, *Appl. Phys. Lett.*, Vol. 45, p. 419, 1983). This W-shaped distribution of deep levels is related to the well-pronounced W-shaped variation of dislocation density caused by the thermal stress across the wafer during the crystal growth (A. S. Jordan, R. Caruso, and A. R. Von Neida, *Bell Syst. Tech. J.*, Vol. 59, p. 593, 1980).

For an extensive discussion of native defects in GaAs and their experimental identification, the reader is referred to an excellent review article by Bourgoin et al. (J. C. Bourgoin, H. J. von Bardeleban, and D. Stiévenard, *J. Appl. Phys.*, Vol. 64, p. R65, November 1, 1988). In addition to vacancies and interstitials, antisites are intrinsic defects in compound semiconductors. The EL2 defect is thought to be due to an As_{Ga} antisite complexed with an As_i (interstitial). The bond between As_i and As_{Ga} is an admixture of a p-p bond (discussed in Section 3.7) and a p-sp^3 bond. Since the requirements on bond angle and bond length for the two bonds are different from that of the normal sp^3 bond, lattice relaxation (or distortion in the local atomic arrangement) results. A diagram relating the relevant energy levels to the displacement of an atom from a reference position is called the configuration-coordinate diagram. Such a diagram for the KCl:Tl phosphor is shown in Fig. 14.1b. For the GaAs:EL2 system, several changes must be made. The $^1S_0$ level and $^1P_1$ level are replaced by E_v and E_c, respectively, and the $^3P_1$ level is replaced by the EL2 level. The energy scale, of course, is changed. Due to the difference in the bond requirement, the minimum of the EL2 level is located at a position different from the minimum position of E_c and E_v in the configuration coordinate. Since the energy separation is a function of the configuration coordinate, different experiments (e.g., DLTS and photoluminescence experiments) may exhibit different activation energies for the EL2 level.)

The As_{Ga} antisite is a double donor possessing two ionized states, D^+ and D^{2+}. Two energy levels, one at $E_c - 0.76eV$ and the other at $E_v + 0.5eV$, have been identified as being associated with the EL2 defect. Semi-insulating property of GaAs results from charge compensation either between the double donor and a shallow acceptor or between the double donor of EL2 and the double acceptor ascribed to Ga_{As}. The fact that the EL2 density and hence the SI property are sensitive to heat treatment can be understood as follows. Dislocations are known to be impurity getters. Dislocation-associated As precipitates may either form or dissociate as the dislocation density changes. It has been suggested that thermal annealing tends to deplete the EL2 traps in SI substrates, causing surface-leakage and backgating problems. This model of surface conversion from i into p^- type has been confirmed experimentally by Lo et al., using the electro-optic probing technique (Y. H. Lo, Z. H. Zhu, C. L. Pan, S. Y. Wang, and S. Wang, *Appl. Phys. Lett.*, Vol. 50, p. 1123, 1987).

Anti-site defects and deep impurities trap charges in order to fulfill the covalent-bond requirement; therefore, a spatial variation of these defects produces a corresponding variation in the electric potential caused by the local space charge. To minimize the effects of substrate-related defects, GaAs FETs are generally fabricated on films grown either by MBE (molecular beam epitaxy) or OMCVD (organometallic chemical vapor deposition) on SI (semi-insulating) GaAs substrates. Figure 13.32 shows the electron mobility and areal electron concentration measured at 77 K in MBE grown films as reported by Abe et al. (M. Abe. T. Timura, K. Nishiuchi, A. Shibatoni, and M. Kobayashi, *IEEE J. Quantum Electron.*, Vol. QE-22, p. 1870, 1986). The spatial variation of both quantities was estimated to be less than $\pm 1\%$ over the entire wafer of 3-in. diameter. High-electron-mobility transistors (HEMTs) or modulation-doped FETs (MODFETs) fabricated on MBE-grown films showed a standard deviation of about 20

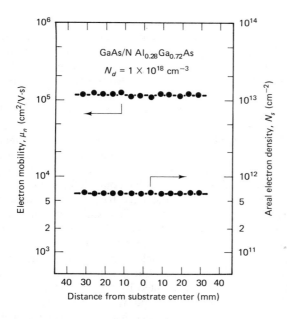

Figure 13.32 Electron mobility and areal electron concentration measured at 77 K in a modulation-doped GaAs–GaAlAs structure grown by MBE on GaAs substrate. (After M. Abe, T. Timura, K. Nishiuchi, A. Shibatoni, and M. Koybayashi, "Recent Advances in Ultra-High-Speed HEMT Technology," *IEEE J. Quantum Electron.,* Vol. QE-22, p. 1870, 1986, © 1986 IEEE.)

mV and a maximum variation of 100 mV in the threshold voltage, similar to the value obtained in FETs fabricated on In-doped dislocation-free LEC substrates. The residual variation in V_{th} is due to thickness variation of the doped (AlGa)As layer. The sensitivity of the threshold voltage was estimated to be on the order of 70 mV/nm at a threshold voltage $V_{th} = 0.13$ V. Therefore, a standard deviation of 100 mV corresponds to a thickness variation of only 1.5 nm over a 2-in wafer. We should mention, however, one serious problem caused by surface defects generally known as *oval defects,* which have a density in the range 300 to 5000 cm^{-2} and a dimension from submicrometer to several micrometers. If an oval defect happens to be in the gate region of a MODFET, it can cause anomalous behavior in the drain *I–V* characteristics (T. Mimura, M. Abe, and M. Kobayashi, *Fujitsu Sci. Tech. J.,* Vol. 21, p. 370, 1985). Oval defects have been suggested as being caused by Ga sputtering associated with gallium oxides, and films grown by OMCVD generally have a much lower oval-defect density than do films grown by MBE.

Now let us turn our attention to silicon. The rapid development of Si IC technology is made possible by advances in crystal-growing and wafer-processing methodologies. Even though melt-grown silicon ingots can be free of dislocations under the condition that the speed of crystal growth (typically about 2×10^{-3} cm/s) far exceeds the dislocation speed (about 10^{-5} cm/s), they are not free of point defects. While vacancies are the dominant point defect in metals (Section 2.10), self-interstitials are dominant in silicon with an estimated concentration of 10^{16} cm^{-3}. Also, the liquid silicon corrodes the quartz crucible that holds the molten silicon. Oxygen and carbon are incorporated into the grown crystal in the form of interstitial oxygen with a concentration from 10^{17} to 10^{18} cm^{-3} and substitutional carbon with a concentration about 10^{16} cm^{-3}. Oxygen incorporation is known to have increased the mechanical strength and the resistance to thermal warping of a silicon wafer by pinning dislocations and retarding slips. Oxygen is also effective in gettering impurities by precipitating oxide and the associated dislocation network. On the other hand, if the initial oxygen concentration is too high, the oxide precipitation results in the formation of stacking faults. Therefore, controlling oxygen in silicon is extremely important to achieving high yields in wafer processing.

For extensive discussions of silicon material and processing technologies, the reader is referred to the monograph series edited by Einspruch (*VLSI Electronics—Microstructure Science,* N. G. Einspruch, ed., Academic Press, New York, 1983) and the symposium series edited by Huff et al. (*Semiconductor Silicon/1981,* H. R. Huff, R. J. Kriegler, and Y. Takeishi, eds., Electrochemical Society Symposium Series, Pennington, N. J.). Here we give the highlights of defect engineering in Si IC technology. A silicon wafer, prior to device fabrication, consists of (1) a denuded zone of about 25 μm thickness which is almost free of point defects and defect clusters, (2) a gettering zone which serves as an infinitive sink for unwanted point defects and possibly for unwanted impurities as well, and (3) a heavily damaged back side for impurity gettering. When a silicon wafer is heated to 1050 to 1200° C, interstitial oxygen atoms diffuse out, leaving a denuded zone near the surface of the wafer. Then the wafer is heated to 650 to 800° C for nucleation of oxygen clusters at high-energy lattice sites such as point defects and dislocations. These sites are future nucleation sites for oxide precipitates. During device fabrication at 900° to 1250° C, mobile impurities, including interstitial silicon, are precipitated at dislocation sites generated by the excessive strain field due to oxide precipitates (T. Y. Tan, E. E. Gardner, and W. K. Tice, *Appl. Phys. Lett.,* Vol. 30, p. 175, 1977). Therefore, the gettering zone has a high density of defects in the form of precipitate-dislocation complexes.

As the Si IC technology moves from large-scale integration (LSI) to very large scale integration (VLSI), the requirement on high chemical purity and crystal perfection will become increasingly stringent. Localized defect clusters, such as interstitial oxygen, left in the denuded zone may become the limiting factor to the achievement of superior device performance in VLSI environment. Note that oxygen clusters of subcritical size formed during the nucleation cycle may not precipitate as oxide. Even though these clusters will redissolve and diffuse out from the wafer surface upon subsequent heating to high temperatures ($> 1100°C$), some oxygen clusters may remain in the denuded zone. Understanding the interactive effects of processing steps on defect generation and defect interactions is of paramount importance for the development of optimal processing methodologies to achieve VLSI circuitry.

Graded Drain Structure

In terms of device-packing density, the Si-based IC (integrated circuits) technology is far more advanced than the GaAs-based IC technology. While GaAs 16K RAM (random-access memory) is still being developed, Si 256K RAM is commercially available. The rapid development of Si IC technology is due in large part to two factors: the availability of high-purity large-dimension Si crystals and the existence of a stable native oxide that can be thermally grown at low cost. The Si–SiO$_2$ interface and the SiO$_2$ film are not perfect. Since the quality of a SiO$_2$ film is a key element in Si IC technology, the integrity of SiO$_2$ films has been a subject of interest for the development of submicrometer technology. Solutions to this problem have been studied in terms of improved device structure and improved SiO$_2$ quality. In Section 10.7 we discussed hot-electron injection as a possible mechanism for oxide degradation. Generation of hot carriers by impact ionization occurs near the drain end of the channel in a MOSFET. To reduce the electric field near the drain, a number of graded-drain structures have been proposed.

They are an As-P (n^+-n^-) diffused drain, an offset gate (E. Takeda, H. Kume, T. Toyabe, and S. Asai, *IEEE Trans. Electron Devices,* Vol. ED-29, p. 611, 1982), and a lightly doped drain (LDD) (S. Ogura, P. J. Tsang, W. W. Walker, J. F. Shepard, and D. L. Critchlow, *Tech. Dig.,* International Electron Devices Meeting (IEDM), p. 651, 1981; *IEEE Trans. Electron Devices,* Vol. ED-29, p. 590, 1982). As first proposed by Takeda et al. (E. Takeda, H. Kume, T. Toyabe, and S. Asai, *IEEE Trans. Electron Devices,* Vol. ED-29, p. 611, 1982) and subsequently analyzed and used by Hu et al. (C. Hu, S. C. Tam, F. C. Hsu, P. K. Ko, T. Y. Chan, and K. W. Terrill, *IEEE Trans. Electron Devices,* Vol. ED-32, p. 375, 1985), substrate and gate currents can be used to monitor the hot-carrier generation in a channel. The use of the various modified MOSFET structures has been proved effective in reducing substrate and gate currents.

The LDD drain structure has been optimized and studied by Matsumoto et al. for 1-μm NMOSFETs (Y. Matsumoto, T. Higuchi, T. Mizuno, S. Sawada. S. Shinozaki, and O. Ozawa, *IEEE Trans. Electron Devices,* Vol. ED-32, p. 429, 1985). Reduction of gate and substrate currents has indeed been observed. However, in terms of MOSFET reliability, the results are still inconclusive. While the results reported by Hsu and coworkers show either inferior degradation or little net advantage of nonconventional MOSFET structures (F. C. Hsu and H. R. Grinolds, *IEEE Electron Device Lett.,* Vol. EDL-5, p. 71, 1984; F. C. Hsu and K. Y. Chiu, *IEEE Electron Device Lett.,* Vol. EDL-5, p. 162, 1984), the results reported by Matsumoto et al. show a much reduced threshold-voltage shift ΔV_{th} at the expense of the current drive in the form of a 25% reduction in I_D in LDD MOSFETs having a surface concentration $n^- = 1 \times 10^{17}$ cm^{-3} region. The difference in the reliability results is due primarily to the difference in the stress condition. The current drive in the experiments of Hsu and coworkers was maintained constant for nonconventional and conventional MOSFETs, whereas the bias voltages V_D and V_G were kept the same in the experiments of Matsumoto et al. Therefore, further investigations are required to determine the effectiveness of the LDD structure in improving MOSFET reliability. Furthermore, the use of a LDD structure inevitably increases the drain resistance and hence lowers the high-frequency response of the device.

Oxide Degradation

Concerning the oxide itself, the experimental findings reported by Yamabe and Taniguchi are worth noting. (K. Yamabe and K. Taniguchi, *IEEE Trans. Electron Devices,* Vol. ED-32, p. 423, 1985). The time-dependent dielectric breakdown (TDDB) can be divided into three categories: A, B, and C modes. The A mode can be attributed to pinholes in the gate oxide, causing an initial short. The defect density responsible for the A mode increases with decreasing oxide thickness. The B mode is caused by weak spots in the oxide, and mode B failure results in the A mode in subsequent measurements. The defect density responsible for the B mode decreases with decreasing oxide thickness. The C mode is for defect-free oxide, in which the breakdown is caused by the Fowler–Nordheim tunneling current into the gate oxide. Three groups of Si wafers were used in the study of Yamabe and Taniguchi: (1) wafers as received from vendor, (2) wafers subjected to preoxidation annealing at 1200°C for 1 h, and (3) wafers subjected to phosphorus diffusion at 1000°C for 30 min from the backside after preoxidation annealing. Figure 13.33 shows the histograms of breakdown field for the three groups. While preoxidation annealing removes oxygen precipitates from the silicon surface,

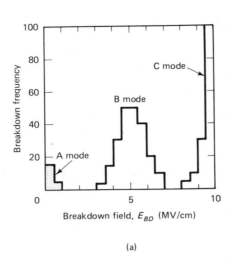

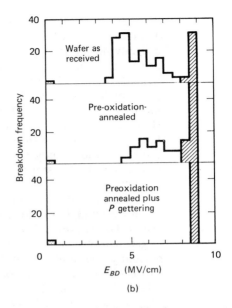

Figure 13.33 Histograms showing the breakdown frequency as a function of breakdown field: (a) for three modes of oxide failure and (b) for wafers under different treatments. (After K. Yamabe and K. Taniguchi, "Time Dependent Dielectric Breakdown of Thin Thermally Grown SiO₂ Films," *IEEE Trans. Electron Devices*, Vol. ED-32, p. 423, 1985, © 1985 IEEE.)

phosphorus diffusion is effective in impurity gettering. The most likely metal impurities in gate oxide, as suggested by Yamabe and Taniguchi, are Cu, Ni, and Na. Figure 13.33 also shows that after removal of the B-mode type of defects, the breakdown field should approach 9×10^6 V/cm in a thermally grown oxide of $t_{ox} = 439$ Å.

In summary, we have discussed various factors important to the development of ultrahigh-speed and ultrasmall-dimension integrated circuits. The need for a self-alignment process to control the gate length is common to MODFETs, MESFETs, and MOSFETs. The recent advances in thin-film growth technology have greatly improved the uniformity of threshold voltage for both MODFETs and MESFETs to a quality level comparable to that for MOSFETs, even though GaAs bulk crystals have a much higher dislocation density and a more pronounced spatial defect-density variation than Si bulk crystals. In other words, the use of MBE-grown or OMCVD-grown films has largely overcome the problem of inferior quality associated with GaAs crystals. However, the problem with the semi-insulating (SI) substrate remains. Although the SI substrate is extremely useful for building millimeter-wave integrated circuits (MMICs), it has a serious drawback for building high-density integrated circuits. A semi-insulating substrate is not an ideal insulator. The filling of traps in a SI substrate can cause a significant increase in the substrate current and hence a significant change in the potential distribution in the substrate. One way to overcome the backgating effect is to use junction isolation for electrical separation of the conduction channel from the SI substrate. For MOSFETs, the integrity of the gate insulator SiO₂ is a major concern. Although hot carriers are present in MOSFETs as well as in MODFETs and MESFETs, they are especially serious in MOSFETs because their injection into SiO_2 can cause degradation of SiO_2 in creating oxide and interface traps.

13.12 MESFET AND MODFET

In Sections 13.10 and 13.11 we discussed the basic requirements for high-speed operation and large-scale integration of FETs. Two different aspects were emphasized: one concerning their response as an individual circuit element and the other concerning their ability to function in an integrated-circuit environment. Our discussion on the former was based on circuit analysis and that on the latter was related to the state of material quality and processing technology. In this section and Section 13.13 we discuss aspects relevant to the physics of FETs of ultrasmall dimensions. Since the current–voltage characteristics of MOSFET and MESFET have been treated extensively in Sections 9.10, 9.11, and 9.13, only those aspects pertaining to the physics of ultrasmall devices are emphasized.

Figure 13.34 shows the structures of (a) a MESFET with self-aligned gate and (b) a MODFET. The MESFET structure consists of a thin epitaxial layer of n-GaAs grown on a semi-insulating (SI) GaAs substrate. Typical values for thickness and doping are 1000 Å and 2×10^{17} cm^{-3}, as indicated. To reduce the parasitic source and drain resistances R_S and R_D, silicon ions are implanted in the source and drain regions, with the gate metal serving as a mask. The self-aligned n^+ regions prevent depletion layers from forming at the surface and reduce the source-to-gate and gate-to-drain distances (Fig. 9.33) practically to zero. In addition to a considerable reduction in R_S and R_D, self-aligned MESFETs are amenable to high integration density because of their simple planar geometry and noncritical gate alignment. The MODFET structure consists of epitaxial layers of undoped GaAs, n-AlGaAs, and n-GaAs, both Si-doped. The two-dimensional electron gas (2DEG; refer to discussion in Section 11.7 in connection with Fig. 11.14) formed in the undoped GaAs at the interface serves as the conduction channel. However, the MODFET processing technology is more complicated than the MESFET processing technology.

The complication in MODFET technology arises from the following considerations. First, ion implantation generally requires subsequent thermal annealing at sufficiently high temperatures, typically around 800°C for 20 min, to remove implant-caused damages. However, during the annealing process, implanted Si ions also diffuse into

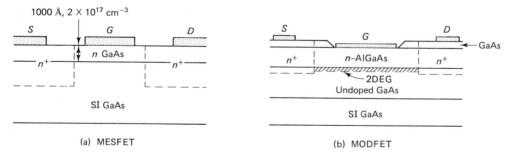

Figure 13.34 Schematic diagram showing (a) the structure of a self-aligned MESFET and (b) the structure of a MODFET.

the undoped 2DEG conduction channel, lowering the electron mobility. Silicon diffusion can cause disordering of GaAs–AlGaAs interface at sufficiently high Si concentration. Therefore, rapid thermal annealing and only moderately high Si doping, typically below 2×10^{18} cm^{-3}, are used to minimize Si diffusion into the 2DEG channel. Second, a recessed gate is generally used, for reasons to be discussed shortly. The recess leaves a finite space between the source and the gate and between the gate and the drain. Therefore, the parasitic source and drain resistances R_S and R_D in MODFETs are not as small as in self-aligned MESFETs. For microwave and analog applications, MESFETs are favored because of simplicity in the processing steps and low parasitic resistances.

Figure 13.35 shows the basic building block for logic circuits. It consists of an enhancement-mode (E-mode) MODFET and a depletion-mode (D-mode) MODFET arranged to function as an inverter. As discussed in Section 11.7, the existence of a 2DEG at a GaAs–AlGaAs interface depends on the thickness of the doped AlGaAs layer (Fig. 11.14). A MODFET with an interface shown in Fig. 11.14a is normally on and thus operates in the D mode. The n-GaAs layer grown on n-AlGaAs serves two purposes. It increases the total dopants in the combined GaAs and AlGaAs layers so that a 2DEG forms. It also facilitates the formation of an ohmic contact. The metal alloy used for ohmic contacts to n-type GaAs is AuGe–Ni, while that for Schottky-barrier contact and for interconnect is TiAl. The thicknesses of and the doping concentrations in the n-GaAs and n-AlGaAs layers are so chosen that with both layers, a MODFET operates in the D mode and with the n-GaAs layer removed, a MODFET operates in the E mode. The energy-band diagrams for the two cases are shown in Fig. 13.36, with (a) corresponding to Fig. 11.14a and (b) corresponding to Fig. 11.14b.

Now let us discuss briefly the mechanism controlling the current in MESFETs and MODFETs. As discussed in Section 9.13, the gate metal forms a Schottky barrier with

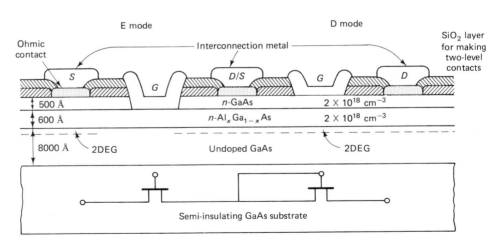

Figure 13.35 Cross-sectional view of a 2-MODFET inverter, with the E-MODFET serving as a switching element and the D-MODFET serving as a load. The metal for ohmic contacts to GaAs is generally Au/Ge–Ni and the metal for gate and interconnect is Ti/Pt/Au or TiAl.

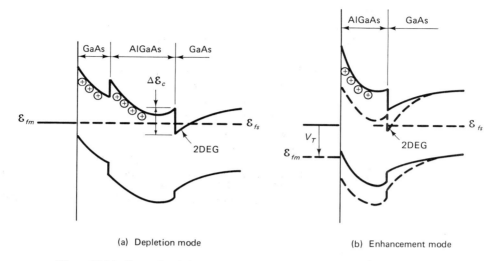

GaAs | AlGaAs | GaAs

$\Delta \mathcal{E}_c$

$\mathcal{E}_{fm}$

$\mathcal{E}_{fs}$

2DEG

(a) Depletion mode

AlGaAs | GaAs

V_T

$\mathcal{E}_{fm}$

$\mathcal{E}_{fs}$

2DEG

(b) Enhancement mode

Figure 13.36 Energy-band diagram for (a) a D-MODFET and (b) an E-MODFET with dashed lines representing the situation under a gate voltage $V_G = V_T$. Without bias, the band bending in case (b) is insufficient to cause the formation of a 2DEG in GaAs.

n-GaAs in a MESFET. When a negative bias is applied to the gate electrode, the width of the space-charge region is increased and the width of the conduction channel is decreased. The situation is very similar to that in a junction FET discussed in Section 9.6, with the Schottky barrier contact serving the function of a p-n junction. The saturation source-to-drain current is given by Eq. (9.13.11). A MODFET, on the other hand, behaves like a MOSFET. Referring to Fig. 11.16, we see that the areal electron density N_s in the 2DEG conduction channel is controlled by the gate voltage. Because the GaAlAs layers in Fig. 13.36 are depleted of free carriers, they can serve the function of an insulator. Therefore, the saturation source-to-drain current I_{DS} is expected to obey the relation

$$I_{DS} = K(V_G - V_T)^2 \qquad (13.12.1)$$

as in Eq. (9.11.14), where $K = \epsilon Z \mu / (2d \, L_G)$ is called the K factor, V_T is the threshold voltage, Z is the gate width, L_G is the gate length, and d is the thickness of the GaAlAs layer. Figure 13.37 shows the measured I_{DS} as a function of gate voltage for both D- and E-MODFETs reported by Abe et al. (M. Abe, T. Mimura, and N. Yokoyama, in A. Goetzberger and M. Zerbst, eds., *Solid State Devices,* Verlag Chemie GmbH, Weinheim, West Germany, 1983, p. 35). Both curves show a K value of 34 mA/V^2. Using an average ϵ value of 10^{-12} F/cm, a total d of 0.11 μm, $Z = 300$ μm, and $L_G = 2$ μm, we find $\mu \sim 6000$ cm^2/V-s for the D-MODFET of Fig. 13.35. We should point out that Eq. (13.12.1) assumes a constant mobility and hence it is valid only for long-channel FETs where a linear velocity-field relation is a reasonable approximation. Therefore, the μ value in the K factor is expected to be lower than the low-field mobility value of 8000 cm^2/V-s.

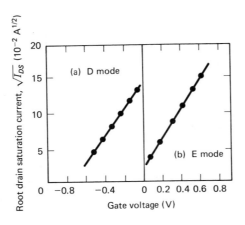

Figure 13.37 Characteristics of the square root of the drain saturation current $\sqrt{I_{DS}}$ versus gate voltage V_G for (a) the D-MODFET and (b) the E-MODFET.

Analysis of Short-Channel MODFETs

Now we consider limitations imposed by device physics and analyze the performance of MODFETs. One important consideration is the short-channel effect ($L_G < 1 \ \mu m$). The drain–source resistance R_{DS} shown in Fig. 13.23 lowers f_{max}, as can be seen from Eq. (13.10.10). We also note from Eq. (9.10.7) that I_D is inversely proportional to the gate length L_G. In deriving the equation, we assumed that L_G was not affected by the extension of the space-charge region at the drain end (Fig. 9.27). This approximation is a good one for a relatively long channel ($L_G > 3 \ \mu m$). When L_G is decreased, the spreading of this space-charge region with increased V_D may occupy a significant part of L_G, thus reducing the effective gate length from L_G to L_G', which is considerably shorter than L_G. The reduced R_{DS} due to nonsaturation of I_{DS} is commonly observed in short-channel FETs. A similar situation occurs in a MODFET when the point of velocity saturation moves into the 2DEG conduction channel. However, there is another phenomenon which is more pronounced in MODFETs than in MESFETs and MOSFETs in the form of a reduced and compressed transconductance g_m. Figure 13.38 shows the variation of g_m computed for a GaAs MODFET at 300 K by Hida et al. (H. Hida, T. Itoh, and K. Ohata, *IEEE Trans. Electron Devices*, Vol. ED-33, p. 1580, 1986). Two possible explanations for the compressed g_m are (1) conduction in the (GaAl)As barrier, due to incomplete depletion of free electrons, in parallel with that in the GaAs channel (K. Lee, M. S. Shur, T. J. Drummond, and H. Morkoç, *IEEE Trans. Electron Devices*, Vol. ED-31, p. 29, 1984) and (2) nonlinear velocity-field relation in the channel and source-to-gate regions (Hida et al.). Here we follow the latter model.

Figure 13.39 shows the source resistance R_S and the average field in the channel $\overline{E}_{CH}$ (dashed curve) and the average field in the source-to-gate region $\overline{E}_{SG}$ (solid curve) as functions of the drain voltage V_D. The curves were computed by Hida et al. for a D-MODFET with a source-to-gate spacing $L_S = 0.5 \ \mu m$, a gate length $L_G = 1 \ \mu m$, a gate-to-drain spacing $L_D = 0.5 \ \mu m$, and a gate width $Z = 200 \ \mu m$. The curves computed for $V_G = 0$ V correspond to the g_m compression region and those for

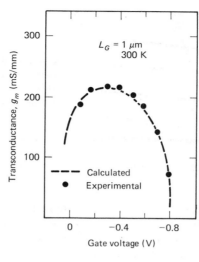

Figure 13.38 Transconductance as a function of gate voltage with dots representing the experimental results and dashed curve representing the theoretical result. (After H. Hida, T. Itoh, and K. Ohata, "A Novel 2DEGFET Model Based on the Parabolic Velocity-Field Curve Approximation," *IEEE Trans. Electron Devices*, Vol. ED-33, p. 1580, 1986, © 1986 IEEE.)

$V_G = 0.4$ V correspond to the normal g_m region in Fig. 13.38. In the analysis by Hida et al., a parabolic velocity relation is used with

$$v(E) = \frac{v_s}{E_s}\left(2 - \frac{E}{E_s}\right)E \qquad \text{for} \quad 0 < E < E_s \qquad (13.12.2)$$

where v_s is the saturation velocity and $v(E) = v_s$ for $E > E_s$. The parameter E_s is the field at and above which $v = v_s$. If we define an effective mobility $\mu_{\text{eff}} = v/E$, then $\mu_{\text{eff}}(0) = 2v_s/E_s$ in the low-field region. Figure 13.40 shows the measured (dashed curves) and calculated (solid curves) drain-current versus drain-voltage characteristics of the 1-μm gate MODFET. We see that good agreement is obtained between the experimental and theoretical curves.

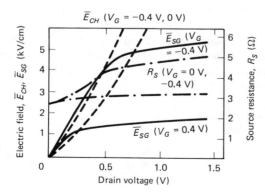

Figure 13.39 Average electric field $\overline{E}_{CH}$ in the channel (dashed lines) and that $\overline{E}_{SG}$ in the source-to-gate region, and source-to-gate resistance (solid lines) computed for a 1-μm gate MODFET. (After H. Hida, T. Itoh, and K. Ohata, "A Novel 2DEGFET Model Based on the Parabolic Velocity-Field Curve Approximation," *IEEE Trans. Electron Devices*, Vol. ED-33, p. 1580, 1986, © 1986 IEEE.)

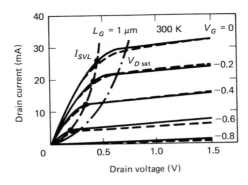

Figure 13.40 Drain-current versus drain-voltage characteristics for the 1-μm gate MODFET: experimental curves (dashed) and calculated characteristics (solid). The curve with x and that with ● represent, respectively, the condition for velocity saturation and that for current saturation, using Eqs. (13.12.10) and $V_{DS} = V_G - V_T$ and assuming $V_S = 0$.

Computing the theoretical curves would require numerical methods. To gain physical insight into the problem, it is best that we use the theoretical model to check the consistency of the computed results instead of trying to reproduce these results. In the source-to-gate region, the drain–source current I_D is

$$I_D = eN_sZv \qquad (13.12.3)$$

Substituting Eq. (13.12.2) in Eq. (13.12.3) and solving for E, we obtain

$$E = E_s\left[1 - \sqrt{1 - \frac{I_D}{eN_sZv_s}}\right] \qquad (13.12.4)$$

where N_s is the areal electron density and Z is the width of the gate. Thus the source resistance R_S is

$$R_S = \frac{E_sL_S}{I_D}\left[1 - \sqrt{1 - \frac{I_D}{eN_sZv_s}}\right] \qquad (13.12.5)$$

which lies between $R_S = R_S(0) = L_S/eN_sZ\mu_{eff}(0)$ in the low-field region and $R_S = 2R_S(0) = E_sL_S/eN_sZv_s$ in the velocity-saturation region. The following values were used in the paper by Hida et al.: $v_s = 1.6 \times 10^7$ cm/s, $\mu_{eff}(0) = 5000$ cm^2/V-s, and $R_S(0) = 3.5$ Ω, yielding $N_s = 9 \times 10^{11}$ cm^{-2} and $E_s = 6.4 \times 10^3$ V/cm. At $V_G = 0$ V and $V_D = 0.5$ V, we find from Fig. 13.40 $I_D = 26$ mA and then from Eq. (13.12.5), $R_S = 4.3$ Ω, in agreement with the value from Fig. 13.39.

We note from Fig. 13.40 that I_D does not saturate but increases gradually even for $V_D > V_{Dsat}$. When the point of $E = E_s$ moves into the 2DEG channel, the gate length is effectively reduced, causing nonsaturation of I_D. In the theoretical model the channel is divided into a velocity-nonsaturation region with a length L_G' and a velocity-saturation region with a length ΔL such that $L_G' + \Delta L = L_G$. In the velocity-nonsaturation region, the current–voltage relation is given by

$$E_sL_G' = \frac{\beta}{8}\left\{\ln\left[\frac{1 - t_1}{1 - t_2}\frac{1 + t_2}{1 + t_1}\right] + 2\left[\left(1 - t_1\right)^{-2} - \left(1 - t_2\right)^{-2} + \left(1 + t_1\right)^{-1} - \left(1 + t_2\right)^{-1}\right]\right\} \qquad (13.12.6)$$

The quantities t_1 and t_2 are given by

$$t_1 = \sqrt{1 - \frac{\beta}{U_S}}, \qquad t_2 = \sqrt{1 - \frac{\beta}{U_D}} \qquad (13.12.7)$$

where $\beta = I_D/C_G Z v_s$, $U_S = V_G - V_T - V_S$, $U_D = V_G - V_T - V_D^*$, C_G is the gate capacitance per unit area, V_S is the source voltage, and V_D^* is the potential at a point in the channel where $E(x) = E_s$. In the velocity-saturation region, the potential drop is given by

$$V_D - V_D^* = \left(\frac{e}{2\epsilon}\right)\left(N_a + \frac{N_s}{d_{2D}}\right)(\Delta L)^2 + E_s \, \Delta L \qquad (13.12.8)$$

where N_a represents the background acceptor concentration, due mostly to unintentionally doped carbon, $N_s = I_D/eZv_s$ is the areal carrier density, and d_{2D} is the depth of the 2DEG conduction channel.

If no point in the channel has reached the state of velocity saturation, $L_G' = L_G$ and $V_D^* = V_D$. Furthermore, if we assume β/U to be a small quantity, $1 - t$ can be approximated by $\beta/2U$. If we keep only the $(1 - t)^{-2}$ terms, Eq. (13.12.8) can be rearranged to yield

$$I_{DS} = \frac{v_s}{E_s}\frac{Z}{L_G} C_G(2V_G - 2V_T - V_D - V_S)(V_D - V_S) \qquad (13.12.9)$$

which is identical to Eq. (9.11.12) by noting that $C_G = \epsilon'/d$, $2v_s/E_s = \mu_{eff}(0)$ by definition, and $V_S = 0$. Therefore, the condition $\beta \ll U$ is an important one and can be rewritten as

$$I_D \ll I_{SVL} = C_G Z v_s [V_G - V_T - V(x)] \qquad (13.12.10)$$

where I_{SVL} is a current parameter, and C_G has a value of 2.2×10^{-7} F/cm^2.

Note that $C_G[V_G - V_T - V(x)]$ is the charge concentration per unit area (C/cm^2) in the channel. If they move with a saturation velocity v_s, they contribute a current I_{SVL} which we call the saturation-velocity-limited current. We also note from Fig. 13.38 that the device has a threshold voltage $V_T = -0.83$ V. Therefore, at $V_G = 0$ V and $V_D = 0.45$ V, we find $I_{SVL} = 26.4$ mA, which is very close to the value of $I_{DS} = 26$ mA found in Fig. 13.40. The line marked by $\times$ is the I_{SVL} curve as a function of V_D. The voltage drop in the source-to-gate region is $V_S = (R_S + R_C)I_D = \overline{E_{SG}}L_S$, including the drop at contact resistance R_C. Using $\overline{E_{SG}} = 4.4 \times 10^3$ V/cm, we find $V_S = 0.22$ V. From Eq. (13.12.9), we obtain $I_D = 66$ mA if $\mu_{eff} = 5000$ cm^2/V-s is used and $I_D = 33$ mA if $\mu_{eff}(E_s) = 2500$ cm^2/V-s is used. It is obvious that Eq. (13.12.9) is no longer useful even if we replace $2v_s/E_s$ by an effective mobility μ_{eff}. The validity of Eq. (13.12.9) is based on the condition $\beta \ll U$. In a short-channel MODFET, this condition is violated because velocity saturation occurs even at low and moderate V_D.

To the right of the I_{SVL} curve in Fig. 13.40 a point exists in the channel at which $E = E_s$. The I_D–V_D characteristics in this region must be calculated from Eqs. (13.12.6) and (13.12.8). At $V_G = -0.4$ V and $V_D = 1.0$ V we find $I_D = 14$ mA from Fig. 13.40. Since the case is already in the velocity-saturation region, $I_D = I_{SVL}$ at $V_D = V_D^*$ in Eq. (13.12.10), yielding $V_D^* = 0.25$ V and $N_s = 2.8 \times 10^{11}$ cm^{-2}. From Fig. 11.10 we see that the electron wave function in a 2DEG spreads over 60 Å at

$N_s = 10^{12}$ cm^{-2}. A smaller areal density means a shallower potential and thus a wider spread. We assume a spread of 80 Å. Thus we use a value of 3.5×10^{17} cm^{-3} for N_s/d_{2D}. This high value makes it the dominant term in Eq. (13.12.8). Using a value $V_D - V_D^* = 0.75$ V, we find $\Delta L = 0.052$ μm and $L_G' = 0.948$ μm in Eq. (13.12.6). The value of I_D also yields a value of 0.175 for $\beta = I_D/C_G Z v_s$. The point at which $E = E_s$ is the point where $V = V_D^*$, by definition. It is also the point where $t_2 = 0$ in Eq. (13.12.7). The reader is asked to show that Eq. (13.12.6) is satisfied with $t_1 = 0.74$. This value of t_1 yields a value of 0.04 V for V_S versus a value of 0.06 V from $\overline{E}_{SG}$ in Fig. 13.40.

In summary, in short-channel MODFETs, velocity saturation plays an important role in determining the I_D versus V_D characteristics. In Fig. 13.40 we show two curves: the I_{SVL} curve, which is the locus for the condition $E = E_s$, and the V_{Dsat} curve, which is the locus for $N_s = 0$, both at the drain end of the gate. For the 1-μm gate MODFET, the I_{SVL} curve occurs at a lower drain voltage V_D than the V_{Dsat} curve, meaning that velocity saturation occurs before drain current saturates. The nonlinear dependence of electron velocity on field is expected to be even more important in MODFETs of shorter gate length. The finite slope of the I_D versus V_D curve in the saturation region is partially due to the fact that the point $E = E_s$ moves inside the 2DEG conduction channel, thus shortening the effective gate length L_G'. The slope becomes larger and hence the resistance R_{DS} in Fig. 13.23 becomes smaller in MODFETs of shorter gate length. Although Eq. (13.12.9) is not accurate for quantitative analysis, it is still useful qualitatively. The g_m compression shown in Fig. 13.38 is due to a substantial increase in V_S due to an increased I_D. According to Eqs. (9.11.15) and (9.13.12), the transconductance g_m is inversely proportional to the gate length. This relation holds only in long-channel FETs where the voltage drop across parasitic elements, such as V_S across R_S, can be neglected. Figure 13.41 shows the measured (dots) and computed (dashed line) g_m for D-MODFETs with different gate lengths. The g_m dependence on L_G is much weaker than the theoretical L_G^{-1} dependence, especially in short-channel MODFETs. Reducing R_S is important not only for reducing noise but also for increasing g_m.

Model for Device Simulation

The analytical model presented in the preceding subsection provides physical insight into the short-channel effect. However, the model is too complicated to use for device simulation even though the I_D–V_D characteristics of a MODFET can be predicted, in

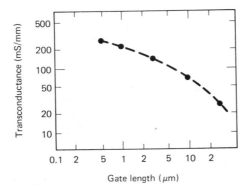

Figure 13.41 Maximum transconductance as a function of gate length of various D-MODFETs: experimental data (dots) and calculated curves (dashed). (After H. Hida, T. Itoh, and K. Ohata, "A Novel 2DEGFET Model Based on the Parabolic Velocity-Field Curve Approximation," *IEEE Trans. Electron Devices*, Vol. ED-33, p. 1580, 1986, © 1986 IEEE.)

principle, from Eqs. (13.12.6) and (13.12.8). Therefore, a semiempirical approach is often adopted. One important physical quantity in device modeling is the velocity dependence on applied field. In the empirical model used by Hida et al. (H. Hida, T. Itoh, and K. Ohata, *IEEE Electron Device Lett.*, Vol. EDL-7, p. 393, 1986), the following velocity expression is used:

$$v(E) = \frac{\mu_0 E}{1 + \alpha(E/E_C) + \beta(E_{av}/E_C)^2} \tag{13.12.11}$$

Using Eq. (13.12.11) in the current equation

$$I_D = \frac{Z\epsilon}{d}[V_G - V_T - V(x)]v(E) \tag{13.12.12}$$

and integrating the resultant equation, we obtain

$$I_D = \frac{\epsilon\mu_0 Z}{dL_G} \frac{(V_G - V_T - V_D/2)V_D - (V_G - V_T - V_0)^2/2}{1 + \alpha(V_D/E_C L_G) + \beta(V_D/E_C L_G)^2} \tag{13.12.13}$$

The quantities ϵ, d, Z, and L_G have the same meaning as those in Eq. (13.12.1). The quantities μ_0, α, β, E_C, E_{av}, and V_0 are parameters used to fit the empirical model to the I_D versus V_D characteristics.

We should point out that the I_D versus V_D relation derived in Sections 9.11 and 9.13 represents the terminal characteristic of a FET. Its derivation does not require precise knowledge of the field variation along the length of the channel. In the same spirit, Eq. (13.12.11) is intended to describe the dependence of an overall velocity v on the drain voltage V_D and not to fit the v_d versus E curve of Fig. 10.25. In the overall picture, the negative differential mobility part $(dv_d/dE = -)$ of Fig. 10.25 is not explicitly represented in Eq. (13.12.11), and its effect on $v(E)$ is incorporated in Eq. (13.12.11) by using a value for the low-field mobility μ_0 considerably smaller than the initial slope (8000 cm^2/V-s) of the v_d versus E curve in Fig. 10.25. The quantity E_C is defined as $E_C = v_s/\mu_0$ and the quantity E_{av} as $E_{av} = V_D/L_G$, where v_s is the saturation velocity.

The modeling of a MODFET is complicated by the fact that when the gate voltage V_G exceeds a certain value, a conduction channel forms in the AlGaAs layer in parallel with the 2DEG. Referring to Fig. 13.36, we see that applying a positive V_G reduces the separation between the conduction-band edge in AlGaAs and the Fermi level. A situation eventually develops in which not only will there be electrons in the AlGaAs layer but also most induced electrons go there, because of their close proximity to the gate. The gradual saturation of the areal electron density in the 2DEG and the onset of the areal electron density in the AlGaAs layer are shown in Fig. 11.16 as curve (a) and curve (b), respectively. The quantity V_0 in Eq. (13.12.13) is the value of $V_G - V_T$ at the onset of parallel conduction in the AlGaAs layer.

The drain-to-source current I_D in Eq. (13.12.13) represents only the current carried in the 2DEG, to which the current in the AlGaAs layer is added to yield the total current. The following values were used by Hida et al. for the various parameters: $v_s = 1.5 \times 10^7$ cm/s, $\mu_0 = 3000$ cm^2/V-s, $\alpha = 0.21$ and $\beta = 0.26$ for the 2DEG, and $v_s = 0.6 \times 10^7$ cm/s and $\mu_0 = 400$ cm^2/V-s in the AlGaAs layer to fit the experimental I_D versus V_D curves of an E-MODFET with $d = 300$ Å and $V_T = 0.06$ V from $V_G = 0.2$ V to $V_G = 1$ V. Excellent agreement between the measured and calculated curves was obtained. We emphasize again that the main purpose of the empirical model is to facilitate modeling the device without resorting to the use of excessive computer time. However, success with Eq. (13.12.11) does not mean that it is the only reasonable way to model the behavior of $v(E)$.

The following dependence

$$v(E) = \frac{\mu_0 E + v_{seff}(E/E_C)^4}{1 + (E/E_C)^4} \qquad (13.12.14)$$

with $\mu_0 = 2800$ cm^2/V-s, $v_{seff} = 2.5 \times 10^7$ cm/s, and $E_C = v_{seff}/\mu_0$ was used by Yamasaki and Hirayama in their simulation study of a self-aligned MESFET, and excellent agreement with the measured I–V characteristics was also obtained (K. Yamasaki and M. Hirayama, *IEEE Trans. Electron Devices,* Vol. ED-33, p. 1652, 1986). The reader is also referred to the article by Snowden and Loret on a two-dimensional numerical simulation of a short-gate GaAs MESFET. The analysis involves the numerical solution of the current-continuity, momentum-conservation, and energy-conservation equations (Section 10.9) derived from the Boltzmann transport equation, and the two-dimensional mapping of the field based on Poisson's equation (C. M. Snowden and C. Loret, *IEEE Trans. Electron Devices,* Vol. ED-34, p. 212, 1987). Therefore, device modeling can range from completely numerical to partly analytical and partly numerical analyses of the basic device equations. The former approach is favored in understanding device physics, while the latter approach is preferred in circuit analysis.

Nonstationary Carrier Dynamics—Velocity Overshoot

The drift velocity versus field curve (dashed) shown in Fig. 10.25 is obtained from the steady-state solution of the Boltzmann transport equation. The steady-state characteristic is applicable to device analysis as long as the transit time τ_t is longer than the relaxation time τ. As the gate length of a FET is reduced, the ratio τ_t/τ is also reduced. It is interesting, therefore, for us to examine whether the state-steady solution is appropriate for submicrometer devices. Here we follow the procedure used by Carnez et al. to set up the time-dependent equations for momentum conservation and energy conservation (B. Carnez, A. Cappy, A. Kaszynski, E. Constant, and G. Salmer, *J. Appl. Phys.,* Vol. 51, p. 784, 1980).

We start with the momentum equation

$$\frac{d(m^*v)}{dt} = eE - \frac{m^*v}{\tau} \qquad (13.12.15)$$

and the energy equation

$$\frac{d\mathscr{E}}{dt} = eEv - \frac{\mathscr{E} - \mathscr{E}_0}{\tau_\mathscr{E}} \qquad (13.12.16)$$

where $\mathscr{E}_0$ is the thermal-equilibrium energy and $\tau_\mathscr{E}$ is the energy relaxation time. The quantities m^*, τ, and $\tau_\mathscr{E}$ all depend on the energy. Since the steady-state solutions for v and $\mathscr{E}$ are known functions of E (Figs. 10.25 and 10.24) and the steady-state value of m^* can be computed from the $n_1/(n_1 + n_2)$ curve of Fig. 10.24, it is natural that we express τ and $\tau_\mathscr{E}$ in terms of v_{ss}, $\mathscr{E}_{ss}$, and m^*_{ss}, where the subscript ss refers to the steady-state value at a given E. Figure 13.42 shows the values of v_{ss}, $\mathscr{E}_{ss}$, and m^*_{ss} as functions of E.

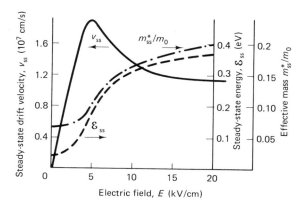

Figure 13.42 Curves showing the steady-state values of the drift velocity v_{ss}, the energy $\mathscr{E}_{ss}$, and the effective mass m_{ss}^* as functions of applied field E in GaAs. The changes are caused by a redistribution of electrons among the central and satellite valleys, that is by decreasing ratio of $n_1/(n_1 + n_2)$ in Figure 10.24. (After B. Carnez, A. Cappy, A. Kaszynski, E. Constant, and G. Salmer, *J. Appl. Phys.*, Vol. 51, p. 784, 1980.)

The steady-state solutions of Eqs. (13.12.15) and (13.12.16) are

$$\tau = \frac{m_{ss}^* v_{ss}}{e E_{ss}}, \qquad \tau_{\mathscr{E}} = \frac{\mathscr{E}'}{e E_{ss} v_{ss}} \qquad (13.12.17)$$

Using Eq. (13.12.17) in Eqs. (13.12.15) and (13.12.16), we find that

$$\frac{d(m^* v)}{dt} = e\left(E - E_{ss}\frac{m^* v}{m_{ss}^* v_{ss}}\right) \qquad (13.12.18)$$

$$\frac{d\mathscr{E}'}{dt} = e(Ev - E_{ss} v_{ss}) \qquad (13.12.19)$$

where $\mathscr{E}' = \mathscr{E} - \mathscr{E}_0$. Note that the quantities v and m^* are functions of E because the ratio $n_1/(n_1 + n_2)$, that is, the fraction of electrons remaining in the central valley, changes with $\mathscr{E}$, which is a function of E. It is obvious that energy $\mathscr{E}$ is the main quantity of concern. Both v_{ss} and m_{ss}^* change with $\mathscr{E}$ and so does $E_{ss}(\mathscr{E})$, the steady-state value of E to reach a given $\mathscr{E}$.

Figure 13.43 shows the time response of the drift velocity to a pulse field E,

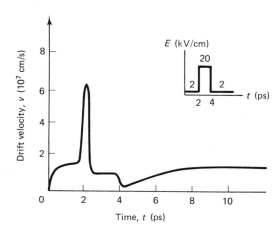

Figure 13.43 Time variation of the drift velocity in GaAs as computed from Eqs. (13.12.20) and (13.12.21) in response to an electric field shown in the inset. (After B. Carnez, A. Cappy, A. Kaszynski, E. Constant, and G. Salmer, *J. Appl. Phys.*, Vol. 51, p. 784, 1980.)

Sec. 13.12   MESFET and MODFET

obtained from a numerical solution of Eqs. (13.12.18) and (13.12.19). Here we give a qualitative explanation of the computed curve. Note that the first $E_1 = 2$ kV/cm pulse is in the low-field region, and no electron transfer occurs during this part of the field. Since m^* is a constant, the solution of Eq. (13.12.15) is simply

$$v = \frac{e\tau E_1}{m^*}\left[1 - \exp\left(\frac{-t}{\tau}\right)\right] \tag{13.12.20}$$

Therefore, v approaches the steady-state value of 1.7×10^6 cm/s with a time constant $\tau = 3.3 \times 10^{-13}$ s. At $t = 2 \times 10^{-12}$ s, v is already at the steady-state value.

The initial response to the $E_2 = 20$ kV/cm pulse can be described as

$$v = \frac{e\tau E_1}{m^*} + \frac{e\tau(E_2 - E_1)}{m^*}\left[1 - \exp\left(-\frac{t}{\tau}\right)\right]\exp\left(-\frac{t}{\tau_{12}}\right) \tag{13.12.21}$$

where τ_{12} is the intervally scattering time from the central valley to the satellite valleys. Since the mobility of the satellite valley is much lower than that of the central valley, we may ignore their contribution to v. In this approximation, m^* is entirely constituted by the mass of the central-valley electrons and thus can be treated as a constant. However, the population of the central valley is reduced by scattering into the satellite valley. This fact is accounted for by the factor $\exp(-t/\tau_{12})$. The velocity v of Eq. (13.12.21) reaches a maximum value

$$v_{\text{max}} = \frac{e\tau}{m^*}\left[E_1 + (E_2 - E_1)\frac{\alpha}{(1 + \alpha)^{1 + 1/\alpha}}\right] \tag{13.12.22}$$

at a time $t_{\text{max}} = \tau \ln(1 + \alpha)$, where $\alpha = \tau_{12}/\tau$. From fitting v_{max} and t_{max} to the values obtained from Fig. 13.43, we find $\alpha = 1.3$ yielding a value of $\tau_{12} = 4.5 \times 10^{-13}$ s. If we use $m^* = 0.55 m_0$ as the effective density-of-state mass for the satellite valley, the ratio B of the density of states N_{c2}/N_{c1} is 22 and the scattering time $\tau_{21} = B\tau_{12}$ for satellite valley electrons to be scattered into the central valley is 9.9×10^{-12} s, in reasonable agreement with the generally accepted value.

We should again point out that Eq. (13.12.21) is useful only for a qualitative description of the nonstationary behavior of hot electrons under a high-field pulse. Numerical solutions of Eqs. (13.12.18) and (13.12.19) are needed to obtain quantitative results. Excellent agreement was reported by Carnez et al. between computed results based on the two equations and simulation results of $v(t)$ based on Monte Carlo method. Note that during the initial part (about 2×10^{-13} s) of the 20 kV/cm pulse, the drift velocity far exceeds its stationary value of about 10^7 cm/s. This phenomenon is known as *velocity overshoot*. The drift-velocity versus time curve can be translated into a drift-velocity versus distance curve which can be used to generate device *I–V* characteristics. The simulation results of Carnez et al. indicate a significant increase of I_D due to velocity overshoot in FETs of submicrometer dimensions. Earlier, Kratzer and Frey computed the carrier velocity as a function of distance by Monte Carlo method. As carriers travel in GaAs under a constant, high field, they start from thermal velocity, and then gain velocity in excess of the stationary value before they settle with the stationary velocity. The distance over which velocity overshoot is significant is again of submicrometer dimensions (S. Kratzer and J. Frey, *J. Appl. Phys.*, Vol. 49, p. 4064, 1978). Recently, Yamasaki and Hirayama reported experimental determination of v_s in FET structures with different gate length and found v_s to vary from 1.7×10^7 cm/s for $L_G = 1$ μm to 2.7×10^7 cm/s for $L_G = 0.3$ μm. Therefore, velocity overshoot is expected to enhance the high-speed performance of GaAs devices of submicrometer dimensions.

Inverted Interface and DX Center

In Section 13.12 we have discussed two effects, the threshold-voltage variation and the backgating effect, which will limit the scale of integration for both MESFETs and MODFETs. Here we mention two additional problems or phenomena, the origin of which is still unresolved. These problems are important to MODFETs because they are related to (GaAl)As. One problem concerns the inverted interface between a (GaAl)As film and a GaAs film grown on it. Inverted modulation doped heterostructures have low electron mobility (4×10^4 cm²/V-s at 10 K as compared to 10^6 cm²/V-s in normal MOD heterostructure) and MODFETs with inverted interface have low mutual transconductance (70 mS/mm versus 300 mS/mm in normal MODFETs) (H. Morkoç, T. J. Drummond, R. E. Thorne, and W. Kopp, *Jpn. J. Appl. Phys.*, Vol. 20, p. L913, 1981; R. E. Thorne, R. Fischer, S. L. Su, W. Kopp, T. J. Drummond, and H. Morkoç, *Jpn. J. Appl. Phys.*, Vol. 21, p. L223, 1982). The origin of this inferior performance is not understood, although the use of a thin superlattice did raise the electron mobility (T. J. Drummond, J. Kelm, D. Arnold, R. Fischer, R. E. Thorne, W. G. Lyons, and H. Morkoç, *Appl. Phys. Lett.*, Vol. 42, p. 615, 1983). The other problem is related to deep traps in (AlGa)As, and the effect is known as *persistent photoconductivity*. Because of an exceedingly small electron-capture cross section ($\sim 10^{-20}$ cm²) at low temperatures (<100 K), optically excited electrons from these traps are not readily recaptured by the traps and hence the photoconductive effect persists for a long time (D. V. Lang, R. A. Logan, and M. Jaros, *Phys. Rev.*, Vol. B19, p. 1015, 1979; B. L. Zhou, K. Ploog, E. Gmelin, X. Q. Zheng, and M. Shulz, *Appl. Phys.*, Vol. A28, p. 223, 1982). The phenomenon is believed to be caused by a donor-related deep trap in $Al_xGa_{1-x}As$ if x exceeds a certain percentage. Such a trap is known as the DX center (H. Kunzel, K. Ploog, K. Kunstel, and B. L. Zhou, *J. Electron. Mater.*, Vol. 13, p. 281, 1984). One possibility is that Si forms a donor complex in association with a deep-level impurity or a lattice defect. Another possibility is the result of changing donor wave function. A donor, such as Si and Sn, possesses two energy states: $\mathscr{E}_1(\Gamma)$, associated with the Γ minimum, and $\mathscr{E}_1(X, L)$, associated with the X or L minima. As the Al content x in $Ga_{1-x}Al_xAs$ increases, the conduction-band minimum changes from Γ minimum to X minima (Fig. 11.8). What is unexpected, however, is that the donor ionization energy does not change monotonically from $\mathscr{E}_1(\Gamma)$ to $\mathscr{E}_1(X)$.

Figure 13.44 shows the calculated ionization energies (solid curves) $\mathscr{E}_1(\Gamma)$, $\mathscr{E}_1(X)$, and $\mathscr{E}_1(L)$ for donors and $\mathscr{E}_1(V)$ for acceptors from the hydrogenic model and the measured donor ionization energy from photoluminescence as compared by Adachi (S. Adachi, *J. Appl. Phys.*, Vol. 58, p. R1, August 1985). The abnormal behavior between $x = 0.3$ to $x = 0.7$ cannot be explained by the effective-mass theory alone. The range of x corresponds to the composition region where crossover of the conduction-band minima occurs, first between Γ and L minima and then between L and X minima. An explanation of the behavior of donor ionization energy in the crossover region would require an appreciable coupling between the Γ, L, and X bands (G. G. Kleiman and M. Francostoro-Decker, *Phys. Rev.*, Vol. B21, p. 3478, 1980; A. K. Saxena, *Appl. Phys.*

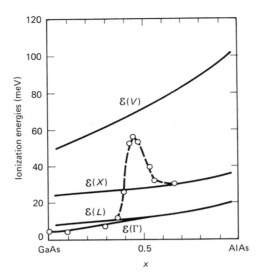

Figure 13.44 Computed curves of the ionization energies for donors, $\mathscr{E}(\Gamma)$, $\mathscr{E}(X)$, and $\mathscr{E}(L)$ and for acceptors $\mathscr{E}(V)$ as functions of the composition x in $Al_xGa_{1-x}As$. The open circles represent the donor-ionization energy deduced from photoluminescence data. (After S. Adachi, *J. Appl. Phys.*, Vol. 58, p. R1, August 1985.)

Lett., Vol. 86, p. 79, 1980). Applying the tight-binding approximation to a deep-level donor, Saxena has obtained the following expression for the donor energy $\mathscr{E}_{DX}$:

$$\mathscr{E}_{DX} = \tfrac{1}{2}(\mathscr{E}_L + \mathscr{E}_X) - \tfrac{1}{2}[(\mathscr{E}_L - \mathscr{E}_X)^2 + 4V^2]^{1/2} \qquad (13.12.23)$$

where $\mathscr{E}_L$ and $\mathscr{E}_X$ represent the band-extrema energies measured from the top of the valence band, and V is the interaction potential between the L and X minima. In the crossover region where the difference $\mathscr{E}_\Gamma - \mathscr{E}_X$ is smaller than V, the donor energy is decreased by an amount V. This result is in qualitative agreement with the behavior of $\mathscr{E}_1(\Gamma) = \mathscr{E}_{DX} - \mathscr{E}_\Gamma$, and so on, shown in Fig. 13.44. If not resolved, problems with inverted interface and with the DX center will impose restrictions on the structure and performance of MODFETs.

13.13 MOSFET

Since its inception, the integrated-circuit (IC) technology has been continually making remarkable progress in reducing the physical dimensions of IC devices. Much of the success in device miniaturization is derived from the scalability of device structures to small dimensions. The planar geometry of MOSFETs makes scaling rules particularly simple, which in turn render possible rapid progress toward device miniaturization. In this section we present scaling rules for Si-MOSFETs and examine effects and issues relevant to devices of ultrasmall physical dimensions. In setting up scaling rules, special considerations are given to the electric field and the voltage swing, the former for minimizing hot-carrier effects and the latter for having a safe margin for errorless digital operation. Specific subjects to be discussed include short-channel and narrow-channel effects, high-field effects, and channel mobility.

Scaling Rules for MOSFET Miniaturization

In the last decade we have witnessed a continual and dramatic increase in the packing density of devices in integrated circuits (IC) as a result of device miniaturization. A set of rules, generally known as *scaling rules,* has been formulated to guide us in the choice of material parameters as well as physical dimensions of IC devices. The scaling analysis serves two important purposes: to ensure that the underlying device physics is not significantly changed, and to foresee problems that we may encounter in further device miniaturization. In the following discussion we first present the linear scaling rules to physical quantities that scale linearly with potential and dimension, and then discuss the effects on device performance resulting from physical quantities that do not follow the linear scaling rules.

The two basic equations used in analyzing MOSFET physics are Poisson's equation and the current-density equation:

$$\frac{\partial^2 \phi}{\partial x^2} + \frac{\partial^2 \phi}{\partial y^2} + \frac{\partial^2 \phi}{\partial z^2} = \frac{e}{\epsilon_s}(N_d - N_a + p - n) \tag{13.13.1}$$

$$J_n = -e\mu_n n \text{ grad } \phi + eD_n \text{ grad } n \tag{13.13.2}$$

where ϕ is the electrostatic potential. First let us consider the simplified case for which n as a result of J_n is much smaller than the impurity concentration, so Eq. (13.13.1) can be decoupled from Eq. (13.13.2). If we reduce the device dimension by a scaling factor λ and the potential by another scaling factor κ by letting

$$(x', y', z') = \frac{x, y, z}{\lambda} \quad \text{and} \quad \phi' = \frac{\phi}{\kappa} \tag{13.13.3}$$

then in terms of scaled variables, Eq. (13.13.1) becomes

$$\frac{\partial^2 \phi'}{\partial x'^2} + \frac{\partial^2 \phi'}{\partial y'^2} + \frac{\partial^2 \phi'}{\partial z'^2} = \frac{e}{\epsilon_s}(N_d - N_a + p - n)\frac{\lambda^2}{\kappa} \tag{13.13.4}$$

Equation (13.13.4) is identical to Eq. (13.13.1) if we make

$$(N_d' - N_a' + p' - n') = \frac{(N_d - N_a + p - n)\lambda^2}{\kappa} \tag{13.13.5}$$

Note that for $\lambda = \kappa$, the electric field $\mathbf{E} = -\nabla\phi$ is kept the same in the scaled MOSFET as in the unscaled MOSFET if the impurity concentration is increased by κ. By so doing, we can keep hot-carrier effects, such as velocity saturation and carrier multiplication, in check in the scaled-down device. Constant-field scaling, proposed by Dennard et al., has been used in establishing the 1-μm FET technology from the 5-μm FET technology (R. H. Dennard, F. H. Gaensslen, H. N. Yu, V. L. Rideout, E. Bassous, and A. LeBlanc, *IEEE J. Solid-State Circuits,* Vol. SC-9, p. 256, 1974; H. N. Yu, A. Riesman, C. M. Osburn, and D. L. Critchlow, *IEEE Trans. Electron Devices,* Vol. ED-26, p. 318, 1979; R. H. Dennard, F. H. Gaensslen, E. J. Walker, and P. W. Cook, *IEEE J. Solid-State Circuits,* Vol. SC-14, p. 247, 1979). However, further miniaturization of device dimension to $\frac{1}{4}$-μm gate length requires a relaxation of the constant-field requirement. In an analysis by Baccarani et al., two different scaling factors, λ for the linear dimension and κ for the potential, were proposed (G. Baccarani, M. R. Wordeman, and R. H. Dennard, *IEEE Trans. Electron Devices,* Vol. ED-31, p. 452, 1984). Table 13.2 lists the scaling factors associated with the most important physical quantities for the general case $\lambda \neq \kappa$. We should point out that the drain current of Eq. (9.11.12) is based on the channel-charge density of Eq. (9.11.10). In deriving Eq.

TABLE 13.2 SCALING FACTORS FOR IMPORTANT PHYSICAL QUANTITIES IN MOSFETS

Quantity[a]	Symbol	Scaling factor
Linear dimension	L_G, Z, d_{ox}	$1/\lambda$
Potential	ϕ, V_S, V_G, V_D	$1/\kappa$
Impurity concentration	N_a, N_d	λ^2/κ
Electric field	E	λ/κ
Capacitance	C_{ox}	$1/\lambda$
Current*	I_{DS} of Eq. (9.11.12)	λ/κ^2
Power*	$I_D V_{DD} \sim P$	λ/κ^3
Power density*	$I_D V_{DD}/A$	λ^3/κ^3
Power delay*	τ_d of Eq. (13.10.11)	κ/λ^2
Power–delay product	$P\tau_d$ of Eq. (13.10.12)	$1/\lambda\kappa^2$
Line resistance	R	λ
Time constant	RC	1

[a]The current expression given in Eq. (9.11.14) assumes a constant mobility and hence the drift velocity scales as λ/κ. The scaling factor for quantities with an asterisk should be multiplied by κ/λ, changing for example to $1/\kappa$ for current, if a MOSFET operates in the regime of velocity saturation which may happen at 77 K. In the latter case, Eq. (9.12.2) is used for I_D instead of Eq. (9.11.12).

(9.11.10) from Eq. (9.11.8), we have assumed a constant inversion-layer thickness. Figure 13.45 shows a comparison of numerically calculated (solid curve) and experimentally measured (dots) values of the areal channel-charge density as a function of the gate voltage for a MOSFET with an oxide thickness $d_{ox} = 10.3$ nm, as reported by Baccarani et al. The experimental and theoretical values agree well not only with each other but also with the approximate linear relation of ρ_c with V_G as stated in Eq. (9.11.7). The good agreement assures us the applicability of the analysis presented in Section 9.11 to MOSFETs scaled down to submicrometer gate length. The scaling factor λ/κ^2 for current applies only to the operation region with biases above threshold. It further assumes nonsaturation of carrier velocity, that is, a constant carrier mobility in Eq. (9.11.12).

Short-Channel and Narrow-Channel Effects on Threshold Voltage

In the analysis presented in Section 9.11, we have assumed a MOSFET with well-defined channel length L_G and width Z. As a result of this assumption, the threshold voltage V_T of Eq. (9.11.9) depends only on the material properties, such as the flat-band voltage V_{FB}, the surface-inversion potential ϕ_{SI}, and the substrate doping N_d for P-MOS or N_a for N-MOS. As we reduce the physical dimensions of a MOSFET, the assumption of constant L_G and Z is no longer valid. Figure 13.46 illustrates (a) the short-channel effect and (b) the narrow-channel effect. Because of the space-charge region extending into the channel region in the y direction and outside the channel region in the z region, the effective gate length L_G is shortened from L to $L_1 = (L + L')/2$ and the effective gate width is increased from Z to $Z_1 = Z + \pi x_d/2$. To

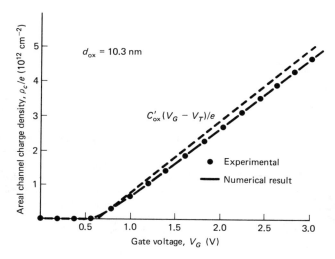

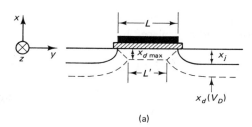

Figure 13.45 Comparison of experimental and computed values of areal channel-charge density ρ_c as a function of the gate voltage for a MOSFET with an oxide thickness $d_{ox} = 10.3$ nm. The dashed line represents the approximate linear relation of Eq. (9.11.7), from which the current I_D of Eq. (9.11.12) is derived. (G. Baccarani, M. R. Wordeman, and R. H. Dennard, "Generalized Scaling Theory and Its Application to a ¼ Micrometer MOSFET Design," *IEEE Trans. Electron Devices,* Vol. ED-31, p. 452, 1984, © 1984 IEEE.) Note that the curves are for a N-MOSFET, whereas the analysis given in Sec. 9.11 is for a P-MOSFET. Therefore, appropriate changes in the equations must be made.

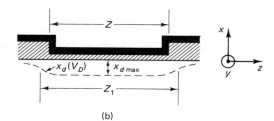

Figure 13.46 Schematic diagrams illustrating (a) short-channel effect and (b) narrow-channel effect as a result of space-charge region extending into and outside the channel region, respectively. The symbols x_j and x_d denote, respectively, the junction depth and depletion width.

reach the condition for surface inversion, the free carriers in the volume $x_{dmax} \times Z_1 \times (L + L')/2$ must be removed from the interface region, where x_{dmax} is given by

$$x_{dmax} = \left(\frac{2\epsilon\phi_{SI}}{eN_{a,d}}\right)^{1/2} \tag{13.13.6}$$

For a MOSFET of a constant gate length L and a constant gate width Z, the charges removed from the interface region is $eN_{a,d}x_{dmax}LZ$, resulting in a surface density $\rho_s = (2e\epsilon N_{a,d}\phi_{SI})^{1/2}$, which appears as the last term in V_T of

Sec. 13.13   MOSFET

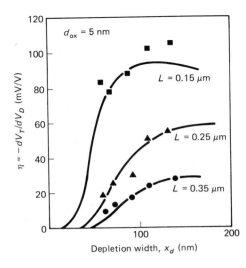

Figure 13.47 Computed values of drain-induced barrier-lowering coefficient, $\eta = -dV_T/dV_d$, as a function of depletion width x_d. (G. Baccarani, M. R. Wordeman, and R. H. Dennard, "Generalized Scaling Theory and Its Application to a ¼ Micrometer MOSFET Design," *IEEE Trans. Electron Devices*, Vol. ED-31, p. 452, 1984, © 1984 IEEE.)

Because of the presence of the drain junction, part of the induced electric flux terminates at the drain. As a result, Eq. (9.11.6) should be replaced by

$$V_T = V_{\mathrm{FB}} - \phi_{\mathrm{SI}} - \frac{1}{C'_{\mathrm{ox}}} \frac{L_1 Z_1}{LZ} (2e\epsilon N_{a,d}\phi_{\mathrm{SI}})^{1/2} \qquad (13.13.7)$$

Note that both L_1 and Z_1 have a slight dependence on drain voltage V_D through the depletion width x_d of the drain junction. A quantity η that measures the threshold-voltage dependence on drain voltage according to

$$\eta = -\frac{dV_T}{dV_D} \qquad (13.13.8)$$

is generally known as a drain-induced barrier-lowering coefficient. Since Z is much bigger than L, the short-channel effect is more important than the narrow-channel effect.

Figure 13.47 shows the calculated values of η for MOSFETs with an oxide thickness $d_{\mathrm{ox}} = 5$ nm and different gate lengths $L = 0.15$ μm, 0.25 μm, and 0.35 μm as a function of the depletion width x_d. For the ¼-μm device, the value of η can be kept below 30 mV/V if x_d is chosen to be below 70 nm. However, a smaller depletion width means a higher impurity concentration $N_{a,d}$, which leads to a more substrate sensitivity of threshold voltage. Therefore, the choice of the depletion width x_d is a trade-off between substrate sensitivity of the threshold voltage and the short-channel effect. The value of depletion width can be calculated from

$$x_d = \left[\frac{2\epsilon(V_D - V_B + V_d)}{eN_{a,d}} \right]^{1/2} \qquad (13.13.9)$$

where V_D, V_B, and V_d are, respectively, the drain voltage, the substrate bias, and the built-in potential. Thus for $\epsilon = 1.04 \times 10^{-10}$ F/m and $N_{a,d} = 10^{17}$ cm^{-3}, the value of x_d is found to be 110 nm at $V_D - V_B + V_d = 2$ V. Therefore, keeping x_d below 100 nm will impose a severe restriction on the maximum value of the drain voltage.

Proposed Design for $\frac{1}{4}$-μm MOSFET

Table 13.3 compares important parameters for a typical one-micrometer MOSFET (H. N. Yu, A. Riesman, C. M. Osburn, and D. L. Critchlow, *IEEE Trans. Electron Devices,* Vol. ED-26, p. 318, 1979; R. H. Dennard, F. H. Gaensslen, E. J. Walker, and P. W. Cook, *IEEE J. Solid-State Circuits,* Vol. SC-14, p. 247, 1979) and the proposed $\frac{1}{4}$-μm MOSFET as reported by Baccarani et al. (G. Baccarani, M. R. Wordeman, and R. H. Dennard, *IEEE Trans. Electron Devices,* Vol. ED-31, p. 452, 1984). It is instructive to review the considerations given in the choice of physical parameters. In terms of dimension scaling, one important parameter is oxide thickness. As discussed in Section 13.11, weak spots in oxides (mode B failure in Fig. 13.33) can be removed by a combination of preoxidation annealing and phosphorus diffusion. Therefore, growing 5-nm oxides of device quality appears technically feasible even though controlling oxide thickness and uniformity will be a challenging task. A thickness of 5 nm contains fewer than 10 layers of SiO_2 molecules. The spatial variation in film thickness is expected to become progressively more important in contributing to standard deviation in V_T in MOSFETs of smaller physical dimensions. We also note from Table 13.2 that the line resistance R scales as λ. The decrease in the junction depth x_j (Fig. 13.46a) will lead to a corresponding increase in the source and drain resistances R_S and R_D in Fig. 13.23b. The value of $x_j = 0.14$ μm is proposed in Table 13.3 to lessen the increase in R_S and R_D.

The scaling factor κ for the potential is based on several considerations: to keep the hot-carrier effect in check and the power dissipation under control, on the one hand, and to provide a safe noise margin, on the other hand. According to Table 13.2, the field scales as λ/κ and the power density scales as λ^3/κ^3. From the view points of hot-carrier effect and heat dissipation, it is desirable to keep κ as close to λ as possible. In the meantime, it is also important to have a threshold voltage V_T within a predicted range. In the design of N-MOSFETs of small dimensions, boron implantation in the channel region is often performed to adjust V_T to the desired value. However, process tolerances and temperature variations ($dV_T/dT \sim 1$ mV/°C) cause V_T to vary. The threshold voltage V_T also changes with V_D (Fig. 13.47). The choice of $V_T = 250$ mV for the $\frac{1}{4}$-μm FET in Table 13.3 represents the lowest possible value for V_T taking into account the effects above. The supply voltage V_{DD} is set at $V_{DD} = 4V_T$. It is obvious from Table 13.3 that $\kappa < \lambda$ is chosen. To provide a safe noise margin for V_T, we are forced to use scaling rules that will lead to a moderate increase in the electric field and to a significant

TABLE 13.3 PROPOSED DESIGN OF $\frac{1}{4}$-μm MOSFET FOR ROOM-TEMPERATURE OPERATION AS COMPARED TO THAT ADOPTED FOR 1-μm MOSFET

Physical parameter	1-μm	$\frac{1}{4}$-μm	Scaling factor
Channel length, L_G (μm)	1.3	0.25	5.2
Oxide thickness, d_{ox} (nm)	25	5	5.0
Junction depth, x_j (μm)	0.35	0.07–0.14	5.0–2.5
Supply voltage, V_{DD} (V)	2.5	1.0	2.5
Threshold voltage, V_T (V)	0.6	0.25	2.4
Impurity concentration $N_{a,d}$ (cm^{-3})	3×10^{15}	3×10^{16}	0.1

increase in the density of heat dissipation. The latter problem may become a fundamental obstacle to developing room-temperature submicrometer-device technology for integrated circuits.

High-Field Effects

In Section 10.7 we discussed hot-carrier effects in MOSFETs. The degradation of device performance caused by hot carriers is manifested in a number of device characteristics, such as shift in the threshold voltage and reduction in the transconductance. The hot carriers are generated near the drain by impact ionization. Referring to Fig. 10.16 for a N-MOSFET, we see that hot electrons tunneling through the gate oxide and hot holes moving toward the substrate electrode contribute, respectively, to the gate current I_G and the substrate current I_{SUB}. These two currents are given phenomenologically by Eqs. (10.7.3) and (10.7.2). A correlation between I_G/I_D and I_{SUB}/I_D is shown in Fig. 10.18 for a N-MOSFET with a gate length $L_G = 1.9$ μm. As a MOSFET is scaled down to smaller dimensions, the hot-carrier problem is expected to become significant at smaller drain voltages or to become more pronounced under the same bias conditions.

Figure 13.48 shows the ratio I_G/I_D as a function of effective gate length L_G measured at 300 K and 77 K under the bias conditions specified in the figure caption, as reported by Sangiorgi et al. (E. Sangiorgi, E. A. Hofstatter, R. K. Smith, P. F. Bechtold, and W. Fichtner, *IEEE Electron Device Lett.*, Vol. EDL-7, p. 115, 1986). We see that the ratio I_G/I_D increases rapidly as the gate length is reduced. The value of I_D is on the order of 10 mA for a FET with a gate width $Z = 30$ μm. From the ratio I_G/I_D given in Fig. 13.48, we obtain values for I_G ranging from 10^{-13} to 10^{-9} A. Extrapolation of the measured I_G as a function of V_D with $V_G = V_D + 0.5$ V, as reported by Sangiorgi et al., indicates a maximum power-supply voltage of 3 V at 300 K and 2 V at 77 K in order to keep the gate current $I_G < 10^{-15}$ A in a scaled-

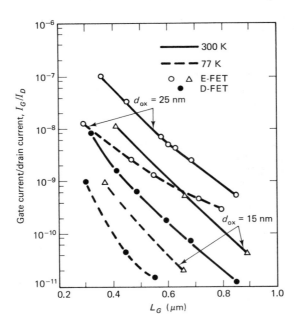

Figure 13.48 Ratio of gate current to drain current, I_G/I_D, as a function of effective gate length L_G measured in MOSFETs at 300 K with $V_D = 5$ V and $V_G = 5.5$ V and at 77 K with $V_D = 4$ V and $V_G = 4.5$ V: enhancement FETs with $d_{ox} = 25$ nm (open circles) and $d_{ox} = 15$ nm (open triangles) and depletion FETs with $d_{ox} = 25$ nm and 18 nm (both in dots). (E. Sangiorgi, E. A. Hofstatter, R. K. Smith, P. F. Bechtold, and W. Fitchner, "Scale Issues Related to High Field Phenomena in Submicrometer MOSFETs," *IEEE Electron Device Lett.*, Vol. EDL-7, p. 115, 1986, © 1986 IEEE.)

down MOSFET with $L_G = 0.6$ μm, $Z = 30$ μm, and $d_{ox} = 18$ nm. The value of 10^{-15} A is chosen as a safe level for I_G to prevent performance degradation.

To help us visualize the situation as described heuristically in Fig. 10.16, we present the two-dimensional simulation results of the lateral electric field near the drain junction at a depth of 50 Å below the interface of MOSFETs under $V_D = 5$ V and $V_G = 5.5$ V for two oxide thicknesses as reported by Sangiorgi et al. (see Fig. 13.49). Note that a high-field region exists with E from 1 to 3×10^5 V/cm. From Fig. 10.9 the ionization coefficient for electrons α_e is found to be about 1.4×10^4 cm^{-1} at $E = 3 \times 10^5$ V/cm in Si. Based on the lateral field alone, significant generation of electron–hole pairs by energetic electrons can be expected near the drain junction. Since α_e decreases rapidly with decreasing E, a reduction of power-supply voltage from $V_{DD} = 5$ V to 3 V should drastically reduce the impact-ionization rate and hence the gate current. As the dimension is scaled down further, the power-supply voltage should be reduced accordingly. The proposed reduction of power-supply voltage to 1 V for the $\frac{1}{4}$-μm MOSFETs appears not only reasonable but may be necessary.

Now let us turn our attention to P-MOSFETs. Figure 13.50 shows the effect of stress on the punch-through voltage V_{PT}, as reported by Koyanagi et al. (M. Koyanagi, A. G. Lewis, R. A. Martin, T. Y. Huang, and J. Y. Chen, *IEEE Trans. Electron Devices*, Vol. ED-34, p. 839, 1987). By *stress* we mean biasing the drain of a MOS-FET to a preset value for a finite time, on the order of 1 h. Under transient conditions, the drain voltage may exceed temporarily the power-supply voltage; therefore, a drain voltage about 50% higher than the power-supply voltage is generally used. For Fig. 13.50 the stress bias was $V_D = -8$ V for 100 min with the P-MOSFET operating under subthreshold condition. As discussed in Section 10.7, stress experiments, such as those shown in Fig. 10.22, provide useful information about the operating lifetime, such as τ of Eq. (10.7.8), of a MOSFET.

For Fig. 13.50, the punch-through voltage is defined as the value of V_D at which the drain current reaches a value of 1 μA. The lateral shift of the curve as a result of stressing has been attributed to a shortening of the effective gate length, as illustrated in Fig. 13.51. A stress bias of 8 V results in an average field of 1.3×10^5 V/cm in a channel of length 0.6μm and a peak field several times that near the drain. Hot electrons and holes are generated by impact ionization. In the model proposed by Koyanagi et al.

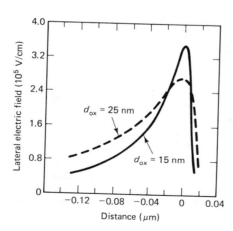

Figure 13.49 Lateral electric field near the drain junction at a depth of 50 Å below the interface, based on two-dimensional device-simulation results of Sangiorgi et al. for MOSFETs with $d_{ox} = 25$ nm (dashed curve) and $d_{ox} = 15$ nm (solid curve). Other parameters are $L_G = 0.8$ μm, $V_D = 5$ V, and $V_G = 5.5$ V. The abscissa indicates the distance from the (simulated) drain junction. (E. Sangiorgi, E. A. Hofstatter, R. K. Smith, P. F. Bechtold, and W. Fitchner, "Scale Issues Related to High Field Phenomena in Submicrometer MOSFETs," *IEEE Electron Device Lett.*, Vol. EDL-7, p. 115, 1986, © 1986 IEEE.)

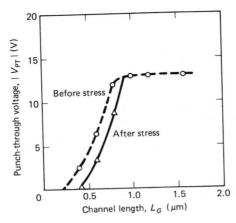

Figure 13.50 Relation between punch-through voltage and channel (gate) length before and after stressing observed in P-MOSFETs. Stress condition was $V_D = -8$ V for 100 min under subthreshold condition. The punch-through voltage is defined as the value of drain voltage that gives rise to a drain current of 1 μA at zero gate voltage. (M. Koyanagi, A. G. Lewis, R. A. Martin, T. Y. Huang, and J. Y. Chen, "Hot-Electron-Induced Punchthrough (HEIP) Effect in Submicrometer PMOS-FETs," *IEEE Trans. Electron Devices*, Vol. ED-34, p. 839, 1987, © 1987 IEEE.)

hot electrons under acceleration toward the gate electrode by the gate field become trapped in the gate oxide. These trapped electrons result in an increased hole concentration near the drain, effectively extending the p^+ drain region and thus shortening the channel region. Consequently, the punch-through voltage is reduced.

As a result of the trapped electrons, the transistor section near the drain will have a threshold voltage V_T shifted in the positive direction. This positive shift of threshold voltage can also provide information about the operating lifetime of a MOSFET. The P-MOSFET had an initial threshold voltage $V_T = -0.55$ V before stressing, defined as the gate voltage to produce a drain current of 10 nA at a drain voltage of -5 V. The operating lifetime of a device is defined in the paper by Koyanagi et al. as the stress time required to produce a shift $\Delta V_T = 100$ mV or a new threshold voltage $V_T' = -0.45$ V. Figure 13.52 shows the measured lifetime τ in devices with $L_G = 0.8$ μm and $L_G = 0.6$ μm, as a function of reciprocal stressing drain voltage. Two curves for

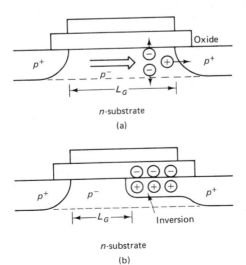

Figure 13.51 Schematic diagrams illustrating (a) hot-electron generation and (b) electron trapping in oxide of P-MOSFET. The model is used by Koyanagi et al. to explain the observed change in punch-through voltage. (M. Koyanagi, A. G. Lewis, R. A. Martin, T. Y. Huang, and J. Y. Chen, "Hot-Electron-Induced Punchthrough (HEIP) Effect in Submicrometer PMOSFETs," *IEEE Trans. Electron Devices*, Vol. ED-34, p. 839, 1987, © 1987 IEEE.)

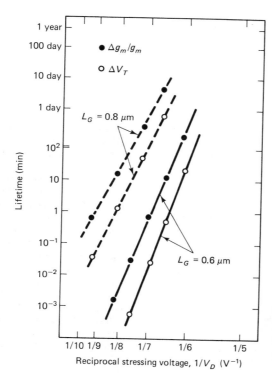

Figure 13.52 Device lifetime deduced from ΔV_T and $\Delta g_m/g_m$ as a function of reciprocal stressing voltage for P-MOSFETs with $L_G = 0.8$ μm and 0.6 μm. (M. Koyanagi, A. G. Lewis, R. A. Martin, T. Y. Huang, and J. Y. Chen, "Hot-Electron-Induced Punchthrough (HEIP) Effect in Submicrometer PMOSFETs," *IEEE Trans. Electron Devices*, Vol. ED-34, p. 839, 1987, © 1987 IEEE.)

each device are given: one based on $\Delta V_T = 100$ mV (open circles) and the other based on a 10% change in g_m (dots). Based on Fig. 13.52 the device lifetime is projected to be about 1 year for the $L_G = 0.80$ μm device and to be less than 10 days for the $L_G = 0.60$ μm device with a stressing voltage $V_D = 5$ V + 10% variation. The projected lifetime suggests a power-supply voltage of 3 V for the 0.8 μm P-MOSFET, compared to a value of 2.5 V for the 1-μm N-MOSFET (Table 13.3).

Channel Mobility

The carrier mobility in a scaled-down MOSFET is expected to be considerably lower than that in a bulk semiconductor because of the existence of a high electric field normal to the channel and the increased impurity concentration. Figure 13.53a shows the measured electron mobility as a function of the average normal field in a N-MOSFET. The mobility drops to 700 cm²/V-s at $E_{av} = 10^5$ V/cm from a value of about 1000 cm²/V-s at $E_{av} = 10^4$ V/cm because of increased surface scattering (F. F. Fang and A. B. Fowler, *Phys. Rev.*, Vol. 169, p. 619, 1968). Figure 13.53b shows the variation of carrier mobility as a function of impurity concentration N_I. The electron mobility drops to 700 cm²/V-s at $N_I = 10^{17}$ cm^{-3} from a value of 1200 cm²/V-s at $N_I = 10^{16}$ cm^{-3} due to increased impurity scattering. Both drops of carrier mobility have not been taken into account in the scaling theory given in Table 13.2.

To gain some physical insight into the expected mobility behavior, we refer to Fig. 11.11b. At high temperatures (around 300 K) and under low-bias voltages, μ has a temperature dependence between T^{-1} and $T^{-1.5}$, indicating the dominance of lattice

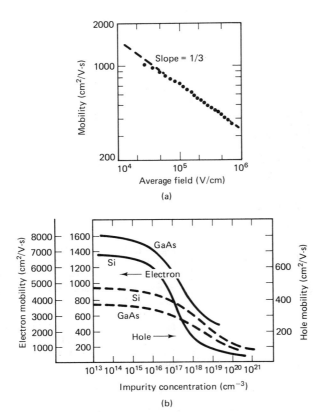

Figure 13.53 Curves showing (a) the variation of electron mobility in N-MOS-FETs as a function of average field normal to the interface and (b) the variation of carrier mobility with impurity concentration in Si and GaAs. The hole mobility is taken as 0.8 μ_H (Hall mobility) for holes in both Si and GaAs. The electron mobility is taken as μ_H in Si and is the drift mobility in GaAs. (a) G. Baccarani, M. R. Wordeman, and R. H. Dennard, "Generalized Scaling Theory and Its Application to a ¼ Micrometer MOSFET Design," *IEEE Trans. Electron Devices*, Vol. ED-31, p. 452, 1984. (b) G. Masetti, M. Severi, and S. Solmi, "Modeling of Carrier Mobility against Carrier Concentration in Arsenic-, Phosphorous-, and Boron-Doped Silicon," *IEEE Trans. Electron Devices*, Vol. ED-30, p. 764, 1983, © 1983 IEEE; J. S. Blakemore, *J. Appl. Phys.*, Vol. 53, p. R123, October 1982.)

scattering. At low temperatures (below 70 K) and under low-bias voltage, μ has a temperature dependence between T and $T^{1.5}$, signifying the importance of impurity scattering. Under high-bias voltages ($V_G' > 12$ V), the mobility stays almost constant with respect to temperature but decreases with increased bias voltage. This behavior can be attributed to surface scattering. The transport properties of electrons in a surface potential well has been studied extensively. Based on the first analysis of the problem by Schrieffer, the surface-scattering-controlled mobility μ_S is given by (J. R. Schrieffer, *Phys. Rev.*, Vol. 97, p. 641, 1955)

$$\frac{\mu_S}{\mu_B} = \frac{(kTm^*)^{1/2}}{e\tau_B E_\perp} \qquad (13.13.10)$$

where μ_B and τ_B are, respectively, the mobility and the scattering time in the bulk and $E_\perp$ is the field normal to the interface. From Fig. 13.53a it is obvious that Eq. (13.13.10) overestimates the importance of $E_\perp$.

To make proper modifications of Eq. (13.13.10), we make following observations. First, we let

$$\mu^{-1} = \mu_S^{-1} + \mu_B^{-1} \qquad (13.13.11)$$

to be a combination of surface and bulk scattering. Next, we assume that

$$\mu_S = AE_\perp^{-q} \tag{13.13.12}$$

where A and q are two parameters to be determined. We note that the constant $(kTm^*)^{1/2}$ in Eq. (13.13.10) is proportional to $(kT)^{1/2}$ because of increased thermal velocity at elevated temperatures. For a degenerate electron gas, kT should be replaced by $\mathscr{E}_f - \mathscr{E}_c$, which is proportional to $n^{2/3}$. Since n is proportional to $E_\perp$, the constant A in Eq. (13.13.12) increases as $E_\perp^{1/3}$. Therefore, Eq. (13.13.12) can be rewritten as

$$\mu_S = BE_\perp^{-2/3} \tag{13.13.13}$$

The data shown in Fig. 13.53a can be fit reasonably well by Eq. (13.13.11) if we let $\mu_B = 1200$ cm^2/V-s and

$$\mu_S = 1900 \left(\frac{10^5}{E_\perp}\right)^{2/3} \qquad \text{cm}^2\text{/V-s} \tag{13.13.14}$$

where $E_\perp$ is expressed in V/cm.

With the background noted above, the behavior of μ at low temperatures in Fig. 11.11b can be explained as follows. The curve for $V_G' = 1$ V is dominated by impurity scattering. The curves for 2 V $< V_G' <$ 7 V are a result of mixed impurity and surface scattering. As V_G' increases, the areal density N_s increases. The increased carriers not only provide an effective screening of ionized impurities but also raise the constant A. The two factors combined are responsible for the initial rise of μ with increased V_G'. After N_s reaches a certain level, the carrier system becomes degenerate and Eq. (13.13.14) should be used. The curves for $V_G' = 12$ V to 32 V in Fig. 11.11b can also be fit by Eq. (13.13.14) with a slightly larger constant 2500 instead of 1900. The establishment of an equation for μ_S is very useful in projecting the performance of MOSFETs at cryogenic temperatures.

In summary, we have provided in this section the background on important issues for further miniaturization of MOSFETs for IC applications. With reduced power-supply voltage, the $\frac{1}{4}$-μm MOSFET appears quite feasible. However, it is generally believed that reduction of feature size below 0.25 μm will require breakthroughs in technology. Even if it is technologically feasible, we must weigh the potential benefits, such as improved performance, against the anticipated problems, such as reliability and cost. It is important for us to investigate approaches and to examine the issues in a broadened perspective. These are subjects for discussion in Section 13.14.

13.14 ALTERNATIVE APPROACHES FOR PERFORMANCE ENHANCEMENT

In Section 13.12 we discussed MESFETs and MODFETs based mainly on GaAs technology, and in Section 13.13 we discussed MOSFETs based on Si technology. So far, the IC technology is dominated by Si-MOSFETs. The main driving force for MOSFET technology is the scaling of device dimensions. However, we are approaching the limits of miniaturization from the viewpoints of device performance and reliability. In this section we present alternative approaches for performance enhancement and examine the relevant issues. The approaches to be discussed include (1) scaling, (2) GaAs versus Si, (3) cryogenic operation, and (4) three-dimensional architecture.

Scaling

According to the scaling rules presented in Table 13.2, the transconductance g_m (mS/mm) is expected to scale as λ^2/κ. For a 1-μm MOSFET, g_m has a typical value around 100 mS/mm. Therefore, a value for g_m around 1000 ms/mm should be expected in a scaled 0.1-μm MOSFET, as compared to the experimental value of about 500 mS/mm (G. A. Sai-Halasz, M. R. Wordeman, D. P. Kern, E. Ganin, S. Richton, D. S. Zicherman, H. Schmid, M. R. Polari, H. Y. Ng, P. J. Restle, T. H. P. Chang, and R. H. Dennard, *IEEE Electron Device Lett.*, Vol. EDL-8, p. 463, 1987). If scaling rules are strictly observed, velocity saturation is not expected. However, the impurity concentration scales as λ^2/κ to a value about 7.5×10^{16} cm^{-3} for $L_G = 0.1$ μm from a value of 3×10^{16} cm^{-3} for $L_G = 0.25$ μm. Note from Fig. 13.53b that μ_n drops appreciably with increased impurity concentration N_I for $N_I > 10^{16}$ cm^{-3}. Therefore, we do not expect a proportionate increase in g_m as L_G is reduced from 0.5 μm to 0.1 μm as from 2.5 μm to 1 μm.

Another important consideration is power density, which scales as $(\lambda/\kappa)^3$. Note that the scaling factor κ for potential is only half the scaling factor λ for dimension listed in Table 13.2, in order to provide a safe noise margin for V_T due to process tolerances. Therefore, the heat-dissipation problem is expected to become more severe as the device dimension is scaled down further. With $\lambda > \kappa$, the electric field will be higher and hot-carrier problems will be more pronounced. An associated problem with impact ionization is latch-up in complementary MOS (C-MOS) circuits. Referring to Fig. 9.36c for a C-MOSFET, we see that there are two parasitic *n-p-n* and *p-n-p* transistors in the formation of a *n-p-n-p* device with the P well and N well serving as the common base and collector. As discussed in Section 9.5 with reference to Eq. (9.5.4), switching a *n-p-n-p* device from the off state to the on state can be triggered by impact ionization by making $(\alpha_1 + \alpha_2) M > 1$. By *latch-up* we mean the appearance of a low-resistance circuit between the power supply (the drain) and the ground (the source). In a scaled-down C-MOSFET, the sum $\alpha_1 + \alpha_2$ becomes higher because of the shorter base region and the multiplication factor becomes larger on account of higher fields. Therefore, latch-up is expected to be a more challenging problem in VLSI circuitry.

Aside from device physics, reliability problems for scaled-down devices become more stringent. The problems include reduction of particle counts in processing environment, contact electromigration, oxide failure, and α-particle-induced errors. These problems are aggravated by the ultrasmall device dimensions. The reader is referred to an excellent review article by Woods for a discussion of these problems (M. H. Woods, *Proc. IEEE*, Vol. 74, p. 1715, 1987). Our ability for continued scaling of VLSI (very-large-scale integration) devices depends on advances in processing and packaging technology, on the one hand, and on understanding of failure mechanisms, on the other. A precise knowledge of reliability mechanisms is needed in foreseeing the limits of scaling. Because the economic stakes are high, the potential benefits must also be weighed against mounting costs.

GaAs versus Si

Even though silicon-based devices have dominated IC technology, gallium-arsenide-based devices are rapidly being developed. Now that Si-based VLSI technology is approaching the limit of further scaling, it is time to compare the performance and physics of GaAs and Si submicrometer FETs. Referring to Fig. 13.22, we see that GaAs

High-Frequency and High-Speed Devices   Chap. 13

MODFETs have a speed advantage over Si MOSFETs by a factor of about 10 for $L_G = 1$ μm and about 4 for $L_G = 0.5$ μm. Therefore, based on speed alone, GaAs seems to be the obvious choice. In Table 13.4 we present performance data at 300 K of a 0.25-μm GaAs MODFET and a 0.10-μm Si MOSFET reported, respectively, by Awano et al. and by Sai-Halasz et al. (Y. Awano, M. Kosugi, T. Mimura, and M. Abe, *IEEE Electron Device Lett.*, Vol. EDL-8, p. 451, 1987; G. A. Sai-Halasz et al., *IEEE Electron Device Lett.*, Vol. EDL-8, p. 463, 1987 cited earlier in this section). Obviously, the characteristics are comparable. The MODFET was reported to have a measured propagation delay time of 9.2 ps per gate and a predicted unity-current-gain cutoff frequency of 110 GHz. No corresponding values were given for the MOSFET.

For enhancement of device performance, we have two alternative approaches: to push the scaling MOSFETs to the limit or to develop MODFET technology for VLSI. The approaching scaling limits and the anticipated technological challenges for ultra-small devices make GaAs devices attractive alternatives to Si devices because what we hope to achieve in scaled-down Si devices can be accomplished in GaAs devices of considerably larger dimensions. The increasing activities in GaAs device research and development reflect recognition of this fact.

The development of GaAs VLSI technology is hindered by the lack of large high-quality substrates. Remarkable advances in heteroepitaxy of GaAs films on Si substrates offer hope that this obstacle may be removed. Maximum transconductances of 196 mS/mm for enhancement and 182 mS/mm for depletion MESFETs with $L_G = 1$ μm grown on a 2-in. Si wafer were obtained, with standard deviations about 8% and 16%, respectively (H. Shichijo, L. T. Tran, R. J. Matyi, and J. W. Lee, *Mater. Res. Soc. Symp. Proc.*, Vol. 91, 1987, p. 213). The g_m values are comparable to those obtained from MESFETs grown on semi-insulating GaAs substrates. Because field-effect transistors are majority-carrier devices, their characteristics are less susceptible to defects than are the characteristics of bipolar transistors.

Let us now turn our attention to the devices. Whereas oxide failure is still a problem for submicrometer MOSFETs,(GaAl)As films grown by MBE or MOCVD are perfectly adequate at room temperature as gate insulator in submicrometer MODFETs. The threshold voltage in MOSFETs is adjusted to the desired value by boron implantation, and the threshold voltage in MODFETs is controlled by the thickness and doping in (GaAl)As gate. Process control is more precise in MBE or MOCVD technology than in implantation technology. Finally, the speed advantage is obviously in favor of GaAs devices over Si devices even though it is below the ratio of their low-field mobilities. However, the considerations above must be weighed against a well-established Si technology and the heavy investment in efforts as well as in resources committed to the Si

TABLE 13.4 PERFORMANCE COMPARISON OF GaAs MODFET AND Si MOSFET AT 300 K

	Gate length, L_G (μm)	Transconductance g_m (mS/mm)	I_{DS}/Z^a (μA/μm)	V_T (V)
MODFET	0.25	400	200	0
MOSFET	0.10	500	300	0.1

[a]The values of saturation drain current per gate width I_{DS}/Z are taken at $V_G = 0.8$ V and $V_D = 1.0$ V.

technology. Therefore, GaAs devices need to show a clear and decisive performance/cost advantage over Si devices for GaAs to compete effectively with Si in VLSI applications.

Cryogenic Operation

Cryogenic operation of logic circuits has been suggested as an effective means for performance enhancement. Possible advantages include increased operation speed and reduced threshold voltage. Studies of cryogenic operation of C-MOSFETs were reported by Aoki et al. (M. Aoki, S. Hanamura, T. Masuhara, and K. Yano, *IEEE Trans. Electron Devices*, Vol. ED-34, p. 8, 1987). In Table 13.5 we compare the threshold voltages and transconductances of the N-MOS and P-MOS, and the delay time and power per gate in a C-MOS inverter circuit, at three different temperatures. To explain the device behavior, we again refer to the mobility behavior shown in Fig. 11.11b and to Eqs. (13.13.11) and (13.13.14). At low temperatures and under sufficient gate bias, the channel mobility is strongly influenced, if not controlled, by surface scattering. Decreasing mobility at increased gate bias was indeed observed by Aoki et al. The fact that both transconductance and gate delay hardly changed between 77 and 4.2 K can be interpreted as a manifestation of mobility being dominated by surface scattering.

The recent interest in liquid-nitrogen-temperature (LNT) operation of logic devices is prompted by the realization of increasing difficulty to achieve performance enhancement through feature size reduction below 1 μm. There are several additional advantages for LNT operation, including higher thermal conductivity and smaller subthreshold swing. A higher thermal conductivity will allow a proportionately higher power density. With regard to subthreshold swing, we refer to Fig. 13.26. The threshold voltage V_T is defined by the projection of the linear portion of the I_{DS} current to the V_G axis. Because of soft turn-off of I_{DS}, a significant I_{DS} exists until V_G is smaller than V_T by ΔV_{ST}. The quantity $\Delta V_{ST} = 2.3kT/e$ is known as *subthreshold swing*. In the experiment of Aoki et al., a reduction of subthreshold swing from 80 mV at 300 K to 20 mV at 77 K was observed. Reduction of subthreshold swing will make possible reduction of threshold voltage.

Because interest in LNT operation is relatively recent, many problems need to be investigated. Impurity freeze-out becomes a problem, especially in heavily doped region. Device degradation by hot-carrier effects may become more severe at 77 K than at 300 K. The device design must be optimized for LNT operation. Channel implanta-

TABLE 13.5 PERFORMANCE COMPARISONS OF A C-MOSFET AT DIFFERENT TEMPERATURES[a]

Temperature (K)	V_T (V)	g_m (mS/mm)	Delay (ps/gate)	Power (μW/gate)
300	0.31 (−0.49)	101 (49)	190	12
77	0.68 (−0.85)	170 (61)	117	11
4.2	0.76 (−0.88)	189 (62)	83	11

[a]The values of g_m are taken at $V_D = 5$ V. The effective channel lengths of the N-MOS and P-MOS are, respectively, 0.7 μm and 0.8 μm. The values of V_T and g_m for the P-MOS are given in parentheses.
Source: M. Aoki, S. Hanamura, T. Masuhara, and K. Yano, *IEEE Trans. Electron Devices*, Vol. ED-34, p. 8, 1987.

tion is useful for threshold-voltage adjustment but lowers the carrier mobility significantly at low temperatures. Finally, the effect of surface scattering on carrier mobility needs careful study. Surface scattering and impurity scattering will limit the potential speed enhancement at 77 K.

Three-Dimensional Architecture

Scaling down feature size to submicrometer dimensions not only presents technological challenges of increased reliability problems at reduced performance benefits but also creates traffic jams for data flow between chips. The number of signals to be taken out from a chip is limited by the number of bonding pads or terminal pins on a chip. The maximum number is now between 200 and 250 for chips, with feature size in the micrometer range. One approach to solve this traffic problem is three-dimensional architecture for device integration. The idea is to build multiple layers of two-dimensional functional devices and to interconnect devices at different levels through vertical conducting paths built in the three-dimensional structure between layers at strategic locations.

There are many potential advantages for three-dimensional device integration. They include the possibility of parallel processing, high packing density and multifunctional operation within one chip. However, there are many technological challenges on the road to three-dimensional integration. Obviously, the layers must be isolated electrically from one another. The introduction of an insulator between semiconducting layers means repeated semiconductor–insulator interfaces. One fundamental problem is how to grow semiconductor films of device quality on an insulator. In addition, the surface at each level must be flat enough for photolithography. Therefore, technology for surface planarization must be developed. From the circuit point of view, devices placed in close proximity can interfere with one another through electric fluxes that are not properly terminated. The reader is referred to an excellent review article by Akasaka for discussions of the progress being made as well as anticipated problems in the area of three-dimensional integration (Y. Akasaka, *Proc. IEEE,* Vol. 74, p. 1703, 1986).

In summary, we have discussed various approaches for further performance enhancement of integrated circuits. As we are approaching the limit of miniaturization, in the neighborhood of 0.25 μm feature size, we must look for alternatives to scaling. The three possible approaches, GaAs technology, cryogenic operation, and three-dimensional integration are all based on existing technologies that need to be further developed. The emerging field of heteroepitaxy, of which silicon growth on insulator and gallium-arsenide growth on silicon are examples, will not only aid the development of these technologies but may lead to possibilities of new material combinations and new device structures.

13.15 BIPOLAR AND HETEROSTRUCTURE BIPOLAR TRANSISTORS

Silicon bipolar transistor (BT) technology is more advanced than gallium-arsenide heterostructure bipolar transistor (HBT) technology because much of the processing and fabrication techniques can be borrowed from the well-developed MOSFET technology. The development in BT technology is relatively slow compared with that in FET technology on account of nonscalability, high power requirement, and nonplanar geometry. However, bipolar transistors possess a distinct advantage over field-effect transistors in that control of the critical dimension, the base width, does not depend on submicrometer lithography yet to be developed. In this section we review the basic device physics and relevant performance characteristics of Si BTs and GaAs HBTs.

The junction capacitance C_T of Eq. (8.2.7) is inversely proportional to the depletion width W of Eq. (8.2.13). A reduction in W means an increase in C_T, which is not desirable. Furthermore, a reduction in W requires increases in doping concentrations. For the forward-biased emitter junction, increased doping concentrations mean an increased turn-on voltage due to the increased built-in potential V_d of Eq. (8.1.3). Furthermore, heavy doping concentrations lead to tunneling (Section 13.3). To have a high emitter-injection efficiency γ, the emitter side must be doped more heavily than the base side, according to Eq. (9.1.24). On the other hand, the base doping must be high enough to give an acceptable base resistance. As to the bias voltages, their respective values must be sufficiently large to turn on the emitter–base junction and to keep the collector–base capacitance small. Therefore, our choice of the physical parameters for a bipolar transistor is somewhat limited by these considerations, which make the device nonscalable.

Next we turn our attention to device speed. The expressions for f_t and $f_{\max}$ presented in Section 13.10, specifically Eqs. (13.10.6) and (13.10.10), are used for field-effect transistors. For bipolar transistors, the corresponding equations for f_t and $f_{\max}$ are, respectively,

$$f_t = (2\pi\tau_{EC})^{-1} \tag{13.15.1}$$

$$f_{\max} = f_t(8\pi f_t r_B C_{TE})^{-1/2} \tag{13.15.2}$$

where r_B is the base resistance and C_{TE} is the emitter-junction capacitance. In a common-emitter configuration the total time τ_{EC} for injected carriers to traverse from emitter to collector junction consists of the time $\tau_E = r_E (C_{TE} + C_{TC})$ to charge the emitter–base junction, the time τ_B for carriers to diffuse across the base region, the time $\tau_C = (r_E + r_C)C_{TC}$ to charge the collector junction and the time τ_D for carriers to drift through the depleted collector regions. Therefore, the quantity τ_{EC} is given by

$$\tau_{EC} = \tau_E + \tau_B + \tau_C + \tau_D \tag{13.15.3}$$

(H. F. Cooke, *Proc. IEEE,* Vol. 59, p. 1163, 1971).

Let us pay special attention to the product $r_B C_{TE}$. Note that the base current flows in the lateral direction in Fig. 9.35b. Therefore, if the lateral dimension of a bipolar transistor is reduced, both r_B and C_{TE} become smaller, due to a shorter path for r_B and a smaller area for C_{TE}. If both the lateral and longitudinal dimensions of the transistor are reduced by a factor λ, r_B remains the same but C_{TE} is reduced by a factor λ^2. We also note from Fig. 9.35b that the collector area is much bigger than the emitter area and that there is also a parasitic capacitance C_{CS} between the collector and the substrate. Therefore, it is extremely important to reduce the collector area as much as possible. Figure 13.54 shows (a) the structure of an advanced Si bipolar transistor with deep trench isolation and (b) the structure proposed for a complimentary pair of p^+-n-p and n^+-p-n transistors. The transistor is surrounded by a deep trench filled with silicon dioxide for isolation and for reduction of parasitic capacitances (T. H. Ning and D. D. Tang, *Proc. IEEE,* Vol. 74, p. 1669, 1986).

To see the relative importance of the various terms in Eq. (13.15.3) in devices with different geometries, we choose an n^+-p-n-n^+ transistor with structural properties listed in Table 13.6 as an example. The structure is very similar to the one shown in

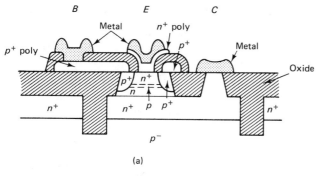

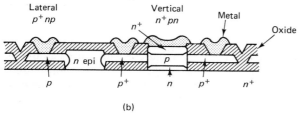

Figure 13.54 Schematic diagrams for (a) a Si bipolar n^+-p-n transistor of advanced design with deep trench isolation and self-aligned base contact and (b) a complementary pair of p^+-n-p and n^+-p-n transistors. (T. H. Ning and D. D. Tang, "Bipolar Trends," *IEEE Proc.,* Vol. 74, p. 1669, 1986, © 1986 IEEE.)

Fig. 9.9 except that the layer thicknesses are much reduced and the doping concentrations are raised. For example, we can start with a wafer with a 1-μm n^+ layer on a p^- substrate, and then grow a thin epitaxial n layer on the wafer. The base and emitter regions are formed first by boron implantation and then by arsenic diffusion. The base region of a bipolar transistor of advanced design consists of two differently doped regions, a moderately doped p region, generally referred to as the *intrinsic region*, directly under the emitter and a heavily doped p^+ region, generally referred to as the *extrinsic region*, from the p base to the base contact (D. D. Tang, T. Z. Chen, C. T. Chuang, G. P. Li, J. M. C. Stork, M. B. Ketchen, E. Hackbarth, and T. H. Ning, *IEEE Electron Device Lett.,* Vol. EDL-8, p. 174, 1987). The sheet resistance for the intrinsic base region is given in parentheses. The variation of mobility (Fig. 13.53b) has been taken into account in the computation of sheet resistances.

The built-in potentials are $V_d = 1.12$ V for the emitter junction and $V_d = 0.84$ V for the collector junction. At zero bias the widths of the space-charge regions are found to be $W_E = 0.04$ μm for the emitter junction and $W'_C = 0.33$ μm based on $N_d = 10^{16}$ cm^{-3} for the collector junction. Since at zero bias the space charge already extends into the heavily doped subcollector region, the width of the collector junction W_C can be taken to be $W_C = 0.20$ μm when the collector junction is reverse biased. As a first design, we choose the emitter area to be 2.5×4 μm^2 and the collector area to be 12×6 μm^2. The following calculation is made for an emitter current $I_E = 1$ mA and a collector bias $V_C = 5$ V. From I_E, the value of injected electron concentration is found to be $n_p = 2.7 \times 10^{16}$ cm^{-3}, requiring an applied bias voltage $V_a = 0.85$ V. From $V_d - V_a = 0.27$ V we find the emitter capacitance to be $C_{TE} = 61$ fF. Since the collector n region is fully depleted at $V_C = 5$ V and since the base and subcollector regions are heavily doped, C_{TC} can be approximated by $\epsilon A/d$ with $d = 0.20$ μm, yielding a value of 37 fF for collector-junction capacitance. At $I_E = 1$ mA, the intrinsic emitter resistance is $r_E = kT/(eI_E) = 26$ Ω. Thus we have the emitter

TABLE 13.6 SUGGESTED STRUCTURAL PROPERTIES AND COMPUTED DEVICE PARAMETERS FOR A SILICON n^+-p-n-n^+ TRANSISTOR

	Doping concentration (cm^{-3})	Layer thickness (μm)	Sheet resistance (Ω/square)	Capacitance per area at zero bias[a] (fF/μm^2)
Emitter	10^{21} (n^{++})	0.10		2.8
Base	10^{18} (p)	0.15	400 (2000)	
Collector	10^{16} (n)	0.20	5000	0.5
Subcollector	10^{19} (n^+)	1.00	20	

[a]The computation of capacitances is based on built-in potential $V_d = 1.12$ V for the emitter junction and $V_d = 0.84$ for the collector junction at which the space-charge region extends into the subcollector.

charging time to be $\tau_E = 2.8$ ps. Note that the base and collector currents flow in the lateral direction. Assuming a flow path of 3 μm length and 8 μm width (from two sides) in the extrinsic region and a flow path of 1.25 μm length and 8 μm width (from both sides) in the intrinsic region, we find a total base resistance $r_B = 460$ Ω. The collector resistance is determined primarily by the resistance of the subcollector region. Assuming a current path of 10-μm length (from one side) and 6-μm width, we have a collector resistance $r_C = 33$ Ω. Therefore, the collector charging time $\tau_C = (r_E + r_C)C_{TC}$ is 2.2 ps.

The base-transit time $\tau_B = W_B^2/2D$ is 2.8 ps and the collector-drift time $\tau_D = W_C/v$ is 2 ps, using saturation velocity of 10^7 cm/s for v. There is evidence from Monte Carlo particle simulation for GaAs HBT that the carrier velocity in the collector region may exceed saturation velocity (P. M. Asbeck, D. L. Miller, R. Asatourian, and C. G. Kirkpatrick, *IEEE Electron Device Lett.*, Vol. EDL-3, p. 403, 1982). Velocity overshoot can be expected also in Si-BT. Therefore, the values of τ_D may have been overestimated. The sum of $\tau_E + \tau_B + \tau_C$ is 9.8 ps. Keeping the doping concentrations unchanged, the values of τ_E and τ_C can be reduced by reducing the junction areas. For an emitter area of 1.25×2 μm^2, the value of n_p must be raised to 1.1×10^{17} cm^{-3} to keep $I_E = 1$ mA. This value of n_p is still smaller than the majority-carrier concentration in the base, and increases emitter capacitance per unit area by only 40%, resulting in the value $C_{TE} = 25$ fF. If we use a collector area of 6×3 μm^2 with the same aspect ratio, we expect r_B and r_C to be unchanged but C_{TC} to be reduced to 9 fF. Therefore, for the second design, the sum $\tau_E + \tau_C + \tau_B + \tau_D$ is expected to be 6.2 ps. If we use $\mu_n = 300$ cm^2/V-s at $N_a = 10^{18}$ cm^{-3} from Fig. 13.53b instead of 1350 cm^2/V-s in the computation of D_n, we find that τ_B has a value of 14.5 ps and becomes the dominant term in τ_{EC} (Problem 13.42). Therefore, reducing basewidth is essential in reducing τ_{EC}. Using photoepitaxy, a base width as thin as 0.065 μm has been reported (T. Sugii, T. Yamazaki, T. Fukano, and T. Ito, *IEEE Electron Device Lett.*, Vol. EDL-8, p. 528, 1987). Reduced base width, reduced device area, and a proper doping profile should enable silicon BTs to have a speed capability comparable to that of silicon MOSFETs.

Heterostructure bipolar transistors were studied for high-frequency operation by Dumke et al. and by Kroemer (W. P. Dumke, J. M. Woodall, and V. L. Rideout, *Solid-State Electron.*, Vol. 15, p. 1339, 1972; H. Kroemer, *Proc. IEEE*, Vol. 70,

p. 13, 1982). Even though silicon technology is at a far more advanced state than gallium-arsenide technology, research in heterostructure bipolar transistors has been more active than that in biplar transistors. To compare the relative advantages of HBT and BT, we use as an example the results on AlGaAs–GaAs HBT reported by Yamauchi and Ishibashi (Y. Yamauchi and T. Ishibashi, *IEEE Electron Device Lett.*, Vol. EDL-7, p. 655, 1986). The structural properties are given in Table 13.7 and the device parameters are given in Table 13.8. We note that the base is more heavily doped than the emitter. Heavy doping of base is avoided in BTs to ensure high emitter efficiency. However, in HBTs, minority-carrier injection into the emitter is greatly reduced by the potential barrier at the heterostructure interface (Fig. 14.24 and Problem 13.43). Heavy base doping in HBT lowers the base resistance r_B, while moderate emitter doping lowers the emitter-junction capacitance per unit area. Both features are advantageous.

Following a procedure similar to that used in the calculation for the Si BT, the various capacitances and resistance can be computed (Problems 13.44 and 13.45). In the calculation, we take note of the following points. First, the values given in Table 13.8 are for $I_C = 10$ mA at a base–emitter voltage $V_{BE} = 1.74$ V in a common-emitter configuration. Second, both V_{BE} and V_{BC} include the potential across the junction and the potential drop in the resistances. Third, the base is degenerately doped, so the built-in potential V_d in the two junctions either exceeds or nearly equals 1.42 V. Fourth, due to the graded $Al_xGa_{1-x}As$ composition, a constant built-in field E_B exists in the base region. The current actually is dominated by the drift term $e\mu E_B n$ instead of the diffu-

TABLE 13.7   STRUCTURAL PROPERTIES OF $Al_xGa_{1-x}As$/GaAs HBT GROWN BY MBE[a]

Layer	Material	Doping (cm^{-3})	Thickness (μm)	Content, x
Cap	n^+-GaAs	5×10^{18}	0.20	0
Emitter	n-AlGaAs	5×10^{17}	0.15	0.30
Base	p-AlGaAs	4×10^{19}	0.12	0.12–0
Collector	n-GaAs	5×10^{16}	0.60	0
Buffer	n^+-GaAs	3×10^{18}	0.50	0

[a]The processed structure has an emitter area $A_E = 5 \times 10$ μm^2 and a collector area $A_C = 8 \times 13$ μm^2.

Source: Y. Yamauchi and T. Ishibashi, *IEEE Electron Device Lett.*, Vol. EDL-7, p. 655, 1986.

TABLE 13.8   CALCULATED DEVICE PARAMETERS FOR THE FABRICATED HBT AT TWO COLLECTOR–EMITTER VOLTAGES V_{CE} IN A COMMON-EMITTER CONFIGURATION FOR A COLLECTOR CURRENT $I_C = 10$ mA[a]

V_{CE} (V)	W_C' (μm)	C_{TE} (fF)	C_{TC} (fF)	r_{EE} (Ω)	r_C (Ω)	r_B (Ω)	τ_E (ps)	τ_C (ps)	τ_B (ps)	τ_D (ps)	f_t (GHz)	f_{max} (GHz)
2	0.22	196	71	6.1	19.3	18.3	0.69	1.81	0.79	0.79	40	36
5	0.37	196	43	6.1	18.9	18.3	0.61	1.07	0.79	3.44	27	37

[a]The quantity W_C' is the width of the space-charge region of the collector junction. The computation is for $I_C = 10$mA and $V_{BE} = 1.74$V.

Source: (Y. Yamauchi and T. Ishibashi, *IEEE Electron Device Lett.*, Vol. EDL-7, p. 655, 1986.)

sion term $eD\,dn/dx$. The base-transmit time is therefore equal to $W_B/\mu E_B$. Note that τ_B in the HBT is much shorter than that in the BT as a result of the built-in field.

The delay in the collector τ_D deserves special attention. Calculations of drift velocity by Monte Carlo particle simulation show that the drift velocity can reach a value around 5 to 7×10^7 cm/s and stay above the saturation velocity 8×10^6 cm/s for a distance about 0.05 to 0.10 μm (P. M. Asbeck, D. L. Miller, R. Asatourian, and C. G. Kirkpartick, *IEEE Electron Device Lett.*, Vol. EDL-3, p. 403, 1982; K. Tomizawa, Y. Awano, and N. Hashizume, *IEEE Electron Device Lett.*, Vol. EDL-5, p. 362, 1984). A two-section model was used by Yamauchi and Ishibashi by assuming a velocity v_{os} in the overshoot region of length d_{os} and a saturation velocity v_s in the remaining collector region of length $W_C' - d_{os}$. Therefore, the collector-delay time is

$$\tau_D = \frac{W_C'}{v_d} = \frac{d_{os}}{v_{os}} + \frac{W_C' - d_{os}}{v_s} \qquad (13.15.4)$$

where W_C' is the width of the collector-depletion region, which is smaller than 0.60 μm given in Table 13.7.

Figure 13.55 shows the measured and calculated values of f_t and $f_{\max}$ as functions of (a) collector current I_C and (b) collector-emitter voltage V_{CE}. For constant V_{CE}, a constant $v_d = 1.5 \times 10^7$ cm/s was used to fit the data in Fig. 13.55a. For different values of V_{CE}, the values of W_C' were different and a best fit to the experimental data shown in Fig. 13.55b was obtained by using $d_{os} = 0.15$ μm, $v_{os} = 5 \times 10^7$ cm/s, and

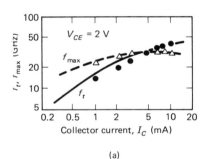

(a)

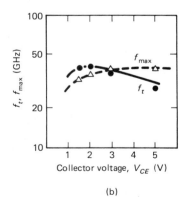

(b)

Figure 13.55 The measured values and computed curves of f_t (solid) and $f_{\max}$ (dashed) for AlGaAs–GaAs HBT as functions of (a) the collector currrent and (b) the collector–emitter voltage. (Y. Yamauchi and T. Ishibashi, "Electron Velocity Overshoot in the Collector Depletion Layer of AlGaAs/GaAs HBTs," *IEEE Electron Device Lett.*, Vol. EDL-7, p. 655, 1986, © 1986 IEEE.)

$v_s = 4 \times 10^6$ cm/s. Velocity overshoot may also have an effect on the base-transit time. The use of $\tau_B = W_B^2/2D$ assumes an average velocity

$$\langle v \rangle = \frac{2D}{W_B} \qquad (13.15.5)$$

for the injected carriers. At a base width $W_B = 0.15$ μm, the value of $\langle v \rangle$ is 4.6×10^6 cm/s for $D_n = 35$ cm^2/V-s in pure Si and 2.7×10^7 cm/s for $D_n = 206$ cm^2/s in pure GaAs. The electron mobility was calculated to be about 2000 and 1000 cm^2/V-s in samples with acceptor concentrations of 10^{18} and 4×10^{19} cm^{-3}, respectively. Using a mobility of 2000 cm^2/V-s and $D = \mu kT/e$ the average electron velocity is 6.8×10^6 cm/s, which is close to saturation velocity. Results on Monte Carlo simulation show that electrons on the average suffer collisions numbering from 9.2 to 2 and attain a velocity from 7.1×10^6 to 1.2×10^7 cm/s in p regions (doped at 3×10^{18} cm^{-3}) of widths from 0.15 to 0.05 μm (C. M. Maziar and M. S. Lundstrom, *IEEE Electron Device Lett.*, Vol. EDL-8, p. 90, 1987).

We should mention that a value of $f_t = 18$ GHz was achieved in InGaAsP–InP HBT having an emitter area of 5×10 μm^2 (R. N. Nottenburg, J. C. Bischoff, M. B. Panish, and H. Temkin, *IEEE Electron Device Lett.*, Vol. EDL-8, p. 282, 1987). Furthermore, work on AlGaAs–GaAs HBTs has shown that both f_t and f_{max} increase inversely with emitter area A_E. A value of $f_t = 20$ GHz in a HBT with $A_E = 4 \times 10$ μm^2 was also below the value reported by Yamauchi and Ishibashi (K. Eda, M. Inada, Y. Ota, A. Nakagawa, T. Hirose, and M. Yanagihara, *IEEE Electron Device Lett.*, Vol. EDL-7, p. 694, 1986). Fully self-aligned AlGaAs–GaAs HBT with submicrometer emitter stripe was also reported (N. Hayama, A. Okamoto, M. Madihian, and K. Honjo, *IEEE Electron Device Lett.*, Vol. EDL-8, p. 246, 1987). With the HBT technology being rapidly developed, a cutoff frequency f_t approaching 100 GHz should be attainable.

In summary, we have reviewed the physics and performance of BTs and HBTs. Because of the potential barrier created by band-edge discontinuities at a heterostructure interface, the emitter-doping concentration can be more moderate and the base-doping concentration can be higher in HBTs than those in BTs, resulting in smaller C_{TE} and r_B. Therefore, a faster speed can be achieved in HBTs with a larger physical dimension, which is an advantage similar to that of MODFETs over MOSFETs. On the other hand, processing technology developed for MOSFETs is readily available and can be a great benefit to developing BT technology. In terms of device physics, velocity overshoot becomes important in both base and collector regions of dimensions smaller than 0.2 μm. With the base being heavily doped, impurity scattering must be taken into account in calculating the mobility and diffusion constant of carriers.

PROBLEMS

13.1. Derive expressions for $Y(\omega_i)$ and $Y(\omega_p)$ similar to Eq. (13.2.6). In your derivation, be sure that the cross products involving two frequencies are treated properly. The power $P(\omega)$ delivered to the component at ω is given by Re $Y(\omega)v(\omega)^2/2$. Show that

$$\frac{P_p(\omega_p)}{\omega_p} + \frac{P_s(\omega_s)}{\omega_s} = 0 \text{ and } \frac{P_p(\omega_p)}{\omega_p} + \frac{P_i(\omega_i)}{\omega_i} = 0$$

The relation is a special case of the Manley–Rowe relation derived from parametric amplifiers and oscillators [J. M. Manley and H. E. Rowe, *Proc. IRE* (now IEEE), Vol. 44, p. 904, 1956; Vol. 47, p. 2115, 1959]. Explain physically why it is impossible for all the *P*'s to have the same sign.

13.2. (a) The choice of an operating point for the varactor diode (Fig. P13.2) is a compromise between a large capacitance change (that means a large interaction parameter) and a high quality factor Q. The value of Q sets a limit as to the maximum swing of v_p to the positive side of the bias. Show that in an abrupt p^+-n narrow-base diode, the value of Q is

$$Q = \frac{\omega C_T}{G} = \frac{N_d W}{e\mu_p n_i^2} \left(\frac{e\epsilon N_d}{2}\right)^{1/2} \frac{\omega}{\sqrt{V_d - V_f}\, \exp(eV_f/kT)}$$

where W is the base width and N_d is the donor concentration in the n region. Since in general the value of V_f is small, only the space-charge capacitance is taken into account.

(b) Consider a Si diode with $W = 5 \times 10^{-4}$ cm, $N_d = 10^{16}$ cm^{-3}, $\epsilon = 12\epsilon_0$, $\mu_p = 480$ cm^2/V-s, $n_i = 1 \times 10^{10}$ cm^{-3}, and $V_d = 0.90$ V. Find the maximum value of V_f allowed in order to give Q a value of 100 at $\omega = 5 \times 10^{10}$ rad/s.

(c) Once the maximum value of V_f is set, we turn our attention to the reverse dc bias V_0. Show that for maximum negative conductance, the value of V_0 should be chosen that

$$V_0 = \frac{V_d - (m + 1)V_f}{m}$$

where $m = \frac{1}{2}$ for an abrupt junction.

13.3. From Fig. 13.1 we see that energy input at ω is converted into energy output at $\omega/2$. This case represents the degenerate operation of a parametric amplifier with $\omega_p = \omega$ and $\omega_s = \omega_i = \omega/2$ in Eq. (13.2.4). Show that for $v_1 = v_p \sin(\omega_p t + \theta) + v_s \sin \omega t$, the admittance seen by an input signal at ω_s is given by

$$Y(\omega_s) = i\omega_s C_0 + \frac{m v_p \omega_s C_0}{2(V_d + V_0)} \exp(+i\theta)$$

Based on Eq. (13.2.1) and Fig. 13.1b, deduce the value of θ for maximum amplification of the input signal.

13.4. (a) For a transmission line loaded with semiconductor diodes at regular intervals, the current and the voltage obey the equations

$$\frac{\partial i}{\partial z} = -(C + fv)\frac{\partial v}{\partial t} \quad \text{and} \quad \frac{\partial v}{\partial z} = -L\frac{\partial i}{\partial t}$$

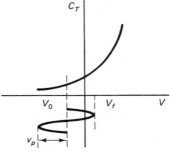

Figure P13.2

where C and L represent the capacitance per unit length and the inductance per unit length of the line, and fv represents the voltage-dependent part of the line capacitance. Letting

$$v_s + v_s^* = a_s \exp\left[i(\omega_s t - k_s z)\right] + \text{complex conjugate}$$

and similar expressions for $v_p + v_p^*$ and $v_i + v_i^*$, show that

$$\frac{\partial^2 v_s}{\partial z^2} = -\omega_s^2 \left(LC + Lf\frac{a_p a_i^*}{a_s} \right) v_s$$

(b) From the equation above and a similar equation for $\partial^2 v_i/\partial z^2$, derive the equations

$$-2ik_s \frac{\partial a_s}{\partial z} = -\omega_s^2 Lf(a_p a_i^*)$$

$$+2ik_i \frac{\partial a_i^*}{\partial z} = -\omega_i^2 Lf(a_p^* a_s)$$

by neglection $\partial^2 a_s/\partial z^2$ and $\partial^2 a_i^*/\partial z^2$ terms. Show that both the signal wave and the idler wave will grow according to $\exp(\alpha z)$, where α is the amplification factor given by

$$\alpha = \frac{f}{2C} (k_s k_i)^{1/2} |a_p|$$

under the phase-match condition $k_p = k_s + k_i$.

13.5. **(a)** Since $m_c^* < m_v^*$ in direct-gap semiconductors such as GaAs, ξ_p is expected to be much smaller than ξ_n in Fig. 13.4. Using Eq. (10.6.14) for P_1 in Eq. (13.3.3), show that the quantity D in Eq. (13.3.7) is given by

$$D = \overline{\mathscr{C}}\xi_p$$

(b) Given: $m_c^* = 0.068m_0$, $m_v^* = 0.64m_0$, $n = 5 \times 10^{18}$ cm^{-3}, and $p = 10^{20}$ cm^{-3}. Find the values of ξ_n, ξ_p, and V_d for the tunnel diode with the aid of Fig. 6.33. Also calculate the average junction field at a forward bias of 0.2 V, and the forward bias at which the tunneling current ceases. In the circulation, ignore the band-shrinkage and band-tailing effects.

13.6. **(a)** Find the values of the junction potential $V_d - V_a$, the depletion width W_d, and the average field $E = (V_d - V_a)/W_d$ for the tunnel diode considered in the chapter in conjunction with Eq. (13.3.8) at a forward bias $V_a = 0.14$ V. Then calculate the tunnel current-density.

(b) Suppose that the tunnel diode has a base width $W_n = W_p = 10^{-5}$ cm and diffusion constants $D_n = 72$ cm^2/s and $D_p = 3.2$ cm^2/s. Calculate the diffusion current density at $V_a = 0.14$ V and the forward bias needed to give a diffusion current density equal to the tunnel current density at $V_a = 0.14$ V.

13.7. We illustrate the principle of resonant tunneling by using a much simplified example. As illustrated in Fig. P13.7, a quantum well is formed by an infinite barrier at right ($V_2 = \infty$ at $x = b$) and a finite barrier (V_1 at $x = a$). If the barrier thickness a is very large, the well is isolated from the contact region (I) and hence the wave functions in the well and barrier regions can be represented by

$$\psi_3 = A_3 \sin k(x - a - b)$$
$$\psi_2 = A_2 \exp \alpha(x - a)$$

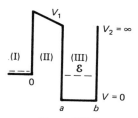

Figure P13.7

(a) Derive the characteristic equation

$$k \cos kb + \alpha \sin kb = 0$$

and express k and α in terms of the electron energy $\mathscr{E}$ and barrier potential V_1, which is assumed constant. The effective masses are assumed equal in all the three regions.

(b) Now we reduce the barrier thickness a, and the problem becomes a three-region problem with the replacement of ψ_2 by

$$\psi_2(\text{new}) = A_2 \exp [\alpha(x - a)] + B_2 \exp [-\alpha(x - a)]$$

and the addition of ψ_1

$$\psi_1 = A_1 \exp (ik_1 x) + B_1 \exp (-ik_1 x)$$

Find the ratio of A_3/A_1.

(c) For practical devices, the value of αa cannot be very small; therefore, the terms with $\exp (\alpha a)$ dependence are more important than the terms with $\exp (-\alpha a)$ dependence. Under the assumption of small perturbation, show that the ratio A_3/A_1 is maximized when the characteristic equation is satisfied.

13.8. (a) Calculate the first quantized energy $\mathscr{E}_0$ for a quantum well of width $b = 60$ Å between two potential barriers of height $\mathscr{E}_b = V_1 = V_2 = 0.2$ eV for two well materials: (1) GaAs with $m_c^* = 0.067m_0$, and (2) Ga$_{0.53}$In$_{0.47}$As with $m_c^* = 0.045m_0$. In this calculation, the electric field is zero.

(b) Now we consider the situation in a resonant tunneling device where a field $E = 5 \times 10^5$ V/cm exists at a bias to produce the peak current. The width of the barrier layer is $a = L_B = 40$ Å. Use Eq. (11.4.3) to find $\mathscr{E}_0$ and explain why the two calculated values of $\mathscr{E}_0$ are different. Also show that the tunneling probability is

$$P = \exp \left[-\frac{4\sqrt{2}}{3} \frac{m_c^* \mathscr{E}_b^{3/2}}{eE\hbar} (1 - \cos^3\theta_1) \right]$$

where $\theta_1 = \sin^{-1}\sqrt{eEa/\mathscr{E}_b}$.

13.9. Draw an energy-band diagram for Ge to show the relative position of the two bands involved in (a) direct transition of Fig. 13.8b and (b) indirect transition of Fig. 13.8a. Also obtain the phonon energies involved in part (b) and check the values thus obtained against the values from the phonon spectrum shown in Fig. 2.38.

13.10. (a) The breakdown voltage of a Si p-n junction shown in Fig. 10.13 can be fit by

$$V_B = 58 \left(\frac{N}{10^{16}} \right)^{-0.715}$$

where N is expressed in cm^{-3} and V_B is expressed in V. Given $N = 10^{17}$ cm^{-3}, calculate the values of V_B, E_{max}, and W, and compare the calculated values of E_{max} and W from Eqs. (13.4.1) and (13.4.2) with the values given in Fig. 13.9.

(b) According to our discussion in Section 10.6, the coefficient for impact ionization can be approximated by $\alpha = aE^m$. From the dependence of V_B on N above, find the value of m and then the position at which α drops to $0.1\alpha_{max}$.

13.11. (a) The second harmonic plays an important role in the operation of TRAPATT diode. Sketch the voltage waveform represented by

$$v = v_1[\sin \omega t + a_v \sin (2\omega t + \theta_v)]$$

for $\theta_v = 0$ and $\theta_v = \pi$. Compare the waveforms with $a_v = 1$ and $a_v = 0.5$. Show that the two waveforms for $\theta_v = \pi$ resemble the waveforms shown in Fig. 13.11e.

(b) The current, however, takes a different waveform. A large current exists only during the interval *abcd*. Sketch the current waveform given by

$$i = i_1[\sin (\omega t + \theta) + a_i \sin (2\omega t + 2\theta + \theta_i)]$$

for $\theta_i = \pi/2$ and $\theta_i = \pi$, and determine which θ_i should be chosen to approximate the desired waveform. Also comment on the role that the phase angle θ may play.

13.12. (a) Analyze the compositions of J_e and J_h in Eq. (13.5.9) at $x = 0$ and $-w_a$, and their relations to J_c and the electron and hole components of J_0. By doing so, verify Eq. (13.5.9).

(b) For an inductive element, the voltage $v = L \, di/dt$ is positive with increasing i. Show that the term involving J_c/τ_a in Eq. (13.5.12) is inductive.

13.13. The coefficient for impact ionization shown in Fig. 10.9 for Si can be approximated by

$$\alpha = \sqrt{\alpha_e \alpha_h} = 2.5 \times 10^4 \left(\frac{E}{4 \times 10^5}\right)^{5.4}$$

Find the value of $\langle \alpha' \rangle$ at $E = 4 \times 10^5$ V/cm. Also calculate the value of ω_r in Eq. (13.5.16) at $J = 10^3$ A/cm^2.

13.14. In the derivation of Eq. (13.5.29), we first let $K_1 = ev_s(p_1 + n_1)$ and $K_2 = ev_s(p_1 - n_1)$. Then we find $\partial K_1/\partial x$ and $\partial K_2/\partial x$ by subtracting and adding Eqs. (13.5.2) and (13.5.3), respectively. Finally, obtain Eq. (13.5.29) by keeping the ac components of $\alpha(n + p)v_s$ and by eliminating K_1 and K_2 from the three equations for $\partial K_1/\partial x$, $\partial K_2/\partial x$, and $\partial E_1/\partial x$.

13.15. (a) For the TRAPATT mode of operation, the current density $J_r = J_{c0}$ in Eq. (13.5.30) is very large, so that the second term becomes the dominant term. Under the circumstances, k can be approximated by $ig + \beta$, where g is the gain constant and β is the phase constant. Show that

$$\beta = \frac{\alpha_0 \omega}{v_s} \left(\frac{\epsilon v_s}{2\langle \alpha' \rangle J_{c0}}\right)^{1/2}$$

(b) An avalanche diode goes into the TRAPATT mode when the phase velocity ω/β of the shock wave (the E field) exceeds the saturation velocity v_s of carriers, and thus the generated carriers become trapped behind the moving avalanche region. Show that the condition for $\omega/\beta > v_s$ is

$$J_{c0} > \frac{\epsilon \alpha_0^2 v_s}{2\langle \alpha' \rangle}$$

(c) Given $\langle \alpha_0 \rangle = 2.5 \times 10^4$ cm^{-1}, $\langle \alpha' \rangle = 0.34$ V^{-1}, and $v_s = 8 \times 10^6$ cm/s, calculate the minimum value of J_{c0} needed.

13.16. The v_d versus E curve for GaAs shown in Fig. 10.25 can be approximated by an empirical relation

$$v_d = \frac{\mu_1 + 1.4(E/5000)^3\mu_2}{1 + 1.4(E/5000)^3} E$$

with $\mu_1 = 8000$ cm^2/V-s and $\mu_2 = 500$ cm^2/V-s. Find the value of dv_d/dE at $E = 5000$ V/cm, and compare the calculated value with the slope of the curve. Given $n = 10^{16}$ cm^{-3} and $\epsilon = 1.14 \times 10^{-12}$ F/cm, calculate the time constant and the sample length for a significant buildup of space-charge waves.

13.17. Show that $\rho = B \exp{(\omega_{ac}x/|v_d|)}$ satisfies Eq. (13.6.3). Also find the value of dv_d/dE from Fig. 10.25 and calculate the value of the gain coefficient at $E = 5000$ V/cm for $n = 10^{16}$ cm^{-3}.

13.18. Starting from Eq. (13.6.12), derive Eq. (13.6.14). Show that $ikl = 2.09 \pm i7.46$ is a solution of Eq. (13.6.14). Also obtain the stability condition of Eq. (13.6.16) from the value of ikl.

13.19. (a) Use Poisson's equation to obtain the field distribution across a depletion ($\Delta n = -$) region and the v_d curve of Fig. 13.16 to show that the depletion charge $\rho = -e\,\Delta n$ will also grow if the sample is biased in the negative differential-mobility region.

(b) Use the velocity curve and the field distribution of Fig. 13.17 to explain that both the accumulation and depletion layers will grow in amplitude by showing that $\Delta n = +$ becomes more positive and $\Delta n = -$ becomes more negative if E_0 is biased in the negative differential-mobility region.

13.20. The derivation of Eq. (13.8.13) from Eq. (13.6.5) involves three steps. First, include the diffusion-current term in Eq. (13.8.11) and show that the integral resulting from it is identically zero. Second, explain the origin of the term env in Eq. (13.8.11) and the physical meaning of v. Show that the second integral is also zero. Finally, state clearly the steps taken in transforming the first integral in Eq. (13.8.11) into the integral in Eq. (13.8.13).

13.21. Give a detailed explanation of why operating point A in Fig. 13.18b is unstable and operating point B is stable by applying the equal-area rule to the two points after they are slightly perturbed from A and B.

13.22. (a) Using the data in Table 13.1, calculate the relative effects of the two terms in parentheses in Eq. (13.10.10). Note that the areal density N_s in a MODFET is limited to a value less than 8×10^{11} cm^{-2} according to Fig. 11.6 and that the doping concentration N_d in a MESFET is limited by considerations of keeping the gate–drain capacitance and the gate current low.

(b) Given: (1) $N_s = 8 \times 10^{11}$ cm^{-2}, d_s = source–gate separation of 0.4 μm, $L_G = 1$μm, and μ_n' = channel mobility of 3000 cm^2/V-s for the MODFET, and (2) $N_d = 5 \times 10^{16}$ cm^{-3}, ϕ_i = Schottky-barrier potential of 0.8 V, a = n-channel thickness of 0.3 μm, the same d_s and L_G, and μ_n' = 4000 cm^2/V-s for the MESFET, calculate the values of source resistance R_S per gate width in Ω-mm for the two devices. Also estimate the gate–drain capacitances C_{GD} (pF/mm) for the two devices if the GaAlAs layer in Fig. 13.29 has a doping concentration 10^{18} cm^{-3} and a thickness 600 Å for the MODFET and if the Schottky gate requires a potential $\phi_i + V = 3$ V to pinch off the channel for the MESFET.

13.23. From Eq. (13.10.10), it is apparent that in order to achieve a high f_{max}, we need to increase R_{DS} and to reduce C_{GD}. Sketch the boundary of the gate–drain junction and its movement as the drain voltage is increased in a MESFET. Explain the physical origin of

R_{DS}. Discuss the effects of increasing N_d in the channel of a MESFET on the values of R_{DS} and C_{GD}.

13.24. The sum of all capacitances in Table 13.1 has a value about 1 pF/mm for $L_G = 1$ μm. Using the value of τ_t shown in Fig. 13.22, calculate the values of the logic delay τ_d for 1-μm MOSFET, MESFET, and MODFET with a gate area 1 μ × 100 μm if $V_{dd} = 5$ V, $\Delta V = 2$ V, and $P = 10^{-3}$ W per gate are assumed for the FETs. Repeat the calculation if $V_{dd} = 3$ V, $\Delta V = 0.5$ V, and a gate area of 0.5 μm × 50 μm are used. Is there a clear speed advantage for GaAs FET over Si FET?

13.25. (a) Consider a Si-implanted GaAs MESFET with a barrier of $\phi_{Bn} = 0.95$ V (in Fig. 9.21b) and an implant depth of $a = 0.15$ μm. Calculate the threshold voltage and indicate whether the MESFET operates in the enhancement or depletion mode for two sheet carrier concentrations $Na = 8.2 \times 10^{11}$ cm^{-2} and 10.1×10^{11} cm^{-2}. Compare the values of V_T thus obtained with the values shown in Fig. 13.28.

   (b) Read the article by Anholt and Sigmon (R. Anholt and T. R. Sigmon, *IEEE Electron Device Lett.*, Vol. EDL-8, p. 16, 1987) and give a concise description of the model used to explain the effect of dislocations on threshold voltage.

13.26. (a) Assuming that the "EL2" deep donor has an energy 0.74 eV below the conduction band edge, calculate the resistivity of a semi-insulating GaAs sample with $N_d = 5 \times 10^{14}$ cm^{-3} and a background shallow-acceptor concentration $N_a = 3 \times 10^{13}$ cm^{-3}. Also find the current in a sample 26 μm in length and 10 μm × 400 μm in cross section at $V_{BG} = 2$ V.

   (b) Assuming further that the contact regions for the backgate, the source, and the drain in Fig. 13.30 have $\mathscr{E}_f = \mathscr{E}_c$, find the width of the two space-charge regions for the n^+-i-n^+ structure. Also calculate the voltage required to extend the acceptor space charge over the entire 26-μm length. Compare the calculated values of I and V with the values shown in Fig. 13.30.

13.27. (a) The threshold voltage V_T for a MODFET is defined as the applied gate voltage to make $\mathscr{E}_f = \mathscr{E}_c$ in GaAs at the interface, and the approximate value of V_T can be found by assuming full ionization in the doped AlGaAs layer and ignoring potential variation in the undoped AlGaAs buffer layer in Fig. 11.14. Given $\phi_B = 1.23$ V for the gate metal, $\Delta\mathscr{E}_c = 0.22$ eV for the conduction-band-edge discontinuity, and $N_d = 2 \times 10^{18}$ cm^{-3} for the doping concentration, find the thickness d for the doped AlGaAs layer to have the MODFET operating in the E-mode with $V_T = 0.13$ V.

   (b) Show that the variation of V_T is related to the variation of d by

$$\frac{\Delta V_T}{\Delta d} = - \left[\frac{2N_d(e\phi_B - eV_T - \Delta\mathscr{E}_c)}{\epsilon} \right]^{1/2}$$

   Calculate the value of $\Delta V_T/\Delta d$ for $V_T = 0.13$ V and that of Δd allowed in order to keep $\Delta V_T < 0.1$ V.

13.28. Repeat Problem 13.27 for a MODFET operating in the D mode with $V_T = -0.93$ V.

13.29. Read the article by Ogura et al. on the design and characteristics of LDD-MOSFET (S. Ogura, P. J. Tsang, W. W. Walker, D. L. Critchlow, and J. F. Shepard, *IEEE Trans. Electron Devices*, Vol. ED-27, p. 1359, 1980). Compare the doping profiles for the LDD and conventional MOSFETs used in the study (Figs. 1 and 6). Discuss the effects of the lightly doped drain on breakdown voltage, threshold voltage, and their variations with channel length (Figs. 5 and 9). Also comment on the effect of the LDD on the transconductance (Fig. 14).

13.30. Read the article by Matsumoto et al. on the choice of n^- surface concentration in LDD-MOSFET (Y. Matsumoto, T. Higuchi, T. Mizuno, S. Sawada, S. Shinozaki, and O. Ozawa, *IEEE Trans. Electron Devices,* Vol. ED-32, p. 429, 1985). Discuss the experimental results, showing the effects of reducing n^- surface concentration on threshold-voltage shift, drain current, and gate current (Figs. 8, 10, and 7). Also comment on why it is desirable to have an offset ΔL between the positions of gate edge and n^- depletion edge.

13.31. The values of N_d for doped AlGaAs and GaAs layers are 2×10^{18} cm^{-3} in Fig. 13.36, and donors are fully ionized. We define a threshold condition such that $\mathscr{E}_f$ starts to rise above $\mathscr{E}_c$ for $V_G > V_T$. Given $d_1 = 110$ Å for GaAs, $d_2 = 260$ Å for AlGaAs, $\phi_B = 1.23$ V for the metal barrier, and $\Delta \mathscr{E}_c = 0.22$ V in Fig. 13.36, calculate the values of threshold voltage V_T for the D mode (Fig. 13.36a) and the E mode (Fig. 13.36b) MOD-FETs. Also find the values of K factor for the two MODFETs if we use $\mu = 4000$ cm^2/V-s, $L_G = 2$ μm, and $Z = 20$ μm.

13.32. **(a)** Refer to the discussion in Section 13.12 regarding Figs. 13.39 and 13.40. Assuming that $d = 450$ Å for the thickness of the AlGaAs layer and $V_T = -0.83$ V for the threshold voltage, calculate the value of the areal carrier density N_s for $V_G = 0$ and $V_D = 0$, and that of the parasitic source resistance R_S.

(b) Note that the curve for V_{DS} in Fig. 13.40 assumes that $V_S = 0$. Use $\overline{E}_{GS}$ given in Fig. 13.39 to obtain a new curve for V_{DS} corrected for V_S. Determine whether the cases (1) $V_G = 0$ and $V_D = 0.5$ V and (2) $V_G = -0.4$ V and $V_D = 0.5$ V fall in the current saturation region. Next use $\overline{E}_{CH}$ given in Fig. 13.39 and Eq. (13.12.2) to calculate the values of average channel mobility $\overline{\mu}_{CH}$ for the two cases. Finally, make appropriate modification of either Eq. (13.12.1) or (13.12.2) to find I_D for the two cases. Compare the calculated values of I_D with the values given in Fig. 13.40.

13.33. Repeat Problem 13.32 for a different case with $V_G = 0$ and $V_D = 0.9$ V. For comparison, calculate the value of I_D based on Eq. (13.12.1) with $V_S = 0$ and $\overline{\mu}_{CH} = 5000$ cm^2/V-s.

13.34. The results shown in Fig. 13.42 and those shown in Fig. 10.24 are all based on Monte Carlo calculation. The two results are somewhat different because of different parameters used in the two calculations, such as deformation potentials for optic and acoustic phonon scattering, and the satellite valley chosen and hence the effective mass m_s^*. Therefore, this problem is intended only for a qualitative comparison. Compare the values of $\mathscr{E}$ from Figs. 10.24 and 13.42 at $E = 10$ kV/cm and 20 kV/cm. From Table 11.2 find m_s^* for the X valley and then calculate the values of m_{ss}^* from $n_1/(n_1 + n_2)$ of Fig. 10.24 and compare the values thus obtained with those of Fig. 13.42. Finally, using Eq. (13.12.17), calculate the values of τ and $\tau_{\mathscr{E}}$ at $E = 10$ kV/cm and 20 kV/cm, and compare the calculated values with the low-field value of τ.

13.35. **(a)** Derive Eq. (13.12.22) from Eq. (13.12.21). Given $\alpha = 1.3$, calculate the values of t_{max} and v_{max} and compare the calculated values with the values obtained from Fig. 13.43.

(b) For a GaAs FET with $L_G = 1$ μm operated under a $V_D = 2$ V, find the steady-state velocity v_{ss} from Fig. 13.42 and the average transient velocity v_{av} from Fig. 13.43. As the gate length L_G is scaled down to 0.5 μm under constant E, do we expect v_{av} to remain unchanged?

13.36. This problem illustrates the sensitivity of the threshold voltage V_T to substrate doping and oxide thickness in a MOSFET designed for 1-μm gate length (Table 13.3). Given ϕ_M(Al) $= 4.10$ V, χ(Si) $= 4.15$ V, ϵ(Si) $= 11.7\epsilon_0$, and ϵ'(SiO$_2$) $= 3.7\epsilon_0$, find the values of V_T for Al–SiO$_2$–Si MOSFETs with $d_{ox} = 200$ Å and various substrate doping concentrations $N_a = 10^{15}$ cm^{-3}, 10^{16} cm^{-3}, and 10^{17} cm^{-3}. Choose the value of N_a to keep $|V_T| < 0.5$ V.

Repeat the calculation for MOSFETs with $N_a = 5 \times 10^{15}$ cm^{-3} and various oxide thicknesses $d_{ox} = 150$ Å, 200 Å, and 250 Å. Find the tolerance on Δd_{ox} to keep $|\Delta V_T| < 0.1$ V. In the calculation, ignore the short-channel effect.

13.37. **(a)** This problem illustrates the sensitivity of the threshold voltage V_T to substrate doping and oxide thickness in a MOSFET designed for a 0.25-μm gate length (Table 13.3). Using the material constants given in Problem 13.36, find the value of V_T for Al–SiO$_2$–Si MOSFETs with $d_{ox} = 50$ Å and various substrate doping concentrations $N_a = 10^{15}$ cm^{-3}, 10^{16} cm^{-3}, and 10^{17} cm^{-3}. Choose the value for N_a to keep $|V_T| < 0.25$ V.

(b) Repeat the calculation for MOSFETs with $N_a = 5 \times 10^{16}$ cm^{-3} and various oxide thicknesses $d_{ox} = 45$ Å, 50 Å, and 55 Å. Find the tolerance on Δd_{ox} to keep $|\Delta V_T| < 0.05$ V. In the calculation, ignore the short-channel effect.

13.38. **(a)** Calculate the contribution of the voltage across oxide, V_{ox}, to the threshold voltage V_T in Eq. (13.13.7) for $d_{ox} = 5$ nm and $N_a = 3 \times 10^{16}$ cm^{-3}.

(b) Approximating the boundaries of the junction and space-charge region in Fig. 13.46a by two circular arcs with a center at a distance x_j away from the gate edge and counting the short-channel effect only at the drain side, show that

$$\frac{L'}{L} = 1 - \frac{x_j}{L}\{[(1 + a_1)^2 - a_2^2]^{1/2} - 1\}$$

where $a_1 = x_d/x_j$, $a_2 = x_{dmax}/x_j$, x_{dmax} is the space-charge width under the channel for surface inversion, and x_d is an extension of the depletion region of the drain junction on the channel side.

(c) Given $V_d = 0.93$ V for the built-in potential of the drain junction and $V_B = 0$ in Eq. (13.13.9), calculate the values of x_d for $V_D = 0$ and 1 V, and thus the threshold voltage shift ΔV_T due to ΔV_D for a MOSFET with $x_j = 0.14$ μm and $L = 0.35$ μm.

13.39. Repeat Problem 13.38 for $N_a = 3 \times 10^{17}$ cm^{-3}. The other change is $V_d = 0.99$ V instead of 0.93 V.

13.40. Figure 13.49 shows the beneficial effect of reduced oxide thickness on gate current from $I_G/I_D = 8 \times 10^{-8}$ for $d_{ox} = 250$ Å to 2×10^{-8} for $d_{ox} = 150$ Å at $L_G = 0.6$ μm. This improvement was attributed to a shift in the lateral field distribution toward the drain (Fig. 13.49). However, the oxide thickness is ultimately limited by tunneling of energetic carriers.

(a) For thermal carriers, the tunneling probability is given by Eq. (10.5.18) with $\mathcal{E}_g$ replaced by an effective tunneling barrier $\mathcal{E}_b$. Using $m = m_0$ and $E = 8.5 \times 10^6$ V/cm from Fig. 13.33 as the breakdown field, and setting $P = \exp(-10)$ as the condition for breakdown, find the value of $\mathcal{E}_b$ for the SiO$_2$-Si interface.

(b) Hot carriers, however, need a minimum energy $3\mathcal{E}_g/2$ to initiate impact ionization. For these electrons, the effective barrier becomes $\mathcal{E}_b - 1.68$ eV. Calculate the attenuation constant for hot carriers in the oxide and the minimum oxide thickness required to keep $P < \exp(-10)$.

13.41. Using a step-function approximation for the Fermi function, that is, n_{D0} of Eq. (6.11.9), and approximating the normal field by

$$E_\perp = \frac{ean}{\epsilon}$$

show that the proportionality constant B in Eq. (13.13.13) is given by

$$B = \left(\frac{5}{9}\right)^{1/2} \frac{h}{m^*} \left(\frac{3\epsilon}{8\pi ae}\right)^{1/3}$$

where a is the extent of the potential well in Fig. 11.10 (about 40 Å for $\mathscr{E}_0$ and considerably larger for higher-energy states). Further, if we express

$$\mu_s = A\left(\frac{10^5}{E_\perp}\right)^{2/3}$$

with $E_\perp$ expressed in V/cm, find the values of A in Si MOSFET and GaAs MODFET with $a = 60$ Å, $m^*(\text{Si}) = 0.26m_0$, and $m^*(\text{GaAs}) = 0.063m_0$.

13.42. Following the procedure outlined in Section 13.15 and using μ_n from Fig. 13.53b, calculate the value of τ_{EC} for the Si BT of Table 13.6 with an emitter area 2.5×4 μm^2 and a collector area 12×6 μm^2.

13.43. Draw the energy diagram for a n-p heterojunction in which the composition in the p region changes from $\text{Al}_{0.3}\text{Ga}_{0.7}\text{As}$ (same as in the n region) to GaAs in a distance 0.12 μm. Assuming that $N_d = 10^{16}$ cm^{-3} in the n region and $N_a = 10^{18}$ cm^{-3} in the p region, and the ratio of $N_c(\text{AlGaAs})/N_c(\text{GaAs}) = 10$, find the built-in potential V_d. Then draw the energy-band diagram under a forward bias to give an injected electron concentration $n_p = 10^{16}$ cm^{-3} in the GaAs region. From the position of the quasi-Fermi level for holes, find the injected hole concentration p_n in the n region. Show that p_n in the heterojunction is reduced by a factor $A\exp\left(-\Delta\mathscr{E}_g/kT\right)$ from p_n in a homojunction, where $\Delta\mathscr{E}_g$ is the difference in gap energy between $\text{Al}_{0.3}\text{Ga}_{0.7}\text{As}$ and GaAs and A is the ratio of N_cN_v in AlGaAs to N_cN_v in GaAs.

13.44. We assume for the emitter–base junction of Table 13.7, $\mathscr{E}_f = \mathscr{E}_{c1}$ in n-(GaAl)As and $\mathscr{E}_f = \mathscr{E}_{v2}$ in p-GaAs.
   (a) Draw energy-band diagram for the heterojunction under thermal equilibrium and show that

$$\mathscr{E}_{c2} - \mathscr{E}_{c1} = -\Delta\mathscr{E}_c + eV_{d1} + eV_{d2}$$

   where V_{d1} and V_{d2} are the built-in potential in each region.
   (b) Show that for $N_a \gg N_d$, $eV_{d1} = \mathscr{E}_{g2}$ (GaAs) $+ \Delta\mathscr{E}_c$.
   (c) Calculate the value of n_p needed to produce $I_E = 10$ mA and then the value of $V_d - V_a$ in (GaAl)As.
   (d) From $V_d - V_a$, calculate the value of C_{TE} and compare the calculated value with the measured value given in Table 13.8.

13.45. Using the structural properties of a HBT in Table 13.7 and letting $V_d - V_a = 0.28$ V (Problem 13.44) for the emitter-base junction, calculate the various device parameters and compare the calculated values with those given in Table 13.8. Take built-in field and mobility dependence on impurity concentration into consideration in your calculation.

14

Optical Devices

14.1 INTRODUCTION

During the past 30 years we have witnessed a revolution in the electronics industry from one based on vacuum tubes to one based on semiconductor devices. The remarkable advances made in semiconductor materials technology have enabled us to explore the use of semiconductors in a different but equally exciting area, the area of optics. Optical devices such as injection lasers, light-emitting diodes (LEDs), and photodetectors, are required and being used in a number of industrial and commercial applications. These include optical fiber communications, optical data storage, robotics, and optical fiber sensors. Potential applications being explored include space communications, and optical interconnects for computer chips, to name a few. It is not an exaggeration to say that we are entering another stage of industrial revolution, the age of optoelectronics, in combining the use of electronic and optical devices for data transmission and information processing.

The purpose of this chapter is to discuss the physical processes involved in the generation and detection of photons in optical devices. The selection and organization of the subject matter for this chapter is based on the following considerations. First, the entire field of photonics includes not only sources and detectors but also modulators, switches, nonlinear devices, and nonreciprocal devices. The field is so vast that a judicious selection of the topics to be discussed is essential. Only the photon generation and detection processes are discussed here because the interaction of radiation with matter is one of the basic components in the treatment of semiconductors. Second, most commonly used semiconductors have band-gap energies in the near-infrared and infrared regions, and involve either band states or shallow impurity states in optical transitions. However, other means are also possible to make a material to respond to photons, for example, by using transitions between the electronic states of a dopant and between the vibrational states of a defect complex. The discussions in Sections 14.2 to 14.4 are presented to illustrate the difference in optical transitions used in various materials ranging from insulators to common semiconductors. The principles of lasers are discussed in Section 14.5 and applied specifically to semiconductor injection lasers in Sections 14.6 to 14.8. Detectors are presented in Sections 14.9 to 14.11.

14.2 LUMINESCENCE IN LARGE-GAP PHOSPHORS

Luminescence refers to any emission of light not directly ascribed to incandescence. Many opto-electronic devices in common usage and of practical importance employ the luminescent property of materials. A cathode-ray tube is a good example. The inside face of the tube is coated with a material known as *phosphor* which possesses luminescent centers. When the fast primary electrons of the cathode ray hit the phosphor, they impart their energy to the luminescent centers by raising the centers from a lower-energy state to a higher-energy state. Upon subsequent return to the lower-energy state, a luminescent center emits electromagnetic radiation (C. C. Klick and J. H. Schulman, "Luminescence in Solids" in H. Ehrenreich, F. Seitz, and D. Turnbull, *Solid State Physics,* Vol. 5, Academic Press, Inc., New York, 1957).

According to the means by which a luminescent material is excited, luminescence is generally categorized into three types: (1) photoluminescence if the excitation is by light or X rays, (2) cathodoluminescence if the excitation is by fast electrons, and (3) electroluminescence if the excitation is by electric field or current. Many household and commercial items use one of the mechanisms as means of excitation for the light-emission process. Here we mention one example for each type: the phosphorescent dial of a clock or watch (photoluminescence), the color (or black and white) TV tube (cathodoluminescence), and the sodium (gas discharge) lamp (electroluminescence). The light-emission mechanisms are similar in these examples even though the luminescent material employed and the method of excitation used are different.

In general, phosphors can be divided into two groups: one for which the host lattice is an insulator and the other for which the host lattice is a semiconductor. In the former case, the photon energy involved in the luminescent transition is determined primarily by the impurity atom, known as the *activator*. Examples of this group are $KCl : Tl^+$, $Al_2O_3 : Cr^{3+}$, and $Y_3Al_5O_{12} : Nd^{3+}$, with Tl^+, Cr^{3+}, and Nd^{3+} ions being the activators. In the latter case, the energy bands of the host crystal as well as the energy levels of the activators are involved in determining the photon energy of the luminescent transition. Important examples of semiconductor phosphors are ZnS, CdS, GaAs, and GaP.

To illustrate the luminescent process in the first group of materials, we consider the case of $KCl : Tl^+$. Figure 14.1a shows the energy band of KCl. In an ionic crystal, the valence band is formed by electrons attached to the negative ions. The transition of an electron from the valence band to the conduction band corresponds to the migration of the electron from the Cl^- ion to the K^+ ion, which thus becomes a neutral atom. Therefore, the conduction band in KCl is ascribed to energy levels of neutral K atoms (N. F. Mott and R. W. Gurney, *Electronic Processes in Ionic Crystals*, Clarendon Press, Oxford, 1948; for further discussion of the energy band of KCl, read L. P. Howland, *Phys. Rev.,* Vol. 109, p. 1927, 1958).

The thallium-activated alkali-halide phosphor is an alkali-halide crystal doped with thallous halide in dilute concentration. The thallous ions (Tl^+) are distributed randomly over lattice sites normally occupied by the alkali ions. For Tl^+ ions to be active luminescent centers, the concentration of thallium must be low so that the thallous ions are essentially isolated from each other. The electron configuration of the ground state of

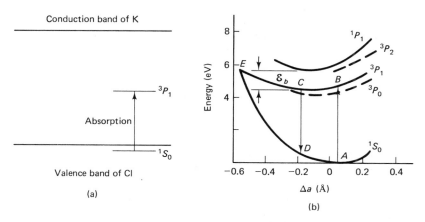

Figure 14.1 (a) Energy-band diagram and (b) configuration model for the ground state ($^1S_0$), the emitting states ($^3P_1$ and $^1P_1$), and the trapping states ($^3P_2$ and $^3P_0$) in KCl : Tl. (F. E. Williams, and P. D. Johnson, *Phys. Rev.*, Vol. 113, p. 97, 1959.)

Tl^+ is $6s^2$ (Table 1.2) with antiparallel spins for the two electrons. The ground state is represented by $^1S_0$. The lowest excited states of a Tl^+ ion have the configuration of $6s6p$. The two electrons in the excited states can have either parallel or antiparallel spins. The excited states are represented by $^3P_0$, $^3P_1$, $^3P_2$, and $^1P_1$. The notation ^xL_y, with x and y being denoted by a number and L by a letter, is called the *spectroscopic notation*.

When an atom has more than two electrons, questions arise as to how the electrons occupy the various quantum states available to them. For an atom having N electrons and an effective nuclear charge ze, the energy (or the quantum mechanical Hamiltonian) can be expressed as

$$\mathcal{H} = \sum_{i=1}^{N} \left[\frac{p_i^2}{2m_0} - \frac{ze^2}{4\pi\epsilon_0 r_i} + \xi(r_i)\mathbf{l}_i \cdot \mathbf{s}_i \right] + \sum_{i \neq j=1}^{N} \frac{e^2}{4\pi\epsilon_0 r_{ij}} \qquad (14.2.1)$$

where the first and second terms represent, respectively, the usual kinetic and potential energy. The third term represents the interaction energy between the spin and the orbital motion of an electron, and the fourth term represents the Coulomb interaction energy between the *i*th and *j*th electrons. In cases where the coulombic interaction is much stronger than the spin-orbit interaction, the spins of different electrons add first and then the orbital angular momenta add. For two electrons, we have

$$\mathbf{S} = \mathbf{s}_1 + \mathbf{s}_2 \qquad (14.2.2)$$

$$\mathbf{L} = \mathbf{l}_1 + \mathbf{l}_2 \qquad (14.2.3)$$

The $6s6p$ electrons have a resultant $S = \frac{1}{2} + \frac{1}{2} = 1$, and a resultant $L = |m_1 + m_2| = 1$ with $m_1 = 0$ and $m_2 = 1$. The $6s^2$ electrons, on the other hand, have $S = 0$ and $L = 0$. The superscript x is related to S by $x = 2S + 1$ and called the *multiplicity number*. The letters S, P, D, G, F indicate the resultant orbital quantum number $L = 0, 1, 2, 3, 4$ for S, P, D, G, F, respectively.

If no external field is applied or if the external field is not strong enough to break the spin-orbit coupling, the spin vector **S** and the orbital momentum vector **L** combine to form a resultant **J** with

$$\mathbf{J} = \mathbf{S} + \mathbf{L} \qquad (14.2.4)$$

expressed in units of $\hbar$. The quantity **J** represents the total angular momentum of the electrons involved, and the value of J is often called the *inner quantum number*. The coupling scheme represented by Eq. (14.2.4) is often called *Russell–Saunders coupling* (E. U. Condon and G. H. Shortley, *The Theory of Atomic Spectra*, Cambridge University Press, Cambridge, 1959; J. H. Van Vleck, *The Theory of Electrical and Magnetic Susceptibilities*, Oxford University Press, Oxford, 1932). The value of J is given by the subscript y. Because the coulombic interaction is the dominant term in Eq. (14.1.1), the state with the least energy (ground state) is the one that has electrons in different orbital states. On the basis of the Pauli exclusion principle, this rule is equivalent to requiring maximization of S, which is known as the *Hund rule* in atomic spectroscopy. If there are several choices of L for the same multiplicity (i.e., the same S value), the state with the largest L value has the lowest energy.

The energies of the various excited states of a Tl^+ ion in KCl have been calculated by Williams and coworkers, and are shown in Fig. 14.1b (P. D. Johnson and F. E. Williams, *J. Chem. Phys.*, Vol. 21, p. 125, 1953; *Phys. Rev.*, Vol. 113, p. 97, 1959; Vol. 117, p. 964, 1960). The abscissa called the *configuration coordinate*, describes the overall geometrical configuration of nuclei surrounding the luminescent center. Each Tl^+ ion occupies the place of a K^+ ion and is symmetrically surrounded by six Cl^- ions. Since only the interaction of the Tl^+ ion with the six nearest Cl^- neighbors is relevant to the energy of the Tl^+ ion, the abscissa can simply be taken as the distance between the Tl^+ ion and its nearest Cl^- neighbors, using the corresponding distance in a perfect KCl crystal as reference. In the ground state, the thallous ion has a configuration coordinate close to point A, the point of lowest potential energy.

There is a quantum mechanical rule called the *selection rule* which governs the transition between the various states, that is, the allowed values for ΔL, ΔS, and ΔJ. For the present case, the principal selection rule is $\Delta J = \Delta y = 0, \pm 1$ except for the $y = 0$ to $y' = 0$ transition, which is forbidden. For the excitation processes in Fig. 14.1b transitions from $^1S_0$ to $^3P_1$ and $^1P_1$ are allowed, and transitions from $^1S_0$ to $^3P_0$ and $^3P_2$ are forbidden. The same rule applies to the luminescent process. Therefore, the states $^3P_1$ and $^1P_1$ act as luminescent centers, whereas the states $^3P_0$ and $^3P_2$ act as electron traps. After being excited to the $^3P_1$ and $^1P_1$ states, electrons may be transferred to the $^3P_0$ and $^3P_2$ states through transitions induced by lattice vibration. Since the trapping states are metastable states by virtue of the selection rule, electrons once trapped can be stored in these states at low temperatures. Upon heating, electrons may return to the emitting states $^3P_1$ and $^1P_1$, and emission of light results. This effect is known as *thermoluminescence*.

One of the characteristics of an alkali-halide phosphor is that the absorption and emission spectra may occur at slightly different frequencies. First we consider the absorption process. The quantum mechanical selection rule only tells us that the final states can be anywhere along the $^3P_1$ line. According to the Franck–Condon principle, the most probable transition is the one that requires the least change in momentum and position. This transition is shown as the absorption process $A \rightarrow B$ in Fig. 14.1b.

Following excitation, the activated ion is no longer in equilibrium with the lattice because the point B does not represent the point of minimum energy on the $^3P_1$ curve. Since the lifetime of the excited state is sufficiently long for equilibrium to be established between the excited Tl^+ ion and its neighbors, the excited state moves to point C before it returns to the ground state. The emission process is shown as transition $C \rightarrow D$ in Fig. 14.1b.

Figure 14.2 shows the experimental absorption and emission spectra of Tl-activated KCl at 298 K and the theoretical curve calculated from the energy-level diagram of Fig. 14.1b. The general agreement gives conclusive evidence that the $^1S_0 \rightleftharpoons {}^3P_1$ transitions are responsible for the 2460-Å absorption and 3050-Å emission bands. The spread in the absorption and emission bands is due to perturbation of the electronic energy levels caused by thermal vibration of the lattice. Therefore, the width of the band should decrease as temperature goes down, in agreement with experiment. It is obvious from Fig. 14.1b that the absorbed light must have a larger quantum energy than the emitted light. One useful application of this phenomenon is in fluorescent lamps. Common fluorescent lamps are filled with a mixture of argon and mercury gas. Such a gas mixture, when excited, emits an appreciable portion of radiation in the ultraviolet region. To convert the ultraviolet radiation into visible light, we commonly coat the wall of the lamp by a properly chosen luminescent material.

In addition to the shift in the emission spectrum, the configuration-coordinate curves (Fig. 14.1b) can be used to explain the rapid decrease in luminescent efficiency at elevated temperatures. As pointed out by Mott and Gurney (R. W. Gurney and N. F. Mott, *Trans. Faraday Soc.*, Vol. 35, p. 69, 1939), a luminescent center in the excited state (point C of Fig. 14.1b), given sufficient thermal energy, may reach a crossover point E and make a transition to the ground state there. Such a transition is called the radiationless transition because no photon is emitted in the process. The probability that an excited center may reach the point E from C is proportional to $\exp(-\mathscr{E}_b/kT)$; hence the rate of the nonradiative transition (the indirect $C \rightarrow D$ via E transition) is

$$R_1 = N_{ex}B \exp\left(-\frac{\mathscr{E}_b}{kT}\right) \qquad (14.2.5)$$

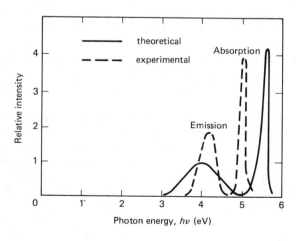

Figure 14.2 Theoretical (solid curve) and experimental (dashed curve) absorption and emission spectra of KCl:Tl. (F. E. Williams, *Phys. Rev.*, Vol. 80, p. 306, 1950.)

Sec. 14.2   Luminescence in Large-Gap Phosphors

where N_{ex} is the density of the excited center, $\mathscr{E}_b$ is the activation energy (Fig. 14.1b), and B is a constant of proportionality. If τ is the lifetime of the center for the radiative process (the direct $C \to D$ transition), the rate for the direct process is $R_2 = AN_{ex}$ with $A = 1/\tau$. Therefore, the efficiency of the luminescent process is

$$\eta = \frac{R_2}{R_1 + R_2} = \frac{A}{A + B \exp(-\mathscr{E}_b/kT)} \tag{14.2.6}$$

For KCl : Tl, $\mathscr{E}_b = 0.69$ eV.

The example KCl : Tl belongs to a class of luminescent materials in which absorption and emission of radiation take place in the same center. In the conventional notation, a phosphor is symbolized by the formula of the host followed by the symbol of the luminescent center. In KCl : Tl, the emission and absorption spectra are characteristic of the luminescent center Tl^+. There is another class of luminescent materials in which the absorption and emission of radiation occur in different atoms. An example of the latter class is $CaSiO_3$: (Pb, Mn). In this material, we have two species, the activator and the sensitizer. The Mn^{2+} ion plays the role of an activator and the Pb^{2+} ion plays the role of a sensitizer.

The fluorescence of divalent manganese has a characteristic spectrum extending from the blue-green to the red. The Mn^{2+} ion has a $3d^5$ electron configuration. The ground state is ^6S, which has been verified by paramagnetic resonance experiments (see, for example, W. D. Hershberger and H. N. Leifer, *Phys. Rev.*, Vol. 88, p. 714, 1952; S. P. Keller, et al., *Phys. Rev.*, Vol. 110, p. 850, 1958; J. Lambe and G. Kikuchi, *Phys. Rev.*, Vol. 119, p. 1256, 1960). The radiative transition is from a group of excited states ^4G to the ground state (A. M. Clogston, *J. Phys. Chem. Solids*, Vol. 7, p. 201, 1958; the calculation was made for ZnS : Mn but is expected to apply to an Mn^{2+} ion in other host lattices). The corresponding characteristic absorption of Mn^{2+} consists of a number of very weak bands which lie in the near ultraviolet and short-wavelength regions of the visible spectrum. In a phosphor such as $CaCO_3$: Mn, ultraviolet radiation around 2500 Å is very ineffective in causing luminescence because of weak absorption in the characteristic Mn^{2+} bands.

The role of a sensitizer is to provide an efficient absorption band so that the activator can be effectively excited. When lead, thallium, or cerium ions are incorporated as sensitizers in $CaCO_3$ or $CaSiO_3$ along with Mn^{2+} ions, the resulting phosphors emit the characteristics orange-red manganese luminescence quite efficiently. Figure 14.3a shows the absorption and emission spectra in $CaSiO_3$: (Pb,Mn). It should be pointed out that each sensitizer introduces a different excitation band, with Pb^{2+} and Tl^+ in the short ultraviolet region (around 2500 Å) and Ce^{3+} in the near ultraviolet region (around 3100 Å).

The mechanism of energy transfer between the sensitizer and the activator has been the subject of many experimental and theoretical investigations (see, for example, F. A. Kröger, *Physica*, Vol. 15, p. 801, 1949; D. L. Dexter, *J. Chem. Phys.*, Vol. 21, p. 836, 1953). In brief, the sensitizer and activator atoms are assumed to oscillate over the range of frequencies corresponding to the broad absorption and emission spectra actually observed. The broadening is mainly the result of the modulation of the electronic energy levels by atomic vibrations. When the broadened spectra of the sensitizer and activator overlap, resonant transfer of energy between them takes place. Such an energy-transfer process is effective over a distance extending from 25 to 50 lattice sites.

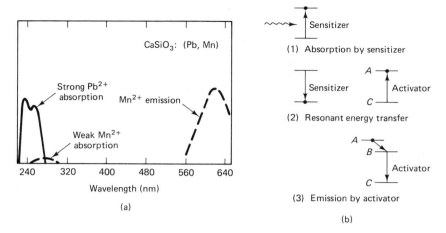

Figure 14.3 (a) Absorption and emission spectra of CaSiO₃: (Pb, Mn) and (b) photoluminescent process involving a sensitizer and an activator.

The Mn^{2+} ion has very weak absorption bands around 2600 Å and 3600 Å. When a proper sensitizer is present, resonant transfer of energy occurs between these levels and the corresponding levels of the sensitizer. This process is illustrated schematically in Fig. 14.3b. The 6000-Å emission band is caused by transitions from the various excited states B to the ground state C of the Mn^{2+} ion.

14.3 LUMINESCENCE IN SULFIDE (MEDIUM-GAP) PHOSPHORS

The group II–VI compounds are an important class of materials for photoconduction and luminescence. Among the group, ZnS and CdS have been most extensively studied, the former for its luminescent properties, the latter for its photoconductive properties. As a general rule, the electrical properties of the group II–VI compounds are between those of the group I–VII and III–V compounds. For example, the energy gaps of ZnS and CdS are, respectively, 3.7 eV and 2.4 eV, compared to 9.4 eV for KCl and 1.4 eV for GaAs. The best estimate of the nature of the binding seems to indicate that the ZnS lattice is about 75% ionic, which is intermediate between predominantly ionic KCl and predominantly covalent GaAs. Because the gap energy in ZnS and CdS is either close to or in the visible region, the materials are useful for generation and detection of visible radiation. The luminescent and photoconductive processes in these materials may involve the band states as well as the impurity states.

In common with most phosphors, the group II–VI compounds are activated by suitable impurities known as activator to produce luminescence in the desired spectral range. To increase the solubility of the activator and to preserve charge balance, it is necessary to introduce impurities of another kind, known as *coactivator*. In Fig. 14.4 the most common activator and coactivator impurities are shown in relation to the ZnS family of phosphors in the periodic table. The role of coactivators can be explained on the basis of charge compensation as follows. To be specific, take ZnS : Cu as an example. The monovalent Cu^+ ion is supposed to enter the ZnS crystal substitutionally and take a normal Zn lattice site. For charge compensation, trivalent ion, say Ga^{3+} ion,

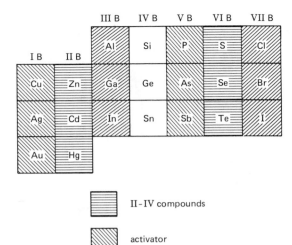

		III B	IV B	V B	VI B	VII B
		Al	Si	P	S	Cl
I B	II B					
Cu	Zn	Ga	Ge	As	Se	Br
Ag	Cd	In	Sn	Sb	Te	I
Au	Hg					

II-IV compounds

activator

coactivator

Figure 14.4 The relation of activator and coactivator impurities to the II–VI compound family of phosphors with regard to their relative positions in the periodic table. (W. W. Piper and F. E. Williams, *Solid State Phys.*, Vol. 6, p. 96, 1958.)

must also replace a Zn^{2+} ion such that one Cu^+ and one Ga^{3+} ion are equivalent to two Zn^{2+} ions. By similar reasoning, a halide negative ion has to take the place of a normal S lattice site to pair with a Cu^+ ion on the normal Zn site.

Since II–VI compounds are situated between the ionic insulators and the predominantly covalent semiconductors, it is natural to adopt the existing and established models for insulators and semiconductors. For covalent semiconductors, the impurity atoms that appear in the periodic table to the left of the original constituent element enter the lattice substitutionally as acceptors; for example, Zn atoms act as acceptors, replacing Ga atoms, in GaAs crystals. By the same token, the activator and coactivator elements become acceptors and donors, respectively, in a II–VI compound. We may look in the other direction and consider ZnS as an ionic crystal made of Zn^{2+} and S^{2-} ions. Thus Cu^{2+} ions replacing Zn^{2+} ions leave the crystal neutral, while Cu^+ ions replacing Zn^{2+} ions create a local region negatively charged. Therefore, Cu^{2+} ions may be considered as neutral or empty acceptor states, while Cu^+ ions represent negatively charged or occupied acceptor states.

One of the most extensively studied phosphors is ZnS : Cu, for its electroluminescent properties. The ZnS : Cu phosphor is suspended in a transparent binder of high dielectric constant. When an alternating voltage is applied to the phosphor, light emission from the phosphor is detected. In ZnS, copper produces two main emission bands, a green band at about 5230 Å and a blue band at about 4450 Å. Figure 14.5 shows the relevant energy levels of the ZnS : Cu system (G. Curie and D. Curie, *J. Phys. Radium,* Vol. 21, p. 127, 1960). The copper impurity acts as a recombination center in both emission processes. Further, the copper level has been identified as due to Cu^+ ion by magnetic susceptibility measurements (R. Bowers and N. T. Melamed, *Phys. Rev.,* Vol. 99, p. 1781, 1955). The identification is based on the fact that the Cu^{2+} ion is paramagnetic, whereas the Cu^+ ion is diamagnetic.

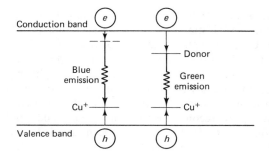

Figure 14.5 Energy levels associated with the characteristic emission bands in Cu-activated ZnS.

Since the process of excitation in an electroluminescent solid is still not well understood, the reader is asked to consult the literature for the proposed theory (for a general view, see H. F. Ivey, "Electroluminescence and Related Effects," *Adv. Electron. Electron Phys. Suppl.*, Vol. 1, 1963). In brief, the following events seem to have taken place. First, electron–hole pairs are generated in ZnS under a high electric field. Second, most luminescent centers have been emptied, and hence become ready to receive an electron. When an electron makes a transition from the donor level to the acceptor (Cu) level, emission of green light results. This model, proposed by Williams and Frener, is known as the *model of associated donor–acceptor centers,* as the green emission needs the simultaneous presence of donors and acceptors (J. S. Frener and F. E. Williams, *J. Phys. Radium,* Vol. 17, p. 667, 1956).

Associated with the green emission in ZnS : Cu, the donors are believed due to halogen or trivalent-metal (Al, Ga, and In) coactivators. In the experiment of Van Gool and Cleiren, equal concentrations of copper and halogen atoms were deliberately introduced into the sulfide, and the intensity of green band was indeed found to be greatly enhanced (W. Van Gool and A. P. Cleiren, *Philips Res. Rep.,* Vol. 15, p. 238, 1960). The copper level also plays the role of a recombination center for the blue emission. However, the origin of the other state represented by the dashed line in Fig. 14.5 is not at all certain. The state could be due to lattice defects or the excited states of copper. It is also possible that the blue emission is the result of a direct transition from the conduction band to the copper acceptor state.

14.4 JUNCTION LUMINESCENCE AND LIGHT-EMITTING DIODES

The electroluminescent process discussed in Section 14.3 requires a high electric field for exciting electrons from the valence band into the conduction band. Another means of excitation is by injection of minority carriers across a semiconductor junction. When a *p-n* junction is biased in the forward direction, excess carriers are generated near the junction (Section 8.3). By *excess carriers* we mean carriers whose concentrations exceed their respective thermal-equilibrium concentrations. In other words, we have more electrons in the conduction band and more holes (or less electrons) in the valence band under a forward-bias condition. The reason for the existence of excess carriers is that electrons are constantly being excited from the valence band to the conduction band in a forward-biased junction.

The injection-excitation process has been applied successfully to III–V compounds. However, the method has only limited use for II–VI compounds because most

II–VI compounds cannot be made to exhibit *p*-type conduction. Figure 14.6 shows (a) the energy-band diagram and (b) the measured emission spectrum of a GaP junction. The junction is doped with Cd as acceptors and with S as donors. The green band is produced by the transition from the S donor level to the Cd acceptor level. If oxygen is also introduced, it forms a Cd-O complex with Cd as its nearest neighbor. The red band denoted by $\mathscr{E}_{\text{pair}}$ is believed to be caused by the recombination of an electron at the Cd–O complex with a hole at an isolated Cd acceptor. The other red emission marked by $\mathscr{E}_{\text{exciton}}$ has been attributed to the recombination of an electron and a hole trapped on a single site of the Cd–O complex (C. H. Henry, P. J. Dean, and J. D. Cuthbert, *Phys. Rev.*, Vol. 166, p. 754, 1968; T. N. Morgan, B. Welber, and R. N. Bhargava, *Phys. Rev.*, Vol. 166, p. 751, 1968; M. Gershenzon, *Bell Syst. Tech. J.*, Vol. 45, p. 1599, 1966).

One of the desirable features of the compound semiconductors is their miscibility. By mixing two compounds of the same family and changing their relative proportions, we may select one particular composition for the desired photon frequency. For example, we may obtain emission spectrum ranging from $h\nu = 0.40$ eV to 1.37 eV from $\text{InAs}_x\text{P}_{1-x}$ (at 77 K) by changing the composition parameter x from 1 to 0. It is also possible to increase the gap energy of a direct-gap semiconductor by mixing it with an indirect-gap semiconductor of a larger gap energy. The variations of the gap energy in GaAs–GaP and GaSb–AlSb are shown in Fig. 14.7. Both GaP and AlSb are indirect-gap semiconductors, having Si-like band structures (Fig. 6.22). The dashed lines represent the positions of the various conduction-band valleys if a linear variation of their position with the composition is assumed. In GaAs, the *L* minima located at *k*[111] actually have a lower energy than the *X* minima (Fig. 6.23), but the variation of *L* minima with composition is not relevant to subsequent discussions and hence not shown in Fig. 14.7.

A light-emitting diode (LED) is made of a *p-n* junction which under a forward bias emits electromagnetic radiation. The use of LEDs can be divided into two broad categories: (1) as a light source for short-distance optical-fiber communications and (2) as a light indicator for display purposes. Because the Rayleigh scattering loss in fibers decreases with wavelength according to λ^{-4}, compound–semiconductor alloys such as $\text{Ga}_{1-x}\text{Al}_x\text{As}$ and $\text{Ga}_{1-x}\text{In}_x\text{As}_{1-y}\text{P}_y$, which cover the wavelength range 0.80 to 1.55 μm,

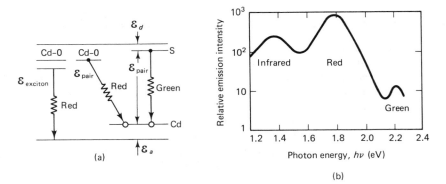

Figure 14.6 (a) Energy-band diagram showing the relevant energies and (b) emission spectrum of a GaP junction.

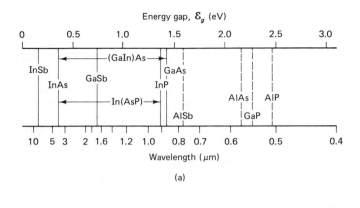

Energy gap, $\mathcal{E}_g$ (eV)

(a)

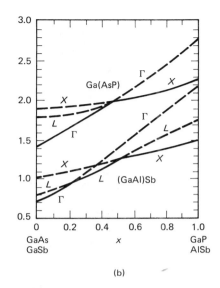

(b)

Figure 14.7 (a) Band-gap energies of III–V compounds and their alloys, and (b) band-gap energies for the $GaAs_{1-x}P_x$ and $Ga_{1-x}Al_xSb$ alloys as functions of composition.

are most suitable for optical fiber communications. Display indicators, on the other hand, require sources emitting in the visible region. Gallium phosphide is most suitable for this purpose because it has emission peaks in the red and green, two of the three primary colors, in addition to the infrared peak. Furthermore, the red or green peak can be enhanced relative to the others by controlling the impurities.

The absorption and fluorescence spectra of $GaAs_{1-x}P_x$ have been studied extensively (D. G. Thomas and J. J. Hopfield, *Phys. Rev.*, Vol. 150, p. 680, 1966; P. J. Dean and D. G. Thomas, *Phys. Rev.*, Vol. 150, p. 690, 1966). For $GaAs_{1-x}P_x$ at 300 K, the direct-gap energy varies with composition as (N. Holonyak, Jr., R. D. Dupuis, H. M. Macksey, M. G. Craford, and W. G. Groves, *J. Appl. Phys.*, Vol. 43, p. 4148, 1972)

$$\mathcal{E}_\Gamma(x) = \mathcal{E}_c[000] - \mathcal{E}_v[000] = 1.441 + 1.091x + 0.210x^2 \qquad (14.4.1)$$

and the (Si-like) indirect-gap energy varies as

$$\mathcal{E}_X(x) = \mathcal{E}_c[100] - \mathcal{E}_v[000] = 1.977 + 0.144x + 0.211x^2 \qquad (14.4.2)$$

The direct gap and indirect gap have a crossover point at $x = 0.45$ with $\mathcal{E}_g = 2.06$ eV. In the indirect gap region, a pure $GaAs_{1-x}P_x$ sample will involve the participation of phonons in photon absorption and emission processes. The various phonon energies identified in GaP are 12.8 meV, 31.3 meV, and 46.5 meV for TA, LA, and TO phonons, respectively (Dean and Thomas). Nitrogen, which is isoelectronic with phosphorus, attracted a great deal of attention because of the role it plays in the absorption and luminescent processes, especially at low temperatures, where phonon-assisted processes are relatively weak. An absorption coefficient as high as 160 cm^{-1} has been reported in GaP with a N concentration of 1.0×10^{19} cm^{-3} (Thomas and Hopfield).

In the study of absorption spectra of GaP : N at 1.6 K by Thomas and Hopfield, many N-related lines were identified. An absorption line labeled as the A line with an energy 0.02 eV below the bandgap energy was attributed to an exciton trapped at an isolated nitrogen atom. Other absorption lines labeled NN_1, NN_2, and so on, were attributed to NN_n pairs, with the NN_1 pair being ascribed to a nitrogen atom on a phosphorus site paired with another nitrogen atom on the nearest shell of phosphorus site and with NN_2 pair having the second nitrogen atom on the next shell of phosphorus site. The situation is rather unusual in that isoelectronic substitution of phosphorus by nitrogen gives rise to discrete states. This is in sharp contrast to a continuous shift of energy gap in the case of isoelectronic substitution of phosphorus by arsenic (Fig. 14.7). The marked difference in the behavior of N and As can be attributed to the fact that N has relatively large electronegativity value (Table 6.6) and small covalent radius compared to P. As a result, a nitrogen atom in GaP will bind an electron and thus act as an electron trap. Once the trap captures an electron, it is ready to capture a hole, thus forming a bound exciton. Furthermore, since the electron is bound to the immediate vicinity of the N atom, the electron wave function must be diffuse in k space due to the uncertainty principle. The diffuse nature of the electron wave vector makes no-phonon radiative transition possible when the bound exciton decays.

The wave-vector dependence of the electron wave function has been analyzed by several workers (R. A. Faulkner, *Phys. Rev.*, Vol. 175, p. 991, 1968; P. J. Dean, *J. Lumin.*, Vol. 1-2, p. 398, 1970; J. C. Campbell, N. Holonyak, Jr., M. G. Craford, and D. L. Keune, *J. Appl. Phys.*, Vol. 45, p. 4543, 1974). Figure 14.8 shows the probability density $|\phi(k)|^2$ in momentum presentation for an electron bound to a shallow donor S in GaP (curve S), to an isoelectronic trap N in GaP (curve N), and to an isoelectronic trap N in $GaAs_{0.55}P_{0.45}$ near the direct–indirect crossover point (curve NC) along the line connecting the Γ point and X point in k space. As mentioned earlier, the relatively large portion of the electron distribution near the electron-attractive N core in real space [as measured by $r\psi(r)$] is responsible for a relatively large value of $|\phi(k)|^2$ in $\mathbf{k}$ space since $\mathbf{k}$ are $\mathbf{r}$ are conjugate quantities. The ratio of the calculated $|\phi(k)|^2$ at $k = 0$ for N and S is consistent with the ratio of the oscillator strength f_N/f_S (Dean). The difference between curves N and NC is caused by band structure enhancement

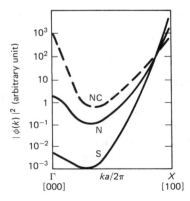

Figure 14.8 Relative magnitude of the probability density $|\phi(k)|^2$ for an electron to be bound to a donor in GaP (curve S), to an isoelectronic trap in GaP (curve N), and to an isoelectronic trap in GaAsP (curve NC). (J. C. Campbell, N. Holonyak, Jr., M. G. Craford, and D. L. Keune, *J. Appl. Phys.*, Vol. 45, p. 4543, 1974.)

(Campbell et al.). For isoelectronic N traps, $|\phi(k)|^2$ has been shown to have the following principal dependence:

$$|\phi(k)|^2 = A[\mathscr{E}(k) - \mathscr{E}_N]^2 \tag{14.4.3}$$

where A is a normalization constant to make $\int |\phi(k)|^2 \, d^3k = 1$, $\mathscr{E}_N$ the isoelectronic trap energy, and $\mathscr{E}(k)$ the energy of the conduction-band electron. Since $\mathscr{E}_N$ is only 10 meV below $\mathscr{E}_X$, the modulus $|\phi(k)|^2$ at $k = 0$ increases rapidly as the composition x approaches the crossover point where $\mathscr{E}_\Gamma = \mathscr{E}_X$.

Figure 14.9 shows the experimental data on quantum efficiency η of $GaAs_{1-x}P_x$ junction luminescence as a function of composition compiled by Campbell et al. for diodes (a) without N doping (square data points) and (b) with N doping (circle data points). The rapid drop of η in case (a) is due to an increasingly large percentage of electrons being in the indirect-gap minima X as the composition x increases. For Boltzmann distribution of electrons between the Γ and X valleys, the electron concentrations n in the two valleys are in the ratio

$$\frac{n_X}{n_\Gamma} = N_X \left(\frac{m_X^*}{m_\Gamma^*}\right)^{3/2} \exp\left(-\frac{\Delta\mathscr{E}}{kT}\right) \tag{14.4.4}$$

where $\Delta\mathscr{E} = \mathscr{E}_X - \mathscr{E}_\Gamma$ and N_X is the number of equivalent X minima. Once n_X becomes comparable to n_Γ, the quantum efficiency η drops rapidly as $\Delta\mathscr{E}$ decreases. The rapid rise of η in case (b) as x moves toward the direct–indirect gap crossover point can be attributed to band-structure enhancement, that is, increase of $|\phi(k)|^2$ at $k = 0$. For extensive discussions of the theory and the practice concerning light-emitting diodes, the reader is referred to an excellent review article on the subject by Bergh and Dean (A. A. Bergh and P. J. Dean, *Proc. IEEE*, Vol. 60, p. 156, 1972).

In Fig. 14.10 we present some pertinent information about light-emitting diodes (LEDs) for use as visible indicators. The wavelengths and photon energies for the three primary colors, red, green and blue, are shown for easy reference. Beside the GaP : N LED, which emits in the green, the $Ga_{0.6}In_{0.4}P$ ternary compound is potentially attrac-

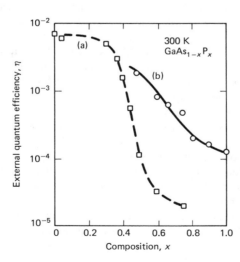

Figure 14.9 External quantum efficiency measured as a function of composition in $GaAs_{1-x}P_x$ light-emitting diodes (a) without N doping (dashed curve) and (b) with N doping (solid curve). (J. C. Campbell, N. Holonyak, Jr., M. G. Craford, and D. L. Kuene, *J. Appl. Phys.*, Vol. 45, p. 4543, 1974.)

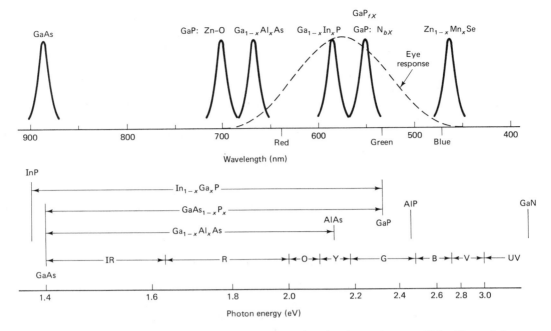

Figure 14.10 Spectral range of various compound semiconductors for use as LEDs. The symbols IR, R, O, Y, G, B, V, and UV, respectively, stand for infrared, red, orange, yellow, green, blue, violet, and ultraviolet.

tive because it emits in the yellow which is in the region of maximum eye response. However, its development was hampered by difficulty in crystal growth and lack of a suitable lattice-match scheme. In the red GaP : Zn–O or GaP : Cd–O LED, the Zn–O and Cd–O complexes play a role similar to N in GaP. The pair with eight valence electrons is isoelectronic with the Ga–P pair except for the increased electron binding energy of about 0.3 eV for the Zn–O complex and about 0.4 eV for the Cd–O complex, compared to a value of 0.01 eV for N. Furthermore, both Zn and Cd are shallow acceptors, with $\mathcal{E}_a = 0.064$ eV for Zn and 0.095 eV for Cd. Therefore, the pair emission from the complex to the acceptor is in the energy range 1.70 to 1.90 eV. The complex trap is also estimated to have an oscillator strength f of about 0.05, comparable to a value of about 0.1 for the N trap.

Several recent developments are worth noting. First, an external efficiency as high as 1.4×10^{-3} has been reported in GaP without isoelectronic doping (T. Kawabata and S. Koike, *Appl. Phys. Lett.*, Vol. 43, p. 490, 1983). The LED is based on phonon-assisted radiative recombination of a free exciton, resulting in an emission peak at 555 nm. This process is dominant in GaP, with N concentration N_N less than 10^{17} cm^{-3}, and still accounts for 40% of the radiative recombination at $N_N = 2 \times 10^{18}$ cm^{-3} and 30% at 1×10^{19} (R. Z. Bachrach and O. G. Lorimor, *Phys. Rev.*, Vol. B7, p. 700, 1973). The slow fall-off of the free-exciton component implies contributions from impurity scattering to the free-exciton recombination in addition to contributions from phonon scattering. However, the formation of free excitons may be hindered by high dopant compensation. Therefore, the success of the GaP free-exciton LED depends on the ability to reduce the background donor concentration in the p region, where the free-exciton recombination takes place. The dopants used in the experiment are S as donors and Zn as acceptors. The second development concerns direct-gap LED. An efficiency as high as 8×10^{-2} in $Ga_{1-x}Al_xAs$ LED of 660 nm has been achieved using double heterostructures to confine injected carriers (H. Ishiguro, K. Sawa, S. Nagao, H. Yamanaka, and S. Koike, *Appl. Phys. Lett.,* Vol. 43, p. 1034, 1983). The wavelength 660 nm is at the point where the electron concentration in the indirect-gap L minima is expected to increase rapidly upon further increase of compositon x. Because of the near lattice match of GaAS and AlAs, heterostructures can easily be made of the ternary compounds of different x. The combined advantages of carrier confinement and operation in the direct-gap region make GaAlAs LED more efficient than GaP : Zn–O LED.

Another interesting and potentially important development is the discovery of $Cd_{1-x}Mn_xTe$ as a dilute magnetic semiconductor (R. R. Galazka, in L. H. Wilson, ed., *Physics of Semiconductors 1978,* Institute of Physics, Bristol, England, 1979, p. 133). Since Mn can be divalent, it replaces Cd without the need of coactivators for charge compensation. For $x < 0.7$, the crystal has the zinc-blende structure. The Mn^{2+} ion has a $3d^5$ configuration; therefore, the ground state is ^6S with a spin parallel arrangement. The excited states belong to ^4G with the spin of one electron being antiparallel to the spins of the other four electrons (J. Lambe and G. Kikuchi, *Phys. Rev.,* Vol. 119, p. 1256, 1960). The spin-spin exchange interaction between the localized magnetic ions and the band electrons leads to giant Faraday rotation in $Cd_{1-x}Mn_xTe$ (A. Gaj, R. R. Galazka, and M. Nawrocki, *Solid State Commun.,* Vol. 25, p. 193, 1978). Under the cubic crystal field, the ground state becomes $^6A_1$ and the excited states split into $^4T_1$, $^4T_2$, $^4A_1$, and ^4E in group-theoretic notation. The photoluminescence peak observed around 2.0 eV has been attributed to intra-ion transitions in Mn^{2+} between $^4T_1$ and $^6A_1$ (R. Y. Tao, M. M. Moriwaki, W. M. Becker, and R. R. Galazka, *J. Appl. Phys.,* Vol. 53, p. 3772, 1982).

Since the initial work of Galazka and coworkers, a number of interesting possibilities have been investigated. These include the observation of photoluminescence of remarkable strength (~ 0.8 mW/sr) of the 2.0-eV peak (~ 600 nm) in $Zn_{0.6}Mn_{0.4}S$ at 300 K (N. V. Joshi, S. Ray, and G. Menk, *Appl. Phys. Lett.,* Vol. 47, p. 1108, 1985) and the observation of stimulated emission in the blue region of the visible spectrum from 451.5 to 455 nm in $ZnSe-Zn_{1-x}Mn_xSe$ multiple quantum wells at temperatures up to 80 K. The energy gap $\mathscr{E}_g$ of $Cd_{1-x}Mn_xTe$ has a value of 1.6 eV for $x = 0$ and 2.7 eV for $x = 0.7$, while that of ZnS is 3.7 eV and that of ZnSe is 2.7 eV at 300 K. Whereas the 2.0-eV emission observed in $Zn_{0.6}Mn_{0.4}S$ is caused by intra-ion transitions between Mn^{2+} levels, the blue emission observed in $ZnSe-Zn_{1-x}Mn_xS$ multiple quantum wells (MQW) is due to free-exciton recombination in ZnSe (R. B. Bylsma, W. M. Becker, T. C. Bonsett, L. A. Kolodziejski, R. L. Gunshor, M. Yamanishi, and S. Datta, *Appl. Phys. Lett.,* Vol. 47, p. 1039, 1985; L. A. Kolodziejski, R. L. Gunshor, T. C. Bonsett, R. Venkatasubramanian, S. Datta, R. B. Bylsma, W. M. Becker, and N. Otsuka, *Appl. Phys. Lett.,* Vol. 47, p. 169, 1985). In dilute magnetic semiconductors, the energy gap generally increases with the addition of Mn. Therefore, in the MQW structure, the excited carriers are confined in the ZnSe wells, where they recombine. The intra-ion transition between Mn^{2+} levels is greatly weakened because of scarcity of carriers in the $Zn_{1-x}Mn_xSe$ barriers.

The development of II–VI compound–semiconductor device technology has been hindered by the problem of self-compensation, which prevents the attainment of either *n*-type or *p*-type conduction. So far, junction luminescence is possible only in hetero-junctions, for example, in *p*ZnTe–*n*CdS (T. Ota, K. Kobayashi, and K. Takahashi, *J. Appl. Phys.*, Vol. 45, p. 1750, 1974). However, lattice mismatch presents a new problem resulting in high interface nonradiative recombination and thus low luminescence efficiency. The advent of advanced thin-film growth technology, namely molecular beam epitaxy (MBE) and metalorganic chemical vapor deposition (MOCVD), will aid our efforts in attaining the stoichiometric composition of and in controlling the dopant types in II–VI compound semiconductors. It is hoped that the recent interests in dilute magnetic semiconductors will stimulate materials research on II–VI compounds. The development of efficient LED in the blue region will require the attainment of lattice-matched heterojunctions with appropriate *p* and *n* regions.

We should point out that it is possible to use multiphoton absorption of radiation from GaAs LED to generate red, green, or blue emission in phosphors containing rare-earth ions (J. E. Geusic, F. W. Ostermeyer, H. M. Marcos, L. G. Van Uitert, and J. P. Van der Ziel, *J. Appl. Phys.*, Vol. 42, p. 1958, 1971; L. F. Johnson, H. J. Guggenheim, T. C. Rich, and F. W. Ostermeyer, *J. Appl. Phys.*, Vol. 43, p. 1125, 1972). For example, 475-nm and 480-nm emissions have been observed in YF_3: Yb^{3+}– Tm^{3+} and BaY_2F_8 : Yb^{3+}–Tm^{3+}, respectively. The Yb^{3+} ion serves as a sensitizer that absorbs the infrared radiation. Through multiple steps of nonresonant energy transfer from Yb^{3+} to Tm^{3+} and multiple-phonon decay between Tm^{3+} levels, Tm^{3+} ions are excited from the ground state $^3H_6$ to an excited state $^1G_4$. The blue emission is caused by transition from $^1G_4$ to $^3H_6$ in Tm^{3+}. We should also mention that GaN films of good optical quality have been grown by MBE on sapphire using AlN as buffer. The close lattice match of GaN and AlN has greatly improved the quality of the GaN film. Cathodoluminescence spectrum shows a strong band-to-band emission at 360 nm corresponding approximately to the gap energy of 3.4 eV in GaN (S. Yoshida, S. Misawa, and S. Gonda, *Appl. Phys. Lett.*, Vol. 42, p. 427, 1983). Since GaN is always *n*-type, metal–insulator–semiconductor (MIS) structures are generally used to observe electroluminescence, with the insulating GaN being highly compensated by Zn doping (G. Jacob and D. Bois, *Appl. Phys. Lett.*, Vol. 30, p. 412, 1977). However, in large-gap *i-n* junctions, a potential barrier is formed in the *i* region when the MIS junction is forward biased. Therefore, the potential barrier and hence the luminescent spectrum will depend on zinc concentration in the intrinsic *i* region.

Finally, we comment on LEDs for optical fiber communications. The ternary $Ga_{1-x}Al_xAs$ LEDs emit at a wavelength between 0.8 and 0.9 μm, while the quaternary $Ga_{1-x}In_xAs_{1-y}P_y$ LEDs cover the wavelength range 0.92 to 1.65 μm, being lattice-matched to the InP substrate. Both types have high quantum efficiency. First, they operate in the direct-gap region. Second, with proper choice of composition, heterostructures with almost perfect lattice match can be made for both optical and carrier confinements. These desirable attributes are discussed further in Section 14.7. In Fig. 14.11, the schematic diagrams for two common LED structures are shown: (a) surface-emitting structure and (b) edge-emitting structure. A hemispherical lens is sometimes formed on the top surface of a surface LED to increase the coupling efficiency to an optical fiber (O. Wada, S. Yamakoshi, M. Abe, Y. Nishitani, and T. Sakurai, *IEEE J. Quantum Electron.*, Vol. QE-17, p. 174, 1981). Using the lensed structure, a maximum power of 200 μW coupled to a 85-μm core, 0.16NA fiber has been reported. This compares with a value of 135 μW coupled into a 50-μm core, 0.2NA fiber for the edge LED (G. H. Olsen, F. Z. Hawrylo, D. J. Chanin, D. Botez, and M. Ettenberg, *IEEE*

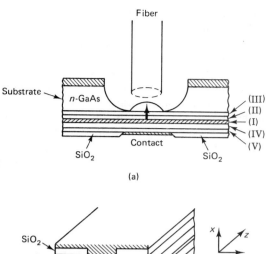

(a)

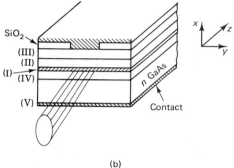

(b)

Figure 14.11 Schematic diagrams showing the structure of (a) surface-emitting and (b) edge-emitting diodes. The various layers are made of (I) GaAs active layer, (II, IV) (AlGa)As cladding layers and (III,V) GaAs contact layer and substrate.

J. Quantum Electron., Vol. QE-17, p. 2130, 1981). Another consideration in the design of a LED is the modulation bandwidth. For diodes with a large bandwidth, the carrier lifetime must be shortened either by decreasing the active-layer thickness or by increasing the doping concentration. However, by so doing the external efficiency of the diode is also reduced, due to increased nonradiative recombination. For doping concentrations $> 10^{18}$ cm^{-3}, the power–bandwidth product appears to be a constant, at least in the case of surface LEDs (D. Botez and G. J. Herskowitz, *Proc. IEEE,* Vol. 68, p. 689, 1980).

14.5 SPONTANEOUS EMISSION AND CARRIER LIFETIME FOR BAND-TO-BAND TRANSITIONS IN DIRECT-GAP SEMICONDUCTORS

In Section 12.9 we presented the Einstein theory of absorption and emission of radiation in terms of two basic processes: the spontaneous process represented by AN_m of Eq. (12.9.2) and the stimulated process represented by $B_{mn}I(\nu)N_m$ of Eq. (12.9.2) and $B_{nm}I(\nu)N_n$ of Eq. (12.9.3). The emission from a LED is derived from the spontaneous process and that from a laser is based on the stimulated process. In this section we apply the Einstein theory to spontaneous emission from direct-gap semiconductor diodes. We also recall that the derivation of the absorption coefficient $\alpha(\nu)$ in semiconductors presented in Section 12.11 is also based on transitions between energy states.

Therefore, a relation must exist between the absorption spectrum and emission spectrum. The purpose of the discussion in this section is twofold: to relate the emission and absorption measurements, and to prepare the background for our subsequent discussion of semiconductor injection lasers.

To apply Eqs. (12.9.2) and (12.9.3) to semiconductors, we make the following observations. First, the upper and lower energy states correspond, respectively, to the conduction- and valence-band states. These states form a continuum and obey Fermi–Dirac statistics. Second, for band-to-band transitions, the k-selection rule of Eq. (12.10.13) is observed. For each $\mathbf{k}_c$ conduction band state there is one and only one corresponding valence-band state with $\mathbf{k}_v = \mathbf{k}_c$. Therefore, the total number of states participating in a spontaneous transition in a volume V is $V\rho(k)\,dk$, where $\rho(k)$ is the density of states in k space. Based on these considerations, the total rate of spontaneous emission is

$$R_{\text{spon}} = V \int A\rho(k)f_c(1 - f_v)\,dk \qquad \text{s}^{-1} \qquad (14.5.1)$$

where f_c and f_v are the two Fermi occupancy factors. The Einstein A coefficient is given by Eq. (12.11.14), or

$$A = \frac{2e^2v}{m_0^2 c^3 \epsilon \hbar} \langle M \rangle^2 \qquad (14.5.2)$$

where c is the velocity of light in the medium and $\langle M \rangle^2$ is the average value of the momentum matrix element squared, that is, $\Sigma \langle p_x \rangle^2/3$. The matrix element $\langle p_x \rangle$ is related to the matrix element $\langle x \rangle$ by (N. F. Mott and I. N. Sneddon, *Wave Mechanics and Its Applications*, Oxford University Press, Oxford, 1948, p. 256)

$$\langle p_x \rangle = m_0 \omega_{cv} \langle x \rangle \qquad (14.5.3)$$

Before we proceed with Eq. (14.5.1), it may be instructive to present the analysis by Roosbroek and Shockley based on the method of detailed balance (W. Van Roosbroeck and W. Shockley, *Phys. Rev.*, Vol. 94, p. 1558, 1954). Let $P(v)$ be the probability per unit time that a photon of energy hv be absorbed. Following our discussion in Section 12.11, we see that the absorption coefficient $\alpha(hv)$ is related to $P(v)$ by

$$\alpha(hv) = \frac{P(v)}{v_g} = \frac{P(v)\bar{n}}{c_0} \qquad (14.5.4)$$

where $\bar{n} = n + dn/dv$ is the group index, n the index of refraction, and c_0 the velocity of light in free space. For a semiconductor at thermal equilibrium, the total rate of recombination $r(hv)$ per unit volume is equal to the total rate of generation $g(hv)$ of electron–hole pairs per unit volume within any frequency interval dv as required by the principle of detailed balance. Under thermal equilibrium, a blackbody has within frequency interval dv a photon density

$$\rho(v)\,dv = \frac{8\pi n^2 v^2 \bar{n}}{c_0^3} \frac{dv}{\exp(hv/kT) - 1} \qquad (14.5.5)$$

Since $g(h\nu) = P(\nu)\rho(\nu)$ and $r(h\nu) = g(h\nu)$, the spontaneous emission should have a spectrum $I(\nu) = h\nu r(h\nu)$, or

$$I(\nu) = \frac{8\pi n^2 \nu^2}{c_0^2} h\nu\alpha(\nu)\exp\left(\frac{-h\nu}{kT}\right) \qquad W/m^3 - Hz \qquad (14.5.6)$$

if $\exp(h\nu/kT) \gg 1$ as it is the case under consideration.

Now we return to Eq. (14.5.1). First, we let $R_{spon} = V \int r_{spon}(\nu) \, d\nu$. Further, we assume A to be independent of $h\nu$ for transitions near the direct band gap. Thus we find that

$$r_{spon}(\nu) = Ah\rho_r(\mathcal{E})f_c(1 - f_v) \qquad (m^3 - Hz - s)^{-1} \qquad (14.5.7)$$

where $\rho_r(\mathcal{E}) = \rho(k)dk/d\mathcal{E}$ is the reduced density of states in energy space. Use of Eq. (12.11.20) yields

$$\rho_r(\mathcal{E}) = \frac{(2m_r)^{3/2}}{2\pi^2\hbar^3} (h\nu - \mathcal{E}_g)^{1/2} \qquad (14.5.8)$$

where m_r is the reduced mass and $h\nu = \mathcal{E}_m - \mathcal{E}_n$ is the photon energy (G. Lasher and F. Stern, *Phys. Rev.*, Vol. 133, p. A553, 1964—equation in cgs units). For each spontaneous emission, a photon of energy $h\nu$ is emitted. Therefore, the spontaneous emission has a power spectrum $I(\nu) = h\nu r_{spon}(\nu)$, that is,

$$I(\nu) = \frac{8\pi e^2 h\nu^4 (2m_r)^{3/2}}{c^3\epsilon\hbar^3} \langle x \rangle^2 (h\nu - \mathcal{E}_g)^{1/2} f_c(1 - f_v) \qquad (14.5.9)$$

which can be expressed in terms of α of Eq. (12.11.25) as

$$I(\nu) = \frac{8\pi\nu^2}{c^2} h\nu\alpha(\nu)f_c(1 - f_v) \qquad (14.5.10)$$

The two Fermi functions can be expressed in terms of two quasi-Fermi levels $\mathcal{F}_n$ and $\mathcal{F}_p$ as follows:

$$f_c = \left[1 + \exp\left(\frac{\mathcal{E}_m - \mathcal{F}_n}{kT}\right)\right]^{-1} \qquad (14.5.11)$$

$$1 - f_v = \left[1 + \exp\left(\frac{\mathcal{F}_p - \mathcal{E}_n}{kT}\right)\right]^{-1} \qquad (14.5.12)$$

For nondegenerate semiconductors, f_c and $(1 - f_v)$ can be approximated by the corresponding Boltzmann factor. Thus we have

$$I(\nu) = \frac{8\pi\nu^2}{c^2} h\nu\alpha(\nu) \exp\left(\frac{\mathcal{F}_n - \mathcal{F}_p}{kT}\right) \exp\left(-\frac{h\nu}{kT}\right) \qquad (14.5.13)$$

In Eq. (14,5,13), $c = c_0/n$ is the velocity of light in the medium. Under thermal equilibrium, $\mathcal{F}_n = \mathcal{F}_p$ and Eq. (14.5.13) reduces to Eq. (14.5.6).

Figure 14.12 shows the measured absorption curve in GaAs doped with $p = 1.2 \times 10^{18}$ cm^{-3} at 77 K and a theoretical fit to the $(h\nu - \mathcal{E}_g)^{1/2}$ dependence of Eq. (12.11.24) with $\mathcal{E}_g = 1.42$ eV and $K = 3.0 \times 10^4$ cm^{-1}(eV)$^{-1/2}$ (H. C. Casey, Jr., and F. Stern, *J. Appl. Phys.*, Vol. 47, p. 631, 1976). We see that a good fit is obtained

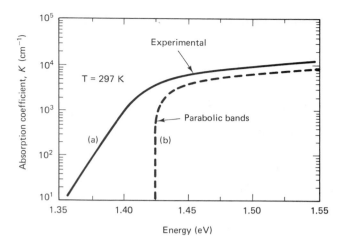

Figure 14.12 Comparison of (a) measured (solid curve) and (b) theoretical (dashed curve) absorption coefficient $\alpha(\nu)$ in GaAs. (H. C. Casey, Jr., and F. Stern, *J. Appl. Phys.*, Vol. 47, p. 631, 1976.)

in the high photon-energy region. In the low-energy region, however, the theoretical $\alpha(\nu)$ curve deviates considerably from the experimental measured absorption curve. The discrepancy is due to two effects at high doping concentrations: (1) formation of band-tail states and (2) shrinkage of energy gap. These two effects are discussed in Section 14.6. Figure 14.13 shows (a) the photoluminescence spectrum $I(\nu)$ measured in the same sample and (b) the calculated spectrum from using the experimental $\alpha(\nu)$ of Fig. 14.12a in Eq. (14.5.6). Even though the two curves have similar shapes, they differ appreciably at higher photon energies.

We should point out that the luminescent spectrum $I_i(\nu)$ calculated from Eq. (14.5.6) is the internal spectrum generated locally inside the sample, while the photoluminescent spectrum $I_e(\nu)$ of Fig. 14.13a is the external spectrum measured outside the sample. A comparison of $I_i(\nu)$ with $I_e(\nu)$ must take the following factors into consideration: (1) spatial decay of the photoexcitation beam, (2) diffusion of the generated carriers, and (3) reabsorption of the photoluminescent signal (C. J. Hwang, *J. Appl. Phys.*,

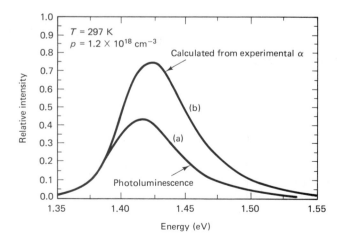

Figure 14.13 Comparison of (a) measured and (b) calculated emission spectra from GaAs at 297 K. (H. C. Casey, Jr., and F. Stern, *J. Appl. Phys.*, Vol. 47, p. 631, 1976.)

Vol. 40, p. 3731, 1969). The effect of internal absorption is also manifested in light-emitting diodes. Figure 14.14 shows the spectral output curves for surface-emitting and edge-emitting InGaAsP LEDs (G. H. Olsen, F. Z. Hawrylo, D. J. Chanin, D. Botez, and M. Ettenberg, *IEEE J. Quantum Electron.*, Vol. QE-17, p. 2130, 1981). In the latter case, the measured spontaneous emission is the part of emission propagating along the length of the waveguide (Fig. 14.11b). It undergoes multiple reflections at the two end mirrors before it transmits through the mirrors. Therefore, the total power density $P(v)$ emitting from one end mirror is

$$P(v) = \int dz \, \beta I(v, z)(1 - r^2)[\exp - \alpha(L - z) +$$
$$r^2 \exp - \alpha(L + z) + 2r \cos 2kz]^2 \, D^{-1} \qquad W/m^2 - Hz \qquad (14.5.14)$$

where $D = [1 - r^2 \exp(-\alpha + i2k)L]$ takes into account the effect of multiple reflections, r is the mirror reflection coefficient, k the phase constant, and β the fraction of spontaneous emission coupled into the waveguide in one direction of propagation. Since the absorption coefficient $\alpha(v)$ increases rapidly with v (Fig. 14.12a), the high-energy components of $I(v)$ is greatly attenuated with respect to the low-energy components. As a result, the emission spectrum $P(v)$ of an edge emitting diode is much narrower than that of a surface emitting diode. Figure 14.14 shows a half-power full-width $\Delta\lambda$ of 700Å for the former and 1350Å for the latter. Because the two diodes are made of InGaAsP of slightly different composition, the two spectra are displaced from each other. The dependence of $P(v)$ on the injection current is discussed in Section 14.7.

Besides spontaneous emission spectrum, another physical quantity of practical interest is the carrier lifetime. For nondegenerate semiconductors under injection, the product of Eqs. (14.5.11) and (14.5.12) can be approximated by

$$f_c(1 - f_v) \cong \exp\left(-\frac{\mathscr{E}_c - \mathscr{F}_n}{kT}\right) \exp\left(-\frac{\mathscr{F}_p - \mathscr{E}_v}{kT}\right) \exp\left(-\frac{\hbar^2 k^2}{2m_r kT}\right) \qquad (14.5.15)$$

Use of Eq. (14.5.15) in Eq. (14.5.1) leads to

$$R_{\text{spon}} = V\frac{A}{2}\left(\frac{h^2 m_r}{2\pi kT m_c m_v}\right)^{3/2} np \qquad (14.5.16)$$

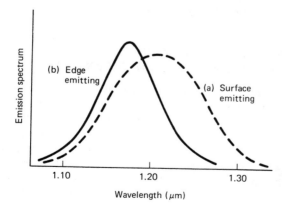

Figure 14.14 Emission spectrum of GaAs diodes of (a) edge-emitting type (solid curve) and (b) surface-emitting type (dashed curve). (G. H. Olsen, F. Z. Hawrylo, D. J. Chanin, D. Botez, and M. Ettenberg, "1.3 μm LPE- and VPE-grown InGaAsP Edge-emitting LEDs," *IEEE J. Quantum Electron.*, Vol. QE-17, p. 2130, 1981, © 1981 IEEE.)

This recombination rate is balanced by a generation rate with $np = n_i^2$ in Eq. (14.5.16). Therefore, if we define a radiative recombination lifetime time τ_r, the total net rate of recombination is $A \Delta n/\tau_r$ where Δn is the excess carrier density. Thus from Eq. (14.5.16) we find that

$$\frac{1}{\tau_r} = \frac{1}{2\tau} \left(\frac{h^2 m_r}{2 \pi kT m_c m_v} \right)^{3/2} (n_0 + p_0) \tag{14.5.17}$$

where $\tau = A^{-1}$ is the spontaneous lifetime. Since in a nondegenerate semiconductor $n_0 < N_c$ and $p_0 < N_v$, τ_r is always longer than τ. The longer carrier lifetime is a direct result of the fact that the conduction band states are sparsely occupied and the valence-band states are heavily populated. Under the circumstance, the probability for an occupied state in the conduction band to recombine with a vacant state in the valence band is greatly reduced.

Another interesting case is one for degenerate semiconductors. We assume a degenerate electron distribution so that we can take $f_c = 1$ and approximate $1 - f_v$ by a Boltzmann distribution in Eq. (14.5.1) for all transitions of interest. Thus we find for the total spontaneous emission rate

$$R_{\text{spon}} = VA \left(\frac{m_r}{m_v} \right)^{3/2} p \tag{14.5.18}$$

and for the hole lifetime,

$$\frac{1}{\tau_r} = \frac{1}{\tau} \left(\frac{m_c}{m_c + m_v} \right)^{3/2} \tag{14.5.19}$$

Similarly, in a degenerate p-type sample the electron lifetime is

$$\frac{1}{\tau_r} = \frac{1}{\tau} \left(\frac{m_v}{m_c + m_v} \right)^{3/2} \tag{14.5.20}$$

We should point out that in the treatment presented above only one valence band with mass m_v is assumed. For III–V compound semiconductors, the valence band actually consists of one heavy-hole band and one light-hole band degenerate at $k = 0$. Therefore, the expressions for the carrier lifetime should be modified to account for the two valence bands. In cases where $m_{vh} >> m_{vl}$, it is sufficient to use heavy-hole mass m_{vh} for m_v because most holes are in the heavy-hole band.

14.6 EFFECTS OF HEAVY DOPANT AND CARRIER CONCENTRATIONS: BAND TAILING AND GAP SHRINKAGE

Before the use of heterostructures in injection lasers, heavily doped p-n homojunctions were needed to establish a gain region inside the junction. Even with heterojunctions, the injected carrier concentrations in the gain region of a diode laser will have a value in the neighborhood of 2×10^{18} cm^{-3}. In this section we consider the effects of heavy doping and carrier concentrations on the energy band structure of a semiconductor. In our previous discussions we have used a parabolic energy dependence for the density of state $\rho(\mathscr{E})$, and assumed a constant energy $\mathscr{E}_g$ for the band gap. Under heavy impurity concentrations, the Coulomb potential of the randomly distributed impurities results in a significant redistribution of energy states near the band edge. This effect is known as *band tailing*. Under high carrier concentrations, the lattice potential, which an electron experiences, is affected by the presence of other electrons, and the behavior of each electron is influenced by all other electrons. The former effect is known as *screening* and the latter as *correlation*. The combination of the two effects results in a lowering of the potential energy for both electrons and holes, which is known as *band-gap shrinkage*.

The band-gap lowering can be estimated by following a procedure similar to that used in the calculation of cohesive energy in metals. In Section 3.4 we used classical treatment and obtained Eq. (3.4.2)

$$\mathscr{E}_p = -\frac{9e^2}{40\pi\epsilon R} \tag{14.6.1}$$

as the amount of energy lowering per atom by forming an electron sea where $R = (3/4\pi n)^{1/3}$ is the average spacing between electrons. A quantum-mechanical treatment yields the following result (M. Gell-Mann and A. R. Brueckner, *Phys. Rev.*, Vol. 106, p. 364, 1957):

$$\mathscr{E}_c^e = \mathscr{E}_p = -2b\frac{a_0}{R}\mathscr{E}_i \tag{14.6.2}$$

where $b = (\pi^2/18)^{1/3}$, $a_0 = 4\pi\hbar^2\epsilon/m_c^*e^2$ is the Bohr radius, and $\mathscr{E}_i = m_c^*e^4/2(4\pi\epsilon\hbar)^2$ is the ionization energy of a hydrogen-atom-like orbit in the semiconductor. The reader can easily show that Eq. (14.6.2) differs from Eq. (14.6.1) only in the numerical factor $b = (\pi^2/18)^{1/3}$ instead of $\frac{9}{10}$. For $m_c^* = 0.07m_0$ and $\epsilon = 13.1\epsilon_0$, $\mathscr{E}_i = 0.006$ eV, and $a_0 = 100$ Å. Therefore, the band-edge lowering caused by exchange interaction denoted by $\mathscr{E}_c^e$ is around 0.010 eV for $R = a_0$ or $n = 2.5 \times 10^{17}$ cm^{-3}. An empirical relation for the dependence of energy gap as a function of hole concentration is found by a best fit between the calculated and measured absorption-coefficient $\alpha(h\nu)$ curves (H. C. Casey, Jr. and F. Stern, *J. Appl. Phys.*, Vol. 47, p. 631, 1976). Assuming equal contributions to gap shrinkage from holes and electrons, the formula for GaAs reads

$$\mathscr{E}_g = 1.424 - 1.6 \times 10^{-8} (p^{1/3} + n^{1/3}) \qquad \text{eV} \tag{14.6.3}$$

where the carrier concentrations p and n are expressed in cm^{-3}. This dependence agrees well with that predicted from Eq. (14.6.2).

The density of states near the band edge, which we refer to as the *band tail*, has been treated by several authors (E. O. Kane, *Phys. Rev.*, Vol. 131, p. 79, 1963; V. L. Bonch-Bruevich, in R. K. Willardson and A. C. Beer, eds., *Semiconductors and Semimetals*, Vol. 1, Academic Press, Inc., New York, 1966, p. 101; B. I. Halperin and M. Lax, *Phys. Rev.*, Vol. 148, p. 722, 1966). Existence of band-tail states has been manifested in a number of experiments. Figure 6.35 shows evidence observed from *I–V* characteristics of heavily doped GaAs tunnel diodes. Figure 14.15 shows the ionization energies of shallow donors (S, Se, and Si) and shallow acceptors (Zn) deduced from Hall measurements (O. V. Emel'yanenko, T. S. Lagunova, D. N. Nasledov, and G. N. Talalakin, *Sov. Phys.-Solid State*, Vol. 7, p. 1063, 1965; F. Ermanis and K. Wolfstirn, *J. Appl. Phys.*, Vol. 37, 1963, 1966). As the impurity concentration increases, the impurity level broadens to form an impurity band, and band-tail states also extend in energy. When the impurity concentration exceeds certain value, the impurity band and band-tail states merge. As a result, the impurity ionization energy goes to zero and the Hall coefficient becomes temperature independent (sample 66, Fig. 6.18). The transition to metallic impurity conduction occurs when the average impurity separation r goes below

$$r \cong 3.0a_0 \tag{14.6.4}$$

where a_0 is the Bohr radius in the semiconductor (N. F. Mott and W. D. Twose, *Advan. Phys.*, Vol. 10, p. 107, 1961). The experimentally measured impurity concentration $N_i = 3/4\pi r^3$ at the transition point (Fig. 14.15) agrees well with Eq. (14.6.4). Additional evidences of band-tail states are apparent from the absorption measurement (Fig. 14.12) and the luminescence experiment (Fig. 14.13).

Figure 14.16 shows a comparison of the band structures of GaAs obtained by Hwang by applying a self-consistent calculation to the models of the Kane and of Halperin and Lax (C. J. Hwang, *Phys. Rev.*, Vol. B2, p. 4117, 1970). Both models are based on the treatment of Wolff, which approximates the many-body Hamiltonian by a one-electron Hamiltonian

$$\mathcal{H} = \frac{p^2}{2m_c} - \mathcal{E}_c^e + \mathcal{E}_c^0 + V(r) \tag{14.6.5}$$

for conduction-band electrons where $\mathcal{E}_c^e$ is the exchange energy given by Eq. (14.6.2) and $\mathcal{E}_c^0$ is the conduction-band edge at low impurity concentrations (P. A. Wolff, *Phys. Rev.*, Vol. 126, p. 405, 1962). The quantity $V(r)$ represents the screened Coulomb potential of an impurity atom

$$V(r) = -\frac{e^2}{4\pi\epsilon r} \exp(-Qr) \tag{14.6.6}$$

where r is the local electron coordinate and $1/Q$ is the screening length. Since $V(r)$ is randomly distributed, its effects on the electrons can be expressed in terms of an average shift $\mathcal{E}_c^c$ of the band edge and an energy spread with a root-mean-square (rms) value V_{rms} where

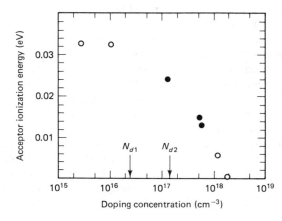

Figure 14.15 Ionization energy of Zn acceptor and formation of (shallow) donor band in GaAs as a function of doping concentration. The dots and open circles represents two sets of experimentally deduced values for Zn. The two arrows indicate, respectively, the onset of donor-band formation at N_{d1} and its merging with conduction band at N_{d2}. (F. Ermanis and K. Wolfstirn, *J. Appl. Phys.*, Vol. 37, p. 1963, 1966; G. Lucovsky and C. J. Repper, *Appl. Phys. Lett.*, Vol. 3, p. 71, 1963.)

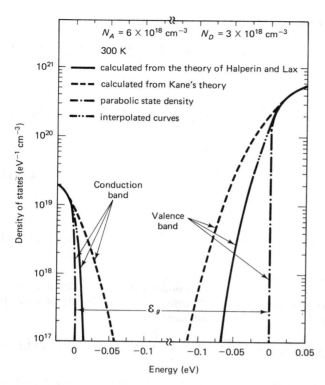

Figure 14.16 Density of states function used by Hwang for conduction band and valence band in heavily doped GaAs by joining calculated results of Kane and of Halperin and Lax. (C. J. Hwang, *Phys. Rev.*, Vol. B2, p. 4117, 1970.)

$$\mathscr{E}_c^c = -\frac{e^2(N_d - N_a)}{\epsilon Q^2} \tag{14.6.7}$$

$$V_{rms}^2 = \frac{e^4(N_d^+ + N_a^-)}{8\pi\epsilon^2 Q} \tag{14.6.8}$$

if $r < a_0$ (C. J. Hwang, *Phys. Rev.*, Vol. B2, p. 4117, 1970) and $\mathscr{E}_c^c$ is related to donor-ionization energy by $\mathscr{E}_c^c = -0.61\,\mathscr{E}_d\sqrt{a_0/r}$ if $r > a_0$ (P. R. Rimbey and G. D. Mahan, *Phys. Rev.*, Vol. B10, p. 3419, 1974).

The screening length Q^{-1} can be obtained by solving Poisson's equation

$$\nabla^2 V = -\frac{\rho}{\epsilon} = -\frac{e(N_d - n)}{\epsilon} \tag{14.6.9}$$

Because the potential energy of an electron is lowered by $V(r)$ near an ionized donor, n varies with V. In a nondegenerate semiconductor, n can be approximated by $n_b \exp(eV/kT)$, where $n_b = N_d$. For $eV/kT < 1$, $\exp(eV/kT) \cong 1 + (eV/kT)$ and Eq. (14.6.9) becomes

$$\nabla^2 V = \frac{e^2 n}{\epsilon kT} V \tag{14.6.10}$$

It can be shown that Eq. (14.6.6) satisfies Eq. (14.6.10) if

$$Q^{-1} = \text{Debye screening length} = \left(\frac{\epsilon kT}{e^2 n}\right)^{1/2} \qquad (14.6.11)$$

At low temperatures, Fermi–Dirac distribution must be used and

$$N_d - n = -\frac{dn}{dV} V = -V \frac{d}{dV} \int \rho(\mathscr{E}) f_c \, d\mathscr{E} \qquad (14.6.12)$$

Since df_c/dV behaves like a delta function, we have

$$\nabla^2 V = \frac{e\rho(\mathscr{E}_f)V}{\epsilon} \qquad (14.6.13)$$

and Q^{-1} becomes

$$Q^{-1} = \left(\frac{\epsilon \pi^2 \hbar^2}{m_c e}\right)^{1/2} (3\pi^2 n)^{-1/6} \qquad (14.6.14)$$

which applies to degenerate semiconductors.

Although the Halperin–Lax model is more accurate, the Kane model is easier to use. A procedure is adopted by Casey and Stern to approximate the density of states by fitting a one-parameter (η_c) density of states of the Kane form

$$\rho_c(\mathscr{E}) = (2\eta_c)^{1/2} \frac{m_{dc}^{3/2}}{\pi^2 \hbar^3} y\left(\frac{\mathscr{E} - \mathscr{E}_c}{\eta_c}\right) \qquad (14.6.15)$$

to the Halperin–Lax form in the band tail, where m_{dc} is the density-of-state effective mass and $\eta_c = \sqrt{2} V_{\text{rms}}$. Figure 14.17 shows the function $y(x)$ as a function of $x = (\mathscr{E} - \mathscr{E}_c)/\eta_c$, which is computed from the integral

$$y(x) = \pi^{-1/2} \int_{-\infty}^{x} (x - z)^{1/2} \exp(-z^2) \, dz \qquad (14.6.16)$$

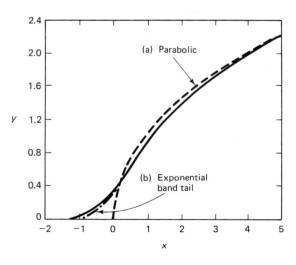

Figure 14.17 Plot of the function $y(x)$, defined in Eq. (14.6.16) and used in Eq. (14.6.15), as a function of $x = (\mathscr{E} - \mathscr{E}_c)/\eta_c$. (H. C. Casey, Jr., and F. Stern, *J. Appl. Phys.*, Vol. 47, p. 631, 1976.) The two dashed curves are for (a) the parabolic band and (b) the band tail.

For comparison, the two dashed curves represent (a) the parabolic band and (b) an exponential band tail, respectively, with

$$y = \begin{cases} x^{1/2} & \text{for } x > 0 & (14.6.17) \\ 0.34 \exp(2.2x) & \text{for } x < 0 & (14.6.18) \end{cases}$$

obviously, high impurity concentration results in a redistribution of energy states. Some states from the parabolic band are moved into the band tail, with the total states remaining unchanged. Table 14.1 shows the value of η computed by Casey and Stern for a number of heavily doped and compensated p GaAs samples with η_v being the band-tail parameter for holes. Obviously, the valence band has a more extended tail than the conduction band.

The absorption and luminescent spectra measured on the low-energy side in Figs. 14.12 and 14.13 can be attributed to transitions involving band-tail states. However, an application of the theory to a quantitative analysis of the data would require our knowledge of (1) the density of states, (2) the energy distribution of impurity states, (3) the matrix elements connecting band-to-band and band-to-impurity states, and (4) the Fermi level. The discussion above covers only the first topic. As regards to the matrix elements, $\langle M_{bb} \rangle^2$ for parabolic band-to-band transition is given by Eq. (12.11.23), while that for band-to-impurity transition has been shown by Eagle and Dumke to be (D. M. Eagle, *J. Phys. Chem. Solids,* Vol. 16, p. 76, 1960; W. P. Dumke, *Phys. Rev.,* Vol. 132, p. 1998, 1963)

$$\langle M_{bi} \rangle^2 = \frac{\langle M_{bb} \rangle^2 \, 64\pi a_0^3}{(1 + a_0^2 k^2)V} \tag{14.6.19}$$

where a_0 is the effective Bohr radius of the impurity state, k the carrier wave number, and V the volume. The corresponding quantity $\langle M_{tt} \rangle^2$ for transitions between band-tail states has been derived by Casey and Stern. Therefore, information is available concerning the third topic. What is lacking is our knowledge about the energy distribution of impurity states. Therefore, it becomes necessary that Hall measurements be made to determine the majority carrier concentration.

From Eqs. (14.6.17) and (14.6.18) we find that the total band-tail states and the parabolic band states within $x = 0$ and 1 are in the ratio

$$r = \frac{\int_{-\infty}^{0} 0.34 \exp(2.2x) \, dx}{\int_{0}^{1} x^{1/2} \, dx} = 0.23 \tag{14.6.20}$$

TABLE 14.1 VALUES OF η_c AND η_v

p	N_a (cm^{-3})	N_d (cm^{-3})	η_c (meV)	η_v (meV)
$1.2 \times 10^{18} p$	1.5×10^{18}	3×10^{17}	10	20
$2.4 \times 10^{18} p$	3×10^{18}	6×10^{17}	9	22
$1.6 \times 10^{19} p$	2×10^{19}	4×10^{18}	5.5	29
p	6×10^{18}	4×10^{18}	23	43
p	1.1×10^{19}	9×10^{18}	37	63

For $\eta = 20$ meV and $m_v = 0.52m_0$, we find that the density of states within $x = 0$ and 1 is 3.7×10^{18} cm^{-3} and hence the total density of band-tail states is only 8.5×10^{17} cm^{-3}, in comparison with the hole concentration $p = 1.2 \times 10^{18}$ cm^{-3} in Table 14.1 and an effective density of states $N_v = 9 \times 10^{18}$ cm^{-3} for the valence band. From these numbers it is obvious that the Fermi level is determined by both the parabolic-band and band-tail states in cases where p is comparable to the total density of band-tail states. Similar calculations can be made for the conduction band. Therefore, determination of majority-carrier concentration from the Hall measurement should enable us to find the Fermi level and hence to fit the theory to the measured absorption and luminescent spectra. Good-to-fair agreement has been obtained (Casey and Frank). However, the theory is still incomplete without the knowledge of energy distribution of the impurities involved, which is needed to find N_d^+ and N_a^-.

14.7 STIMULATED EMISSION: POPULATION INVERSION AND GAIN SPECTRUM

In Section 14.5 we presented Eq. (14.5.7) for the rate of spontaneous emission. Similarly, based on the Einstein theory, the net rate for stimulated emission per energy interval $dh\nu$ per unit volume is

$$r_{stim}(h\nu) = BI(\nu)[f_c(1 - f_v)\rho_r(\mathscr{E}) - f_v(1 - f_c)\rho_r(\mathscr{E})] \quad (\text{m}^3 - \text{eV} - \text{s})^{-1} \quad (14.7.1)$$

The first term corresponds to the stimulated downward-transition (emission) term in Eq. (12.9.2) and the second term corresponds to the stimulated upward-transition (absorption) term in Eq. (12.9.3). The Fermi factors are introduced in Eq. (14.7.1) to account for Pauli's exclusion principle. If we let $S(h\nu)$ be the photon density per energy interval $dh\nu$ and $N(h\nu)$ be the density of electromagnetic modes within energy interval $dh\nu$, then

$$I(\nu) = \frac{I(\mathscr{E})\, d\mathscr{E}}{d\nu} = S(\mathscr{E})h\nu ch \quad (14.7.2)$$

$$N(h\nu) = \frac{N(\nu)\, d\nu}{d\mathscr{E}} = \frac{8\pi\nu^2}{c^3}\frac{1}{h} \quad (14.7.3)$$

Using Eq. (12.9.8) for B in terms of A in Eq. (14.7.1), we obtain

$$r_{stim}(h\nu) = A\frac{S(h\nu)}{N(h\nu)}(f_c - f_v)\rho_r(\mathscr{E}) \quad (14.7.4)$$

which is the rate of net downward transitions per unit volume per unit time within energy interval between $\mathscr{E}$ and $\mathscr{E} + d\mathscr{E}$.

Since each downward transition emits a photon while each upward transition absorbs a photon, a net increase in photon density results if $r_{stim}(h\nu) > 0$, that is,

$$f_c > f_v \quad (14.7.5)$$

Under thermal equilibrium, f_c is always smaller than f_v because the conduction band is higher in energy than the valence band. Therefore, a nonequilibrium situation must be established to achieve $f_c > f_v$. The condition stated in Eq. (14.7.5) is known as the *condition for population inversion*. This condition is satisfied in the space-charge region of a forward-biased degenerate *p-n* junction, as shown schematically in Fig. 14.18a. In

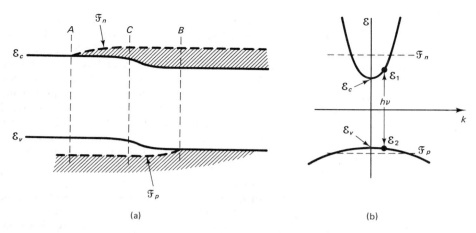

Figure 14.18 Energy-band diagram (a) in real space along the direction of a degenerate p-n junction and (b) in k space at a fixed point in the space-charge region.

the region between A and B, there exist energy states for which $f_c \cong 1$ (represented by the shaded area) and $f_v \cong 0$ (represented by the unshaded area). Expressing the Fermi functions in terms of the relevant energies, we can rewrite Eq. (14.7.5) as

$$hv = \mathscr{E}_1 - \mathscr{E}_2 < \mathscr{F}_n - \mathscr{F}_p \qquad (14.7.6)$$

This condition was first pointed out by Bernard and Duraffourg, (M. G. A. Bernard and G. Duraffourg, *Phys. Status Solidi*, Vol. 1, p. 699, 1961). Figure 14.18b shows the energy band in k space at a given position C inside the space-charge region. The intersections of the vertical line with the bands define two energies $\mathscr{E}_1$ and $\mathscr{E}_2$ of the states involved in an optical transition, as required by the k-selection rule. Gain or positive r_{stim} is possible as long as $\mathscr{E}_1$ is below quasi-Fermi level $\mathscr{F}_n$ and $\mathscr{E}_2$ is above the quasi-Fermi level $\mathscr{F}_p$.

From Eqs. (14.5.7) and (14.7.4) we find that

$$\frac{r_{\text{stim}}(hv)}{r_{\text{spon}}(hv)} = \frac{S(hv)}{N(hv)} \left\{ 1 - \exp\left[\frac{hv - (\mathscr{F}_n - \mathscr{F}_p)}{kT} \right] \right\} \qquad (14.7.7)$$

Furthermore, the gain coefficient $g(hv)$ for an optical beam is

$$g(v) = \frac{1}{S}\frac{dS}{dx} = \frac{r_{\text{stim}}}{Sc} = Bh^2 v \rho_r(\mathscr{E})(f_c - f_v) \qquad \text{m}^{-1} \qquad (14.7.8)$$

which if applied to parabolic band-to-band transition is equal to the negative of the absorption coefficient α of Eq. (12.11.22). For $hv > \mathscr{F}_n - \mathscr{F}_p$, r_{stim} is negative, which means attentuation; and for $hv < \mathscr{F}_n - \mathscr{F}_p$, r_{stim} is positive, which means amplification. Therefore, stimulated emission spectrum will have its low-energy side amplified and the high-energy side attentuated in relation to the spontaneous emission spectrum. We also see from Eq. (14.7.7) that the stimulated process becomes important only when the photon density $S(hv)$ builds up to a value exceeding the density of spontaneous emission modes $N(hv)$. It is obvious that a cavity structure is needed to retain a portion of the emitted photons.

The gain coefficient $g(\nu)$ is an important quantity which is used to characterize operation of an injection laser. Consider a double heterostructure (DH) p-n junction with an intrinsic GaAs active layer sandwiched between two (GaAl)As layers, one n type and the other p type. Without any bias, the GaAs layer has an absorption coefficient $\alpha(\nu)$ given by Eq. (12.11.22) with $f_c = 0$ and $f_v = 1$, which is shown schematically as the dashed curve in Fig. 14.19a. Under a forward bias, electrons are injected from n-(GaAl)As and holes are injected from p-(GaAl)As into the middle GaAs. For ease of discussion, we assume that the temperature is sufficiently low that the Fermi function can be approximated by a step function. As the injection current is raised, the two quasi-Fermi levels rapidly move first toward and then into their respective band, leading to the situation shown in Fig. 14.18b. Within the energy region $h\nu < \mathscr{F}_n - \mathscr{F}_p$, complete population inversion is established at 0 K with $f_c = 1$ and $f_v = 0$. Thus the gain coefficient is exactly equal to the negative of the absorption coefficient and is represented by the solid curve AB_1 in Fig. 14.19a. In the energy region $h\nu > \mathscr{F}_n - \mathscr{F}_p$, the situation becomes normal again with $f_c = 0$ and $f_v = 1$, and we have an absorption coefficient $\alpha(\nu)$ given by Eq. (12.11.22), which is represented by the curve C_1D. At 0 K, the transition from point B_1 (gain) to point C_1 (absorption) is abrupt and hence vertical in Fig. 14.19a. Upon further increase of the injection current, the two quasi-Fermi levels move deeper into their respective band and thus the transition line B_2C_2 moves to a higher-energy $h\nu_2$.

From Fig. 14.19a it is obvious that the spectral gain profile $g(\nu, J)$ changes with the injection current density J. Figure 14.19b shows the variation of $g(\nu, J)$ with J at a fixed photon energy $h\nu$. At a current density J_1, the separation $\mathscr{F}_n - \mathscr{F}_p$ equals $h\nu_1$ and

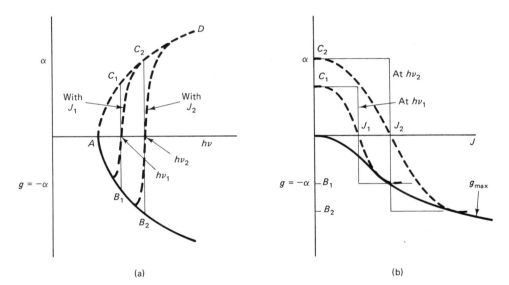

(a)   (b)

Figure 14.19 Schematic diagrams showing variation of absorption coefficient α (a) as a function of photon energy $h\nu$ in a p-n junction under different current densities and (b) as a function of injection-current density J at two different photon energies. The gain coefficient $g(\nu)$ is equal to $-\alpha(\nu)$.

hence an abrupt transition from absorption (point C_1) for $hv > hv_1$ to gain (point B_1) for $hv < hv_1$ occurs at 0 K. Therefore, the $g(J)$ curve for a fixed hv is itself a step function. At a finite temperature, the change in f_c and f_v becomes gradual as a function of energy and $g(J)$ curves for various hv are represented schematically by a family of dashed curves in Fig. 14.19b. One useful quantity in analyzing laser operation is the peak gain, which is represented by the heavy curve in Fig. 14.19b as the envelope of the individual spectral gain curves. We should point out that in obtaining the shapes of the curves shown schematically in Fig. 14.19, the k-selection rule is assumed and the gap-shrinkage effect is ignored. Even so, both figures are useful in gaining an understanding of the behavior of theoretically calculated and experimentally measured spectral gain curves.

Figure 14.20a shows the measured spectral gain curve from stripe geometry DH GaAs injection lasers with an n-doped active region (B. W. Hakki and T. L. Paoli, *J. Appl. Phys.*, Vol. 44, p. 4113, 1973). Referring to Eq. (14.5.14), we see that the power density $P(v)$ undergoes maxima ($2kL = m2\pi$) and minima [$2kL = (2m + 1)\pi$] as a function of v. The undulation is a result of constructive and destructive interference of the electromagnetic waves upon reflections at the two mirrors. The peaks and valleys of $P(v)$, represented by P_p and P_v, respectively, become discernible at injection currents slightly below the laser-threshold current. We let

$$P_p(v) = P_0[1 - r^2 \exp{(-\alpha L)}]^{-2} \tag{14.7.9}$$

$$P_v(v) = P_0[1 + r^2 \exp{(-\alpha L)}]^{-2} \tag{14.7.10}$$

Hence $\alpha(v)$ can be determined from Eqs. (14.7.9) and (14.7.10) in terms of P_p and P_v as

$$\alpha(v) = L^{-1}\ln{\left[\frac{P_p^{1/2} + P_v^{1/2}}{P_p^{1/2} - P_v^{1/2}}\right]} + L^{-1}\ln{r} \tag{14.7.11}$$

where v is the average of v_p and v_v. The curves shown in Fig. 14.20a were computed from the measured P_p and P_v through the use of Eq. (14.7.11). We should point out that the baseline for $g(v) = 0$ should be at $\alpha = 30$ cm^{-1}, which accounts for losses in the laser structure other than that associated with band-to-band and band-to-impurity transitions. Therefore, the gain has a spectral width about 100 Å as compared to a typical width over 200 Å for the spontaneous emission (Fig. 14.14). For comparison, the theoretical curves computed for GaAs samples with $p - n = 4 \times 10^{17}$ cm^{-3} and $N_a + N_d - p = 6 \times 10^{17}$ cm^{-3} are shown in Fig. 14.20b, using the modified Halperin–Lax band-tail model of Eq. (14.6.15) and the gap shrinkage of Eq. (14.6.3) (F. Stern, *J. Appl. Phys.*, Vol. 47, p. 5382, 1976). The position of the peak gain shifts to shorter wavelength at a rate of 1 Å/mA in Fig. 14.20a compared to a value of about 0.5 Å/mA estimated from Fig. 14.20b for a laser with a dimension 380 μm $\times$ 13 μm $\times$ 0.3 μm for the active region.

Figure 14.21 shows the same computed gain coefficient shown in Fig. 14.20b but arranged as a function of the nominal current density $J_{\text{nom}} = J/d$, where d is the thickness of the active layer. The rate of carriers injected into the active region is J/e per unit area per unit time. Assuming uniform recombination over d, the rate of carriers recombined is dn/τ or R_{spon} per unit area per unit time, where τ is the carrier lifetime

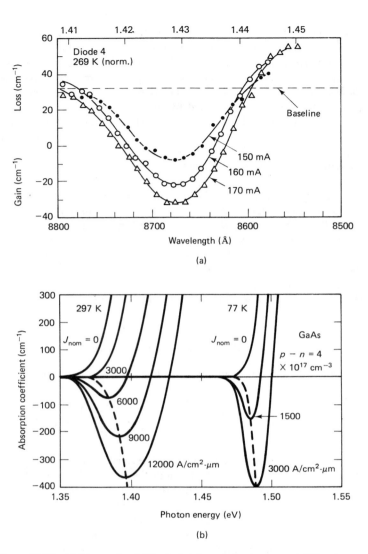

Figure 14.20 (a) Measured and (b) calculated spectral gain curves at different currents in (a) and at different nominal current densities J/d in (b), where d is thickness of the active GaAs layer in μm. The baseline in (a) represents the internal loss coefficient; therefore, the gain coefficient due to band-to-band or band-to-impurity transition is measured from the baseline. (B. W. Hakki and T. L. Paoli, *J. Appl. Phys.*, Vol. 44, p. 4113, 1973; F. Stern, *J. Appl. Phys.*, Vol. 47, p. 5382, 1976.)

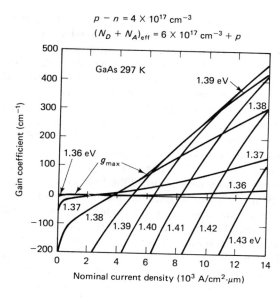

$$p - n = 4 \times 10^{17} \text{ cm}^{-3}$$
$$(N_D + N_A)_{\text{eff}} = 6 \times 10^{17} \text{ cm}^{-3} + p$$

Figure 14.21 Family of curves showing the gain coefficient at a fixed photon energy as a function of nominal current density J/d. The envelope (dashed line) of the curves represents the maximum gain attainable at a given nominal current density and is denoted by g_{max}. (F. Stern, *J. Appl. Phys.*, Vol. 47, p. 5382, 1976.)

and R_{spon} is the spontaneous emission rate given by Eq. (14.5.16). Under steady state, the two rates are equal. Thus we have

$$J_{\text{nom}} = \frac{J}{d} = \frac{en}{\tau_r} = eR_{\text{spon}} \tag{14.7.12}$$

from which the injected carrier density n can be found. Experimentally determined values of the radiative lifetime in p-type GaAs samples range from $\tau_r = 3 \times 10^{-10}$ s for $p_0 = 2 \times 10^{19}$ cm^{-3} to $\tau_r = 2 \times 10^{-9}$ s for $p_0 = 1 \times 10^{18}$ cm^{-3} (H. C. Casey, Jr., and M. B. Panish, *Heterostructure Lasers*, Academic Press, Inc., New York, 1978, Fig. 3.7-5, p. 161). The peak gain g_{max}, which is the envelope of the individual spectral gain curves, is found to be approximately given by a linear relation

$$g_{\text{max}} = \beta(J_{\text{nom}} - J_0) = 0.045J_{\text{nom}} - 190 \qquad \text{cm}^{-1} \tag{14.7.13}$$

in the gain region $50 < g < 100$ cm^{-1}, where J_{nom} is expressed in A/cm^2-μm. The linear dependence of g_{max} on J_{nom} is commonly used in analyzing the dynamic operation of injection lasers even though a quadratic variation $\beta(J_{\text{nom}} - J_0)^2$ is a better approximation in the low-gain region ($g < 50$ cm^{-1}). We should also point out that the values of the two constants β and J_0 will be different in active regions with different material properties such as impurity concentrations.

14.8 INJECTION LASERS

Until recent years the primary interests of electrical engineers were confined to physical electronics. By *physical electronics* we mean those subjects the foundation of which is built on classical physics. Examples of these are vacuum tubes, gas-discharge devices, semiconductor diodes, transistors, and metal–semiconductor–interface devices. The operation of these devices can be described by a set of current–voltage characteristics based on the classical equations of current flow. With the invention of such remarkable devices as the maser and the laser, we find it necessary to extend our knowledge beyond the domain of classical physics. The operation of the maser and the laser is based on the interaction of an electromagnetic radiation with an atomic system, during which energy is released from the atomic system to enhance the electromagnetic radiation. The idea of using the process of induced emission for amplification was conceived by C. H. Townes and independently by N. G. Basov and A. M. Prokhorov. The first successful maser was the ammonia maser developed by Townes and his students for operation in the microwave region (J. P. Gordon, H. J. Zeiger, and C. H. Townes, *Phys. Rev.*, Vol. 95, p. 444, 1954). Similar attempts were also reported by Basov and Prokhorov (N. G. Basov and A. M. Prokhorov, *J. Exp. Theor. Phys. U.S.S.R.*, Vol. 27, p. 431, 1954). Townes, Prokhorov, and Basov were awarded the Nobel Prize in Physics in 1964 for their fundamental work in the field of quantum electronics. An extension of the maser principle to the visible and infrared region of the electromagnetic spectrum (laser) was proposed by Schawlow and Townes, and the first successful laser was the ruby laser developed by Maiman (A. L. Schawlow and C. H. Townes, *Phys. Rev.*, Vol. 112, p. 1940, 1958; T. H. Maiman, *Nature,* Vol. 187, p. 493, 1960). For their work in optical spectroscopy, A. L. Schawlow and N. Bloembergen, shared the Nobel Prize in physics with K. Siegbahn in 1981. With the advent of the maser and the laser, a new field generally referred to as *quantum electronics* was born. The word *quantum* is used because, to help us understand the basic principles governing the operation of such devices, a certain knowledge of quantum mechanics has become essential.

Shortly after the demonstration of ruby laser, spectral line narrowing and a sharp increase in light intensity were observed in GaAs and Ga(AsP) *p-n* homojunctions and reported by several groups.[†] (R. N. Hall, G. E. Fenner, J. D. Kingsley, T. J. Soltys, and R. O. Carlson, *Phys. Rev. Lett.,* Vol. 9, p. 366, 1962; M. I. Nathan, W. P. Dumke, G. Burns, F. H. Dill, Jr., and G. Lasher, *Appl. Phys. Lett.,* Vol. 1, p. 62, 1962; N. Holonyak, Jr., and S. F. Bevacqua, *Appl. Phys. Lett.,* Vol. 1, p. 82, 1962; T. M. Quist, R. H. Rediker, R. J. Keyes, W. E. Krag, B. Lax, A. L. McWhorter, and H. J. Zeiger, *Appl. Phys. Lett.,* Vol. 1, p. 91, 1962). Both characteristics are typical of the onset of stimulated emission, which is a prerequisite to laser action. Since the early work on GaAs, the list of semiconductor laser has continued to grow, including (1) III–V compounds and their alloys, such as (GaAl)As and (GaIn)AsP; (2) II–VI compounds and their alloys, such as CdS and Cd(SSe); and (3) lead salts and mixed IV–VI compounds such as PbS and (PbHg)Te. As pointed out in Section 14.4, one outstanding feature of compound semiconductors is their miscibility to form an alloy, opening up the possibility of covering a wide range of frequency spectrum. For example, we can have a photon energy between 2.50 and 1.80 eV by changing the composition x in CdS_xSe_{1-x} from $x = 1$ to $x = 0$. On the long-wavelength side, laser action has been reported in $Sn_{0.27}Pb_{0.73}Te$ at $\lambda = 27$ μm, and on the short-wavelength side, in ZnS at $\lambda = 3250$ Å. Therefore, semiconductor lasers provide coherent sources from the far-infrared region to the near-ultraviolet region. In Table 14.2 we list the photon energy and the operating condition (including the excitation scheme) of a selected number of semiconductor lasers.

[†]An excellent review of the early work has been given by Casey and Panish (H. C. Casey, Jr., and M. B. Panish, *Heterostructure Lasers,* Academic Press, Inc., New York, 1978, Chap. 1).

Table 14.2 EXAMPLES OF SEMICONDUCTOR LASERS

Material[a]	Wavelength	Threshold[b]	Operating condition
ZnS (platelet)	350 nm	35 MW/cm^2 (2 - p)	Pulsed (300 K)[c]
ZnMnSe/GaAs (MBE)	454 nm	0.2 MW/cm^2 (1 - p)	Pulsed (5.5 K)[d]
CdS (platelet)	Blue-green	12 MW/cm^2 (e - b)	Pulsed (300 K)[e]
CdMnTe/GaAs (MBE)	765 nm	14 kW/cm^2 (1 - p)	Pulsed (25 K)[f]
AlGaInP/GaAs (MOVPE)	584 nm	1.9 kA/cm^2	CW (77 K)[g]
AlGaAs/GaAs (MBE)	680 nm	2.5 kA/cm^2	CW (300 K)[h]
AlGaAs/GaAs (MBE)	880 nm	0.7 kA/cm^2	CW (300 K)[i]
GaInAsP/InP (MOCVD)	1.58 μm	4.3 kA/cm^2	CW (300 K)[j]
InGaAsSb/GaSb (MBE)	2.2 μm	4.2 kA/cm^2	Pulsed (300 K)[k]
AlGaSb/InAsSb/GaSb (MBE)	3.9 μm	4 kA/cm^2 (1 − p)	Pulsed (80 K)[l]
PbSnTe/PbTe (VPE)	6 μm	9 kA/cm^2	CW (130 K)[m]
PbEuSeTe/PbTe (MBE)	6.6–2.6 μm	—	Pulsed (150 K)[n]

[a]In lasers with several materials listed, the last material represents the substrate upon which the laser structure is grown. Also noted is the fact that each laser system may cover a finite range of wavelength by varying the material composition in the active layer or the layer thickness in quantum-well lasers.

[b]The symbols 2 - p, 1 - p, and e - b indicate excitation by two photon absorption, by one-photon absorption, and by electron beam, respectively. Otherwise, the lasers are excited by injection currents.

[c]S. Wang and C. C. Chang, *Appl. Phys. Lett.*, Vol. 12, p. 193, 1968.

[d]R. B. Byslma, W. M. Becker, T. C. Bonsett, L. A. Kolodziejski, R. L. Gunshor, M. Yamanishi, and S. Datta, *Appl. Phys. Lett.*, Vol. 47, p. 1039, 1985.

[e]V. Daneu, D. P. DeGloria, A. Sanchez, F. Tong, and R. M. Osgood, Jr., *Appl. Phys. Lett.*, Vol. 49, p. 546, 1986.

[f]R. N. Bicknell, N. C. Giles-Taylor, J. F. Schetzina, N. G. Anderson, and W. D. Laidig, *Appl. Phys. Lett.*, Vol. 46, p. 238, 1985.

[g]I. Hino, K. Akiko, A. Gomyo, K. Kobayashi, and T. Suzuki, *Appl. Phys. Lett.*, Vol. 48, p. 557, 1986.

[h]T. Hayakawa, T. Suyama, K. Takahashi, M. Kondo, S. Yamamoto, and T. Hijikata, *Appl. Phys. Lett.*, Vol. 49, p. 636, 1986.

[i]W. T. Tsang, *Appl. Phys. Lett.*, Vol. 39, p. 134, 1981.

[j]M. Razeghi, R. Blondeau, K. Kazmierski, M. Krakowski, B. de Cremoux, J. P. Duchemin, and J. C. Bouley, *Appl. Phys. Lett.*, Vol. 45, p. 784, 1984.

[k]T. H. Chiu, W. T. Tsang, J. A. Ditzenberger, and J. P. van der Ziel, *Appl. Phys. Lett.*, Vol. 49, p. 1051, 1986.

[l]J. P. van der Ziel, T. H. Chiu, and W. T. Tsang, *Appl. Phys. Lett.*, Vol. 48, p. 315, 1986.

[m]K. Shinohara, Y. Nishijima, H. Ebe, A. Ishida, and F. Fujiyasu, *Appl. Phys. Lett.*, Vol. 47, p. 1184, 1985.

[n]D. L. Partin and C. M. Thrush, *Appl. Phys. Lett.*, Vol. 45, p. 193, 1984.

In the discussion to follow, we focus our attention on injection lasers even though the general principles to be presented are equally applicable to photo-excited and electron-beam-excited lasers. The structure of a semiconductor laser is made of a semiconductor cleaved by both ends. The change in the index of refraction at semiconductor–air interface provides sufficient reflection required of an optical cavity for laser action. Therefore, a semiconductor laser can be modeled by an amplifying medium bounded at both ends by a set of two parallel mirrors as schematically shown in Fig. 14.22. Con-

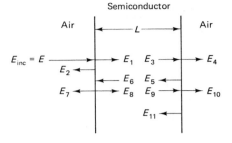

Semiconductor

Air ← L → Air

$E_{inc} = E$

$E_1 = t_2E,\ E_3 = AE_1,\ E_5 = r_1E_3$
$E_2 = r_2E,\ E_4 = t_1E_3,\ E_6 = AE_5$

Figure 14.22 Schematic diagram showing multiple reflections at the air–semiconductor interface of a semiconductor-laser cavity. The total transmitted wave is made of E_4, E_{10}, and so on, whereas the total reflected wave is made of E_2, E_7, and so on.

sider a plane wave incident on the cavity. As shown in Problem 14.22, the transmitted field E_{tran} and the reflected field E_{ref} are related to the incident field E_{inc} by

$$E_{tran} = \frac{t_1 t_2 A}{1 - A^2 r_1^2} E_{inc} \tag{14.8.1}$$

$$E_{ref} = \left(r_2 + \frac{r_1 t_1 t_2 A^2}{1 - A^2 r_1^2} \right) E_{inc} \tag{14.8.2}$$

where the subscripts 1 and 2 apply to transmission and reflection coefficients for waves inside the semiconductor and for waves in the air, and A is the propagation factor inside the gain medium given by

$$A = \exp\left[\left(\frac{g}{2} + ik \right) L \right] \tag{14.8.3}$$

Laser action occurs when nonzero E_{tran} and E_{ref} exist even for zero E_{inc} requiring $r_1^2 A^2 = 1$. This equation simply says that upon making a round trip inside the gain medium, a wave returns to the original value in both amplitude and phase.

The amplitude condition requires that the gain coefficient g_{th} provided by stimulated emission equals the equivalent loss coefficient, or

$$g_{th} = \alpha + L^{-1} \ln r^{-2} \tag{14.8.4}$$

where α is the inherent loss coefficient of the cavity without stimulated emission. The baseline $\alpha = 30 \text{ cm}^{-1}$ shown in Fig. 14.20b, for example, represents this inherent α. The last term in Eq. (14.8.4) represents the equivalent loss coefficient due to transmission loss through the two mirrors. The phase condition of $r_1^2 A^2 = 1$ requires that $2kL$ be integral multiple of 2π, or

$$L = \frac{m\lambda}{2n} = \frac{m\lambda_s}{2} \tag{14.8.5}$$

where $\lambda_s = \lambda/n$ is the wavelength inside a semiconductor, n the index of refraction, and m an integer. Figure 14.23 shows (a) the light intensity versus injection current of a low-threshold injection laser and (b) the lasing spectrum of a multiple-mode injection laser. Assuming a value of 40 cm^{-1} for g_{th}, we find from Fig. 14.21 a value of about

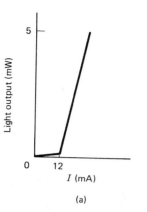

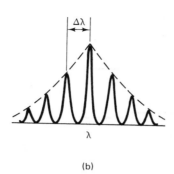

Figure 14.23 Schematic diagrams showing (a) light intensity versus injection current of a low-threshold injection laser and (b) the lasing spectrum of a multiple-mode injection laser.

6×10^3 A/cm²-μm for J_{nom}. Thus in a laser with dimensions of 300 μm × 4 μm × 0.1 μm, the value of the threshold current J_{th} is estimated to be about 7 mA. Effects such as current spreading and nonradiative recombination could raise J_{th} to the value of 12 mA shown in Fig. 14.22a. From Eq. (12.8.5) the spacing $\Delta\lambda$ between two neighboring longitudinal modes m and $m + 1$ is

$$\Delta\lambda = \frac{\lambda^2}{2L\bar{n}} \qquad (14.8.6)$$

where $\bar{n} = n + \lambda\, dn/d\lambda$ is the group index of refraction. For $L = 310$ μm and $\bar{n} = 4.5$, $\Delta\lambda = 2.8$ Å, which is in agreement with experimentally observed value.

The laser action observed in 1962 was achieved in homojunctions under pulsed operation at 77 K. The progress was very slow during ensuing years, in an effort to reduce the threshold current necessary for CW operation at room temperature. Failure to accomplish the goal is due to inherent problems present in a homojunction (Fig. 14.18a). First, besides the current caused by carrier recombination in region AB, a substantial wasteful current exists due to electron diffusion beyond A and hole diffusion beyond B. Second, because of lack of sufficient guiding, the optical field spreads to regions beyond A and beyond B where the field suffers significant losses. These two problems are rectified in a double heterostructure. As shown in Fig. 14.24a, because of the difference in energy gap, discontinuities $\Delta\mathscr{E}_c$ and $\Delta\mathscr{E}_v$ occur in the conduction-band edge and the valence-band edge, respectively. The band-edge discontinuities, $\Delta\mathscr{E}_c$ on the left and $\Delta\mathscr{E}_v$ on the right, do not impede electron injection from n-(GaAl)As to GaAs and hole injection from p-(GaAl)As to GaAs. However, electron diffusion from GaAs to p-(GaAl)As and hole diffusion from GaAs to n-(GaAl)As are blocked, respectively, by $\Delta\mathscr{E}_c$ on the right and by $\Delta\mathscr{E}_v$ on the left. Therefore, injected carriers are confined in the potential well constituted by the energy-gap difference. Consequently, the ratio of the diffusion current (wasteful component) to the recombination current (useful component) is greatly reduced. Based on a split of $\Delta\mathscr{E}_g$ into $\Delta\mathscr{E}_c = 0.7\,\Delta\mathscr{E}_g$ and $\Delta\mathscr{E}_v = 0.3\,\Delta\mathscr{E}_g$, it is estimated (Problem 14.24) that a value greater than 0.3 for the composition x in the outer $Ga_{1-x}Al_xAs$ regions is needed in order that the potential barrier at the interface is sufficiently high to render the diffusion current insignificant.

Another important attribute of a double heterostructure is the provision of a guiding structure with GaAs serving as the guiding layer and (GaAl)As serving as the two

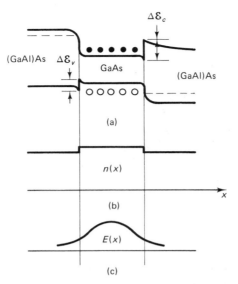

(GaAl)As $\Delta \mathcal{E}_v$ GaAs (GaAl)As

$\Delta \mathcal{E}_c$

(a)

$n(x)$

x

(b)

$E(x)$

(c)

Figure 14.24 (a) Energy-band diagram of an (AlGa)As/GaAs double heterostructure with band-edge discontinuities at the two interfaces, (b) index variation due to compositional change and (c) variation of optical field in the double heterostructure. The potential barriers caused by the band-edge discontinuities confine the injected carriers in the GaAs well.

cladding layers. For compound semiconductors a general tendency exists that larger gap materials have a lower index of refraction (Fig. 14.24b). Therefore, a solution (Fig. 14.24c) of the form $E(x) \exp (i\beta z)$ can be found for the wave equation with

$$E(x) = A \cos (k_x x + \phi) \qquad (14.8.7)$$

being oscillatory in the GaAs region and

$$E(x) = B_{1,2} \exp (\pm p_{1,2} x) \qquad (14.8.8)$$

decaying exponentially in the two (GaAl)As regions, where $p_{1,2}$ is the decay constant. We further note that since (GaAl)As has a higher gap than GaAs, the optical wave generated in the GaAs region will not be reabsorbed in the (GaAl)As regions even though the field spreads into the (GaAl)As regions.

In summary, a double heterostructure (DH) provides a potential well (Fig. 14.24a) to confine injected carriers and a refractive index step (Fig. 14.24b) to confine the optical field. The carrier confinement greatly enhances the utilization of injected carriers. The optical confinement makes the stimulated emission radiate into clearly defined modes of the optical waveguide formed by the index step. The simultaneous confinement of injected carriers and the optical field in the same region greatly increases their interaction and thus the rate of stimulated emission. Furthermore, a double heterostructure eliminates the reabsorption loss inherent in a homojunction. As a result of these factors, the use of double heterostructures in semiconductor lasers has produced substantial reduction of the threshold current and thus made possible CW operation of injection lasers at room temperature (Zh. I. Alferov, V. M. Andreev, E. L. Portnoi, and M. K. Trukan, *Sov. Phys. Semicond.*, Vol. 3, p. 1107, 1969; I Hayashi, M. B. Panish, P. W. Foy, and S. Sumski, *Appl. Phys. Lett.*, Vol. 17, p. 109, 1970).

The advent of DH lasers is a major step in the development of injection lasers. However, a planar DH laser still lacks a proper guiding mechanism in the lateral direction, that is, the direction in the junction plane and perpendicular to the direction of wave propagation (the longitudinal direction). Referring to the laser structure shown in Fig. 14.11b, we see that guiding in the longitudinal or z direction is provided by the two end facets and guiding in the transverse or x direction is provided by the double heterostructure. We further notice that the structure is planar. By definition a planar structure has uniform properties in the horizontal plane. Therefore, nonplanar structures must be introduced to provide guiding in the lateral direction. One obvious way of accomplishing this purpose is to surround the active GaAs region completely by (GaAl)As in the form of a buried heterostructure (BH) (T. Tsukada, *J. Appl. Phys.*, Vol. 45, p. 4899, 1974). Another way is to vary the thickness of the active layer in the lateral direction so that k_x in Eq. (14.8.7) is smaller in the intended guiding region than in the regions outside the guiding region. Figure 14.25 shows one of the guiding structures of the latter type together with the observed intensity distributions of the various lateral modes. The symbol TE$_{mn}$ indicates a TE mode with electric field in the y direction, and the indices m and n denote the mode number in the transverse and lateral direction, respectively (W. T. Tsang and S. Wang, *Appl. Phys. Lett.*, Vol. 28, p. 665, 1976). The thinner GaAs regions have a higher k_x than the thicker GaAs region, and as

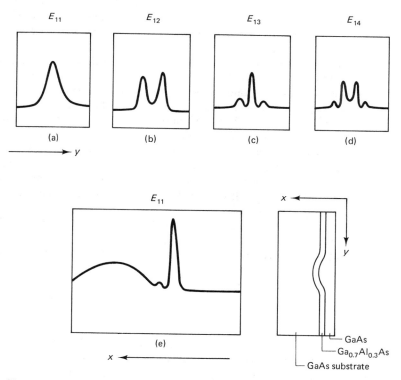

Figure 14.25 An optical waveguide formed by thickness variation in the lateral (y) direction, and observed light-intensity distributions for the various waveguide modes in the lateral (y) direction for (a), (b), (c), and (d), and in the transverse direction for (e). The index m or n in E_{mn} signifies the number q of intensity minima (strictly speaking, intensity zeros in the respective direction with $q = m - 1$ or $n - 1$). (W. T. Tsang and S. Wang, *Appl. Phys. Lett.*, Vol. 28, p. 665, 1976.)

Sec. 14.8   Injection Lasers

a result, have an imaginary k_y (meaning decaying solution) in order to keep $k_x^2 + k_y^2$ the same throughout the GaAs region. The spatial variation in k_x due to the thickness variation causes an optical beam to be confined to the thicker region. A DH laser with lateral guiding has clearly defined modes in all three directions, a prerequisite for the laser to oscillate at a single wavelength. The multiple-wavelength spectrum shown in Fig. 14.23b is from a broad-area planar DH laser which emits in filaments instead of a single beam.

The GaAs–(GaAl)As laser that we use as an example of DH lasers is unique because GaAs and (GaAl)As have nearly equal lattice constants. As discussed in Section 11.1, one of the basic requirements for growth of high-quality heteroepitaxial films is a nearly perfect lattice match. Due to substrate availability heterostructures being extensively studied and rapidly developed are the GaAs–(GaAl)As system matched to GaAs substrate, the (GaIn)(AsP) system matched to InP substrate, and the (GaIn)(AsP) system matched to GaAs substrate. Referring to Fig. 11.1, we see that the first system covers the spectral range from $\lambda = 0.87$ to 0.64 μm, the second from $\lambda = 1.56$ to 0.92 μm, and the third from $\lambda = 0.87$ to 0.66 μm, not taking into consideration the effects of band-tail states and satellite valleys. The long-wavelength lasers (1.3 μm and 1.55 μm) are primary sources for long-distance optical fiber communication systems. The medium-wavelength lasers (0.80 to 0.87 μm) are suitable sources for short-distance fiber systems in local-area-network applications. The short-wavelength lasers (< 0.78 μm) are for optical storage and laser printing applications.

An interesting question arises as to the feasibility of extending the spectral range of injection lasers into the visible (< 0.63 μm) and far-infrared regions. Besides the problems of controlling impurities and crystal imperfections in, and making ohmic contacts to, semiconductors of large gap as mentioned in Section 14.3, there is also one basic problem relating to lasers of short wavelength. Referring to Eq. (14.7.7), we see that the rates of stimulated and spontaneous emissions are proportional to $S(h\nu)/N(h\nu)$. Since the density of electromagnetic modes $N(h\nu)$ increases as λ^{-4}, the photon density required to make $r_{\text{stim}}(h\nu) > r_{\text{spon}}(h\nu)$ also increases as λ^{-4}. This fact imposes stricter conditions on the optical cavity, such as higher mirror reflectivity and lower inherent cavity loss, in order to keep the threshold current reasonably low. As to the long-wavelength lasers, one serious problem is the *Auger recombination*. It involves the simultaneous annihilation of an electron–hole pair and absorption of the released energy by a third particle, which can be either an electron or a hole. The threshold current of an injection laser measured as a function of temperature generally obeys an empirical formula

$$J_{\text{th}}(T + \Delta T) = J_{\text{th}}(T) \exp\left(\frac{\Delta T}{T_0}\right) \tag{14.8.9}$$

Around room temperature, T_0 ranges from 160 to 220 K in GaAs lasers and T_0 is around 60 K in (GaIn)(AsP) lasers. The Auger process is a dominant factor in the rapid increase of J_{th} with temperature in quaternary lasers (G. H. B. Thompson and G. D. Henshall, *Electron. Lett.*, Vol. 16, p. 42, 1980).

Because the Auger process involves three particles, the rate of Auger recombination R_{Aug} can be expressed as

$$R_{\text{Aug}} = Bn^2p \quad \text{or} \quad Bnp^2 \tag{14.8.10}$$

The proportionality constant B has a strong temperature dependence

$$B(T) = B_0 \left(\frac{kT}{\mathscr{E}_g} \right)^{3/2} \exp \left(-\frac{\mathscr{E}_{\text{th}}}{kT} \right) \qquad (14.8.11)$$

where the threshold energy is given by

$$\mathscr{E}_{\text{th}} = \frac{(m_c + 2m_v)\mathscr{E}_g}{m_c + m_v} \qquad (14.8.12)$$

if the process is dominated by electron–electron collision (A. R. Beattie and P. T. Landsberg, *Proc. R. Soc. London,* Vol. 249, p. 16, 1958) and by

$$\mathscr{E}_{\text{th}} = \frac{(2m_{vh} + m_c)(E_g - \Delta)}{2m_{vh} + m_c - m_{vs}} \qquad (14.8.13)$$

if the process is dominated by collisions between a heavy hole with mass m_{vh} and a split-off-band hole with mass m_{vs} (A. Sugimura, *IEEE J. Quantum Electron.,* Vol. QE-17, p. 627, 1981). After the Auger process, the third particle becomes either an energetic electron or energetic hole. The energetic carrier loses its energy through either carrier–carrier collision or optical-phonon emission. The first process will enhance carrier diffusion across the heterojunction while the second process will raise the carrier temperature above the lattice temperature. Therefore, the Auger recombination is a nonradiative process and will have both direct and indirect effects on the operation of injection lasers, especially in the long-wavelength region. A new measurement technique, the recently developed differential carrier lifetime measurement, has made possible the separation of radiative recombination current ($\sim Anp$) and nonradiative recombination current ($\sim Bn^2 p$ or Bnp^2), thus establishing the presence of Auger recombination in quaternary LEDs and lasers (R. Oshlansky, C. B. Su, J. Manning, and W. Powazinik, *IEEE J. Quantum Electron.,* Vol. QE-20, p. 1023, 1984). Since the coefficient $B(T)$ according to Eq. (14.8.11) increases very rapidly with decreasing gap energy $\mathscr{E}_g$, the Auger process may limit the operation of long-wavelength lasers to temperatures below room temperature.

14.9 RATE EQUATIONS

In the preceding section we presented the amplitude and phase conditions [Eqs. (14.8.4) and (14.8.5)] for a laser. In this section we combine the results of our discussion in Sections 14.5 and 14.7 to formulate a set of equations from which important characteristics of an injection laser can be derived. We assume that the impurity concentration in the active region is low so the carrier concentration N is equal to the injected electron or hole concentration $n = p$. The carriers are supplied by the injection current at a rate of J/ed, and depleted through nonradiative and spontaneous recombination at a rate of N/τ and through stimulated emission at a rate of R_{stim} per unit volume per second. Therefore, the net rate of change in N with time is given by

$$\frac{dN}{dt} = \frac{J}{ed} - \frac{N}{\tau} - R_{\text{stim}} \qquad (14.9.1)$$

where $\tau = \tau_r \tau_{nr}/(\tau_r + \tau_{nr})$, τ_r is the radiative lifetime and τ_{nr} is the nonradiative lifetime. The photons, on the other hand, are supplied through stimulated emission at the same rate R_{stim} and through spontaneous emission at rate $\gamma N/\tau_r$, but depleted through cavity losses at a rate S/τ_p. Thus the net rate of change in S with time is given by

$$\frac{dS}{dt} = R_{stim} + \frac{\gamma N}{\tau_r} - \frac{S}{\tau_p}$$ (14.9.2)

where γ represents the fraction of spontaneous emission coupled to a given mode of the cavity and has a value ranging from 10^{-4} to 10^{-5}, depending on wavelength and the cavity volume. The photon lifetime τ_p is given by $\tau_p^{-1} = c(\alpha + L^{-1} \ln r^{-2})$, where c is the velocity of light in the medium.

Since the gain coefficient is defined as $g = S^{-1} dS/dx$, the rate $R_{stim} = dS/dt$ is equal to cgS. Furthermore, g depends on $(f_c - f_v)$, which itself is a function of the carrier density. Using a linear approximation for $g = A(N - N_0)$, we have

$$R_{stim} = cgS = cA(N - N_0)S$$ (14.9.3)

where N_0 is the carrier density required to make the medium lossless ($g = 0$) and $A = dg/dN$ is the linear expansion coefficient. The term R_{stim} represents the carrier–photon interaction, and hence is nonlinear by nature. In the region where the spontaneous emission dominates, we can ignore R_{stim} and thus obtain, under a constant J, the following steady-state solution for N and S:

$$N = \frac{\tau}{ed} J \quad \text{and} \quad S = \frac{\gamma \tau}{\tau_r} \frac{\tau_p}{ed} J$$ (14.9.4)

Since γ is a small number, S increases slowly with increasing J. In the region where the stimulated emission dominates, we can neglect the term $\gamma N/\tau_r$ in Eq. (14.9.2) and thus obtain the following steady-state solution:

$$N = N_1 = N_0 + (cA\tau_p)^{-1}$$ (14.9.5)

which is the carrier density required to produce a gain to balance the cavity loss. Using this value of N in Eq. (14.9.1) and setting $dN/dt = 0$, we find that

$$S = \frac{\tau_p}{ed} (J - J_{th})$$ (14.9.6)

where J_{th} is the threshold current density given by

$$J_{th} = \frac{N_1 ed}{\tau}$$ (14.9.7)

The light output versus current relation in the stimulated-emission-dominated region differs from that in the spontaneous-emission-dominated region in two aspects. It has a much larger slope by a factor of $\tau_r/\tau\gamma$, and it has an intercept on the current axis which defines the threshold current I_{th}. Both features are clearly shown in Fig. 14.23a.

In the discussion above, we tacitly assume that only one dominant cavity mode is involved. Since the linear dimension of a laser cavity is much larger than the wavelength, especially in the longitudinal direction, we must consider the possibility of many cavity modes being excited. We use the subscript m to indicate the main mode with the highest intensity and the subscript s to indicate a secondary mode with diminished intensity. It is generally accepted that the gain spectrum of a semiconductor is homogeneously broadened. In other words, carriers interact with all the cavity modes within the gain spectrum. Under the circumstances, the photon density can be solved separately for each mode (T. P. Lee, C. A. Burrus, C. A. Copeland, A. G. Dentai, and D. Marcuse, *IEEE J. Quantum Electron.*, Vol. QE-18, p. 1101, 1982). Thus we obtain from Eq. (14.9.2)

$$S_m = \frac{\gamma N}{c \tau_r} [g_{th} - g(\lambda_m)]^{-1} \tag{14.9.8}$$

$$S_s = \frac{\gamma N}{c \tau_r} [g_{th} - g(\lambda_s)]^{-1} \tag{14.9.9}$$

where $g_{th} = \alpha + L^{-1} \ln (r^{-2})$. Refer to the gain spectrum shown in Fig. 14.20a. We assume that the wavelength of the peak gain is at λ_m. Further, we expand the spectral dependence of gain around λ_m as

$$g(\lambda_s) = g(\lambda_m) - (\lambda_s - \lambda_m)^2 \frac{g''_\lambda}{2} \tag{14.9.10}$$

where $g''_\lambda = d^2 g/d\lambda^2$. As the injection current is raised, the difference $g_{th} - g(\lambda_m)$ is reduced. As a result, S_m increases indefinitely with increasing J for the main mode. For all other modes, the difference $g_{th} - g(\lambda_s)$ approaches $(\lambda_s - \lambda_m)^2 g''_\lambda/2$ as a limit. Therefore, for secondary modes, the photon density and hence the output intensity tend to saturate. This tendency has indeed been observed (C. A. Burrus, T. P. Lee, and A. G. Dentai, *Electron. Lett.*, Vol. 17, p. 954, 1981).

Now we turn to the temporal behavior of an injection laser. Since the coupled equations are nonlinear, numerical methods are often used. Figure 14.26 shows the computer-simulation results for (a) the transient response to a step driving current from $J = 0$ to $J = 1.5J_{th}$, and (b) the steady-state response to a sinusoidal current $J = J_b + mJ_b \sin \omega t$ with a dc bias current $J_b = 1.5J_{th}$, a modulation frequency $\nu = 1 \times 10^9$ Hz, and a modulation depth $m = 0.4$ or 0.6. The following set of values for the various parameters are used: $N_0 = 1.5 \times 10^{18}$ cm^{-3}, $\tau = \tau_r = 3 \times 10^{-9}$ s, $\tau_p = 6 \times 10^{-12}$ s, $\gamma = 5 \times 10^{-5}$ and $cA = 2.5 \times 10^{-7}$ cm^3/s. Referring to Fig.

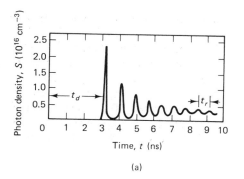

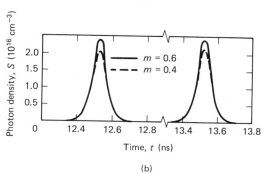

(a)

(b)

Figure 14.26 Computer simulation results of the solutions of Eqs. (14.9.1) and (14.9.2) for (a) the transient response to a step driving current from $J = 0$ to $J = 1.5J_{th}$ and (b) the steady-state response to a sinusoidal driving current superimposed on a dc bias current $J_b = 1.5J_{th}$. (M. Tang and S. Wang, *J. Opt. Commun.*, Vol. 8, p. 82, 1987.) Refer to the text for the parameters used in the simulation.

Sec. 14.9   Rate Equations

14.26, we see that during the initial response to a step excitation, the photon density S has a very large swing. The dynamics of the lasing process can be stated as follows. As soon as the step driving current is on, the carrier density N begins to build up with a time constant τ. During this phase, $S = 0$ and integration of Eq. (14.9.1) yields the time t_d required for the carrier density to build up from an initial value N_i to a final value N_f:

$$t_d = \tau \ln \frac{J - edN_i/\tau}{J - edN_f/\tau} \tag{14.9.11}$$

Once N exceeds N_1 of Eq. (14.9.5), dS/dt becomes positive and the photon density S starts to build up very rapidly, with a time constant on the order of τ_p. Since τ is much longer than τ_p, we may take $N_f = N_1$ in Eq. (14.9.11) and let $N_i = 0$. Thus we have

$$t_d = \tau \ln \frac{J}{J - J_{\text{th}}} \tag{14.9.12}$$

which is the delay time between the application of a current pulse and the initiation of a laser pulse.

Note that the rapid rise of S is indeed confirmed by computer simulation. During this phase the carrier density N depletes very rapidly. When N goes below N_1, dS/dt becomes negative and soon R_{stim} becomes unimportant. When this happens, N starts to build up again from a higher initial value $N_1 > N_i > N_0$ and the process repeats itself with a smaller swing. Finally, the photon density S approaches the steady-state value given by Eq. (14.9.6). We should point out that even though the swing in S is very large, the swing in N is relatively small, on the order of $0.1N_1$ or about 2×10^{17} cm^{-3}. This value yields a total rise and fall time t_{rf} on the order of 1.5×10^{-10} s for the photon density S. In the region where the amplitude swing in S is small, Eq. (14.9.3) can be linearized and the transient solution of Eqs. (14.9.1) and (14.9.2) shows typical relaxation oscillation with a characteristic angular frequency

$$\omega_r = [1 + cg_N'\tau_pN_0]^{1/2}(\tau_r\tau_p)^{-1/2}\left(\frac{J}{J_{\text{th}}} - 1\right)^{1/2} \tag{14.9.13}$$

where $g_N' = dg/dN$.

Finally, we refer to Fig. 14.26b, which simulates the formation of short pulses in an injection laser under high-frequency modulation with large modulation indices. For modulation frequency in the range $t_d^{-1} < \nu < t_{rf}^{-1}$, a single fully developed laser pulse can be expected during each modulation cycle and is confirmed by the simulation results. It appears that the minimum value of the total rise and fall time t_{rf} is relatively insensitive to the modulation depth. However, it depends on τ and τ_p. Shortening τ_p to 2×10^{-12} s and τ to 1×10^{-9} s, one can reduce t_{rf} to around 50 ps. Generation of optical pulses of about 50 ps duration by a microwave signal superimposed on a dc bias has indeed been achieved (H. Ito, H. Yokoyama, S. Murata, and H. Ibana, *IEEE J. Quantum Electron.*, Vol. QE-17, p. 663, 1981).

14.10 PHOTODETECTORS

The central element in the detection of an optical signal is the detector, a device that converts electromagnetic energy into electric energy. Photodetectors can be divided into three basic types: the photoemissive type, the photoconductive type, and the photodiode type. In this section we present a general discussion on the basic principles as well as performance considerations for the various photodetectors. Since the photoconductive and photodiode detectors are compatible with the basic concepts on which modern electronics and optoelectronics are based in terms of physical compactness and functional integrability, they are discussed further in the following sections. Specifically, the recent advances in photoconductors are discussed in Section 14.11. The photodiode detectors can be divided into two distinct types. The recent developments of *p-i-n* and Schottky photodiodes without internal gain are discussed in Section 14.12, while those of photodetectors with internal gain, namely avalanche photodiodes and phototransistors, are discussed in Section 14.13.

A vacuum tube with a photoemissive cathode conducts electric current when photons impinge on, and liberate electrons from, the cathode. Figure 14.27 shows the spectral response of several commonly used photocathodes. As discussed in Sections 4.6 and 4.7, an electron must overcome a potential barrier, called the *work function,* in order to escape from a solid. The photocathodes in Fig. 14.27 have different long-wavelength thresholds because their work functions are different. A photomultiplier tube is a vacuum tube that consists of a photocathode, an anode, and several dynodes. The electrons emitted from the photocathode are accelerated by a voltage and focused on the first dynode, where each incident electron causes the emission of several secondary electrons. The electrons thus emitted are again focused on the second dynode, and so on, until in the last stage the electrons are collected by the positive anode. If each dynode is capable of producing δ secondary electrons upon the impact of a primary electron, the gain factor g of an M-stage photomultiplier tube is $g = \delta^M$. A tube with 10 stages and a value of 4 for δ has a gain of about 10^6. The response time of a photomultiplier tube is about 10^{-8} s, whereas that of a phototube is about 10^{-9} s. Two special types of photomultiplier tube have been developed with a flat response to 4 GHz and a gain of 10^5 (R. C. Miller and N. C. Witwer, *IEEE J. Quantum Electron.,* Vol. QE-1, p. 62, 1965). Because of the large postdetection gain and the low-noise figure, photomultipliers are widely used for the detection of weak signals in the visible range.

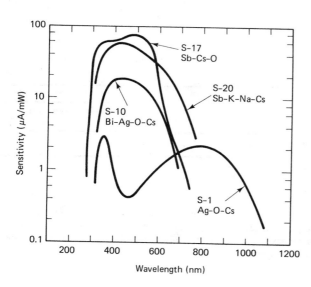

Figure 14.27 Sensitivity of a selected number of photocathodes.

However, no photoemissive material has been found to have a low-enough work function for the detection of infrared signals. On the other hand, photoconductive and photodiode detectors are natural candidates in the infrared region because the technology of III–V compounds is well developed in this spectral range.

When a semiconductor is irradiated with a beam of light having sufficient photon energy, carriers are generated either by band-to-band transitions (intrinsic photoconductor) or by impurity-to-band or band-to-impurity transitions (extrinsic photoconductor). The extra conduction due to photon-generated carriers is called *photoconductance*. Figure 14.28 shows the three basic processes for possible use in photoconductive detectors: (a) band-to-band transition, (b) band-to-impurity transition, and (c) impurity-to-band transition. Here we use the systems Ge:Au and Ge:(Au, Sb) as examples of cases (b) and (c) used in practical photoconductors; case (a) is discussed in detail in Section 14.11. From Fig. 6.21a we see that Au has four levels located at 0.04 and 0.20 eV from the conduction band and at 0.05 and 0.15 eV from the valence band. In Au-doped Ge, photons having energy greater than 0.15 eV (or wavelength shorter than 8.3 μ) free holes from the 0.15-eV level (Fig. 14.28b). The photoconductivity is therefore p type. To make the photoconductor respond to 8.2-μm radiation, it is necessary to cool it to 77 K. In the Ge:(Au, Sb) photoconductor, the 0.20-eV level is used. The optimum doping concentration is the addition of two Sb atoms for every Au atom. Under this condition, the Fermi level lies between the 0.01-eV Sb level and the 0.20-eV Au level. The material becomes photoconductive when a photon excites an electron from the 0.20-eV level to the conduction band (Fig. 14.28c). The Ge:(Au, Sb) photoconductor is n type and operates at 77 K or below. Extrinsic photoconductors are commonly used for the detection of far-infrared signals.

The performance of a photoconductor is measured in terms of two parameters: the photoconductivity gain and the response time. Consider the excitation process illustrated in Fig. 14.28b. Let G be the rate of hole generation per unit volume. If τ is the lifetime of holes, the steady-state condition requires that

$$\frac{dp_1}{dt} = -\frac{p_1}{\tau} + G = 0 \quad \text{or} \quad p_1 = G\tau \tag{14.10.1}$$

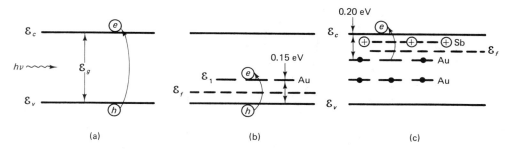

Figure 14.28 Three basic processes used in photoconductive detectors: (a) band-to-band transitions by which electron–hole pairs are generated, (b) band-to-impurity transitions by which holes are generated and electrons are trapped, and (c) impurity-to-band transitions by which electrons are excited from impurities to the conduction band.

where p_1 is the excess hole concentration. Thus the photon-induced current (photocurrent) in a photoconductor of uniform area A is

$$I = e\mu_p p_1 EA \qquad (14.10.2)$$

where E is the applied electric field. In a sample of length L, the transit time τ_t is $L/\mu_p E$. Using this relation and Eq. (14.10.1) in Eq. (14.10.2), we find that

$$I = \frac{\tau}{\tau_t} eGAL \qquad (14.10.3)$$

The product $eGAL$ represents the amount of charges generated per unit time. Therefore, the dimensionless quantity τ/τ_t is the photoconductivity gain. A sensitive photoconductor must have long lifetime and high mobility. The response time of a photoconductor is related to the lifetime τ. For a light source that is modulated sinusoidally, the generation rate G varies as $G = G_1 \exp (i\omega t)$. Accordingly, the excess-hole concentration also varies with time as

$$p_1 = \frac{G_1 \tau}{1 + i\omega\tau} \exp (i\omega t) \qquad (14.10.4)$$

Therefore, the response of a photoconductor has a bandwidth $\Delta\omega \sim \tau^{-1}$, and the gain–bandwidth product is independent of τ but increases with carrier mobility μ.

Besides gain and bandwidth, another important consideration is noise, which ultimately limits the sensitivity of a detector. There are three noise sources: (1) the Johnson noise due to fluctuations caused by thermal energy, (2) the shot noise due to random fluctuations in current-generating events, and (3) the $1/f$ noise of undetermined origin. For high-speed or high-frequency (> 500 MHz) applications, the $1/f$ noise is generally not the limiting factor and thus will not be considered further. The Johnson noise can be represented by a noise-voltage source v_N, which has a mean-square value

$$\langle v_{NJ}^2 \rangle = 4RkTB \qquad (14.10.5)$$

measured within a frequency bandwidth $B = \Delta\nu$, where R is the dark resistance of a sample. The shot noise can be represented by a noise-current source i_N, which has a mean-square value given by

$$\langle i_{NS}^2 \rangle = 2eIB \qquad (14.10.6)$$

within a frequency bandwidth B, where I is the current through a diode. Whereas the Johnson noise is the principal noise source in photoconductors, the shot noise is the main noise source in photodiodes. To reduce the Johnson noise, a photoconductor is often cooled down to low temperatures. In a photodiode, I consists of both the dark current I_{dark} and the photocurrent I_{ph}. To minimize the shot noise, the dark current in a diode must be made small. From Eqs. (14.10.5) and (14.10.6) we see that at room temperature, a resistance R of 10^6 Ω and a current I of 10^{-7} A produce the same amount of noise current.

Figure 14.29 shows the physical structures of two types of photodiodes: (a) the p-i-n photodiode and (b) the avalanche photodiode. Recent advances in laser and optical fiber technology have prompted the research and development of fast-response photodetectors for use in laser and fiber communication receivers. The two types of photo-

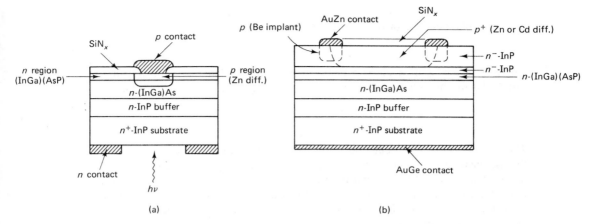

Figure 14.29 Structures of (a) *p-i-n* photodiode in which photons are absorbed in the near-intrinsic *n*-(InGa)As layer of 2 to 4 μm thickness, and (b) avalanche photodiode in which light is absorbed in the *n*-(InGa)(AsP) layer and carriers are multiplied in the *n*-InP layer, where a high field exists.

diodes are designed for detection of light with a modulation frequency in the microwave region. The *p-i-n* junction consists of a thin, heavily doped *n* region, a near-intrinsic *n* region, and a heavily doped *p* region. The middle *n*-(GaIn)As region is comparatively thick (between 2 to 4 μm), so that most incident light is absorbed in the near-intrinsic region. The junction is biased in the reverse direction. The photoexcited electron–hole pairs are swept by the high electric field across the junction with electrons to the *n* side and holes to the *p* side. If the electric field in the depletion region is high enough for carriers to attain a saturation velocity v_s, the transit time is given by

$$\tau_t = \frac{W_d}{v_s} \tag{14.10.7}$$

where W_d is the width of the drift region and equal to the width of the intrinsic region in the case of a *p-i-n* photodiode. For $v_s = 10^7$ cm/s and $W_d = 4$ μm, we have $\tau_r = 40$ ps. Fast responses of *p-i-n* diodes in the GHz region have been demonstrated (R. P. Riesz, *Rev. Sci. Instrum.,* Vol. 33, p. 994, 1962). Excellent reviews of photodetectors for use in laser and fiber communication receivers are now available either in book form or in journal articles. These include a book by Ross and articles by Anderson and McMurtry; Smith; Stillman, Robbins, and Tabatabaie; and Brain and Lee (M. Ross, *Laser Receivers,* John Wiley & Sons, Inc., New York, 1966; M. Brain and T. P. Lee, *IEEE Trans. Electron Devices,* Vol. ED-12, p. 2673, 1985; L. K. Anderson and B. J. McMurtry, *Proc. IEEE,* Vol. 54, p. 1336, 1966; R. G. Smith, *Proc. IEEE,* Vol. 68, p. 1247, 1980; G. E. Stillman, V. M. Robbins, and N. Tabatabaie, *IEEE Trans. Electron Devices,* Vol. ED-31, p. 1643, 1984).

Besides speed, another important consideration for a photodetector is sensitivity, which is related to the efficiency of detection. The number of photon absorbed per incident photon can be found by integrating I of Eq. (12.11.1). The quantum efficiency η is thus given by

$$\eta = (1 - R)[1 - \exp(-\alpha W)] \qquad (14.10.8)$$

where R is the surface reflectance and W is the width of the absorbing layer. A thicker W means a higher η but it also means a slower speed. For direct-gap semiconductors, α is on the order of 10^{-4} cm^{-1} or larger (Fig. 12.19). Therefore, a thickness of a few micrometers is sufficient to achieve over 90% absorption. For silicon, which is an indirect-gap semiconductor, α is in the range 500 to 1000 cm^{-1} and hence W is required to be in the range of a few tens of micrometers, typically 30 to 60 μm. The larger W used in silicon p-i-n diodes results in a much slower response time, typically on the order of 0.5 ns. The larger W also means a higher applied voltage to reach the state of full depletion of carriers in the intrinsic region. Under partial depletion, carriers generated in an undepleted region have to diffuse before they are collected. Therefore, the photoresponse of the diode has a component of a slower response time, governed by the carrier-diffusion time

$$\tau_d = \frac{W_{\mathrm{UD}}^2}{2D} \qquad (14.10.9)$$

where W_{UD} is the width of the undepleted region and D the diffusion constant. The condition for full depletion sets a limit on the impurity concentration N_I for a given bias voltage V_b. In silicon, $N_I \leq 10^{13}$ cm^{-3} is required in the intrinsic region for $V_b \sim 20$ V.

From the discussion above, we see that a compromise between high speed and high efficiency is unavoidable because they have conflicting requirements on W. A III–V compound has several advantages over Si insofar as the p-i-n photodiode is concerned. Because of a much larger α, the speed of a III–V diode is much faster. Second, through the use of heterostructure such as the one shown in Fig. 14.29a, most of the light is absorbed in the intrinsic region and not in the doped regions. The use of heterostructure is not possible for Si diodes. Third, because the substrate is made of a larger-gap material and hence nonabsorbing, the light can be incident from the substrate side. This arrangement makes it possible to reduce the size of the top contact. Finally, and perhaps most important, the III–V p-i-n diode is amenable to integration with fast III–V electronic devices. The subject of optoelectronic integration is discussed in Section 14.12. However, the p-i-n diode is simply a detector; it does not provide any internal amplification of the detected signal. The avalanche photodiode, on the other hand, has this capability. Whereas III–V compounds are the favored material for p-i-n diodes, silicon is most commonly used in avalanche diodes.

In principle, a junction photodiode can be operated in the avalanche region to provide internal current gain. Such a possibility has been investigated and demonstrated by many workers. In practical avalanche diodes, ways must be devised to overcome problems associated with the high electric field and with the avalanche process itself. First, a junction often breaks down in small areas, creating microplasmas. The probability of microplasmas can be reduced by keeping the active area as small and with as uniform properties as possible. Second, breakdown may occur at places where the junction intersects the surface. Surface breakdown can be prevented by the use of a guard ring, the Be-implanted p region in Fig. 14.29b. The doping concentration in the guard-ring region is kept low so that the high electric field is confined to the bulk region (the n-p^+ junction in Fig. 14.29b). With the elimination of these undesirable effects, the true response of an avalanche photodiode (APD) can be determined. Because carriers must first drift to the high-field region and after impact ionization, carriers of the opposite type must drift back, the effective length of the drift region and hence the frequency response of an APD will depend on the dc multiplication factor M_0 as well as the ratio α_e/α_h of the ionizaton coefficient for electrons and holes. A theoretical analysis based on the solution of current transport equations similar to Eqs. (13.6.2) and (13.6.3) but with different α_e and α_h shows that the multiplication factor M has the frequency dependence

$$M(\omega) = \frac{M_0}{(1 + \omega^2 M_0^2 \tau_1^2)^{1/2}} \qquad (14.10.10)$$

where τ_1 is the effective transit time related to the transit time τ_t of Eq. (14.10.7) by

$$\tau_1 = N \left(\frac{\alpha_e}{\alpha_h} \right) \tau_t \qquad (14.10.11)$$

The quantity $N(\alpha_e/\alpha_h)$ has a value between $\frac{1}{3}$ and 2 as the ratio α_e/α_h or α_h/α_e, whichever is smaller, varies from 1 to 10^{-3}. Equation (14.10.10) means physically that the response time of an avalanche diode is slowed down considerably by the multiplication process. It further implies a more-or-less fixed gain–bandwidth product (R. B. Emmons, *J. Appl. Phys.*, Vol. 38, p. 3705, 1967). Experiments on Si APDs indeed confirmed this relation and yielded a response time $\tau = 5 \times 10^{-13} M_0$ s (T. Kaneda, H. Takanashi, H. Matsumoto, and T. Yamaoka, *J. Appl. Phys.*, Vol. 47, p. 4960, 1976).

Compared to the *p-i-n* photodiode, the APD has the advantage of providing internal gain but suffers from having a slow speed. Also associated with gain is the additional noise caused by the random occurrence of the impact-ionization process. An analysis by McIntyre shows that the shot noise of Eq. (14.10.6) is replaced by

$$\langle i_{NS}^2 \rangle = 2eIM^2 F(M)B \qquad (14.10.12)$$

The factor M^2 is expected because the noise current is also amplified by the multiplication process. The quantity $F(M)$ is referred to as the *excess noise factor*. Under a uniform electric field as in a *p-i-n* diode, the factor $F(M)$ is given by

$$F = M \left[1 - (1 - k) \left(\frac{M - 1}{M} \right)^2 \right] \qquad (14.10.13)$$

which for large M can be approximated by

$$F = 2(1 - k) + kM \qquad (14.10.14)$$

where $k = \alpha_h/\alpha_e$ or α_e/α_h, whichever is smaller (R. J. McIntyre, *IEEE Trans. Electron Devices*, Vol. ED-13, p. 164, 1966). For the case of a nonuniform electric field, the reader is referred to an analysis by Webb et al. (P. P. Webb, R. J. McIntyre and J. Conradi, *RCA Rev.*, Vol. 35, p. 234, 1974). We note from Eq. (14.10.14) that for low-noise APDs it is essential that the ionization rate for one type of carrier be much larger than that for the other. Such is the case for silicon, with $k \cong 0.02$ to 0.04. Even so, the term kM becomes important for $M \geq 50$. Because of the rapid increase of shot noise with M, the useful internal gain of junction avalanche photodiodes is limited (H. Melchoir and W. T. Lynch, *IEEE Trans. Electron Devices,* Vol. ED-13, p. 829, 1966; R. J. McIntyre, *IEEE Trans. Electron Devices,* Vol. ED-13, p. 164, 1966). The same limitation applies to Schottky-barrier avalanche diodes.

One useful quantity concerning the performance of a photodetector is the minimum detectable signal. First, we express the response of a detector in terms of the sensitivity S (A/W), defined as the ratio of the signal current (photocurrent) to the incident power. If the detector has a noise current $i'_N = \langle i^2_N \rangle^{1/2}$, its noise equivalent power (NEP) is

$$\text{NEP} = \frac{i'_N}{S} \qquad (14.10.15)$$

At a signal power $P_S = \text{NEP}$, the photocurrent is equal to the noise current, and the signal-to-noise ratio is 1. Thus the quantity NEP is a measure of the minimum detectable signal. The only question about Eq. (14.10.15) is the condition under which i'_N is measured. Figure 14.30 shows the NEP calculated for a silicon *p-i-n* diode as a function of the load resistance R_L for various background currents I_B (G. E. Stillman, V. M. Robbins, and N. Tabatabaie, *IEEE Trans. Electron Devices,* Vol. ED-31, p. 1643, 1984). In the calculation of the photocurrent, an optical source with $\lambda = 770$ nm and a detector efficiency of 75% were assumed. From Fig. 14.30 we find a minimum detectable power of 10^{-9} W with $B = 100$ MHz, $R_L = 10^7$ Ω, and $I_B = 100$ nA.

The quantity NEP defined in Eq. (14.10.15) depends on both the bandwidth B of the detecting system and the area A of the detector. If we divide the detector volume into columns of equal area ΔA, the total noise power P_N is simply the sum of the noise power ΔP_N generated in each element because ΔP_N in one element is statistically independent of ΔP_N generated in another element. Since $\Delta P_N / \Delta A$ is a constant, P_N is proportional to the surface area A of the detector and the noise current i'_N is proportional to $(AB)^{1/2}$. Therefore, we define a quantity D^*, called the *detectivity,* as

$$D^* = \frac{(AB)^{1/2}}{\text{NEP}} = \frac{S(AB)^{1/2}}{i'_N} \qquad (14.10.16)$$

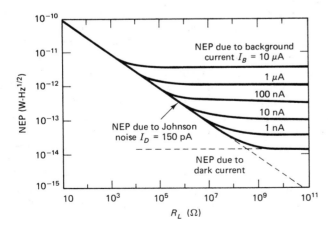

Figure 14.30 Noise equivalent power (NEP) calculated for a silicon *p-i-n* diode as a function of the load resistance R_L. The family of curves is computed for different background currents. (G. E. Stillman, V. M. Robbins, and N. Tabatabaie, "III-V Compound Semiconductor Devices: Optical Detectors," *IEEE Trans. Electron Devices,* Vol. ED-31, p. 1643, 1984 © 1984 IEEE.)

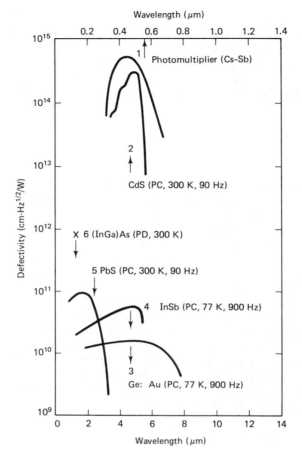

Wavelength (μm)

Figure 14.31 Measured values of detectivity, D^*, defined in Eq. (14.10.16), for a selected number of photodetectors. (O. K. Kim, B. V. Dutt, R. J. McCoy, and J. R. Zuber, "A Low Dark-Current, Planer In GaAs p-i-n Photodiode," *IEEE J. Quantum Electron.*, Vol. QE-21, p. 138, 1985 © 1985 IEEE.)

The detectivity thus defined has a dimension of $cm(Hz)^{1/2} W^{-1}$. The additional factor $(AB)^{1/2}$ makes D^* independent of the area and the bandwidth. The measured values of D^* for a selected number of photodetectors are shown in Fig. 14.31 (P. W. Kruse, L. D. McGlauchlin, and R. B. McQuistan, *Elements of Infrared Technology*, John Wiley & Sons, Inc., New York, 1962). For a PbS detector with a surface area of $1\ cm^2$, the minimum detectable power is about 10^{-11} W measured at $\lambda = 2\ \mu m$ and in a detection system with a bandwidth of $\Delta\nu = 1$ Hz. The point for a planar (InGa)As *p-i-n* diode is calculated by using the NEP figure of Fig. 14.30 and the experimental data of $J_B = 2.2 \times 10^{-6}$ A/cm² and $C/A = 10$ nF/cm² (O. K. Kim, B. V. Dutt, R. J. McCoy, and J. R. Zuber, *IEEE J. Quantum Electron.*, Vol. QE-21, p. 138, 1985). For example, for the value of NEP $= 10^{-13}$ (W-Hz$^{1/2}$), we find that $R_L = 10^6$ Ω and $I_B = 10$ nA. Thus we need an area $A = 5 \times 10^{-3}$ cm^{-2} to achieve a maximum D^* of 7×10^{11} cm-Hz$^{1/2}$/W, but then we have to accept a slow $R_L C$ constant of 50 μs. The high value of D^* for CdS is also achieved at the expense of speed. For optical fiber communications where the signal can be guided and confined in a very small area and where speed is a major consideration, NEP is more appropriate to characterize a detector than D^*. For space communications where the signal is spread over an area much larger than the detector surface, the detectivity D^* is a useful quantity to compare various detectors.

In summary, there are three principal types of photodetectors: the photoemissive type, the photoconductive type, and the photodiode type. In the visible region, high gain and fast response (or large bandwidth) have been made possible in a photomultiplier by combining the traditional art of controlling electron transit time with modern advances in microwave techniques. However, efficient response in available photoemissive material is limited to wavelength shorter than 1 μm. Since the prospect for substantial improvement in the photocathode emission efficiency in the near-infrared region and beyond is very poor, future advances in infrared detection seem most likely to depend on solid-state photodetectors.

Of solid-state detectors, the photoconductive type has a limited gain–bandwidth product, with the gain being equal to τ/τ_t and the bandwidth being of the order of τ^{-1}. Our task is to make τ_t sufficiently short to obtain useful gain at moderately high modulation frequencies. This task proves to be rather difficult even in high-mobility materials because τ_t is controlled by the mobility of low-mobility carriers (holes). The p-i-n photodiodes, on the other hand, have very fast response but no gain. The avalanche photodiodes offer the hope of providing internal gain. However, the gain is limited by noise associated with the avalanche process. Besides zero gain, photodiodes have the disadvantage of narrow apertures. To limit the junction capacitance and the dark current, the junction area must be kept small. However, this requirement is not a limitation on photodiodes for use in optical fiber communications because the signal is guided and confined in a very small area, typically under 10^{-4} cm^2. We should emphasize that there is a trade-off between gain and speed, on the one hand, and between speed and noise, on the other. Therefore, comparison of photodetectors needs to take all these aspects into consideration. The interrelation between these aspects is discussed for each detector type in the sections that follow.

14.11 PHOTOCONDUCTIVE DEVICES

Photoconductive devices are simpler to fabricate and hence more amenable than photodiodes to optoelectronic integration. Although photoconductors are limited in their performance by noise and by gain–bandwidth product, they can be useful for special applications. In this section we present two experimental structures: the photoconductive switch and the interdigital photoconductive detector. Figure 14.32 shows the structures used (a) for autocorrelation measurements and (b) for sampling of electrical signals. The structures are in the form of transmission lines. The dielectric medium in Fig. 14.32a was a 5000-Å film of amorphous silicon (a-Si) deposited on fused-silica substrate by electron-gun evaporation (D. H. Auston, A. M. Johnson, P. R. Smith, and J. C. Bean, *Appl. Phys. Lett.*, Vol. 37, p. 371, 1980). One of the photoconductors has a dc bias V_b. Since the a-Si is an insulator, the gap keeps the charge on the section of the transmission line connected to V_b. When a laser pulse is incident on the gap area, the gap becomes closed by the photo-induced current and is opened again when the photogenerated carriers decay to insignificant values. A second laser pulse closes the second gap. Therefore, a laser pulse serves the function of an optical switch, and the device utilizing a laser-activated switch is called the *Auston switch,* named after the inventor (D. H. Auston, *Appl. Phys. Lett.*, Vol. 26, p. 101, 1975).

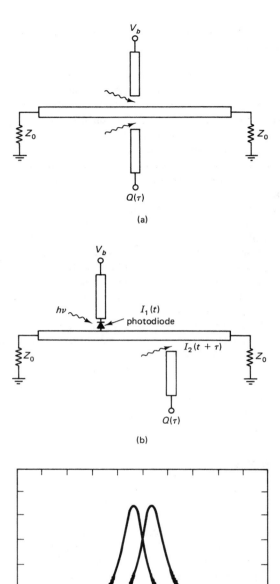

(a)

(b)

10 ps

(c)

Figure 14.32 Structures of photoconductive devices for (a) autocorrelation measurements, and (b) sampling of electrical signals. The two oscilloscope traces in (c) represent autocorrelation measurements taken with the role of bias and signal pulses (optical) interchanged. (D. A. Auston, A. M. Johnson, P. R. Smith, and J. C. Bean, *Appl. Phys. Lett.*, Vol. 37, p. 371, 1980.)

Note that the transfer of charges from the originally charged section of the transmission line (marked by bias voltage V_b) to the load section (marked by Q) takes two steps. If $v(t)$ represents the normalized voltage generated on the originally uncharged transmission line by the action of one laser pulse, the charge Q appearing on the load section of the transmission line is

$$Q(\tau) = B \int_{-\infty}^{\infty} v_1(t)v_2(t + \tau) \, dt \qquad (14.11.1)$$

where τ is the delay between two laser pulses and B a proportionality constant. When the same laser pulse is used at the two gaps, the measured sampling signal $Q(\tau)$ is proportional to the autocorrelation function with $v_1 = v_2$. Figure 14.32c shows the measured autocorrelation function, with the two traces representing the interchange of role of the bias and signal electrodes. Clearly, the technique has the capability of measuring signals with picosecond resolution. Since the pulse of the laser (mode-locked rhodamine 6G dye laser) has a 3.5-ps duration, the measured width is caused by the combined effects of carrier decay time and circuit RC time. Finally, we should mention that the use of the structure of Fig. 14.32c to sample the response of a Si p-i-n photodiode has also been reported, using irradiated semi-insulating GaAs (to shorten carrier lifetime) as the sampling photoconductor (D. H. Auston and P. R. Smith, *Appl. Phys. Lett.*, Vol. 41, p. 599, 1982). The response of the diode was well resolved, with a full width at half maximum (FWHM) of 36 ps versus a FWHM value of 55 ps obtained from a sampling oscilloscope. Many novel electronic devices now have a response time faster than the best sampling oscilloscope. The Auston switch offers a powerful measurement tool to sample by optical pulses those fast electrical signals which cannot be resolved by electronic sampling techniques.

Photoconductive detectors have structures similar to FETs. Therefore, they are easy to fabricate and amenable to integration with FETs. The interdigital photoconductive detector is especially attractive because of the advantage of having a large area for optical absorption, yet keeping a short spacing between electrodes (L. Figueroa and C. W. Slayman, *IEEE Electron Device Lett.*, Vol. EDL-2, p. 208, 1981). Figure 14.33a shows the structure of an interdigital detector having a modulation-doped (AlIn)As (n-type) and (InGa)As (undoped) heterostructure which is lattice-matched to a semi-insulating InP substrate. A two-dimensional electron gas is formed in undoped (InGa)As to minimize the effect of impurity scattering. Figure 14.33b shows the response of the detector to a 100-ps, $\lambda = 1.29$ μm laser pulse. It has a rise time of 80 ps but a fall time of 1.2 ns. The slow fall time (or long tail) is a phenomenon commonly observed in photoconductors. It is caused by either the transit time or the recombination time of photo-generated carriers, whichever is shorter. Although the photo-generated electrons are collected almost immediately at the anode ($\tau_t = 36$ ps for a finger spacing of 3.6 μm), the holes continue to drift in the conducting channel ($\tau_t = 360$ ps for $\mu = 200$ cm^2/V-s and $E = 5 \times 10^3$ V/cm). The charge imbalance sets up a space-charge field that draws electrons to neutralize the excess holes. This process goes on until the excess holes are recombined or collected. Therefore, for a band-to-band photoconductor (Fig. 14.28a), the lifetime τ is that of holes and the transit time τ_t is that of electrons in the gain factor $G = \tau/\tau_t$.

The observed long fall time of 1.2 ns is due primarily to the hole lifetime as a result of either recombination or collection by the electrode. Since electron–hole pairs

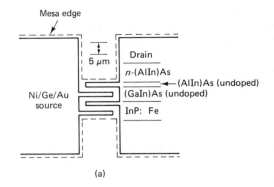

(a)

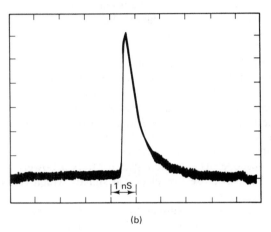

1 nS

(b)

Figure 14.33 (a) Structure and (b) measured response of an interdigital photoconductive detector. (C. Y. Chen, Y. M. Pang, K. Alavi, and P. A. Garbinski, *Appl. Phys. Lett.*, Vol. 44, p. 99, 1984.)

are generated throughout the (InGa)As layer, the distance carriers have to traverse is longer than the interfinger spacing, thus bringing the hole transit time closer to the observed value of 1.2 ns (C. Y. Chen, Y. M. Pang, K. Alavi, and P. A. Garbinski, *Appl. Phys. Lett.*, Vol. 44, p. 99, 1984). On the other hand, experiments on semi-insulating InP photoconductive switch show a fall time of 130 ps. From the applied voltage of 1.5 V across a 25-μm gap, the transit time of holes is estimated to be about 20 ns (P. M. Downey, D. H. Auston, and P. R. Smith, *Appl. Phys. Lett.*, Vol. 42, p. 215, 1983). It appears, therefore, that the fall time of 130 ps observed in the switching experiment was caused by a short carrier lifetime, possibly due to a large surface recombination velocity. The photoconductive (GaIn)As layer in the interdigital photoconductor, on the other hand, has two heterostructures as its boundaries, and hence should have a reasonable surface recombination velocity. Therefore, if we are willing to sacrifice the gain, the fall time can be reduced by shortening the recombination lifetime or transit time of carriers (J. Degani, R. F. Leheny, R. E. Nahory, M. A. Pollack, J. P. Heritage, and J. C. DeWinter, *Appl. Phys. Lett.*, Vol. 38, p. 27, 1981).

Although the interdigital photoconductive detector has a higher responsivity than a *p-i-n* photodiode, its dark current is much higher, contributing to a much larger NEP $\sim 10^{-11}$ W-Hz$^{1/2}$. The dc resistance of a photoconductive device is on the order of 5 kΩ, which means a dark current of 1 mA at a bias of 5 V. This value is to be

Optical Devices   Chap. 14

compared with dark currents of 1 nA in *p-i-n* photodiodes. Both improvement of speed and reduction of dark current are possible. Since the initial work of Chen and coworkers, attempts have been made to shorten the fall time and to reduce the dark current. The best result is obtained in a mesa structure, which has the electrodes directly placed on the photoconductive n^--(GaIn)As layer with an unintentional doping concentration of 1×10^{15} cm^{-3}. Furthermore, the mesa structure contains a n^--p (GaIn)As junction grown on a p^+-InP substrate instead of a semi-insulating InP : Fe substrate. The combination appears effective in reducing the fall time to 90 ps due to a shorter recombination lifetime at the surface and a faster transit time through collection of holes by the contact and by the junction. In addition, the detector has a dark current of 200 nA at a drain–source voltage of 2 V, indicating a 10-MΩ channel resistance (C. Y. Chen, A. G. Dentai, B. L. Kasper, and P. A. Garbinski, *Appl. Phys. Lett.*, Vol. 46, 1164, 1985). These improvements make the photoconductive detector competitive with the *p-i-n* photodiode despite a reduction of ac gain to about 1.3. Photoreceivers consisting of a (GaIn)As photoconductive detector and a GaAs FET have shown a sensitivity of -28.8 dBm at 2 Gb/s and $\lambda = 1.51$ μm. For a discussion of the sensitivity of photoconductive detectors, the reader is referred to an article by Forrest (S. R. Forrest, *IEEE J. Lightwave Technol.*, Vol. LT-3, p. 347, 1985).

14.12 *p-i-n* AND SCHOTTKY PHOTODIODES

The initial *p-i-n* photodiode has a mesa structure, with the *p-n* junction being illuminated via the top layer. The structure shown in Fig. 14.29a is evolved from the original structure and based on considerations of reducing the dark current and improving the speed. The illumination of the junction via the nonabsorbing *n*-InP substrate makes possible the incorporation of a *p* region by *p*-dopant diffusion. The small area of a diffused diode reduces both the junction capacitance and the reverse current of the diode compared to a planar diode. The use of a higher-gap quaternary (GaIn)(AsP) layer as the cap removes the insulator from the photoconductive ternary (GaIn)As layer, thus reducing the contribution to dark current from the insulator–semiconductor interface. Using the structure of Fig. 14.29a, Kim and coworkers have achieved the following results: (1) a lowest dark current of 0.1 nA at -10 V, (2) a typical capacitance of 0.46 pF at -10 V, (3) a rise time of 200 ps, and (4) a fall time of 700 ps. The diode has a diffusion window area of 4.6×10^{-5} cm^2, a *n*-(GaIn)As layer thickness of 3.5 μm, but reduced to 2.5 μm in the zinc-diffused region (O. K. Kim, B. V. Dutt, R. J. McCoy, and J. R. Zuber, *IEEE J. Quantum Electron.*, Vol. QE-21, p. 138, 1985).

We note that the fall time of 700 ps is between the values of 1.2 ns and 80 ps given in Section 14.11 for the photoconductive detector. As we recall, one major difference between the structures for these two values is that the latter structure has contacts directly placed on the photoconductor. Although interface or surface states may raise the dark current, they may also improve the speed. It appears, therefore, that an inverse relation or a conflicting requirement exists between the dark current and the response time of a *p-i-n* diode in a manner similar to the gain and the bandwidth of a photoconductive detector. The slow fall time is due to incomplete depletion of the *n*-(GaIn)As layer. Using a bias voltage of 10 V and a doping concentration $N_d = 10^{16}$ cm^{-3}, the width of the depletion region is only 1.15 μm. Using a value of $D = 11$ cm^2/V-s for holes in Eq. (14.10.9), we find a hole-diffusion time τ_d of 870 ps, which is reasonably close to the measured fall time of 700 ps. If the unintentional doping concentration is reduced to 10^{15} cm^{-3} from 10^{16} cm^{-3}, the photoconductive layer will be fully depleted. Even though Eq. (14.10.7) predicts a hole transit time of 25 ps for $W_d = 2.5$ μm and $v_s = 10^7$ cm/s, such a short fall time has not been achieved in practice. For further discussions of *p-i-n* photodetectors, especially their performance in optical communication systems, the reader is referred to an excellent review article by Brain and Lee (M. Brain and T. P. Lee, *IEEE Trans. Electron Devices,* Vol. ED-32, p. 2673, 1985).

An extremely interesting photodetector is the Schottky-barrier photodiode studied by Wang and coworkers (S. Y. Wang, D. M. Bloom, and D. M. Collis, *Appl. Phys. Lett.,* Vol. 42, p. 190, 1983). It not only has a 3-dB bandwidth exceeding 20 GHz, but also is valuable as an example to illustrate the basic physics governing an ultrafast photodetector. Figure 14.34 shows (a) the basic structure of the device and (b) its equivalent circuit and (c) its impulse response to a 5-ps mode-locked dye-laser pulse reported by Wang (S. Y. Wang, "Ultra High Speed Photodetectors," *Tech. Dig.,* International Electron Devices Meeting (IEDM), p. 712, 1984). The following values were quoted in the paper: C_j = junction capacitance = 20 fF, R_d = diode resistance = 100 MΩ, R_s = spreading resistance = 50 Ω, C_p = parasitic capacitance = 15 fF, L_b = bond-wire inductance = 80 pH, R_L = load resistance = 25 Ω, A = junction area = 25 μm^2, W^+ = n^+-layer thickness = 0.4 μm, and W^- = n^--layer thickness = 0.2 to 0.3 μm. Based on these values, the equivalent circuits have the following time constants: $R_d C_j$ = 2 ps, $R_s C_p$ = 0.75 ps, and L_b/R_L = 3.2 ps. The impulse response of

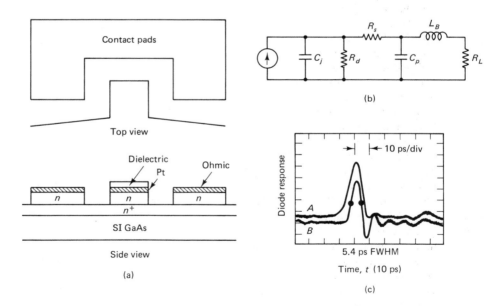

Figure 14.34 (a) Basic structure, (b) equivalent circuit, and (c) impulse response of a Schottky-barrier photodiode (Courtesy of S. Y. Wang). The use of a mesa structure greatly reduces the area of the diode. A semitransparent layer of Pt of 100 Å thickness is used as a Schottky contact, over which a dielectric layer is deposited for scratch protection. (S. Y. Wang, *Techn. Dig.,* International Electron Device Meeting (IEDM), 1984, p. 712.)

Fig. 14.34c shows a response time of 8.8 ps for trace A, including the laser-pulse width contribution, and a deconvolved response time of 5.4 ps for trace B, excluding the laser contribution. A Fourier transform of the impulse response of trace B indicates a 3-dB bandwidth exceeding 100 GHz.

It is interesting to note that the early version of the Schottky-barrier photodiode had a 3-dB bandwidth of 20 GHz. For the 100-GHz detector, the following steps were taken. The wafer was subject to proton bombardment outside the photosensitive and contact regions to minimize parasitic effects. The n^+ contact layer is heavily doped to reduce R_s. The thickness of the n^- photosensitive layer with 10^{15} cm^{-3} silicon doping was reduced from 1.6 μm to 0.2 to 0.3 μm. This reduction has two effects. It ensures full depletion of the n^- layer even at a reverse bias of 1 V. Furthermore, it reduces the transit time. Using $v_s = 6 \times 10^6$ cm/s for holes at $E = 2 \times 10^5$ V/cm, we obtain a τ_t value of 4 ps. Because the n^- epitaxial layer outside the photo-sensitive region was etched away and because the n^+ epitaxial layer was proton-bombarded, photo-generated carriers outside the photosensitive region decayed very rapidly due to damage-caused shortening of recombination lifetime. These steps may well be responsible for the elimination of a long tail generally observed in photoconductors and photodiodes by preventing excess carriers diffusing back to the photosensitive high-field region. We should also mention that a dark current of 2 pA at a reverse bias of 2 V and an external quantum efficiency of 30% were reported for the 20-GHz Schottky photodiode having a junction area of 1×10^{-5} cm^2. Therefore, we expect an even smaller dark current in the 100-GHz photodetector due to the smaller junction area. On the other hand, we also expect a much lower quantum efficiency due to a shorter absorption region.

From our discussions on the photoconductive detector, the p-i-n photodiode and Schottky photodiode, we see that the three principal quantities, the speed, the sensitivity, and the dark current, are all interrelated in the sense that they are influenced by photosensitive layer thickness, photosensitive area, doping concentration, and recombination lifetime. Therefore, the selection of a particular type of photodetector depends on the application. Since they are governed by the same underlying device physics, the difference in the detector performance is due primarily to the difference in the emphasis on a certain aspect of the performance, for example, speed versus minimum detectable signal. Another important aspect which we have not yet discussed concerns integrability. Optoelectronic integration promises the advantages of lower cost, better reliability, faster speed, and smaller physical size. The feasibility of monolithic detector-FET integration was demonstrated in (InGa)As by Leheny et al. in the form of a p-i-n photodiode and a junction FET (R. F. Leheny, R. E. Nahory, M. A. Pollack, A. A. Ballman, E. D. Beebe, J. C. DeWinter, and R. J. Winter, *Electron. Lett.*, Vol. 16, p. 353, 1980). It is the long-wavelength ($\lambda = 1.30$ to 1.56 μm) region where optoelectronic integration is important to the future advance of optical fiber communication systems. However, integration must overcome the conflicting requirements on the devices as well as technological problems.

Photodetectors require a relatively thick (2 to 4 μm) depleted light-absorbing region. Full depletion at a moderate bias further requires a light doping concentration ($\sim 10^{15}$ cm^{-3}). In contrast, FETs with high transconductance require a narrow channel (~ 0.5 μm) of moderate doping concentration ($\sim 10^{17}$ cm^{-3}). Obviously, the epitaxial layers for the detector and for the FET have to be grown separately. Separate growth complicates the fabrication procedure and increases the cost even though we still gain in speed by minimizing the parasitic effects. So far results on monolithic integration on

GaAs are still behind those on hybrid integration, for which the design of the devices can be optimized separately. Therefore, much work is still needed to find a suitable, yet simple scheme for device fabrication. Further hindering the progress on monolithic integration in (GaIn)As or InP-based materials is the lack of a suitable metal to provide a Schottky barrier of sufficient height. The MISFET technology is also unsuitable because of the high interface state density at a SiO_2–or SiN_x–(GaIn)As interface. Therefore, comparison data on the performance of various photodetectors in optical fiber communication systems are available only for hybrid receivers. For further discussions on optoelectronic integration of detectors and transimpedance amplifiers, the reader is referred to excellent review articles on the subject by Brain and Lee and by Forrest (M. Brain and T. P. Lee, *IEEE Trans. Electron. Devices,* Vol. ED-32, p. 2673, 1985; S. R. Forrest, *IEEE Trans. Electron. Devices,* Vol. ED-32, p. 2640, 1985).

14.13 PHOTOTRANSISTORS AND AVALANCHE PHOTODIODES

In this section we turn our attention to photodetectors with gain. Both phototransistors and avalanche photodiodes exhibit internal current gain. The phototransistor studied by Campbell and Ogawa consisted of a n^--InP emitter, a p^--(InGa)As base, and a n^--(InGa)As collector grown on a n-InP substrate (J. C. Campbell and K. Ogawa, *J. Appl. Phys.*, Vol. 53, 1203, 1982). The transistor showed a small-signal current gain h_{fe}

$$h_{fe} = \frac{h\nu}{e\eta} \frac{\Delta I_C}{\Delta P_i}$$

(14.13.1)

of 90 at $I_C = 1$ μA and 300 at $I_c = 100$ μA, where η is the external quantum efficiency, I_C the dc collector current, and P_i the incident light power. A high gain–bandwidth product requires low base–emitter and base–collector capacitances and a low emitter–base resistance. The latter in turn means the need for a dc emitter–base bias current, applied either electrically or optically. Analyses of phototransistor performance have shown that an integrated phototransistor receiver is expected to have a sensitivity at best comparable to that of a *p-i-n*-FET hybrid receiver. The need for an electric or optical bias, however, makes phototransistor less attractive than the *p-i-n*-FET hybrid.

Among the photodetectors, avalanche photodiodes have shown the best sensitivity. However, to achieve the best performance, several steps have been taken to reduce the dark current and to improve the speed. There are three current sources: the usual generation–recombination current in the depletion region, the surface leakage current, and the tunneling current. As mentioned in Section 14.10, the guard ring in Fig. 14.29b is effective in reducing the leakage current. We note from Eq. (10.5.12) that the tunneling probability across the band gap is exponentially dependent in $\mathscr{E}_g^{3/2}$. Therefore, the tunneling current can be greatly reduced if the high-field avalanche region is mainly in a high-gap material. This is accomplished in Fig. 14.29b by placing a n^--InP avalanche region between the n-(InGa)As absorbing region and the n^+-InP contact region. The structure proposed by Susa and coworkers is called a *separate absorption and multiplication* (SAM) *avalanche photodiode* (APD) (N. Susa, H. Nakagome, H. Ando, and H. Kanbe, *IEEE J. Quantum Electron.*, Vol. QE-17, p. 243, 1981). However, the response time of a SAM APD is limited by the accumulation of holes at the InP–(InGa)As interface due to a large valence-band discontinuity. A solution to this problem has been proposed by adding a graded quaternary (GaIn)(AsP) layer to make a smooth transition from the n^--InP layer to the n-(InGa)As layer. Such a photodetector is called the *separate-absorption-graded-multiplication* (SAGM) APD (S. R. Forrest, O. K. Kim, and R. G. Smith, *Appl. Phys. Lett.*, Vol. 40, p. 95, 1982).

Typical SAM and SAGM APDs have the following values for the physical parameters used in the various layers: (1) the n-(GaIn)As absorption layer, with thickness ranging from 2 to 4 μm and doping concentration below 1×10^{16} cm^{-3}; (2) the n^--InP avalanche layer, with thickness ranging from 1.5 to 2 μm and doping concentration between 8×10^{15} and 2×10^{16} cm^{-3}, and (3) the graded (GaIn)(AsP) layer, with thickness about 0.3 μm and doping concentration in the range 1 to 2×10^{16} cm^{-3}. The best values reported so far for the dark current are 3×10^{-6} A/cm^2 for the SAM APD and 5×10^{-4} A/cm^2 for the SAGM APD. Because there is considerable variation in the reported values of the dark current in SAM APD from a high value of 2.2×10^{-3} A/cm^2 to the best value of 3×10^{-6} A/cm^2 and because there is scarcity of experimental data on the SAGM APD, the difference between the two best values may not be due entirely to the introduction of the quaternary (GaIn)(AsP) layer, which has a smaller band gap than the InP layer. The SAGM APD was reported to have a dark current of 20 nA and a gain–bandwidth product of 17 GHz (J. C. Campbell, W. S. Holden, G. J. Qua, and A. G. Dentai, *IEEE J. Quantum Electron.*, Vol. QE-21, p. 1743, 1985). The dark current of 20 nA is to be compared with typical values of 1 nA in p-i-n photodiodes (J. C. Campbell, A. G. Dentai, W. S. Holden, and B. L. Kasper, *Electron. Lett.*, Vol. 19, p. 818, 1983).

We should point out that in SAGM APDs two components of photocurrent were generally observed: the fast component, with a response time about 200 ps, and the slow component, with a response time ranging from 10 to 100 ns. The elimination of the slow component should bring the response time of a SAGM APD comparable to that observed in p-i-n photodiode. Although the SAGM APD is much more difficult to fabricate and more complicated to operate than the p-i-n photodiode, it provides the best performance in optical fiber communication systems based on direct modulation and detection. A value of -32.6 dBm for the minimum detectable optical power has been quoted at 4 Gbits/s (M. Brain and T. P. Lee, *IEEE Trans. Electron. Devices,* Vol. ED-32, p. 2673, 1985). In Fig. 14.35 we show the computed values and the observed data on the time-averaged mean minimum detected power for p-i-n, avalanche, and photo-

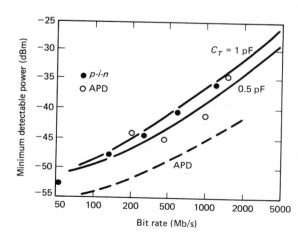

Figure 14.35 Data on minimum detectable power observed in p-i-n diodes (dots) and APDs (open circles) and results computed for p-i-n diodes (solid curves) and APDs (dashed curve). Both experimental data and calculated results are plotted as functions of bit rate. (S. R. Forrest, "Optical Detectors: Three Contenders," *IEEE Spectrum,* Vol. 23 (5) p. 76, 1986, © 1986 IEEE.)

conductive photodetectors as reported by Forrest (S. R. Forrest, *IEEE Lightwave Technol.*, Vol. 3, p. 347, 1985; *IEEE Spectrum*, Vol. 23, p. 76, May 1986). The curves were computed for detectors operating at $\lambda = 1.3$ μm with a maximum bit error rate (BER) of 10^{-9}. Also, a total capacitance of 1 pF and a dark current of 50 nA were assumed. Based on Fig. 14.35, APDs are expected to show a better performance than *p-i-n* and photoconductive (PC) detectors. However, APDs require a high reverse bias, exceeding 50 V, which is much higher than the standard power-supply voltage for integrated circuits. The low voltage requirement, the simplicity of the fabrication process, and the comparative ease for optoelectronic integration make the *p-i-n*-FET and PC-FET combination attractive in certain applications. For further comparison of detector performance, the reader is referred to the review article by Brain and Lee.

A discussion of APDs is incomplete without commenting on the ionization coefficients. For some time, many III–V compound semiconductors, notably those with direct band gap, such as GaAs and InP, were thought to have equal ionization coefficients for electrons and holes. This assumption would make the excess noise unacceptably high. However, this assumption has been proven incorrect because it was based on experimental data which included mixed injection in the high-field region due to the Franz–Keldysh effect. In experimental structures that eliminated this effect, unequal ionization coefficients have been established. The ratio of α's has the following values: (1) in GaAs, $\alpha_e/\alpha_h = 2.2$ at $E = 2.5 \times 10^5$ V/cm, which decreases to 1.4 at $E = 5.0 \times 10^5$ V/cm; (2) in InP, $\alpha_h/\alpha_e = 4$ at $E = 2.4 \times 10^5$ V/cm to 1.3 at $E = 7.7 \times 10^5$ V/cm; and (3) in $Ga_{0.47}In_{0.53}As$, $\alpha_e/\alpha_h = 2$ in the field range 2×10^5 to 2.5×10^5 V/cm. The unequal coefficients have also been confirmed by noise measurements (G. E. Stillman, V. M. Robbins, and N. Tabatabie, *IEEE Trans. Electron Devices*, Vol. ED-31, p. 1643, 1984). The possibility of using the SAM structure makes the (InGa)As/InP APD more attractive than Ge APD. Finally, we should mention that the ratio α_e/α_h or α_h/α_e can be greatly enhanced in a superlattice, where the ionization coefficient of the carriers with a larger band-edge discontinuity will be increased relative to that of the carriers with a smaller band-edge discontinuity. Such a scheme was proposed by Chin et al. and has been demonstrated by Capasso et al. (R. Chin, N. Holonyak, Jr., G. E. Stillman, Y. Y. Tang, and K. Hess, *Electron. Lett.*, Vol. 17, p. 467, 1980; F. Capasso, W. T. Tsang, A. L. Hutchinson, and G. F. Williams, *Appl. Phys. Lett.*, Vol. 40, p. 38, 1982). Much work is still needed, however, to realize a practical superlattice APD with performance characteristics comparable to those of existing APDs.

PROBLEMS

14.1. One important development in infrared lasers is the advent of F_2^+-center laser in alkali-halide crystals, pioneered by Mollenauer (L. F. Mollenauer, *Opt. Lett.*, Vol. 1, p. 164, 1977). The F_2^+ center consists of a single electron shared by two anion (halide ion) vacancies along the [110] axis. Therefore, it resembles a H_2^+ molecule ion discussed in Section 3.2 with specific reference to Fig. 3.5 except that it is embedded in a dielectric continuum. The absorption spectra have been attributed to $1S\sigma_g \rightarrow 2P\sigma_u$ in the infrared and $1S\sigma_g \rightarrow 2P\pi_u$ in the visible. The emission band is shifted to longer wavelength with respect to the absorption band.

Referring to the NaCl structure, draw a diagram showing the charge locations in a F_2^+ center. Explain the spectroscopic notation and discuss why the transition is allowed. Also explain why the emission spectrum is shifted (J. C. Slater, *Quantum Theory of Matter*, McGraw-Hill Book Company, New York, 1968, p. 370 and Figs. 19.5 and 19.8; M. A. Aegerter and F. Lüty, *Phys. Status Solidi*, Vol. 43, p. 244, 1971).

14.2. Read Sections 5 and 8 in the article by D. L. Dexter, "Theory of the Optical Properties of Imperfections in Nonmetals" (*Solid State Phys.*, Vol. 6, pp. 367, 387, 1958). Explain why lattice vibrations play a role in the luminescent process of KCl : Tl. State the Franck–Condon principle in reference to the change of vibrational modes during the luminescent process.

14.3. Read the paper by J. J. Hopfield, "A Theory of Edge Emission in CdS, ZnS, and ZnO" (*J. Phys. Chem. Solids*, Vol. 10, p. 110, 1959). Describe the process of edge emission on the basis of Hopfield's analysis.

14.4. Read the article by F. E. Williams, "Theory of the Energy Levels of Donor–Acceptor Pairs" (*J. Phys. Chem. Solids*, Vol. 12, p. 265, 1960). Present the theory and experimental evidence supporting the associated donor–acceptor model.

14.5. Read the articles by R. Bowers and N. T. Melamed, "Luminescent Centers in ZnS (Cu, Cl) Phosphors" (*Phys. Rev.*, Vol. 99, p. 1781, 1955) and by J. Lambe and G. Kikuchi, "Paramagnetic Resonance of CdTe (Mn) and CdS(Mn)" (*Phys. Rev.*, Vol. 119, p. 1256, 1960). Discuss experimental evidence from which the ionized states of Cu and Mn can be deduced.

14.6. Read Sections 4, 5, and 6 in the article by W. W. Piper and F. E. Williams, "Electroluminescence" (*Solid State Phys.*, Vol. 6, p. 95, 1958). Discuss the processes of excitation, energy transfer, and emission in an electroluminescent material.

14.7. The absorption coefficient α involving phonon emission obey the following relations:

$$\alpha = B(h\nu - \mathscr{E}_g - \mathscr{E}_x - \hbar\omega_{op})^{1/2}$$

for parity-allowed indirect transitions to free exciton states and

$$\alpha = C(h\nu - \mathscr{E}_g - \hbar\omega_{op})^2$$

for transitions to free electron–hole band states. Read the articles by M. Gershenzon, "Radiative Recombination in the III–V Compounds" (*Semicond. Semimet.*, Vol. 2, p. 289, 1966), by Y. P. Varshni, "Band-to-Band Radiative Recombination in Groups IV, VI, and III–V Semiconductors" (*Phys. Status Solidi*, Vol. 19, p. 459, 1967; Vol. 20, p. 9, 1967), and by P. J. Dean and D. G. Thomas, "Intrinsic Absorption-Edge Spectrum of Gallium Phosphide" (*Phys. Rev.*, Vol. 150, p. 690, 1966). Comment on how the two relations are obtained and how they can be applied to find the phonon energies $\hbar\omega_{op}$.

14.8. Read Section 4A in the article by A. A. Bergh and P. J. Dean, "Light-Emitting Diodes" (*Proc. IEEE*, Vol. 60, p. 156, 1972). Discuss how the absorption spectrum at the fundamental absorption edge is affected by indirect transitions involving free exciton states.

Show that the quantity $A = e^2(2\mu)^{3/2}f/(ncm\hbar^2)$ defined in Eq. (43) and used in the expression $\alpha = A(h\nu - \mathscr{E}_g)^{1/2}$ is identical to K in $\alpha = K (h\nu - \mathscr{E}_g)^{1/2}$ of Eq. (12.11.24).

14.9. Derive Eqs. (14.5.8) and (14.5.9), and explain the physical meaning of $\rho_r(\mathscr{E})$ and $I(\nu)$.

14.10. Using the experimental curve of Fig. 14.12 for $\alpha(\nu)$ in Eq. (14.5.6), calculate spontaneous emission spectrum for GaAs. Explain why the measured curve in Fig. 14.13 differs considerably from the calculated curve on the high-energy side.

14.11. Experimental data on emission spectra show that the emission intensity drops by a factor about 4.8 in Ge from $h\nu = 0.882$ eV to 0.892 eV at 77 K. Show that the behavior is expected in the spectral region $\alpha\, d > 1$ where d is the sample thickness (A. A. Bergh and P. J. Dean, *Proc. IEEE,* Vol. 60, p. 156, Fig. 25, 1972).

14.12. Apply Eq. (14.5.1) to transitions which obey the k-selection rule in a nondegenerate semiconductor. Derive Eq. (14.5.17) and calculate the spontaneous lifetime in GaAs samples with $n_0 = 10^{16}$ cm^{-3} and 10^{17} cm^{-3}.

14.13. Referring to Figs. 11.1 and 12.28 showing the band-gap energy and lattice constant of various III–V and II–VI compounds, discuss possible candidates for LEDs in the yellow, green, and blue regions of the visible spectrum. Also comment whether the alloy can be lattice-matched to a suitable substrate, such as Si, GaAs, or GaP.

14.14. Compare Eqs. (14.6.1) and (14.6.2). Express Eq. (14.6.2) in the form $\mathscr{E}_c^e = A n^{1/3}$, and calculate the value of A for GaAs.

14.15. Show that Eq. (14.6.6) satisfies Eq. (14.6.10) if Q is given by Eq. (14.6.11). By comparing Eqs. (14.6.13) and (14.6.10), show that Q should be given by Eq. (14.6.14) to satisfy Eq. (14.6.6) for degenerate semiconductors.

14.16. Calculate the screening length Q^{-1}, the band-edge shift, and the band-edge spread in GaAs samples **(a)** with $N_d = 5 \times 10^{17}$ cm^{-3} and $N_a = 3 \times 10^{17}$ cm^{-3} and **(b)** with $N_d = 10^{16}$ cm^{-3} and $N_a = 10^{15}$ cm^{-3}.

14.17. Consider a GaAs sample with $p = 1.6 \times 10^{19}$ cm^{-3}. Given $m_{dv} = 0.52\, m_0$ and $\eta_v = 29$ meV, calculate the total density of states N_{bt} in the band tail from Eqs. (14.6.15) and (14.6.18). Since $p > N_{bt}$, we may assume N_{bt} to be completely occupied by holes. Then use Fig. 6.33 corrected for m_d to find the Fermi level, that is, the value of $\mathscr{E}_v - \mathscr{E}_f$ where $\mathscr{E}_v$ is the normal valence band edge.

14.18. Using Eqs. (14.6.15) and (14.6.17) with proper modifications, find the expression of p for both nondegenerate and degenerate p-type semiconductors.

For a GaAs sample with $p = 1.2 \times 10^{18}$ cm^{-3} find the Fermi level, that is, $\mathscr{E}_f - \mathscr{E}_v$. In the calculations, use the step-function approximation for the Fermi factor where appropriate.

14.19. Given $\mathscr{E}_g = 1.02$ eV, use Eq. (14.5.9) to predict the shape of $I(\nu)$. Then convert $I(\nu)$ into $I(\lambda)$, and compare the calculated $I(\lambda)$ with the emission spectrum shown in Fig. 14.14 from a surface-emitting LED. Explain why Eq. (14.5.9) is not useful to predict the behavior of $I(\lambda)$ for $\lambda > 1.22$ μm. Discuss the effects of increased densities of injected carriers on the intensity and the spectral shape of spontaneous emission.

14.20. Sketch the spontaneous emission spectrum $I(\nu)$ expected from Eq. (14.5.13) and the gain spectrum $g(\nu)$ expected from Eq. (14.7.8) on the same graph. Discuss why the maximum of $g(\nu)$ occurs on the lower photon-energy side of the maximum of $I(\nu)$. Also derive Eq. (14.7.7) from Eqs. (14.5.7) and (14.7.4).

14.21. Given $\tau_r = 3 \times 10^{-9}$ s, $m_{dc}^* = 0.063\, m_0$, and $m_{dv}^* = 0.52 m_0$ in GaAlAs, find the positions of the quasi-Fermi levels in Eq. (14.7.7) at two nominal current densities $J = 6000$ and 9000 A/cm$^2 - \mu$m. Then from Eq. (14.7.8), estimate the maximum gain, g_{max}, at the two current densities.

14.22. Derive Eqs. (14.8.1) and (14.8.2) by adding all the E fields in Fig. 14.22 for the transmitted and reflected wave, respectively.

14.23. Derive Eq. (14.8.6) from Eq. (14.8.5). Given $\bar{n} = 4.5$, $n = 3.59$, $\alpha = 10$ cm^{-1}, and $L = 250$ μm for a GaAs laser, determine the threshold gain g_{th} and the mode spacing $\Delta\lambda$.

14.24. Assuming a constant potential well and a constant potential barrier in Fig. 14.24, find an expression for the electron concentration in the p-(GaAl)As region at the interface boundary. Given $\tau_r = 3 \times 10^{-9}$ s, $d = 0.1$ μm, and $m_c = 0.063 m_0$ in GaAs and $m_c =$

$0.50m_0$, $d = 2$ μm, and $\mu_n = 1000$ cm²/V-s in (GaAl)As, find the quasi-Fermi level $\mathscr{F}_n - \mathscr{E}_c$ in GaAs at $n = 2 \times 10^{18}$ cm⁻³ and the band-edge discontinuity $\Delta\mathscr{E}_c$ needed at the interface to limit the diffusion current $I_{\text{diff}} = 0.02I_{\text{rec}}$, where I_{rec} is the recombination current.

14.25. The Auger recombination dominated by electron–electron collision can be described symbolically as follows:

$$e_1 + h_1 \to h\nu, \qquad e_2 + h\nu \to e_3$$

The first equation represents the recombination of an electron–hole ($e_1 - h_1$) pair, and the second equation represents the absorption of the released photon energy $h\nu$ by another electron (e_2), which consequently becomes an energetic electron (e_3). Draw an energy-band diagram to illustrate the Auger recombination process. Applying momentum and energy conservation to the process, find the minimum energy for the electron e_3. Show that this energy is given by Eq. (14.8.12).

14.26. The Auger recombination represented by Eq. (14.8.13) involves the recombination of an electron–hole ($e_1 - h_1$) pair and the absorption of the released photon energy by a split-off band electron (h_2) to become a heavy-hole band electron (h_3). Draw an energy-band diagram to illustrate the process, and set up momentum- and energy-conservation equations for the process. Given $m_c = 0.066m_0$, $m_{vh} = 0.68m_0$, $m_{vs} = 0.24m_0$, and $\Delta = 0.38$ eV in a $In_{1-x}Ga_xAs_yP_{1-y}$ compound with $\mathscr{E}_g = 1.03$ eV for $x = 0.266$ and $y = 0.578$, calculate the ratio B_{e-e}/B_{h-s} in Eq. (14.8.11) for the two Auger recombination processes involving electron–electron collision and a heavy hole and split-off-band hole, respectively. Which process is more important in controlling the nonradiative radiation lifetime in the laser?

14.27. (a) Derive Eqs. (14.9.5) and (14.9.6) from Eqs. (14.9.1) and (14.9.2). Calculate the photon lifetime τ_p for a cavity with an internal loss coefficient $\alpha = 30$ cm⁻¹, a mirror reflectivity $r^2 = 0.32$, and a length $L = 250$ μm.
(b) Convert Eq. (14.7.13) into the form $g = A(N - N_0)$ and find the values for A and N_0 if N and N_0 are expressed in cm⁻³ instead of cm⁻²μm⁻¹. Given $\epsilon = 10.9\epsilon_0$ and $\tau = \tau_r = 3 \times 10^{-9}$ s, calculate the values of N_1 and J at threshold for a GaAs laser with an active-layer thickness $d = 0.1$ μm.

14.28. The gain $g(\lambda)$ in a semiconductor laser near gain-peak wavelength λ_0 can be approximated by

$$g(\lambda) = g(\lambda_0) - g(\lambda_0)\left(\frac{\lambda - \lambda_0}{\Delta\lambda}\right)^2$$

(a) Suppose that $g(\lambda_0)$ is maintained at a value

$$g(\lambda_0) = g_{\text{th}} - 10^{-4} \text{ cm}^{-1}$$

with $g_{\text{th}} = 100$ cm⁻¹. We further suppose that at 300 K, λ_0 coincides with the cavity-resonance wavelength λ_m of the mth longitudinal mode. Find the ratios for S_s/S_m with $s = m \pm 1$ and $s = m \pm 2$ if $\lambda_s - \lambda_m = (s - m)3$ Å and $\Delta\lambda = 100$ Å.
(b) It is known that λ_0 changes at a rate of $+3$ Å/K and λ_m changes at a rate of $+0.8$ Å/K. Find the ratios $S_m(T)/S_m(300 \text{ K})$ and $S_s(T)/S_m(300 \text{ K})$ at $T = 301$ K.

14.29. (a) Using exp (t/τ) as an integration factor, derive Eq. (14.9.11). Is this equation valid above the laser threshold?
(b) Using $d = 0.1$ μm and the data given in the text for the computation of the curve in Fig. 14.26a, find the value of J_{th}. Also calculate the time delay τ_d for the onset of

laser oscillation with $J = 1.6 \times 10^3$ A/cm^2. Compare the calculated value with the value obtained from Fig. 14.26a.

14.30. **(a)** Applying small-signal analysis to Eqs. (14.9.1) and (14.9.2), show that the carrier-photon system has a natural resonance at ω_r given by Eq. (14.9.13). In the derivation, take note of the fact that above laser threshold, the steady-state value of N is held at N_{th}. In other words, of the total injection current density J, the part J_{th} is used to maintain N_{th} while the part $J - J_{\text{th}}$ is converted into laser output.

(b) Using the data given in the text for the computation of the curve in Fig. 14.26a, calculate the value of ω_r for $J = 1.5J_{\text{th}}$ and compare the calculated value with the value obtained from Fig. 14.26a.

14.31. A Si *p-i-n* photodetector has an absorption coefficient $\alpha = 500$ cm^{-1} at $h\nu = 1.27$ eV and an intrinsic region of $W = 40$ μm in thickness with a background $N_d = 10^{14}$ cm^{-3}. The corresponding numbers for a Ga$_{0.53}$In$_{0.47}$As photodetector are $\alpha = 10^4$ cm^{-1} at 0.93 eV and $W = 2$ μm with 10^{14} cm^{-3}. Given $\mathscr{E}_g = 1.12$ eV in Si and 0.73 eV in Ga$_{0.53}$In$_{0.47}$As, find the limits of wavelength for the two detectors. Also calculate the efficiency of detection at 1.27 eV for the former and at 0.93 eV for the latter. Given $v_s = 10^7$ cm/s, find the fast response time and the voltage required for complete depletion in the two devices.

14.32. **(a)** The bit-error rate (BER) for a digital system depends on the signal-to-noise amplitude ratio S/N as follows:

$$\text{BER} = \frac{1}{2} \text{erfc} \left(\frac{S}{2\sqrt{2} N} \right)$$

Using Eq. (7.7.17a), calculate the value of BER for $S/N = 12$ or 21.6 dB.

(b) Using the data given in Problem 14.31 for the (GaIn)As detector, find the photocurrent as a function of incident light power P at $h\nu = 0.93$ eV. Supposing that the detector has an input resistance of 10^6 Ω and a dark current of 10^{-8} A, calculate the incident power (in terms of dBm) needed to achieve a BER of 10^{-9} for a detection bandwidth B = 1kHz. The unit dBm stands for decibels below 1 mW of power.

14.33. **(a)** Calculate the value of NEP for a (InGa)As *p-i-n* detector with $R_L = 10^7$ Ω and $I_B = 100$ nA used for detection of 770 nm radiation at $\eta = 75\%$ efficiency. Check the calculated value with that given in Fig. 14.30. Repeat the calculation if the wavelength is changed to 1.3 μm.

(b) Using $J_B = 2.2 \times 10^{-6}$ A/cm^2 and $C/A = 10$ nF/cm^2 given in the text, find the junction area to achieve maximum D^* and the $R_L C$ time constant.

14.34. Read the article by Auston and Smith (*Appl. Phys. Lett.*, Vol. 41, p. 599, 1982) regarding the use of irradiated semi-insulating GaAs to sample the response of a Si *p-i-n* photodiode. Describe the sampling arrangement and device structure used in the experiment. Explain the purpose of irradiation and the effect of deep traps on the speed and sensitivity of the measurement.

Appendix 1

Boltzmann Statistics

Consider an ideal gas consisting of N free, noninteracting particles. The energy of a particle is

$$\mathscr{E} = \frac{p^2}{2m} \tag{A1.1}$$

where p is the momentum of the particle. According to our discussion in Section 4.4,

$$p = \hbar k = \frac{2\pi\hbar}{L}(q_1^2 + q_2^2 + q_3^2)^{1/2} \tag{A1.2}$$

where q_1, q_2, and q_3 are positive integers (0, 1, 2, . . .). In Fig. A1.1 we draw a vertical line to represent an energy state, and divide the energy spectrum into groups of levels. We call each group of levels a cell, and designate the number of levels in the ith cell by g_i. We further let n_i be the number of particles in the ith cell and $\mathscr{E}_i$ be the average energy of the levels in the ith cell. By definition, the total number N of particles in the system is

$$N = \sum_i n_i \tag{A1.3}$$

and the total energy U of the system is

$$U = \sum_i n_i \mathscr{E}_i \tag{A1.4}$$

As stated in Section 4.5, the occurrence of events in a system involving a large number of particles is accurately predicted by statistical methods. It is our task to find the average value of n_i. First, let us calculate the number of various possible ways to arrive at a given distribution, for example, n_1, n_2, . . . , n_i, . . . , n_l particles in cells 1, 2, . . . i, . . . , l. There are $N!/(n_1! \cdots n_l!)$ ways of arranging N particles into the l cells in this manner. Next, we must consider the distributions within a given cell. There are g_1 levels in the first cell. This means that each particle can occupy the levels

Figure A1.1 Schematic diagram used to find the distribution of particles among various energy states. Each energy state is represented by a vertical line, and states with the same energy are grouped together.

in the first cell in g_1 ways. Thus, for the first cell, there are $[g_1]^{n_1}$ ways in which n_1 particles can occupy the g_1 levels. A similar procedure applies to the other cells. Therefore, the total number of ways to get $n_1, n_2, \ldots, n_i, \ldots, n_l$ particles, respectively, into cells $1, 2, \ldots, i, \ldots, l$ is

$$W = \frac{N!}{n_1! \, n_2! \cdots n_l!} \, g_1^{n_1} \, g_2^{n_2} \cdots g_l^{n_l} \tag{A1.5}$$

Using the Stirling approximation for large n,

$$\log n! \cong n(\log n) - n \tag{A1.6}†$$

we obtain

$$\log W = N \log N - \sum_i n_i \log n_i + \sum_i n_i \log g_i \tag{A1.7}$$

To find the equilibrium distribution (i.e., the statistical average value of n_i), we vary the numbers $n_1, n_2, \ldots, n_i, \ldots, n_l$ in Eq. (A1.7) until $\log W$ reaches a maximum. This maximization process is subject to the condition that N of Eq. (A1.3) and U of Eq. (A1.4) remain unchanged. Using the method of Lagrangian multipliers, we obtain the following condition for maximum W:

$$\delta(\log W) - \delta\left(\alpha \sum_i n_i + \beta \sum_i n_i \mathscr{E}_i\right) = 0 \tag{A1.8}$$

where α and β are two Lagrangian multipliers. Substituting Eq. (A1.7) into Eq. (A1.8), we obtain

$$\sum_i [-(1 + \log n_i) + \log g_i - \alpha - \beta \mathscr{E}_i] = 0 \tag{A1.9}$$

Since we have already used Lagrangian multipliers to conserve N and U, the n_i's in Eq. (A1.9) are considered independent of one another. Thus maximization of W requires each coefficient of δn_i to be zero, or

$$\log \frac{n_i}{g_i} = -1 - \alpha - \beta \mathscr{E}_i \tag{A1.10}$$

†In this appendix, log is used for natural logarithm to the base 2.71828.

Let us now apply Eq. (A1.10) to a physical system. As discussed in Sections 4.3 and 4.4, for a laboratory-size system ($L \sim 1$ mm), the spacing between adjacent energy levels is so close that the energy levels can be treated as if they were continuous. For a continuous distribution, the number of levels within an elementary volume in momentum space $d^3p = dp_x dp_y\, dp_z$ is proportional to d^3p. In other words, $g = A\, d^3p$, where A is a proportionality constant. Using this g in Eq. (A1.10), we obtain

$$dN = C[\exp(-\beta\mathscr{E})]\, dp_x\, dp_y\, dp_z \qquad (A1.11)$$

In Eq. (A1.11), $C = A/\exp(1 + \alpha)$. Since the energy $\mathscr{E}$ varies continuously, we no longer need the subscript i in $\mathscr{E}_i$. For the same reason, the continuous variable dN is used instead of n_i to denote the number of particles in d^3p.

To find the constants C and β in Eq. (A1.11), we use monatomic gas as an example. Realizing that in spherical coordinates $d^3p = 4\pi p^2\, dp$, we integrate Eq. (A1.11) to obtain

$$N = \int dN = C \int_0^\infty 4\pi p^2 \exp\left(-\frac{\beta p^2}{2m}\right) dp = C\left(\frac{2\pi m}{\beta}\right)^{3/2} \qquad (A1.12)$$

Next, we calculate the total energy U of the system. For a continuous distribution, Eq. (A1.4) is replaced by

$$U = \int \mathscr{E}\, dN = C \int_0^\infty 4\pi \frac{p^4}{2m} \exp\left(\frac{\beta p^2}{2m}\right) dp = C\frac{3}{2\beta}\left(\frac{2\pi m}{\beta}\right)^{3/2} \qquad (A1.13)$$

It is well known from experiments that the specific heat for a monatomic gas is $3Nk/2$, where k is the Boltzmann constant. This means that

$$U = \tfrac{3}{2} NkT \qquad (A1.14)$$

Thus, from Eqs. (A1.12) and (A1.13), we find that

$$\beta = \frac{1}{kT} \quad \text{and} \quad C = N\left(\frac{1}{2\pi mkT}\right)^{3/2} \qquad (A1.15)$$

This distribution function for an ideal gas is

$$D(p) = \frac{dN}{dp} = 4\pi N p^2\left(\frac{1}{2\pi mkT}\right)^{3/2} \exp\left(-\frac{\mathscr{E}}{kT}\right) \qquad (A1.16)$$

in the momentum space or

$$D(v) = \frac{dN}{dv} = 4\pi N v^2\left(\frac{m}{2\pi kT}\right)^{3/2} \exp\left(-\frac{\mathscr{E}}{kT}\right) \qquad (A1.17)$$

in the velocity space. Equation (A1.17) is called the Maxwell velocity distribution.

Since the value of β is found, we return to the case of discrete levels. From Eq. (A1.10) we obtain

$$n_i = g_i G \exp\left(-\frac{\mathscr{E}}{kT}\right) \qquad (A1.18)$$

where $G = \exp(-1 - \alpha)$. Equation (A1.18) is useful in calculating the distribution in a system that has discrete energy levels. The atomic system is such a system. For

Boltzmann Statistics

example, at high temperatures, an appreciable percentage of hydrogen atoms may have its electron occupying one of its excited levels ($2s$, $2p$, . . .) even though the lowest-energy level is the $1s$ level. From Eq. (A1.18) the population of the atoms in the various energy levels is in the ratio

$$\frac{n_i}{n_j} = \frac{g_i}{g_j} \exp\left(-\frac{\mathscr{E}_i - \mathscr{E}_j}{kT}\right)$$ (A1.19)

In Eq. (A1.19), the g_i represents the number of quantum states available to each level. If we treat the $2s$ states as a group and the $2p$ states as a group, then the ratio $g(2p)/g(2s)$ is equal to 3 for hydrogen atoms.

We should mention one important fact about statistics of random variables. By their very nature, any number $\bar{n}$ predicted by statistical methods is the mean value of the number n which is averaged either over a long period of observation time or over a large number of samples. At any given instant or for a given sample, there is bound to be deviation from the mean average. The quantity $\overline{\delta n^2} = \overline{(\bar{n} - n)^2}$ measures the mean-square fluctuation from the mean value. It can be shown from the statistics of random variables that

$$\frac{\overline{\delta n^2}}{\overline{n}^2} = \frac{1}{\overline{n}}$$ (A1.20)

for large n.

Equation (A1.20) shows that as the average number $\bar{n}$ increases, the ratio of the deviation from the mean value to the mean value decreases. An example familiar to electrical engineers is the shot noise caused by the fluctuation in the number of electrons emitted. The power associated with the shot noise is proportional to the current I_0, whereas the useful power delivered to the load is proportional to I_0^2. Since I_0 is proportional to the average number $\bar{n}$ of electrons emitted, the ratio of the noise power to the useful power is proportional to $1/\bar{n}$. The result is what we expect from Eq. (A1.20).

In regard to the three statistical distributions—the Boltzmann, the Fermi–Dirac, and the Bose-Einstein—we should emphasize the fact that the number of particles involved in a laboratory-size specimen is enormous. Therefore, any deviation from the predicted average value is insignificantly small as compared to the predicted average value itself. In other words, the most probable distribution given by Eq. (A1.10) which is derived by maximizing W represents the actual distribution in the system.

Appendix 2

Review of Statistical Thermodynamics: The Boltzmann Relation and Thermodynamic Energy Functions

Consider a thermal system under constant temperature T and constant pressure p. The internal energy U of the system is changed if heat Q is supplied to the system or if work $-p\,dV$ is done by the system. According to the first law of thermodynamics,

$$dU = dQ - p\,dV \tag{A2.1}$$

where V is the volume. On the other hand, the internal energy is a sum of the energies of individual particles. From Eq. (A1.4),

$$dU = \sum \mathcal{E}_i\,dn_i + \sum n_i\,d\mathcal{E}_i \tag{A2.2}$$

where $\mathcal{E}$ represents the energy and n represents the number of particles (Appendix A1). The connection between the two expressions can be seen from statistical thermodynamics.

Under constant temperature and pressure, the change in $\mathcal{E}_i$ can be related to the change in volume by

$$d\mathcal{E}_i = \frac{\partial \mathcal{E}_i}{\partial V}\,dV \tag{A2.3}$$

Using Eq. (A2.3) in Eq. (A2.2) and comparing Eq. (A2.2) with Eq. (A2.1), we obtain

$$p = -\sum n_i \frac{\partial \mathcal{E}_i}{\partial V} \tag{A2.4}$$

$$dQ = \sum \mathcal{E}_i\,dn_i \tag{A2.5}$$

The term $\partial \mathcal{E}_i / \partial V$ represents the partial force (per unit area) acting on the particles in the different energy states. Equation (A2.4) simply says that the pressure p being exerted on the system is equal to the total force (per unit area) acting on all the particles. The physical meaning of Eq. (A2.5) is also simple. Any heat exchange between the system and its environment will result in a redistribution of the particles among the various energy states.

In classical thermodynamics, the differential heat input to a system maintained at thermal equilibrium is equal to

$$dQ = T \, dS \tag{A2.6}$$

where S is the entropy of the system. From Eqs. (A2.5) and (A2.6), we find

$$T \, dS = \sum \mathcal{E}_i \, dn_i \tag{A2.7}$$

Therefore, the entropy S is a function of the distribution of particles. In statistical thermodynamics, the relation between S and the various n_i's is made explicit. Referring to our discussion in Appendix A1 and specifically to Eq. (A1.8), we see that under a redistribution of particles

$$d(\log W) = \alpha \sum dn_i + \beta \sum \mathcal{E}_i \, dn_i \tag{A2.8}^\dagger$$

Since $\sum dn_i = 0$, we have

$$\sum \mathcal{E}_i \, dn_i = \frac{1}{\beta} \, d(\log W) \tag{A2.9}$$

Using Eq. (A2.9) in Eq. (A2.7), we obtain

$$T \, dS = \frac{1}{\beta} \, d(\log W) \tag{A2.10}$$

where $\beta = 1/kT$ from Eq. (A1.15). After integration, Eq. (A2.10) becomes

$$S = k \log W \tag{A2.11}$$

where W is given by Eq. (A1.5). Equation (A2.11) is known as the Boltzmann relation.

Equation (A2.6) applies only to a system under thermal equilibrium. In general, we have

$$dQ \leq T \, dS \tag{A2.12}$$

$$dU \leq T \, dS - p \, dV \tag{A2.13}$$

Equations (A2.12) and (A2.13) are a statement of the second law of thermodynamics in its general form. For $dQ = 0$, $T \, dS \geq 0$. Thus the entropy can only remain constant ($dS = 0$) or increase ($dS > 0$). In equilibrium, the entropy of a system reaches its maximum value. The condition for maximum entropy is, therefore, the condition for thermal equilibrium if $dQ = 0$.

---

†In this appendix, log is used for natural logarithm to the base 2.71828.

In many real cases, however, dQ is nonzero. Different thermodynamic energy functions are used under different circumstances. The frequently used ones are the enthalpy H, the Helmholtz free energy F, and the Gibbs free energy G, defined by

$$H = U + pV \tag{A2.14}$$

$$F = U - TS \tag{A2.15}$$

$$G = U + pV - TS \tag{A2.16}$$

Under constant volume and temperature, the correct function to use is F. For $dV = 0$, Eq. (A2.13) yields $T\,dS \geq dU$. For $dT = 0$, Eq. (A2.15) becomes

$$dF = dU - T\,dS - S\,dT = dU - T\,dS \leq 0 \tag{A2.17}$$

Therefore, the Helmholtz free energy can only remain constant ($dF = 0$) or decrease ($dF < 0$). In equilibrium, F of a system reaches a minimum value. This condition is used, for example, in finding the concentration of Schottky defects in Section 2.10.

Under constant temperature and pressure, the correct function to use is G. For $dT = 0$ and $dp = 0$, differentiation of Eq. (A2.16) gives

$$dG = dU + p\,dV - T\,dS \tag{A2.18}$$

In view of Eq. (A2.13), we have

$$dG \leq 0 \tag{A2.19}$$

Therefore, the Gibbs free energy can only remain constant ($dG = 0$) or decrease ($dG < 0$). Under thermal equilibrium, G of a system reaches a minimum value ($dG = 0$). This condition is used, for example, in establishing the relation between the Fermi levels of two metals in contact (Section 4.6).

The discussion above can be extended to a complex mixture consisting of any collection of electrons, molecules, ions, or solids. As pointed out earlier in our discussion of the Boltzmann relation, the entropy S is a function of the distribution of particles among the energy states. For a pure (homogeneous) system, a thermodynamic state is specified by its energy and volume. Under thermal equilibrium, we can rewrite Eq. (A2.13) as

$$dS = \frac{1}{T}\,dU + \frac{p}{T}\,dV \tag{A2.20}$$

For dS to be a total differential, we must have

$$\frac{1}{T} = \left(\frac{\partial S}{\partial U}\right)_V \quad \text{and} \quad \frac{p}{T} = \left(\frac{\partial S}{\partial V}\right)_U \tag{A2.21}$$

In other words, the entropy S is a function of U and V only.

In a mixture (heterogeneous system), a thermodynamic state is determined by its energy, volume, and composition. The composition is specified by a set of numbers $n_1, \ldots, n_n$ where n_i represents the number of particles in the ith component. Thus the entropy S can be expressed functionally as

$$S = S(U, V, n_1, \ldots, n_n) \tag{A2.22}$$

A quantity known as the electrochemical potential, or sometimes the chemical potential, is defined as

$$g_i = -T\left(\frac{\partial S}{\partial n_i}\right)_{U,V,n_1,\,\ldots,n_{i-1},n_{i+1},\cdots,n_n} \qquad (A2.23)$$

Differentiating Eq. (A2.22) yields

$$dS = \frac{1}{T}\,dU + \frac{p}{T}\,dV - \sum_i \frac{g_i}{T}\,dn_i \qquad (A2.24)$$

which is known as the Gibbs equation.

Next, we differentiate Eq. (A2.16) and obtain

$$dG = dU + p\,dV + V\,dp - T\,dS - S\,dT \qquad (A2.25)$$

Substituting Eq. (A2.24) in Eq. (A2.25), we find

$$dG = V\,dp - S\,dT + \sum_i g_i\,dn_i \qquad (A2.26)$$

Under constant p and T, eq. (A2.26) becomes

$$dG = \sum g_i\,dn_i \qquad (A2.27)$$

For a system of particles obeying Fermi–Dirac statistics, the entropy S in Eq. (A2.11) is given by

$$S = k \sum_i \left[n_i \log\left(\frac{z_i}{n_i} - 1\right) - z_i \log\left(1 - \frac{n_i}{z_i}\right) \right] \qquad (A2.28)$$

where z_i is the number of quantum states having the same energy and

$$\frac{n_i}{z_i} = \frac{1}{1 + \exp\left[(\mathscr{E}_i - \mathscr{E}_{fi})/kT\right]} \qquad (A2.29)$$

Therefore, the chemical potential g_i can be evaluated from Eq. (A2.29) through the use of Eq. (A2.23). The quantity $\mathscr{E}_f$ is known as the Fermi energy and its physical significance is discussed in Section 4.5.

Now we apply Eq. (A2.27) to the problem of contact between two metals. Consider the transfer of dn electrons having the same energy from metal 1 to metal 2. Thus, in Eqs. (A2.27) and (A2.29), we have

$$dn_2 = -dn_1 = dn, \qquad \mathscr{E}_1 = \mathscr{E}_2 = \mathscr{E} \qquad (A2.30)$$

From Eqs. (A2.23) and (A2.28), it can be shown that

$$g_2 - g_1 = \mathscr{E}_{f2} - \mathscr{E}_{f1} \qquad (A2.31)$$

Thus Eq. (A2.27) becomes

$$dG = (g_2 - g_1)\,dn = (\mathscr{E}_{f2} - \mathscr{E}_{f1})\,dn \qquad (A2.32)$$

which is used in Section 4.6 as Eq. (4.6.5). For a detailed discussion of the chemical potential and for a derivation of Eq. (A2.28), the reader is referred to a text on thermodynamics or statistical mechanics (see, for example, W. C. Reynolds, *Thermodynamics,* McGraw-Hill Book Company, New York, 1965, Chaps. 10 and 12).

Appendix 3

Useful Integrals for the Evaluation of Conductive Properties of Metals and Semiconductors

For particles obeying Boltzmann statistics, one useful integral is

$$\int_{-\infty}^{\infty} \exp\,(-\alpha p^2)\,dp = A = \left(\frac{\pi}{\alpha}\right)^{1/2} \tag{A3.1}$$

Equation (A3.1) can be derived in the following manner. Consider two such integrals, one with variable p_x and the other with variable p_y. Let

$$A^2 = \int_{-\infty}^{\infty} \exp\,(-\alpha p_x^2 - \alpha p_y^2)\,dp_x\,dp_y \tag{A3.2a}$$

In cylindrical coordinates, $p_x^2 + p_y^2 = p_r^2$ and $dp_x\,dp_y = 2\pi p_r\,dp_r$. Thus Eq. (A3.2a) becomes

$$A^2 = \pi \int_0^{\infty} \exp\,(-\alpha p_r^2)\,dp_r^2 = \frac{\pi}{\alpha} \tag{A3.2b}$$

Other useful integrals are

$$\int_{-\infty}^{\infty} p^{2n} \exp\,(-\alpha p^2)dp = (-1)^n\,\frac{\partial^n A}{\partial \alpha^n} \tag{A3.3a}$$

$$\int_{-\infty}^{\infty} p^{2n+1} \exp\,(-\alpha p^2)dp = \frac{n!}{\alpha^n}\,\frac{1}{2\alpha} \tag{A3.3b}$$

For particles obeying Fermi-Dirac statistics, consider the following integral:

$$I_q = \int_0^\infty \mathscr{E}^q D(\mathscr{E}) f(\mathscr{E}) \, d\mathscr{E} = G \int_0^\infty \mathscr{E}^p f(\mathscr{E}) \, d\mathscr{E} \tag{A3.4}$$

where $G = 4\pi(2m/h^2)^{3/2}$ and $p = q + \frac{1}{2}$ according to Eq. (4.4.15). Integrating Eq. (A3.4) by parts, we obtain

$$I_q = \frac{G}{p+1} \left[\mathscr{E}^{p+1} f(\mathscr{E}) \Big|_0^\infty - \int_0^\infty \mathscr{E}^{p+1} \frac{\partial f(\mathscr{E})}{\partial \mathscr{E}} \, d\mathscr{E} \right] \tag{A3.5}$$

$$= \frac{G}{(p+1)kT} \int_0^\infty \frac{\exp x}{(1 + \exp x)^2} \mathscr{E}^{p+1} \, d\mathscr{E}$$

where $x = (\mathscr{E} - \mathscr{E}_f)/kT$. The first term in I_q is zero because $\mathscr{E}^{p+1} = 0$ at $\mathscr{E} = 0$ and $f(\mathscr{E}) = 0$ at $\mathscr{E} = \infty$.

Next, we expand $\mathscr{E}^{p+1}$ around $\mathscr{E}_f$ and obtain

$$\mathscr{E}^{p+1} = (\mathscr{E} - \mathscr{E}_f + \mathscr{E}_f)^{p+1} \tag{A3.6}$$

$$= \mathscr{E}_f^{p+1} \left[1 + (p+1) \frac{\mathscr{E}_1}{\mathscr{E}_f} + \frac{(p+1)p}{2} \frac{\mathscr{E}_1^2}{\mathscr{E}_f^2} + \cdots \right]$$

where $\mathscr{E}_1 = \mathscr{E} - \mathscr{E}_f$. The reason for doing this is that most significant change in $f(\mathscr{E})$ occurs around $\mathscr{E}_f$ as Fig. 4.17 shows. Substituting Eq. (A3.6) in Eq. (A3.5), we have

$$I_q = \frac{G}{p+1} \mathscr{E}_f^{p+1} \int_0^\infty \frac{\exp x}{(1 + \exp x)^2} \, dx \tag{A3.7}$$

$$+ \frac{G}{2} p \mathscr{E}_f^{p+1} \left(\frac{kT}{\mathscr{E}_f} \right)^2 \int_0^\infty \frac{\exp x}{(1 + \exp x)^2} x^2 \, dx$$

The second term $\mathscr{E}_1/\mathscr{E}_f$ is an odd function. Since the function $\exp x/(1 \exp x)^2$ is an even function, the product of the two is an odd function and hence does not contribute to I_q.

Using the definite integrals

$$\int_0^\infty \frac{\exp x}{(1 + \exp x)^2} \, dx = 1 \tag{A3.8}$$

$$\int_0^\infty \frac{\exp x}{(1 + \exp x)^2} x^2 \, dx = \frac{\pi^2}{3} \tag{A3.9}$$

in Eq. (A3.7), we obtain

$$I_q = G\mathscr{E}_f^{p+1} \left[\frac{1}{p+1} + \frac{p\pi^2}{6} \left(\frac{kT}{\mathscr{E}_f} \right)^2 \right] \tag{A3.10}$$

For $q = 1$, which corresponds to the situation in Eq. (4.9.2),

$$I_1 = G\mathscr{E}_f^{5/2} \left[\frac{2}{5} + \frac{\pi^2}{4} \left(\frac{kT}{\mathscr{E}_f} \right)^2 \right] \tag{A3.11}$$

For $q = 0$, I_0 is equal to the total number of electrons in a volume V. Thus we have

$$I_0 = n = G\mathscr{E}_f^{3/2}\left[\frac{2}{3} + \frac{\pi^2}{12}\left(\frac{kT}{\mathscr{E}_f}\right)^2\right] \tag{A3.12}$$

Eliminating n from Eqs. (4.6.3) and (A3.12), we obtain

$$\mathscr{E}_{f0}^{3/2} = \mathscr{E}_f^{3/2}\left[1 + \frac{\pi^2}{8}\left(\frac{kT}{\mathscr{E}_f}\right)^2\right] \tag{A3.13}$$

Since change in $\mathscr{E}_f$ is small, we approximate Eq. (A3.13) by

$$\mathscr{E}_f = \mathscr{E}_{f0}\left[1 + \frac{\pi^2}{8}\left(\frac{kT}{\mathscr{E}_{f0}}\right)^2\right]^{-2/3} \cong \mathscr{E}_{f0}\left[1 - \frac{\pi^2}{12}\left(\frac{kT}{\mathscr{E}_{f0}}\right)^2\right] \tag{A3.14}$$

Appendix 4

Momentum, Velocity, and Current in Bloch-Wave Formulation

Consider a Bloch wave

$$\psi(\mathbf{r}, \mathbf{k}) = u(\mathbf{r}, \mathbf{k}) \exp(i\mathbf{k} \cdot \mathbf{r}) \qquad (A4.1)$$

which satisfies the Schrödinger equation

$$\left[-\frac{\hbar^2}{2m} \nabla^2 + V(\mathbf{r}) - \mathcal{E}(\mathbf{k}) \right] \psi(\mathbf{r}, \mathbf{k}) = (\mathcal{H} - \mathcal{E})\psi(\mathbf{r}, \mathbf{k}) = 0 \qquad (A4.2)$$

Our first task is to show that the expectation value of momentum $\langle \mathbf{p} \rangle$ for a Bloch wave is indeed given by Eq. (5.5.3).

To evaluate p_x, we differentiate Eq. (A4.2) with respect to k_x, then multiply it with $\psi^*(\mathbf{r}, \mathbf{k})$, and integrate the resultant equation. Thus we obtain

$$\iiint \psi^*(\mathbf{r}, \mathbf{k}) \frac{\partial}{\partial k_x} (\mathcal{H} - \mathcal{E})\psi(\mathbf{r}, \mathbf{k}) \, dx \, dy \, dz = 0 \qquad (A4.3)$$

Note that

$$\frac{\partial}{\partial k_x} (\mathcal{H} - \mathcal{E})\psi(\mathbf{r}, \mathbf{k}) = -\frac{\partial \mathcal{E}}{\partial k_x}\psi(\mathbf{r}, \mathbf{k}) + (\mathcal{H} - \mathcal{E})\frac{\partial}{\partial k_x}\psi(\mathbf{r}, \mathbf{k}) \qquad (A4.4)$$

Substituting Eq. (A4.4) in Eq. (A4.3), we find that

$$\frac{\partial \mathcal{E}(\mathbf{k})}{\partial k_x} = \iiint \psi^*(\mathbf{r}, \mathbf{k}) \, (\mathcal{H} - \mathcal{E})\frac{\partial}{\partial k_x} \psi^*(\mathbf{r}, \mathbf{k}) \, dx \, dy \, dz \qquad (A4.5)$$

We further note that

$$\frac{\partial}{\partial k_x}\psi(\mathbf{r}, \mathbf{k}) = \left(ixu + \frac{\partial u}{\partial k_x} \right) \exp(i\mathbf{k} \cdot \mathbf{r}) \qquad (A4.6)$$

$$(\mathcal{H} - \mathcal{E})ix\psi(\mathbf{r, k}) = \left[ix(\mathcal{H} - \mathcal{E}) + \frac{\hbar^2}{mi}\frac{\partial}{\partial x}\right]\psi(\mathbf{r, k}) = \frac{\hbar^2}{mi}\frac{\partial}{\partial x}\psi(\mathbf{r, k}) \quad \text{(A4.7)}$$

Using Eqs. (A4.6) and (A4.7) in Eq. (A4.5) yields

$$\frac{\partial \mathcal{E}(\mathbf{k})}{\partial k_x} = \iiint \psi^*(\mathbf{r, k})(\mathcal{H} - \mathcal{E})\frac{\partial u}{\partial k_x} \exp{(i\mathbf{k} \cdot \mathbf{r})}\ dx\ dy\ dz$$

$$+ \iiint \psi^*(\mathbf{r, k})\frac{\hbar^2}{mi}\frac{\partial}{\partial x}\psi(\mathbf{r, k})\ dx\ dy\ dz \quad \text{(A4.8)}$$

In quantum mechanics, there is a general property of a Hermitian operator L that

$$\iiint \psi_1^* L\psi_2\ dx\ dy\ dz = \iiint \psi_2 L^*\psi_1^*\ dx\ dy\ dz \quad \text{(A4.9)}$$

through integration by parts. However, in the derivation of Eq. (A4.9), we use the fact that both wave functions ψ_1 and ψ_2 vanish at infinity. Here we utilize the periodic property of the Bloch wave, that is,

$$\psi_1^*(x_1)\psi_2(x_1) = \psi_1^*(x_2)\psi_2(x_2)$$

$$\psi_1^*(x_1)\frac{\partial}{\partial x}\psi_2(x_1) = \psi_1^*(x_2)\frac{\partial}{\partial x}\psi_2(x_2)$$

and so on, where x_1 and x_2 represent two boundaries of the unit cell.

Applying the Hermitian property of Eq. (A4.9) to $\mathcal{H}$ of Eq. (A4.8) and letting $\psi_1 = \psi(\mathbf{r, k})$ and $\psi_2 = (\partial u/\partial k_x) \exp{(i\mathbf{k} \cdot \mathbf{r})}$, we find that

$$\iiint \psi^*(\mathbf{r, k})(\mathcal{H} - \mathcal{E})\frac{\partial u}{\partial k_x} \exp{(i\mathbf{k} \cdot \mathbf{r})}\ dx\ dy\ dz = 0 \quad \text{(A4.10)}$$

because $(\mathcal{H} - \mathcal{E})^*\psi^* = 0$. Therefore, Eq. (A4.8) becomes

$$p_x = \frac{m}{\hbar}\frac{\partial \mathcal{E}(\mathbf{k})}{\partial k_x} \quad \text{or} \quad v_x = \frac{1}{\hbar}\frac{\partial \mathcal{E}(\mathbf{k})}{\partial k_x} \quad \text{(A4.11)}$$

To find an expression for the current density, we start with the charge-conservation equation

$$\frac{\partial \rho}{\partial t} = -\nabla \cdot \mathbf{J} \quad \text{(A4.12)}$$

where $\rho = -e\psi^*\psi$ is the quantum-mechanical charge density. From the time-dependent Schrödinger equation

$$\mathcal{H}\psi = i\hbar\frac{\partial \psi}{\partial t}, \quad \mathcal{H}^*\psi^* = -i\hbar\frac{\partial \psi^*}{\partial t} \quad \text{(A4.13)}$$

we obtain

$$\frac{\partial \psi^*\psi}{\partial t} = \frac{1}{i\hbar}(\psi^*\mathcal{H}\psi - \psi\mathcal{H}^*\psi^*) = \frac{\hbar}{-2mi}\nabla \cdot (\psi^* \nabla\psi - \psi \nabla\psi^*) \quad \text{(A4.14)}$$

Momentum, Velocity, and Current in Bloch-Wave Formulation

Comparing Eq. (A4.14) with Eq. (A4.12) yields

$$\mathbf{J} = -\frac{\hbar}{2i}\frac{e}{m}(\psi^* \, \boldsymbol{\nabla}\psi - \psi \, \boldsymbol{\nabla}\psi^*) \qquad (A4.15)$$

The expectation value of the current density denoted by J_{se} is the spatial average of J over a unit cell, or

$$\mathbf{J}_{se} = \frac{1}{V_{\text{cell}}}\frac{-e}{2m} \iiint\limits_{\text{cell}} (\psi^*\mathbf{p}\psi + \psi\mathbf{p}^*\psi^*) \, dx \, dy \, dz \qquad (A4.16)$$

The subscript se denotes a single electron. Using the Hermitian property of the momentum operator and summing over all electrons in the cell, we obtain

$$\mathbf{J}_e = -\frac{e}{m}\frac{1}{V_{\text{cell}}} \sum \mathbf{p} = -\frac{e}{V_{\text{cell}}} \sum \mathbf{v} \qquad (A4.17)$$

Appendix 5

Boltzmann Equation

The drift current and the diffusion current are the results of carrier transport. In treating such problems, we are dealing with an enormous number of charged particles. Obviously, it is impossible to follow the motion of each particle. We presented in Section 4.8 a phenomenological treatment of the drift current and in Section 7.1 that of the diffusion current. Such treatments, although conceptually simple, are limited in their usefulness because they cannot be generalized to deal with complex problems. A formal treatment of carrier-transport problems must be based on the statistical distribution of carriers.

In Chapter 5, our derivation of carrier concentrations is based on the statistical distribution of electrons among the available energy states in k-space. Let us define a general distribution function $f(\mathbf{r}, \mathbf{v}, \mathbf{t})$ in the six-dimensional space (x, y, z, v_x, v_y, v_z), called the phase space such that the number of particles having spatial coordinates between $\mathbf{r}$ and $\mathbf{r} + d\mathbf{r}$ and velocity coordinates between $\mathbf{v}$ and $\mathbf{v} + d\mathbf{v}$ is given

$$dN = f(\mathbf{r}, \mathbf{v}, t)\, d\mathbf{r}d\mathbf{v} \qquad (A5.1)$$

where $d\mathbf{r} = dx\, dy\, dz$ and $d\mathbf{v} = dv_x\, dv_y\, dv_z$. Realizing that $q\, dN/d\mathbf{r}$ is simply the charge density with each particle carrying a charge q, we find the current density to be

$$J_x = \iiint\limits_{-\infty}^{\infty} (qv_x)f(\mathbf{r}, \mathbf{v}, t)\, d\mathbf{v} \qquad (A5.2)$$

Since electrons in motion carry a kinetic energy, accompanying any electron flow there must be a corresponding energy flow. The rate of energy flow per unit area is given by

$$C_x = \iiint\limits_{-\infty}^{\infty} v_x \mathscr{E}f(\mathbf{r}, \mathbf{v}, t)\, d\mathbf{v} \qquad (A5.3)$$

Under thermal equilibrium, the distribution function is related to either the Fermi–Dirac distribution or the Boltzmann distribution. Both distributions are even functions of v_x, v_y, and v_z; hence J_x and C_x are identically zero. Under a nonequilibrium situation, we must know $f(\mathbf{r}, \mathbf{v}, t)$ in order to find J_x and C_x. Therefore, finding the distribution function for a nonequilibrium state is of central importance in determining the macroscopic behavior of carriers. The equation that governs the change in the distribution function in response to external force is called the Boltzmann equation.

Consider an elementary volume $d\mathbf{r} \, d\mathbf{v}$ in the six-dimensional phase space in which the number of particles at a given time t is given by Eq. (A5.1). Because of the motion of the particles, the number of particles in this elementary volume must change. The particles originally in the box will move to another box in the phase space and will be replaced by particles coming from a box that has coordinates

$$\mathbf{r}' = \mathbf{r} - \mathbf{v} \, \Delta t \quad \text{and} \quad \mathbf{v}' = \mathbf{v} - \mathbf{a} \, \Delta t \tag{A5.4}$$

where $\mathbf{a}$ is the acceleration experienced by the particle. Therefore, the change in $f(\mathbf{r}, \mathbf{v}, t)$ due to particle motion is given by

$$\Delta f_{\text{motion}} = f(\mathbf{r}', \mathbf{v}') - f(\mathbf{r}, \mathbf{v})$$

$$= -\sum \frac{\partial f}{\partial x} v_x \, \Delta t - \sum \frac{\partial f}{\partial v_x} a_x \, \Delta t \tag{A5.5}$$

The two summations are over x, y, and z and over v_x, v_y, and v_z, respectively. It is to be noted that the changes in $f(\mathbf{r}, \mathbf{v}, t)$ caused by the particle motion are smooth and continuous. Therefore, the steps taken in expanding $f(\mathbf{r}', \mathbf{v}', t)$ into a Taylor's series is valid.

The distribution function can also be changed during a collision process. However, here only the velocity of the particle is changed, and the change is abrupt, that is, from $\mathbf{v}$ to $\mathbf{v}'$ instantaneously. For these reasons, the collision process must be treated differently. Let $S(\mathbf{v}, \mathbf{v}')$ be the probability per unit time that a state $(\mathbf{r}, \mathbf{v})$ is scattered into $(\mathbf{r}, \mathbf{v}')$, and $S(\mathbf{v}', \mathbf{v})$ be the probability per unit time for the reverse process. Obviously, the time rate of change in $f(\mathbf{r}, \mathbf{v}, t)$ is the difference of the rates for the two processes, or

$$\left. \frac{\partial f(\mathbf{r}, \mathbf{v}, t)}{\partial t} \right|_{\text{collision}} = \sum_{\mathbf{v}'} \left[S(\mathbf{v}', \mathbf{v}) f(\mathbf{r}, \mathbf{v}') - S(\mathbf{v}, \mathbf{v}') f(\mathbf{r}, \mathbf{v}) \right] \tag{A5.6}$$

The total rate of change in $f(\mathbf{r}, \mathbf{v}, t)$ can be obtained by summing the rates from Eqs. (A5.5) and (A5.6), or

$$\frac{\partial f}{\partial t} = -\mathbf{a} \cdot \nabla_v f - \mathbf{v} \cdot \nabla_r f + \left. \frac{\partial f}{\partial t} \right|_{\text{collision}} \tag{A5.7}$$

where $f = f(\mathbf{r}, \mathbf{v}, t)$, ∇_r is the conventional gradient operator in coordinate space (x, y, z), and ∇_v is the gradient operator in velocity space (v_x, v_y, v_z). Both gradient operators are vector operators. Equation (A5.7) is the Boltzmann equation, from which the transport properties of carriers are derived. In our derivation of the Eq. (6.6.11), the collision term is represented phenomenologically by

$$\left. \frac{\partial f}{\partial t} \right|_{\text{collision}} = -\frac{f - f_0}{\tau} \tag{A5.8}$$

where f_0 is the distribution function under thermal equilibrium and τ is the mean free time. Substituting Eq. (A5.8) into Eq. (A5.7), we see that in a homogeneous sample (i.e., $\nabla_r f = 0$), the distribution function f relaxes to f_0 with a characteristic time constant τ as soon as the force is removed.

The simplest case is one for which $\nabla_r f = 0$ and $df/dt = 0$. This case represents the dc condition in a homogeneous sample. Under these conditions, use of Eq. (A5.8) in Eq. (A5.7) yields

$$f - f_0 = -\tau \mathbf{a} \cdot \nabla_v f \tag{A5.9}$$

For simplicity, we assume the applied field to be in the x direction and the effective mass to be isotropic. Thus Eq. (A5.9) reduces to

$$f - f_0 = -\tau \frac{qE_x}{m^*} \frac{\partial f}{\partial v_x} \tag{A5.10}$$

For small disturbances such that $f - f_0 \ll f_0$, Eq. (A5.10) can be solved by the standard method of successive approximation commonly known as iteration. To the first-order approximation, the function on the right-hand side of Eq. (A5.10) can be replaced by f_0. Thus we have

$$f = f_0 - \tau \frac{qE_x}{m^*} \frac{\partial f_0}{\partial v_x} \tag{A5.11}$$

Substituting Eq. (A5.11) into Eq. (A5.2) and realizing that $\partial f_0/\partial v_x = m^* v_x/kT$, we have

$$J_x = \frac{e^2}{kT} E_x \int\int\int_{-\infty}^{\infty} \tau v_x^2 f_0 \, dv_x \, dv_y \, dv_z \tag{A5.12}$$

because $q^2 = e^2$ irrespective of the carrier type. The total carrier density is, by definition, given by

$$n = \int\int\int f_0 \, dv_x \, dv_y \, dv_z \tag{A5.13}$$

We define an assemble average $\langle g \rangle$ for a physical quantity g as

$$\langle g \rangle = n^{-1} \int\int\int_{-\infty}^{\infty} g f_0 \, dv_x \, dv_y \, dv_z \tag{A5.14}$$

Note that the average kinetic energy is

$$\frac{3kT}{2} = n^{-1} \int\int\int_{-\infty}^{\infty} \frac{1}{2} m^* v^2 f_0 \, dv_x \, dv_y \, dv_z \tag{A5.15}$$

Boltzman Equation

821

Using Eq. (A5.15) in Eq. (A5.12), we obtain

$$\mu = \frac{J_x}{E_x} = \frac{e}{m^*} \frac{\langle v^2 \tau \rangle}{\langle v^2 \rangle} \qquad (A5.16)$$

which is used in Eq. (6.6.11). Note that the distribution function is unchanged if we interchange v_x with v_y or v_z for carriers with isotropic effective mass. Therefore we can replace v_x^2 by $v^2/3$ in Eq. (A5.12).

The expressions for the Hall coefficient and for the magnetoresistance are derived in Appendix A6. The derivations of equations used in analyzing hot electron effects are given in Appendix A9.

Appendix 6

Calculation of Hall Coefficient and Magnetoresistance by Boltzmann Transport Equation

In this appendix we set up the mathematical formulation based on the classical Boltzmann transport equation for analyzing galvanomagnetic effects.

In the presence of electric field $\mathbf{E}$ and magnetic flux density $\mathbf{B}$, the acceleration $\mathbf{a}$ of an electron moving with a velocity $\mathbf{v}$ is given by

$$\mathbf{a} = -\frac{e}{m^*}(\mathbf{E} + \mathbf{v} \times \mathbf{B}) \tag{A6.1}$$

If we assign $\mathbf{B}$ to be in the z direction and $\mathbf{E}$ to be in the xy plane, then the solution of the Boltzmann transport equation

$$f - f_0 = -\tau \sum_{x,y,z} a_x \frac{\partial f}{\partial v_x} \tag{A6.2}$$

takes the following form:

$$f = f_0 + v_x g_1 + v_y g_2 \tag{A6.3}$$

Substituting Eq. (A6.3) into (A6.2) and equating the coefficients of v_x and v_y, we obtain

$$g_1 + \frac{e\tau B}{m^*} g_2 = e\tau E_x \frac{\partial f_0}{\partial \mathscr{E}} \tag{A6.4}$$

$$-\frac{e\tau B}{m^*} g_1 + g_2 = e\tau E_y \frac{\partial f_0}{\partial \mathscr{E}} \tag{A6.5}$$

823

In Eq. (A6.2), f is the distribution function of electrons in k space such that the electron volume density is $n = \int f_0 \, dk_x \, dk_y \, dk_z$ (m^{-3}) for a three-dimensional system.

The solution of Eqs. (A6.4) and (A6.5) is

$$g_1 = e\tau \frac{\partial f_0}{\partial \mathcal{E}} \frac{E_x - sE_y}{1 + s^2} \tag{A6.6}$$

$$g_2 = e\tau \frac{\partial f_0}{\partial \mathcal{E}} \frac{sE_x + E_y}{1 + s^2} \tag{A6.7}$$

where $s = e\tau B/m^*$. For the three-dimensional system, the current density is given by $\mathbf{J} = \int -ev f \, dk_x \, dk_y \, dk_z$. thus we have

$$J_x = \sigma_{xx} E_x - \sigma_{xy} E_y \tag{A6.8}$$

$$J_y = \sigma_{yx} E_x + \sigma_{yy} E_y \tag{A6.9}$$

where the coeffficients $\sigma_{xx} = \sigma_{yy}$ and $\sigma_{xy} = \sigma_{yx}$ are given by

$$\sigma_{xx} = \int -\frac{\partial f_0}{\partial \mathcal{E}} \frac{e^2 \tau s}{1 + s^2} v_x^2 \, dk_x \, dk_y \, dk_z \tag{A6.10}$$

$$\sigma_{xy} = \int -\frac{\partial f_0}{\partial \mathcal{E}} \frac{e^2 \tau s}{1 + s^2} v_x^2 \, dk_x \, dk_y \, dk_z \tag{A6.11}$$

Note that the terms involving $v_x v_y$ are odd functions and $\partial f_0/\partial \mathcal{E}$ is an even function of k_x and k_z. Therefore, any integral involving $(\partial f_0/\partial \mathcal{E})v_x v_y$ is zero. The Hall coefficient can be found from Eqs. (A6.8) and (A6.9) by setting $J_y = 0$. Thus we find that

$$J_x = \frac{\sigma_{xx}^2 + \sigma_{xy}^2}{\sigma_{xx}} E_x \tag{A6.12}$$

$$R_H = \frac{E_y}{J_x B} = \frac{-\sigma_{xy}}{\sigma_{xx}^2 + \sigma_{xy}^2} \frac{1}{B} \tag{A6.13}$$

The change of longitudinal resistivity E_x/J_x with B under $J_y = 0$ is known as magnetoresistance.

Note that in Eq. (A6.1) only one effective mass m^* is used. Therefore, Eqs. (A6.10) and (A6.11) apply only to electrons in GaAs-like semiconductors. For electrons in Si-like and Ge-like semiconductors, we must start from equations of motion for electrons in one constant-energy ellipsoid and choose a coordinate system with the principal axes of the ellipsoid as its coordinate axes x, y, and z. In this case we must replace Eq. (A6.3) by

$$f = f_0 + \sum f_\mu^{(10)} E_\mu + \sum f_{\mu\nu}^{(11)} E_\mu B_\nu + \sum f_{\mu\nu\rho}^{(12)} E_\mu B_\nu B_\rho \tag{A6.14}$$

where the superscripts $(1n)$ represent the terms first-order in E and nth order in B. The subscripts μ, ν, and ρ represent the components of E and B along the chosen axes. Then we must sum the contributions from all ellipsoids. Therefore, we expect a general expression for J as follows:

$$J_\lambda = \sigma_{\lambda\mu} E_\mu + \sigma_{\lambda\mu\nu} E_\mu B_\nu + \sigma_{\lambda\mu\nu\rho} E_\mu B_\nu B_\rho \tag{A6.15}$$

Obviously, a complete analysis of the galvanomagnetic effects in multivalley semiconductors is very complicated. The reader is referred to an article by Pearson and Suhl (G. L. Pearson and H. Suhl, *Phys. Rev.*, Vol. 83, p. 768, 1951) for experimental results on Ge and articles by Meiboom and Abeles and by Shibuya (S. Meiboom and B. Abeles, *Phys. Rev.*, Vol. 95, p. 31, 1954; M. Shibuya, *Phys. Rev.*, Vol. 95, p. 1388, 1954) for theoretical analyses of the galvanomagnetic effects. An excellent review of the mathematical formulation and the theoretical results can be found in books by Smith and by Anselm (R. A. Smith, *Semiconductors,* 2nd ed., Cambridge University Press, Cambridge, 1978, pp. 122–142; A. Anselm, *Introduction to Semiconductor Theory* (English translation), MIR Publishers, Moscow, and Prentice-Hall, Inc., Englewood Cliffs, N.J., 1981 , pp. 538–559).

Magnetoresistance data were used early to deduce the band structure of semiconductors. However, cyclotron-resonance experiments are now used extensively for this purpose because they give more direct and easy interpretation of the experimental results than does magnetoresistance measurement. In the following discussion, we present selected theoretical results on the Hall coefficient to indicate the relation between Hall and conductivity mobility and on the magnetoresistance to illustrate its sensitivity to band structure.

Hall Coefficient

In general, the low-field ($\mu B \ll 1$) Hall coefficient is given by

$$R_0 = -\frac{1}{ne} \frac{\langle v^2\tau^2\rangle\langle v^2\rangle}{\langle v^2\tau\rangle^2} \frac{3r(r+2)}{(2r+1)^2} \tag{A6.16}$$

where $r = m_l/m_t$ is the mass ratio and $\langle v^2\tau^n\rangle$ is given by

$$\langle v^2\tau^n\rangle = \iiint f_0 v^2\tau^n \, dk_x \, dk_y \, dk_z \tag{A6.17}$$

The ratio of Hall to conductivity mobility is

$$\frac{\mu_H}{\mu} = \frac{\langle v^2\tau^2\rangle \langle v^2\rangle}{\langle v^2\tau\rangle^2} \frac{3r(2r+1)}{(r+2)^2} = \xi F(r) \tag{A6.18}$$

In GaAs, electron mobility is dominated by polar scattering and $\xi = 1.17$ at 300 K. Since $r = 1$, $\mu_H = 1.17\,\mu$. At low temperatures, lattice scattering and impurity scattering become important. The value of ξ varies between 1.02 around 50 K and 1.24 around 200 K (J. S. Blakemore, *J. Appl. Phys.*, Vol. 53, p. R123, October 1982). In Si, $\xi = 3\pi/8$ at 300 K and $r = 5.17$, yielding $\mu_H = 1.02\,\mu$. Therefore, μ_H can be used for μ in Si around room temperature.

For holes, acoustic and nonpolar optical scattering processes are of comparable importance for $T > 100$ K in GaAs and Ge, and for $T > 300$ K in Si. An analysis of a two-carrier system yields results similar to those of Eqs. (6.6.2) to (6.6.7) except that proper change in the sign must be made due to the change from electron to hole. Incorporating this change in Eq. (6.6.8) and adding contributions from heavy and light holes (indicated by subscript h and l), we obtain

$$R_H = \frac{1}{ep_h} \xi \frac{1 + (p_l/p_h)(\mu_l/\mu_h)^2}{[1 + (p_l/p_h)(\mu_l/\mu_h)]^2} \tag{A6.19}$$

Values of 1.25 for GaAs and 1.18 for Si have been used for the ratio μ_H/μ around 300 K.

Magnetoresistance

For semiconductors with an isotropic effective mass, we can use Eq. (A6.12) to obtain $\rho = E_x/J_x$, which is also isotropic. Magnetoresistance is defined as $\Delta\rho/\rho$ as a result of the applied B. The low-field ($\mu B << 1$) magnetoresistance can be evaluated by approximating $(1 + s^2)^{-1}$ by $1 - s^2$ in Eqs. (A6.10) and (A6.11), and by keeping only terms up to B^2. Thus we find that

$$\frac{\Delta\rho}{\rho_0} = -\frac{\Delta\sigma}{\sigma_0} = \left(\frac{eB}{m^*}\right)^2 (\eta - 1)\xi \tag{A6.20}$$

where η depends on the scattering mechanism as

$$\eta = \frac{\langle v^2\tau^3\rangle\langle v^2\tau\rangle}{\langle v^2\tau^2\rangle^2} \quad \text{and} \quad \xi = \frac{\langle v^2\tau^2\rangle^2}{\langle v^2\tau\rangle^2} \tag{A6.21}$$

If τ is independent of v^2, we have $\eta = 1$ and $\Delta\rho/\rho_0 = 0$. If $v\tau$ is independent of v^2, as is the case with acoustic phonon scattering, then we have

$$\frac{\Delta\rho}{\rho_0} = \frac{9\pi}{16}\left(1 - \frac{\pi}{4}\right)(\mu B)^2 \tag{A6.22}$$

For semiconductors with Si-like or Ge-like conduction bands, $\Delta\rho/\rho_0$ depends on the relative orientation of $\mathbf{J}$ and $\mathbf{B}$. For crystals with cubic symmetry, $\mathbf{J}$ has a general dependence on $\mathbf{E}$ and $\mathbf{B}$ as follows (F. Seitz, *Phys. Rev.*, Vol. 79, p. 372, 1950):

$$\mathbf{J} = \sigma_0\mathbf{E} + \alpha\mathbf{E} \times \mathbf{B} + \beta B^2\mathbf{E} + \gamma B(\mathbf{E} \cdot \mathbf{B}) + \delta\overset{\leftrightarrow}{T}\mathbf{E} \tag{A6.23}$$

where $\overset{\leftrightarrow}{T}$ is a diagonal tensor with elements B_x^2, B_y^2, and B_z^2. The coefficients, α, β, γ, and δ as evaluated by Shibuya are given in Table A6.1.

TABLE A6.1 PARAMETERS α, β, γ, AND δ

	Si-like	Ge-like
$\dfrac{\alpha}{\sigma_0^2 R_0}$	1	1
$\dfrac{\beta}{\sigma_0^3 R_0^2\eta}$	$\dfrac{-(r^2 + r + 1)(2r + 1)}{r(r + 2)^2}$	$\dfrac{-(2r + 1)^2}{3r(r + 2)}$
$\dfrac{\gamma}{\sigma_0^3 R_0^2\eta}$	$\dfrac{3r(2r + 1)}{(r + 2)^2}$	$\dfrac{(2r + 1)^2}{3r(r + 2)}$
$\dfrac{\delta}{\sigma_0^3 R_0^2\eta}$	$\dfrac{(r - 1)^2(2r + 1)}{r(r + 2)^2}$	$\dfrac{-2(r - 1)^2(2r + 1)}{3r(r + 2)^2}$

The quantity R_0 is the low-field Hall coefficient given by Eq. (A6.16) and η is the scattering factor given by Eq. (A6.21).

Take Si as an example. With applied $\mathbf{E}$ in the x direction and $\mathbf{B}$ in the z direction, we have

$$J_x = \sigma_0 E_x + \alpha E_y B_z + \beta E_x B^2 \tag{A6.24}$$

$$J_y = \sigma_0 E_y - \alpha E_x B_y \tag{A6.25}$$

where E_y is the Hall field. Eliminating E_y, we obtain $\sigma = J_x/E_x$ and the magnetoresistance

$$\frac{\Delta\rho}{\rho_0} = -\frac{\Delta\sigma}{\sigma_0} = R_0^2 \sigma_0^2 [\eta F(r) - 1] \tag{A6.26}$$

where $F(r)$ is equal to $\beta/\sigma_0^3 R_0^2 \eta$ given in Table A6.1. For different orientations of $\mathbf{J}$ (or $\mathbf{E}$) and $\mathbf{B}$, the function $F(r)$ will have different dependence on $r = m_l/m_t$.

High-Field Case

Earlier we treated the weak-field case ($\mu B << 1$). Here we comment on the high-field case ($\mu B >> 1$) and denote the high-field quantities by the subscript ∞. For the Hall coefficient

$$R_\infty = \frac{-1}{ne} \quad \text{or} \quad \frac{1}{pe} \tag{A6.27}$$

The magnetoresistance is much more complicated. For electrons in GaAs-like semiconductors, the transverse effect is given by

$$\frac{\rho_\infty - \rho_0}{\rho_0} = \frac{\langle v^2/\tau \rangle \langle v^2 \tau \rangle}{\langle v^2 \rangle^2} - 1 \tag{A6.28}$$

which yields $\Delta\rho/\rho_0 = (32/9\pi) - 1$ in the lattice-scattering dominated region and $\Delta\rho/\rho_0 = (32/3\pi) - 1$ in the impurity-scattering dominated region. In Si-like and Ge-like semiconductors, $\Delta\rho/\rho_0$ also tends to approach a saturated value which depends on the relative orientations of $\mathbf{B}$ and $\mathbf{J}$.

Calculation of Hall Coefficient and Magnetoresistance

Appendix 7

Solutions of the Time-Dependent Diffusion Equation

Response to a Pulse

A simple way of solving Eq. (7.7.11) is through Laplace transformation. Let

$$\bar{p} = \mathscr{L}p' = \int_0^\infty \exp{(-st_1)}p' \, dt_1 \tag{A7.1}$$

After taking the Laplace transform, Eq. (7.7.11) becomes

$$D_p \frac{\partial^2 \bar{p}}{\partial x_1^2} = s\bar{p} - p_1(t = 0) - \mathscr{L}f(x_1, t_1) \tag{A7.2}$$

where the symbol $\mathscr{L}$ represents the Laplace transform. For a unit pulse injected at $x = 0$ and at $t = 0$,

$$f(x_1, t_1) = \delta(x)\delta(t) \tag{A7.3}$$

In Eq. (A7.3), $\delta(x)$ and $\delta(t)$ are the Dirac delta functions. A Dirac delta function $\delta(y)$ has the following properties:

$$\delta(y) = \begin{cases} \infty, & \text{at } y = 0 \\ 0 & \text{elsewhere} \end{cases} \tag{A7.4}$$

$$\int_{-\infty}^{\infty} \delta(y) \, dy = 1 \tag{A7.5}$$

$$\int_{-\infty}^{\infty} F(y)\delta(y) \, dy = F(y = 0) \tag{A7.6}$$

Using Eq. (A7.6), we find the Laplace transform of $f(x_1, t_1)$ to be

$$\mathscr{L}f(x_1, t_1) = \delta(x) \tag{A7.7}$$

If we are given a physical problem, the problem is uniquely defined in mathematical language by a definite set of initial and boundary conditions, that is, a definite set of functions for $p_1(x, t = 0)$ and $p_1 (x = a, t)$. As we can see from the example above, the initial condition is taken into account in the Laplace transform by the function $p_1(t = 0) = p_1(x, t = 0)$ in Eq. (A7.2). For the case presented in Section 7.7, we assume that no initial charge distribution exists, that is, $p_1(t = 0) = 0$. Thus Eq. (A7.2) becomes

$$D_p \frac{d^2\bar{p}}{dx_1^2} = s\bar{p} - \delta(x_1) \tag{A7.8}$$

Therefore, the step that we took through Laplace transformation is to remove the time dependence in the time-dependent diffusion equation. We regain the time dependence of p_1 when we take the inverse Laplace transformation later in the discussion.

Equation (A7.8) is an ordinary differential equation. In the theory of linear differential equations, the solution of a given differential equation,

$$f(D)y = \frac{d^n y}{dx^n} + a_1 \frac{d^{n-1}y}{dx^{n-1}} + \cdots + a_{n-1}\frac{dy}{dx} + a_n y = F(x) \tag{A7.9}$$

can be separated into two parts: the complementary function y_c, which satisfies the homogeneous equation

$$f(D)y_c = 0 \tag{A7.10}$$

and the particular solution, which satisfies the equation

$$f(D)y_p = F(x) \tag{A7.11}$$

The complete solution of Eq. (A7.9) is

$$y = y_c + y_p \tag{A7.12}$$

The complementary solution of Eq. (A7.8) is

$$\bar{p} = B \exp(-\alpha x_1) + C \exp(\alpha x_1) \tag{A7.13}$$

where $\alpha = (s/D_p)^{1/2}$. To keep $\bar{p}$ finite at $x_1 = \pm\infty$, only one of the two terms should be used. For $x_1 > 0$,

$$\bar{p} = B \exp(-\alpha x_1) \tag{A7.14}$$

and for $x_1 < 0$,

$$\bar{p} = C \exp(\alpha x_1) \tag{A7.15}$$

The Dirac delta function $\delta(x_1)$ in Eq. (A7.8) is taken into account in the following manner. If Eq. (A7.8) is integrated with respect to x_1, we find that

$$D_p \frac{\partial\bar{p}}{\partial x_1}\bigg|_{x_1=0+} - D_p \frac{\partial\bar{p}}{\partial x_1}\bigg|_{x_1=0-} = -\int_{0-}^{0+} \delta(x_1)\, dx_1 = -1 \tag{A7.16}$$

The symbols 0_+ and 0_- represent, respectively, points just right and just left of the source. In other words, because of the source, there is a discontinuity in the derivative as given by Eq. (A7.16). Equation (A7.16) now replaces Eq. (A7.11).

Solutions of the Time-Dependent Diffusion Equation **829**

At $x_1 = 0$, $\bar{p}$ must be continuous, that is, $B = C$ from Eqs. (A7.14) and (A7.15). The derivatives of $\bar{p}$ at $x_1 = 0$ are related through Eq. (A7.16), yielding $(B + C)\alpha D_p = 1$. Combining the results, we find that

$$\bar{p} = \begin{cases} \dfrac{\exp{(-\alpha x_1)}}{2\alpha D_p} & \text{for } x_1 > 0 \\[4mm] \dfrac{\exp{(\alpha x_1)}}{2\alpha D_p} & \text{for } x_1 < 0 \end{cases} \tag{A7.17}$$

The inverse Laplace transform of $\bar{p}$ is

$$p' = \frac{1}{(4\pi D_p t_1)^{1/2}} \exp\left(-\frac{x_1^2}{4D_p t_1}\right) \tag{A7.18}$$

which is known as the Gaussian distribution function. In terms of the original independent variables x and t, the excess-carrier concentration p_1 can be expressed as

$$p_1 = \frac{\exp{(-t/\tau_0)}}{(4\pi D_p t)^{1/2}} \exp\left[-\frac{(x - \mu_p E_x t)^2}{4D_p t}\right] \tag{A7.19}$$

Response to a Step Function

Now let us consider the case in which a source is applied at $x = 0$ for $t > 0$ so as to maintain a unity concentration at the point $x = 0$. Noting that the Laplace transform of unity is $1/s$, the boundary condition becomes

$$\bar{p} = \mathscr{L}p_1 = \frac{1}{s} \quad \text{at } x = 0 \tag{A7.20}$$

The initial condition in Eq. (A7.2) is

$$p_1(t = 0) = 0 \tag{A7.21}$$

Since the source is accounted for by Eq. (A7.20), the term $\mathscr{L}f(x_1, t_1)$ should be ignored in Eq. (A7.2). The solution of Eq. (A7.2) is, therefore, given by

$$\bar{p} = \begin{cases} \dfrac{\exp{(-\alpha x)}}{s} & \text{for } x > 0 \\[4mm] \dfrac{\exp{(\alpha x)}}{s} & \text{for } x < 0 \end{cases} \tag{A7.22}$$

where $\alpha = (s/D_p)^{1/2}$. The inverse Laplace transform of $\bar{p}$ is

$$p_1(x, t) = \text{erfc}\left(\frac{x}{\sqrt{4Dt}}\right) = \frac{2}{\sqrt{\pi}} \int_{\frac{x}{\sqrt{4Dt}}}^{\infty} \exp{(-\xi^2)}\, d\xi \tag{A7.23}$$

where the symbol erfc stands for the complementary error function.

When recombination of excess carriers is no longer negligible, p' of Eq. (7.7.10) must be used in Eq. (A7.20) instead of p_1. However, it is p_1 that is unity at $x = 0$ for $t > 0$, not p'. Therefore, Eq. (A7.10) must be replaced by

$$\bar{p} = \mathscr{L}p' = \mathscr{L}\left[p_1 \exp\left(\frac{t}{\tau_0}\right)\right] = \frac{1}{s - \beta} \qquad \text{at } x = 0 \qquad \text{(A7.24)}$$

where $\beta = 1/\tau_0$. Similarly, Eq. (A7.22) becomes

$$\bar{p} = \begin{cases} \dfrac{\exp(-\alpha x)}{s - \beta} & \text{for } x > 0 \\[2ex] \dfrac{\exp(\alpha x)}{s - \beta} & \text{for } x < 0 \end{cases} \qquad \text{(A7.25)}$$

After taking the inverse Laplace transform of $\bar{p}$ and multiplying p' by $\exp(-t/\tau_0)$, we find that

$$p_1(x, t) = \frac{1}{2}\left[\exp\left(-\frac{x}{\sqrt{D\tau_0}}\right) \text{erfc}\left(\frac{x}{\sqrt{4Dt}} - \sqrt{\frac{t}{\tau_0}}\right)\right.$$

$$\text{(A7.26)}$$

$$\left. + \exp\left(\frac{x}{\sqrt{D\tau_0}}\right) \text{erfc}\left(\frac{x}{\sqrt{4Dt}} + \sqrt{\frac{t}{\tau_0}}\right)\right]$$

Appendix 8

Transient Response of Junction Diodes

The following transient analysis of the hole distribution is given for a wide-base diode in which the base width W is much larger than the hole diffusion length L ($= \sqrt{D_p\tau_0}$). The time-dependent diffusion equation for excess-hole concentration p_1,

$$\frac{\partial p_1}{\partial t} = D_p \frac{\partial^2 p_1}{\partial x^2} - \frac{p_1}{\tau_0} \tag{A8.1}$$

takes the form

$$D_p \frac{\partial^2 \bar{p}}{\partial x^2} = -p'(0) + s\bar{p} \tag{A8.2}$$

after the Laplace transform $\bar{p} = \mathscr{L}p'$ and the substitution $p_1 = p' \exp(-t/\tau_0)$. In Eq. (A8.2), $p'(0)$ represents the initial distribution of excess holes.

For the phase I operation,

$$p'(0) = p_{1f} \exp\left(-\frac{x}{L}\right) = p_{n0}\left[\exp\left(\frac{eV_{f0}}{kT}\right) - 1\right]\exp\left(-\frac{x}{L}\right) \tag{A8.3}$$

where p_{1f} is the excess-hole concentration at the edge of the space-charge region under a forward bias V_{f0} applied across the junction. Equation (A8.3) is the same as Eq. (8.3.11) except that the origin for x in Eq. (A8.3) is taken right at the edge of the space-charge region. The general solution of Eq. (A8.2) is given by

$$\bar{p} = B \exp(-\alpha x) + \frac{p'(0)}{s - \beta} \tag{A8.4}$$

where $\alpha = \sqrt{s/D_p}$ and $\beta = 1/\tau_0$. Using Eq. (A8.3) for $p'(0)$ and realizing that $D_p \, \partial^2 p'(0)/\partial x^2 = \beta p'(0)$, the reader can easily show that the last term in Eq. (A8.4) is the particular solution of Eq. (A8.2).

As discussed in the text, the current is equal to I_r during phase I operation. This condition for constant current requires that the concentration gradient of p_1 at $x = 0$ be such that

$$\frac{-I_r}{eA} = -D_p \left.\frac{\partial p_1}{\partial x}\right|_{x=0} = -D_p \exp\left(-\frac{t}{\tau_0}\right) \left.\frac{\partial p'}{\partial x}\right|_{x=0}$$

In terms of $\bar{p}$, the equation above becomes

$$D_p \left.\frac{\partial \bar{p}}{\partial x}\right|_{x=0} = \frac{I_r}{eA} \mathcal{L}\left[\exp\left(\frac{t}{\tau_0}\right)\right] = \frac{I_r}{eA(s - \beta)} \qquad (A8.5)$$

where A is the cross-sectional area of the junction and $\mathcal{L}$ stands for Laplace transform. Applying the boundary condition of Eq. (A8.5) to Eq. (A8.4) yields

$$\bar{p} = \frac{L}{eAD_p} \frac{I_f}{s - \beta} \exp\left(-\frac{x}{L}\right) - \frac{I_r + I_f \exp(-\alpha x)}{eAD_p} \frac{1}{\alpha(s - \beta)} \qquad (A8.6)$$

where $I_f = eAD_p p_{1f}/L$ is the forward junction current and $L = \sqrt{D_p\tau_0}$. Inverse Laplace transform of Eq. (A8.6) results in the following:

$$p_1 = \frac{L}{2eAD_p}\left\{ 2I_f \exp(-v) - (I_r + I_f)\left[\exp(-v)\,\mathrm{erfc}\left(\frac{v}{2u} - u\right)\right.\right.$$

$$\left.\left. - \exp(v)\,\mathrm{erfc}\left(\frac{v}{2u} + u\right)\right]\right\} \qquad (A8.7)$$

In Eq. (A8.7), erfc is the complementary error function and $v = x/L$ and $u = \sqrt{t/\tau_0}$.

For phase II operation, because the boundary condition expressed in Eq. (A8.5) no longer applies, we must look for a new solution. To do this exactly, we should set, in Eq. (A8.2), the initial excess-hole distribution in phase I operation at $t = t_I$. We immediately see that it would be very difficult to find the particular solution of Eq. (A8.2) if p_1 evaluated at $t = t_I$ from Eq. (A8.7) is taken as $p'(0)$ in Eq. (A8.2). At this point, a reasonable approximation must be made.

The initial excess-hole distribution is chosen to be the same as that in phase I; that is, at $t = 0$, $p'(0)$ is given by Eq. (A8.3). However, the boundary condition at $x = 0$ is that for a reverse-biased diode, that is,

$$\bar{p} = \mathcal{L}\, p_1 \exp\left(\frac{t}{\tau_0}\right) \cong 0 \qquad \text{at } x = 0 \qquad (A8.8)$$

after the use of Eq. (8.5.10) for p_1. The solution of Eq. (A8.2) now reads

$$\bar{p} = \frac{p_{1f}[\exp(-x/L) - \exp(-\alpha x)]}{s - \beta} \qquad (A8.9)$$

As in Eq. (A8.4), the term $p_{1f}[\exp(-x/L)]/(s - \beta)$ represents the particular solution. What is different from Eq. (A8.4) is the coefficient in front of the complementary solution, which is determined by the boundary condition. The coefficient in Eq. (A8.7) is so chosen that $\bar{p} = 0$ at $x = 0$. Inverse Laplace transform of $\bar{p}$ gives

$$p_1 = \frac{LI_f}{2eAD_p} \left[2 \exp{(-v)} - \exp{(-v)} \operatorname{erfc}\left(\frac{v}{2u} - u\right) \right.$$
$$\left. - \exp{(v)} \operatorname{erfc}\left(\frac{v}{2u} + u\right) \right] \qquad (A8.10)$$

The first term in Eq. (A8.10) represents the initial excess-hole distribution, and is the same as the first term in Eq. (A8.7). The remaining terms in Eqs. (A8.7) and (A8.10) are different because of the different boundary conditions.

Appendix 9

Transport Equations for One-Valley Semiconductors

The Boltzmann equation presented in Appendix A5 is the foundation for analyzing problems of carrier transport. The four basic equations are: (1) the continuity equation, (2) the momentum (or current-density) equation, (3) the energy equation, and (4) the energy-flow equation. In this appendix we present the derivation of these four equations for one-valley semiconductors where only intravalley scattering is considered. This is in contrast to two-valley semiconductors discussed in Section 10.9, where the intervalley scattering becomes important and is treated. Our starting point is the Boltzmann equation, Eq. (A5.7),

$$\frac{\partial f}{\partial t} = -\mathbf{a} \cdot \nabla_v f - \mathbf{v} \cdot \nabla_r f + \frac{\partial f}{\partial t}\bigg|_{\text{collision}} \tag{A9.1}$$

and its auxiliary equation

$$\frac{\partial f}{\partial t}\bigg|_{\text{collision}} = \int\!\!\!\int\!\!\!\int_{-\infty}^{\infty} [S(\mathbf{v}', \mathbf{v})f(\mathbf{r}, \mathbf{v}') - S(\mathbf{v}, \mathbf{v}')f(\mathbf{r}, \mathbf{v})] \, d\mathbf{v}' \tag{A9.2}$$

where $f(\mathbf{r}, \mathbf{v}, t)$ is the time-dependent distribution function of carriers in the six-dimensional phase space (x, y, z, v_x, v_y, v_z) and $d\mathbf{v} = dv_x \, dv_y \, dv_z$.

The first two terms on the right-hand side of Eq. (A9.1) represent the rates of change in the distribution function caused by the motion of carriers. The quantities $S(\mathbf{v}', \mathbf{v})$ and $S(\mathbf{v}, \mathbf{v}')$ represent the scattering matrix elements or the probabilities per unit time for a carrier to be scattered in the phase space from $\mathbf{v}'$ to $\mathbf{v}$ and from $\mathbf{v}$ to $\mathbf{v}'$, respectively. During a scattering event, the position $\mathbf{r}$ of the carrier remains unchanged. By definition, the carrier concentration n is given by

$$n = \int\!\!\!\int\!\!\!\int_{-\infty}^{\infty} f \, d\mathbf{v} = \int\!\!\!\int\!\!\!\int_{-\infty}^{\infty} f \, dv_x \, dv_y \, dv_z \tag{A9.3}$$

and the assembled average $\langle u \rangle$ of a physical quantity u is given by Eq. (A5.14), or

$$\langle u \rangle = n^{-1} \int\!\!\!\int\!\!\!\int_{-\infty}^{\infty} uf \, d\mathbf{v} \tag{A9.4}$$

where $d\mathbf{v} = dv_x dv_y dv_z$. The distribution function f is given by either the Boltzmann distribution or the Fermi–Dirac distribution except for a proportional constant.

The continuity equation can be derived from Eq. (A9.1) by multiplying it by $d\mathbf{v}$ and then integrating the resultant equation over the velocity space. Thus we obtain

$$\frac{\partial n}{\partial t} = -\nabla \cdot (n\langle \mathbf{v} \rangle) + \left. \frac{\partial n}{\partial t} \right|_{\text{collision}} \tag{A9.5}$$

One useful identity exists:

$$\int\!\!\!\int\!\!\!\int_{-\infty}^{\infty} G \frac{\partial f}{\partial v_x} \, dv_x \, dv_y \, dv_z = \left. Gf \right|_{-\infty}^{\infty} - \int\!\!\!\int\!\!\!\int_{-\infty}^{\infty} f \frac{\partial G}{\partial v_x} \, dv_x \, dv_y \, dv_z \tag{A9.6}$$

Because the distribution function must go to zero at $v_x = -\infty$ and $v_x = +\infty$, the first term on the right-hand side of Eq. (A9.6) is identically zero. Since we simply integrate Eq. (A9.1), $G = 1$ and hence $\partial G / \partial v_x = 0$ in Eq. (A9.6). Therefore, the contribution from the $\mathbf{a} \cdot \nabla_v f$ term in Eq. (A9.1) to Eq. (A9.5) is identically zero. We calculate the contribution from $\mathbf{v} \cdot \nabla_r f$ by making the following observation. The spatial dependence of this term is contained in the distribution function f. Therefore, we can take the following step:

$$\int\!\!\!\int\!\!\!\int_{-\infty}^{\infty} Gv_x \frac{\partial f}{\partial x} \, dv_x \, dv_y \, dv_z = \frac{\partial}{\partial x} \int\!\!\!\int\!\!\!\int_{-\infty}^{\infty} Gv_x f \, dv_x \, dv_y \, dv_z \tag{A9.7}$$

Letting $G = 1$ in Eq. (A9.7) and using the definition of $\langle \mathbf{v} \rangle$ according to Eq. (A9.4), we obtain the term $\nabla \cdot (n\langle \mathbf{v} \rangle)$ in Eq. (A9.5). Of course, the term $\partial n / \partial t$ comes from the $\partial f / \partial t$ term in Eq. (A9.1).

The treatment that we are about to present is called the method of moments first developed by Blotekjaer in his analysis of the electron-transfer effect in GaAs (K. Blotekjaer, *IEEE Trans. Electron Devices,* Vol. ED-17, p. 38, 1970). It involves successive multiplications of the Boltzmann equation by 1 (zeroth moment), $m^*\mathbf{v}$ (first moment), $m^*v^2/2$ (second moment), and $\mathbf{v}m^*v^2/2$ (third moment) to derive, respectively, the continuity equation, the momentum equation, the energy equation, and the energy-flow equation. The order of the moment refers to the power of velocity in the multiplication factor. The method of moments was later used by Hänsch and Miura-Mittausch in their analysis of velocity saturation in Si (W. Hänsch and M. Miura-Mittausch, *J. Appl. Phys.,* Vol. 60, p. 650, 1986). In this section we limit our discussion to one-valley semiconductors by ignoring intervalley scattering.

For one-valley semiconductors, scattering of electrons takes place within the same valley, and hence does not change the electron concentration. In other words, the term

$\partial n/\partial t$ due to collisions is identically zero. Furthermore, the current density $\mathbf{J}_e$ due to electrons is $\mathbf{J}_e = -en\langle\mathbf{v}\rangle$. Therefore, Eq. (A9.5) reduces to

$$\frac{\partial \rho_e}{\partial t} = -\nabla \cdot \mathbf{J}_e \tag{A9.8}$$

where $\rho_e = -en$ is the charge density associated with electrons. Obviously, Eq. (A9.8) is the well-known continuity equation for electrons.

To derive the momentum equation, we multiply both sides of Eq. (A9.1) by $m^*\mathbf{v}$ and then integrate the resultant equation over the velocity space. First, we deal with the collision term. We note that in the conventional treatment of the Boltzmann equation as outlined in Appendix A5, we use Eq. (A5.8) or $-(f - f_0)/\tau$ for the collision term in Eq. (A9.1). We further note that integrating $m^*\mathbf{v}f$ over the velocity space is simply $m^*\mathbf{J}/(-e)$ and integrating $m^*\mathbf{v}f_0$ over the velocity space is zero. Therefore, we let

$$\left.\frac{\partial nm^*\langle\mathbf{v}\rangle}{\partial t}\right|_{\text{collision}} = \frac{m^*\mathbf{J}_e}{e\tau_m} \tag{A9.9}$$

Since there are two relaxation times, one for momentum and the other for energy, we use τ_m to represent the momentum relaxation time.

Performing the same operation on Eq. (A9.2) and interchanging $\mathbf{v}'$ and $\mathbf{v}$ for the second term on the right-hand side of Eq. (A9.2) after multiplication of $\mathbf{v}d\mathbf{v}$, we have

$$\left.\frac{\partial nm^*\langle\mathbf{v}\rangle}{\partial t}\right|_{\text{collision}} = m^* \iint_{-\infty}^{\infty} (\mathbf{v} - \mathbf{v}')S(\mathbf{v}', \mathbf{v})f(\mathbf{r}, \mathbf{v}') \, d\mathbf{v} \, d\mathbf{v}' \tag{A9.10}$$

From Eqs. (A9.9) and (A9.10), we obtain

$$\tau_m^{-1} = \left(\frac{e}{J_e^2}\right)\mathbf{J}_e \cdot \iint_{-\infty}^{\infty} (\mathbf{v} - \mathbf{v}')S(\mathbf{v}', \mathbf{v})f(\mathbf{r}, \mathbf{v}') \, d\mathbf{v} \, d\mathbf{v}' \tag{A9.11}$$

from which the momentum relaxation time can be evaluated from the scattering matrix $S(\mathbf{v}', \mathbf{v})$. The integration in Eq. (A9.11) involves two triple integrals, noting that $d\mathbf{v} = dv_x \, dv_y \, dv_z$ and $d\mathbf{v}' = dv_x' \, dv_y' \, dv_z'$.

Now we calculate the contributions from the $\mathbf{a} \cdot \nabla_v f - \mathbf{v} \nabla_r f$ terms. Letting $G = m^*v_x$ and $F_x = m^*a_x$ in Eqs. (A9.6) and (A9.7), we obtain, respectively,

$$\iiint_{-\infty}^{\infty} F_x v_x \frac{\partial f}{\partial v_x} \, dv_x \, dv_y \, dv_z = -F_x n \tag{A9.12}$$

$$\iiint_{-\infty}^{\infty} m^*v_x^2 \frac{\partial f}{\partial x} \, dv_x \, dv_y \, dv_z = \frac{\partial}{\partial x}(nm^*\langle v_x^2\rangle) \tag{A9.13}$$

The results for $G = m^*v_y$ and $G = m^*v_z$ are similar to those given in Eqs. (A9.12) and (A9.13) except that F_x is replaced by F_y and F_z and $\langle v_x^2\rangle$ is replaced by $\langle v_y^2\rangle$ and $\langle v_z^2\rangle$, respectively. We note that terms having dependences such as $F_y v_x \partial/\partial v_y$ and $v_x v_y \partial/$

Transport Equations for One-Valley Semiconductors

∂x will have zero contributions to the triple integrals in Eqs. (A9.12) and (A9.13). Using Eqs. (A9.9), (A9.12), (A9.13) and similar equations, we find that

$$\frac{\partial nm^*\langle \mathbf{v}\rangle}{\partial t} = -e\mathbf{E}n - \frac{2}{3}\nabla\mathscr{E} + \frac{\mathbf{J}_e}{\mu} \tag{A9.14}$$

where $\mathscr{E} = n\langle m^*v^2/2\rangle$ is the energy density (joule/m^3) and

$$\mu = \frac{e\tau_m}{m^*} \quad \text{m}^2/\text{V-s} \tag{A9.15}$$

is the electron mobility. Equation (A9.14) is known as the momentum equation.

To obtain energy equation, we multiply Eq. (A9.1) by $m^*v^2/2$ and integrate the resultant equation over the velocity space. Letting $G = m^*v^2/2$ in Eqs. (A9.6) and (A9.7), respectively, yields

$$\iiint\limits_{-\infty} \frac{1}{2}m^*v^2\frac{\partial f}{\partial v_x}\, dv_x\, dv_y\, dv_z = \frac{m^*}{e}J_x \tag{A9.16}$$

$$\iiint\limits_{-\infty}^{\infty} v_x\frac{1}{2}m^*v^2\frac{\partial f}{\partial x}\, dv_x\, dv_y\, dv_z = \frac{\partial}{\partial x}(C_{\mathscr{E}x}) \tag{A9.17}$$

where $C_{\mathscr{E}x}$ (W/m^2) is the x-component of power density defined as

$$\mathbf{C}_{\mathscr{E}} = \iiint\limits_{-\infty}^{\infty} \mathbf{v}\left(\frac{1}{2}m^*v^2\right)f\, dv_x\, dv_y\, dv_z \tag{A9.18}$$

If we use Eq. (A5.8) or $-(f - f_0)/\tau$ for the collision term in Eq. (A9.1), we have

$$\left.\frac{\partial\mathscr{E}}{\partial t}\right|_{\text{collision}} = -\frac{\mathscr{E} - \mathscr{E}_0}{\tau_{\mathscr{E}}} \tag{A9.19}$$

On the other hand, use of Eq. (A9.2) yields

$$\left.\frac{\partial\mathscr{E}}{\partial t}\right|_{\text{collision}} = \iint\limits_{-\infty}^{\infty} \frac{m^*}{2}(v^2 - v'^2)S(\mathbf{v}', \mathbf{v})f(\mathbf{r}, \mathbf{v}')\, d\mathbf{v}\, d\mathbf{v}' \tag{A9.20}$$

following a procedure similar to the one used in deriving Eq. (A9.10). The integration again involves two triple integrals with $d\mathbf{v} = dv_x\, dv_y\, dv_z$ and $d\mathbf{v}' = dv_x'\, dv_y'\, dv_z'$.

Comparing Eqs. (A9.19) and (A9.20) yields

$$\tau_{\mathscr{E}}^{-1} = (\mathscr{E} - \mathscr{E}_0)^{-1}\iint\limits_{-\infty}^{\infty} \frac{m^*}{2}(v'^2 - v^2)S(\mathbf{v}', \mathbf{v})f(\mathbf{r}, \mathbf{v}')\, d\mathbf{v}\, d\mathbf{v}' \tag{A9.21}$$

Using the results given in Eqs. (A9.16), (A9.17), and (A9.19) we have

$$\frac{\partial\mathscr{E}}{\partial t} = \mathbf{J}_e \cdot \mathbf{E} - \nabla \cdot (\mathbf{C}_{\mathscr{E}}) - \frac{\mathscr{E} - \mathscr{E}_0}{\tau_{\mathscr{E}}} \tag{A9.22}$$

Equation (A9.22) is known as the energy-conservation equation. Note that power density is the product of energy density and flow rate, and has a direction determined by the direction of energy flow. Therefore, the quantity $\mathbf{C}_\mathscr{E}$ is a vector.

To obtain the energy-flow equation, we multiply Eq. (A9.1) by $vm^*v^2/2$ and then integrate the resultant equation over the velocity space. Letting $G = vm^*v^2/2$ in Eq. (A9.6), we find that

$$a_x \int\!\!\!\int\!\!\!\int_{-\infty}^{\infty} v_x \frac{m^*v^2}{2} \frac{\partial f}{\partial v_x}\, dv_x\, dv_y\, dv_z = a_x\left(\frac{5\mathscr{E}}{3}\right) \tag{A9.23}$$

In obtaining Eq. (A9.23) we assume that $\langle m^*v_x^2\rangle = \langle m^*v^2/3\rangle = 2\mathscr{E}/3$. To find the contribution from the $\mathbf{v} \cdot \nabla_r f$ term, we assume that the solution of the Boltzmann equation takes the form

$$f(\mathbf{r},\, \mathbf{v},\, t) = f_0(\mathbf{r},\, \mathbf{v},\, t) + \mathbf{E} \cdot \mathbf{v} f_1(\mathbf{r},\, \mathbf{v},\, t) \tag{A9.24}$$

Letting $G = vm^*v^2/2$, using Eq. (A9.24) in Eq. (A9.7), and keeping only terms up to third moment, we find that

$$\int\!\!\!\int\!\!\!\int_{-\infty}^{\infty} v_x \frac{m^*v^2}{2} \frac{\partial f}{\partial x}\, dv_x\, dv_y\, dv_z \cong \frac{\partial}{\partial x} \int\!\!\!\int\!\!\!\int_{-\infty}^{\infty} f_0 v_x \frac{m^*v^2}{2}\, dv_x\, dv_y\, dv_z = 0 \tag{A9.25}$$

because the integrand is an odd function.

To find the contribution from the collision term in Eq. (A9.1), we again use $-(f - f_0)/\tau_m$ and thus obtain

$$\left.\frac{\partial \mathbf{C}_\mathscr{E}}{\partial t}\right|_{\text{collision}} = -\frac{\mathbf{C}_\mathscr{E}}{\tau_m} \tag{A9.26}$$

Adding all the contributions from Eqs. (A9.23), (A9.25), and (A9.26), we have

$$\frac{\partial \mathbf{C}_\mathscr{E}}{\partial t} = \frac{-5}{3} \frac{e\mathbf{E}\mathscr{E}}{m^*} - \frac{\mathbf{C}_\mathscr{E}}{\tau_m} \tag{A9.27}$$

Equation (A9.27) is known as the energy-flow equation and is approximate in the sense that the contribution from the fourth moment is ignored.

Appendix 10

Transition Probability and Time-Dependent Perturbation Calculation

In Sections 12.8 and 12.10, we discussed optical transitions in solids and used transition probability w as a fundamental physical quantity to describe quantitatively the rate of transition between two states. In this appendix we develop a general mathematical scheme for perturbation calculation necessary for the derivation of transition probability. The fundamental question concerns the effect of the perturbation Hamiltonian $\mathcal{H}_1$ of Eq. (12.8.13) on the wave functions.

Consider an atomic system whose eigenstates satisfy

$$\mathcal{H}_0 \psi_i = i\hbar \frac{\partial \psi_i}{\partial t} = \mathcal{E}_i \psi_i \tag{A10.1}$$

where the subscript i refers to one of the states. In the presence of a radiation field, the Hamiltonian becomes $\mathcal{H}_0 + \mathcal{H}_1$, and as a consequence, a new wave function ψ must replace ψ_i in Eq. (A10.1). The new ψ satisfies the new time-dependent Schrödinger equation:

$$(\mathcal{H}_0 + \mathcal{H}_1)\psi = i\hbar \frac{\partial \psi}{\partial t} \tag{A10.2}$$

In general, Eq. (A10.2) cannot be solved exactly; therefore, an appropriate approximation method must be developed. The method to be presented here is called the perturbation method.

The perturbation scheme is applicable only if $\mathcal{H}_1$ is small compared to $\mathcal{H}_0$. This condition is satisfied even for intense laser beams. Under the circumstance of small perturbation, we may express the new wave function ψ as a linear combination of the eigenstates of $\mathcal{H}_0$ as follows:

$$\psi = \sum_i C_i(t)\psi_i \tag{A10.3}$$

Such a linear expansion is possible because the Schrödinger equation, Eq. (A10.1), is a linear differential equation in ψ. Substituting Eq. (A10.3) into Eq. (A10.1), we find that

$$\sum_i (C_i \mathcal{H}_0 \psi_i + C_i \mathcal{H}_1 \psi_i) = i\hbar \sum_i \left(\frac{dC_i}{dt} \psi_i + C_i \frac{\partial \psi_i}{\partial t} \right) \tag{A10.4}$$

Because of Eq. (A10.1), the terms $\mathcal{H}_0 \psi_i$ and $i\hbar\, \partial\psi_i/\partial t$ in Eq. (A10.4) cancel out. Thus we obtain

$$i\hbar \sum_i \psi_i \frac{dC_i}{dt} = \sum_i \mathcal{H}_1 \psi_i C_i \tag{A10.5}$$

The eigenfunctions of $\mathcal{H}_0$ form an orthonormal set, and hence the relation

$$\int_{-\infty}^{\infty} \psi_i^* \psi_j \, d\tau = \delta_{ij} \tag{A10.6}$$

exists between two eigenfunctions ψ_i and ψ_j. In Eq. (A10.6), $d\tau = dx\,dy\,dz$, and δ_{ij} is the Kronecker delta with $\delta_{ij} = 1$ for $i = j$ and $\delta_{ij} = 0$ for $i \neq j$. Multiplying both sides of Eq. (A10.5) by ψ_j^* and then integrating the resultant equation over $d\tau$, we obtain

$$i\hbar \frac{dC_j}{dt} = \sum_i \int_{-\infty}^{\infty} C_i \psi_j^* \mathcal{H}_1 \psi_i \, d\tau \tag{A10.7}$$

Out of all the terms in the summation on the left-hand side of Eq. (A10.5), only the term ψ_j survives the orthonormality condition of Eq. (A10.6) and contributes to the left-hand side of Eq. (A10.7).

According to our discussion in Section 1.3, the probability dW of finding an electron in a volume element $d\tau = dx\,dy\,dz$ is equal to

$$dW = |\psi|^2 \, d\tau = \psi\psi^* \, d\tau \tag{A10.8}$$

Since in general ψ is complex, the asterisk indicates the complex conjugate. The above probabilistic interpretation of $|\psi|^2$ was originally advanced by Born. A generalization of the statistical interpretation of $|\psi|^2$ permits us to make other predictions. Substituting Eq. (A10.3) in Eq. (A10.8), we find that

$$\int_{-\infty}^{\infty} |\psi|^2 \, d\tau = \sum_i C_i(t)C_i^*(t) = \sum_i |C_i(t)|^2 = 1 \tag{A10.9}$$

after the use of the orthonormality condition of Eq. (A10.6). Since the total probability of finding an electron must be unity, the summation in Eq. (A10.9) must be equal to 1. Now the physical meaning of the coefficients $C_i(t)$ in Eq. (A10.3) is clear. The quantity $|C_i(t)|^2$ represents the probability of finding the electron in the state i. Information concerning the time variation of $C_i(t)$ is given by Eq. (A10.7), which is the quantum-mechanical rate equation. We should emphasize, however, that the quantity $|C_i(t)|^2$, but not $C_i(t)$ or $C_i^*(t)$ alone, corresponds to the classical population N_i.

Both the perturbation Hamiltonian and the wave functions have a time-dependent part and a spatial-dependent part. Thus we express $\mathcal{H}_1$ and ψ as

$$\mathcal{H}_1 = \hbar \exp(i\omega t) + \hbar \exp(-i\omega t) \tag{A10.10}$$

$$\psi_i(r,\ t) = \psi_i(r) \exp \left(-\frac{i\mathscr{E}_i t}{\hbar} \right) \qquad \text{(A10.11)}$$

By letting

$$\mathscr{H}_{ji} = \langle j|\hbar|i \rangle = \int_{-\infty}^{\infty} \psi_j^*(r)\hbar\psi_i(r)\ d\tau \qquad \text{(A10.12)}$$

$$\omega_{ji} = \frac{\mathscr{E}_j - \mathscr{E}_i}{\hbar} \qquad \text{(A10.13)}$$

we rewrite Eq. (A10.7) as

$$i\hbar \frac{dC_j}{dt} = \sum_i C_i \mathscr{H}_{ji} \{ \exp\ [i(\omega_{ji} - \omega)t] + \exp\ [i(\omega_{ji} + \omega)t] \} \qquad \text{(A10.14)}$$

Note that Eq. (A10.14) is a linear equation in C's. This property permits us to use the method of superposition. We assume that the electron is initially in the state i. In other words, $C_i = 1$ and other C's are zero on the right-hand side of Eq. (A10.14). Using this information and integrating Eq. (A10.14) with respect to t, we obtain

$$C_j(t) = -\frac{t\mathscr{H}_{ji}}{\hbar} \left[\frac{\exp\ (i\theta_1) - 1}{\theta_1} + \frac{\exp\ (i\theta_2) - 1}{\theta_2} \right] \qquad \text{(A10.15)}$$

The two angles θ_1 and θ_2 are

$$\theta_1 = (\omega_{ji} - \omega)t \quad \text{and} \quad \theta_2 = (\omega_{ji} + \omega)t \qquad \text{(A10.16)}$$

and the two terms in Eq. (A10.15) represent the absorption and emission process, respectively.

To obtain the rate of atomic transition, we apply the probabilistic interpretation of wave function to Eq. (A10.15). According to Eq. (A10.9), the probability of finding an electron in the state j is

$$W_j = \frac{|C_j(t)|^2}{\Sigma_i |C_i(t)|^2} = |C_j(t)|^2 \qquad \text{(A10.17)}$$

For the emission process, only the term containing θ_2 in Eq. (A10.15) is important. Substituting Eq. (A10.15) in Eq. (A10.17) and letting $j = n$ and $i = m$, we find that

$$W_n = |C_n(t)|^2 = \frac{4}{\hbar^2} |\mathscr{H}_{nm}|^2 \frac{\sin^2\ (\theta_2/2)}{(\theta_2/t)^2} \qquad \text{(A10.18)}$$

The other terms having θ_1^2 and $\theta_1\theta_2$ in the denominator are insignificant and hence neglected.

We should point out that depending on whether we are interested in the transient solution or the steady-state solution, the treatment of Eq. (A10.18) is different. In the present treatment, we are interested in the steady-state solution. In the limit $t \rightarrow \infty$, the function $F = \sin^2\ (\theta/2)/(\theta/t)^2$ is vanishingly small everywhere except near the origin, and hence the function F behaves like a Dirac delta function. It can be shown that

$$\int_{-\infty}^{\infty} \frac{\sin^2\ (\theta/2)}{\theta^2}\ d\theta = \frac{\pi}{2} \qquad \text{(A10.19)}$$

Changing the variable from θ to $\omega_{nm} + \omega$, we have

$$\int_{-\infty}^{\infty} F(\omega_{nm} + \omega)d(\omega_{nm} + \omega) = \frac{\pi t}{2} \qquad \text{(A10.20)}$$

Therefore, we can replace $F(\omega_{nm} + \omega)$ by $(\pi t/2)\delta(\omega_{nm} + \omega)$ in Eq. (A10.18), and thus obtain

$$W_n = \frac{2\pi t}{\hbar^2} |\mathcal{H}_{nm}|^2 \delta(\omega_{nm} + \omega) \qquad \text{(A10.21)}$$

A Dirac delta function $\delta(x)$ is nonzero only when its argument x is zero. This condition says that W_n is nonzero only when $\omega = -\omega_{nm}$, and this condition leads to the energy-conservation condition. The transition probability w is defined as the probability of transition per unit time. Thus from Eq. (A10.21), we have

$$w = \frac{W_n}{t} = \frac{2\pi}{\hbar} |\mathcal{H}_{nm}|^2 \delta(\mathscr{E}_m - \mathscr{E}_n - \hbar\omega) \qquad \text{(A10.22)}$$

The change from $\hbar^2$ in Eq. (A10.21) to $\hbar$ in Eq. (A10.22) is due to the change in the variable from ω to $\hbar\omega$ in the Dirac delta function.

From the one-photon absorption process, we can follow a similar procedure as outlined above except that the roles of states m and n should be interchanged. That means that $i = n$ and $j = m$ in Eq. (A10.15). Since θ_1 is zero, only the first term in Eq. (A10.15) needs to be considered. Therefore, for the upward transition, the transition probability w is given by

$$w = \frac{W_m}{t} = \frac{2\pi}{\hbar} |\mathcal{H}_{mn}|^2 \delta(\mathscr{E}_m - \mathscr{E}_n - \hbar\omega) \qquad \text{(A10.23)}$$

The quantities $\mathcal{H}_{nm}$ and $\mathcal{H}_{mn}$ in Eqs. (A10.22) and (A10.23) are called the matrix elements which connect two states m and n through the perturbation Hamiltonian $\mathcal{H}_1$. The transition is said to be allowed if the matrix element is nonzero, and forbidden if the matrix element is zero.

Index

Abadalla, M., 621
Abe, M., 447, 677, 690, 691, 697, 721, 754
Abeles, B., 825
Abramowitz, M., 226, 531, 619
Abrupt junction, 334–38
Absorption coefficient, 236, 603–7
Acceptor states, 202–4
Acoustic-phonon branch, 69, 466–69, 534
Activator and sensitizer, 740
Adachi, S., 525, 707
Adachi, Y., 688
Akasaka, Y., 723
Akiba, S., 519
Alavi, K., 794
Alexandre, F., 621
Alferov, Zh. I., 519, 776
Allen, J. W., 508
Allowed energy bands, 112–13
Alloying process, 354, 355–56
Altarelli, M., 557
Amorphous solid, 29
Amphoteric impurities, 230
Anderson, L. K., 786
Anderson, N. G., 624, 626
Anderson, R. L., 516
Ando, H., 798

Ando, T., 531, 557
Andreev, V. M., 776
Anholt, R., 685
Anselm, A., 608, 609, 610, 618, 825
Antypas, G. A., 519
Aoki, M., 722
Arcentev, I. N., 519
Arnold, D. J., 529, 678, 707
Arthur, J. R., 366
Asai, S., 693
Asatourian, R., 726, 728
Asbeck, F. M., 726, 728
Aspnes, D. E., 246, 615
Associated Laguerre polynomial, 13, 14
Associated Legendre polynomial, 13, 14
Atomic bonding. *See* Bonding
Atomic number, 3, 5
Atomic packing, 37–38
Atomic Physics (Born), 1, 2
Atomic structure
  Bohr postulates, 6
  of elements, 19–26
  hydrogen atom, 10–17
  Rutherford atom, 1–5
  Schrödinger wave equation, 7–10
  spin and exclusion principle, 17–19

Wilson-Sommerfeld rule of quantization, 6–7
Atomic Theory (Hume-Rothery), 162
Atom-size ratio in various packing forms, 37–38
Auger recombination, 778–79
Augustyniak, W. M., 318
Austin, E. J., 624
Auston, D. H., 791, 793, 794
Auston switch, 791, 793
Avalanche
  breakdown, 383, 419, 478, 486, 487–88, 492
  diodes, 649–54
  photodiodes, 785–86, 798–800
Avalanche-Diode Microwave Oscillators (Gibbons), 654
Awano, Y., 721, 728

Baccarani, G., 709, 713
Bachenstoss, G., 374
Bachrach, R. Z., 752
Backgating effect, 685, 686–88
Backward diode, 647
Bakanowski, A. E., 352
Ballistic electrons, 552–55
Ballman, A. A., 797
Baltensperger, W., 254

Band(s)
  allowed, 112–13
  conduction, 114
  -gap shrinkage, 255, 760–66
  -structure diagram, 175
  tailing, 253, 760–66
  valence, 114
  See also Energy bands; Energy-band structure
Band-edge discontinuities, 526–30
Band-to-band transition, 275–77
  carrier lifetime in, 755–60
Baraff, G. A., 480
Barber, M. R., 664
Bardeen, J., 216, 390
Barrier height, metal-semiconductor contact and, 422–23
Bartelink, D. J., 654
Base-width modulation effect, 394
Basinski, J., 426
Basov, N. G., 772
Bassous, E., 709
Bastard, G., 557, 559, 567, 568, 623
Batdorf, R. L., 650
Batey, J., 525, 526
Bauer, R. S., 524
Bean, J. C., 791
Beattie, A. R., 779
Bechtold, P. F., 682, 714
Becker, W. M., 625, 626, 753
Beebe, E. D., 516, 797
Beer, A. C., 246, 612, 615, 761
Berenz, J. J., 678
Bergh, A. A., 751
Bergstresser, T. K., 238
Bernard, M. G. A., 767
Bethe, H. A., 427
Bevacqua, S. F., 772
Bhargava, R. N., 748
Bhattacharya, P. K., 564
Bicknell, R. N., 624, 626
Biersack, J. P., 371
Bipolar transistors, 390–97, 723–29

current gain, 392
diffused, 397
heterostructure, 723–29
microwave, 406
power, 403
switching, 408
transit time in, 724
Birman, J. L., 224
Birrittella, M. S., 686
Bischoff, J. C., 729
Blackbody radiation, 597–98
Blakemore, J. S., 252, 558, 562, 611, 612, 825
Blewer, R. S., 444, 452
Bloch, F., 152
Bloch theorem, 151–54
Bloch wave, 154
  momentum, velocity, and current of, 816–18
Bloem, J., 232
Bloembergen, N., 772
Bloom, D. M., 621, 796
Blotekjaer, K., 836
Body-centered cubic (bcc) structure, 34, 37
Bohr magneton, 626
Bohr postulates, 6
Bohr radius, 7
Bois, D., 754
Boltzmann
  constant, 127, 133
  equation, 819–22
  Maxwell-Boltzmann distribution function, 127, 128, 129
  relation and thermodynamic energy functions, 809–12
  statistics, 805–8
  transport equation, 500, 823–25
Bombardment of particles, excess carrier lifetime and, 319–22
Bonch-Bruevich, V. L., 761
Bonding/bond
  classification of bond types, 79–87
  covalent, 80–84
  electron-pair, 81

interatomic forces, 77–79
  in ionic crystals, 80
  in metallic crystals, 84
  in molecular crystals, 84–87
  primary, 79
  saturation and directional properties of covalent bonds, 96–101
  semiconductors and chemical, 221–24
  strength of, 79
  structure and, in covalent crystals, 101–5
  van der Waals, 84–87
Bonsett, T. C., 626, 753
Born, M., 1, 2
Bose-Einstein distribution function, 127
Botez, D., 517, 754, 755, 759
Bott, I. B., 504
Bottka, N., 246, 615
Bourdman, A. D., 509
Bourgoin, J. C., 690
Bowers, R., 746
Bowing or nonlinear parameter, 524
Bragg law of diffraction, 54, 147, 158–60
Brain, M., 786, 796, 798, 799
Braittain, W. H., 390
Bravais lattices, 31–34
Brenac, A., 621
Brillouin zones, 158–62
  reciprocal-lattice vectors, 159
Broido, D. A., 557
Brooks, H., 247
Brown, E. R., 647
Brown, G. A., 254
Browne, P. F., 224
Brozel, M. R., 689
Brueckner, A. R., 761
Brum, J. A., 557, 559, 623
Buget, U., 428
Bujatti, M., 448
Bulman, G. E., 478
Bulman, P. G., 509
Burns, G., 772
Burrus, C. A., 622, 623, 781
Butcher, P. N., 504, 508, 668

Butler, J. K., 520, 613
Bylsma, R. B., 626, 753

Caine, E. J., 567
Callaway, J., 619
Campbell, J. C., 750, 798, 799
Capacitance
  carrier storage and diffusion,
    343–46
  diffused diodes and transi-
    tion-region, 375–77
  space-charge region and junc-
    tion, 333–38
  -voltage in SIS structures,
    526–30
Capasso, F., 800
Cappy, A., 704
Carbon, energy bands in, 114,
  115
Cardona, M., 246, 615, 625
Carenco, A., 621
Carlson, R. O., 772
Carnez, B., 704
Carrier concentration
  band tailing and shrinkage
    and, 760–66
  in degenerate semiconduc-
    tors, 250–55
  in nondegenerate semicon-
    ductors, 204–9
  in intrinsic semiconductors,
    199
  and Hall measurement, 180–
    2, 209–14
Carrier dynamics
  energy relaxation time, 332
  momentum relaxation time,
    332, 507
  nonstationary, 704–7
Carrier lifetime, 279–83,
  755–60
Carrier mobility
  impurity scattering, 214–16
  lattice scattering, 216–21
  optical phonon scattering,
    219
  piezoelectric scattering, 228
  polar scattering, 225

values in Ge, Si, and GaAs,
  222
Carrier multiplication
  avalanche breakdown, 383,
    419, 478, 486, 487–88,
    492
  impact ionization and,
    474–81
Carriers, drift velocity of,
  464–70
  saturation velocity, 470, 500
  velocity overshoot, 701
Carriers, excess
  charge-neutrality condition,
    272–74
  continuity equation, 283–86
  definition of, 265
  diffusion current, 265–66
  factors that affect the lifetime
    of, 312–22
  Haynes-Shockley experiment,
    295–301
  lifetime of, 279–83, 755–60
  quasi-Fermi levels and,
    269–72
  recombination processes,
    274–83
  storage and diffusion capaci-
    tance, 343–46
  surface-recombination veloc-
    ity, 309–12
  time-dependent diffusion
    equations, 283–95
Carrier waves in two-valley
  semiconductors, growth
  and propagation of,
  661–64
Carroll, J. E., 481, 654, 664,
  675
Carslaw, H. S., 290
Caruso, R., 448, 690
Casey, H. C., Jr., 358, 517,
  519, 520, 521, 612, 613,
  757, 761, 771
Cave, E. F., 365
Cesium chloride (CsCl) structure, 36–37
Chan, T. Y., 493, 693
Chang, A., 546

Chang, C. A., 567
Chang, K. K. N., 650
Chang, L. L., 566, 567
Chang, T. H. P., 720
Chang, T. Y., 440
Chang, Y. C., 567
Chanin, D. J., 754, 759
Channeling, 369
Channel mobility, 717–19
Chao, P. C., 678
Chapin, P. C., 646
Charge-control model, 413–16
Charge-neutrality condition,
  272–74
Chemical potential, 127
Chemla, D. S., 615, 616, 617,
  622, 623
Chen, C. Y., 794, 795
Chen, J. Y., 715
Chen, R. T., 686, 689
Chen, T. Z., 725
Chen, Y. S., 508
Chin, A. K., 448
Chin, R., 800
Chiu, K. Y., 693
Chloride vapor-phase epitaxy,
  365
Cho, A. Y., 366, 367, 567
Chow, W. F., 647
Christel, L. A., 370
Chuang, C. T., 725
Chynoweth, A. G., 478, 508,
  648, 663, 668
Circuit integration, 449–53
Circulator, 647
Classical Electrodynamics
  (Jackson), 577, 578, 588,
  610
Classical mechanics, 7–8
Clausius-Mossotti equation, 593
Cleiren, A. P., 747
Cleland, J. W., 320
Clogston, A. M., 744
Clorfeine, A. S., 653
Coactivator, 745
Cohen, B. G., 650
Cohen, M. L., 238
Cohesive energy, 79, 80
  in a metallic crystal, 89–92

Coleman, P. D., 646
Collector of transistor, 391
Collis, D. M., 796
Compensated semiconductors, 208–9
Complementary error function, 292
II-VI Compounds (Thomas, ed.), 238
II-VI compounds, diluted magnetic semiconductors and, 624–27
Condon, E. U., 13, 742
Conduction band, 114
Conduction of Heat in Solids (Carslaw and Yaeger), 290
Configuration coordinate, 742
Conley, J. W., 255
Conradi, J., 788
Constant, E., 704
Constructive interference, 48
Contact potential
  Fermi energy and, 129–31
  metal-semiconductor contact and, 422–23
Contact resistance, specific, 442
Continuity equation, 283–86
Conwell, E. M., 209, 225–26, 232, 469, 470, 505, 547, 554
Cook, P. W., 492, 709, 713
Cooke, H. F., 724
Coordination number, 34
Copeland, C. A., 781
Copeland, J. A., 672, 675
Core repulsive energy, 78
Coriell, A. S., 508
Correlation, 761
Coulomb scattering, 534
Covalent bonds, 80–84
  saturation and directional properties of, 96–101
Covalent crystals
  structure and bonding of, 101–5
  structure of, 38–41
Craford, M. G., 749, 750
Crasemann, B., 482, 599
Crawford, J. H., Jr., 320
Credé, R. H., 654

Critchlow, D. L., 693, 709, 713
Crowell, C. R., 424, 426, 428, 453, 493, 523
Cryogenic operation, 722–23
Crystal diffraction, 52–57
  Laue method, 55
  powder method, 54–55
  rotating-crystal method, 55
Crystal imperfections, 61–65
Crystallographic notation, 41–44
Crystallography, definition of, 30
Crystal momentum, 158
Crystal structure
  Bravais lattices, 31–34
  coordination number, 34
  crystalline state, 29–31
  lattice vibration, 65–71
  reciprocal-lattice vectors and plane normal, 44–46
  rotational symmetry and material-parameter tensor, 57–61
Crystal structure, simple
  arrangement of atoms in a, 34–41
  atom-size ratio in various packing forms, 37–38
  body-centered cubic (bcc), 34, 37
  in covalent crystals, 38–41
  diamond structure, 38–40
  face-centered cubic (fcc), 34, 35, 37
  hexagonal close-packed (hcp), 34, 35, 36, 37
  in ionic crystals, 36–37
  in metals, 34–36
  octahedral packing, 37–38
  tetrahedral packing, 37–38
  zinc-blende (sphalerite) structure, 40
Crystal systems, 31, 32
  symmetry operations in, 57–61
Cubic crystal system, 58, 59, 60
Curie, D., 746

Curie, G., 746
Current conduction in
  metal-semiconductor barrier, 426–28
  SIS GaAs/AlGaAs/GaAs structure, 526–30
Current gain, 392
  cutoff frequency, 677
Current Injection in Solids (Lampert and Mark), 687
Curtis, O. L., Jr., 320
Cuthbert, J. D., 748
Cutoff frequency, 406–7
  power-gain, 679–80
Cyclotron frequency, 182, 239, 537
Cyclotron-resonance experiment, 182–86, 239–42
Czorny, D., 365

Damen, T. C., 622, 623
Dangling bonds, 496–97
Dapkus, P. D., 368
Darwin, C. G., 594
Dash, W. C., 236
Das Sarma, S., 541–42
Datta, S., 624, 625, 626, 627, 753
Davisson-Germer experiment, 51
Deal, B. E., 434
Dean, P. J., 748, 749, 750, 751
de Broglie, L., 51
Debye, P. P., 209, 585
Debye theory of specific heat, 140
Deep-lying impurities, 230
Defects in Crystalline Solids (Bloem, Kroger, Vink), 232
Deformation potential, 217
Degani, J., 794
Degenerate semiconductors, 197, 250–55, 641
DeGrefte, H. A. M., 370
Dekker, A. J., 64
Delagebaudeuf, D., 540
DeLoach, B. C., 650, 654
Dennard, R. H., 492, 709, 713, 720

Density-of-state, 124–26
mobility effective masses
and, 242–43
Dentai, A. G., 781, 795, 799
Depolarization field, 587–90
Derick, L., 372
Detectivity, 789
Device-fabrication technology.
See Planar structures, fab-
rication techniques for; p-n
junction formation
DeWinter, J. C., 516, 794, 797
Dexter, D. L., 744
Dhar, S., 564
Diamond structure, 38–40
energy bands of carbon in,
114, 115
*Dielectric Behavior and Struc-
ture* (Smyth), 586
Dielectric constant, polarizabil-
ity and, 591–94
Dielectric properties of materi-
als
chemical bond and, 586
Faraday rotation, 627
Franz-Keldysh effect, 618
Kramers-Kronig relation,
609–10
Lyddane-Sachs-Teller relation, 607–9
Pockels effect, 620
polarization and, 586–87
Stark effect, 623
Dielectric-relaxation time, 273,
661
Diffraction experiments, 46
Bragg law of diffraction, 54
crystal, 52–57
neutron, 51–52
X-ray, 51–52
Diffused transistors, 397–403
Diffusion
coefficient of impurities,
364
constant of carriers, 266,
301
current, 265–66
fabrication techniques for
planar structures and,
371–75

p-n junction formation and,
354, 359–65
Diffusion capacitance, carrier
storage and, 343–46
Diffusion equations, time-
dependent, 283–95, 828–
31
Dill, F. H., Jr., 772
Diluted magnetic semiconductors, II-VI
compounds and, 624–27
Dimmock, J. O., 542, 563, 612
Dingle, R., 524, 542
Diodes
backward, 647
Esaki, 641–49
IMPATT, 462, 647, 649–61
junction luminescence and
light-emitting, 747–55
photo, 785–86
transient response of junction, 832–34
TRAPATT, 649–54
varactor, and parametric interaction, 636–40
Diodes, p-n junction, 330–33
ideal characteristics, 340–43
narrow-base, 340–41, 353,
354
spreading resistance of, 345
transition-region capacitance
of diffused, 375–77
wide-based, 340–41, 353–54
Dipole domains, 664, 668–72
Dipole moment, polarization
and, 577–79
Dipole Moments (LeFèvre), 586
Dirac, P. A. M., 18
Direct-gap energy, 236, 238
Direct-gap semiconductors, 233
spontaneous emission and
carrier lifetime in, 755–60
Direct optical transition, 236
Direct radiative recombination
processes, 276
Direct tunneling, 484, 486–87
Dislocations, 62, 64–65
Dislocations (Friedel), 64
Dixon, R. W., 252, 516
Döhler, G. H., 559

Domain(s)
dynamics of dipole, 668–72
formation, 664–68
Donor states, 199–202
Dopant, band tailing and
shrinkage and, 760–66
Dorda, G., 535, 537
Downey, P. M., 794
Drain conductance, 422
Dresselhaus, G., 184, 185, 562
Drift mobility, 114, 213, 300–
301
Drift velocity, 464–70
Drummond, T. J., 549, 677,
680, 698, 707
Duchemin, J. P., 368
Duke, C. B., 255
Dumke, W. P., 277, 726, 765,
772
Dunlap, W. C., Jr., 516
Dupuis, R. D., 368, 749
Duraffourg, G., 767
Dushman equation, 9
-Richardson equation, 133,
527, 529
Dutt, B. V., 790, 795
DX center, 707–8
Dynamic characteristic of dipole
domain, 670

Eagle, D. M., 765
Early, J. M., 394
Early effect, 394
Eastman, D. E., 615
Eastman, L. F., 678
Ebers, J. J., 412
-Moll model, 411–13
Eda, K., 729
Effective diffusion constant,
285
Effective lifetime of excess car-
riers, 315
Effective mass(es)
density-of-state and mobility,
242
experimental determination,
238–50
of a wave packet, 157–58
relation among, 562

Effective mobility of excess
  carriers, 285
Ehrenreich, H., 200, 229, 614,
  740
Eilenberger, D. J., 615
Einspruch, N. G., 692
Einstein relation, 266
Eisberg, R. M., 2
Ekenberg, U., 557
*Electrical Properties of Semi-
  conductor Surfaces*
  (Frankl), 434
Electric dipole, 78
  moment, 85, 577–79
  transition, 596
Electric-quadrupole transition,
  596
Electric susceptibility, 579
Electrode interconnections,
  444
Electromagnetic theory, 5
Electromagnetic waves and
  matter waves, 48–52
Electromechanical coupling
  constant, 228
Electron affinity, 423
Electronegativity, 223–24
Electronic conduction, energy
  bands and
  Bloch theorem, 151–54
  Brillouin zones, 158–62
  cyclotron resonance experi-
    ment, 182–86
  effective mass, 157–58
  Hall measurement, 180–82
  hole conduction, 176–80
  Kronig-Penney model,
    147–51
  wave packet and group ve-
    locity, 154–57
  zone theory, 162–67
Electronic conductivity and
  mean free time, 133–37
Electronic contribution
  to dipole moment, 579
  to heat capacity, 137–38
  to thermal conductivity,
    138–40
Electronic polarizability,
  580–82

*Electronic Principles: Physics,
  Models, and Circuits* (Gray
  and Searle), 416
*Electronic Processes in Ionic
  Crystals* (Mott and
  Gurney), 740
Electronic Semiconductors
  (Spenke), 157
Electron-pair bond, 81
Electrons
  angular momentum of, 17–19
  diffraction experiment and,
    51, 52
  hole conduction and, 176–80
  hot and ballistic electrons in
    heterostructures, 547–55
  motion of, in atoms, 5–7
  motion of, in a hydrogen
    atom, 10–17
  Pauli exclusion principle,
    18–19
  spin of, 18
  transfer of, in two-valley
    semiconductors, 503–9
  two-dimensional electron
    (hole) gas, 530–34
  See also Free-electrons
*Electrons, Atoms, Metals and
  Alloys* (Hume-Rothery), 36
*Electrons and Holes in Semi-
  conductors* (Shockley),
  157, 213
Electrons and Phonons
  (Ziman), 182, 220, 226
Electron-transfer effect, 470–74
Electroreflectance experiment,
  246
Elements
  atomic structure of, 19–26
  periodic table of chemical, 3,
    4, 19
*Elements of Infrared Technol-
  ogy* (Kruse, McGlauchlin,
  McQuistan), 790
Elliot, K. R., 686
Emel'yanenko, O. V., 761
Emitter efficiency, 392
Emmons, R. B., 788
Energy
  cohesive, 79, 80

crystal types and, 77–79
  direct-gap, 236
  Fermi, 90
  forbidden energy gap, 113
  indirect-gap, 236
  thermal gap, 210
Energy bands
  in carbon, 114, 115
  in germanium, 115
  in ionic crystals, 116–17
  in metals and insulators,
    113–19
  in semiconductors, 118–19,
    232–38
  in semimetals, 119
  in silicon, 115
  in sodium, 115, 116
  in solids, 108–13
  See also Electronic conduc-
    tion, energy bands and
*Energy Bands in Semiconduc-
  tors* (Long), 517
Energy-band structure
  examples of direct-gap and
    indirect-gap semiconduc-
    tors, 233
  experimental studies of en-
    ergy-band structure and ef-
    fective masses, 238–50
  free-electron approximation,
    167–70
  tight-binding approximation,
    171–76
Energy discontinuity, zone theory and, 165–67
Eng, S. T., 647
Epitaxy, 354
  chloride vapor-phase, 365
  liquid-phase, 356–59
  molecular-beam, 358,
    366–67
Equal-area rule, 668
Erbaduk, M., 615
Ermanis, F., 761
Esaki, L., 559, 566, 567, 641,
  642
Esaki diodes, 641–49
Etch pit technique, 65
Ettenberg, M., 516, 754, 759
Eutectic point, 355

Evans, W. J., 653
Excess carriers. see Carriers, excess
Excess noise factor, 788
Excitons, 615
Expectation value, 10
Extrinsic recombination processes, 277–83

Face-centered cubic (fcc) structure, 34, 35, 37
Fang, F. F., 717
Faraday rotation, 626
Fast surface states, 301
and surface-recombination velocity, 309–12
Faulkner, R. A., 750
Faurie, J. P., 567, 570
Fawcett, W., 504, 508, 509
Fay, B., 664
F center, 63
Feldman, W. L., 508
Fenner, G. E., 772
Fermi-Dirac distribution function, 126
carrier concentrations and Fermi level, 204–9
description of, 126–29
Pauli exclusion principle and, 128–29
semiconductors and, 196–99, 204–9, 250–55
thermionic emission and, 131–33
Fermi energy, 90
contact potential and, 129–31
metals and values of, 130
Fermi level(s), 127
evaluation of, 204
excess carriers and quasi-, 269–72
p-n junctions and quasi-, 338–40
in semiconductors, 196–99, 204–9, 250–55
thermal equilibrium and, 268
Fern, R. E., 523
Fichtner, W., 682, 714

Field-effect experiments, 303–9
Field-effect transistors
current gain cut-off frequency, 677–78
depletion mode, 437, 696
enhancement mode, 437, 696
flat band voltage, 694
junction, 420
metal-oxide-semiconductor, 432–39
metal-semiconductor, 439, 441–42, 444–48
power delay product, 677
reliability, 681–3
self-aligned gate, 684
threshold voltage, 435, 697
transconductance, 422, 433
ultra-high speed, 676
ultra-small dimension, 683
Fields and Waves in Communication Electronics (Ramo, Whinnery, Van Duzer), 588, 596
Figueroa, L., 793
Fischer, R., 525, 529, 646, 678, 680, 707
Flat-band voltage, 434
Floquet theorem, 152
Forbidden energy gap, 113
Forbidden transitions, 601
Forbidden zones, 151
Fordemwalt, J. N., 451
Forrest, S. R., 795, 798, 799, 800
Forster, J. H., 352
Fowler, A. B., 531, 717
Fowler, R. H., 424
Foy, P. W., 776
Foyt, A. G., 509, 664
Francostoro-Decker, M., 707
Frankl, D. R., 434
Franz, W., 618
Franz-Keldysh effect, 618–20
Free-electron(s)
approximation, 167–70
in a box, 119–23
definition of, 119
electronic conductivity and mean free time, 133–37

electronic contribution to heat capacity and thermal conductivity, 137–40
energy bands in solids, 108–13
Fermi-Dirac distribution function, 126–29
in metals, insulators, semiconductors, and semimetals, 113–19
model and density of states, 123–26
theory, 121
thermionic emission, 131–33
Free holes, 119
Freeouf, J. L., 615
Frener, J. S., 747
Frenkel defect, 63
Frey, J., 706
Friedel, J., 64
Fritzsche, H., 648
Fröhlich, H., 508
Fröhlich coupling constant, 225
Frosch, C. J., 371–72
Fujii, T., 645
Fukano, T., 726
Fukukawa, Y., 650, 677
Fundamentals of Modern Physics (Eisberg), 2
Furdyna, J. K., 626

(GaAl)As
/GaAs system, 521–26
properties of, 521–26, 561–65
GaAs
dielectric properties of, 593
electric properties of, 222, 235, 243
(GaAl)As/ system, 521–26
impurity states in, 281–82
intrinsic carrier concentration, 341
intrinsic defects in, 688–91
ionicity, 224
optical properties of, 610–17
properties of, 521, 523, 561–65
versus Si, 720–22

Gaensslen, F. H., 709, 713
Gain spectrum, population inversion and, 766–71
Gaj, A., 753
Gakbuzov, D. Z., 519
Galazka, R. R., 753
Galt, J. K., 185
Galvanomagnetic Effects in Semiconductors (Beer), 246
Ganin, E., 720
Gannon, J. J., 516
Garbinski, P. A., 794, 795
Gardner, E. E., 692
Gate structures, 439
   junction field-effect (unipolar) transistor and, 420–22
   MESFET, 439, 441–42
   MOS, 439–40, 441
   MOSFET, 439–40
   self-aligned, 684–85
Gatos, H. C., 689
Gaussian distribution, 291
Gell-Mann, M., 761
Generation and recombination current, 377–81
Gentry, F. E., 420
Germanium (Ge)
   dielectric property of, 593
   drift mobility in, 300–301
   electrical property of, 222, 234, 243
   energy bands in, 115
   impurities in, 281–82, 359
Gershenzon, M., 748
Geusic, J. E., 754
Ghijsen, J., 625
Giaever, I., 641
Gibbons, G., 384, 480, 488, 654
Gibbons, J. F., 369, 370, 371, 419
Gibbs equation, 812
Gibbs free energy, 130
Gibson, A. F., 246, 247
Gilden, M., 659
Giles-Taylor, N. C., 624, 626
Ginzberg, A. S., 254
Glicksman, M., 246, 647
Gmelin, E., 707

Goetzberger, A., 432, 677, 697
Golio, J. M., 554
Gonda, S., 754
Goodhue, W. D., 641, 647
Goodman, A. M., 426
Gordon, J. P., 772
Goronkin, H., 686
Gossard, A. C., 539, 542, 546, 547, 549, 552, 615, 616, 617, 622, 623
Gossick, B. R., 354
Goudsmit, S., 18
Graded-drain structures, 692–93
Graded junction, 334
Grant, I., 689
Gray, P. E., 416
Grinolds, H. R., 693
Grobman, W. D., 615
Group Theory and Quantum Mechanics (Timken), 517
Group velocity, wave packet and, 154–57
Grove, A. S., 373, 429, 432, 434
Groves, W. G., 749
Gubanov, A. J., 516
Guggenheim, H. J., 754
Guldner, Y., 567, 568
Gummel, H. K., 661
Gunn, J. B., 462, 470, 503
Gunn-effect devices, 232, 462, 503–9, 647
Gunn oscillators, 273, 504, 649
Gunshor, R. L., 624, 625, 626, 627, 753
Gurney, R. W., 740, 743
Gutzwieler, F. W., 420

Hackbarth, E., 725
Hakki, B. W., 769
Hall, R. N., 277, 516, 772
Hall coefficient, 180, 825–26
Hall effect
   measurement of, 180–82, 209–14
   quantum, 535–39
Hall mobility, 212–13
Halperin, B. I., 255, 761
Hamiltonian, perturbation, 599

Hanamura, S., 722
Handbook of Mathematical Functions (Abramowitz and Stegun), 226, 531, 619
Hannay, N. B., 223, 226
Hänsch, W., 500–503, 836
Harmonic oscillator, 27
Harrison, W. A., 517
Hashizume, N., 728
Hauser, J. R., 554
Hawrylo, F. Z., 754, 759
Hayama, N., 729
Hayashi, I., 367, 776
Hayes, J. R., 553
Haynes, J. R., 298
   -Shockley experiment, 295–301
Heat capacity, electronic contribution to, 137–38
Hefetz, Y., 625
Heiblum, M., 552, 554
Heilmeier, G. H., 463
Heisenberg's uncertainty principle, 16–17, 51–52
Heitler and London method, 93, 94, 95
Hendel, R. H., 549, 552
Henderson, T., 529, 678
Henisch, H. K., 428
Henry, C. H., 524, 748
Henshall, G. D., 778
Heritage, J. P., 794
Herman, F., 238
Herman, M. A., 570
Hermann, G., 131
Hermitian operators, 599
Herring, C., 131
Hershberger, W. D., 744
Herskowitz, G. J., 517, 755
Hess, K., 549, 800
Heterostructure bipolar transistors, 723–29
Heterostructure devices, 462–63
Heterostructure Lasers (Casey and Panish), 358, 517, 519, 520, 521, 771
Heterostructures, electronic properties of

band-edge discontinuities, 526–30
(GaAl)As/GaAs system, 521–26
(GaIn)(AsP)/InP system, 521–26
hot and ballistic electrons in, 547–55
interface, 539–42
magnetoresistance and quantum Hall effect, 535–39
material requirements, 516–21
modulation-doped, 542–46
quantum wells and, 555–58
superlattices and, 558–60, 565–70
two-dimensional electron (hole) gas, 530–34
See also Optical devices
Hexagonal close-packed (hcp) structure, 34, 35, 36, 37
Hexagonal crystal system, 59, 60–61
Hickmott, T. W., 525, 529
Hida, H., 698, 703
High Field Transport in Semiconductors (Conwell), 225–26
Higuchi, T., 693
Hilsum, C., 470
Hines, M. F., 659
Hirachi, Y., 650
Hirao, T., 370
Hirayama, M., 685, 704
Hirose, T., 729
Hisatsugu, T., 677
Hiyamizu, S., 553, 645
Hobson, G. S., 509
Hoefflinger, B., 660
Hofker, W. K., 370
Hofstatter, E. A., 682, 714
Hogarth, C. A., 229
Hök, B., 626
Hojo, A., 525, 529
Holden, W. S., 799
Hole(s)
conduction, 176–80, 195
diffusion length of, 289
two-dimensional electron

(hole) gas, 530–34
Holmes, D. E., 686, 689
Holonyak, N. H., 420
Holonyak, N., Jr., 749, 750, 772, 800
Hong, W. P., 564
Honjo, K., 729
Hooke's law, 66
Hopfield, J. J., 749
Horio, K., 687
Hoshikawa, K., 685
Hot-electron effects in MOS-FET, 492–500
Hot Electron Microwave Generators (Carroll), 481, 654, 664, 675
Hot electrons in heterostructures, 547–55
Howarth, D. J., 220, 225
Howarth, L. E., 424
Howland, L. P., 740
Hsieh, J. J., 366, 517
Hsu, F. C., 493, 693
Hu, C., 493, 693
Hu, T., 678
Huang, T. Y., 715
Huff, H. R., 692
Hume-Rothery, W., 36, 162
Hund rule, 742
Hutchinson, A. L., 800
Hutson, A. R., 226, 228, 508
Huygens' wave theory, 50
Hwang, C. J., 758, 762, 763
Hwang, J. C. M., 546, 678
Hydrogen atom, electron motion in, 10–17
Hydrogen molecule, methods used in studying
Heitler and London, 93, 94, 95
molecular orbitals, 93–94
Schrödinger equation for a, 92
spin wave function for a, 94–95
Hyuga, F., 685

Ibana, H., 782
ideality factor, 414
Ihantola, H. K., 434

Ikola, R. J., 653
Ikoma, T., 687, 688, 689
Illegems, M., 367
Imamura, K., 553
Impact ionization process, 382–84, 419
and carrier multiplication, 474–81
ionization coefficient, 475, 800
multiplication factor, 477, 788
IMPATT diodes, 462, 647, 649–54
transit-time effects, 654–61
Implantation, 354, 369–71
Impurities, recombination via, 277–83
Impurity band, 253
Impurity effect on carrier lifetime, 319–22
Impurity scattering, 214–16
Impurity states, 229–32
diffusion coefficients in GE and SI, 364
distribution coefficients in GE and Si, 359
Inada, M., 729
InAs, properties of, 234, 521, 523
Inata, F., 645
Indirect-gap energy, 236, 238
Indirect-gap semiconductors, 233, 236
Indirect optical transition, 236, 238
Indirect radiative recombination processes, 276
Injection Currents in Solids (Lampert), 428
Injection lasers, 771–79
rate equations and, 779–82
Inner quantum number, 742
Inoue, K., 370
InP, properties of, 521, 523
Insulated-gate field-effect transistor (IGFET), 432
Insulators, energy bands in, 113–19
Integrated circuits, 449–53

Integrated Circuits (Warner and Fordemwalt), 451
Integrated Electronics (Lin), 451
Interband magnetooptic effect, 244–45
Interference phenomenon, 46–48
Int. J. Electron., 654
Interstitials, 62, 63
Intrinsic carrier concentration, 199
  values in Ge, Si, and GaAs, 341
Intrinsic recombination process, 275–77
Introduction to Quantum Mechanics (Pauling and Wilson), 13
Introduction to Semiconductor Theory (Anselm), 608, 610, 618, 825
Introduction to Solid-State Physics (Kittel), 64, 162, 220
Ionic contribution to dipole moment, 579
Ionic crystals
  bonding in, 80
  energy bands in, 116–17
  physical models for, 87–89
  structures of, 36–37
  transition probability and selection rules and, 600–602
Ionicity, 223–24
Ionic polarizability, 583–84
Ion implantation, 354, 369–71
Ionization coefficient, 475
Ionization process, impact, 382–84, 419
  and carrier multiplication, 474–81
Irvin, J. C., 232, 374
Ishibashi, T., 727
Ishida, S., 689
Ishiguro, H., 753
Ishikawa, H., 677
Ito, H., 782
Itoh, T., 698, 703, 726

Ivey, H. F., 747
Iwata, S., 444

Jackson, J. D., 577, 578, 588, 610
Jackson, T. N., 682
Jacob, G., 754
Jaggi, R., 503
James, H. M., 254
James, L. W., 519
Jaros, M., 624, 707
Jayaraman, A., 508
Jeppson, K. O., 497, 498
Johnson, A. M., 791
Johnson, E. O., 408
Johnson, L. F., 754
Johnson, P. D., 742
Johnson, R. L., 625
Johnson, V. A., 249
Johnson noise, 785
Johnston, R. L., 650, 654
Jones, H., 173
Jordan, A. S., 690
Josephson, B. D., 641
Joshi, N. V., 753
Joyce, W. B., 252
Juang, F. Y., 564
Junction breakdown, 381, 486–92
  avalanche breakdown, 383, 419, 478, 486, 487–88, 492
  Zener breakdown, 486–87, 488–92
Junction diodes, transient response of, 832–34
Junction field-effect (unipolar) transistors, 420–22
Jylha, O., 625

Kaiser, W., 602, 604
Kamejima, T., 689
Kaminska, M., 689
Kaminsky, G., 650
Kanbe, H., 798
Kane, E. O., 238, 255, 486, 561, 606, 643, 761
Kaneda, T., 788
Karlovsky, J., 647

Kasper, B. L., 795, 799
Kastalsky, A., 549, 552
Kaszynski, A., 704
Kato, N., 448, 686
Kawabata, T., 752
Kawasaki, Y., 686
Kazarinov, R., 539
Keever, M., 549
Keldysh, L. V., 618
Keller, S. P., 744
Kelley, J., 434
Kelm, J., 707
Kennedy, D. P., 384
Kern, D. P., 720
Ketchen, M. B., 725
Kettersen, A., 529
Keune, D. L., 750
Keyes, R. J., 772
Keyes, R. W., 247
Kikuchi, G., 744, 753
Kim, B., 677
Kim, O. K., 790, 795, 799
Kingsley, J. D., 772
Kingston, R. H., 349
Kino, G. S., 504, 508, 664
Kip, A. F., 184, 185, 562
Kirchner, P. D., 682
Kirkpatrick, C. G., 686, 726, 728
Kiss, Z. J., 602
Kittel, C., 64, 162, 184, 185, 220, 562
Kleiman, G. G., 707
Kleinman, D. A., 223
Kleinman, L., 238
Klem, J., 529
Klick, C. C., 740
Knoedler, C. M., 552, 554
Ko, P. K., 493, 693
Ko, W. H., 434
Kobayashi, K., 754
Kobayashi, M., 690, 691
Kobayashi, N., 444
Koeman, N. J., 370
Kohn, W., 200
Koike, S., 752, 753
Kolner, B. H., 621
Kolodziejski, L. A., 624, 625, 626, 753

Kopp, W. F., 549, 646, 678, 707
Koshikawa, K., 448
Kosugi, M., 721
Koval'chik, T. L., 232
Koyanagi, M., 715
Krag, W. E., 772
Kramers, H. A., 610
  -Kronig relation, 609–10
Kratzer, S., 706
Kressel, H., 520
Kriegler, R. J., 692
Kroemer, H., 508, 516, 567, 666, 727
Kroger, F. A., 232, 744
Kronig, R. L., 148, 610
  -Penney model, 147–51, 558–59, 567–68
Kruse, P. W., 790
Kuhn, M., 498
Kukimoto, H., 565
Kullendorff, N., 626
Kume, H., 444, 693
Kunstel, K., 707
Kunzel, H., 707
Kuper, A. B., 434

LaBate, E. E., 424
Lagowski, J., 689
Lagrangian multipliers, 806
Lagunova, T. S., 761
Laidig, W. D., 624, 626
Lambe, J., 744, 753
Lamond, P., 434
Lampert, M. A., 428, 687
Landau levels, 244, 537–38
Landsberg, P. T., 779
Lang, D. V., 565, 707
Langevin function, 585
Lark-Horowitz, K., 249
Laser Receivers (Ross), 786
Lasers, injection, 771–79
  Auger recombination in, 778
  carrier confinement in, 776
  delay time of, 782
  gain spectrum of, 770
  optical guiding in, 775
  population-inversion condition, 766

rate equations for, 779
relaxation-oscillation frequency, 782
threshold current density, 780
Lasher, G., 757, 772
Lattice(s)
  Bravais, 31–34
  Brillouin zones, 158–62
  definition of, 30
  heterostructures and super, 558–60, 565–70
  periodicity, 147–51
  reciprocal-lattice vectors and plane normal, 44–46
  recombination process via imperfections in, 277–83
  scattering, 216–21
  vibration, 65–71
Laue method of crystal diffraction, 55
Laughlin, R., 539
Lavirain, J. P., 368
Law of mass action, 127
Lawrence, H., 377
Lax, B., 245, 351, 772
Lax, M., 254, 255, 761
Le, H. Q., 647
LeBlanc, A., 709
Lee, C. A., 650
Lee, C. P., 687
Lee, D. H., 369, 650
Lee, H. P., 547, 678
Lee, J. W., 721
Lee, K., 698
Lee, S. J., 687
Lee, T. P., 781, 786, 796, 798, 799
LeFèvre, R. J. W., 586
Leheny, R. F., 440, 794, 797
Lehovec, K., 429, 445
Leifer, H. N., 744
Leighton, R. B., 215
Lepselter, M. P., 434
Levi, A. F. J., 553
Lewis, A. G., 715
Ley, L., 625
Li, G. P., 725
Liao, A. S. H., 440
Liechti, C. A., 679

Lien, W. H., 137
Lifetime
  carrier, 279–283, 755–60
  factors that affect the lifetime of excess carriers, 312–22
  spontaneous, 277, 604–5
Light-emitting diodes, junction luminescence and, 747–55
Lin, H. C., 451
Lin, P. J., 238
Linear electrooptic effect, 621
Linear Stark effect, 621
Line defects, 62, 64–65
Linh, N. T., 540
Liquid-phase epitaxy (LPE), 356–59
Littlejohn, M. A., 554
Littmark, U., 371
Liu, S. G., 654
Lo, Y. H., 547, 687, 689, 690
Logan, R. A., 565, 648, 707
Long, D., 517
Lorentz field, 591
Lorenz number for metals and semiconductors, 140
Loret, C., 704
Lorimor, O. G., 752
Ludeke, R., 567
Ludwig, G. W., 300
Lukes, Z., 434
Luminescence
  in large-gap phosphors, 740–45
  light-emitting diodes and junction, 747–55
  in sulfide (medium-gap) phosphors, 745–47
Lundstrom, M. S., 729
Luryi, S., 539, 549, 552
Lyddane, R. H., 608
  -Sachs-Teller relation, 607–9
Lynch, W. T., 789
Lyons, W. G., 707

McCoy, R. J., 790, 795
McCumber, D. E., 663, 668
MacDonald, D. K. C., 247
McGill, T. C., 570
McGlauchlin, L. D., 790

McIntyre, R. J., 788, 789
Macksey, H. M., 749
McMurtry, B. J., 786
McQuistan, R. B., 790
McWhorter, A. L., 509, 664, 772
Madelung constant, 88
Madihian, M., 729
Magnetic-dipole transition, 596
Magnetoresistance
    calculation of, 826–27
    measurement of, 246–47
    and quantum hall effect, 535–39
Mahan, G. D., 255, 763
Maiman, T. H., 772
Maita, J. P., 218
Manasevit, H. M., 367
Manning, J., 779
Marcos, H. M., 754
Marcuse, D., 781
Mark, P., 687
Mars, D., 687
Martin, R. A., 715
Masers, 772
Mashita, M., 525, 529
Masking and photoetching procedures, 371–73
Maslakovets, Y. P., 232
Masselink, W. T., 677, 680
Masuhara, T., 722
Materials Used in Semiconductor Devices (Hogarth), 229
Matsui, J., 689
Matsumoto, H., 788
Matsumoto, Y., 689, 693
Matsuoka, Y., 686
Matter (corpuscle) waves, 48–52
Matyi, R. J., 721
Maxwell-Boltzmann distribution function, 127, 128, 129
Maxwell's equations, 48–49
Maxwell velocity distribution, 807
Mayer, J. W., 369
Maziar, C. M., 729
Mead, C. A., 426, 448, 452
Mean free path, 136
Mean free time, 136

electronic conductivity and, 133–37
mean free path and, 136
Measurements
    carrier concentration and mobility, 209–14
    Hall, 180–82, 209–14
    magnetoresistance and piezoresistance, 246–47
Meiboom, S., 825
Melamed, N. T., 746
Melchoir, H., 788–89
Menigaux, L., 621
Menk, G., 753
Merz, J. L., 567
Metal-insulator-semiconductor (MIS) structure, 428–32
Metallic crystals
    bonding in, 84
    cohesive energy in, 89–92
    physical models for, 89–92
Metalorganic chemical-vapor deposition (MOCVD), 367–69
Metal-oxide-semiconductor (MOS), 428
Metal-oxide-semiconductor field-effect transistor (MOSFET)
    channel mobility, 717–19
    comparison of design parameters for submicrometer, 713–14
    cryogenic operation, 722
    degradation, 496, 695
    device parameters, 434–39
    graded drain structure, 692
    high-field effects, 714–17
    hot-electron effects in, 492–500
    power-gain cutoff frequency, 679–80
    scaling rules for miniaturization, 708–10
    short- and narrow-channel effects on threshold voltage, 710–12
    structure, 432–34
    three-dimensional architecture, 723

Metals
    energy bands in, 113–19
    integrals for the evaluation of conductive properties of, 813–15
    Lorenz number for, 140
    thermal data for, 140
    values of Fermi energy for, 130
    work functions of, 131
Metal-semiconductor barrier/contact
    barrier height, 422–23
    current conduction in, 426–28
Metal-semiconductor field-effect transistors (MESFET)
    gates, 439, 441–42
    model for device simulation, 704
    power-gain cutoff frequency, 679–80
    short channel effect, 697
    structure, 444–48, 695–97
Method of image, 295
Method of molecular orbital, 93
Method of superposition, 294, 411
Meyerhofer, D., 254
Microwave Semiconductor Devices and Their Circuit Application (Watson), 676
Microwave Solid-State Engineering (Nergaard and Glicksman), 647
Microwave transistors, 406–8
Miller, D. A. B., 615, 616, 617, 622, 623
Miller, D. L., 726, 728
Miller, J., 687
Miller, R. C., 783
Miller, S. L., 383, 478, 488
Miller indices of a plane, 41–43
Mimura, T., 677, 691, 697, 721
Minomura, S., 565
Minority-carrier injection, 270, 338–40
Misawa, S., 754

Misawa, T., 659, 660
Miura-Mittausch, M., 500–503, 836
Miyazaki, H., 444
Miyazawa, S., 685, 689
Mizuno, T., 693
Mizuta, M., 565
Mobility
  channel, 717–19
  drift, 114, 213, 300–301
  Hall, 212–13
  See also carrier mobility
Mobility effective masses, density-of-state and, 242–43
Mobility measurements, carrier concentration and, 209–14
Model of associated donor-acceptor centers, 747
Modern Physics (Sproull), 1
Modulation-doped field-effect transistor (MODFET)
  analysis of short-channel, 698–702
  depletion mode, 696
  DX center, 707–8
  enhancement mode, 696
  inverted interface and, 707–8
  model for device simulation, 702–4
  power-gain cutoff frequency, 679–80
  structure of, 695–97
  velocity overshot, 704
  velocity saturation, 697
Modulation-doped heterostructure, 542–46
Modulation Spectroscopy (Cardona), 615
Molecular-beam epitaxy (MBE), 358, 366–67
Molecular crystals, 84–87
Molecular orbital, 82, 93–94
Molecular structure of materials, polarization and, 586–87
Moll, J. L., 412, 434, 453, 481
  Ebers-Moll model, 411–13
Monoclinic crystal system, 58, 59, 61
Moon, R. L., 519

Morgan, T. N., 748
Mori, N., 444
Morin, F. J., 218
Moriwaki, M. M., 753
Morkoç, H., 525, 529, 549, 646, 677, 678, 680, 698, 707
Moss, T. S., 236, 609
Mott, N. F., 173, 581, 599, 740, 743, 756, 762
Mott barrier, 428
Moudy, L. A., 368
Multiple quantum wells (MQW), 556
Murase, K., 448
Murata, S., 782
Muto, S., 553, 645
Mylroie, S., 370

Nagao, S., 753
Nahory, R. E., 516, 794, 797
Nakagami, T., 650
Nakagawa, A., 729
Nakagome, H., 798
Nakahara, J., 625
Nakamura, T., 677
Nakanishi, H., 448
Nakanisi, T., 525, 529
Nanishi, Y., 689
Napoli, L. S., 653
Nasledov, D.N., 761
Nathan, M. I., 552, 554, 772
Nawrocki, M., 753
Negative differential conductance, 641–49
Nergaard, L. S., 647
Neustadter, S. F., 351
Neutrons, 1
  diffraction experiment and, 51, 52
Newman, R., 236
Newton's corpuscle theory, 50
Ng, H. Y., 720
Nichols, M. H., 131
Ning, T. H., 492, 493, 724, 725
Nishi, H., 553
Nishimoto, Y., 564
Nishitani, Y., 754

Nishiuchi, K., 677, 690
Nondegenerate semiconductors, 197
Nottenburg, R. N., 729
Noyce, R. N., 379
Nozières, P., 591, 594
Nucleus of an atom, 1
Nuese, C., 516
Nurmikko, A. V., 625

O'Brien, R. R., 384
Octahedral packing, 37–38
Odani, K., 447
Ogawa, K., 798
Ogura, S., 693
Ohata, K., 698, 703
Ohmic contact, 442–44
Ohm's law, 114
Ohnishi, T., 447
Ohwada, K., 677
Okada, T., 368
Okamoto, A., 729
Olsen, G. H., 516, 754, 759
Omura, Y., 677
One-valley semiconductors, transport equations for, 500–503, 835–39
Onodera, H., 447
Onsager, L., 594
Onton, A., 523
Oosthoek, D. P., 370
Optical devices, photon generation and detection processes in, 739
  effects of heavy dopant and carrier concentrations, 760–66
  injection lasers, 771–79
  junction luminescence and light-emitting diodes, 747–55
  luminescence in large-gap phosphors, 740–45
  luminescence in sulfide (medium-gap) phosphors, 745–47
  photoconductive devices, 791–95
  photodetectors, 782–91

Optical devices (cont.)
  phototransistors and ava-
    lanche photodiodes, 798–
    800
  p-i-n and Schottky photo-
    diodes, 795–98
  population inversion and gain
    spectrum, 766–71
  rate equations, 779–82
  spontaneous emission and
    carrier lifetime, 755–60
  stimulated emission, 748
Optically biaxial crystals, 61
Optically uniaxial crystals, 61
Optical-phonon branch, 70,
  466, 469–70
Optical properties of GaAs,
  610–17
Optical Properties of Semiconductors (Moss), 609
Optical transitions in solids,
  594–97
  absorption coefficient, 603–7
  Franz-Keldysh effect, 618–20
  oscillator strength, 582, 604–5
  selection rules, 601, 602, 742
  spontaneous lifetime, 277,
    604–5
  spontaneous and stimulated
    processes, 597
  transition probability, 598
Orientational polarization, 579,
  584–86
Orthorhombic crystal system,
  58, 59, 60, 61
Osaka, J., 685
Osborn, J. A., 589
Osburn, C. M., 492, 493, 709,
  713
Oscillator strength, 582, 604–5
Oshlansky, R., 779
Ostermeyer, F. W., 754
Ota, T., 754
Otsuka, N., 625, 753
Oval defects, 691
Oxide Coated Cathode, The
  (Hermann and Wagener),
  131
Oxide degradation, 693–95
Ozawa, O., 693

Paige, E. G. S., 619
Pan, C. L., 690
Pang, Y. M., 794
Panish, M. B., 358, 368, 517,
  519, 520, 521, 729, 776
Paoli, T. L., 769
Papkoff, M., 434
Parajape, B. V., 508
Parametric interaction, varactor
  diodes and, 636–40
  Manley-Rowe relation, 730
Parker, C. D., 641
Parsey, J. M., 689
Paul, W., 247
Pauli exclusion principle, 18–
  19, 95, 116
  Fermi-Dirac distribution
    function and, 128–29
Pauling, L., 13, 223, 224
Pearson, G. L., 508, 825
Peck, D. D., 641
Peltier effect, 247
Penney, W. G., 148
Pepper, M., 535, 537
Periodic boundary condition,
  163
Periodic table of elements, 3,
  4, 19
Perkins, W. H., 678
Persistent photoconductivity,
  707
Perturbation calculation, time-
  dependent, 840–43
Perturbation Hamiltonian, 599
Pessa, M., 570, 625
Pettit, G. D., 682
Phase velocity, 156, 157
Philipp, H. R., 591, 614
Phillips, J. C., 223, 224, 246,
  254
Phillips, N. E., 137
Phonons, 465
  acoustic-phonon branch,
    466–69, 534
  optical-phonon branch, 466,
    469–70
  tunneling and, 648
Phosphors
  luminescence in large-gap,
    740–45

luminescence in sulfide (me-
  dium-gap), 745–47
Photoconductance, 784–85
Photoconductive devices,
  791–95
Photodetectors, 782–91
Photodiodes, 785–86
  avalanche, 798–800
  p-i-n and Schottky, 795–98
Photoetching procedures,
  371–73
Photons. See Optical devices,
  photon generation and de-
  tection processes in
Phototransistors, 798–800
Physical electronics, 771
Physics and Technology of
  Semiconductor Devices
  (Grove), 373, 429
Physics of Semiconductor De-
  vices (Sze), 429
Physics of Semiconductor Laser
  Devices (Thompson), 610
Physics of Semiconductors
  (Galazka and Wilson), 753
Physics of Semiconductors
  (Moll), 481
Picus, G., 222
Piezoelectric scattering, 228–29
Piezoresistance measurements,
  246–47
Pigg, J. C., 320
Pinczuk, W., 547
Pine, D., 591, 594
p-i-n photodiodes, 795–98
Planar structures, fabrication
  techniques for, 371–75
Planck law of blackbody radia-
  tion, 598
Planck's constant, 6
Plane defects, 62
Plasma frequency, 137
Ploog, K., 707
p-n junction
  breakdown, 381–84
  built-in or diffusion potential
    in, 269
  carrier storage and diffusion
    capacitance, 343–46
  diodes, 330–33

generation and recombination current, 377–81
graded junction, 334
ideal diode characteristics, 340–43
minority-carrier injection, 338–40
potential barrier in, 330–33
quasi-Fermi levels and, 338–40
reverse saturation current, 333
space-charge region and junction capacitance, 333–38
step/abrupt junction, 334–38
switching response and recovery time, 346–54
transition-region capacitance of diffused diodes, 375–77
p-n junction formation
alloying, 354, 355–56
chloride vapor-phase epitaxy, 365
diffusion, 354, 359–65
epitaxy, 354, 356–59, 365–67
implantation, 354, 369–71
liquid-phase epitaxy, 356–59
metalorganic chemical-vapor deposition, 367–69
molecular-beam epitaxy, 358, 366–67
p-n-p-n structure, 416–20
Pockels effect, 619–21
Point defects, 61–62
Poisson, M. A., 368
Poisson's equation, 272
Polari, M. R., 720
Polarizability/polarization
definition of, 85, 579
depolarization field, 587–90
dielectric constant and, 591–94
dipole moment and, 577–79
electronic, 580–82
ionic, 583–84
Lorentz field, 591
orientational, 579, 584–86
Polar Molecules (Debye), 209, 585

Polar scattering, 224–28
Pollack, M. A., 516, 794, 797
Pollak, F. H., 246
Population inversion and gain spectrum, 766–71
Portnoi, E. L., 776
Potential well, electrons in a, 121–22
two-dimensional electron (hole) gas, 530–34
Powazinik, W., 779
Powder method of crystal diffraction, 54–55
Powell, J. L., 482, 599
Power-delay product, 681–83
Power-gain cutoff frequency, 679–80
Power transistors, 403–6
Prager, H. J., 650
Primary bonds, 79
Primitive cell, 31
Primitive lattice vectors, 32
Prince, M. B., 300
Principles of Modern Physics (Leighton), 215
Principles of the Theory of Solids (Ziman), 176, 185
Principles of Tunnel Diode Circuits (Chow), 647
Prokhorov, A. M., 772
Putley, E. H., 229

Qua, G. J., 799
Quantization, Wilson-Sommerfeld rule of, 6–7
Quantum electronics, 772
Quantum mechanics, 7–8, 15–17, 51
Quantum Mechanics (Powell and Crasemann), 482, 599
Quantum Mechanics and Its Applications (Mott and Sneddon), 599
Quantum numbers, 14, 18, 741
Quantum wells
heterostructures and, 555–58
Stark effect and, 621–24
Quist, T. M., 772

Radiation, absorption and emission of, 597–98
Ramo, S., 588, 596
Rate equations, 779–82
Ray, S., 753
Razeghi, M., 368
Read, W. T., 277, 462, 649–50
Read diode, 462
Reciprocal-lattice vectors and plane normal, 44–46, 159
Recombination centers, 274–75
Recombination current, generation and, 377–81
Recombination processes, 274
extrinsic, 277–83
intrinsic, 275–77
Rectifying Semiconductor Contacts (Henisch), 428
Rediker, R. H., 772
Reduced zone scheme, 165
Reed, L. J., 689
Rees, H. D., 619
Reflection experiment, 245–46
Resistivity of different materials, 108, 109
Resonant-tunneling barrier structure, 641–49
Restle, P. J., 720
Restrahlen absorption, 611
Reverse saturation current, 333, 381
Reynolds, F. W., 136
Reynolds, W. C., 812
Rich, T. C., 754
Richton, S., 720
Rideout, V. L., 709, 726
Ridley, B. K., 470, 663–64, 666
Riesman, A., 709, 713
Riesz, R. P., 786
Rimbey, P. R., 763
Robbins, V. M., 478, 786, 789, 800
Robson, P. N., 664
Ross, M., 786
Rotating-crystal method of crystal diffraction, 55
Rotational symmetry operations in a crystal, 57–61

Ruby laser, 772
Ruch, J. G., 504, 508, 509
Rumyantsev, V. D., 519
Russell-Saunders coupling, 742
Rutherford, E., 215
Rutherford atom, 1–5
Rutherford scattering law, 3, 5,
    215

Sachs, R. G., 608
Sah, C. T., 379, 434
Sai-Halasz, G. A., 567, 720,
    721
Sakai, K., 519
Sakaki, H., 567
Sakamoto, M., 368
Sakamoto, T., 624
Saksena, B. D., 224
Sakurai, T., 754
Salmer, G., 704
Sangiorgi, E., 682, 714
Sans, E., 677
Sarace, J. C., 426
Saturation intensity, 616
Saturation velocity, 465, 677,
    699
Sawa, K., 753
Sawada, S., 693
Saxena, A. K., 707
Scaling, 720
    rules for MOSFET miniatur-
      ization, 708–10
Scattering
    acoustic-phonon, 466–69,
      534
    Coulomb, 534
    impurity, 214–16
    inter- and intravalley pro-
      cesses, 219, 227
    lattice, 216–21
    optical-phonon, 466, 469–70
    polar and piezoelectric,
      224–29
    Rutherford scattering law, 3,
      5, 215
    Si-SiO$_2$ interface-roughness,
      534
    surface, 434, 534
Scharfetter, D. L., 653, 654,
    661

Schawlow, A. L., 772
Schetzina, J. F., 624, 626
Schmid, H., 720
Schottky, W., 428
Schottky defect, 62, 63
Schottky photodiodes, 795–98
Schrieffer, J. R., 434, 718
Schrödinger wave equation, 7–10
    for a hydrogen molecule, 92
Schulman, J. H., 740
Schulman, J. N., 567, 570
Schuster, S. E., 492
Schwarzchild, B., 539
Screening, 761
Searle, C. L., 416
Seebeck effect, 247–50
Seeger, K., 548
Seelbach, W. C., 686
Seitz, F., 200, 246, 247, 740,
    826
Selection rule, 742
Self-aligned gate, 684–85
Sell, D. D., 612, 613
Semiconductor controlled recti-
    fier, 416
*Semiconductor Controlled Rec-
    tifiers* (Gentry), 420
Semiconductor-insulator-semi-
    conductor (SIS) structures,
    band-edge discontinuities
    in, 526–30
*Semiconductor Lasers and Het-
    erojunction LEDs* (Kressel
    and Butler), 520
Semiconductor Physics
    (Seeger), 548
Semiconductors
    band tailing, 761
    band-gap shrinkage, 760
    carrier concentration and mo-
      bility measurements,
      209–14
    carrier concentrations and
      fermi level according to
      temperature regions,
      204–9
    chemical bond and, 221–24
    common properties of, 221,
      222
    compensated, 208–9

Debye screening length, 763
degenerate, 197, 250–55,
    641
diluted magnetic, 624–27
direct-gap, 233, 755–60
direct-gap energy, 236, 238
donor and acceptor states,
    199–204
electronegativity, 223–24
electron transfer in two-val-
    ley, 503–9
energy bands in, 118–19,
    232–38
experimental studies of en-
    ergy-band structure and ef-
    fective masses, 238–50
Fermi level and, 196–99,
    204–9, 250–55
growth and propagation of
    carrier waves in two-val-
    ley, 661–64
-heterostructure interface,
    539–42
impurity scattering, 214–16
impurity states, 229–32
indirect-gap, 233, 236
indirect-gap energy, 236, 238
integrals for the evaluation of
    conductive properties of,
    813–15
intrinsic and extrinsic,
    193–96
intrinsic carrier concentration,
    199
ionicity of compound semi-
    conductors, 223–24
lattice scattering, 216–21
Lorenz number for, 140
materials, 221–32
metal-semiconductor barrier
    and current conduction,
    426–28
metal-semiconductor contact
    and barrier height, 422–23
metal-semiconductor field-
    effect transistors, 439,
    441–42, 444–48
nondegenerate, 197
polar and piezoelectric scat-
    tering, 224–29

surface states, 301–12
thermal data for, 140
transition probability and selection rules and, 602–3
transport equations for one-valley, 500–503, 835–39
work functions of, 131
Semiconductors (Hannay), 223, 226
Semiconductors (Smith), 535, 825
Semiconductors and Semimetals, 612, 615, 761
Semiconductor Silicon/1981, 692
Semiconductor Statistics (Blakemore), 252
Semimetals, energy bands in, 119
Senechal, R. R., 426
Separate absorption and multiplication (SAM) avalanche photodiode, 798
Separate-absorption-graded-multiplication (SAGM) avalanche photodiode, 799
Seraphin, B. O., 246, 615
Shah, J., 547
Shaklee, K. L., 246
Shallow impurities, 204–9
Shallow states, 230
Sham, L. J., 557
Shaver, D. C., 677
Shay, J. S., 516
Shepard, J. F., 693
Shepherd, W. H., 373
Shewchuk, T. J., 646
Shewchun, J., 516
Shibatoni, A., 690
Shibayama, H., 677
Shibuya, M., 825
Shichijo, H., 721
Shih, H. D., 678
Shinado, M., 616
Shinozaki, S., 693
Shirayama, M., 448
Shockley, W., 157, 213, 216, 277, 379, 390, 465, 469, 649, 756

Haynes-Shockley experiment, 295–301
Shortley, G. H., 13, 742
Shot noise, 785
Shulz, M., 707
Shur, M. S., 698
Siedel, T. E., 653
Siegbahn, K., 772
Sigmon, T. R., 685
Silicon (Si)
dielectric property of, 593
drift mobility in, 300–301
electric property of, 222, 234, 243
energy bands in, 115
GaAs versus, 720–22
impurities in, 281–82, 359
intrinsic carrier concentration, 341
intrinsic defects in, 691–92
ionization energy of donors and acceptors, 199
-SiO$_2$ interface-roughness scattering, 534
Simpson, W. I., 368
Sinnott, M. J., 162
Skolnick, M. S., 689
Slayman, C. W., 793
Slip, 64
Slobodskoy, A., 429
Slow surface states, 301–9
Smith, P. M., 678
Smith, P. R., 791, 793, 794
Smith, P. W., 615, 616, 617
Smith, R. A., 535, 825
Smith, R. G., 786, 799
Smith, R. K., 682, 714
Smits, F. M., 374
Smyth, C. P., 586
Sneddon, I. N., 581, 599, 756
Snow, E. H., 434
Snowden, C. M., 704
Sodium, energy bands in, 115, 116
Sodium chloride (NaCl) structure, 36, 37
Solids
energy bands in, 108–13
optical transitions in, 594–97
Solid State Devices (Goetzberger and Zerbst), 677, 697
Solid State Electronics (Wang), 207, 517, 535
Solid State for Engineers, The, (Sinnott), 162
Solid-State Physics (Dekker), 64
Solid State Theory (Harrison), 517
Sollner, T. C. L. G., 641, 647
Solomon, P. M., 525, 529
Soltys, T. J., 772
Sommers, H. S., Jr., 254
Sondheimer, E. H., 220, 225
Space-charge region and junction capacitance, 333–38
Sparks, M., 390
Spectroscopic notation, 741
Spenke, E., 157
Spin, 18
Spitzer, W. G., 223, 424, 426
Spontaneous emission process, 597
in band-to-band transitions and direct-gap semiconductors, 755–60
Spontaneous lifetime, 277, 604–5
Sproull, R. L., 1
sp^3 hybrid orbitals, 100, 102
Stark effect, 621–24
Staymake, N. A., 654
Stegun, I. A., 226, 531, 619
Step junction, 334–38
Stern, F., 531, 541, 543, 546, 613, 757, 761, 769
Stern-Gerlach experiment, 17
Sterzer, F., 647
Stiévenard, D., 690
Stillman, G. E., 478, 542, 563, 612, 786, 789, 800
Stimulated emission process, 597
Stirland, D. J., 689
Stiwell, G. R., 136
Stoner, E. C., 589
Stork, J. M. C., 725
Störmer, H. L., 539, 542, 546, 547
Straggling, 369

Stratton, R., 469, 470, 508
Strazalkowski, I., 523
Streetman, B. G., 549
Sturge, M. D., 612
Su, C. B., 779
Su, S. L., 707
Subbanna, S., 567
Subthreshold current, 439
Successive approximation, 19
Sugano, S., 616
Sugii, T., 726
Sugimura, A., 779
Suhl, H., 825
Sulfide (medium-gap) phos-
    phors, luminescene in,
    745–47
Sumski, S., 368, 776
Superlattices, heterostructures
    and, 558–60, 565–70
Surface accumulation and inver-
    sion, 303
Surface-recombination process,
    excess carrier lifetime and,
    312–19
Surface-recombination velocity,
    fast states and, 309–12
Surface scattering, 434, 534
Surface states
    fast states and surface-recom-
    bination velocity, 309–12
    slow states and field-effect
    experiments, 301–9
Susa, N., 798
Sustaining voltage, 673
Suyama, S., 677
Suzuki, M., 448, 686
Svensson, C. M., 497, 498
Swain, S., 509
Swan, C. W., 653
Switching delay, 677
Switching transistors, 408–11
Symmetry operations in a crys-
    tal, 57–61
Sze, S. M., 232, 383, 384,
    426, 428, 429, 434, 453,
    480, 488, 493
Szigeti, B., 221–22

Tabatabaie, N., 478, 786, 789,
    800

Tachikawa, M., 565
Taft, E. A., 591
Takahashi, K., 754
Takanashi, H., 788
Takayanagi, S., 370
Takeda, E., 693
Takeishi, Y., 692
Talalakin, G. N., 761
Tam, S. C., 493, 693
Tan, T. Y., 692
Tang, D. D., 724, 725
Tang, Y. Y., 800
Taniguchi, K., 693
Taniguchi, M., 625, 688–89
Tannenwald, P. E., 641
Tao, R. Y., 753
Taylor, B. C., 509
Teal, G. K., 390
Tell, B., 440
Teller, E., 608
Temkin, H., 729
Tensor, symmetry operations in
    material-parameter, 60–61
Terman, L. M., 429
Terrill, K. W., 493, 693
Tetragonal crystal system, 58,
    59, 60–61
Tetrahedral packing, 37–38
Theory of Atomic Spectra, The
    (Condon and Shortley), 13,
    742
Theory of Electrical and Mag-
    netic Susceptibilities, The
    (Van Vleck), 742
Theory of Metals (A. H.
    Wilson), 152, 162, 173
Theory of Metals and Alloys
    (Mott and Jones), 173
Thermal conductivity, electronic
    contribution to, 138–40
Thermal data for metals and
    semiconductors, 140
Thermal equilibrium condition,
    266–69
Thermal gap energy, 210
Thermionic emission, 131–33
Thermodynamics, review of
    statistical, 809–12
Thermodynamics (Reynolds),
    812

Thermoelectric effects, 213–14,
    247–50
Thermoelectricity: An Introduc-
    tion to the Principles
    (MacDonald), 247
Thermoluminescence, 742
Theuerer, H. C., 373
Thim, H. W., 664
Thomas, D. C., 552, 554
Thomas, D. E., 648
Thomas, D. G., 238, 749
Thompson, G. H. B., 610, 778
Thorne, R. E., 707
Three-dimensional architecture,
    723
Threshold voltage, 435
    short- and narrow-channel ef-
    fects on, 710–12
    variation, 685–86
Thyristor, 416
Tice, W. K., 692
Tiemann, J. J., 255, 648
Tight-binding approximation,
    171–76
Time-dependent
    diffusion equations, 283–95,
    828–31
    perturbation calculation,
    840–43
Timken, M., 517
Timura, T., 690
Togashi, M., 448, 685–86
Tomizawa, K., 728
Townes, C. H., 772
Toyabe, T., 693
Toyama, Y., 650
Tran, L. T., 721
Transferred-electron devices,
    672–76
    dipole domain, 664, 668–72
    equal area rule, 668
    Gunn oscillators, 273, 504,
    649
    LSA and other modes of op-
    eration, 673
Transferred Electron Devices
    (Bulman), 509
Transistors
    bipolar, 390–97, 723–29

charge-control model, 413–16
comparison of p-n-p and n-p-n, 390–97
diffused, 397–403
Ebers-Moll model, 411–13
field-effect, 676–83
heterostructure bipolar, 723–29
junction field-effect (unipolar), 420–22
large-signal analytical models, 411–16
metal-insulator-semiconductor structure, 428–32
metal-oxide-semiconductor, 428
metal-oxide-semiconductor field-effect, 432–39
microwave, 406–8
modulation-doped field-effect, 679–80, 695–702
p-n-p-n structure, 416–20
point-contact, 390
power, 403–6
switching, 408–11
Transition, spontaneous emission and carrier lifetime in band-to-band, 755–60
Transition probability
and selection rules, 598–603
and time-dependent perturbation calculation, 840–43
Transitions, atomic, 596
forbidden, 601
Transit-time effect, 463–64
IMPATT, 654–61
Transport efficiency, 392
Transport equations for one-valley semiconductors, 500–503, 835–39
TRAPATT diodes, 649–54
Trew, R. W., 554
Triclinic crystal system, 58, 59, 61
Trigonal crystal system, 59, 60–61
Trukan, M. K., 776
Trumbore, F. A., 360
Tsang, P. J., 693

Tsang, W. T., 368, 615–16, 777, 800
Tserng, H. Q., 677
Tsu, R., 559, 641
Tsui, D. C., 539, 546
Tsukada, T., 777
Tunneling, 382, 442, 481–86
Esaki diodes, 641–9
phonons and, 648
probability, 484, 646
resonant-tunneling barrier structure, 641–49
WKB approximation, 483
Zener breakdown, 486–87, 488–92
Turnbull, D., 200, 246, 247, 740
Turner, A. E., 626
Two-dimensional electron (hole) gas, 530–34
quantum Hall effect, 535–39
Twose, W. D., 762
Two-valley semiconductors
growth and propagation of carrier waves in, 661–64
transfer of electrons in, 503–9

Uenohara, M., 640, 675
Uhlenbeck, G. E., 18
Uncertainty principle, 16–17, 51–52
Unipolar transistors, 420–22
Unit cells, 31

Vacancies, 62–63
Vakhshoori, D., 547, 554, 646
Valence band, 114
Van der Waals
bond, 84–87
energy, 78
interaction, 86
Van der Ziel, J. P., 754
Van Duzer, T., 588, 596
Van Gool, W., 747
Van Roosbroeck, W., 277, 756
Van Uitert, L. G., 754
Van Vleck, J. H., 742
Van Zastrow, E. E., 420

Varactor diodes and parametric interaction, 636–40
Vassel, M. O., 505
V center, 63
Velocity
drift, 464–70
overshoot, 704–7
saturation by transport equations, 500–503
Velocity of a wave packet
group, 154–57
phase 156, 157
Venkatasubramanian, R., 753
Vibration, lattice, 65–71
Vink, H. J., 232
Vinter, B., 543
Von Bardeleban, H. J., 690
Von Klitzing, K., 535, 537, 539
Von Neida, A. R., 448, 690
Voos, M., 567, 568
Voulgaris, N. C., 417
Vrehen, Q. H. F., 562

Wada, K., 685, 689
Wada, O., 754
Wagener, S., 131
Walker, E. J., 709, 713
Walker, W. W., 693
Walukiewicz, W., 689
Wang, F. C., 448
Wang, H. W., Jr., 524
Wang, S., 207, 517, 535, 547, 554, 621, 646, 687, 689, 690, 777
Wang, S. Y., 621, 687, 689, 690, 796
Ware, R. M., 689
Warner, R. M., 377
Warner, R. M., Jr., 451
Watanabe, H., 689
Watanabe, K., 448, 685
Watanabe, M. O., 525, 529
Watkins, T. B., 470
Watson, H. A., 676
Watters, R. L., 300
Wave(s)
Bloch, 154
electromagnetic and matter, 48–52

Wave functions
  normalized, 15
  orthogonal, 15
Wave Mechanics and Its Applications (Mott and
  Sneddon), 581, 756
Wave packet, 52
  group velocity and, 154–57
  motion and effective mass of,
  157–58
Wayna, S., 678
Webb, P. P., 788
Weber, J. P., 621
Wecht, K. W., 612, 613
Wei, L. Y., 516
Weidemann-Franz law, 140
Weimer, P. K., 452
Weiner, J. S., 622, 623
Weisbrod, J., 650
Welber, B., 748
Welch, B. M., 687
Well
  electrons in a potential,
  121–22
  quantum wells and hetero-
  structures, 555–58
  quantum wells and Stark ef-
  fect, 621–24
Weller, K. P., 650, 678
Wertheim, G. K., 318, 320,
  321
Whinnery, J. R., 588, 596
Wiegmann, W., 524, 542, 546,
  547, 553, 616, 617, 622,
  623, 650
Willardson, R. K., 612, 615,
  761
Williams, F. E., 742, 747
Williams, G. F., 800
Wilson, A. H., 152, 162, 173
Wilson, E. B., 13

Wilson, L. H., 753
Wilson-Sommerfeld rule of
  quantization, 6–7
Winter, R. J., 797
Witwer, N. C., 783
WKB (Wentzel-Kramers-
  Brillouin) approximation,
  483
Wolfe, C. M., 542, 563, 612
Wolff, P. A., 762
Wolfstirn, K., 761
Wood, D. L., 602, 604
Wood, T. H., 622, 623
Woodall, J. M., 682, 726
Woods, M. H., 720
Wordeman, M. R., 709, 713,
  720
Work function(s), 121–22, 783
  of metals and semiconduc-
  tors, 131
Wright, G. T., 428
Wright, S. L., 525, 526

Yaegashi, Y., 370
Yaeger, J. C., 290
Yagi, K., 444
Yamabe, K., 693
Yamada, Y., 448
Yamakoshi, S., 754
Yamamoto, N., 444
Yamamoto, T., 519
Yamanaka, H., 753
Yamanishi, M., 626, 753
Yamaoka, T., 788
Yamasaki, K., 704
Yamauchi, Y., 727
Yamazaki, T., 726
Yanai, H., 687
Yang, E. S., 417
Yang, J. J., 367, 689
Yangihara, M., 729

Yano, K., 722
Ying, R. S., 650
Yokoyama, H., 782
Yokoyama, N., 447, 553, 677,
  697
Yon, E., 434
Yoshida, J., 525, 529
Yoshida, S., 754
Young, T., 50–51
Young, L., 504, 640, 647
Young, M. S. S., 448
Yu, H. N., 492, 493, 709, 713
Yu, P. W., 686

Zamerowski, T. J., 516
Zawadski, W., 566
Zeiger, H. J., 772
Zener, C., 487
Zener breakdown, 486–87,
  488–92
Zerbst, M., 677, 697
Zheng, X. Q., 707
Zhou, B. L., 707
Zhu, H. Z., 690
Zhu, Z. H., 621, 688, 689, 690
Zicherman, D. S., 720
Ziegler, J. F., 371
Ziman, J. M., 176, 182, 185,
  220, 226
Zisman, W. A., 131
Zone purification, 359
Zone theory
  density of states and, 162–65
  energy discontinuity and,
  165–67
  reduced zone scheme, 165
Zoroofchi, J., 613
Zuber, J. R., 790, 795
Zuleeg, R., 445
Zurcher, P., 524
Zwerdling, S., 245

SYMBOLS

Script and Greek (scalar quantities)

$\mathscr{E}$	energy
$\mathscr{E}_f$	Fermi level
$\mathscr{E}_g$	gap energy
$\mathscr{F}$	quasi-Fermi level
$\mathscr{H}$	Hamiltonian
$\mathscr{E}_{fi}$	intrinsic Fermi level
$\mathscr{E}_w$	work function
α (alpha)	crystal angle; spin-up state; impact-ionization coefficient; current-amplification factor; polarizability; absorption coefficient; loss coefficient
β (beta)	crystal angle; spin-down state; phase angle
γ (gamma)	crystal angle; force constant; emitter efficiency
δ (delta)	skin depth; strain; dilation (Δ); split-off band energy (Δ)
ϵ (epsilon)	permittivity or dielectric constant
ζ (zeta)	wavefunction
θ (theta)	coordinate angle; phase angle; equivalent temperature; wave function (Θ)
κ (kappa)	scaling factor; imaginary part of refractive index
λ (lambda)	wavelength; scaling factor; split-off band energy
μ (mu)	mobility; permeability
ν (nu)	frequency
ξ (xi)	Fermi energy relative to band edge
ρ (rho)	mass density; charge density; resistivity
σ (sigma)	conductivity; scattering or capture cross section
τ (tau)	relaxation or mean free time; carrier lifetime
ϕ (phi)	coordinate angle; surface potential (ϕ_s); barrier potential (ϕ_B); built-in potential (ϕ_{bi}); surface-inversion potential (ϕ_{SI}); wavefunction (Φ)
χ (chi)	electron affinity; electrical susceptibility
ψ (psi)	wavefunction; potential energy; wavefunction (Ψ)
ω (omega)	angular frequency

Roman (Symbols set in boldface type represent vector quantities.)

a	Bohr radius; lattice constant
$\mathbf{B}, B$	magnetic induction or flux density
c	velocity of light; elastic constant (c_{ij})
C	heat capacity; concentration; capacitance; junction capacitance (C_T); diffusion capacitance (C_D)
$\mathbf{D}, D$	electric displacement or flux density
D	density-of-state function; diffusion constant
e	electronic charge
$\mathbf{E}, E$	electric field
f	Fermi function, oscillator strength
$\mathbf{F}, F$	force; free energy
g	conductance; gain coefficient
h	Planck's constant; $\hbar = h/2\pi$
$\mathbf{H}, H$	magnetic field
i	$= \sqrt{-1}$; ac current
I	electric current; drain current (I_D); drain saturation current (I_{DS}); light intensity
j	$= \sqrt{-1}$; ac current density